OXFORD IB DIPLOMA PROGRAMME

IN COOPERATION WITH ib

2014 EDITION

PHYSICS

COURSE COMPANION

David Homer
Michael Bowen-Jones

OXFORD UNIVERSITY PRESS

OXFORD
UNIVERSITY PRESS

Great Clarendon Street, Oxford, OX2 6DP, United Kingdom

Oxford University Press is a department of the University of Oxford. It furthers the University's objective of excellence in research, scholarship, and education by publishing worldwide. Oxford is a registered trade mark of Oxford University Press in the UK and in certain other countries

First published in 2014

British Library Cataloguing in Publication Data
Data available

978-0-19-839213-2

14

Paper used in the production of this book is a natural, recyclable product made from wood grown in sustainable forests. The manufacturing process conforms to the environmental regulations of the country of origin.

Printed in India by Multivista Global Pvt. Ltd

Acknowledgements

The publishers would like to thank the following for permissions to use their photographs:

Cover image: © James Brittain/Corbis; p1: Shutterstock; p3: ANDREW BROOKES, NATIONAL PHYSICAL LABORATORY/SCIENCE PHOTO LIBRARY; p4: Victor Habbick/Shutterstock; p9: Shutterstock; p10a: Pavel Mitrofanov/ Shutterstock; p10b: Yury Kosourov/Shutterstock; p10c: ANDREW LAMBERT PHOTOGRAPHY/SCIENCE PHOTO LIBRARY; p27: OUP; p49: Shutterstock; p63: EMILIO SEGRE VISUAL ARCHIVES/AMERICAN INSTITUTE OF PHYSICS/ SCIENCE PHOTO LIBRARY; p66: Georgios Kollidas/Shutterstock; p74: Liviu Ionut Pantelimon/Shutterstock; p80: Helen H. Richardson/The Denver Post/Getty Images; p81: JOHN HESELTINE/SCIENCE PHOTO LIBRARY; p82: MATTEIS/LOOK AT SCIENCES/SCIENCE PHOTO LIBRARY; p85: Autoguide; p91: Rainer Albiez/Shutterstock; p92: MAURICIO ANTON/SCIENCE PHOTO LIBRARY; p101: SCIENCE PHOTO LIBRARY; p106: ANDREW MCCLENAGHAN/SCIENCE PHOTO LIBRARY; p115: BERENICE ABBOTT/ SCIENCE PHOTO LIBRARY; p115: Bill McMullen/Getty Images; p116: Shutterstock; p124a: F1 Online/REX; p124b: © sciencephotos/Alamy; p131a: Shutterstock; p131b: FRIEDRICH SAURER/SCIENCE PHOTO LIBRARY; p136: Harvard.Edu; p146: ANDREW LAMBERT PHOTOGRAPHY/SCIENCE PHOTO LIBRARY; p154: GIPHOTOSTOCK/SCIENCE PHOTO LIBRARY; p160: Jerry Di Marco, Montana State Univ. Physics Dept; p169: Volodymyr Krasyuk/ Shutterstock; p195: TREVOR CLIFFORD PHOTOGRAPHY/SCIENCE PHOTO LIBRARY; p229: MARTYN F. CHILLMAID/SCIENCE PHOTO LIBRARY; p245: Shutterstock; p246: Henri Silberman/Getty Images; p252: DAVID DUCROS, CNES/SCIENCE PHOTO LIBRARY; p253: FPG/Hulton Archive/ Getty Images; p254: ASSOCIATED PRESS; p255: Patty Lagera/Getty Images; p258a: Portrait of Nicolaus Copernicus (1473-1543) (oil on canvas), Pomeranian School, (16th century) / Nicolaus Copernicus Museum, Frombork, Poland / Giraudon / The Bridgeman Art Library; p258b: Tycho Brahe, Planella Coromina, Josep or Jose (1804-90) / Private Collection / © Look and Learn / The Bridgeman Art Library; p258c: Leemage/Getty Images; p258d: Portrait of Isaac Newton (1642-1727) 1702 (oil on canvas), Kneller, Sir Godfrey (1646-1723) / National Portrait Gallery, London, UK / The Bridgeman Art Library; p258e: DEA PICTURE LIBRARY/Getty Images; p265: Kevin Clogstoun/Getty Images; p267: GORONWY TUDOR JONES, UNIVERSITY OF BIRMINGHAM/SCIENCE PHOTO LIBRARY; p271: SCIENCE PHOTO LIBRARY; p272a: Phil Degginger/Alamy; p272a: © Phil Degginger/ Alamy; p272b: Deutsche Bundespost/NobbiP/Wikipedia; p278: Dr Steven Murray/Shutterstock; p279: Bromsgrove School (W. Dainty/C. Shakespear); p280: Bromsgrove School (W. Dainty/C. Shakespear); p296: GORONWY TUDOR JONES, UNIVERSITY OF BIRMINGHAM/SCIENCE PHOTO LIBRARY; p301: Photo courtesy of Berkeley Lab; p303: MissMJ/Wikipedia; p307: Shawn Hempel/Shutterstock; p316: Public Domain/Wikipedia; p322a: Ramon grosso dolarea/Shutterstock; p322b: Worldpics/Shutterstock; p325a: Markuso/Shutterstock; p325b: www.Quebecgetaways.com; p327: Shutterstock; p337: Shutterstock; p348: Public Domain/Wikipedia; p353: Shutterstock; p366: Dietrich Zawischa; p368: GIPHOTOSTOCK/SCIENCE PHOTO LIBRARY; p370: labman.phys.utk.edu; p372: GIPHOTOSTOCK/ SCIENCE PHOTO LIBRARY; p375: CHARLES D. WINTERS/SCIENCE PHOTO LIBRARY; p385: www.astro.cornell.edu; p386: NOAA; p391: Robert Adrian Hillman/Shutterstock; p420: © sciencephotos / Alamy; p427: Zigzag Mountain Art/Shutterstock; p445: Ziga Cetrtic/Shutterstock; p461: Shutterstock; p475: Kamioka Observatory, ICRR (Institute for Cosmic Ray Research), The University of Tokyo; p481: ANDREW LAMBERT PHOTOGRAPHY/SCIENCE PHOTO LIBRARY; p486: Paul Noth/Condenast Cartoons; p499: Kamioka Observatory, ICRR (Institute for Cosmic Ray Research), The University of Tokyo; p507: MIKKEL JUUL JENSEN / SCIENCE PHOTO LIBRARY; p540: NASA, ESA, and STScI; p549: PERY BURGE/SCIENCE PHOTO LIBRARY; p552: Ricardo high speed carbon-fibre flywheel; p570: Shutterstock; p584: iStock; p587: Travelpix Ltd/Getty Images; p593: Allison Herreid/Shutterstock; p594: GIPhotoStock/Science Source; p599a: DAVID PARKER/SCIENCE PHOTO LIBRARY; p599b: DAVID PARKER/SCIENCE PHOTO LIBRARY; p612: Donald Joski/Shutterstock; p614: Image courtesy of Celestron; p616: Israel Pabon/Shutterstock; p617: PETER BASSETT/ SCIENCE PHOTO LIBRARY; p620: Edward Kinsman/Getty Images; p623: US AIR FORCE/SCIENCE PHOTO LIBRARY; p627: iStock; p631: GUSTOIMAGES/ SCIENCE PHOTO LIBRARY; p641: ROBERT GENDLER/SCIENCE PHOTO LIBRARY; p642: ALMA/NAOJ/NRAO/EUROPEAN SOUTHERN OBSERVATORY/ NASA/ESA HUBBLE SPACE TELESCOPE/SCIENCE PHOTO LIBRARY; p643: © Alan Dyer, Inc/Visuals Unlimited/Corbis; p644a: CHRIS COOK/SCIENCE PHOTO LIBRARY; p644b: ESO/NASA; p644c: GALEX, JPL-Caltech/NASA; p646: C. CARREAU/EUROPEAN SPACE AGENCY/SCIENCE PHOTO LIBRARY; p654: ROYAL ASTRONOMICAL SOCIETY/SCIENCE PHOTO LIBRARY; p657: NASA/ CXC/ SAO/NASA; p658: ROYAL OBSERVATORY, EDINBURGH/SCIENCE PHOTO LIBRARY; p664: ESA and the Planck Collaboration; p665: GSFC/ NASA; p667: ESO/VISTA/J. Emerson; p676: GSFC/NASA; p677: NASA/WMAP Science Team; p677: NATIONAL OPTICAL ASTRONOMY OBSERVATORIES/ SCIENCE PHOTO LIBRARY; p680: NASA/JPL-Caltech/S.Willner (Harvard-Smithsonian CfA); p683: NASA/SCIENCE PHOTO LIBRARY; p689: Dorling Kindersley/Getty Images

Artwork by Six Red Marbles and OUP

The authors and publisher are grateful for permission to reprint extracts from the following copyright material:

P651 Nick Strobel, table 'Main Sequence Star Properties' from www.astronomynotes.com, reprinted by permission.

P499 John Updike, 'Telephone Poles and Other Poems' from Cosmic Gall, (Deutsch, 1963), copyright © 1959, 1963 by John Updike, reprinted by permission of Alfred A. Knopf, an imprint of the Knopf Doubleday Publishing Group, a division of Random House LLC, and Penguin Books Ltd, all rights reserved.

Sources:

P472 Albert Einstein 'Considerations concerning the fundaments of theoretical physics', Science 91:492 (1940)

p112 'Neutrino 'faster than light' scientist resigns', BBC News © 2013 BBC

Contents

Course book definition

The IB Diploma Programme course books are resource materials designed to support students throughout their two-year Diploma Programme course of study in a particular subject. They will help students gain an understanding of what is expected from the study of an IB Diploma Programme subject while presenting content in a way that illustrates the purpose and aims of the IB. They reflect the philosophy and approach of the IB and encourage a deep understanding of each subject by making connections to wider issues and providing opportunities for critical thinking.

The books mirror the IB philosophy of viewing the curriculum in terms of a whole-course approach; the use of a wide range of resources, international mindedness, the IB learner profile and the IB Diploma Programme core requirements, theory of knowledge, the extended essay, and creativity, action, service (CAS).

Each book can be used in conjunction with other materials and indeed, students of the IB are required and encouraged to draw conclusions from a variety of resources. Suggestions for additional and further reading are given in each book and suggestions for how to extend research are provided.

In addition, the course books provide advice and guidance on the specific course assessment requirements and on academic honesty protocol. They are distinctive and authoritative without being prescriptive.

IB mission statement

The International Baccalaureate aims to develop inquiring, knowledgeable and caring young people who help to create a better and more peaceful world through intercultural understanding and respect.

To this end the organization works with schools, governments and international organizations to develop challenging programmes of international education and rigorous assessment.

These programmes encourage students across the world to become active, compassionate and lifelong learners who understand that other people, with their differences, can also be right.

The IB Learner Profile

The aim of all IB programmes to develop internationally minded people who work to create a better and more peaceful world. The aim of the programme is to develop this person through ten learner attributes, as described below.

Inquirers: They develop their natural curiosity. They acquire the skills necessary to conduct inquiry and research and snow independence in learning. They actively enjoy learning and this love of learning will be sustained throughout their lives.

Knowledgeable: They explore concepts, ideas, and issues that have local and global significance. In so doing, they acquire in-depth knowledge and develop understanding across a broad and balanced range of disciplines.

Thinkers: They exercise initiative in applying thinking skills critically and creatively to recognize and approach complex problems, and make reasoned, ethical decisions.

Communicators: They understand and express ideas and information confidently and creatively in more than one language and in a variety of modes of communication. They work effectively and willingly in collaboration with others.

Principled: They act with integrity and honesty, with a strong sense of fairness, justice and respect for the dignity of the individual, groups and communities. They take responsibility for their own action and the consequences that accompany them.

Open-minded: They understand and appreciate their own cultures and personal histories, and are open to the perspectives, values and traditions of other individuals and communities. They are accustomed to seeking and evaluating a range of points of view, and are willing to grow from the experience.

Caring: They show empathy, compassion and respect towards the needs and feelings of others. They have a personal commitment to service, and to act to make a positive difference to the lives of others and to the environment.

Risk-takers: They approach unfamiliar situations and uncertainty with courage and forethought, and have the independence of spirit to explore new roles, ideas, and strategies. They are brave and articulate in defending their beliefs.

Balanced: They understand the importance of intellectual, physical and emotional balance to achieve personal well-being for themselves and others.

Reflective: They give thoughtful consideration to their own learning and experience. They are able to assess and understand their strengths and limitations in order to support their learning and personal development.

A note on academic honesty

It is of vital importance to acknowledge and appropriately credit the owners of information when that information is used in your work. After all, owners of ideas (intellectual property) have property rights. To have an authentic piece of work, it must be based on your individual and original ideas with the work of others fully acknowledged. Therefore, all assignments, written or oral, completed for assessment must use your own language and expression. Where sources are used or referred to, whether in the form of direct quotation or paraphrase, such sources must be appropriately acknowledged.

How do I acknowledge the work of others?

The way that you acknowledge that you have used the ideas of other people is through the use of footnotes and bibliographies.

Footnotes (placed at the bottom of a page) or endnotes (placed at the end of a document) are to be provided when you quote or paraphrase from another document, or closely summarize the information provided in another document. You do not need to provide a footnote for information that is part of a 'body of knowledge'. That is, definitions do not need to be footnoted as they are part of the assumed knowledge.

Bibliographies should include a formal list of the resources that you used in your work. 'Formal' means that you should use one of the several accepted forms of presentation. This usually involves separating the resources that you use into different categories (e.g. books, magazines, newspaper articles, internet-based resources, CDs and works of art) and providing full information as to how a reader or viewer of your work can find the same information. A bibliography is compulsory in the Extended Essay.

What constitutes malpractice?

Malpractice is behaviour that results in, or may result in, you or any student gaining an unfair advantage in one or more assessment component. Malpractice includes plagiarism and collusion.

Plagiarism is defined as the representation of the ideas or work of another person as your own. The following are some of the ways to avoid plagiarism:

- words and ideas of another person to support one's arguments must be acknowledged
- passages that are quoted verbatim must be enclosed within quotation marks and acknowledged
- CD-Roms, email messages, web sites on the Internet and any other electronic media must be treated in the same way as books and journals
- the sources of all photographs, maps, illustrations, computer programs, data, graphs, audio-visual and similar material must be acknowledged if they are not your own work
- works of art, whether music, film dance, theatre arts or visual arts and where the creative use of a part of a work takes place, the original artist must be acknowledged.

Collusion is defined as supporting malpractice by another student. This includes:

- allowing your work to be copied or submitted for assessment by another student
- duplicating work for different assessment components and/or diploma requirements.

Other forms of malpractice include any action that gives you an unfair advantage or affects the results of another student. Examples include, taking unauthorized material into an examination room, misconduct during an examination and falsifying a CAS record.

Using your IB Physics kerboodle Online Resources

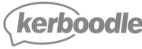

What is Kerboodle?

Kerboodle is an online learning platform. If your school has a subscription to IB Physics Kerboodle Online Resources you will be able to access a bank of resources and assessments to guide you through this course.

What is in your Kerboodle Online Resources?

There are three main areas on the IB Physics Kerboodle: planning, resources, and assessment.

Resources

There a hundreds of extra resources available on the IB Physics Kerboodle Online. You can use these at home or in the classroom to develop your skills and knowledge as you progress through the course.

- Hundreds of worksheets – read articles, perform experiments and simulations, practice your skills, or use your knowledge to answer questions.
- Find out more by looking at links to recommended sites on the Internet, answer questions, or do more research.
- Plus more to come in regular updates to Kerboodle!

Planning

This area is for your teacher so you won't have access to material in here.

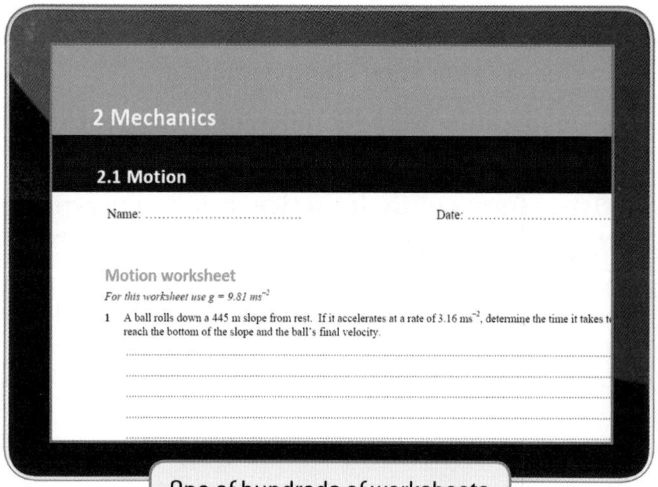

One of hundreds of worksheets

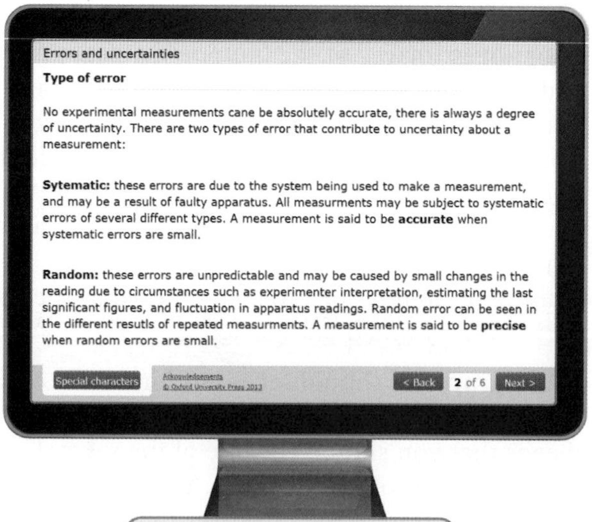

Practical skills presentation

Assessment

Click on the assessment tab to check your knowledge or revise for your examinations. Here you will find lots of interactive quizzes and exam-style practice questions.

- Formative tests: use these to check your comprehension. Evaluate how confident you feel about a sub-topic, then complete the test. You will have two attempts at each question and get feedback after every question. The marks are automatically reported in the markbook, so you can see how you progress throughout the year.
- Summative tests: use these to practice for your exams or as revision. Work through the test as if it were an examination – go back and change any questions you aren't sure about until you are happy, then submit the test for a final mark. The marks are automatically reported in the markbook, so you can see where you may need more practice.
- Assessment practice: use these to practice answering the longer written questions you will come across when you are examined. These worksheets can be printed out and performed as a timed test.

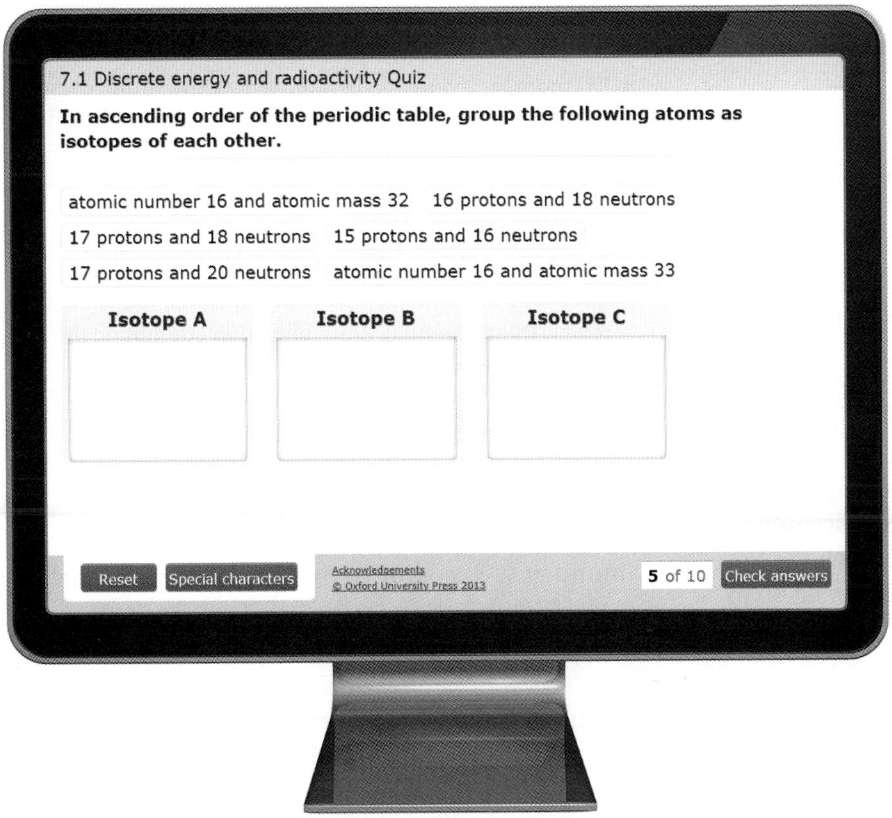

Don't forget!

You can also find all of the textbook answers on our free website **www.oxfordsecondary.co.uk/ib-physics**

Introduction

Physics is one of the earliest academic disciplines known – if you include observational astronomy, possibly the oldest. In physics we analyse the natural world to develop the best understanding we can of how the universe and its constituent parts interrelate. Our aim as physicists is to develop models that correspond to what is observed in the laboratory and beyond. These models come in many forms: some may be quantitative and based on mathematics; some may be qualitative and give a verbal description of the world around us. But, whatever form the models take, physicists must all agree on their validity before they can be accepted as part of our physical description of the universe.

Models used by physicists are linked by a coherent set of principles known as concepts. These are over-arching ideas that link the development of the subject not only within a particular physical topic (for example, forces in mechanics) but also between topics (for example, the common mathematics that links radioactive decay and capacitor discharge). In studying physics, take every opportunity to understand a new concept when you meet it. When the concept occurs elsewhere your prior knowledge will make the later learning easier.

This book is designed to support your learning of physics within group 4 of the IB Diploma Programme. Like all the disciplines represented in this subject group it has a thorough basis in the facts and concepts of science, but it also draws out the nature of science. This is to give you a better understanding of what it means to be a scientist, so that you can, for example, identify shortcomings in scientific topics presented to you in the media or elsewhere. Not everyone taking IB Physics will want to go on to be a physicist or engineer, but all citizens need to have an awareness of the importance of science in modern society.

The structure of this book needs an explanation; all of the topics include the following elements:

Understanding

The specifics of the content requirements for each sub-topic are covered in detail. Concepts are presented in ways that will promote enduring understanding.

Investigate!

These sections describe practical work you can undertake. You may need to modify these experiments slightly to suit the apparatus in your school. These are a valuable opportunity to build the skills that are assessed in IA (see page 687).

Nature of science

These sections help you to develop your understanding by studying a specific illustrative example or learning about a significant experiment in the history of physics.

Here you can explore the methods of science and some of the knowledge issues that are associated with scientific endeavour. This is done using carefully selected examples, including research that led to paradigm shifts in our understanding of the natural world.

Theory of Knowledge

These short sections have articles on scientific questions that arise from Theory of knowledge. We encourage you draw on these examples of knowledge issues in your TOK essays. Of course, much of the material elsewhere in the book, particularly in the nature of science sections, can be used to prompt TOK discussions.

Worked example

These are step-by-step examples of how to answer questions or how to complete calculations. You should review them carefully, preferably after attempting the question yourself.

End -of-Topic Questions

At the end of each topic you will find a range of questions, including both past IB Physics exam questions and new questions. Answers can be found at www.oxfordsecondary.co.uk/ib-physics

Authors do not write in isolation. In particular, our ways of describing and explaining physics have been honed by the students we have been privileged to teach over the years, and by colleagues who have challenged our ways of thinking about the subject. Our thanks go to them all. More specifically, we thank Jean Godin for much sound advice during the preparation of this text. Any errors are, of course, our responsibility.

Last but in no sense least, we thank our wives, Adele and Brenda, for their full support during the preparation of this book. We could not have completed it without their understanding and enormous patience.

M Bowen-Jones
D Homer

1 MEASUREMENTS AND UNCERTAINTIES

Introduction

This topic is different from other topics in the course book. The content discussed here will be used in most aspects of your studies in physics. You will come across many aspects of this work in the context of other subject matter. Although you may wish to do so, you would not be expected to read this topic in one go, rather you would return to it as and when it is relevant.

1.1 Measurements in physics

Understanding

- → Fundamental and derived SI units
- → Scientific notation and metric multipliers
- → Significant figures
- → Orders of magnitude
- → Estimation

Applications and skills

- → Using SI units in the correct format for all required measurements, final answers to calculations and presentation of raw and processed data
- → Using scientific notation and metric multipliers
- → Quoting and comparing ratios, values, and approximations to the nearest order of magnitude
- → Estimating quantities to an appropriate number of significant figures

Nature of science

In physics you will deal with the qualitative and the quantitative, that is, descriptions of phenomena using words and descriptions using numbers. When we use words we need to interpret the meaning and one person's interpretation will not necessarily be the same as another's. When we deal with numbers (or equations), providing we have learned the rules, there is no mistaking someone else's meaning. It is likely that some readers will be more comfortable with words than symbols and vice-versa. It is impossible to avoid either methodology on the IB Diploma course and you must learn to be careful with both your numbers and your words. In examinations you are likely to be penalized by writing contradictory statements or mathematically incorrect ones. At the outset of the course you should make sure that you understand the mathematical skills that will make you into a good physicist.

Quantities and units

Physicists deal with **physical quantities**, which are those things that are *measureable* such as mass, length, time, electrical current, etc. Quantities are related to one another by equations such as $\rho = \frac{m}{V}$ which is the symbolic form of saying that density is the ratio of the mass of an object to its volume. Note that the symbols in the equation are all written in italic (sloping) fonts – this is how we can be sure that the symbols represent quantities. Units are always written in Roman (upright) font because they sometimes share the same symbol with a quantity. So "*m*" represents the quantity "mass" but "m" represents the unit "metre". We will use this convention throughout the course book, and it is also the convention used by the IB.

Nature of science

The use of symbols

The use of Greek letters such as rho (ρ) is very common in physics. There are so many quantities that, even using the 52 Arabic letters (lower case and capitals), we soon run out of unique symbols. Sometimes symbols such as *d* and *x* have multiple uses, meaning that Greek letters have become just one way of trying to tie a symbol to a quantity uniquely. Of course, we must consider what happens when we run out of Greek letters too – we then use Russian ones from the Cyrillic alphabet.

Greek

Α	α	alpha	Ν	ν	nu
Β	β	beta	Ξ	ξ	ksi
Γ	γ	gamma	Ο	ο	omicron
Δ	δ	delta	Π	π	pi
Ε	ε	epsilon	Ρ	ρ	rho
Ζ	ζ	zeta	Σ	σ	sigma
Η	η	eta	Τ	τ	tau
Θ	θ	theta	Υ	υ	upsilon
Ι	ι	iota	Φ	φ	phi
Κ	κ	kappa	Χ	χ	chi
Λ	λ	lambda	Ψ	ψ	psi
Μ	μ	mu	Ω	ω	omega

Russian

Аа Әә Бб Вв Гг Дд Ее
Жж Зз Ҙҙ Ии Йй Кк
Лл Мм Нн Ңң Оо Пп
Рр Сс Тт Уу Ўў Фф
Хх Цц Чч Шш Ыы
Юю Яя

Fundamental quantities are those quantities that are considered to be so basic that all other quantities need to be expressed in terms of them. In the density equation $\rho = \frac{m}{V}$ only mass is chosen to be fundamental (volume being the product of three lengths), density and volume are said to be **derived quantities**.

It is essential that all measurements made by one person are understood by others. To achieve this we use units that are understood to have unambiguous meaning. The worldwide standard for units is known as SI – *Système international d'unités*. This system has been developed from the metric system of units and means that, when values of scientific quantities are communicated between people, there should never be any confusion. The SI defines both units and prefixes – letters used to form decimal multiples or sub-multiples of the units. The units themselves are classified as being either fundamental (or *base*), derived, and supplementary.

There are only two supplementary units in SI and you will meet only one of these during the Diploma course, so we might as well mention them first. The two supplementary units are the radian (rad) – the unit of angular measurement and the steradian (sr) – the unit of "solid angle". The radian is a useful alternative to the degree and is defined as *the angle subtended by an arc of a circle having the same length as the radius,* as shown in figure 1. We will look at the radian in more detail in Sub-topic 6.1. The steradian is the three-dimensional equivalent of the radian and uses the idea of mapping a circle on to the surface of a sphere.

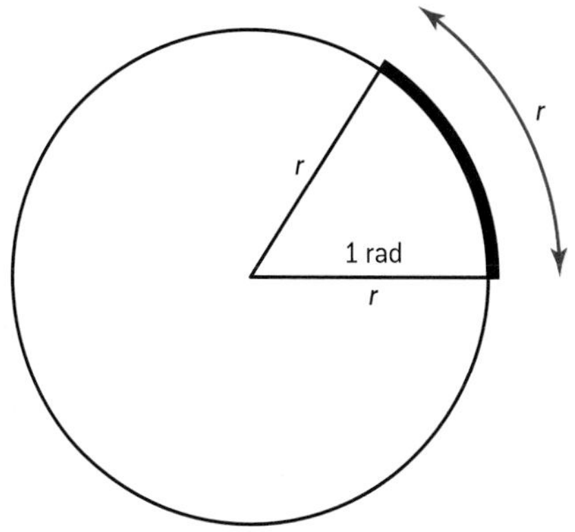

▲ Figure 1 Definition of the radian.

Fundamental and derived units

In SI there are seven **fundamental units** and you will use six of these on the Diploma course (the seventh, the candela, is included here for completeness). The fundamental quantities are length, mass, time, electric current, thermodynamic temperature, amount of substance, and luminous intensity. The units for these quantities have exact definitions and are precisely reproducible, given the right equipment. This means that any quantity can, in theory, be compared with the fundamental measurement to ensure that a measurement of that quantity is accurate. In practice, most measurements are made against more easily achieved standards so, for example, length will usually be compared with a standard metre rather than the distance travelled by light in a vacuum. You will not be expected to know the definitions of the fundamental quantities, but they are provided here to allow you to see just how precise they are.

metre (m): the length of the path travelled by light in a vacuum during a time interval of $\frac{1}{299\,792\,458}$ of a second.

kilogram (kg): mass equal to the mass of the international prototype of the kilogram kept at the Bureau International des Poids et Mesures at Sèvres, near Paris.

▲ Figure 2 The international prototype kilogram.

second (s): the duration of 9 192 631 770 periods of the radiation corresponding to the transition between the two hyperfine levels of the ground state of the caesium-133 atom.

ampere (A): that constant current which, if maintained in two straight parallel conductors of infinite length, negligible circular cross-section, and placed 1 m apart in vacuum, would produce between these conductors a force equal to 2×10^{-7} newtons per metre of length.

kelvin (K): the fraction $\frac{1}{273.16}$ of the thermodynamic temperature of the triple point of water.

mole (mol): the amount of substance of a system that contains as many elementary entities as there are atoms in 0.012 kg of carbon–12. When the mole is used, the elementary entities must be specified and may be atoms, molecules, ions, electrons, other particles, or specified groups of such particles.

candela (cd): the luminous intensity, in a given direction, of a source that emits monochromatic radiation of frequency 540×10^{12} hertz and that has a radiant intensity in that direction of $\frac{1}{683}$ watt per steradian.

All quantities that are not fundamental are known as *derived* and these can always be expressed in terms of the fundamental quantities through a relevant equation. For example, speed is the rate of change of distance with respect to time or in equation form $v = \frac{\Delta s}{\Delta t}$ (where Δs means the change in distance and Δt means the change in time). As both distance (and length) and time are fundamental quantities, speed is a derived quantity.

TOK

Deciding on what is fundamental

Who has made the decision that the fundamental quantities are those of mass, length, time, electrical current, temperature, luminous intensity, and amount of substance? In an alternative universe it may be that the fundamental quantities are based on force, volume, frequency, potential difference, specific heat capacity, and brightness. Would that be a drawback or would it have meant that "humanity" would have progressed at a faster rate?

Note

If you are reading this at the start of the course, it may seem that there are so many things that you might not know; but, take heart, "Rome was not built in a day" and soon much will come as second nature. When we write units as $\mathrm{m\,s^{-1}}$ and $\mathrm{m\,s^{-2}}$ it is a more effective and preferable way to writing what you may have written in the past as m/s and m/s^2; both forms are still read as "metres per second" and "metres per second squared."

The units used for fundamental quantities are unsurprisingly known as fundamental units and those for derived quantities are known as derived units. It is a straightforward approach to be able to express the unit of any quantity in terms of its fundamental units, provided you know the equation relating the quantities. Nineteen fundamental quantities have their own unit but it is also valid, if cumbersome, to express this in terms of fundamental units. For example, the SI unit of pressure is the pascal (Pa), which is expressed in fundamental units as $\mathrm{m^{-1}\,kg\,s^{-2}}$.

Nature of science

Capitals or lower case?

Notice that when we write the unit newton in full, we use a lower case n but we use a capital N for the symbol for the unit – unfortunately some word processors have default setting to correct this so take care! All units written in full should start with a lower case letter, but those that have been derived in honour of a scientist will have a symbol that is a capital letter. In this way there is no confusion between the scientist and the unit: "Newton" refers to Sir Isaac Newton but "newton" means the unit. Sometimes units are abbreviations of the scientist's surname, so amp (which is a shortened form of ampère anyway) is named after Ampère, the volt after Volta, the farad, Faraday, etc.

▲ Figure 3 Choosing fundamental units in an alternative universe.

Example of how to relate fundamental and derived units

The unit of force is the newton (N). This is a derived unit and can be expressed in terms of fundamental units as $\mathrm{kg\ m\ s^{-2}}$. The reason for this is that force can be defined as being the product of mass and acceleration or $F = ma$. Mass is a fundamental quantity but acceleration is not. Acceleration is the rate of change of velocity or $a = \frac{\Delta v}{\Delta t}$ where Δv represents the change in velocity and Δt the change in time. Although time is a fundamental quantity, velocity is not so we need to take another step in defining velocity in fundamental quantities. Velocity is the rate of change of displacement (a quantity that we will discuss later in the topic but, for now, it simply means distance in a given direction). So the equation for velocity is $v = \frac{\Delta s}{\Delta t}$ with Δs being the change in displacement and Δt again being the change in time. Displacement (a length) and time are both fundamental, so we are now in a position to put N into fundamental units. The unit of velocity is $\mathrm{m\,s^{-1}}$ and these are already fundamental – there is no shortened form of this. The units of acceleration will therefore be those of velocity divided by time and so will be $\frac{\mathrm{m\,s^{-1}}}{\mathrm{s}}$ which is written as $\mathrm{m\,s^{-2}}$. So the unit of force will be the unit of mass multiplied by the unit of acceleration and, therefore, be $\mathrm{kg\,m\,s^{-2}}$. This is such a common unit that it has its own name, the newton, ($\mathrm{N \equiv kg\,m\,s^{-2}}$ – a mathematical way of expressing that the two units are identical). So if you are in an examination and forget the unit of force you could always write $\mathrm{kg\,m\,s^{-2}}$ (if you have time to work it out!).

Significant figures

Calculators usually give you many digits in an answer. How do you decide how many digits to write down for the final answer?

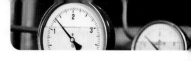

Scientists use a method of rounding to a certain number of significant figures (often abbreviated to s.f.). "Significant" here means meaningful.

Consider the number 84 072, the 8 is the most significant digit, because it tells us that the number is eighty thousand and something. The 4 is the next most significant telling us that there are also four thousand and something. Even though it is a zero, the next digit, the 0, is the third most significant digit here.

When we face a decimal number such as 0.00245, the 2 is the most significant digit because it tells us that the number is two thousandth and something. The 4 is the next most significant, showing that there are four ten thousandths and something.

If we wish to express this number to two significant figures we need to round the number from three to two digits. If the last number had been 0.00244 we would have rounded down to 0.0024 and if it had been 0.00246 we would have rounded up to 0.0025. However, it is a 5 so what do we do? In this case there is equal justification for rounding up and down, so all you really need to be is consistent with your choice for a set of figures – you can choose to round up or down. Often you will have further digits to help you, so if the number had been 0.002451 and you wanted it rounded to two significant figures it would be rounded up to 0.0025.

Some rules for using significant figures

- A digit that is not a zero will always be significant – 345 is three significant figures (3 s.f.).
- Zeros that occur sandwiched between non-zero digits are always significant – 3405 (4 s.f.); 10.3405 (6 s.f.).
- Non-sandwiched zeros that occur to the left of a non-zero digit are not significant – 0.345 (3 s.f); 0.034 (2 s.f.).
- Zeros that occur to the right of the decimal point are significant, provided that they are to the right of a non-zero digit – 1.034 (4 s.f.); 1.00 (3 s.f.); 0.34500 (5 s.f.); 0.003 (1 s.f.).
- When there is no decimal point, trailing zeros are not significant (to make them significant there needs to be a decimal point) – 400 (1 s.f.); 400. (3 s.f.) – but this is rarely written.

Scientific notation

One of the fascinations for physicists is dealing with the very large (e.g. the universe) and the very small (e.g. electrons). Many physical constants (quantities that do not change) are also very large or very small. This presents a problem: how can writing many digits be avoided? The answer is to use scientific notation.

The speed of light has a value of 299 792 458 m s^{-1}. This can be rounded to three significant figures as 300 000 000 m s^{-1}. There are a lot of zeros in this and it would be easy to miss one out or add another. In scientific notation this number is written as $3.00 \times 10^8\ \text{m s}^{-1}$ (to three significant figures).

Let us analyse writing another large number in scientific notation. The mass of the Sun to four significant figures is 1 989 000 000 000 000 000 000 000 000 000 kg (that is 1989 and twenty-seven zeros). To convert this into scientific notation we write it as 1.989 and then we imagine moving the decimal point 30 places to the left (remember we can write as many trailing zeros as we like to a decimal number without changing it). This brings our number back to the original number and so it gives the mass of the Sun as 1.989×10^{30} kg.

A similar idea is applied to very small numbers such as the charge on the electron, which has an accepted value of approximately 0.000 000 000 000 000 000 1602 coulombs. Again we write the coefficient as 1.602 and we must move the decimal point 19 places to the right in order to bring 0.000 000 000 000 000 000 1602 into this form. The base is always 10 and moving our decimal point to the right means the exponent is negative. We can write this number as 1.602×10^{-19} C.

Apart from avoiding making mistakes, there is a second reason why scientific notation is preferable to writing numbers in longhand. This is when we are dealing with several numbers in an equation. In writing the value of the speed of light as $3.00 \times 10^8\ \text{m s}^{-1}$, 3.00 is called the "coefficient" of the number and it will always be a number between 1 and 10. The 10 is called the "base" and the 8 is the "exponent".

There are some simple rules to apply:

- When adding or subtracting numbers the exponent must be the same or made to be the same.
- When multiplying numbers we add the exponents.
- When dividing numbers we subtract one exponent from the other.
- When raising a number to a power we raise the coefficient to the power and multiply the exponent by the power.

Worked examples

In these examples we are going to evaluate each of the calculations.

1 $1.40 \times 10^6 + 3.5 \times 10^5$

Solution

These must be written as $1.40 \times 10^6 + 0.35 \times 10^6$ so that both numbers have the same exponents.

They can now be added directly to give 1.75×10^6

2 $3.7 \times 10^5 \times 2.1 \times 10^8$

Solution

The coefficients are multiplied and the exponents are added, so we have: $3.7 \times 2.1 = 7.77$ (which we round to 7.8 to be in line with the data – something we will discuss later in this topic) and: $5 + 8 = 13$

So we write this product as: 7.8×10^{13}

3 $3.7 \times 10^5 \times 2.1 \times 10^{-8}$

Solution

Again the coefficients are multiplied and the exponents are added, so we have: $3.7 \times 2.1 = 7.8$

Here the exponents are subtracted (since the 8 is negative) to give: $5 - 8 = -3$

So we write this product as: 7.8×10^{-3}

4 $\dfrac{4.8 \times 10^5}{3.1 \times 10^2}$

Solution

The coefficients are divided and the exponents are subtracted so we have: $4.8 \div 3.1 = 1.548$ (which we round to 1.5)

And $5 - 2 = 3$

This makes the result of the division 1.5×10^3

5 $(3.6 \times 10^7)^3$

Solution

We cube 3.6 and $3.6^3 = 46.7$

And multiply 7 by 3 to give 21

This gives 46.7×10^{21}, which should become 4.7×10^{22} in scientific notation.

Metric multipliers (prefixes)

Scientists have a second way of abbreviating units: by using metric multipliers (usually called "prefixes"). An SI prefix is a name or associated symbol that is written before a unit to indicate the appropriate power of 10. So instead of writing 2.5×10^{12} J we could alternatively write this as 2.5 TJ (terajoule). Figure 4 gives the 20 SI prefixes – these are provided for you as part of the data booklet used in examinations.

Factor	Name	Symbol
10^{24}	yotta	Y
10^{21}	zetta	Z
10^{18}	exa	E
10^{15}	peta	P
10^{12}	tera	T
10^{9}	giga	G
10^{6}	mega	M
10^{3}	kilo	k
10^{2}	hecto	h
10^{1}	deka	da
10^{-1}	deci	d
10^{-2}	centi	c
10^{-3}	milli	m
10^{-6}	micro	μ
10^{-9}	nano	n
10^{-12}	pico	p
10^{-15}	femto	f
10^{-18}	atto	a
10^{-21}	zepto	z
10^{-24}	yocto	y

▲ Figure 4 SI metric multipliers.

Orders of magnitude

An important skill for physicists is to understand whether or not the physics being considered is *sensible*. When performing a calculation in which someone's mass was calculated to be 5000 kg, this should ring alarm bells. Since average adult masses ("weights") will usually be 60–90 kg, a value of 5000 kg is an impossibility.

A number rounded to the nearest power of 10 is called an **order of magnitude**. For example, when considering the average adult human mass: 60–80 kg is closer to 100 kg than 10 kg, making the order of magnitude 10^2 and not 10^1. Of course, we are not saying that all adult humans have a mass of 100 kg, simply that their average mass is closer to 100 than 10. In a similar way, the mass of a sheet of A4 paper may be 3.8 g which, expressed in kg, will be 3.8×10^{-3} kg. Since 3.8 is closer to 1 than to 10, this makes the order of magnitude of its mass 10^{-3} kg. This suggests that the ratio of adult mass to the mass of a piece of paper (should you wish to make this comparison)$= \frac{10^2}{10^{-3}} = 10^{2-(-3)} = 10^5 = 100\,000$. In other words, an adult human is 5 orders of magnitude (5 powers of 10) heavier than a sheet of A4 paper.

Estimation

Estimation is a skill that is used by scientists and others in order to produce a value that is a useable approximation to a true value. Estimation is closely related to finding an order of magnitude, but may result in a value that is more precise than the nearest power of 10.

Whenever you measure a length with a ruler calibrated in millimetres you can usually see the whole number of millimetres but will need to estimate to the next $\frac{1}{10}$ mm – you may need a magnifying glass to help you to do this. The same thing is true with most non-digital measuring instruments.

Similarly, when you need to find the area under a non-regular curve, you cannot truly work out the actual area so you will need to find the area of a rectangle and estimate how many rectangles there are. Figure 5 shows a graph of how the force applied to an object varies with time. The area under the graph gives the impulse (as you will see in Topic 2). There are 26 complete or nearly complete yellow squares under the curve and there are further partial squares totalling about four full squares in all. This gives about 30 full squares under the curve. Each curve has an area equivalent to $2\text{ N} \times 1\text{ s} = 2\text{ N s}$. This gives an estimate of about 60 N s for the total impulse.

In an examination, estimation questions will always have a tolerance given with the accepted answer, so in this case it might be (60 ± 2) N s.

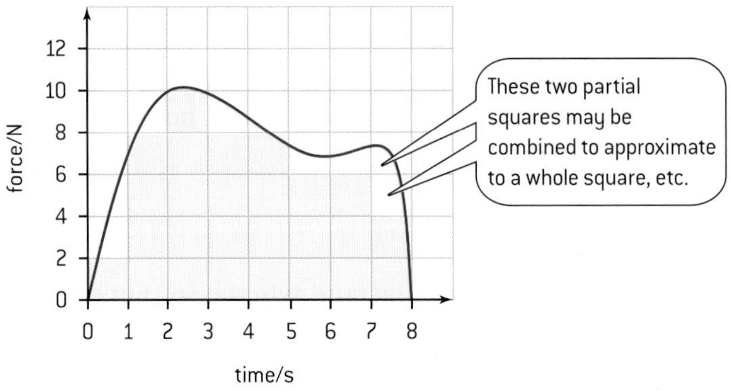
force/N
time/s
These two partial squares may be combined to approximate to a whole square, etc.

▲ Figure 5

1.2 Uncertainties and errors

Understanding

- → Random and systematic errors
- → Absolute, fractional, and percentage uncertainties
- → Error bars
- → Uncertainty of gradient and intercepts

Nature of science

In Sub-topic 1.1 we looked at the how we define the fundamental physical quantities. Each of these is measured on a scale by comparing the quantity with something that is "precisely reproducible". By precisely reproducible do we mean "exact"? The answer to this is no. If we think about the definition of the ampere, we will measure a force of 2×10^7 N. If we measure it to be 2.1×10^7 it doesn't invalidate the measurement since the definition is given to just one significant figure. All measurements have their limitations or uncertainties and it is important that both the measurer and the person working with the measurement understand what the limitations are. This is why we must always consider the uncertainty in any measurement of a physical quantity.

Applications and skills

- → Explaining how random and systematic errors can be identified and reduced
- → Collecting data that include absolute and/or fractional uncertainties and stating these as an uncertainty range (using $\pm$)
- → Propagating uncertainties through calculations involving addition, subtraction, multiplication, division, and raising to a power
- → Using error bars to calculate the uncertainty in gradients and intercepts

Equations

Propagation of uncertainties:

If: $y = a \pm b$

then: $\Delta y = \Delta a + \Delta b$

If: $y = \frac{ab}{c}$

then: $\frac{\Delta y}{y} = \frac{\Delta a}{a} + \frac{\Delta b}{b} + \frac{\Delta c}{c}$

If: $y = a^n$

then: $\frac{\Delta y}{y} = \left| n\frac{\Delta a}{a} \right|$

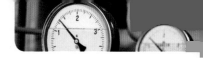

Uncertainties in measurement

Introduction

No experimental quantity can be absolutely accurate when measured – it is always subject to some degree of uncertainty. We will look at the reasons for this in this section.

There are two types of error that contribute to our uncertainty about a reading – systematic and random.

Systematic errors

As the name suggests, these types of errors are due to the system being used to make the measurement. This may be due to faulty apparatus. For example, a scale may be incorrectly calibrated either during manufacture of the equipment, or because it has changed over a period of time. Rulers warp and, as a result, the divisions are no longer symmetrical.

A timer can run slowly if its quartz crystal becomes damaged (not because the battery voltage has fallen – when the timer simply stops).

▲ Figure 1 Zero error on digital calliper.

When measuring distances from sealed radioactive sources or light-dependent resistors (LDRs), it is hard to know where the source is actually positioned or where the active surface of the LDR is.

The zero setting on apparatus can drift, due to usage, so that it no longer reads zero when it should – this is called a **zero error**.

Figure 1 shows a digital calliper with the jaws closed. This should read 0.000 mm but there is a zero error and it reads 0.01 mm. This means that all readings will be 0.01 mm bigger than they should be. The calliper can be reset to zero or 0.01 mm could be subtracted from any readings made.

Often it is not possible to spot a systematic error and experimenters have to accept the reading on their instruments, or else spend significant effort in making sure that they are re-calibrated by checking the scale against a standard scale. Repeating a reading never removes the systematic error. The real problem with systematic errors is that it is only possible to check them by performing the same task with another apparatus. If the two sets give the same results, the likelihood is that they are both performing well; however, if there is disagreement in the results a third set may be needed to resolve any difference.

In general we deal with zero errors as well as we can and then move on with our experimentation. **When systematic errors are small, a measurement is said to be accurate**.

Nature of science

Systematic errors

Uncertainty when using a 300 mm ruler may be quoted to ±0.5 mm or ±1 mm depending on your view of how precisely you can gauge the reading. To be on the safe side you might wish to use the larger uncertainty and then you will be sure that the reading lies within your bounds.

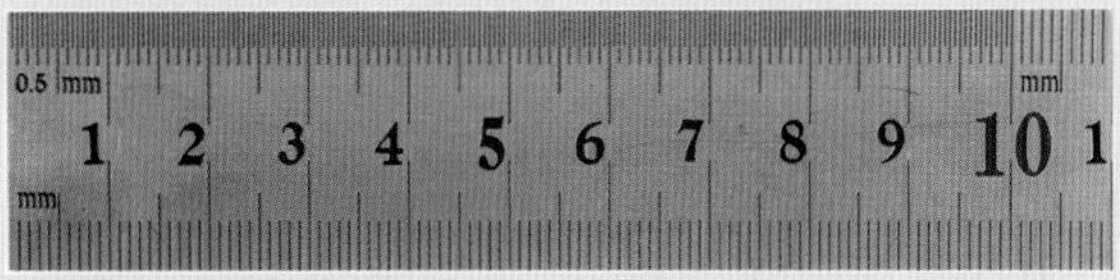

▲ Figure 2 Millimeter (mm) scale on ruler.

You should make sure you observe the scale from directly above and at right angles to the plane of the ruler in order to avoid parallax errors.

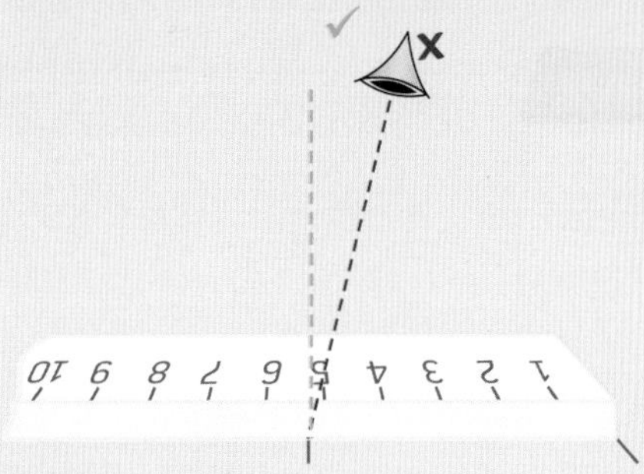

▲ Figure 3 Parallax error.

The meter in figure 4 shows an analogue ammeter with a fairly large scale – there is justification in giving this reading as being (40 ± 5) A.

▲ Figure 4 Analogue scale.

▲ Figure5 Digital scale.

The digital ammeter in figure 5 gives a value of 0.27 A which should be recorded as (0.27 ± 0.01) A.

In each of these examples the uncertainty is quoted to the same **precision** (number of decimal places) as the reading – it is essential to do this as the number of decimal places is always indicative of precision. When we write an energy value as being 8 J we are implying that it is (8 ± 1) J and if we write it as 8.0 J it implies a precision of ±0.1 J.

Random errors

Random errors can occur in any measurement, but crop up most frequently when the experimenter has to estimate the last significant figure when reading a scale. If an instrument is insensitive then it may be difficult to judge whether a reading would have changed in different circumstances. For a single reading the uncertainty could well be better than the smallest scale division available. But, since you are determining the maximum possible range of values, it is a sensible precaution to use this larger precision. Dealing with digital scales is a problem – the likelihood is that you have really no idea how precisely the scales are calibrated. Choosing the least significant digit on the scale may severely underestimate the uncertainty but, unless you know the manufacturer's data regarding calibration, it is probably the best you can do.

When measuring a time manually it is inappropriate to use the precision of the timer as the uncertainty in a reading, since your reaction time is likely to be far greater than this. For example, if you timed twenty oscillations of a pendulum to take 16.27 s this should be recorded as being (16.3 ± 0.1) s. This is because your reaction time dominates the precision of the timer. If you know that your reaction time is greater than 0.1 s then you should quote that value instead of 0.1 s.

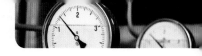

The best way of handling random errors is to take a series of repeat readings and find the average of each set of data. Half the range of the values will give a value that is a good approximation to the statistical value that more advanced error analysis provides. The range is the largest value minus the smallest value.

Readings with small random errors are said to be precise (this does not mean they are accurate, however).

Worked examples

1 In measuring the angle of refraction at an air-glass interface for a constant angle of incidence the following results were obtained (using a protractor with a precision of $\pm 1°$):

$45°, 47°, 46°, 45°, 44°$

How should we express the angle of refraction?

Solution

The mean of these values is 45.4° and the range is $(47° - 44°) = 3°$.

Half the range is 1.5°.

How then do we record our overall value for the angle of refraction?

Since the precision of the protractor is $\pm 1°$, we should quote our mean to a whole number (integral) value and it will round down to 45°. We should not minimize our uncertainty unrealistically and so we should round this up to 2°. This means that the angle of refraction should be recorded as $45 \pm 2°$.

2 The diagram below shows the position of the meniscus of the mercury in a mercury-in-glass thermometer.

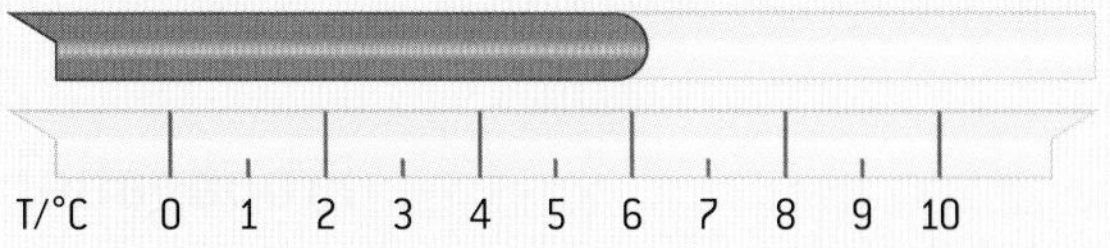

Express the temperature and its uncertainty to an appropriate number of significant figures.

Solution

The scale is calibrated in degrees but they are quite clear here, so it is reasonable to expect a precision of ±0.5 °C. The meniscus is closer to 6 than to 6.5 (although that is a judgement decision) so the values should be recorded as (6.0 ± 0.5) °C. Remember the measurement and the uncertainty should be to the same number of decimal places.

3 A student takes a series of measurements of a certain quantity. He then averages his measurements. What aspects of systematic and random uncertainties is he addressing by taking repeats and averages?

Solution

Systematic errors are not dealt with by means of repeat readings, but taking repeat readings and averaging them should cause the average value to be closer to the true value than a randomly chosen individual measurement.

Absolute and fractional uncertainties

The values of uncertainties that we have been looking at are called **absolute uncertainties**. These values have the same units as the quantity and should be written to the same number of decimal places.

Dividing the uncertainty by the value itself leaves a dimensionless quantity (one with no units) and gives us the **fractional uncertainty**. Percentaging the fractional uncertainty gives the **percentage uncertainty**.

Worked example

Calculate the absolute, fractional, and percentage uncertainties for the following measurements of a force, F:

2.5 N, 2.8 N, 2.6 N

Solution

mean value $= \frac{2.5\text{ N} + 2.8\text{ N} + 2.6\text{ N}}{3} = 2.63$ N, this is rounded down to 2.6 N

range $= (2.8 - 2.5)$ N $= 0.3$ N, giving an absolute uncertainty of 0.15 N that rounds up to 0.2 N

We would write our value for F as (2.6 ± 0.2) N

the fractional uncertainty is $\frac{0.2}{2.6} = 0.077$ and the percentage uncertainty will be $0.077 \times 100\% = 7.7\%$

Propagation of uncertainties

Often we measure quantities and then use our measurements to calculate other quantities with an equation. The uncertainty in the calculated value will be determined from a combination of the uncertainties in the quantities that we have used to calculate the value from. This is known as **propagation of uncertainties**.

There are some simple rules that we can apply when we are propagating uncertainties. In more advanced treatment of this topic we would demonstrate how these rules are developed, but we are going to focus on your application of these rules here (since you will never be asked to prove them and you can look them up in a text book or on the Internet if you want further information).

In the uncertainty equations discussed next, a, b, c, etc. are the quantities and Δa, Δb, Δc, etc. are the absolute uncertainties in these quantities.

Addition and subtraction

This is the easiest of the rules because when we add or subtract quantities we always *add* their absolute uncertainties.

When $a = b + c$ or $a = b - c$ then $\Delta a = \Delta b + \Delta c$

In order to use these relationships don't forget that the quantities being added or subtracted must have the same units.

So if we are combining two masses m_1 and m_2 then the total mass m will be the sum of the other two masses.

$m_1 = (200 \pm 10)$ g and $m_2 = (100 \pm 10)$ g so $m = 300$ g and $\Delta m = 20$ g meaning we should write this as:

$m = (300 \pm 20)$ g

We use subtraction more often than we realise when we are measuring lengths. When we set the zero of our ruler against one end of an object we are making a judgement of where the zero is positioned and this really means that the value is (0.0 ± 0.5) mm.

A ruler is used to measure a metal rod as shown in figure 6. The length is found by subtracting the smaller measurement from the larger one. The uncertainty for each measurement is ± 0.5 mm.

Larger measurement $= 195.0$ mm

Smaller measurement $= 118.5$ mm

Length $= (76.5 \pm 1.0)$ mm as the uncertainty is 0.5 mm + 0.5 mm

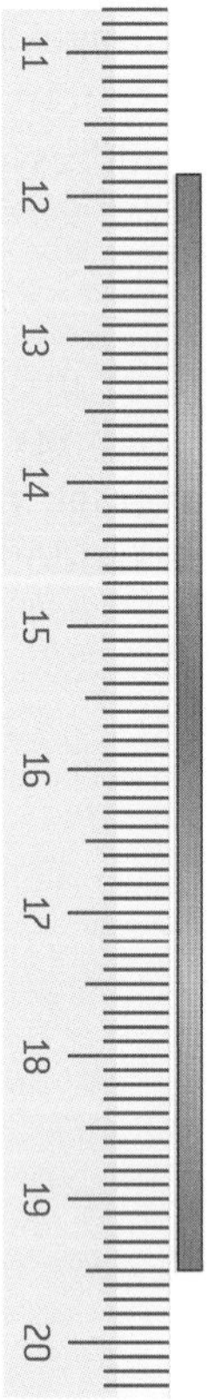

▲ Figure 6 Measuring a length.

Nature of science

Subtracting values

When subtraction is involved in a relationship you need to be particularly careful. The resulting quantity becomes smaller in size (because of subtraction), while the absolute uncertainty becomes larger (because of addition). Imagine two values that are subtracted: $b = 4.0 \pm 0.1$ and $c = 3.0 \pm 0.1$.

We won't concern ourselves with what these quantities actually are here.

If $a = b - c$ then $a = 1.0$ and since $\Delta a = \Delta b + \Delta c$ then $\Delta a = 0.2$

We have gone from two values in which the percentage uncertainty is 2.5% and 3.3% respectively to a calculated value with uncertainty of 20%. Now that really is propagation of uncertainties!

Multiplication and division

When we multiply or divide quantities we *add* their fractional or percentage uncertainties, so:

when $a = bc$ or $a = \frac{b}{c}$ or $a = \frac{c}{b}$

then $\frac{\Delta a}{a} = \frac{\Delta b}{b} + \frac{\Delta c}{c}$

There are very few relationships in physics that do not include some form of multiplication or division.

We have seen that density ρ is given by the expression $\rho = \frac{m}{V}$ where m is the mass of a sample of the substance and V is its volume. For a particular sample, the percentage uncertainty in the mass is 5% and for the volume is 12%.

The percentage uncertainty in the calculated value of the density will therefore be $\pm 17\%$.

If the sample had been cubical in shape and the uncertainty in each of the sides was 4% we can see how this brings about a volume with uncertainty of 12%:

For a cube the volume is the cube of the side length ($V = l^3 = l \times l \times l$)

so $\frac{\Delta V}{V} = \frac{\Delta l}{l} + \frac{\Delta l}{l} + \frac{\Delta l}{l} = 4\% + 4\% + 4\% = 12\%$

This example leads us to:

Raising a quantity to a power

From the cube example you might have spotted that $\frac{\Delta V}{V} = 3\frac{\Delta l}{l}$

This result can be generalized so that when $a = b^n$ (where n can be a positive or negative whole, integral, or decimal number)

then $\frac{\Delta a}{a} = \left| n\frac{\Delta b}{b} \right|$

The modulus sign is included as an alternative way of telling us that the uncertainty can be either positive or negative.

Worked example

The period T of oscillation of a mass m on a spring, having spring constant k is $T = 2\pi\sqrt{\frac{m}{k}}$

Don't worry about what these quantities actually mean at this stage.

The uncertainty in k is 11% and the uncertainty in m is 5%.

Calculate the approximate uncertainty in a value for T of 1.20 s.

Solution

First let's adjust the equation a little – we can write it as

$T = 2\pi\left(\frac{m}{k}\right)^{\frac{1}{2}}$ which is of the form $a = 2\pi\left(\frac{b}{c}\right)^n$

Although we will truncate π, we can really write it to as many significant figures as we wish and so the percentage uncertainty in π, as in 2 will be zero.

Using the division and power relationships:

$\frac{\Delta a}{a} = n\frac{\Delta b}{b} + n\frac{\Delta c}{c}$ or here $\frac{\Delta T}{T} = \frac{1}{2}\frac{\Delta m}{m} + \frac{1}{2}\frac{\Delta k}{k}$

so the percentage uncertainty in T will be half that in m + half that in k.

This means that the percentage error in $T = 0.5 \times 5\% + 0.5 \times 11\% = 8\%$

If the measured value of T is 1.20 s then the absolute uncertainty is $1.20 \times \frac{8}{100} = 0.096$ This rounds up to 0.10 and so we quote T as being (1.20 ± 0.10) s.

Remember that the quantity and the uncertainty must be to the same number of decimal places and so the zeros are important, as they give us the precision in the value.

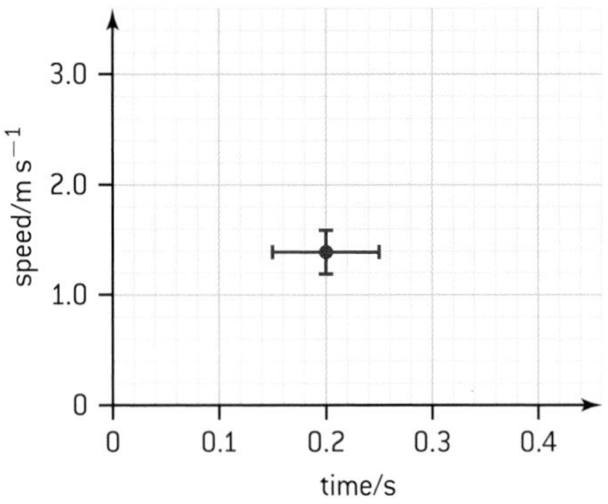

Figure 7 Error bars.

Drawing graphs

An important justification for experimental work is to investigate the relationship between physical quantities. One set of values is rarely very revealing even if it can be used to calculate a physical constant, such as dividing the potential difference across a resistor by the current in the resistor to find the resistance. Although the calculation does tell you the resistance for one value of current, it says nothing about whether the resistance depends upon the current. Taking a series of values would tell you if the resistance was constant but, with the expected random errors, it would still not be definitive. By plotting a graph and drawing the line of best fit the pattern of results is far easier to spot, whether it is linear or some other relationship.

Error bars

In plotting a point on a graph, uncertainties are recognized by adding error bars. These are vertical and horizontal lines that indicate the possible range of the quantity being measured. Suppose at a time of (0.2 ± 0.05) s the speed of an object was (1.2 ± 0.2) m s^{-1} this would be plotted as shown in figure 7.

This means that the value could possibly be within the rectangle that touches the ends of the error bars as shown in figure 8. This is the **zone of uncertainty** for the data point. A line of best fit should be one that spreads the points so that they are evenly distributed on both sides of the line and also passes through the error bars.

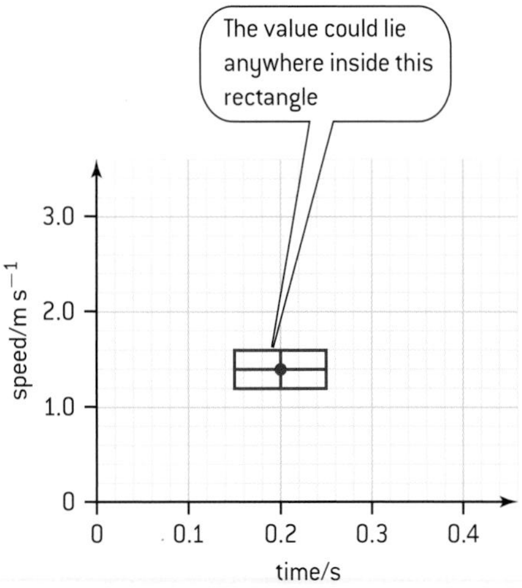

Figure 8 Zone of uncertainty.

Uncertainties with gradients

Using a computer application, such as a spreadsheet, can allow you to plot a graph with data points and error bars. You can then read off the gradient and the intercepts from a linear graph directly. The application will automatically draw the best trend line. You can then add the trend lines with the steepest and shallowest gradients that are just possible – while still passing through all the error bars. Students quite commonly, but incorrectly, use the extremes of the error bars that are furthest apart on the graphs. Although these could be appropriate, it is essential that all the trend lines you draw pass through all of the error bars.

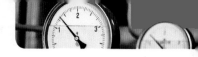

In an experiment to measure the electromotive force (emf) and internal resistance of a cell, a series of resistors are connected across the cell. The currents in and potential differences across the resistors are then measured. A graph of potential difference, V, against current, I, should give a straight line of negative gradient. As you will see in Topic 5 the emf of a cell is related to the internal resistance r by the equation:

$$\varepsilon = I(R + r) = V + Ir$$

This can be rearranged to give $V = \varepsilon - Ir$

So a graph of V against I is of gradient $-r$ (the internal resistance) and intercept ε (the emf of the cell).

The table on the right shows a set of results from this experiment. With a milliammeter and voltmeter of low precision the repeat values are identical to the measurements given in the table.

The graph of figure 9 shows the line of best fit together with two lines that are just possible.

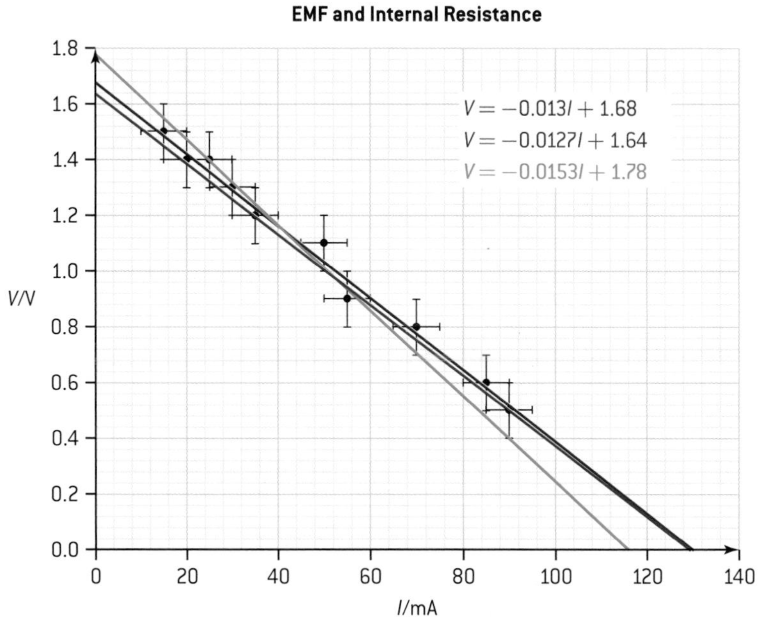

Figure 9 A graph of potential difference, V, against current, I, for a cell.

$I \pm 5$/mA	$V \pm 0.1$/V
15	1.5
20	1.4
25	1.4
30	1.3
35	1.2
50	1.1
55	0.9
70	0.8
85	0.6
90	0.5

Converting from milliamps to amps, the equations of these lines suggest that the internal resistance (the gradient) is 13.0 Ω and the range is from 12.7 Ω to 15.3 Ω (= 2.6 Ω) meaning that half the range = 1.3 Ω.

This leads to a value for $r = (13.0 \pm 1.3)\ \Omega$.

The intercept on the V axis of the line of best fit = 1.68 which rounds to 1.7 V (since the data is essentially to 2 significant figures). The range of the just possible lines gives 1.6 to 1.8 V (when rounded to two significant figures). This means that $\varepsilon = (1.6 \pm 0.1)$ V.

Nature of science

Drawing graphs manually

- One of the skills expected of physicists is to draw graphs by hand and you may well be tested on this in the data analysis question in Paper 3 of the IB Diploma Programme physics examination. You are also likely to need to draw graphs for your internal assessment.
- Try to look at your extreme values so that you have an idea of what scales to use. You will need a minimum of six points to give you a reasonable chance of drawing a valid line.
- Use scales that will allow you to spread your points out as much as possible (you should fill your page, but not overspill onto a second sheet as that would damage your line quality and lose you marks). You can always calculate an intercept if you need one; when you don't include the origin, your axes give you a **false origin** (which is fine).
- Use sensible scales that will make both plotting and your calculations clear-cut (avoid scales that are multiples of 3, 4, or 7 – stick to 2, 5, and 10).
- Try to plot your graph as you are doing the experiment – if apparently unusual values crop up, you will see them and can check that they are correct.
- Before you draw your line of best fit, you need to consider whether or not it is straight or a curve. There may well be anomalous points (outliers) that you can ignore, but if there is a definite trend to the curve then you should opt for a smooth curve drawn with a single line and not "sketched" artistically!

Figure 10 shows some of the key elements of a good hand-drawn graph. Calculating the gradients **on** the graph is very useful when checking values.

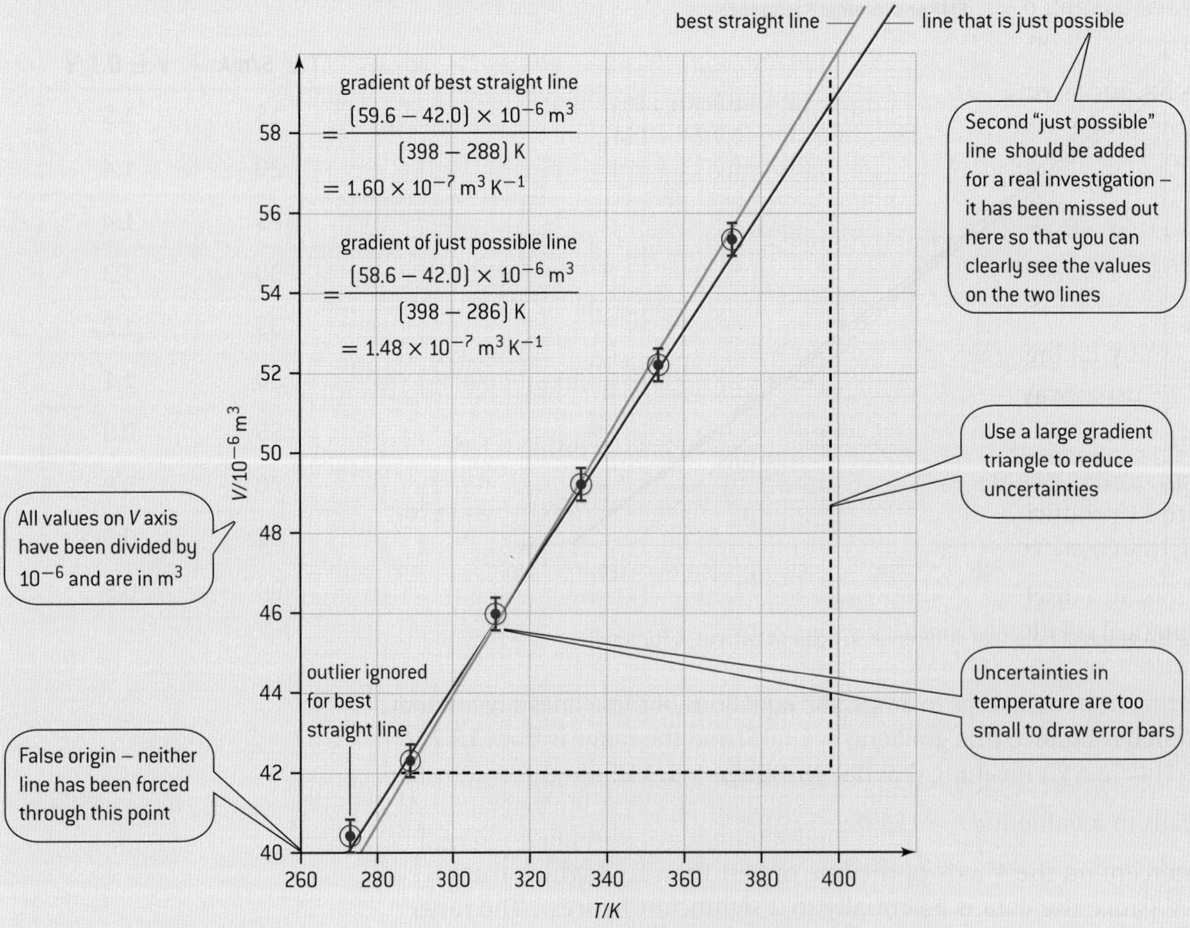

▲ Figure 10 Hand-drawn graph.

Linearizing graphs

Many relationships between physical quantities are not directly proportional and a straight line cannot be obtained simply by plotting one quantity against the other. There are two approaches to dealing with non-proportional relationships: when we know the form of the relationship and when we do not.

If we do know the form of the relationship such as $p \propto \frac{1}{V}$ (for a gas held at constant temperature) or $T \propto \sqrt{l}$ (for a simple pendulum) we can plot a graph of one quantity against the power of the other quantity to obtain a straight-line origin graph. An alternative for the simple pendulum is to plot a graph of T^2 against l which will give the same result.

You should think about the propagation of errors when you consider the relative merits of plotting T against $\sqrt{l}$ or T^2 against l.

The following discussion applies to HL examinations, but offers such a useful technique that SL students may wish to utilize it when completing IAs or if they undertake an Extended Essay in a science subject.

If we don't know the actual power involved in a relationship, but we suspect that one quantity is related to the other, we can write a general relationship in the form $y = kx^n$ where k and n are constants.

By taking logs of this equation we obtain $\log y = \log k + n\log x$ which we can arrange into $\log y = n\log x + \log k$ and is of the form $y = mx + c$. This means that a graph of $\log y$ against $\log x$ will be linear of gradient n and have an intercept on the $\log y$ axis of $\log k$.

This technique is very useful in carrying out investigations when a relationship between two quantities really is not known. The technique is also a useful way of dealing with exponential relationships by taking logs to base e, instead of base 10. For example, radioactive nuclides decay so that either the activity or the number of nuclei remaining falls according to the same general form. Writing the decay equation for the number of nuclei remaining gives:

$N = N_0 e^{-\lambda t}$ by taking logs to the base e we get $\ln N = \ln N_0 - \lambda t$ (where $\ln N$ is the usual way of writing $\log_e N$).

By plotting a graph of $\ln N$ against t the gradient will be $-\lambda t$ and the intercept on the $\ln N$ axis will be $\ln N_0$. This linearizes the graph shown in figure 11 producing the graph of figure 12. A linear graph is easier to analyse than a curve.

Capacitors also discharge through resistors using the same general mathematical relationship as that used for radioactive decay.

Note

Relationships can never be "indirectly proportional" – this is a meaningless term since it is too vague. Consequently the term "proportional" means the same as "directly proportional".

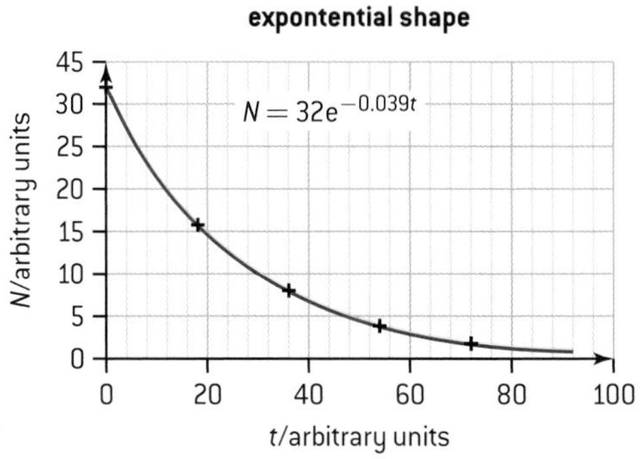

Figure 11

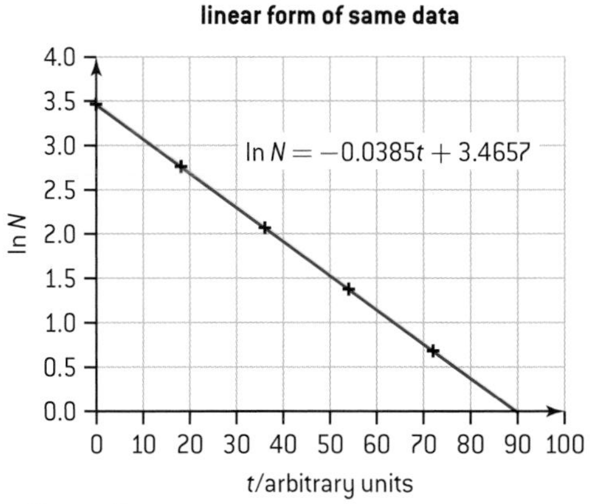

Figure 12

1.3 Vectors and scalars

Understanding

- → Vector and scalar quantities
- → Combination and resolution of vectors

Applications and skills

- → Solving vector problems graphically and algebraically.

Equations

The horizontal and vertical components of vector A:

- → $A_H = A\cos\theta$
- → $A_V = A\sin\theta$

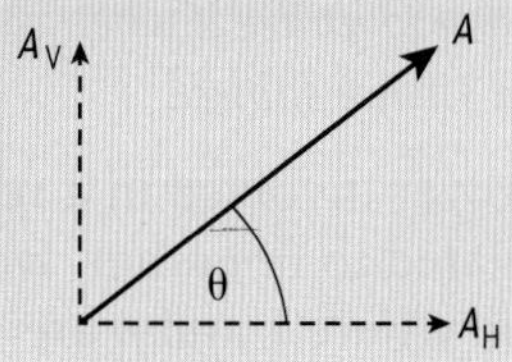

Nature of science

All physical quantities that you will meet on the course are classified as being vectors or scalars. It is important to know whether any quantity is a vector or a scalar since this will affect how the quantity is treated mathematically. Although the concept of adding forces is an intuitive application of vectors that has probably been used by sailors for millennia, the analytical aspect of it is a recent development. In the *Philosophiæ Naturalis Principia Mathematica*, published in 1687, Newton used quantities which we now call vectors, but never generalized this to deal with the concepts of vectors. At the start of the 19th century vectors became an indispensible tool for representing three-dimensional space and complex numbers. Vectors are now used as a matter of course by physicists and mathematicians alike.

Vector and scalar quantities

Scalar quantities are those that have magnitude (or *size*) but no direction. We treat scalar quantities as numbers (albeit with units) and use the rules of algebra when dealing with them. Distance and time are both scalars, as is speed. The average speed is simply the distance divided by the time, so if you travel 80 m in 10 s the speed will always be 8 m s^{-1}. There are no surprises.

Vector quantities are those which have both magnitude and direction. We must use vector algebra when dealing with vectors since we must take into account direction. The vector equivalent of distance is called displacement (i.e., it is a distance in a specified direction). The vector equivalent of speed is velocity (i.e., it is the speed in a specified direction). Time, as we have seen, is a scalar. Average velocity is defined as being displacement divided by time.

Dividing a vector by a scalar is the easiest operation that we need to do involving a vector. To continue with the example that we looked at with scalars, suppose the displacement was 80 m due north and the time was, again, 10 s. The average velocity would be 8 m s^{-1} due north. So, to generalize, when we divide a vector by a scalar we end up with a new vector that has the direction of the original one, but which will be of magnitude equal to that of the vector divided by that of the scalar.

Commonly used vectors and scalars		
Vectors	**Scalars**	**Comments**
force (F)	mass (m)	$\vec{F}$
displacement (s)	length/distance (s, d, etc.)	displacement used to be called "space" – now that means something else!
velocity (v or u)	time (t)	
momentum (p)	volume (V)	
acceleration (a)	temperature (T)	
gravitational field strength (g)	speed (v or u)	velocity and speed often have the same symbol
electric field strength (E)	density (ρ)	the symbol for density is the Greek "rho" not the letter "p"
magnetic field strength (B)	pressure (p)	
area (A)	energy/work (W, etc.)	the direction of an area is taken as being at right angles to the surface
	power (P)	
	current (I)	with current having direction you might think that it should be a vector but it is not (it is the ratio of two scalars, charge and time, so it cannot be a vector). In more advanced work you might come across *current density* which is a vector.
	resistance (R)	
	gravitational potential (V_G)	the subscripts tell us whether it is gravitational or electrical
	electric potential (V_E)	
	magnetic flux (Φ)	flux is often thought as having a direction – it doesn't!

Representing vector quantities

A vector quantity is represented by a line with an arrow.

- The direction the arrow points represents the direction of the vector.
- The length of the line represents the magnitude of the vector to a chosen scale.

When we are dealing with vectors that act in one dimension it is a simple matter to assign one direction as being positive and the opposite direction as being negative. Which direction is positive and which negative really doesn't matter as long as you are consistent. So, if one force acts upwards on an object and another force acts downwards, it is a simple matter to find the resultant by subtracting one from the other.

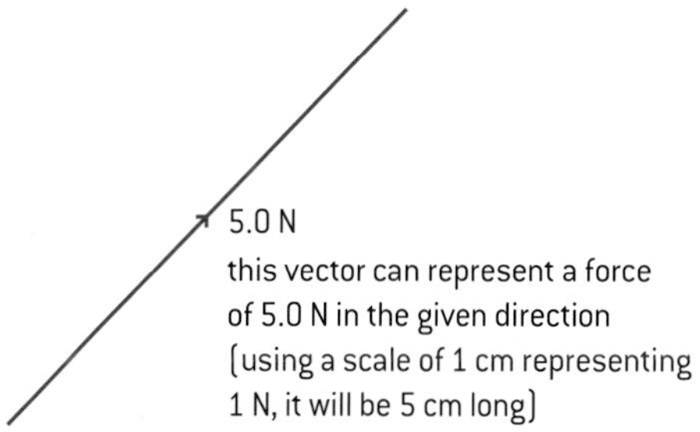

5.0 N
this vector can represent a force of 5.0 N in the given direction (using a scale of 1 cm representing 1 N, it will be 5 cm long)

▲ Figure 1 Representing a vector.

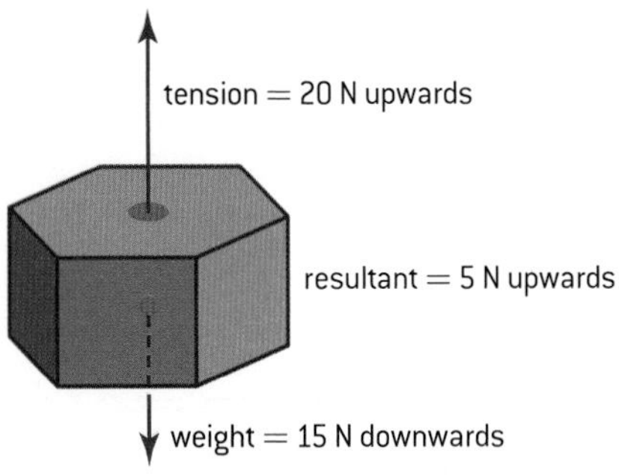

▲ Figure 2 Two vectors acting on an object.

Figure 2 shows an upward tension and downward weight acting on an object – the upward line is longer than the downward line since the object is not in equilibrium and has an upward resultant.

Adding and subtracting vectors

When adding and subtracting vectors, account has to be taken of their direction. This can be done either by a scale drawing (graphically) or algebraically.

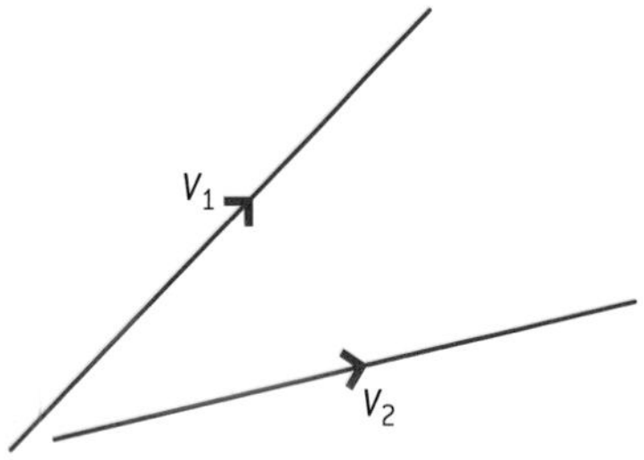

▲ Figure 3 Two vectors to be added.

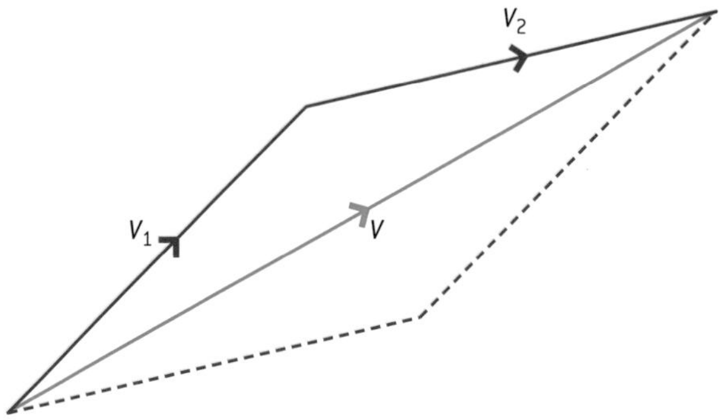

▲ Figure 4 Adding the vectors.

Scale drawing (graphical) approach

Adding two vectors V_1 and V_2 which are not in the same direction can be done by forming a parallelogram to scale.

- Make a rough sketch of how the vectors are going to add together to give you an idea of how large your scale needs to be in order to fill the space available to you. This is a good idea when you are adding the vectors mathematically too.
- Having chosen a suitable scale, draw the scaled lines in the direction of V_1 and V_2 (so that they form two adjacent sides of the parallelogram).
- Complete the parallelogram by drawing in the remaining two sides.
- The blue diagonal represents the **resultant vector** in both magnitude and direction.

Worked example

Two forces of magnitude 4.0 N and 6.0 N act on a single point. The forces make an angle of 60° with each other. Using a scale diagram, determine the resultant force.

Solution

Don't forget that the vector must have a magnitude and a direction; this means that the angle is just as important as the size of the force.

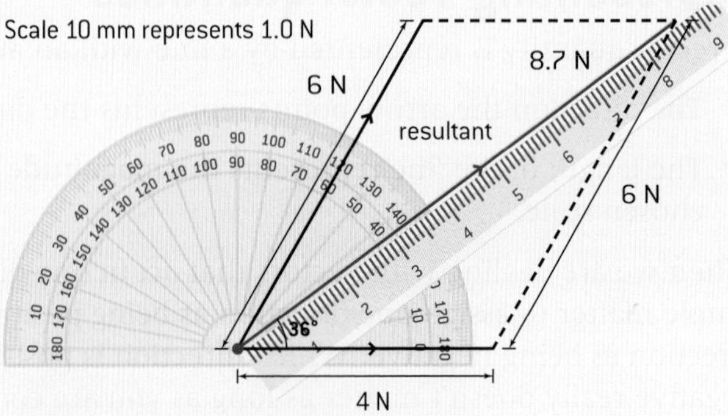

length of resultant = 87 mm so the force = 8.7 N
angle resultant makes with 4 N force = 36°

Algebraic approach

Vectors can act at any angle to each other but the most common situation that you are going to deal with is when they are at right angles to each other. We will deal with this first.

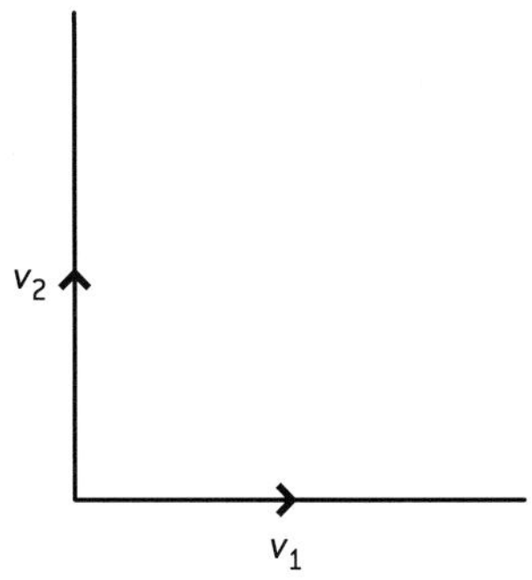

Adding vector quantities at right angles

Pythagoras' theorem can be used to calculate a resultant vector when two perpendicular vectors are added (or subtracted). Assuming that the two vector quantities are horizontal and vertical *but the principle is the same as long as they are perpendicular.*

Figure 5 shows two perpendicular velocities v_1 and v_2; they form a parallelogram that is a rectangle.

The magnitude of the resultant velocity $= \sqrt{v_1^2 + v_2^2}$

- The resultant velocity makes an angle θ to the horizontal given by

$$\tan\theta = \left(\frac{v_1}{v_2}\right) \text{ so that } \theta = \tan^{-1}\left(\frac{v_1}{v_2}\right)$$

- Notice that the order of adding the two vectors makes no difference to the length or the direction of the resultant.

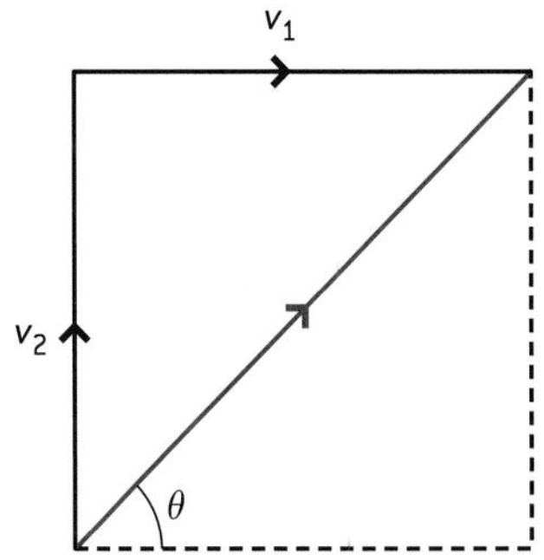

▲ Figure 5 Adding two perpendicular vectors.

Worked example

A walker walks 4.0 km due west from his starting point. He then stops before walking 3.0 km due north. At the end of his journey, how far is the walker from his starting point?

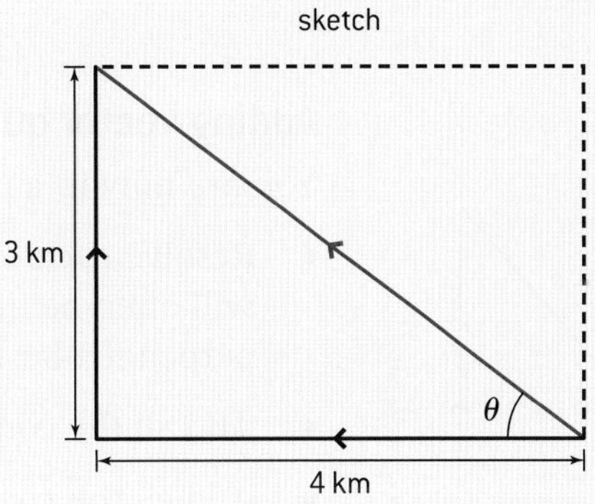

Solution

$$\textit{resultant} = \sqrt{4^2 + 3^2} = 5 \text{ km}$$

$$\text{angle } \theta = \tan^{-1}\left(\frac{3}{4}\right) = 36.9°$$

Before we look at adding vectors that are not perpendicular, we need to see how to resolve a vector – i.e. split it into two components.

Resolving vectors

We have seen that adding two vectors together produces a **resultant vector**. It is sensible, therefore, to imagine that we could split the resultant into the two vectors from which it was formed. In fact this is true for any vector – it can be divided into components which, added together, make the resultant vector. There is no limit to the number of vectors that can be added together and, consequently, there is no limit to the number of components that a vector can be divided into. However, we most commonly divide a vector into two components that are perpendicular to one another. The reason for doing this is that perpendicular vectors have no affect on each other as we will see when we look at projectiles in Topic 2.

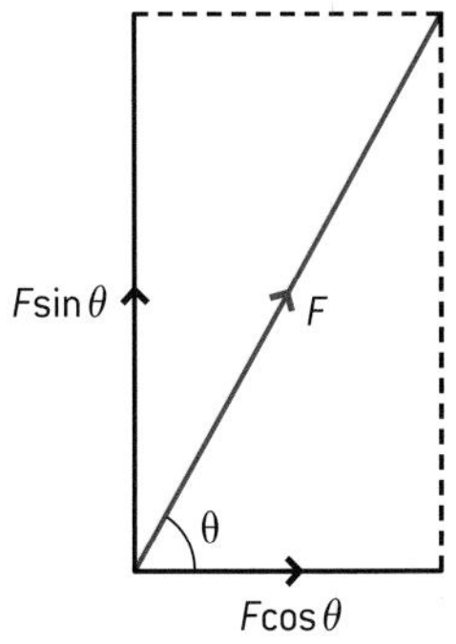

▲ Figure 6 Resolving a force.

The force F in figure 6 has been resolved into the horizontal component equal to $F\cos\theta$ and a vertical component equal to $F\sin\theta$. (The component opposite to the angle used is always the sine component.)

Worked example

An ice-hockey puck is struck at a constant speed of 40 m s^{-1} at an angle of 60° to the longer side of an ice rink. How far will the puck have travelled in directions **a)** parallel and **b)** perpendicular to the long side after 0.5 s?

Solution

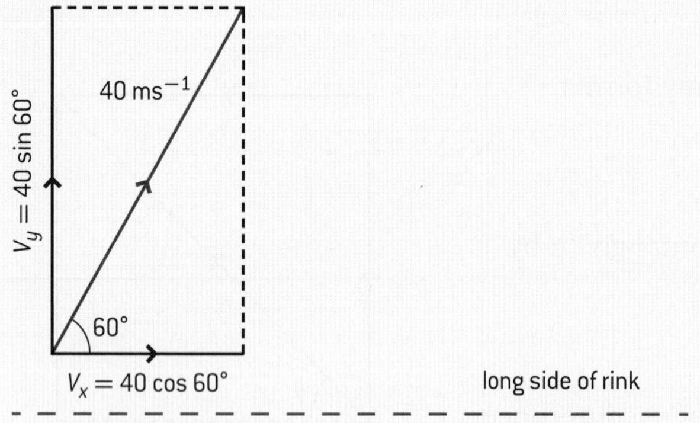

a) Resolving parallel to longer side:

$v_x = v \cos 60°$

$v_x = 40 \cos 60° = 20$ m s^{-1}

distance travelled $(x) = v_x t = 20 \times 0.5$
$= 10$ m

b) Resolving parallel to shorter side:

$v_x = v \sin 60°$

$v_y = 40 \cos 60° = 34.6$ m s^{-1}

distance travelled $(y) = v_y t = 34.6 \times 0.5$
$= 17$ m

Just to demonstrate that resolving is the reverse of adding the components we can use Pythagoras' theorem to add together our two components giving:

total speed $= \sqrt{20^2 + 34.6^2} = 39.96$ m s^{-1}

as the value for v_y was rounded this gives the expected 40 m s^{-1}

Adding vector quantities that are not at right angles

You are now in a position to add any vectors.

- Resolve each of the vectors in two directions at right angles – this will often be horizontally and vertically, but may be parallel and perpendicular to a surface.
- Add all the components in one direction to give a single component.
- Add all the components in the perpendicular direction to give a second single component.
- Combine the two components using Pythagoras' theorem, as for two vector quantities at right angles.

V_1 and V_2 are the vectors to be added.

Each vector is resolved into components in the *x* and *y* directions. Note that since the *x* component of V_2 is to the left it is treated as being negative (the *y* component of each vector is in the same direction... so upwards is treated as positive).

Total *x* component $V_x = V_{1x} + V_{2x} = V_1 \cos \theta_1 - V_2 \cos \theta_2$

Total *y* component $V_y = V_{1y} + V_{2y} = V_1 \sin \theta_1 + V_2 \sin \theta_2$

Having calculated V_x and V_y we can find the resultant by using Pythagoras so $V = \sqrt{V_x^2 + V_y^2}$

and the angle θ made with the horizontal $= \tan^{-1}\left(\frac{V_y}{V_x}\right)$.

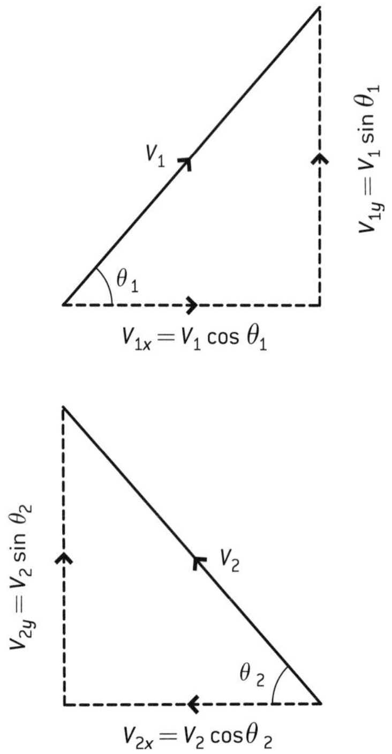

Figure 7 Finding the resultant of two vectors that are not perpendicular.

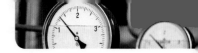

Worked example

Magnetic fields have strength 200 mT and 150 mT respectively. The fields act at 27° to one another as shown in the diagram.

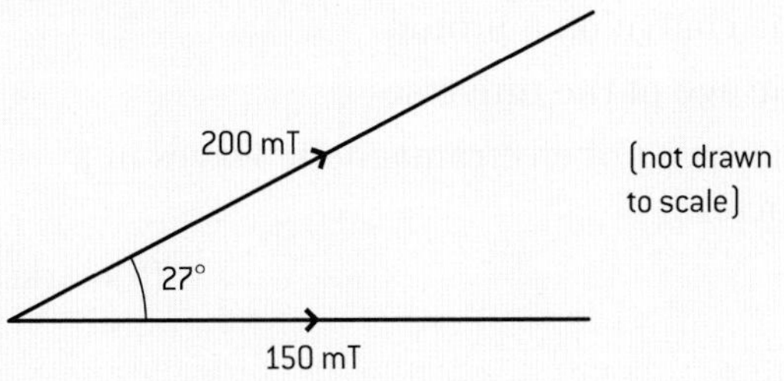

Calculate the resultant magnetic field strength.

Solution

The 150 mT field is horizontal and so has no vertical component.

Vertical component of the 200 mT field = 200 sin 27° = 90.8 mT

This makes the total vertical component of the resultant field.

Horizontal component of the 200 mT field = 200 cos 27° = 178.2 mT

Total horizontal component of resultant field = (150.0 + 178.2) mT = 328.2 mT

Resultant field strength $\sqrt{90.8^2 + 328.2^2} = 340$ mT

The resultant field makes an angle of:

$\tan^{-1}\left(\frac{90.8}{328.2}\right) = 15°$ with the150 mT field.

Subtraction of vectors

Subtracting one vector from another is very simple – you just form the negative of the vector to be subtracted and add this to the other vector. The negative of a vector has the same magnitude but the opposite direction. Let's look at an example to see how this works:

Suppose we wish to find the difference between two velocities v_1 and v_2 shown in figure 8.

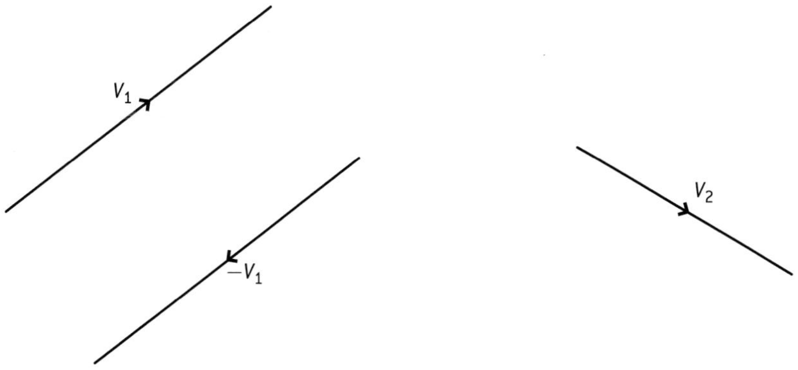

▲ Figure 8 Positive and negative vectors.

In finding the difference between two values we subtract the first value from the second; so we need $-v_1$. We then add $-v_1$ to v_2 as shown in figure 9 to give the red resultant.

The order of combining the two vectors doesn't matter as can be seen from the two versions in figure 9. In each case the resultant is the same – it doesn't matter where the resultant is positioned as long as it has the same length and direction it is the same vector.

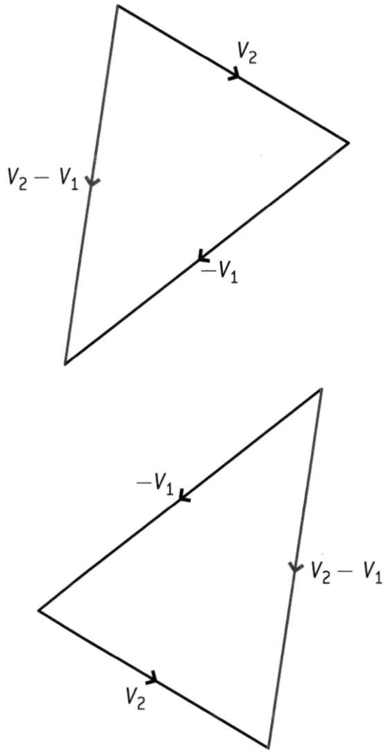

▲ Figure 9 Subtracting vectors.

Questions

1 Express the following units in terms of the SI fundamental units.

a) newton (N)
b) watt (W)
c) pascal (Pa)
d) coulomb (C)
e) volt (V)

(5 marks)

2 Express the following numbers to three significant figures.

a) 257.52
b) 0.002 347
c) 0.1783
d) 7873
e) 1.997

(5 marks)

3 Complete the following calculations and express your answers to the most appropriate number of significant figures.

a) 1.34×3.2
b) $\dfrac{1.34 \times 10^2}{2.1 \times 10^3}$
c) $1.87 \times 10^2 + 1.97 \times 10^3$
d) $(1.97 \times 10^5) \times (1.0 \times 10^4)$
e) $(9.47 \times 10^{-2}) \times (4.0 \times 10^3)$

(5 marks)

4 Use the appropriate metric multiplier instead of a power of ten in the following.

a) 1.1×10^4 V
b) 4.22×10^{-4} m
c) 8.5×10^{10} W
d) 4.22×10^{-7} m
e) 3.5×10^{-13} C

(5 marks)

5 Write down the order of magnitude of the following (you may need to do some research).

a) the length of a human foot
b) the mass of a fly
c) the charge on a proton
d) the age of the universe
e) the speed of electromagnetic waves in a vacuum

(5 marks)

6 a) Without using a calculator estimate to one significant figure the value of $\dfrac{2\pi 4.9}{480}$.

b) When a wire is stretched, the area under the line of a graph of force against extension of the wire gives the elastic potential energy stored in the wire. Estimate the energy stored in the wire with the following characteristic:

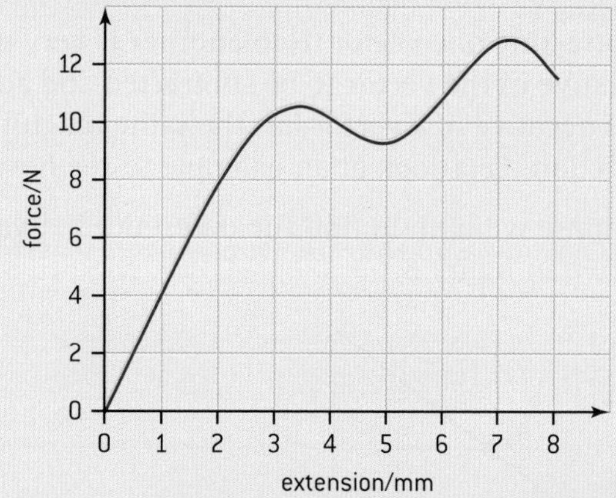

(4 marks)

7 The grid below shows one data point and its associated error bar on a graph. The *x*-axis is not shown. State the *y*-value of the data point together with its absolute and percentage uncertainty.

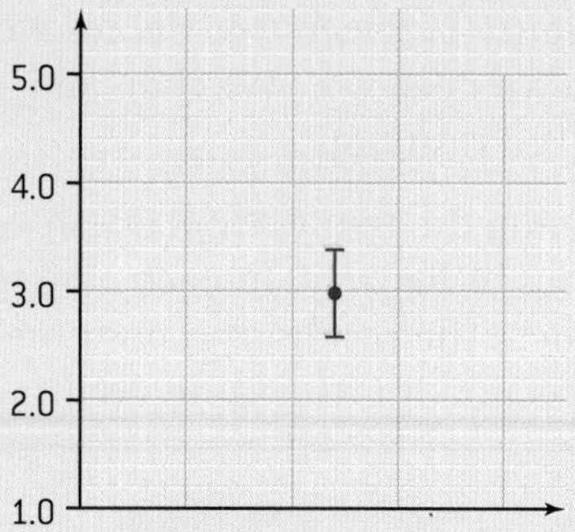

(3 marks)

8 A ball falls freely from rest with an acceleration g. The variation with time t of its displacement s is given by $s = \frac{1}{2}gt^2$. The percentage uncertainty in the value of t is $\pm 3\%$ and that in the value of g is $\pm 2\%$. Calculate the percentage uncertainty in the value of s.

(2 marks)

9 The volume V of a cylinder of height h and radius r is given by the expression $V = \pi r^2 h$. In a particular experiment, r is to be determined from measurements of V and h. The percentage uncertainty in V is $\pm 5\%$ and that in h is $\pm 2\%$. Calculate the percentage uncertainty in r.

(3 marks)

10 (*IB*)

At high pressures, a real gas does not behave as an ideal gas. For a certain range of pressures, it is suggested that for one mole of a real gas at constant temperature the relation between the pressure p and volume V is given by the equation

$pV = A + Bp$ where A and B are constants.

In an experiment, 1 mole of nitrogen gas was compressed at a constant temperature of 150 K. The volume V of the gas was measured for different values of the pressure p. A graph of the product pV against p is shown in the diagram below.

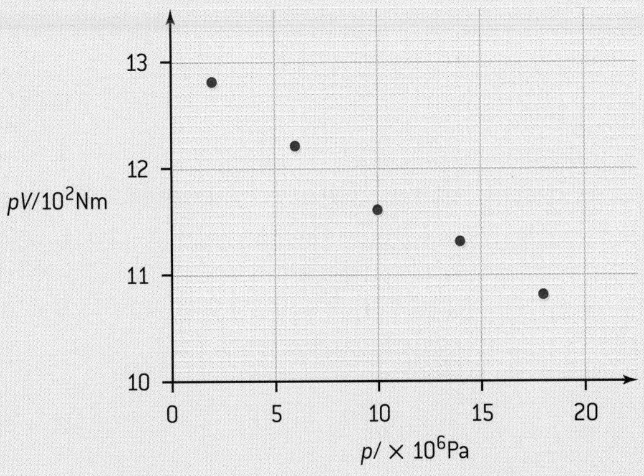

a) Copy the graph and draw a line of best fit for the data points.

b) Use your graph to determine the values of the constants A and B in the equation $pV = A + Bp$

c) p was measured to an accuracy of 5% and V was measured to an accuracy of 2%. Determine the absolute error in the value of the constant A.

(6 marks)

11 (*IB*)

An experiment was carried out to measure the extension x of a thread of a spider's web when a load F is applied to it.

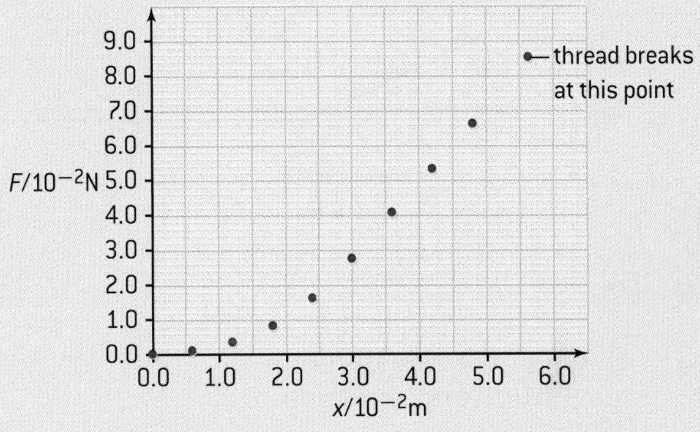

a) Copy the graph and draw a best-fit line for the data points.

b) The relationship between F and x is of the form

$$F = kx^n$$

State and explain the graph you would plot in order to determine the value n.

c) When a load is applied to a material, it is said to be *under stress*. The magnitude p of the stress is given by

$$p = \frac{F}{A}$$

where A is the cross-sectional area of the sample of the material.

Use the graph and the data below to deduce that the thread used in the experiment has a greater breaking stress than steel.

Breaking stress of steel $= 1.0 \times 10^9$ N m^{-2}

Radius of spider web thread $= 4.5 \times 10^{-6}$ m

d) The uncertainty in the measurement of the radius of the thread is $\pm 0.1 \times 10^{-6}$ m. Determine the percentage uncertainty in the value of the area of the thread.

(9 marks)

12 A cyclist travels a distance of 1200 m due north before going 2000 m due east followed by 500 m south-west. Draw a scale diagram to calculate the cyclist's final displacement from her initial position.

(4 marks)

13 The diagram shows three forces *P*, *Q*, and *R* in equilibrium. *P* acts horizontally and *Q* vertically.

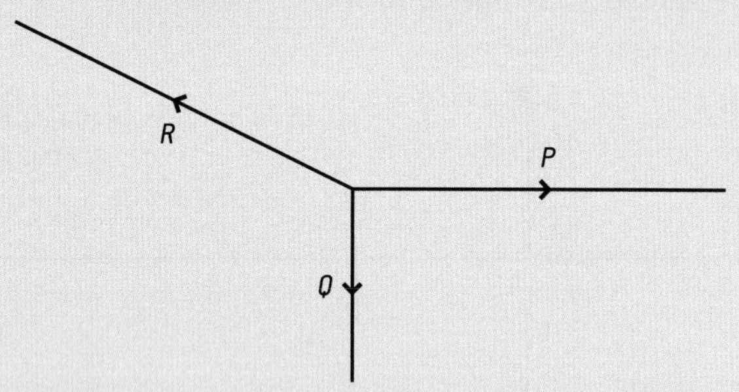

When $P = 5.0\,\text{N}$ and $Q = 3.0\,\text{N}$, calculate the magnitude and direction of *R*.

(3 marks)

14 A boat, starting on one bank of a river, heads due south with a speed of $1.5\ \text{m s}^{-1}$. The river flows due east at $0.8\ \text{m s}^{-1}$.

a) Calculate the resultant velocity of the boat relative to the bank of the river.

b) The river is 50 m wide. Calculate the displacement from its initial position when the boat reaches the opposite bank.

(7 marks)

15 A car of mass 850 kg rests on a slope at 25° to the horizontal. Calculate the magnitude of the component of the car's weight which acts parallel to the slope.

(3 marks)

2 MECHANICS

Introduction

Everything moves. The Earth revolves on its axis as it travels around the Sun. The Sun orbits within the Milky Way. Galaxies move apart.

Motion and its causes are important to a study of physics. We begin by defining the meaning of everyday terms such as distance, velocity, acceleration and go on to develop models for motion that will allow us to predict the future motion of an object.

2.1 Motion

Understanding

- → Distance and displacement
- → Speed and velocity
- → Acceleration
- → Graphs describing motion
- → Equations of motion for uniform acceleration
- → Projectile motion
- → Fluid resistance and terminal speed

Nature of science

An understanding of motion lies at the heart of physics. The areas of the subject are linked by the concepts of movement and the forces that produce motion. Links are used in a creative way by scientists to illuminate one part of the subject by reference to insights developed for other topics. The study of motion also relies on careful observation of the world so that accurate models of motion can be developed.

Applications and skills

- → Determining instantaneous and average values for velocity, speed, and acceleration
- → Solving problems using equations of motion for uniform acceleration
- → Sketching and interpreting motion graphs
- → Determining the acceleration of free-fall experimentally
- → Analysing projectile motion, including the resolution of vertical and horizontal components of acceleration, velocity, and displacement
- → Qualitatively describing the effect of fluid resistance on falling objects or projectiles, including reaching terminal speed

Equations

Kinematic equations of motion:

- → $v = u + at$
- → $s = ut + \frac{1}{2}at^2$
- → $v^2 = u^2 + 2as$
- → $s = \frac{(v + u)t}{2}$

Distance and displacement

Your journey to school is unlikely to take a completely straight line from home to classroom. How do we describe journeys when we turn a corner and change direction?

The map shows the journey a student makes to get to school together with the times of arrival at various points on the way. To keep life simple, the journey has no hills.

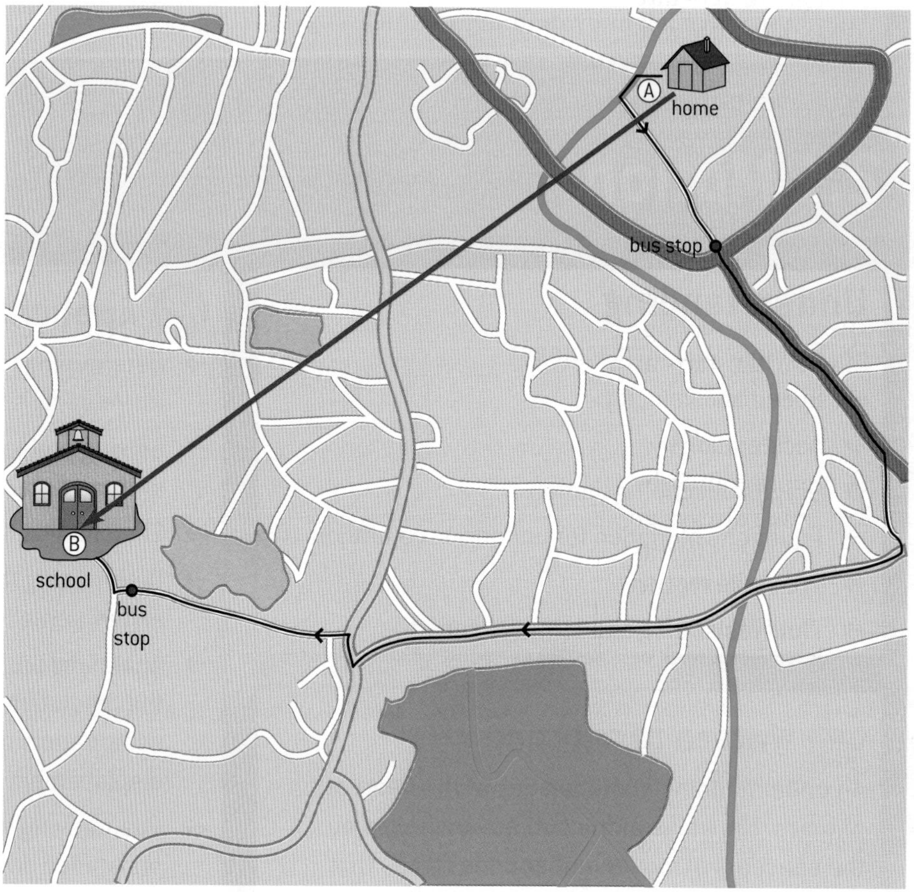

▲ Figure 1 Journey to school.

journey leg	time	distance for leg / m
leave home	08.10.00	0
walk to bus stop	08.20.15	800
bus arrives at stop	08.24.30	0
bus arrives near school	08.31.10	2400
walk from bus to school	08.34.00	200

The total length of this journey is 3.4 km including all the twists and turns. This is the distance travelled. As we saw in Topic 1, distance is a scalar quantity; it has no direction. If the student walks home by exactly the same route, then the distance travelled going home is the same as going to school.

Nature of science

How long is a piece of string?

Distance can be measured in any appropriate unit of length: metres, miles, and millimetres are all common. Some countries and some professions use alternatives. Surveyors use chains, the English-speaking world once used measures of length called rods, poles, and perches – all related to agricultural measurements. Astronomers use light years (a unit of length, not time) and a measure called simply the "astronomical unit". Sometimes in everyday life it is a question of using a convenient unit rather than the correct metric version. In your exam, however, lengths will be in multiples and sub-multiples of the metre or in a well-recognised scientific unit such as the light year.

Journeys can be defined by the starting point (A) and finishing point (B) without saying anything about the intermediate route. A vector is drawn that starts at A and finishes at B. This straight line from start to finish has, as a vector, *magnitude*, and *direction*. This vector measurement is known as the **displacement**.

As you saw in Sub-topic 1.3, when you give the details of a displacement you must always give two pieces of information: the magnitude (or size) of the quantity together with its unit, and also the direction of the vector. This direction can be written in a number of ways. One way is as a heading such as N35°E meaning that the finishing direction is at an angle of 35° clockwise from north. If you are a sailor or a keen orienteer, you may have other ways of measuring the direction of displacement.

The displacement of the student's journey to school (AB) is also shown on the diagram. The displacement is the vector that connects home to school. This time the journey from school to home is not the same as the trip to school. It has the same vector length, but the direction is the opposite to that of the outward route.

Nature of science

Moving in 3D

Displacement has been described here in terms of a journey in a flat landscape. Does a change in level alter things? Only one thing changes, and that is the number of pieces of information required to specify the final position relative to the start. Three pieces of information are now required: the magnitude plus its unit, the heading, and the *overall change in height* during the journey. Specifying motion in three dimensions requires three numbers or coordinates, you are already familiar with the idea of a coordinate from drawing and using graphs.

There is flexibility in how the three numbers can be chosen. You may have seen three-dimensional graphs with three axes each at 90° to the others, in this case coordinate numbers are given that relate to the distance along each axis. Another option is to use polar coordinates (figure 2): where a distance and two angles are required, one angle is the bearing from North, the other the angle up or down from the horizontal needed to look directly at the object above (or below) us.

In some circumstances, even the distance itself may not be required. Sailors use latitude and longitude when they are navigating. They stay on the surface of the sea and this is effectively a constant distance from the centre of the Earth.

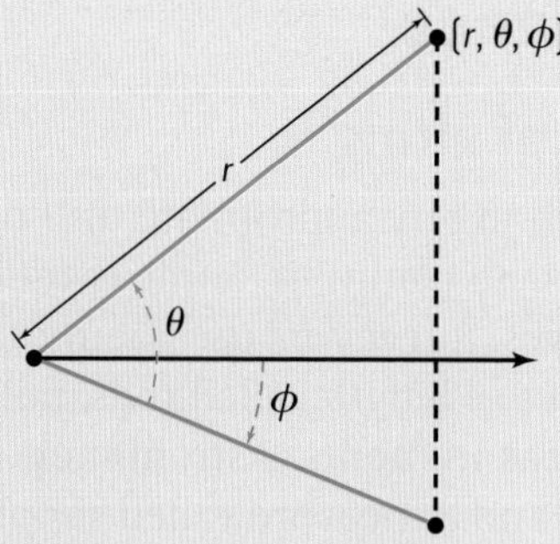

▲ Figure 2 Polar coordinates.

Worked examples

1 A cyclist travels 16 km in 70 minutes. Calculate, in m s^{-1}, the speed of the cyclist.

Solution

70 minutes is 60 × 70 = 4200 s, 16 km is 16 000 m. The quantities are now in the units required by the question.

The speed of the cyclist is $\frac{16\,000}{4200} = 3.8$ m s^{-1}.

2 The speed of light in a vacuum is 3.0×10^8 m s^{-1}. A star is 22 light years from Earth (1 light year is the distance travelled by light in one year). Calculate the distance of the star from Earth in kilometres.

Solution

Light travels 3.0×10^8 m in 1 s. So in a year it travels $3.0 \times 10^8 \times 365 \times 24 \times 60 \times 60 = 9.5 \times 10^{15}$ m. The distance of the star from the Earth is $22 \times 9.5 \times 10^{15} = 2.1 \times 10^{17}$ m.

The question asks for an answer in kilometres, so the distance is 2.1×10^{14} km.

Speed and velocity

In just the same way that there are scalar and vector measures of the length of a journey, so there are two ways of measuring how quickly we cover the ground.

The first of these is a scalar quantity **speed**, which is defined as $\frac{\text{distance travelled on the journey}}{\text{time taken for the journey}}$. Units for speed, familiar to you already, may include metre per second (m s^{-1}) and kilometres per hour (km h^{-1}), but any accepted distance unit can be combined with any time unit to specify speed.

Velocity is the vector term and, just as for displacement, a magnitude and direction are required. Examples might be "4.2 m s^{-1} due north" or "55 km h^{-1} at N22.5°E".

Nature of science

Measuring speed?

To measure speed it is necessary to measure the the distance travelled (using a "ruler") and time taken (using a "clock"). The trick is in a good choice of "ruler", "clock" and method for recording the measurements! A 30 cm ruler and a wrist watch will be fine for a biologist measuring the speed at which an earthworm moves. But if we want to measure the speed of a 100 m sprinter, then a measured distance on the ground, a good stop watch and a human observer is barely good enough. Even then the observer has to be careful to watch the smoke from the starting pistol and not to wait for the sound of the gun.

If you need to measure the speed of a soccer ball during a penalty kick, then stop watch-plus-human is no longer adequate to make a valid measurement. Perhaps a video camera that takes frames at a known rate (the clock) and a scale near the path of the ball visible on the picture (the ruler) is needed now?

Move up to measuring the speed of a jet aircraft and the equipment needs to change again.

Choosing the right equipment for the task in hand is all part of the job of the working science student.

Describing motion with a graph – I

The use of two variables, distance and time, to calculate speeds and velocities means that things become more complicated. Figure 1 showed a map of the journey the student makes to school. Part of this journey is on foot, part is by bus. It is unlikely that the student will travel at the same speed all the time, as the bus will travel faster than the student walks.

If we want to display data in a visual way, a graph of distance against time is one of the most common approaches.

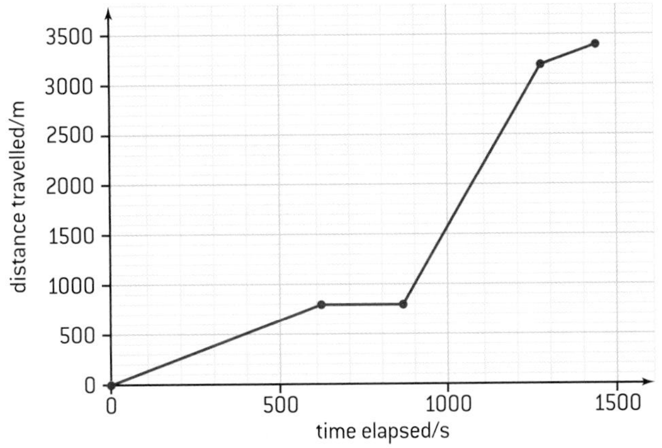

▲ Figure 3 Distance–time graph.

The distance travelled by the student is plotted on the vertical axis while the time since the beginning of the journey is plotted horizontally (the clock times of the journey have been translated into elapsed times since the start of the journey for this graph). The different regions of the graph are identified and you should confirm that they are matched correctly. Notice how the gradient of the graph changes for the different parts of the journey: small values of the gradient for the walking sections of the journey, horizontal (zero gradient) for stationary at the bus stop, and steep for the bus journey. What will the graph for the journey home from school look like, assuming that the time for each segment of the journey is the same as in the morning?

Information can easily be extracted from this graph. The gradient of the graph is the speed. Add the overall direction to this speed and we have the velocity too.

For the first walk to the bus stop the distance was 800 m and the time taken was 615 s. The constant walking speed was therefore $\frac{800}{615}$, which is 1.3 m s^{-1}.

The gradient of the bus journey is $\frac{2400}{400}$ (this is marked on the graph) and so the speed was 6.0 m s^{-1}.

Of course, even this journey with its changes is simplistic. Real journeys have few straight lines, so we must introduce some ways to handle rapidly varying speeds and velocities.

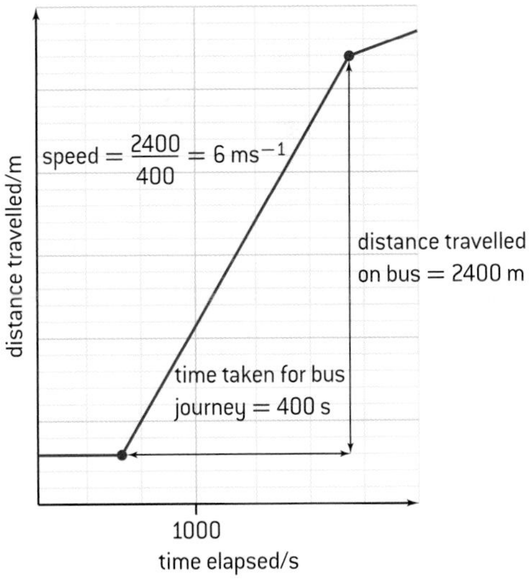

▲ Figure 4 Distance–time graph from figure 3 with gradients.

Instantaneous and average values

The bus driver knows how fast the bus in the student's journey is travelling because it is displayed on the speedometer. This is the **instantaneous speed**, as it gives the value of the speed at the moment in time at which the speed is determined.

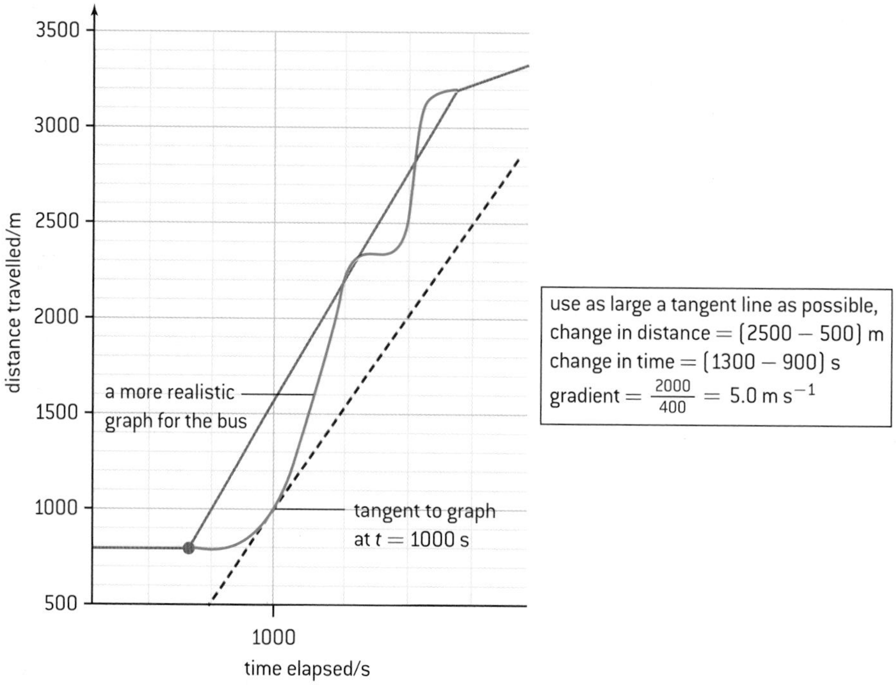

▲ Figure 5 Instantaneous speed.

Nature of science

Calculating a gradient of a graph

Taking the gradient of a graph is actually a way of averaging results. As discussed in Topic 1, a straight-line graph drawn through points that have some scatter makes this obvious. The value of the gradient is the $\frac{\text{change in the value on the } y\text{-axis}}{\text{change in the value on the } x\text{-axis}}$. The unit associated with the gradient is the $\frac{y\text{-axis unit}}{x\text{-axis unit}}$. Don't forget to quote the unit every time you write down the value of the gradient.

The instantaneous speed is also the gradient of the distance–time graph at the instant concerned. Figure 5 shows how this is calculated when 1000 s into the whole journey to school. The original red line for the bus from figure has been replaced by a green line that is more realistic for the motion of a real bus – the speed varies as the driver negotiates the traffic. You will frequently be asked to calculate gradients on physics graphs. Make sure that you can do this accurately by using a transparent ruler.

From a mathematical point of view we can describe the instantaneous speed as the **rate of change of position with respect to time**.
A mathematician will write this as $\frac{ds}{dt}$, where s is the distance and t is the time. You may also have seen $\frac{\Delta s}{\Delta t}$, where the symbol Δ means *change in*.
So $\frac{\Delta s}{\Delta t}$ is just shorthand for $\frac{\textit{change in distance}}{\textit{change in time}}$.

There is however another useful measure of speed. This is the **average speed** and is the speed calculated over the whole the journey without regard to variations in speed. So as an equation this is

$$\textit{average speed} = \frac{\text{distance travelled over whole journey}}{\text{time taken for whole journey}}$$

In terms of the distance–time graph, the average speed is equal to the gradient of the straight line that joins the beginning and the end of the time interval concerned. So, for the part of the student's journey up to the moment when the bus arrives at the stop, the distance travelled is 800 m, the time taken is 870 s (*including* the wait at the stop) so the average speed is 0.92 m s^{-1}.

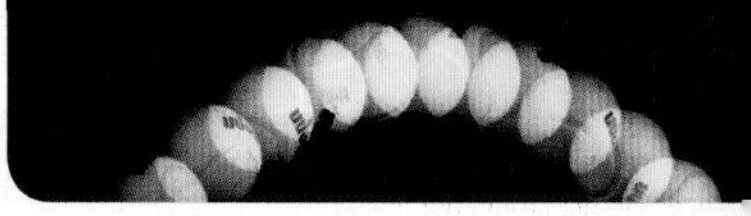

Everything said here about average and instantaneous speeds can also refer to average and instantaneous velocities. Remember, of course, to include the directions when quoting these measurements.

Worked example

The graph shows how the distance run by a boy varies with the time since he began to run.

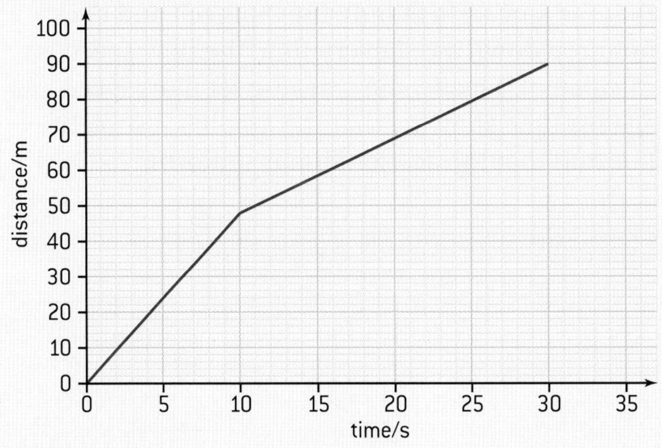

Calculate:

a) the instantaneous speed at
(i) 5.0 s (ii) 20 s

b) the average speed for the whole 30 s run.

Solution

a) (i) The boy runs at constant speed so the graph is straight from 0 – 10 s.

The gradient of the straight line is $\frac{48}{10} = 4.8\ \text{m s}^{-1}$.

So the instantaneous speed at 5.0 s is $4.8\ \text{m s}^{-1}$.

(ii) Again the boy is running at a constant speed, but this time slower than in the first 10 s.

The speed from 10 s to 30 s is:

$$\frac{(90 - 48)}{(30 - 10)} = \frac{42}{20} = 2.1\ \text{m s}^{-1}.$$

The instantaneous speed at 20 s is $2.1\ \text{m s}^{-1}$.

b) The total distance travelled in 30 s is 90 m, so the average speed is $\frac{90}{30} = 3.0\ \text{m s}^{-1}$.

Acceleration

In real journeys, instantaneous speeds and velocities change frequently. Again we need to develop a mathematical language that will help us to understand the changes.

The quantity we use is **acceleration**. Acceleration is taken to be a vector, but sometimes we are not interested in the direction and then write the "magnitude of the acceleration" meaning the size of the acceleration ignoring direction.

The definition is: $\textit{acceleration} = \frac{\text{change in velocity}}{\text{time taken for the change}}$

and this means that the units of acceleration are $\frac{\text{m s}^{-1}}{\text{s}}$ which is usually written as m s^{-2} (or sometimes you will see m/s^2).

It is important to understand what acceleration means, not just to be able to use it in an equation. If an object has an acceleration of $5\ \text{m s}^{-2}$ then for every second it travels, its velocity increases in magnitude by $5\ \text{m s}^{-1}$ in the direction of the acceleration vector.

As an example: the Japanese N700 train has a quoted acceleration of $0.72\ \text{m s}^{-2}$. Assume that this is a constant value (very unlikely). One second after starting from rest, the speed of the train will be $0.72\ \text{m s}^{-1}$. One second later (at 2 s from the start) the speed will be $0.72 + 0.72 = 1.44\ \text{m s}^{-1}$. At 3 s it will be $2.16\ \text{m s}^{-1}$ and so on. Each second the speed is $0.72\ \text{m s}^{-1}$ more.

Worked example

How many seconds will it take the N700 to reach its maximum speed of $300\ \text{km h}^{-1}$ on the Sanyo Shinkansen route?

Solution

$300\ \text{km h}^{-1} \equiv = 83.3\ \text{m s}^{-1}$

Time taken to reach the maximum speed:

$\frac{83.3}{0.72} = 116$ s,

just under 2 minutes.

Nature of science

Spreadsheet models

One powerful way to think about acceleration (and other quantities that change in a predictable way) is to model them using a spreadsheet. The examples here use a version of Microsoft Excel© but any computer spreadsheet can be used for this.

This is a spreadsheet model for the N700 train. The value of the acceleration is in cell B1. Cells A4 to A29 give the time in increments of 5 s; the computed speed at each of these times is in cells B4 to B29. The speed is calculated by taking the *change* in time between the present cell and the one above it, and then multiplying by the acceleration (the acceleration is written as \$B\$1 so that the spreadsheet only uses this cell and does not drop down a cell every time the new speed is calculated). Finally, the spreadsheet plots speed against time showing that the graph is a straight line and that the acceleration is uniform.

	A	B
1	acceleration/s	0.72
2		
3	time/s	speed/ms^{-1}
4	0	0
5	5	3.6
6	10	7.2
7	15	10.8
8	20	14.4
9	25	18.0
10	30	21.6
11	35	25.2
12	40	28.8
13	45	32.4
14	50	36.0
15	55	39.6
16	60	43.2
17	65	46.8
18	70	50.4
19	75	54.0
20	80	57.6
21	85	61.2
22	90	64.8
23	95	68.4
24	100	72.0
25	105	75.6
26	110	79.2
27	115	82.8
28	120	86.4
29	125	90.0

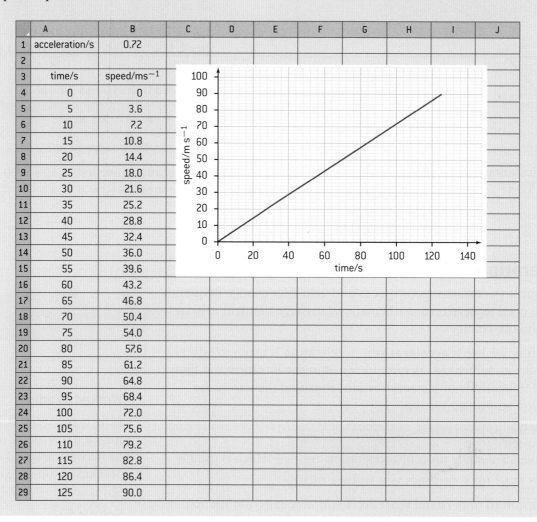

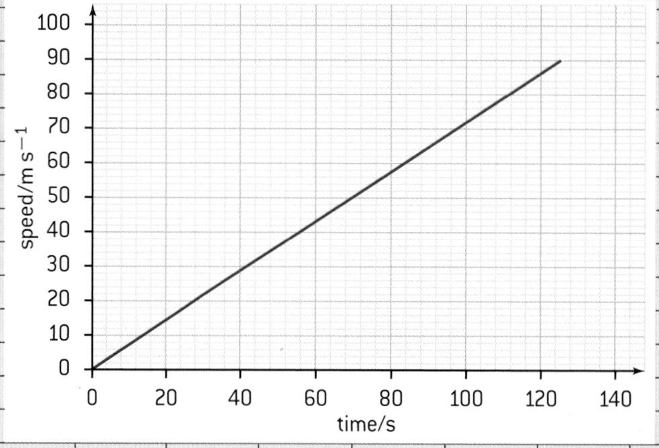

Describing motion with a graph – II

Distance–time plots lead to a convenient display of speed and velocity changes. Plots of speed (or velocity) against time also help to display and visualize acceleration.

The data table and the graph show a journey with various stages on a bicycle. From the start until 10 s has elapsed, the bicycle accelerates at a uniform rate to a velocity of +4 m s^{-1}. The positive sign means that the velocity is directed to the right.

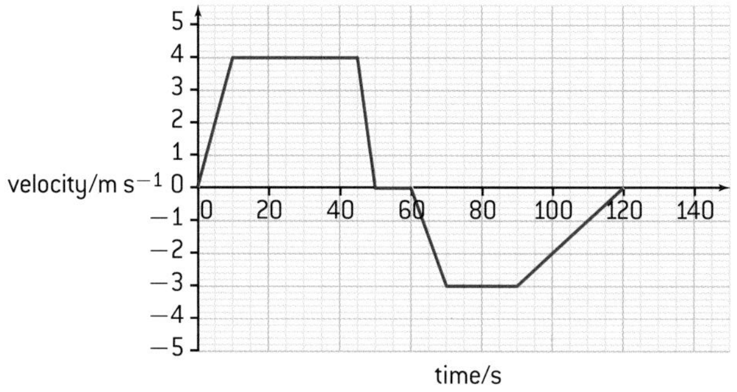

▲ Figure 6 Velocity–time graph for the bicycle.

From 10 s to 45 s the cycle moves at a constant velocity of +4 m s^{-1} and at 45 s the cyclist applies the brakes so that the cycle stops in 5 s. The cycle is then stationary for 10 s.

From then on the velocity is negative meaning that the cycle is travelling in the opposite direction. The pattern is similar, an acceleration to −3 m s^{-1}, a period of constant velocity and a deceleration to a stop at 120 s.

As before, the gradient of the graph has a meaning. The gradient of this velocity–time graph gives the magnitude (size) of the acceleration and its sign (direction) as well.

From 45 s to 50 s the velocity goes from 4 m s^{-1} to 0 and so the acceleration is $\frac{\text{final speed} - \text{initial speed}}{\text{time taken}}$ $\frac{(0-4)}{5} = -0.8$ m s^{-2}. From 90 s to 120 s the magnitude of the acceleration is $\frac{3}{30} = 0.1$ m s^{-2}. We need to take care with the sign of the acceleration here. Because the cycle is moving in the negative direction and is slowing down, the acceleration is positive (as is the gradient on the graph) – this simply means that there is a force acting to the right, that is, in the positive direction which is slowing the cycle down. We will discuss how force leads to acceleration later in this topic.

The area under a velocity–time graph gives yet more information. It tells us the total displacement of the moving object. The way to see this is to realize that the product of *velocity* × *time* is a displacement (and that product of *speed* × *time* is a distance). The units tell you this too: when the units are multiplied the seconds in $\frac{\text{metre}}{\text{second}}$ × ~~second~~ cancel to leave metre only.

In the case of a graph with uniform acceleration, the areas, and hence the displacements (distances) are straightforward to calculate. Divide the graph into right-angled triangles and rectangles and then work out the areas for each individual part. This is shown in figure 7.

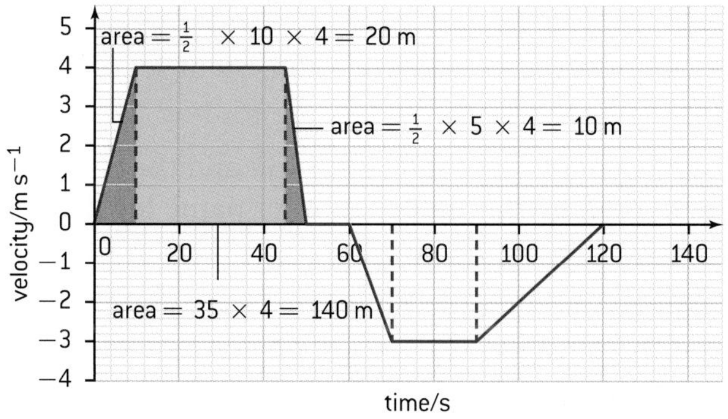

▲ Figure 7 Velocity-time graph broken down into areas.

The individual areas are shown on the diagram and for the motion up to a time of 60 s the area is 170 m; the area from the 60 s time to the end is −120 m. As usual, the negative sign indicates motion in the opposite direction to the original.

As discussed in Topic 1, when the velocity–time graph is curved, you will need to:

(i) estimate the number of squares

(ii) assess the area (distance) for one square, and finally

(iii) multiply the number of squares by the area of one square.

This will usually give you an estimate of the overall distance.

Figure 8 gives an example of how this is done.

There are about 85 squares between the *x*-axis and the line. (You may disagree slightly with this estimate, but that is fine – there is always an allowance made for this.)

Each of the squares is 2 s along the time axis and 0.5 m along the speed axis. So the area of one square is equivalent to 1.0 m of distance. The total distance travelled is 85 m (or, at least, somewhere between 80 and 90 m).

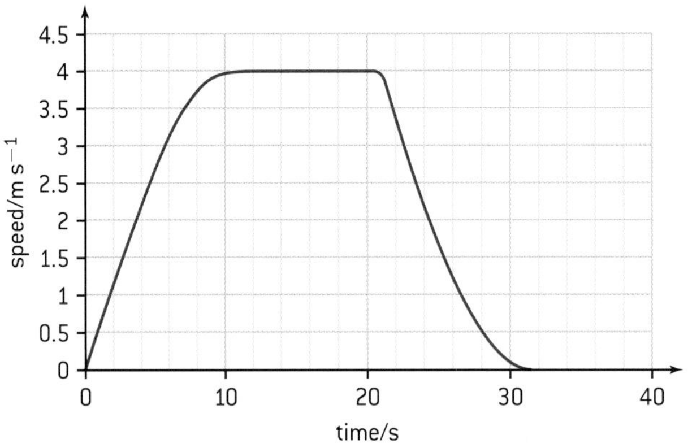

▲ Figure 8 A more difficult area to estimate.

The "*suvat*" equations of motion

The graphs of distance–time and speed–time lead to a set of equations that can be used to predict the value of the parameters in motion. They also help you to understand the connection between the various quantities in your study of motion.

A note about symbols: from now on we will use a consistent set of symbols for the quantities. The table gives the list.

symbol	quantity
s	displacement/distance
u	initial (starting) velocity/speed
v	final velocity/speed
a	acceleration
t	time taken to travel the distance *s*

If you read down the list of symbols they spell out *suvat* and the equations are sometimes known by this name, another name for them is the *kinematic* equations of motion.

The derivation of the equations of uniformly accelerated motion begins from a simple graph of speed against time for a constant acceleration from velocity *u* to velocity *v* in a time *t*.

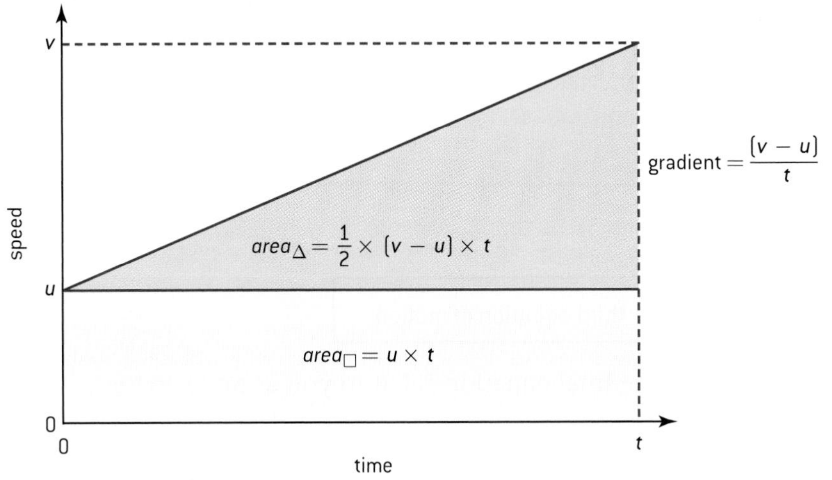

▲ Figure 9 Deriving the first two equations of motion.

The acceleration is the gradient of the graph:

$$\frac{\text{change in speed}}{\text{time taken for change}}$$

The change in speed is $v - u$, the time taken is t.

Therefore,

$$a = \frac{v-u}{t}$$

and re arranging gives

$v = u + at$	first equation of motion

The area under the speed–time graph from 0 to t is made up of two parts, the lower rectangle, $area_{\square}$ and the upper right-angled triangle, $area_{\Delta}$.

$$area_{\Delta} = \frac{1}{2}\,base \times height = \frac{1}{2}t(v-u)$$

$$area_{\square} = base \times height = ut$$

$$s = \text{total area} = area_{\Delta} + area_{\square} = ut + \frac{1}{2}(v-u)t = ut + \frac{1}{2}(at)t$$

So

$s = ut + \frac{1}{2}at^2$	second equation of motion

The first equation has no s in it; the second has no v. There are three more equations, one with a missing t and one with a missing a. There is one equation that has a missing u, but this is not often used.

To eliminate t from the first and second equations, re arrange the first in terms of t:

$$t = \frac{v-u}{a}$$

This can be substituted into the second equation:

$$s = u\frac{v-u}{a} + \frac{1}{2}a\left(\frac{v-u}{a}\right)^2$$

and

$$as = u(v-u) + \frac{1}{2}(v-u)^2 = uv - u^2 + \frac{1}{2}v^2 + \frac{1}{2}u^2 - \frac{1}{2}2uv$$

which gives

$$2as = v^2 - u^2$$

or

$v^2 = u^2 + 2as$	third equation of motion

The derivation of the final equation is left to you as an exercise:

$s = \left(\frac{v+u}{2}\right)t$	fourth equation of motion

Remember!

These equations only apply if the acceleration is *uniform*. In other words, acceleration must not change during the motion.

There are two ways to approach this proof. One way is to think about the meaning of the speed that corresponds to $\frac{v+u}{2}$ and then to recognize the time at which this speed occurs in the motion. The second way is to take the third equation and amalgamate it with the first.

You will not be expected to remember these proofs or the equations themselves (which appear in the data booklet), but they do illustrate how useful graphs and equations can be when solving problems in kinematics.

Worked examples

1 A driver of a car travelling at 25 m s^{-1} along a road applies the brakes. The car comes to a stop in 150 m with a uniform deceleration. Calculate **a)** the time the car takes to stop, and **b)** the deceleration of the car.

Solution

a) One way to answer kinematic equation questions is to begin by writing down what you do and don't know from the question.

$s = 150$ m; $u = 25$ m s^{-1}; $v = 0$; $a = ?$; $t = ?$

To work out t, the fourth equation is required: $s = \left(\frac{v+u}{2}\right)t$

which rearranges to $t = \left(\frac{2}{v+u}\right)s$

Substituting the values gives

$t = \left(\frac{2}{25}\right) 150 = 2 \times 6 = 12$ s

b) To find a the equation $v^2 = u^2 + 2as$ is best.

Substituting:

$0 = 25^2 + 2 \times a \times 150$

$a = -\frac{25 \times 25}{300} = -\frac{25}{12} = -2.1$ m s^{-2}.

The minus sign shows that the car is decelerating rather than accelerating.

2 A cyclist slows uniformly from a speed of 7.5 m s^{-1} to a speed of 2.5 m s^{-1} in a time of 5.0 s.

Calculate **a)** the acceleration, and **b)** the distance moved in the 5.0 s.

Solution

a) $s = ?$; $u = 7.5$ m s^{-1}; $v = 2.5$ m s^{-1}; $a = ?$; $t = 5.0$ s

Use $v = u + at$ and therefore $2.5 = 7.5 + a \times 5.0$

so, $a = -\frac{5.0}{5.0} = -1.0$ m s^{-2}

The negative sign shows that this is a deceleration.

b) $s = ut + \frac{1}{2}at^2$ so

$s = 7.5 \times 5.0 - \frac{1}{2} \times 1.0 \times 5.0^2$

$= 37.5 - 12.5$

$= 25$ m

Projectile motion

Falling freely

When an object is released close to the Earth's surface, it accelerates downwards. We say that the force of gravity acts on the object, meaning that it is pulled towards the centre of the Earth. Equally the object pulls with the same force on the Earth in the opposite direction. Not surprisingly, with small objects, the effect of the force on the Earth is so small that we do not notice it.

In Topic 6, we shall look in more detail at the effects of gravity but for the moment we assume that there is a constant acceleration that acts on all bodies close to the surface of the Earth.

The acceleration due to gravity at the Earth's surface is given the symbol g. The accepted value varies from place to place on the surface, so that at Kuala Lumpur g is 9.776 m s^{-2} whereas at Stockholm it is 9.818 m s^{-2}. Reasons for the variation include variations in the shape of the Earth (it is not a perfect sphere, being slightly flattened at the poles) and effects that are due to the densities of the rocks in different locations. The different tangential speeds of the Earth at different latitudes also have an effect. It is better to buy gold by the newton at the equator and sell it at the North Pole than the other way round!

Investigate!

Measuring g

Alternative 1

- There are a number of ways to measure g. This method uses a data logger to collect data. One of the problems with measuring g "by hand" is that the experiment happens quickly. Manual collection of the data is difficult.
- An ultrasound sensor is mounted so that it senses objects below it.
- Set the logging system up to measure the speed of the object over a time of about 1 s. Set a sensible interval between measurements.
- Switch on the logger system, and drop an object vertically that is large enough for the sensor to detect.
- The output from the system should be a speed–time graph that is a straight line; it may be that the logger's software can calculate the gradient for you.
- You could extend this experiment by testing objects of different mass but similar size and shape to confirm a suggestion by Galileo that such differences do not affect the drop.

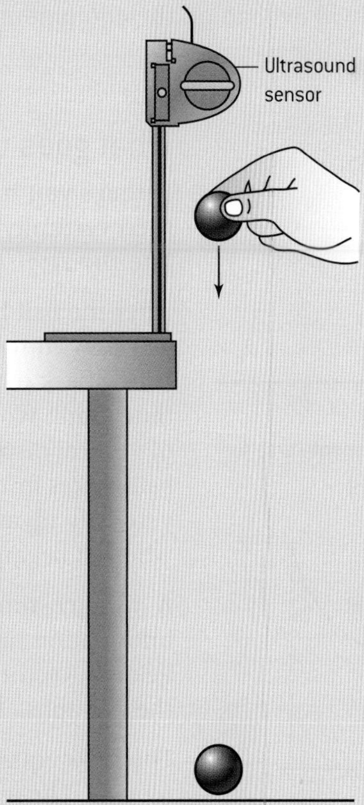

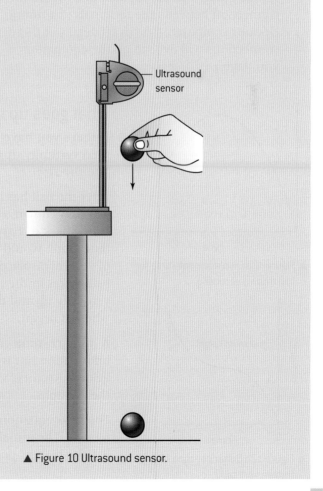

▲ Figure 10 Ultrasound sensor.

Alternative 2

There are other options that do not involve a data logger.

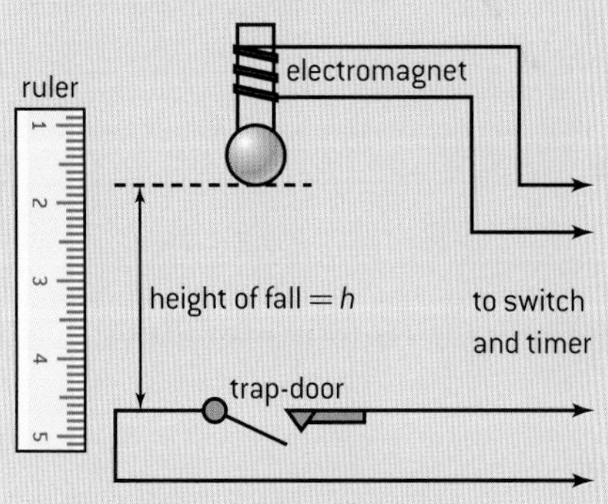

▲ Figure 11 Trap door method.

- A magnetic field holds a small steel sphere (such as a ball bearing) between two metal contacts. The magnetic field is produced by a coil of wire with an electric current in it. When the current is switched off, the field disappears and the sphere is released to fall vertically.
- As the sphere leaves the metal contacts, a clock starts. The clock stops when the sphere opens a small trapdoor and breaks the connection between the terminals of a timing clock or computer. (The exact details of these connections will depend on the equipment you have.)
- This system measures the time of flight t of the sphere from the contacts to the trapdoor.
- Measure the distance h from the bottom of the sphere to the top of the trapdoor (you might think about why these are the appropriate measurement points).
- A possible way to carry out the experiment is to measure t for one value of h – with, of course, a few repeat measurements for the same h. Then use $h = ut + \frac{1}{2}gt^2$ with $u = 0$ to calculate g.
- This is a one-off measurement that is prone to error. Can you think of some reasons why? One way to reduce the errors is to change the vertical distance h between the sphere and trapdoor and to plot a graph of h against t^2. The gradient of the graph is $\frac{g}{2}$. If you observe an intercept on the h-axis, what do you think it represents?

Alternative 3

- There is a further method that involves taking a video or a multiflash image of a falling object and analysing the images to measure g. You will see an example of such an analysis later in this sub-topic.

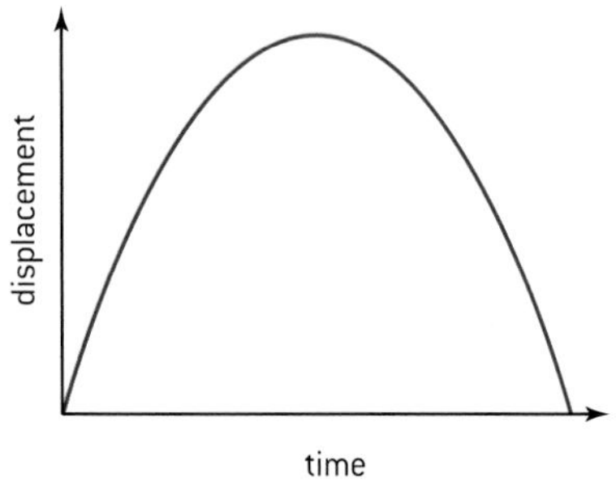

▲ Figure 12a Displacement–time for ball thrown vertically.

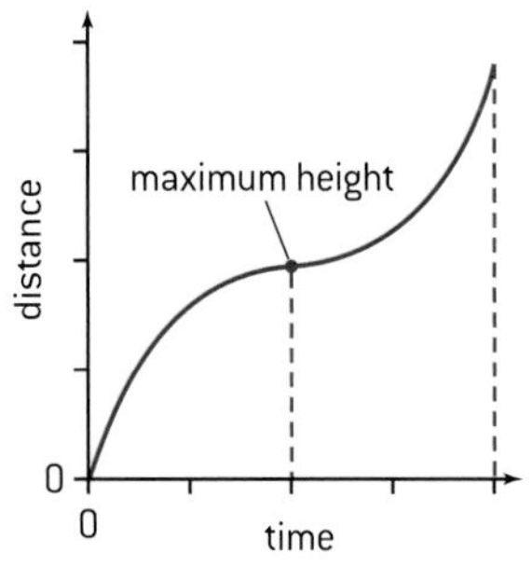

▲ Figure 12b Distance–time for ball thrown vertically.

What goes up must come down

Perhaps you have seen a toy rocket filled with water under pressure and fired vertically upwards? Or you may have thrown a ball vertically, high into the air?

After the ball has been released, as the pull of gravity takes effect, the ball slows down, eventually stopping at the top of its motion and then falling back to Earth. If there was no air resistance the displacement–time graph would look like figure 12a. Remember that this is a graph of vertical displacement against time, not the shape of the path the ball makes in the air – which is called the **trajectory**. The ball is going vertically up and then vertically down to land in the same spot from where it began.

A distance–time graph would look different (figure 12b), it gives similar information but without the direction part of the displacement and velocity vectors. Make sure that you understand the difference between these graphs.

The *suvat* equations introduced earlier can be used to analyse this motion. The initial vertical speed is u, the time to reach the highest point is t, the highest point is h and the acceleration of the rocket is $-g$. The sign of g is negative because upwards is the positive direction. As the acceleration due to gravity is downwards, g must have the opposite sign. The kinematic equations are printed again but with differences to reflect the vertical motion to the highest point:

$0 = u - gt$ which comes from $v = u + at$

$h = ut - \frac{1}{2}gt^2$ which comes from $s = ut + \frac{1}{2}at^2$

$0 = u^2 - 2gh$ which comes from $v^2 = u^2 + 2as$

If you want to find out the time for the entire motion (that is, up to the highest point and then back to Earth again), it is simply $2t$.

Reminder!

Try to remember this crucial point about the signs in the equations when you answer questions on vertical motion: upwards is +ve and downwards is −ve. Something else that students forget is that at the top of the motion the vertical speed of the rocket is zero.

Worked examples

1 A student drops a stone from rest at the top of a well. She hears the stone splash into the water at the bottom of the well 2.3 s after releasing the stone. Ignore the time taken for the sound to reach the student from the bottom of the well.

a) Calculate the depth of the well.

b) Calculate the speed at which the stone hits the water surface.

c) Explain why the time taken for the sound to reach the student can be ignored.

Solution

a) The acceleration due to gravity g is 9.8 m s^{-2}.

$u = 0$; $t = 2.3$ s.

$s = ut + \frac{1}{2}at^2$ and therefore

$s = 0 + \frac{1}{2} \times 9.8 \times 2.3^2$

$= 26$ m

b) $v = u + at$; $v = 0 + 9.8 \times 2.3 = 23$ m s^{-1}

c) The speed of sound is about 300 m s^{-1} and so the time to travel about 25 m is about 0.08 s. Only about 4% of the time taken for the stone to fall.

2 A hot-air balloon is rising vertically at a constant speed of 5.0 m s^{-1}. A small object is released from rest relative to the balloon when the balloon is 30 m above the ground.

a) Calculate the maximum height of the object above the ground.

b) Calculate the time taken to reach the maximum height.

c) Calculate the total time taken for the object to reach the ground.

Solution

a) The object is moving upwards at +5.0 m s^{-1} when it is released. The acceleration due to gravity is −9.8 m s^{-2}.

When the object is released it will continue to travel upwards but this upward speed will decrease under the influence of gravity. When it reaches its maximum height it will stop moving and then begin to fall.

$v^2 = u^2 + 2as$ and $s = \frac{0 - 5^2}{-2 \times 9.8} = +1.3$ m

This shows that the object rises a further 1.3 m above its release point, and is therefore 31.3 m above the ground at the maximum height.

b) $v = u + at$; $t = \frac{0 - 5}{-9.8} = +0.51$ s (the plus sign shows that this is 0.51 s after release)

c) After reaching the maximum height (at which point the speed is zero) the object falls with the acceleration due to gravity.

$s = 31.3$ m ; $u = 0$; $v = ?$; $a = -9.8$ m s^{-2}; $t = ?$

Using $s = ut + \frac{1}{2}at^2$, $-31.3 = 0 - 0.5 \times 9.8 \times t^2$.

(Notice that s is −31.3 m as it is in the opposite direction to the upwards + direction.)

This gives a value for t of ±2.53 s. The positive value is the one to use. Think about what the negative value stands for.

So the total time is the 0.51 s to get to the maximum height together with the 2.53 s to fall back to Earth.

This gives a total of 3.04 s which rounds to 3.0 s.

Notice that, in this example, if you carry the signs through consistently, they give you information about the motion of the object.

Moving horizontally

There is little that is new here. We are going to assume that the surface of the Earth is large enough for its surface to be considered flat (Topic 6 will go into more details of what happens in reality) and that there is no friction.

Gravity acts vertically and not in the horizontal direction. This will be important when we combine the horizontal and vertical motions later.

Because the horizontal acceleration is zero, the *suvat* equations are simple.

For horizontal motion:

- the horizontal velocity does not change
- the horizontal distance travelled is *horizontal speed* × *time for the motion*.

Putting it all together

A student throws a ball horizontally. Figure 13 shows multiple images of the ball every 0.10 s as it moves through the air. The picture also shows, for comparison, the image of a similar ball dropped vertically at the same moment as the ball is thrown. What do you notice about both images?

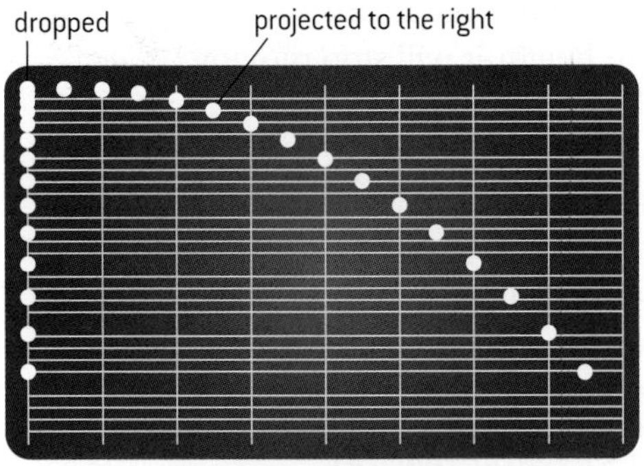

▲ Figure 13 Multi-flash images of two falling objects.

It is obvious which of the two balls was thrown horizontally. Careful examination of the images of this ball should convince you that the *horizontal* distance between them is constant. Knowing the time interval and the distance scale on the picture means that you can work out the initial (and unchanging) horizontal speed.

The images tell us about the vertical speeds too. The clue here is to concentrate on the ball that was dropped vertically. The distance between images (strictly, between the same point on the ball in each image) is increasing. The distance s travelled varies with time t from release as $s \propto t^2$. If t doubles then s should increase by factor of 4. Does it look as though this is what happens? Careful measurements from this figure followed by a plot of s against t^2 could help you to confirm this.

The two motions, horizontal and vertical, are completely independent of each other. The horizontal speed continues unchanged (we assumed no air resistance) while the vertical speed changes as gravity acts on the ball. This independence allows a straightforward analysis of the motion. The horizontal and the vertical parts of the motion can be split up and treated separately and then re-combined to answer questions about the velocity and the displacement for the whole of the motion.

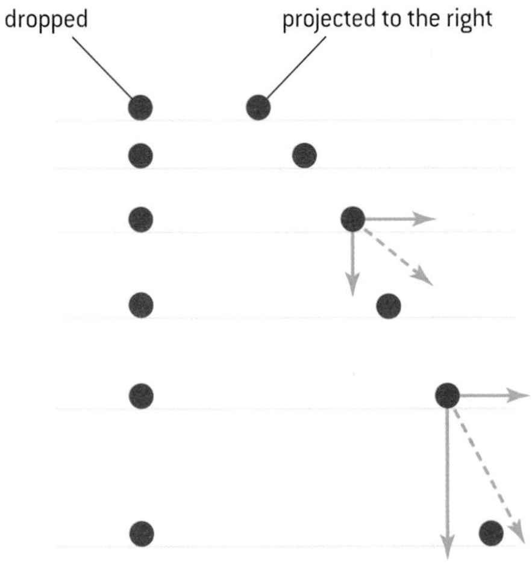

▲ Figure 14 Horizontal and vertical speed components.

The position is summed up in figure 14. At two positions along the trajectory, the separate components of velocity are shown and the resultant (the actual velocity including its direction) is drawn.

Real-life situations do not always begin with horizontal motion. A well-aimed throw will project the ball upwards into the air to achieve the best range (overall distance travelled). But the general principles above still allow the situation to be analysed.

at maximum height:
$$0 = u \sin\theta - gt$$
$$h = (u \sin\theta)t - \frac{gt^2}{2} \text{ and}$$
$$0 = u^2 \sin^2\theta - 2gh$$
range:
$$= 2t \, (u \cos\theta)$$
t is the time to maximum height

initial velocity u at θ to horizontal
initial horizontal component $= u \cos\theta$
initial vertical component $= u \sin\theta$

time for whole motion is **twice** time to maximum height

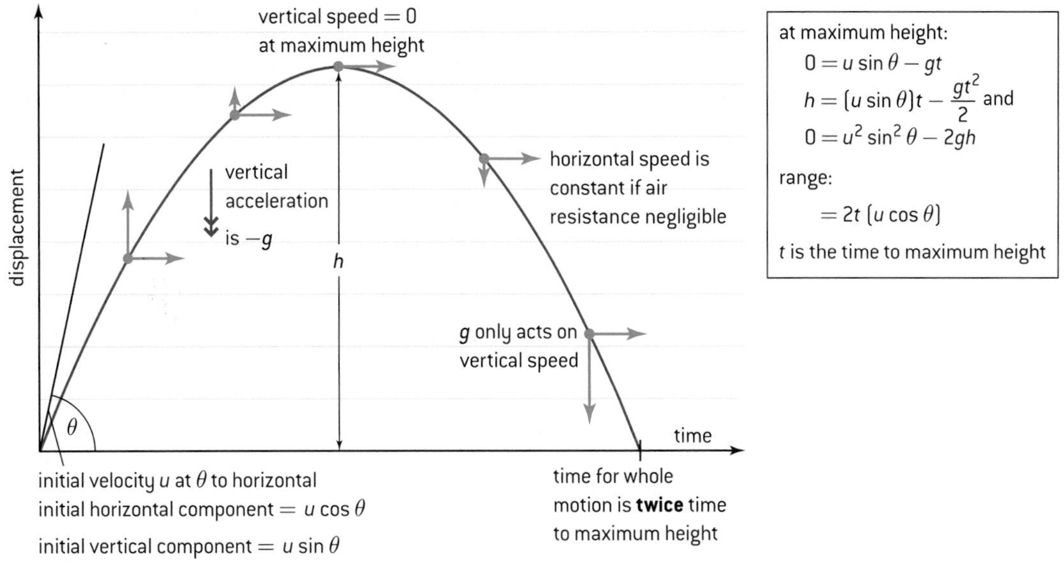

▲ Figure 15 Projectile motion.

Study figure 15 carefully and apply the ideas in it to any projectile problems you need to solve.

Worked examples

1 An arrow is fired horizontally from the top of a tower 35 m above the ground. The initial horizontal speed is 30 m s⁻¹. Assume that air resistance is negligible.

Calculate:

a) the time for which the arrow is in the air

b) the distance from the foot of the tower at which the arrow strikes the ground

c) the velocity at which the arrow strikes the ground.

Solution

a) The time taken to reach the ground depends on the vertical motion of the arrow.

At the instant when the arrow is fired, the vertical speed is zero.

The time to reach the ground can be found using $s = ut + \frac{1}{2}at^2$

$t^2 = \frac{2 \times 35}{9.8}$, so t = 2.67 s or 2.7 s to 2 s.f.

b) The distance from the foot of the tower depends only on the horizontal speed.

$s = ut = 30 \times 2.67 = 80.1$ m = 80 m.

c) To calculate the velocity, the horizontal and vertical components are required. The horizontal component remains at 30 m s⁻¹. The vertical speed is calculated using $v = u + at$ and is $0 + 9.8 \times 2.67 = 26.2$ m s⁻¹.

The speed is $\sqrt{30^2 + 26.2^2} = 39.8$ m s⁻¹ which rounds to 40 m s⁻¹.

The angle at which the arrow strikes the ground is $\tan\frac{26.2}{30} = 41°$

2 An object is thrown horizontally from a ship and strikes the sea 1.6 s later at a distance of 37 m from the ship.

Calculate:

a) the initial horizontal speed of the object

b) the height of the object above the sea when it was fired thrown.

Solution

a) The object travelled 37 m in 1.6 s and the horizontal speed was $\frac{37}{1.6} = 23$ m s⁻¹.

b) Use $s = ut + \frac{1}{2}at^2$ to calculate s above the sea.

$s = 0 + 0.5 \times 9.8 \times 1.6^2 = 12.5$ m

2.2 Forces

Understanding

- Objects as point particles
- Free-body diagrams
- Translational equilibrium
- Newton's laws of motion
- Solid friction

Nature of science

The application of mathematics enhances our understanding of force and motion. Isaac Newton was able to use the work of earlier scientists and to formalize it through the use of the calculus, a form of which he developed for this purpose. He made many insights during his scientific thinking within the topic of force and motion, but also beyond it. The story of the falling apple, whether true or not, illustrates the importance of serendipity in science and the requirement that the creative scientist can form links that go beyond what exists already.

Applications and skills

- Representing forces as vectors
- Sketching and interpreting free-body diagrams
- Describing the consequences of Newton's first law for translational equilibrium
- Using Newton's second law quantitatively and qualitatively
- Identifying force pairs in the context of Newton's third law
- Solving problems involving forces and determining resultant force
- Describing solid friction (static and dynamic) by coefficients of friction

Equations

- Newton's second law: $F = ma$
- static friction equation: $F_f \leq \mu_s R$
- dynamic friction equation: $F_f = \mu_d R$

Introduction

We depend on forces and their effects for all aspects of our life. Forces are often taught in most elementary physics courses as though they are "pushes or pulls", but forces go well beyond this simple description. Forces can change the motion of a body and they can deform the shapes of bodies. Forces can act at a distance so that there is no contact between objects or between a system that produces a force and the object on which it acts.

TOK

Aristotle and the concept of force

Discussions about the concept of what is meant by a force go back to the dawn of scientific thought. Aristotle, a Greek philosopher who lived about 2300 years ago, had an overarching view of the world (called an Aristotelian cosmology) and he can be regarded as being an important factor in the development of science. The German philosopher Heidegger wrote that there would have been no Galileo without Aristotle before him.

Despite his importance to us, however, we would not have regarded Aristotle as a scientist in any modern sense. For one thing he is not known to have performed experiments to verify his ideas; some of his ideas seem very odd to us today. Aristotle believed in the "nature" of all objects including living things. He believed that all objects had a natural state which was to be motionless on the surface of the Earth and that all objects, if left alone, would try to attain this state. Then he distinguished between "natural motion" in which, for example, heavy objects fall downwards and "unnatural" or "forced" motion in which the objects need to have a force continually applied if they are to remain anywhere other than their natural state.

Unfortunately for those learning physics this is a very persuasive idea because we know intuitively that if we want to hold something in our hand, our muscles have to keep "working" in order to do this.

There are many other examples of Aristotelian thought and how later scientists moved our thinking forward. But it is important to remember the contribution that Aristotle made to science, even if some of his ideas are now overturned. What do you think students in 50, 100, or even a 1000 years will make of the physics of our century?

Newton's laws of motion

Newton's first law

By the time of Galileo, scientists had begun to realise that things were not as simple as the Greek philosophers such as Aristotle had thought. They were coming to the view that moving objects have **inertia**, meaning a resistance to stopping and that, once in motion, objects continue to move.

Galileo carried out an experiment with inclined planes and spheres. In fact, this may have been a thought experiment – this was often the way forward in those days – but in any event it is easy to see what Galileo was trying to suggest.

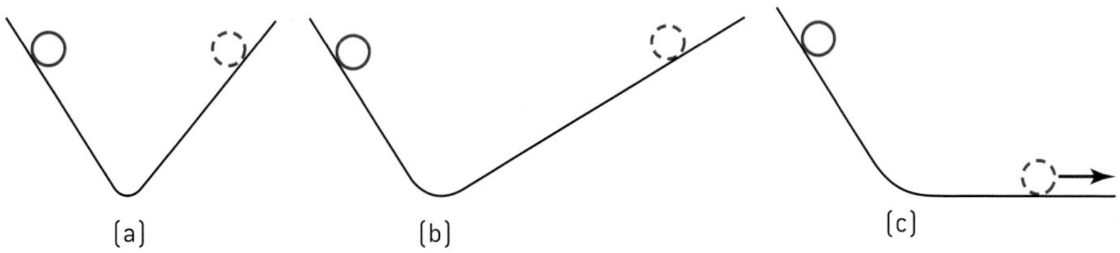

▲ Figure 1 Galileo's thought experiment.

In the first experiment (a), the two arms of the inclined plane are at the same angle and the sphere rolls the same distance up the slope as it rolled down (assuming no energy losses). In the second experiment (b), the second arm is at a lower angle than before but the sphere rolls up this plane to the same height as that from which it was released. Galileo then concluded that if the second plane is horizontal (c), the sphere will go on rolling for ever because it will never be able to climb to the original release height.

Newton included this idea in his **first law of motion** that says:

An object continues to remain stationary or to move at a constant velocity unless an external force acts on it.

Galileo had suggested that the sphere on his horizontal plane went on forever, but it was Newton who realised that there is more to say than this. Unless something from outside applies a force to change it, the velocity of an object (both its speed and its direction) must remain the same.

This was directly opposed to Aristotle's view that a force had to keep pushing constantly at a moving object for the speed to remain the same.

Newton's second law

The next step is to ask: if a force does act on an object, in what way does the velocity change?

Newton proposed that there was a fundamental equation that connected force and rate of change of velocity. This is contained in his second law of motion. This law can be written in two ways, one way is more complex than the other and we will look at the more complicated form of the law later in this topic.

Newton's second law, in its simpler form, says that

Force = mass × acceleration

As an equation this can be written

$$F = ma$$

The appropriate SI units are force in newtons (N), mass in kilograms (kg), and acceleration in metres per second^2 (m s^{-2}). As discussed in Topic 1 the newton is a derived unit in SI. We can represent it in terms of fundamental units alone as kg m s^{-2}.

Two things arise from this equation:

- Mass is a scalar, so there will not be a change in the direction of the acceleration if we multiply acceleration by the mass. The direction of the force and the direction of the acceleration must be the same. So, applying a force to a mass will change the velocity *in the same direction* as that of the force.
- One way to think about the mass in this equation is that it is the ratio of the force required per unit of acceleration for a given object. This helps us to standardize our units of force. If an object of mass 1 kg is observed to accelerate with an acceleration of 1 m s^{-2} then one unit of force (1 N) must have acted on it.

Nature of science

Inertial and gravitational mass

What exactly do we mean by *mass*? The mass that we use in Newton's second law of motion is inertial mass. Inertial mass is the property that permits an object to resist the effects of a force that is trying to change its motion.
In other parts of this book, mass is used in a different context. We talk about the weight of an object and we know that this weight arises from the gravitational attraction between the mass of the object and the mass of the Earth. The two masses in this case are gravitational masses and are the response of matter to the effects of gravity. It has been experimentally verified that weight is proportional to mass to better than 3 parts in 10^{11}. It is a postulate of Einstein's general theory of relativity that inertial mass and gravitational mass are proportional.

Investigate!

Force, mass, and acceleration

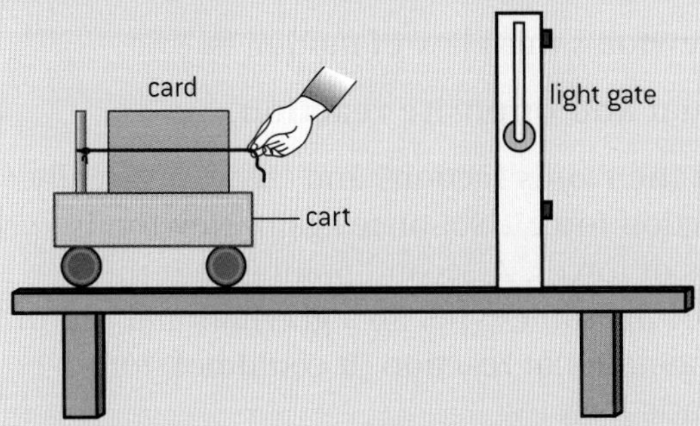

Experiment 1: force and acceleration

- The idea in this experiment is to measure the acceleration of the cart (of constant mass) when it is towed by different numbers of elastic threads each one extended by the same amount.
- The timing can be done using light gates and an electronic timer as shown here. Alternatively it can be done using a data logger with, say, an ultrasound sensor and the cart moving away from the sensor.
- Practise accelerating the cart with one elastic thread attached to the rear of the cart. The thread(s) will have to be extended by the same amount for each run. A convenient point to judge is the forward end of the cart. Make sure that your hand clears the light gates if you are using this method. Your hand needs to move at the same speed as the cart so that the thread is the same length throughout the run.
- The card on the cart needs to be of a known length so that you can use the time taken to break the light beam to calculate, first, the average speed at each gate and then (using the distance apart of the gates), the acceleration of the cart. The kinematic equations are used for this. (This is why you must pull the cart with a constant force so that the acceleration is uniform.)
- Repeat the experiment with two, three, and possibly four elastic threads, all identical, all extended by the same amount. This means that you will be using one unit of force (with one thread), two units (with two threads) and so on.
- Plot a graph of calculated acceleration against number of force units. Is your graph straight? Does it go through the origin? Remember that this experiment has a number of possible uncertainties in judging the best-fit line.

Experiment 2: mass and acceleration

- The setup is essentially the same as for Experiment 1, except this time you will use a constant force (possibly two elastic threads is appropriate).
- Change the mass of your cart (some laboratory carts are specially designed to stack, one on top of another) and measure the acceleration with a constant force and varying numbers of carts.
- This time Newton's second law predicts that mass should be inversely proportional to acceleration. Plot a graph of acceleration against $\frac{1}{\text{mass}}$. Is it a straight line?

Worked examples

1 A car with a mass of 1500 kg accelerates uniformly from rest to a speed of 28 m s^{-1} (about 100 km h^{-1}) in a time of 11 s. Calculate the average force that acts on the car to produce this acceleration.

Solution

The acceleration $= a = \frac{v-u}{t} = \frac{28}{11} = 2.54$ m s^{-2}.

Force $= ma = 1500 \times 2.54 = 3.8$ kN.

2 An aircraft of mass 3.3×10^5 kg takes off from rest in a distance of 1.7 km. The maximum thrust of the engines is 830 kN. Calculate the take-off speed.

The acceleration of the aircraft is $\frac{8.3 \times 10^5}{3.3 \times 10^5} =$ 2.51 m s^{-2}.

$v^2 = u^2 + 2as$ so $v = 0 + 2 \times 2.51 \times 1700$ which leads to $v = 92.4$ m s^{-1}.

TOK

But are they really laws?

The essential question is: can Newton's laws of motion be proved? The answer is that they cannot; strictly speaking they are assertions as Newton himself recognized. In his famous *Principia* (written in Latin as was a custom in those days) he writes *Axiomata sive leges motus*, or "the axioms or laws of motion".

However, they appear to be an excellent set of rules that allow us to predict most of the motion that we undertake. They remained unchallenged as a theory for about 200 years until the two theories of relativity were formulated by Einstein at the turn of the twentieth century. Essentially (said Einstein) the rules that Newton proposed do not always apply to, for example, motion that is very fast. However, for the modest speeds at which humans travel, the rules are reliable to a high degree and are certainly good enough for our needs most of the time.

Newton's third law

Newton's third law of motion is sometimes written in such a way that the true meaning of the law has to be teased out. The law can be expressed in a number of equivalent ways.

One common way to write Newton's third law of motion is:

Every action has an equal and opposite reaction.

The first point to make is that the words "action" and "reaction" really mean "action force" and "reaction force". So, once again, Newton is referring directly to the effects of forces. A second point is that the action–reaction pair must be of the same type. So a gravitational action force must correspond to a gravitational reaction. It could not, for example, refer to an electrostatic force.

The law suggests that forces must appear in pairs, but in thinking about a particular situation it is important to identify all possible force pairs and then to pair them up correctly. Take, as an example, the situation of a rubber ball resting on a table.

At first glance the obvious action force here is the weight of the ball, that is, the gravitational pull the Earth exerts on the sphere. This force acts downwards and – if the table were not there – the ball would accelerate downwards according to Newton's second law. It would fall to the floor.

What is the reaction force here? Given that action force and reaction force must pair up like for like, the reaction must be the gravitational force that the ball exerts on the Earth. This is of exactly the same size as the pull of the Earth on the ball, but is in the opposite direction.

What prevents the ball accelerating downwards to the floor? A force must be exerted by the table on the ball – and if there are no accelerations happening, then this force is equal and opposite to the downwards gravitational pull of the Earth on the ball. But the upwards table force is *not* the reaction force to the ball's weight – we have already seen that this is the gravitation pull on the Earth. The origin of the table force is the electrostatic forces between atoms. As the ball lies on the table, it deforms the horizontal surface very slightly, rather like what happens when you push downwards with a finger on a metre ruler suspended by supports at its end – the ruler bends in a spring-like way to provide a resistance to the force acting downwards. The dent in the table surface is the response of the atoms in the table to the weight lying on it. Remove the ball and the surface will return to being flat. This upwards force that returns the table to the horizontal is pushing upwards on the ball. There is a corresponding downwards force from the deformed ball (the ball will become slightly flattened as a response to the gravitational pull). So here is the second action–reaction pair, between two forces that are electrostatic in origin.

In summary, there are four forces in this situation: the weight of the ball, the upwards "spring-like" force of the table, and the reactions to these forces which are the pull of the ball on the Earth and the downwards push of the deformed ball on the table.

To (literally) get a feeling for this, take a one-metre laboratory ruler and suspend it between two lab stools as suggested earlier. Press down gently with your finger in the centre of the ruler so that it becomes curved. The ruler will bend, you will be able to feel it resisting your efforts to deform it too far. Remove your finger and the ruler will return to its original shape.

In explaining Newton's third law in a particular example, you must remember to emphasise the nature, the size, and the direction of the force you are describing. Another common example is that of a rocket in space. Students sometimes write that "...by Newton's third law, the rocket pushes on the atmosphere to accelerate" but this shows a poor understanding of how the propulsion actually works.

First, of course, the rocket does not "push" on anything. The fact that a rocket can accelerate in space where there is no atmosphere proves this.

What happens inside the rocket is that chemicals react together producing a gas with a very high temperature and pressure. The rocket has exhaust nozzles through which this gas escapes from the combustion chamber. At one end of the chamber inside the rocket the gas molecules rebound off the end wall and exert a force on it, as a result they reverse their direction. In principle, the rebounding molecules could then travel down the rocket and leave through the nozzles. So there is an action-reaction pair here, the force forwards that the gas molecules exert on the chamber (and therefore the rocket) and the force that the chamber exerts on the gas molecules. It is the first of these two forces that accelerates the rocket. If the chamber were completely sealed and the gas could exert an equal and opposite force at the back of the rocket, then the forward force would be exactly countered by the backwards force and no acceleration would occur.

▲ Figure 2 Chemical rockets in action.

Later we shall interpret this acceleration in a different way. But the explanation given here will still be correct at a microscopic level.

Think about the following situations and discuss them with fellow students.

- A fireman has to exert considerable force on a fire hose to keep it pointing in the direction that will send water to the correct place.
- There is a suggestion to power space travel to deep space by ejecting ions from a spaceship.
- A sailing dinghy moves forward when the wind blows into the sails.

Free-body force diagrams

As a vector quantity, a force can be represented by an arrow that gives both the scaled length and the direction of the force. In simple cases where few forces act this works well, but as the situations become more complex, diagrams that show all the arrows can become complicated.

One way to avoid this problem is to use a **free-body force diagram** to illustrate what is happening.

The rules for a free-body diagram for a body are simple, as follows.

- The diagram is for one body only and the force vectors are represented as arrows.
- Only the forces acting on the body are considered.
- The force (vector) arrows are drawn to scale originating at a point that represents the centre of mass of the body.
- All forces have a clear label.

Procedure for drawing and using a free-body diagram

- Begin by sketching the general situation with all the bodies that interact in the situation.
- Select the body of interest and draw it again removed from the situation.
- Draw, to scale, and label all the forces that act on this body due to the other bodies and forces.
- Add the force vectors together (either by drawing or calculation) to give the net force acting on the body. The sum can be used later to draw other conclusions about the motion of the object.

Examples of free-body diagrams

1 *A ball falling freely under gravity with no air resistance.*

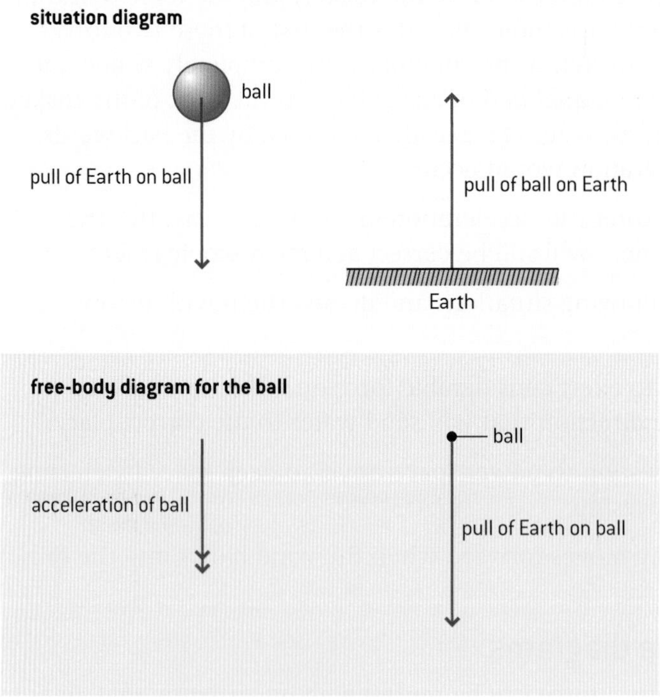

▲ Figure 3 Ball falling freely under gravity.

This straightforward situation speaks for itself. There are two forces acting: the Earth pulling on the ball and the ball pulling on the Earth. For the free-body diagram of the ball we are only interested in the first of these. So the free-body diagram is a particularly simple one, showing the object and one force. Notice that the ball is not represented as a real object, but as a point that refers to the centre of mass of the object.

2 *The same ball resting on the ground.*

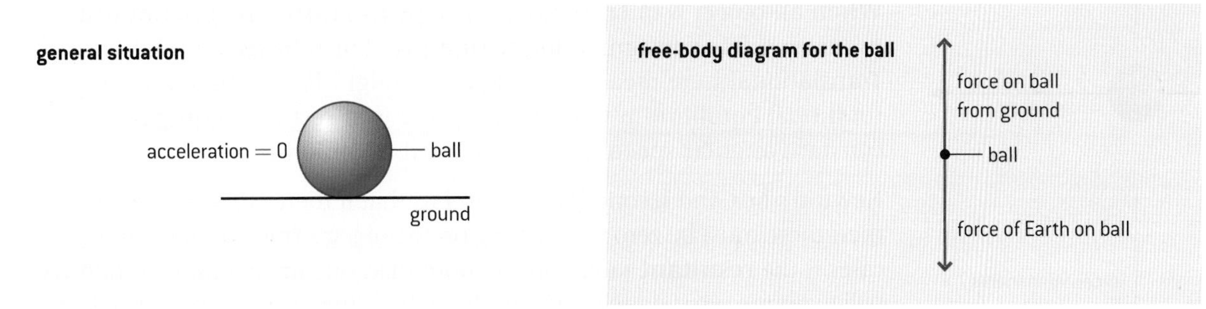

▲ Figure 4 Ball resting on the ground.

As we saw earlier, four forces act in this case:

- the weight of the ball downwards
- the reaction of the Earth to this weight
- the upwards force from the ground
- the reaction of the ball to this "spring-like" force.

A free-body diagram (figure 4) definitely helps here because, by restricting ourselves to the forces acting on the ball, the four forces reduce to two: the weight of the ball and the upwards force on it due to the deformation of the ground. These two forces are equal and opposite. The net resultant (vector sum) of the forces is zero and there is therefore no acceleration.

3 *An object accelerating upwards in a lift.*

The weight of the object is downwards and the size of this force is the same as though the object had been stationary on the Earth's surface. However, the upwards force of the floor of the lift on the object is now larger than the weight and the resultant force of the two has a net upwards component. The object is accelerated upwards as you would expect.

For the lift, there is an upward force in the lift cable and a downwards weight of the lift is downwards together with the weight of the object. The resultant force is upwards and is equal to the force in the cable less the weight forces of lift and object.

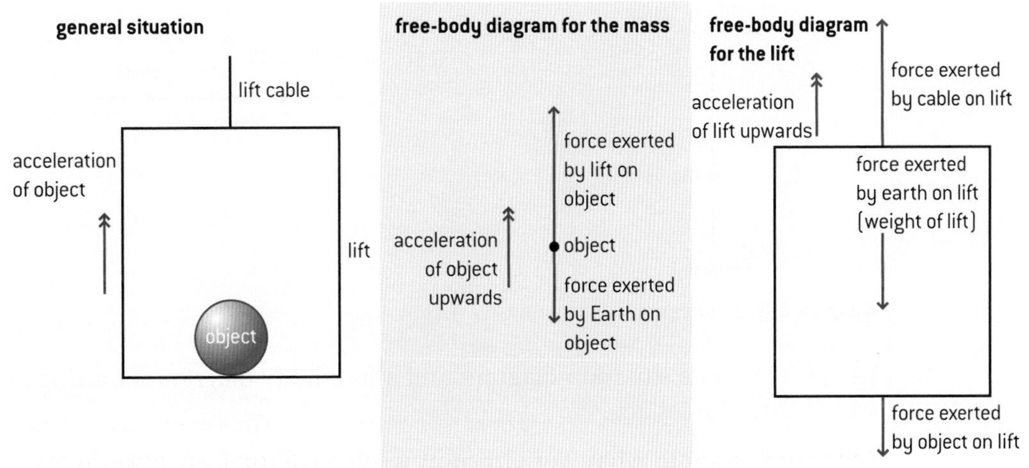

▲ Figure 5 Mass in a lift accelerating upwards.

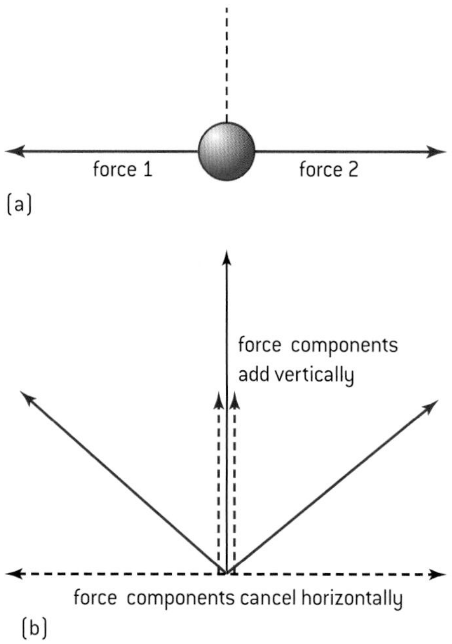

▲ Figure 6 Two equal forces in different directions.

Translational equilibrium

When an object is in **translational equilibrium** it is either at rest or moving at a constant velocity (not just constant *speed* in this case). Translational here means moving in a straight line. (There is also rotational equilibrium where something is at rest or rotating at a constant angular speed; this is discussed in option B).

Newton's first and second laws remind us that if there is no change of velocity then there must be zero force acting on the object. This zero force in many cases is the **resultant** (addition) of more than one force. In this section we will examine what equilibrium implies when there is more than one force.

The simplest case is that of two forces. If they are equal in size and opposite in direction, they will cancel out and be in equilibrium (Figure 6(a)).

If the forces are equal in size but not in the same direction then equilibrium is not possible. Horizontally, in figure 6(b), the two components of the force vectors are still equal and opposite and we could expect that there would be no change in this direction. But vertically the two vector components point in the same direction so that overall there will be an unbalanced vertical force and an acceleration acting on the object on which these forces act.

This gives a clue as to how we should proceed when there are three or more forces.

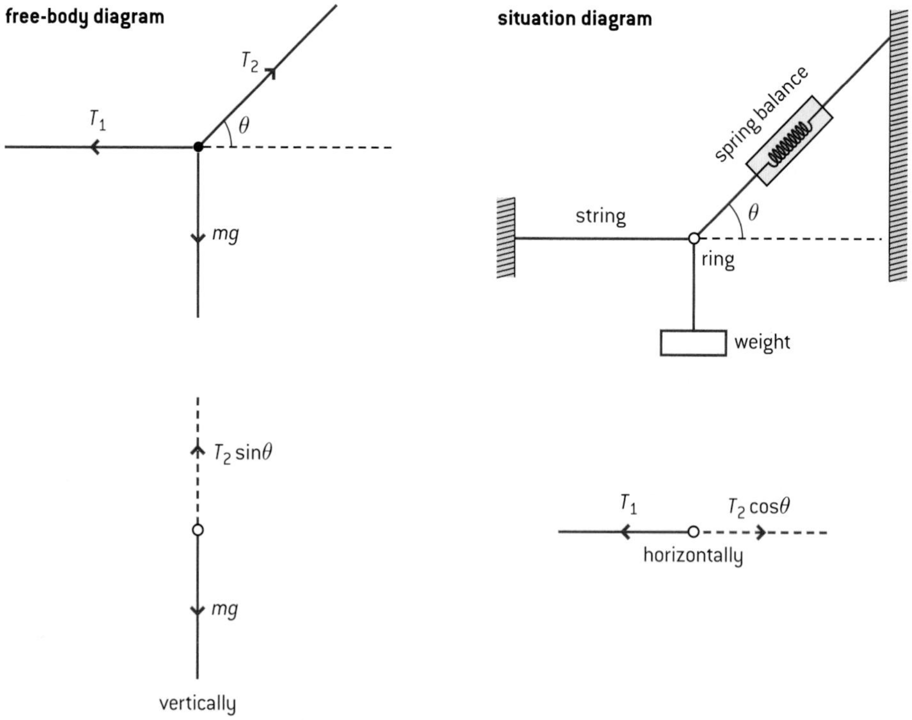

▲ Figure 7 Three forces acting on a ring.

Figure 7 shows a situation diagram and a free-body diagram for a ring on which three forces act.

For equilibrium, in whatever direction we resolve the forces, all three components must add up to zero in this direction.

Figure 7 shows that horizontal and vertical are two good directions with which to begin, because two forces are aligned with these directions and one disappears in each direction chosen. Thus, vertical force mg has no component in the horizontal direction and horizontal force T_1 has no part to play vertically. But whichever direction is chosen, if there is to be no resultant force and consequently no acceleration, then all the forces must cancel.

There is one more consequence of this idea. Figure 8 shows the forces drawn, as usual, to scale and in the correct direction (in red). The forces can be moved, as shown by the green arrows, into a new arrangement (shown in black). What is special here is that the three forces form a closed triangle where all the arrows meet.

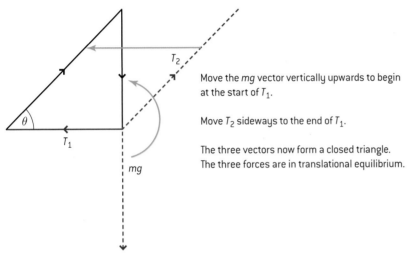

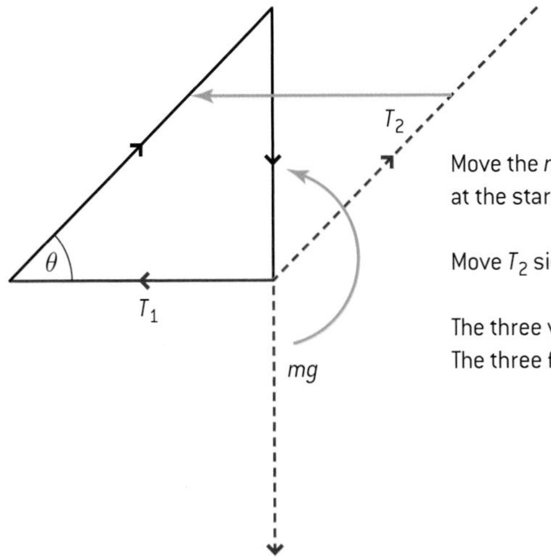
T_2
θ
T_1
mg
Move the mg vector vertically upwards to begin at the start of T_1.
Move T_2 sideways to the end of T_1.
The three vectors now form a closed triangle. The three forces are in translational equilibrium.

▲ Figure 8 Making a triangle for forces.

Algebraically, this must be true because we know (ignoring the directions of the forces) that

$$T_1 = T_2 \cos \theta \text{ (horizontally) and } mg = T_2 \sin \theta \text{ (vertically)}$$

So

$$T_1^2 = T_2^2 \cos^2\theta \text{ and } (mg)^2 = T_2^2 \sin^2\theta$$

Adding these together gives

$$T_1^2 + (mg)^2 = T_2^2 (\sin^2\theta + \cos^2\theta)$$

and therefore as $\sin^2 \theta + \cos^2 \theta = 1$

$$T_1^2 + (mg)^2 = T_2^2$$

This is equivalent to Pythagoras' theorem where $x = T_1^2$, $y = (mg)^2$ and $z = T_2^2$ so that $x^2 + y^2 = z^2$.

So, because the sum of the squares of two vector lengths equals the square of the third vector length, the three must fit together as a right-angled triangle. We have a right-angled triangle formed by the scaled lengths and directions of the three force vectors. This important figure is known as **a triangle of forces**. If you can draw the vectors for a system in this way, then the system must be in translational equilibrium.

Investigate!

Three forces in equilibrium

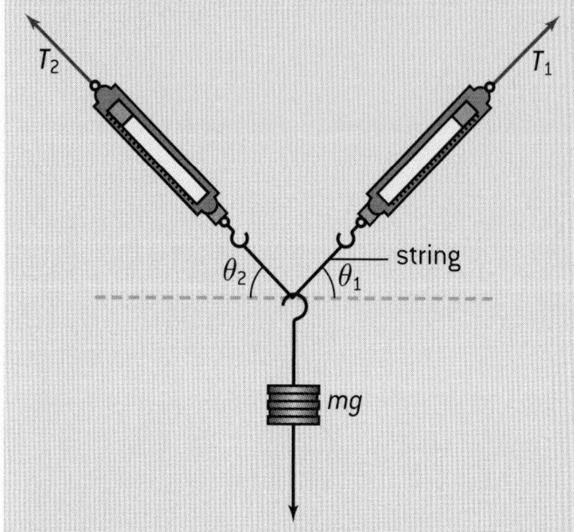

- If a point is in equilibrium under the influence of three forces, then the three forces must form a closed triangle.
- To verify this, use a mass and two spring balances with a knot halfway along the string between the two spring balances to act as the point object.
- Your arrangement for hanging the spring balances will depend on the resources in your school.
- Select a mass to suit the sensitivity of your spring balances.
- Place a vertical drawing board with a large sheet of paper pinned to it behind the balances so that you can mark the position of the strings. When these are marked, use a protractor to measure the angles.
- Begin with a simple case, say when T_1 is horizontal so that $\theta_1 = 90°$.
- The knot must be stationary and when this is true:
  - $T_1 \cos \theta_1 = T_2 \cos \theta_2$ (resolving horizontally)
  - $T_1 \sin \theta_1 + T_2 \sin \theta_2 = mg$ (resolving vertically)
- Construct this table to enable you to verify that both equations are true for every case you set up.

	T_1/N	θ_1/°	T_2/N	θ_2/°	mg/N	$T_1 \cos \theta_1$/N	$T_1 \sin \theta_1$/N	$T_2 \cos \theta_2$/N	$T_2 \sin \theta_2$/N
Case 1									

For each case check that the equations are correct within experimental error.

Solid friction

Friction is the force that occurs between two surfaces in contact. If you live in a part of the world where there is snow and ice then you will know that when the friction between your shoes and the ice disappears it can be a good thing (for skiing) or a bad thing (for falling over).

Investigate!

Solid friction

In this experiment you use a spring balance to measure the force that acts between two surfaces. It gives some unexpected results.

- Set up the platform on rollers with a suitable weight and a spring balance. A string connects the spring balance to the weight. Ensure that the weight is attached securely to the block.
- Pull the platform using the winch system with an increasing force observing, at the same time, the reading on the balance.

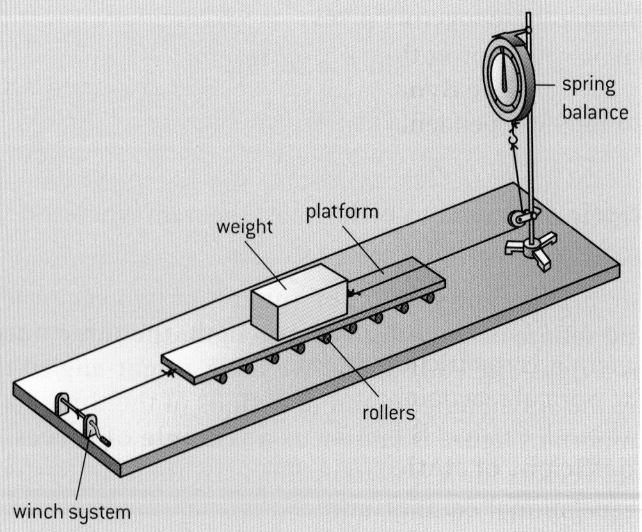

- Note the maximum value of the force as measured on the balance.
- Note the way the size of the force changes when the platform begins to move.
- Is this force constant? Or does it fluctuate?
- Change the size of the weights on the platform. How does the total weight affect the force measured on the balance?
- Change the type of surface at the bottom of the weights. You might use abrasive (emery) paper or cloth in different abrasive grades pinned to the platform.
- Investigate other changes: does lubrication at the bottom of the weights affect the friction?

If you carried out the *Investigate!* Solid friction experiment you may have found that:

- The force on the balance increases as you pull harder and harder, but the platform does not begin to move relative to the weights immediately.
- Eventually the platform suddenly begins to move at a particular value of force and at this instant the force, shown by the balance, drops to a new lower value.
- This new value is then maintained as the platform moves steadily.
- You may observe "stick-slip" behaviour where the platform alternately sticks and then jumps to a new sticking position. This behaviour is associated with two values of friction, but this may be too difficult to observe unless you get very suitable surfaces.
- The friction forces depend on the magnitude of the weights.

The frictional forces that occur in this experiment are described empirically as:

- **static friction** (when there is no relative movement between the surfaces)
- **dynamic friction** (when there is relative movement).

As the pulling force increases but without any slip happening, the friction is said to be **static** (because there is no motion). Eventually however, the pulling force will exceed the value of the static friction and the surfaces will start to move. As the movement continues, the friction force drops to a new value, lower than the maximum static value. This new lower value is known as the **dynamic friction**. Both static and dynamic friction forces are highly dependent on the two surfaces concerned.

Static friction

The static friction force F_f is found to be given empirically by

$$F_f \leq \mu_s R$$

where F_f is the frictional force exerted by the surface on the block. R is the normal reaction of the surface on the block, this equals the weight of the block as there is no vertical acceleration. The symbol μ_s refers to the **coefficient of static friction**.

The "less than or equal to" symbol indicates that the static friction force can vary from zero up to a maximum value. Between these limits F_f is equal to the pull on the block. Once the pull on the block is equal to F_f, then the block is just about to move. Once the pull exceeds F_f then the block begins to slide (figure 9). For these larger forces, the friction operating is in the dynamic regime.

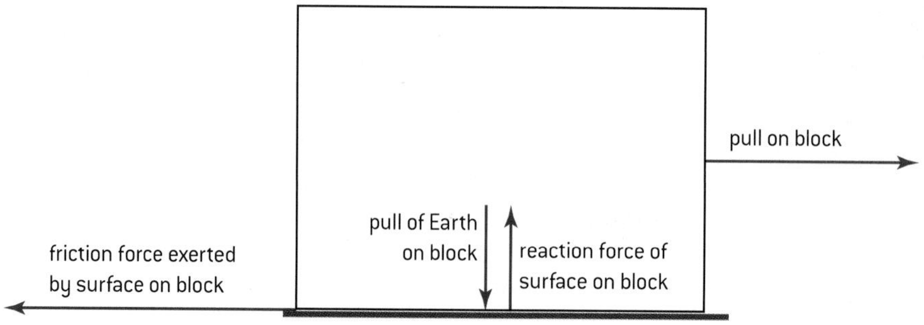

▲ Figure 9 Friction force acting on an object.

Dynamic friction

Dynamic friction only applies when the surfaces move relative to each other. The friction drops from its maximum static value (there is an explanation for this below) and remains at a constant value. This value depends on the total reaction force acting on the surface but (according to simple theory) is not thought to depend on the relative speed between the two surfaces. So, for dynamic friction

$$F_f = \mu_d R$$

where μ_d is the **coefficient of dynamic friction**.

The values of μ_s and μ_d vary greatly depending on the pair of surfaces being used and also the condition of the surfaces (for example, whether lubricated or not). A few typical values are given in the table. If you want to investigate a wider range of surfaces, there are many sources of the coefficient values on the Internet – search for "Coefficients of friction".

Each friction coefficient is a ratio of two forces (F_f and R) and so has no units.

Surface 1	Surface 2	μ_s	μ_d
glass	metal	0.7	0.6
rubber	concrete	1.0	0.8
rubber	wet tarmac	0.6	0.4
rubber	ice	0.3	0.2
metal	metal (lubricated)	0.15	0.06

Values of μ_s and μ_d for pairs of surfaces

It is possible for the coefficients to be greater than 1 for some surface pairs. This reflects the fact that for these surfaces the friction is very strong and greater than the weight of the block. Remember that the surfaces are being pulled sideways by a horizontal force whereas the reaction force is vertical so we are not really comparing like with like in these empirical rules for friction.

Investigate!

Friction between a block and a ramp

One way to measure the static coefficient of friction between two surfaces in a school laboratory is to use the two surfaces as part of a ramp system.

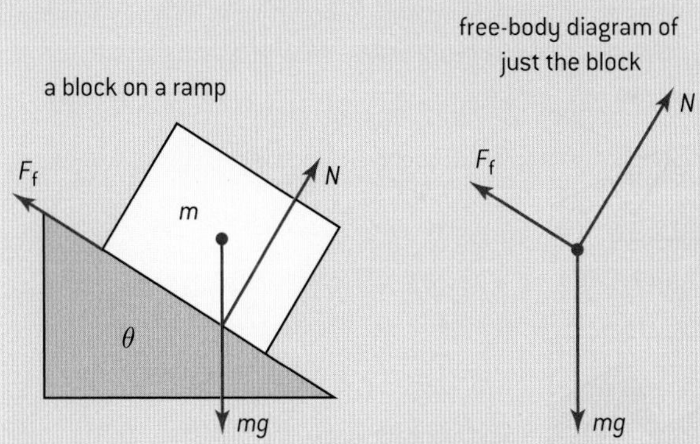

- One surface is the top of the ramp, the other is the base of the block.
- Resolving at 90° to plane, $N = mg\cos\theta$; resolving along the plane, $F = mg\sin\theta$
- So $\tan\theta = \frac{F_f}{N} = \mu_s$
- Start with the ramp horizontal, and then gradually raise one end until the block starts to slip.
- At this moment of slip, measure the angle of the ramp. The tangent of this angle is equal to the coefficient of static friction.

Worked examples

1 A box is pushed across a level floor at a constant speed with a force of 280 N at 45° to the floor. The mass of the box is 50 kg.

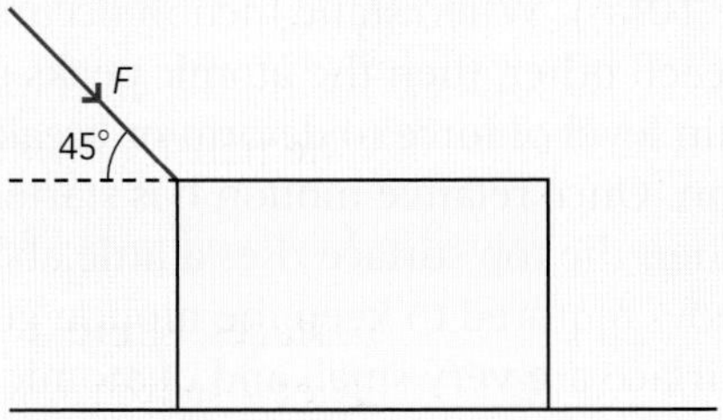

Calculate:

a) the vertical component of the force

b) the weight of the box

c) the horizontal component of the force

d) the coefficient of dynamic friction between the box and the floor.

Solution

a) the vertical component is 280 sin 45° = 198 N

b) the weight of the box = $mg = 50 \times 9.8 = 490$ N

c) the horizontal component of the force = 280 cos 45° = 198 N

d) the vertical component of the force exerted by the floor on the box = 490 + 198 N = 688 N

the friction force = the horizontal component (the box is travelling at a steady speed), so

$$\mu_d = \frac{198}{688} = 0.29$$

2 A skier places a pair of skis on a snow slope that is at an angle of 1.7° to the horizontal. The coefficient of static friction between the skis and the snow is 0.025.

Determine whether the skis will slide away by themselves.

Solution

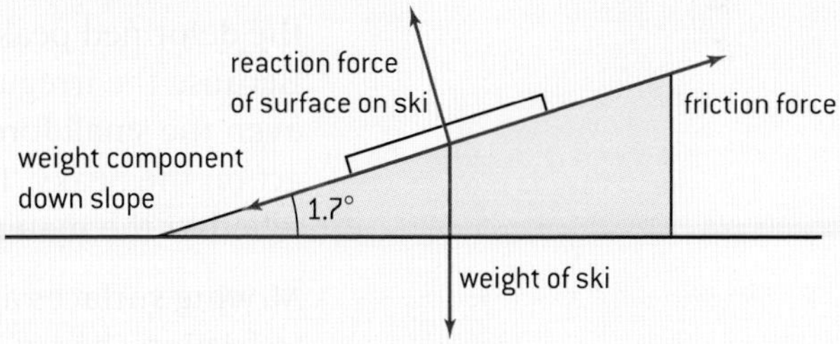

Call the weight of the skis W.

The component of weight down the slope = $W\sin 1.7°$

The reaction force of the surface on the ski = $W\cos 1.7°$.

Therefore, the maximum friction force up slope = $\mu_S W\cos 1.7°$.

The skis will slide if

$\mu_S W\cos 1.7° < W\sin 1.7°$

in other words, if $\mu_S < \tan 1.7°$.
The value of tan 1.7° is 0.0296 and this is greater than the value of μ_S, which is 0.025, so the skis will slide away.

Origins of friction

Friction originates at the interface between the two materials, in other words at the surface where they meet. The actual causes of friction are still being investigated today and the explanation given here is highly simplified and a historical one. Leonardo da Vinci mentions friction in his notebooks and some of the next scientific writings about friction appear in works by Guillaume Amontons from around 1700.

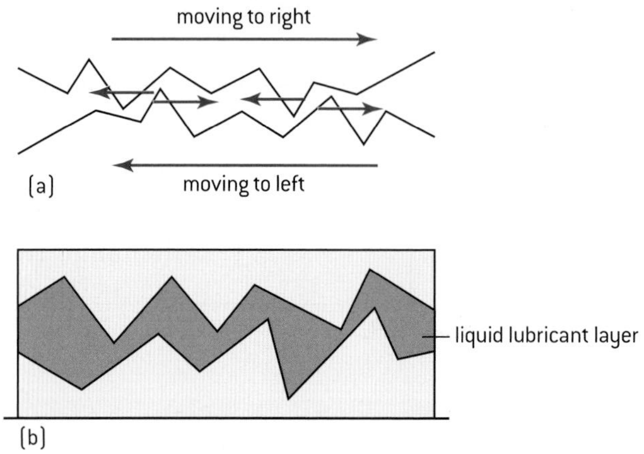

▲ Figure 10 How solid friction arises.

One model for solid friction suggests that what are very smooth surfaces to us are not smooth at all. At the atomic level, they are actually full of peaks and troughs of atoms (figure 10(a)). When static friction occurs and two surfaces are at rest relative to each other, then the atomic peaks rest in the troughs, and it needs a certain level of force to deform or break the peaks sufficiently for sliding to begin. Once relative motion has started (so that dynamic friction occurs), then the top surface rises a little above the deformed peaks. Less force is now required to keep the motion going. Because the irregularities on the surface are very small and of atomic size, even the small forces applied in our lab experiments cause large stresses to act on the peaks. The peaks then deform like a soft plastic irrespective of whether the material is hard steel or something much softer.

Moving surfaces are often coated with a lubricant to reduce wear due to friction (figure 10(b)). The lubricant fills the space between the two surfaces and either prevents the peaks and troughs of atoms from touching or reduces the amount of contact. In either event, the atoms from the surfaces do not interact as much as before and the friction force and the coefficient are reduced.

The origins of these friction forces are bound up in the complex electronic properties of the materials that make up the surfaces. However, this simple theory should give you some understanding of friction as well as an awareness that the bulk materials we perceive on the macroscopic scale arise from microscopic properties that are operating at the atomic level.

What is important to realize is that the two equations for static and dynamic friction do not arise from a study of the interatomic forces between the surfaces; they are derived purely from experiments with bulk materials. The results are empirical not theoretical.

Fluid resistance and terminal speed

The assumption that air resistance is negligible is often unrealistic. An object that travels through a fluid (a liquid or a gas) is subject to a complex process in which, as the object travels through the material, the fluid is "stirred up" and becomes subject to a **drag force**. The process is complex, so even after introducing some simple assumptions about the resistance of a fluid, we will still not be able to give a complete analysis.

In energy terms, the action of **air resistance** is to transfer some of the energy of the moving body into the fluid through which it is moving. Some fluids absorb this energy better than others: swimming through water is much more tiring than running in the air.

For your IB Diploma Programme physics exams, you only need to describe the effects of fluid resistance without going into the mathematics.

Skydiving

In 2012, Felix Baumgartner jumped safely from a height of 39 km above New Mexico to reach a top speed of 1342 km h^{-1} – faster than the speed of sound.

A skydive from more usual heights will not take place at such high speeds, usually up to about 200 km h^{-1}. The difference is due to the variation in the resistance of the air at different heights above the Earth.

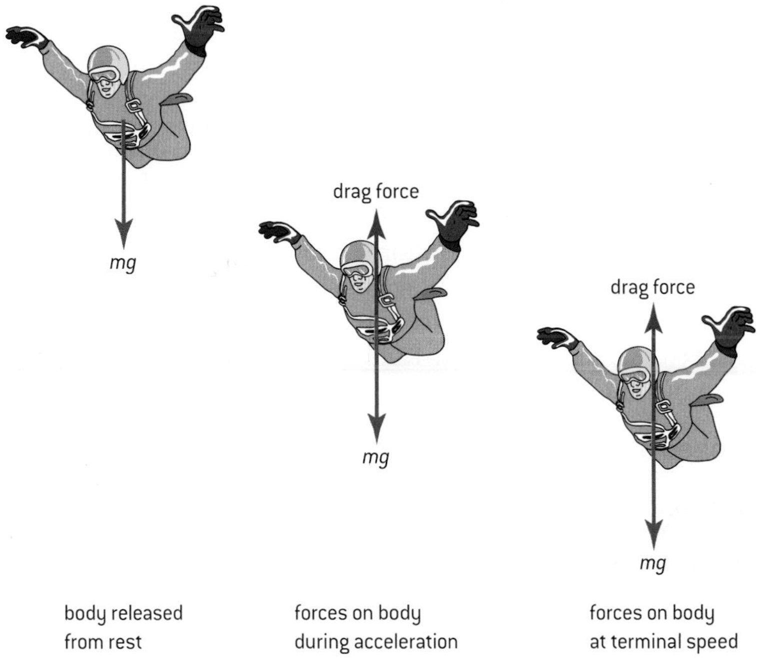

▲ Figure 11 Forces acting on skydiver.

The diagram shows the forces acting on a skydiver. The weight of the diver acts vertically downwards and is effectively constant (because there is little change in the Earth's gravitational field strength at the height of the dive). The air resistance force acts in the opposite direction to the motion of the diver and for a diver falling vertically, this will be vertical too. Other forces acting on the diver include the upwards buoyancy caused by the displacement of air by the diver.

When the skydiver initially leaves the aircraft, the diver's weight acts downwards and because the vertical speed is almost zero, there is almost no air resistance. Air resistance increases as the speed increases so that as the diver goes faster and faster the resistance force becomes larger and larger. The net force therefore decreases and consequently the acceleration of the diver downwards also decreases. Eventually the weight force downwards and the resistance force upwards are equal in magnitude and, of course, opposite in direction. At this point there is no longer any acceleration and the diver has reached a constant rate of fall known as the **terminal speed**.

The graph of vertical speed against time is as shown in figure 12.

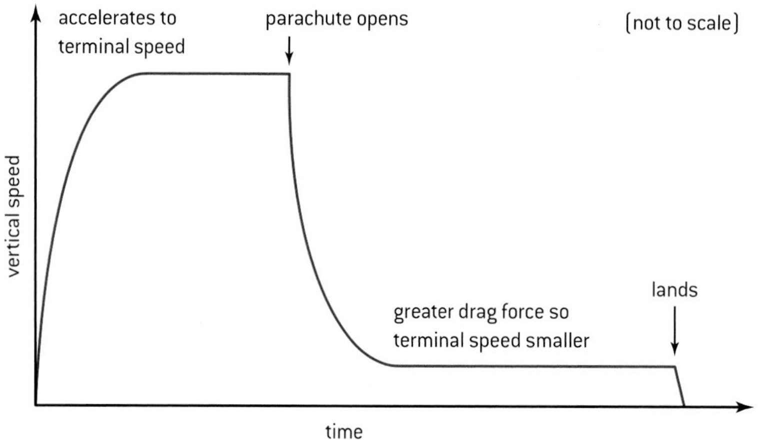

▲ Figure 12 Speed–time for a parachute jump.

Eventually the skydiver opens the parachute. Now, because of the large surface area of the parachute envelope, the upwards resistive force is much larger than before and is greater than the weight. As a result, the directions of the net force and acceleration are also upwards; the vertical velocity decreases in magnitude. Once again a balance will be reached where the upward and downward forces are equal and opposite – but at a much lower speed than before (about 12 m s^{-1} for a landing) and the diver reaches the ground safely.

Maximum speed of a car

The top (maximum) speed of a motor car is determined by a number of factors. The most obvious of these is the maximum force that the engine can exert through the tyres on the road surface. But, as with the skydiver, this is not the only force acting. There is a considerable drag on the vehicle due to the air and this drag force increases markedly as the speed of the car becomes larger. Typically, when the speed doubles the drag force will increase by at least a factor of four.

There is a maximum power that the car engine can produce. When the car accelerates and the speed increases towards the maximum, the power dissipated in friction also increases. When the maximum energy output of the engine every second is completely used in overcoming the energy losses, then the car cannot accelerate further and has reached its maximum speed.

Worked example

1 Calculate the upward force acting on a skydiver of mass 80 kg who is falling at a constant speed.

Solution

The weight of the skydiver is $80 \times 9.8 = 784$ N. Because the skydiver is falling at constant speed (i.e. terminal speed) the upwards drag force is equal to the downwards weight.

The upward force is 784 N.

2 P and Q are two boxes that are pushed across a rough surface at a constant velocity with a horizontal force of 30 N. The mass of P is 2.0 kg and the mass of Q is 4.0 kg.

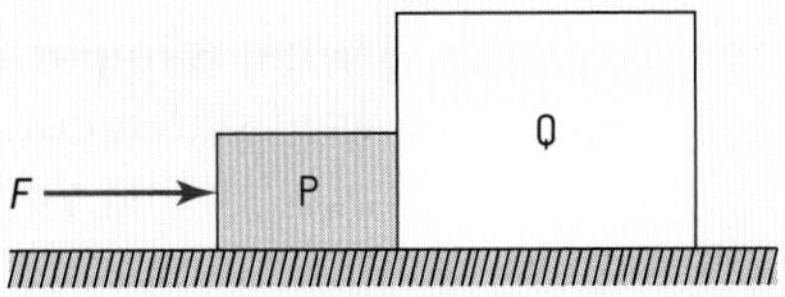

State the resultant force on box Q.

Solution

Q is moving at constant speed. So the resultant force on this box must be 0 N.

2.3 Work, energy, and power

Understanding

- Principle of conservation of energy
- Kinetic energy
- Gravitational potential energy
- Elastic potential energy
- Work done as energy transfer
- Power as rate of energy transfer
- Efficiency

Nature of science

The theory of conservation of energy allows many areas of science to be understood at a fundamental level. It allows the explanation of natural phenomena but also means that scientists can predict the outcome of a previously unknown effect. The conservation of energy also demonstrates that paradigm shifts occur in science: the interchangeability of mass and energy as predicted by Einstein is an example of this.

Applications and skills

- Discussing the conservation of total energy within energy transformations
- Sketching and interpreting force – distance graphs
- Determining work done including cases where a resistive force acts
- Solving problems involving power
- Quantitatively describing efficiency in energy transfers

Equations

- work: $W = Fs\cos\theta$
- kinetic energy: $E_K = \frac{1}{2}mv^2$
- elastic potential energy: $E_p = \frac{1}{2}k(\Delta x)^2$
- change in gravitational potential energy: $\Delta E_p = mg\Delta h$
- power $= Fv$
- efficiency $= \frac{\text{useful workout}}{\text{total work in}} = \frac{\text{useful power out}}{\text{total power in}}$

Introduction

This sub-topic looks at the physics of energy transfers. The importance of energy is best appreciated when it moves or transfers between different forms. Then we can make it do a useful job for us.

What is important is that you learn to look below the surface of the bald statement: "electrical energy is converted into internal energy" and ensure that you can talk about the detailed changes that are taking place.

Energy forms and transfers

Energy can be stored in many different forms. Some of the important ones are listed in the table below.

Energy	Nature of energy associated with...	Notes
kinetic	the motion of a mass	
(gravitational) potential	the position of a mass in a gravitational field	sometimes the word "gravitational" is not used
electric/magnetic	charge flowing	
chemical	atoms and their molecular arrangements	
nuclear	the nucleus of an atom	related to a mass change by $\Delta E = \Delta mc^2$
elastic (potential)	an object being deformed	The word "potential" is not always used
thermal (heat)	a change in temperature or a change of state	A change of state is a change of a substance between phases, i.e. solid to liquid, or liquid to gas. This is referred to as "energy transferred as a result of temperature difference" in line with the IB Guide. The colloquial term "heat" is usually acceptable when referring to situations involving conservation of energy situations.
mass	conversion to binding (nuclear) energy when nuclear changes occur	
vibration (sound)	mechanical waves in solids, liquids, or gases	the amount of sound energy transferred is almost always negligible when compared with other energy forms
light	photons of light	sometimes called "radiant energy" another form of electric/magnetic

Energy can be transferred between any of its forms and it is during such transfers we see the effects of energy. For example, water can fall vertically to turn the turbine of a hydroelectric power station and drive a generator. Lots of things are happening here. The water molecules are attracted by the Earth and accelerate downwards through the pipe. Their momentum is transferred to the blades of the turbine which rotates and turns the coils in the generator. As a result of the coils turning, electrons are forced to move and there is an electric current. Overall, this chain of physical processes can be summed up as the conversion of gravitational potential energy of the water into an electrical form. Another example of an energy transfer is that of an animal converting stored chemical energy in the muscles leading to kinetic and gravitational potential energy forms with some of the chemical energy also appearing as frictional losses.

Learn to recognise the physical (and sometimes chemical) processes that are going on in the system. It is easy to describe the changes in fairly broad energy transfer terms. Always try, however, to explain the effects in terms of microscopic or macroscopic interactions. Whatever the form of the energy transfer, we use a unit of energy called the joule (J). This is in honour of James Joule (1818–1889) the English scientist who devoted his scientific efforts to studying energy and its transfers.

One joule is the energy required when a force of one newton acts through a distance of one metre.

When energy changes from one form to another we find that nothing is lost (providing that we take care to include every single form of energy that is included in the changes).

This is known as the **principle of conservation of energy** which says that energy cannot be created or destroyed.

Since Einstein's work at the turn of the twentieth century, we now recognise that mass must be included in our table of energy forms. For most changes the mass difference is insignificant, but in nuclear changes it makes a major contribution.

In some applications, such as when discussing the output of power stations, the joule is too small a unit so you will frequently see energies expressed in megajoules (MJ or 10^6 J) or even gigajoules (GJ or 10^9 J). Get used to working in large powers of ten and with prefixes when dealing with energies.

In some parts of physics, different energy units are used. These have usually arisen historically. An example of this is the electronvolt (eV), which is the energy gained by an electron when it is accelerated through a potential difference of one volt. Another example is the calorie used by dieticians; this is an old unit that has lingered in the public domain for perhaps longer than it should! One calorie (cal) is 4.2 J. You will learn the detail about any special units used in the course in the appropriate place in this book.

Worked example

Describe the mechanisms associated with the energy changes that occur when a balloon is blown up.

Solution

Air molecules gain kinetic energy that is used to store elastic potential energy in the skin of the balloon and to make changes in the energy of the air. The air inside is able to exert more force (pressure) outwards on the skin of the balloon until a new equilibrium is established between the tension in the skin and the atmospheric pressure.

Doing work

In 1826, Gaspard-Gustave Coriolis was studying the engineering involved in raising water from a flooded underground mine. He realized that energy was being transferred when the steam engines were pumping the water vertically. He described this energy transfer as "work done", and he recognized that the energy transferred when the pumping engines exerted a force on a particular mass of water and lifted it from the bottom of the mine to the surface.

In other words

work done (in J) = *force exerted* (in N) × *distance moved in the direction of the force* (in m)

So, when a weight of 5 N of water is lifted vertically through a height of 150 m, then the work done by the engine on the water is $5 \times 150 = 750$ J.

In the example of the mine, the force and the distance moved are in the same direction (vertically upwards), but in many cases this will not

▲ Figure 1 Mine steam engine.

be the case. A good example of this is a sand yacht, where the force F from the wind acts in one direction and the sail is set so that the yacht moves through a displacement s that is at an angle θ to the wind.

In this case, the distances moved in the direction of the force and by the yacht are not the same. You must use the component of force in the direction of movement.

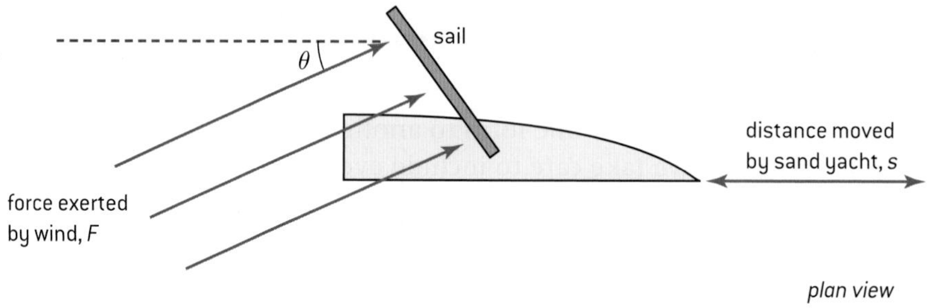

▲ Figure 2 Force on sail.

In this case,

work done $= F\cos\theta \times s$

Work done against a resistive force

Work is done when a resistive force is operating too. Consider a box being pushed at a constant speed in a horizontal straight line. For the speed to be constant, friction forces must be overcome. The force that overcomes the friction may not act in the direction of movement. Again, the work that will be done by the force (or the force provider) is *force acting* × *distance travelled* × $\cos\theta$ (figure 3).

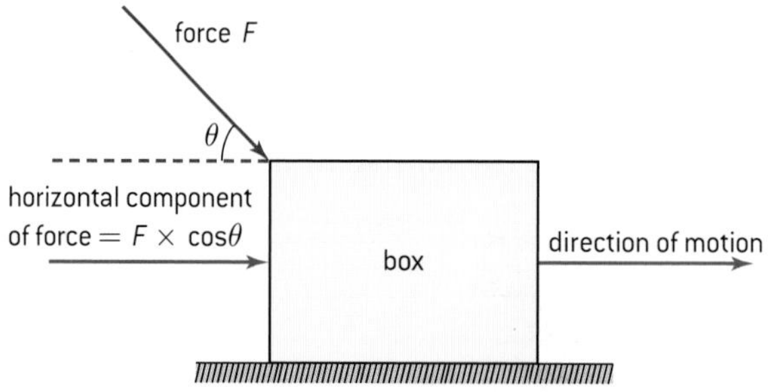

▲ Figure 3 Resistive force.

Worked examples

1 The thrust (driving force) of a microlight aircraft engine is 3.5×10^3 N. Calculate the work done by the thrust when the aircraft travels a distance of 15 km.

Solution

Work done = force × distance = 3500 × 15 000 = 5.3×10^7 J ≡ 53 MJ.

2 A large box is pulled a distance of 8.5 m along a rough horizontal surface by a force of 55 N that acts at 50° to the horizontal. Calculate the work done in moving the box 8.5 m.

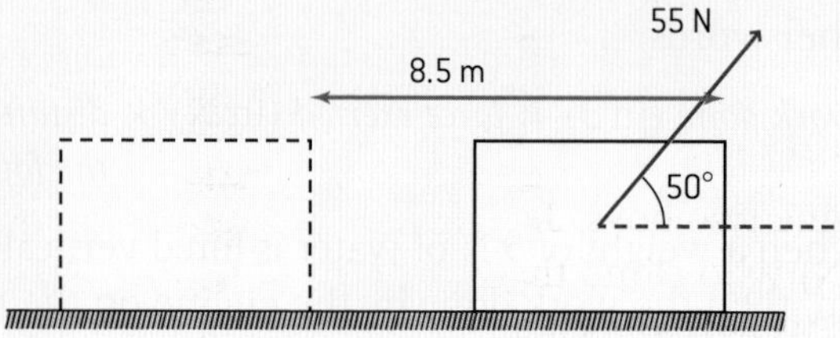

Solution

The component of force in the direction of travel is $55 \times \cos 50 = 35.4$ N.

The work done = this force component × distance travelled $= 35.4 \times 8.5 = 301$ J.

3 An object moving in a straight line has an initial kinetic energy of 24 J. Calculate the distance in which the object will come to rest if a net force of 4.0 N opposes the motion.

Solution

There must be 24 J of work done to stop the motion of the object. The force acting is 4.0 so $\frac{24}{4} = 6$ m.

Force–distance graphs

In practice, it is rare for the force acting on a moving object to be constant. Real railway trucks or sand yachts have air resistance and other forces that lead to energy losses that vary with the speed of the object or the surface over which it runs.

We can deal with this if we know how the force varies with distance. Some examples are shown in figure 4.

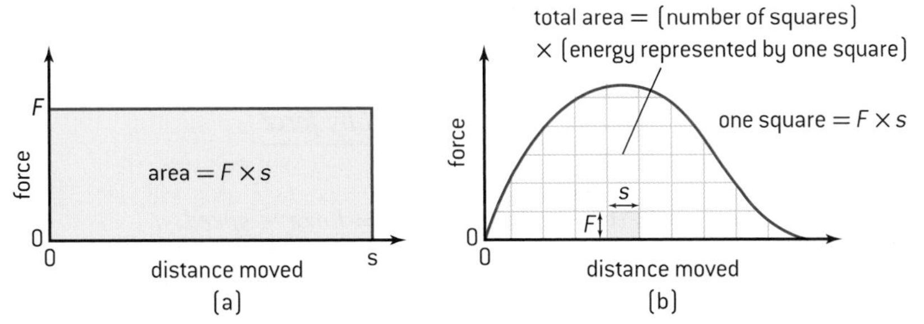

▲ Figure 4 Force–distance graphs.

For a constant force (figure 4(a)), the graph of force against distance will be a straight line parallel to the *x*-axis.

The *work done* is the product of *force* × *distance* (we are assuming that $\theta = 90°$ in this case); this corresponds to the area under the graph of force against distance.

When the force is not constant with distance moved (figure 4(b)) the work done is still the area under the line, but this time you have to work a little harder by estimating the number of squares under the graph and equating each square to the energy that it represents. The product of (*energy for one square*) × (*number of squares*) will then give you the overall work done.

There is a further example of this type of calculation later in the topic on elastic potential energy.

Power

Imagine two boys, Jean and Phillipe with the same weight (let's say 650 N) who climb the same hill (70 m high). Because they have the same weight and climb the same vertical distance they both gain the same amount of gravitational potential energy. This is at the expense of the chemical energy reserves in their bodies. However, suppose Jean climbs the hill in 150 s whereas Phillipe takes 300 s.

Worked example

The graph shows the variation with displacement *d* of a force *F* that is applied to a toy car. Calculate the work done by *F* in moving the toy through a distance of 4.0 cm.

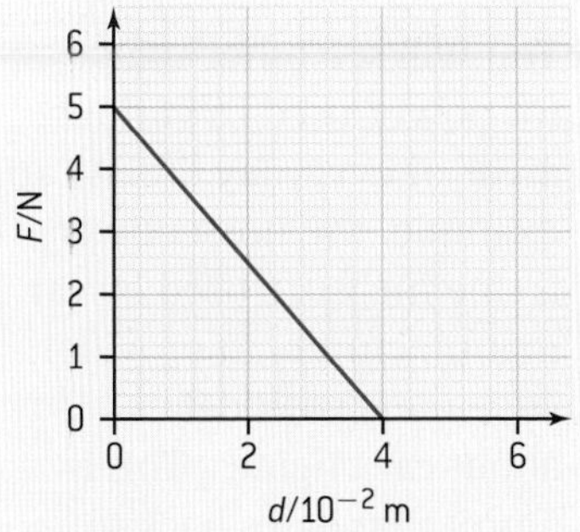

Solution

The work done is equal to the area of the triangle enclosed by the graph and the axes. This is $\frac{1}{2}$ × base of the triangle × height of triangle $= \frac{1}{2} \times 4.0 \times 10^{-2} \times 5.0 = 0.10$ J

James Watt was a mechanical engineer who worked at the end of the 18th century and the beginning of the 19th. He did major work in improving steam engine design and even developed a way of making copies of paper documents that was used in offices until early in the 20th century.

The obvious difference here is that Jean is gaining potential energy twice as fast as Phillipe because Jean's time to make the climb is half that of Phillipe. This difference is important when we want to compare two machines taking different times to carry out the same amount of work.

The quantity **power** is used to measure the **rate of doing work**, in other words it is the number of joules that can be converted every second. So, power is defined as:

$$\text{power} = \frac{\textit{energy transferred}}{\textit{time taken for transfer}}$$

Using the correct quantities is important here. If the energy change is in joules, and the time for the transfer measured in seconds then the power is in watts (W).

$$1\ \text{W} \equiv 1\ \text{J s}^{-1}$$

In the example of Jean and Phillipe above, both did 45.5 kJ of work, but Jean's power in climbing the hill was $\frac{650 \times 70}{150} = 303$ W and Phillipe's was 152 W because he took twice as long in the climb.

The equation *work done* = *force* × *distance* can be rearranged to give another useful expression for power.

$$\textit{power} = \frac{\textit{work done}}{\textit{time}} = \textit{force} \times \frac{\textit{distance moved by force}}{\textit{time}}$$

It is easy to see that this is the same as *power* = *force* × *speed*.

So the power required to move an object travelling at a speed v with a force F is Fv.

Nature of science

You will see other units for power used in various areas of everyday life. As an example, the horsepower is one measure used by car manufacturers to provide customers with data about car engines. This is a unit that dates back to the beginning of the 18th century and was used to compare the power of an average horse with steam engines. (1 horsepower is equal to about 750 W in modern units.)

Kinetic energy, KE

Kinetic energy is the energy an object has because of its motion; kinetic energy (sometimes abbreviated to KE) has its own symbol, E_K. Objects gain kinetic energy when their speed increases.

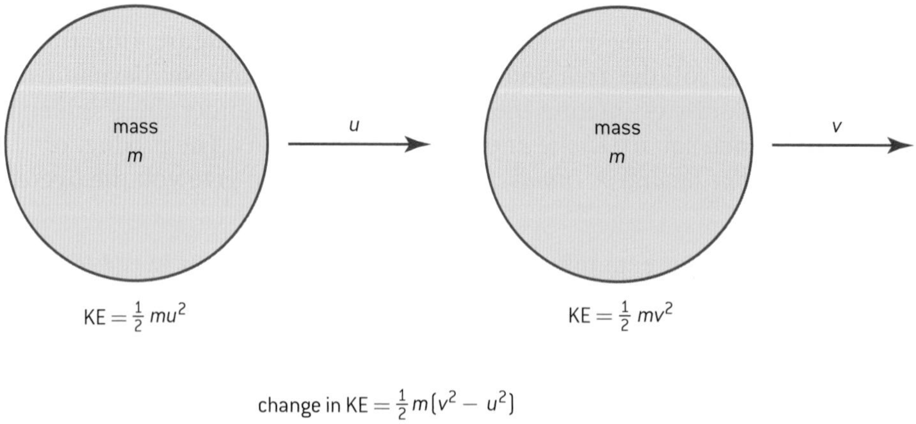

▲ Figure 5 Kinetic energy of mass.

Imagine an object of mass m which is at rest at time $t = 0$ and which is accelerated by a force F for a time T. The kinematic equations and Newton's second law allow us to work out the speed of the object v at time $t = T$.

The acceleration a is $\frac{F}{m}$ (using Newton's second law of motion).

Therefore $v = 0 + \frac{F}{m}T$ (because the initial speed is zero). So $F = \frac{mv}{T}$

The work done on the mass is the gain in its kinetic energy, E_K, and is $F \times s$ where s is the distance travelled and therefore

$$E_K = F \times s = \frac{mv}{T} \times \frac{vT}{2}$$

because $s = \frac{(v + 0)T}{2}$

The work done by the force is equal to the gain in kinetic energy and is

$$E_K = \frac{1}{2}mv^2$$

Remember that this is the case where the initial speed was 0. If the object is already moving at an initial speed u then the change in kinetic energy ΔE_K will be $\frac{1}{2}m(v^2 - u^2)$.

There is a subtle piece of notation here. When we talk about a value of kinetic energy, we write E_K, but when we are talking about a change in kinetic energy from one value to another then we should write ΔE_K where Δ, as usual, means "the change in".

Worked examples

1 A vehicle is being designed to capture the world land speed record. It has a maximum design speed of 1700 km h^{-1} and a fully fuelled mass of 7800 kg.

Calculate the maximum kinetic energy of the vehicle.

Solution

1700 km h^{-1} $\equiv$ 470 m s^{-1} (this is greater than the speed of sound in air!)

$E_K = \frac{1}{2}mv^2 = 0.5 \times 7800 \times 470^2 = 8.6 \times 10^8$ J

$\equiv 0.86$ GJ

2 A car of mass 1.3×10^3 kg accelerates from a speed of 12 m s^{-1} to a speed of 20 m s^{-1}. Calculate the change in kinetic energy of the car.

Solution

$\Delta E_K = \frac{1}{2}m(v^2 - u^2) = 0.5 \times 1300 \times (20^2 - 12^2)$

$= 650 \times (400 - 144) = 1.7 \times 10^5$ J

Worked example

A car is travelling at a constant speed of 25 m s^{-1} and its engine is producing a useful power output of 20 kW. Calculate the driving force required to maintain this speed.

Solution

driving force $= \frac{\text{power}}{\text{speed}}$

$= \frac{20\,000}{25} = 800$ N

Examiner's tip

This equation for ΔE_K is one that needs a little care. Notice where the squares are; they are attached to each individual speed. This equation is not the same as $\frac{1}{2}m(v - u)^2$.

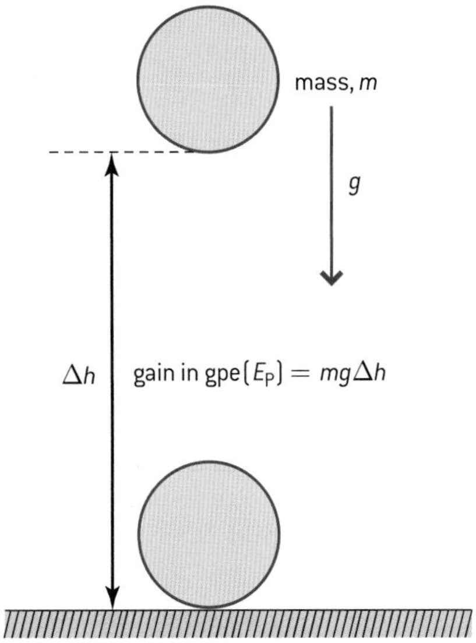

▲ Figure 6 Gravitational potential energy.

Gravitational potential energy, GPE

Gravitational potential energy is the energy an object has because of its position in a gravitational field. When a mass is moved vertically up or down in the gravity field of the Earth, it gains or loses gravitational potential energy (GPE). The symbol assigned to GPE is E_p. Only the initial and final positions relative to the surface determine the change of GPE (assuming that there is no air resistance on the way). For this reason, gravitational force is among the group of forces said to be **conservative**, because they conserve energy (see the Nature of Science box below).

The work done when an object is raised at constant speed through a change in height Δh is, as usual, equal to *force* × *distance* moved. In this case the force required is mg and work done, $E_p = mg \times \Delta h$.

As usual, the value of g is 9.8 m s^{-2} close to the Earth's surface, but this value becomes smaller when we move away from the Earth.

Nature of science

Conservative forces

There is an important difference between forces such as gravity and frictional forces. When only gravity acts, the energy change depends on the start height and end height of the motion but not on the route taken by an object to get from start to finish. Horizontal movement does not have to be counted if there is no friction acting. This type of force is said to be *conservative;* in other words, it conserves energy. We could recover all the energy by moving the object back to the start. Contrast this with the friction force that acts between a book and a table as the book is moved around the table's surface from one point to another. If the book goes directly from start to finish a certain amount of energy will be used up to overcome the friction. But if the book goes by a longer route, more energy is needed (*work done = friction force × distance travelled*). When it is necessary to know the exact route before we can calculate the total energy conversion, the force is said to be *non-conservative*. If we move the book back again, we cannot recover the energy in the way that we could when only gravity acted.

You will not need to write about the meaning of the term "conservative force" in the IB Diploma Programme physics examination.

Energy moving between GPE and KE

Sometimes the use of gravitational potential energy and kinetic energy together provides a neat way to solve a problem.

A snowboarder is moving down a curved slope starting from rest ($u = 0$). The vertical change in height of the slope is $\Delta h = 50$ m.

What is the speed of the snowboarder at the bottom of the slope? Assume we can ignore friction at the base of the board and the air resistance.

What we must *not* do in this example is to use the kinematic (*suvat*) equations. They cannot be applied in this case because the acceleration of the snowboarder will not be constant – *suvat* only works when we are certain that a does not change.

Although the *suvat* equations give the correct answer here, they should not be used because the physics is incorrect. The final answer happens to be correct only because we use the start and end points and also as a consequence of the conservative nature of gravity. We are also using an average value for acceleration down the slope by assuming that the angle to the horizontal is constant. The equations would not give the correct answer if we brought friction forces into the calculation.

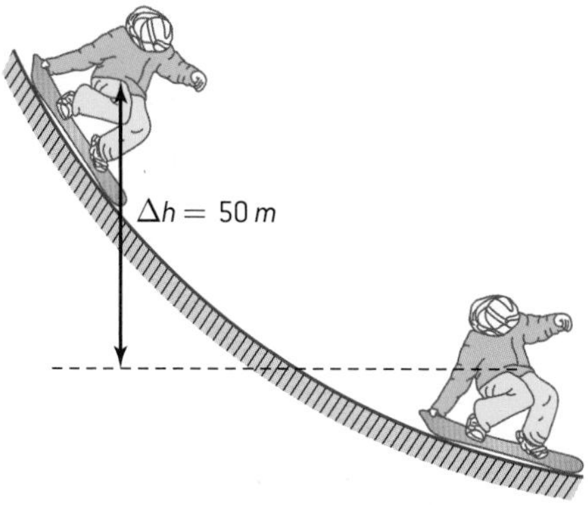

▲ Figure 7 Mechanics of snowboarding.

Conservation of energy comes to our aid because (as the friction losses are negligible) we know that the loss of gravitational potential energy as the boarder goes down the slope is equal to the gain in kinetic energy over the length of the slope. Because we know that the GPE change only depends on the initial and final positions then we do not need to worry at all about what is going on at the base of a board during the ski run.

So $\Delta E_P = mg\Delta h = \frac{1}{2}mv^2$ and $v = \sqrt{2g\Delta h}$,

in this case $v = \sqrt{2 \times 9.8 \times 50} = 31 \text{ m s}^{-1}$

(This is a speed of about 110 km h^{-1} which tells you that the assumption about no air resistance and no friction is a poor one, as any snowboarder will tell you!)

Notice that the answer does not depend on the mass of the boarder; the mass term cancels out in the equations.

Investigate!

Converting GPE to KE

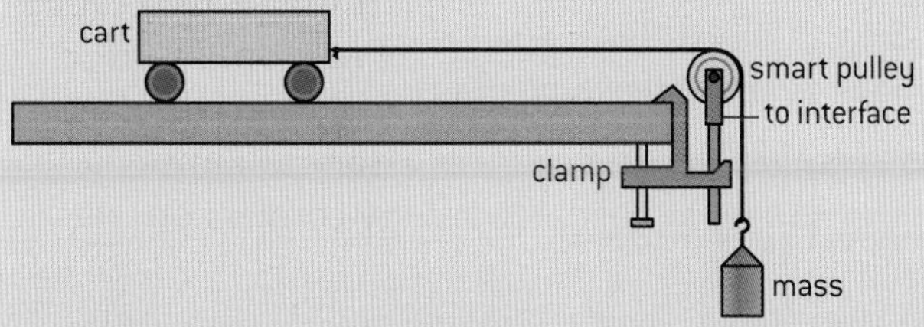

This experiment will help you to understand the conversion between KE and GPE.

- Arrange a cart on a track and compensate the track for friction. This is done by raising the left-hand end of the track through a small distance so that the cart neither gains nor loses speed when travelling down the track without the string attached.
- Have a string passing over a pulley at the end of the track and tie a weight of known mass to the other end of the track.
- Measure the mass of the cart too.
- Devise a way to measure the speed of the cart. You could use a "smart pulley" that can measure the speed as the string turns the pulley wheel, or an ultrasound sensor, or a data logger with light gates.
- When the mass is released, the cart will gain speed, as the gravitational potential energy of the weight changes.
- Make measurements to assess the gravitational potential energy lost (you will need to know the vertical height through which the mass falls) and the kinetic energy gained (you will need the final speed). Notice that only the falling mass is losing GPE but both the mass and the cart are gaining kinetic energy.
- Compare the two energies in the light of the likely errors in the experiment. Is the energy conserved? Where do you expect energy losses in the experiment to occur?

Worked examples

1 A ball of mass 0.35 kg is thrown vertically upwards at a speed of 8.0 m s^{-1}. Calculate

a) the initial kinetic energy

b) the maximum gravitational potential energy

c) the maximum height reached.

Solution

a) $E_K = \frac{1}{2}mv^2 = \frac{1}{2} \times 0.35 \times 8^2 = 11.2$ J

b) At the maximum height all the initial kinetic energy will have been converted to gravitational potential energy so the maximum value of the GPE is also 11.2 J.

c) The maximum GPE is 11.2 J and this is equal to $mg\Delta h$, so $\Delta h = \frac{11.2}{mg} = \frac{11.2}{0.35 \times 9.8} = 3.3$ m.

2 A pendulum bob is released from rest 0.15 m above its rest position. Calculate the speed as it passes through the rest position.

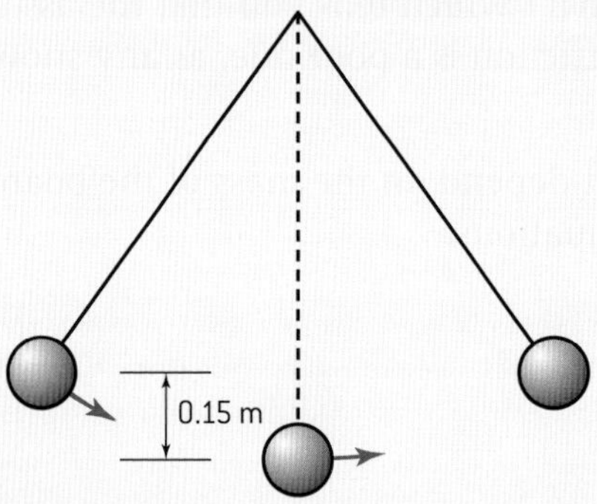

Solution

E_K at the rest position = E_P at the release position;

$\frac{1}{2}mv^2 = mg\Delta h$ which rearranges to

$v = \sqrt{2gh} = \sqrt{2 \times 9.8 \times 0.15} = 1.7$ m s^{-1}.

Elastic potential energy

The shape of a solid can be changed by applying a force to it. Different materials will respond to a given force in different ways; some materials will be able to return the energy that has been stored in them when the force is removed. A metal spring is a good example of this; most springs are designed to store energy in this way in many different contexts. The materials that can return energy in this way have stored **elastic potential energy**.

To discuss elastic potential energy we need to know something about the properties of springs. This is an area of physics that has been studied for a long time. Robert Hooke, a contemporary of Newton, published a rule about springs that has become known as **Hooke's law**.

Investigate!

Investigating Hooke's law

Robert Hooke realized that there was a relationship between the load on a spring and the extension of the spring.

- Arrange a spring of known unstretched length with a weight hanging on the end of the spring.
- You will need to devise a way to measure the **extension** (change in length from the original unstretched length) of the string for each of a number of different increasing weights hanging on the end of the spring.
- Repeat the measurements as you remove the weights as a check.
- Plot a graph of force (weight) acting on the spring (y-axis) against the extension (x-axis). This is not the obvious way to draw the graph, but it is the way normally used. Normally, we would plot the dependent variable (the extension in this case) on the y-axis.

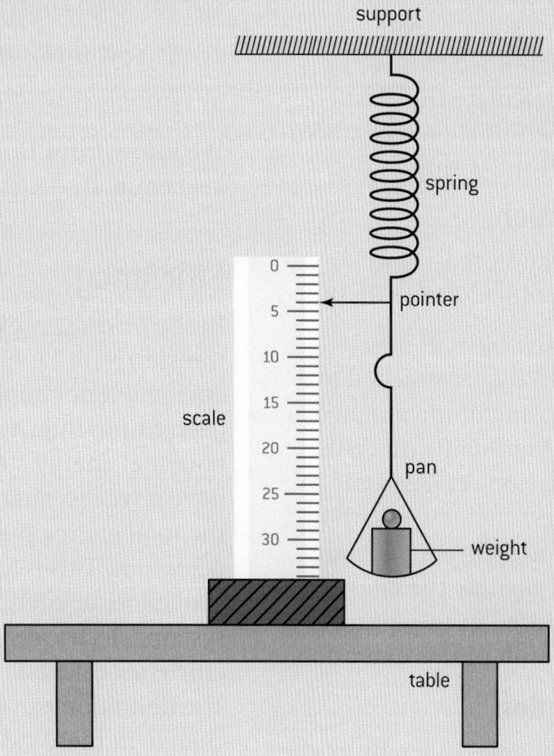

For small loads acting on the spring, Hooke showed that the extension of the spring is directly proportional to the load. (For larger loads, this relationship breaks down as the microscopic arrangement of atoms in the spring changes. We shall not consider this part of the deformation, however.)

Hooke's rule means that the graph of force F against extension (Δx) is a straight line going through the origin.

In symbols the rule is $F \propto \Delta x$, or $F = k\Delta x$ where k is a constant known as the **spring constant**, which has units of N m^{-1}. The gradient of the graph is equal to k (this is why the graph is plotted the "wrong way round").

We can now relate this graph to the work done in stretching the spring. The force is not constant (the bigger the extension, the bigger the force

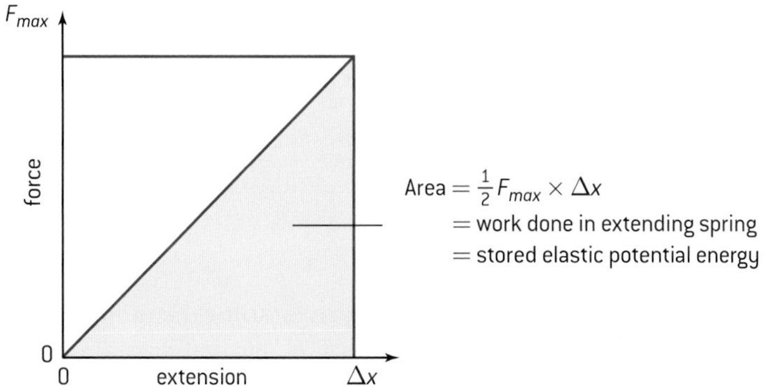

▲ Figure 8 Work done in stretching spring.

Worked examples

1 A spring, of spring constant 48 N m^{-1}, is extended by 0.40 m. Calculate the elastic potential energy stored in the spring.

Solution

Energy stored $= \frac{1}{2}kx^2 = 0.5 \times 48 \times 0.40^2 = 0.38$ J

2 An object of mass 0.78 kg is attached to a spring of unstretched length 560 mm. When the object has come to rest the new length of the spring is 620 mm. Calculate the energy stored in the spring as a result of this extension.

Solution

The change in length of the spring $\Delta x = 620 - 560 = 60$ mm

The tension in the spring will be equal to the weight of the object $= mg = 0.78 \times 9.8 = 7.64$ N

The energy stored in the spring $= \frac{1}{2}F\Delta x = 0.5 \times 7.64 \times 0.06 = 0.23$ J

required) but we now know how to deal with this. The work done on the spring is the area under the graph of force against extension.

The area is a right-angled triangle and is equal to

$\frac{1}{2} \times$ *force to stretch spring* $\times$ *extension*,

in symbols $E_P = \frac{1}{2} F_{max} \Delta x$.

We know from Hooke's law that $F = k\Delta x$, so $k = \frac{F}{\Delta x}$ and therefore E_P is also equal to $\frac{1}{2} k(\Delta x)^2$.

Efficiency

In the *Investigate!* experiment where gravitational potential energy was transferred to kinetic energy, it is likely that some gravitational potential energy did not appear in the motion of the cart. Some energy will have been lost to internal energy as a consequence of friction and to elastic potential energy stored when the string changes its shape. We need to have a way to quantify these losses. One way to do this is to compare the total energy put into a system with the useful energy that can be taken out. This is known as the **efficiency** of the transfer and can be applied to all energy transfers, whether carried out in a mechanical system, electrical system or other type of transfer. You can expect to meet efficiency calculations in any area of physics where energy transfers occur. The definition can also be applied to power transfers, because the energy change in these cases takes place in the same time for the total energy in and the useful work out.

$$\text{efficiency} = \frac{\text{useful work out}}{\text{total energy in}} = \frac{\text{useful power out}}{\text{total power}}$$

Worked example

1 An electric motor raises a weight of 150 N through a height of 7.2 m. The energy supplied to the motor during this process is 3.5×10^4 J. Calculate:

a) the increase in gravitational potential energy

b) the efficiency of the process.

Solution

a) $\Delta E_P = 150 \times 7.2 = 1080$ J

b) Efficiency $= \frac{\text{useful work out}}{\text{energy in}} = \frac{1080}{3500} = 0.31$ or 31%

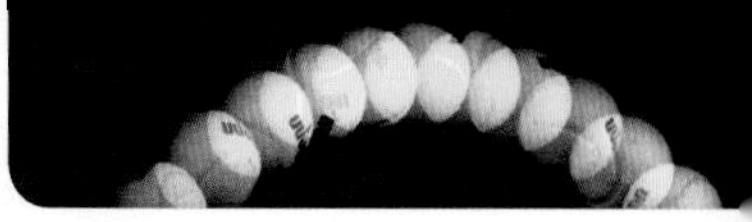

2.4 Momentum

Understanding

- Newton's second law expressed as a rate of change of momentum
- Impulse and force–time graphs
- Conservation of linear momentum
- Elastic collisions, inelastic collisions, and explosions

Nature of science

The concept of momentum has arisen as a result of evidence from observations carried out over many centuries. The principle of conservation of momentum is an example of a law that has universal applicability. It allows the prediction of the outcomes of physical interactions at both the macroscopic and the microscopic level. Many areas of physics are informed by it, from the kinetic theory of gases to nuclear interactions.

Applications and skills

- Applying conservation of momentum in simple isolated systems including (but not limited to) collisions, explosions, or water jets
- Using Newton's second law quantitatively and qualitatively in cases where mass is not constant
- Sketching and interpreting force–time graphs
- Determining impulse in various contexts including (but not limited to) car safety and sports
- Qualitatively and quantitatively comparing situations involving elastic collisions, inelastic collisions, and explosions

Equations

- momentum: $p = mv$
- Newton's second law (momentum version): $F = \frac{\Delta p}{\Delta t}$
- kinetic energy: $E_K = \frac{p^2}{2m}$
- impulse: $= F\Delta t = \Delta p$

Introduction

Many sports involve throwing or catching a ball. Compare catching a table-tennis (ping-pong) ball with catching a baseball travelling at the same speed. One of these may be a more painful experience than the other! What is different in these two cases is the mass of the object. The velocity may be the same in both cases but the combination of velocity and mass makes a substantial difference. Equally, comparing the experience of catching a baseball when gently tossed from one person to another with catching a firm hit from a good player should tell you that changes in velocity make a difference too.

Momentum

We call the product of the mass m of an object and its instantaneous velocity v the **momentum p of the object** ($p = m \times v$) and this quantity turns out to have far-reaching importance in physics.

First, here are some basics ideas about momentum.

- Momentum is mass × velocity *never* mass × speed.
- Momentum has direction. The mass is a scalar but velocity is a vector. When mass and velocity are multiplied together, the momentum is also a vector with the same direction as the velocity. Think of the mass as "scaling" the velocity – in other words, just making it bigger by a factor equal to the mass of the object.
- It follows that the unit of momentum is the product of the units of mass and velocity, in other words **kg m s^{-1}**. There is a shorter alternative to this that we shall see later.
- If velocity or mass is changing then the momentum must also be changing. We shall be looking at the two cases where one quantity changes while the other is held constant.
- When a net resultant force acts on an object, the object accelerates and the velocity must change. This means a change in momentum too. So a net force leads to a change in momentum.

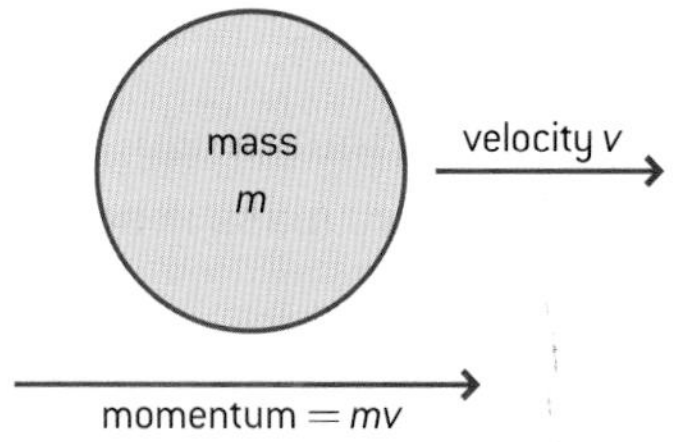

▲ Figure 1 Momentum and velocity.

Worked examples

1 A ball of mass 0.25 kg is moving to the right at a speed of 7.4 m s^{-1}. Calculate the momentum of the ball.

Solution

$p = mv = 0.25 \times 7.4 = 1.85$ kg m s^{-1} to the right.

2 A ball of mass 0.25 kg is moving to the right at a speed of 7.4 m s^{-1}. It strikes a wall at 90° and rebounds from the wall leaving it with a speed of 5.8 m s^{-1} moving to the left. Calculate the change in momentum.

Solution

From the previous example the initial momentum is 1.85 kg m s^{-1} to the right.

The final momentum is $0.25 \times 5.8 = 1.45$ kg m s^{-1} to the left.

So taking the direction to the right as being positive, the change in momentum $= -1.45 - (+1.85) = -3.3$ kg m s^{-1} to the right (or alternatively $+3.3$ kg m s^{-1} to the left).

▲ Figure 2 Newton's cradle.

Collisions and changing momentum

You may have seen a "Newton's cradle". Newton did not invent this device (it was developed in the twentieth century as an executive toy), but it helps us to visualize some important rules relating to his laws of motion, and it certainly seems appropriate to associate his name with it.

One of the balls (the right-hand one in figure 2) is moved up and to the right, away from the remaining four. When released the ball falls back and hits the second ball from the right. The right-hand ball then stops moving and the left-most ball moves off to the left. It is as though the motion of the original ball transfers through the middle three – which remain stationary – and appears at the left-hand end.

You can analyse this in terms of physics that you already know.

The right-hand ball gains gravitational potential energy when it is raised. When it is released the potential energy is converted into kinetic energy as the ball gains speed. Eventually the ball strikes the next stationary one along, and a pair of action–reaction forces acts at the surfaces of the two balls (you might like to consider what they are). The stationary ball is compressed slightly by the force acting on it and this compression moves as a wave through the middle three balls. When the compression reaches the left-hand ball the elastic potential energy associated with the compression wave is converted into kinetic energy (because there is a net force on this sphere) and work begins to be done against the Earth's gravity. The left-hand ball starts to move to the left. When the speed of the ball becomes zero, it is at its highest point on the left-hand side and the motion repeats in the reverse direction. The overall result is that the end spheres appear to perform an alternate series of half oscillations.

Another way to view this sequence is as a transfer of momentum. Think about a simpler case where only two spheres are in contact (in the toy, three balls can be lifted out of the way). The right-hand sphere gains momentum as it falls from the top of its swing. When it collides with the other sphere, the momentum appears to be transferred to the second sphere. The first sphere now has zero momentum (it is stationary) and the second has gained momentum. What rules govern this transfer of momentum?

These interactions between the balls in the Newton's cradle are called **collisions**. This is the term given to any interaction where momentum transfers. Examples of collisions include firing a gun, hitting a ball with a bat in sport, two toy cars running into each other, and a pile driver sinking vertical cylinders into the ground on a construction site. There are many more.

Investigate!

Is momentum conserved?

The exact details of this experiment will depend on the apparatus you have in your school.

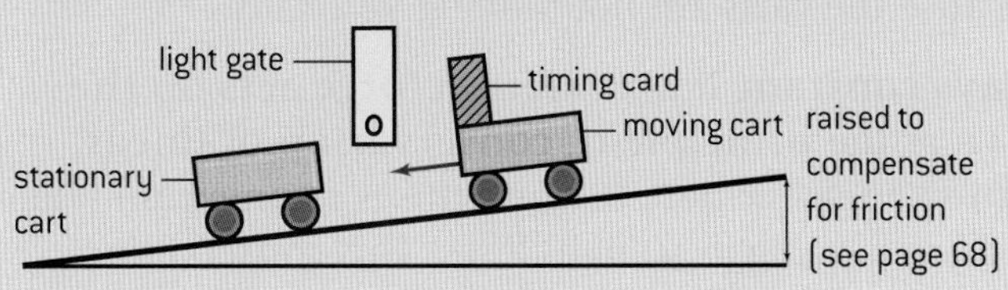

- The experiment consists of measuring the speed of a cart of known mass and then launching it at another cart also of known mass that is initially stationary. Often there will be carts available to you of almost identical mass, as this is a particularly easy case to begin with.
- You will need a way to measure the velocity of the carts just before and just after the collision. This could be done in various ways:
  - using a data logger with motion sensors
  - using a paper tape system where a tape attached to the cart is pulled through a device that makes dots at regular time intervals on the tape
  - using a video camera and computer software
  - using a stop watch to measure the time taken to cover a short, known distance before and after the collision.
- If your carts run on a track, you can allow for the friction at the cart axles and the air resistance. This is done by raising the end of the track a little so that, when pushed, a cart runs at a constant speed. The friction at the bearings and the air resistance will be exactly compensated by the component of the cart weight down the track.
- For the first part of the experiment, begin by arranging that the two carts will stick together after colliding. This can be done in a number of ways, including using modelling clay, or a pin attached to one cart entering a piece of cork in the other, or two magnets (one on each cart) that attract.
- Make the first (moving) cart collide with and stick to the second (stationary) cart.
- Measure the speed of the first cart before the collision and the combined speed of the carts after the collision.
- Repeat the experiment a number of times and think carefully about the likely errors in the results.

The initial momentum is *mass of first cart* × *velocity of first cart*. The final momentum is (*mass of first cart* + *mass of second cart*) × *combined velocity of both carts*.

- What can you say about the total momentum of the system before the collision compared with the total momentum after the collision? You should consider the experimental errors in the experiment before making your judgement.
- If you have done the experiment carefully, you should find that the momentum before and the momentum after the collision are approximately equal.
- Now extend your experiment to different cases:
  - where the carts do not stick together
  - where they are both moving before the collision
  - where the masses are not the same, and so on.
- You may need to alter how you measure the velocity to cope with the different cases.

If you carry out an experiment that compares the momentum before with the momentum after a collision then you should find (within the experimental uncertainty of your measurements) that the total momentum in the system does not change.

An important point here is that there must be no force from outside the system acting on the objects taking part in the collision. As we saw earlier, external forces produce accelerations, and this changes the velocity and hence the momentum.

You might argue that gravitational force is acting – and gravity certainly is acting on the carts in the experiment – but because the gravitational force is not being allowed to do any work, the force does not contribute to the interaction because the carts are not moving vertically.

Momentum is always constant if no external force acts on the system. This is known as the **principle of conservation of linear momentum**, the word "linear" is here because this is momentum when the objects concerned are moving in straight lines. This conservation rule has never been observed to be broken, and is one of the important conservation rules that (as far as we know) are true throughout the universe. In nuclear physics, particles have been proposed in order to conserve momentum in cases where it was apparently not being conserved. These particles were subsequently found to exist.

Momentum conservation is such an important rule that it is worth us considering a few different situations to see how momentum conservation works.

In each of these cases we will assume that the centres of the objects lie on a straight line so that the collision happens in one dimension.

Two objects with the same mass, one initially stationary, when no energy is lost

This is known as an **elastic collision** and for it to happen no permanent deformation must occur in the colliding objects and no energy can be released as internal energy (through friction), sound, or any other way. The spheres in the Newton's cradle lose only a little energy every time they collide and this is why this is a reasonable demonstration of momentum effects.

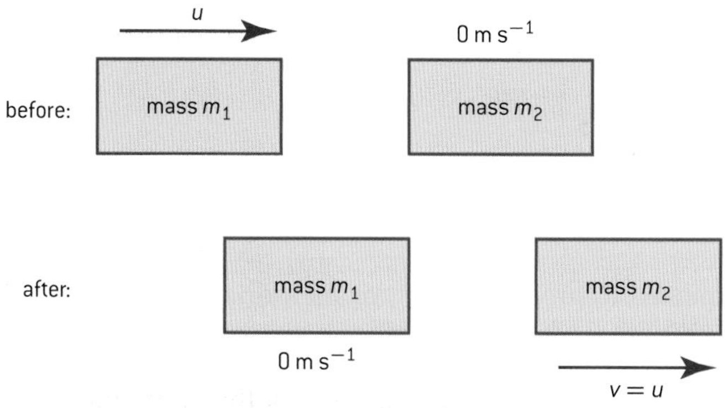

▲ Figure 3 Elastic collision between two identical masses.

The first object collides with the second stationary object. The first object stops and remains at rest while the second moves off at the speed that the first object had before the collision. (Try flicking a coin across a smooth table to hit an identical coin head-on.) In this case momentum is conserved because (using an obvious set of symbols for the mass m of the objects and their velocities u and v)

$$m_1 u = m_2 v$$

Because $m_1 = m_2$ then $u = v$ so the velocity of one mass before the collision is equal to the velocity of the second mass afterwards. The kinetic energy of the moving mass (whichever mass is moving) is $\frac{1}{2}mu^2$ and it does not change either.

Two objects with different masses when no energy is lost

This time (as you may have seen in an experiment or demonstration) the situation is more complicated.

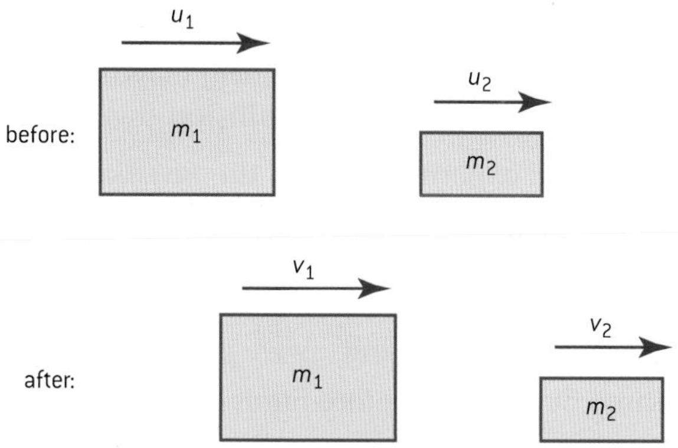

▲ Figure 4 Two moving objects with different mass in an elastic collision.

Again

$$m_1 u_1 + m_2 u_2 = m_1 v_1 + m_2 v_2$$

but this time we cannot eliminate the mass terms so easily. What we do know, though, is that kinetic energy is conserved. So the kinetic energy before the collision must equal the kinetic energy after the collision (because no energy is lost).

This means that (summing the kinetic energies before and after the collision)

$$\frac{1}{2}m_1 u_1^2 + \frac{1}{2}m_2 u_2^2 = \frac{1}{2}m_1 v_1^2 + \frac{1}{2}m_2 v_2^2$$

TOK

Popper and falsifiability

No-one has yet observed a case where the momentum in an isolated system is not conserved, but we should continue to look! Karl Popper, a philosopher of the twentieth century, argued that the test of whether a theory was truly scientific was that it was capable of being falsified. By this he meant that there has to be an experiment that could in principle contradict the hypothesis being tested. Popper argued that psychoanalysis was not a science because it could not be falsified by experiment.

Popper also applied his ideas to scientific induction. He said that, although we cannot prove that the Sun will rise tomorrow, because it always has we can use the theory that the Sun rises in the morning until the day when it fails to do so. At that point we must revise our theory. What should happen if momentum appears not to be conserved in an experiment? Is it more sensible to look for a new theory or to suggest that something has been overlooked?

The momentum and kinetic energy equations can be solved to show that

$$v_1 = \left(\frac{m_1 - m_2}{m_1 + m_2}\right) u_1 + \left(\frac{2m_2}{m_1 + m_2}\right) u_2$$

and

$$v_2 = \left(\frac{m_2 - m_1}{m_1 + m_2}\right) u_2 + \left(\frac{2m_1}{m_1 + m_2}\right) u_1$$

(You will not be expected to prove these in, or memorize them for, an examination.)

There are some interesting cases when mass m_2 is stationary and is struck by m_1:

- Case 1: m_1 is much smaller than m_2. Look at the v_1 equation. If $m_1 \ll m_2$ then the first term becomes roughly $\left(\frac{-m_2}{m_2}\right) u_1$ which is $-u_1$; the second term is zero because $u_2 = 0$. So the small mass "bounces" off the large mass, reversing its direction (shown by the minus sign). The large mass gains speed in the forward direction (the original direction of the small mass). The magnitude of the speed of the larger mass is roughly $\left(\frac{2m_1}{m_2}\right) u_1$ and this is a small fraction of the original speed of the small mass because $\left(\frac{2m_1}{m_2}\right)$ is much smaller than 1.
- Case 2: m_1 is much greater than m_2. Here the original mass loses hardly any speed (though it must lose a little). The momentum lost by m_1 is given to m_2 which moves off in the same direction, but at about twice the original speed of m_1. Look at the v_1 and v_2 equations and satisfy yourself that this is true.

Two objects colliding when energy is lost

When a moving object collides with a stationary one and the two objects stick together, then some of the initial kinetic energy is lost. After the collision, there is a single object with an increased (combined) mass and a single common velocity. This is known as an **inelastic collision** (figure 5).

This is a case you may have studied experimentally in the *Investigate!*

The momentum equation this time is

$$m_1 u_1 = (m_1 + m_2) v_1$$

A rearrangement shows that $v_1 = \frac{m_1}{(m_1 + m_2)} u_1$ and, as we might expect, the final velocity is in the same direction as before but is always smaller than the initial velocity.

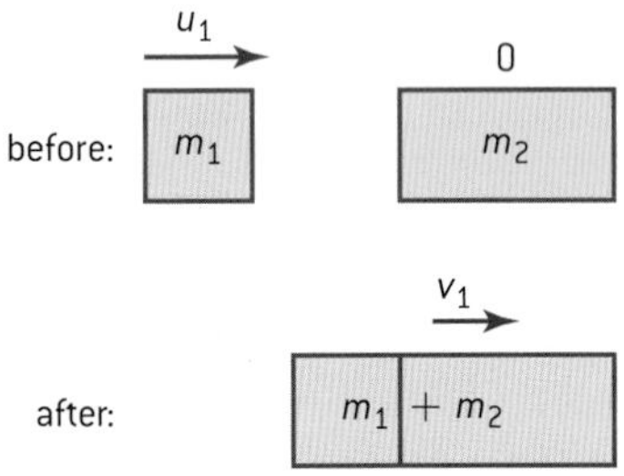

▲ Figure 5 Inelastic collision between two masses.

As for energy loss, the incoming kinetic energy is $\frac{1}{2} m_1 u_1^2$ and the final kinetic energy is (substituting for v_1)

$$\frac{1}{2}(m_1 + m_2) \frac{m_1^2}{(m_1 + m_2)^2} u_1^2$$

This is

$$\frac{1}{2} \frac{m_1^2}{(m_1 + m_2)} u_1^2.$$

The ratio $\frac{\text{initial kinetic energy}}{\text{final kinetic energy}}$ is $\frac{(m_1 + m_2)}{m_1}$

Two objects when energy is gained

There are many occasions when two initially stationary masses gain energy in some way. Some laboratory dynamics carts have a way to show this. An easy way is to attach two small strong magnets to the front of the carts so that when the carts are released after being held together with like poles of the magnets facing each other, the magnets repel and drive the carts apart.

The analysis is quite straightforward in this case.

The initial momentum is zero as neither object is moving.

After the collision the momentum is $m_1v_1 + m_2v_2 = 0$.

So $m_1v_1 = -m_2v_2$

The objects move apart in opposite directions. If the masses are equal the speeds are the same, with one velocity the negative of the other. If the masses are not equal then

$$\frac{m_1}{m_2} = \frac{-v_2}{v_1}$$

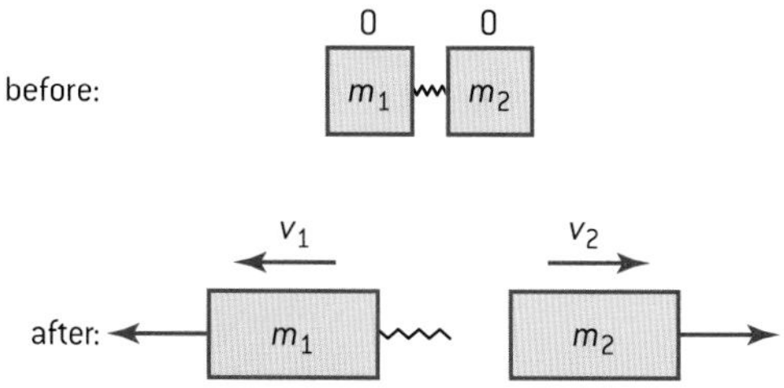

▲ Figure 6 Energy gained in a collision.

TOK

Helpful or not?

Are conservation laws helpful to scientists? On the one hand they allow the prediction of as yet untested cases, but on the other hand they may restrict the progress of science. This can happen if scientists are not prepared to challenge the status quo.

In 2012 the results of an experiment suggested that neutrinos could travel faster than the speed of light. This flew in the face of the accepted science originally proposed by Einstein. Later investigations showed that small errors in the timings had occurred in the experiments, and that there was no evidence for faster-than-light travel. Were the scientists right to publish their results so that others could test the new proposals?

Worked examples

1 A rail truck of mass 4500 kg moving at a speed of 1.8 m s^{-1} collides with a stationary truck of mass 1500 kg. The two trucks couple together. Calculate the speed of the trucks immediately after the collision.

Solution

The intial momentum = 4500 × 1.8 kg m s^{-1}.

The final momentum = (4500 + 1500) × v, where v is the final speed.

Momentum is conserved and so $v = \frac{4500 \times 1.8}{(4500 + 1500)} = \frac{4.5 \times 1.8}{6.0} = 1.3(5) = 1.4$ m s^{-1}.

2 Stone A of mass 0.5 kg travelling at 3.8 m s^{-1} across the surface of a frozen pond collides with stationary stone B of mass 3.0 kg. Stone B moves off at a speed of 0.65 m s^{-1} in the same original direction as stone A.

Calculate the final velocity of stone A.

Solution

The initial momentum is 0.5 × 3.8 = 1.9 kg m s^{-1}

The final momentum is 3.0 × 0.65 + 0.5v_A and this is equal to 1.9 kg m s^{-1}

$$v_A = \frac{1.9 - (3.0 \times 0.65)}{0.5} = -\frac{0.05}{0.5} = -0.1 \text{ kg m s}^{-1}$$

The final velocity of stone A is 0.1 kg m s^{-1} in the opposite direction to its original motion.

3

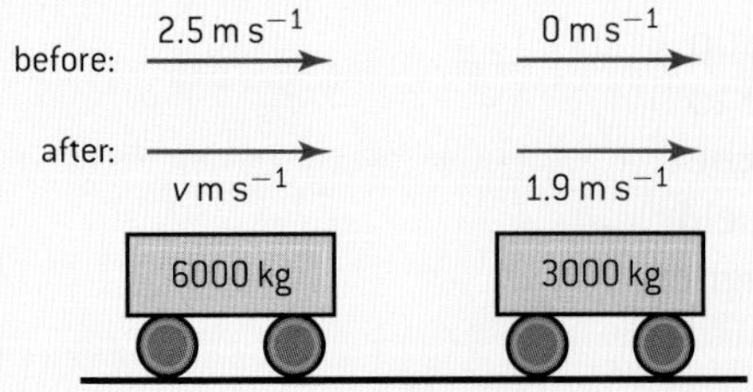

A railway truck of mass 6000 kg collides with a stationary truck of mass 3000 kg. The first truck moves with an initial speed of 2.5 m s^{-1} and the second truck moves off with a speed of 1.9 m s^{-1} in the same direction.

Calculate:

a) the velocity of the first truck immediately after the collision

b) the loss in kinetic energy as a result of the collision.

Solution

a) The initial momentum is $6000 \times 2.5 = 15\,000$ kg m s^{-1}.

The final momentum is $6000v + 3000 \times 1.9$

Equating these values: $15\,000 = 6000v + 5700$, so $6000v = 9300$ and $v = 1.5(5)$ kg m s^{-1}. This is positive so the first truck continues to move to the right after the collision at a speed of 1.6 kg m s^{-1}.

b) The initial kinetic energy $= \frac{1}{2} \times 6000 \times 2.5^2 = 18\,750$ J.

After the collision the total kinetic energy $= \frac{1}{2} \times 6000 \times 1.55^2 + \frac{1}{2} \times 3000 \times 1.9^2$

$= 7208 + 5415 = 12\,623$ or to 2 s.f. 13 000 J.

So the loss in kinetic energy is 18 750 – 13 000 = 5800 J.

Energy and momentum

There is a convenient link between kinetic energy and momentum.

Kinetic energy, $E_K = \frac{1}{2}mu^2$; momentum, $p = mu$ and therefore $p^2 = m^2u^2$. So

$$E_K = \frac{p^2}{2m}$$

This is often of use in calculations.

Applications of momentum conservation

Recoil of a gun

Figure 7 shows a gun being fired to trigger a snow fall. This prevents a more dangerous avalanche. When the gun fires its shell, the gun moves backwards in the opposite direction to that in which the shell goes. You should be able to explain this in terms of **momentum conservation**.

▲ Figure 7 Firing a gun to cause an avalanche.

Initially, both gun and shell are stationary; the initial total momentum is zero. The shell is propelled in the forwards direction through the gun by the expansion of gas following the detonation of the explosive in the shell. So this is one of the cases discussed earlier, one in which energy is gained. The explosion is a force internal to the system. Gas at high pressure is generated by the explosion in the chamber behind the shell. The gas exerts a force on the interior of the shell chamber and hence a force on the gun as well. The explosive releases energy and this is transferred into kinetic energies of both the shell and the gun.

The initial linear momentum was zero and no *external* force has acted on the system. The momentum must continue to be zero and this can only be so if the gun and the shell move in opposite directions with the same magnitude of momentum. The shell will go fast because it has a small mass compared to the gun; the gun moves relatively slowly.

Water hoses

Watch a fire being extinguished by firemen using a high-pressure hose and you will see the effect of water leaving the system. Often two or more firemen are needed to keep the hose on target because there is a large force on the hose in the opposite direction to that of the water. This can be seen in a more modest form when a garden hose connected to a tap starts to shoot backwards in unpredictable directions if it is not held when the tap is turned on.

▲ Figure 8 Fire fighting.

The cross-sectional area of the hose is greater than that of the nozzle through which the water emerges. The mass of water flowing past a point in the hose every second is the same as the mass that emerges from the nozzle every second. So the speed of the water emerging from the nozzle must be greater than the water speed along the hose itself. The water gains momentum as it leaves the hose because of this increase in exit speed compared with the flow speed in the hose.

The momentum of the system has to be constant and so there must be a force backwards on the end of the hose which needs to be countered by the efforts of the firemen. The kinetic energy and the momentum are being supplied by the water pump that feeds water to the fire hose or whatever originally created the pressure in the supply to the garden tap.

The momentum lost by the system per second is *(mass of water leaving per second) × (speed at which water leaves the nozzle – speed in the hose).*

The mass of water lost per second is $\frac{\Delta m}{\Delta t}$ and so the momentum lost per second $\dot{p}$ is $\frac{\Delta m}{\Delta t}(v - u)$ where v is the speed of water as it leaves the nozzle and u is the speed of the water in the hose.

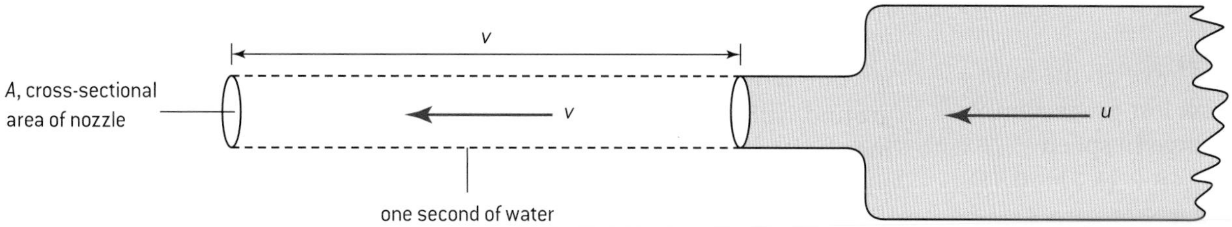

▲ Figure 9 Water leaving the hose.

If we know the cross-sectional area A of the nozzle of the hose and the density of the water, ρ, then $\frac{\Delta m}{\Delta t}$ can be determined. Figure 9 shows what happens inside the hose during a one-second time interval. Every second you can imagine a cylinder of water leaving the hose; this cylinder is v long and has an area A. The volume leaving per second is therefore Av

The mass of the water leaving in one second $\frac{\Delta m}{\Delta t}$

$= \textit{(density of water)} \times \textit{(volume of cylinder)} = \rho Av.$

Because the mass entering and leaving the nozzle per second must be the same, this means that the change of momentum in one second $= \frac{\Delta m}{\Delta t} \times (v - u) = \rho Av(v - u)$. If $u \ll v$ then the expression simplifies to ρAv^2.

The hose is an example of where care is needed when looking at the whole system. Consider what happens if the water is directed at a vertical wall. The water strikes the wall, loses all its horizontal momentum, and trickles vertically down the wall. The momentum must

Worked example

A mass of 0.48 kg of water leaves a garden hose every second. The nozzle of the hose has a cross-sectional area of 8.4×10^{-5} m^2. The water flows in the hose at a speed of 0.71 m s^{-1}. The density of water is 1000 kg m^{-3}.

Calculate:

a) the speed at which water leaves the hose

b) the force on the hose.

Solution

a) 0.48 kg of water in one second corresponds to a volume of $\frac{0.48}{1000}$ m^{-3} leaving the hose per second. This leaves through a nozzle of area 8.4×10^{-5} m^2 so the speed v must be $\frac{0.48}{1000 \times 8.4 \times 10^{-5}} =$ 5.71 m s^{-1}.

b) The force on the hose = mass lost per second × change in speed
$= 0.48 \times (5.71 - 0.71)$
$= 2.4$ N

have gone into the wall, its foundations and, therefore, the ground. So we might conclude that the Earth itself has gained momentum and that we can speed up the Earth's rotation by using a garden hose. But this is not true, because the water had to be given momentum originally by a pump. This gain in momentum at the pump must have given some momentum to the Earth too. The amount of momentum the Earth gained at the pump is equal and opposite to the momentum gained by the Earth when the water strikes the wall.

Rocketry

Earlier in this topic we discussed the acceleration of a rocket and looked at the situation from the perspective of Newton's second and third laws. A similar analysis is possible in terms of momentum conservation.

Rockets operate effectively in the absence of an atmosphere because momentum is conserved. All rockets release a liquid or gas at high speed. The fluid can be a very hot gas generated in the combustion of a solid chemical (as in a domestic rocket) or from the chemical reaction when two gases are mixed and burnt. Or it can be a fluid stored inside the rocket under pressure. In each case, the fluid escapes from the combustion/storage chamber through nozzles at the base of the rocket. As a result the rocket accelerates in the opposite direction to the direction in which the fluid is ejected. Momentum is conserved. The rate of loss of momentum from the rocket in the form of high-speed fluid must be equal to the rate of gain in momentum of the rocket. We shall look again at the mathematics of the rocket later in this topic.

Helicopters

Helicopters are aircraft that can take off and land vertically and also can hover motionless above a point on the ground. Vertical flight is said to have been invented in China about 2500 years before the present, but anyone who has seen the seeds from certain trees spiral down to the ground will realize where the original design came from. There were many attempts to build flying machines on the helicopter principle over the centuries, but the first commercial aircraft flew in the 1930s.

A helicopter uses the principle of conservation of linear momentum in order to hover. The rotating blades exert a force on originally stationary air causing it to move downwards towards the ground gaining momentum in the process. No external force acts and as a result there is an upward force on the helicopter through the rotors.

▲ Figure 10 Helicopter.

Impulse

Earlier we used Newton's second law of motion: $F = ma$ where the symbols have their usual meaning.

We can rearrange this equation using one of the kinematic equations:

$$a = \frac{(v - u)}{t}$$

Eliminating a gives $F = \frac{m(v - u)}{t}$

which means that $\textit{force} = \frac{\textit{change in momentum}}{\textit{time taken for change}}$

or

$$\textbf{\textit{force}} \times \textbf{\textit{time}} = \textbf{\textit{change in momentum}}$$

This equation gives a relationship between force and momentum and provides a further clue to the real meaning of momentum itself. The equation shows that we can change momentum (in other words, accelerate an object) by exerting a large force for a short time or by exerting a small force for a long time. A small number of people can get a heavy vehicle moving at a reasonable speed, but they have to push for a much longer time than the vehicle itself would take if powered by its own engine (which produces a larger force).

The product of ***force*** and ***time*** is given the name **impulse**, its units are newton seconds (N s). Impulse is the same as change in momentum, and N s gives us an alternative to kg m s^{-1} as a unit for momentum. You can use whichever you prefer.

There is a short-hand way to write the $force = \frac{change\ in\ momentum}{time}$ equation. We have already used the convention that "Δ" means "change in". So in symbols the equation becomes

$$F = \frac{\Delta p}{\Delta t}$$

where p is the symbol for momentum and t (as usual) means time.

(If you are familiar with differential calculus you may also come across this expression in the form $F = \frac{dp}{dt}$, but using the equation in this form will take us too far away from IB Diploma Programme physics.)

Worked examples

1 An impulse of 85 N s acts on a body of mass 5.0 kg that is initially at rest. Calculate the distance moved by the body in 2.0 s after the impulse has been delivered.

Solution

The change in momentum is 85 kg m s^{-1} so that the final speed is $\frac{85}{5} = 17$ m s^{-1}. In 2.0 s the distance travelled is 34 m.

2 An impulse I acts on an object of mass m initially at rest. Determine the kinetic energy gained by the object.

Solution

The change in speed of the object is $\frac{I}{m}$. The object is initially at rest and the gain in kinetic energy is $\frac{1}{2}mv^2$ which is $\frac{1}{2}m\left(\frac{I}{m}\right)^2 = \frac{I^2}{2m}$

Force–time graphs

Up till now we have usually assumed that forces are constant and do not change with time. This is rarely the case in real life and we need a way to cope with changes in momentum when the force is not constant. The equation $F = \frac{\Delta p}{\Delta t}$ helps here because it suggests that we use a force–time graph.

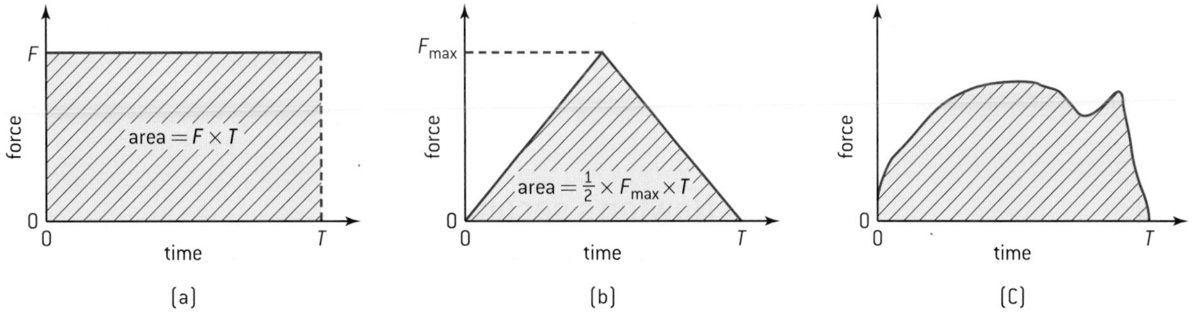

▲ Figure 11 Force–time graphs.

If a constant force acts on a mass, then the graph of force, F, against time, t, will look like figure 11(a). The change in momentum is $F \times T$ and this is the shaded area below the line. The area below the line in a force–time graph is equal to the change in momentum.

Another straightforward case that is more plausible than a constant force is the graph shown in figure 11(b) where the force rises to a maximum (F_{max}) and then falls back to zero in a total time T. The area under the graph this time is $\frac{1}{2}F_{max}\,T$ and, again, this is the change in momentum.

The final case (figure 11(c)) is one where there is no obvious mathematical relationship between F and t, but nevertheless we have a graph of how F varies with t. This time you will need to estimate the number of squares under the graph and use the area of one square to evaluate the momentum change.

Worked examples

1 The sketch graph shows how the force acting on an object varies with time.

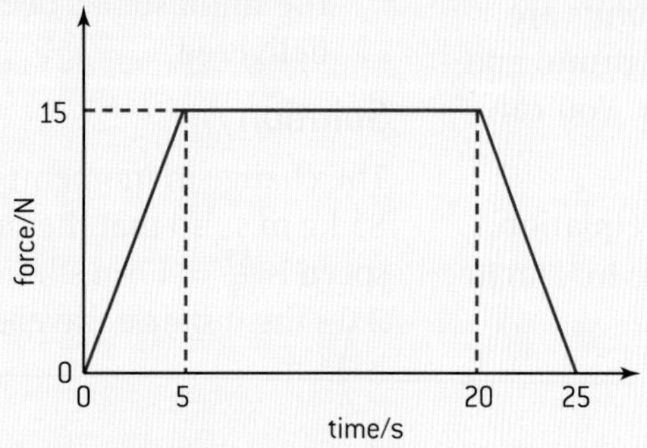

The mass of the object is 50 kg and its initial speed is zero.

Calculate the final speed of the object.

Solution

The total area under the $F - t$ graph is $2 \times \left(\frac{1}{2} \times 15 \times 5\right) + (15 \times 15) = 75 + 225 = 300$ N s

This is the change in momentum.

So the final speed is $\frac{300}{50} = 6.0$ m s^{-1}

2 The graph shows how the momentum of an object of mass 40 kg varies with time.

Calculate, for the object:

a) the force acting on it

b) the change in kinetic energy over the 10 s of the motion.

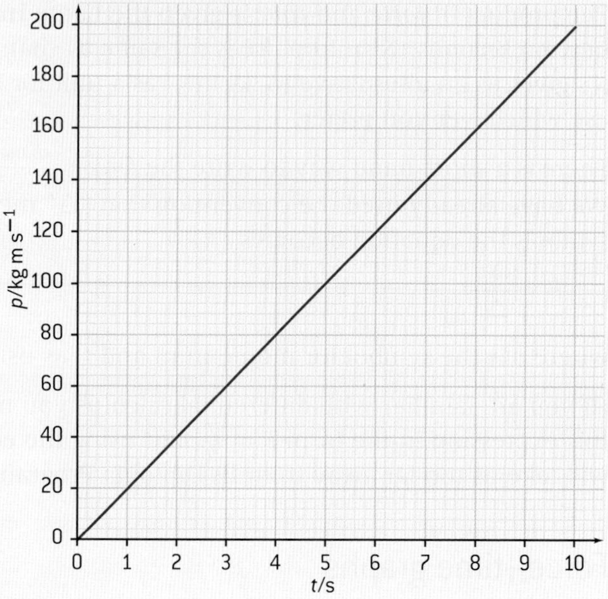

Solution

a) The gradient of the graph is $\frac{\Delta p}{\Delta t}$ and this is equal to the force.

The magnitude of the gradient is $\frac{200}{10}$ and this gives a value for force of 20 N.

b) $\Delta E_K = \frac{p^2}{2m} = \frac{200^2}{2 \times 40} = 500$ J

Revisiting Newton's second law

In the last sub-topic we used $F = ma$ to show that F also equals $\frac{\Delta p}{\Delta t}$. Using the full expression for momentum p gives

$$F = \frac{\Delta(mv)}{\Delta t}$$

This can be written as (using the product rule)

$$F = m\frac{\Delta v}{\Delta t} + v\frac{\Delta m}{\Delta t}$$

You may have to take this algebra on trust but, by thinking through what the two terms stand for, you should be able to understand the physics that they represent.

The first term on the left-hand side is just *mass* $\times$ $\frac{\textit{change in velocity}}{\textit{change in time}}$ which we know as *mass* $\times$ *acceleration* – our original form of Newton's second law of motion. The second term on the right-hand side is something new, it is the *instantaneous velocity* $\times$ $\frac{\textit{change in mass}}{\textit{change in time}}$.

Our first version of Newton's second law was a simpler form of the law than the full "change in momentum" version. The new extra term takes account of what happens when the mass of the accelerating object is changing.

Rockets and helicopters again

Rockets

We showed that rockets are an excellent example of momentum conservation in action. The difficulty in analysing the rocket example is that the rocket is always losing mass (in the form of gas or liquid propellant), so m is not constant.

$F = \frac{m\Delta v}{\Delta t} + \frac{v\Delta m}{\Delta t}$ is our new version of Newton's second law of motion and, in this case, $F = 0$ because there is no external force acting on the system.

So $\frac{m\Delta v}{\Delta t} = -\frac{v\Delta m}{\Delta t}$. The two terms have quite separate meanings: $\frac{m\Delta v}{\Delta t}$ refers to the *instantaneous* mass of the rocket (including the remaining fuel) and to the acceleration of this total mass $\frac{\Delta v}{\Delta t}$. The other term is the ejection speed of the fuel and the rate at which mass is lost from the system $\frac{\Delta m}{\Delta t}$. So at one instant in time, the acceleration of the rocket

$$a = \frac{\Delta v}{\Delta t} = -\frac{v\Delta m}{m\Delta t}.$$

The negative sign reminds us that the rocket is *losing* mass while *gaining* speed.

Helicopters

With the helicopter hovering, there is a weight force downwards and so the equation now becomes

$$Mg = \frac{m\Delta v}{\Delta t} + \frac{v\Delta m}{\Delta t} \text{ where } M \text{ is the mass of the helicopter}$$

There is no change in the speed of the helicopter (it is hovering) so $\Delta v = 0$, there is however momentum gained by the air that moves downwards. This is the speed, v, the air gains multiplied by the mass of the air accelerated every second so $Mg = v\frac{\Delta m}{\Delta t}$ or

weight of helicopter (in N) = *mass of air pushed downwards per second* (in kg s^{-1}) × *speed of air downwards* (in m s^{-1})

Momentum and safety

Many countries have made it compulsory to wear seat belts in a moving vehicle. Likewise, many modern cars have airbags that inflate very quickly if the car is involved in a collision.

Obviously, both these devices restrain the occupants of the car, preventing them from striking the windscreen or the hard areas around it. But there is more to the physics of the air bag and the seat belt than this.

On the face of it, someone in a car has to lose the same amount of kinetic energy and momentum whether they are stopped by the windscreen or restrained by the seat belt. What differs in these two

Worked example

A small firework rocket has a mass of 65 g. The initial rate at which hot gas is lost from the firework has been lit is 3.5 g s^{-1} and the speed of release of this gas from the rear of the rocket is 130 m s^{-1}.

Calculate the initial acceleration of the rocket.

Solution

$a = \frac{\Delta v}{\Delta t} = -\frac{v\Delta m}{m\Delta t}$ where v is the release speed of the gas, $\frac{\Delta m}{\Delta t}$ is the rate of loss of gas, and m is the mass of the rocket

$a = \frac{130 \times 3.5}{65} = 7.0 \text{ m s}^{-2}$

▲ Figure 12 Car safety: seat belts and air bags.

cases is the time during which the loss of energy and momentum occur. Unrestrained, the time to stop will be very short and the deceleration will therefore be very large. A large deceleration means a large force and it is the magnitude of the force that determines the amount of damage.

Seat belts and air bags dramatically increase the time taken by the occupants of the car to stop and as *force* $\times$ *time* = *momentum change*, for a constant change in momentum, a long stopping time will imply a smaller, and less damaging, force.

Momentum and sport

This sub-topic began with a suggestion that it was less painful to catch a table-tennis ball than a baseball. You should now be able to understand the reason for the difference. You should also realize why good technique in many sports hinges on the application of momentum change.

Many sports in which an object – usually a ball – is struck by hand, foot or bat rely on the efficient transfer of momentum. This transfer is often enhanced by a "follow through", which increases the contact time between bat and ball. The player maintains the same force but for a longer time, so the impulse on the ball will increase, increasing the momentum change as well.

Think about your sport and how effective use of momentum change can help you.

Investigate!

Estimating the force on a soccer ball

This experiment will allow you to estimate the force used to kick a soccer ball. It uses many of the ideas contained in this topic and is a good place to conclude our look at IB mechanics!

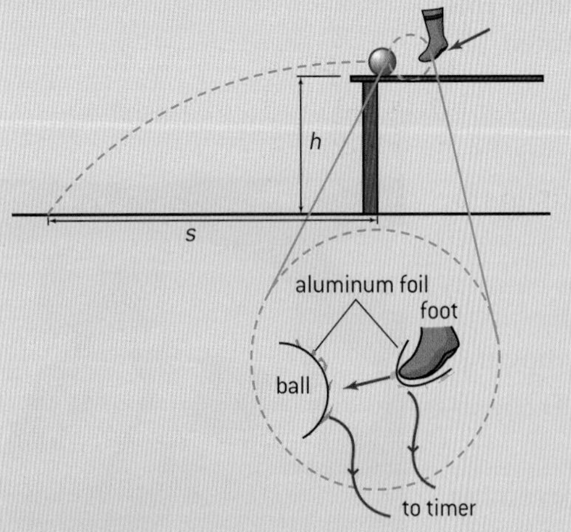

- The basis of the method is to measure the contact time between the foot and the ball and the subsequent change in momentum of the ball. The use of *force* $\times$ *contact time* = *change in momentum* allows the force to be calculated.
- To measure the contact time: stick some aluminium foil to the shoe of the kicker and to the soccer ball. Set up a data logger or fast-timer so that it will measure the time T for which the two pieces of foil are in contact.
- To measure the change in momentum: the ball starts from rest so all you need is the magnitude of the final momentum. Get the kicker to kick the ball horizontally from a lab bench. Measure the distance s from where the ball is kicked to where it lands. Measure the distance h from the bottom of the ball on the bench to the floor. Use the ideas of projectile motion to calculate (i) the time t taken for the ball to reach the floor $\left[h = \frac{1}{2}gt^2 \text{ and so } t = \sqrt{\frac{2h}{g}}\right]$. Then using this t, the initial speed u of the ball can be estimated as $\frac{s}{t}$. Measure the mass of the ball M and therefore the change in momentum is Mu which is equal to the *force on the ball* $\times$ T.
- This method can be modified for many sports including hockey, baseball, and golf.

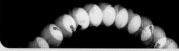

Questions

1 *(IB)*

Christina stands close to the edge of a vertical cliff and throws a stone at 15 m s^{-1} at an angle of 45° to the horizontal. Air resistance is negligible.

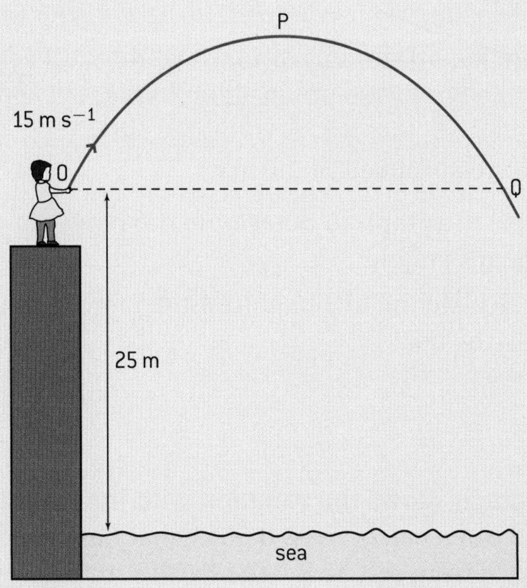

Point P on the diagram is the highest point reached by the stone and point Q is at the same height above sea level as point O. Christina's hand is at a height of 25 m above sea level.

a) At point P on a copy of the diagram above draw arrows to represent:

(i) the acceleration of the stone (label this A)

(ii) the velocity of the stone (label this V).

b) Determine the speed with which the stone hits the sea. (8 marks)

2 *(IB)*

Antonia stands at the edge of a vertical cliff and throws a stone vertically upwards.

The stone leaves Antonia's hand with a speed $v = 8.0$ m s^{-1}. The time between the stone leaving Antonia's hand and hitting the sea is 3.0 s. Assume air resistance is negligible.

a) Calculate:

(i) the maximum height reached by the stone

(ii) the time taken by the stone to reach its maximum height.

b) Determine the height of the cliff. (6 marks)

3 *(IB)*

A marble is projected horizontally from the edge of a wall 1.8 m high with an initial speed *V*.

A series of flash photographs are taken of the marble and combined as shown below. The images of the marble are superimposed on a grid that shows the horizontal distance *x* and vertical distance *y* travelled by the marble.

The time interval between each image of the marble is 0.10 s.

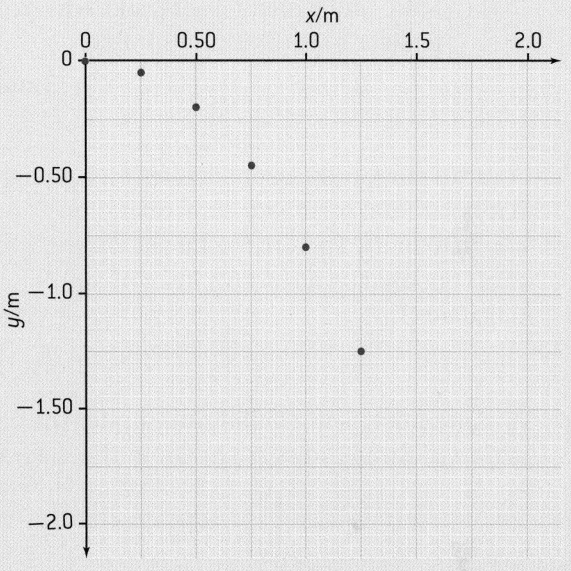

Use data from the photograph to calculate a value of the acceleration of free fall. (3 marks)

4 *(IB)*

A cyclist and his bicycle travel at a constant velocity along a horizontal road.

a) (i) State the value of the resultant force acting on the cyclist.

(ii) Copy the diagram and draw labelled arrows to represent the vertical forces acting on the bicycle.

(iii) Explain why the cyclist and bicycle are travelling at constant velocity.

b) The total mass of the cyclist and bicycle is 70 kg and the total resistive force acting on them is 40 N. The initial speed of the cycle is 8.0 m s^{-1}. The cyclist stops pedalling and the bicycle comes to rest.

(i) Calculate the magnitude of the initial acceleration of the bicycle and rider.

(ii) Estimate the distance taken by the bicycle to come to rest from the time the cyclist stops pedalling.

(iii) State and explain *one* reason why your answer to b)(ii) is an estimate.

(13 marks)

5 A car of mass 1000 kg accelerates on a straight flat horizontal road with an acceleration $a = 0.30$ m s^{-2}. The driving force T on the car is opposed by a resistive force of 500 N. Calculate T. (3 marks)

6 A crane hook is in equilibirium under the action of three forces as shown in the diagram.

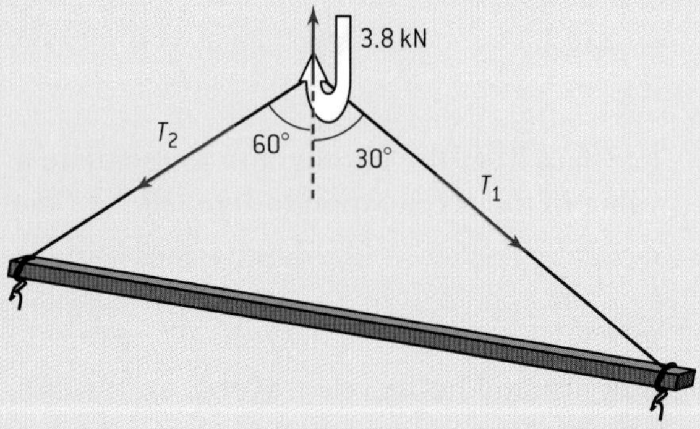

Calculate T_1 and T_2. (4 marks)

7 *(IB)*

A small boat is powered by an outboard motor of variable power P. The graph below shows the variation with speed v of the power P for a particular load.

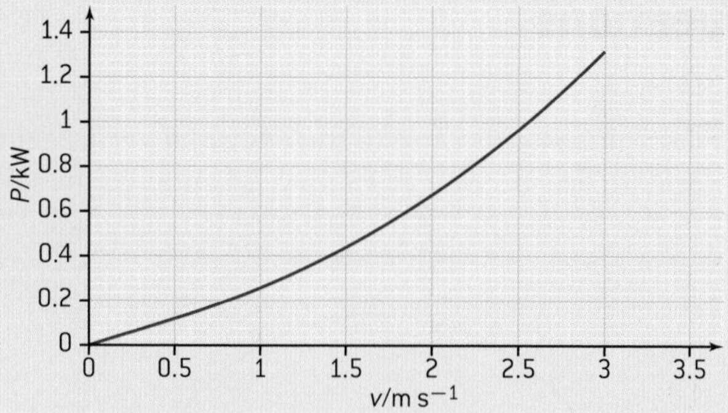

For a steady speed of 2.0 m s^{-1}:

a) use the graph to determine the power of the boat's engine

b) calculate the frictional (resistive) force acting on the boat. (3 marks)

8 *(IB)*

The graph shows the variation with time t of the speed v of a ball of mass 0.50 kg that has been released from rest above the Earth's surface.

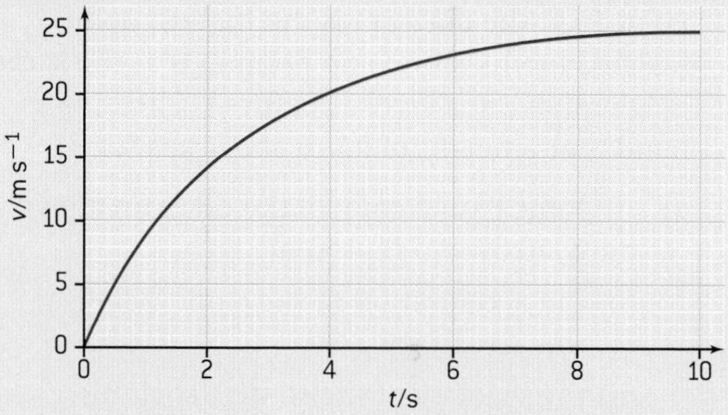

The force of air resistance is *not* negligible.

a) State, without any calculations, how the graph could be used to determine the distance fallen.

b) (i) Copy the diagram and draw and label arrows to represent the forces on the ball at 2.0 s.

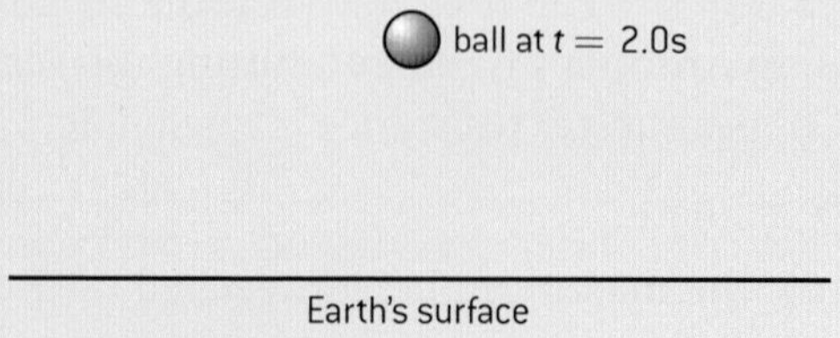

(ii) Use the graph to show that the acceleration of the ball at 2.0 s is approximately 4 m s^{-2}.

(iii) Calculate the magnitude of the force of air resistance on the ball at 2.0 s.

(iv) State and explain whether the air resistance on the ball at $t = 5.0$ s is smaller than, equal to, or greater than the air resistance at $t = 2.0$ s.

c) After 10 s the ball has fallen 190 m.

(i) Show that the sum of the potential and kinetic energies of the ball has decreased by about 800 J. (14 marks)

9 *(IB)*

A bus is travelling at a constant speed of $6.2\ \mathrm{m\,s^{-1}}$ along a section of road that is inclined at an angle of 6.0° to the horizontal.

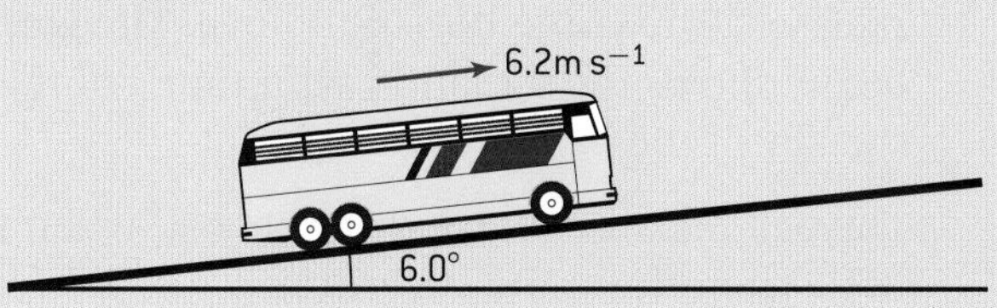

a) (i) Draw a labelled sketch to represent the forces acting on the bus.

(ii) State the value of net force acting on the bus.

b) The total output power of the engine of the bus is 70 kW and the efficiency of the engine is 35%.

Calculate the input power to the engine.

c) The mass of the bus is 8.5×10^3 kg.

Determine the rate of increase of gravitational potential energy of the bus.

d) Using your answer to c (and the data in b), estimate the magnitude of the resistive forces acting on the bus. (12 marks)

10 *(IB)*

Railway truck A moves along a horizontal track and collides with a stationary truck B. The two join together in the collision. Immediately before the collision, truck A has a speed of $5.0\ \mathrm{m\,s^{-1}}$. Immediately after collision, the speed of the trucks is v.

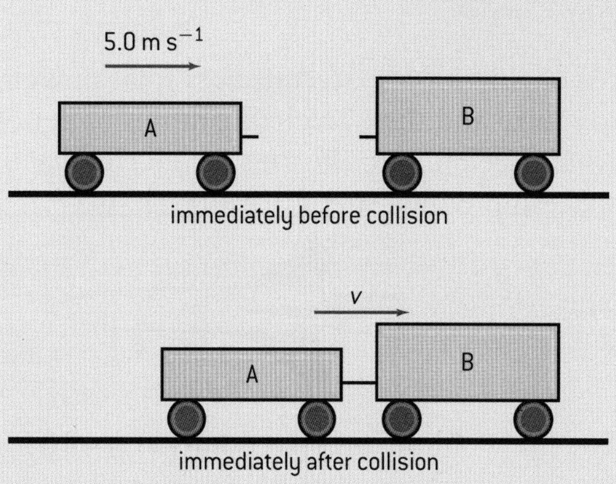

The mass of truck A is 800 kg and the mass of truck B is 1200 kg.

a) (i) Calculate v.

(ii) Calculate the total kinetic energy lost during the collision.

b) Suggest where the lost kinetic energy has gone. (6 marks)

11 *(IB)*

Large metal bars are driven into the ground using a heavy falling object.

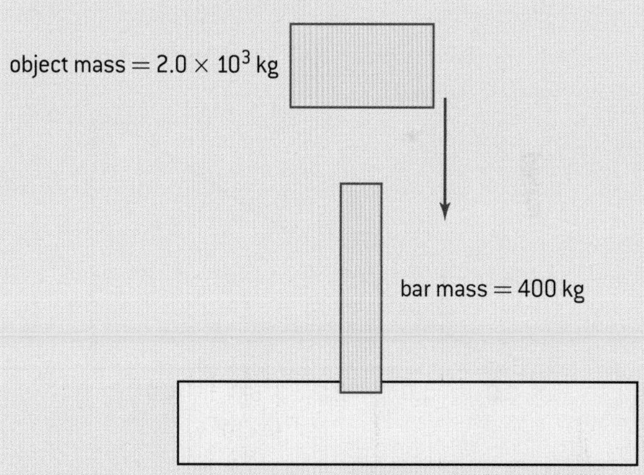

The falling object has a mass 2000 kg and the metal bar has a mass of 400 kg.

The object strikes the bar at a speed of $6.0\ \mathrm{m\,s^{-1}}$. It comes to rest on the bar without bouncing. As a result of the collision, the bar is driven into the ground to a depth of 0.75 m.

a) Determine the speed of the bar immediately after the object strikes it.

b) Determine the average frictional force exerted by the ground on the bar. (7 marks)

12 *(IB)*

An engine for a spacecraft uses solar power to ionize and then accelerate atoms of xenon. After acceleration, the ions are ejected from the spaceship with a speed of $3.0 \times 10^4\,\text{m s}^{-1}$.

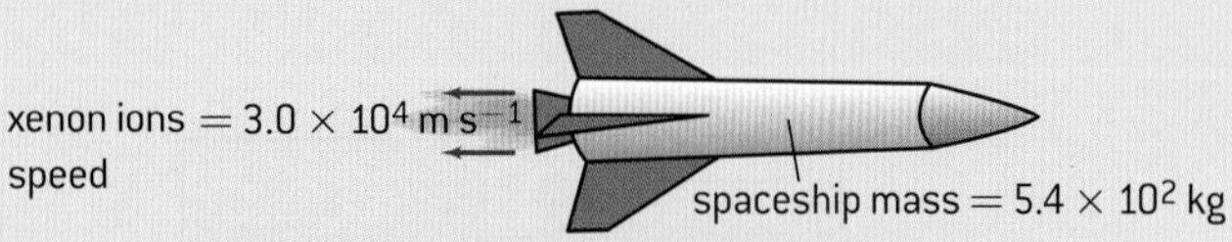

The mass of one ion of xenon is 2.2×10^{-25} kg

a) The original mass of the fuel is 81 kg. Determine how long the fuel will last if the engine ejects 77×10^{18} xenon ions every second.

b) The mass of the spaceship is 5.4×10^2 kg. Determine the initial acceleration of the spaceship.

The graph below shows the variation with time t of the acceleration a of the spaceship. The solar propulsion engine is switched on at time $t = 0$ when the speed of the spaceship is $1.2 \times 10^3\ \text{m s}^{-1}$.

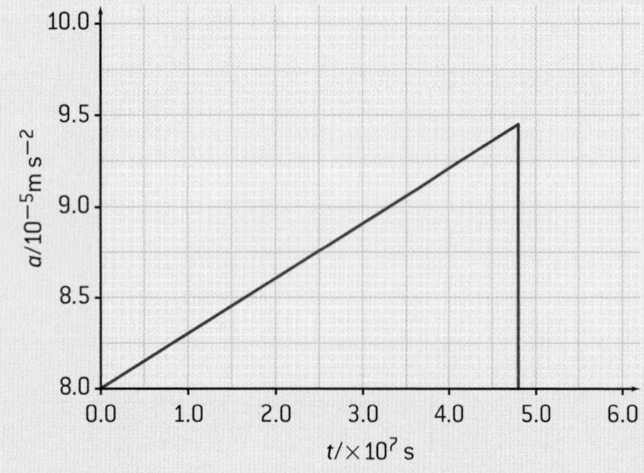

c) Explain why the acceleration of the spaceship increases with time.

d) Using data from the graph, calculate the speed of the spaceship at the time when the xenon fuel has all been used. (15 marks)

13 *(IB)*

A large metal ball is hung from a crane by means of a cable of length 5.8 m as shown below.

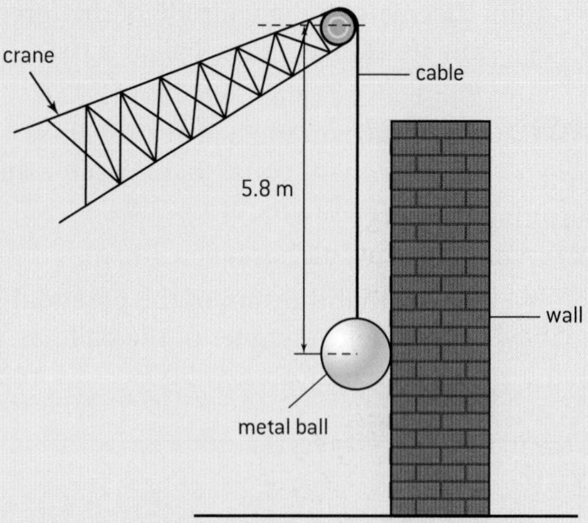

To knock a wall down, the metal ball of mass 350 kg is pulled away from the wall and then released. The crane does not move. The graph below shows the variation with time t of the speed v of the ball after release.

The ball makes contact with the wall when the cable from the crane is vertical.

a) For the ball just before it hits the wall use the graph, to estimate the tension in the cable. The acceleration of free fall is $9.8\,\text{m s}^{-2}$.

b) Determine the distance moved by the ball after coming into contact with the wall.

c) Calculate the total change in momentum of the ball during the collision of the ball with the wall. (7 marks)

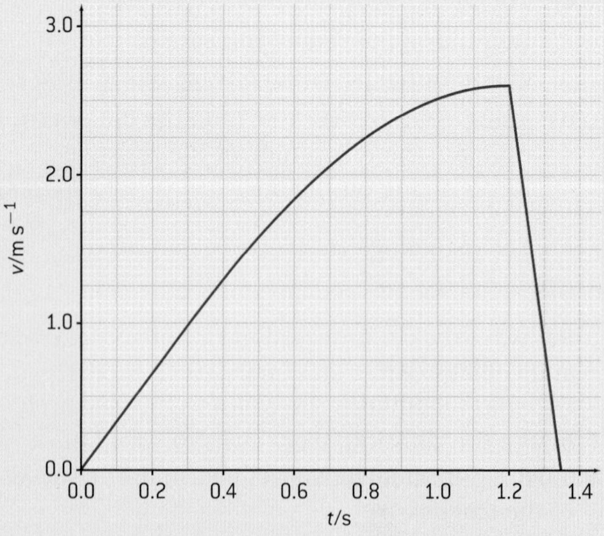

3 THERMAL PHYSICS

Introduction

In this topic we look at at thermal processes resulting in energy transfer between objects at different temperatures. We consider how the energy transfer brings about further temperature changes and/or changes of state or phase.

We then go on to look at the effect of energy changes in gases and use the kinetic theory to explain macroscopic properties of gases in terms of the behaviour of gas molecules.

3.1 Temperature and energy changes

Understanding

- → Temperature and absolute temperature
- → Internal energy
- → Specific heat capacity
- → Phase change
- → Specific latent heat

Nature of science

Evidence through experimentation

Since early humans began to control the use of fire, energy transfer because of temperature differences has had significant impact on society. By controlling the energy flow by using insulators and conductors we stay warm or cool off, and we prepare life-sustaining food to provide our energy needs (see figure 2). Despite the importance of energy and temperature to our everyday lives, confusion regarding the difference between "thermal energy" or "heat" and temperature is commonplace. The word "heat" is used colloquially to mean energy transferred because of a temperature difference, but this is a throwback to the days in which scientists thought that heat was a substance and different from energy.

Applications and skills

- → The three states of matter are solid, liquid, and gas
- → Solids have fixed shape and volume and comprise of particles that vibrate with respect to each other
- → Liquids have no fixed shape but a fixed volume and comprise of particles that both vibrate and move in straight lines before colliding with other particles
- → Gases (or vapours) have no fixed volume or shape and move in straight lines before colliding with other particles – this is an ideal gas
- → Thermal energy is often misnamed "heat" or "heat energy" and is the energy that is transferred from an object at a higher temperature by conduction, convection, and thermal radiation

Equations

- → Conversion from Celsius to kelvin: $T(\mathrm{K}) = \theta(^{\circ}\mathrm{C}) + 273$
- → Specific heat capacity relationship: $Q = mc\Delta T$
- → Specific latent heat relationship: $Q = mL$

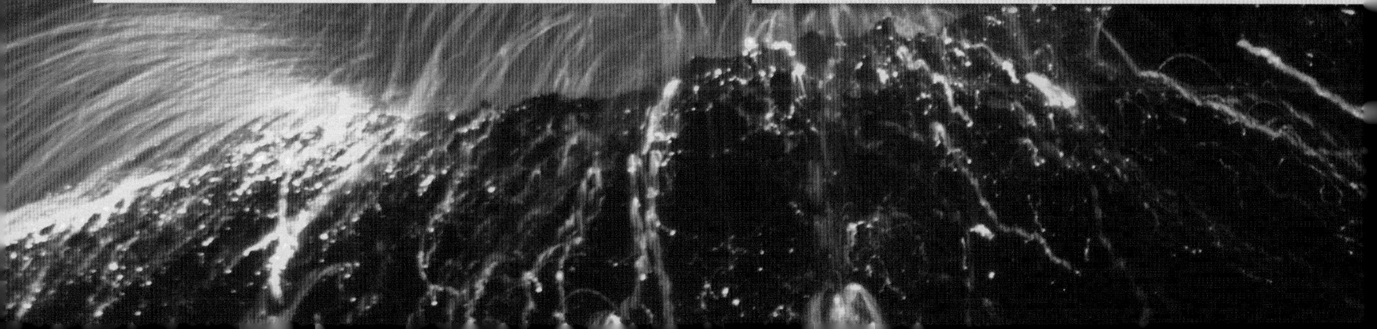

Introduction

From the 17th century until the end of the 19th century scientists believed that "heat" was a substance that flowed between hot and cold objects. This substance travelling between hot and cold objects was known as "phlogiston" or "caloric" and there were even advocates of a substance called "frigoric" that flowed from cold bodies to hot ones. In the 1840s James Joule showed that the temperature of a substance could be increased by doing work on that substance and that doing work was equivalent to heating. In his paddle wheel experiment (see figure 1) he dropped masses attached to a mechanism connected to a paddle wheel; the wheel churned water in a container and the temperature of the water was found to increase. Although the caloric theory continued to have its supporters it was eventually universally abandoned. The unit "calorie" which is sometimes used relating to food energy is a residual of the caloric theory, as is the use of the word heat as a noun.

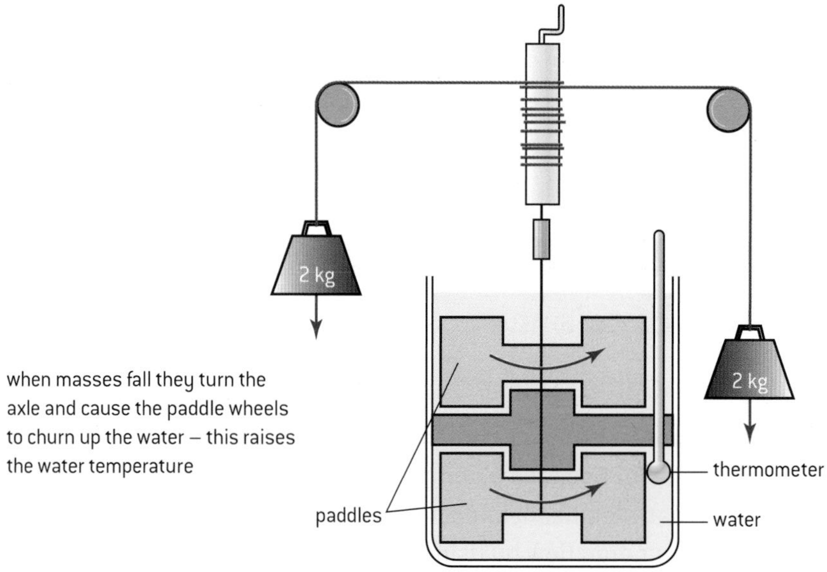
2 kg
2 kg
when masses fall they turn the axle and cause the paddle wheels to churn up the water – this raises the water temperature
thermometer
paddles
water

▲ Figure 1 Joule's paddle wheel experiment.

▲ Figure 2 Dinner is served!

Temperature and energy transfer

You may have come across temperature described as the "degree of hotness" of an object. This is a good starting point since it relates to our senses. A pot of boiling water feels very hot to the touch and we know instinctively that the water and the pot are at a higher temperature than the cold water taken from a refrigerator. The relative temperature of two objects determines the direction in which energy passes from one object to the other; energy will tend to pass from the hotter object to the colder object until they are both at the same temperature (or in **thermal equilibrium**). The energy flowing as a result of conduction, convection, and thermal radiation is what is often called "heat".

Temperature is a scalar quantity and is measured in units of degrees Celsius (°C) or kelvin (K) using a thermometer.

Nature of science

Thermometers

Most people are familiar with liquid-in-glass thermometers in which the movement of a column of liquid along a scale is used to measure temperature. Thermometers are not just restricted to this liquid-in-glass type; others use the expansion of a gas, the change in electrical resistance of a metal wire, or the change in emf (electromotive force) at the junction of two metal wires of different materials. A thermometer can be constructed from any object that has a property that varies with temperature (a **thermometric property**). In the case of a liquid-in-glass thermometer the thermometric property is the expansion of the liquid along a glass capillary tube. The liquid is contained in the bulb which is a reservoir; when the bulb is heated the liquid expands, travelling along the capillary. Since the bore of the capillary is assumed to be constant, as the volume of the liquid increases with temperature so does the length of the liquid. Such thermometers are simple but not particularly accurate. Reasons for the inaccuracy could be that the capillary may not be uniform, or its cross-sectional area may vary with temperature, or it is difficult to make sure that all the liquid is at the temperature of the object being investigated. Glass is also a relatively good insulator and it takes time for thermal energy to conduct through the glass to the liquid, making the thermometer slow to respond to rapid changes in temperature.

Liquid-in-glass thermometers are often calibrated in degrees Celsius, a scale based on the scale reading at two **fixed points**, the ice point and the steam point. These two temperatures are defined to be 0 °C and 100 °C respectively (although Celsius actually used the steam point for 0 and ice point for 100). The manufacturer of the thermometer assumes that the length of the liquid in the capillary changes linearly with temperature between these two points, even though it may not actually do so. This is a fundamental assumption made for all thermometers, so that thermometers only agree with each other at the fixed points – between these points they could well give different values for the same actual temperature.

Digital thermometers or temperature sensors have significant advantages over liquid-in-glass thermometers and have largely taken over from liquid-in-glass thermometers in many walks of life. The heart of such devices is usually a thermistor. The resistance of most thermistors falls with temperature (they are known as "ntc" or "negative temperature coefficient of resistance" thermistors). Since the thermistor is usually quite small, it responds very quickly to temperature changes. Thermistor thermometers are usually far more robust than liquid-in-glass ones.

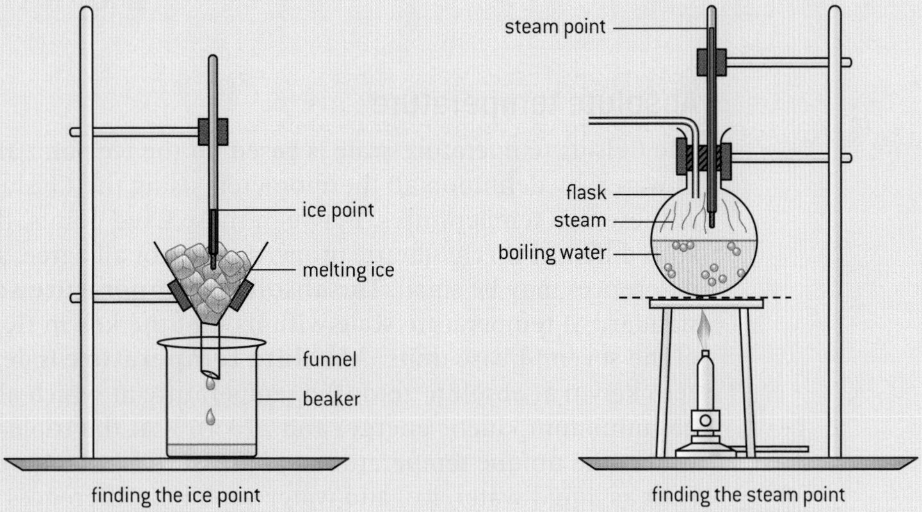

▲ Figure 3 Calibration of a liquid-in glass thermometer at the ice point and steam point.

Investigate!

Calibrating a thermistor against an alcohol-in-glass thermometer

This investigation can be performed by taking readings manually or by using a data logger. Here we describe the manual way of performing the experiment. The temperature can be adjusted either using a heating coil or adding water at different temperatures (including some iced water, perhaps).

- A multimeter set to "ohms" or an ohm-meter is connected across the thermistor (an ammeter/voltmeter method would be a suitable alternative but resistance would need to be calculated).
- The thermistor is clamped so that it lies below the water surface in a Styrofoam (expanded polystyrene) container next to an alcohol-in-glass thermometer.
- Obtain pairs of values of readings on the multimeter and the thermometer and record these in a table.
- Plot a graph of resistance/Ω against temperature/°C (since this is a calibration curve there can be no systematic uncertainties and any uncertainties will be random – being sufficiently small to be ignored in the context of this investigation).
- The graph shown is typical for a "ntc" thermistor.
- Suggest why, on a calibration curve, systematic uncertainties are not appropriate.
- Would the designers of a digital thermometer assume a linear relationship between resistance and temperature?
- Would you expect the reading on a digital thermometer to correspond to that on a liquid-in-glass thermometer?

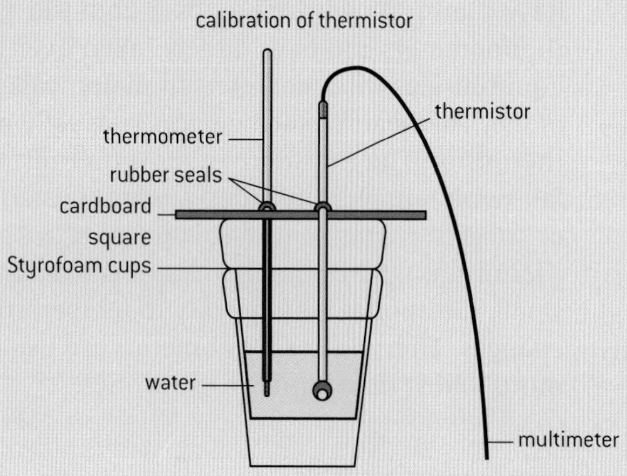

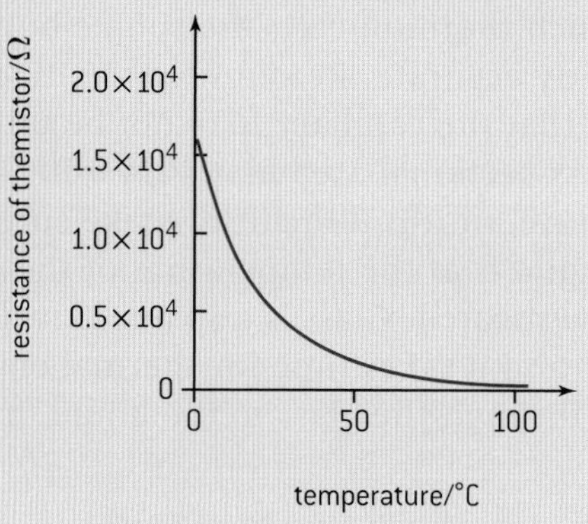

Absolute temperature

The Celsius temperature scale is based on the ice point and steam point of water; by definition all thermometers using the Celsius scale agree at these two temperatures. Between these fixed points, thermometers with different thermometric properties do not all agree, although differences may be small. The **absolute temperature scale** is the standard SI temperature scale with its unit the kelvin (K) being one of the seven SI base units. **Absolute temperature** is defined to be zero kelvin at absolute zero (the temperature at which all matter has minimum kinetic energy) and 273.16 K at the triple point of water (the unique temperature and pressure at which water can exist as liquid water, ice, and water vapour). Differences in absolute temperatures exactly correspond to those in Celsius temperatures (with a temperature difference of 1 °C being identical to 1 K). For this reason it is usual to write the units of temperature difference as K but not °C (although you are unlikely to lose marks in an examination for making this slip). To convert from temperatures in degrees Celsius to absolute temperatures the following relationship is used:

$$T(\text{K}) = \theta(^\circ\text{C}) + 273$$

where T represents the absolute temperatures and θ the temperature in degrees Celsius.

Other aspects of the absolute temperature scale will be considered in Sub-topic 3.2.

Worked example

A temperature of 73 K is equivalent to a temperature of

A. −346 °C. B. −200 °C.
C. +73 °C. D. +200 °C.

Solution

Substituting values into the conversion equation:
$73 = \theta + 273$
so $\theta = 73 - 273 = -200$
making the correct option B.

Internal energy

Substances consist of particles (e.g., molecules) in constant random motion. As energy is transferred to a substance the separation of the particles could increase and they could move faster. When particles move further apart (or closer to other neighbouring particles) the potential energy of the particles increases. As they move faster their random kinetic energy increases. **The internal energy of a substance is the total of the potential energy and the random kinetic energy of all the particles in the substance.** For a solid, these two forms of energy are present in roughly equal amounts; however in a gas the forces between the particles are so small that the internal energy is almost totally kinetic. We will discuss this further in the Sub-topic 3.2.

Note: Students often confuse the terms internal energy and thermal energy. Avoid using the term thermal energy – talk about internal energy and the energy transferred because of temperature differences, which is what most people mean by thermal energy.

Specific heat capacity

When two different objects receive the same amount of energy they are most unlikely to undergo the same temperature change. For example, when 1000 J of energy is transferred to 2 kg of water or to 1 kg of copper the temperature of each mass changes by different amounts. The temperature of the water would be expected to increase by about 0.12 K while that of copper by 2.6 K. The two masses have different **heat capacities**. You may think that it is hardly a fair comparison since there is 2 kg of water and 1 kg of copper; and you would be right to think this! If we chose equal masses of the two substances we would discover that 1 kg of water would increase by 0.24 K under the same conditions.

In order to be able to compare substances more closely we define **the specific heat capacity (c) of a substance as the energy transferred to 1 kg of the substance causing its temperature to increase by 1 K.**

The defining equation for this is

$$c = \frac{Q}{m\Delta T}$$

where Q is the amount of energy supplied to the object of mass m and causing its temperature to rise by ΔT. When Q is in J, m in kg, and ΔT in K, c will be in units of J kg^{-1} K^{-1}.

You should check that the data provided show that water has a specific heat capacity of approximately 4.2×10^3 J kg^{-1} K^{-1} and copper a value of approximately 380 J kg^{-1} K^{-1}.

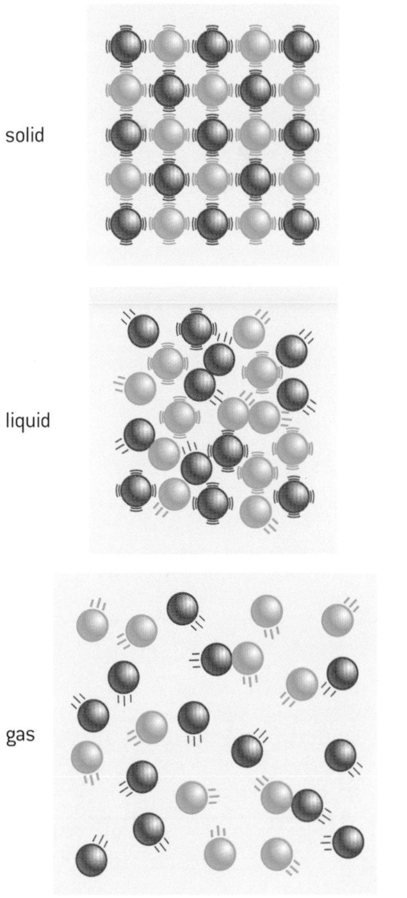

▲ Figure 4 Motion of molecules in solids, liquids, and gases.

Worked example

A piece of iron of mass 0.133 kg is placed in a kiln until it reaches the temperature θ of the kiln. The iron is then quickly transferred to 0.476 kg of water held in a thermally insulated container. The water is stirred until it reaches a steady temperature. The following data are available.

Specific heat capacity of iron $= 450\ \text{J kg}^{-1}\ \text{K}^{-1}$

Specific heat capacity of water $= 4.2 \times 10^3\ \text{J kg}^{-1}\ \text{K}^{-1}$

Initial temperature of the water $= 16\ °\text{C}$

Final temperature of the water $= 45\ °\text{C}$

The specific heat capacity of the container and insulation is negligible.

a) State an expression, in terms of θ and the above data, for the energy transfer of the iron in cooling from the temperature of the kiln to the final temperature of the water.

b) Calculate the increase in internal energy of the water as the iron cools in the water.

c) Use your answers to b) and c) to determine θ.

Solution

a) Using $Q = mc\Delta T$ for the iron
$Q = 0.133 \times 450 \times (\theta - 45) = 60 \times (\theta - 45)$

b) Using $Q = mc\Delta T$ for the water
$Q = 0.476 \times 4200 \times (45 - 16) = 5.8 \times 10^4\ \text{J}$

c) Equating these and assuming no energy is transferred to the surroundings
$60 \times (\theta - 45) = 5.8 \times 10^4$

$\therefore 60\theta - 2700 = 5.8 \times 10^4$

$60\theta = 6.1 \times 10^4$

$\therefore \theta = \dfrac{6.1 \times 10^4}{60}$

$\theta = 1000\ °\text{C}$ (or $1010\ °\text{C}$ to 3 s.f.)

Investigate!

Estimating the specific heat capacity of a metal

This is another investigation that can be performed manually or by using a data logger. Here we describe the manual way of performing the experiment. A metal block of known mass is heated directly by a heating coil connected to a low-voltage electrical supply. Channels in the block allow the coil to be inserted and also a thermometer or temperature probe. The power supplied to the block is calculated from the product of the current (I) in the heating coil and the potential difference (V) across it. The supply is switched on for a measured time (t) so that the temperature changes by at least 10 K. When the temperature has risen sufficiently the power is switched off, but the temperature continues to be monitored to find the maximum temperature rise (it takes time for the thermometer to reach the same temperature as the surrounding block).

- Thermal energy transferred to the block in time $t = VIt$
- The heater itself, the temperature probe and the insulation will all have heat capacities that will mean that there is a further unknown term to the relationship: which should become $VIt = (mc + C)\Delta T$ where C is the heat capacity of everything except the block – which will undergo the same temperature change as the block since they are all in good **thermal contact**. This term may be ignored if the mass of the block is high (say, 1 kg or so).
- As the block is heated it will also be losing thermal energy to the surroundings and so make the actual rise in temperature lower

that might be predicted. You are not expected to calculate "cooling corrections" in your IB calculations but it is important to recognize that this happens. A simple way of compensating for this is to cool the apparatus by, say 5 K, below room temperature before switching on the power supply. If you then allow its temperature to rise to 5 K above room temperature, then you can assume that the thermal energy gained from the room in reaching room temperature is equal to that lost to the room when the apparatus is above room temperature.

- This method can be adapted to measure the specific heat capacity of a liquid, but you must remember that the temperature of the container will also rise.

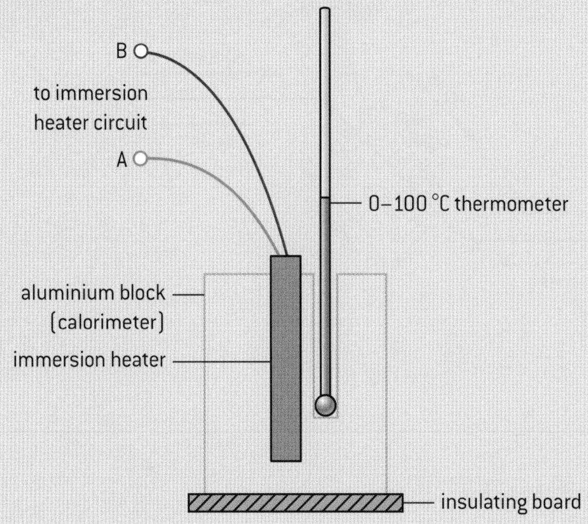

Specific latent heat

The temperature of a block of ice in a freezer is likely to be well below 0 °C. If we measure the temperature of the ice from the time it is taken from the freezer until it has all melted and reached room temperature we would note a number of things. Initially the temperature of the ice would rise; it would then stay constant until all of it had melted then it would rise again but at a different rate from before. If we now put the water into a pan and heat it we would see its temperature rise quickly until it was boiling and then stay constant until all of the water had boiled away. These observations are typical of most substances which are heated sufficiently to make them melt and then boil. Figure 5 shows how the temperature changes with time for energy being supplied. Energy is continually being supplied to the ice but there is no temperature change occurring during melting or boiling. The energy required to achieve the change of phase is called **latent heat**; the word *latent* meaning "hidden". In a similar way to how specific heat capacity was defined in order to compare equal masses of substances, physicists define:

- **specific latent heat of fusion** (melting) as the energy required to change the phase of 1 kg of substance from a solid to a liquid without any temperature change

TOK

The development of understanding of energy transfer due to temperature differences

The term specific heat capacity is a throwback to the caloric days when energy was thought to be a substance. A more appropriate term would be to call this quantity *specific energy capacity* but specific heat capacity is a term that has stuck even though scientists now recognize its inappropriateness.

Are there other cases where we still use an old-fashioned term or concept because it is simpler to do so than to change all the books? Has the paradigm shift believing that heating was a transfer of substance into being a transfer of energy been taken on by society or is there still confusion regarding this?

- **specific latent heat of vaporization** (boiling) as the energy required to change the phase of 1 kg of liquid into a gas without any temperature change.

The equations for these take the form

$$L = \frac{Q}{m}$$

where Q is the energy supplied to the object of mass which causes its phase to change without any temperature change. When Q is in J, m in kg, L will be in units of J kg^{-1}.

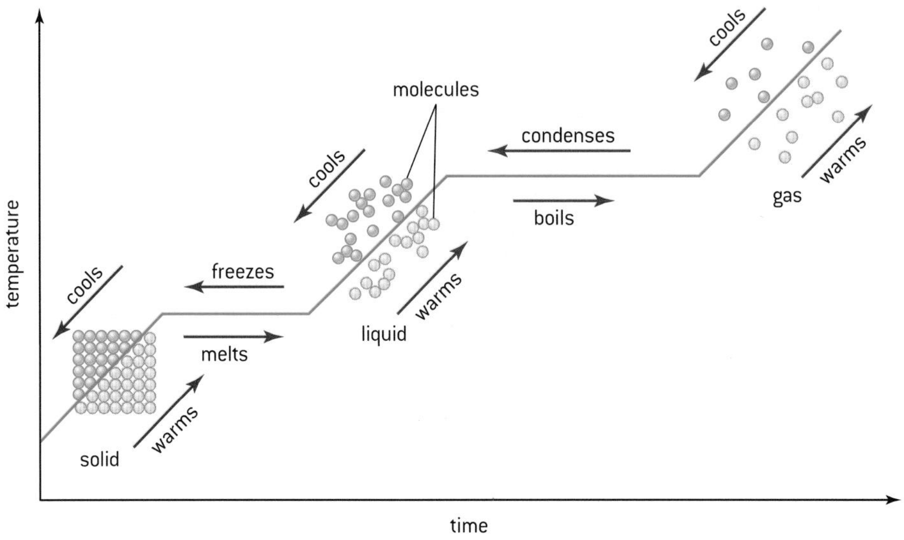

▲ Figure 5 Change of phase (not to scale).

The graph shows how the temperature of a substance changes with time. The portions of the graph that are flat indicate when there is no temperature change and so the substance is changing phase.

- This graph assumes that energy is supplied by a constant power source and so, since energy is the product of power and time, a graph of temperature against energy transferred to the substance will take the same shape as this.
- If you are told the rate at which energy is supplied (i.e. the power supplied) then from the gradients of the temperature–time graph you can work out the heat capacity of the liquid, the solid and the gas.
- From the time of each horizontal section of the graph you can also calculate the specific heat capacity of the solid and liquid phases.
- When a gas transfers energy to another object at a constant rate we see the temperature of the gas drop until it reaches the boiling point (condensing point is the same temperature as this) at which time the graph becomes flat. Again as the liquid transfers energy away the temperature will fall until the melting (or solidifying) temperature is reached when the graph again becomes flat until all the substance has solidified. After this it will cool again.
- Cooling is the term generally used to mean falling temperature rather than loss of energy, so it is important not to say the substance is cooling when the temperature is constant (flat parts of the graph).

Worked example

A heater is used to boil a liquid in a pan for a measured time.

The following data are available.

Power rating of heater = 25 W

Time for which liquid is boiled = 6.2×10^2 s

Mass of liquid boiled away = 4.1×10^{-2} kg

Use the data to determine the specific latent heat of vaporization of the liquid.

Solution

Using $Q = Pt$ for heater: $Q = 25 \times 6.2 \times 10^2 = 15\,500$ J

Using $Q = mL$ for the liquid: $15\,500 = 4.1 \times 10^{-2} \times L$

$$L = \frac{15\,500}{4.1 \times 10^{-2}} = 3.8 \times 10^5 \text{ J kg}^{-1}$$

Molecular explanation of phase change

The transfer of energy to a solid increases its internal energy – this means that the mean random kinetic energy of the molecules increases (as the vibrations and speeds increase) and the intermolecular potential energy increases (as the molecules move further apart). Eventually some groups of molecules move far enough away from their neighbours for the influence of their neighbours to be reduced – chemists would often use the model of the intermolecular bonds being broken. When this is happening the energy supplied does not increase the mean random kinetic energy of the molecules but instead increases the potential energy of the molecules.

Eventually the groups of molecules are sufficiently free so that the solid has melted. The mean speed of the groups of molecules now increases and the temperature rises once more – the molecules are breaking away from each other and joining together at a constant rate. The potential energy is not changing on average. As the liquid reaches its boiling point the molecules start moving away from each other within their groups. Individual molecules break away and the potential energy once more increases as energy is supplied. Again a stage is reached in which the mean kinetic energy remains constant until all the molecules are separated from each other. We see a constant temperature (although much higher than during melting). The energy is vaporizing the liquid and only when the liquid has all vaporized is the energy re-applied to raising the gas temperature.

3.2 Modelling a gas

Understanding

- Pressure
- Equation of state for an ideal gas
- Kinetic model of an ideal gas
- Mole, molar mass, and the Avogadro constant
- Differences between real and ideal gases

Applications and skills

- Equation of state for an ideal gas
- Kinetic model of an ideal gas
- Boltzmann equation
- Mole, molar mass, and the Avogadro constant
- Differences between real and ideal gases

Equations

- Pressure: $p = \frac{F}{A}$
- Number of moles of a gas as the ratio of number of molecules to Avogadro's number: $n = \frac{N}{N_A}$
- Equation of state for an ideal gas: $pV = nRT$
- Pressure and mean square velocity of molecules of an ideal gas: $p = \frac{1}{3}\rho\overline{c^2}$
- The mean kinetic energy of ideal gas molecules: $E_{K_{mean}} = \frac{3}{2}k_BT = \frac{3}{2}\frac{R}{N_A}T$

Nature of science

Progress towards understanding by collaboration and modelling

Much of this sub-topic relates to dealing with how scientists collaborated with each other and gradually revised their ideas in the light of the work of others. Repeating and often improving the original experiment using more reliable instrumentation meant that generalizations could be made. Modelling the behaviour of a real gas by an ideal gas and the use of statistical methods was groundbreaking in science and has had a major impact on modern approaches to science including the quantum theory where *certainty* must be replaced by *probability*.

Introduction

One of the triumphs of the use of mathematics in physics was the successful modelling of the microscopic behaviour of atoms to give an understanding of the macroscopic properties of a gas. Although the Swiss mathematician Daniel Bernoulli had suggested that the motion of atoms was responsible for gas pressure in the mid-1700s, the kinetic theory of gases was not widely accepted until the work of the Scottish physicist, James Clerk-Maxwell and the Austrian, Ludwig Boltzmann working independently over a hundred years later.

The gas laws

The gas laws were developed independently experimentally between the mid-seventeenth century and the start of the nineteenth century. An ideal gas can be defined as one that obeys the gas laws under all conditions. Real gases do not do this but they approximate well to an ideal gas so long as the pressure is little more than normal atmospheric pressure. With the modern apparatus available to us it is not difficult to verify the gas laws experimentally.

As a consequence of developing the vacuum pump and thus the hermetically-sealed thermometer, the Irish physicist Robert Boyle was able to show in 1662 that the pressure of a gas was inversely proportional to its volume. Boyle's experiment is now easily repeatable with modern apparatus. **Boyle's law** states that **for a fixed mass of gas at constant temperature the pressure is inversely proportional to the volume**. The French experimenter Edmé Mariotte independently performed a similar experiment to that of Boyle and was responsible for recognizing that the relationship only holds at a constant temperature – so with modern statements of the law, there is much justification for calling this law Mariotte's law.

Mathematically this can be written as $p \propto \frac{1}{V}$ (at constant temperature) or $pV =$ constant (at constant temperature).

A graph of pressure against volume at constant temperature (i.e., a Boyle's law graph) is known as an isothermal curve ("iso" meaning the same and "thermo" relating to temperature). Isothermal curves are shown in figure 2(b) on p104.

TOK

Boyle's impact on scientific method

Robert Boyle was a scientific giant and experimental science has much to thank him for in addition to his eponymous law. In the mid-1600s where philosophical reasoning was preferred to experimentation, Boyle championed performing experiments. Boyle was most careful to describe his experimental techniques to allow them to be reproduced by others – this gave reliability to experimental results and their interpretation. At this time, many experimenters were working independently. The confusion in correct attribution of gas laws to their discoverer may have been less involved had others followed Boyle's lead. His rapid and clear reporting of his experimental work and data avoided the secrecy that was common at the time; this was advantageous to the progress of other workers.

Although Boyle was apparently not involved in the ensuing quarrels, there was great debate regarding whether Boyle or the German chemist Henning Brand discovered the element phosphorus. This is largely because Brand kept his discovery secret while he pursued the "philosopher's stone" in an attempt to convert base metals such as lead into gold. Is rapid or frequent publication usually a case of a scientist's self-promotion or is the practice more often altruistic?

▲ Figure 1 Robert Boyle.

The French physicist Jacques Charles around 1787 repeated the experiments of his compatriot Guillaume Amontons to show that all gases expanded by equal amounts when subjected to equal temperature changes. Charles showed that the volume changed by $\frac{1}{273}$ of the volume at 0 °C for each 1 K temperature change. This implied that at −273 °C, the volume of a gas becomes zero. Charles's work went unpublished and was repeated in 1802 by another French experimenter, Joseph Gay-Lussac. The law that is usually attributed to **Charles** is now stated as **for a fixed mass of gas at constant pressure the volume is directly proportional to the absolute temperature**.

Mathematically this law can be written as $V \propto T$ (at constant pressure) or $\frac{V}{T} =$ constant (at constant pressure).

Amontons investigated the relationship between pressure and temperature but used relatively insensitive equipment. He was able to show that the pressure increases when temperature increases but failed to quantify this completely. The **third gas** law is stated as **for a gas of fixed mass and volume, the pressure is directly proportional to the absolute temperature.** This law is sometimes attributed to Amontons and often (incorrectly) to Gay-Lussac or Avogadro and sometimes it is simply called the *pressure law*. Maybe it is safest to refer to this as the third gas law!

Mathematically this law can be written as $p \propto T$ (at constant volume) or $\frac{p}{T} =$ constant(at constant volume).

The Italian physicist, Count Amadeo Avogadro used the discovery that gases expand by equal amounts for equal temperature rises to support a hypothesis that all gases at the same temperature and pressure contain equal numbers of particles per unit volume. This was published in a paper in 1811. We now state this as **the number of particles in a gas at constant temperature and pressure is directly proportional to the volume of the gas**.

Mathematically this law can be written as $n \propto V$ (at constant pressure and temperature) or $\frac{n}{V} =$ constant (at constant pressure and temperature).

Since each of the gas laws applies under different conditions the constants are not the same in the four relationships. The four equations can be combined to give a single constant, the ideal gas constant, R. Combining the four equations gives what is known as the **equation of state of an ideal gas**:

$$\frac{pV}{nT} = R \quad \text{or} \quad pV = nRT$$

When pressure is measured in pascal (Pa), volume in cubic metre (m^3), temperature in kelvin (K), and n is the number of moles in the gas, then R has the value 8.31 J K^{-1} mol^{-1}.

The mole and the Avogadro constant

The mole, which is given the symbol "mol", is a measure of the amount of substance that something has. It is one of the seven SI base units and is defined as being **the amount of substance having the same**

number of particles as there are neutral atoms in 12 grams of carbon-12. One mole of a gas contains 6.02×10^{23} atoms or molecules. This number is the Avogadro constant N_A. In the same way that you can refer to a dozen roses (and everyone agrees that a dozen is the name for 12) you can talk about three moles of nitrogen gas which will mean that you have 18.06×10^{23} gas molecules. The mole is used as an alternative to expressing quantities in volumes or masses.

Molar mass

As we have just seen, the mole is simply a number that can be used to count atoms, molecules, ions, electrons, or roses (if you wanted). In order to calculate the molar mass of a substance (which differs from substance to substance) we need to know the chemical formula of a compound (or whether a molecule of an element is made up from one or more atoms). Nitrogen gas is normally **diatomic** (its molecules have two atoms), so we write it as N_2. One mole of nitrogen gas will contain 6.02×10^{23} molecules but 12.04×10^{23} atoms. As one mole of nitrogen atoms has a mass of approximately 14.01 g then a mole of nitrogen molecules will have a mass of 28.02 g – this is its **molar mass**. The chemical formula for water is, of course, H_2O, so one mole of water molecules contains two moles of hydrogen atoms and one mole of oxygen atoms. 1 mol of hydrogen atoms has a mass of 1.00 g and 1 mol of oxygen atoms has a mass of 16.00 g, so 1 mol of water has a mass of $(2 \times 1.00\text{ g}) + 16.00\text{ g} = 18.00\text{ g}$. The mass of water is 18.00 g mol^{-1}.

Worked example

Calculate the percentage change in volume of a fixed mass of an ideal gas when its pressure is increased by a factor of 2 and its temperature increases from 30 °C to 120 °C.

Solution

Since n is constant the equation of state can be written

$$\frac{p_1V_1}{T_1} = \frac{p_2V_2}{T_2}$$

We are trying to obtain the ratio $\frac{V_2}{V_1}$ and so the equation can be rearranged into the form $\frac{V_2}{V_1} = \frac{p_1T_2}{p_2T_1}$

$p_2 = 2p_1$ and $T_1 = 303\text{ K}$ and $T_2 = 393\text{ K}$, so $\frac{V_2}{V_1}$

$$= \frac{393}{2 \times 303} = 0.65 \text{ or } 65\%$$

This means that there is a 35% reduction in the volume of the gas.

Investigate!

Verifying the gas laws experimentally

Boyle's law

- Pressure is changed using the pump.
- This pushes different amounts of oil into the vertical tube, which changes the pressure on the gas (air).
- Changes are carried out slowly to ensure there is no temperature change.
- A graph of pressure against volume gives a curve known as an **isothermal** (line at constant temperature).

- Isothermals can be plotted over a range of different temperatures to give a series of curves such as those in figure 2(b).
- A graph of pressure against the reciprocal of volume $\frac{1}{V}$ should give a straight line passing through the origin.
- This investigation could be monitored automatically using a data logger with a pressure sensor.

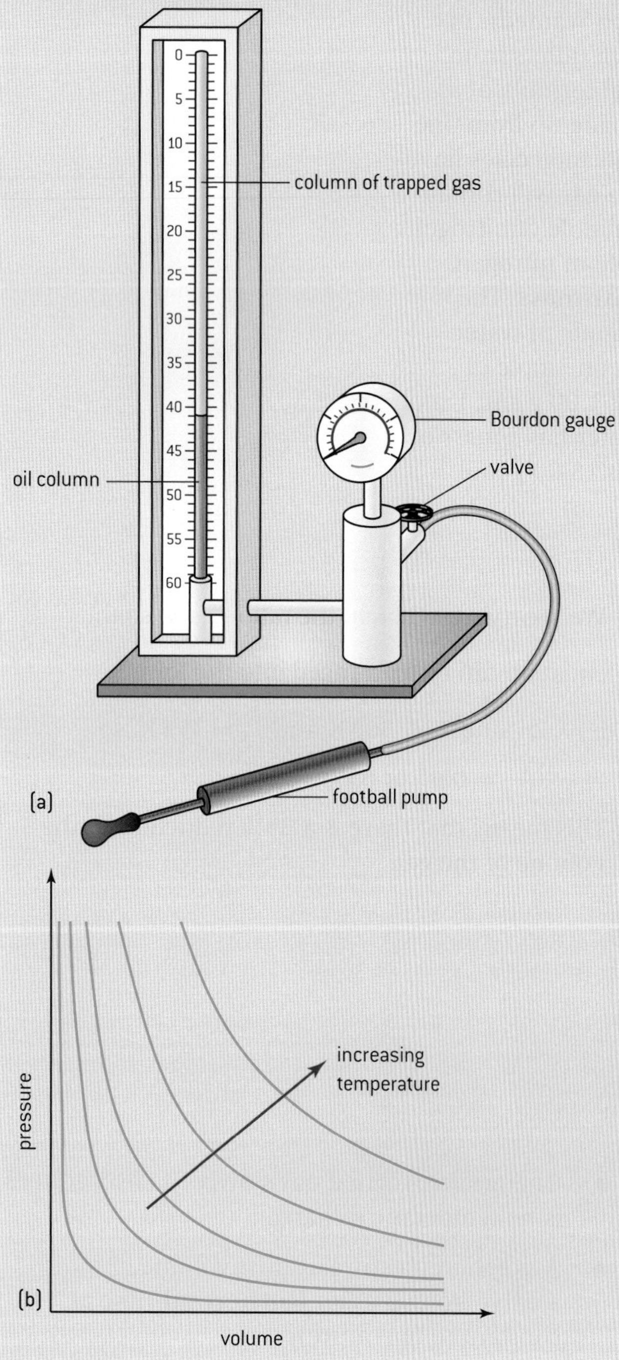

▲ Figure 2 (a) Boyle's law apparatus and (b) graph.

Charles's law

- Pressure is kept constant because the capillary tube is open to the surrounding atmosphere (provided the atmospheric pressure does not change).
- Change the temperature by heating (or cooling/icing) the water.
- Assuming that the capillary is totally uniform (which is unlikely) the length of the column of air trapped in the tube will be proportional to its volume.
- Stir the water thoroughly to ensure that it is at a constant temperature throughout.

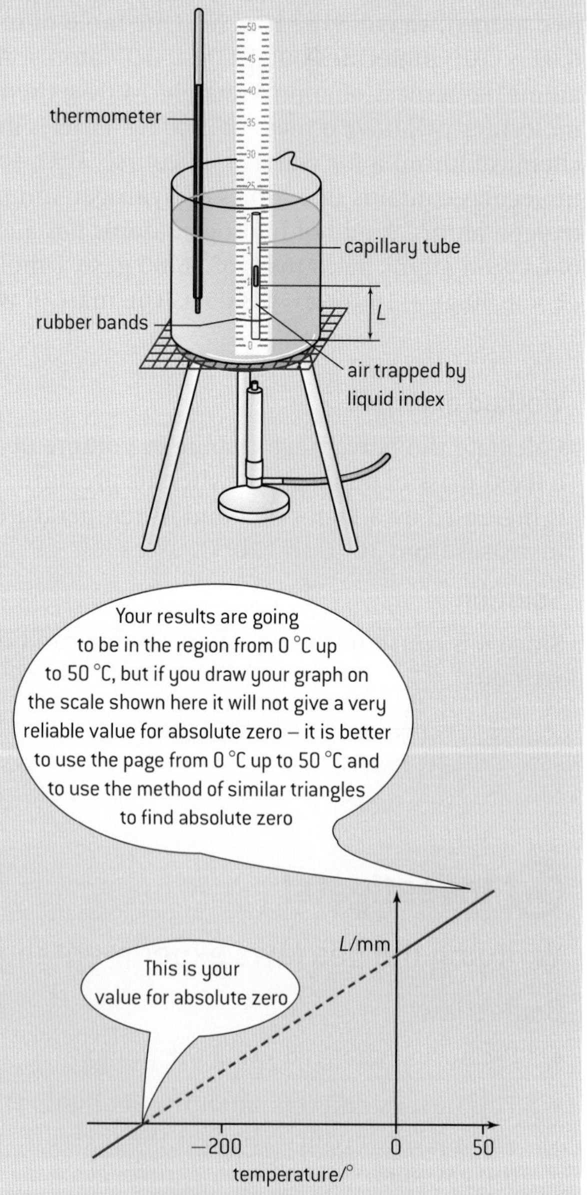

▲ Figure 3 Charles' law apparatus and graph.

- Leave a reasonable length of time in between readings to allow energy to conduct to the air in the capillary – when this happens the length of the capillary no longer changes.
- A graph of length of the air column against the temperature in °C should give a straight line, which can be extrapolated back to absolute zero.
- To choose larger (and therefore better) scales you can use similar triangles to calculate absolute zero.

The third gas law

- There are many variants of this apparatus which could, yet again, be performed using a data logger with temperature and pressure sensors.
- The gas (air) is trapped in the glass bulb and the temperature of the water bath is changed.
- The pressure is read from the Bourdon gauge (or manometer or pressure sensor, etc.).
- Leave a reasonable length of time in between readings to allow energy to conduct to the air in the glass bulb – when this happens the pressure no longer changes.
- A graph of pressure against the temperature in °C should give a straight line, which can be extrapolated back to absolute zero.
- To choose larger (and therefore better) scales you can use similar triangles to calculate absolute zero.

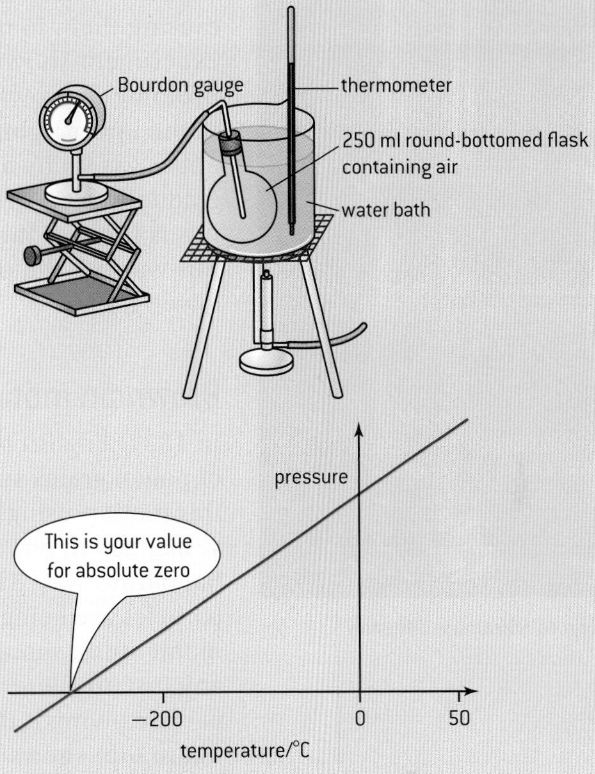

▲ Figure 4 Third gas law apparatus and graph.

Nature of science

Evidence for atoms

The concept of *atoms* is not a new one but it has only been accepted universally by scientists over the past century. Around 400 BCE the Greek philosopher Democritus theorized the existence of *atoms* – named from ancient Greek: "ατομοσ" – meaning without division. He suggested that matter consisted of tiny indivisible but discrete particles that determined the nature of the matter of which they comprised. Experimental science did not become fashionable until the 17th century and so Democritus' theory remained unproven. Sir Isaac Newton had a limited view of the atomic nature of matter but believed that all matter was ultimately made of the same substance. In the 19th century the chemist, John Dalton, was the first to suggest that the individual elements were made of different atoms and could be combined in fixed ratios. Yet even Dalton and his successors were only able to infer the presence of atoms from chemical reactions.

Less explicit evidence (yet still indirect) came from experiments with diffusion and "Brownian motion".

▲ Figure 5 Bromine diffusion.

The microscopic interpretation of gases

Diffusion

The smell of cooked food wafting from a barbecue is something that triggers the flow of stomach juices in many people. This can happen on a windless day and is an example of diffusion of gases. The atomic vapours of the cooking food are being bombarded by air molecules causing them to move through the air randomly. The bromine vapour experiment shown in figure 5 is a classic demonstration of diffusion. Bromine (a brown vapour) is denser than air and sinks to the bottom of the lower right-hand gas jar. This is initially separated from the upper gas jar by a cover slide. As the slide is removed the gas is gradually seen to fill the upper gas jar (as seen in the jars on the left); this is because the air molecules collide with the bromine atoms. If this was not the case we would expect the bromine to remain in the lower gas jar.

Brownian motion

In 1827, English botanist, Robert Brown, first observed the motion of pollen grains suspended in water. Today we often demonstrate this motion using a microscope to see the motion of smoke particles suspended in air. The smoke particles are seen to move around in a haphazard way. This is because the relatively big and heavy smoke particles are being bombarded by air molecules. The air molecules have momentum, some of which is transferred to the smoke particles. At any instant there will be an imbalance of forces acting on each smoke particle giving it the random motion observed. The experiment of figure 6(a) shows how smoke in an air cell is illuminated from the side. Each smoke particle scatters light in all directions and so some reaches the microscope. The observer sees the motion as tiny specks of bright light that wobble around unpredictably as shown in figure 6(b).

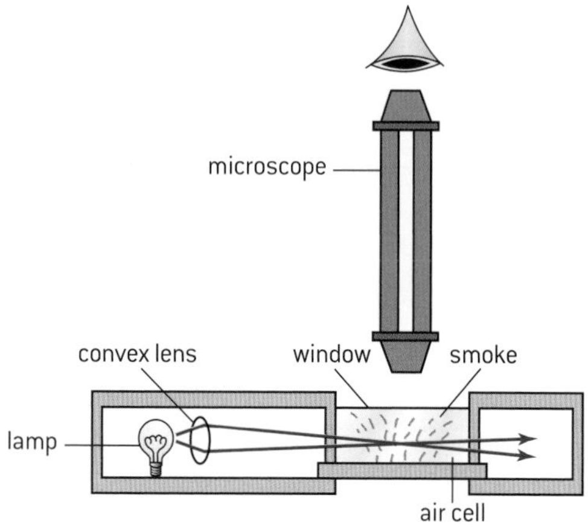

(a)

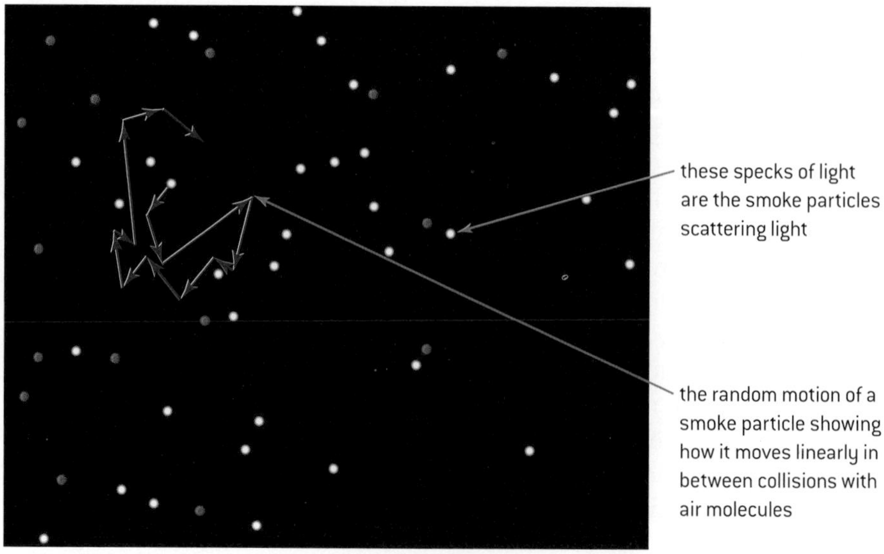

these specks of light are the smoke particles scattering light
the random motion of a smoke particle showing how it moves linearly in between collisions with air molecules

(b)

▲ Figure 6 The smoke cell.

Kinetic model of an ideal gas

The **kinetic theory of gases** is a statistical treatment of the movement of gas molecules in which macroscopic properties such as pressure are interpreted by considering molecular movement. The key assumptions of the kinetic theory are:

- A gas consists of a large number of identical tiny particles called molecules; they are in constant random motion.
- The number is large enough for statistical averages to be made.
- Each molecule has negligible volume when compared with the volume of the gas as a whole.
- At any instant as many molecules are moving in one direction as in any other direction.
- The molecules undergo perfectly elastic collisions between themselves and also with the walls of their containing vessel; during collisions each momentum of each molecule is reversed.
- There are no intermolecular forces between the molecules between collisions (energy is entirely kinetic).
- The duration of a collision is negligible compared with the time between collisions.
- Each molecule produces a force on the wall of the container.
- The forces of individual molecules will average out to produce a uniform pressure throughout the gas (ignoring the effect of gravity).

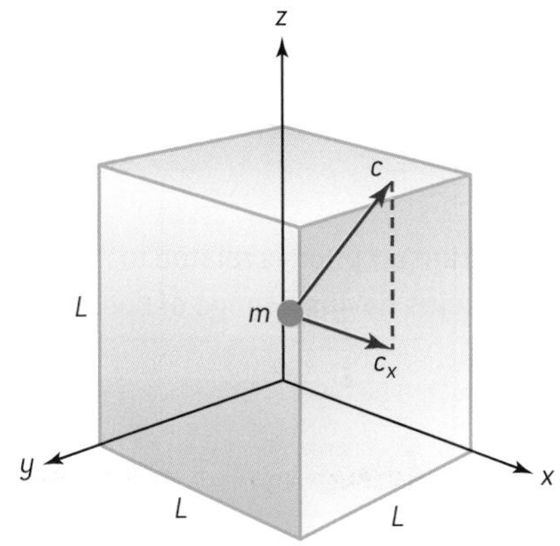

▲ Figure 7 Focusing on the *x*-component of the velocity.

Using these assumptions and a little algebra it is possible to derive the **ideal gas equation**:

In figure 7 we see one of N molecules each of mass m moving in a box of volume V. The box is a cube with edges of length L. We consider the collision of one molecule moving with a velocity c towards the right-hand wall of the box. The components of the molecule's velocity in the x, y, and z directions are c_x, c_y, and c_z respectively. As the molecule collides with the wall elastically, its x component of velocity is reversed, while its y and z components remain unchanged.

The x component of the momentum of the molecule is mc_x before the collision and $-mc_x$ after the collision.

The change in momentum of the molecule is therefore $-mc_x - (mc_x) = -2mc_x$

As force is the rate of change of momentum, the force F_x that the right-hand wall of the box exerts on our molecule will be $\frac{-2mc_x}{t}$ where t is the time taken by the molecule to travel from the right-hand wall of the box to the opposite side and back again (in other words it is the time between collisions with the right-hand wall of the box).

$$\text{Thus } t = \frac{2L}{c_x} \text{ and so } F_m = \frac{-2mc_x}{\frac{2L}{c_x}} = \frac{-mc_x^2}{L}$$

What quantity is equivalent to the reciprocal of *t*?

Using Newton's third law of motion we see that the molecule must exert a force of $\frac{mc_x^2}{L}$ on the right-hand wall of the box (i.e. a force equal in magnitude but opposite in direction to the force exerted by the right-hand wall on the molecule).

Since there are N molecules in total and they will have a range of speeds, the total force exerted on the right-hand wall of the box

$$F_x = \frac{m}{L}\left(c_{x_1}^2 + c_{x_2}^2 + c_{x_3}^2 + \ldots + c_{x_N}^2\right)$$

Where c_{x_1} is the x component of velocity of the first molecule, c_{x_2} that of the second molecule, etc.

Which assumption is related to molecules having a range of speeds?

With so many molecules in even a small volume of gas, the forces average out to give a constant force.

The mean value of the square of the velocities is given by

$$\overline{c_x^2} = \frac{\left(c_{x_1}^2 + c_{x_2}^2 + c_{x_3}^2 + \ldots + c_{x_N}^2\right)}{N}$$

This means that the total force on the right-hand wall of the box is given by $F_x = \frac{Nm}{L}\overline{c_x^2}$.

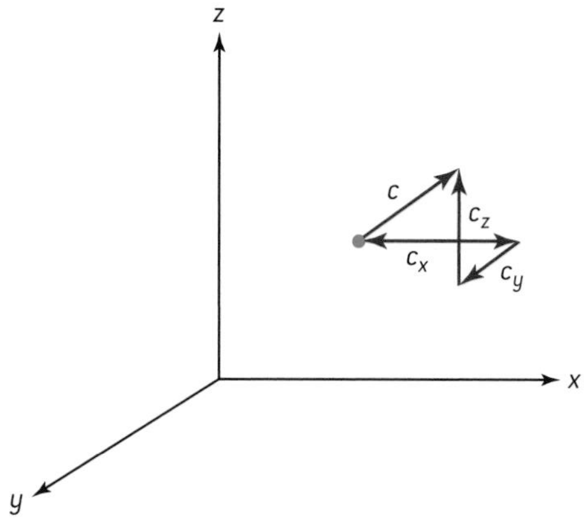

▲ Figure 8 Resolving the velocity of a molecule.

Figure 8 shows a velocity vector c being resolved into components c_x, c_y, and c_z.

Using Pythagoras' theorem we see that $c^2 = c_x^2 + c_y^2 + c_z^2$

It follows that the mean values of velocity can be resolved into the mean values of its components such that

$$\overline{c^2} = \overline{c_x^2} + \overline{c_y^2} + \overline{c_z^2}$$

$\overline{c^2}$ is called the mean square speed of the molecules.

Which assumption is being used here?

On average there is an equal likelihood of a molecule moving in any direction as in any other direction so the magnitude of the mean components of the velocity will be the same, i.e.,

$$\overline{c_x^2} = \overline{c_y^2} = \overline{c_z^2}\text{, so } \overline{c^2} = 3\,\overline{c_x^2} \text{ or } \overline{c_x^2} = \tfrac{1}{3}\,\overline{c^2}$$

So our equation for the total force on the right-hand wall becomes $F = \frac{1}{3}\frac{Nm}{L}\overline{c^2}$

The pressure on this wall (which has an area A) is given by $p_x = \frac{F_x}{A}$ and $A = L^2$ so that

$$p_x = \frac{1}{3}\frac{Nm}{L^3}\overline{c^2} \text{ or } p_x = \frac{1}{3}\frac{Nm}{V}\overline{c^2}$$

Since pressure at a point in a fluid (gas or liquid) acts equally in all directions we can write this equation as

$$p = \frac{1}{3}\frac{Nm}{V}\overline{c^2}$$

As Nm is the total mass of the gas and the density ρ (Greek, rho) is the total mass per unit volume, our equation simplifies to

$$p = \frac{1}{3}\rho\overline{c^2}$$

Do collisions between molecules affect this result?

Be careful not to confuse p, the pressure, with ρ, the density.

You should note that the guidance in the IB Diploma Programme physics guide says that you should understand this proof – that does not mean that you need to learn it line by line but that each of the steps should make sense to you. You should then be able to answer any question asked in examinations.

Molecular interpretation of temperature

Returning to the equation of an ideal gas in the form

$$p = \frac{1}{3}\frac{Nm}{V}\overline{c^2}$$

we get

$$pV = \frac{Nm}{3}\overline{c^2}$$

Multiplying each side by $\frac{3}{2}$ gives

$$\frac{3}{2}pV = \left(\frac{3}{2}\frac{Nm}{3}\overline{c^2}\right) = N \times \frac{1}{2}m\overline{c^2}$$

Comparing this with the equation of state for an ideal gas $pV = nRT$

we can see that $\frac{3}{2}nRT = N \times \frac{1}{2}m\overline{c^2}$

This may look a little confusing with both n and N in the equation so let's recap:

n is the number of moles of the gas and N is the number of molecules so let's combine them to give a simpler equation:

$$\frac{3}{2}\frac{nRT}{N} = \frac{1}{2}m\overline{c^2}$$

But $\frac{N}{N_A} = n$ so $\frac{N}{n}$ is the number of molecules per mole (the Avogadro number, N_A) making the equation:

$$\frac{3}{2}\frac{RT}{N_A} = \frac{1}{2}m\overline{c^2}$$

Things get even easier when we define a new constant to be $\frac{R}{N_A}$; this constant is given the symbol k_B and is called the **Boltzmann constant**.

So the equation now becomes:

$$\frac{3}{2}k_B T = \frac{1}{2}m\overline{c^2}$$

Now $\frac{1}{2}m\overline{c^2}$ should remind you of the equation for kinetic energy and, indeed, it represents the mean translational kinetic energy of the gas molecules. We can see from this equation that the mean kinetic energy gives a measure of the absolute temperature of the gas molecules.

The kinetic theory has linked the temperature (a macroscopic property) to the microscopic energies of the gas molecules. In an ideal gas there are no long-range intermolecular forces and therefore no potential energy components; the internal energy of an ideal gas is entirely kinetic. This means that the total internal energy of an ideal gas is found by multiplying the number of molecules by the mean kinetic energy of the molecules

total internal energy of an ideal gas $= \frac{3}{2}Nk_B T$

We have only considered the translational aspects of our gas molecules in this derivation and this is fine for atomic gases (gases with only one atom in their molecules); when more complex molecules are considered this equation is slightly adapted using a principle called the equipartition of energies (something not included in the IB Diploma Programme Physics guide).

Note

When you look at the unit for Boltzmann's constant you should see a similarity with specific heat capacity. This is like the specific heat capacity for one mole of **every** monatomic gas

Alternative equation of state of an ideal gas

The gas laws produced the equation of state in the form of $pV = nRT$. Now we have met the Boltzmann constant we can use an alternative form of this equation. As we have seen n represents the number of moles of our ideal gas and R is the universal molar gas constant. If we want to work with the number of molecules in the gas (as physicists often do) the equation of state can be written in the form of $pV = Nk_BT$ where N represents the number of molecules and k_B is the Boltzmann constant. Thus $nR = Nk_B$ and if we consider 1 mol of gas then $n = 1$ and $N = N_A = 6.02 \times 10^{23}$. We see from this why $k_B = \frac{R}{N_A}$ as was used previously.

With $R = 8.3\ \text{J K}^{-1}\ \text{mol}^{-1}$ k_B must $= \frac{8.3\ \text{J mol}^{-1}\ \text{K}^{-1}}{6.02 \times 10^{23}\ \text{mol}^{-1}} = 1.38 \times 10^{-23}\ \text{J K}^{-1}$

Worked example

Nitrogen gas is sealed in a container at a temperature of 320 K and a pressure of 1.01×10^5 Pa.

a) Calculate the mean square speed of the molecules.

b) Calculate the temperature at which the mean square speed of the molecules reduces to 50% of that in a).

mean density of nitrogen gas over the temperatures considered = 1.2 kg m^{-3}

Solution

a) $p = \frac{1}{3}\rho\overline{c^2}$ so $\overline{c^2} = \frac{3p}{\rho} = \frac{3 \times 1.01 \times 10^5}{1.2}$

$= 2.53 \times 10^5\ \text{m}^2\ \text{s}^{-2}$ (notice the unit here?)

b) $\overline{c^2} \propto T$ so the temperature would need to be 50% of the original value (i.e. it would be 160 K).

Nature of science

Maxwell–Boltzmann distribution

In the 1850s James Clerk Maxwell realized that a gas had too many molecules to have any chance of being analysed using Newton's laws (even though this could be done in principle). With no real necessity to consider the motion of individual molecules he realized that averaging techniques could be used to link the motion of the molecules with their macroscopic properties. He recognized that he needed to know the distribution of molecules having different speeds. Ludwig Boltzmann had proposed a general idea about how energy was distributed among systems consisting of many particles and Maxwell developed Boltzmann's distribution to show how many particles have a particular speed in the gas. Figure 9 shows the three typical distributions for the same number of gas molecules at three different temperatures. At higher temperatures the most probable speed increases but overall there are less molecules travelling at this speed since there are more molecules travelling at higher speeds. The speeds of molecules at these temperatures can be greater than 1 km s^{-1} (but they don't travel very far before colliding with another molecule!).

The Maxwell–Boltzmann distribution was first verified experimentally between 1930 and 1934 by the American physicist I F Zartman using methods devised by the German-American Otto Stern in the 1920s. Zartman measured the speed of molecules emitted from an oven by collecting them on the inner surface of a rotating cylindrical drum.

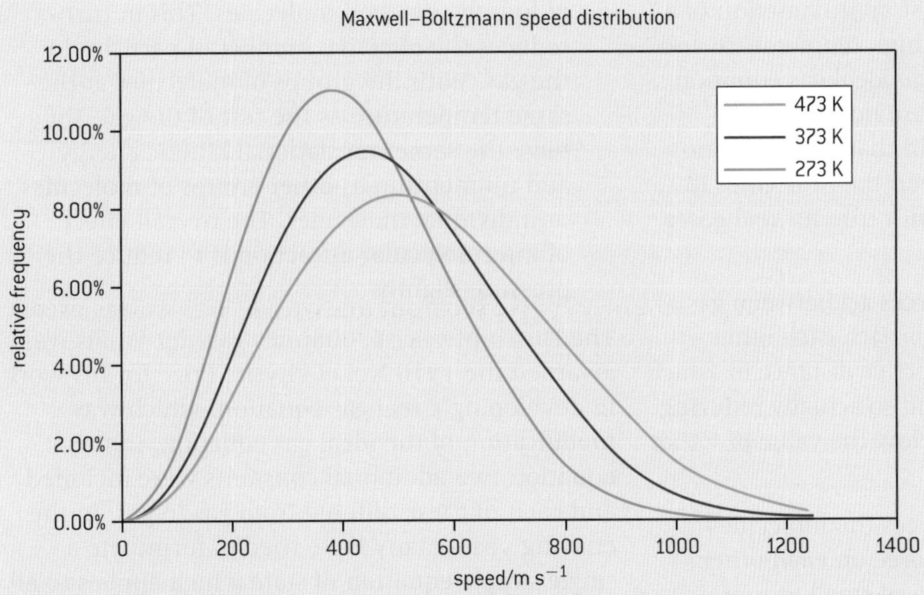

▲ Figure 9 Maxwell–Boltzmann distribution for the speed of gas molecules.

Nature of science

Real gases

An ideal gas is one that would obey the gas laws and the ideal gas equation under all conditions – so **ideal gases cannot be liquefied**. In 1863, the Irish physician and chemist Thomas Andrews succeeded in plotting a series of p–V curves for carbon dioxide; these curves deviated from the Boyle's law curves at high pressures and low temperatures. Until this time it had been believed that certain gases could never be liquefied. Andrews showed that there was a critical temperature above which the gas could not be liquefied by simply increasing the pressure. He demonstrated that for carbon dioxide the critical temperature is approximately 31 °C.

The difference between ideal gases and real gases

The 19th century experimenters showed that ideal gases are just that – ideal – and that real gases do not behave as ideal gases. Under high pressures and low temperatures all gases can be liquefied and thus become almost incompressible. This can be seen if we plot a graph of $\frac{pV}{RT}$ against p for one mole of a real gas – we would expect this plot to give a horizontal straight line of value of 1.00 for an ideal gas. Figure 10 is such a plot for nitrogen. The pressure axis is in atmospheres – this is a non-SI unit that is appropriate for high pressures and quite commonly used by experimenters (e.g. 900 atm converts to 900 × 1.01 × 10^5 Pa or 9.1 × 10^7 Pa). The ideal gas line is a better fit to the real gas line at 1000 K and better still at low pressure.

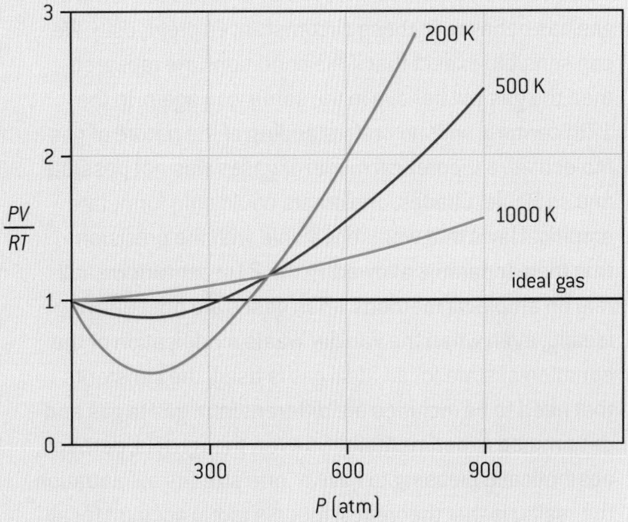

▲ Figure 10 Deviation of a real gas from an ideal gas situation.

Thus we can see that the best approximation of a real gas to an ideal gas is at high temperature and low pressures. In deriving the ideal gas equation we made assumptions that are not true for a real gas. In particular we said that "there are no intermolecular forces between the molecules in between collisions." This is not true for real gases on two accounts:

- Short-range repulsive forces act between gas molecules when they approach each other – thereby reducing the effective distance in which they can move freely (and so actually reducing the useful gas volume below the value of V that we considered).
- At slightly greater distances the molecules will exert an attractive force on each other – this causes the formation of small groups of loosely attached molecules. This in turn reduces the effective number of particles in the gas. With the groups of molecules at the same temperature as the rest of the gas, they have the same translational kinetic energy and momentum as other groups of molecules or individual molecules. The overall effect of intermolecular attraction is to reduce the pressure slightly.

The Dutch physicist Johannes van der Waals was awarded the 1910 Nobel Physics Prize for his work in developing a real gas equation, which was a modification of the ideal gas equation. In this equation two additional constants were included and each of these differed from gas to gas. In the ensuing years many have tried to formulate a single simple equation of state which applies to all gases, no one has been successful to date.

TOK

Empirical versus theoretical models

Theoretical models such as the kinetic theory of gas are based on certain assumptions. If these assumptions are not met in the real world, the model may fail to predict a correct outcome as is the case with real gases at high pressure and low temperature.

Despite the best efforts of van der Waals and others, there is no theorectical model that gives the right answer for all gases in all states. This, however, does not mean that we cannot make correct predictions about a particular gas – we can use an empirical model that is based on how the gas has behaved in these circumstances previously. We can sensibly expect that if the conditions are repeated then the gas will behave in the same way again. In the 17th century, with no understanding of the nature of gas molecules, a theoretical model of gases was not possible and so Boyle, Charles, and others could only formulate empirical laws that were repeatable with the precision that their apparatus allowed. In the 21st century we still rely on empirical methods when gases are not behaving ideally; even when the van der Waals modification of the equation of state for an ideal gas is used, the constants that need to be included are different from gas to gas and determined experimentally. So even though it is, perhaps, aesthetically pleasing to have a "one size fits all" equation, the reality is that theoretical models rarely account for all circumstances. Even Newton's laws of motion must be modified when objects are moving at speeds approaching that of light. Theoretical models are not inherently better than empirical models, or vice versa. The best model, whether it is theoretical or empirical, is the model that gives the best predictions for a particular set of circumstances.

- When a set of experimental results are not in line with theory does this mean that the theory must be abandoned?
- In March 2012 the following report appeared on the BBC website:

> "The head of an experiment that appeared to show subatomic particles travelling faster than the speed of light has resigned from his post..... Earlier in March, a repeat experiment found that the particles, known as neutrinos, did not exceed light speed. When the results from the Opera group at the Gran Sasso underground laboratory in Italy were first published last year, they shocked the world, threatening to up-end a century of physics as well as relativity theory – which holds the speed of light to be the Universe's absolute speed limit."

The media were very quick to applaud the experiment when the initial results were published, but the scientific community was less quick to abandon a well-established "law". Which of the two groups was showing bias in this case?

Questions

1 Two objects are in thermal contact. State and explain which of the following quantities will determine the direction of the transfer of energy between these objects?

a) The mass of each object.

b) The area of contact between the objects.

c) The specific heat capacity of each object.

d) The temperature of each object. (4 marks)

2 Two objects are at the same temperature. Explain why they must have the same internal energy. (2 marks)

3 The internal energy of a piece of copper is increased by heating.

a) Explain what is meant, in this context, by internal energy and heating.

b) The piece of copper has mass 0.25 kg. The increase in internal energy of the copper is 1.2×10^3 J and its increase in temperature is 20 K. Estimate the specific heat capacity of copper. (4 marks)

4 Calculate the amount of energy needed to raise the temperature of 3.0 kg of steel from 20 °C to 120 °C. The specific heat capacity of steel is 490 J kg^{-1} K^{-1}. (2 marks)

5 Calculate the energy supplied to 0.070 kg of water contained in a copper cup of mass 0.080 kg. The temperature of the water and the cup increases from 17 °C to 25 °C.

Specific heat capacity of water = 4200 J $kg^{-1}K^{-1}$

Specific heat capacity of copper = 390 J kg^{-1} K^{-1} (2 marks)

6 The temperature difference between the inlet and the outlet of an air-cooled engine is 30.0 K. The engine generates 7.0 kW of waste power that the air extracts from the engine. Calculate the rate of flow of air (in kg s^{-1}) needed to extract this power.

Specific heat capacity of air (at constant pressure) = 1.01×10^3 J kg^{-1} K^{-1} (3 marks)

7 2.0 kg of water at 0 °C is to be changed into ice at this temperature. The same mass of water now at 100 °C is to be changed into steam at this temperature.

Specific latent heat of fusion of water = 3.34×10^5 J kg^{-1}

Specific latent heat of vaporization of water = 2.26×10^6 J kg^{-1}

a) Calculate the amount of energy needed to be removed from the water to freeze it.

b) Calculate the amount of energy required by the water to vaporize it.

c) Explain the difference between the values calculated in a) and b). (6 marks)

8 (*IB*) A container holds 20 g of neon and also 8 g of helium. The molar mass of neon is 20 g and that of helium is 4 g.

Calculate the ratio of the number of atoms of neon to the number of atoms of helium. (2 marks)

9 A fixed mass of an ideal gas is heated at constant volume. Sketch a graph to show the variation with Celsius temperature t with pressure p of the gas? (3 marks)

10 Under what conditions does the equation of state for an ideal gas, $pV = nRT$, apply to a real gas? (2 marks)

11 a) (i) Explain the difference between an ideal gas and a real gas.

(ii) Explain why the internal energy of an ideal gas comprises of kinetic energy only.

b) A fixed mass of an ideal gas has a volume of 870 cm^3 at a pressure of 1.00×10^5 Pa and a temperature of 20.0 °C. The gas is heated at constant pressure to a temperature of 21.0 °C.

Calculate the change in volume of the gas. (6 marks)

12 *(IB)*

a) The pressure p of a fixed mass of an ideal gas is directly proportional to the kelvin temperature T of the gas, that is, $p \propto T$.

(i) State the relation between the pressure p and the volume V for a change at constant temperature.

(ii) State the relation between the volume V and kelvin temperature T for a change at a constant pressure.

b) The ideal gas is held in a cylinder by a moveable piston. The pressure of the gas is p_1, its volume is V_1 and its kelvin temperature is T_1.

The pressure, volume and temperature are changed to p_2, V_2 and T_2 respectively. The change is brought about as illustrated below.

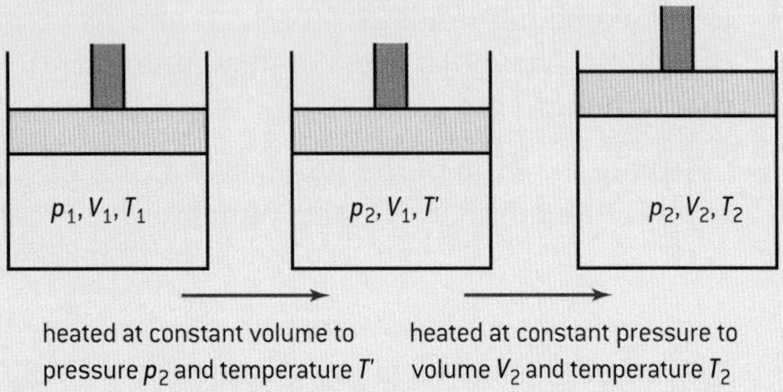

State the relation between

(i) p_1, p_2, T_1 and T'

(ii) V_1, V_2, T' and T_2

c) Use your answers to b) to deduce, that for an ideal gas

$pV = KT$,

where K is a constant. (6 marks)

13 A helium balloon has a volume of 0.25 m^3 when it is released at ground level. The temperature is 30 °C and the pressure 1.01×10^5 Pa. The balloon reaches a height such that its temperature has fallen to −10 °C and its pressure to 0.65×10^5 Pa.

a) Calculate the new volume of the balloon.

b) State two assumptions that must be made about the helium in the balloon.

c) Calculate the number of moles of helium in the balloon. (6 marks)

4 OSCILLATIONS AND WAVES

Introduction

All motion is either *periodic* or *non-periodic*. In **periodic motion** an object repeats its pattern of motion at fixed intervals of time: it is regular and repeated. Wave motion is also periodic and there are many similarities between oscillations and waves; in this topic we will consider the common features but also see that there are differences.

4.1 Oscillations

Understanding

- → Simple harmonic oscillations
- → Time period, frequency, amplitude, displacement, and phase difference
- → Conditions for simple harmonic motion

Applications and skills

- → Qualitatively describing the energy changes taking place during one cycle of an oscillation
- → Sketching and interpreting graphs of simple harmonic motion examples

Equations

- → Period-frequency relationship: $T = \frac{1}{f}$
- → Proportionality between acceleration and displacement: $a \propto -x$

Nature of science

Oscillations in nature

Naturally occurring oscillations are very common, although they are often enormously complex. When analysed in detail, using a slow-motion camera, a hummingbird can be seen to flap its wings at a frequency of around 20 beats per second as it hovers, drinking nectar. Electrocardiographs are used to monitor heartbeats as hearts pulsate, pushing blood around our bodies at about one per second when we are resting and maybe two or three times this rate as we exert ourselves. Stroboscopes can be used to freeze the motion in engines and motors where periodic motion is essential, but too strong vibrations can be potentially very destructive. The practical techniques that have been developed combined with the mathematical modelling that is used to interpret oscillations are very powerful tools; they can help us to understand and make predictions about many natural phenomena.

▲ Figure 1 Hummingbird hovering over flower.

▲ Figure 2 Pocket watch under strobe light.

Isochronous oscillations

A very common type of oscillation is known as **isochronous** (taking the same time). These are oscillations that repeat in the same time period, maintaining this constant time property no matter what amplitude changes due to damping occur. It is the isochronicity of oscillations such as that of a simple pendulum that has made the pendulum such an important element of the clock. If we use a stroboscope (or strobe) we can freeze an isochronous oscillation so that it appears to be stationary; the strobe is made to flash at a regular interval, which can be matched to the oscillation. If the frequency of the strobe matches that of the oscillating object then, when the object is in a certain position, the strobe flashes and illuminates it. It is then dark while the object completes an oscillation and returns to the same position when the strobe flashes again. In this way the object appears motionless. The lowest frequency of the strobe that gives the object the appearance of being static will be the frequency of the object. If you now double the strobe frequency the object appears to be in two places. Figure 2 shows a swinging pocket watch which is illuminated with strobe light at 4 Hz, so the time interval between the images is 0.25 s. The watch speeds up in the middle and slows down at the edges of the oscillation.

Describing periodic motion

The graph in figure 3 is an electrocardiograph display showing the rhythm of the heart of a healthy 48-year-old male pulsing at 65 beats per minute. The pattern repeats regularly with the repeated pattern being called the *cycle* of the motion. The time duration of the cycle is called the **time period** or the **period (T)** of the motion. Thus, a person with a pulse of 65 beats per minute has an average heartbeat period of 0.92 s.

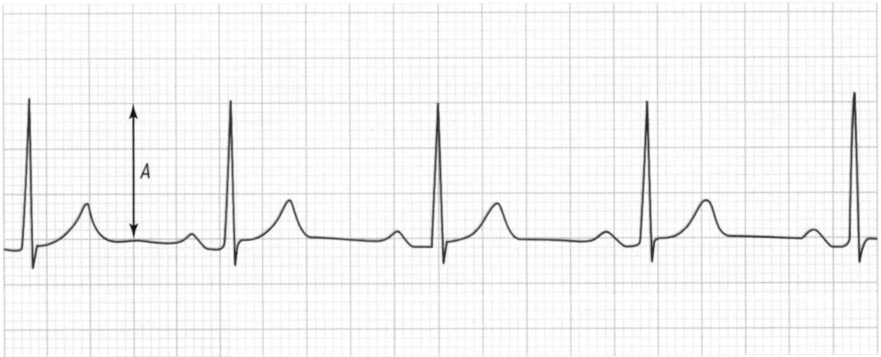

▲ Figure 3 Normal adult male heart rhythm.

Let us now compare the pattern of figure 3 with that of figure 4, which is a data-logged graph of a loaded spring. The apparatus used for the datalogging is shown in figure 5 and the techniques discussed in the *Investigate!* section on p117. The graph for the loaded spring is a more straightforward example of periodic motion, obtained as the mass suspended on a spring oscillates up and down, above and below its normal rest position. There are just over 12 complete oscillations occurring in 20 seconds giving a period of approximately 1.7 s. Although the two graphs are very different the period is calculated in the same way.

As the mass passes its rest or **equilibrium position** its displacement (x) is zero. We have seen in Topics 1 and 2 that displacement is a vector

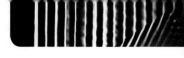

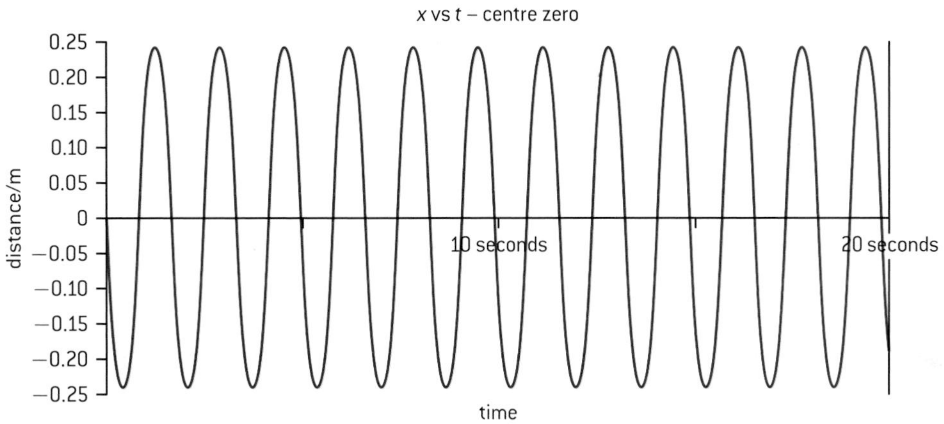

▲ Figure 4 Loaded spring.

quantity and, therefore, must have a direction. In this case, since the motion is linear or one dimensional, it is sufficient to specify the direction as simply being positive or negative. The choice of whether above or below the rest position is positive is arbitrary but, once decided, this must be used consistently. In the graph of figure 4 the data logger has been triggered to start as the mass passes through the rest position. In this case positive displacements are above the rest position – so the mass is moving downwards when timing is started.

The maximum value of the displacement is called the **amplitude** (x_0). In the case of the loaded spring the amplitude is a little smaller than 0.25 m. It is much more difficult to measure the amplitude for the heart rhythm (figure 3) because of determining the rest position – additionally with no calibration of the scale it is impossible to give any absolute values. The amplitude is marked as A which is approximately 3 cm on this scale.

The **frequency** (f) of an oscillation is the number of oscillations completed per unit time. When the time is in seconds the frequency will be in hertz (Hz). Scientists and engineers regularly deal with high frequencies and so kHz, MHz, and GHz are commonly seen. The vibrations producing sound waves range from about 20 Hz to 20 kHz while radio waves range from about 100 kHz to 100 MHz.

With **frequency** being the number of oscillations per second and **period** the time for one oscillation you may have already spotted the relationship between these two quantities:

$$T = \frac{1}{f}$$

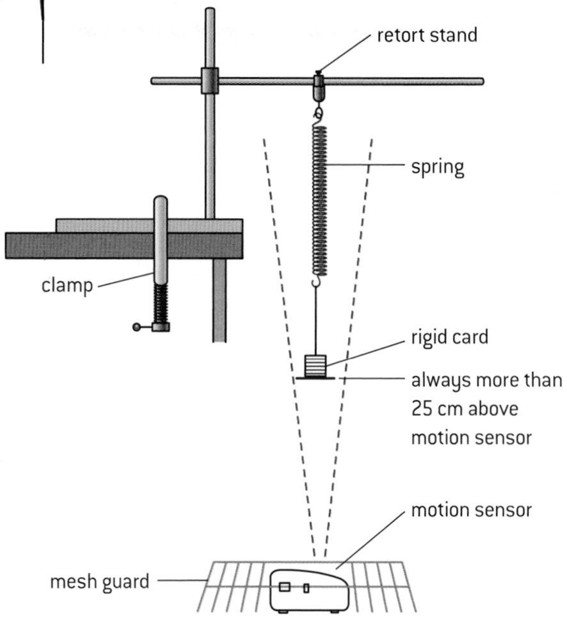

▲ Figure 5 Data logger arrangement.

Worked example

A child's swing oscillates simple harmonically with a period of 3.2 s. What is the frequency of the swing?

Solution

$T = \frac{1}{f}$ so $f = \frac{1}{T}$

$= \frac{1}{3.2} = 0.31$ Hz

Investigate!

In this investigation, a motion sensor is used to monitor the position of a mass suspended from the end of a long spring. The data logger software processes the data to produce a graph showing the variation of displacement with time. The apparatus is arranged as shown in figure 5 with the spring having a period of at least 1 second.

- The apparatus is set up as shown above.
- The mass is put into oscillation by displacing and releasing it.
- The details of the data logger will determine the setting values but data loggers can usually be triggered to start reading at a given displacement value.

- Motion sensors use ultrasound, so reflections from surroundings should be avoided.
- The software can normally be used to plot graphs of velocity and acceleration (in addition to displacement) against time.
- Similar investigations can be performed with other oscillations.

Nature of Science

"Simple" for physicists

"Simple" does not mean "easy" in the context of physics! In fact, mathematically (as discussed in Sub-topic 9.1) simple harmonic motion (SHM) is not "simple". Simple may be better thought of as *simplified* in that we make life easier for ourselves by either limiting or ignoring some of the forces that act on a body. A simple pendulum consists of a mass suspended from a string. In dealing with the simple pendulum the frictional forces acting on it (air resistance and friction between the string and the suspension) and the fact that the mass will slightly stretch the string are ignored without having any disastrous effect on the equations produced. It also means that smooth graphs are produced, such as for the loaded spring.

Simple harmonic motion

We saw that the graph produced by the mass oscillating on the spring was much less complex than the output of a human heart. The mass is said to be undergoing simple harmonic motion or SHM.

In order to perform SHM an object must have a restoring force acting on it.

- The magnitude of the force (and therefore the acceleration) is proportional to the displacement of the body from a fixed point.
- The direction of the force (and therefore the acceleration) is always towards that fixed point.

Focusing on the loaded spring, the "fixed point" in the above definition is the equilibrium position of the mass – where it was before it was pulled down. The forces acting on the mass are the tension in the spring and the pull of gravity (the weight). In the equilibrium position the tension will equal the weight but above the equilibrium the tension will be less and the weight will pull the mass downward; below the equilibrium position the tension will be greater than the weight and this will tend to pull the mass upwards. The difference between the tension and the weight provides the restoring force – the one that tends to return the mass to its equilibrium position.

We can express the relationship between acceleration a and displacement x as:

$$a \propto -x$$

which is equivalent to

$$a = -kx$$

This equation makes sense if we think about the loaded spring. When the spring is stretched further the displacement increases and the tension increases. Because *force* = *mass* × *acceleration*, increasing the force increases the acceleration. The same thing applies when the mass is raised but, in this case, the tension decreases, meaning that the weight dominates and again the acceleration will increase. So although we cannot prove the proportionality (which we leave for HL) the relationship makes sense.

The minus sign is explained by the second bullet point "the acceleration is always in the opposite direction to the displacement". So choosing upwards as positive when the mass is above the equilibrium position, the acceleration will be downwards (negative). When the mass is below the equilibrium position (negative), the (net) force and acceleration are upwards (positive). The force decelerates the mass as it goes up and then accelerates it downwards after it stopped.

Figure 6 is a graph of a against x. We can see that $a = -kx$. takes the form $y = mx + c$ with the gradient being a negative constant.

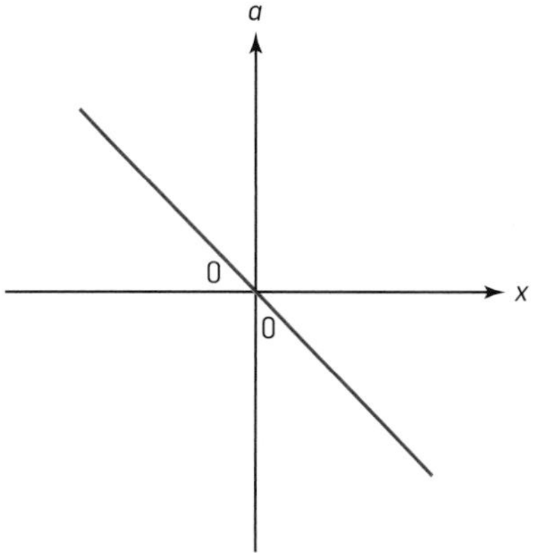

▲ Figure 6 Acceleration–displacement graph.

Graphing SHM

From Topic 2 we know that the gradient of a displacement–time graph gives the velocity and the gradient of the velocity–time graph gives the acceleration at any given time. This remains true for any motion. Let's look at the displacement graph for a typical SHM.

In figure 7 the graph is a sine curve (but if we chose to start measuring the displacement from any other time it could just as easily be a cosine, or a negative sine, or any other sinusoidally shaped graph). If we want to find the velocity at any particular time we simply need to find the gradient of the displacement–time graph at that time. With a curved graph we must draw a tangent to the curve (at our chosen time) in order to find a gradient. The blue tangent at 0 s has a gradient of +2.0 cm s^{-1}, which gives the maximum velocity (this is the steepest tangent and so we see the velocity is a maximum at this time). Looking at the gradient at around 1.6 s or 4.7 s and you will see from the symmetry that it will give a velocity of −2.0 cm s^{-1}, in other words it will be the minimum velocity (i.e. the biggest negative velocity). At around 0.8, 2.4, 3.9, and 5.5 s the gradient is zero so the velocity is zero.

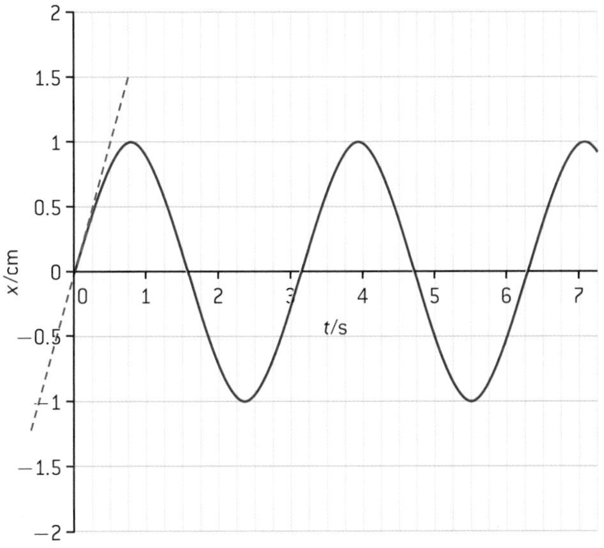

▲ Figure 7 The displacement graph for a typical SHM.

Directions with SHM

In many situations it is helpful to think about the magnitude of the velocity i. e. speed.

For SHM organize a table for an object oscillating between $+A$ and $-A$. In your table show where the magnitudes of the position, velocity, acceleration and force are (a) zero and (b) maximum. Once you have done that mark in the directions of these these vectors at $+A$, $-A$ and zero. By doing this you should be thinking physics before algebra.

Using these values you will see that, overall, the velocity is going to change as shown in figure 8(b); it is a cosine graph in this instance.

You may remember that the gradient of the velocity–time graph gives the acceleration and so if we repeat what we did for displacement to velocity, we get figure 8(c) for the acceleration graph. You will notice that this is a reflection of the displacement–time graph in the time axis i.e. a negative sine curve. This should be no surprise, since from our definition of simple harmonic motion we expect the acceleration to be a negative constant multiplied by the displacement.

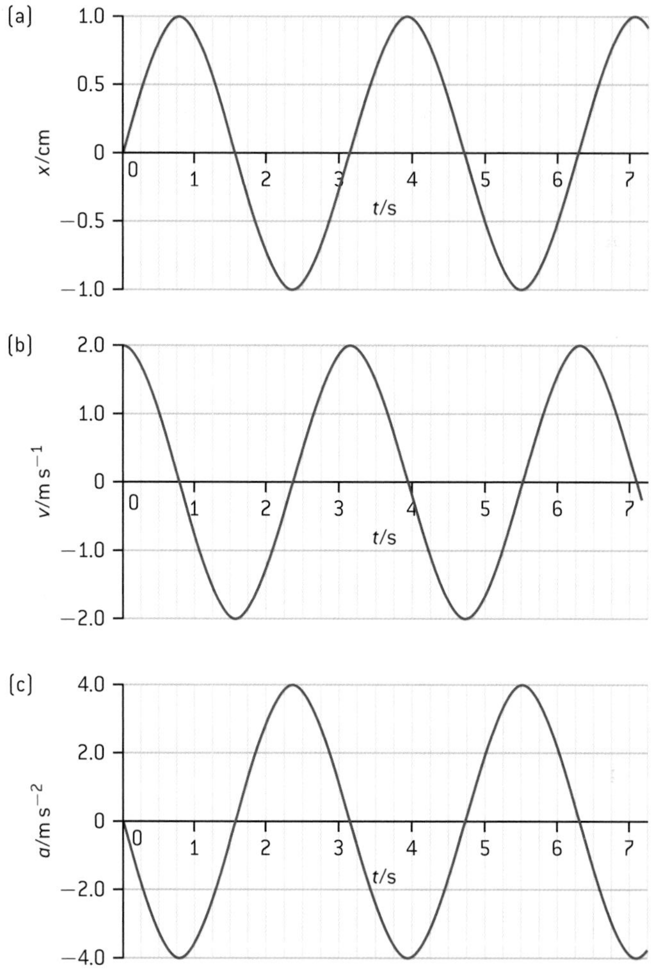

▲ Figure 8 The variation with time of displacement, velocity, and acceleration.

Worked examples

1 On a sheet of graph paper sketch two cycles of the displacement–time $(x - t)$ relationship for a simple pendulum. Assume that its displacement is a maximum at time 0 seconds. Mark on the graph a time for which the velocity is maximum (labeled **A**), a time for which the velocity is zero (labeled **B**) and a time for which the acceleration is a maximum (labeled **C**).

Solution

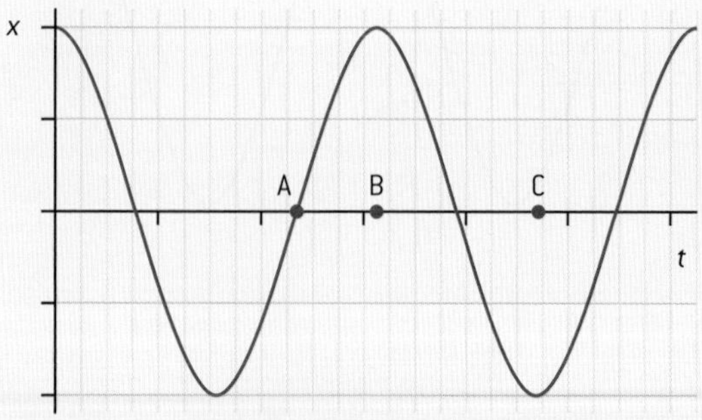

Note

- This is a sketch graph so no units are needed – they are arbitrary.
- **A**, **B**, and **C** are each labelled on the time axis – as this is what the question asks.
- There are just two cycles (two complete periods) marked – since this is what the question asks.
- **A** is where the gradient of the displacement–time graph is a maximum.
- **B** is where the gradient of the displacement–time graph is a zero.
- **C** is where the displacement is a minimum – because $a = -kx$ this means that the acceleration will be a maximum.

2 The equation defining simple harmonic motion is $a = -kx$.

a) What are the units of the constant k?

b) Two similar systems oscillate with simple harmonic motion. The constant for system S_1 is k, while that for system S_2 is $4k$. Explain the difference between the oscillations of the two systems.

Solution

a) Rearranging the equation we obtain

$$k = -\frac{a}{x}$$

Substituting the units for the quantities gives unit of $k = -\frac{\text{ms}^{-2}}{\text{m}} = \text{s}^{-2}$

So the units of k are s^{-2} or per second squared – this is the same as frequency squared (and could be written as Hz^2).

b) Referring to the solution to (a) we can see that for S_2 the square of the frequency would be 4 times that of S_1. This means that S_2 has $\sqrt{4}$ or twice the frequency (or half the period) of S. That is the difference.

Phase and phase difference

Referring back to figures 8 a, b, and c we can see that there is a big similarity in how the shapes of the displacement, velocity, and acceleration graphs change with time. The three graphs are all sinusoidal – they take the same shape as a sine curve. The difference between them is that the graphs all start at different points on the sine curve and continue like this. The graphs are said to have a **phase difference**.

When timing an oscillation it really doesn't matter when we start timing – we could choose to start at the extremes of the oscillation or the middle. In doing this the shape of the displacement graph would not change but would look like the velocity graph (quarter of a period later) or acceleration graph (half a period later). From this we can see that the phase difference between the displacement–time graph and the velocity–time graph is equivalent to quarter of a period or $\frac{T}{4}$ (we could say that velocity leads displacement by quarter of a period). The phase difference between the displacement–time graph and the acceleration–time graph is equivalent to half a period or $\frac{T}{2}$ (we could say that acceleration leads displacement by half a period or that displacement leads acceleration by half a period – it makes no difference which).

Although we have discussed phase difference in term of periods here, it is more common to use angles. However, transferring between period and angle is not difficult:

Period T is equivalent to 360° or 2π radians,

so $\frac{T}{2}$ is equivalent to 180° or π radians

and $\frac{T}{4}$ is equivalent to 90° or $\frac{\pi}{2}$ radians.

When the phase difference is 0 or T then two systems are said to be oscillating **in phase**.

Worked example

1 Calculate the phase difference between the two displacement–time graphs shown in figure 9. Give your answers in a) seconds b) radians c) degrees.

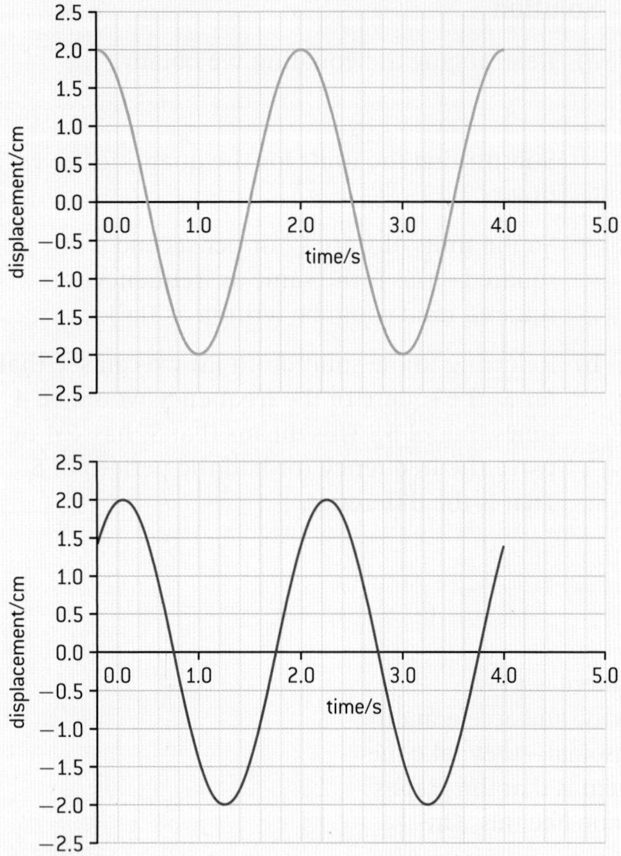

Figure 9

Solution

a) It can be seen from either graph that the period is 2.0 s.

Drawing vertical lines through the peaks of the two curves shows the phase difference. This appears to be between 0.25–0.26 s so we will go with (0.25 ± 0.01) s.

b) The period (2.0 s) is equivalent to 2π radians, so taking ratios:

$\frac{2.0}{2\pi} = \frac{0.25}{\phi}$ (where ϕ is the phase difference in radians)

This gives a value for ϕ of 0.79 radian.

c) The period (2.0 s) is equivalent to 360°, so taking ratios:

$\frac{2.0}{360} = \frac{0.25}{\phi}$ (where ϕ is the phase difference in degrees)

This gives a value for ϕ of 45°.

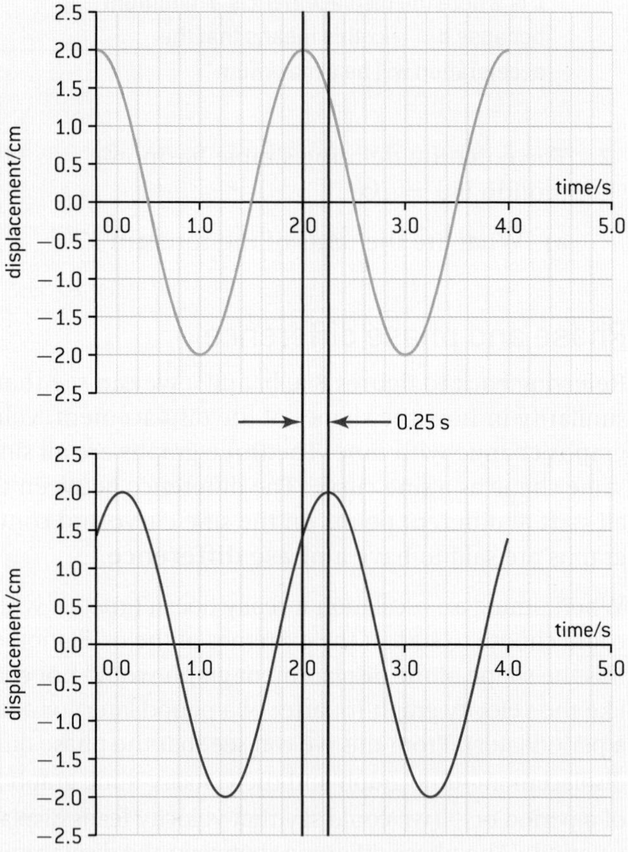

Figure 10

Energy changes in SHM

Let's think of the motion of the simple pendulum of figure 11 as an example of a system which undergoes simple harmonic motion. For a pendulum to oscillate simple harmonically the string needs to be long and to make small angle swings (less than 10°). The diagram is, therefore, not drawn to scale. Let's imagine that the bob just brushes along the ground when it is at its lowest position. When the bob is pulled to position A, it is at its highest point and has a maximum gravitational potential energy

(GPE). As the bob passes through the rest position B, it loses all the GPE and gains a maximum kinetic energy (KE). The bob now starts to slow down and move towards position C when it briefly stops, having regained all its GPE. In between A and B and between B and C the bob has a combination of KE and GPE.

In a damped system over a long period of time the maximum height of the bob and its maximum speed will gradually decay. The energy gradually transfers into the internal energy of the bob and the air around it. Damping will be discussed in Option B.4.

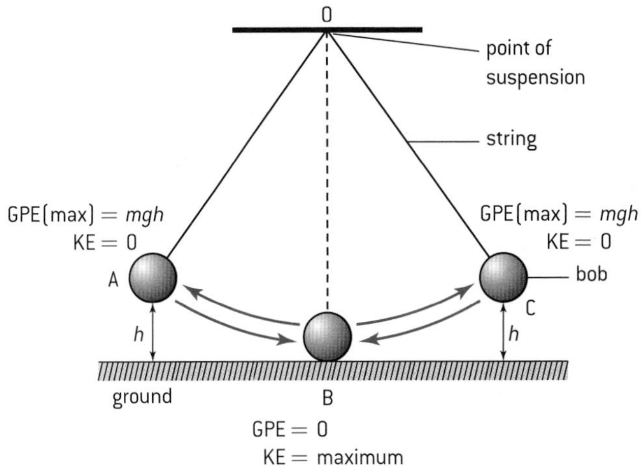

▲ Figure 11 simple pendulum.

4.2 Travelling waves

Understanding

- → Travelling waves
- → Wavelength, frequency, period, and wave speed
- → Transverse and longitudinal waves
- → The nature of electromagnetic waves
- → The nature of sound waves

Applications and skills

- → Explaining the motion of particles of a medium when a wave passes through it for both transverse and longitudinal cases
- → Sketching and interpreting displacement – distance graphs and displacement – time graphs for transverse and longitudinal waves
- → Solving problems involving wave speed, frequency, and wavelength
- → Investigating the speed of sound experimentally

Equation

- → The wave equation: $c = f\lambda$

Nature of science

Patterns and trends in physics

One of the aspects of waves that makes it an interesting topic is that there are so many similarities with just enough twists to make a physicist think. Transverse waves have many similarities to longitudinal waves, but there are equally many differences. An electromagnetic wave requires no medium through which to travel, but a mechanical wave such a sound does need a medium to carry it; yet the intensity of each depends upon the square of the amplitude and the two waves use the same wave equation. Physicists enjoy patterns but at the same time they enjoy the places where patterns and trends change – the similarities give confidence but the differences can be thought-provoking.

▲ Figure 1 Ripples spreading out from a stone thrown into a pond.

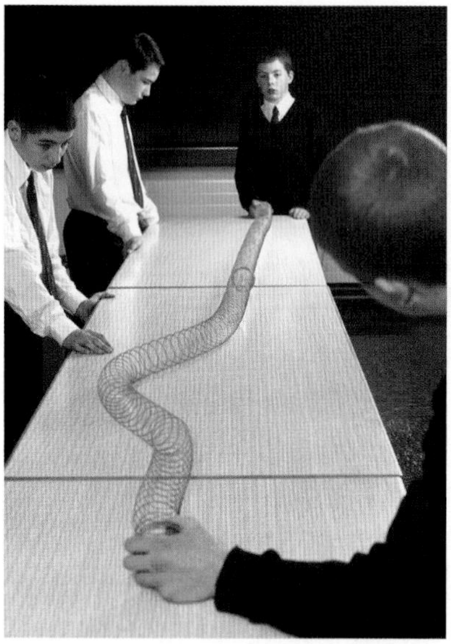

▲ Figure 2 Slinky spring being used to demonstrate travelling waves.

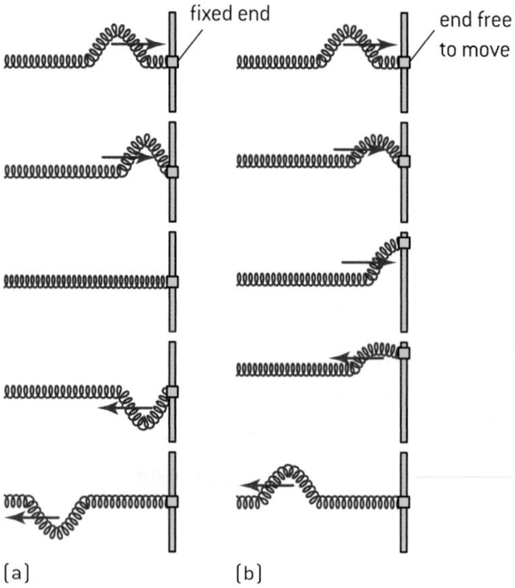

▲ Figure 3 Reflection of wave pulses.

Introduction

When you consider waves you probably immediately think of the ripples on the surface of a lake or the sea. It may be difficult to think of many examples of waves, but this topic is integral to our ability to communicate and, since life on Earth is dependent on the radiation that arrives from the Sun, humankind could not exist without waves.

Waves are of two fundamental types:

- **mechanical waves**, which require a material medium through which to travel
- **electromagnetic waves**, which can travel through a vacuum.

Both types of wave motion can be treated analytically by equations of the same form. Wave motion occurs in several branches of physics and an understanding of the general principles underlying their behaviour is very important. Modelling waves can help us to understand the properties of light, radio, sound ... even aspects of the behaviour of electrons.

Travelling waves

Figure 2 shows a slinky spring being used to demonstrate some of the properties of travelling waves.

With one end of the stretched spring fixed (as in figure 3(a)), moving the other end sharply upwards will send a pulse along the spring. The pulse can be seen to travel along the spring until it reaches the fixed end; then it reflects and returns along the spring. On reflection the pulse changes sides and what was an upward pulse becomes a downwards pulse. The pulse has undergone a phase change of 180° or π radians on reflection. When the end that was fixed is allowed to move there is no phase change on reflection as shown in figure 3(b) and the pulse travels back on the same side that it went out.

> **Note**
>
> You should discuss with fellow students how Newton's second and third laws of motion apply to the motion of the far end of the spring.

Observing the motion of a slinky can give much insight into the movement of travelling waves. To help focus on the motion of the spring, it is useful if one of the slinky coils is painted to make it stand out from the others. The coils are being used to model the particles of the medium through which the wave travels – they could be air or water molecules. If a table tennis ball is placed beside the spring, it shoots off at right angles to the direction of the wave pulse. This shows that this is a **transverse wave** pulse – the direction of the pulse being perpendicular to the direction in which the pulse travels along the slinky. When the slinky is vibrated continuously from side to side a transverse wave is sent along the spring as shown in figure 4 overleaf.

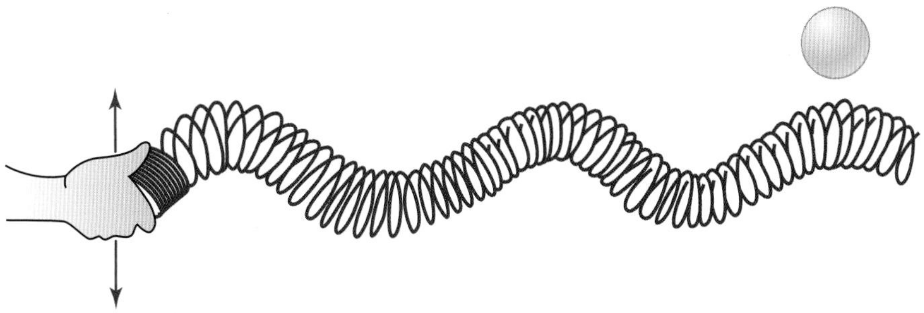

▲ Figure 4 Transverse waves on a slinky spring.

From observing the wave on the spring and water waves on a pond we can draw the following conclusions:

- A wave is initiated by a vibrating object (the source) and it travels away from the object.
- The particles of the medium vibrate about their rest position at the same frequency as the source.
- The wave transfers energy from one place to another.

A slinky can generate a different type of wave called a **longitudinal wave**. In this case the free end of a slinky must be vibrated back and forth, rather than from side to side. As a result of this the coils of the spring will vibrate about their rest position and energy will travel along the spring in a direction parallel to that of the spring's vibration as shown in figure 5.

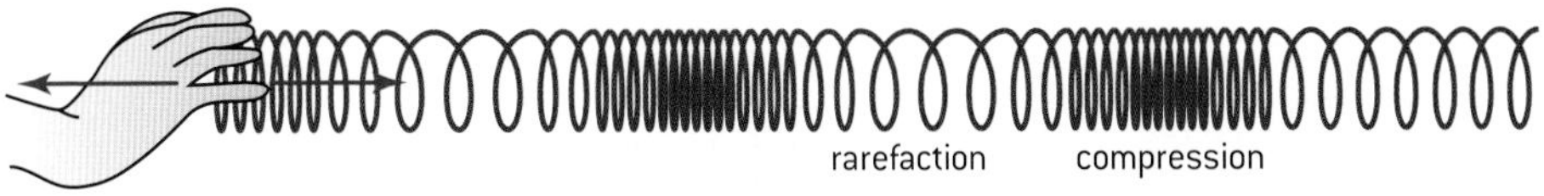

▲ Figure 5 Longitudinal waves on a slinky spring.

Describing waves

When we describe wave properties we need specialist vocabulary to help us:

- **Wavelength** λ is the shortest distance between two points that are in phase on a wave, i.e. two consecutive **crests** or two consecutive **troughs**.
- **Frequency** f is the number of vibrations per second performed by the source of the waves and so is equivalent to the number of crests passing a fixed point per second.
- **Period** T is the time that it takes for one complete wavelength to pass a fixed point or for a particle to undergo one complete oscillation.
- **Amplitude** A is the maximum displacement of a wave from its rest position.

These definitions must be learned, but it is often easier to describe waves graphically. There are two types of graph that are generally used when describing waves: displacement–distance and displacement–time graphs. Such graphs are applicable to every type of wave.

For example *displacement* could represent:

- the displacement of the water surface from its normal flat position for water waves
- the displacement of air molecules from their normal position for sound waves
- the value of the electric field strength vector for electromagnetic waves.

Displacement–distance graphs

This type of graph is sometimes called a wave profile and represents the displacement of many wave particles at a particular instant of time. On figure 6 the two axes are perpendicular but, in reality, this is only the case when we describe transverse waves (when the graph looks like a photograph of the wave taken at a particular instant). For longitudinal waves the actual displacement and distance are parallel. Figure 7 shows how the particles of the medium are displaced from the equilibrium position when the longitudinal wave travels through the medium forming a series of **compressions** (where the particles are more bunched up than normal) and **rarefactions** (where the particles are more spread out than normal). In this case the particles that are displaced to the left of their equilibrium position are given a negative displacement, while those to the right are allocated a positive displacement.

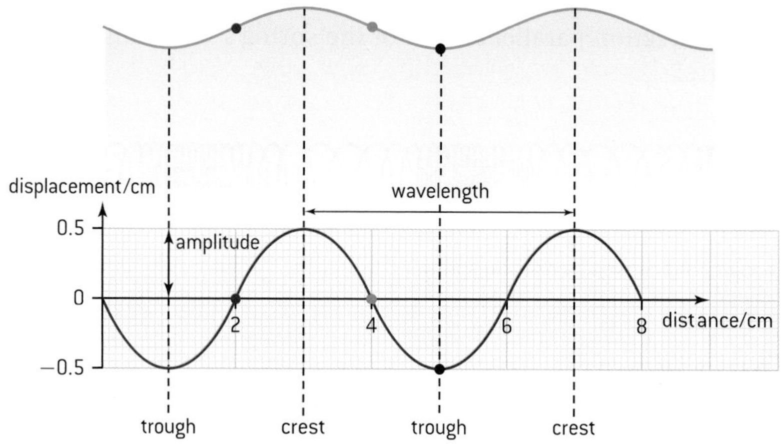

▲ Figure 6 Transverse wave profile.

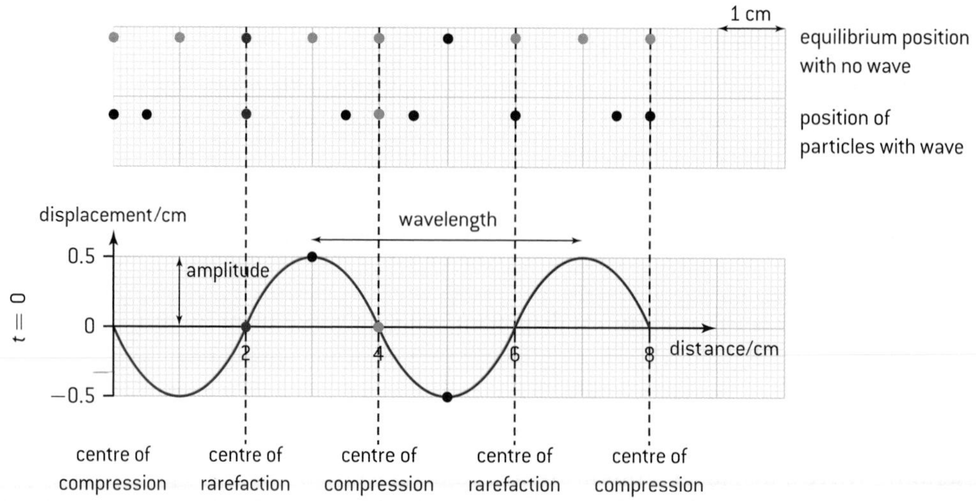

▲ Figure 7 Longitudinal wave profile.

It is very easy to read the amplitude and wavelength directly from a displacement–distance graph. For the longitudinal wave the wavelength is both the crest-to-crest distance and the distance between two consecutive compressions or rarefactions.

Using a sequence of displacement–distance graphs can provide a good understanding of how the position of an individual particle changes with time in both transverse and longitudinal waves (figure 8). On this diagram the wave profile is shown at a quarter period $\left(\frac{T}{4}\right)$ intervals; diagrams like this are very useful in spotting the phase difference between the particles. **P** and **Q** are in anti-phase here (i.e. 180° or π radians out of phase) and **Q** leading **R** by 90° or $\frac{\pi}{2}$ radians. You may wish to consider the phase difference between **P** and **R**.

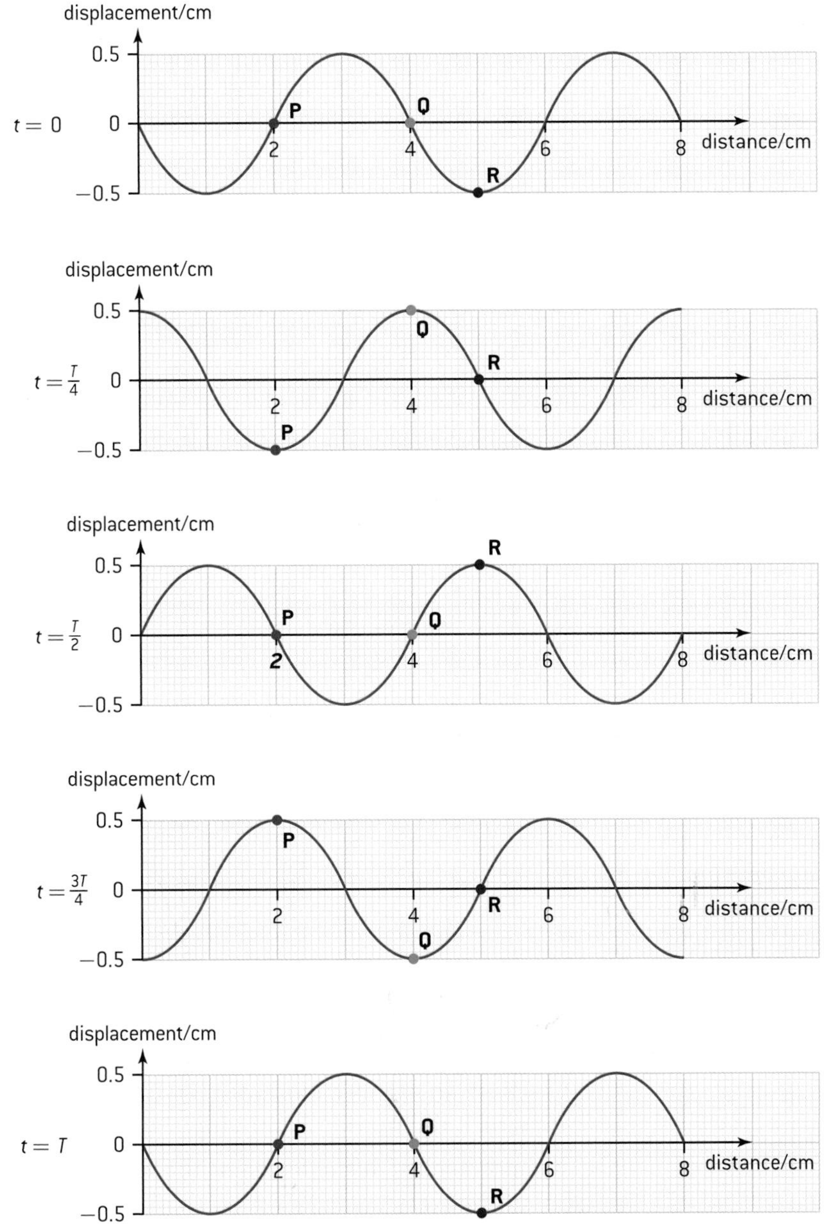

▲ Figure 8 Sequence of displacement-distance graphs at $\frac{T}{4}$ time intervals.

Displacement–time graphs

A displacement–time graph describes the displacement of one particle at a certain position during a continuous range of times. Figure 9 shows the variation with time of the displacement of a single particle. Each particle along the wave will undergo this change (although with phase difference between individual particles).

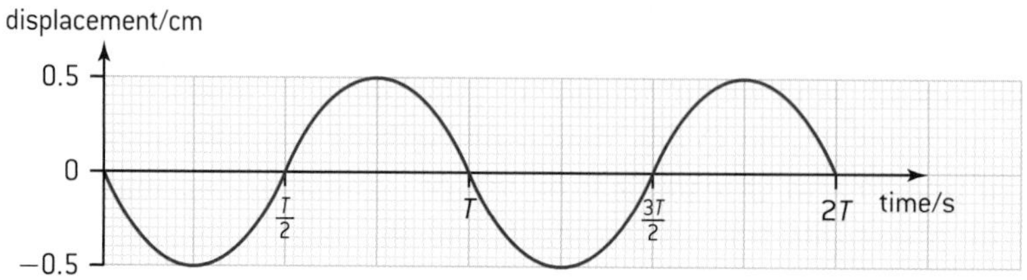

▲ Figure 9 Displacement–time graphs.

> **Note**
>
> It is common to confuse displacement–distance graphs (where crest-to-crest gives the wavelength) with displacement–time graphs (where crest-to-crest gives the period).

This graph makes it is very easy to spot the period and the amplitude of the wave.

The wave equation

When a source of a wave undergoes one complete oscillation the wave it produces moves forward by one wavelength (λ). Since there are f oscillations per second, the wave progresses by $f\lambda$ during this time and, therefore, the velocity (c) of the wave is given by $c = f\lambda$.

With f in hertz and λ in metres, c will have units of hertz metres or (more usually) metres per second. You do need to learn this derivation and it is probably the easiest that you will come across in IB Physics.

Worked example

The diagram below represents the direction of oscillation of a disturbance that gives rise to a wave.

a) Draw two copies of the diagram and add arrows to show the direction of wave energy transfer to illustrate the difference between (i) a transverse wave and (ii) a longitudinal wave.

b) A wave travels along a stretched string. The diagram to the right shows the variation with distance along the string of the displacement of the string at a particular instant in time. A small marker is attached to the string at the point labelled M. The undisturbed position of the string is shown as a dotted line.

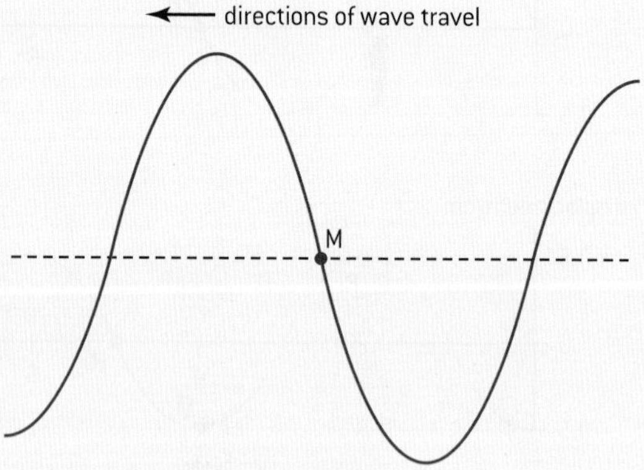

On a copy of the diagram:

(i) Draw an arrow to indicate the direction in which the marker is moving.

(ii) Indicate, with the letter A, the amplitude of the wave.

(iii) Indicate, with the letter λ, the wavelength of the wave.

(iv) Draw the displacement of the string a time $T/4$ later, where T is the period of oscillation of the wave.

(v) Indicate, with the letter **N**, the new position of the marker.

c) The wavelength of the wave is 5.0 cm and its speed is 10 cm s^{-1}.

Determine:

(i) the frequency of the wave

(ii) how far the wave has moved in a quarter of a period.

Solution

a) (i) The energy in a transverse wave travels in a direction perpendicular to the direction of vibration of the medium:

(ii) The energy in a longitudinal wave travels in a direction parallel to the direction of vibration of the medium:

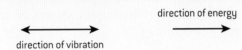

b)

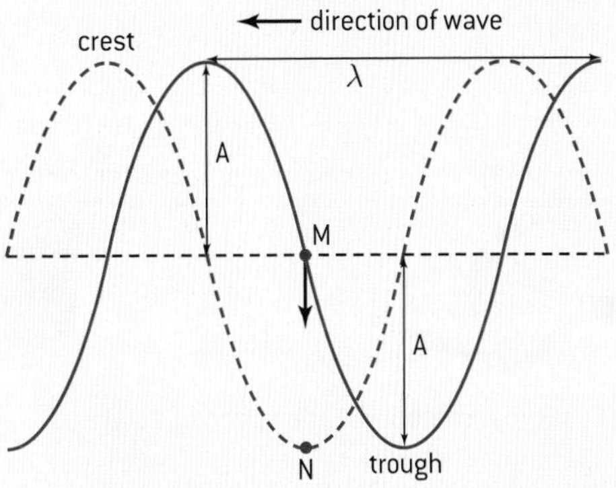

(i) With the wave travelling to the left, the trough shown will move to the left and that must mean that M moves downwards.

(ii) The amplitude (A) is the height of a crest or depth of a trough.

(iii) The wavelength (λ) is the distance equivalent to crest to next crest or trough to next trough.

(iv) In quarter of a period the wave will have moved quarter of a wavelength to the left (broken line curve).

(v) After quarter of a period the marker is now at the position of the trough (N).

c) (i) $f = \frac{c}{\lambda} = \frac{10}{5} = 2.0$ Hz (since both wavelength and speed are in cm, there is no need to convert the units)

(ii) $T = \frac{1}{f} = 0.5$ s

Because the wave moves at a constant speed $s = ct = c\frac{T}{4} = 10 \times 0.125 = 1.25$ cm

Investigate!

Measuring the speed of sound

Here are two of the many ways of measuring the speed of sound in free air (i.e. not trapped in a tube).

Method one – using a fast timer

This is a very simple method of measuring the speed of sound:

- Two microphones are connected to a fast timer (one which can measure the nearest millisecond or even microsecond).
- The first microphone triggers the timer to start.
- The second microphone triggers the time to stop.
- When the hammer is made to strike the plate the sound wave travels to the two microphones triggering the nearer microphone first and the further microphone second.
- By separating the microphones by 1 m, the time delay is around 3.2 ms.

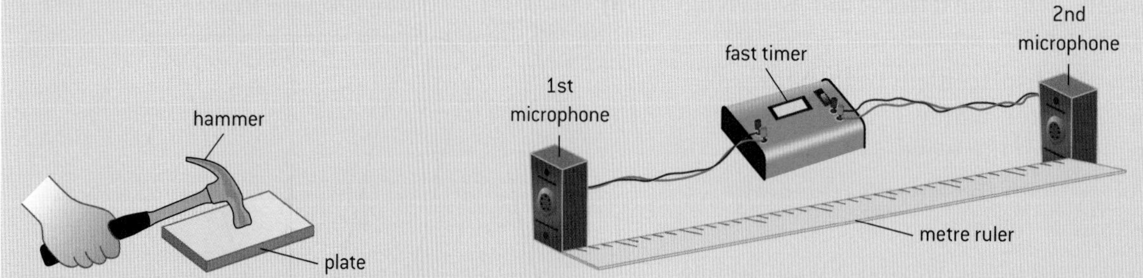

▲ Figure 10 Fast timer method for the speed of sound.

- This gives a value for the speed of sound to be $c = \frac{s}{t} = \frac{1.0}{3.2 \times 10^{-3}} \approx 310 \text{ m s}^{-1}$
- This should be repeated a few times and an average value obtained.
- You might think about how you could develop this experiment to measure the speed of sound in different media. Sound travels faster in solids and liquids than in gases – can you think why this should be the case? If the temperature in your country varies significantly over the year you might try to perform the experiment in a hot or cold corridor and compare your results.

Method two – using a double beam oscilloscope

- Connect two microphones to the inputs of a double-beam oscilloscope (figure 11).
- Connect a signal generator to a loudspeaker and set the frequency to between 500 Hz and 2.0 kHz.
- One of the microphones needs to be close to the loudspeaker, with the second a metre or so further away.
- Compare the two traces as you move the second microphone back and forth in line with the first microphone and the speaker.
- Use a ruler to measure the distance that you need to move the second microphone for the traces to change from being in phase to changing to antiphase and then back in phase again.
- The distance moved between the microphones being in consecutive phases will be the wavelength of the wave.
- The speed is then found by multiplying the wavelength by the frequency shown on the signal generator.

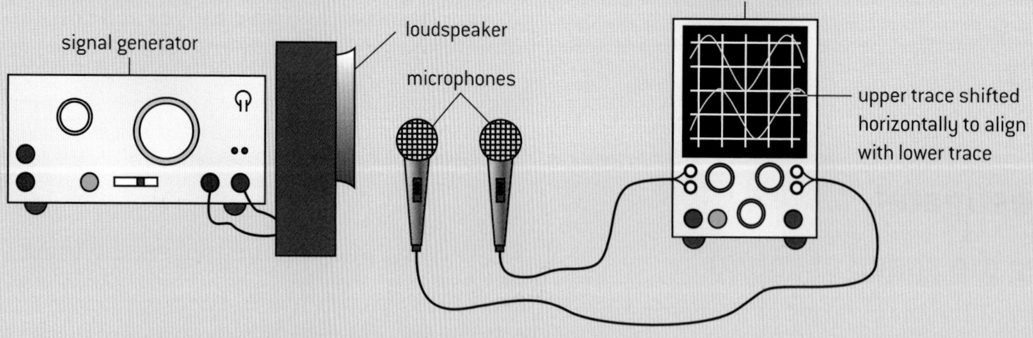

▲ Figure 11 Double beam oscilloscope method for the speed of sound.

Nature of Science

Analogies can slow down progress

You may have seen a demonstration, called the "bell jar" experiment, to show that sound needs a material medium through which to travel.

An electric bell is suspended from the mouth of a sealed bell jar and set ringing (figure 12). A vacuum pump is used to evacuate the jar. When there is no longer any air present in the jar, no sound can be heard. This experiment was first performed by Robert Boyle in 1660 although, long before this time, the Ancient Greeks understood that sound needed something to travel through. The first measurements of the speed of sound were made over four hundred years ago. These experiments were based on measuring the time delay either between a sound being produced and its echo reflecting from a distant surface or between seeing the flash of a cannon when fired and hearing the bang. From the middle to the end of the seventeenth century physicists such as Robert Hooke and Christiaan Huygens proposed a wave theory of light. In line with the model for sound, Huygens suggested that light was carried by a medium called the luminiferous ether. It was not until 1887, when Michelson and Morley devised an experiment in an attempt to detect the ether wind, that people began to realize that electromagnetic waves were very different from sound waves.

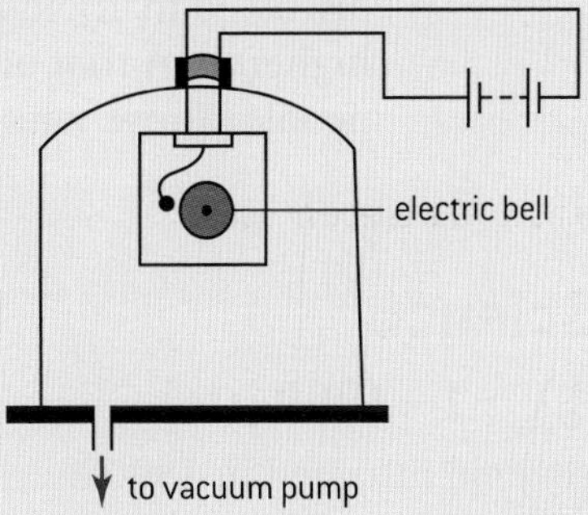

▲ Figure 12 The bell jar experiment.

Electromagnetic waves

You may have seen the experiment shown in figure 13 in which a beam of *white* light is dispersed into the colours of the visible spectrum by passing it through a prism. The fact that what appears to be a single "colour" actually consists of multiple colours triggers the question "what happens beyond the red and blue ends of the spectrum?" The two colours represent the limit of vision of the human eye but not the limit of detection of electromagnetic radiation by the human body. Holding the back of your hand towards the Sun allows you to feel the warmth of the Sun's infra-red radiation. In the longer term, your hand will be subjected to sunburn and even skin cancer caused by the higher energy ultraviolet radiation. Figure 14 shows the full electromagnetic spectrum with the *atmospheric windows*; these are the ranges of electromagnetic waves that can pass through the layers of the atmosphere.

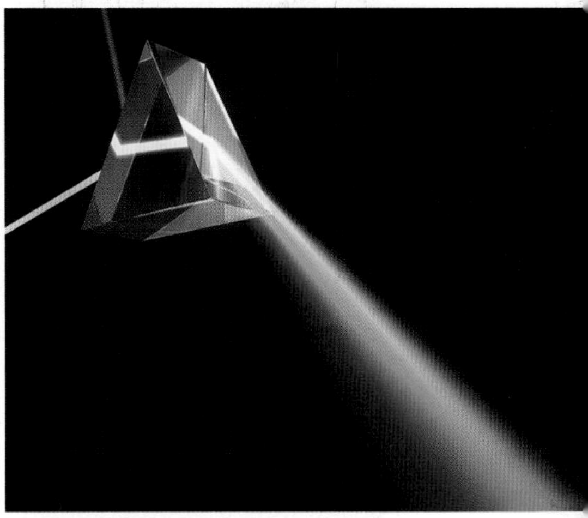

▲ Figure 13 Dispersion of white light using a glass prism.

All electromagnetic waves are transverse, carry energy, and exhibit the full range of wave properties. They travel at 300 million metres per second (3.00×10^8 m s^{-1}) in a vacuum. All electromagnetic waves (except gamma rays) are produced when electrons undergo an energy change, even though the mechanisms might differ. For example, radio waves are emitted when electrons are accelerated in an aerial or antenna. Gamma rays are different, they are emitted by a nucleus or by means of other particle decays or annihilation events. All electromagnetic waves consist of a time-varying electric field with an associated time-varying magnetic field. As the human eye is sensitive to the electric component, the amplitude of an electromagnetic wave is usually taken as the wave's maximum electric field strength. By graphing the electric field strength on the *y*-axes, we can use displacement–distance and displacement–time graphs to represent electromagnetic waves. In the following discussion of the different areas of the electromagnetic spectrum the range of the wavelength is given but these values are not hard and fast. There are overlaps of wavelength when radiation of the same wavelength is emitted by different

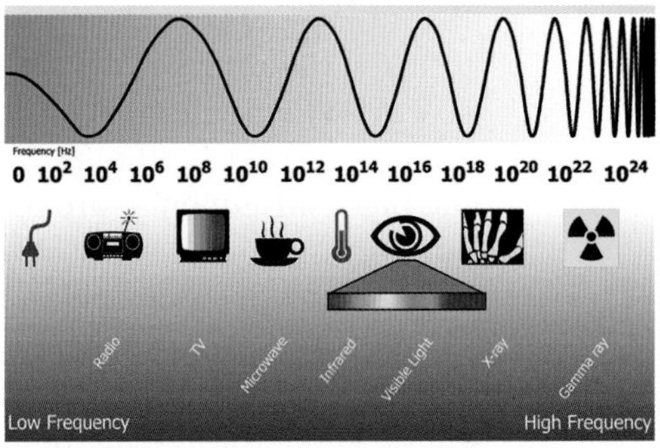

▲ Figure 14 The electromagnetic spectrum.

mechanisms (most notably X-rays and gamma rays). In addition to comparing the wavelengths of the radiations it is sometimes more appropriate to compare their frequencies (for example, radio waves) or the energy of the wave photon (for example, with X-rays and gamma rays). The frequency range is easy to calculate from wavelengths using the equation $c = f\lambda$ with $c = 3.00 \times 10^8$ m s^{-1}. Photon energies will be discussed in Topic 7. Those electromagnetic waves with frequencies higher than that of visible light ionize atoms – and are thus harmful to people. Those with lower frequencies are generally believed to be safe.

Nature of Science

Infra-red radiation ($\lambda_{ir} \sim 1 - 1000$ μm)

William Herschel discovered infra-red in 1800 by placing a thermometer just beyond the red end of the spectrum formed by a prism. Today we might do the same experiment but using an infra-red detector as a modern thermometer. Objects that are hot but not glowing, i.e. below 500 °C, emit infra-red only. At this temperature objects become red-hot and emit red light in addition to infra-red. At around 1000 °C objects become white hot and emit the full visible spectrum colours. Remote controllers for multimedia devices utilize infra-red as do thermal imagers used for night vision. Infra-red astronomy is used to "see" through dense regions of gas and dust in space with less scattering and absorption than is exhibited by visible light.

Ultraviolet radiation ($\lambda_{uv} \sim 100 - 400$ nm)

In 1801, Johann Ritter detected ultraviolet by positioning a photographic plate beyond the violet part of the spectrum formed by a prism. It can also be detected using fluorescent paints and inks – these absorb the ultraviolet (and shorter wavelengths) and re-emit the radiation as visible light. Absorption of ultraviolet produces important vitamins in the skin but an overdose can be harmful, especially to the eyes. The Sun emits the full range of ultraviolet: UV-A, UV-B, and UV-C. These classifications are made in terms of the range of the wavelengths emitted (UV-A ~315–400 nm, UV-B ~280–315 nm, and UV-C ~ 100–280 nm). UV-C rays, having the highest frequency, are the most harmful but, fortunately, they are almost completely absorbed by the atmosphere. UV-B rays cause sunburn and increase the risk of DNA and other cellular damage in living organisms; luckily only about 5% of this radiation passes through the ionosphere (this consists of layers of electrically charged gases between 80 and 400 km above the Earth). Fluorescent tubes, used for lighting, contain mercury vapour and their inner surfaces are coated with powders. The mercury vapour and powders fluoresce when radiated with ultraviolet light. With the atmosphere absorbing much of the ultraviolet spectrum, satellites must be positioned above the atmosphere in order to utilize ultraviolet astronomy, which is very useful in observing the structure and evolution of galaxies.

Radio waves ($\lambda_{radio} \sim 1$ mm $- 100$ km)

James Clerk Maxwell predicted the existence of radio waves in 1864. Between 1885 and 1889 Heinrich Hertz produced electromagnetic waves in the laboratory, confirming that light waves are electromagnetic radiation obeying Maxwell's equations. Radio waves are used to transmit radio and television signals. VHF (or FM) radio waves have shorter wavelengths than those of AM. VHF and television signals have wavelengths of a few metres and travel in straight lines (or **rays**) from the transmitter to the receiver. Long-wave radio relies on reflections from the ionosphere and diffracts around obstacles on the Earth's surface. Satellite communication requires signals with wavelengths less than 10 m in order to penetrate the ionosphere. Radio telescopes are used by astronomers to observe the composition, structure and motion of astronomic bodies. Such telescopes are physically large, for example the Very Large Array (VLA) radio telescope in New Mexico which consists of 27 antennas arranged in a "Y" pattern up to 36 km across.

Microwaves ($\lambda_{micro} \sim 1$ mm $- 30$ cm)

Microwaves are short wavelength radio waves that have been used so extensively with radar, microwave cooking, global navigation satellite systems and astronomy that they deserve their own category. In 1940, Sir John Randall and Dr H A Boot invented the magnetron which produced microwaves that could be used in **radar** (an acronym for *radio detection and ranging*) to locate aircraft on bombing missions. The microwave oven is now a common kitchen appliance; the waves are tuned to frequencies that can be absorbed by the water and fat molecules in food, causing these molecules to vibrate. This increases the internal energy of the food – the container holding the food absorbs an insignificant amount of energy and stays much cooler. Longer wavelength microwaves pass through the Earth's atmosphere more effectively than those of shorter wavelength. The **Cosmic Background Radiation** is the elemental radiation field that fills the universe, having been created in the form of gamma rays at the time of the Big Bang. With the universe now cooled to a temperature of 2.73 K the peak wavelength is approximately 1.1 mm (in the microwave region of the spectrum).

X-rays ($\lambda_X \sim 30$ pm – 3 nm)

X-rays were first produced and detected in 1895 by the German physicist Wilhelm Conrad Roentgen. An X-ray tube works by firing a beam of electrons at a metal target. If the electrons have sufficient energy, X-rays will emitted by the target. X-rays are well known for obtaining images of broken bones. They are also used in hospitals to destroy cancer cells. Since they can also damage healthy cells, using lead shielding in an X-ray tube is imperative. Less energetic and less invasive X-rays have longer wavelengths and penetrate flesh but not bone: such X-rays are used in dental surgery. In industry they are used to examine welded metal joints and castings for faults. X-rays are emitted by astronomical objects having temperatures of millions of kelvin, including pulsars, galactic supernovae remnants, and the accretion disk of black holes. The measurement of X-rays can provide information about the composition, temperature, and density of distant galaxies.

Gamma rays ($\lambda_\gamma < 1$ pm)

Gamma rays were discovered by the French scientist Paul Villard in 1900. Gamma rays have the shortest wavelength and the highest frequency of all electromagnetic radiation. They are generated, amongst other mechanisms, by naturally occurring radioactive nuclei in nuclear explosions and by neutron stars and pulsars. Only extra-terrestrial gamma rays of the very highest energies can reach the surface of the Earth – the rest being absorbed by ozone in the Earth's upper atmosphere.

Nature of Science

Night vision

Humans eyes are sensitive to the electric component of a portion of the electromagnetic spectrum known as the visible spectrum; this is what we call *sight*. Some animals, in particular insects and birds, are able to see using the ultraviolet part of spectrum. However, ultraviolet consists of relatively short wavelength radiation that can damage animal tissue, yet these animals appear to be immune to these dangers. It has been speculated whether any animal eye might be adapted to be able to use the infra-red part of the electromagnetic spectrum. As these long wavelengths have low energy it might be far-fetched to believe that they can be detected visually. There are animals that have evolved ways of sensing infra-red that are similar to the processes occurring in the eye. For example, the brains of some snakes are able to interpret the infra-red radiation in a way that can be combined with other sensory information to enable them to have a better understanding of surrounding danger or food sources.

4.3 Wave characteristics

Understanding

- Wavefronts and rays
- Amplitude and intensity
- Superposition
- Polarization

Applications and skills

- Sketching and interpreting diagrams involving wavefronts and rays
- Solving problems involving amplitude and intensity
- Sketching and interpreting the superposition of pulses and waves
- Describing methods of polarization
- Sketching and interpreting diagrams illustrating polarized reflected and transmitted beams
- Solving problems involving Malus's law

Equations

- Relationship between intensity and amplitude: $I \propto A^2$
- Malus's law: $I = I_0 \cos^2 \theta$

Nature of science

Imagination and physics

"Imagination ... is more important than knowledge. Knowledge is limited. Imagination encircles the world[1]."

Einstein was famous for his *gedanken* "thought" experiments and one of the qualities that makes a great physicist is surely a hunger to ask the question "what if ... ?". Mathematics is a crucial tool for the physicist and it is central to what a physicist does, to be able to quantify an argument. However, imagination can also play a major role in interpreting the results. Without imagination it is hard to believe that we would have Huygens' principle, Newton's law of gravitation, or Einstein's theory of special relativity. To visualize the abstract and apply theory to a practical situation is a flair that is fundamental to being an extraordinary physicist.

[1]Viereck, George Sylvester (October 26, 1929). 'What life means to Einstein: an interview'. *The Saturday Evening Post*.

Introduction

Wavefronts and rays are visualizations that help our understanding of how waves behave in different circumstances. By drawing ray or wave diagrams using simple rules we can predict how waves will behave when they encounter obstacles or a different material medium.

Definitions of these two quantities are:

- A **wavefront** is a surface that travels with a wave and is perpendicular to the direction in which the wave travels – the ray.
- A **ray** is a line showing the direction in which a wave transfers energy and is, of course, perpendicular to a wavefront.

The distance between two consecutive wavefronts is one wavelength (λ).

The motion of wavefronts

One of the simplest ways to demonstrate the motion of wavefronts is to use a ripple tank such as that shown in figure 1. This is simply a glass-bottomed tank that contains water, illuminated from above. Any waves on the water focus the light onto a screen often placed below the tank. The bright patches result from the crests focusing the light and the dark patches from the troughs defocusing the light. Using a vibrating dipper, plane or circular waves can be produced allowing us to see what happens to wavefronts in different situations.

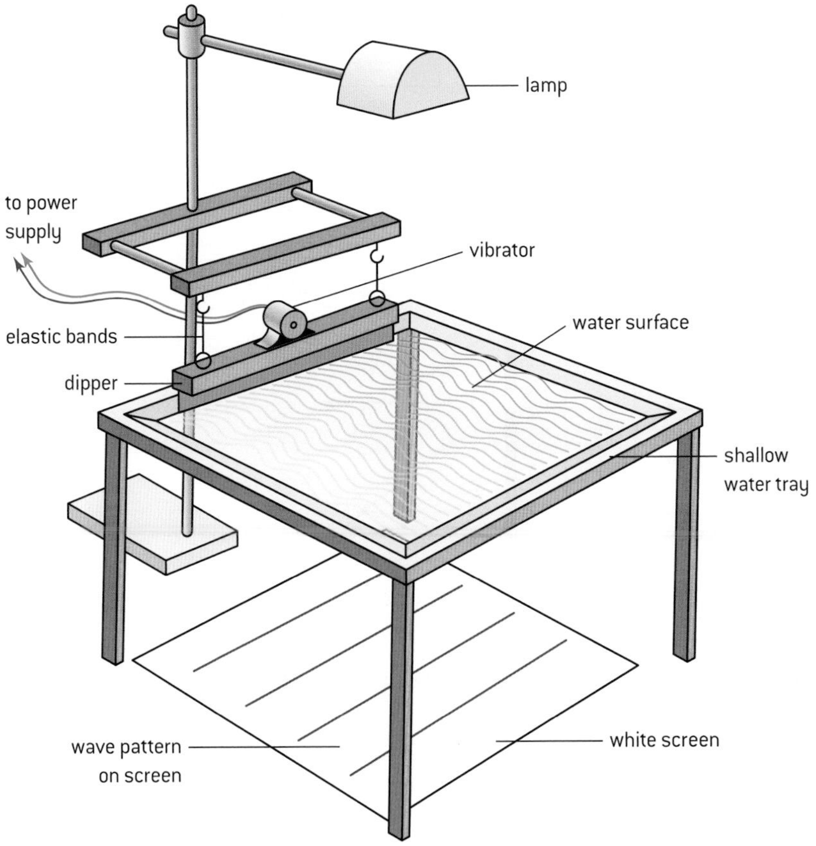

▲ Figure 1 A ripple tank.

The images shown in figure 2 show the effect on the wavefronts as they meet (a) a plane barrier, (b) a shallower region over a prism-shaped glass and (c) a single narrow slit.

Using wavefront or ray diagrams we can illustrate how waves behave when they are reflected, refracted, and diffracted. We will refer to these diagrams in the sections of this topic following on from this.

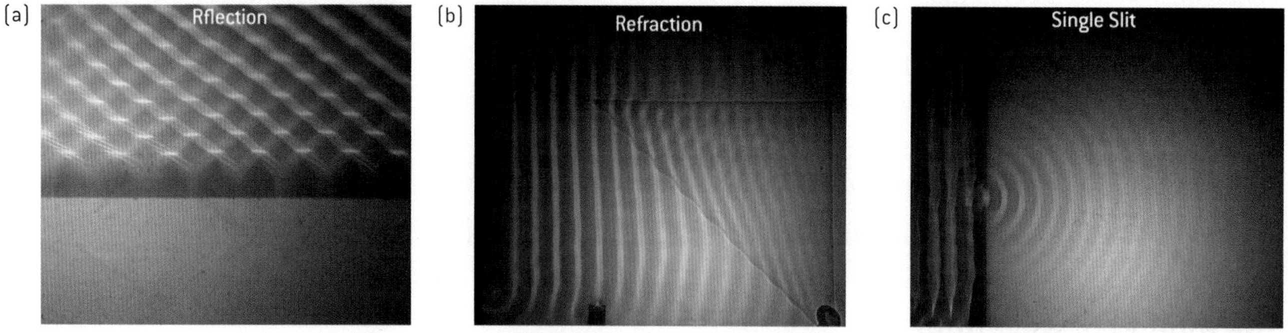

▲ Figure 2 Reflection, refraction, and diffraction of waves in a ripple tank.

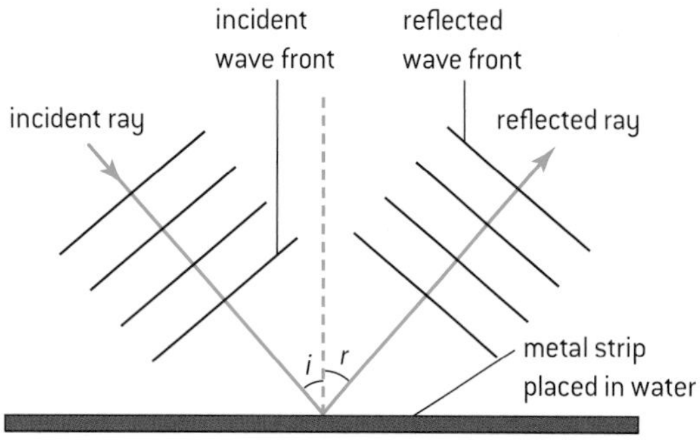

▲ Figure 3a Wavefront and ray diagrams for reflection of waves.

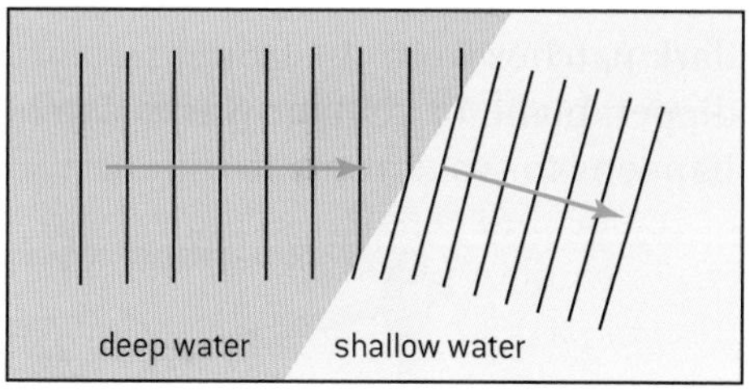

▲ Figure 3b Wavefront and ray diagrams for refraction of waves.

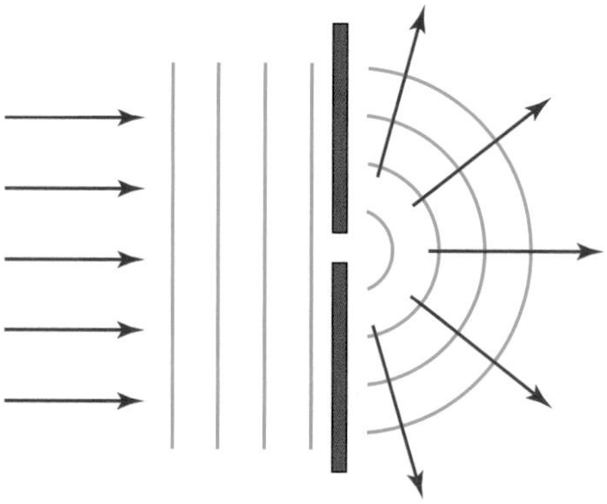

▲ Figure 3c Wavefront and ray diagrams for diffraction of waves by a single slit.

In figure 3a we can see that there is no change of wavelength and that the angle of incidence (i) is equal to the angle of reflection (r).

In figure 3b we see the wave slowing down and bending as it enters the denser medium. The wavelength of the wave in the denser medium is shorter than in the less dense medium – but the frequency remains unchanged (although it is not possible to tell this from the the pattern shown in the ripple tank).

Figure 3c shows diffraction where the wave spreads out on passing through the slit but there is no change in the wavelength.

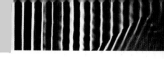

Nature of Science

Huygens' principle

One of the ways to predict what will happen to wavefronts under different circumstances is to use Huygens' principle. This was suggested in 1678 by the Dutch physicist Christiaan Huygens: the wavefront of a travelling wave at a particular instant consists of the tangent to circular wavelets given out by each point on the previous wavefront as shown in figure 4. In this way the wavefront travels forward with a velocity c.

Huygens was able to derive the laws of reflection and refraction from his principle. What the principle does not explain is why an expanding circular (but really spherical) wave continues to expand outwards from its source rather than travel back and focus on the source. The French physicist Augustin Fresnel (1788–1827) adapted Huygens' principle to explain diffraction by proposing the principle of superposition, discussed in Sub-topic 4.4. The Huygens–Fresnel principle (as the overall principle should more accurately be called) is useful in explaining many wave phenomena.

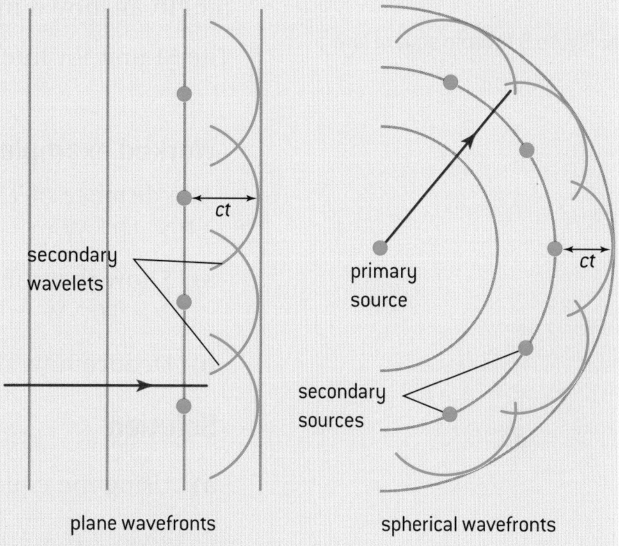

▲ Figure 4 Huygens' principle.

The intensity of waves

The loudness of a sound wave or the brightness of a light depends on the amount of energy that is received by an observer. For example, when a guitar string is plucked more forcefully the string does more work on the air and so there will be more energy in the sound wave. In a similar way, to make a filament lamp glow more brightly requires more electrical energy. The energy E is found to be proportional to the square of the amplitude A:

$$E \propto A^2$$

So doubling the amplitude increases the energy by a factor of four; tripling the amplitude increases the energy by a factor of nine, etc.

Loudness is the observer's perception of the intensity of a sound and brightness that of light; loudness and brightness are each affected by frequency.

If we picture waves being emitted by a point source, S, they will spread out in all directions. This will mean that the total energy emitted will be spread increasingly thinly the further we go from the source. Figure 5 shows how the energy spreads out over the surface area of a sphere.

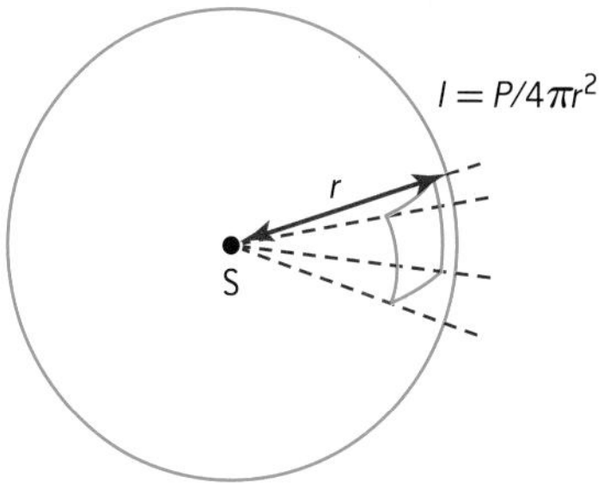

▲ Figure 5 Energy spreading out from a point source.

In order to make intensity comparisons more straightforward it is usual to use the idea of the energy transferred per second – this is the *power* (P) of the source. This means that the intensity (I) at a distance (r)

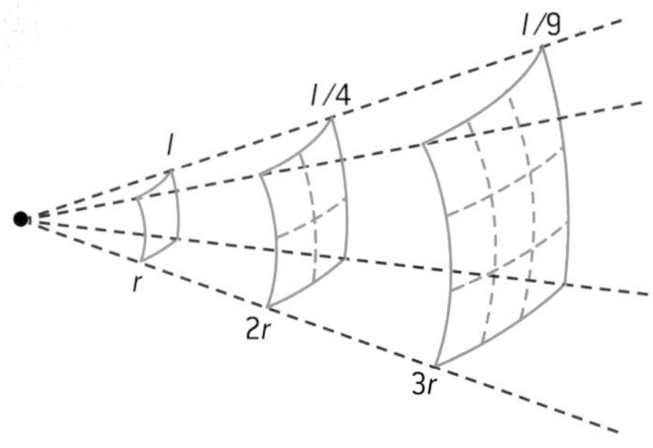

▲ Figure 6 Inverse square law.

from a point source is given by the power divided by the surface area of the sphere at that radius:

$$I = \frac{P}{4\pi r^2}$$

This equation shows that intensity has an **inverse-square relationship** with distance from the point source; this means that, as the distance doubles, the intensity falls to a quarter of the previous value and when the distance the is tripled, the intensity falls to one nineth as shown in figure 6.

The SI unit for intensity is W m^{-2}.

Worked example

At a distance of 15 m from the source, the intensity of a loud sound is 2.0×10^{-4} W m^{-2}.

a) Show that the intensity at 120 m from the source is approximately 3×10^{-6} W m^{-2}.

b) Deduce how the amplitude of the wave changes.

Solution

a) Using the equation $I = \frac{P}{4\pi r^2}$, $\frac{P}{4\pi}$ remains constant so $I_1 r_1^2 = I_2 r_2^2$ this gives $2.0 \times 10^{-4} \times 15^2 = I_2 \times 120^2$

$I_2 = 2.0 \times 10^{-4} \times \frac{15^2}{120^2} = 3.1 \times 10^{-6}$ W $m^{-2} \approx$ 3×10^{-6} W m^{-2}

Note

In "show that" questions there is an expectation that you will give a detailed answer, showing all your working and that you will give a final answer to more significant figures than the data in the question – actually this is good practice for any answer!

b) With the intensity changing there must be a change of amplitude. The intensity is proportional to the square of the amplitude so:

$$\frac{I_2}{I_2} = \frac{A_1^2}{A_2^2} \text{ or } \frac{A_1}{A_2} = \sqrt{\frac{I_2}{I_2}} = \sqrt{\frac{2.0 \times 10^{-4}}{3.1 \times 10^{-6}}} = 8.0$$

Thus the amplitude at 15 m from the source is 8.0 times that at 120 m from the source. Another way of looking at this is to say that A is proportional to $\frac{1}{r}$.

Note

Because the previous part was a "show that" question you had all the data needed to answer this question. In questions where you need to calculate data you will never be penalized for using incorrect data that you have calculated previously. In a question that has several parts you might fail to gain a sensible answer to one of the parts. When a subsequent part requires the use of your answer as data – don't give up, invent a sensible value (and say that is what you are doing). You should then gain any marks available for the correct method.

The principle of superposition

Unlike when solid objects collide, when two or more waves meet the total displacement is the vector sum of their individual displacements. Having interacted, the waves continue on their way as if they had never met at all. This principle is used to explain interference and standing waves in Sub-topics 4.4 and 4.5 respectively. For a consistent pattern, waves need to be of the same type and have the same frequency and speed; the best patterns are achieved when the waves have the same or very similar amplitudes. Figure 7 shows two pulses approaching and passing through each other. When they meet, the resultant amplitude is the algebraic sum of the two amplitudes of the individual pulses.

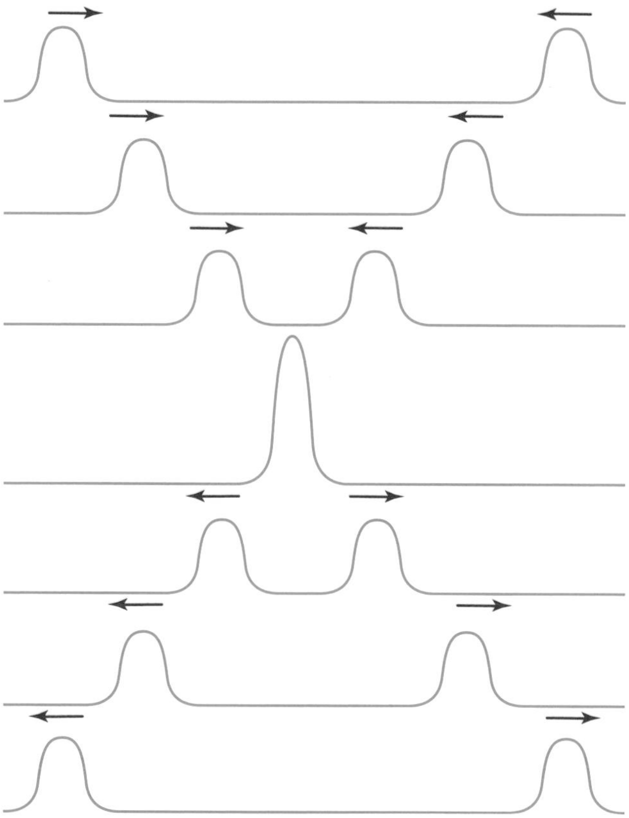

▲ Figure 7 Superposition of pulses.

This principle applies equally well to complete waves as to pulses – this is shown in figure 8. You should convince yourself that the green wave is the vector sum of the red and blue waves at every instant. It is equally valid to use the principle of superposition with displacement–distance graphs.

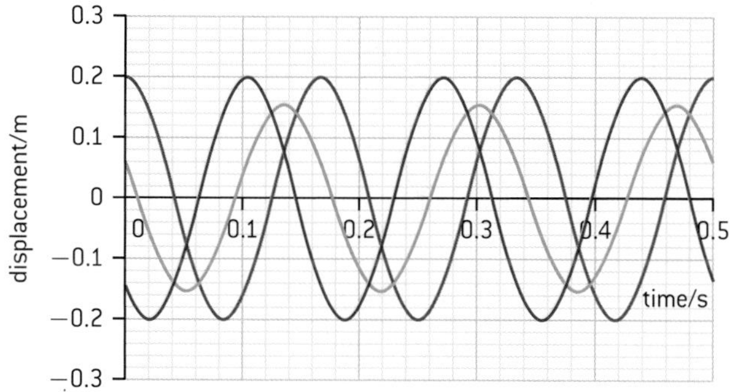

▲ Figure 8 Displacement–time graph showing the superposition of waves.

Worked examples

1 Two identical triangular pulses of amplitude A travel towards each other along a rubber cord. At the instant shown on the diagram below, point M is midway between the two pulses.

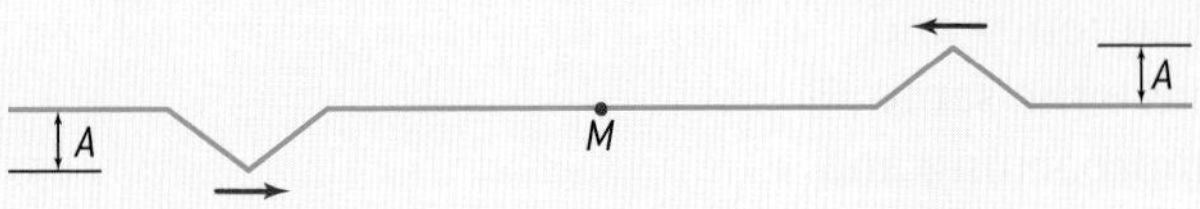

What is the amplitude of the disturbance in the string as the pulses move through M?

Solution

The pulses are symmetrical, so when they meet they completely cancel out giving zero amplitude at M.

2 **a)** For a travelling wave, distinguish between a *ray* and a *wavefront*.

The diagram below shows three wavefronts incident on a boundary between medium I and medium R. Wavefront CD is shown crossing the boundary. Wavefront EF is incomplete.

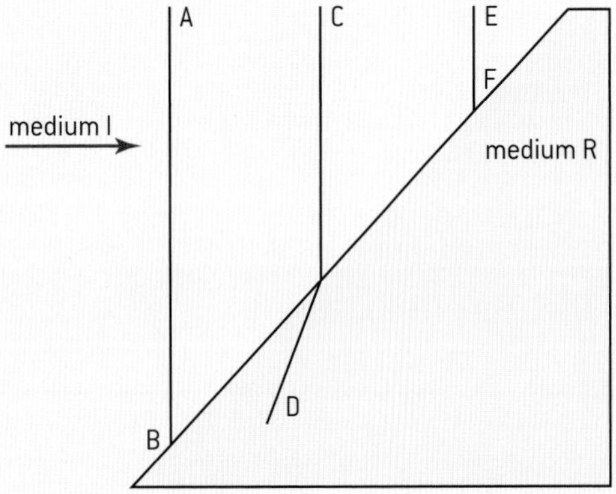

b) (i) On a copy of the diagram above, draw a line to complete the wavefront EF.

(ii) Explain in which medium, I or R, the wave has the higher speed.

(iii) By taking appropriate measurements from the diagram, determine the ratio of the speeds of the wave travelling from medium I to medium R.

Solution

a) A ray is a line that shows the direction of propagation of a wave. Wavefronts are lines connecting points on the wave that are in phase, such as a crest or a trough. The distance between wavefronts is one wavelength and wavefronts are always perpendicular to rays.

b) (i)

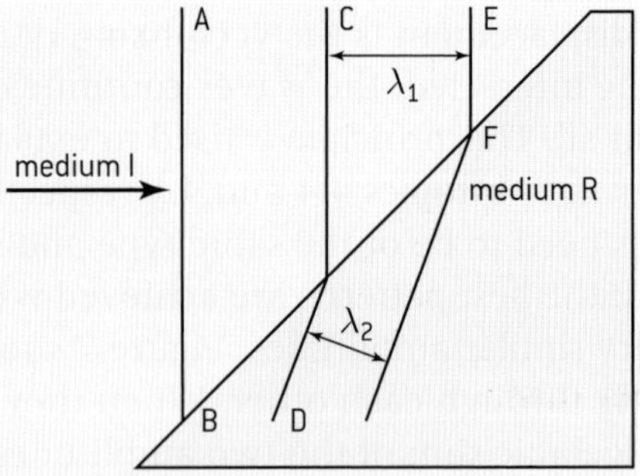

On entering the new medium the waves refract by the same amount to give parallel wavefronts.

(ii) As the wavefronts are closer in the medium R the waves are travelling more slowly.

(iii) The frequency of a wave does not change when the wave moves from one medium to another. As $c = f\lambda$ then $\frac{c}{\lambda}$ = constant and so:

$$\frac{c_1}{\lambda_1} = \frac{c_2}{\lambda_2} \therefore \frac{c_1}{c_2} = \frac{\lambda_1}{\lambda_2}$$

3 The graphs below show the variation with time of the individual displacements of two waves as they pass through the same point.

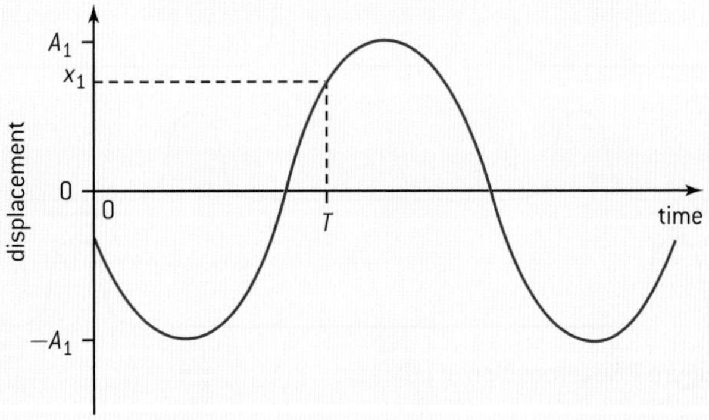

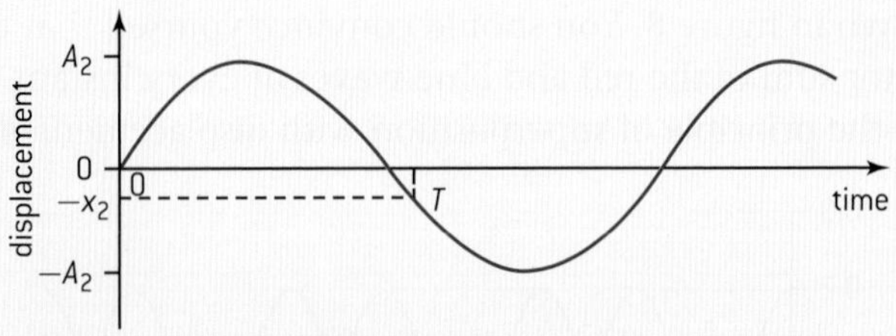

What is the total displacement of the resultant wave at the point at time T?

Solution

The total displacement will be the vector sum of the individual displacements. As they are in opposite directions the vector sum will be their difference, i.e. $x_1 - x_2$

Polarization

Although transverse and longitudinal waves have common properties – they reflect, refract, diffract and superpose – the difference between them can be seen by the property of polarization. Polarization of a transverse wave restricts the direction of oscillation to a plane perpendicular to the direction of propagation. Longitudinal waves, such as sound waves, do not exhibit polarization because, for these waves, the direction of oscillation is parallel to the direction of propagation. Figure 9 shows a demonstration of the polarization of a transverse wave on a rubber tube.

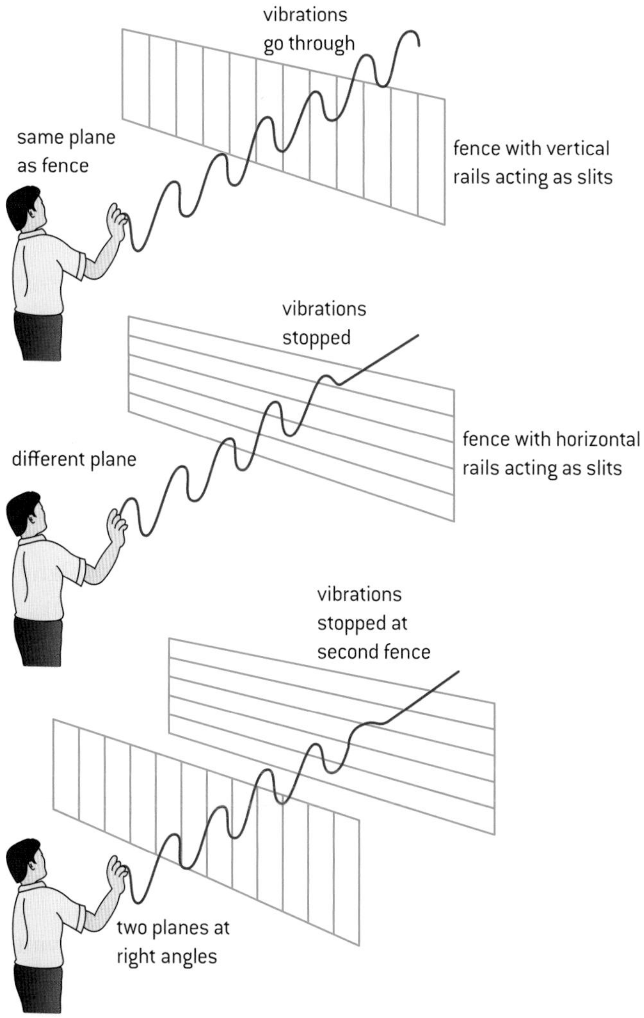

▲ Figure 9 Polarization demonstration.

Most naturally occurring electromagnetic waves are completely unpolarized; this means the electric field vector (and therefore the magnetic field vector perpendicular to it) vibrate in random directions but in a plane always at right angles to the direction of propagation of the wave. When the direction of vibration stays constant over time, the wave is said to be **plane polarized** in the direction of vibration – this is the case with many radio waves which are polarized as a result of the orientation of the transmitting aerial (antenna). **Partial polarization** is when there is some restriction to direction of vibration but not 100%. There is a further type of polarization when the direction of vibration rotates at the same

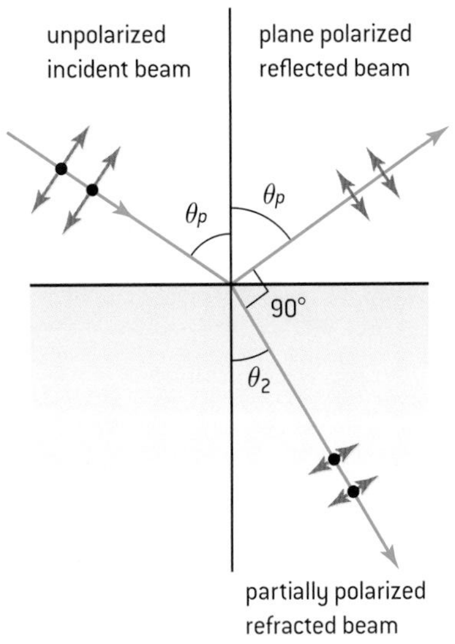

▲ Figure 11 Polarization of reflected light.

frequency as the wave – this called circular or elliptical polarization and is caused when a wave is in a strong magnetic field – but you will not be examined on this. Figure 10 shows how we represent polarized and unpolarized light diagramatically; the double-headed arrow represents a polarized wave, showing the plane of polarization of the wave. The crossed arrows show that the vibration has an electric field vector in all planes and these are resolved into the two perpendicular planes shown (you may see this marked as four double-headed arrows in some texts). These diagrams can become a little confusing when rays are added to show the direction in which the waves are travelling. Figure 11 shows some examples of this.

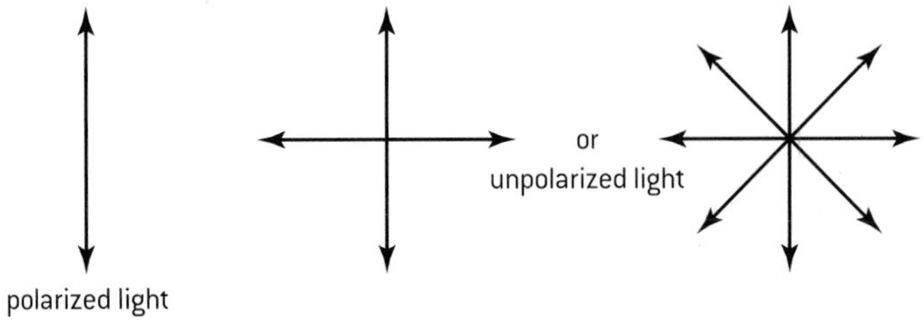

▲ Figure 10 Representing polarized and unpolarized waves.

Polarization of light

In 1809 the French experimenter Étienne-Louis Malus showed that when unpolarized light reflected off a glass plate it could be polarized depending upon the angle of incidence – the plane of polarization being that of the flat surface reflecting the light. In 1812 the Scottish physicist Sir David Brewster showed that when unpolarized light incident on the surface of an optically denser material (such as glass), at an angle called the polarizing angle, the reflected ray would be completely plane polarized. At this angle the reflected ray and refracted ray are at right angles as shown in figure 11.

Today the most common method of producing polarized light is to use a polarizing filter (usually called **Polaroid**). These filters are made from chains of microcrystals of iodoquinine sulfate embedded in a transparent cellulose nitrate film. The crystals are aligned during manufacture and electric field vibration components, parallel to the direction of alignment, become absorbed. The electric field vector causes the electrons in the crystal chains to oscillate and thus removes energy from the wave. The direction perpendicular to the chains allows the electric field to pass through. The reason for this is that the limited width of the molecules restricts the motion of the electrons, meaning that they cannot absorb the wave energy. When a pair of Polaroids are oriented to be at 90° to each other, or "crossed", no light is able to pass through. The first Polaroid restricts the electric field to the direction perpendicular to the crystal chains; the second Polaroid has its crystals aligned in this direction and so absorbs the remaining energy. The first of the two Polaroids is called the **polarizer** and the second is called the **analyser**.

Malus's law

When totally plane-polarized light (from a polarizer) is incident on an analyser, the intensity I of the light transmitted by the analyser is directly proportional to the square of the cosine of angle between the transmission axes of the analyser and the polarizer.

Figure 12 shows polarized light with the electric field vector of amplitude E_0 incident on an analyser. The axis of transmission of the analyser makes an angle θ with the incident light. The electric field vector E_0 can be resolved into two perpendicular components $E_0 \cos\theta$ and $E_0 \sin\theta$. The analyser transmits the component that is parallel to its transmission axis, which is $E_0 \cos\theta$.

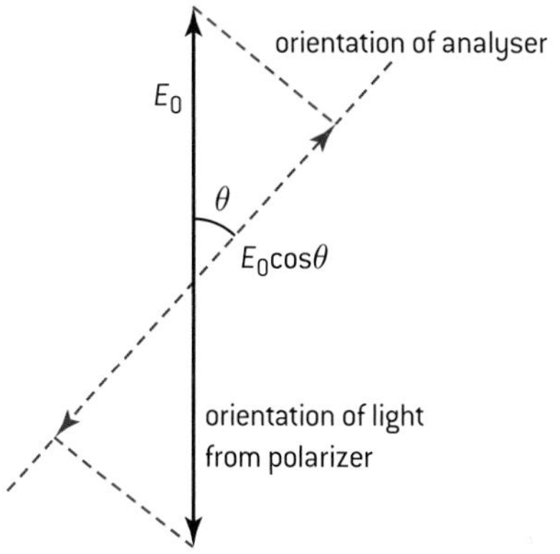

▲ Figure 12 Analysing polarized light.

We have seen that intensity is proportional to the square of the amplitude of a wave so $I_0 \propto E_0^2$ the transmitted intensity will be I which is proportional to $(E_0 \cos\theta)^2$ or $I \propto E_0^2 \cos^2\theta$

Taking ratios we have $\frac{I}{I_0} = \frac{E_0^{\,2}\cos^2\theta}{E_0^{\,2}}$ cancelling $E_0^{\,2}$ gives

$$I = I_0 \cos^2\theta$$

When $\theta = 0°$ (or 180°) $I = I_0$ (since $\cos 0° = 1$); this means that the intensity of light transmitted by the analyser is maximum when the transmission axes of the two Polaroids are parallel.

When $\theta = 90°$, $I = I_0 \cos^2 90° = 0$; this means that no light is transmitted by the analyser when the Polaroids are crossed.

Figure 13 shows a pair of Polaroids. The left-hand image shows that, when their transmission axes are aligned, the same proportion of light passing through one Polaroid passes through both. The central image shows that when one of the Polaroids is rotated slightly, less light passes through their region of overlap. The right-hand image shows that where the Polaroids are crossed no light is transmitted.

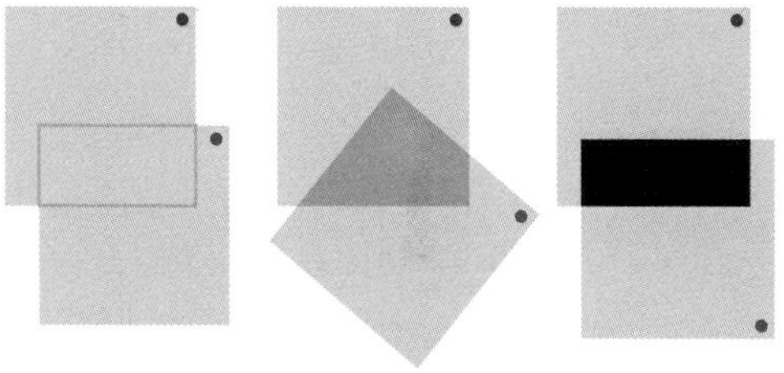
▲ Figure 13 Two Polaroids.

Nature of science

Uses of polarization of light

Polaroid sunglasses are used to reduce the glare coming from the light scattered by surfaces such as the sea or a swimming pool. In industry, stress analysis can be performed on models made of transparent plastic by placing the model in between a pair of crossed Polaroids. As white light passes through a plastic, each colour of the spectrum is polarized with a unique orientation. There is high stress in the regions where the colours are most concentrated – this is where the model (or real object being modelled) is most likely to break when it is put under stress.

Certain asymmetric molecules (called *chiral molecules*) are **optically active** – this is the ability to rotate the plane of plane-polarized light. The angle that the light is rotated through is measured using a **polarimeter**, which consists of a light source, and a pair of Polaroids. The light passes through the first Polaroid (the polarizer) and, initially with no sample present, the second Polaroid (the analyser) is aligned so that no light passes through. With a sample placed between the Polaroids, light does pass through because the sample has rotated the plane of polarization. The analyser is now rotated so that again no light passes through. The angle of rotation is measured and from this the concentration of a solution, for example, can be found. An easily produced, optically active substance is a sugar solution.

Polarization can also be used in the recording and projection of 3D films. These consist of two films projected at the same time. Each of the films is recorded from a slightly different camera angle and the projectors are also set up in this way. The two films are projected through polarizing filters – one with its axis of transmission horizontal and one with it vertical. By wearing polarized eye glasses with one lens horizontally polarized and one vertically polarized, the viewer's left eye only sees the light from the left projector and the right eye light from the right projector – thus giving the viewer the perception of depth. In cinemas the screen needs to be metallic as non-metallic flat surfaces have a polarizing effect.

Worked examples

1 Unpolarized light of intensity I_0 is incident on a polarizer. The transmitted light is then incident on an analyser. The axis of the analyser makes an angle of 60° to the axis of the polarizer.

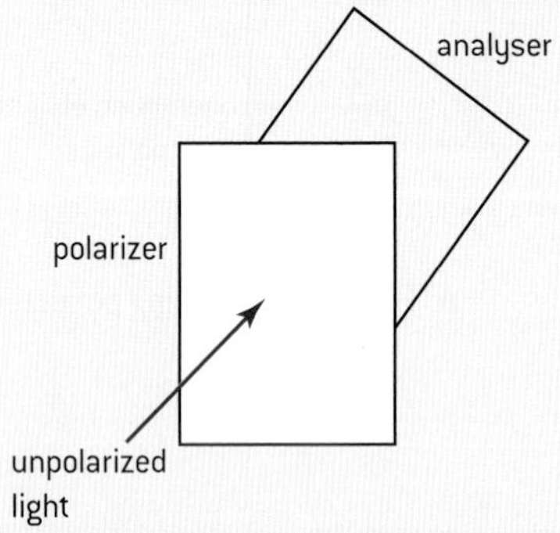

Calculate the intensity emitted by the analyser.

Solution

The first polarizer restricts the intensity to $\frac{I_0}{2}$. Using Malus's law $I = I_0 \cos^2 \theta$

$\cos 60° = 0.5$ so $\cos^2 60° = 0.25$

thus $I = 0.25\frac{I_0}{2} = 0.125I_0$

2 **a)** Distinguish between polarized and unpolarized light.

b) A beam of plane-polarized light of intensity I_0 is incident on an analyser.

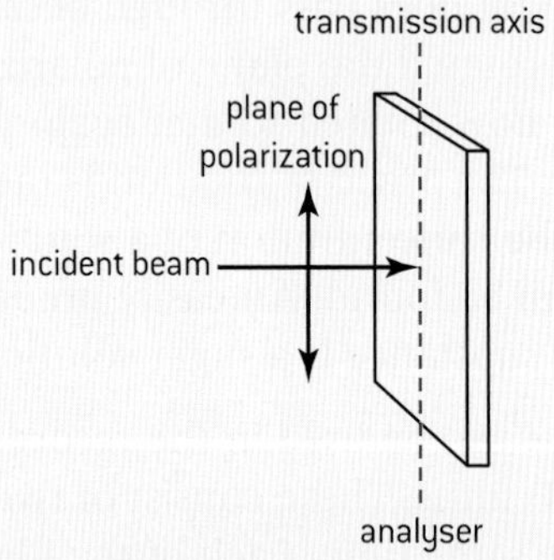

The angle between the transmission axis of the analyser and the plane of polarization of the light θ can be varied by rotating the analyser about an axis parallel to the direction of the incident beam. In the position shown, the transmission axis of the analyser is parallel to the plane of polarization of the light ($\theta = 0°$).

Sketch a graph to show how the intensity I of the transmitted light varies with θ as the analyser is rotated through 180°.

Solution

a) In unpolarized light the **electric field vector** vibrates randomly in any plane (perpendicular to the direction of propagation). In polarized light this vector is restricted to just one plane.

b) This is an application of Malus's law $I = I_0 \cos^2 \theta$

When $\theta = 0$ or 180°, $\cos \theta = 1$ and so $\cos^2 \theta = 1$ and $I = I_0$

When $\theta = 90°$, $\cos \theta = 0$ and so $\cos^2 \theta = 0$ and $I = 0$

These are the key points to focus on. Note that $\cos^2 \theta$ will never become negative. There is no need to include a unit for intensity as this is a sketch graph.

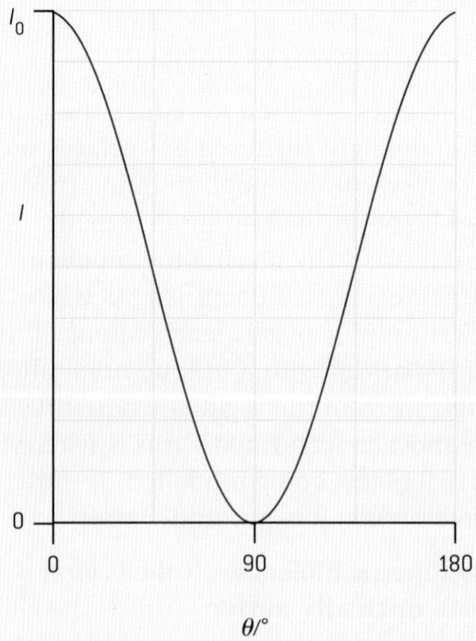

4.4 Wave behaviour

Understanding

- Reflection and refraction
- Snell's law, critical angle, and total internal reflection
- Diffraction through a single-slit and around objects
- Interference patterns
- Double-slit interference
- Path difference

Applications and Skills

- Sketching and interpreting incident, reflected and transmitted waves at boundaries between media
- Solving problems involving reflection at a plane interface
- Solving problems involving Snell's law, critical angle, and total internal reflection
- Determining refractive index experimentally
- Qualitatively describing the diffraction pattern formed when plane waves are incident normally on a single-slit
- Quantitatively describing double-slit interference intensity patterns

Equations

- Snell's law: $\frac{n_1}{n_2} = \frac{\sin\theta_2}{\sin\theta_1} = \frac{v_2}{v_1}$

  Interference at a double slit: $s = \frac{\lambda D}{d}$

Nature of science

Wave or particle?

In the late seventeenth century two rival theories of the nature of light were proposed by Newton and Huygens. Newton believed light to be particulate and supported his view by the facts that it apparently travels in straight lines and can travel through a vacuum; at this time it was a strongly held belief that waves needed a medium through which to travel. Huygens' wave model was supported by the work of Grimaldi who had shown that light diffracts around small objects and through narrow openings. He was also able to argue that when a wave meets a boundary the total incident energy is shared by the reflected and transmitted waves; Newton's argument for this was based on the particles themselves deciding whether or not to reflect or transmit – this was not a strong argument and the wave theory of light became predominant. In the 21st century light is treated as both a wave and a particle in order to explain the full range of its properties.

Introduction

Now we have looked at how to describe waves, we are in the position to look at wave properties – or, as it is described in IB Physics, wave behaviour. You are likely to have come across some of this topic if you

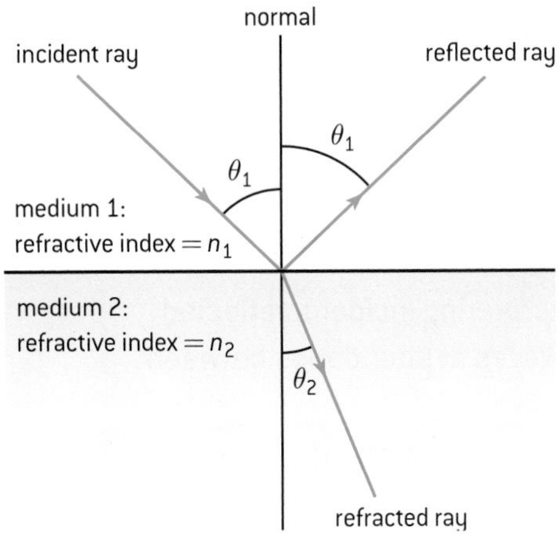

▲ Figure 1 Reflection and refraction of waves.

have studied physics before starting the IB Diploma Programme and we have used some of the ideas in the previous sub-topic. In Sub-topic 4.3 we studied polarization – a property that is restricted to transverse waves alone. The ideas examined in this sub-topic apply to transverse waves (both mechanical and electromagnetic) and longitudinal waves. There are many demonstrations of wave properties utilizing sound and light; however, microwaves are commonly used for demonstrating these wave properties too.

Reflection and refraction of waves

We have looked at much of the content of figure 1 when we considered polarization. We will now focus on what is happening to the rays – remember we could always add wavefronts at right angles to the rays drawn on these diagrams. What the ray diagrams do not show is what is happening to the wavelength of the waves – we will return to this in due course.

The laws of reflection and refraction can be summarized in three laws as follows:

1 **The reflected and refracted rays are in the same plane as the incident ray and the normal.**
This means that the event of reflection or refraction does not alter the plane in which the light ray travels – this is not obvious because we draw ray diagrams in two dimensions but, when we use ray boxes to perform experiments with light beams, we can confirm that this is the case.

2 **The angle of incidence equals the angle of reflection.**
The angle of incidence is the angle between the incident ray and the normal and the angle of reflection is the angle between the reflected ray and the normal. The normal is a line perpendicular to a surface at any chosen point. The angle of incidence and reflection are both labelled as θ_1 on figure 1.

3 **For waves of a particular frequency and for a chosen pair of media the ratio of the sine of the angle of incidence to the sine of the angle of refraction is a constant called the (relative) refractive index.**
This is called Snell's law (or Descartes's law in the French speaking world). The angle of refraction is the angle between the refracted ray and the normal. Snell's law can be written as

$$\frac{\sin \theta_1}{\sin \theta_2} = {}_1n_2$$

For light going from medium 1 to medium 2 – this way of writing Snell's law has several variants and the IB course uses one that we will look at soon.

When light is normal on a surface Snell's law breaks down because the light passes directly through the surface.

▲ Figure 2 Use of ray box to demonstrate the laws of reflection and refraction.

Nature of science

What happens to light at an interface between two media?

This is a complex process but in general terms, when charges are accelerated, for example when they are vibrated, they can emit energy as an electromagnetic wave. In moving through a vacuum the electromagnetic wave travels with a velocity of 3.00×10^8 m s^{-1}. When the wave reaches an atom, energy is absorbed and causes electrons within the atom to vibrate. All particles have frequencies at which they tend to vibrate most efficiently – called the **natural frequency**. When the frequency of the electromagnetic wave does not match the natural frequency of vibration of the electron, then the energy will be re-emitted as an electromagnetic wave. This new electromagnetic wave has the same frequency as the original wave and will travel at the usual speed in the vacuum between atoms. This process continues to be repeated as the new wave comes into contact with further atoms of unmatched natural frequency. With the wave travelling at 3.00×10^8 m s^{-1} in space but being delayed by the absorption–re-emission process, the overall speed of the wave will be reduced. In general, the more atoms per unit volume in the material, the slower the radiation will travel. When the frequency of the light does match that of the atom's electrons the re-emission process is occurs in all directions and the atom gains energy, increasing the internal energy of the material.

Refractive index and Snell's law

The **absolute refractive index (*n*)** of a medium is defined in terms of the speed of electromagnetic waves as:

$$n = \frac{\text{speed of electromagnetic waves in a vacuum}}{\text{speed of electromagnetic waves in the medium}} = \frac{c}{v}$$

The refractive index depends on the frequency of the electromagnetic radiation and, since the speed of light in a vacuum is the limit of speed, the absolute refractive index is always greater than 1 (although there are circumstances when this is not true – but that is well beyond the expectation of your IB Physics course). For all practical purposes the absolute refractive index of air is 1 so it is not necessary to perform refractive index experiments in a vacuum.

Worked example

Calculate the angle of refraction when the angle of incidence at a glass surface is 55° (refractive index of the glass = 1.48).

Solution

As we are dealing with air and glass there is no difference between absolute refractive index and relative refractive index.

Snell's law gives $\frac{\sin\theta_1}{\sin\theta_2} = n_{\text{glass}}$

$$= \frac{\sin 55°}{\sin\theta_2} = 1.48$$

$$\sin\theta_2 = \frac{\sin 55°}{1.48} = \frac{0.819}{1.48} = 0.553$$

$$\theta = \sin^{-1}(0.553) = 33.6 \approx 34°$$

TOK

Conservation of energy and waves

The wave and ray diagram for reflection and refraction tells just part of the story. As with many areas of physics, returning to the conservation of energy is important. For the total energy incident on the interface between the two media the energy is shared between the reflected wave, the transmitted wave and the energy that is ***absorbed*** – the further the wave passes through the second medium, the more of the energy is likely to be absorbed. The conservation of mass/energy is a principle in physics which, to date, has not let physicists down. Does this mean that we have proved that the principle of conservation of energy is infallible?

Reversibility of light

You should be able to prove to yourself that rays are reversible. Place a ray box and a glass block on a piece of paper. Mark, on the paper, the path of the beam of light emitted by the ray box as it approaches and leaves the glass block. Then place the ray box on the other side of the block and you will see that the light travels along the same path in the opposite direction.

We have seen that for light travelling from medium 1 to medium 2 Snell's law can be written as

$$\frac{\sin\theta_1}{\sin\theta_2} = {}_1n_2$$

here ${}_1n_2$ means the relative refractive index going from medium 1 to medium 2. For light travelling in the opposite direction and since light is reversible we have

$$\frac{\sin\theta_2}{\sin\theta_1} = {}_2n_1$$

It should be clear from this that ${}_1n_2 = \frac{1}{{}_2n_1}$

Worked example

The (absolute) refractive index of water is 1.3 and that of glass is 1.5.

a) Calculate the relative refractive index from glass to water.

b) Explain what this implies regarding the refraction of light rays.

c) Draw a wavefront diagram to show how light travels through a plane interface from glass to water.

Solution

a) $n_{water} = {}_{vac}n_{water} = 1.3$ and $n_{glass} = {}_{vac}n_{glass} = 1.5$

We are calculating ${}_{glass}n_{water}$

$$\frac{\sin\theta_{vac}}{\sin\theta_{water}} = {}_{vac}n_{water} \text{ and } \frac{\sin\theta_{vac}}{\sin\theta_{glass}} = {}_{vac}n_{glass}$$

$$\frac{\sin\theta_{glass}}{\sin\theta_{vac}} \times \frac{\sin\theta_{vac}}{\sin\theta_{water}} = \frac{\sin\theta_{glass}}{\sin\theta_{water}} = {}_{glass}n_{water}$$

This means that ${}_{glass}n_{water} = \frac{1}{1.5} \times 1.3 = 0.87$

b) With a relative refractive index less than one this means that the light travels faster in the water than the glass and therefore bends away from the normal.

c)

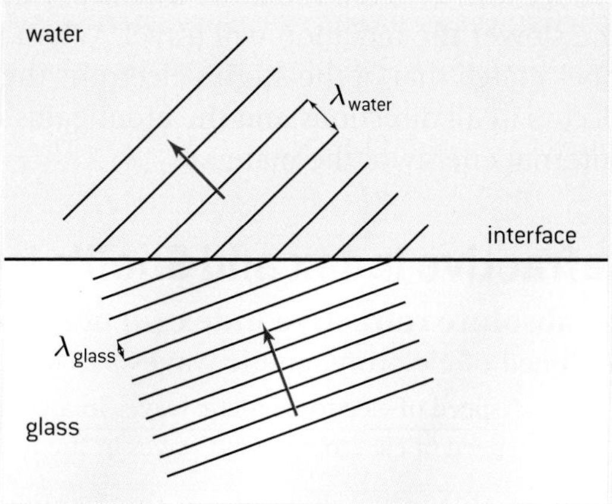

As the refractive index of water is lower than that of glass the light wave speeds up on entering the water. The frequency is constant and, as $v = f\lambda$, the wavelength will be greater in water – meaning that the wavefronts are further apart. The fact that the frequency remains constant is a consequence of Maxwell's electromagnetic equations – something not covered in IB Physics.

The critical angle and total internal reflection

When a light wave reaches an interface travelling from a higher optically dense medium to a lower one the wave speeds up. This means that the wavelength of the wave increases (frequency being constant) and the direction of the wave moves away from the normal – the angle of refraction being greater than the angle of incidence so as the angle of incidence increases the angle of refraction will approach 90°. **Optical density** is not the same as physical density, i.e. mass per unit volume, it is measured in terms of refractive index – higher refractive index material having higher optical density.

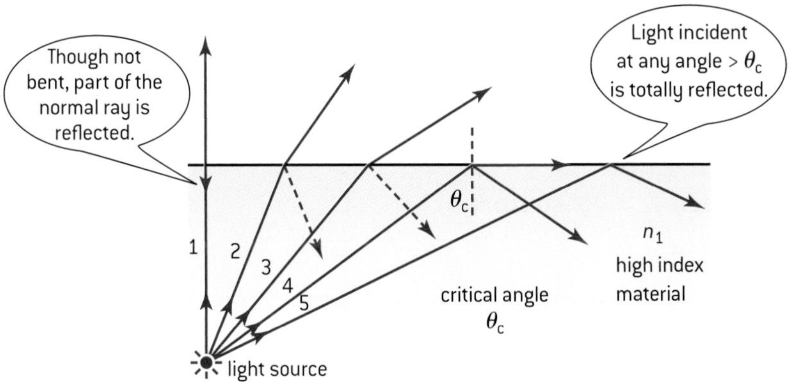

▲ Figure 3 Light passing from more optically dense medium to less optically dense medium.

Ray **1** in figure 3 shows a ray passing from a more optically dense medium to a less optically dense medium normal to an interface. Most of the light passes though the interface but a portion is reflected back into the original medium. Increasing the angle of incidence (as for rays **2** and **3**) will increase the angle of refraction and ray **4** shows an angle of incidence when the angle of refraction is 90°. The angle of incidence at this value is called the **critical angle (θ_c)**. Ray **5** shows that, when the angle of incidence is larger than the critical angle, the light wave does not move into the new medium at all but is reflected back into the original medium. This process is called **total internal reflection**.

It should be noted that for angles smaller than the critical angle there will always be a reflected ray; although this will carry only a small portion of the incident energy.

A ray box and semicircular glass block can be used to measure the critical angle for glass as shown in figure 4. When the beam is incident on the curved face of the block making it travel towards the centre of the flat face, it is acting along a radius and so enters the block normally and, therefore, without bending. By moving the ray box, different angles of incidence can be obtained. Both the critical angle and total internal reflection can be seen.

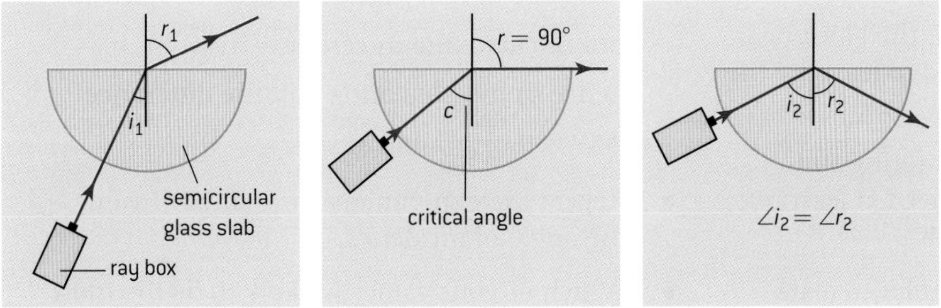

▲ Figure 4 Use of ray box to investigate critical angle and total internal reflection.

Worked example

Calculate the (average) critical angle for a material of (average) absolute refractive index 1.2.

Solution

The word "average" is included because the refractive index and critical angle would each be different for different colours. Don't be surprised if it is left out in some questions – it is implied by talking about single values.

$sin\ \theta_c = \frac{1}{n_1} = \frac{1}{1.2} = 0.833$

$\theta_c = \sin^{-1} 0.833 = 56.4° \approx 56°$

When a ray box emitting white light is used, the light emerging through the glass block is seen to **disperse** into the colours of the rainbow. This is due to each of the colours, of which white light is comprised, having a different frequency. The refractive index for each of the colours is different.

Calculating the critical angle

Snell's law gives $\frac{\sin\theta_1}{\sin\theta_2} = {}_1n_2$.

In order to obtain a critical angle, medium 1 must be more optically dense than medium 2.

When $\theta_1 = \theta_c$ then $\theta_2 = 90°$ so $\sin\theta_2 = 1$

This gives $\sin\theta_c = {}_1n_2$

$${}_1n_2 = \frac{n_2}{n_1}$$

When the less dense medium (medium 2) is a vacuum or air then $n_2 = 1$

So $\sin\theta_c = \frac{1}{n_1}$

Investigate!

Measuring the refractive index

There are several possible experiments that you could do to measure the refractive index depending on whether a substance is a liquid or a solid (we assume the refractive index of gases is 1 – although mirages are a good example of how variations in air density can affect the refraction of light). You could research how to use real and apparent depth measurements to measure the refractive index of a liquid. The investigation outlined below will be to trace some rays (or beams of light, really) through a glass block of rectangular cross-section.

- The arrangement is similar to that shown in figure 2 on p146.
- Place the block on a piece of white paper and mark its position by drawing around the edges.
- Direct a beam of light to enter the block near the centre of a longer side and to leave by the opposite side.
- Mark the path of the beam entering and leaving the block – you will need at least two points on each beam to do this.
- Remove the block and use a ruler to mark the path of the beam on either side of the block and then inside. Add arrows to these to remind you that they represent rays and to indicate which is the incident beam and which is the refracted beam.
- Using a protractor, mark in and draw normals for the beam entering and leaving the block.
- Remembering that light is reversible and the beam is symmetrical, measure two values for each of θ_1 and θ_2.
- Calculate the refractive index of the block using Snell's law.
- Estimate the experimental uncertainty on your measurements of θ_1 and θ_2.
- Note that the uncertainty in θ_1 and θ_2 is not the same as that in $\sin\theta_1$ and $\sin\theta_2$ but you can calculate the uncertainty in $\frac{\sin\theta_1}{\sin\theta_2}$ (and hence n) by calculating half the difference between $\frac{\sin\theta_{1max}}{\sin\theta_{2min}}$ and $\frac{\sin\theta_{1min}}{\sin\theta_{2max}}$
- Repeat the experiment for a range of values of the angle of incidence.
- Which of your values is likely to be the most reliable?

Diffraction

The first detailed observation and description of the phenomenon that was named **diffraction** was made by the Italian priest Francesco Grimaldi. His work was published in 1665, two years after his death. He found that when waves pass through a narrow gap or slit (called an aperture), or when their path is partly blocked by an object, the waves spread out into what we would expect to be the shadow region. This is illustrated by figure 5 and can be demonstrated in a ripple tank. He noted that close to the edges the shadows were bordered by alternating bright and dark fringes. Given the limited apparatus available to Grimaldi his observations were quite extraordinary.

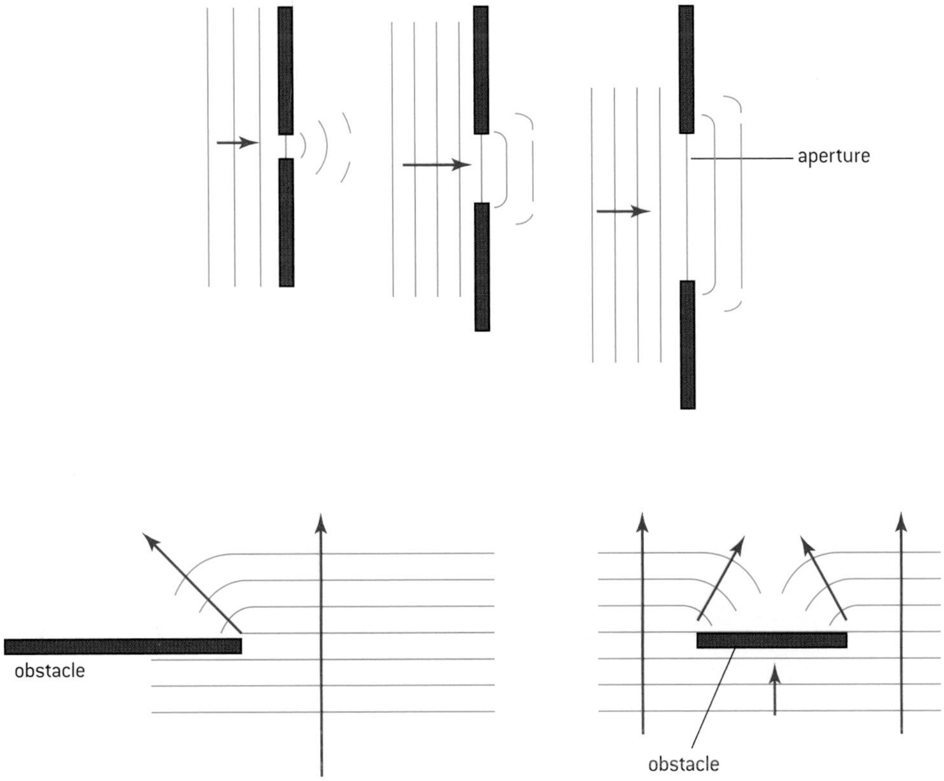

▲ Figure 5 Diffraction.

Further observation of diffraction has shown:

- the frequency, wavelength, and speed of the waves each remains the same after diffraction
- the direction of propagation and the pattern of the waves change
- the effect of diffraction is most obvious when the aperture width is approximately equal to the wavelength of the waves.
- the amplitude of the diffracted wave is less than that of the incident wave because the energy is distributed over a larger area.

Explaining diffraction by a single slit is complex and you will only be asked for a qualitative description of single-slit diffraction at SL – however, as mentioned in Sub-topic 4.3, the Huygens–Fresnel principle gives a good insight into how the single-slit diffraction pattern comes about (see figure 6).

Worked example

Complete the following diagrams to show the wavefronts after they have passed through the gaps.

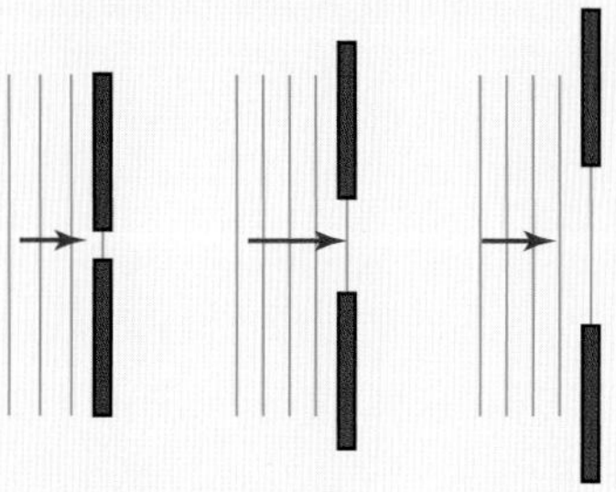

Solution

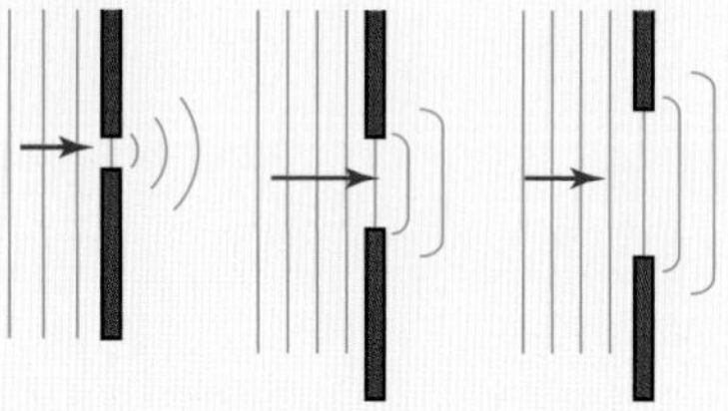

When the width of the slit is less than or equal to the wavelength λ of the wave, the waves emerge from the slit as circular wavefronts. As the slit width is increased, the spreading of the waves only occurs at the edges and the diffraction is less noticeable.

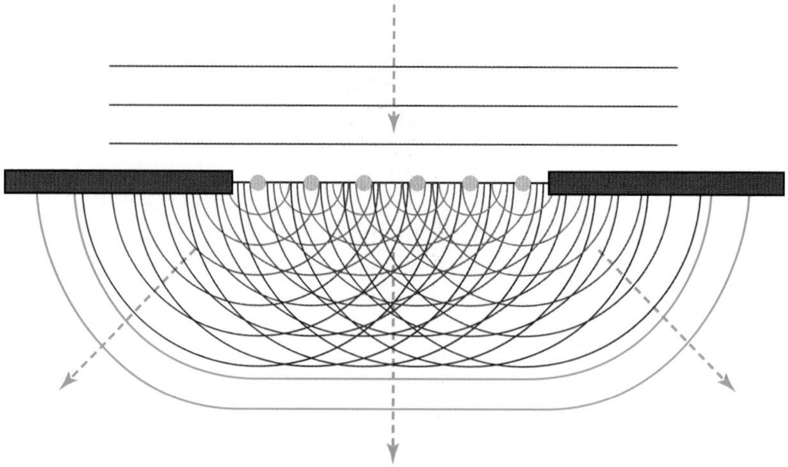

▲ Figure 6 Huygens–Fresnel explanation of diffraction.

Plane waves travelling towards the slit behave as if they were sources of secondary wavelets. The orange dots in figure 6 show these "secondary sources" within the slit. These "sources" each spread out as circular waves. The tangents to these waves will now become the new wavefront. The central image is bright and wide, beyond it are further narrower bright images separated by darkness. Single-slit diffraction is further explored for those studying HL Physics in Topic 9.

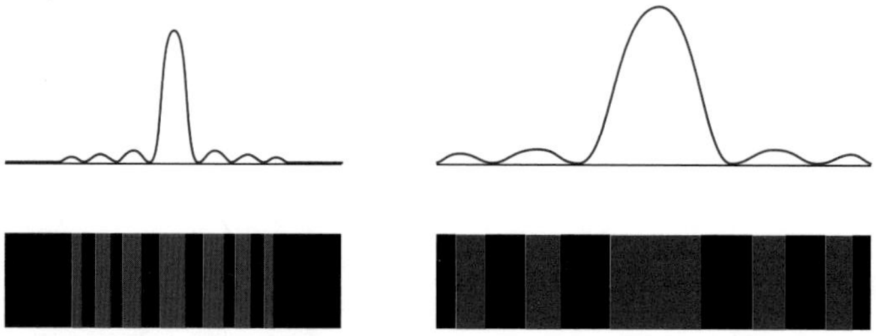

▲ Figure 7 Single-slit diffraction pattern.

Double-slit interference

We briefly discussed the principle of superposition in Sub-topic 4.3. Interference is one application of this principle.

When two or more waves meet they combine to produce a new wave – this is called **interference**. When the resultant wave has larger amplitude than any of the individual waves the interference is said to be **constructive**; when the resultant has smaller amplitude the interference is **destructive**. Interference can be achieved by using two similar sources of all types of wave. It is usually only observable if the two sources have a constant phase relationship – this means that although they need not emit the two sets of waves in phase, the phase of the waves cannot alter relative to one another. Such sources will have the same frequency and are said to be **coherent**.

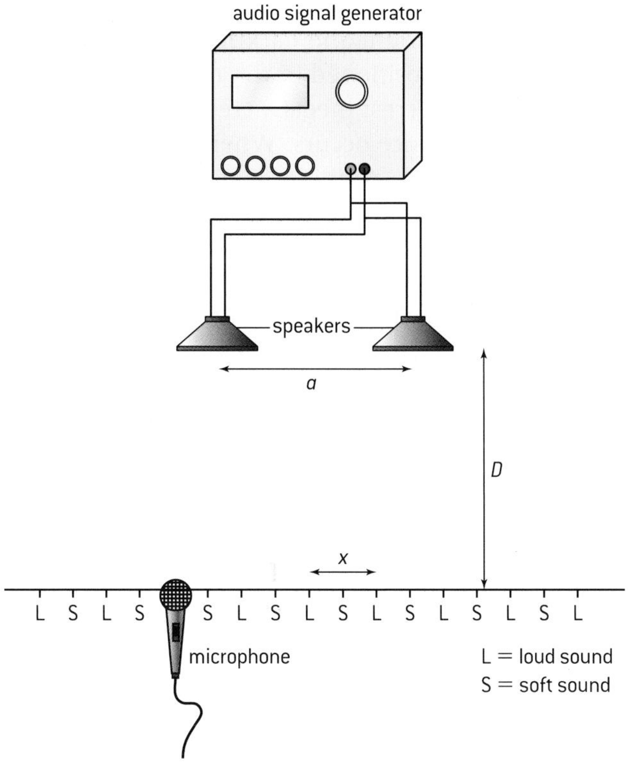

Figure 8 Interference of sound waves.

Interference of sound waves is easy to demonstrate using two loudspeakers connected to the same audio frequency oscillator as shown in figure 8. Moving the microphone (connected to an oscilloscope) in a line perpendicular to the direction in which the waves are travelling allows an equally spaced loud–soft sequence to be detected. More simply, the effect can be demonstrated by the observer walking along the loud–soft line while listening to the loudness.

When a coherent beam of light is incident on two narrow slits very close together the beam is diffracted at each slit and, in the region

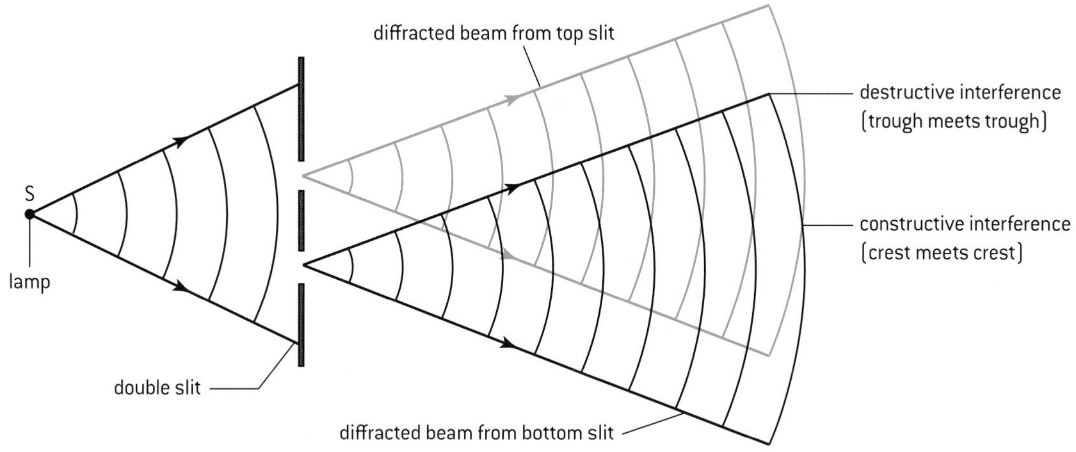

Figure 9 Interference of light waves.

▲ Figure 10 Fringes produced by a double-slit.

where the two diffracted beams cross, interference occurs as seen in figure 9. A pattern of equally spaced bright and dark fringes (shown in figure 10) is obtained on a screen positioned in the region where the diffracted beams overlap. When a crest meets a crest (or a trough meets a trough) constructive interference occurs. When a crest meets a trough destructive interference occurs. A similar experiment to this was performed by the talented English physicist (and later physician) Thomas Young in 1801. The coherent beam is achieved by placing a single slit close to the source of light – this means that the wavefronts spreading from the single slit each reach the double slit with the same phase relationship and so the secondary waves coming from the double slit retain their constant phase relationship.

Path difference and the double-slit equation

You will never be asked to derive this equation but the ideas regarding path difference are vital to your understanding of interference.

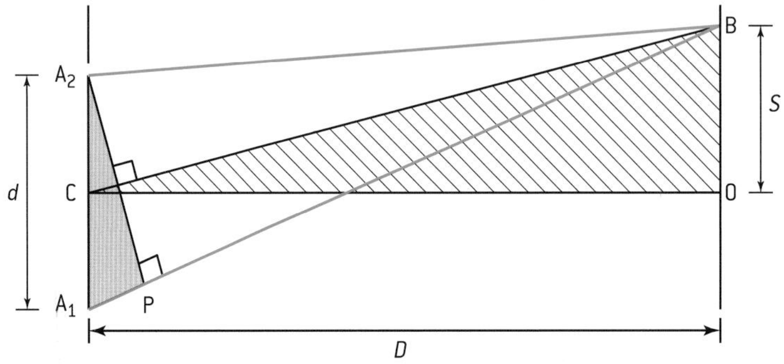

▲ Figure 11 The double-slit geometry.

Figure 11 shows two slits (apertures) A_1 and A_2 distance *d* apart. The double slit is at distance *D* from a screen. O is the position of the central bright fringe (arising from constructive interference). B is the position of the next bright fringe above O; the distance OB is the fringe spacing *s*. There will be another bright fringe distance *s* below O. The beams from A_1 and A_2 to O will travel equal distances and so will meet with the same phase relationship that they had at A_1 and A_2 – they have zero path difference. At B the beam from A_1 will travel an extra wavelength compared with the beam from A_2 – the path distance (= A_1P) equals λ. Because of the short wavelength of light and the fact that *D* is very much larger than *d*, the line A_2P is effectively perpendicular to lines A_1B *and* CB (C being the midpoint of A_1 A_2). This means that the triangles A_1A_2P and CBO are similar triangles.

$$\text{Taking ratios } \frac{BO}{CO} = \frac{A_1P}{A_1A_2} \text{ or } \frac{s}{D} = \frac{\lambda}{d}$$

Rearranging gives $s = \dfrac{\lambda D}{d}$

This gives the separation of successive bright fringes (or bands of loud sound for a sound experiment).

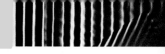

In general for two coherent beams starting in phase, if the path difference is a whole number of wavelengths we get constructive interference and if it is an odd number of half wavelengths we get destructive interference. It must be an odd number of half wavelengths for destructive interference because an even number would give a whole number (integer) and that is constructive interference!

Summarizing this:

For constructive interference the path difference must $= n\lambda$ where $n = 0,1,2...$

For destructive interference the path difference must $= \left(n + \frac{1}{2}\right)\lambda$ where $n = 0,1,2...$

n is known as the order of the fringe, $n = 0$ being the zeroth order, $n = 1$ the first order, etc.

Worked examples

1 In a double-slit experiment using coherent light of wavelength λ, the central bright fringe is observed on a screen at point O, as shown below.

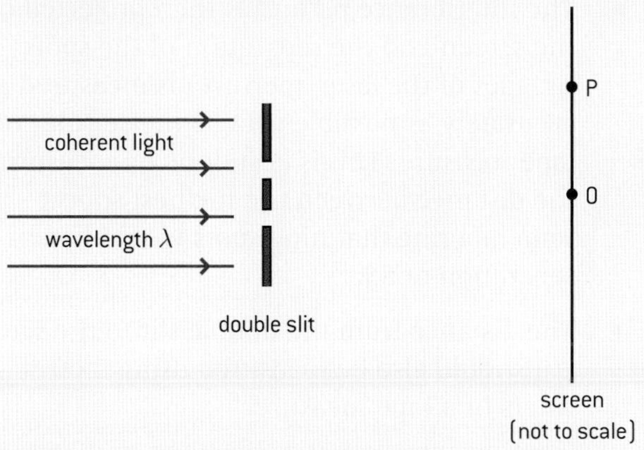

At point P, the path difference between light arriving at P from the two slits is 7λ.

a) Explain the nature of the fringe at P.

b) State and explain the number of dark fringes between O and P.

Solution

a) As the path difference is an integral number of wavelengths there will be a bright fringe at P.

b) For destructive interference the path difference must be an odd number of half wavelengths, so there will be dark fringes when the path difference is $\frac{\lambda}{2}, \frac{3\lambda}{2}, \frac{5\lambda}{2}, \frac{7\lambda}{2}, \frac{9\lambda}{2}, \frac{11\lambda}{2}, \frac{13\lambda}{2}$ giving a total of 7 dark fringes.

2 Two coherent point sources S_1 and S_2 oscillate in a ripple tank and send out a series of coherent wavefronts as shown in the diagram.

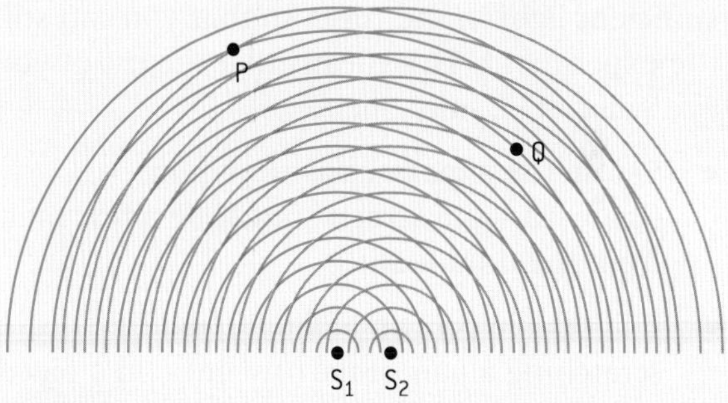

State and explain the intensity of the waves at P and Q?

Solution

Considering each wavefront to be a crest; at point P two crests meet and so superpose constructively giving an amplitude which is twice that of one wave (assuming the wavefronts each have the same amplitude) – the intensity is proportional to the square of this so will be four times the intensity of either of the waves alone. At Q a blue crest meets a red trough and so there is cancellation occurring and there will be zero intensity.

Investigate!

Measuring the wavelength of laser light using a double slit

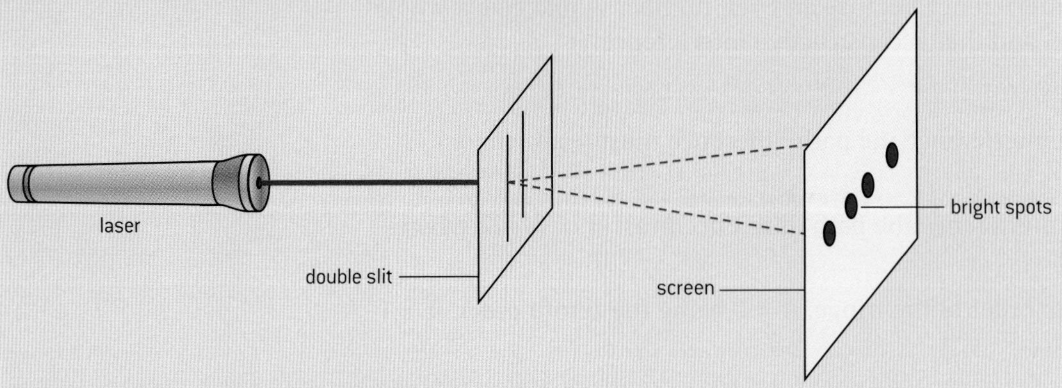

▲ Figure 12 Measuring the wavelength of laser light using a double slit.

This experiment, using a gas laser (or laser pointer), is a modern version of the one performed by Young. A laser emits a highly coherent beam of light ideal for performing this experiment. **Care should be taken not to shine the laser beam or its reflection into your eye** – should this happen, look away immediately to avoid the risk of permanent damage to your eye. One way to minimise eye damage is to keep sufficient light in the room so that you can still do the experiment. This means that the iris of your eye will not be fully open.

- Set up the apparatus as shown in figure 12 – the screen should be a few metres from the double slit.
- The double slit can be homemade by scratching a pair of narrow lines on a piece of glass painted with a blackened material or it could be a ready prepared slide.
- The slit separation (d) can be measured using a travelling microscope (however, d is likely to be provided by a manufacturer).
- Light from the laser beam diffracts through the slits and emerges as two separate coherent waves. Both slits must be illuminated by the narrow laser beam; sometimes it may be necessary to use a diverging (concave) lens to achieve this.
- The interference pattern is then projected onto the screen and the separation of the spots (images of the laser aperture) is measured as accurately as possible using a metre ruler or tape measure. This is best done by measuring the distance between the furthest spots, remembering that nine spots would have a separation of $8S$.
- The distance from the double slit to the screen (D) should also be measured using metre rulers or a tape measure.
- Once your readings are taken the wavelength of the light can be calculated from $\lambda = \frac{Sd}{D}$.

When red light is used as the source, the bright fringes are all red. When blue light is used as a source, the bright fringes are all blue. As blue light has a shorter wavelength than red light the blue fringes are closer together than the red fringes. When the source is white light the zeroth order (central) bright fringe is white but the other fringes are coloured with the blue edges closer to the centre and the red edges furthest from the centre. Can you explain this?

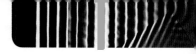

Investigate!

Measuring the wavelength of microwaves using a double slit

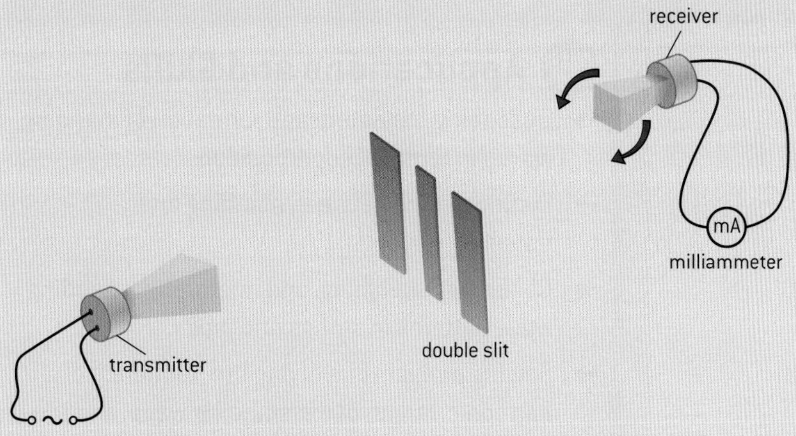

▲ Figure 13 Microwave arrangement.

- The double slit is adjusted so that the slits are around 3 cm apart – for maximum diffraction.
- Arrange both transmitter and receiver about half a metre from the slit.
- Alter the position of the receiver until the received signal is at its strongest.
- Now slowly rotate the receiver until the signal is weakest.
- Cover up one of the slits with a book and explain the result.
- Remove the book so that two slits are again available and attempt to discover if "fringes" occur as in Young's double-slit interference experiment with light. Look for fringes on both sides of the maximum.
- Measure as accurately as you can the values for D, d, and S.
- Calculate the wavelength of the microwaves.
- Repeat for other distances of transmitter and receiver from the slits.

These investigations can be repeated with sound waves or radio waves given appropriate transmitters, receivers and slits of the correct dimension.

4.5 Standing waves

Understanding

- ➔ The nature of standing waves
- ➔ Boundary conditions
- ➔ Nodes and antinodes

Applications and Skills

- ➔ Describing the nature and formation of standing waves in terms of superposition
- ➔ Distinguishing between standing and travelling waves
- ➔ Observing, sketching, and interpreting standing wave patterns in strings and pipes
- ➔ Solving problems involving the frequency of a harmonic, length of the standing wave, and the speed of the wave

Nature of science

Fourier synthesis

Synthesizers are used to generate a copy of the sounds naturally produced by a wide range of musical instruments. Such devices use the principle of superposition to join together a range of harmonics that are able to emulate the sound of the natural instrument. Fourier synthesis works by combining a sine-wave signal with sine-wave or cosine-wave harmonics of correctly chosen amplitude. The process is named after the French mathematician and physicist Jean Baptiste Joseph, Baron de Fourier who, in the early part of the nineteenth century, developed the mathematical principles on which synthesis of music is based. In many ways Fourier synthesis is a visualization of music.

Introduction

We have seen in Sub-topic 4.2 how travelling waves transfer energy from the source to the surroundings. In a travelling wave the position of the crests and troughs changes with time. Under the right circumstances, waves can be formed in which the positions of the crests and troughs do not change – in such a case the wave is called a **standing wave**. When two travelling waves of equal amplitude and equal frequency travelling with the same speed in opposite directions are superposed, a standing wave is formed.

Standing waves on strings

Figure 1 shows two travelling waves (coloured green and blue) moving towards each other at four consecutive times t_1, t_2, t_3, and t_4. The green and blue waves superpose to give the red standing wave. As the green and blue waves move forward there are points where the total displacement (seen on the red wave) always remains zero – these are called **nodes**. At other places the displacement varies between a maximum in one direction and a maximum in the other direction – these are called **antinodes**.

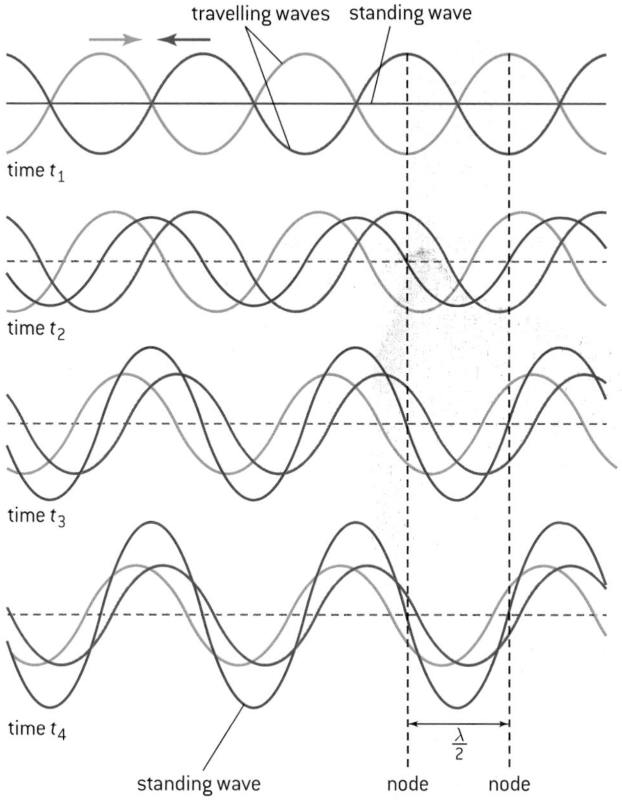

▲ Figure 1 The formation of a standing wave.

At any instant the displacement of the standing wave will vary at all positions other than the nodes. Thus a single frame shot of a standing wave would look like a progressive wave – as can be seen from the red wave in figure 1. When representing the standing wave graphically it is usual to show the extremes of standing waves, but over a complete time period of the oscillation the wave will occupy a variety of positions as shown by the arrow in the loop of figure 2.

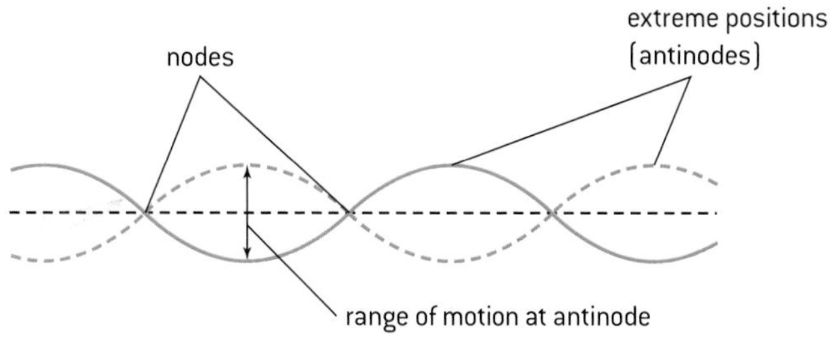

▲ Figure 2 Nodes and antinodes on a string.

Melde's string

The apparatus shown in figure 3 is useful in demonstrating standing waves on a string. A variant of this apparatus was first used in the late nineteenth century by the German physicist Franz Melde. A string is strung between a vibration generator and a fixed end. When the vibration generator is

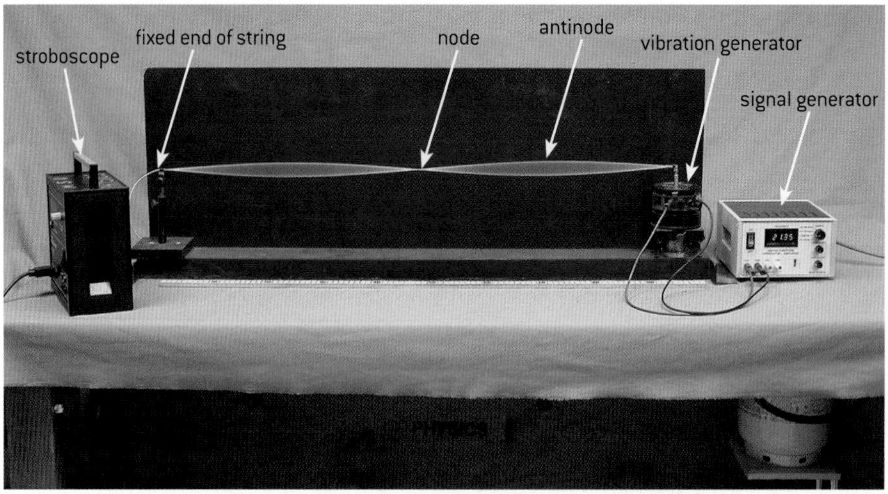

▲ Figure 3 Melde's string.

connected to an audio frequency oscillator, the end of the string attached to the vibration generator oscillates vertically. A wave travels down the string before undergoing a phase change of 180° when it reflects at the fixed end. The reflected wave superposes with the incident wave and (at certain frequencies) a standing wave is formed. The frequency of the audio frequency generator is slowly increased from zero and eventually a frequency is reached at which the string vibrates with large amplitude in the form of a single loop – **the first harmonic**. If the frequency is further increased, the amplitude of the vibrations dies away until a frequency of twice the first harmonic frequency is achieved – in this case two loops are formed and we have the **second harmonic** frequency shown in figure 3. This is an example of resonance – the string vibrates with large amplitude only when the applied frequency is an integral multiple of the natural frequency of the string. We return to resonance in more detail in Option B. Using a stroboscope to freeze the string reveals detail about a standing wave. For example, when the flash frequency is slightly out of synch with the vibration frequency, it is possible to see the variation with time of the string's displacement; this will be zero at a node but a maximum at an antinode.

Note

- Although we often treat the point of attachment of the string to the vibration generator as being a node, this is not really correct. The generator vibrates the string to set up the wave and therefore the nearest node to the generator will be a short distance from the vibrator. We call this inaccuracy an "end correction" but often draw a diagram showing the node at the vibrator.
- Because some of the travelling wave energy is transmitted or absorbed by whatever is clamping the fixed end, the reflected waves will be slightly "weaker" than the incident waves. This means that cancellation is not complete and there will be some slight displacement at the nodes.
- Within each loop all parts of the string vibrate together in phase, but loops next to each other are constistently 180° out of phase.
- Although strings are often used to produce sound waves, it is not a sound wave travelling along the string – it is a transverse wave. This wave travels at a speed which is determined by the characteristics of the string.

Harmonics on strings

We have seen from Melde's string that a string has a number of frequencies at which it will naturally vibrate. These natural frequencies are known as the **harmonics of string**. The natural frequency at which a string vibrates depends upon the tension of the string, the mass per unit length and the length of the string. With a stringed musical instrument each end of a string is fixed, meaning there will be a node at either end. The first harmonic is the lowest frequency by which a standing wave will be set up and consists of a single loop. Doubling the frequency of vibration halves the wavelength and means that two loops are formed – this is called the second harmonic; three times the fundamental frequency gives the third harmonic (see figure 4).

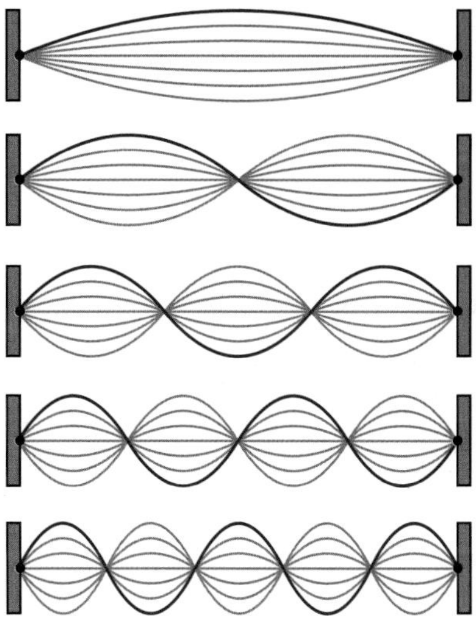

▲ Figure 4 Harmonics on a string.

You will see from figure 1 that the distance between two consecutive nodes is equal to half a wavelength – in other words each loop in a standing wave is equivalent to $\frac{\lambda}{2}$. Figure 4 shows the first five harmonics of a wave on a string of a fixed length. With the speed of the wave along the string being constant, halving the wavelength doubles the frequency; reducing the wavelength by a factor of three triples the frequency, etc. (the wave equation $c = f\lambda$ applies to all travelling waves). The different harmonics can be achieved either by vibrating the string at the appropriate frequency or by plucking, striking or bowing the string at a different position – although plucking at the centre is likely to produce the first, third and fifth harmonics. By pinching the string at different places a node is produced so, for example, pinching at the centre of the string would produce the even harmonics. In a musical instrument several harmonics occur at the same time, giving the instrument its rich sound.

Worked example

A string is attached between two rigid supports and is made to vibrate at its first harmonic frequency *f*.

The diagram shows the displacement of the string at $t = 0$.

(a) Draw the displacement of the string at time

(i) $t = \frac{1}{4f}$ (ii) $t = \frac{1}{2f}$

(b) The distance between the supports is 1.0 m. A wave in the string travels at a speed of 240 m s^{-1}. Calculate the frequency of the vibration of the string.

Solution

(a) (i) We must remember the relationship between period *T* and frequency *f* is $T = \frac{1}{f}$.

This means that $t = \frac{T}{4} = \frac{1}{4f}$ so a quarter of a period has elapsed and the string has gone through quarter of a cycle to give:

(ii) In this case the wave has gone through half a period $\left(t = \frac{T}{2} = \frac{1}{2f}\right)$ and so will have moved from a crest to a trough:

(b) The string is vibrating in first harmonic mode and so the distance between the fixed ends is half a wavelength $\left(\frac{\lambda}{2}\right)$.

So $\frac{\lambda}{2} = 1.0$ m and $\lambda = 2.0$ m.

Using $c = f\lambda$ $\quad f = \frac{c}{\lambda} = \frac{240}{2.0}$

$= 120$ Hz

Standing waves in pipes

Standing waves in a pipe differ from standing waves on a string. In pipes the wave medium is (usually) air and the waves themselves are longitudinal. Pipes can have two closed ends, two open ends, or one open and one closed; the latter two being shown in figure 5. The sound waves are reflected at both ends of the pipe irrespective of whether they are open or closed. The variation of displacement of the air molecules in the pipe determines how we graphically represent standing waves in pipes. There is a displacement antinode at an open end (since the molecules can be displaced by the largest amount here) and there will be a displacement node at the closed end (because the molecules next to a closed end are unable to be displaced). Similar ideas to strings apply to the harmonics in pipes. Strictly speaking, the displacement antinode forms just beyond an open end of the pipe but this end effect can be ignored for most purposes.

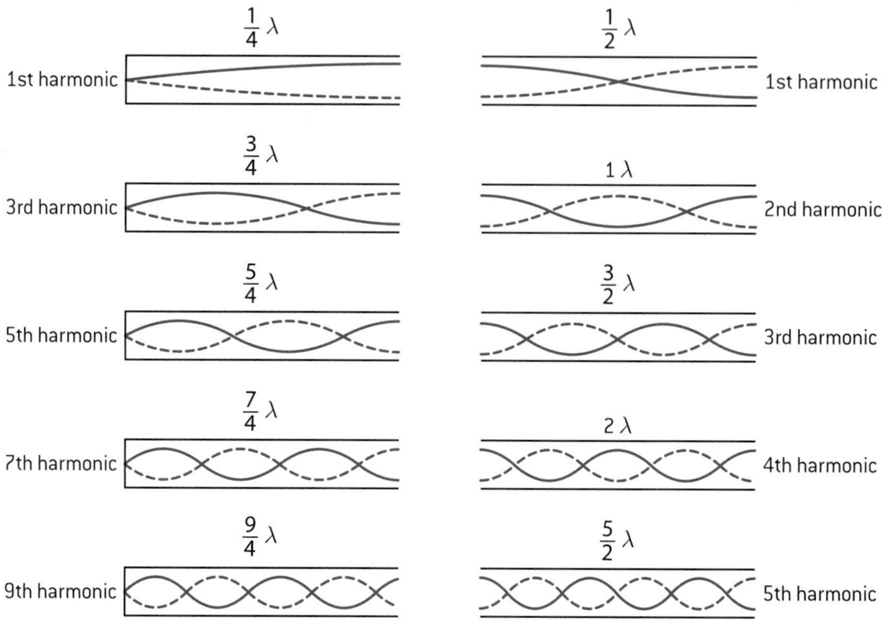

▲ Figure 5 Harmonics in a pipe.

Harmonics in pipes

The harmonic in a pipe depends on whether or not the ends of a pipe are open or closed. For a pipe of fixed length with one open and one closed end there must always be a node at the closed end and an antinode at the open end. This means that only odd harmonics are available (the number of the harmonic is the number of half loops in this type of pipe). For a pipe with two open ends there must always be an antinode at each end and this means that all harmonics are achieveable (in this case the number of loops gives the number of the harmonic).

Let's compare the frequencies of the harmonics for the "one open end" pipe. Suppose the pipe has a length L. The wavelength (λ) of the first harmonic would be $4L$ and, from $c = f\lambda$, the frequency would be $\frac{c}{4L}$. c is the speed of sound in the pipe.

For the third harmonic $L = \frac{3}{4}\lambda$, so $\lambda = \frac{4}{3}L$ and the frequency $= \frac{3c}{4L}$ (or three times that of the first harmonic).

A harmonic is named by the ratio of its frequency to that of the first harmonic.

Worked examples

1 The first harmonic frequency for a particular organ pipe is 330 Hz. The pipe is closed at one end but open at the other. What is the frequency of its third harmonic?

Solution

A named harmonic is the ratio of its frequency to that of the first harmonic – so in this case the third harmonic will be 990 Hz since the third harmonic has three times the frequency of the first harmonic.

2 The first harmonic frequency of the note emitted by an organ pipe which is closed at one end is f. What is the first harmonic frequency of the note emitted by an organ pipe of the same length that is open at both ends?

Solution

The length of a pipe closed at one end in first harmonic mode is $\frac{\lambda}{4}$ $(= L)$ so $\lambda = 4L$

This length will be half the wavelength of a pipe open at both ends so $L = \frac{\lambda}{2}$ so $\lambda = 2L$

Since the wavelength has halved the frequency must double so the new frequency will be $2f$.

Boundary conditions

In considering both pipes and strings we have assumed reflections at the ends or boundaries. In meeting the boundary of a string the wave reflects (or at least partially reflects) – this is known as a **fixed boundary**; there will be the usual phase change of 180° at a fixed boundary meaning that the reflected wave cancels the incident wave and so forms a node. The closed ends of pipes and edges of a drumhead also have fixed boundaries. In the case of an open-ended pipe there is still a reflection of the wave at the boundary but no phase change, so the reflected wave does not cancel the incident wave and there is an antinode formed – the same idea applies to strips of metal vibrated at the centre, xylophones, and vibrating tuning forks. This type of boundary is called a **free boundary**.

Comparison of travelling waves and stationary waves

The following table summarizes the similarities and differences between travelling waves and standing waves:

Property	Travelling wave	Standing wave
energy transfer	energy is transferred in the direction of propagation	no energy is transferred by the wave although there is interchange of kinetic and potential energy within the standing wave
amplitude	all particles have the same amplitude	amplitude varies within a loop – maximum occurs at an antinode and zero at a node
phase	within a wavelength the phase is different for each particle	all particles within a "loop" are in phase and are antiphase (180° out of phase) with the particles in adjacent "loops"
wave profile (shape)	propagates in the direction of the wave at the speed of the wave	stays in the same position
wavelength	the distance between adjacent particles which are in phase	twice the distance between adjacent nodes (or adjacent antinodes)
frequency	all particles vibrate with same frequency.	all particles vibrate with same frequency except at nodes (which are stationary)

TOK

Pitch and frequency

Musical pitch is closely linked to frequency but also has a psychological component in relation to music. We think of pitch as being someone's perception of frequency. Musical notes of certain pitches, when heard together, will produce a pleasant sensation and are said to be *consonant* or *harmonic*. These sound waves form the basis of a *musical interval*. For example, any two musical notes of frequency ratio 2:1 are said to be separated by an *octave* and result in a particularly pleasing sensation when heard. Similarly, two notes of frequency ratio of 5:4 are said to be separated by an interval of a *third* (or strictly a *pure* third); again this interval sounds pleasing. Has music always been thought of in this way? Is the concept of consonance accepted by all societies?

Questions

1 *(IB)*

a) A pendulum consists of a bob suspended by a light inextensible string from a rigid support. The pendulum bob is moved to one side and then released. The sketch graph shows how the displacement of the pendulum bob undergoing simple harmonic motion varies with time over one time period.

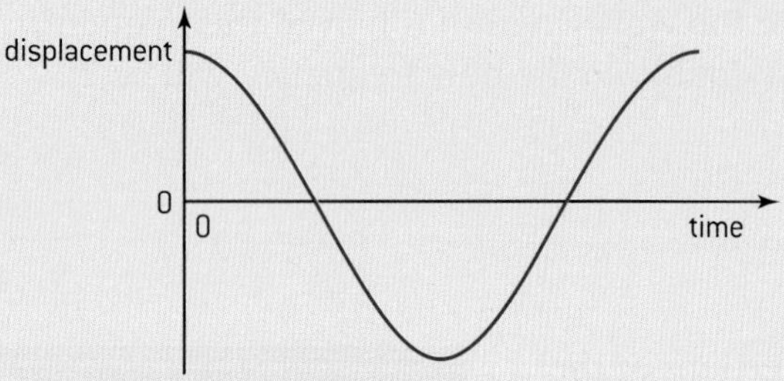

On a copy of the sketch graph:

(i) Label, with the letter A, a point at which the acceleration of the pendulum bob is a maximum.

(ii) Label, with the letter V, a point at which the speed of the pendulum bob is a maximum.

b) Explain why the magnitude of the tension in the string at the midpoint of the oscillation is greater than the weight of the pendulum bob.

2 *(IB)*

The graph below shows how the displacement x of a particle undergoing simple harmonic motion varies with time t. The motion is undamped.

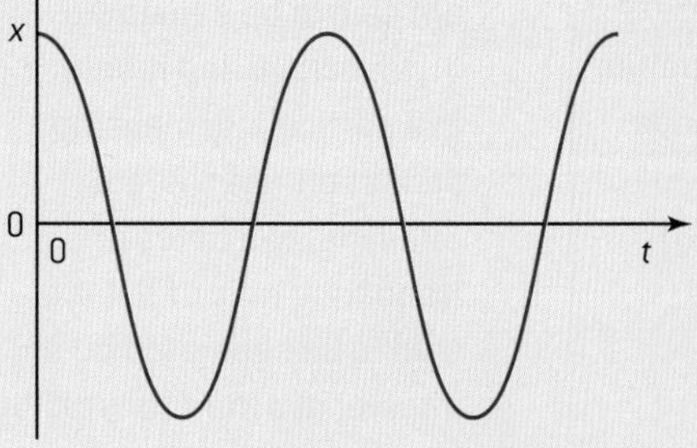

a) Sketch a graph showing how the velocity v of the particle varies with time.

b) Explain why the graph takes this form.

(4 marks)

3 *(IB)*

a) In terms of the acceleration, state *two* conditions necessary for a system to perform simple harmonic motion.

b) A tuning fork is sounded and it is assumed that each tip vibrates with simple harmonic motion.

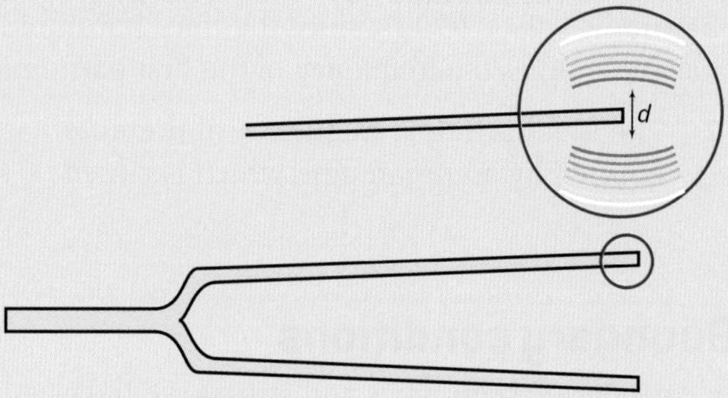

The extreme positions of the oscillating tip of one fork are separated by a distance d.

(i) State, in terms of d, the amplitude of vibration.

(ii) Sketch a graph to show how the displacement of one tip of the tuning fork varies with time.

(iii) On your graph, label the time period T and the amplitude A.

(8 marks)

4 *(IB)*

a) Graph 1 below shows the variation with time t of the displacement d of a travelling (progressive) wave. Graph 2 shows the variation with distance x along the same wave of its displacement d.

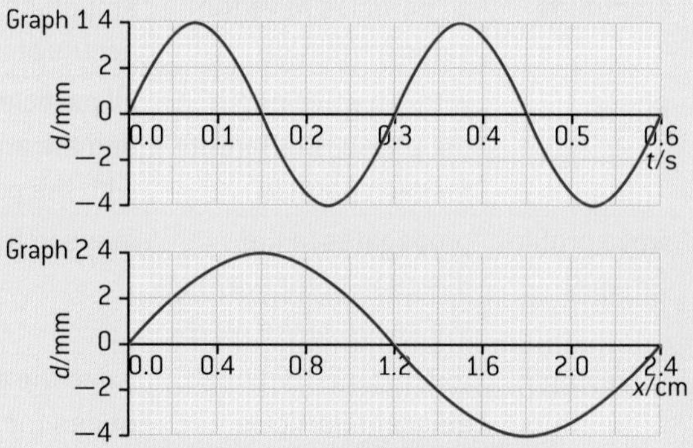

a) State what is meant by a *travelling wave*.

b) Use the graphs to determine the amplitude, wavelength, frequency and speed of the wave.

(5 marks)

5 *(IB)*

a) With reference to the direction of energy transfer through a medium, distinguish between a transverse wave and a longitudinal wave.

b) A wave is travelling along the surface of some shallow water in the *x*-direction. The graph shows the variation with time *t* of the displacement *d* of a particle of water.

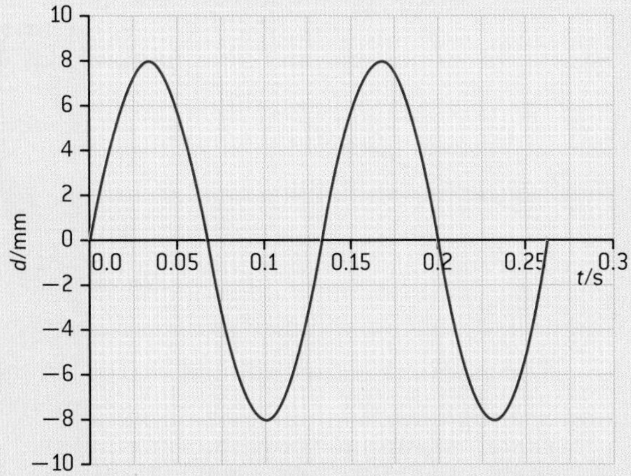

Use the graph to determine the frequency and the amplitude of the wave.

c) The speed of the wave in b) is 15 cm s^{-1}. Deduce that the wavelength of this wave is 2.0 cm.

d) The graph in b) shows the displacement of a particle at the position $x = 0$.

Draw a graph to show the variation with distance *x* along the water surface of the displacement *d* of the water surface at time $t = 0.070$ s.

(11 marks)

6 *(IB)*

a) By referring to the energy of a travelling wave, explain what is meant by:

(i) a ray

(ii) wave speed.

b) The following graph shows the variation with time *t* of the displacement x_A of wave A as it passes through a point P.

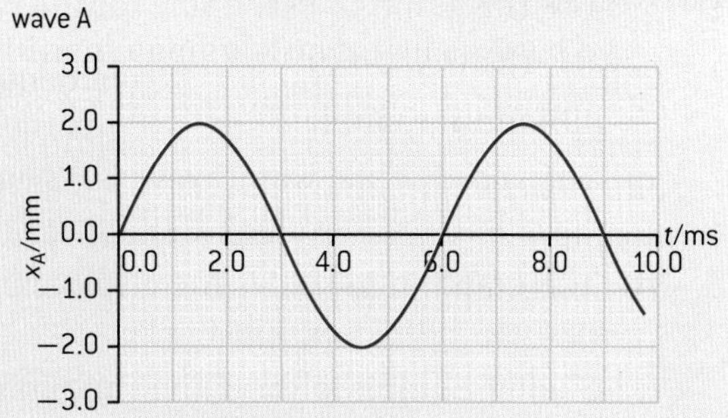

The graph below shows the variation with time *t* of the displacement x_B of wave B as it passes through point P. The waves have equal frequencies.

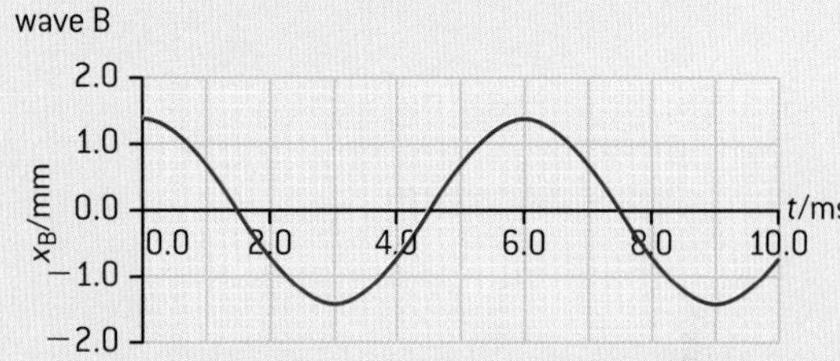

(i) Calculate the frequency of the waves.

(ii) The waves pass simultaneously through point P. Use the graphs to determine the resultant displacement at point P of the two waves at time $t = 1.0$ ms and at time $t = 8.0$ ms.

(6 marks)

7 *(IB)*

a) With reference to the direction of energy transfer through a medium, distinguish between a transverse wave and a longitudinal wave.

b) The graph shows the variation with time *t* of the displacement *d* of a particular water particle as a surface water wave travels through it.

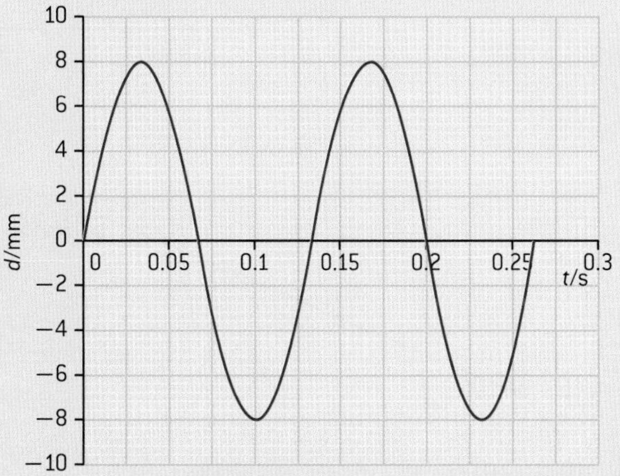

Use the graph to determine for the wave:

(i) the frequency

(ii) the amplitude.

c) The speed of the water wave is 12 cm s^{-1}. Calculate the wavelength of the wave.

d) The graph in b) shows the displacement of a particle at the position $x = 0$.

Sketch a graph to show the variation with distance x along the water surface of the displacement d of the water surface at time $t = 0.20$ s.

e) The wave meets a shelf that reduces the depth of the water.

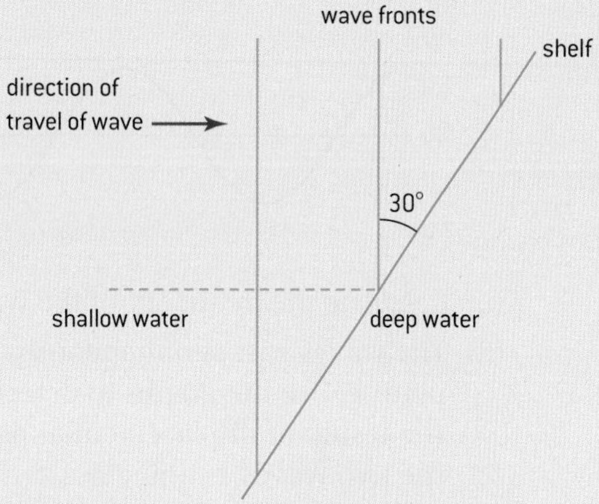

The angle between the wavefronts in the shallow water and the shelf is 30°. The speed of the wave in the shallow water is 12 cm s^{-1} and in the deeper water is 18 cm s^{-1}. For the wave in the deeper water, determine the angle between the normal to the wavefronts and the shelf.

(12 marks)

8 *(IB)*

a) A beam of unpolarized light of intensity I_0 is incident on a polarizer. The polarization axis of the polarizer is initially vertical as shown.

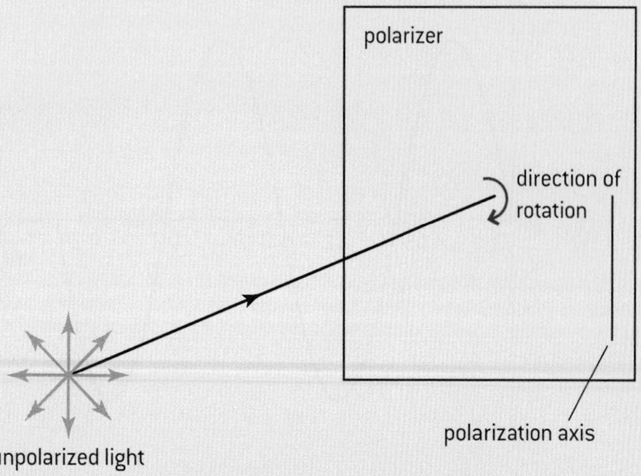

The polarizer is then rotated by 180° in the direction shown. Sketch a graph to show the variation with the rotation angle θ, of the transmitted light intensity I, as θ varies from 0° to 180°. Label your sketch-graph with the letter U.

b) The beam in a) is now replaced with a polarized beam of light of the same intensity.

The plane of polarization of the light is initially parallel to the polarization axis of the polarizer.

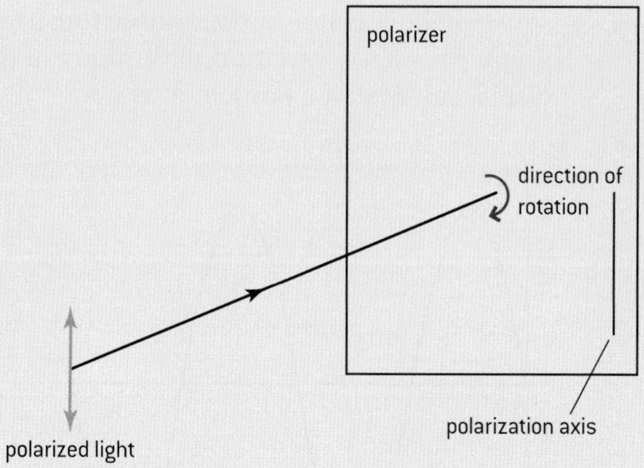

The polarizer is then rotated by 180° in the direction shown. On the same axes in a), sketch a graph to show the variation with the rotation angle θ, of the transmitted light intensity I, as θ varies from 0° to 180°.

(5 marks)

9 *(IB)*

An orchestra playing on boat X can be heard by tourists on boat Y, which is situated out of sight of boat X around a headland.

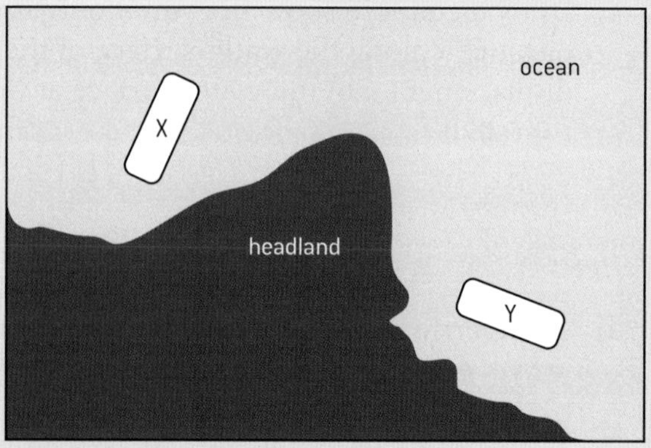

The sound from X can be heard on Y due to

A. refraction

B. reflection

C. diffraction

D. transmission.

10 *(IB)*

A small sphere, mounted at the end of a vertical rod, dips below the surface of shallow water in a tray. The sphere is driven vertically up and down by a motor attached to the rod.

The oscillations of the sphere produce travelling waves on the surface of the water.

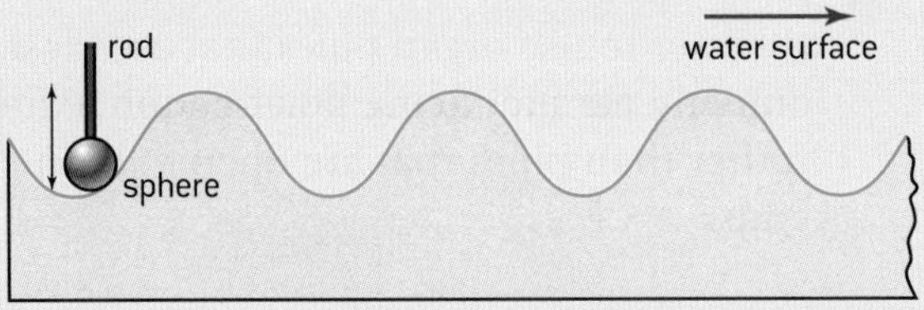

a) The diagram shows how the displacement of the water surface at a particular instant in time varies with distance from the sphere. The period of oscillation of the sphere is 0.027 s.

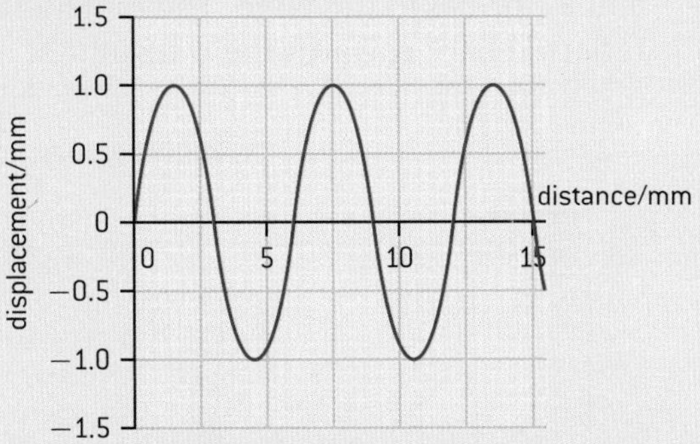

Use the diagram to calculate, for the wave:

(i) the amplitude

(ii) the wavelength

(iii) the frequency

(iv) the speed.

b) The wave moves from region A into a region B of shallower water. The waves move more slowly in region B. The diagram (not to scale) shows some of the wavefronts in region A.

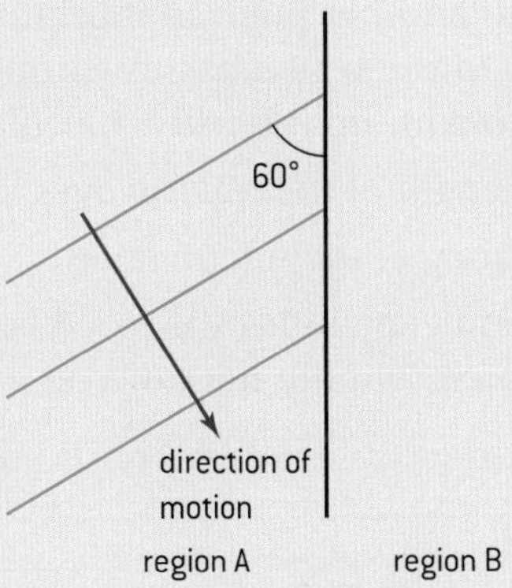

(i) With reference to a wave, distinguish between a ray and a wavefront.

(ii) The angle between the wavefronts and the interface in region A is 60°. The refractive index ${}_{A}n_{B}$ is 1.4. Determine the angle between the wavefronts and the interface in region B.

(iii) On the diagram above, construct *three* lines to show the position of three wavefronts in region B.

c) Another sphere is dipped into the water. The spheres oscillate in phase. The diagram shows some lines in region A along which the disturbance of the water surface is a minimum.

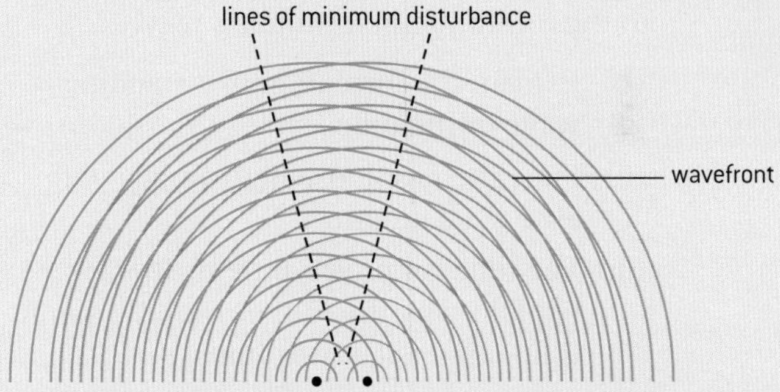

(i) Outline how the regions of minimum disturbance occur on the surface.

(ii) The frequency of oscillation of the spheres is increased. State *and* explain how this will affect the positions of minimum disturbance.

(15 marks)

11 *(IB)*

a) Describe *two* ways in which standing waves differ from travelling waves.

b) An experiment is carried out to measure the speed of sound in air, using the apparatus shown below.

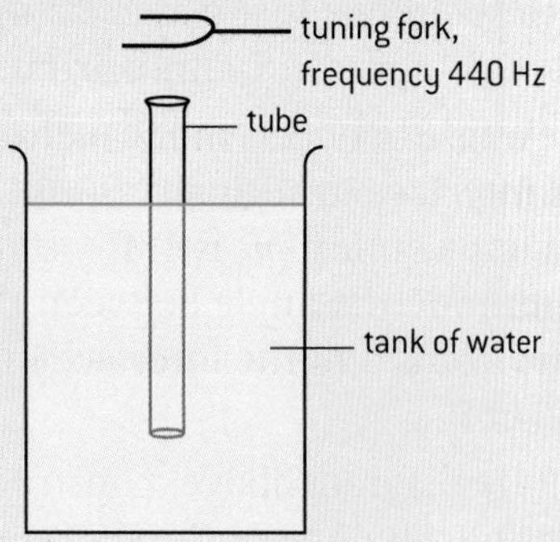

A tube that is open at both ends is placed vertically in a tank of water until the top of the tube is just at the surface of the water. A tuning fork of frequency 440 Hz is sounded above the tube. The tube is slowly raised out of the water until the loudness of the sound reaches a maximum for the first time, due to the formation of a standing wave.

(i) Explain the formation of a standing wave in the tube.

(ii) State the position where a node will always be produced.

(iii) The tube is raised a little further. Explain why the loudness of the sound is no longer at a maximum.

c) The tube is raised until the loudness of the sound reaches a maximum for a second time.

Between the two positions of maximum loudness the tube has been raised by 36.8 cm.

The frequency of the sound is 440 Hz. Estimate the speed of sound in air.

(10 marks)

12 *(IB)*

a) State *two* properties of a standing (stationary) wave.

b) The diagram shows an organ pipe that is open at one end.

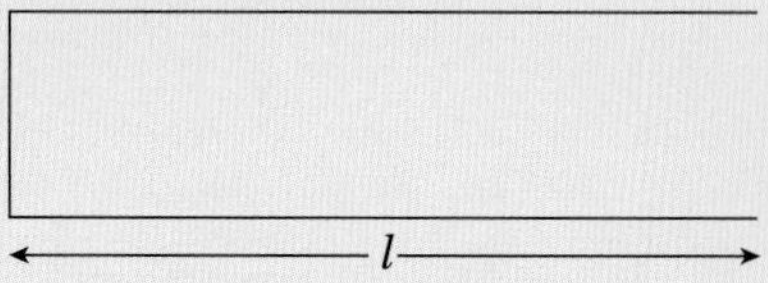

The length of the pipe is l. The frequency of the fundamental (first harmonic) note emitted by the pipe is 16 Hz.

(i) On a copy of the diagram, label with the letter P the position along the pipe where the amplitude of oscillation of the air molecules is the largest.

(ii) The speed of sound in the air in the pipe is 330 m s^{-1}. Calculate the length l.

c) Use your answer to b) (ii) to suggest why it is better to use organ pipes that are closed at one end for producing low frequency notes rather than pipes that are open at both ends.

(8 marks)

13 *(IB)*

A microwave transmitter emits radiation of a single wavelength towards a metal plate along a line normal to the plate. The radiation is reflected back towards the transmitter.

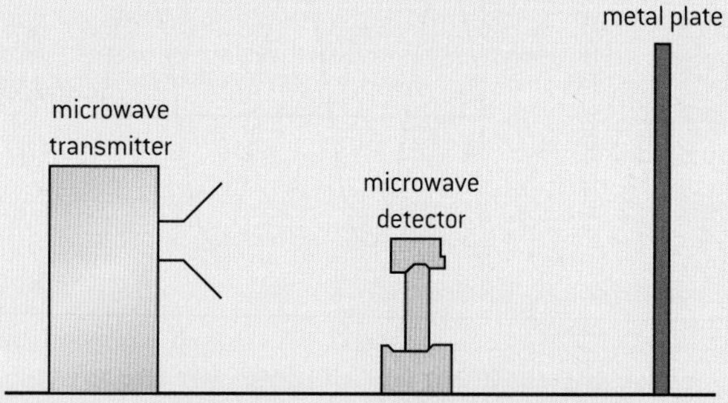

A microwave detector is moved along a line normal to the microwave transmitter and the metal plate. The detector records a sequence of equally spaced maxima and minima of intensity.

a) Explain how these maxima and minima are formed.

b) The microwave detector is moved through 130 mm from one point of minimum intensity to another point of minimum intensity. On the way it passes through nine points of maximum intensity. Calculate the

(i) wavelength of the microwaves.

(ii) frequency of the microwaves.

c) Describe and explain how it could be demonstrated that the microwaves are polarized.

(11 marks)

5 ELECTRICITY AND MAGNETISM

Introduction

Modern society uses a whole range of electrical devices from the simplest heated metal filaments that provide light, through to the most sophisticated medical instruments and computers. Devices of increasing technical complexity are developed every day.

In this topic we look at the phenomenon of electricity, and what is meant by charge and electric current. We consider the three effects that can be observed when charge flows in an electric circuit.

5.1 Electric fields

Understanding

- Charge
- Electric field
- Coulomb's law
- Electric current
- Direct current (dc)
- Potential difference (pd)

Nature of science

Electrical theory resembles the kinetic theory of gases in that a theory of the microscopic was developed to explain the macroscopic observations that had been made over centuries. The development of this subject and some of the byways that were taken make this a fascinating study. We should remember the many scientists who were involved. It is a tribute to them that they could make so much progress when the details of the microscopic nature of electronic charge were unknown to them.

Applications and skills

- Identifying two species of charge and the direction of the forces between them
- Solving problems involving electric fields and Coulomb's law
- Calculating work done when charge moves in an electric field in both joules and electronvolts
- Identifying sign and nature of charge carriers in a metal
- Identifying drift speed of charge carriers
- Solving problems using the drift-speed equation
- Solving problems involving current, potential difference, and charge

Equations

- current-charge relationship: $I = \frac{\Delta q}{\Delta t}$
- Coulomb's law: $F = k\frac{q_1 q_2}{r^2}$
- the coulomb constant: $k = \frac{1}{4\pi\varepsilon_0}$
- potential difference definition: $V = \frac{W}{q}$
- conversion of energy in joule to electron-volt: $W(\text{J}) \equiv \frac{W(eV)}{e}$
- electric field strength: $E = \frac{F}{q}$
- drift speed: $I = nAvq$

Charge and field

Simple beginnings

Take a plastic comb and pull it through your hair. Afterwards, the comb may be able to pick up small pieces of paper. Look closely and you may see the paper being thrown off shortly after touching the comb.

Similar observations were made early in the history of science. The discovery that objects can be charged by friction (you were doing this when you drew the comb through your hair) is attributed to the Greek scientist Thales who lived about 2600 years ago. In those days, silk was spun on amber spindles and as the amber rotated and rubbed against its bearings, the silk was attracted to the amber. The ancient Greek word for amber is ηλεκτρον (electron).

In the 1700s, du Fay found that both conductors and insulators could be "electrified" (the term used then) and that there were two opposite kinds of "electrification". However, he was unable to provide any explanation for these effects. Gradually, scientists developed the idea that there were two separate types of charge: positive and negative. The American physicist, Benjamin Franklin, carrying out a famous series of experiments flying kites during thunderstorms, named the charge on a glass rod rubbed with silk as "positive electricity". The charge on materials similar to ebonite (a very hard form of rubber) rubbed with animal fur was called "negative". One of Franklin's other discoveries was that a charged conducting sphere has no electric field inside it (the field and the charges always being outside the sphere). Joseph Priestley was able to deduce from this that the force between two charges is inversely proportional to the square of the distance between the charges.

At the end of the nineteenth century J. J. Thomson detected the presence of a small particle that he called the electron. Experiments showed that all electrons have the same small quantity of charge and that electrons are present in all atoms. Atoms were found to have protons that have the same magnitude of electronic charge as, but have opposite charge to, the electron. In a neutral atom or material the number of electrons and the number of protons are equal. We now assign a negative charge to the electron and positive to the proton. There are only these two species of charge.

Explaining electrostatics

Experiments also show that positively charged objects are attracted to negatively charged objects but repelled by any other positively charged object. The possible cases are summed up in figure 1. There can also be an attraction between charged and uncharged objects due to the separation of charge in the uncharged object. In figure 1(c) the electrons in the uncharged sphere are attracted to the positively charged sphere (sphere A) and move towards it. The electrons in sphere B are now closer to the positives in sphere A than the fixed positive charges on B. So the overall force is towards sphere A as the force between two charges increases as the distance between them decreases.

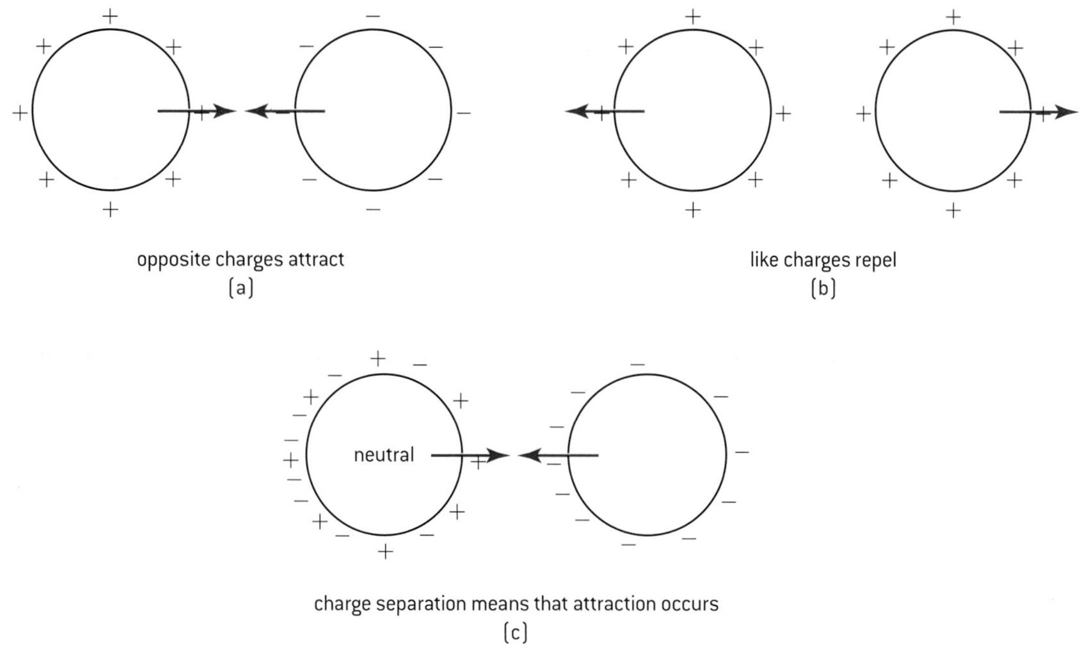
opposite charges attract
(a)
like charges repel
(b)
neutral
charge separation means that attraction occurs
(c)

▲ Figure 1 Attractions between charges.

We now know that the simple electrostatic effects early scientists observed are due only to the movement of the negatively charged electrons. An object with no *observed* charge has an exact balance between the electrons and the positively charged protons; it is said to be **neutral**. Some electrons in conducting materials are loosely attached to their respective atoms and can leave the atoms to move from one object to another. This leaves the object that lost electrons with an overall positive charge. The electrons transferred give the second object an overall negative charge. Notice that electrons are not lost in these transfers. If 1000 electrons are removed from a rod when the rod is charged by a cloth, the cloth will be left with 1000 extra electrons at the end of the process. **Charge is conserved**; the law of conservation of charge states that in a closed system the amount of charge is constant.

When explaining the effects described here, always use the idea of surplus of electrons for a negative charge, and describe positive charge in terms of a lack (or deficit) of electrons. Figure 2 shows how an experiment to charge a metal sphere by induction is explained.

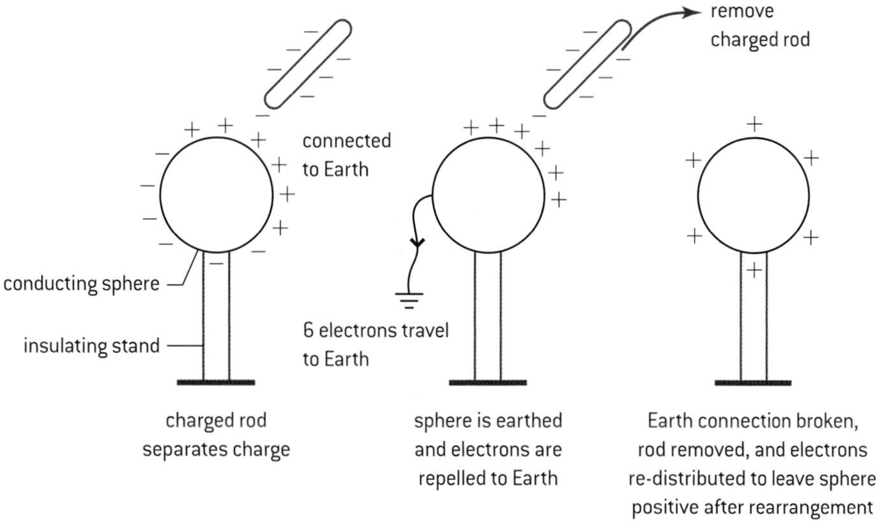
remove
charged rod
connected
to Earth
conducting sphere
insulating stand
6 electrons travel
to Earth
charged rod
separates charge
sphere is earthed
and electrons are
repelled to Earth
Earth connection broken,
rod removed, and electrons
re-distributed to leave sphere
positive after rearrangement

▲ Figure 2 Charging by induction.

TOK

Inverse-square laws

Forces between charged objects is one of several examples of inverse-square laws that you meet in this course. They are of great importance in physics. Inverse-square laws model a characteristic property of some fields, which is that as distance doubles, observed effects go down by one quarter.

Mathematics helps you to learn and conceptualize your ideas about the subject. When you have learnt the physics of one situation (here, electrostatics) then you will be able to apply the same rules to new situations (for example, gravitation, in the next topic).

Is the idea of field a human construct or does it reflect the reality of the universe?

Measuring and defining charge

The unit of charge is the coulomb (abbreviated to C). Charge is a scalar quantity.

The coulomb is defined as the charge transported by a current of one ampere in one second.

Measurements show that all electrons are identical, with each one having a charge equal to -1.6×10^{-19} C; this fundamental amount of charge is known as the **electronic (or elementary) charge** and given the symbol e.

Charges smaller than the electronic charge are not observed in nature. (Quarks have fractional charges that appear as $\pm\frac{1}{3}e$ or $\pm\frac{2}{3}e$; however, they are never observed outside their nucleons.)

In terms of the experiments described here, the coulomb is a very large unit. When a comb runs through your hair, there might be a charge of somewhere between 1 pC and 1 nC transferred to it.

Forces between charged objects

In 1785, Coulomb published the first of several *Mémoires* in which he reported a series of experiments that he had carried out to investigate the effects of forces arising from charges.

He found, experimentally, that the force between two point charges separated by distance r is proportional to $\frac{1}{r^2}$ thus confirming the earlier theory of Priestley. Such a relationship is known as an inverse-square law.

Investigate!

Forces between charges

- These are sensitive experiments that need care and a dry atmosphere to achieve a result.
- Take two small polystyrene spheres and paint them with a metal paint or colloidal graphite, or cover with aluminum foil. Suspend one from an insulating rod using an insulating (perhaps nylon) thread. Mount the other on top of a sensitive top-pan balance, again using an insulating rod.
- Charge both spheres by induction when they are apart from each other. An alternative charging method is to use a laboratory high voltage power supply. Take care, your teacher will want to give you instructions about this.
- Bring the spheres together as shown in figure 3 and observe changes in the reading on the balance.

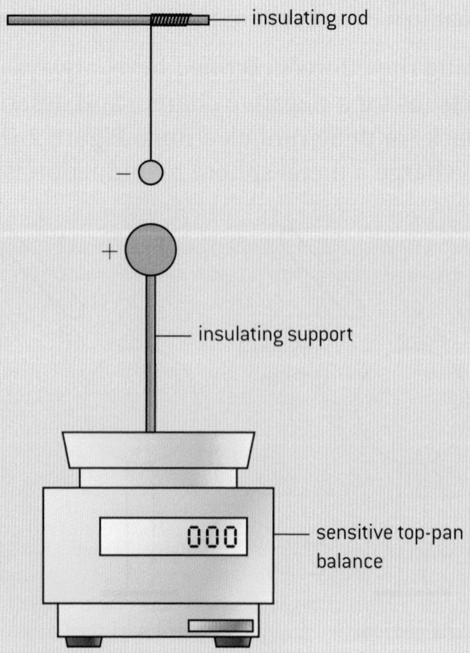

▲ Figure 3

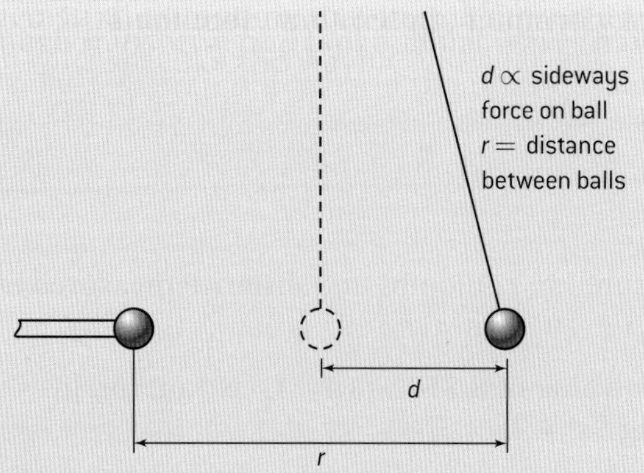

▲ Figure 4

- Another method is to bring both charged spheres together as shown in figure 4.
- The distance d moved by the sphere depends on the force between the charged spheres. The distance r is the distance between the centres of the spheres.
- Vary d and r making careful measurements of them both.
- Plot a graph of d against $\frac{1}{r^2}$. An experiment performed with care can give a straight-line graph.
- For small deflections, d is a measure of the force between the spheres (the larger the force the greater the distance that the sphere is moved) whereas r is the distance between sphere centres.

Nature of science

Scientists in Coulomb's day published their work in a very different way from scientists today. Coulomb wrote his results in a series of books called *Mémoires*. Part of Coulomb's original *Mémoire* in which he states the result is shown in figure 5.

Loi fondamentale de l'Éleċtricité

La force répulfive de deux petits globes éleċtrifés de la viéme nature d'éleċtricité, eft en raifon inverfe du carré de la diftance du centre des deux globes.

▲ Figure 5

Later experiments confirmed that the force is proportional to the product of the size of the point charges q_1 and q_2.

Combining Coulomb's results together with these gives

$$F \propto \frac{q_1 q_2}{r^2}$$

where the symbol $\propto$ means "is proportional to".

The magnitude of the force F between two point charges of charge q_1 and q_2 separated by distance r in a vacuum is given by

$$F = \frac{k q_1 q_2}{r^2}$$

where k is the constant of proportionality and is known as Coulomb's constant.

In fact we do not always quote the law in quite this mathematical form. The constant is frequently quoted differently as

$$k = \frac{1}{4\pi\varepsilon_0}$$

The new constant ε_0 is called the permittivity of free space (free space is an older term for "vacuum". The 4π is added to rationalize electric and magnetic equations – in other words, to give them a similar shape and to retain an important relationship between them (see the TOK section on page 177).

So the equation becomes

$$F = \frac{1}{4\pi\varepsilon_0}\frac{q_1 q_2}{r^2}$$

When using charge measured in coulombs and distance measured in metres, the value of k is 9×10^9 N m^2 C^{-2}.

This means that ε_0 takes a value of 8.854×10^{-12} C^2 N^{-1} m^{-2} or, in fundamental units, m^{-3} kg^{-1} s^4 A^2.

The equation as it stands applies only for charges that are in a vacuum. If the charges are immersed in a different medium (say, air or water) then the value of the permittivity is different. It is usual to amend the equation slightly too, k becomes $\frac{1}{4\pi\varepsilon}$ as the "0" subscript in ε_0 should only be used for the vacuum case. For example, the permittivity of water is 7.8×10^{-10} C^2 N^{-1} m^{-2} and the permittivity of air is 8.8549×10^{-12} C^2 N^{-1} m^{-2}. The value for air is so close to the free-space value that we normally use 8.85×10^{-12} C^2 N^{-1} m^{-2} for both. The table gives a number of permittivity values for different materials.

Material	Permittivity / 10^{-12} C^2 N^{-1} m^{-2}
paper	34
rubber	62
water	779
graphite	106
diamond	71

This equation appears to say nothing about the direction of the force between the charged objects. Forces are vectors, but charge and (distance)^2 are scalars. There are mathematical ways to cope with this, but for point charges the equation gives an excellent clue when the signs of the charges are included.

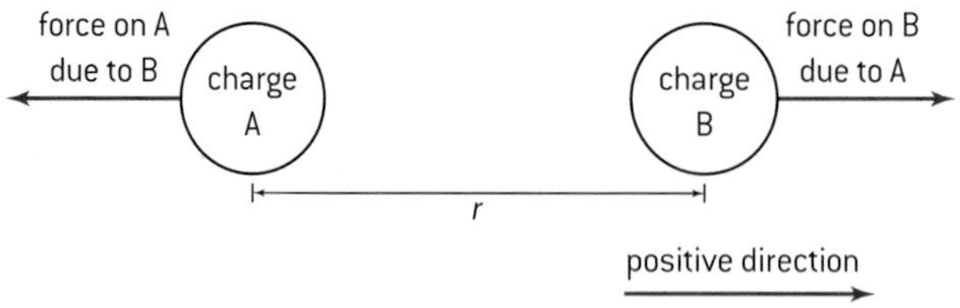

▲ Figure 6 Force directions.

We take the positive direction to be from charge A to charge B; in figure 6 that is from left to right. Let's begin with both charge A and charge B being positive. When two positive charges are multiplied together in $\frac{q_1 q_2}{r^2}$, the resulting sign of the force acting on charge B due to charge A is also positive. This means that the direction of the force will be assigned the positive direction (from charge A to charge B): in other words, left to right. This agrees with the physics because the charges are repelled. If both charges are negative then the answer is the same, the charges are repelled and the force is to the right.

If, however, one of the charges is positive and the other negative, then the product of the charges is negative and the force direction will be opposite to the left-to-right positive direction. So the force on charge 2 is now to the left. Again, this agrees with what we expect, that the charges attract because they have opposite signs.

Worked examples

1 Two point charges of +10 nC and −10 nC in air are separated by a distance of 15 mm.

a) Calculate the force acting between the two charges.

b) Comment on whether this force can lift a small piece of paper about 2 mm × 2 mm in area.

Solution

a) It is important to take great care with the prefixes and the powers of ten in electrostatic calculations.

The charges are: $+10 \times 10^{-9}$ C and -10×10^{-9} C. The separation distance is 1.5×10^{-2} m (notice how the distance is converted right at the outset into consistent units).

$$\text{So } F = \frac{(+1 \times 10^{-8}) \times (-1 \times 10^{-8})}{4\pi\varepsilon_0 (1.5 \times 10^{-2})^2}$$

$$= 4.0 \times 10^{-3}\ \text{N}$$

The charges are attracted along the line joining them. (Do not forget that force is a vector and needs both magnitude and direction for a complete answer.)

b) A sheet of thin A4 paper of dimensions 210 mm by 297 mm has a mass of about 2 g. So the small area of paper has a mass of about 1.3×10^{-7} kg and therefore a weight of 1.3×10^{-6} N. The electrostatic force could lift this paper easily.

2 Two point charges of magnitude +5 μC and +3 μC are 1.5 m apart in a liquid that has a permittivity of $2.3 \times 10^{-11}\ \text{C}^2\ \text{N}^{-1}\ \text{m}^{-2}$.

Calculate the force between the point charges.

Solution

$$F = \frac{(+5 \times 10^{-6}) \times (+3 \times 10^{-6})}{4\pi \times 2.3 \times 10^{-11} \times (1.5)^2} = 23\ \text{mN};$$

a repulsive force acting along the line joining the charges.

Electric fields

Sometimes the origin of a force between two objects is obvious, an example is the friction pad in a brake rubbing on the rim of a bicycle wheel to slow the cycle down. In other cases there is no physical contact between two objects yet a force exists between them. Examples of this include the magnetic force between two magnets and the electrostatic force between two charged objects. Such forces are said to "act at a distance".

The term **field** is used in physics for cases where two separated objects exert forces on each other. We say that in the case of the comb picking up the paper, the paper is sitting in the **electric field** due to the comb. The concept of the field is an extremely powerful one in physics not least because there are many ideas common to all fields. As well as the magnetic and electrostatic fields already mentioned, gravity fields also obey the same rules. Learn the underlying ideas for one type of field and you have learnt them all.

Mapping fields

Investigate!

Plotting electric fields

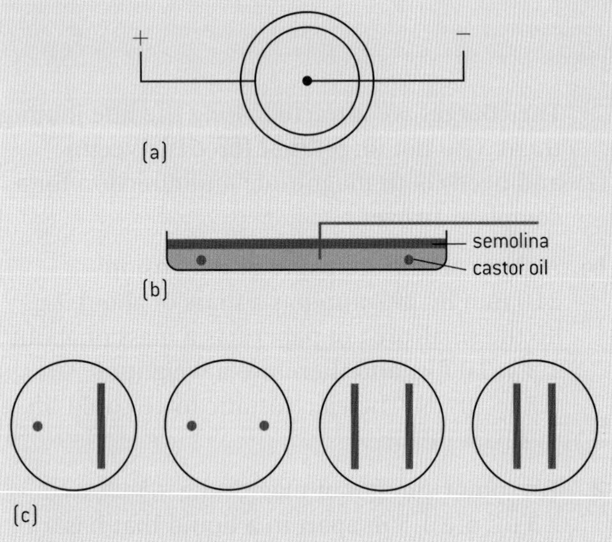

▲ Figure 7

- At an earlier stage in your school career you may have plotted magnetic field patterns using iron filings (if you have not done this there is an Investigate! in Sub-topic 5.4 to illustrate the method). This experiment allows patterns to be observed for electric fields.
- Put some castor oil in a Petri dish and sprinkle some grains of semolina (or grits) onto the oil. Alternatives for the semolina include grass seed and hairs cut about 1 mm long from an artist paint brush.
- Take two copper wires and bend one of them to form a circle just a little smaller than the internal diameter of the Petri dish. Place the end of the other wire in the centre of the Petri dish.
- Connect a 5 kV power supply to the wires – take care with the power supply!
- Observe the grains slowly lining up in the electric field.
- Sketch the pattern of the grains that is produced.
- Repeat with other wire shapes such as the four shown in figure 7(c).

In the plotting experiment, the grains line up in the field that is produced between the wires. The patterns observed resemble those in figure 8. The experiment cannot easily show the patterns for charges with the same sign.

The idea of field lines was first introduced by Michael Faraday (his original idea was of an elastic tube that repelled other tubes). Although field lines are imaginary, they are useful for illustrating and understanding the nature of a particular field. There are some conventions for drawing these electric field patterns.

- The lines start and end on charges of opposite sign.
- An arrow is essential to show the direction in which a positive charge would move (i.e. away from the positive charge and towards the negative charge).
- Where the field is strong the lines are close together. The lines act to repel each other.
- The lines never cross.
- The lines meet a conducting surface at 90°.

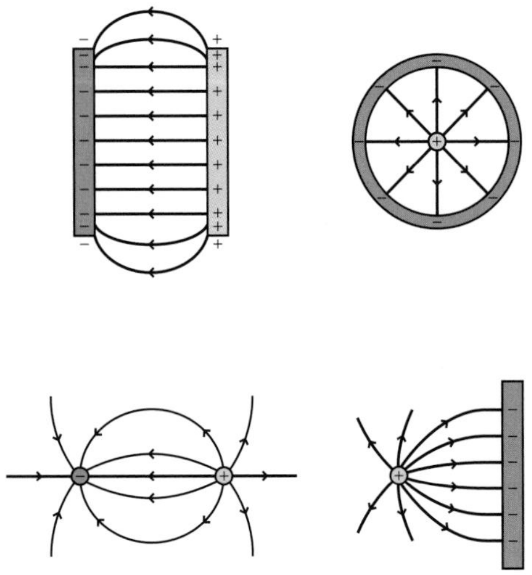

▲ Figure 8 Electric field patterns.

Electric field strength

As well as understanding the field pattern, we need to be able to measure the strength of the electric field. The **electric field strength** is defined using the concept of a **positive test charge**. Imagine an isolated charge Q sitting in space. We wish to know what the strength of the field is at a point P, a distance r away from the isolated charge. We put another charge, a positive test charge of size q, at P and measure the force F that acts on the test charge due to Q. Then the magnitude of the electric field strength is defined to be

$$E = \frac{F}{q}$$

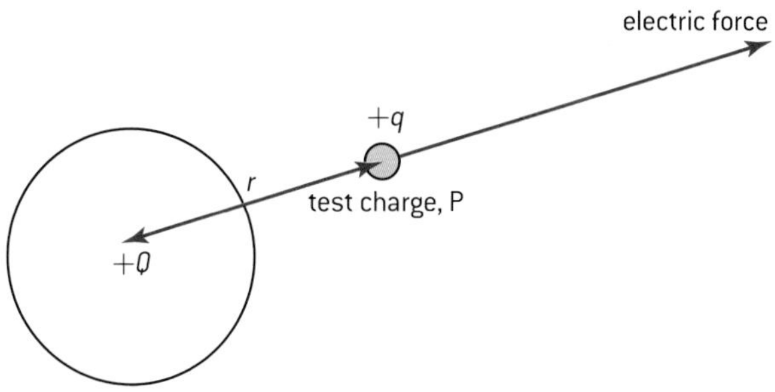

▲ Figure 9 Definition of electric field strength.

The units of electric field strength are N C^{-1}. (Alternative units, that have the same meaning, are V m^{-1} and will be discussed in Topic 11.) Electric field strength is a vector, it has the same direction as the force F (this is because the charge is a scalar which only "scales" the value of F up or down). **A formal definition for electric field strength at a point is the force per unit charge experienced by a small positive point charge placed at that point.**

Coulomb's law can be used to find how the electric field strength varies with distance for a point charge.

Q is the isolated point charge and q is the test charge, so

$$F = \frac{1}{4\pi\varepsilon_0}\frac{Qq}{r^2}$$

$$E = \frac{F}{q}$$

Therefore

$$E = \frac{1}{4\pi\varepsilon_0}\frac{Qq}{r^2} \times \frac{1}{q}$$

so

$$E = \frac{1}{4\pi\varepsilon_0}\frac{Q}{r^2}$$

The electric field strength of the charge at a point is proportional to the charge and inversely proportional to the square of the distance from the charge.

If Q is a positive charge then E is also positive. Applying the rule that r is measured from the charge to the test charge, then if E is positive it acts outwards away from the charge. This is what we expect as both charge and test charge are positive. When Q is negative, E acts towards the charge Q.

The field shape for a point charge is known as a **radial field**. The field lines radiate away (positive) or towards (negative) the point charge as shown in figure 10.

TOK

So why not use *k*?

James Maxwell, working in the middle of the nineteenth century, realized that there was an important connection between electricity, magnetism and the speed of light. In particular he was able to show that the permittivity of free space ε_0 (which relates to electrostatics) and the permeability of free space μ_0 (which relates to electromagnetism) are themselves connected to the speed of light c:

$$\frac{1}{\varepsilon_0\mu_0} = c^2$$

This proves to be an important equation. So much so that, in the set of equations that arise from the SI units we use, we choose to use ε_0 in all the electrostatics equations and μ_0 in all the magnetic equations.

However, not all unit systems choose to do this. There is another common system, the cgs system (based on the centimetre, the gram and the second, rather than the metre, kilogram and second). In cgs, the value of the constant k in Coulomb's law is chosen to be 1 and the equation appears as $F = \frac{q_1 q_2}{r^2}$. If the numbers are different, is the physics the same?

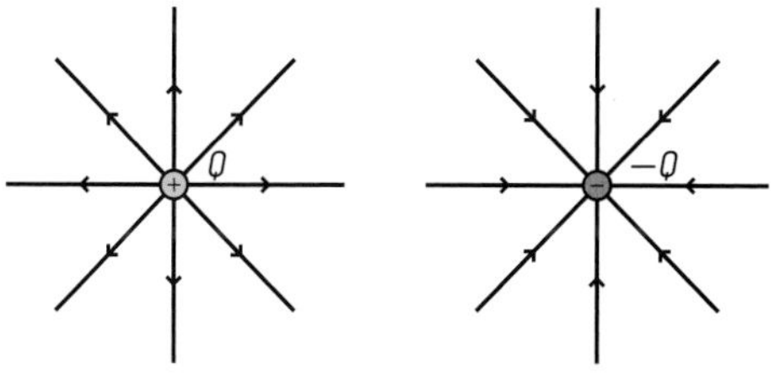

▲ Figure 10 Radial fields for positive and negative point charges.

Worked examples

1 Calculate the electric field strengths in a vacuum

a) 1.5 cm from a +10 μC charge

b) 2.5 m from a –0.85 mC charge

Solution

a) Begin by putting the quantities into consistent units: $r = 1.5 \times 10^{-2}$ m and $Q = 1.0 \times 10^{-5}$ C.

Then $E = \frac{1.0 \times 10^{-5}}{4\pi\varepsilon_0 (1.5 \times 10^{-2})^2} = +\ 4.0 \times 10^{8}$ N C^{-1}.

The field direction is away from the positive charge.

b) $E = \frac{-8.5 \times 10^{-4}}{4\pi\varepsilon_0 (2.5)^2} = -1.2 \times 10^{6}$ N C^{-1}

The field direction is towards the negative charge.

2 An oxygen nucleus has a charge of $+8e$. Calculate the electric field strength at a distance of 0.68 nm from the nucleus.

Solution

The charge on the oxygen nucleus is $8 \times 1.6 \times 10^{-19}$ C; the distance is 6.8×10^{-10} m.

$E = \frac{1.3 \times 10^{-18}}{4\pi\varepsilon_0 \times (6.8 \times 10^{-10})^2} = -2.5 \times 10^{10}$ N C^{-1} away from the nucleus.

The electric field strengths can be added using either a calculation or a scale diagram as outlined in Topic 1.

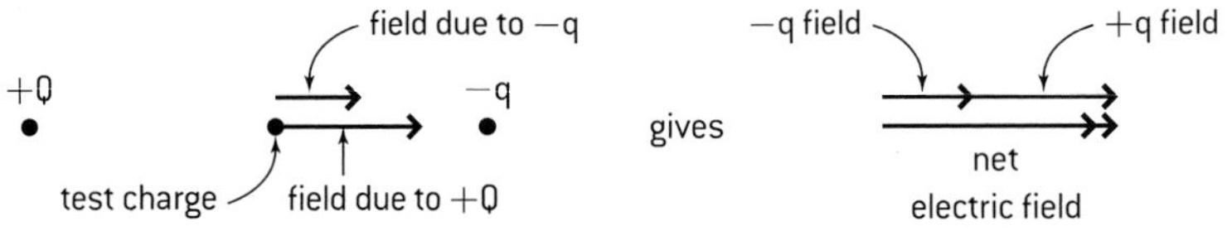

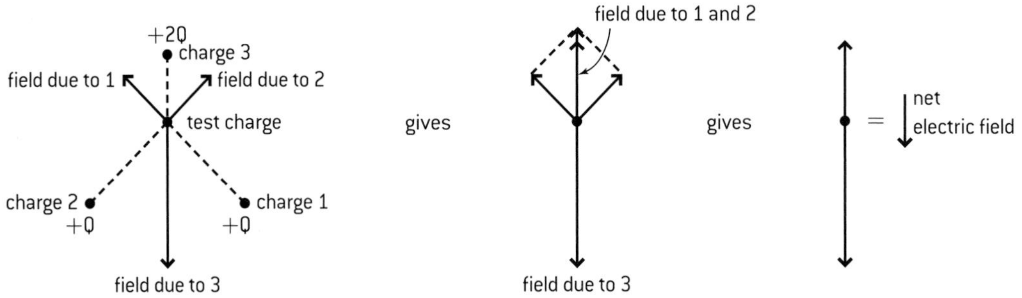

▲ Figure 11 Vector addition of electric fields.

This vector addition of field strengths (figure 11) can give us an insight into electric fields that arise from charge configurations that are more complex than a single point charge.

Close to a conductor

Imagine going very close to the surface of a conductor. Figure 12 shows what you might see. First of all, if we are close enough then the surface will appear flat (in just the same way that we are not aware of the curvature of the Earth until we go up in an aircraft). Secondly we would see that all the free electrons are equally spaced. There is a good reason for this: any one electron has forces acting on it from the other electrons. The electron will accelerate until all these forces balance out and it is in equilibrium, for this to happen they must be equally spaced.

Now look at the field strength vectors radiating out from these individual electrons. Parallel to the surface, these all cancel out with each other so there is no electric field in this direction. (This is the same as saying that any one electron will not accelerate as the horizontal field strength is

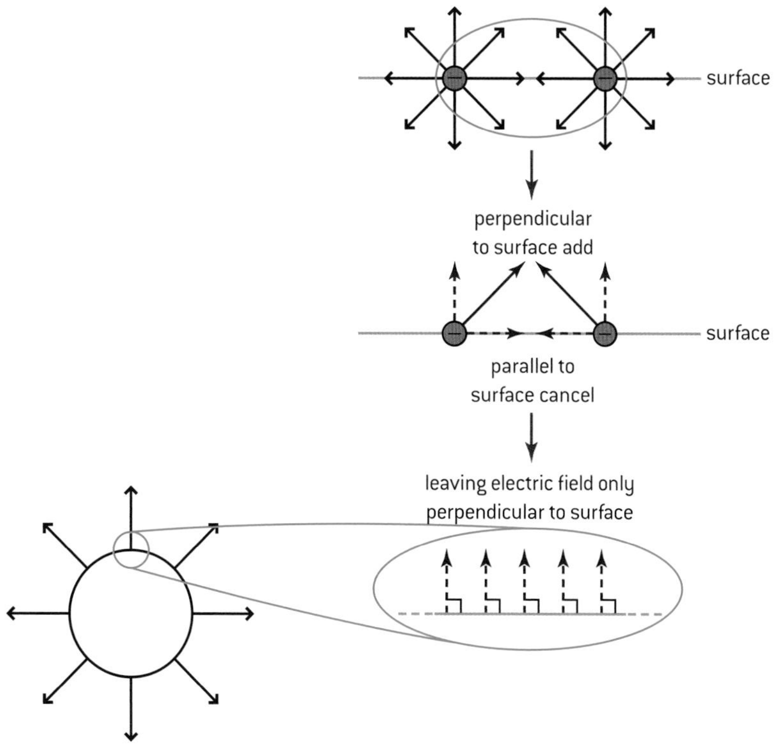

▲ Figure 12 Close to conducting surfaces.

zero). Perpendicular to the surface, however, things are different. Now the field vectors all add up, and because there is no field component parallel to the surface, the local field must act at 90° to it.

So, close to a conducting surface, the electric field is at 90° to the surface.

Conducting sphere

This can be taken a step further for a conducting sphere, whether it is hollow or solid. Again, the free electrons at the surface are equally spaced and all the field lines at the surface of the sphere are at 90° to it. The consequence is that the field must be radial, just like the field of an isolated point charge. So, to a test charge outside the sphere, the field of the sphere appears exactly the same as that of the point charge. Mathematical analysis shows that outside a sphere the field indeed behaves as though it came from a point charge placed at the centre of the sphere with a charge equal to the total charge spread over the sphere.

(Inside the sphere is a different matter, it turns out that there is no electric field inside a sphere, hollow or solid, a result that was experimentally determined by Franklin.)

TOK

But does the test charge affect the original field?

If we wanted to use the definition of the strength of an electric field practically by using a test charge, we would have to take care. Just as a thermometer alters the temperature of the object it measures, so the test charge will exert a force on the original charge (the one with the field we are trying to measure) and may accelerate it, or disturb the field lines. This means in practice that a test charge is really an imaginary construct that we use to help our understanding. In practice we prefer to measure electric field strengths using the idea that it is the potential gradient, in other words the change in the voltage divided by the change in distance: the larger this value, the greater the field strength. We will leave further discussion of this point to Topic 11 but it is food for thought from a theory of knowledge perspective: a measurement that we can think about but not, in practice, carry out. The German language has a name for it: *gedankenexperiment* "thought experiment". How can a practical subject such as a science have such a thing?

Worked example

Two point charges, a +25 nC charge X and a +15 nC charge Y are separated by a distance of 0.5 m.

a) Calculate the resultant electric field strength at the midpoint between the charges.

b) Calculate the distance from X at which the electric field strength is zero.

c) Calculate the magnitude of the electric field strength at the point P on the diagram. X and Y are 0.4 m and 0.3 m from P respectively.

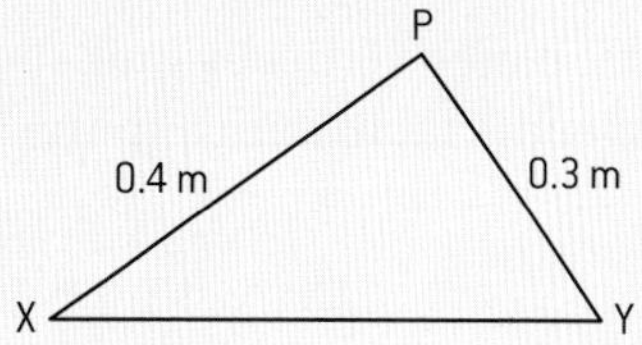

Solution

a) $E_A = \dfrac{2.5 \times 10^{-8}}{4\pi\varepsilon_0 \times 0.25^2} = 3600 \text{ N C}^{-1}$

$E_B = \dfrac{1.5 \times 10^{-8}}{4\pi\varepsilon_0 \times 0.25^2} = 2200 \text{ N C}^{-1}$

The field strengths act in opposite directions, so the net electric field strength is $(3600 - 2200) = 1400 \text{ N C}^{-1}$; this is directed away from X towards Y.

b) For E to be zero, $E_A = -E_B$ and so

$$\frac{2.5 \times 10^{-8}}{4\pi\varepsilon_0 \times d^2} = \frac{1.5 \times 10^{-8}}{4\pi\varepsilon_0 \times (0.5 - d)^2}$$

thus

$$\frac{d^2}{(0.5 - d)^2} = \frac{2.5}{1.5}$$

or

$$\frac{d}{(0.5 - d)} = \sqrt{\frac{2.5}{1.5}} = 1.3$$

$$d = 0.65 - 1.3d$$

$$2.3d = 0.65$$

$$d = 0.28 \text{ m}$$

c) PX = 0.4 m so E_X at P is $\dfrac{2.5 \times 10^{-8}}{4\pi\varepsilon_0 \times 0.4^2}$

$= 1400 \text{ N C}^{-1}$ along XP in the direction away from X.

PY = 0.3 m so E_Y at P is $\dfrac{1.5 \times 10^{-8}}{4\pi\varepsilon_0 \times 0.3^2}$

$= 1500 \text{ N C}^{-1}$ along PY in the direction towards Y.

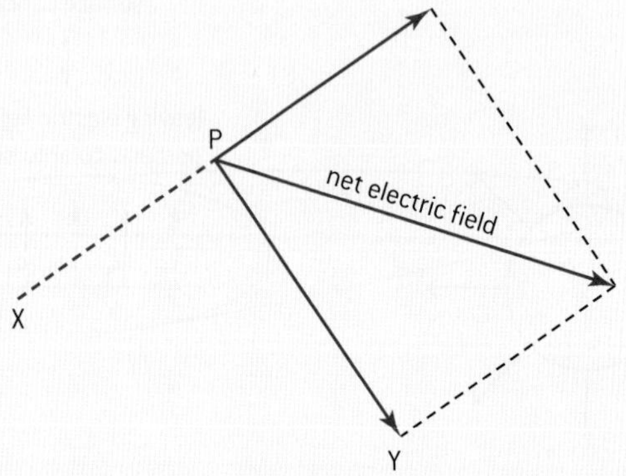

The magnitude of the resultant electric field strength is $\sqrt{1400^2 + 1500^2} = 2100 \text{ N C}^{-1}$

(The calculation of the angles was not required in the question and is left for the reader.)

Moving charge

Investigate!

Moving charge around

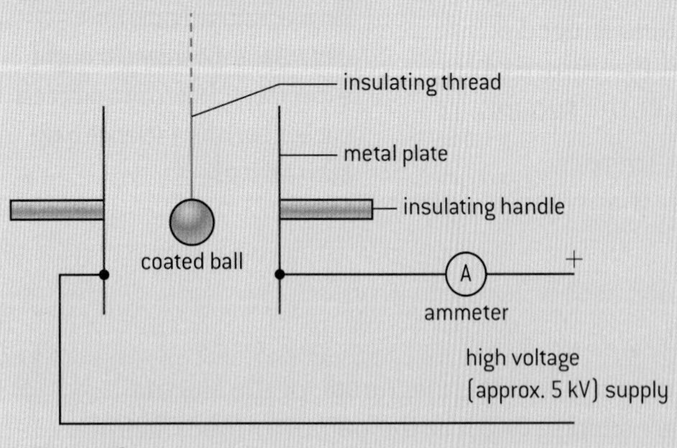

▲ Figure 13

- Arrange two metal plates vertically and about 20 cm apart. Suspend a table tennis (ping-pong) ball with a painted metal surface from a long insulating thread so that it is midway between the plates.
- Connect the plates to a high-voltage power supply with a sensitive ammeter or light-beam galvanometer in the circuit.
- When the supply is turned on, the ball should "shuttle" between the plates (it may need to be kickstarted by making it touch one of the plates).
- Notice what happens on the scale of the galvanometer. Every time the ball touches a plate there is a deflection on the meter. Look at the direction and size of the deflections – do they vary?

In the previous *Investigate!* the power supply is connected to the plates with conducting leads. Electrons move easily along these leads. When the supply is turned on, the electrons soon distribute themselves so that the plate connected to the negative supply has surplus electrons, and the other plate has a deficit of electrons, becoming positive. As the ball touches one of the plates it loses or gains some of the electrons. The charge gained by the ball will have the same sign as the plate and as a result the ball is almost immediately repelled. A force acts on the ball because it is in the electric field that is acting between the plates. The ball accelerates towards the other plate where it transfers all its charge to the new plate and gains more charge. This time the charge gained has the sign of the new plate. The process repeats itself with the ball transferring charge from plate to plate.

The meter is a sensitive ammeter, so when it deflects it shows that there is current in the wires leading to the plates. Charge is moving along these wires, so this is evidence that:

- an electric current results when charge moves
- the charge is moved by the presence of an electric field.

A mechanism for electric current

The shuttling ball and its charge show clearly what is moving in the space between the plates. However, the microscopic mechanisms that are operating in wires and cells are not so obvious. This was one of the major historical problems in explaining the physics of electricity.

Electrical conduction is possible in gases, liquids, solids, and a vacuum. Of particular importance to us is the electrical conduction that takes place in metals.

Conduction in metals

The metal atoms in a solid are bound together by the **metallic bond**. The full details of the bonding are complex, but a simple model of what happens is as follows.

When a metal solidifies from a liquid, its atoms form a regular lattice arrangement. The shape of the lattice varies from metal to metal but the common feature of metals is that as the bonding happens, electrons are donated from the outer shells of the atoms to a common sea of electrons that occupies the entire volume of the metal.

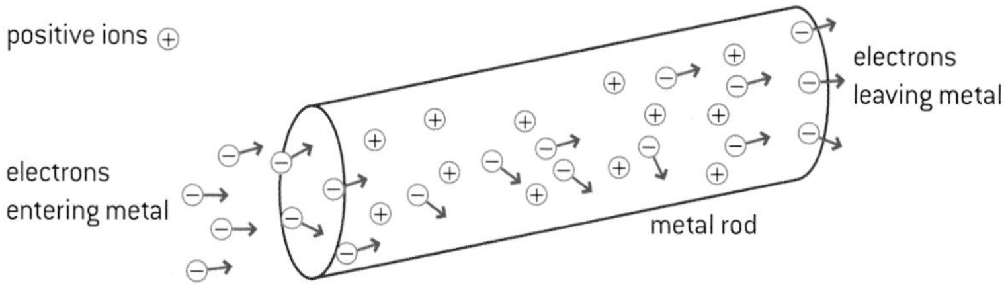

▲ Figure 14 Conduction by free electrons in a metal.

Figure 14 shows the model. The positive ions sit in fixed positions on the lattice. There are ions at each lattice site because each atom has now lost an electron. Of course, at all temperatures above absolute zero they vibrate in these positions. Most of the electrons are still bound to them but around the ions is the sea of free electrons or **conduction electrons**; these are responsible for the electrical conduction.

Although the conduction electrons have been released from the atoms, this does not mean that there is no interaction between ions and electrons. The electrons interact with the vibrating ions and transfer their kinetic energy to them. It is this transfer of energy from electrons to ions that accounts for the phenomenon that we will call "resistance" in the next sub-topic.

The energy transfer in a conductor arises as follows:

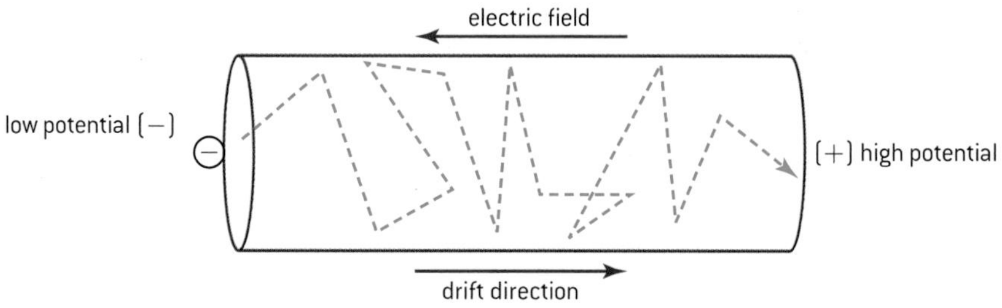

▲ Figure 15

In a metal in the absence of an electric field, the free electrons are moving and interacting with the ions in the lattice, but they do so at random and at average speeds close to the speed of sound in the material. Nothing in the material makes an electron move in any particular direction.

However, when an electric field is present, then an electric force will act on the electrons with their negative charge. The definition of electric field direction reminds us that the electric field is the direction in which a positive charge moves, so the force on the electrons will be in the opposite direction to the electric field in the metal (figure 15).

In the presence of an electric field, the negatively charged electrons drift along the conductor. The electrons are known as **charge carriers**. Their movement is like the random motion of a colony of ants carried along a moving walkway.

Conduction in gases and liquids

Electrical conduction is possible in other materials too. Some gases and liquids contain free ions as a consequence of their chemistry. When an electric field is applied to these materials the ions will move, positive in the direction of the field, negative the opposite way. When this happens an electric current is observed.

If the electric field is strong enough it can, itself, lead to the creation of ions in a gas or liquid. This is known as **electrical breakdown**. It is a common effect during electrical storms when lightning moves between a charged cloud and the Earth. You will have seen such conduction in neon display tubes or fluorescent tubes use for lighting.

Nature of science

Models of conduction

The model here is of a simple flow of free electrons through a solid, a liquid, or a gas. But this is not the end of the story. There are other, more sophisticated models of conduction in solids that can explain the differences between conductors, semiconductors and insulators better than the flow model here. These involve the electronic band theory which arises from the interactions between the electrons within individual atoms and between the atoms themselves.

Essentially, this band model proposes that electrons have to adopt different energies within the substance and that some groups of energy levels (called band gaps) are not permitted to the electrons. Where there are wide band gaps, electrons cannot easily move from one set of levels to another and this makes the substance an insulator. Where the band gap is narrow, adding energy to the atomic structure allows electrons to jump across the band gap and conduct more freely – this makes a semiconductor, and you will later see that one of the semiconductor properties is that adding internal energy allows them to conduct better. In conductors the band gap is of less relevance because the electrons have many available energy states and so conduction happens very easily indeed.

Full details of this theory are beyond IB Physics, but if you have an interest in taking this further, you can find many references to the theory on the Internet.

Electric current

When charge flows in a conductor we say that there is an electric current in the conductor. Current is measured in **ampères**, the symbol for the unit is A. Often, in the English speaking world, the accent is omitted.

Current is linked to flow of charge in a simple way.

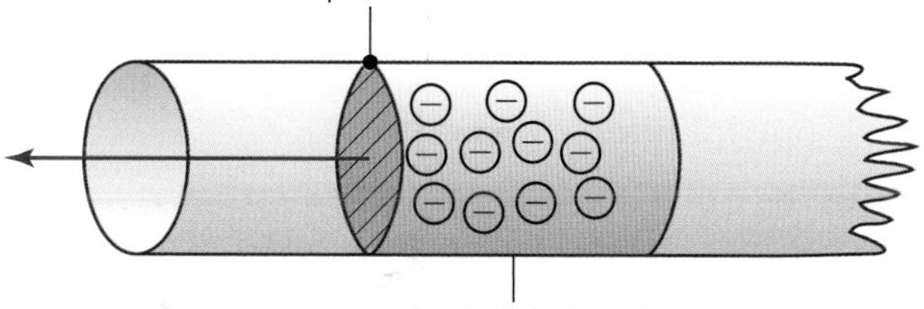

▲ Figure 16 Charge flow leading to current.

Imagine a block of electrons with a total charge of one coulomb moving along a conductor.

An observer at point P is watching these electrons move along the conductor. If all the electrons in the block move past the point in one second then, the current is one ampere.

If it takes twice as long (2 s) for the block to pass, then the current is half and is 0.5 A.

If the block takes 0.1 s to pass the observer, then the current is 10 A.

Mathematically

$$\text{electric current, } I = \frac{\textit{total charge that moved past a point}}{\textit{time taken for charge to move past the point}}$$

Tip

It is not good practice to write or say that "current flows", what is flowing in the circuit is the electric charge. The movement of this charge is what we call current and it is best to write that "there is a current in the circuit, or in a component" as appropriate.

or in symbols

$$I = \frac{\Delta Q}{\Delta t}$$

The ampere is a **fundamental unit** defined as part of the SI. Although it is *explained* here in terms of the flow of charge, this is not how it is *defined*. The SI definition is based on ideas from magnetism and is covered in Sub-topics 1.1 and 5.4.

Nature of science

Another physics link

This link between flowing charge and current is a crucial one. Electrical current is a macroscopic quantity, transfer of charge by electrons is a microscopic phenomenon in every sense of the word. This is another example of a link in physics between macroscopic observations and inferences about what is happening on the smallest scales.

It was the lack of knowledge of what happens inside conductors at the atomic scale that forced scientists, up to the end of the nineteenth century, to develop concepts such as current and field to explain the effects they observed. It also, as we shall see, led to a crucial mistake.

Worked examples

1 In the shuttling ball experiment, the ball moves between the two charged plates at a frequency of 0.67 Hz. The ball carries a charge of magnitude 72 nC each time it crosses from one plate to the other.

Calculate:

a) the average current in the circuit

b) the number of electrons transferred each time the ball touches one of the plates.

Solution

a) The time between the ball being at the same plate $= \frac{1}{f} = \frac{1}{0.67} = 1.5$ s. The time to transfer 72 nC is therefore 0.75 s.

Current $= \frac{7.2 \times 10^{-8}}{0.75} = 96$ nA

b) The charge transferred is 72 nC $= 7.2 \times 10^{-8}$ C

Each electron has a charge of -1.6×10^{-19} C, so the number of electrons involved in the transfer is $\frac{7.2 \times 10^{-8}}{1.6 \times 10^{-19}} = 4.5 \times 10^{11}$

2 **a)** Calculate the current in a wire through which a charge of 25 C passes in 1500 s.

b) The current in a wire is 36 mA. Calculate the charge that flows along the wire in one minute.

Solution

a) $I = \frac{\Delta Q}{\Delta t}$, so the current $= \frac{25}{1500} = 17$ mA

b) $\Delta Q = I\Delta t$ and $\Delta t = 60$ s. Thus charge that flows $= 3.6 \times 10^{-2} \times 60 = 2.2$ C

Charge carrier drift speed

Turn on a lighting circuit at home and the lamp lights almost immediately. Does this give us a clue to the speed at which the electrons in the wires move? In the *Investigate!* experiment on page 185, the stain indicating the position of the ions moves at no more than a few millimetres per second. The lower the value of the current, the slower the rate at which the ions move. This slow speed at which the ions move along the conductor is known as the **drift speed**.

We need a mathematical model to confirm this observation.

Imagine a cylindrical conductor that is carrying an electric current I. The cross-sectional area of the conductor is A and it contains charge carriers each with charge q. We assume that each carrier has a speed v and that there are n charge carriers in 1 m^3 of conductor – this quantity is known as the **charge density**.

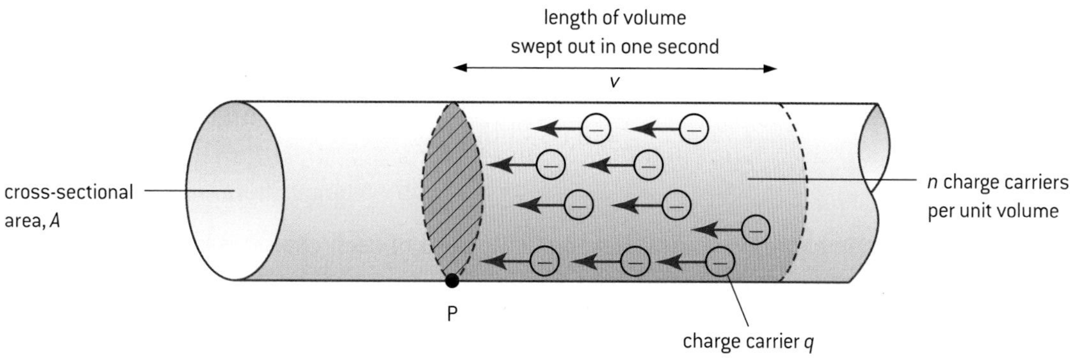

▲ Figure 18 A model for conduction.

Figure 18 shows charge carriers, each of charge q, moving past point P at a speed v.

In one second, a volume Av of charge carriers passes P.

The total number of charge carriers in this volume is nAv and therefore the total charge in the volume is $nAvq$.

However, this is the charge that passes point P in one second, which is what we mean by the electric current. So

$$I = nAvq$$

Investigate!

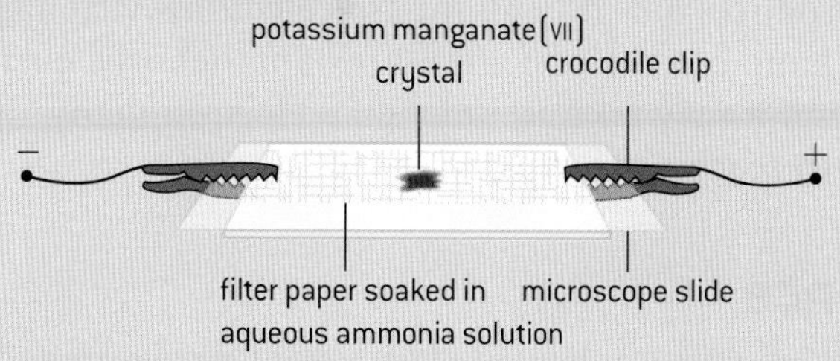

▲ Figure 17

- The speeds with which electrons move in a metal conductor during conduction are difficult to observe, but the progress of conducting ions in a liquid can be inferred by the trace they leave.
- You should wear eye protection during this experiment.
- Fold a piece of filter paper around a microscope slide and fix it with two crocodile clips at the ends of the slide. The crocodile clips should be attached to leads that are connected to a low voltage power supply (no more than 25 V is required in this experiment).
- Wet the filter paper with aqueous ammonia solution.
- Take a small crystal of potassium manganate(VII) and place it in the centre of the filter paper. Ensure that the slide is horizontal.
- Turn on the current and watch the crystal. You should see a stain on the paper moving away from the crystal.
- Reverse the current direction to check that the effect is not due to the slide not being horizontal.
- How fast is the stain moving?

Worked example

1 A copper wire of diameter 0.65 mm carries a current of 0.25 A. There are 8.5×10^{28} charge carriers in each cubic metre of copper; the charge on each charge carrier (electron) is 1.6×10^{-19} C. Calculate the drift speed of the charge carriers.

Solution

Rearranging the equation

$$v = \frac{I}{nAQ}$$

and the area A of the wire is $\pi\left(\frac{0.65 \times 10^{-3}}{2}\right)^2$ $= 3.3 \times 10^{-7}$ m^2.

So $v = \frac{0.25}{8.5 \times 10^{28} \times 3.3 \times 10^{-7} \times 1.6 \times 10^{-19}}$

$= 0.055$ mm s^{-1}

The example shows that the drift speed of each charge is less than one-tenth of a millimetre each second. You may well be surprised by this result – but it is probably of the same order as the speed you observed in the experiment with the potassium manganate(VII). Although the electron charge is very small, the speed can also be small because there are very large numbers of free electrons available for conduction in the metal.

To see how sensitive the drift speed is to changes in the charge carrier density, we can compare the drift speed in copper with the drift speed in a semiconductor called germanium. The number of charge carriers in one cubic metre of germanium is about 10^9 less than in copper. So to sustain the same current in a germanium sample would require a drift speed 10^9 times greater than in the copper or a cross-sectional area 10^9 as large.

The slow drift speed in conductors for substantial currents poses the question of how a lamp can turn on when there may be a significant run of cable between switch and lamp. The charge carriers in the cable are drifting slowly around the cable. However, the information that the charge carriers are to begin to move when the switch is closed travels much more quickly – close to the speed of light in fact. The information is transferred when an electromagnetic wave propagates around the cable and produces a drift in all the free electrons virtually simultaneously. So the lamp can turn on almost instantaneously, even though, for direct current, it may take an individual electron many minutes or even hours to reach the lamp itself.

Potential difference

Free electrons move in a conductor when an electric field acts on the conductor. Later we shall see how devices such as electric cells and power supplies provide this electric field. At the same time the power supplies transfer energy to the electrons. As the electrons move through the conductors, they collide with the positive ions in the lattice and transfer the energy gained from the field to the ions.

In situations where fields act, physicists use two quantities called potential and **potential difference** when dealing with energy transfers. Potential difference (often abbreviated to "pd") is a measure of the electrical potential energy transferred from an electron when it is moving between two points in a circuit. However, given the very small amount of charge possessed by each electron this amount of energy is also very small. It is better to use the much larger quantity represented by one coulomb of charge.

Potential difference between two points is defined as the work done (energy transferred) W when one unit of charge Q moves between the points.

$$\text{potential difference} = \frac{W}{Q}$$

The symbol given to potential difference is V; its unit is the J C^{-1} and is named the volt (symbol: V) after the Italian scientist Alessandro Volta who was born in the middle of the eighteenth century and who worked on the development of electricity.

The potential difference between two points is one volt if one joule of energy is transferred per coulomb of charge passing between the two points.

A simple circuit will illustrate these ideas:

An electric cell is connected to a lamp via a switch and three leads. Figure 19 shows a picture of the circuit as it would look set up on the bench.

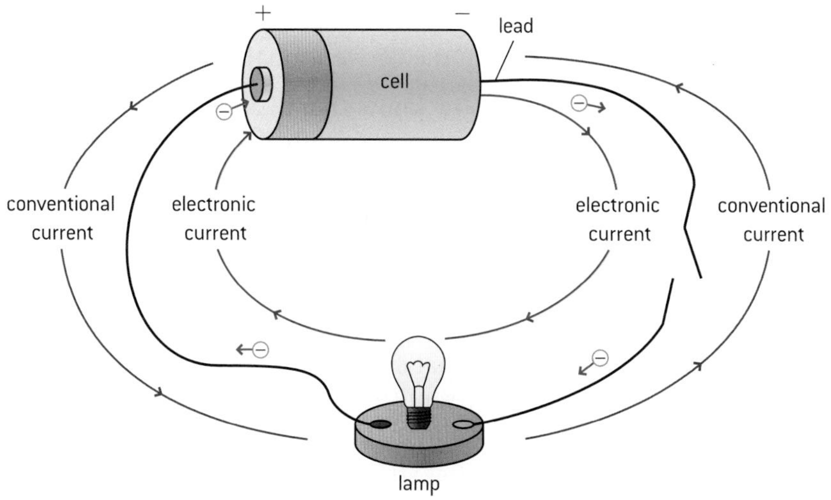

▲ Figure 19 Conventional and electronic current in a circuit.

When the switch is closed, electrons flow round the circuit. Notice the direction in which the electrons move and also that the diagram shows the direction of a **conventional current**. The two directions are opposite; in this case, clockwise for the electron flow and anti-clockwise for the conventional current. The reason for this difference is explained in a later Nature of science section. You need to take care with this difference, particularly when using some of the direction rules that are introduced later in this topic.

What happens to an electron as it goes round the circuit once? The electron gains electric potential energy as it moves through the cell (this will be covered in Sub-topic 5.3). The electron then leaves the cell and begins to move through the connecting lead. Leads are designed so that they do not require much energy transfer to allow the electrons through (we say that they have a low electrical resistance) and so the potential difference from one end of the lead to the other is small. The electron moves through the switch which also gains little energy from the charge carrier.

After moving through another lead the electron reaches the lamp. This component is different from others in the circuit, it is deliberately designed so that it can gain much of the electrical potential energy from the electrons as they pass through it. The metal lattice in the filament gains energy and as a result the ions vibrate at greater speeds and with greater amplitudes. At these high temperatures the filament in the lamp will glow brightly; the lamp will be lit.

In potential difference terms, the pds across the leads and the switch are small because the passage of one coulomb of charge through them will not result in much energy transfer to the lattice. The pd across the lamp will be large because, for each coulomb going through it, large amounts of energy are transferred from electrons to the lattice ions in the filament raising its temperature.

Nature of science

Conventional and electron currents

In early studies of current electricity, the idea emerged that there was a flow of "electrical fluid" in wires and that this flow was responsible for the observed effects of electricity. At first the suggestion was that there were two types of fluid known as "vitreous" and "resinous". Benjamin Franklin (the same man who helped draft the US Declaration of Independence) proposed that there was only one fluid but that it behaved differently depending on the circumstances. He was also the first scientist to use the terms "positive" and "negative".

What then happened was that scientists assigned a positive charge to the "fluid" thought to be moving in the wires. This positive charge was said to flow out of the positive terminal of a power supply (because the charge was repelled) and went around the circuit re-entering the power supply through the negative terminal. This is what we now term the conventional current.

In fact, we now know that in a metal the charge carriers are electrons and that they move in the opposite direction, leaving a power supply at the negative terminal. This is termed the **electronic current**. You should take care with these two currents and not confuse them.

You may ask: why do we now not simply drop the conventional current and talk only about the electronic current? The answer is that other rules in electricity and magnetism were set up on the assumption that charge carriers are positive. All these rules would need to be reversed to take account of our later knowledge. It is better to leave things as they are.

Worked examples

1 A high efficiency LED lamp is lit for 2 hours. Calculate the energy transfer to the lamp when the pd across it is 240 V and the current in it is 50 mA.

Solution

2 hours is $2 \times 60 \times 60 = 7200$ s.

The charge transferred is $I\Delta t$
$= 7.2 \times 10^3 \times 50 \times 10^{-3} = 360$ C

Work done = *charge* × *pd* = $360 \times 240 = 86\,400$ J

2 A cell has a terminal voltage of 1.5 V and can deliver a charge of 460 C before it becomes discharged.

a) Calculate the maximum energy the cell can deliver.

b) The current in the cell never exceeds 5 mA. Estimate the lifetime of the cell.

Solution

a) Potential difference, $V = \frac{W}{q}$
so $W = qV = 460 \times 1.5 = 690$ J

b) The current of 5 mA means that no more than 5 mC flows through the cell at any time.
So $\frac{460}{0.005} = 92\,000$ s (which is about 25 hours)

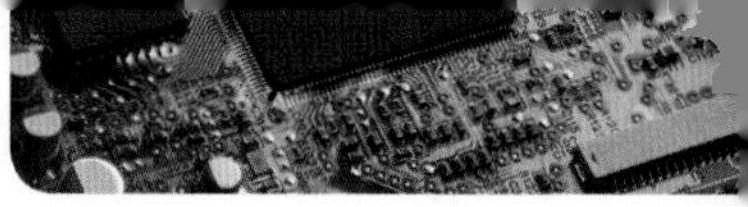

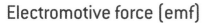

Electromotive force (emf)

Another important term used in electric theory is **electromotive force** (usually written as "emf" for brevity). This term seems to imply that there is a force involved in the movement of charge, but the real meaning of emf is connected to the energy changes in the circuit. When charge flows electrical energy can go *into* another form such as internal energy (through the heating, or Joule, effect), or it can be converted *from* another form (for example, light (radiant energy) in solar (photovoltaic) cells). The term emf will be used in this course when energy is transferred *to* the electrons in, for example, a battery. (Other devices can also convert energy into an electrical form. Examples include microphones and dynamos.)

The term *potential difference* will be used when the energy is transferred *from* the electrical form. So, examples of this would be electrical into heat and light, or electrical into motion energy.

The table shows some of the devices that transfer electrical energy and it gives the term that is most appropriate to use for each one.

Device					pd or emf?
Cell		chemical		electrical	emf
Resistor		electrical		internal	pd
Microphone		sound		electrical	emf
Loudspeaker	***converts energy from***	electrical	***into***	sound	pd
Lamp		electrical		light (and internal)	pd
Photovoltaic cell		light		electrical	emf
Dynamo		kinetic		electrical	emf
Electric motor		electrical		kinetic	pd

Power, current, and pd

We can now answer the question of how much energy is delivered to a conductor by the electrons as they move through it.

Suppose there is a conductor with a potential difference V between its ends when a current I is in the conductor.

In time Δt the charge Q that moves through the conductor is equal to $I\Delta t$.

The energy W transferred to the conductor from the electrons is QV which is $(I\Delta t)V$.

So the energy transferred in time Δt is

$$W = IV\Delta t$$

The electrical power being supplied to the conductor is $\frac{\text{energy}}{\text{time}} = \frac{W}{\Delta t}$ and therefore

$$\text{electrical power } P = IV$$

Alternative forms of this expression that you will find useful are $I = \frac{P}{V}$ and $V = \frac{P}{I}$

The **unit of power is the watt** (W) – 1 watt (1 W) is the power developed when 1 J is converted in 1 s, the same in both mechanics and electricity. So another way to think of the volt is as the power transferred per unit current in a conductor.

Worked examples

1 A 3 V, 1.5 W filament lamp is connected to a 3 V battery. Calculate:

a) the current in the lamp

b) the energy transferred in 2400 s.

Solution

a) Electrical power, $P = IV$, so $I = \frac{P}{V} = \frac{1.5}{3} = 0.5$ A

b) The energy transferred every second is 1.5 J so in 2400 s, 3600 J.

2 An electric motor that is connected to a 12 V supply is able to raise a 0.01 kg load through a distance of 1.5 m in 7 s. The motor is 40% efficient. Calculate the average current in the motor while the load is being raised.

Solution

The energy gained is $mg\Delta h$ $= 0.01 \times g \times 1.5 = 0.147$ J

The power output of the motor must be $= \frac{0.147}{7} = 0.021$ W

The current in the motor $= \frac{P}{V} = \frac{0.021}{12} = 1.8$ mA. Since the motor is 40% efficient the current will be 4.5 mA.

Nature of science

A word about potential

The use of the term "potential difference" implies that there is something called potential which can differ from point to point. This is indeed the case.

An isolated positive point charge will have field lines that radiate away from it. A small positive test charge in this field will have a force exerted on it in the field line direction and, if free to do so, will accelerate away from the original charge. When the test charge is close, there is energy stored in the system and we say that the system (the two charges interacting) has a high potential. When the test charge is further away, there is still energy stored, but it is smaller because the system has converted energy into the kinetic energy of the charge – it has "done work" on the charge. So to move the test charge away from the original charge transfers some of the original stored energy; this is described as a loss of potential. Positive charges move from points of high potential to low potential if they are free to do so.

Negative charges on the other hand move from points of low potential to high. You can work this through yourself by imagining a negative test charge near a positive original charge. This time the two charges are attracted and to move the negative charge away we have to do work on the system. This increases the potential of the system. If charges are free they will fall towards each other losing potential energy. In what form does this energy re-appear?

We shall return to a discussion of potential in Topic 10. From now on Topic 5 only refers to potential differences.

The electronvolt

Earlier we said that the energy possessed by individual electrons is very small. If a single electron is moved through a potential difference of equal to 15 V then as $V = \frac{W}{Q}$, so $W = QV$ and the energy gained by this electron is $15 \times 1.6 \times 10^{-19}$ J $= 2.4 \times 10^{-18}$ J. This is a very small amount and involves us in large negative powers of ten. It is more convenient to define a new energy unit called the **electronvolt** (symbol eV).

This is defined as the energy gained by an electron when it moves through a potential difference of one volt. An energy of 1 eV is equivalent to 1.6×10^{-19} J. The electronvolt is used extensively in nuclear and particle physics.

Be careful, although the unit sounds as though it might be connected to potential difference, like the joule it is a unit of energy.

Worked examples

1 An electron, initially at rest, is accelerated through a potential difference of 180 V. Calculate, for the electron:

a) the gain in kinetic energy

b) the final speed.

Solution

(a) The electron gains 180 eV of energy during its acceleration.

$1 \text{ eV} \equiv 1.6 \times 10^{-19}$ J so $180 \text{ eV} \equiv 2.9 \times 10^{-17}$ J

(b) The kinetic energy of the election $= \frac{1}{2}mv^2$ and the mass of the electron is 9.1×10^{-31} kg.

So $v = \sqrt{\dfrac{2E_{ke}}{m_e}} = \sqrt{\dfrac{2 \times 2.9 \times 10^{-17}}{9.1 \times 10^{-31}}} = 8.0 \times 10^6 \text{ m s}^{-1}$.

2 In a nuclear accelerator a proton is accelerated from rest gaining an energy of 250 MeV. Estimate the final speed of the particle and comment on the result.

Solution

The energy gained by the proton, in joules, is 4.0×10^{-11} J.

As before, $v = \sqrt{\dfrac{2E_{ke}}{m_p}} = \sqrt{\dfrac{2 \times 4.0 \times 10^{-12}}{1.7 \times 10^{-27}}}$, but using a value for the mass of the proton this time.

The numerical answer for $v = 2.2 \times 10^8 \text{ m s}^{-1}$. This is a large speed, 70% of the speed of light. In fact the speed will be less than this as some of the energy goes into increasing the mass of the proton through relativistic effects rather than into the speed of the proton.

5.2 Heating effect of an electric current

Understanding

- Circuit diagrams
- Kirchhoff's laws
- Heating effect of an electric current and its consequences
- Resistance
- Ohm's law
- Resistivity
- Power dissipation

Nature of science

Peer review – a process in which scientists repeat and criticize the work of other scientists – is an important part of the modern scientific method. It was not always so. The work of Ohm was neglected in England at first as Barlow, a much-respected figure in his day, had published contradictory material to that of Ohm. In present-day science the need for repeatability in data collection is paramount. If experiments or other findings cannot be repeated, or if they contradict other scientists' work, then a close look is paid to them before they are generally accepted.

Applications and skills

- Drawing and interpreting circuit diagrams
- Indentifying ohmic and non-ohmic conductors through a consideration of the V–I characteristic graph
- Investigating combinations of resistors in parallel and series circuits
- Describing ideal and non-ideal ammeters and voltmeters
- Describing practical uses of potential divider circuits, including the advantages of a potential divider over a series variable resistor in controlling a simple circuit
- Investigating one or more of the factors that affect resistivity
- Solving problems involving current, charge, potential difference, Kirchhoff's laws, power, resistance and resistivity

Equations

- resistance definition: $R = \frac{V}{I}$
- electrical power: $P = VI = I^2 R = \frac{V^2}{R}$
- combining resistors:
  in series $R_{total} = R_1 + R_2 + R_3 \ldots$
- in parallel $\frac{1}{R_{total}} = \frac{1}{R_1} + \frac{1}{R_2} + \frac{1}{R_3} + \ldots$
- resisitivity definition: $\rho = \frac{RA}{l}$

Effects of electric current

Introduction

This is the first of three sub-topics that discusses some of the effects that occur when charge flows in a circuit. The three effects are:

- **heating effect**, when energy is transferred to a resistor as internal energy
- **chemical effect**, when chemicals react together to alter the energy of electrons and to cause them to move, or when electric current in a material causes chemical changes (Sub-topic 5.3)
- **magnetic effect**, when a current produces a magnetic field, or when magnetic fields change near conductors and induce an emf in the conductor (Sub-topic 5.4).

This sub-topic deals with the heating effect of a current after giving you some advice on setting up and drawing electric circuits.

Drawing and using circuit diagrams

At some stage in your study of electricity you need to learn how to construct electrical circuits to carry out practical tasks.

This section deals with drawing, interpreting, and using circuit diagrams and is designed to stand alone so that you can refer to it whenever you are working with diagrams and real circuits.

You may not have met all the components discussed here yet. They will be introduced as they are needed.

Circuit symbols

A set of agreed electrical symbols has been devised so that all physicists understand what is represented in a circuit diagram. The agreed symbols that are used in the IB Diploma Programme are shown in figure 1. Most of these are straightforward and obvious; some may be familiar to you already. Ensure that you can draw and identify all of them accurately.

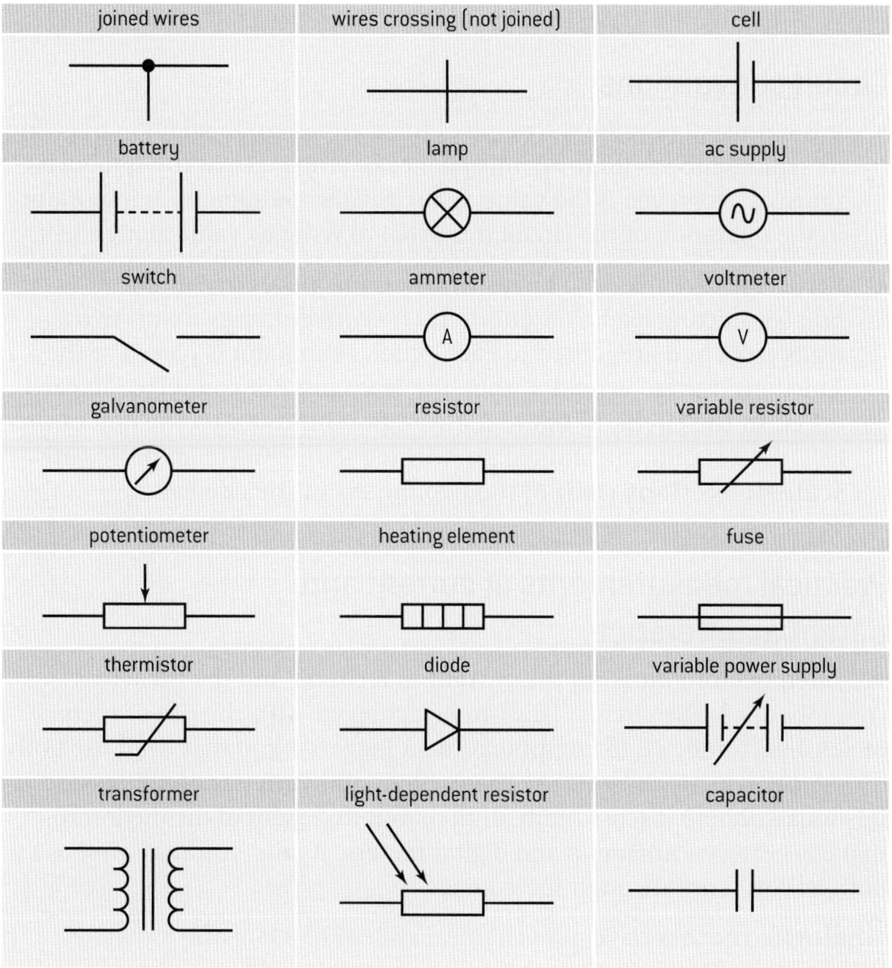

▲ Figure 1 Circuit symbols.

There are points to make about the symbols.

- Some of the symbols here are intended for **direct current (dc)** circuits (cells and batteries, for example). Direct current refers to a circuit in which the charge flows in one direction. Typical examples of this in use would be a low-voltage flashlight or a mobile phone. Other types of electrical circuits use **alternating current (ac)** in which the current direction is first one way around the circuit and then the opposite. The time between changes is typically about 1/100th of a second. Common standards for the frequencies around the world include 50 Hz and 60 Hz. Alternating current is used in high-voltage devices (typically in the home and industry), where large amounts of energy transfer are required: kettles, washing machines, powerful electric motors, and so on. Alternating supplies can be easily transformed from one pd to another, whereas this is more difficult (though not impossible) for dc.
- There are separate symbols for cells and batteries. Most people use these two terms interchangeably, but there is a difference: a battery is a collection of cells arranged positive terminal to negative – the diagram for the battery shows how they are connected. A cell only contains one source of emf. Sub-topic 5.3 goes into more detail about cells and batteries.

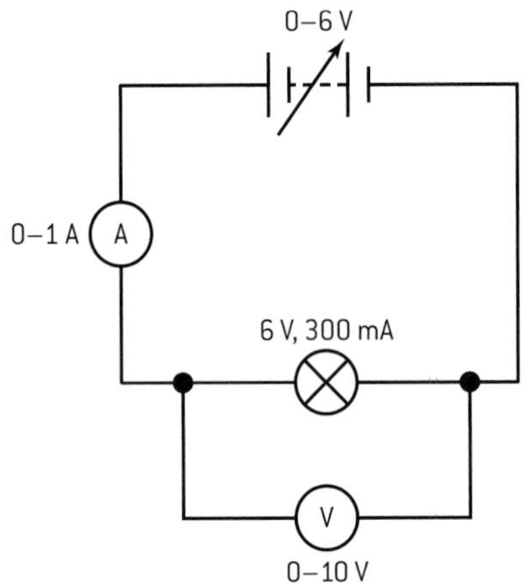

▲ Figure 2 Circuit diagram.

Circuit conventions

- It is not usual to write the name of the component in addition to giving its symbol, unless there is some chance of ambiguity or the symbol is unusual. However, if the value of a particular component is important in the operation of the circuit it is usual to write its value alongside it.
- Particular care needs to be taken when it is necessary to draw one connecting lead over another. The convention is that if two leads cross and are joined to each other, then a dot is placed at the junction. If there is no dot, then the leads are not connected to each other. In the circuit in figure 2 it is important to know the emf of the cell, the data for the lamp at its working temperature, and the **full scale deflections (fsd)** of the two meters in the circuit.

Practical measurements of current and potential difference

We often need to measure the current in a circuit and the pd across components in the circuit. This can be achieved with the use of meters or sensors connected to computers (data loggers). You may well use both during the course. Schools use many types and varieties of meters and it is impossible to discuss them all here, but an essential distinction to make is between analogue and digital meters. Again, you may well use both as you work through the course.

Analogue meters have a mechanical system of a coil and a magnet. When charge flows through the coil, a magnetic field is produced that interacts with the field of the magnet and the coil swings round against a spring. The position reached by the pointer attached to the coil is a measure of the current in the meter.

Digital meters sample the potential difference across the terminals of the meter (or, for current, the pd across a known resistor inside the meter) and then convert the answer into a form suitable for display on the meter.

Ammeters measure the current in the circuit. As we want to know the size of the current in a component it is clear that the ammeter must have the same current. The ammeter needs to be in **series** with the circuit or component. An ideal ammeter will not take any energy from the electrons as they flow through it, otherwise it would disturb the circuit it is trying to measure. Figure 2 shows where the ammeter is placed to measure the current.

Voltmeters measure the energy converted per unit charge that flows in a component or components. You can think of a voltmeter as needing to compare the energy in the electrons before they enter a component to when they leave it, rather like the turnstiles (baffle gates) to a rail station that count the number of people (charges) going through as they give a set amount of money (energy) to the rail company. To do this the voltmeter must be placed across the terminals of the component or components whose pd is being measured. This arrangement is called **parallel**. Again, figure 2 shows this for the voltmeter.

Constructing practical circuits from a diagram

Wiring a circuit is an important skill for anyone studying physics. If you are careful and work in an organized way then you should have no problems with any circuit no matter how complex.

As an example, this is how you might set up one of the more difficult circuits in this course.

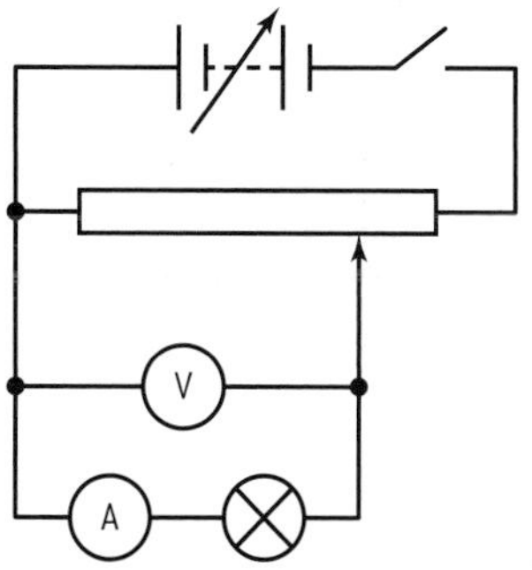

(a)

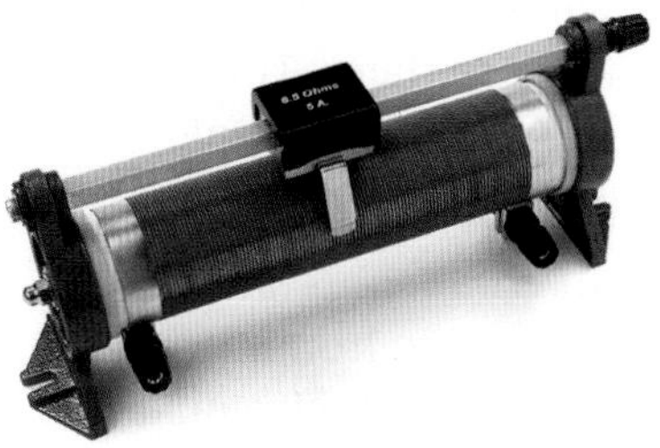

(b)

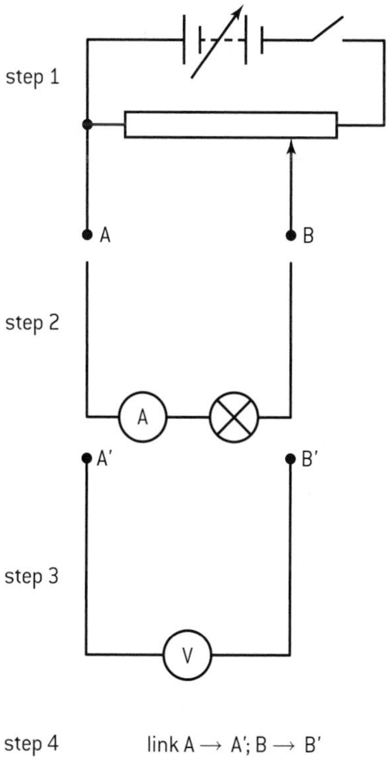

(c)

▲ Figure 3 Variable resistors.

This is a potential divider circuit that it is used to vary the pd across a component, in this case a lamp.

The most awkward component to use here is the potential divider itself (a form of variable resistor, that is sometimes known as a **potentiometer**). In one form it has three terminals (figure 3(b)), in another type it has three in a rotary format. The three-terminal linear device has a terminal at one end of a rod with a wiper that touches the resistance windings, and another two terminals one at each end of the resistance winding itself.

Begin by looking carefully at the diagram figure 3(a). Notice that it is really two smaller circuits that are linked together: the top sub-circuit with the cell, and the bottom sub-circuit with the lamp and the two meters. The bottom circuit itself consists of two parts: the lamp/ammeter link together with the voltmeter loop.

The rules for setting up a circuit like this are:

- If you do not already have one, draw a circuit diagram. Get your teacher to check it if you are not sure that it is correct.
- Before starting to plug leads in, lay out the circuit components on the bench in the same position as they appear on the diagram.
- Connect up one loop of the circuit at a time.
- Ensure that components are set to give minimum or zero current when the circuit is switched on.
- Do not switch the circuit on until you have checked everything.

Figure 3(c) shows a sequence for setting up the circuit step-by-step.

Another skill you will need is that of troubleshooting circuits – this is an art in itself and comes with experience. A possible sequence is:

- Check the circuit – is it really set up as in your diagram?
- Check the power supply (try it with another single component such as a lamp that you know is working properly).
- Check that all the leads are correctly inserted and that there are no loose wires inside the connectors.
- Check that the individual components are working by substituting them into an alternative circuit known to be working.

Resistance

We saw in Sub-topic 5.1 that, as electrons move through a metal, they interact with the positive ions and transfer energy to them. This energy appears as kinetic energy of the lattice, in other words, as internal energy: the metal wire carrying the current heats up.

However, simple comparison between different conductors shows that the amount of energy transferred can vary greatly from metal to metal. When there is the same current in wires of similar size made of tungsten or copper, the tungsten wire will heat up more than the copper. We need to take account of the fact that some conductors can achieve the energy transfer better than others. The concept of **electrical resistance** is used for this.

The resistance of a component is defined as

$$\frac{\text{potential difference across the component}}{\text{current in the component}}$$

The symbol for resistance is R and the definition leads to a well-known equation

$$R = \frac{V}{I}$$

The unit of resistance is the ohm (symbol Ω; named after Georg Simon Ohm, a German physicist). In terms of its fundamental units, 1 Ω is 1 kg m^2 s^{-3} A^{-2}; using the ohm as a unit is much more convenient!

Alternative forms of the equation are: $V = IR$ and $I = \frac{V}{R}$

Investigate!

Resistance of a metal wire

- Take a piece of metal wire (an alloy called constantan is a good one to choose) and connect it in the circuit shown. Use a power supply with a variable output so that you can alter the pd across the wire easily.

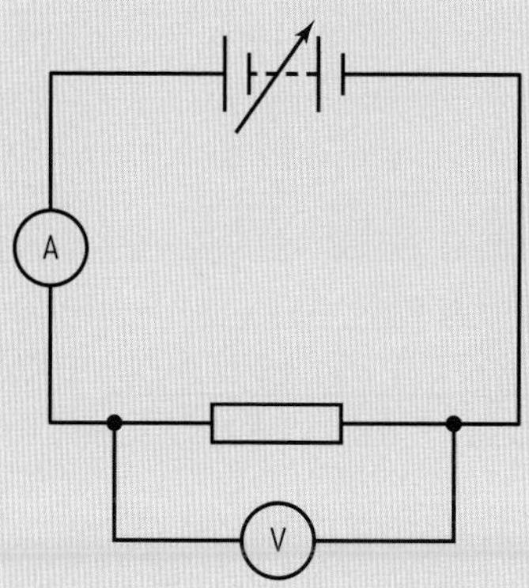

▲ Figure 4

- If your wire is long, coil it around an insulator (perhaps a pencil) and ensure that the coils do not touch.
- Take readings of the current in the wire and the pd across it for a range of currents. Your teacher will tell you an appropriate range to use to avoid changing the temperature of the wire.
- For each pair of readings divide the pd by the current to obtain the resistance of the wire in ohms.

A table of results is shown in figure 5 for a metal wire 1 m in length with a diameter of 0.50 mm. For each pair of readings the resistance of the wire has been calculated by dividing the pd by the current. Although the resistance values are not identical (it is an experiment with real errors, after all), they do give an average value for the resistance of 2.54 Ω. This should be rounded to 2.5 Ω given the signficant figures in the data.

current/A	pd/V	resistance/Ω
0.05	0.13	2.55
0.10	0.26	2.60
0.20	0.50	2.50
0.30	0.76	2.53
0.40	1.01	2.53
0.50	1.27	2.54
0.60	1.53	2.55
0.70	1.78	2.54

▲ Figure 5 Variation of pd with current for a conductor.

Nature of science

Edison and his lamp

The conversion of electrical energy into internal energy was one of the first uses of distributed electricity. Thomas Edison an inventor and entrepreneur, who worked in the US at around the end of the nineteenth century, was a pioneer of electric lighting. The earliest forms of light were provided by producing a current in a metal or carbon filament. These filaments heated up until they glowed. Early lamps were primitive but produced a revolution in the way that homes and public spaces were lit. The development continues today as inventors and manufacturers strive to find more and more efficient electric lamps such as the light-emitting diodes (LED). More developments will undoubtedly occur during the lifetime of this book.

Ohm's law

Figure 6 shows the results from the metal wire when they are plotted as a graph of V against I.

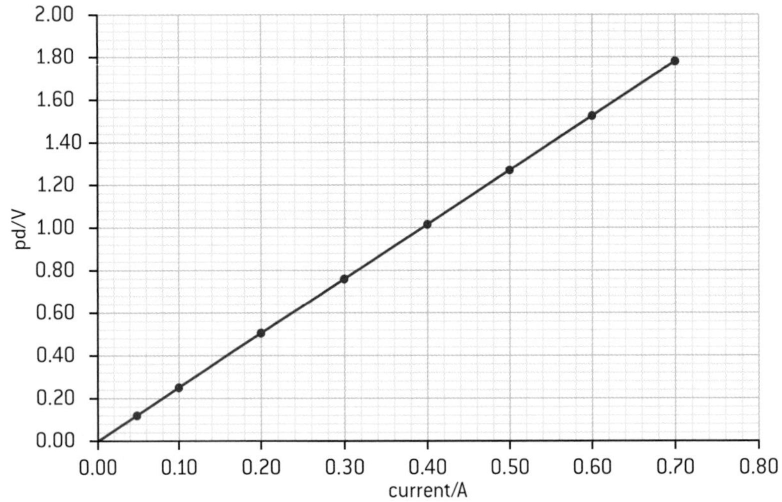

▲ Figure 6 pd against current from the table.

A best straight line has been drawn through the data points. For this wire, the resistance is the same for all values of current measured. Such a resistor is known as **ohmic**. An equivalent way to say this is that the potential difference and the current are proportional (the line is straight and goes through the origin). In the experiment carried out to obtain these data, the temperature of the wire did not change.

This behaviour of metallic wires was first observed by Georg Simon Ohm in 1826. It leads to a rule known as **Ohm's law**.

Ohm's law states that the potential difference across a metallic conductor is directly proportional to the current in the conductor providing that the physical conditions of the conductor do not change.

By physical conditions we mean the temperature (the most important factor as we shall see) and all other factors about the wire. But the temperature factor is so important that the law is sometimes stated replacing the term "physical conditions" with the word "temperature".

Worked example

The current in a component is 5.0 mA when the pd across it is 6.0 V.

Calculate:

a) the resistance of the component

b) the pd across the component when the current in it is 150 μA.

Solution

a) $R = \frac{V}{I} = \frac{6}{5 \times 10^{-3}} = 1.2\ \text{k}\Omega$

b) $V = IR = 1.5 \times 10^{-4} \times 1.2 \times 10^{3} = 0.18\ \text{A}$

Nature of science

Ohm and Barlow

Ohm's law has its limitations because it only tells us about a material when the physical conditions do not change. However, it was a remarkable piece of work that did not find immediate favour. Barlow was an English scientist who was held in high respect for his earlier work and had recently published an alternative theory on conduction. People simply did not believe that Barlow could be wrong.

This immediate acceptance of one scientist's work over another would not necessarily happen today. Scientists use a system of peer review. Work published by one scientist or scientific group must be set out in such a way that other scientists can repeat the experiments or collect the same data to check that there are no errors in the original work. Only if the scientific community as a whole can verify the data is the new work accepted as scientific "fact".

Towards the end of the 20th century a research group thought that it had found evidence that nuclear fusion could occur at low temperatures (so-called "cold fusion"). Repeated attempts by other research groups to replicate the original results failed, and the cold fusion ideas were discarded.

Investigate!

Variation of resistance of a lamp filament

- Use the circuit you used in the investigation on resistance of a metal wire, but repeat the experiment with a filament lamp instead of the wire.

  Your teacher will advise you of the range of currents and pds to use.

- This time do the experiment twice, the second time with the charge flowing through the lamp in the opposite direction to the first. There are two ways to achieve this: the first is to reverse the connections to the power supply, also reversing the connections to the ammeter and voltmeter (if the meters are analogue). The second way is somewhat easier, simply reverse the lamp and call all the readings negative because they are in the opposite direction through the lamp.
- Plot a graph of V (y-axis) against I (x-axis) with the origin (0,0) in the centre of the paper. Figure 7 shows an example of a V–I graph for a lamp.

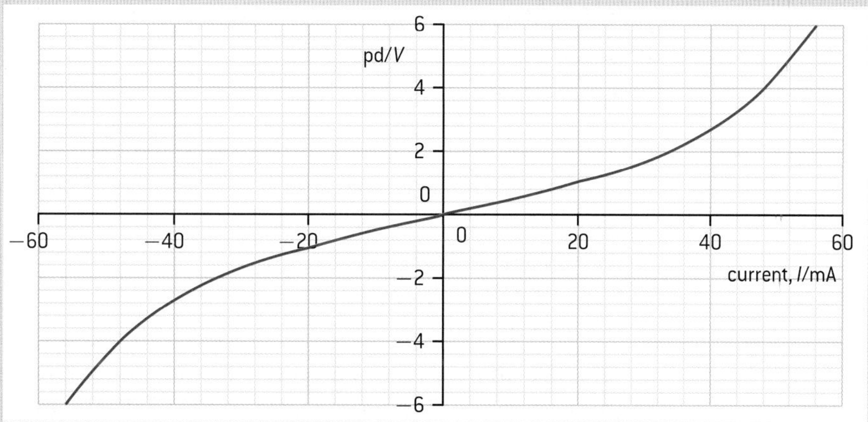

▲ Figure 7

TOK

But is it a law?

This rule of Ohm is always called a law – but is it? In reality it is an experimental description of how a group of materials behave under rather restricted conditions. Does that make it a law? You decide.

There is also another aspect to the law that is often misunderstood.

Our definition of resistance is that $R = \frac{V}{I}$ or $V = IR$. Ohm's law states that

$$V \propto I$$

and, including the constant of proportionality k,

$$V = kI$$

We therefore define R to be the same as k, but the definition of resistance does not correspond to Ohm's law (which talks only about a proportionality). $V = IR$ is emphatically not a statement of Ohm's law and if you write this in an examination as a statement of the law you will lose marks.

The graph is not straight (although it goes through the origin) so V and I are not proportional to each other. The lamp does not obey Ohm's law and it is said to be **non-ohmic**. However, this is not a fair test of ohmic behaviour because the filament is not held at a constant temperature. If it were then, as a metal, it would probably obey the law.

When the resistance is calculated for some of the data points it is not constant either. The table shows the resistance values at each of the positive current points.

Current/mA	Resistance/Ω
20	50
34	59
41	73
47	85
52	96
55	109

Tip

Notice that these resistances were calculated for each individual data point using $\frac{V}{I}$. They were not evaluated from the tangent to the graph at the current concerned. The definition of resistance is in terms of $\frac{V}{I}$ *not* in terms of $\frac{\Delta V}{\Delta I}$ which is what the tangent would give.

The data show that resistance of the lamp increases as the current increases. At large currents, it takes greater changes in pd to change the current by a fixed amount. This is exactly what you might have predicted. As the current increases, more energy is transferred from the electrons every second because more electrons flow at higher currents. The energy goes into increasing the kinetic energy of the lattice ions and therefore the temperature of the bulk material. But the more the ions vibrate in the lattice, the more the electrons can collide with them so at higher temperatures even more energy is transferred to the lattice by the moving charges.

Other non-ohmic conductors include semiconducting diodes and thermistors; these are devices made from a group of materials known as semiconductors.

Investigate!

Diodes and thermistors

- You can easily extend the *Investigate!* lamp experiment to include these two types of device. There are some extra practical points however:
- The diode will require a protecting resistor in series with it (100 Ω is usually appropriate) as the current can become very large at even quite small potential differences; this will cause the diode to melt as large amounts of energy are transferred to it. The resistor limits the current so that the diode is not damaged.
- The thermistor also needs care, because as it heats up its resistance decreases and if too much current is used, the thermistor can also be destroyed.

The results of these experiments and some others are summarized in the *I–V* graphs in figure 8.

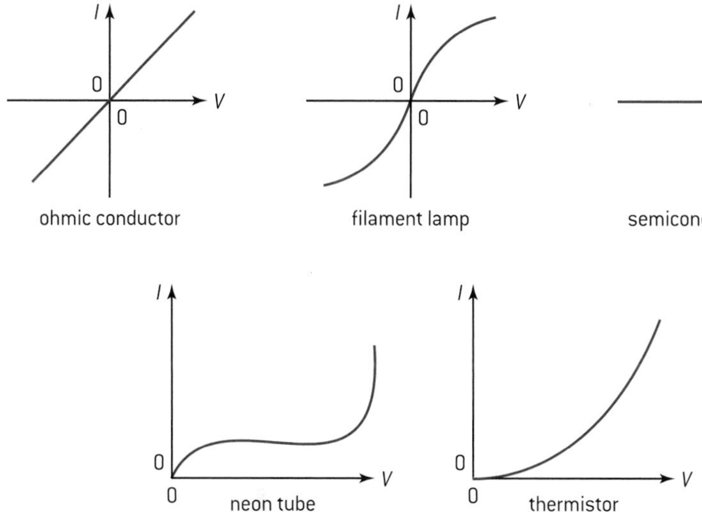

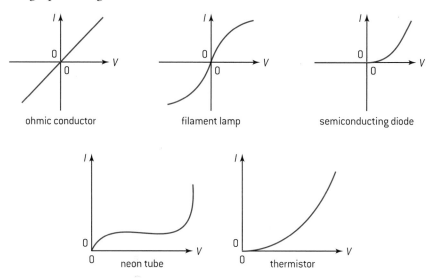

▲ Figure 8 *I–V* graphs for various conductors.

Semiconducting diodes

Semiconducting diodes are designed only to allow charges to flow through them in one direction. This is seen clearly in the graph. For negative values of *V* there is actually a very small current flowing in the negative direction but it is far less than the forward current. The nature of semiconductor material also means that there is no significant current in the forward direction until a certain forward pd is exceeded.

Thermistors

Thermistors are made from one of the two elements that are electrical semiconductors: silicon and germanium. There are several types of thermistor, but we will only consider the negative temperature coefficient type (ntc). As the temperature of an ntc thermistor increases, its resistance falls. This is the opposite behaviour to that of a metal.

Semiconductors have many fewer free electrons per cubic metre compared with metals. Their resistances are typically 10^5 times greater than similar samples of metals. However, unlike in a metal, the charge density in semiconductors depends strongly on the temperature. The higher the temperature of the semiconductor, the more charge carriers are made available in the material.

As the temperature rises in the germanium:

- The lattice ions vibrate more and impede the movement of the charge carriers. This is exactly the same as in metals and also leads to an *increase* in resistance.
- More and more charge carriers become available to conduct because the increase in temperature provides them with enough energy to break away from their atoms. This leads to a large *decrease* in resistance.
- The second effect is much greater than the first and so the net effect is that conduction increases (resistance falls) as the temperature of the semiconductor rises.

Investigate!

How resistance depends on size and shape

- You will need a set of wires made from the same metal or metal alloy. The wires should have a circular cross-section and be available in a range of different diameters. You will also need to devise a way to vary the length of one of the wires in the circuit.
- Use the circuit in figure 4 on p197 to answer the following questions:
  - How does the resistance R of one of the wires vary with length l?
  - How does the resistance R of the wires vary with diameter d when the wires all have the same length?
- Try to make things easy for your analysis. Double and halve the values of length to see what difference this makes to the resistance – is there an obvious relationship? The diameter of the wire may be more difficult to test in this way, but a graph of resistance against diameter may give you a clue.
- Once you have an idea what is going on, you may decide that the best way to answer these questions is to plot graphs of

  R against l, and

  R against $\frac{1}{d^2}$

Resistivity

The resistance of a sample of a material depends not only on what it is made of, but also on the physical dimensions (the size and shape) of the sample itself.

The graphs suggested in the *Investigate!* should give straight lines that go through the origin and can be summed up in the following rule.

The resistance of a conductor is:

- proportional to its length l
- inversely proportional to its cross-sectional area A (which is itself proportional to d^2).

So

$$R \propto \frac{l}{A}$$

This leads to a definition of a new quantity called **resistivity**.

Resistivity ρ is defined by

$$\rho = \frac{RA}{l}$$

The unit of resistivity is the ohm-metre (symbol Ω m). Take care here: this is ohm metre. It is not ohm metre^{-1} – a mistake frequently made by students in examinations. The meaning of ohm metre^{-1} is the resistance of one metre length of a particular conductor, which is a relevant quantity to know, but is not the same as resistivity.

Resistivity is a quantity of considerable use. The resistance of a material depends on the shape of the sample as well as what it is made from. Even a constant volume of a material will have values of resistance that depend on the shape. However, the value of the resistivity is the same for all samples of the material. Resistivity is independent of shape or size just like quantities such as *density* (where the size of a material has been removed by the use of volume) or *specific latent heat* (where the value is related to unit mass of the material).

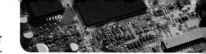

Worked examples

1 A uniform wire has a radius of 0.16 mm and a length of 7.5 m. Calculate the resistance of the wire if its resistivity is $7.0 \times 10^{-7}\ \Omega$ m.

Solution

Unless told otherwise, assume that the wire has a circular cross-section.

area of wire $= \pi(1.6 \times 10^{-4})^2 = 8.04 \times 10^{-8}\ \text{m}^2$

$$\rho = \frac{RA}{l}$$

and $R = \dfrac{\rho l}{A} = \dfrac{7.0 \times 10^{-7} \times 7.5}{8.04 \times 10^{-8}} = 65\ \Omega$

2 Calculate the resistance of a block of copper that has a length of 0.012 m with a width of 0.75 mm and a thickness of 12 mm. The resistivity of copper is $1.7 \times 10^{-8}\ \Omega$ m.

Solution

The cross-sectional area of the block is $7.5 \times 10^{-4} \times 1.2 \times 10^{-2} = 9.0 \times 10^{-6}\ \text{m}^2$

The relevant dimension for the length is 0.012 m, so

$$R = \frac{\rho l}{A} = \frac{1.7 \times 10^{-8} \times 0.012}{9.0 \times 10^{-6}} = 0.023\ \text{m}\Omega$$

Investigate!

Resistivity of pencil lead

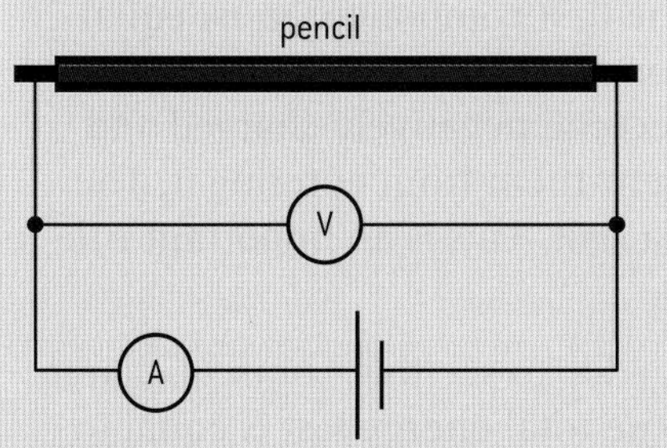

▲ Figure 9

- Graphite is a semi-metallic conductor and is a constituent of the lead in a pencil. Another constituent in the pencil lead is clay. It is the ratio of graphite to clay that determines the "hardness" of the pencil. This experiment enables you to estimate the resistivity of the graphite.
- Take a B grade pencil (sometimes known as #1 grade in the US) and remove about 1.5 cm of the wood from each end leaving a cylinder of the lead exposed. Attach a crocodile (alligator) clip firmly to each end. Use leads attached to the crocodile clips to connect the pencil into a circuit in order to measure the resistance of the lead. Expect the resistance of the pencil lead to be about 1 Ω for the purpose of choosing the power supply and meters.
- Determine the resistance of the lead.
- Measure the length of pencil lead between the crocodile clips.
- Measure the diameter of the lead using a micrometer screw gauge or digital callipers and calculate the area of the lead.
- Using the data, calculate the resistivity of the lead. The accepted value of the resistivity of graphite is about $3 \times 10^{-5}\ \Omega$ m but you will not expect to get this value given the presence of clay in the lead as well.
- Take the experiment one step further with a challenge. Use your pencil to uniformly shade a 10 cm by 2 cm area on a piece of graph paper. This will make a graphite resistance film on the paper. Attach the graphite film to a suitable circuit and measure the resistance of the film. Knowing the resistance and the dimensions of your shaded area should enable you to work out how thick the film is. (*Hint:* in the resistivity equation, the length is the *distance across the film*, and the area is *width of the film* × *thickness of the film*.)

Practical resistors

Resistors are of great importance in the electronics and electrical industries. They can be single value (fixed) devices or they can be variable. They can be manufactured in bulk and are readily and cheaply available.

Resistors come in different sizes. Small resistors can have a large resistance but only be able to dissipate (lose) a modest amount of energy every second. If the power that is being generated in the resistor is too large, then its temperature will increase and in a worst case, it could become a thermal fuse! Resistors are rated by their manufacturers so that, for example, a resistor could have a resistance of 270 Ω with a power rating of 0.5 W. This would mean that the maximum current that the resistor can safely carry is $\sqrt{\frac{0.5}{270}} = 43$ mA.

Combining resistors

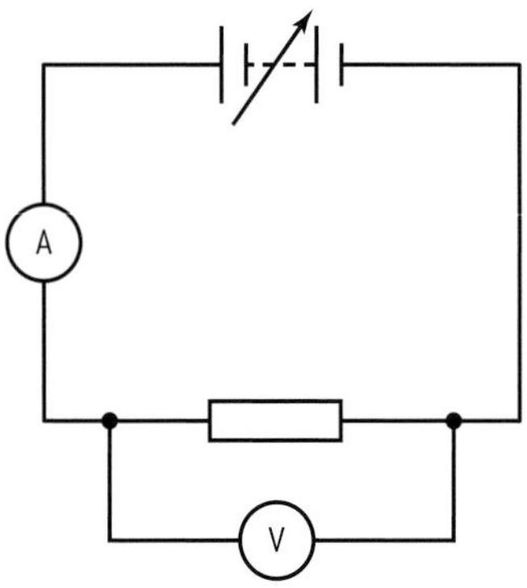

▲ Figure 10

Electrical components can be linked together in two ways in an electrical circuit: **in series**, where the components are joined one after another as the ammeter, the cell and the resistor in figure 10, or **in parallel** as are the resistor and the voltmeter in the same figure.

Two components connected in series have the same current in each. The number of free electrons leaving the first component must equal the number entering the second component; if electrons were to stay in the first component then it would become negatively charged and would repel further electrons and prevent them from entering it. The flow of charges would rapidly grind to a halt. This is an important rule to understand.

In series the potential differences (pds) add. To see this, think of charge as it travels through two components, the total energy lost is equal to the sum of the two separate amounts of energy in the components. Because the charge is the same in both cases, therefore, the sum of the pds is equal to the total pd dropped across them.

Components in parallel, on the other hand, have the same pd across them, but the currents in the components differ when the resistances of the components differ. Consider two resistors of different resistance values, in parallel with each other and connected to a cell. If one of the resistors is temporarily disconnected then the current in the remaining resistor is given by the emf of the cell divided by the resistance. This will also be true for the other resistor if it is connected alone. If both resistors are now connected in parallel with each other, both resistors have the same pd across them because a terminal of each resistor is connected to one of the terminals of the cell. The cell will have to supply more current than if either resistor were there alone – to be precise, it has to supply the sum of the separate currents.

To sum up

	Currents ...	**Potential differences ...**
In series	...are the same	...add
In parallel	...add	...are the same

Investigate!

Resistors in series and parallel

- For this experiment you need six resistors, each one with a tolerance of ±5%. Two of these resistors should be the same. The tolerance figure means that the manufacturer only guarantees the value to be within 5% of the nominal value – "nominal" means the printed value. Your resistors may have the nominal value written on them or you may have to use the colour code printed on them. The code is easy to decipher.

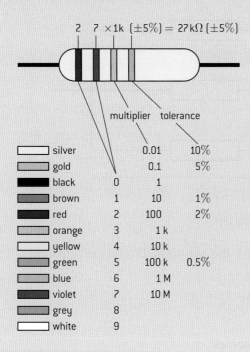

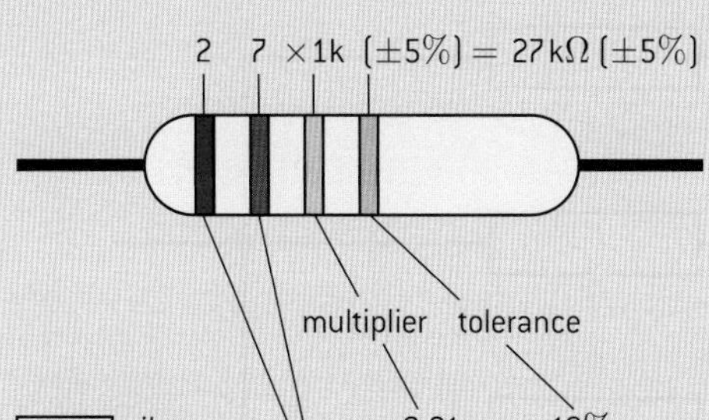

		multiplier	tolerance
silver		0.01	10%
gold		0.1	5%
black	0	1	
brown	1	10	1%
red	2	100	2%
orange	3	1 k	
yellow	4	10 k	
green	5	100 k	0.5%
blue	6	1 M	
violet	7	10 M	
grey	8		
white	9		

- You also need a multimeter set to measure resistance directly and a way to join the resistors together and to connect them to the multimeter.
- First, measure the resistance of each resistor and record this in a table.
- Take the two resistors that have the same nominal value and connect them in series. Measure the resistance of the combination. Can you see a rule straight away for the combined resistance of two resistors?
- Repeat with five of the possible combinations for connecting resistors in series.
- Now measure the combined resistance of the two resistors with the same nominal value when they are in parallel. Is there an obvious rule this time?
- One way to express the rule for combining two resistors R_1 and R_2 in parallel is as $\frac{R_1 R_2}{R_1 + R_2}$. Test this relationship for five combinations of parallel resistors.
- Test your two rules together by forming combinations of three resistors such as:

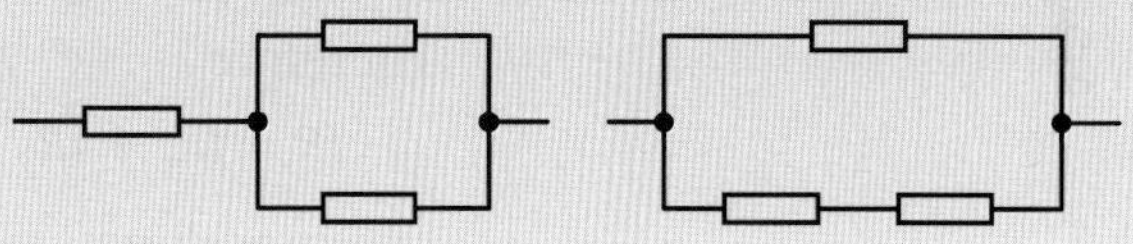

These ideas provide a set of rules for the combination of resistors in various arrangements:

Resistors in series

Suppose that there are three resistors in series: R_1, R_2 and R_3 (figure 11(a)). What is the resistance of the single resistor that could replace them so that the resistance of the single resistor is equivalent to the combination of three?

The resistors are in series (figure 11(a)) and therefore the current I is the same in each resistor.

Using the definition of resistance, the pd across each resistor, V_1, V_2 and V_3 is

$$V_1 = IR_1 \qquad V_2 = IR_2 \qquad \text{and} \qquad V_3 = IR_3$$

The single resistor with a resistance R has to be indistinguishable from the three in series. In other words, when the current through this single resistor is the same as that through the three, then it must have a pd V across it such that

$$V = IR$$

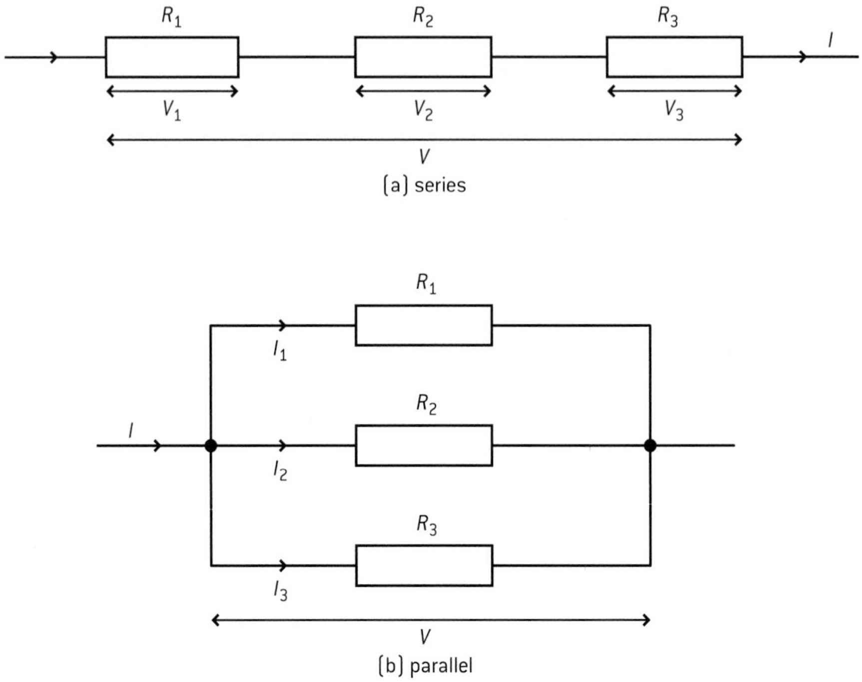

▲ Figure 11 Resistors in series and parallel.

As this is a series combination, the potential differences add, so

$$V = V_1 + V_2 + V_3$$

Therefore

$$IR = IR_1 + IR_2 + IR_3$$

and cancelling I leads to, in **series**

$$\boldsymbol{R = R_1 + R_2 + R_3}$$

When resistors are combined in series, the resistances add to give the total resistance.

Resistors in parallel

Three resistors in parallel (figure 11(b)) have the same pd V across them (figure 11(b)). The currents in the three separate resistors add to give the current in the connecting leads to and from all three resistors, therefore

$$I = I_1 + I_2 + I_3$$

Each current can be written in terms of V and R using the definition of resistance:

$$\frac{V}{R} = \frac{V}{R_1} + \frac{V}{R_2} + \frac{V}{R_3}$$

This time V cancels out and, in **parallel**,

$$\frac{1}{R} = \frac{1}{R_1} + \frac{1}{R_2} + \frac{1}{R_3}$$

In parallel combinations of resistors, the reciprocal of the total resistance is equal to the sum of the reciprocals of the individual resistances.

The parallel equation needs some care in calculations. The steps are:

- calculate the reciprocals of each individual resistor
- add these reciprocals together
- take the reciprocal of the answer.

It is common to see the last step ignored so that the answer is incorrect; a worked example below shows the correct approach.

More complicated networks

When the networks of resistors are more complicated, then the individual parts of the network need to be broken down into the simplest form.

Figure 12 shows the order in which a complex resistor network could be worked out.

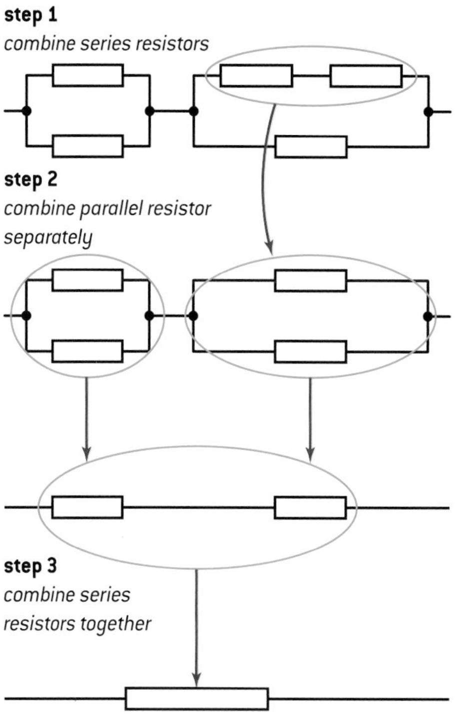
step 1
combine series resistors
step 2
combine parallel resistor separately
step 3
combine series resistors together

▲ Figure 12

Worked examples

1 Three resistors of resistance, 2.0 Ω, 4.0 Ω, and 6.0 Ω are connected. Calculate the total resistance of the three resistors when they are connected:

a) in series

b) in parallel.

Solution

a) In series, the resistances are added together, so $2 + 4 + 6 = 12\ \Omega$

b) In parallel, the reciprocals are used:

$\frac{1}{R} = \frac{1}{R_1} + \frac{1}{R_2} + \frac{1}{R_3} = \frac{1}{2} + \frac{1}{4} + \frac{1}{6} = \frac{6+3+2}{12} = \frac{11}{12}$

The final step is to take the reciprocal of the sum, $R = \frac{12}{11} = 1.1\ \Omega$

2 2.0 Ω, 4.0 Ω, and 8.0 Ω resistors are connected as shown. Calculate the total resistance of this combination.

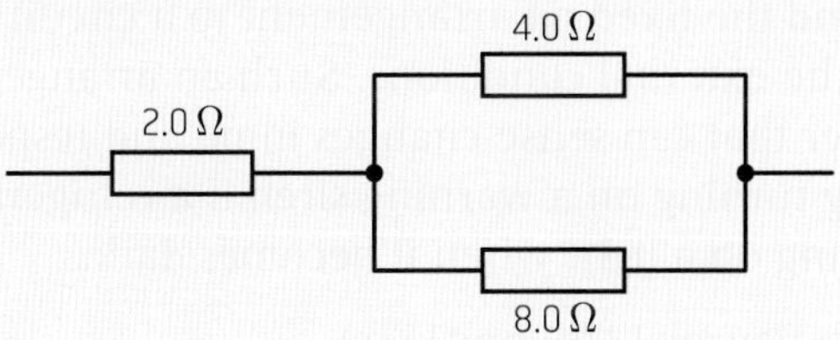
4.0 Ω
2.0 Ω
8.0 Ω

Solution

The two resistors in parallel have a combined resistance of $\frac{1}{R} = \frac{1}{R_1} + \frac{1}{R_2} = \frac{1}{4} + \frac{1}{8} = \frac{3}{8}$

$$R = \frac{8}{3} = 2.67\ \Omega$$

This 2.67 Ω resistor is in series with 2.0 Ω, so the total combined resistance is $2.67 + 2.0 = 4.7\ \Omega$.

3 Four resistors each of resistance 1.5 Ω are connected as shown. Calculate the combined resistance of these resistors.

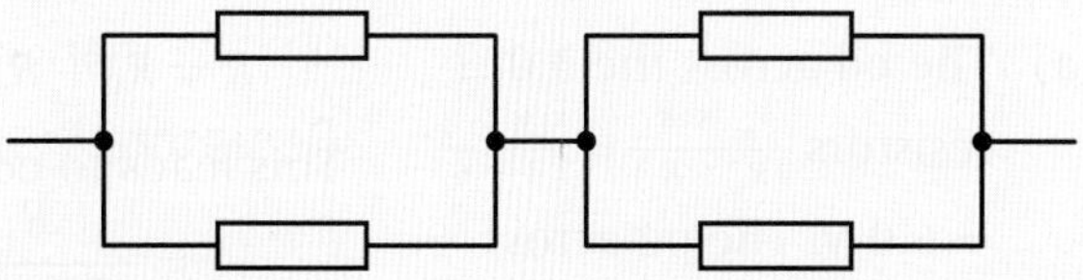

Solution

Two 1.5 Ω resistors in parallel have a resistance given by $\frac{1}{R} = \frac{1}{1.5} + \frac{1}{1.5} = \frac{2}{1.5}$. So $R = 0.75\ \Omega$.

Two 0.75 Ω resistors in series have a combined resistance of $0.75 + 0.75 = 1.5\ \Omega$

Potential divider

In the section on circuit diagrams, a potential divider circuit was shown. This is a circuit commonly used with sensors and also to produce variable potential differences. It has some advantages over the simpler variable resistor circuit even though it is more complicated to set up.

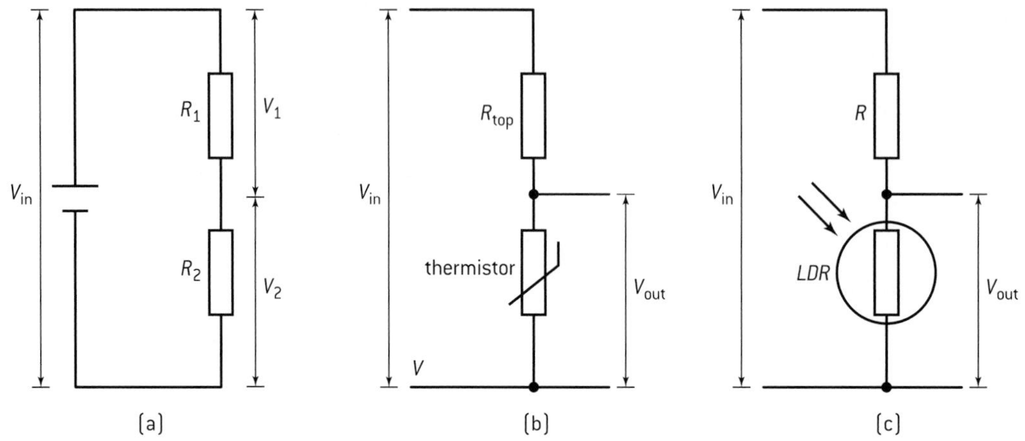

▲ Figure 13 The potential divider.

The most basic potential divider consists of two resistors with resistances R_1 and R_2 in series with a power supply. This arrangement is used to provide a fixed pd at a value somewhere between zero and the emf of the power supply. Figure 13(a) shows the arrangement.

The two resistors have the same current in them, and the sum of the pds across the resistors is equal to the source emf.

So

The current I in both resistors is

$$I = \frac{\textit{total pd across both resistors}}{\textit{total resistance}} = \frac{V}{R}$$

As usual,

$$V_1 = IR_1 \text{ and } V_2 = IR_2$$

The resistors are in series so

$$R = R_1 + R_2$$

This leads to equations for the pd across each resistor

$$V_1 = \frac{R_1}{(R_1 + R_2)} V_{in} \qquad \text{and} \qquad V_2 = \frac{R_2}{(R_1 + R_2)} V_{in}$$

Using a potential divider with sensors

It is a simple matter to extend the fixed pd arrangement to a circuit that will respond to changes in the external conditions. Such an arrangement might be used by a computer that can sense changes in pd and respond accordingly, for example, by turning on a warning siren if a refrigerator becomes too warm, or turning on a light when it becomes dark.

Typical circuits are shown in figures 13(b) and13(c).

In figure 13(b) a thermistor is used instead of one of the fixed resistors.

Worked example

1 A potential divider consists of two resistors in series with a battery of 18 V. The resistors have resistances 3.0 Ω and 6.0 Ω. Calculate, for each resistor:

a) the pd across it

b) the current in it.

Solution

a) The pd across the 3.0 Ω resistors $\frac{V_{in} \times R_1}{(R_1 + R_2)} = \frac{18 \times 3}{3 + 6}$ = 6.0 V. The pd across the 6.0 Ω is then 18 – 6.0 V = 12 V.

b) The current is the same in both resistors because they are in series. The total resistance is 9.0 Ω and the emf of the battery is 18 V. The current = $\frac{V}{R} = \frac{18}{9} = 2.0$ A.

Recall that when a thermistor is at a high temperature its resistance is small, and that the resistance increases when the temperature falls.

Rather than calculating the values, for this example we will use our knowledge of how pd and current are related to work out the behaviour of the circuit from first principles.

Suppose the temperature is low and that the thermistor resistance is high relative to that of the fixed-value resistor. Most of the pd will be dropped across the thermistor and very little across the fixed resistor. If you cannot see this straight away, remember the equations from the previous section:

$$V_{\text{thermistor}} = \frac{V_{\text{in}} R_{\text{thermistor}}}{(R_{\text{thermistor}} + R_{\text{resistor}})} \text{ and } V_{\text{resistor}} = \frac{V_{\text{in}} R_{\text{resistor}}}{(R_{\text{thermistor}} + R_{\text{resistor}})}$$

If $R_{\text{thermistor}} \gg R_{\text{resistor}}$ ($\gg$ means "much greater than") then $R_{\text{thermistor}} \approx (R_{\text{thermistor}} + R_{\text{resistor}})$ and $V_{\text{thermistor}} \approx V_{\text{in}}$.

So the larger resistance (the thermistor at low temperatures) has the larger pd across it.

If the thermistor temperature now increases, then the thermistor resistance will fall. Now the fixed-value resistor will have the larger resistance and the pds will be reversed with the thermistor having the small voltage drop across it. A voltage sensor connected to a computer can be set to detect this voltage change and can activate an alarm if the thermistor has too high a temperature.

You may be asking what the resistance of the fixed resistor should be. The answer is that it is normally set equal to that of the thermistor when it is at its optimum (average) temperature. Then any deviation from the average will change the potential difference and trigger the appropriate change in the sensing circuit.

The same principle can be applied to another sensor device, a light-dependent resistor (LDR). This, like the thermistor, is made of semiconducting material but this time it is sensitive to photons incident on it. When the light intensity is large, charge carriers are released in the LDR and thus the resistance falls. When the intensity is low, the resistance is high as the charge carriers now re-combine with their atoms.

Look at figure 13(c). You should be able to explain that when the LDR is in bright conditions then the pd across the fixed resistor is high.

Using a potential divider to give a variable pd

A variable resistor circuit is shown in figure 14(a). It consists of a power supply, an ammeter, a variable resistor and a resistor. The value of each component is given on the diagram.

We can predict the way this circuit will behave.

When the variable resistor is set to its minimum value, 0 Ω, then there will be a pd of 2 V across the resistor and a current of 0.2 A in the circuit.

When the variable resistor is set to its maximum value, 10 Ω, then the total resistance in the circuit is 20 Ω, and the current is 0.1 A.

This means that with 0.1 A in the 10 Ω fixed resistor, only 1 V is dropped across it. Therefore the range of pd across the fixed resistor can only vary from 1 V to 2 V – half of the available pd that the power supply can in principle provide. You should now be able to predict the range across the variable resistor.

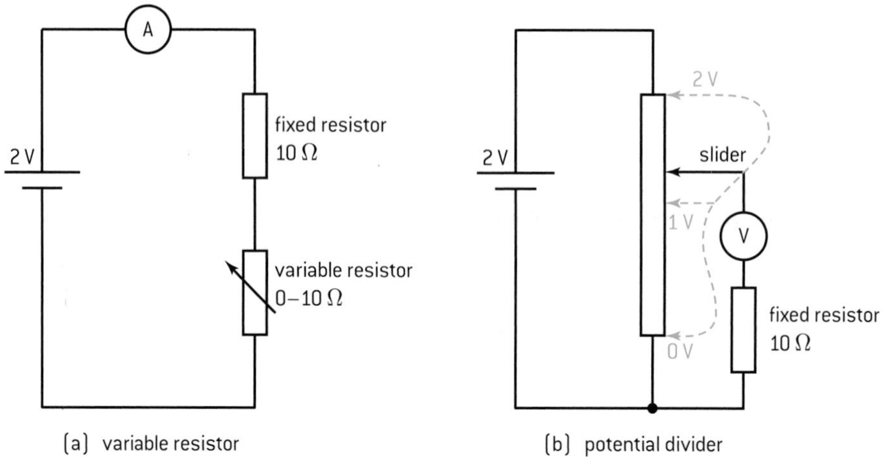

▲ Figure 14

The limited range is a significant limitation in the use of the variable resistor. To achieve a better range, we could use a variable resistor with a much higher range of resistance. To get a pd of 0.1 V across the fixed resistor the resistance of the variable resistor has to be about 200 Ω. If the fixed resistor had a much greater resistance, then the variable resistor would need an even higher value too and this would limit the current.

The potential divider arrangement (figure 14(b)) allows a much greater range of pd to the component under test than does a variable resistor in series with the component. In a potential divider, the same variable resistor can be used but the set up is different and involves the use of the three terminals on the variable resistor. (A variable resistor is also sometimes called a **rheostat**.) One terminal is connected to one side of the cell, and the other end of the rheostat resistor is connected to the other terminal of the cell.

The potential at any point along the resistance winding depends on the position of the slider (or wiper) that can be swept across the windings from one end to the other. Typical values for the potentials at various points on the windings are shown for the three blue slider positions on figure 14(b). The component that is under test (again, a resistor in this case) is connected in a secondary circuit between one terminal of the resistance winding and the slider on the rheostat. When the slider is positioned at one end, the full 2 V from the cell is available to the resistor under test. When at the other end, the pd between the ends of the resistor is 0 V (the two leads to the resistor are effectively connected directly to each other at the variable resistor).

You should know how to set this arrangement up and also how to draw the circuit and explain its use.

Worked examples

1 A light sensor consists of a 6.0 V battery, a 1800 Ω resistor and a light-dependent resistor in series. When the LDR is in darkness the pd across the resistor is 1.2 V.

a) Calculate the resistance of the LDR when it is in darkness.

b) When the sensor is in the light, its resistance falls to 2400 Ω. Calculate the pd across the LDR.

Solution

a) As the pd across the resistor is 1.2 V, the pd across the LDR must be 6 – 1.2 = 4.8 V.

The current in the circuit is $I = \frac{V}{R} = \frac{1.2}{1800} = 0.67$ mA.
The resistance of the LDR is $\frac{V}{I} = \frac{4.8}{0.67 \times 10^{-3}} = 7200\ \Omega$.

b) The ratio of $\frac{\textit{resistance across LDR}}{\textit{resistance across }1800\ \Omega} = \frac{2400}{1800} = 1.33$.

This is the same value as $\frac{\textit{pd across LDR}}{\textit{pd across }1800\ \Omega}$. For the ratio of pds to be 1.33, the pds must be 2.6 V and 3.4 V with the 3.4 V across the LDR.

2 A thermistor is connected in series with a fixed resistor and a battery. Describe and explain how the pd across the thermistor varies with temperature.

Solution

As the temperature of the thermistor rises, its resistance falls. The ratio of the pd across the fixed resistor to the pd across the thermistor rises too because the thermistor resistance is dropping. As the pd across the fixed resistor and thermistor is constant, the pd across the thermistor must fall.

The change in resistance in the thermistor occurs because more charge carriers are released as the temperature rises. Even though the movement of the charge carriers is impeded at higher temperatures, the release of extra carriers means that the resistance of the material decreases.

Investigate!

Variable resistor or potential divider?

- Set up the two circuits shown in figure 14. Match the value of the fixed resistor to the variable resistor – they do not need to be exactly the same but should be reasonably close. Add a voltmeter connected across the fixed resistor to check the pd that is available across it.
- Make sure that the maximum current rating for the fixed resistor and the variable resistor cannot be exceeded.
- Check the pd available in the two cases and convince yourself that the potential divider gives a wider range of voltages.

Heating effect equations

We saw earlier that the power P dissipated in a component is related to the pd V across the component and the current I in it:

$$P = IV$$

The energy E converted in time Δt is

$$E = IV\Delta t$$

When either V or I are unknown, then two more equations become available:

$$P = IV = I^2R = \frac{V^2}{R}$$

and

$$E = IV\Delta t = I^2R\Delta t = \frac{V^2}{R}\Delta t$$

These equations will allow you to calculate the energy converted in electrical heaters and lamps and so on. Applications that you may come across include heating calculations, and determining the consumption of energy in domestic and industrial situations.

Worked examples

1 Calculate the power dissipated in a 250 Ω resistor when the pd across it is 10 V.

Solution

$$P = \frac{V^2}{R} = \frac{10^2}{250} = 0.40 \text{ W}$$

2 A 9.0 kW electrical heater for a shower is designed for use on a 250 V mains supply. Calculate the current in the heater.

Solution

$$P = IV \text{ so } I = \frac{P}{V} = \frac{9000}{250} = 36 \text{ A}$$

3 Calculate the resistance of the heating element in a 2.0 kW electric heater that is designed for a 110 V mains supply.

Solution

$$P = \frac{V^2}{R};\ R = \frac{V^2}{P} = \frac{110^2}{2000} = 6.1 \text{ A.}$$

Kirchhoff's first and second laws

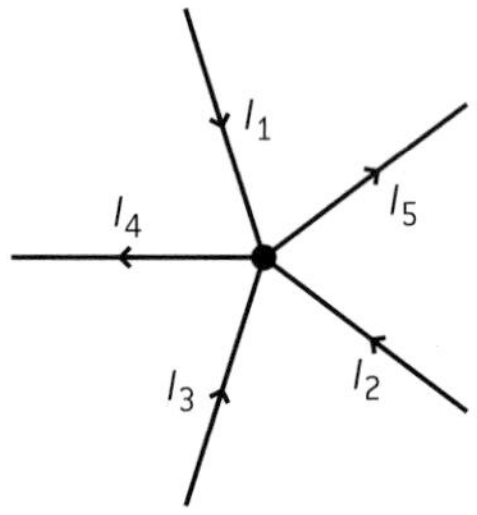

▲ Figure 15 Kirchhoff's first law.

This section contains no new physics, but it will consolidate your knowledge and put the electrical theory you have learnt so far into a wider physical context.

We saw that the charge carriers in a conductor move into and out of the conductor at equal rates. If 10^6 flow into a conductor in one second, then 10^6 must flow out during the same time to avoid the buildup of a static charge.

We also considered what happens when current splits into two or more parts at the junction where a parallel circuit begins. This can be taken one step further to a situation where there is more than one incoming current at the junction too.

Figure 15 shows a junction with three incoming currents and two outgoing ones. Our rule about the incoming charge equating to the outgoing charge must apply here too, so algebraically:

$$I_1 + I_2 + I_3 = I_4 + I_5$$

A general way to write this is to use the Σ sign (meaning "add up everything"), so that

for any junction $\sum I = 0$

When using the Σ notation, remember to get the signs of the currents correct. Call any current in *positive*, and any current out *negative*.

In words this can be written as

the sum of the currents into a junction equals the sum of the currents away from a junction
or
total charge flowing into a junction equals the total charge flowing away from the junction.

This important rule was first quoted by the Prussian physicist Gustav Kirchhoff in 1845. It is known as **Kirchhoff's first law**.

It is equivalent to a statement of **conservation of charge**.

Kirchhoff devised a second law which is also a conservation rule this time of energy.

In any electric circuit there are sources of emf (often a cell or a battery of cells for dc) and sinks of pd (typically, lamps, heating coils, resistors, and thermistors). A general rule in physics is that energy is conserved. Electrical components have to obey this too. So, in any electrical circuit, the energy being converted into electrical energy (in the sources of emf) must be equal to the energy being transferred from electrical to internal, by the sinks of pd.

This is **Kirchhoff's second law** equivalent to **conservation of energy**.

This second law applies to all closed circuits – both simple and complex.

In words, Kirchhoff's second law can be written as

in a complete circuit loop, the sum of the emfs in the loop is equal to the sum of the potential differences in the loop
or
the sum of all variations of potential in a closed loop equals zero

In symbols

for any closed loop in a circuit $\sum \varepsilon = \sum IR$

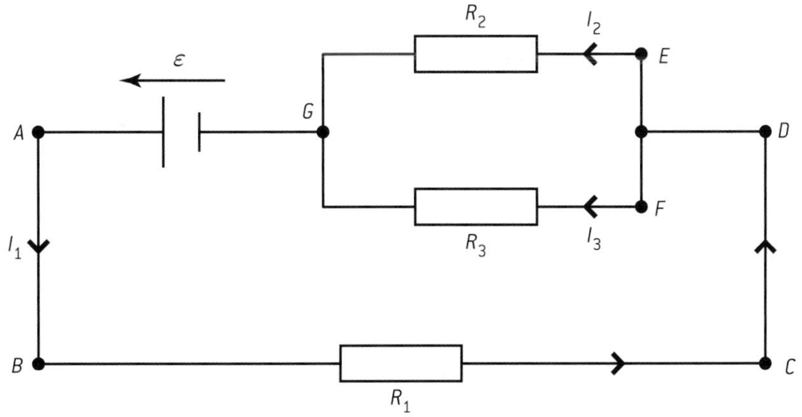

▲ Figure 16 Kirchhoff's second law.

Look at the circuit in figure 16, it has parallel and series elements in it. A number of possible loops are drawn and analysed:

Loop GABCDEG travelling anticlockwise round the loop

This loop begins at the cell and goes around the circuit, through resistor R_1 and resistor R_2 finally ending at the cell again. In this loop there is one source of emf and two sinks of pd (ignoring the leads, which we assume have zero resistance).

So

ε (*the emf of the cell*) $= I_1R_1 + I_2R_2$ (*the total pds aross the resistors*)

The direction of loop travel and the current direction are in all cases the same. We give a positive sign to the currents when this is the case. The emf of the cell is driving in the same direction as the loop travel direction; it gets a positive sign as well. If the loop direction and the current or emf were to be opposed then they would be given a negative sign.

loop EFGE travelling clockwise round the loop

This loop goes first through resistor R_3 and the loop direction is in the same direction as the conventional current. Next the loop goes through resistor R_1 but this time the current direction and the loop are different so there has to be a negative sign. There is no source of emf in the loop so the Kirchhoff equation becomes

$$0 = I_3R_3 - I_2R_2$$

Kirchhoff's first law can be applied at point G. The total current into point G is $I_2 + I_3$; the total out is I_1. The application of the law is $I_1 = I_2 + I_3$.

There are now three separate equations with three unknowns and these equations can be solved to work out the currents in each part of the circuit assuming that we know the value for the emf of the cell and the values of the resistances in the circuit.

By setting up a series of loops it is possible to work out the currents and pds for complicated resistor networks, more complicated than could be done using the resistor series and parallel rules alone.

Ideal and non-ideal meters

So far in this topic we have assumed (without mentioning it!) that the meters used in the circuits were ideal. Ideal meters have no effect on the circuits that they are measuring. We would always want this to be true but, unfortunately, the meters we use in real circuits in the laboratory are not so obliging and we need to know how to allow for this.

Ammeters are placed in series with components so the ammeter has the same current as the components. It is undesirable for the ammeter to change the current in a circuit but, if the ammeter has a resistance of its own, then this is what will happen. An ideal ammeter has zero resistance – clearly not attainable in practice as the coils or circuits inside the ammeter have resistance.

Voltmeters are placed in parallel with the device or parts of a circuit they are measuring. In an ideal world, the voltmeter will not require any energy for its coil to move (if it is a moving-coil type) or for its analogue to digital conversion (if it is a digital meter). The way to avoid current in the voltmeter is for the meter to have an infinite resistance – again, not an attainable situation in practice.

Some modern digital meters can get very close to these ideals of zero ohm for ammeters and infinite ohms for voltmeters. Digital meters are used more and more in modern science.

Worked examples

1 Calculate the unknown branch current in the following junctions.

a)

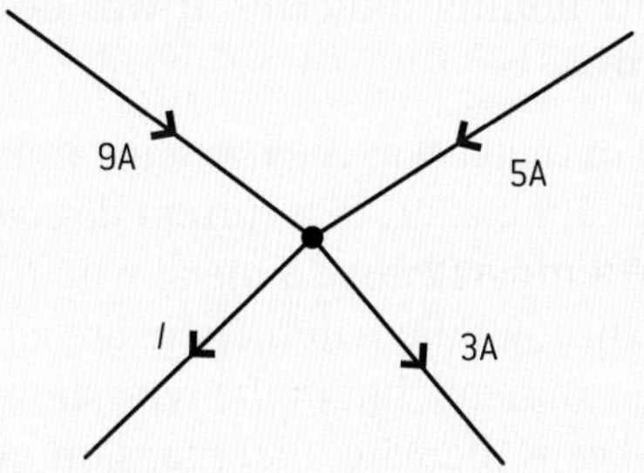

b)

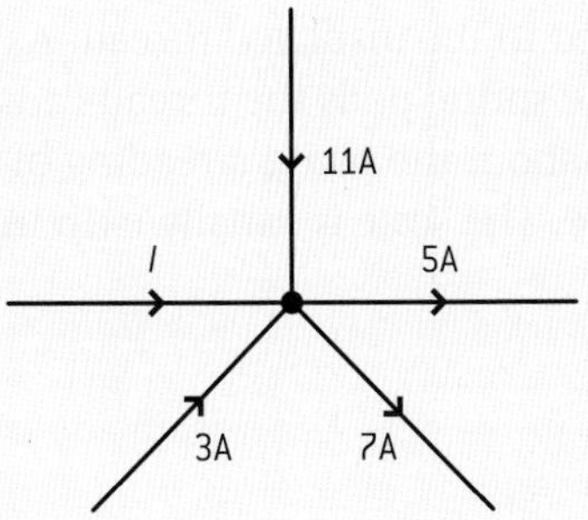

Solution

a) Current into the junction = 9 + 5A

Current out of the junction = 3 + *I*A

$I = 9 + 5 - 3 = 11$ A out of the junction.

b) Current into the junction = $I + 11 + 3 = I + 14$A

Current out of the junction = 5 + 7 = 12A

$I = 12 - 14 = -2$ A, so the current I is directed away from the junction.

2 Calculate the currents in the circuit shown.

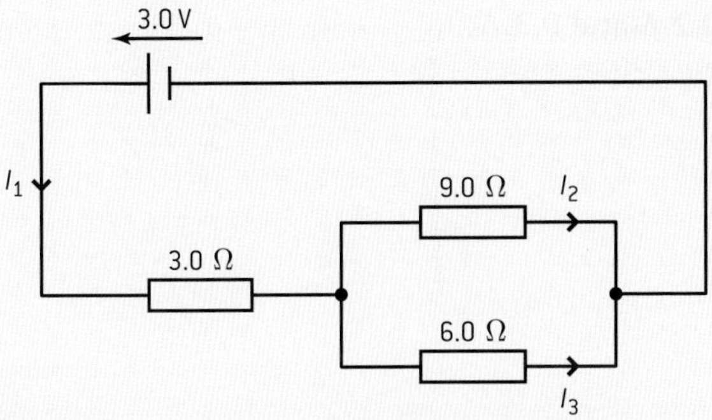

Solution

(This circuit can be analysed using the resistors in series and parallel equations. It is given here to show the technique and to convince you that it works.)

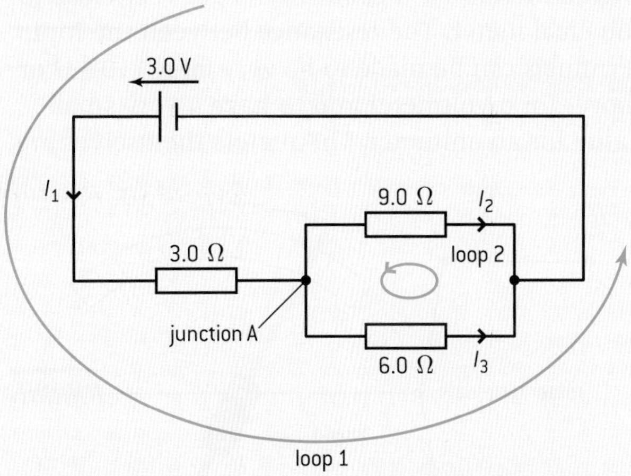

Current directions have been assigned and two loops 1 and 2 and junction A defined in the diagram.

For loop 1

$$3 = 3I_1 + 9I_2 \qquad \textit{[equation 1]}$$

(the emf in the loop is 3 V)

for loop 2

$$0 = 6I_3 - 9I_2 \qquad \textit{[equation 2]}$$

(there is no source of emf in this loop, current I_2 is in the opposite direction to the loop direction 0.

For junction A

$$I_1 = I_2 + I_3$$

so

$$0 = 9I_2 - 6I_1 + 6I_2$$

$$0 = 15I_2 - 6I_1$$

And from equation 1

$$6 = 6I_1 + 18I_2$$

Adding the equations

$$6 = 33I_2$$

$$I_2 = 0.18 \text{ A}$$

Substituting gives:

$$I_1 = 0.45 \text{ A}$$

and

$$I_3 = 0.27 \text{ A}$$

Nature of science

How times change!

Digital meters are a good way to get very close to the ideal meter. The resistance between the input terminals can be made to be very large (10^{12} Ω or more) for a voltmeter and to have a very small value for an ammeter. The meters themselves are easy to read without any judgements required about what the pointer indicates. It was not always like this.

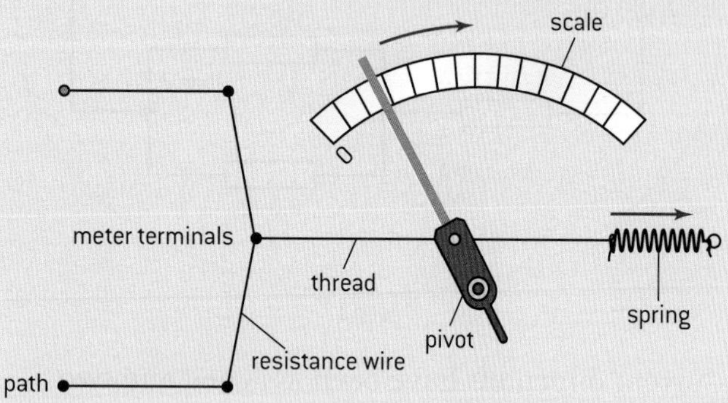

▲ Figure 17

An early form of meter was the hot-wire ammeter. This uses the heating effect of a current directly to increase the temperature of a metal wire (the current flows through the wire). There are a number of forms of the ammeter, but in the type shown here, a spring keeps the wire under tension. As the wire expands with the increase in temperature, any point on the wire moves to the right. A pivoted pointer is attached to the insulated thread. As the wire expands, the spring pulls the thread causing the pointer to rotate about the pivot; a reading of current can now be made. The scale is usually extremely non-linear.

Worked examples

1 A 250 Ω resistor is connected in series with a 500 Ω resistor and a 6.0 V battery.

a) Calculate the pd across the 250 Ω resistor.

b) Calculate the pd that will be measured across the 250 Ω resistor if a voltmeter of resistance 1000 Ω is connected in parallel with it.

Solution

a) The pd across the 250 Ω resistor

$$= \frac{V_{in} \times R_1}{(R_1 + R_2)} \; \frac{6 \times 250}{(250 + 500)} = 2.0 \text{ V.}$$

b) When the voltmeter is connected, the resistance of the parallel combination is

$$R = \frac{R_1 R_2}{(R_1 + R_2)} = \frac{250 \times 1000}{1250} = 200\ \Omega$$

$$V = \frac{200 \times 6}{700} = 1.7 \text{ V}$$

2 An ammeter with a resistance of 5.0 Ω is connected in series with a 3.0 V cell and a lamp rated at 300 mA, 3 V. Calculate the current that the ammeter will measure.

Solution

Resistance of lamp $= \frac{V}{I} = \frac{3}{0.3} = 10\ \Omega$. Total resistance in circuit $= 10 + 5 = 15\ \Omega$. So current in circuit $= \frac{V}{R} = \frac{3}{15} = 200$ mA. This assumes that the resistance of the lamp does not vary between 0.2 A and 0.3 A

5.3 Electric cells

Understanding

- → Cells
- → Primary and secondary cells
- → Terminal potential difference
- → Electromotive force emf
- → Internal resistance

Applications and skills

- → Investigating practical electric cells (primary and secondary)
- → Describing the discharge characteristic of a simple cell (variation of terminal potential difference with time)
- → Identifying the direction current flow required to recharge a cell
- → Determining internal resistance experimentally
- → Solving problems involving emf, internal resistance, and other electrical quantities

Equation

- → emf of a cell: $\varepsilon = I(R + r)$

Nature of science

Scientists need to balance their research into more and more efficient electric cells with the long-term risks associated with the disposal of the cells. Some modern cells have extremely poisonous contents. There will be serious consequences if these chemicals are allowed to enter the water supply or the food chain. Can we afford the risk of using these toxic substances, and what steps should the scientists take when carrying forward the research?

Introduction

Electric currents can produce a **chemical effect**. This has great importance in chemical industries as it can be a method for extracting ores or purifying materials. However, in this course we do not investigate this aspect of the chemical effect. Our emphasis is on the use of an electric cell to store energy in a chemical form and then release it as electrical energy to perform work elsewhere.

Cells

Cells operate as direct-current (dc) devices meaning that the cell drives charge in one direction. The electron charge carriers leave the negative terminal of the cell. After passing around the circuit, the electrons re-enter the cell at the *positive* terminal. The positive terminal has a higher potential than the negative terminal – so electrons appear to gain energy (whereas positive charge carriers would appear to lose it).

The chemicals in the cell are reacting while current flows and as a result of this reaction the electrons gain energy and continue their journey around the circuit.

Nature of science

Anodes and cathodes

You will meet a naming system for anode and cathode which seems different for other cases in physics. In fact it is consistent, but you need to think carefully about it. For example, in a cathode-ray tube where the electrons in the tube are emitted from a hot metal filament, the filament is called the cathode and is at a negative voltage. This appears to be the opposite notation from that used for the electric cell. The reason is that the notation refers to what is happening inside the cell, not to the external circuit. The chemical reaction in the cell leads to positive ions being generated at one of the electrodes and then flowing away into the bulk of the liquid. So as far as the interior of the cell is concerned this is an anode, because it is an (internal) source of positive ions. Of course, as far as the external circuit is concerned, the movement of positive charges away from this anode leaves it negative and the electrode will repel electrons in the external circuit. As external observers, we call this the negative terminal even though (as far as the interior of the cell is concerned) it is an anode.

Primary and secondary cells

Many of the portable devices we use today: torches, music players, computers, can operate with internal cells, either singly or in batteries. In some cases, the cells are used until they are exhausted and then thrown away. These are called **primary cells**. The original chemicals have completely reacted and been used up, and they cannot be recharged. Examples include AA cells (properly called dry cells) and button mercury cells as used in clocks and other small low-current devices.

On the other hand, some devices use rechargeable cells so that when the chemical reactions have finished, the cells can be connected to a charger. Then the chemical reaction is reversed and the original chemicals form again. When as much of the re-conversion as is possible has been achieved, the cell is again available as a chemical energy store. Rechargeable cells are known as **secondary cells**.

There many varieties of cell, here are two examples of the chemistry of a primary cell and that of a secondary cell.

The **Leclanché cell** was invented by Georges Leclanché in 1886. It is a primary cell and is the basis of many domestic torch and radio batteries.

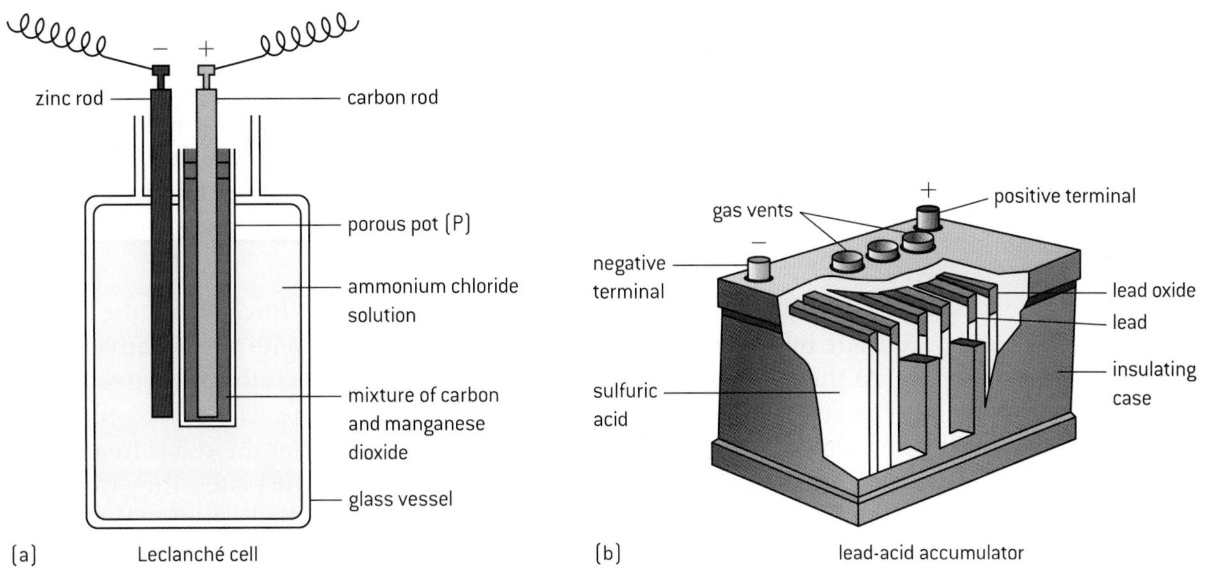

▲ Figure 1 Leclanché cell and lead-acid accumulator.

Figure 1(a) shows one practical arrangement of the cell. It consists of a zinc rod that forms the negative terminal. Inside the zinc is a paste of ammonium chloride separated from manganese dioxide (the cathode) by a porous membrane. In the centre of the manganese dioxide is a carbon rod that acts as the positive terminal for the cell.

Zinc atoms on the inside surface of the case oxidize to become positive ions. They then begin to move away from the inside of the case through the chloride paste leaving the case negatively charged. When the cell is connected to an external circuit, these electrons move around the circuit eventually reaching the carbon rod. A reaction inside the cell uses these electrons together with the components of the cell eventually forming the "waste products" of the cell, which include ammonia, manganese oxide, and manganese hydroxide.

Secondary cells are very important today. They include the lead–acid accumulator, together with more modern developments such as the nickel–cadmium (NiCd), nickel–metal-hydride (NiMH), and lithium ion cells. All these types can be recharged many times and even though they may have a high initial cost (compared to primary cells of an equivalent emf and capacity), the recharge is cheap so that, during the projected lifetime of the cell, the overall cost is lower than that of primary cells.

Although the lead-acid accumulator is one of the earliest examples of a secondary cell, it remains important ("accumulator" is an old word that implies the accumulation or collection of energy into a store). It is capable of delivering the high currents needed to start internal combustion and diesel engines, and it is reliable in the long term to maintain the current to important computer servers if the mains supply fails.

The **lead–acid cell** (figure 1(b)) is slightly older than the Leclanché cell, and was invented in 1859 by Gaston Planté.

In its charged state the cell consists of two plates, one of metallic lead, the other of lead(IV) oxide (PbO_2) immersed in a bath of dilute sulfuric acid. During discharge when the cell is supplying current to an external circuit, the lead plate reacts with the acid to form lead(II) sulfate and the production of two free electrons. The liquid surrounding the plates loses acid and becomes more dilute. At the oxide plate, electrons are gained (from the external circuit) and lead(II) sulfate is formed, also with the removal of acid from the liquid. So the net result of discharging is that the plates convert to identical lead sulfates and the liquid surrounding the plates becomes more dilute.

Recharging reverses these changes: electrons are forced from the positive plate by an external charging circuit and forced onto the negative plate. At the negative plate, atomic lead forms, at the positive, lead oxide is created, which restores the cell to its original state. The charge–recharge cycle can be carried out many times providing that the cell is treated carefully. However, if lead or lead oxide fall off the plates, they can no longer be used in the process and will collect at the bottom of the cell container. In the worst case the lead will short out the plates and the cell will stop operating. If the cell is overcharged (when no more sulfate can be converted into lead and lead oxide) the charging current will begin to split the water in the cell into hydrogen and oxygen gas which are given off from the cell. The total amount of liquid in the cell will decrease, meaning that the plates may not be fully covered by the acid. These parts of the plates will then not take part in the reaction and the ability of the cell to store energy will decrease.

Much industrial research effort is concentrated on the development of rechargeable cells. An important consideration for many manufacturers of electronic devices is the energy density for the battery (the energy stored per unit volume) as this often determines the overall design and mass of an electronic device. Also a larger energy density may well provide a longer lifetime for the device.

Capacity of a cell

Two cells with the same chemistry will generate the same electromotive force (emf) as each other. However if one of the cells has larger plates than the other and contains larger volumes of chemicals, then it will be able to supply energy for longer when both cells carry the same current.

The **capacity** of a cell is the quantity used to measure the ability of a cell to release charge. If a cell is discharged at a high rate then it will not be long before the cell is exhausted or needs recharging, if the discharge current is low then the cell will supply energy for longer times. The capacity of a cell or battery is the constant current that it can supply for a given discharge. So, if a cell can supply a constant current of 2 A for 20 hours then it said to have a capacity of 40 amp-hours (abbreviated as 40 Ah).

The implication is that this cell could supply 1 A for 40 hours, or 0.1 A for 400 hours, or 10 A for 4 hours. However, practical cells do not necessarily discharge in such a linear way and this cell may be able to provide a small discharge current of a few milliamps for much

Worked examples

1 Explain the difference between a primary and a secondary cell.

Solution

A primary cell is one that can convert chemical energy to electrical energy until the chemicals in the cell are exhausted. Recharging the cell is not possible.

A secondary cell can be recharged and the chemicals in it are converted back to their original form so that electrical energy can be supplied by the cell again.

2 A cell has a capacity of 1400 mA h. Calculate the number of hours for which it can supply a current of 1.8 mA.

Solution

Cell capacity = current × time and therefore $1400 = 1.8t$. In this case $t = 780$ hours.

longer than expected from the value of its capacity. Conversely, the cell may be able to discharge at 10 A for much less than 4 hours. An extreme example of this is the lead–acid batteries used to start cars and commercial vehicles. A typical car battery may have a capacity of 100 A h. The current demand for starting the car on a cold winter's day can easily approach 150 A. But the battery will only be able to turn the engine for a few minutes rather than the two-thirds of an hour we might expect.

Discharging cells

Investigate!

Discharge of a cell

- Some cells have high capacity, and studying the discharge of such a cell can take a long time. This is a good example of how electronic data collection (data-logging) can help an experimenter.

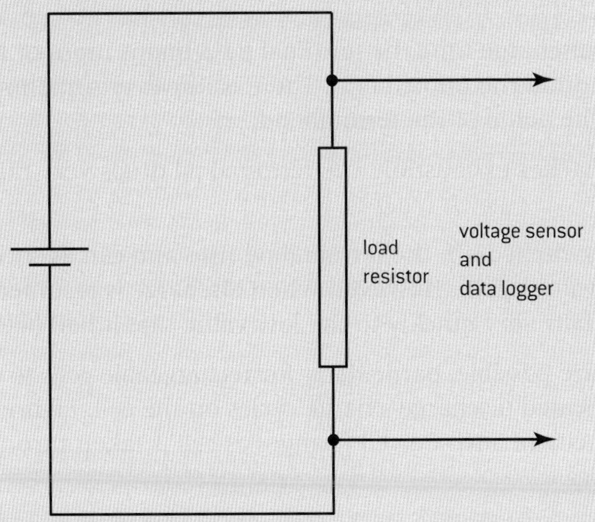

▲ Figure 2

- The aim is to collect data to find how the terminal pd across the cell varies with time from the start of the discharge. The basic circuit is shown in figure 2. Use a new cell to provide the current so that the initial behaviour of the cell can be seen too.
- Many data loggers need to be "told" how often to take measurements (the sampling rate). You will need to make some judgements about the overall length of time for which the experiment is likely to operate. Suppose you have a 1.5 V cell rated at 1500 mA h (you can check the figure by checking the cell specification on the manufacturer's website). It would seem reasonable that if the cell is going to supply a current of 250 mA, then it would discharge in a time of between 4–10 hours. You will need to set up the data logger accordingly with a suitable time between readings (the sampling interval). Do not, however, exceed the maximum discharge current of the cell.
- When the computer is set up, turn on the current in the discharge circuit and start the logging. Eventually the cell will have discharged and you can display the data as a graph. What are the important features of your graph?
- Test other types of cell too, at least one primary and one secondary cell. Are there differences in the way in which they discharge?
- For at least one cell, towards the end of the discharge, switch off the discharge current while still continuing to monitor the terminal pd. Does the value of the pd stay the same or does it recover? When the discharge is resumed what happens to the pd?

The results you obtain will depend on the types and qualities of the cells you test. A typical discharge curve looks like that in figure 3 on p222 in which the terminal potential difference of the cell is plotted against time since the discharge current began.

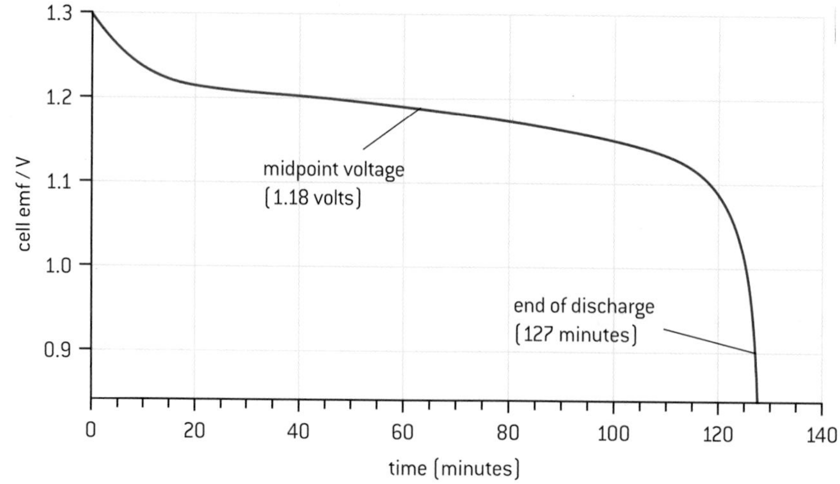

▲ Figure 3 Typical discharge–time graph for a cell.

Important features of this graph are that:

- The initial terminal pd is higher than the quoted emf (the value the manufacturer prints on the case), but the initial value quickly drops to the rated emf (approximately).
- For most of the discharge time the terminal pd remains more or less constant at or around the quoted emf. There is however sometimes a slow decline in the value of the terminal pd.
- As the cell approaches exhaustion, the terminal pd drops very rapidly to a low level.
- If the current is switched off, the terminal pd rises and can eventually reach the rated value again. However, when discharge is resumed, the terminal pd falls very quickly to the low value that it had before.

Other experiments are possible, particularly for rechargeable cells to see the effects of repeated discharge–charge cycles on the cell. Figure 4 shows how the capacity of a Ni–Cd cell changes as cell is taken through more and more cycles.

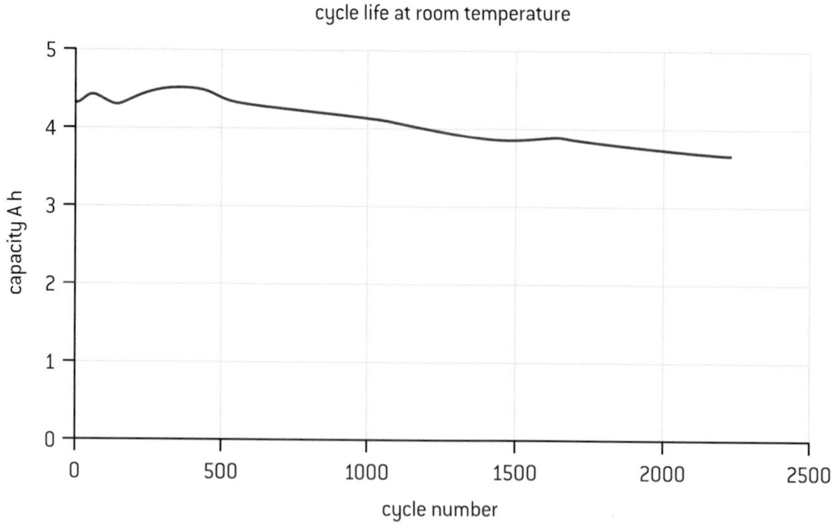

▲ Figure 4 Cycle life of a rechargeable cell.

Recharging secondary cells

The clue to recharging a secondary cell is in the earlier descriptions of cell chemistry. The chemicals produce an excess of electrons at the negative terminal. During discharge these electrons then move through the external circuit transferring their energy as they go. When the electron arrives at the positive terminal, all of its energy will have been transferred to other forms and it will need to gain more from the chemical store in the cell.

To reverse the chemical processes we need to return energy to the cell using electrons as the agents, so that the chemical action can be reversed. When charging, the electrons need to travel in the reverse direction to that of the discharge current and you can imagine that the charger has to force the electrons the "wrong" way through the cell.

Part of a possible circuit to charge a 6-cell lead–acid battery is shown in figure 5.

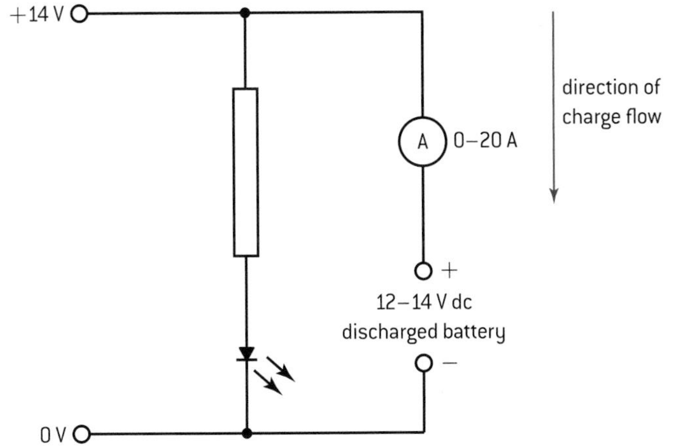

▲ Figure 5 Charging circuit for a cell.

The direction of the charging current is shown in the diagram. It is in the opposite direction to that of the cell when it is supplying energy. An input pd of 14 V is needed, notice the polarity of this input. The light-emitting diode (LED) and its series resistor indicate that the circuit is switched on. The ammeter shows the progress of charging. When the battery is completely discharged, the charging current will be high, but as the charging level (and therefore the emf) of the battery rises, this current gradually falls. During the charging process the terminal voltage will be greater than the emf of the cell. At full charge the emf of the battery is 13.8 V. When the current in the meter is zero, the battery is fully charged.

Internal resistance and emf of a cell

The materials from which the cells are made have electrical resistance in just the same way as the metals in the external circuit. This **internal resistance** has an important effect on the total resistance and current in the circuit.

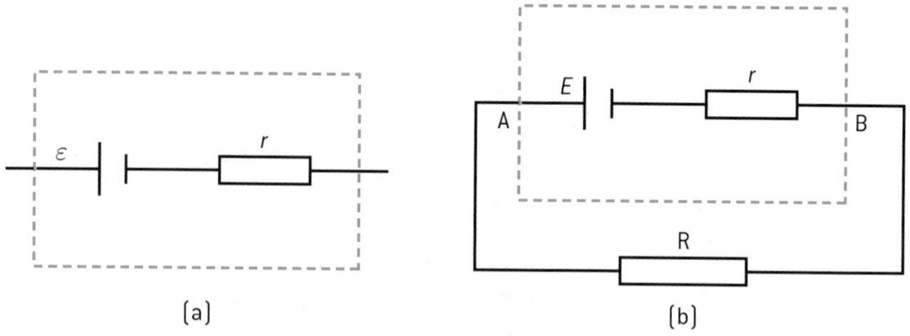

▲ Figure 6

Figure 6(a) shows a simple model for a cell. Inside the dotted box is an "ideal" cell that has no resistance of its own. Also inside the box is a resistance symbol that represents the internal resistance of the cell. The two together make up our model for a real cell. The model assumes that the internal resistance is constant (for a practical cell it varies with the state of discharge) and that the emf is constant (which also varies with discharge current).

The model cell has an emf ε and an internal resistance r in series with an external resistance R (figure 6(b)). The current in the circuit is I.

We can apply Kirchhoff's second law to this circuit:

the emf of the cell supplying energy to the circuit $= \varepsilon$

the sum of the pds $= IR + Ir$

So

$$\varepsilon = IR + Ir$$

If the pd across the external resistor is V, then

$$\varepsilon = V + Ir$$

or

$$V = \varepsilon - Ir$$

It is important to realize that V, which is the pd across the external resistance, is equal to the terminal pd across the ideal cell and its internal resistance (in other words between A and B). **The emf is the open circuit pd across the terminals of a power source – in other words, the terminal pd when no current is supplied.** The pd between A and B is less than the emf unless the current in the circuit is zero. The difference between the emf and the **terminal pd** (the measured pd across the terminal of the cell) is sometimes referred to as the "lost pd" or the "lost volts". These lost volts represent the energy required to drive the electron charge carriers through the cell itself. Once the energy has been used in the cell in this way, it cannot be available for conversion in the external circuit. You may have noticed that when a secondary cell is being charged, or any cell is discharging at a high current, the cell becomes warm. The energy required to raise the temperature of the cell has been converted in the internal resistance.

TOK

Simple assumptions

The model of a cell with a fixed internal resistance and a constant emf is an example of modelling. In this case the model is a simple one that cannot be realized in practice. Another model we used earlier was to suggest that there are ideal ammeters with zero resistance and ideal voltmeters with infinite resistance.

Scientists frequently begin with a very simple model of a system and then explore the possibilities that this model can offer in terms of analysis and behaviour. The next step is to make the model more complicated (but without being too intricate!) and to see how much complexity is needed before the model resembles the real system that is being modelled.

Do the simplifications and assumptions of ideal behaviour form a suitable basis for modelling?

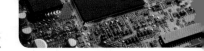

Investigate!

Measuring the internal resistance of a fruit cell

The method given here works for any type of electric cell, but here we use a citrus fruit cell (orange, lemon, lime, etc.) or even a potato for the measurement. The ions in the flesh of the fruit or the potato react with two different metals to produce an emf. With an external circuit that only requires a small current, the fruit cell can discharge over surprisingly long times.

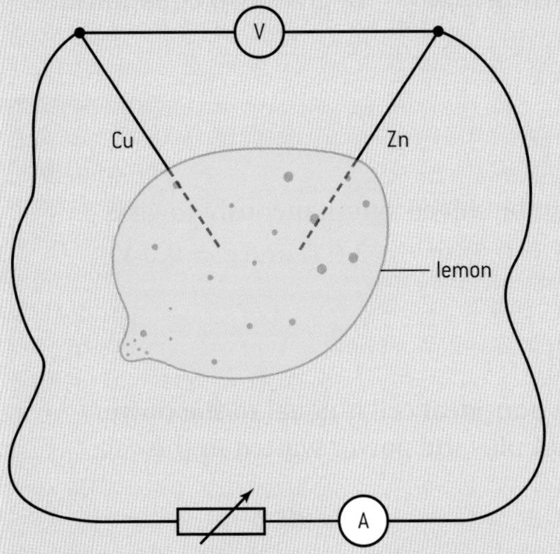

▲ Figure 7

- To make the cell, take a strip of copper foil and a strip of zinc foil, both about 1 cm by 5 cm and insert these, about 5 cm apart deep into the fruit. You may need to use a knife to make an incision unless the foil is stiff enough.
- Connect up the circuit shown in figure 7 using a suitable variable resistor.
- Measure the terminal pd across the fruit cell and the current in it for the largest range of pd you can manage.

  Compare the equation

  $$V = \varepsilon - Ir$$

  with the equation for a straight line

  $$y = c + mx$$

  A plot of V on the y-axis against I on the x-axis should give a straight line with a gradient of $-r$ and an intercept on the V-axis of ε.
- A set of results for a cell (not a fruit cell in this case) is shown together with the corresponding graph of V against I. The intercept for this graph is 1.25 V and the gradient is $-2.4\ \text{V A}^{-1}$ giving an internal resistance value of 2.4 Ω.

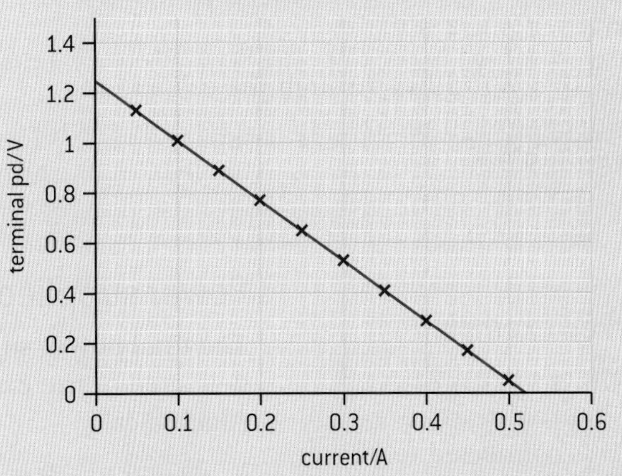

Terminal pd / V	Current / A
1.13	0.05
1.01	0.10
0.89	0.15
0.77	0.20
0.65	0.25
0.53	0.30
0.41	0.35
0.29	0.40
0.17	0.45
0.05	0.50

▲ Figure 8

Worked example

1 A cell of emf 6.0 V and internal resistance 2.5 Ω is connected to a 7.5 Ω resistor.

Calculate:

a) the current in the cell

b) the terminal pd across the cell

c) the energy lost in the cell when charge flows for 10 s.

Solution

a) Total resistance in the circuit is 10 Ω, so current in circuit $= \frac{6}{10} = 0.60$ A

b) The terminal pd is the pd across cell $= IR =$

$0.6 \times 7.5 = 4.5$ V

c) In 10 s, 6 C flows through the cell and the energy lost in the cell is 1.5 J C^{-1}. The energy lost is 9.0 J.

2 A battery is connected in series with an ammeter and a variable resistor R. When $R = 6.0$ Ω, the current in the ammeter is 1.0 A. When $R = 3.0$ Ω, the current is 1.5 A. Calculate the emf and the internal resistance of the battery.

Solution

Using

$$V = \varepsilon - Ir$$

and knowing that $V = IR$ gives two equations:

$$6 \times 1 = \varepsilon - 1 \times r$$

and

$$3 \times 1.5 = \varepsilon - 1.5 \times r$$

These can be solved simultaneously to give $(6 - 4.5) = 0.5r$ or $r = 3.0$ Ω and $\varepsilon = 9.0$ V.

Worked example

A battery of emf 9.0 V and internal resistance 3.0 Ω is connected to a load resistor of resistance 6.0 Ω. Calculate the power delivered to the external load.

Solution

Using the equation $\frac{\varepsilon^2}{(R+r)^2} R$ leads to $\frac{9^2}{(6+3)^2} 6 = 6.0$ W

Power supplied by a cell

The total power supplied by a non-ideal cell is equal to the power delivered to the external circuit plus the power wasted in the cell. Algebraically,

$$P = I^2 R + I^2 r$$

using the notation used earlier.

The power delivered to the external resistance is

$$\frac{\varepsilon^2}{(R+r)^2} R$$

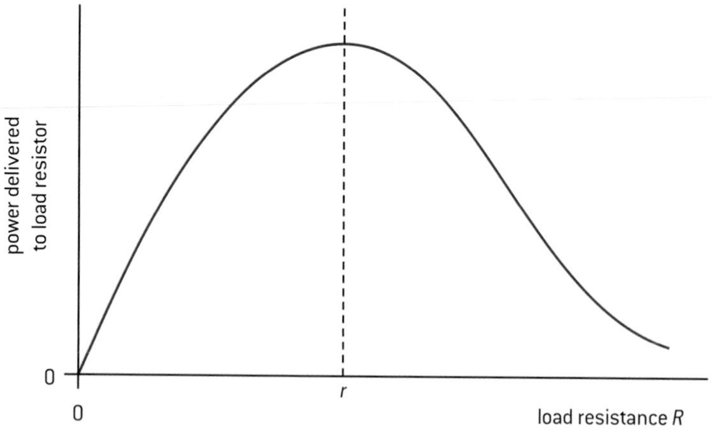

▲ Figure 9 Power delivered to a resistor.

Figure 9 shows how the power that is delivered to the external circuit varies with R. The peak of this curve is when $r = R$, in other words, when the internal resistance of the power supply is equal to the resistance of the external circuit. The load and the supply are "matched" when the resistances are equal in this way. This matching of supply and circuit is important in a number of areas of electronics.

5.4 Magnetic effects of electric currents

Understanding

- Magnetic fields
- Magnetic force

Applications and skills

- Sketching and interpreting magnetic field patterns
- Determining the direction of the magnetic field based on current direction
- Determining the direction of force on a charge moving in a magnetic field
- Determining the direction of force on a current-carrying conductor in a magnetic field
- Solving problems involving magnetic forces, fields, current and charges

Equations

- force on a charge moving in a magnetic field: $F = qvB \sin\theta$
- force on a current-carrying conductor in a magnetic field: $F = ILB \sin\theta$

Nature of science

Sometimes visualization aids our understanding. Magnetic field lines are one of the best examples of this. Scientists began by visualizing the shape and strength of a magnetic field through the position and direction of the fictitious lines of force (or field lines). The image proved to be so powerful that the technique was subsequently used with electric and gravitational fields too.

Introduction

Electromagnetism is the third effect observed when charge moves in a circuit – the electric current gives rise to a magnetic field. But it was not the observation of a magnetic effect arising from a current in a wire that began the ancient study of magnetism. Early navigators knew that some rocks are magnetic. As with the nature of electric charge, the true origins of magnetic effects remained obscure for many centuries. Only comparatively recently has an understanding of the microscopic aspects of materials allowed a full understanding of the origins of magnetism. As with electrostatics, scientists began by using the concept of the field to describe the behaviour of interacting magnets. However, magnetic fields differ fundamentally in character from electrostatic fields and are in some ways more complex.

Magnetic field patterns

The repulsion between the like poles of two bar magnets is familiar to us. The forces between magnets of even quite modest strength are impressive. Modern magnetic alloys can be used to produce tiny magnets (less than 1 cm in diameter and a few millimetres thick) that can easily attract another ferromagnetic material through significant thicknesses of a non-magnetic substance.

At the beginning of a study of magnetism, it is usual to describe the forces in terms of fields and field lines. You may have met this concept before. There is said to be a **magnetic field** at a point if a force acts on a magnetic pole (in practice, a pair of poles) at that point. Magnetic fields are visualized through the construction of field lines.

Magnetic field lines have very similar (but not identical) properties to those of the electric field lines in Sub-topic 5.1. In summary these are:

- Magnetic field lines are conventionally drawn from the north-seeking pole to the south-seeking pole, they represent the direction in which a north-seeking pole at that point would move.
- The strength of the field is shown by the density of the field lines, closer lines mean a stronger field.
- The field lines never cross.
- The field lines act as though made from an elastic thread, they tend to be as short as possible.

Nature of science

Talking about poles

A word about notation: there is a real possibility of confusion when talking about magnetic poles. This partly arises from the origin of our observations of magnetism. When we write "magnetic north pole" what we really mean is "the magnetic pole that seeks the geographic north pole" (figure 1(a)). We often talk loosely about a magnetic north pole pointing to the north pole. Misunderstandings can occur here because we also know that like poles repel and unlike poles attract. So we end up with the situation that a magnetic north pole is attracted to the "geographic north pole" – which seems wrong in the context of two poles repelling. In this book we will talk about north-seeking poles meaning "geographic north-seeking" and south-seeking poles meaning "geographic south seeking". On the diagrams, N will mean geographic north seeking, S will mean geographic south seeking.

Figure 1(b) shows the patterns for a single bar magnet and (c) and (d) two arrangements of a pair of bar magnets of equal strength.

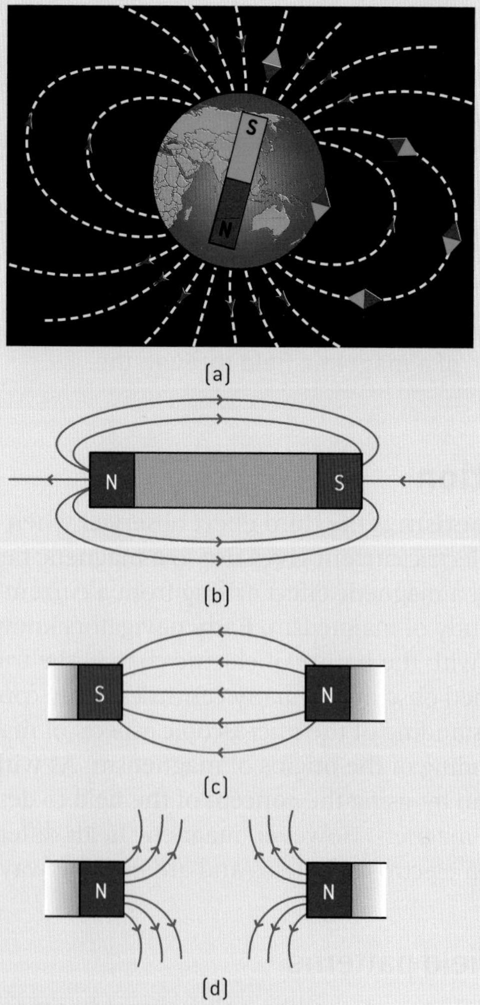

▲ Figure 1 Magnetic field patterns.

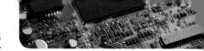

In the pairs, notice the characteristic field pattern when the two opposite poles are close (figure 1(b)) and when the two north-seeking poles are close (figure 1(c)). When two north-seeking poles are close (or two south-seeking poles), there is a position where the field is zero between the magnets (called a null point). When two opposite poles are close, the field lines connect the two magnets. In this situation the magnets will be attracted, so this particular field pattern implies an attraction between poles.

Investigate!

Observing field patterns of permanent magnets and electric currents

- There are a number of ways to carry out this experiment. They can involve the scattering of small iron filings, observation of suspensions of magnetized particles in a special liquid, or other techniques. This *Investigate!* is based on the iron-filing experiment but the details will be similar if you have access to other methods.
- Take a bar magnet and place a piece of rigid white card on top of it. You may need to support the card along its sides. Choose a non-magnetic material for the support.
- Take some iron filings in a shaker (a pepper pot is ideal) and, from a height above the card of about 20 cm sprinkle filings onto the card. It is helpful to tap the card gently as the filings fall onto it.
- You should see the field pattern forming as the magnetic filings fall through the air and come under the influence of the magnetic field. Sketch or photograph the arrangement.
- The iron filings give no indication of field direction. The way to observe this is to use a plotting compass – a small magnetic compass a few centimetres in diameter – that indicates the direction to which a north-seeking pole sets itself. Place one or more of these compasses on the card and note the direction in which the north-seeking pole points.
- Repeat with two magnets in a number of configurations; try at least the two in figure 1.
- Electric currents also give rise to a magnetic field. However, currents small enough to be safe will only give weak fields, not strong enough to affect the filings as they fall.

▲ Figure 2

- To improve the effect: cut a small hole in the centre of the card and run a long lead through it (the lead will need to be a few metres long). Loop the lead in the same direction through a number of times (about ten turns if possible). This trick enables one current to contribute many times to the same field pattern. You may need at least 25 A in total (2.5 A in the lead) to see an effect.

Magnetic field due to a current in conductors

The magnetic field pattern due to a current in a long straight wire is a circular pattern centred on the wire. This seems odd to anyone only used to the bar magnet pattern. Measurements show that the field is strong close to the wire but becomes weaker further away from it. This should be clear in your drawings of the long wire field pattern when the lines of force are drawn at increased spacing as you move further from the wire.

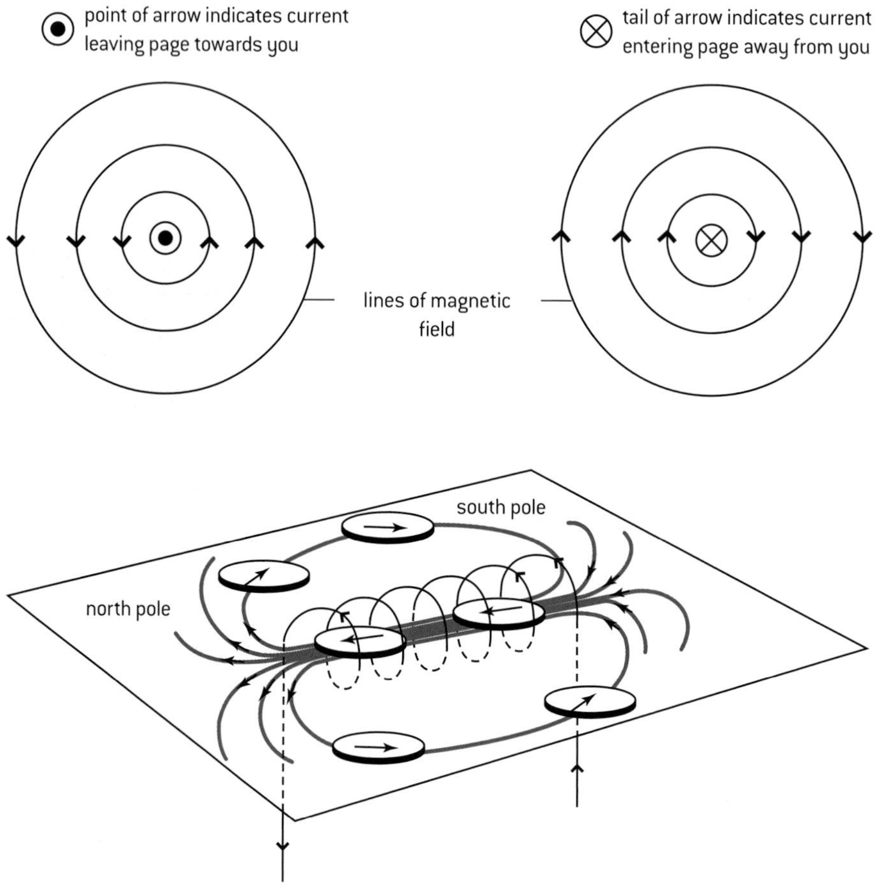

▲ Figure 3 Magnetic field patterns around current-carrying conductors.

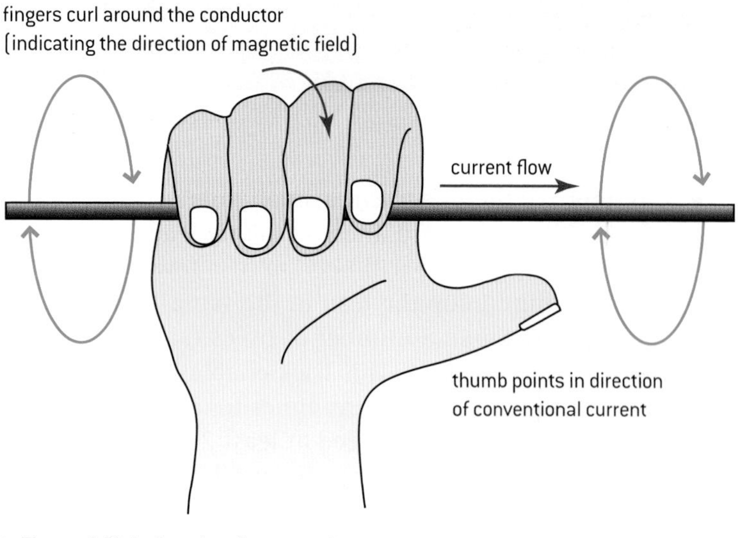

▲ Figure 4 Right-hand corkscrew rule.

Your observations using the plotting compasses should have shown that the direction of the field depends on the direction of the current.

Using the conventional current (i.e. the direction that positive charges are moving in the wire) the relationship between the current and the magnetic field direction obeys a **right-hand corkscrew rule relationship**.

To remember this, hold your right hand with the fingers curled into the palm and the thumb extended away from the fingers (see figure 4). The thumb represents the direction of the conventional current and the fingers represent the direction of the field. Another way to think of the current–field relationship is in terms of a screwdriver being used to insert a screw. The screwdriver has to turn a right-handed screw clockwise to insert it and drive the screw forwards. The direction in which the imaginary screw moves is that of the conventional current, and the direction in which the screwdriver turns is that of the field.

Use whichever **direction rule** you prefer, but use it consistently and remember that the rule works for conventional current.

The strength of the magnetic field can be increased by increasing the current in the conductor.

The magnetic field due to the solenoid is familiar to you already. To understand how it arises, you need to imagine the long straight wire being coiled up into the solenoid shape taking its circular field with it as the coiling takes place. With current in the wire, the circular field is set up in each wire. This circular field adds together with fields from neighbouring turns in the solenoid. Figure 5 shows this; look closely at what happens close to the individual wires. The black lines show the field near the wires, the blue lines show how the fields begin to combine, the red line shows the combined field in the centre of the solenoid.

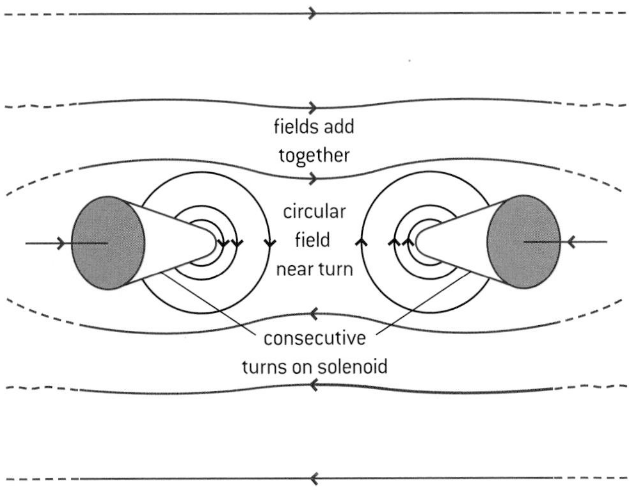

▲ Figure 5 Building up the field pattern.

A field runs along the hollow centre of the solenoid and then outside around the solenoid (figure 3). Outside it is identical to the bar magnet pattern to the extent that we can assign north- and south-seeking poles to the solenoid. Again there is an easy way to remember this using an N and S to show the north-seeking and south-seeking poles. The arrows on the N and S show the current direction when looking into the solenoid from outside. If you look into the solenoid and the conventional current is anticlockwise at the end of the solenoid closer to you then that is the end that is north-seeking. If the current is clockwise then that is the south-seeking end.

The strength of the magnetic field in a solenoid can be increased by:

- increasing the current in the wire
- increasing the number of turns per unit length of the solenoid
- adding an iron core inside the solenoid.

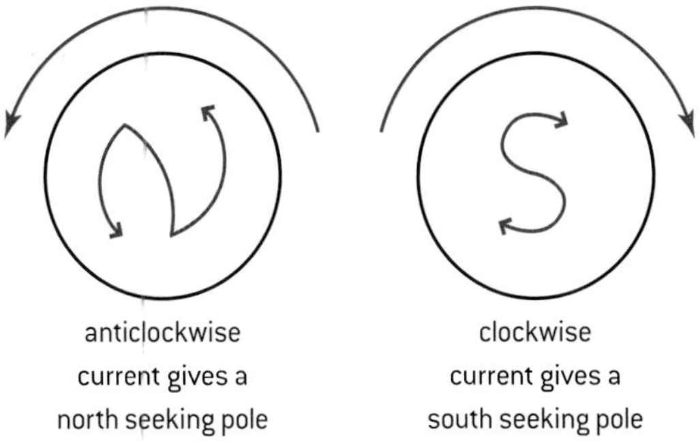

anticlockwise current gives a north seeking pole
clockwise current gives a south seeking pole

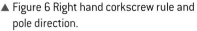

▲ Figure 6 Right hand corkscrew rule and pole direction.

Worked example

Four long straight wires are placed perpendicular to the plane of the paper at the edges of a square.

The same current is in each wire in the direction shown in the diagram. Deduce the direction of the magnetic field at point P in the centre of the square.

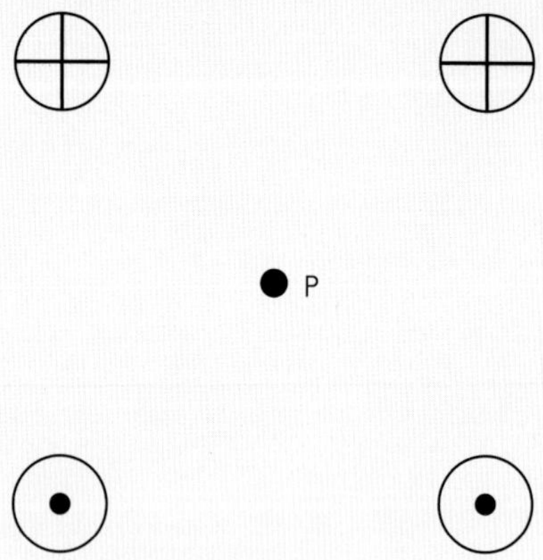

Solution

The four field directions are shown in the diagram. The sum of these four vectors is another vector directed from point P to the left.

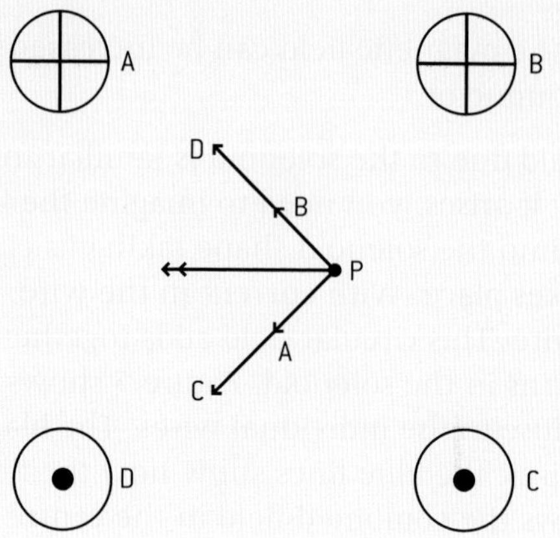

Nature of science

But why permanent magnets?

We have said nothing about the puzzle of permanent magnets or the magnetism of the Earth (Figure 1(a)). The suggestion is that magnetism arises between charges when the charges move relative to each other. Can this idea be extended to help explain the reasons for permanent magnetism (known as ferromagnetism)? In fact, permanent magnetism is comparatively rare in the periodic table, only iron, nickel, and cobalt and alloys of these metals show it.

The reason is due to the arrangement of the electrons in the atoms of these substances. Electrons are now known to have the property of *spin* which can be imagined as an orbiting motion around the atom. In iron, cobalt and nickel, there is a particular arrangement that involves an unpaired electron. This is the atomic origin of the moving charge that is needed for a magnetic field to appear.

The second reason why iron, nickel, and cobalt are strong permanent magnets is that neighbouring atoms can co-operate and line up the spins of their unpaired electrons in the same direction. So, many electrons are all spinning in the same direction and giving rise to a strong magnetic field.

Deep in the centre of the Earth it is thought that a liquid-like metallic core containing free electrons is rotating relative to the rest of the planet. Again, these are conditions that can lead to a magnetic field. In which direction do you predict that the electrons are moving? However, this phenomenon is not well understood and is still the subject of research interest. Why, for example, does the magnetic field of the Earth flip every few thousand years? There is much evidence for this including the magnetic "striping" in the undersea rocks of the mid-Atlantic ridge and in the anomalous magnetism found in some ancient cooking hearths of the aboriginal peoples of Australia.

Forces on moving charges

Force between two current-carrying wires

We have used field ideas to begin our study of magnetic effects, but these conceal from us the underlying physics of magnetism. To begin this study we look at the interactions that arise between conductors when they are carrying electric current.

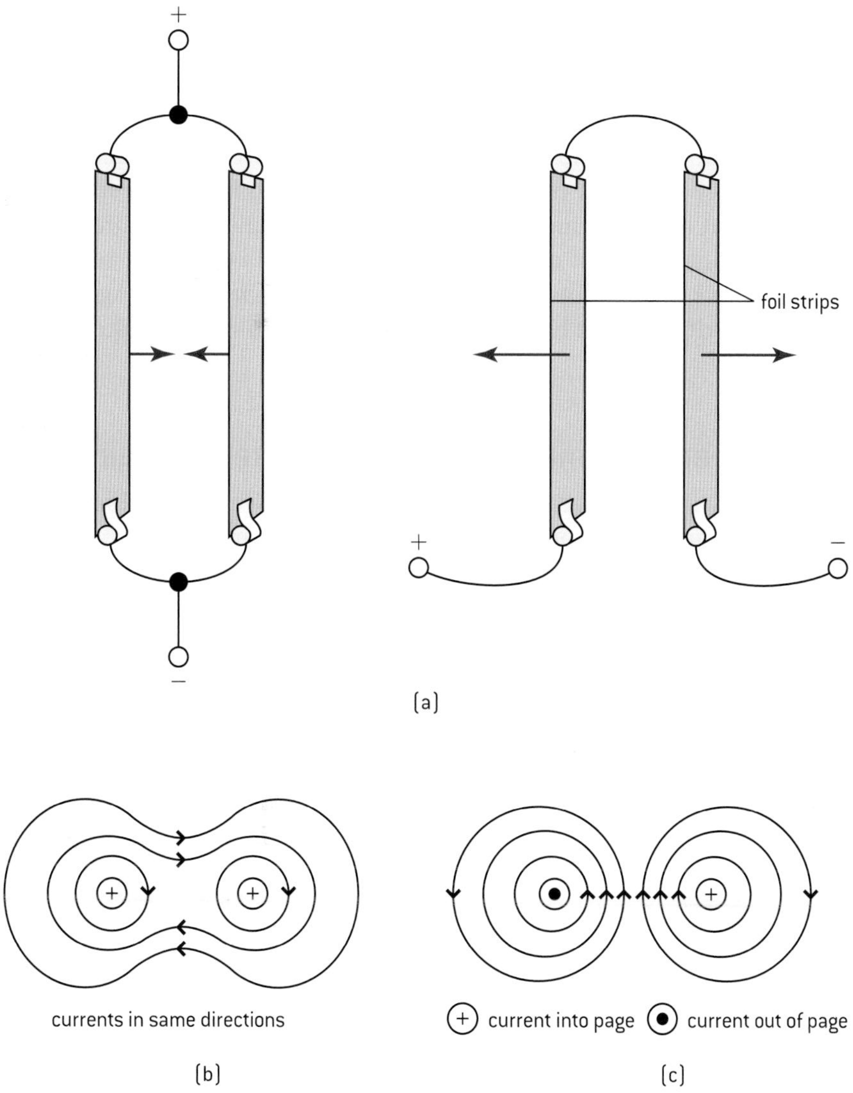

▲ Figure 7

Figure 7(a) shows two foil strips hanging vertically. The current directions in the foils can be the same or opposite directions. When the currents are in the same direction, the strips move together due to the force on one foil strip as it sits in the magnetic field of the other strip. When the currents are in opposing directions, the strips are seen to move apart.

You can set this experiment up for yourself by using two pieces of aluminium foil about 3 cm wide and about 70 cm long for each conductor. The power supply should be capable of providing up to 25 A so take care! Connections are made to the foils using crocodile (alligator) clips.

The forces on the foil can be explained in terms of the interactions between the fields as shown in the figure 7(b). When the currents are in the same direction, the field lines from the foils combine to give a pattern in which field lines loop around both foils. The notation used to show the direction of conventional current in the foil is explained on the diagram. Look back at figure 1(d) which shows the field pattern for two bar magnets with the opposite poles close. You know that the bar magnets are attracted to each other in this situation. The field pattern for the foils is similar and also leads to attraction. Think of the field lines as trying to be as short as possible. They become shorter if the foils are able to move closer together.

When the currents are in opposite directions, the field pattern changes (figure 7(c)). Now the field lines between the foils are close together and in the same direction thus representing a strong field. It seems reasonable that a strong magnetic field (like a strong electrostatic field) represents a large amount of stored energy. This energy can be reduced if the foils move apart allowing the field lines to separate too. Again, this has similarities to the bar magnet case but this time with like poles close together.

Force between a bar magnet field and a current-carrying wire

One extremely important case is the interaction between a uniform magnetic field and that produced by a current in a wire.

Again we can start with the field between two bar magnets with unlike poles close. In the centre of the region between the magnets, the field is uniform because the field lines are parallel and equally spaced.

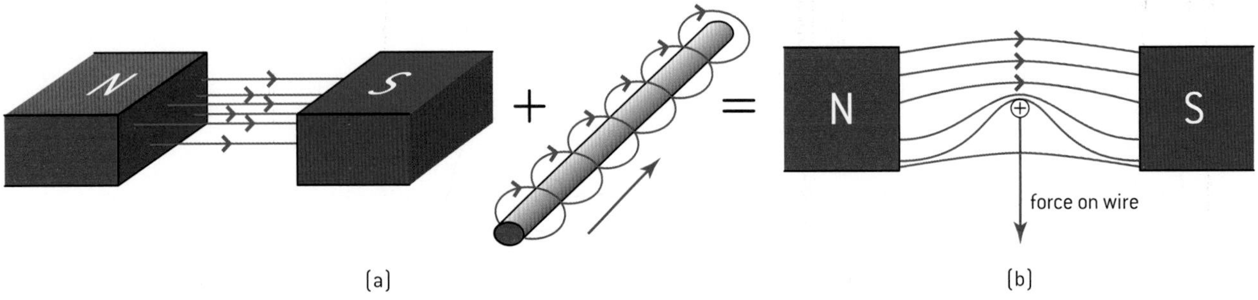

▲ Figure 8 The catapult field.

Suppose a wire carrying a current sits in this field. Figure 8(a) shows the arrangement and the directions of the uniform magnetic field and the field due to the current. A force acts downwards on the wire.

The effect can again be explained in terms of the interaction between the two magnetic fields. The circular field due to the wire adds to the uniform field due to the magnets to produce a more complicated field. This is shown in figure 8(b). Overall, the field is weaker below the wire than above it. Using our ideas of the field lines, it is clear that the system (the wire and the uniform field) can overcome this difference in field strength, either by attempting to move the wire downwards or by moving the magnets that cause the uniform field upwards. This effect is sometimes called the **catapult field** because the field lines above the wire resemble the stretched elastic cord of a catapult just before the object in the catapult is fired.

This effect is of great importance to us. It is the basis for the conversion of electrical energy into kinetic energy. It is used in electric motors, loudspeakers, and other devices where we need to produce movement from an electrical power source. It is called the **motor effect**.

It is possible to predict the direction of motion of the wire by drawing the field lines on each occasion when required but this is tedious. There are a number of direction rules that are used to remember the direction of the force easily. One of the best known of these is due to the English physicist Fleming and is known as **Fleming's left-hand rule**.

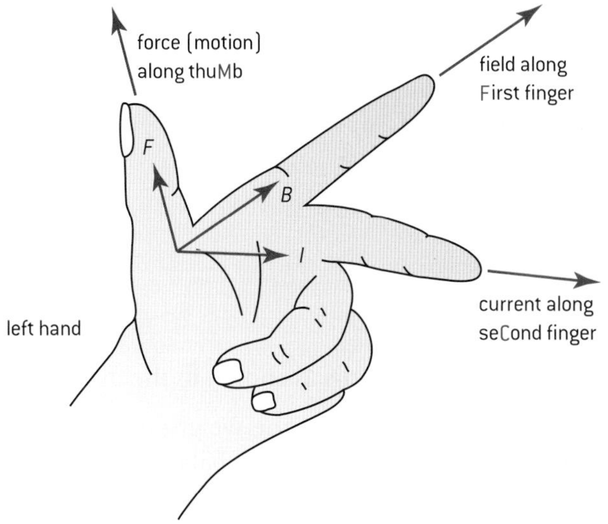

▲ Figure 9 Fleming's left-hand rule.

To use the rule, extend your left hand as shown in figure 9. Your first (index) finger points in the direction of the uniform magnetic field and your second finger points in the direction of the conventional current in the wire, then your thumb gives the direction of the force acting on the wire.

The motor effect

The explanation for the motor effect has been given so far in terms of field lines. This is, of course, not the complete story. We should be looking for explanations that involve interactions between individual charges, both those that produce the uniform magnetic field and those that arise from the current in the wire. Electrostatic effects (as their name implies) arise between charges that are not moving. Magnetic effects arise because the sets of charges that produce the fields are moving relative to each other and are said to be in different frames of reference. There have been no electrostatic effects because we have dealt with conductors in which there is an exact balance of positive and negative charges. Magnetism can be thought of as the residual effect that arises when charges are moving with respect to each other.

Worked example

Wires P and R are equidistant from wire Q.

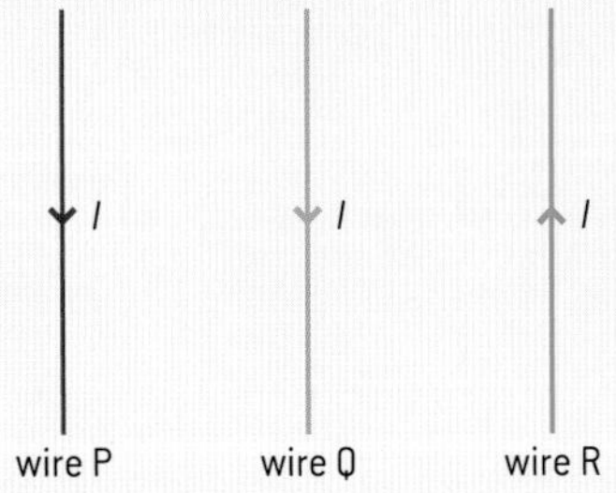

Each wire carries a current of the same magnitude and the currents are in the directions shown.

Describe the direction of the force acting on wire Q due to wires P and R.

Solution

Using the right-hand corkscrew rule, the field due to wire P at wire Q is out of the plane of the paper, and the field due to wire R at wire Q is also out of the plane of the paper.

Using Fleming's left-hand rule, the force on wire Q is in the plane of the paper and to the left.

The magnitude of the magnetic force

Investigate!

Force on a current-carrying conductor

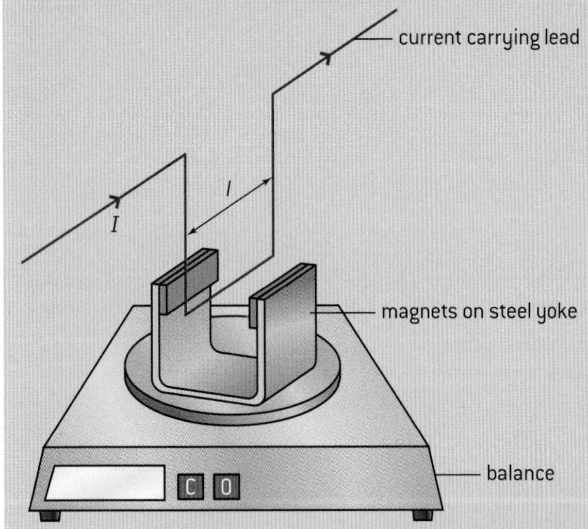

▲ Figure 10

- This is an experiment that will give you an idea of the size of the magnetic force that acts on typical laboratory currents. If carried out carefully it will also allow you to see how the force varies with the length of the conductor and the size of the current.
- You will need some pairs of flat magnets (known as "magnadur" magnets, a sensitive top-pan balance, a power supply and a suitable long straight lead to carry the current.
- Arrange the apparatus as shown in the diagram.
- Zero the balance so that the weight of the magnets is removed from the balance reading.
- The experiment is in two parts:
  - First, vary the current in the wire and collect data for the force acting on the magnets, and therefore the balance, as a result. A trick to improve precision is to reverse the current in the wire and take balance readings for both directions. Then add the two together (ignoring the negative sign of one reading) to give double the answer. Draw a graph to display your data.
  - Second, use two pairs of magnets side-by-side to double the length of the field. Take care that the poles match, otherwise the forces will cancel out. This is not likely to work so well as you need to assume that the magnet pairs have the same strength. This will probably not be true. But, roughly speaking, does doubling the length of wire in the field double the force?

The result of experiments like the one above is that the force acting on the wire due to the field is proportional to:

- the length (l) of the wire
- the current (I) in the wire.

This leads us to a definition of magnetic field strength rather different from that of electric field strength and gravitational field strength.

We cannot define the magnetic field strength in terms of

$$\frac{\text{force}}{\text{a single quantity}}$$

because the force depends on two quantities: current and length.

Instead we define

magnetic field strength

$$= \frac{\text{force acting on a current element}}{\text{current in the element} \times \text{length of the element}}$$

By "element" we mean a short section of a wire that carries a current. If a force F acts on the element of length l when the current in it is I, then the magnetic field strength B is defined by

$$B = \frac{F}{Il}$$

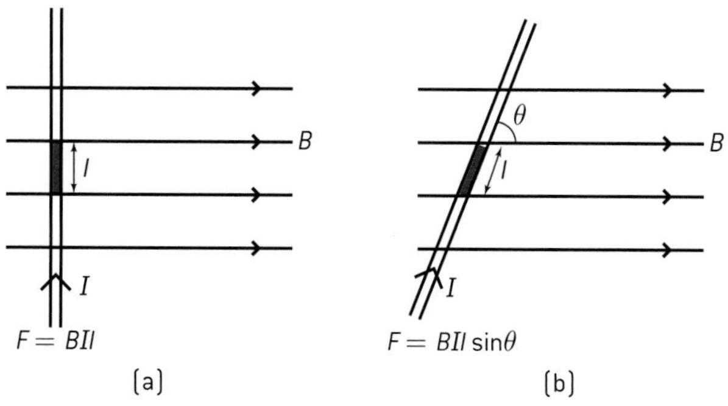

▲ Figure 11

The unit of magnetic field strength is the **tesla**, abbreviated T and the tesla is equivalent to the fundamental units kg s^{-2} A^{-1}. When a 1 metre long current element is in a magnetic field and has a current of 1 A in it, if a magnetic force of 1 N acts on it, then the magnetic field strength is defined to be 1 T. The tesla can also be thought of as a shortened form of N A^{-1} m^{-1}. The tesla turns out to be a very large unit indeed. The largest magnetic field strengths created in a laboratory are a few kT and the magnetic field of the Earth is roughly 10^{-4} T. The very largest fields are associated with some neutron stars. The field strength can be order of 100 GT in such stars.

Having defined B we can go on to rearrange the equation to give the force that acts on a wire:

$$F = BIl$$

This applies when the field lines, the current and the wire are all at 90° to each other as they are when you use the Fleming left-hand rule. If this is not the case (see figure 11(b)) and the wire is at an angle θ to the lines then we need to take the appropriate component of I or l that is at 90° to the field.

In terms of the way the angle is defined in figure 11(b), the equation becomes

$$F = BIl \sin\theta$$

This equation is written in terms of the current in the wire. Of course, the current is, as usual, the result of moving charge carriers. The equation can be changed to reflect this.

From Sub-topic 5.1 we know that the current I is given by

$$I = \frac{Q}{t}$$

where Q is the charge that flows through the current element taking a time t to do it.

Substituting

$$F = B\left(\frac{Q}{t}\right) l \sin\theta$$

$$= BQ\left(\frac{l}{t}\right) \sin\theta$$

The term $\left(\frac{l}{t}\right)$ is the **drift speed v** of the charge carriers and making this substitution gives the expression

$$F = BQv \sin\theta$$

for the force acting on a charge Q moving at speed v at an angle θ to a magnetic field of strength B.

Notice the way that the angle θ is defined in the diagrams. It is the angle between the direction in which the charge is moving (or the current direction – the same thing) and the field lines. Don't get this wrong and use cosine instead of sine in your calculations.

TOK

Why direction rules?

During this sub-topic we have introduced two direction rules: the corkscrew rule for magnetic fields around wires and Fleming's rule for the motor effect. What is the status of these rules? Are they implicit in the way the universe operates, or are they simply indications that depend on the way we define current direction and other fundamental quantities in physics? To begin to answer this, consider what would have happened if Franklin had known that charge carriers in metals were electrons with a negative charge. What would this have changed (if anything)?

Nature of science

The definition of the ampere

In Topic 1 you learnt that the ampere was defined in terms of the force between two current-carrying wires. You may have thought this odd both then and when the relationship between current and charge was developed earlier in this topic. Perhaps it is clearer now.

A precise measurement of charge is quite difficult. At one time it could only be achieved through chemical measurements of electrolysis – and the levels of precision were not great enough to give a good value at that time.

It is much easier to set up an experiment that measures the force between two wires. The experiment can be done to a high precision using a device called a **current balance** (as in the force *Investigate!* above) in which the magnetic force can be measured in terms of the gravitational force needed to cancel it out. This is rather like old-fashioned kitchen scales where the quantity being measured is judged against a standard mass acted on by gravity.

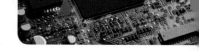

Worked example

1 When a charged particle of mass m, charge q moves at speed v in a uniform magnetic field then a magnetic force F acts on it. Deduce the force acting on a particle mass m of charge $2q$ and speed $2v$ travelling in the same direction in the same magnetic field.

Solution

The equation for the force is $F = BQv \sin \theta$

In the equation, $\sin \theta$ and B do not change but every other quantity does, so the force is $4F$.

2 Electric currents of magnitude I and $3I$ are in wires 1 and 2 as shown.

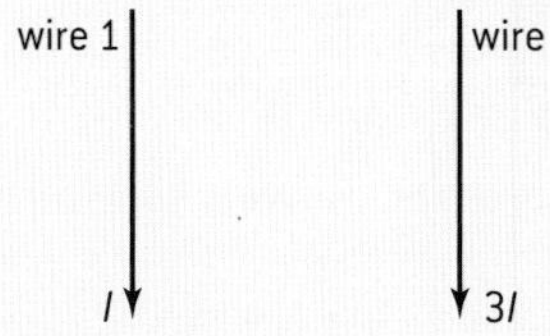

A force F acts on wire 2 due to the current in wire 1. Deduce the magnitude of the force on wire 1 due to the current in wire 2.

Solution

This problem can be solved by reference to Newton's laws of motion, but an alternative is to consider the changes in magnetic field. If the magnetic field at wire 2 due to wire 1 is B then the magnetic field at wire 1 due to wire 2 is $3B$. However, the current in wire 1 is one-third of that in wire 2.

So force at wire 1 due to wire 2 is

$F = 3B \times \frac{I}{3} \times l =$ force in wire 2 due to wire 1.

The forces are the same.

Nature of science

Vectors and their products

You may have formed the view that vectors are just used for scale diagrams in physics. This is not the case, vectors come into their own in mathematical descriptions of magnetic force.

There is only one way to multiply scalars together: run a relay with four stages each of distance 100 m and the total distance travelled by the athletes is 400 m. Multiplying two scalars only gives another scalar.

Vectors, because of the added direction, can be multiplied together in two ways:

- to form a scalar product (sometimes called "dot" product) where, for example, force and displacement are multiplied together to give work done (a scalar) that has no direction. In vector notation this is written as $\boldsymbol{F} \cdot \boldsymbol{s} = W$. The multiplication sign is the dot (hence the name) the vectors are written with a bold font, the scalar in ordinary font.
- to form a vector product (sometimes called "cross" product) where two vectors are multiplied together to give a third vector, thus $\boldsymbol{a} \times \boldsymbol{b} = c$

The multiplication of qv and B to form the vector force F is a vector product. The charge q is a scalar but everything else in the equation is a vector. A mathematician would write

$$\boldsymbol{F} = q\,(\boldsymbol{v} \times \boldsymbol{B})$$

to show that the vector velocity and the vector magnetic field strength are multiplied together. The order of $\boldsymbol{v}$ and $\boldsymbol{B}$ is important. There is a vector rule for the direction of $\boldsymbol{F}$ that is consistent with our observations earlier and the $\sin \theta$ appears when the vector multiplication is worked out in terms of the separate components of the vector.

Vector notation turns out to be an essential language of physics, because it allows a concise notation and because it contains all the direction information within the equations rather than forcing us to use direction rules. However we do not pursue the full theory of vectors in IB Physics.

Questions

1 (*IB*) Four point charges of equal magnitude, are held at the corners of a square as shown below.

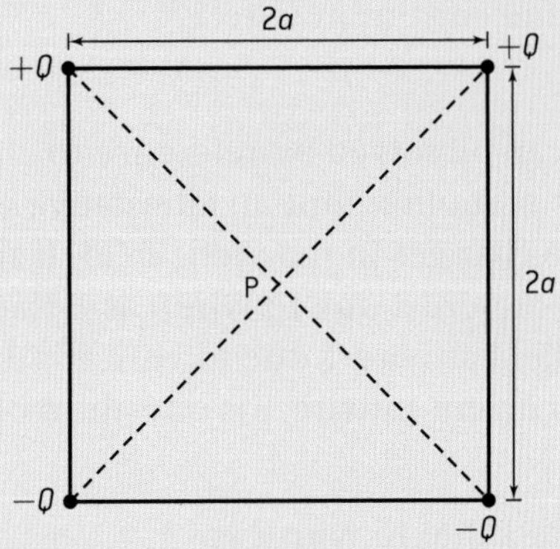

The length of each side of the square is $2a$ and the sign of the charges is as shown. The point P is at the centre of the square.

a) (i) Determine the magnitude of the electric field strength at point P due to one of the point charges.

(ii) On a copy of the diagram above, draw an arrow to represent the direction of the resultant electric field at point P.

(iii) Determine, in terms of Q, a and k, the magnitude of the electric field strength at point P. (7 marks)

2 (*IB*) Two point charges of magnitude $+2Q$ and $-Q$ are fixed at the positions shown below. Discuss the direction of the electric field due to the two charges between A and B. Suggest at which point the electric field is most likely to be zero. (3 marks)

3 (*IB*) Two identical spherical conductors X and Y are mounted on insulated stands. X carries a charge of +6.0 nC and Y carries a charge of −2.0 nC.

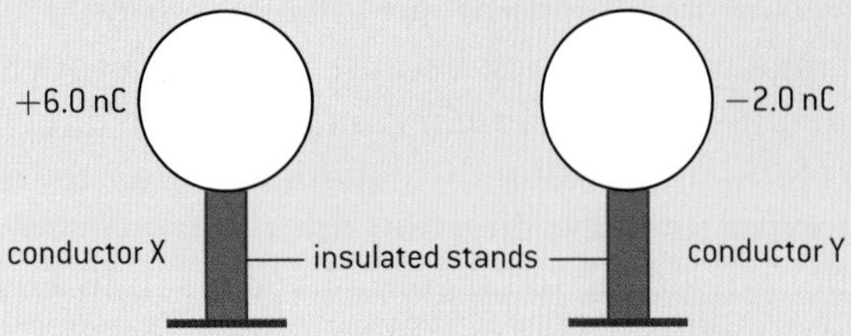

The electric force between them is $+F$ (i.e. attractive). The spheres are touched together and are then returned to their original separation.

a) Calculate the charge on X and the charge on Y.

b) Calculate the value of the electric force between them after being returned to their original separation. (7 marks)

4 (*IB*) Two charged plastic spheres are separated by a distance d in a vertical insulating tube, as shown.

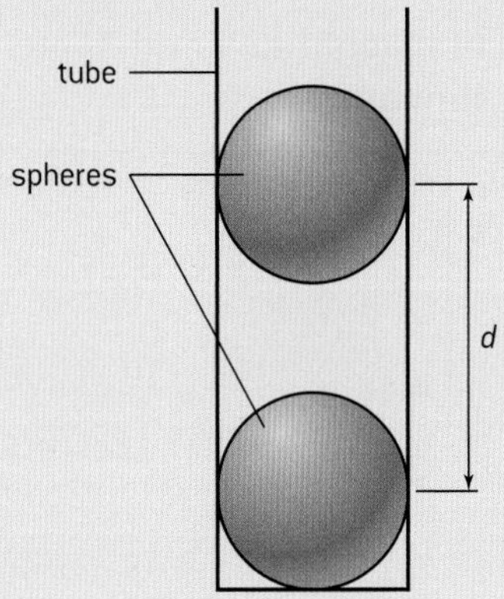

The charge on each sphere is doubled. Friction with the walls of the tube is negligible.

Deduce the new separation of the spheres.

(5 marks)

5 (*IB*)

a) Electric fields may be represented by lines of force. The diagram below shows some lines of force.

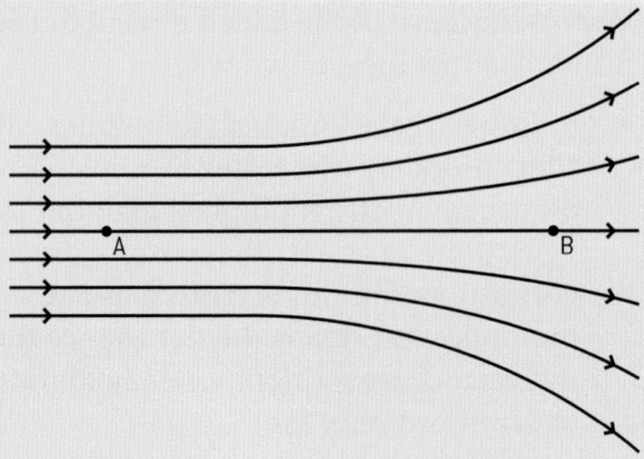

(i) State whether the field strength at A and at B is constant, increasing or decreasing, when measured in the direction from A towards B.

(ii) Explain why field lines can never touch or cross.

b) The diagram below shows two insulated metal spheres. Each sphere has the same positive charge.

Copy the diagram and in the shaded area between the spheres, draw the electric field pattern due to the two spheres. (8 marks)

6 (*IB*) A lamp is at normal brightness when there is a potential difference of 12 V across its filament and a current in the filament of 0.50 A.

a) For the lamp at normal brightness, calculate:

(i) the power dissipated in the filament

(ii) the resistance of the filament.

b) In order to measure the voltage–current (*V–I*) characteristics of the lamp, a student sets up the following electrical circuit.

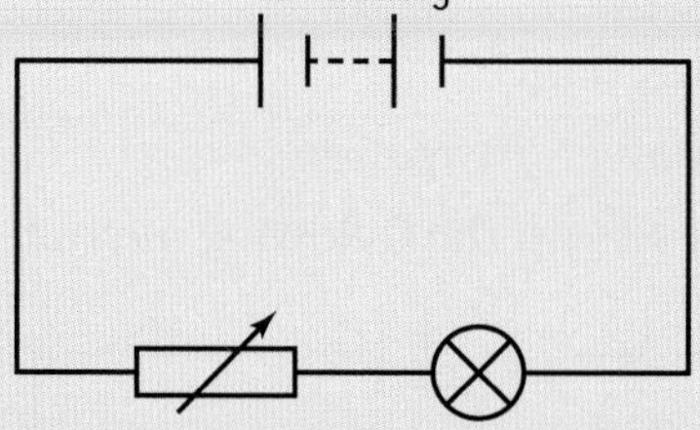

State the correct positions of an ideal ammeter and an ideal voltmeter for the characteristics of the lamp to be measured.

c) The voltmeter and the ammeter are connected correctly in the previous circuit. Explain why the potential difference across the lamp

(i) cannot be increased to 12 V

(ii) cannot be reduced to zero.

d) An alternative circuit for measuring the *V–I* characteristic uses a *potential divider*.

(i) Draw a circuit that uses a potential divider to enable the *V–I* characteristics of the filament to be found.

(ii) Explain why this circuit enables the potential difference across the lamp to be reduced to 0 V. (13 marks)

7 (*IB*) The graph below shows the *V–I* characteristic for two 12 V filament lamps A and B.

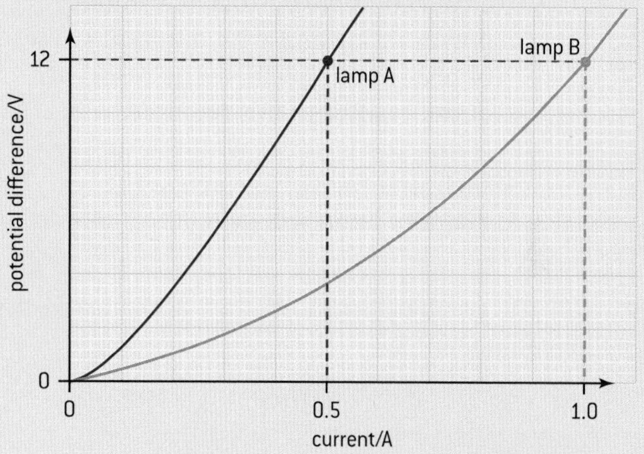

a) (i) Explain why the graphs indicate that these lamps do not obey Ohm's law.

(ii) State and explain which lamp has the greater power dissipation for a potential difference of 12 V.

The two lamps are now connected in series with a 12 V battery as shown below.

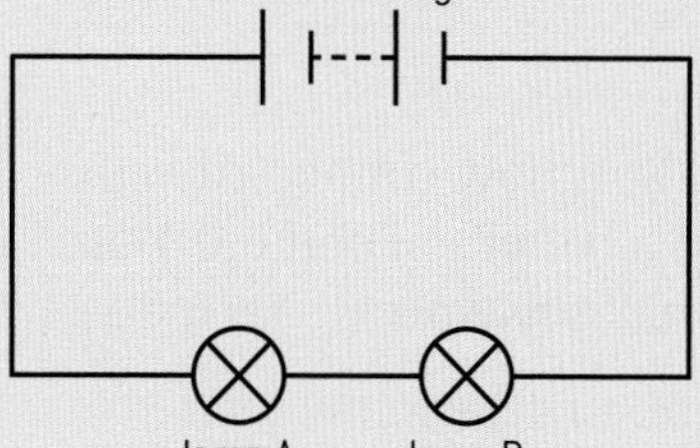

b) (i) State how the current in lamp A compares with that in lamp B.

(ii) Use the *V–I* characteristics of the lamps to deduce the total current from the battery.

(iii) Compare the power dissipated by the two lamps. (11 marks)

8 (*IB*)

a) Explain how the resistance of the filament in a filament lamp can be determined from the *V*–*I* characteristic of the lamp.

b) A filament lamp operates at maximum brightness when connected to a 6.0 V supply. At maximum brightness, the current in the filament is 120 mA.

(i) Calculate the resistance of the filament when it is operating at maximum brightness.

(ii) You have available a 24 V supply and a collection of resistors of a suitable power rating and with different values of resistance. Calculate the resistance of the resistor that is required to be connected in series with the supply such that the voltage across the filament lamp will be 6.0 V. (4 marks)

9 (*IB*) The graph below shows the *I*–*V* characteristics for component X.

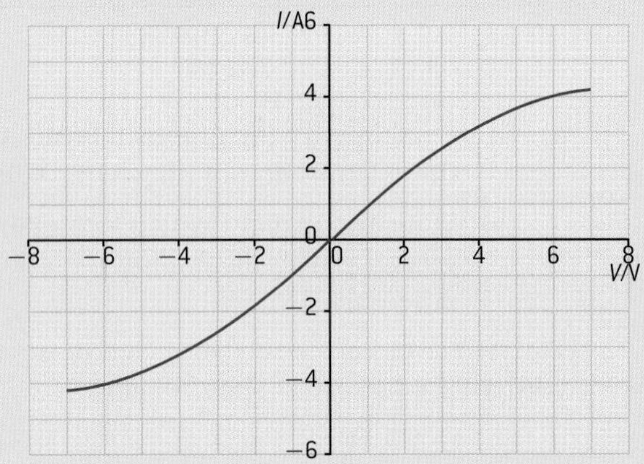

The component X is now connected across the terminals of a battery of emf 6.0 *V* and negligible internal resistance.

a) Use the graph to determine:

(i) the current in component X

(ii) the resistance of component X.

b) A resistor R of constant resistance 2.0 Ω is now connected in series with component X as shown below.

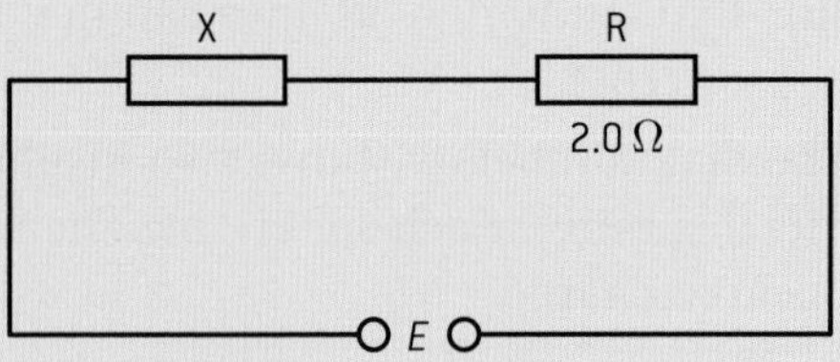

(i) Copy the graph in (a), and draw the *I*–*V* characteristics for the resistor R.

(ii) Determine the total potential difference *E* that must be applied across component X and across resistor R such that the current through X and R is 3.0 A. (7 marks)

10 (*IB*) A student is to measure the current–voltage (*I*–*V*) characteristics of a filament lamp. The following equipment and information are available.

	Information
Battery	emf = 3.0 V, negligible internal resistance
Filament lamp	marked "3 V, 0.2 A"
Voltmeter	resistance = 30 kΩ, reads values between 0.0 and 3.0 V
Ammeter	resistance = 0.1 Ω, reads values between 0.0 and 0.5 A
Potentiometer	resistance = 100 Ω

a) For the filament lamp operating at normal brightness, calculate:

(i) its resistance

(ii) its power dissipation.

The student sets up the following *incorrect* circuit.

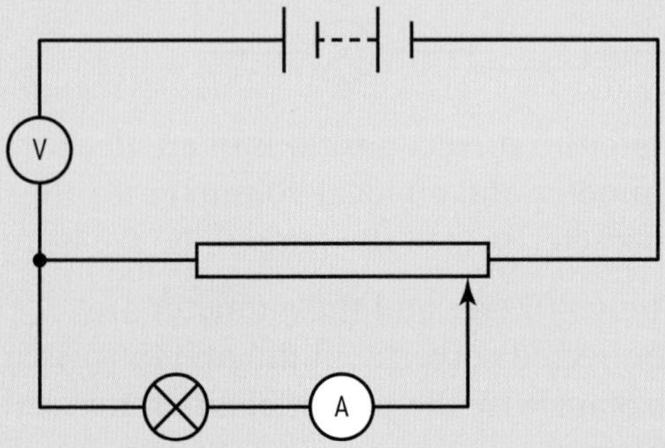

b) (i) Explain why the lamp will not light.

(ii) State the approximate reading on the voltmeter. Explain your answer.

(6 marks)

11 (*IB*) A particular filament lamp is rated at 12 V, 6.0 mA. It just lights when the potential difference across the filament is 6.0 V.

A student sets up an electric circuit to measure the *I–V* characteristics of the filament lamp.

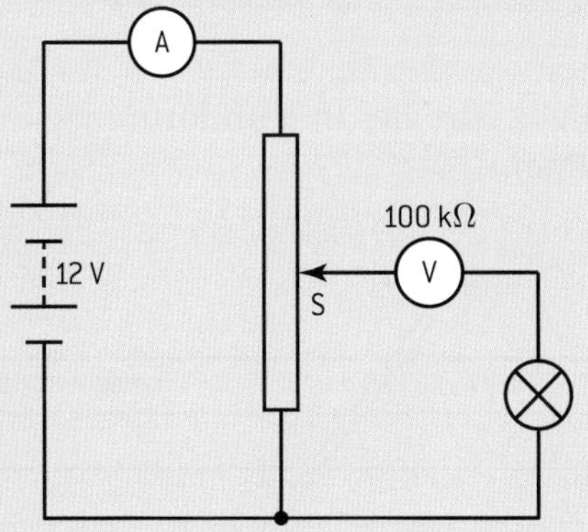

In the circuit, shown below, the student has connected the voltmeter and the ammeter into the circuit *incorrectly*.

The battery has emf 12 V and negligible internal resistance. The ammeter has negligible resistance and the resistance of the voltmeter is 100 kΩ.

The maximum resistance of the variable resistor is 15 Ω.

a) Explain, without doing any calculations, whether there is a position of the slider S at which the lamp will be lit.

b) Estimate the maximum reading of the ammeter. (5 marks)

12 (*IB*) The graph below shows the current–voltage (*I–V*) characteristics of two different conductors X and Y.

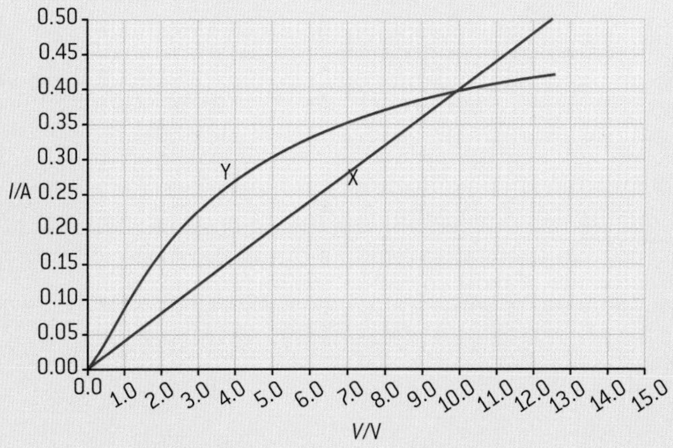

a) (i) State the value of the current for which the resistance of X is the same as the resistance of Y and determine the value of this resistance.

(ii) Describe and suggest an explanation for the *I–V* characteristic of conductor Y.

b) The two conductors X and Y are connected in series with a cell of negligible internal resistance. The current in the conductors is 0.20 A.

Use the graph to determine:

(i) the resistance of Y for this value of current

(ii) the emf of the cell. (8 marks)

13 (*IB*) A cell of electromotive force (emf) *E* and internal resistance *r* is connected in series with a resistor *R*, as shown below.

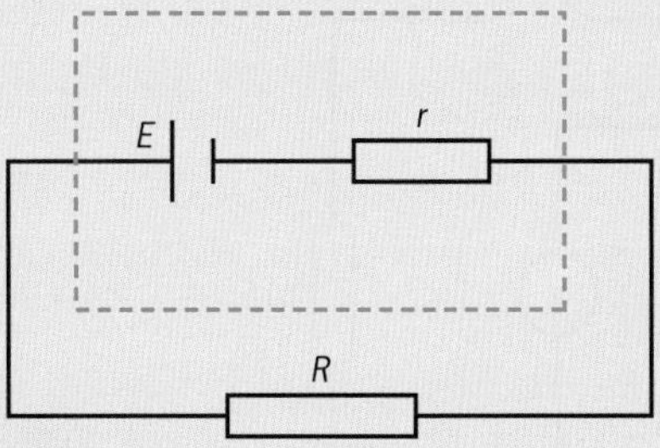

The cell supplies 8.1×10^3 J of energy when 5.8×10^3 C of charge moves completely round the circuit. The current in the circuit is constant.

a) Calculate the emf *E* of the cell.

b) The resistor *R* has resistance 6.0 Ω. The potential difference between its terminals is 1.2 V. Determine the internal resistance *r* of the cell.

c) Calculate the total energy transfer in *R*.

d) Describe, in terms of a simple model of electrical conduction, the mechanism by which the energy transfer in *R* takes place. (12 marks)

14 (*IB*) A battery is connected in series with a resistor R. The battery transfers 2000 C of charge completely round the circuit. During this process, 2500 J of energy is dissipated R and 1500 J is expended in the battery. Calculate the emf of the battery. (3 marks)

15 (*IB*) A student connects a cell in series with a variable resistor and measures the terminal pd V of the cell for a series of currents I in the circuit. The data are shown in the table.

V/V	I/mA
1.50	120
1.10	280
0.85	380
0.75	420
0.60	480
0.50	520

Use the data to determine the emf and internal resistance of the cell. (5 marks)

16 (*IB*) A battery is connected to a resistor as shown.

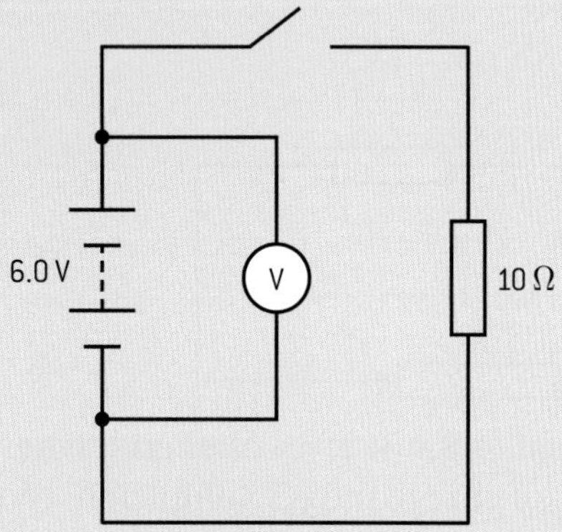

When the switch is open the voltmeter reads 12 V, when the switch is closed it reads 11.6 V.

a) Explain why the readings differ.

b) (i) State the emf of the battery.

(ii) Calculate the internal resistance of the battery.

c) Calculate the power dissipated in the battery. (6 marks)

17 (*IB*) An electron enters a pair of electric and magnetic fields in a vacuum as shown in the diagram.

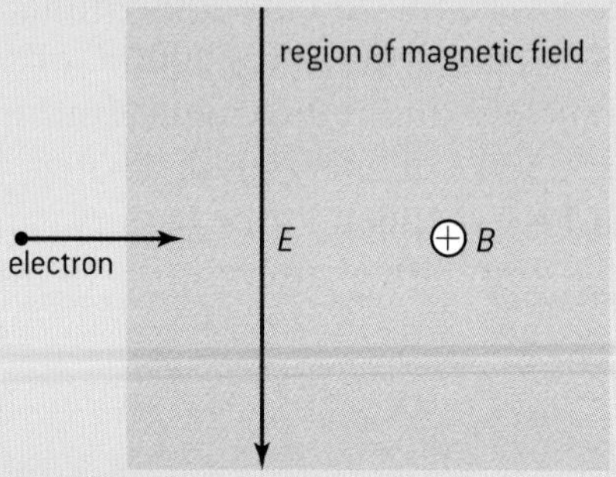

The electric field strength is 3.8×10^5 V m^{-1} and the magnetic field strength is 2.5×10^{-2} T. Calculate the speed of the electron if the net force acting on it due to the fields is zero. (3 marks)

18 (*IB*) A straight wire lies in a uniform magnetic field as shown.

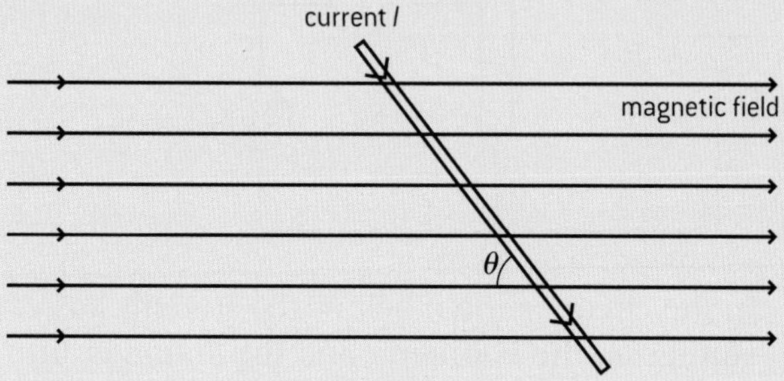

The current in the wire is I and the wire is at an angle of θ to the magnetic field. The force per unit length on the conductor is F. Determine the magnetic field strength. (2 marks)

19 (*IB*) A straight wire of length 0.75 m carries a current of 35 A. The wire is at right angles to a magnetic field of strength 0.058 T. Calculate the force on the wire. (2 marks)

20 (*IB*) An ion with a charge of $+3.2 \times 10^{-19}$ C and a mass of 2.7×10^{-26} kg is moving due south at a speed of 4.8×10^3 m s^{-1}. It enters a uniform magnetic field of strength 4.6×10^{-4} T directed downwards towards the ground. Determine the force acting on the ion. (4 marks)

6 CIRCULAR MOTION AND GRAVITATION

Introduction

Two apparently distinct areas of physics are linked in this topic: motion in a circle and the basic ideas of gravitation. But of course they are not distinct at all. The motion of a satellite about its planet involves both a consideration of the gravitational force and the mechanics of motion in a circle. Man cannot travel beyond the Earth without a knowledge of both these aspects of Physics.

6.1 Circular motion

Understanding

- → Period, frequency, angular displacement, and angular velocity
- → Centripetal force
- → Centripetal acceleration

Nature of science

The drive to develop ideas about circular motion came from observations of the universe. How was it that astronomical objects could move in circular or elliptical orbits? What kept them in place in their motion? Scientists were able to deduce that there must be a force acting radially inwards for every case of circular motion that is observed. Whether it is a bicycle going around a corner or a planet orbiting its star, the physics is the same.

Applications and skills

- → Identifying the forces providing the centripetal forces such as tension, friction, gravitational, electrical, or magnetic
- → Solving problems involving centripetal force, centripetal acceleration, period, frequency, angular displacement, linear speed, and angular velocity
- → Qualitatively and quantitatively describing examples of circular motion including cases of vertical and horizontal circular motion

Equations

- → speed–angular speed relationship: $v = \omega r$
- → centripetal acceleration: $a = \frac{v^2}{r} = \frac{4\pi^2 r}{T^2}$
- → centripetal force: $F = \frac{mv^2}{r} = m\omega^2 r$

▲ Figure 1 A fairground carousel.

Moving in a circle

Most children take great delight in an object on a string whirling in a circle – though they may be less happy with the consequences when the string breaks and the object hits a window! Rides at a theme park and trains on a railway are yet more examples of movement in a circle. What is needed to keep something rotating at constant speed?

The choice of term (as usual in physics!) is very deliberate. In circular motion we say that the "speed is constant" but not the "velocity is constant".

Velocity, as a vector quantity, has both magnitude *and* direction. The object on the string has a constant speed but the direction in which the object is moving is changing all the time. The velocity has a constant *magnitude* but a changing *direction*. If either of the two parts that make up a vector change, then the vector is no longer constant. Whenever velocity changes (even if it is only the direction) then the object is accelerated.

Understanding the physics of this acceleration is the key to understanding circular motion. But before looking at how the acceleration arises we need a language to describe the motion.

Angular displacement

The angle moved around the circle by an object from where its circular motion starts is known as the **angular displacement**. Unlike the linear displacement used in Topic 2, angular displacement will not be considered to be a vector in IB Physics. Angular displacement is the angle through which the object moves and it can be measured in degrees (°) or in radians (rad). Radians are more commonly used than degrees in this branch of physics. If you have not met radians before, read about the differences between radians and degrees.

Nature of science

Radians or degrees

Calculations of circular motion involve the use of angles. In any science you studied before starting this course you will almost certainly have measured all angles in degrees.

1° (degree) is defined to be $\frac{1}{360}$th of the way around a circle.

In some other areas of physics (including circular motion) there is an alternative measure of angle that is much more convenient, the **radian**. Radians are based on the geometry of the arc of a circle.

1 radian (abbreviated as rad) is defined as the angle equal to the circumference of an arc of a circle divided by the radius of the circle. In symbols

$$\theta = \frac{s}{r}$$

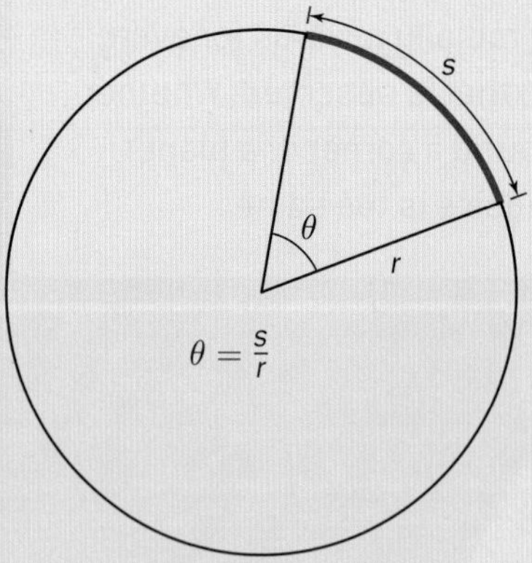

▲ Figure 2 Definition of radian.

Going around the circle once means travelling around the circumference; this is a distance of $2\pi r$. The angle θ in radians subtended by the whole circle is $\frac{2\pi r}{r} = 2\pi$ rad.

So $360° = 2\pi = 6.28$ rad

and $1 \text{ rad} = 57.3°$

Sometimes, the radian numbers are left as fractions, so

$90° = \frac{\pi}{2}\left(\frac{1}{4} \text{ round the circle}\right),$

$30° = \frac{\pi}{6}\left(\frac{1}{12} \text{ round the circle}\right)$

and so on.

To convert other values for yourself, use the equation $\frac{\text{angle in degree}}{360} = \frac{\text{angle in radians}}{2\pi}$

There are some similarities between the sine of an angle and the angle in radians. The two quantities are compared in this Nature of Science box which shows sin θ and θ in radians. Notice that, as θ becomes smaller, sin (θ) and θ become closer together. From angles of 10° down to 0, the differences between sin θ and θ are very small and in some calculations and proofs we treat sin θ and θ as being equal (this is known as "the small angle approximation"). For small angles cos θ approximates to zero radians.

To illustrate this, here are the values of sin θ and θ in radians for four angles: 90°, 45°, 10°, and 5°. Notice how similar the sine values and the radians are for 10° and 5°.

$\sin(90°) = 1.000;$ $\quad \frac{\pi}{2}\text{rad} = 1.571 \text{ rad}$

$\sin(45°) = 0.707;$ $\quad \frac{\pi}{4}\text{rad} = 0.785 \text{ rad}$

$\sin(10°) = 0.175;$ $\quad \frac{\pi}{18}\text{rad} = 0.174 \text{ rad}$

$\sin(5°) = 0.087;$ $\quad \frac{\pi}{36}\text{rad} = 0.087 \text{ rad}$

Finally, a practical point: Scientific and graphic calculators work happily in either degrees or radians (and sometimes in another type of angular measure known as "grad" too). But the calculator has to be "told" what to expect! Always check that your calculator is set to work in radians if that is what you want, or in degrees if those are the units you are using. You will lose calculation marks in an examination if you confuse the calculator!

Angular speed

In Topic 2 we used the term *speed* to mean "linear speed". When the motion is in a circle there is an alternative: **angular speed**, this is given the symbol ω (the lower-case Greek letter, omega).

$$\textit{average angular speed} = \frac{\textit{angular displacement}}{\textit{time for the angular displacement to take place}}$$

Figure 3 shows how things are defined and you will see that in symbols the definition becomes

$$\omega = \frac{\theta}{t}$$

where θ is the angular displacement and t is the time taken for the angular displacement.

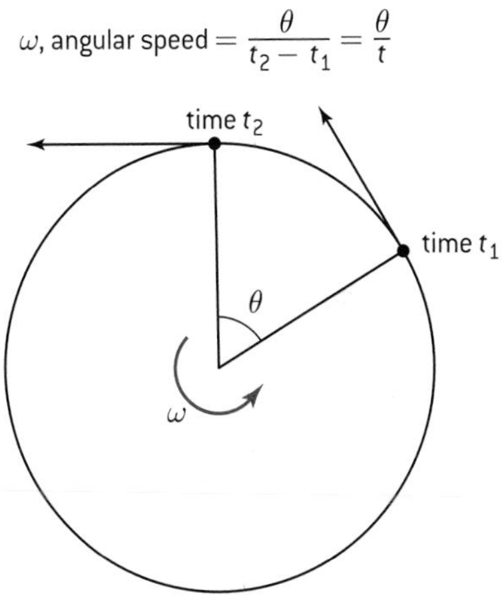

▲ Figure 3 Angular speed.

Nature of science

Angular speed or angular velocity?

You may be wondering about the distinction between angular speed and angular velocity, and whether angular velocity is a vector similar to linear velocity.

The answer is that angular velocity is a vector but an unusual one. It has a magnitude equal to the angular speed, but its direction is surprising! The direction is along the axis of rotation, in other

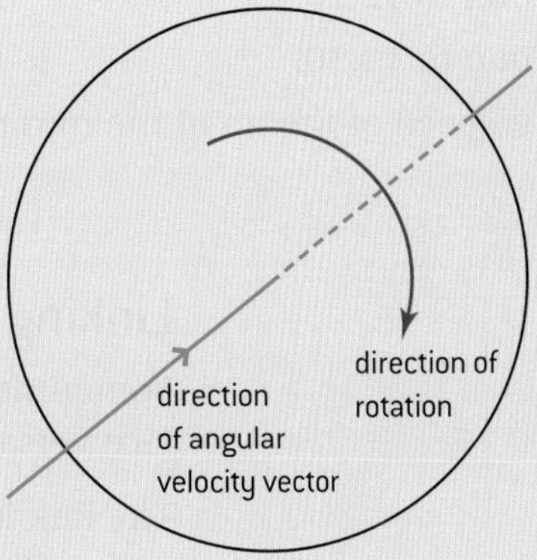

▲ Figure 4 Angular velocity direction.

words, through the centre of the circle around which the object is moving and perpendicular to the plane of the rotation.

The direction follows a clockwise corkscrew rule so that in this example the direction of the angular velocity vector is into the plane of the paper.

In the IB course, only the angular speed – the scalar quantity – is used.

Period and frequency

The time taken for the object to go round the circle once is known as the **periodic time** or simply the **period** of the motion, it has the symbol T. In one period, the angular distance travelled is 2π rad. So,

$$T = \frac{2\pi}{\omega}$$

When T is in seconds the units of ω are radians per second, abbreviated to rad s^{-1}.

If you have already studied waves in this course, you might have met the idea of **time period** – the time for one cycle. Another quantity that is associated with T is **frequency**. Frequency is the number of times an object goes round a circle in unit time (usually taken to be 1 second), so one way to express the unit of frequency would be in "per second" or s^{-1}. However, the unit of frequency is re-named after the 19th century physicist Heinrich Hertz and is abbreviated to Hz. There is a link between T and f so that:

$$T = \frac{1}{f}$$

This leads to a link between ω and f

$$\omega = 2\pi f$$

Worked example

A large clock on a building has a minute hand that is 4.2 m long.

Calculate:

a) the angular speed of the minute hand

b) the angular displacement, in radians, in the time periods

(i) 12 noon to 12.20

(ii) 12 noon to 14.30.

c) the linear speed of the tip of the minute hand.

Solution

a) The minute hand goes round once (2π rad) every hour.

One hour is 3600 s

$$\text{angular speed} = \frac{\textit{angular displacement}}{\textit{time taken}}$$

$$= \frac{2\pi}{3600} = 0.001\,75 \text{ rad s}^{-1}$$

b) (i) 20 minutes is $\frac{1}{3}$ of 2π, so $\frac{2\pi}{3}$ rad

(ii) 2.5 h is $2\pi \times 2.5 = 5\pi$ rad

c) $v = r\omega = 4.2 \times 0.001\,75 = 0.007\,33 \text{ m s}^{-1}$
$= 7.3 \text{ mm s}^{-1}$

Linking angular and linear speeds

Sometimes we know the linear speed and need the angular speed or vice versa.

The link is straightforward: When the circle has a radius r the circumference is $2\pi r$, and T, is the time taken to go around once. So the linear speed of the object along the edge of the circle v is

$$v = \frac{2\pi r}{T}$$

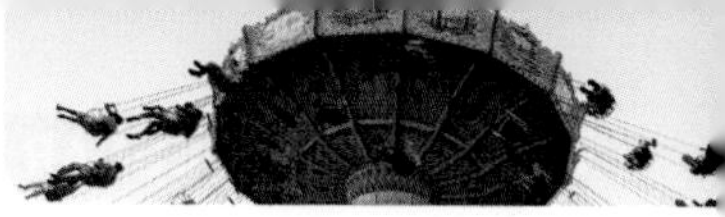

Rearranging the equation gives

$$T = \frac{2\pi r}{v}$$

We have just seen that

$$T = \frac{2\pi}{\omega}$$

so equating the two equations for T gives

$$\frac{2\pi r}{v} = \frac{2\pi}{\omega}$$

Cancelling the 2π and rearranging gives

$$v = \omega r$$

Notice that, in both this equation and in the earlier equation $s = \theta r$, the radius r multiplies the angular term to obtain the linear term. This is a consequence of the definition of the angular measure.

Centripetal acceleration

Earlier we showed that an object moving at a constant angular speed in a circle is being accelerated. Newton's first law tells us that, for any object in which the direction of motion or the speed is changing, there must be an external force acting. In circular motion the direction is constantly changing and so the object accelerates and there must be a force acting on it to cause this to happen. In which direction do the force and the acceleration act, and what are their sizes?

The diagram shows two points P_1 and P_2 on the circle together with the velocity vectors v_{old} and v_{new} at these points. The vectors are the same length as each other because the speed is constant. However, v_{old} and v_{new} point in different directions because the object has moved round the circle by an angular distance $\Delta\theta$ between P_1 and P_2. Acceleration is, as usual,

$$\frac{\text{change of velocity}}{\text{time taken for the change}}$$

The change in velocity is the change-of-velocity vector Δv that has to be added to v_{old} in order to make it become the same length and direction as v_{new}. Identify these vectors on the diagram.

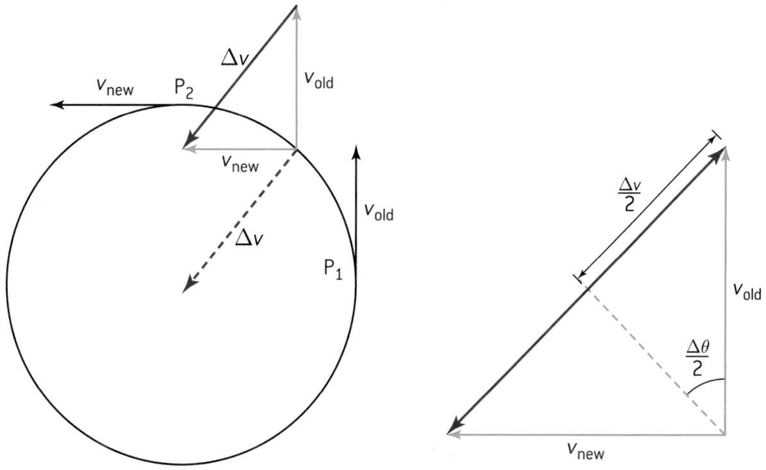

▲ Figure 5 Proof of centripetal acceleration direction.

Notice that v_{old} and v_{new} slide round the circle to meet. Where does the new vector Δv point? The answer is: to the centre of the circle. This is an "averaging" process to find out what the difference is between v_{old} and v_{new} half-way between the two points.

This averaging can be taken further. The time, Δt, to go between P_1 and P_2, and the linear distance around the circle between P_1 and P_2 (which is $r\theta$) are related by

$$\Delta t = \frac{r\Delta\theta}{v}$$

Using some trigonometry on the diagram shows that

$$\frac{\Delta v}{2} = v \sin\left(\frac{\Delta\theta}{2}\right)$$

The size of the average acceleration $\boldsymbol{a}$ that is directed towards the centre of the circle is

$$\boldsymbol{a} = \frac{\Delta v}{\Delta t} = \frac{2v \sin\left(\frac{\Delta\theta}{2}\right)}{\frac{2r}{v}\frac{\Delta\theta}{2}}$$

This can be written as

$$\boldsymbol{a} = \frac{v^2}{r}\frac{\sin\left(\frac{\Delta\theta}{2}\right)}{\frac{\Delta\theta}{2}}$$

When $\Delta\theta$ is very small, the ratio $\frac{\sin\left(\frac{\Delta\theta}{2}\right)}{\frac{\Delta\theta}{2}}$ is almost exactly equal to 1 and so the instantaneous acceleration a when P_1 and P_2 are very close together is

$a = \frac{v^2}{r} = \omega^2 r = v\omega$ directed to the centre of the circle.

This acceleration is at 90° to the velocity vector and it points inwards to the centre of the circle.

The force that acts to keep the object moving in a circle is called the **centripetal force** and this force leads to a **centripetal acceleration**. (The origin of the word centripetal comes from two Latin words *centrum* and *petere* – literally "to lead to the centre".)

Nature of science

Linking it together

Notice that some of these equations have interesting links elsewhere: $mv\omega$ is, for example, the magnitude of the linear momentum multiplied by ω. Try to be alert for these links as they will help you to piece your physics together.

Centripetal force

Newton's second law of motion in its simpler form tells us that $F = ma$ using the usual symbols.

The second law applies to the force that provides the centripetal acceleration, so the magnitude of the force $= m\frac{v^2}{r} = m\omega^2 r = mv\omega$. The question we need to ask for each situation is: what force provides the centripetal force for that situation? The direction of this force must be along the radial line between the object and the centre of the circle.

Investigate!

Investigating how F varies with m, v and r

This experiment tests the relationship

$$m\frac{v^2}{r} = Mg$$

- To do this a bung is whirled in a horizontal circle with a weight hanging from one end of a string and mass (rubber bung) on the other end.

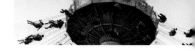

- A paper clip is attached to the string below a glass tube. The clip is used to ensure that the radius of rotation of the bung is constant – the bung should be rotated at a speed so that the paper clip just stays below the glass tube.

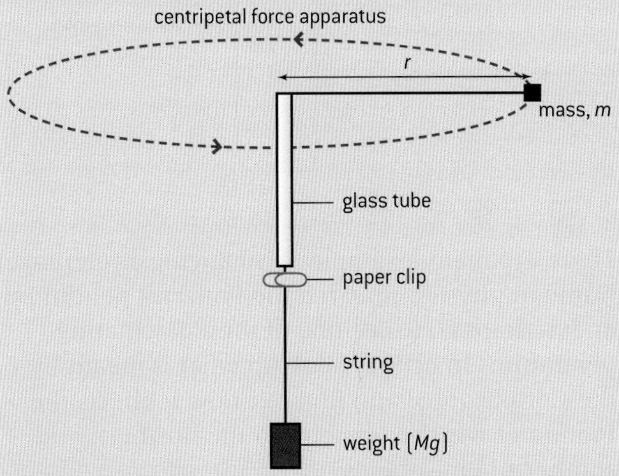

▲ Figure 6 Centripetal force, mass, and speed.

- The tension in the string is the same everywhere (whether below the glass tube or above in the horizontal part). This tension is *F* in the equation and is equal to *Mg* where *M* is the mass of the weight (hanging vertically).
- Use a speed at which you can count the number of rotations of *m* in a particular time and from this work out the linear speed *v* of mass *m*.

To verify the equation you need to test each variable against the others. There are a number of possible experiments in each of which one variable is held constant (a control variable), one is varied (the independent variable), and the third (the dependent variable) is measured. One example is:

Variation of v with r

- In this experiment, *m* and *M* must be unchanged. Move the clip to change *r*, and for each value of *r*, measure *v* using the method given above.
- Analysis:

$$\frac{v^2}{r} = constant$$

- A graph of v^2 against *r* ought to be a straight line passing through the origin. Alternatively you could, for each experimental run, simply divide v^2 by *r* and look critically at the answer (which should be the same each time) to see if the value is really constant. If going down this route, you ought to assess the errors in the experiment and put error limits on your $\frac{v^2}{r}$ value.
- What are the other possible experimental tests?
- In practice the string cannot rotate in the horizontal plane because of its own weight. How can you improve the experiment or the analysis to allow for this?

Nature of science

Centripetal or centrifugal?

When discussing circular motion, you will almost certainly have heard the term "centrifugal force" – probably everywhere except in a physics laboratory! In this course we have spoken exclusively about "centripetal force". Why are there two terms in use?

It should now be clear to you how circular motion arises: a force acts to the centre of the circle around which the object is moving. The alternative idea of centrifugal force comes from common experience. Imagine you are in a car going round a circle at high speed. You will undoubtedly feel as if you are being "flung outwards".

One way to explain this is to imagine the situation from the vantage point of a helicopter hovering

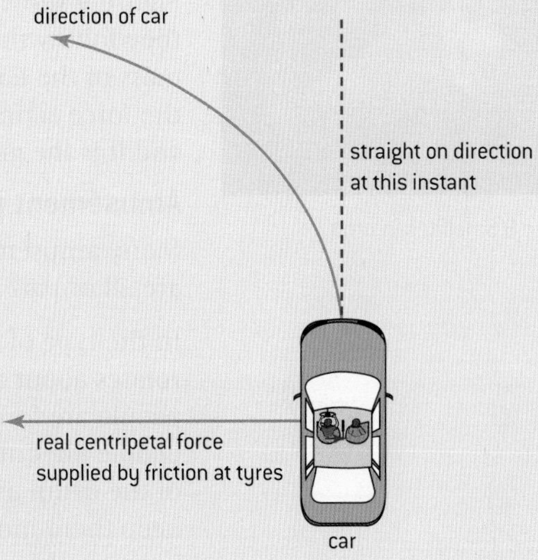

▲ Figure 7 Centripetal forces in a car seen from above.

stationary above the circle around which the car is moving. From the helicopter you will see the passenger attempting to go in a straight line (Newton's first law), but the passenger is forced to move in a circle through friction forces between passenger and seat. If the passenger were sitting on a friction-less seat and not wearing a seat belt, then he or she will not get the "message" that the car is turning. The passenger continues to move in a straight line eventually meeting the door that is turning with the car. If there were no door, what direction will the passenger take?

Another way to explain this is to imagine yourself in the car as it rotates. This is a rotating frame of reference that is accelerating and as such cannot obey Newton's laws of motion. You instinctively think that the rotating frame is actually stationary. Therefore your tendency to go in what you believe to be a straight line actually feels like an outward force away from the centre of the circle (remember the rest of the world now rotates round you, and your straight line is actually part of a circle). Think about a cup of coffee sitting on the floor of the car. If there is insufficient friction at the base of the cup, the cup will slide to the side of the car. In the inertial frame of reference (the Earth) the cup is trying to go in a straight line. In your rotating frame of reference you have to "invent" a force acting outwards from the centre of the circle to explain the motion of the cup.

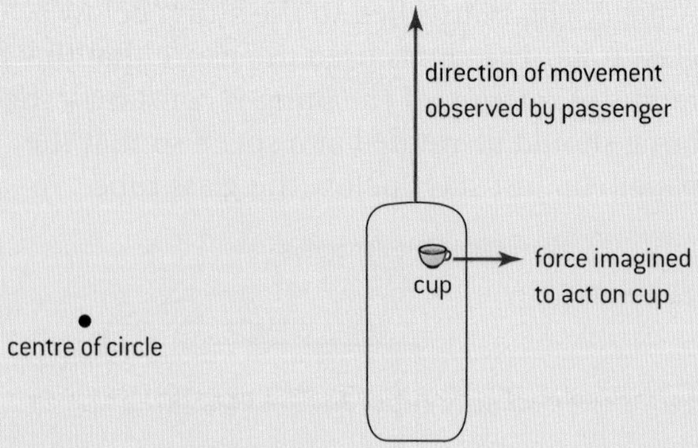

▲ Figure 8 Rotation forces.

There are many examples of changing a reference frame in physics: research the Foucault pendulum and perhaps go to see one of these fascinating pendulums in action. Look up what is meant by the Coriolis force and find out how it affects the motion of weather systems in the northern and southern hemispheres.

One of the tricks that physicists often use is to change reference frames – it's all part of the nature of science to adopt alternative frames of reference to make explanations and theories more accessible.

One last tip: Don't use explanations based on centrifugal force in an IB examination. The real force is centripetal; centrifugal force was invented to satisfy Newton's second law in an accelerated frame of reference.

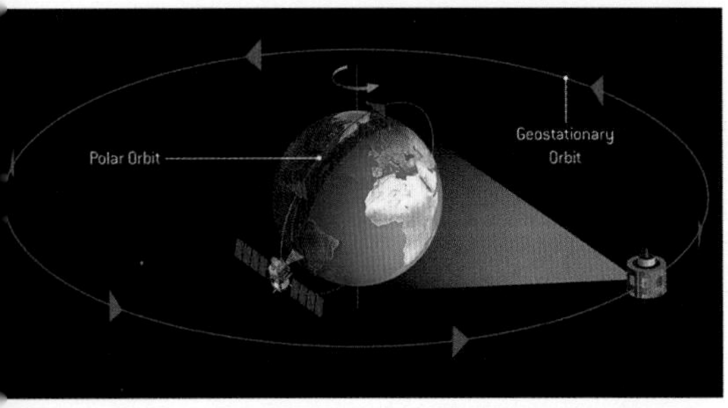

▲ Figure 9 Satellites in orbit.

Centripetal accelerations and forces in action

Satellites in orbit

Figure 9 shows satellites in a circular orbit around the Earth. Why do they follow these paths? Gravitational forces act between the centre of mass of the Earth and the centre of mass of the satellite. The direction of the force acting on the satellite is always towards the centre of the planet and it is the gravity that supplies the centripetal force.

Amusement park rides

Many amusement park rides take their passengers in curved paths that are all or part of a circle. How does circular motion provide a thrill?

In the type of ride shown in figure 10, the people are inside a drum that rotates about a vertical axis. When the rotation speed is large enough the people are forced to the sides of the drum and the floor drops away. The people are quite safe however because they are "held" against the inside of the drum as the reaction at the wall provides the centripetal force to keep them moving in the circle. The people in the ride feel the reaction between their spine and the wall. Friction between the rider and the wall prevents the rider from slipping down the wall.

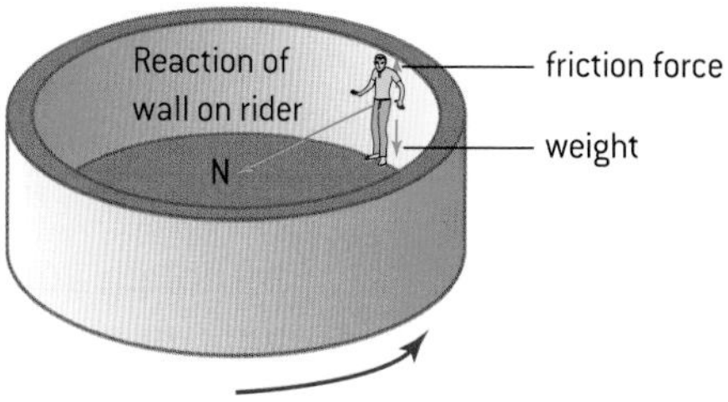

▲ Figure 10 The rotor in action.

Turning and banking

When a driver wants to make a car turn a corner, a resultant force must act towards the centre of the circle to provide a centripetal force. The car is in vertical equilibrium (the driving surface is horizontal) but not in horizontal equilibrium.

Turning on a horizontal road

For a horizontal road surface, the friction acting between the tyres and the road becomes the centripetal force. The friction force is related to the coefficient of friction and the normal reaction at the surface where friction occurs.

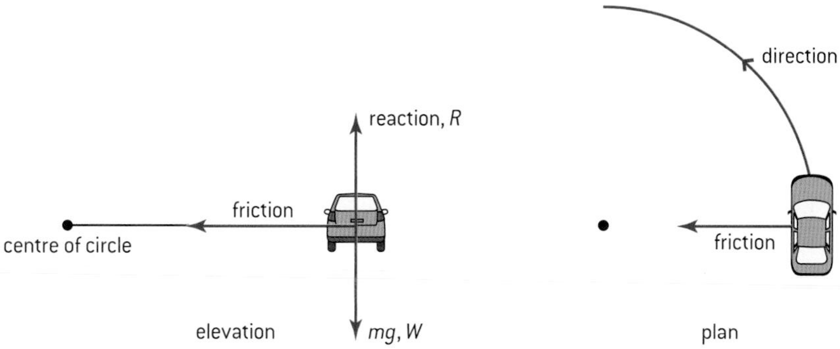

▲ Figure 11 Car moving in a circle.

If the car is not to skid, the centripetal force required has to be less than the frictional force

$$m\frac{v^2}{r} < \mu_s mg$$

where μ_s is the static coefficient of friction. Note that when the vehicle is already skidding the "less than" sign becomes an equality and the dynamic coefficient of friction should be used.

This rearranges to give a maximum speed of $v_{max} = \sqrt{\mu_d rg}$ for a circle of radius r.

Banking

Tracks for motor or cycle racing, and even ordinary roads for cars are sometimes **banked** (figures 12 and 13). The curve of the banked road surface is inclined at an angle so that the normal reaction force contributes to the centripetal force that is needed for the vehicle to go round the track

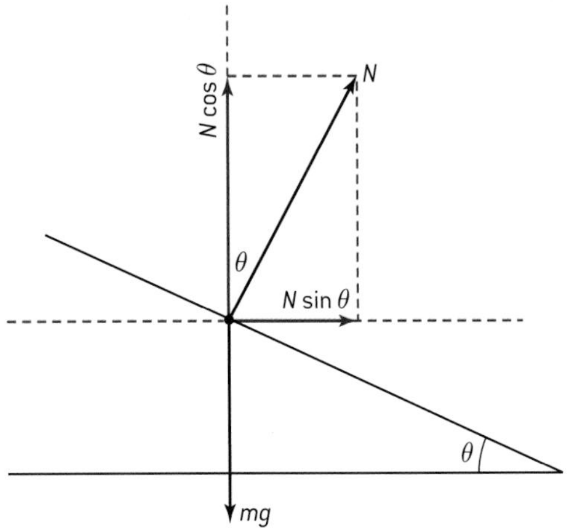

▲ Figure 12 Forces in banking.

at a particular speed. Bicycles and motorcycles can achieve the same effect on a level road surface by "leaning in" to the curve. Tyres do not need to provide so much friction on a banked track compared to a horizontal road; this reduces the risk of skidding and increases safety.

Although you will not be asked to solve mathematical problems on this topic in your IB Physics examination, you do need to understand the principles that underpin banking.

Figure 12 shows forces acting on a small sphere rolling round a track. This is simplified to a point object moving in a circle to remove the complications of two or four wheels. A horizontal centripetal force directed towards the centre of the circle is needed for the rotation. The other forces that act on the ball are the force normal to the surface (which is at the banking angle θ) and its weight acting vertically down. The vector sum of the horizontal components of the weight and the normal force must equal the centripetal force.

Looking at this another way, if N is the normal force then the centripetal force is equal to

$$N \sin\theta$$

The normal force resolved vertically is $N\cos\theta$ and is, of course, equal and opposite to mg. So $F_{\text{centripetal}} = \left(\frac{mg}{\cos\theta}\right)\sin\theta = mg\tan\theta$

$$F_{\text{centripetal}} = \frac{mv^2}{r} \text{ and therefore } \tan\theta = \frac{v^2}{gr}$$

The banking angle is correct at a particular speed and radius. Notice that it does not depend on the mass of the vehicle so a banked road works for a cyclist and a car, provided that they are going at the same speed.

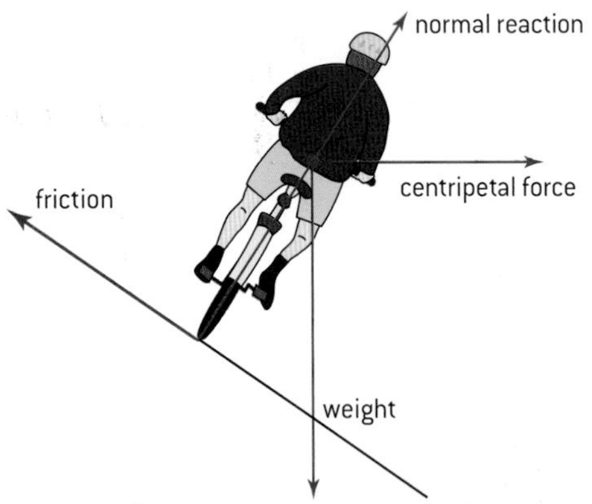

Some more examples of banking:

- Commercial airline pilots fly around a banked curve to change the direction of a passenger jet. If the angle is correct, the passengers will not feel the turn, simply a marginal increase in weight pressing down on their seat).
- Some high-speed trains tilt as they go around curves so that the passengers feel more comfortable.

▲ Figure 13 Cycle velodrome.

Moving in a vertical circle

So far the examples have been of motion around a horizontal circle. People will queue for a long time to experience moderate fear on a fairground attraction like the rollercoaster in figure 14. The amount of thrill from the ride depends on its height, speed, and also the forces that act on the riders.

How is the horizontal situation modified when the circular motion of the mass is in a vertical plane?

1 What are the forces acting when the motion is in a vertical circle?

Imagine a mass on the end of a string that is moving in a vertical circle at constant speed.

Look carefully at figure 15 and notice the way the tension in the string changes as the mass goes around.

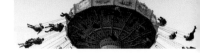

▲ Figure 14 Theme park ride.

Begin with the case when the string is horizontal, at point A. The weight acts downwards and the tension in the string is the horizontal centripetal force towards the centre of the circle.

The mass continues to move upwards and reaches the top of the circle at B. At this point the tension in the string and the weight both act downwards. Thus:

$$T_{down} + mg = m\frac{v^2}{r}$$

and therefore

$$T_{down} = m\frac{v^2}{r} - mg$$

The weight of the mass combines with the tension to provide the centripetal force and so the tension required is less than the tension T when the string is horizontal.

At C, the bottom of the circle, the tension and the weight both act vertically but in opposite directions and so

$$T_{up} = m\frac{v^2}{r} + mg$$

At the bottom, the string tension must overcome weight and also provide the required centripetal force.

As the mass moves around the circle, the tension in the string varies continuously. It has a minimum value at the top of the circle and a maximum at the bottom. The bottom of the circle is the point where the string is most likely to break. If the maximum breaking tension of the string is T_{break}, then, for the string to remain intact,

$$T_{break} > m\frac{v^2}{r} + mg$$

and the linear speed at the bottom of the circle must be less than

$$\sqrt{\frac{r}{m}(T_{break} - mg)}$$

If this seems to you to be a very theoretical idea without much practical value, think about a car going over a bridge. If you assume that the shape of the bridge is part of a circle, then there is a radius of curvature r. What is the speed at which the car will lose contact with the bridge?

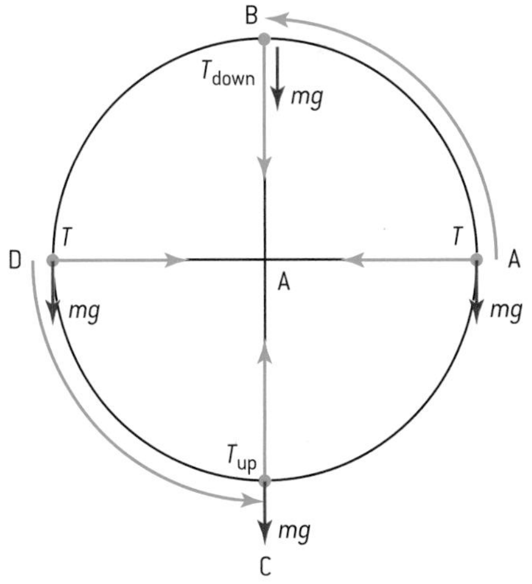

▲ Figure 15 Forces in circular motion in a vertical plane.

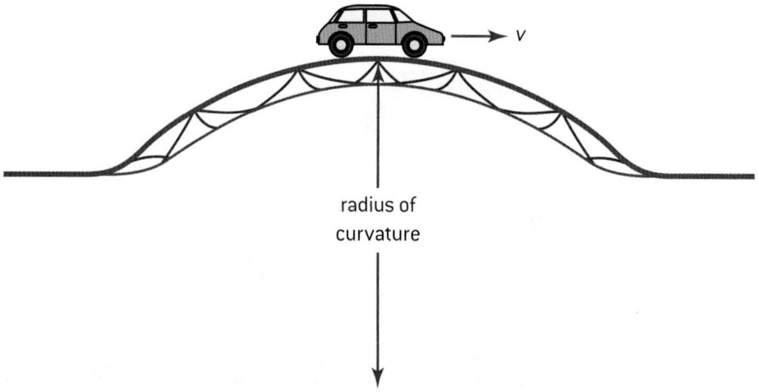

▲ Figure 16 Car going over a bridge.

This is the case considered above, where the object, in this case the car, is at the top of the circle. What is the "tension" (in this case the force between car and road) if the car wheels are to lose contact with the bridge? To answer this question, you might begin with a free-body diagram. You should be able to show that the car loses contact at a speed equal to $\sqrt{gr}$.

2 How does speed change when motion is in a vertical circle?

Not all circular motion in a vertical circle is at a constant speed. As a mass moves upwards it slows as kinetic energy is transferred to gravitational potential energy (if there is nothing to keep it moving at constant speed). At the top of the motion the mass must not stop moving or even go too slowly, because if it did then the string would lose its tension. The motion would no longer be in a circle.

The centripetal force F_c needed to maintain the motion is $F_c = m\frac{v^2}{r}$ as usual, at the top of the circle, if F_c is supplied entirely by gravity then

$$F_c = mg = m\frac{v^2}{r}$$

Just for an instant, the object is in free-fall.

The equation can be rearranged to give

$$v_{top} = \sqrt{gr}$$

and this is the minimum speed at the top of the circle for which the motion will still be circular. The minimum speed does not depend on mass.

Energy is conserved assuming that there are no losses (for example, to internal energy as a result of air resistance as the mass goes round). Equating the energies:

> kinetic energy at top + gravitational potential energy difference between top and bottom = kinetic energy at the bottom

and

$$\frac{1}{2}mv_{top}^2 + mg(2r) = \frac{1}{2}mv_{bottom}^2$$

By substituting for both tensions, T_{bottom} and T_{top}, it is possible to show that

$$T_{bottom} = T_{top} + 6mg$$

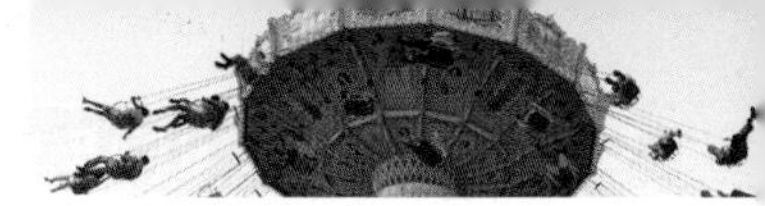

Worked examples

1 A hammer thrower in an athletics competition swings the hammer on its chain round 7.5 times in 5.2 s before releasing it. The hammer describes a circle of radius 4.2 m and has a mass of 4.0 kg. Assume that the hammer is swung in a horizontal circle and that the chain is horizontal.

a) Calculate, for the rotation:

(i) the average angular speed of the hammer

(ii) the average tension in the chain.

b) Comment on the assumptions made in this question.

Solution

a) (i) 7.5 revolutions $= 15\pi$ rad

$$\text{angular speed} = \frac{15\pi}{5.2} = 9.1 \text{ rad s}^{-1}$$

(ii) Tension in the chain = centripetal force required for rotation centripetal force $= mr\omega^2 = 4.0 \times 4.2 \times 9.1^2 = 1400$ N

b) The thrower usually inclines the plane of the circle at about 45° to the horizontal in order to achieve maximum range. Even if the plane were horizontal, then the weight of the hammer would contribute to the system so that a component of the tension in the chain must allow for this. Both assumptions are unlikely.

6.2 Newton's law of gravitation

Understanding

→ Newton's law of gravitation

→ Gravitational field strength

Nature of science

Newton's insights into mechanics and gravitation led him to develop laws of motion and a law of gravitation. One of his motion laws and the law of gravitation are mathematical in nature, two of the motion laws are descriptive. None of these laws can be proved and there is no attempt in them to explain why the masses are accelerated under the influence of a force, or why two masses are attracted by the force of gravity. Newton's ideas about motion have been subsequently modified by the work of Einstein. The questioning and insight that leads to the development of laws are fundamental to the nature of science.

Applications and skills

→ Describing the relationship between gravitational force and centripetal force

→ Applying Newton's law of gravitation to the motion of an object in circular orbit around a point mass

→ Solving problems involving gravitational force, gravitational field strength, orbital speed, and orbital period

→ Determining the resultant gravitational field strength due to two bodies

Equations

→ Newton's law of gravitation: $F = G\frac{Mm}{r^2}$

→ gravitational field strength: $g = \frac{F}{m}$

→ gravitational field strength and the gravitational constant: $g = G\frac{M}{r^2}$

Nature of science

Scientists from the past

Three of the great names from the history of astronomy and physics were Copernicus, Tycho Brahe, and Kepler. Their contributions were linked and ultimately led to the important gravitational work of Newton. Try to find out something about these astronomers. During the lifetimes of these scientists, science was carried out in a very different way from today.

The realization that the Earth orbits the Sun rather than the Sun orbiting the Earth was one of the great developments in scientific understanding. Galileo Galilei and other scientists of the 16th century overcame cultural, philosophical and religious prejudices, and some even suffered persecution for the scientific truths they had discovered. As we study the work of these pioneers we should remember that scientists in past times were not always as free as those today.

Research the life of Galileo (we often drop the second name) and explore why he and others came into conflict with the Roman Catholic Church over their scientific beliefs.

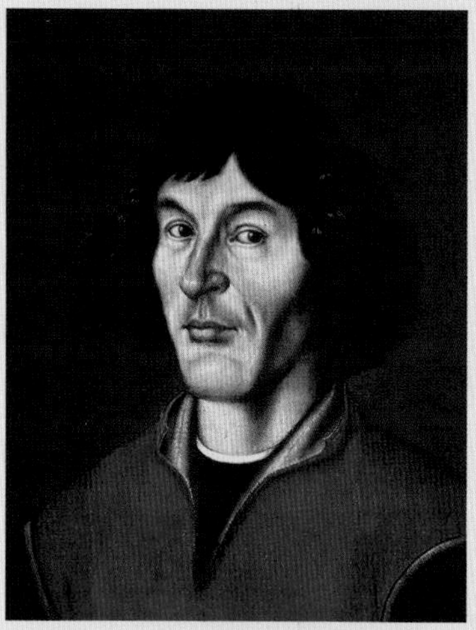

Copernicus (Mikolaj Kopernik)

Tycho Brahe

Johannes Kepler

Isaac Newton

Galileo Galilei

▲ Figure 1 An astronomers' portrait gallery.

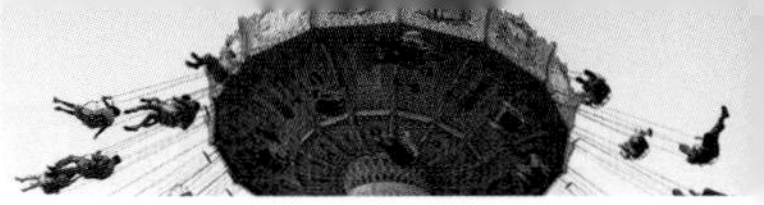

Gravitational field strength

Like electrostatics, gravity acts at a distance and is an example of a force that has an associated force field. Imagine two masses in deep space with no other masses close enough to influence them. One mass (call it A) is in the force field due to the second mass (B) and a force acts on A. B is in the gravitational field of A and also experiences a force. These two forces have an equal magnitude (even though the masses may be different) but act in opposite directions.

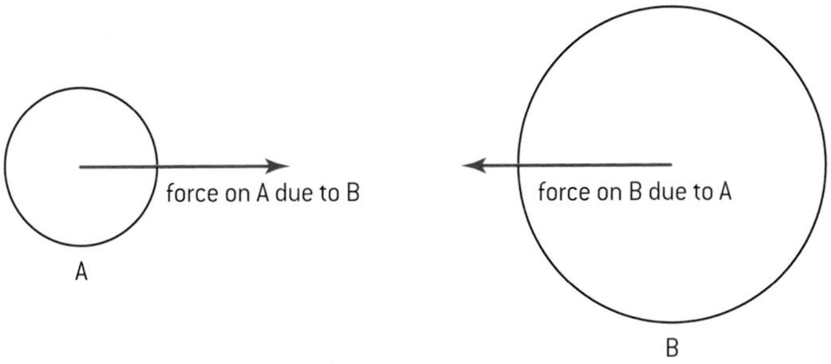

▲ Figure 2 Gravitational forces between two masses.

If both masses are small, the size of a human, say, the force of gravity is extremely weak. Only when one of the masses is as large as a planet does the force become noticeable to us. However, whatever the size of the mass we need a way to measure the strength of the field to which it gives rise. Gravitational forces are the weakest of all the forces in nature and so require large amounts of mass for the force to be felt.

The strength of a gravitational field is defined using the idea of a **small test mass**. This test mass has to be so small that it does not disturb the field being measured. If the test mass is large then it will exert a force of its own on the mass that produces the field being measured. The test mass would then accelerate the other mass and alter the arrangement that is being measured.

The gravitational force that acts on the small test mass has both magnitude and direction. These are shown in the diagram. The test mass will accelerate in this direction if it is free to move.

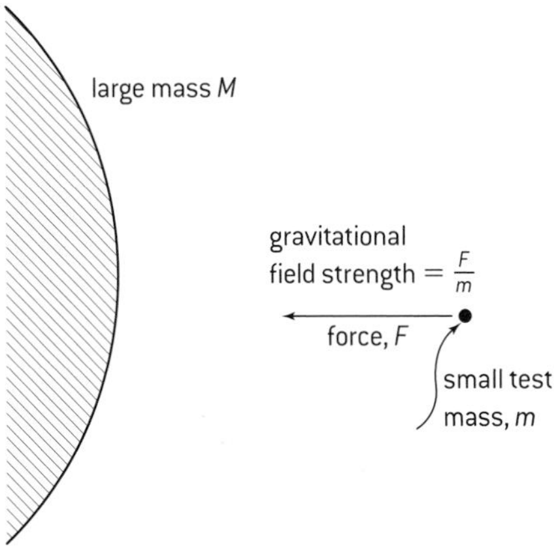

▲ Figure 3 Definition of gravitational field strength.

Defining gravitational field strength

The concept of the small test mass leads to a definition of the strength of the gravitational field.

If the mass of the test mass is m, and the field is producing a gravitational force of F on the test mass, then the **gravitational field strength** (given the symbol g) is defined as

$$g = \frac{F}{m}$$

The units of gravitational field strength are N kg^{-1}.

Since F is a vector and m a scalar, it follows that g is a vector quantity and that its direction is that of F.

A formal definition in words is that **gravitational field strength at a point is the force per unit mass experienced by a small point mass placed at that point.** This definition requires that the test mass is not just small but is also an (infinitesimally) small point in space. You might want to consider what the effect might be if the test mass has a shape and extends in space.

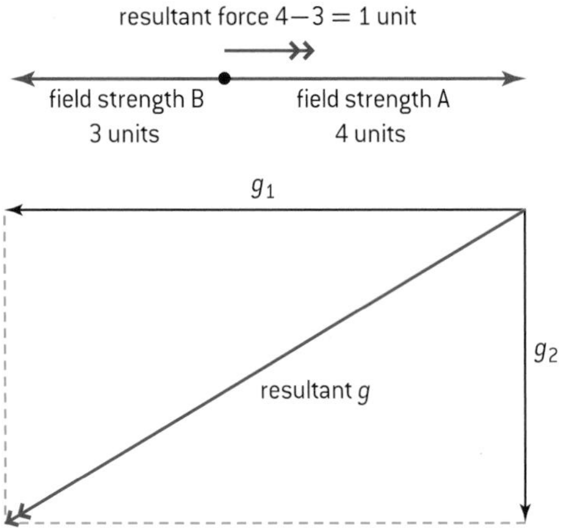

▲ Figure 4 Adding field strengths vectorially.

So far the discussion has been limited to the gravitational field produced by one point mass. How does the situation change if there is more than one mass, excluding the test mass itself?

Field strength is independent of the magnitude of the point test mass (we divided F by m to achieve this). So the vector field strengths can be added together (figure 4).

In IB Physics examinations you will only be asked to add the field strengths of masses that all lie on the same straight line. But even in two dimensions, the addition is straightforward using the ideas of vector addition by drawing or by calculation.

g and the acceleration due to gravity

Sometimes students are surprised that the symbol g is used for the gravitational field strength, they think there might be a risk of confusion with g the acceleration due to gravity! However this does not happen.

At the Earth's surface (using Newton's second law)

acceleration due to gravity at the surface

$$= \frac{\text{force on a mass at the surface due to gravity}}{\text{size of the mass}}$$

So the acceleration $= \frac{F}{m}$ but $\frac{F}{m}$ is also the definition of gravitational field strength so the acceleration due to gravity = gravitational field strength = g.

The magnitude of the gravitational field strength (measured in N kg^{-1}) is equal to the value of the acceleration due to gravity (measured in m s^{-2}). You should be able to show that N $kg^{-1} \equiv$ m s^{-2}. (The symbol $\equiv$ means "is equivalent to".)

Newton's law of gravitation – an inverse-square law

Isaac Newton (1642–1727) was a British scientist who consolidated the work of others and who added important insights of his own. During his life he contributed to the study of optics, mechanics and gravitation. One of his greatest triumphs was work on gravity that he developed using the data and ideas of the German astronomer Johannes Kepler and others about the motion of the planets.

Newton realized that the gravitational force F between two objects with masses M and m whose centres are separated by distance r is:

- always attractive
- proportional to $\frac{1}{r^2}$
- proportional to M and m.

This can be summed up in the equation

$$F \propto \frac{Mm}{r^2}$$

Laws that depend on $\frac{1}{r^2}$ are known as inverse-square. If the distance between the two masses is **doubled** without changing mass, then the force between the masses goes down to **one-quarter** of its original value.

To use this equation numerically a constant of proportionality is needed and is given the symbol G,

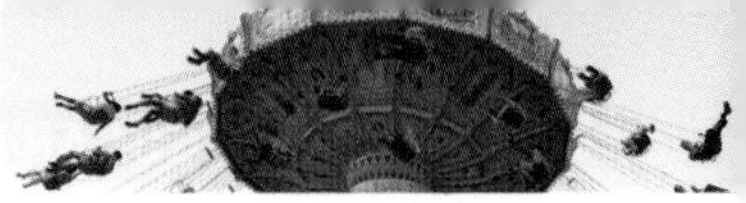

$$F = \frac{GMm}{r^2}$$

G is known as the **universal gravitational constant** and it has an accepted value of 6.67×10^{-11} N m^2 kg^{-2}. Gravity is always attractive so if the distance is measured *from* the centre of mass M *to* mass m then the force on m due to M is *towards* M. In other words, the force is in the opposite direction to the direction in which the distance is measured. You may see some books where a negative sign is introduced to predict this direction. The IB Diploma Programme physics course does not attribute negative signs to attractive forces, the responsibility of keeping track of the force direction is with you.

Nature of science

A universal constant?

What does it mean to say that G is a universal gravitational constant? Newton did not know what the size of the constant was (as the value was determined over a hundred years later by Cavendish). In a sense, it did not matter. Newton realized that all objects are attracted to the Earth – the apocryphal story of him seeing a falling apple reminds us that he knew this. The insight that Newton had was to realize that the force of gravity went on beyond the apple tree and stretched up into the sky, to the Moon, and beyond. He realized that the Moon was falling continuously towards the Earth under the influence of gravity and because it was also moving "horizontally" it was in continuous orbit.

Worked examples

1 Calculate the force of attraction between an apple of mass 100 g and the Earth.
Mass of Earth = 6.0×10^{24} kg.
Radius of Earth = 6.4 Mm

Solution

$$F = \frac{GMm}{r^2} = \frac{6.7 \times 10^{-11} \times 0.1 \times 6.0 \times 10^{24}}{(6.4 \times 10^6)^2}$$

$$= 1.1\text{ N}$$

2 Calculate the force of attraction between a proton of mass 1.7×10^{-27} kg and an electron of mass 9.1×10^{-31} kg a when they are at a distance of 1.5×10^{-10} m apart.

Solution

$$F = \frac{GMm}{r^2}$$

$$= \frac{6.7 \times 10^{-11} \times 1.7 \times 10^{-27} \times 9.1 \times 10^{-31}}{(1.5 \times 10^{-10})^2}$$

$$= 4.6 \times 10^{-48}\text{ N}$$

(Compare this with the electrostatic attraction of the electron and proton at this separation of about 10^{-8} N.)

Gravitational field strengths re-visited

Knowledge of Newton's law of gravitation means that the definition of gravitational field strength can be taken further to help our study of real situations.

The field strength at a distance r from a point mass, M

The simplest case is that of a single point mass M placed, as usual, a long way from any other mass. In practice we say that the point mass is at an infinite distance from any other mass because if r is very large then $\frac{1}{r^2}$ is extremely small (mathematicians say that as r tends to infinity, $\frac{1}{r^2}$ tends to zero).

As usual, the mass of our small test object is m. This means that the *magnitude* of the force F between the two masses M and m is

$$F = \frac{GMm}{r^2}$$

so that the gravitational field strength g is

$$g = \frac{F}{m} = \frac{GM}{r^2}$$

As before, the direction is measured outwards *from* M but the force on the point mass is in the opposite direction *towards* M.

The field strength at a distance r from the centre outside a sphere of mass M

It turns out that the answer for g outside a spherical planet is exactly the same as g for the point mass just quoted:

$$g = \frac{GM}{r^2}$$

If we are outside the sphere, all the mass acts as though it is a point mass of size M positioned at the centre of the sphere (the centre of mass). We only need one equation for both point masses and spheres.

Nature of science

And inside the Earth...?

Science is all about asking questions. Outside the Earth, the gravitational field strength varies as $\frac{1}{r^2}$. But what happens to gravity inside the Earth? Is it zero? Is it a constant? Does it become larger and larger, reaching infinity, as we get closer to the centre? You might, at first sight, expect this from Newton's law.

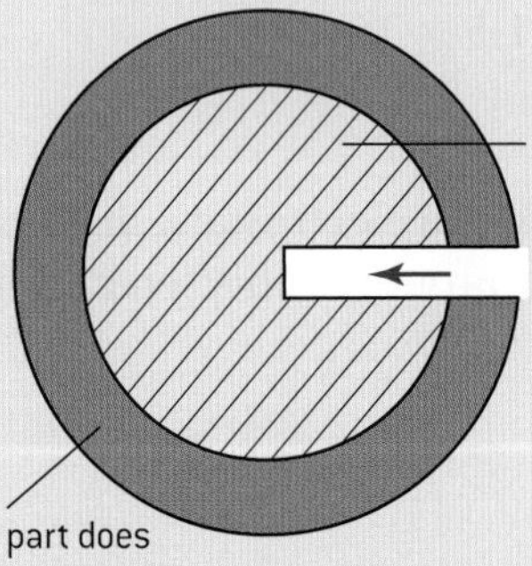

▲ Figure 5 Journey to the centre of the Earth.

In *Journey to the centre of the Earth* the novelist Jules Verne imagined going through a volcanic tunnel to the centre of the planet. Visualize his travellers halfway down the tunnel as they stand on the surface of a "smaller" Earth defined by their present distance from the centre.

Two remarkable things happen: The first is that the contribution from the shell of Earth "above" the travellers makes no contribution to the gravitational field; all the different parts of the outer shell cancel out. The second thing that happens is that only the mass inside the "smaller" Earth contributes to the gravitational pull. The mass of this "smaller" Earth varies with r^3 assuming a constant density for the Earth. Because the gravitational force varies with $\frac{1}{r^2}$, together these give an overall dependence of r. The whole graph for the variation of g for a planet (inside and out) is given in figure 6, g_s is the gravitational field strength at the surface

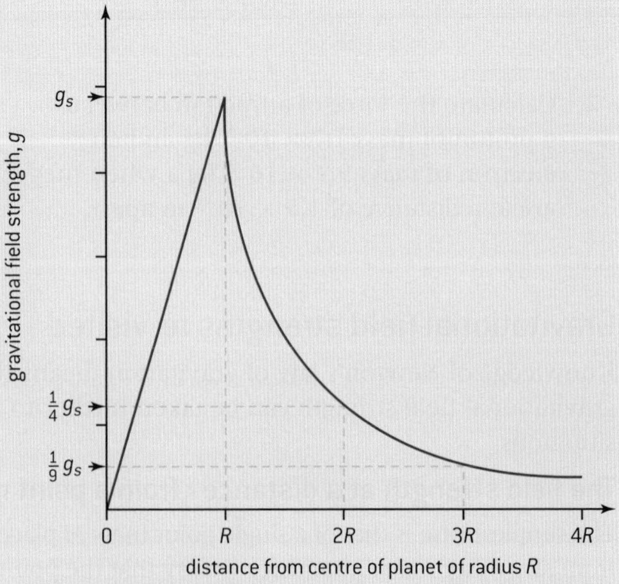

▲ Figure 6 Gravitational field strength inside and outside the Earth.

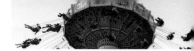

Linking orbits and gravity

The gravitational force of a planet provides the centripetal force to keep a satellite in orbit.

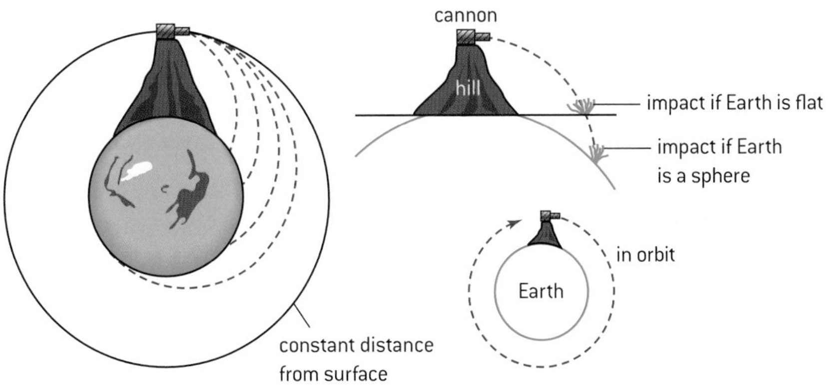

▲ Figure 7 Newton's cannon – how he thought about orbits.

Newton had an insight into this too. He used the example of a cannon on a high mountain (figure 7). The cannon fires its cannonball horizontally and it accelerates vertically downwards. On a flat Earth, it will eventually hit the ground. Newton knew that the Earth was a sphere and that the curvature of the earth allowed the ball to travel further before hitting the ground.

He then imagined the ball being fired at larger and larger initial speeds. Eventually the shell will travel "horizontally" at such a high speed that the curvature of the Earth and the curve of the trajectory will be exactly the same. When this happens the distance between shell and surface is constant and the shell is in orbit around the Earth.

What do you expect to happen to the trajectory of the cannon ball if it is fired at even greater speeds? To check your conclusions, find an applet on the Internet that will allow you to vary the firing speed of Newton's cannon. A good starting point for the search is "applet Newton gun".

This motion of a satellite around the Earth can be analysed by combining the ideas of centripetal force and gravitational attraction. The gravitational attraction F_G provides the centripetal force F_c, so (ignoring signs)

$$F_c = F_G = \frac{mv^2}{r} = \frac{GM_E m}{r^2}$$

where M_E is the mass of the Earth, r is the distance from the satellite to the centre of the Earth, v is the linear speed of the satellite, and m is its mass.

These equations can be simplified to

$$v = \sqrt{\frac{GM_E}{r}}$$

This equation predicts the speed of the satellite at a particular radius. Notice that the speed does not depend on the mass of the satellite. All satellites travelling round the Earth at the same distance above the surface have the same speed.

As an exercise, use the equations on page 250 to show that

$$\omega^2 = \frac{GM_E}{r^3}$$

Worked examples

1 Calculate the gravitational field strength at the surface of the Moon.

(Mass of Moon $= 7.3 \times 10^{22}$ kg; radius of the Moon $= 1.7 \times 10^6$ m)

Solution

$$g = \frac{GM}{r^2}$$

$$= \frac{6.7 \times 10^{-11} \times 7.3 \times 10^{22}}{(1.7 \times 10^6)^2}$$

$$= 1.7 \text{ N kg}^{-1}$$

2 Calculate the gravitational field strength of the Sun at the position of the Earth.

(Mass of Sun $= 2.0 \times 10^{30}$ kg; Earth–Sun distance $= 1.5 \times 10^{11}$ m.)

Solution

$$g = \frac{GM}{r^2}$$

$$= \frac{6.7 \times 10^{-11} \times 2.0 \times 10^{30}}{(1.5 \times 10^{11})^2}$$

$$= 6.0 \text{ mN kg}^{-1}$$

where ω is the angular speed, and that

$$T^2 = \frac{4\pi^2 r^3}{GM_E}$$

where T is the orbital period (time for one orbit) of the satellite. This result:

(orbital period of a satellite)$^2 \propto$ *(orbital radius)*^{3}

is known as **Kepler's third law**. It is one of the three laws that he discovered when he analysed Tycho Brahe's data.

Worked examples

1 Calculate the orbital period of Jupiter about the Sun. Mass of Sun = 2.0×10^{30} kg; radius of Jupiter's orbit = 7.8×10^{11} m)

Solution

$$T^2 = \frac{4\pi^2 r^3}{GM_S}$$

$T = 3.7 \times 10^8$ s (about 12 years)

2 The orbital time period of the Earth about the Sun is 3.2×10^7 s. Calculate the orbital period of Mars.

(radius of Earth orbit = 1.5×10^{11} m; radius of Mars orbit = 2.3×10^{11} m)

Solution

$$\frac{T_M^2}{T_E^2} = \frac{r_M^3}{r_E^3}$$

$$T_M = T_E\sqrt{\frac{r_M^3}{r_E^3}} = 6.1 \times 10^7 \text{ s (about 1.9 years)}$$

Nature of science

What Newton knew …

Several decades before Newton began working on his ideas of gravitation, Kepler had used his own and others' astronomical data to show that, for the planets,

(radius of the orbit)$^3 \propto$ *(time period of orbit)*^{2}

This was an *empirical* result (meaning that it came from experimental data).

Newton was able to show that this proportionality arose from his own theory; this was a *theoretical* result (meaning that it came from a theory, not data).

Here are some data for the satellites of Jupiter.

Moon	Distance from centre of Jupiter/10^3 km	Orbital period/ days
Io	420	1.8
Europa	670	3.6
Ganymede	1070	7.2
Callisto	1890	16.7

Use these data to show that

(radius of the orbit)$^3 \propto$ (time period of orbit)^2

Or, put another way, that $\frac{\text{(radius of the orbit)}^3}{\text{(time period of orbit)}^2}$ is a constant.

Questions

1 A particle P is moving in a circle with uniform speed. Draw a diagram to show the direction of the acceleration a and velocity v of the particle at one instant of time.

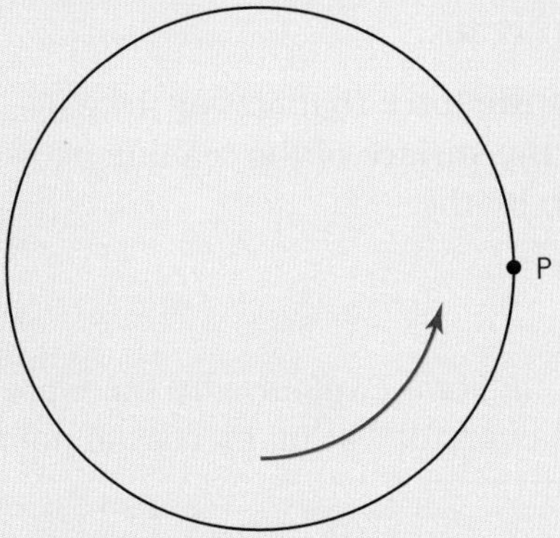

2 State what provides the centripetal force that causes a car to go round a bend.

3 State the centripetal force that acts on a particle of mass m when it is travelling with linear speed v along the arc of a circle of radius r.

4 (*IB*) At time $t = 0$ a car moves off from rest in a straight line. Oil drips from the engine of the car with one drop every 0.80 s. The position of the oil drops on the road are drawn to scale on the grid below such that 1.0 cm represents 4.0 m. The grid starts at time $t = 0$.

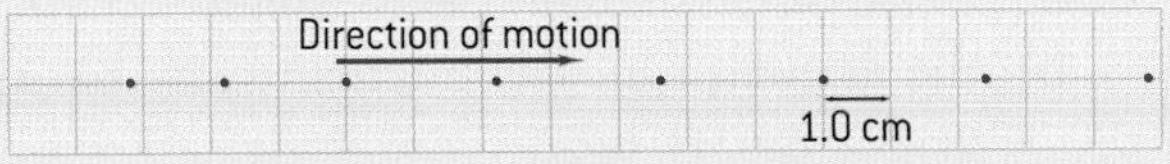

a) (i) State the feature of the diagram that indicates that the car accelerates at the start of the motion.

(ii) Determine the distance moved by the car during the first 5.6 s of its motion.

b) The car then turns a corner at constant speed. Passengers in the car who were sitting upright feel as if their upper bodies are being "thrown outwards".

(i) Identify the force acting on the car, and its line of action, that enables the car to turn the corner.

(ii) Explain why the passengers feel as if they are being thrown outwards.

5 The Singapore Flyer is a large Ferris wheel of radius 85 m that rotates once every 30 minutes.

a) Calculate the linear speed of a point on the rim of the wheel of the Flyer.

b) (i) Determine the fractional change in the weight of a passenger on the Flyer at the top of the ride.

(ii) Explain whether the passenger has a larger or smaller apparent weight at the top of the ride.

c) The capsules need to rotate to keep the floor of the cabin in the correct place.

Calculate the angular speed of the capsule about its central axis.

6 The radius of the Earth is 6400 km. Determine the linear speed of a point on the ground at the following places on Earth:

a) Quito in Ecuador (14 minutes of arc south of the Equator)

b) Geneva in Switzerland (46° north of the Equator)

c) the South Pole.

7 A school bus of total mass 6500 kg is carrying some children to school.

a) During the journey the bus needs to travel round in a horizontal curve of radius 150 m. The dynamic coefficient of friction between the tyres and the road surface is 0.7. Estimate the maximum speed at which the driver should attempt the turn.

b) Later in the journey the driver needs to drive across a curved bridge with a radius of curvature of 75 m. Estimate the maximum speed if the bus is to remain in contact with the road.

8 A velodrome used for bicycle races has a banking angle that varies continuously from 0° to 60°. Explain how the racing cyclists use this variation in angle to their advantage in a race.

Data needed for these questions:

Radius of Earth = 6.4 Mm;
Mass of Earth = 6.0×10^{24} kg;
Mass of Moon = 7.3×10^{22} kg;
Mass of Sun = 2.0×10^{30} kg;
Earth–Moon distance = 3.8×10^{8} m;
Sun–Earth distance = 1.5×10^{11} m;
$G = 6.67 \times 10^{-11}$ N m^2 kg^{-2}

9 Deduce how the radius R of the circular orbit of a planet around a star of mass m_s relates to the period T of the orbit.

10 A satellite orbits the Earth at constant speed as shown below.

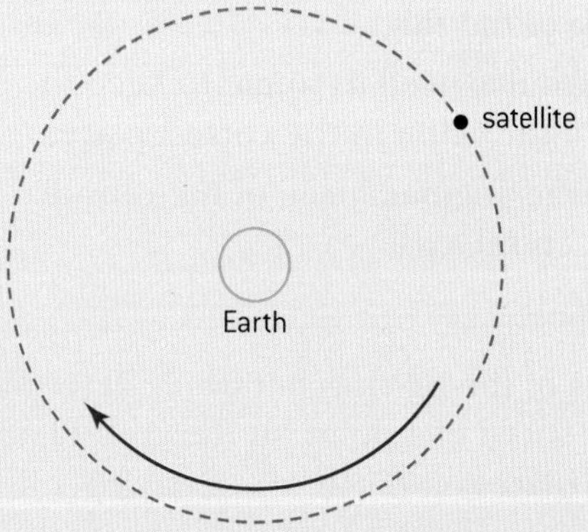

a) Explain why, although the speed of the satellite is constant, the satellite is accelerating.

b) Discuss whether or not the gravitational force does work on the satellite.

11 Determine the distance from the centre of the Earth to the point at which the gravitational field strength of the Earth equals that of the Moon.

12 The ocean tides on the Earth are caused by the tidal attraction of the Moon and the Sun on the water in the oceans.

a) Calculate the force that acts on 1 kg of water at the surface of the sea due its attraction by the

(i) Moon

(ii) Sun.

b) *Optional – difficult.* Explain why there are two tides every day at many coastal points on the Earth.

[Hint: there are two parts to the answer, why a tide at all, and why two every day.]

13 There are two types of communication satellite. One type of communication satellite orbits over the poles at a distance from the centre of the Earth of 7400 km; the other type is geostationary with an orbital radius of 36 000 km. Geostationary satellites stay above one point on the equator whereas polar-orbit satellites have an orbital time of 100 minutes.

Calculate:

a) the gravitational field strength at the position of the polar-orbit satellite

b) the angular speed of a satellite in geostationary orbit

c) the centripetal force acting on a geostationary satellite of mass 1.8×10^{3} kg.

7 ATOMIC, NUCLEAR, AND PARTICLE PHYSICS

Introduction

In this topic we consider the composition of atoms. We look at extra-nuclear electrons and the nucleus and the particles of which the nucleus is composed. We see the vast array of particles that are now known to exist and how these particles can be classified and grouped. As is often the case, energy plays in important role in the atom and fundamental to this is the tendency for particles to be most stable when their energy is minimized.

7.1 Discrete energy and radioactivity

Understanding

- → Discrete energy and discrete energy levels
- → Transitions between energy levels
- → Radioactive decay
- → Alpha particles, beta particles, and gamma rays
- → Half-life
- → Constant decay probability
- → Absorption characteristics of decay particles
- → Background radiation

Applications and skills

- → Describing the emission and absorption spectra of common gases
- → Solving problems involving atomic spectra, including calculating the wavelength of photons emitted during atomic transitions
- → Completing decay equations for alpha and beta decay
- → Determining the half-life of a nuclide from a decay curve
- → Investigating half-life experimentally (or by simulation)

Equations

- → Photon energy–frequency relationship: $E = hf$
- → Planck relationship for wavelength: $\lambda = \frac{hc}{E}$

Nature of science

Unintentional discoveries

Physics is scattered with examples of experimenters making significant discoveries unintentionally. In the case of radioactivity the French physicist Henri Becquerel, in 1896, was investigating a possible link between X-rays and phosphorescence. He had stored a sample of uranium salt (which releases light over a period of time – having being exposed to light) in a drawer with a photographic plate (the forerunner to film) wrapped in an opaque covering. Becquerel discovered that the plate had become exposed even though it had been kept in the dark. Rather than ignore this unpredicted outcome, Becquerel sought an explanation and found other uranium salts produced the same result – suggesting that it was the uranium atom that was responsible. He went on to show that the emissions from uranium ionize gases but differ from X-rays in that they are deflected by electric and magnetic fields. For his work, which developed as the result of a lucky accident, Becquerel was awarded a share of the Nobel Prize for Physics in 1903 (with the Curies for their work also on radioactivity).

Introduction

Although many texts take a historical tour of the development of atomic and nuclear physics, the IB Physics syllabus does not lend itself to such a treatment. Instead, there is a focus on the common aspects of the constituents of the atom. It is likely that you are already familiar with the structure of an atom as consisting of a positively charged nucleus surrounded by negatively charged electrons in fixed orbits. You are probably also be aware that the nucleus consists of positively-charged protons and uncharged neutrons. We will take these aspects as read, but we will also revisit them throughout the sections of Topic 7 (and, for those students studying HL Physics, Topic 12 too).

TOK

Assumptions in physics

Topic 5 showed that protons and electrons are charged. A bigger question is "why do charges behave in this way?" The simple answer to this is that we don't know – we do know that some objects have a property that we call charge and that charges interact at a distance through an electromagnetic field. Yet we don't know what charges are and what the difference between positive and negative actually is. This may seem a huge gap in our understanding. However, we use the same concept of the implied labelling of objects or properties – for example, when we ask the question "what is a chair?", our answers will inevitably give us a description of a chair along the lines of "something we sit on", "having four legs, it has a seat and back" etc. Therefore, what we have done is described the properties and characteristics of chairs – it remains the case that a chair is a chair. So, although we only have a fuzzy idea of what electrons and charges actually are, does this prevent us from developing effective theories that are based on poorly understood concepts?

Energy levels

Returning to the nuclear model of an atom, the orbiting electrons cannot occupy any possible orbit around the nucleus. Different orbits correspond to different amounts of energy, or energy levels, and the electrons are restricted to orbits with specific energies. Electrons change energy so that they can jump from one energy level to another, but they can only occupy allowed energy levels. Although you may ask "what makes an allowed orbit or allowed energy level?" it takes an understanding of the wave nature of electrons and quantum mechanics to attempt to explain this consistently. We return to this for HL students in Topic 12, but for now it is maybe better to accept that this is how nature operates.

Hydrogen is the simplest of all elements, with the hydrogen atom normally consisting of a proton bound to a single electron by the electromagnetic force (an attractive force between oppositely charged particles). There are other **isotopes** of hydrogen, with a nucleus including one or more neutrons. If a hydrogen atom completely loses the electron it becomes a positively-charged hydrogen ion. It is the single proton that defines this nucleus to be hydrogen – any atom with a single proton must be hydrogen. In the same way any atom having two protons is an isotope of helium, and any with three protons will be lithium, etc.

A common visualization of the atom is the "planetary model" in which the electron orbits the proton mimicking the Earth orbiting the Sun. Despite weaknesses in this model, it still gives a good introduction to energy levels in the atom. The energy levels for the hydrogen atom are shown in the table below:

Energy level (n)	Energy/eV
1	−13.58
2	−3.39
3	−1.51
4	−0.85
5	−0.54

The energy levels are usually shown diagrammatically as in figure 1. This diagram shows the energy level and the energy of an electron that occupies this level. An electron in the ground state of the hydrogen atom must have exactly −13.58 eV of energy; one in the second energy level must have −3.39 eV, etc. An atom with an electron occupying an energy level higher than the ground state is said to be **excited**; so an electron in level 2 is in the first excited state, one in level 3 is in the second excited state, etc. The energy levels in atoms are said to be **quantized** – which means they must have discrete finite values. An electron in a hydrogen atom **cannot** have −5.92 eV or −10.21 eV or any other value between −13.58 and −3.39 eV; it **must** have one of the values in the table.

Transitions between energy levels

When an electron in the hydrogen atom jumps from the ground state to the first excited state it must gain some energy; this cannot be any randomly chosen amount of energy – it must be exactly the right amount. The electron needs to gain (−3.39) − (−13.58) = 10.19 eV of energy. The electron does not physically pass through the space between the two energy levels. Before gaining the energy it must have −13.58 eV and immediately after gaining the energy it must have −3.39 eV. The energy has to be transferred to the electron in one discrete amount; it cannot be gradually built up over time.

The energy needed to excite an atom can come from absorption of light by the atom. In order to understand this we must consider light to be a packet or **quantum** called a **photon** – the rationale for this is discussed in Topic 12. The energy, E, carried by a photon is related to the frequency of the radiation by the equation $E = hf$ where h is the Planck constant ($= 6.63 \times 10^{-34}$ J s) and f is the frequency in Hz. The wave equation $c = f\lambda$ (where c is the speed of electromagnetic waves in a vacuum) also applies to the photons and, by combining the two equations, we can relate the energy to the wavelength $E = hf = h\frac{c}{\lambda}$ or $\lambda = h\frac{c}{E}$. We have seen that an electron needs to absorb 10.19 eV in order to jump from the ground state to the first excited state. What, then, will be the wavelength of the electromagnetic radiation needed to do this?

First we need to convert 10.19 eV into joules. This is simply a matter of multiplying the energy in eV by the charge on an electron, so $10.19 \text{ eV} = 10.19 \times 1.6 \times 10^{-19} \text{ J} \sim 1.6 \times 10^{-18} \text{ J}$.

Now using $\lambda = h\frac{c}{E}$ we get $\lambda = 6.63 \times 10^{-34} \text{ J} \times \frac{3.00 \times 10^{8}}{1.6 \times 10^{-18}} = 1.2 \times 10^{-7}$ m or 120 nm.

This radiation in the ultraviolet part of the spectrum.

When an electron has been excited in this way it is likely to be quite unstable and will very quickly (in about a nanosecond) return to a lower energy level. In order to do this it must lose this same amount of energy. This means that a photon of energy 10.19 eV and wavelength 120 nm must be emitted by the atom. The electron will then jump to the lower energy level.

For the transition from the second energy level to the third to take place only 1.88 eV needs to be absorbed by the atom and 0.66 eV to jump from the third level to the fourth. When the electron is given the full 13.58 eV it

Note

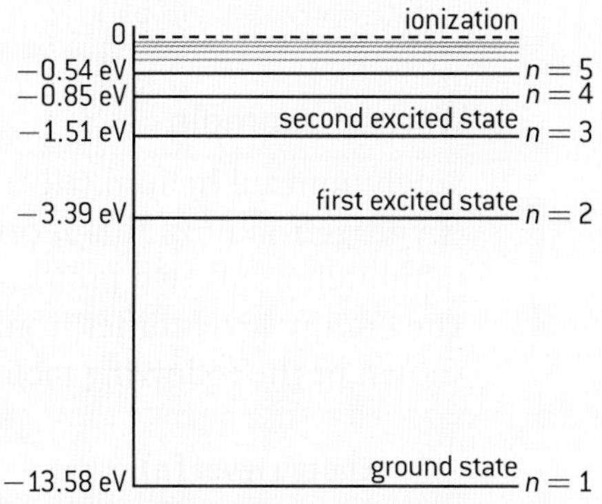

Figure 1 Energy level diagram for hydrogen.

- The energies are given in units of electron volt ($1 \text{ eV} = 1.6 \times 10^{-19}$ J). This is a very common unit (along with the multipliers keV and MeV) for atomic physics and it avoids having to use very small numbers. However, you might find that a question is set in joule.
- The energies are all negative values – this is because the potential energy of two objects is zero only when they are at infinite separation. Because there is an attractive force between a proton and an electron the electron has been moved from infinity until it is in its orbit. The proton–electron system has lost some energy and, since it was zero to start with, it is now negative.
- Energy level 1 is the lowest energy level – having the most negative and, therefore, smallest amount of energy. It is the most stable state and is called the *ground state*.
- n is known as the principal quantum number and we will return to its values in Topic 12.

is completely removed from the nucleus (or *taken to infinity*) and the atom has become ionized. The energy supplied to the electron in the ground state to take it to infinity is called the (first) *ionization energy*.

Worked example

Singly-ionized helium (He^+)is said to be "hydrogen-like" in that it only has one electron (although it has two protons and two neutrons). The energy levels differ from those of hydrogen as shown in the following table:

Energy level (n)	Energy/eV
1	−54.4
2	−13.6
3	−6.0
4	−3.4
5	−2.2

a) Explain why the ground state has a much higher energy level than that of hydrogen (−13.6 eV).

b) (i) Determine the frequency of the photon emitted by an electron transition from energy level 4 to energy level 2.

(ii) State which region of the electromagnetic spectrum the emitted photon belongs to.

Solution

a) The hydrogen atom has a single proton in the nucleus but the helium ion has two. This means the attractive force between the nucleus and an electron is greater for the helium nucleus. The helium electron is more tightly bound to the nucleus and, therefore, requires more energy to remove it to infinity.

b) (i) $\Delta E = (-13.6) - (-3.4) = -10.2$ eV (the minus sign tells us that there is a loss of energy during the emission of photons).

$$10.2 \text{ eV} = 10.2 \times 1.6 \times 10^{-19} \text{ J} = 1.6 \times 10^{-18} \text{ J}$$

$$E = hf \therefore f = \frac{E}{h} = \frac{1.6 \times 10^{-18}}{6.63 \times 10^{-34}} = 2.4 \times 10^{15} \text{ Hz}$$

(ii) We have seen this energy for the hydrogen atom previously and so know that this frequency corresponds to ultraviolet. You can calculate the wavelength by dividing the speed of electromagnetic waves by your known frequency.

Nature of science

Patterns in physics

You may have noticed that the calculation for the energy of photon emitted by the helium ion in the worked example above is identical to an energy value for the hydrogen atom. This sort of coincidence is not unusual in physics and suggests that there may be a link between the energy differences. In fact in 1888 the Swedish physicist Johannes Rydberg proposed a formula for "hydrogen-like" atoms: singly ionized helium, doubly ionized lithium, triply ionized beryllium, etc. based on observation of spectral line patterns emitted by these elements. Rydberg's theory, based on his practical work, utilized the idea of "wave number" (the reciprocal of the wavelength) – this was of fundamental importance to the Danish physicist Niels Bohr in explaining quantization of energy levels in 1913.

Emission spectra

When energy is supplied to a gas of atoms at low pressure the atoms emit electromagnetic radiation. In the laboratory the energy is usually supplied by an electrical discharge (an electrical current passing through the gas when a high voltage is set up between two electrodes across the gas). If the radiation emitted by the gas is incident on the collimating slit of a spectrometer, it can then be dispersed by passing it though a diffraction grating or a glass prism (see figure 5). Observing the spectrum through a telescope, or projecting it onto a screen, will give a series of discrete lines similar to those shown in figure 2.

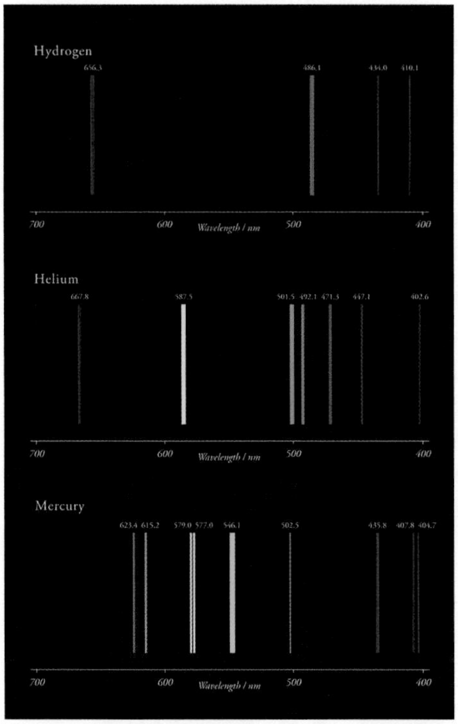

▲ Figure 2 Line emission spectra.

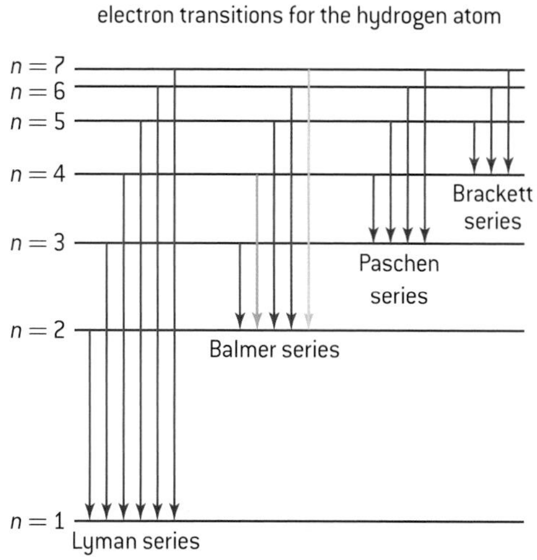

▲ Figure 3 Electron transitions for hydrogen.

Spectral lines appear in series in the different regions of the electromagnetic spectrum (infra-red, visible, and ultraviolet). Each series of lines is dependent on the energy level that the electrons fall to. In the case of hydrogen the series are named after the first person to discover them: in the Lyman series the electrons fall to the ground state ($n = 1$) – this series is in the ultraviolet region of the electromagnetic spectrum; in the Balmer series the electrons fall to the first excited level ($n = 2$) – this is in the visible region; the Paschen series relates to $n = 3$ and is in the infrared region.

Absorption spectra

The electrons in solids, liquids, and dense gases can also be excited – they tend to glow when heated to a high temperature. When the emitted light is observed it is seen to consist of a spectrum of bands of colours rather than lines. In the case of solids this will give a continuous spectrum in which the colours are merged into each other and so are not discrete. This is typical of matter in which the atoms are closely packed. The neighbouring atoms are very close and this causes the energy levels in the atoms to change value. When there are many atoms, the overall energy levels combine to form a series of very similar, but different, energies which makes up an energy band.

When the object emitting a continuous spectrum is surrounded by a cool gas, then the continuous spectrum is modified by the surrounding gas. The continuous spectrum is "streaked" by a number of dark lines. When a heated tungsten filament is viewed through hydrogen gas the absorption spectrum

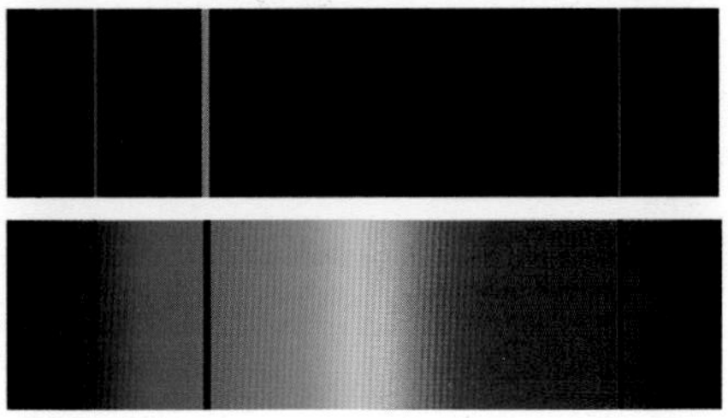

▲ Figure 4 Emission and absorption spectra for hydrogen.

shown in the lower part of figure 4 can be seen. Further inspection shows that the black lines occur at exactly the same positions as the lines of the hydrogen emission spectrum. This pattern cannot simply be a coincidence.

Absorption occurs when an electron in an atom of the absorbing material absorbs a photon. The energy of this photon must be identical to the difference between the energy levels. The material removes photons of this frequency from the continuous range of energies emitted by the light source. Naturally, this will make the absorber's atoms become unstable and they will revert to a lower energy level by emitting photons – these will include the absorbed frequencies but they will be emitted in random directions – and not necessarily in the original direction. This will reduce the intensity of those specific frequencies in the original direction giving the black lines seen crossing the continuous spectrum in figure 4.

Absorption spectra for sodium can be demonstrated with the apparatus shown in figure 5. A white light source emits light which is incident on a diffraction grating on the turntable of the spectrometer. A continuous spectrum is seen through the telescope or displayed on a computer monitor (using an appropriate sensor and software). A burner is used to heat sodium chloride (table salt). The black absorption lines of sodium should be detected in the yellow region of the continuous spectrum.

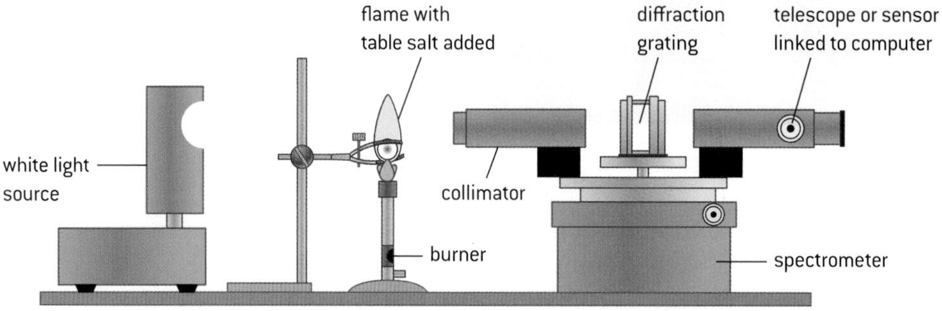

▲ Figure 5 Demonstration of sodium absorption spectra.

Nature of science

Fraunhofer lines

▲ Figure 6 Stamp commemorating the life and work of Fraunhofer.

The English scientist William Woolaston originally observed absorption lines in the Sun's spectrum in 1802. However it was the German physicist, Joseph von Fraunhofer, who, in 1814, built a spectrometer and invented the diffraction grating with which he was able to observe and analyse these absorption lines – now known as *Fraunhofer lines*. These lines were the first lines in a spectrum to be observed. Fraunhofer labelled the most prominent of the lines as A–K. These lines provide astronomers with a means to study the composition of a star.

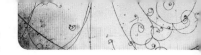

Worked example

a) The element helium was first identified from the *absorption spectrum* of the Sun.

(i) Explain what is meant by the term *absorption spectrum*.

(ii) Outline how this spectrum may be experimentally observed.

b) One of the wavelengths in the absorption spectrum of helium occurs at 588 nm.

(i) Show that the energy of a photon of wavelength 588 nm is 3.38×10^{-19}J.

(ii) The diagram below represents some of the energy levels of the helium atom. Use the information in the diagram to explain how absorption at 588 nm arises.

(iii) Mark this transition on a copy of the energy level diagram below.

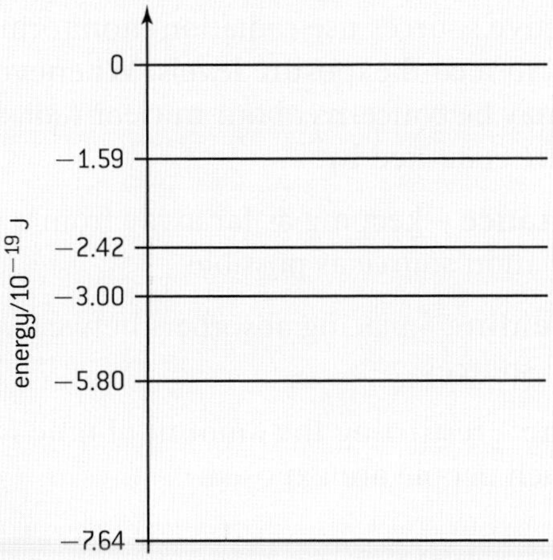

Solution

a) **(i)** An absorption spectrum consists of a continuous spectrum that has a number of absorption lines crossing it. These lines correspond to the frequencies of the light in the emission spectrum of the elements within the substance that is absorbing the light.

(ii) The light from the Sun can be projected onto a screen after passing through a diffraction grating or a prism in order to disperse it into its component wavelengths. This gives evidence about the elements in the gases in the outer part of the Sun.

b) **(i)** $E = \frac{hc}{\lambda} = \frac{6.63 \times 10^{-34} \times 3.00 \times 10^{8}}{588 \times 10^{-9}}$

$= 3.38 \times 10^{-19}$ J

(ii) As this is absorption, the electron is being raised to a higher energy level. The difference between the levels must be equal to 3.38×10^{-19} J and so this is between $(-5.80 \times 10^{-19}$ J) and $(-2.42 \times 10^{-19}$ J) levels, i.e.,
$(-2.42 \times 10^{-19}\text{ J}) - (-5.80 \times 10^{-19}\text{ J})$
$= (+)3.38 \times 10^{-19}$ J

Note that the energy levels are given in joules and so there is no need to do a conversion from electronvolts in this question.

(iii) The transition is marked in red.

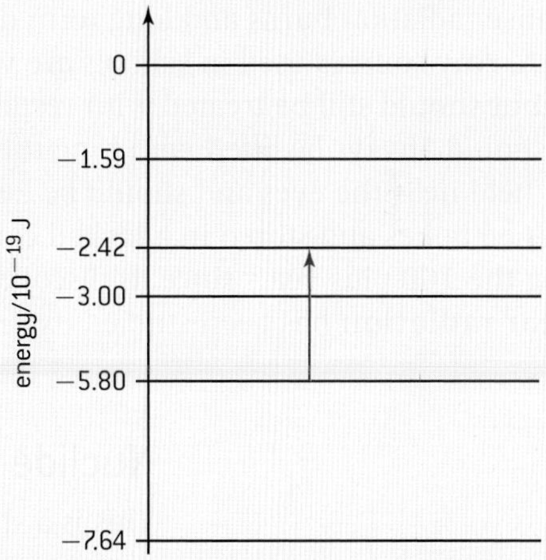

Note that light from the Sun needs to be filtered and should not be observed directly through a telescope.

Radioactive decay

Radioactive decay is a naturally occurring process in which the nucleus of an unstable atom will spontaneously change into a different nuclear configuration by the emission of combinations of alpha particles, beta particles, and gamma radiation. There are fewer than four hundred naturally occurring nuclides (nuclei with a particular number of protons and neutrons) but only about 60 of these are radioactive. The trend is for these nuclei to become more stable – although this may take a very long time. In radioactive decay the nuclide decaying is referred to as the **parent** and the nuclide(s) formed as the **daughter(s)**.

As we have discussed, the nucleus contains protons and neutrons – these are jointly called *nucleons*. These are bound together by the strong nuclear force, which must overcome the electrostatic repulsion between the positively charged protons. The presence of the neutrons moderates this repulsion – the strong nuclear force (which has a very short range $\approx 10^{-15}$ m) acts equally on both the protons and neutrons. For the nuclei with few nucleons, having approximately equal numbers of protons and neutrons corresponds to nuclides being stable and not radioactive. As we will soon see, heavier nuclei need a greater proportion of neutrons in order to be stable.

Nature of science

Nuclear radiation and safety

Many people are apprehensive about any exposure to radiation from radioactive sources. The danger from alpha particles is small unless the source is ingested into the body. Beta particles and gamma rays are much more penetrating and can cause radiation burns and long term damage to DNA. Any sources used in schools are very weak but should still be treated with respect. They should always be lifted with long tongs, never held near the eyes and should be kept in lead-lined boxes and stored in a locked container. During the 100 plus years since its discovery **nuclear radiation** has been studied extensively and can be safely controlled. Those working with radioactive sources use radiation monitoring devices to record exposure levels. Whenever there may be concerns about nuclear radiation, exposure is limited by:

- **distance** – keeping as far away from a radiation source as possible
- **shielding** – placing absorbers between people and sources
- **time** – restricting the amount of time for which people are exposed.

Nuclide nomenclature

This is a shorthand way of describing the composition of a nuclide. The element is described by its chemical symbol – H, He, Li, Be, B, etc. The structure of the nucleus is denoted by showing both the proton number Z (otherwise called the "atomic number") and the nucleon number *A* (or "mass number"). This always takes the form: ${}^{A}_{Z}X$ so, for example, the isotopes of carbon, – carbon-12 and carbon-14, – are written as ${}^{12}_{6}C$ and ${}^{14}_{6}C$. Carbon-12 has 6 protons and 12 nucleons and so it has 6 neutrons; carbon-14 has 6 protons and 14 nucleons making 8 neutrons. Isotopes are nuclides of the same element (and so have the same number of protons) having different numbers of neutrons.

Alpha (α) decay

In alpha-particle decay, an unstable nuclide emits a particle of the same configuration as a helium nucleus ${}^{4}_{2}He$ (having two protons and two neutrons). Many nuclides of heavy elements decay primarily by alpha-particle emission.

Examples:

$^{238}_{92}\mathrm{U} \rightarrow {}^{4}_{2}\mathrm{He} + {}^{234}_{90}\mathrm{Th}$

$^{234}_{90}\mathrm{Th} \rightarrow {}^{4}_{2}\mathrm{He} + {}^{230}_{88}\mathrm{Ra}$

As a consequence of the conservation of charge and mass–energy, the equation must balance so that there are equal numbers of protons and nucleons on either side of it.

The alpha particle is sometimes written as $^{4}_{2}\alpha$ instead of $^{4}_{2}\mathrm{He}$.

Negative beta (β^-) decay

In negative beta-particle emission, an unstable nuclide emits an electron. The emission of a beta particle does not change the nucleon number of the parent nuclide. A neutron is converted to a proton and an electron is ejected. This decay occurs for those nuclides with too high a neutron–proton ratio. The decay is accompanied by an electron antineutrino which we will discuss in Sub-topic 7.3.

Examples:

$^{131}_{53}\mathrm{I} \rightarrow {}^{0}_{-1}\mathrm{e} + {}^{131}_{54}\mathrm{Xe} + \overline{\nu}$

$^{234}_{90}\mathrm{Th} \rightarrow {}^{0}_{-1}\mathrm{e} + {}^{234}_{91}\mathrm{Pa} + \overline{\nu}$

The negative beta particle is sometimes written as $^{0}_{-1}\beta$ instead of $^{0}_{-1}\mathrm{e}$.

Again, the equation must balance. The daughter nuclide has the same nucleon number as the parent but has one extra proton meaning that its proton number increases by 1. The antineutrino has no proton or nucleon number and is often written as $^{0}_{0}\overline{\nu}$.

Positron (β^+) decay

In positron or positive beta-particle emission, an unstable nuclide emits a positron. This is the antiparticle of the electron, having the same characteristics but a positive charge instead of a negative one. The emission of the positron does not change the nucleon number of the parent nuclide. A proton is converted to a neutron and a positron is ejected. This decay occurs for those nuclides with too high a proton–neutron ratio. The decay is accompanied by an electron neutrino which, again, we will discuss in Sub-topic 7.3.

Examples:

$^{11}_{6}\mathrm{C} \rightarrow {}^{0}_{+1}\mathrm{e} + {}^{11}_{5}\mathrm{B} + \nu$

$^{21}_{11}\mathrm{Na} \rightarrow {}^{0}_{+1}\mathrm{e} + {}^{21}_{10}\mathrm{Ne} + \nu$

The positron is sometimes written as $^{0}_{+1}\beta$ instead of $^{0}_{+1}\mathrm{e}$.

Again, the equation balances. The daughter nuclide has the same nucleon number as the parent but has one less proton, meaning that its proton number decreases by 1. The neutrino has no proton or nucleon number and is alternatively written as $^{0}_{0}\nu$.

Gamma ray emission

Gamma rays are high-energy photons often accompanying other decay mechanism. Having emitted an alpha or beta particle the daughter nucleus is often left in an excited state. It stabilizes by emitting gamma photon(s) thus losing its excess energy.

Examples:

$$^{60}_{27}\text{Co} \rightarrow {}^{0}_{-1}\text{e} + {}^{60}_{28}\text{Ni}^{*} + \overline{\nu} + \gamma$$

$$^{60}_{28}\text{Ni}^{*} \rightarrow {}^{60}_{28}\text{Ni} + \gamma$$

Here the cobalt-60 decays by beta emission into an excited nickel-60 nuclide $^{60}_{28}\text{Ni}^{*}$. This is accompanied by a gamma photon. The nickel-60 de-excites by emitting a second gamma photon (of different energy from the original photon). Gamma photons have no proton or nucleon number and are sometimes written as $^{0}_{0}\gamma$.

If we had wished to summarize the cobalt decay in one equation we could write it as:

$$^{60}_{27}\text{Co} \rightarrow {}^{0}_{-1}\text{e} + {}^{60}_{28}\text{Ni} + \overline{\nu} + 2\gamma$$

This tells us that overall there are two gamma photons emitted.

Worked examples

1 A nucleus of strontium-90 ($^{90}_{38}\text{Sr}$) decays into an isotope of yttrium (Y) by negative beta emission. Write down the nuclear equation for this decay.

Solution

This is a normal beta negative decay so the equation will be:

$$^{90}_{38}\text{Sr} \rightarrow {}^{0}_{-1}\text{e} + {}^{90}_{39}\text{Y} + {}^{0}_{0}\overline{\nu}$$

2 Under certain circumstances a nucleus can capture an electron from the innermost shell of electrons surrounding the nucleus. When the iron-55 ($^{55}_{26}\text{Fe}$) nucleus captures an electron in this way the nucleus changes into a manganese (Mn) nucleus. Write a nuclear equation to summarise this.

Solution

This is a case of balancing a nuclear equation with the electron being identical to a negative beta particle. The electron is present before the interaction and so appears on the left-hand side of the equation giving:

$$^{55}_{26}\text{Fe} + {}^{0}_{-1}\text{e} \rightarrow {}^{55}_{25}\text{Mn}$$

Half-life

Radioactive decay is a continuous but random process – there is no way of predicting which particular nucleus in a radioactive sample will decay next. However, statistically, with a large sample of nuclei it is highly probable that in a given time interval a predictable number of nuclei will decay, even if we do not know exactly which particular ones. We say that the nuclide has a constant **probability of decay** and this does not depend upon the size of the sample of a substance we have. Another way to look at this is in terms of **half-life**. The half-life is the time taken for half the total number of nuclei initially in a sample to decay *or* for the initial activity of a sample to fall by half. This varies significantly from nuclide to nuclide: the half-life of uranium-238 is 4.5×10^{9} years, while that of phosphorus-30 is 2.5 minutes, and that of the artificially produced nuclide ununoctium-294 is 5 milliseconds.

The nucleus of an atom has a diameter of the order of 10^{-15} m and is essentially isolated from its surroundings , as atoms themselves are separated by distances of about 10^{-10} m. This means that the decay of a nucleus is independent of the physical state of the nuclide (whether it is solid, liquid, or gas or in a chemical compound) and the physical conditions – such as temperature and pressure. Only nuclear interactions such as a collision with a particle in a particle accelerator can influence the half-life of a nuclide.

Figure 7 shows how the parent nuclei decay during a time equal to four half-lives. During this time the percentage of the parent nuclide present has fallen to 6.25%. This graph shows an exponential decay – the shape of this decay curve is common to all radioactive isotopes. The curve approaches the time axis but never intercepts it (it is said to be "asymptotic").

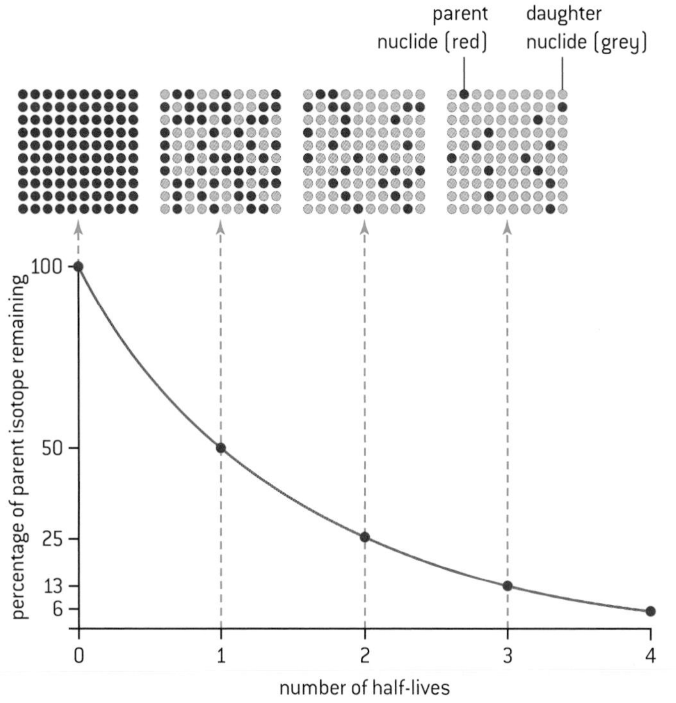

▲ Figure 7 Radioactive half-life.

Consider a sample of radioactive material with N_0 undecayed nuclei at time $t = 0$.

In one half-life this would become $\frac{N_0}{2}$

In two half-lives it would become $\frac{N_0}{4}$

In three half-lives it would be $\frac{N_0}{8}$

And in four half-lives it would become $\frac{N_0}{16}$ (= 6.25% of N_0)

Alternatively you could take half to the power of four to give $\left(\frac{1}{2}\right)^4 = \frac{1}{16}$ or 6.25%

At SL you will only be given calculations which have integral numbers of half-lives, those of you studying HL will see how we deal with non-integral half-lives in Sub-topic 12.2.

Investigate!

Modelling radioactive decay

Different countries have varying regulations regarding the use of radioactive sources. It may be that you have the opportunity to investigate half-life experimentally, for example, using a thorium generator or by the decay of protactinium. If this is possible there is merit, as always, in performing a real experiment. If this is not possible then you should use software or dice to simulate radioactive decay.

It is possible to carry out a very simple modelling experiment using 100 dice (but the more the merrier, if you have lots of time).

- Throw the dice and remove all those which show one.

▲ Figure 8 Dice can be used to model radioactive decay.

- Make a note of the number of dice that you have removed.
- Repeat this until you have just a few dice left.

throw (t)	N (student A)	N (student B)	N (student C)	N (student D)	N (all students)	uncertainty
0	100	100	100	100	400	0
1	92	88	79	88	347	7
2	84	79	71	80	314	7
3	67	59	61	70	257	6
4	52	51	56	57	216	3
5	48	41	36	50	175	7
6	40	36	33	42	151	5
7	36	33	28	31	128	4
8	32	28	24	21	105	6
9	28	20	16	20	84	6
10	23	18	15	19	75	4
11	17	15	11	17	60	3
12	13	15	10	12	50	3
13	10	10	6	11	37	3
14	9	9	6	9	33	2
15	9	9	3	7	28	3
16	8	8	3	7	26	3
17	8	7	2	4	21	3
18	7	7	1	4	19	3
19	6	7	1	4	18	3
20	5	5	0	4	14	3
21	5	4	0	4	13	3
22	5	2	0	3	10	3
23	4	2	0	0	6	2
24	4	2	0	0	6	2
25	4	2	0	0	6	2

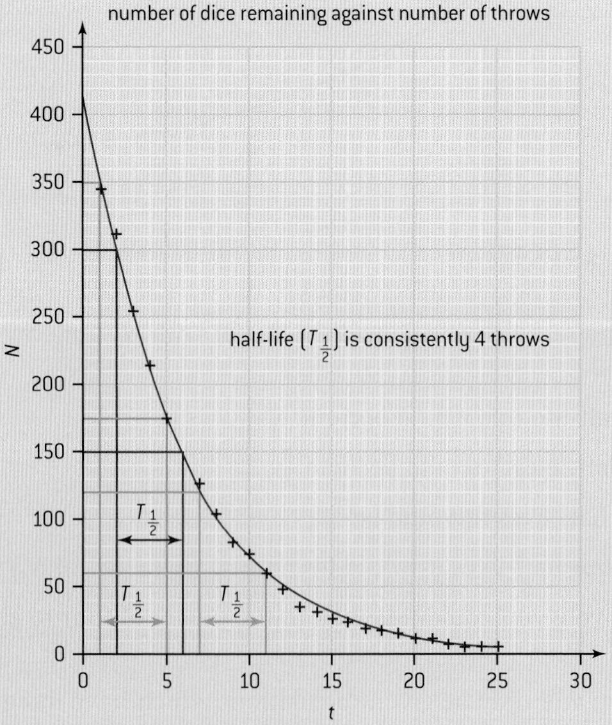

▲ Figure 9 Modelling radioactive decay using a spreadsheet.

- Record the remaining number of dice on a spreadsheet.
- Use the spreadsheet software to draw a line graph (or scatter chart) showing the number of dice remaining against the number of throws.
- Draw a trend line for the points.
- You will see that, with a small sample of just 100 dice, there will be quite a large variation from point to point.
- If several groups of students repeat this experiment you will be able to include error bars on your graph.
- From the trend line you should calculate an average value for the half-life (see figure 9).

Measuring radioactive decay

For beta and gamma radiation the count rate near to a source is measured using a Geiger counter. Strictly, this should be called a Geiger–Müller (G–M) tube and counter. The G–M tube is a metal cylinder filled with a low-pressure gas. At one end of the tube is a thin mica window (that allows radiation to enter the tube). A high voltage is connected across the casing of the tube and the central electrode, as shown in figure 10. The beta or gamma radiation entering the tube ionizes the gas. The ions and electrons released are drawn to the electrodes, thus producing a pulse of current that can be measured by a counting circuit. Alpha particles will be absorbed by the window of a G–M tube but can be detected using a spark counter such as that shown in figure 11. A very high voltage is connected across the gauze on the top and a filament positioned a few millimetres under the gauze. When the alpha particles ionize the air a spark jumps between the gauze and the filament.

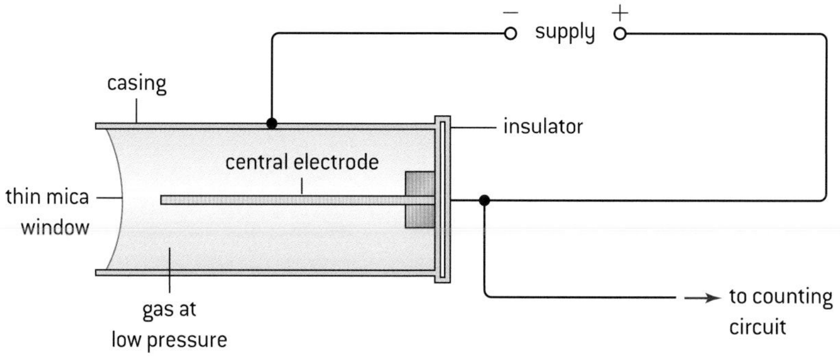

▲ Figure 10 Geiger–Müller tube.

Note

- When you use a graph, in order to calculate the half-life, you should make at least three calculations and take the average of these. Students often halve a value and halve that again and again; as a consequence of doing this the changes are getting very small and will not be reliable. It is good practice to calculate three half-lives, starting from different values, before averaging as shown in figure 9.
- In the example shown the half-life consistently takes four throws the number of throws would be equivalent to a measurement of time in a real experiment with a decaying source.
- For clarity, the error bars have been omitted from this graph.

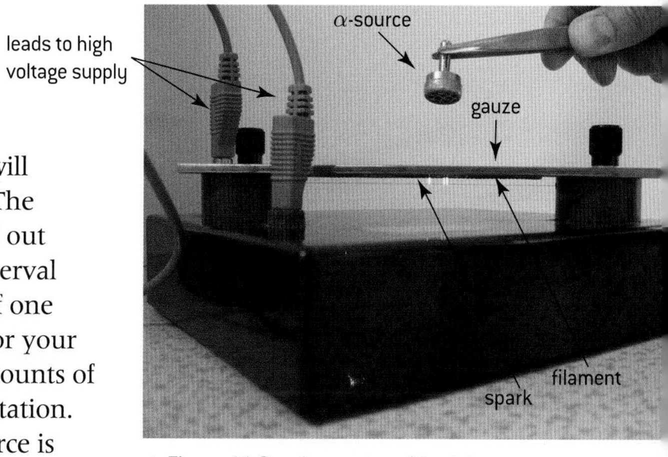

▲ Figure 11 Spark counter with alpha source.

Background count

When a G–M tube is connected to its counter and switched on it will give a reading even when a source of radioactivity is not present. The device will be measuring the **background count**. If this is carried out over a a series of equal, short time intervals, the count for each interval will vary. However, when measurements are taken over periods of one minute several times, you should achieve a fairly constant value for your location. Radioactive material is found everywhere. Detectable amounts of radiation occur naturally in the air, rocks and soil, water, and vegetation. The count rate varies from place to place but the largest single source is from the radioactive gas radon that can accumulate in homes and in the

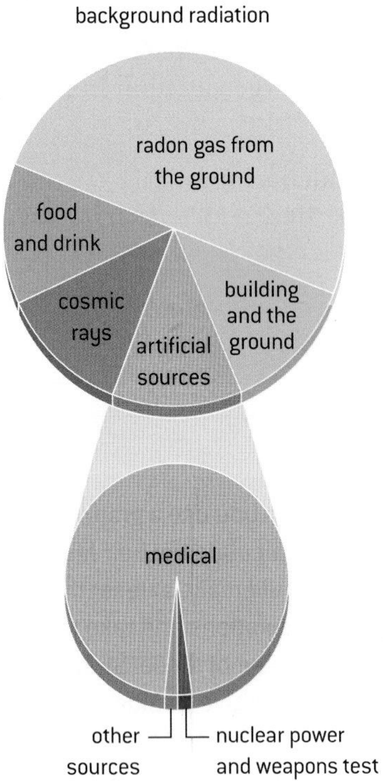

▲ Figure 12 Sources of background radiation.

workplace. Figure 12 is a pie chart that shows the contribution of different sources of background radiation to the total. The sievert (Sv) is a radiation unit that takes the ionizing effects of different radiations into account. Most people absorb between 1.5 and 3.5 millisievert per year largely from background radiation. There are places where the background dose of radiation is in excess of 30 mSv yr^{-1}. However, there is no evidence of increased cancers or other health problems arising from these high natural levels. The sievert will not be examined on the IB Diploma Programme physics course.

Absorption of radiation

The different radioactive emissions interact with materials according to their ionizing ability. Alpha particles ionize gases very strongly, have a very short range in air and are absorbed by thin paper; this is because they are relatively massive and have a charge of $+2e$. Beta particles are poorer ionizers but have a range of several centimetres in air and require a few millimetres of aluminium to absorb them. They are much lighter than alpha particles and have a charge of $-1e$. Gamma rays, being electromagnetic waves, barely interact with matter and it takes many metres of air or several centimetres of lead to be able to absorb them. The apparatus shown in figure 13 is a simple arrangement for measuring the thickness of materials needed to absorb different types of radiation. The source is positioned opposite the window of the G–M tube so that as high a reading as possible is achieved. Different thicknesses of materials are then inserted between the source and the G–M tube until the count rate is brought back in line with the background count. When this happens, the absorption thickness of the named material is found. The following table summarizes the properties of α, β, and γ sources. The information represents "rule of thumb" values and there are exceptions to the suggested ranges.

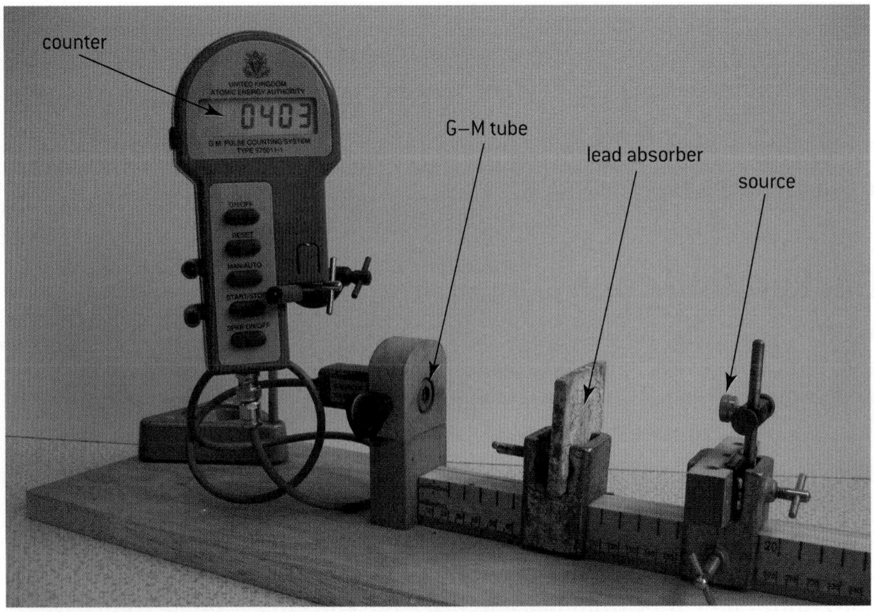

▲ Figure 13 Apparatus to measure absorption of nuclear radiation.

Emission	Composition	Range	Ionizing ability
α	a helium nucleus (2 protons and 2 neutrons)	low penetration, biggest mass and charge, absorbed by a few centimetres of air, skin or thin sheet of paper	very highly ionizing
β	high energy electrons	moderate penetration, most are absorbed by 25 cm of air, a few centimetres of body tissue or a few millimetres of metals such as aluminium	moderately highly ionizing
γ	very high frequency electromagnetic radiation	highly penetrating, most photons are absorbed by a few cm of lead or several metres of concrete few photons will be absorbed by human bodies	poorly ionizing – usually secondary ionization by electrons that the photons can eject from metals

Worked example

Actinium-227 ($^{227}_{89}Ac$), decays into thorium-227 ($^{227}_{90}Th$). Thorium-227 has a half-life of 18 days and undergoes α-decay to the nuclide radium-223. On a particular detector a sample of thorium-227 has an initial count rate of 32 counts per second.

a) Define the terms **(i)** nuclide and **(ii)** half-life.

b) Copy and complete the following reaction equation. $^{227}_{89}Ac \rightarrow {}^{227}_{90}Th + \dots\dots\dots + \dots\dots\dots$

c) **(i)** Draw a graph to show the variation with time t (for $t = 0$ to $t = 72$ days) of the number of nuclei in a sample of thorium-227

(ii) Determine, from your graph, the count rate of thorium after 30 days.

(iii) Outline the experimental procedure to measure the count rate of thorium-227.

Solution

a) **(i)** A nuclide is a nucleus with a particular number of protons and neutrons.

(ii) Half-life is the time for the count rate to halve in value **OR** the time for half the number of nuclei to decay into nuclei of another element.

b) The proton number has increased by one so this must be negative beta decay.

$$^{227}_{89}Ac \rightarrow {}^{227}_{90}Th + {}^{0}_{-1}e + {}^{0}_{0}\bar{\nu}$$

c) **(i)**

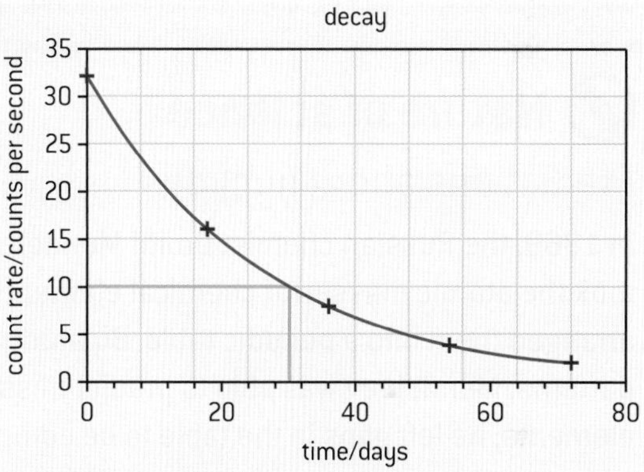

(ii) Marked on the graph in green – the activity is 10 units. *There is likely to be a little tolerance on this type of question in an examination.*

(iii) You would not be expected to give great detail in this sort of question. You should include the following points:

- Use of a G–M tube as detector.
- Measuring the average background count rate in counts per second (this can be done by timing for 100 seconds with no source nearby). This should be done three times, averaged and divided by 100 to give the counts per second.
- Measuring the count rate three times with the source close to the detector.
- Correcting the count rate by subtracting the background count rate from the count rate with the source in position.

7.2 Nuclear reactions

Understanding

- → The unified atomic mass unit
- → Mass defect and nuclear binding energy
- → Nuclear fission and nuclear fusion

Applications and skills

- → Solving problems involving mass defect and binding energy
- → Solving problems involving the energy released in radioactive decay, nuclear fission, and nuclear fusion
- → Sketching and interpreting the general shape of the curve of average binding energy per nucleon against nucleon number

Equations

- → The mass–energy relationship: $\Delta E = \Delta mc^2$

Nature of science

Predictions about nuclides

In 1869, the Russian chemist Dmitri Mendeleev took the atomic masses of chemical elements and arranged them into a periodic table. By studying his patterns, Mendeleev was able to predict missing elements; he left gaps in the table to be completed when the elements were discovered. Since the advent of the periodic table, patterns of atoms have allowed scientists to make predictions about the nuclides of which the elements are composed.

For nuclear physicists the graph of proton number against neutron number tells them whether a nuclide is likely to be unstable and, if it is, by which mechanism it will decay. The graph of nuclear binding energy per nucleon against nucleon number tells them how much energy can be generated by a particular fission or fusion reaction. These patterns have taken the guesswork out of physics and have proved to be an invaluable tool.

Patterns for stability in nuclides

When a graph of the variation of the neutron number with proton number is plotted for the stable nuclei, a clear pattern is formed. This is known as the **zone of stability**. Nuclides lying within the zone are stable, while those outside it are unstable and will spontaneously decay into a nuclide tending towards the stability zone. In this way it is possible to predict the mechanism for the decay: α, β^-, β^+ (or electron capture).

Nuclides having low proton numbers are most stable when the neutron–proton ratio is approximately one. In moving to heavier stable nuclides the neutron–proton ratio gradually increases with the heaviest stable nuclide, bismuth 209, having a ratio of 1.52.

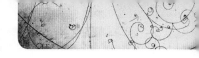

Figure 1 shows this plot of neutron number against proton number.

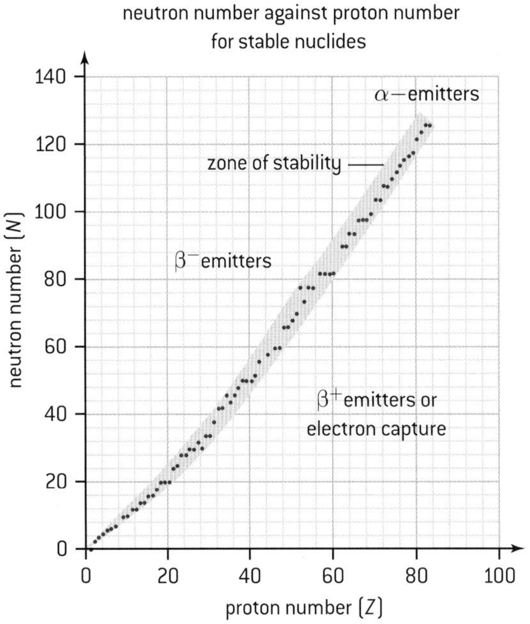

▲ Figure 1 Graph of neutron number against proton number for some stable nuclides.

Unstable nuclides lying to the left of the zone of stability are *neutron rich* and decay by β^- emission. Those nuclides to the right of the zone of stability are *proton rich* and decay by β^+ emission or else by electron capture (this is, as the name suggests, when a nucleus captures an electron and changes a proton into a neutron as a result – thus increasing the neutron–proton ratio). The heaviest nuclides are alpha emitters since emission of both two protons and two neutrons reduces the neutron–proton ratio and brings the overall mass down.

A second pattern that is seen to affect the stability of a nucleus is whether the number of protons and neutrons are even. Almost half the known stable nuclides have both even numbers of protons and neutrons, while only five of the stable nuclides have odd numbers of both protons and neutrons. The elements with even numbers of protons tend to be the most abundant in the universe.

The third stability pattern is when either the number of protons or the number of neutrons is equal to one of the even numbers 2, 8, 20, 28, 50, 82, or 126. Nuclides with proton numbers or neutron numbers equal to one of these **magic numbers** are usually stable. The nuclides, where both the proton number and the neutron number are magic numbers, are highly stable and highly abundant in the universe.

The unified atomic mass unit

The **unified atomic mass unit (u)** is a convenient unit for masses measured on an atomic scale. It is defined as **one-twelfth of the rest mass of an unbound atom of carbon-12 in its nuclear and electronic ground state, having a value of 1.661 × 10^{-27} kg**. Carbon was chosen because it is abundant and present in many different compounds hence making it useful for precise measurements. With carbon-12 having six protons and six neutrons, this unit is the average mass of nucleons and is, therefore, approximately equal to the mass of either a proton or a neutron.

In some areas of science (most notably chemistry) the term "unified atomic mass unit" has been replaced by the term "dalton" (Da). This is an alternative name for the same unit and is gaining in popularity among scientists. For the current IB Diploma Programme Physics syllabus the term "unified atomic mass unit" will be used.

Binding energy

As discussed in Sub-topic 7.1 the strong nuclear force acts between neighbouring nucleons within the nucleus. It has a very short range $\approx 10^{-15}$ m or 1 fm. The stability of many nuclei provides evidence for the strong nuclear force.

In order to completely dismantle a nucleus into all of its constituent nucleons work must be done to separate the nucleons and overcome the strong nuclear force acting between them. This work is known as the *nuclear binding energy*.

Suppose we could reverse the process and construct a nucleus from a group of individual nucleons. We would expect there to be energy released as the strong nuclear force pulls them together. This would be equal to the nuclear binding energy needed to separate them.

This implies that energy is needed to deconstruct a nucleus from nucleons and is given out when we construct a nucleus.

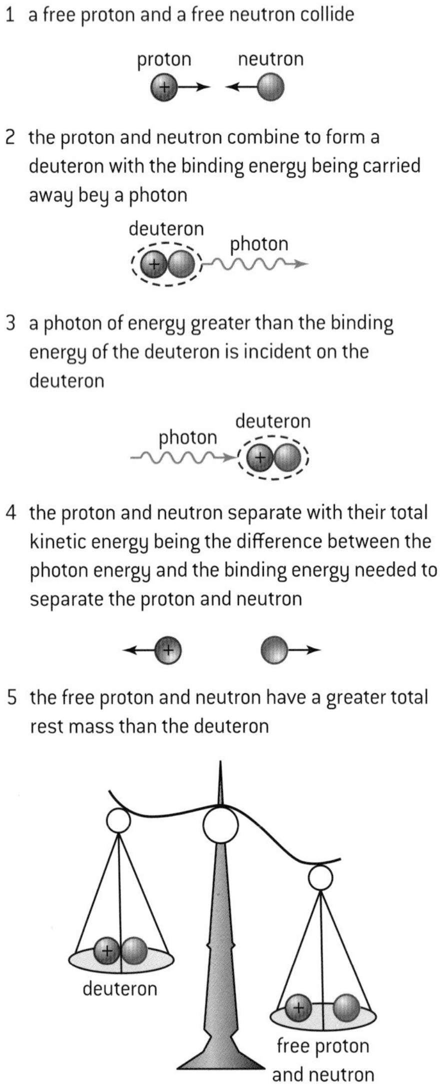

1 a free proton and a free neutron collide
proton
neutron
2 the proton and neutron combine to form a deuteron with the binding energy being carried away bey a photon
deuteron
photon
3 a photon of energy greater than the binding energy of the deuteron is incident on the deuteron
photon
deuteron
4 the proton and neutron separate with their total kinetic energy being the difference between the photon energy and the binding energy needed to separate the proton and neutron
5 the free proton and neutron have a greater total rest mass than the deuteron
deuteron
free proton and neutron

▲ Figure 2 Binding energy of a deuteron.

Mass defect and nuclear binding energy

Energy and mass are different aspects of the same quantity and are shown to be interchangeable through perhaps the most famous equation in physics (Einstein's mass–energy relationship):

$$E = mc^2$$

where E is the energy, m the equivalent mass and c the speed of electromagnetic waves in a vacuum.

This equation has huge implications for physics on an atomic scale.

When work is done on a system so that its energy increases by an amount $+\Delta E$ then its mass will increase by an amount $+\Delta m$ given by:

$$\Delta m = \frac{\Delta E}{c^2}$$

Alternatively when work is done by a system resulting in its energy decreasing by an amount $-\Delta E$ then its mass will decrease by an amount $-\Delta m$ given by:

$$-\Delta m = \frac{-\Delta E}{c^2}$$

These relationships are universal but are only significant on an atomic scale. When energy is supplied to accelerate a rocket, there will be an increase in the mass of the rocket. In an exothermic chemical reaction there will be a decrease in the mass of the reactants. However, these are insignificant amounts and can be ignored without jeopardizing the calculations. It is only with atomic and nuclear changes that the percentage mass change becomes significant.

It follows from Einstein's relationship that *the total mass of the individual nucleons making up a nucleus must be greater than the mass of that nucleus* – since work needs to be done in order to break the nucleus into its component parts. This difference is known as the **mass defect** – which is the "mass equivalent" of the nuclear binding energy.

Mass and energy units for nuclear changes

We have seen in the last sub-topic that the electronvolt (eV) is commonly used as the unit of energy on the atomic and nuclear scale. Nuclear energy changes usually involve much more energy than that needed for electron energy changes and so MeV ($= \text{eV} \times 10^6$) is the usual multiplier. In the mass–energy relation $E = mc^2$, when the energy is measured in MeV, the mass is often quoted in the unit MeV c^{-2}. This is not an SI unit but is very convenient in that it avoids having to convert to kg when the energy is in electronvolts. On this basis, the unified atomic mass unit can also be written as 931.5 MeV c^{-2}. To find the equivalent energy (in MeV) we simply have to multiply by c^2!

Worked example

The nuclide ${}^{24}_{11}\text{Na}$ decays into the stable nuclide ${}^{24}_{12}\text{Mg}$.

a) **(i)** Identify this type of radioactive decay.

(ii) Use the data below to determine the rest mass in unified atomic mass units of the particle emitted in the decay of a sodium-24 nucleus ${}^{24}_{11}\text{Na}$.

rest mass of ${}^{24}_{11}\text{Na} = 23.990\ 96\text{u}$

rest mass of ${}^{24}_{12}\text{Mg} = 23.985\ 04\text{u}$

energy released in decay = 5.002 160 MeV

b) The isotope sodium-24 is radioactive but the isotope sodium-23 is stable. Suggest which of these isotopes has the greater nuclear binding energy.

Solution

a) **(i)** This is an example of negative beta decay since the daughter product has an extra proton (a neutron has decayed into a proton and an electron – with the electron being emitted as a negative beta particle).

(ii) The energy released is equivalent to a mass of 5.002 160 MeV c^{-2}.

1 *u* is equivalent to 931.5 MeV c^{-2}

So the energy is equivalent to a mass of $\frac{5.002160}{931.5} = 0.005\ 37\text{u}$

The energy mass of the electron must therefore be

$$23.990\ 96\text{u} - (23.985\ 04\text{u} + 0.005\ 37\text{u}) = 0.000\ 55\text{u}$$

This is consistent with the value for the mass of an electron (given as 0.000 549u in the IB Physics data booklet).

b) As sodium-24 has 24 nucleons and sodium-23 has 23 nucleons, the total binding energy for sodium 24 is going to be greater than that of sodium 23.

Variation of nuclear binding energy per nucleon

The nuclear binding energy of large nuclei tends to be larger than that for smaller nuclei. This is because, with a greater number of nucleons, there are more opportunities for the strong force to act between nucleons. This means more energy is needed to dismantle a nucleus into its component nucleons. In order to compare nuclei it is usual to plot the average binding energy per nucleon of nuclides against the nucleon number as shown in figure 3.

The average binding energy per nucleon is found by dividing the total binding energy for a nucleus by the number of nucleons in the nucleus. Although there is a general pattern, with most nuclides having a binding energy of around 8 MeV per nucleon, there are wide differences from this.

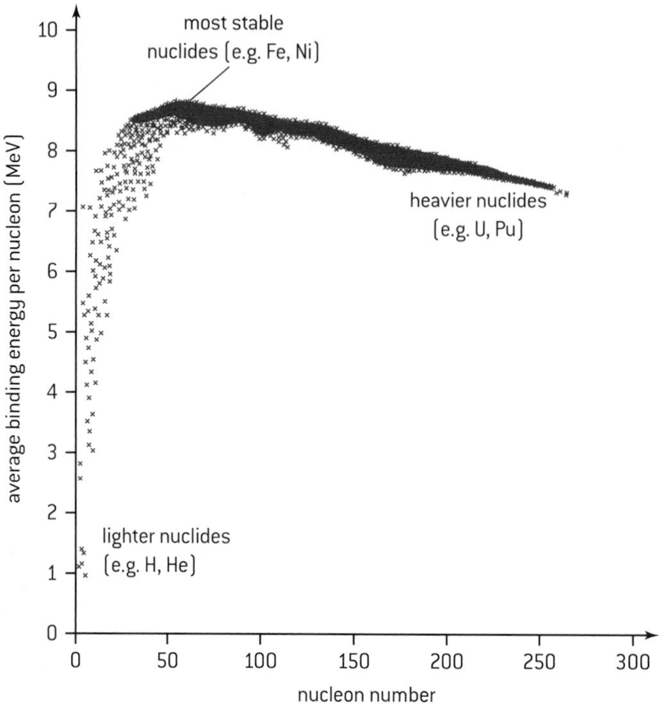

▲ Figure 3 Plot of average binding energy per nucleon against the nucleon number.

On the left of the plot, those nuclides of low nucleon number, such as hydrogen-2 and helium, are less tightly bound than the more massive nuclides (hydrogen-1 or *normal hydrogen* consists of a single proton and has no nuclear binding energy). As we move to the central region of the plot we reach the maximum binding energy per nucleon with nuclides such as iron-58 and nickel-62. These nuclides, having the highest nuclear binding energies per nucleon, are the most stable nuclei and therefore are abundant in the universe. Further to the right than these nuclei, the pattern reverses and the heavier nuclei are less tightly bound than lighter ones. Towards the extreme right we reach the heavy elements of uranium and plutonium – in these the binding energy per nucleon is about 1 MeV less than those in the central region.

Nuclear fusion

The joining together (or **fusing**) of small nuclei to give larger ones releases energy. This is because the total nuclear binding energy of the fused nuclei is larger than the sum of total nuclear binding energies of the component nuclei. The difference in the binding energies between the fusing nuclei and the nucleus produced is emitted as the kinetic energy of the fusion products.

Although it doesn't actually happen in this way, it is useful to think of the energy being released as the difference between the energy emitted in constructing the fused nucleus and the energy required in deconstructing the two nuclei.

When two nuclei of masses m_1 and m_2 fuse to form a nucleus of mass m_3 the masses do not add up as we might expect and $m_1 + m_2 > m_3$

Since the total number of nucleons is conserved this means that m_3 has a smaller mass than the total for $m_1 + m_2$. This loss of mass is emitted as the kinetic energy of the fusion products – in other words: $\Delta E = ((m_1 + m_2) - m_3)c^2$

Let's look at an example:

A helium-4 nucleus is composed of two protons and two neutrons. Let us consider the mass defect of helium-4 compared with its constituent particles.

The mass of helium-4 nucleus = 4.002 602u

The mass of 2 (individual) protons = 2 × 1.007 276u = 2.014 552u

The mass of 2 (individual) neutrons = 2 × 1.008 665u = 2.017 33u

The total mass of all the individual nucleons = 4.031 882u

So the mass defect is 4.031 882u − 4.002 602u = 0.029 28u

This is equivalent to an increased binding energy of 0.029 28 × 931.5 MeV or 27.3 MeV – this energy is given out when the nucleus is formed from the individual nucleons.

You can probably see that the potential for generating energy is immense, but being able to produce the conditions for fusion anywhere except in a star is a significant problem. Initially, the repulsion between the protons means that energy must be supplied to the system in order to allow the strong nuclear force to do its work. In reality, joining together two protons and two neutrons is not a simple task to achieve on Earth. It seems that Earth-based fusion reactions are more likely to occur between the nuclei of deuterium (hydrogen-2) and tritium (hydrogen-3).

Nuclear fission

The concept of nuclear **fission** is not quite as straight-forward as fusion – although fission has been used practically for over 70 years. If we are able to take a large nucleus and split it into two smaller ones, the binding energy per nucleon will increase as we move from the right-hand side to the centre of figure 3. This means that energy must be given out in the form of the kinetic energy of the fission products.

So, in nuclear fission, the energy released is equivalent to the difference between the energy needed to deconstruct a large nucleus and that emitted when two smaller nuclei are constructed from its components.

In terms of masses, the total mass of the two smaller nuclei will be less than that of the parent nucleus and the difference is emitted as the kinetic energy of the fission products. So we see that as $m_1 + m_2 < m_3$ then $\Delta E = (m_3 - (m_1 + m_2))c^2$

There are some nuclei that undergo fission spontaneously but these are largely nuclides of very high nucleon number made synthetically. Uranium-235 and uranium-238 do undergo spontaneous fission but this has a low probability of occurring. Uranium-236 however is not a common naturally occurring isotope of uranium but it does undergo

TOK

The use of fission and fusion

In the last twenty years there has been a development in techniques to extract fossil fuels from locations that were not either physically or financially viable previously. Given the debate relating to nuclear energy, which is the more moral stance to take: to use nuclear fuel with its inherent dangers or to risk damage to landscapes and habitats when extracting fossil fuels?

spontaneous fission. Uranium-236 can be produced from uranium-235 when it absorbs a low energy neutron.

Uranium-236 then undergoes spontaneous fission to split into two lighter nuclides and at the same time emits two or three further neutrons:

$$^{235}_{92}U + ^{1}_{0}n \rightarrow ^{236}_{92}U \rightarrow ^{141}_{56}Ba + ^{92}_{36}Kr + 3^{1}_{0}n$$

These isotopes of barium and krypton are just one of several possible pairs of fission products of uranium-236.

Nature of science

Fission chain reaction

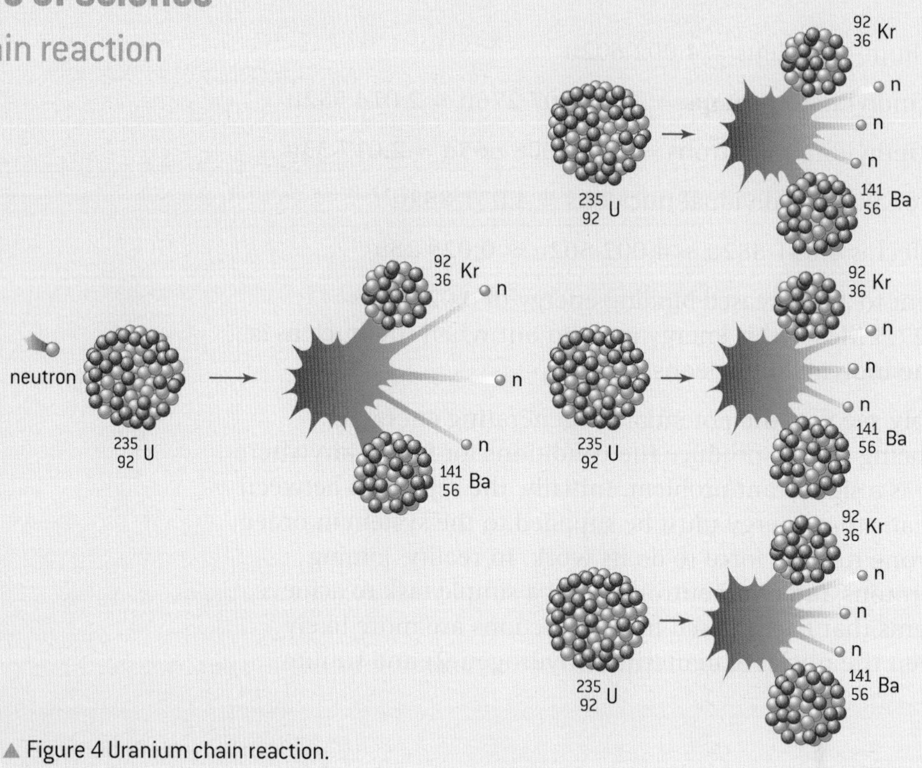

▲ Figure 4 Uranium chain reaction.

The use of nuclear fission in nuclear power stations is discussed in Topic 8. One of the key aspects of a continuous power production plant is that the nuclear fuel is able to fission in a controlled chain reaction. The neutrons produced in the reaction above must be capable of producing more nuclear fission events by encountering further uranium nuclei. When there is sufficient mass of uranium-235 the reaction becomes self-sustaining, producing a great deal of kinetic energy that can be transformed into electricity in the power station.

Worked example

a) Compare the process of nuclear fission with nuclear fusion.

b) Helium-4 ($^{4}_{2}He$) and a neutron are the products of a nuclear fusion reaction between deuterium ($^{2}_{1}H$) and tritium ($^{3}_{1}H$).

$$^{2}_{1}H + ^{3}_{1}H \rightarrow ^{4}_{2}He + ^{1}_{0}n + \text{energy}$$

The masses of these nuclides are as follows:

$^{2}_{1}H$	2.014 102u
$^{3}_{1}H$	3.016 050u
$^{4}_{2}He$	4.002 603u

Show that the energy liberated in each reaction is approximately 2.8×10^{-12} J.

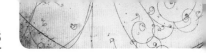

Solution

a) Nuclear fusion involves the joining together of light nuclei while nuclear fission involves the splitting up of a heavy nucleus. In each case the total nuclear binding energy of the product(s) is greater than that of the initial nuclei or nucleus. The difference in binding energy is emitted as the kinetic energy of the product(s). In relation to the plot of nuclear binding energy per nucleon against nucleon number, fission moves nuclei from the far right towards the centre whereas fusion moves nuclei from the far left towards the centre – both processes involve a move up the slopes towards higher values.

b) *To simplify the calculation you should break it down into several steps. Remember to include all of the steps – with so many significant figures, short cuts could cost you marks in an exam.*

Total mass on left-hand side of equation = 2.014 102u + 3.016 050u = 5.030 152u

Looking up the mass of the neutron in the data booklet (= 1.008 665u)

Total mass on right-hand side of equation = 4.002 603u + 1.008 665u = 5.011 268u

Thus there is a loss of mass on the right-hand side (as the binding energy of the helium nucleus is higher than that of the deuterium and tritium nuclei)

Mass difference (Δm) = 5.030 152u − 5.011 268u = 0.018 884u

As $u = 931.5$ MeV c^{-2},
$\Delta m = 0.018\,884 \times 931.5$ MeV c^{-2}
$= 17.59$ MeV c^{-2}

$\Delta E = \frac{\Delta m}{c^2} = 17.59$ MeV

$= 17.59 \times 10^6 \times 1.60 \times 10^{-19}$ J

$\Delta E = 2.81 \times 10^{-12}$ J

Nature of science

Alternative nuclear binding energy plot

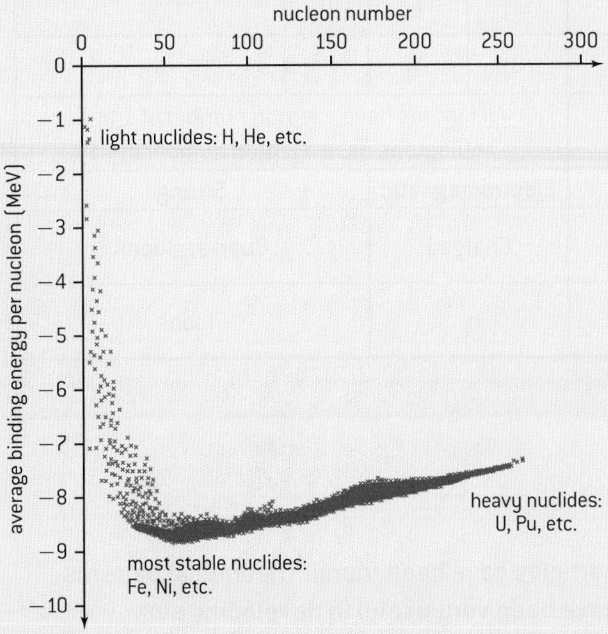

▲ Figure 5 Alternative plot of binding energy per nucleon against the nucleon number.

It is quite usual to see the binding energy plot, shown in figure 3, drawn "upside down" as shown in figure 5. The reason for plotting this "inverted" graph is that when the nucleons are at infinite separation they have no mutual potential energy but, because the strong nuclear force attracts them, they lose energy as they become closer meaning that their potential energy becomes negative. This implies that the more stable nuclei are in a potential valley. Fission and fusion changes that bring about stability, will always move to lower (more negative) energies. This convention is consistent with gravitational and electrical potential energies which are discussed in Topic 10. In line with most text books at IB Diploma level we will continue to use the plot shown in figure 3.

7.3 The structure of matter

Understanding

- Quarks, leptons, and their antiparticles
- Hadrons, baryons, and mesons
- The conservation laws of charge, baryon number, lepton number, and strangeness
- The nature and range of the strong nuclear force, weak nuclear force, and electromagnetic force
- Exchange particles
- Feynman diagrams
- Confinement
- The Higgs boson

Applications and skills

- Describing the Rutherford–Geiger–Marsden experiment that led to the discovery of the nucleus
- Applying conservation laws in particle reactions
- Describing protons and neutrons in terms of quarks
- Comparing the interaction strengths of the fundamental forces, including gravity
- Describing the mediation of the fundamental forces through exchange particles
- Sketching and interpreting simple Feynman diagrams
- Describing why free quarks are not observed

Particle properties

Charge	Quarks			Baryon number		Charge	Leptons		
$\frac{2}{3}e$	u	c	t	$\frac{1}{3}$		−1	e	μ	τ
$-\frac{1}{3}e$	d	s	b	$\frac{1}{3}$		0	υ_e	υ_μ	υ_τ
All quarks have a strangeness number of 0 except the strange quark that has a strangeness number of −1						All leptons have a lepton number of 1 and antileptons have a lepton number of −1			

	Gravitational	Weak	Electromagnetic	Strong
Particles experiencing	All	Quarks, leptons	Charged	Quarks, gluons
Particles mediating	Graviton	W^+, W^-, Z^0	γ	Gluons

Nature of science

Symmetry and physics

Symmetry has played a major part in the development of particle physics. Mathematical symmetry has been responsible for the prediction of particles. By searching the bubble chamber tracks produced by cosmic rays or generated by particle accelerators many of the predicted particles have been found. Theoretical patterns have been very useful in developing our current understanding of particle physics and, increasingly, experiments are confirming that these patterns are valid even though their profound signification is yet to be determined.

Introduction

At the end of the nineteenth century, physicists experimented with electrical discharges through gases at low pressure (see figure 1).

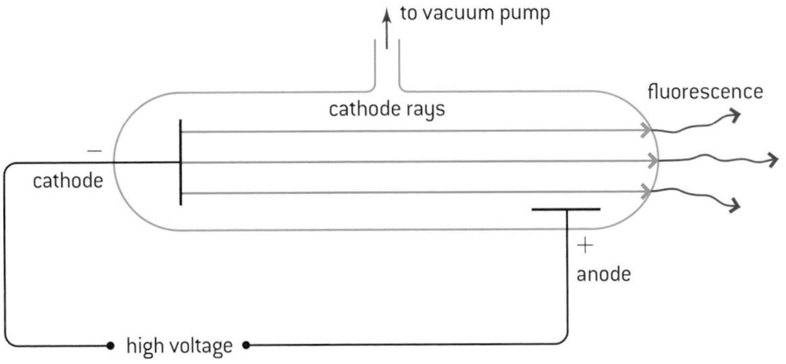

▲ Figure 1 Discharge tube.

In 1869, the German physicist, Johann Hittorf, observed a glow coming from the end of a discharge tube, opposite the cathode. He suggested that radiation was being emitted from the cathode and this caused the tube to fluoresce. The radiation causing the effect was later called a beam of "cathode rays". Eight years later the British physicist, Sir Joseph Thomson, discovered that cathode rays could be deflected by both electric and magnetic fields. Thomson's experiment showed that cathode rays were charged. After further experiments with hydrogen gas in the tube, Thomson concluded that cathode rays were beams of particles coming from atoms. With atoms being neutral, Thompson deduced that the atom must also carry positive charge. Figure 2 shows Thomson's "plum pudding" or "current bun" model of the atom, consisting of a number of electrons buried in a cloud of positive charge.

Although Thomson's model was short-lived, it was the first direct evidence that atoms have structure and are not the most elementary building blocks of matter as had been previously thought.

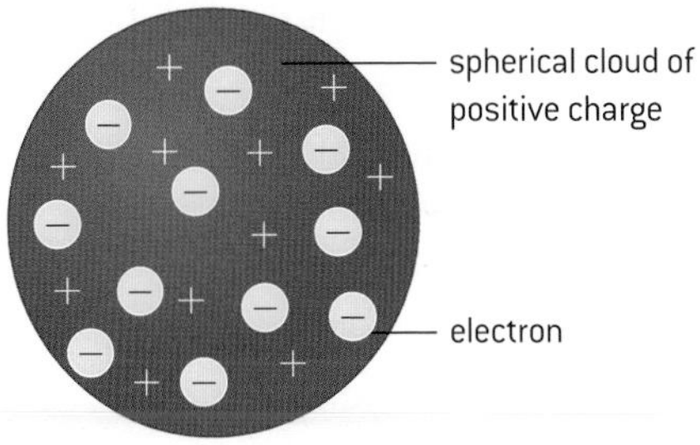

▲ Figure 2 Thomson's "plum pudding" model of the atom.

The scattering of alpha particles

In 1909, the German, Johannes Geiger, and the English–New Zealander, Ernest Marsden, were studying at Manchester University in England. Ernest Rutherford (another New Zealander) was supervising their research. The students were investigating the scattering of alpha particles by a thin gold foil. They used a microscope in a darkened room to detect flashes of light emitted when a alpha particles collided with a zinc sulfide screen surrounding the apparatus. Geiger and Marsden were expecting the alpha particles to be deflected by a very small amount as a result of the electrostatic effects of the charges in the atom. At Rutherford's suggestion they moved their detector to the same side of the foil as the alpha source and were astonished to find that approximately one in every eight thousand alpha particles appeared to be reflected (or "back-scattered") by the thin foil. With the alpha particles travelling at about 3% of the speed of light, reflection by electrons surrounded by a cloud of positive charge was unthinkable and Rutherford described the effect as:

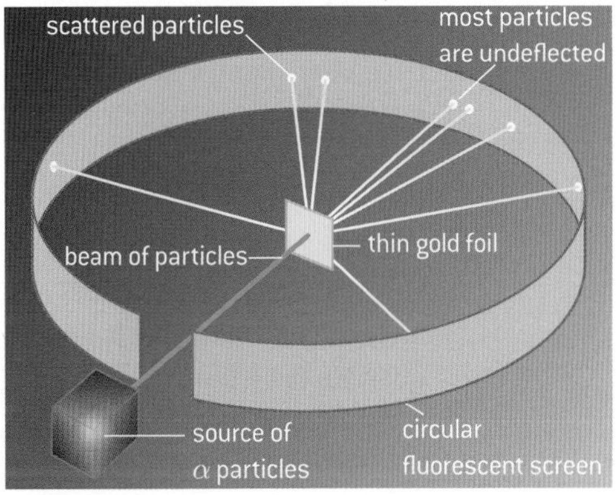

▲ Figure 3 The Rutherford–Geiger–Marsden apparatus.

> *... quite the most incredible event that has ever happened to me in my life. It was almost as incredible as if you fired a 15-inch shell at a piece of tissue paper and it came back and hit you. On consideration, I realized that this scattering backward must be the result of a single collision, and when I made calculations I saw that it was impossible to get anything of that order of magnitude unless you took a system in which the greater part of the mass of the atom was concentrated in a minute nucleus. It was then that I had the idea of an atom with a minute massive centre, carrying a charge.*

Rutherford proposed that, as the alpha particles carry a positive charge, back-scattering could only occur if the massive part of the atom was also positively charged (for the mechanics of this see Sub-topic 2.4). Thus the atom would consist of a small, dense positive nucleus with electrons well outside the nucleus. Rutherford's calculations showed that the diameter of the nucleus would be of the order of 10^{-15} m while that of the atom as a whole would be of the order of 10^{-10} m; this meant that the atom was almost entirely empty space.

As will be discussed in Topic 12, for electrons to be able to occupy the orbits required by later models of the atom, a theory that contradicts classical physics is required. The electrons that are accelerating because of their circular motion should emit electromagnetic radiation and spiral into the nucleus. The consequence of this would be that all atoms should collapse to the size of the nucleus and matter could not exist. Since it is obvious that matter does exist, it is clear that Rutherford's model is not the whole story.

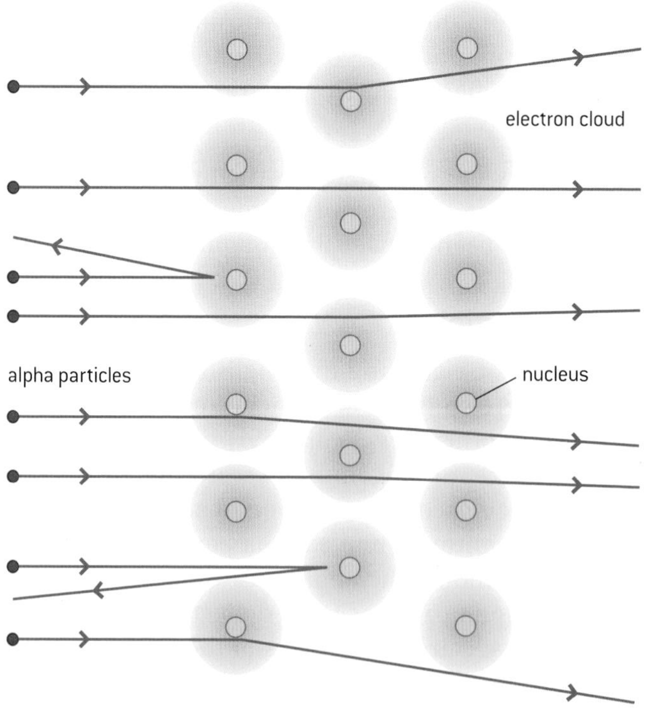

▲ Figure 4 Paths of several alpha particles in the scattering experiment.

TOK

Physicists and knowing

"Now I know what the atom looks like."

— a quote from Sir Ernest Rutherford, after publishing the results of his alpha scattering experiment.

When scientists devote time to investigating a specific area of research, they are often keen to publish their findings. Although peer appraisal is an important aspect of research there have been many instances in which a scientist adheres to one chosen model without being objective about the work of others. Many scientists pride themselves on their objectivity, but is it possible to be truly objective when those who are deemed to be successful are rewarded with prestigious awards and funding?

The particle explosion

In 1928, the British physicist, Paul Dirac, predicted an antiparticle of the electron (the **positron**). Particles and their associated antiparticles have identical rest mass but have reversed charges, spins, baryon numbers, lepton numbers, and strangeness (see later for these). The positron was discovered in cosmic rays by the American, Carl Anderson, in 1932. Cosmic rays consist mainly of high-energy protons and atomic nuclei ejected from the supernovae of massive stars. Anderson used a strong magnetic field to deflect the particles and found that their tracks were identical to those produced by electrons, but they curved in the opposite direction.

When an electron collides with a positron the two particles **annihilate** and their total mass is converted into a pair of photons of identical energy emitted at right angles to each other. The inverse of this process is called **pair production**. This is when a photon interacts with a nucleus and produces a particle and its antiparticle; for this event to occur the photon must have a minimum energy equal to the total rest mass of the particle and the antiparticle.

1932 was a significant year in the development of our understanding of the atom with other developments in addition to the positron discovery. In 1930, Bothe and Becker, working in Germany, had found that an unknown particle was ejected when beryllium is bombarded by alpha particles. James Chadwick at Cambridge University proved experimentally that this particle was what we now know to be the **neutron**.

In 1936, Anderson and his Ph.D. student Seth Neddermeyer at the California Institute of Technology went on to discover the **muon**, using cloud chamber measurements of cosmic rays. Although at the time they believed this particle to be the theoretically predicted pion, this was the start of the recognition that there were more particles than protons, neutrons, electrons and positrons.

The understanding that electrons had antiparticles suggested to Dirac that the same should be true for protons and, in 1955, the antiproton was discovered at the University of California, Berkeley by Emilio Segrè and Owen Chamberlain. In their experiment protons were accelerated to an energy of approximately 6 MeV before colliding with further protons in a stationary target. The reaction is summarized as:

$$\mathrm{p} + \mathrm{p} \rightarrow \mathrm{p} + \mathrm{p} + \mathrm{p} + \overline{\mathrm{p}}$$

Here p represents a proton and $\overline{\mathrm{p}}$ an antiproton. The kinetic energy of the colliding protons (left-hand side of the equation) is sufficient to produce a further proton and an antiproton – using $E = mc^2$.

We have encountered the neutrino and antineutrino in beta decay. According to the standard Big Bang model, these particles are thought to be the most numerous in the universe. The **electron neutrino** was first proposed in 1930 by the German physicist, Wolfgang Pauli, as an explanation of why electrons emitted in beta decay did not have quantized energies as is the case with alpha particles. Pauli suggested that a further particle additional to the electron should be emitted in beta decay so that energy and momentum are conserved. It was the Italian, Enrico Fermi, who named the particle "little neutral one" or neutrino and further developed Pauli's theory. As the neutrino has no charge and almost no mass it interacts only minimally with matter and was not discovered experimentally until 1956.

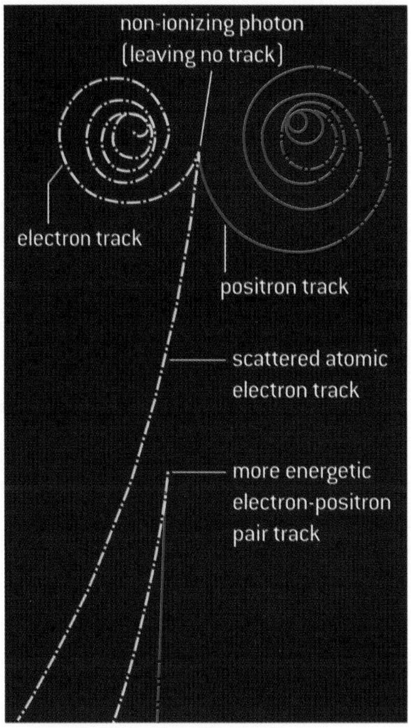

▲ Figure 5 Electron and positron tracks in a cloud chamber.

By the end of the 1960s over 300 particles had been discovered and physicists were starting to classify them. A full theory explaining the structure of matter and the nature of the forces that hold particles together has become a major goal of scientific research. The current theory is that the universe is made up of a relatively small number of fundamental particles.

Classification of particles – the standard model

The Standard Model came about through a combination of experimental discoveries and theoretical developments. Although formulated towards the end of the twentieth century, it is supported by the experimental discoveries of the bottom quark in 1977, the top quark in 1995, the tau particle in 2001, and the Higgs boson in 2012. The Standard Model has been described as being the theory of (almost) everything. On the whole the model has been very successful, but it fails to fully incorporate relativistic gravitation or to predict the accelerating expansion of the universe. The model suggests that the only fundamental particles are leptons, quarks, and gauge bosons. All other particles are believed to consist of combinations of quarks and antiquarks.

Leptons

The leptons are members of the electron family and consist of the electron (e^-), the muon (μ), the tau (τ), their antiparticles plus three neutrinos associated with each of the particles and three neutrinos associated with the antiparticles. Electrons, muons, and taus are all negatively charged and their antiparticles are positively charged. Neutrinos and antineutrinos are electrically uncharged.

The electron is known to have a mass of about $\frac{1}{1800}$th of the mass of a proton, making it a very light particle. The muon is also light, having a mass of about 200 times that of the electron. The tau is heavier and has a mass similar to that of a proton. For reasons that will be shown later, each lepton is given a lepton number. Leptons have lepton number +1 and antileptons −1.

Leptons				Charge/e	Lepton number (L)
Particle	e	μ	τ	−1	+1
Antiparticles	$\bar{e}$	$\bar{\mu}$	$\bar{\tau}$	+1	−1
Neutrinos	ν_e	ν_μ	ν_τ	0	+1
Antineutrinos	$\bar{\nu}_e$	$\bar{\nu}_\mu$	$\bar{\nu}_\tau$	0	−1

Quarks

Speculation about the existence of quarks began after scattering experiments were performed with accelerated electrons at the Stanford Linear Accelerator Center (SLAC) at Stanford University in the USA between 1967 and 1973. The electrons were accelerated up to energies of 6 GeV before colliding with nuclei. In a similar way to alpha scattering, some of the electrons were scattered through large angles by the nucleons – which suggested that the nucleons are not of uniform density but have discrete charges within them. These experiments supported the theories put forward independently by the Russian–American physicist George Zweig and the American Murray Gell-Mann in 1964. Gell-Mann proposed that the charges within nucleons were grouped in threes and coined the term "quarks" (which he pronounced "qworks") – in deference to a quote from Finnegan's Wake by the Irish novelist James Joyce:

> *'Three quarks for Muster Mark! Sure he hasn't got much of a bark.*
> *And sure any he has it's all beside the mark.'*

Zweig referred to quarks as "aces" and believed (incorrectly, as it turned out) there were four of them – as the aces in a pack of cards.

As with leptons there are six quarks and six antiquarks. The quarks are labelled by their "flavour" – which has no physical significance apart from identifying the quark. These flavours are called up (u), down (d), strange (s), charm (c), bottom (b), and top (t). Quarks each carry a charge of either $+\frac{2}{3}$e or $-\frac{1}{3}$e, and antiquarks each carry a charge of either $-\frac{2}{3}$e or $+\frac{1}{3}$e.

Often the e is omitted from charges and they are written as "relative charge" $+\frac{1}{3}$, $-\frac{2}{3}$, etc.

These quarks are split into three generations of increasing mass. The first generation contains the up and down quarks, which are the lightest quarks. The second contains the strange and charm quarks, and the third the bottom and top quarks – the heaviest quarks. The up, down, and strange quarks were the first to be discovered, with the up and down quarks combining to form nucleons.

Quarks with charge $+\frac{2}{3}$e	Quarks with charge $-\frac{1}{3}$e
u	d
c	s
t	b

Antiquarks carry the opposite charge and are denoted by $\bar{u}$, $\bar{d}$, $\bar{c}$, etc.

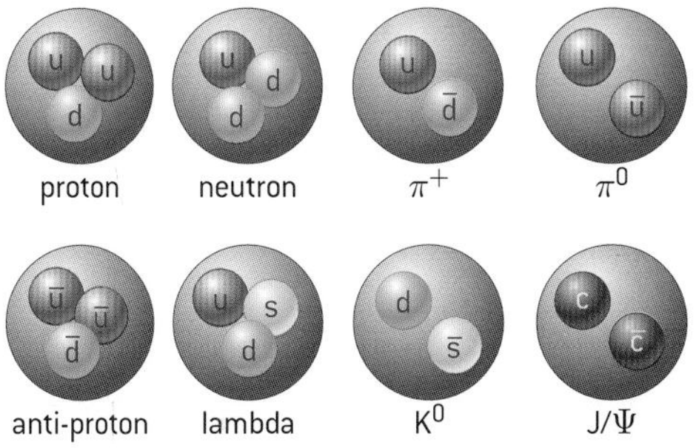

▲ Figure 6 Three quark and two quark hadrons.

Quark confinement

It is thought that quarks never exist on their own but exist in groups within hadrons. Hadrons are formed from a combination of two or three quarks (called **mesons** and **baryons**) – this is known as **quark confinement**. The theory that explains quark confinement is known as quantum chromodynamics (QCD) something that is not included in the IB Diploma Programme Physics syllabus. To hold the quarks in place they exchange gluons (the exchange particle for the strong nuclear force – see later in this sub-topic). Moving a quark away from its neighbours in the baryon or meson stores more energy in the interaction between the quarks and therefore requires increasing amounts of energy to increase their separation. If more and more energy is fed into the system, instead of breaking the force between quarks and separating them (which is what classical physics might suggest), more quarks are produced – this leaves the original quarks unchanged but creates a new meson or baryon by the mass–energy relationship $E = mc^2$.

Baryons qqq and antibaryons $\bar{q}\bar{q}\bar{q}$
These are a few of the many types of baryons.

Symbol	Name	Quark content	Electric charge
p	proton	uud	+1
$\bar{p}$	antiproton	$\bar{u}\bar{u}\bar{d}$	−1
n	neutron	udd	0
Λ	lambda	uds	0
Ω^-	omega	sss	−1

Mesons $q\bar{q}$
These are a few of the many types of mesons.

Symbol	Name	Quark content	Electric charge
π^+	pion	$u\bar{d}$	+1
K^-	kaon	$s\bar{u}$	−1
ρ^+	rho	$u\bar{d}$	+1
B^0	B-zero	$d\bar{b}$	0
η_c	eta-c	$c\bar{c}$	0

▲ Figure 7 Baryons and mesons.

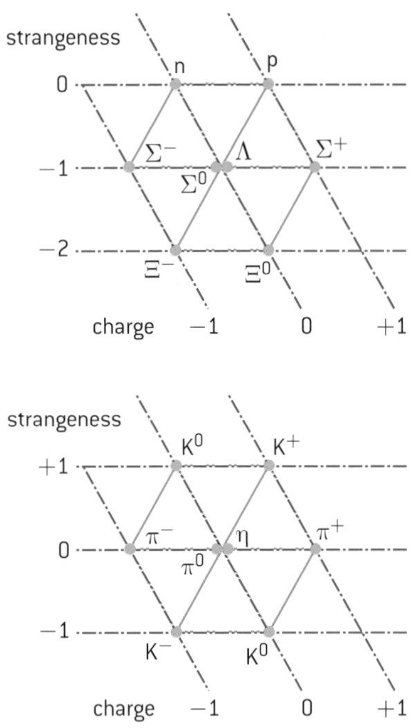

▲ Figure 8 Example patterns for some baryons and mesons (The Eightfold Way).

Hadrons

Hadrons are particles composed of quarks and include baryons (which are made up of three quarks) or mesons (which comprise of quark-antiquark pairs). The strong interaction (see later) acts on all hadrons but not on leptons while the weak interaction acts on both leptons and hadrons. Some particle physicists have hypothesized "pentaquarks" consisting of four quarks and one antiquarks – although experiments at the start of the century looked promising, these have yet to be confirmed experimentally.

In order to explain which particles can exist and to explain the outcome of observed interactions between particles, the quarks are assigned properties described by a numerical value. The quark is given a baryon number (B) of $\frac{1}{3}$ and for an antiquark the baryon number is $-\frac{1}{3}$.

Strangeness (S) is a property that was initially defined to explain the behaviour of massive particles such as kaons and hyperons. These particles are created in pairs in collisions and were thought to be "strange" because they have a surprisingly long lifetime of 10^{-10} s instead of the expected 10^{-23} s. A strange quark has a strangeness of −1, and a strange antiquark has a strangeness of +1. This property of strangeness is conserved when strange particles are created but it is not conserved when they subsequently decay.

Figure 8 shows eight baryons each with a baryon number +1 and eight mesons each with baryon number 0. Gell-Mann used these diagrams as a means of organizing baryons and mesons. Particles on the same horizontal level have the same strangeness, while those on the same diagonal have the same charge. Using this system, Gell-Mann predicted the eta (η) particle in 1961, which was discovered experimentally a few months later. The name "Eightfold Way" refers to Noble Eightfold Path – a way towards enlightenment in Buddhism.

Nature of science

Symmetry and physics

In the same way that the periodic table gives the pattern of the elements based on their electron structure, the **Standard Model** gives the pattern for the fundamental particles in nature. It is believed that there are six leptons and six quarks and these occur in pairs. A further symmetry is seen by each particle having its own antiparticle (which will have the same rest mass as the particle but other properties such as charge and spin are reversed). When an electron collides with its antiparticle the positron, the two particles annihilate producing pairs of photons that travel in opposite directions (thereby conserving momentum and energy). It is not possible for annihilating particles to produce a single photon.

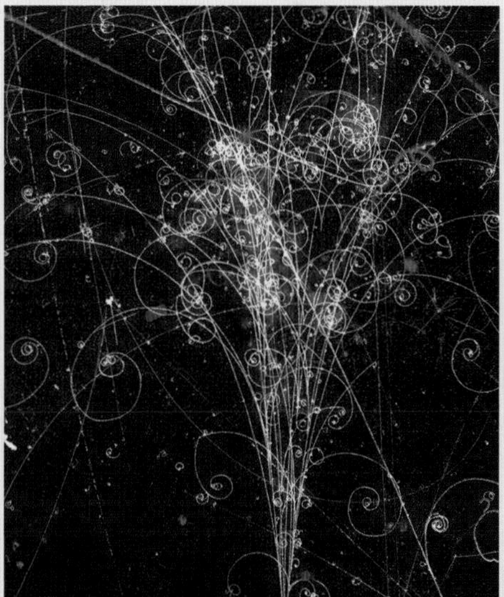

▲ Figure 9 Particle collisions shown in a bubble chamber.

Examples of baryons

Protons and neutrons are important in atoms and you should know their quark composition and be able to work out that of their antiparticles. Since they are both baryons, they each consist of three quarks and have a baryon number +1.

The proton consists of two up quarks and one down quark (uud), which means it has charge

$$\left(+\frac{2}{3}+\frac{2}{3}-\frac{1}{3}\right)=+1$$

The neutron consists of two down quarks and one up quark (ddu), which gives it a charge of

$$\left(-\frac{1}{3}-\frac{1}{3}+\frac{2}{3}\right)=0$$

An antiproton has a baryon number of −1 and consists of two antiup quarks and one antidown quark ($\bar{u}\bar{u}\bar{d}$). This gives a charge of $\left(-\frac{2}{3}-\frac{2}{3}+\frac{1}{3}\right)=-1$

You may wish to prove to yourself that the antineutron has a charge of zero (and baryon number of −1).

Since none of these particles has any strange quarks their strangeness is 0.

Examples of mesons

In questions you will always be given the quark composition of mesons, so there is no need to try to remember these or the composition of any baryons apart from the proton and neutron.

A π^+ meson is also called a positive pion and consists of an up quark and an antidown quark ($u\bar{d}$). The positive pion has charge $\left(+\frac{2}{3}+\frac{1}{3}\right)=+1$. As it is not a baryon, its baryon number is 0 and, as it has no strange quarks, its strangeness is 0 too.

A K^+ meson or positive kaon is the lightest strange meson and consists of an up quark and an antistrange quark ($u\,\bar{s}$). The positive kaon has charge $\left(+\frac{2}{3}+\frac{1}{3}\right)=+1$. Again, a meson is not a baryon and so it has baryon number 0. However, this particle does contain an antistrange quark and so it has strangeness of +1.

Conservation rules

When considering interactions between particles or their decay, the equivalence of mass and energy must be taken into account. Mass may become some form of energy and vice versa using the equation $E = mc^2$.

In addition to the expected conservation of mass–energy and momentum, no interaction that disobeys the conservation of charge has ever been observed; the same is true for baryon number (B) and lepton number (L). **All leptons have a lepton number of +1 and antileptons have a lepton number of −1.**

As mentioned before, the conservation rules for strangeness are different and we will consider them later.

You may be asked to decide whether an interaction or decay is feasible on the basis of the conservation rules.

Worked examples

1 a) Show that, when a proton collides with a negative pion ($\bar{u}d$), the collision products can be a neutron and an uncharged pion.

b) Deduce the quark composition of the uncharged pion.

Solution

a) The equation for the interaction is

$p + \pi^- \rightarrow n + \pi^0$

Q: $+1 - 1 \rightarrow 0 + 0$ ✓

B: $+1 + 0 \rightarrow +1 + 0$ ✓

L: $0 + 0 \rightarrow 0 + 0$ ✓

This interaction is possible on the basis of conservation of charge, baryon number and lepton number.

b) Writing the equation in terms of quarks:

$uud + \bar{u}d \rightarrow ddu + ??$

$?? = u\bar{u}$ in order to balance this equation.

This suggests that the neutral pion is very short lived – since the combination $u\bar{u}$ would mutually annihilate. In fact this particle has a lifetime of about 8×10^{-17} s and annihilates to form two gamma ray photons or, very occasionally, a gamma ray photon, an electron and a positron.

2 Explain whether a collision between two protons could produce two protons and a neutron.

Solution

Writing the equation for the baryons:

$p + p \rightarrow p + p + n$

Q: $+1 + 1 \rightarrow +1 + 1 + 0$ ✓

L: $0 + 0 \rightarrow 0 + 0 + 0$ ✓

B: $+1 + 1 \rightarrow +1 + 1 + 1$ ✗

So this interaction fails on the basis of baryon number.

Fundamental forces

Current theories suggest that there are just four fundamental forces in nature. In the very early universe, when the temperatures were very high, it is possible that at least three of these four forces originated as a single unified force.

The four fundamental forces are:

- The **gravitational force** is weak, has an infinite range and acts on all particles. It is always attractive and over astronomic distances it is the dominant force – on an atomic or sub-atomic scale it is negligible.
- The **electromagnetic force** causes electric and magnetic effects such as the forces between electrical charges or bar magnets. Like gravity, the electromagnetic force has an infinite range but it is much stronger at short distances, holding atoms and molecules together. It can be attractive or repulsive and acts between all charged particles.
- The **strong nuclear force** or **strong interaction** is very strong, but has very short-range. It acts only over ranges of $\approx 10^{-15}$ m and acts between hadrons but not leptons. At this range the force is attractive but it becomes strongly repulsive at distances any smaller than this.
- The **weak nuclear force** or **weak interaction** is responsible for radioactive decay and neutrino interactions. Without the weak interaction stars could not undergo fusion and heavy nuclei could not be built up. It acts only over very short ranges of $\approx 10^{-18}$ m and acts between all particles.

As these four fundamental forces have different ranges it is impossible to generalize their relative strengths for all situations. The table below shows a comparison of the effects that the four forces have on a pair of protons in a nucleus (since all four forces will act on protons at that range).

Force	Range	Relative strength	Roles played by these forces in the universe
Gravitational	∞	1	binding planets, solar system, sun, stars, galaxies, clusters of galaxies
Weak nuclear	$\approx 10^{-18}$ m	10^{24}	(W^+, W^-): transmutation of elements (W^0): breaking up of stars (supernovae)
Electromagnetic	∞	10^{35}	binding atoms, creation of magnetic fields
Strong nuclear	$\approx 10^{-15}$ m	10^{37}	binding atomic nuclei, fusion processes in stars

Exchange particles

A very successful model that has been used to explain the mechanism of the fundamental forces was suggested by the Japanese physicist, Hideki Yukawa, in 1935. Yukawa proposed that the force between a pair of particles is mediated (or transmitted) by particles called **gauge bosons**. The four fundamental forces have different ranges and a different boson is responsible for each force. The mass of the boson establishes the range of the force. The bosons carry the force between particles.

Figure 10 shows how an electron can exchange a photon with a neighbouring electron, leading to electromagnetic repulsion. The exchange particle is said to be a "virtual" particle because it is not detected during the exchange. The exchange particle cannot be detected during its transfer between the particles because detection would mean that it would no longer be acting as the mediator of the force between the particles. The larger the rest mass of the exchange particle is, the lower the time it can be in flight without it being detected and, therefore, the lower the range of the force.

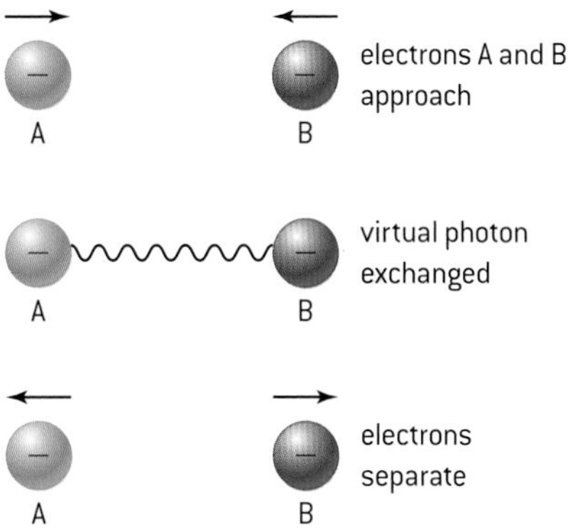

▲ Figure 10 Exchange of virtual photon between two electrons.

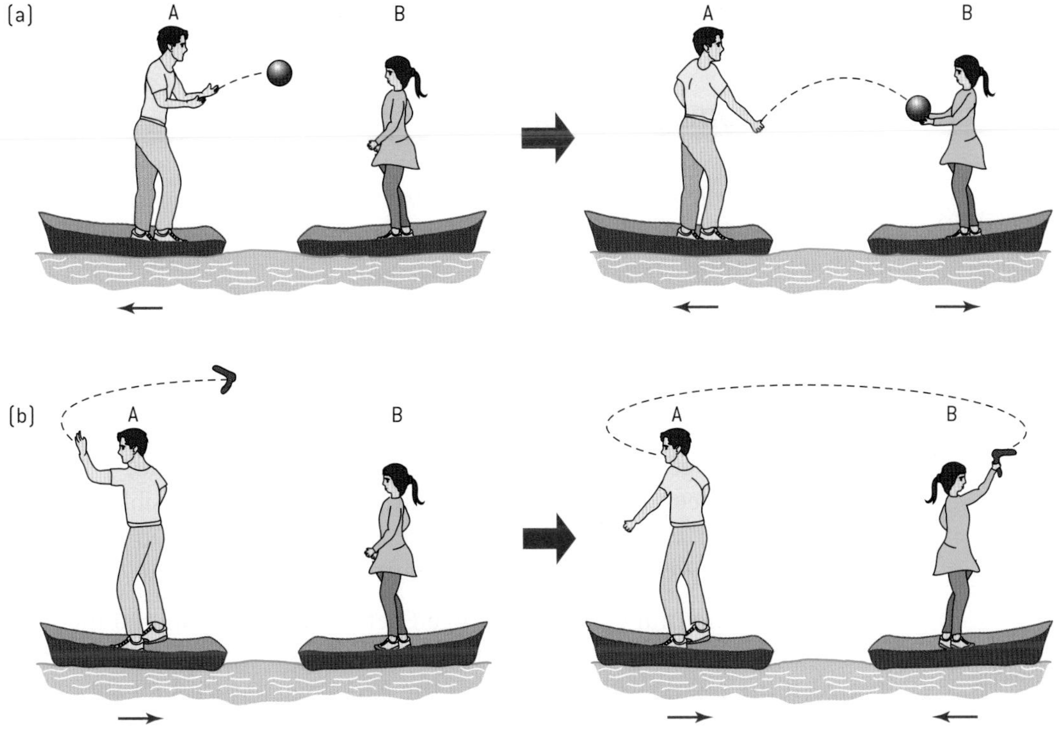

▲ Figure 11 Analogy of exchange particles.

Nature of science

Feynman diagrams and cloud chamber tracks

Don't be confused by thinking that Feynman diagrams are like the pictures of cloud or bubble chamber interactions – they are not and they really should not be thought of as "spacetime" diagrams! The diagrams help theoretical physicists work out the probability of interactions occurring, but for the IB course we treat them as illustrations of the interactions between complex particle events.

The simplest analogy to explain how a repulsive force can be produced by transfer of a particle is to picture what happens when a heavy ball is thrown backwards and forwards between people in two boats (see figure 11). The momentum changes as the ball is thrown and caught and to someone who cannot seen the ball travelling, a repulsive force seems to exist as they move apart. In order to explain the attractive force we need to imagine that a boomerang is being thrown between the people in the boats – in this case the change of momentum brings the boats together. To understand this more fully requires the use of the uncertainty principle discussed in Topic 12.

The table shows the exchange particles for the four fundamental forces.

Force	Exchange particle	Acts on
Gravitational	gravitons (undiscovered)	all particles
Weak nuclear	W^+, W^- and Z^0 bosons	quarks and leptons
Electromagnetic	photons	electrically charged particles
Strong nuclear	gluons (and mesons)	quarks and gluons (and hadrons)

Feynman diagrams

These are graphical visualizations, developed by the American physicist Richard Feynman, that represent interactions between particles. These diagrams are sometimes known as "spacetime diagrams". They have the time axis going upwards and the space or position axis to the right (*although many particle physicists draw these axes with space going upwards and time to the right – so be careful when you are researching interactions*). Straight lines represent particles and upwards arrows show particles moving forwards in time (downward arrows indicate an antiparticle – also moving forwards in time). Wavy or broken lines that have no arrows represent exchange particles. Points at which lines come together are called vertices (plural of vertex) and, at each vertex, conservation of charge, lepton number and baryon number must be applied.

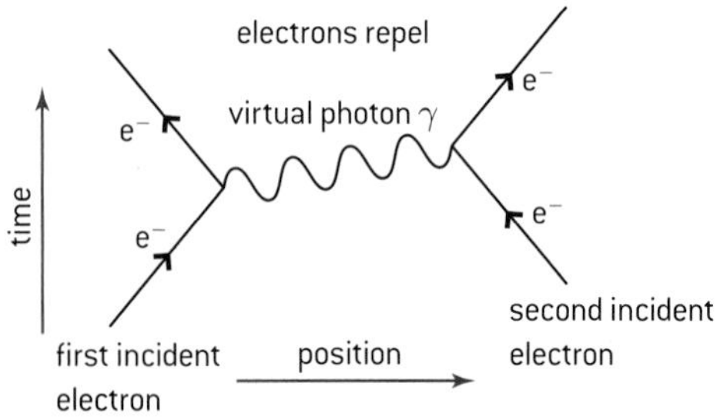

▲ Figure 12 Feynman diagram of the electromagnetic force between two electrons.

The electromagnetic force

Figure 12 shows a Feynman diagram for the electromagnetic force between two electrons. The exchange particle that gives rise to the force is the photon. Photons have no mass and this equates to the force having an infinite range.

This diagram shows two electrons moving closer and interacting by the exchange of a virtual photon before moving apart.

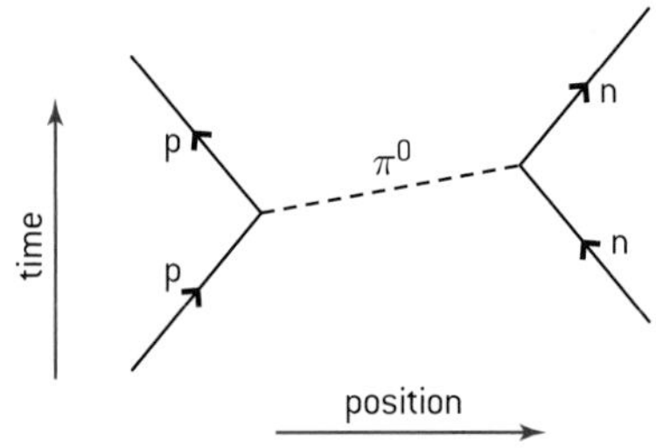

neutral pion (π^0) is exchanged between a proton (p) and a neutron (n) mediating the strong nuclear force between these particles in the nucleus

▲ Figure 13 Feynman diagram of the strong force between a proton and a neutron.

The strong force

This is the strongest of the forces and acts between quarks and, therefore, between nucleons. The exchange particles responsible are pions (π^+, π^- or π^0). Figure 13 shows the Feynman diagram for the strong force between a proton and a neutron.

In this case a neutral pion is exchanged between the proton and the neutron that ties them together. In hadrons, the pion carries gluons between the quarks – the gluons are the exchange particles for the colour force acting between quarks (**colour is not included on the IB Diploma Programme Physics syllabus and so questions will be limited to the exchange of mesons between hadrons**).

Weak nuclear force

This is responsible for radioactive decay by beta emission. In negative beta decay a neutron decays to a proton. In this process a W boson is exchanged as a quark changes from down to up. The W boson then immediately decays into an electron and an electron antineutrino.

The W and Z particles responsible for weak interactions are massive. The weak interaction is the only mechanism by which a quark can change into another quark, or a lepton into another lepton.

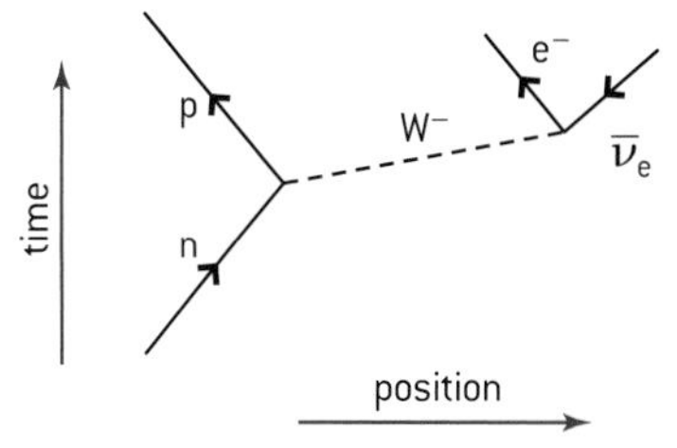

A neutron decays into a proton by the emission of an electron and an electron antineutrino (shown with the arrow moving downwards because it is an antiparticle). The decay is mediated by the negative W boson.

▲ Figure 14 Feynman diagrams for negative beta decay.

> **Note**
>
> The shorter the range of the exchange force the more massive the exchange particle – so the exchange particles for gravitation and the electromagnetic interaction (both of infinite range) must have zero rest masses. The weak interaction will have the heaviest boson because it range is the shortest; the strong interaction has an exchange particle of intermediate mass.

Conservation of strangeness

We are now in a position to discuss conservation of strangeness. The strange quark has a strangeness of -1 and particles containing a single strange quark will also have a strangeness of -1. The antistrange quark has a strangeness of $+1$ and particles containing a single antistrange quark will also have a strangeness of $+1$. One type of K^0 (neutral kaon) has a strangeness of $+1$ and so it contains an antistrange quark. A particle containing two strange quarks would have a strangeness of -2 and one containing two antistrange quarks would have a strangeness of $+2$, etc.

Strangeness is not conserved when strange particles decay through the weak interaction.

For example, strangeness is not conserved when a strange quark decays into an up quark.

Strangeness is conserved when there is a strong interaction.

This is why strange particles are always produced in pairs. If two particles interact to produce a strange particle then a strange antiparticle must also appear.

Figure 15 shows a stationary proton interacting through the strong interaction with a negative pion (short green line) at the bottom right-hand side of the image. These particles create a neutral kaon (K^0) and a second neutral particle called a lambda particle (Λ^0). As these particles are neutral they produce no tracks in a bubble chamber (a device that forms trails of bubbles along the path of a charged particle). The K^0 track is shown as a purple broken line and the Λ^0 as a blue broken line. The reaction is given by:

$$p + \pi^- \rightarrow K^0 + \Lambda^0$$

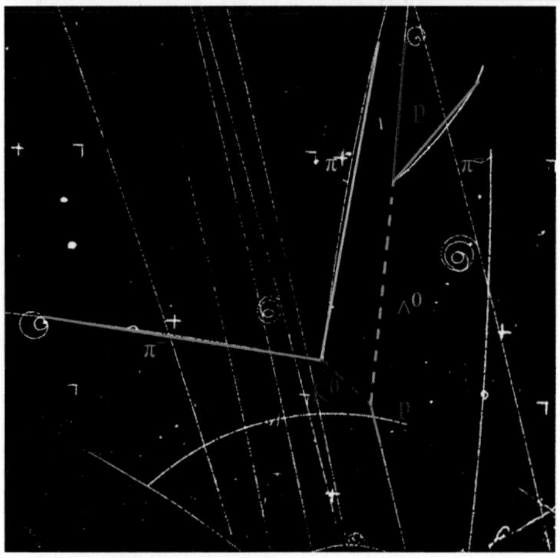

▲ Figure 15 Proton-pion interaction.

There are no strange particles on the left-hand side of this equation. The K^0 has a strangeness of $+1$ so the Λ^0 must have a strangeness of -1 for the reaction to be viable by the strong interaction. In terms of quarks the equation for the reaction is

$$uud + \bar{u}d \rightarrow d\bar{s} + uds$$

Let's check this to see that charge, lepton number, baryon number, and strangeness are conserved:

Q: left-hand side $\left(+\frac{2}{3}+\frac{2}{3}-\frac{1}{3}\right)+\left(-\frac{2}{3}-\frac{1}{3}\right)=0$ and

right hand side $\left(-\frac{1}{3}+\frac{1}{3}\right)+\left(+\frac{2}{3}-\frac{1}{3}-\frac{1}{3}\right)=0$

As $0 = 0$, charge is conserved.

L: as none of the particles are leptons, the lepton numbers on both sides are zero meaning lepton number is conserved.

B: left-hand side $\left(+\frac{1}{3}+\frac{1}{3}+\frac{1}{3}\right)+\left(+\frac{1}{3}+\frac{1}{3}\right)=\frac{5}{3}$ and

right hand side $\left(+\frac{1}{3}+\frac{1}{3}\right)+\left(+\frac{1}{3}+\frac{1}{3}+\frac{1}{3}\right)=\frac{5}{3}$

As $\frac{5}{3}=\frac{5}{3}$, baryon number is conserved (as it will always be if there are the same number of quarks on both sides of the equation.

S: there are no strange quarks on the left-hand side so the strangeness = 0; on the right-hand side there is one strange and one antistrange quark so the total strangeness $= -1 + 1 = 0$. So strangeness is also conserved.

With all four quantities conserved the interaction is viable.

You should check each of these equations to show that charge, baryon number, and lepton number are conserved. Strangeness should not be conserved because each of the two interactions involves the weak nuclear force.

The kaon subsequently decays into positive and negative pions and the lambda particle decays into a negative pion and a proton. As each of these decays is a weak interaction strangeness is **not** conserved.

These reactions are given by $K^0 \rightarrow \pi^+ + \pi^-$
or in terms of quarks $d\bar{s} \rightarrow u\bar{d} + \bar{u}d$
and
$\Lambda^0 \rightarrow \pi^- + p$
or in terms of quarks $uds \rightarrow \bar{u}d + uud$

Worked example

Draw Feynman diagrams to show the following interaction:

a) positive beta (positron) decay: $p \rightarrow n + e^+ + \nu_e$

b) proton–electron collision: $p + e^- \rightarrow n + \nu_e$

c) the two types of neutron–electron neutrino collision: $n + \nu_e \rightarrow \nu_e + n$ and $n + \nu_e \rightarrow p + e^-$

Solution

a)

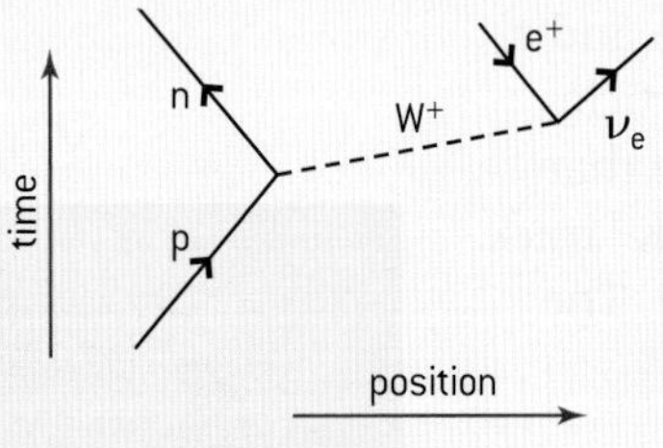

Figure 16 Feynman diagram for positive beta (positron) decay.

In this decay the proton decays into a neutron and emits a positron and electron neutrino. The decay is mediated by the positive W boson (W^+).

b)

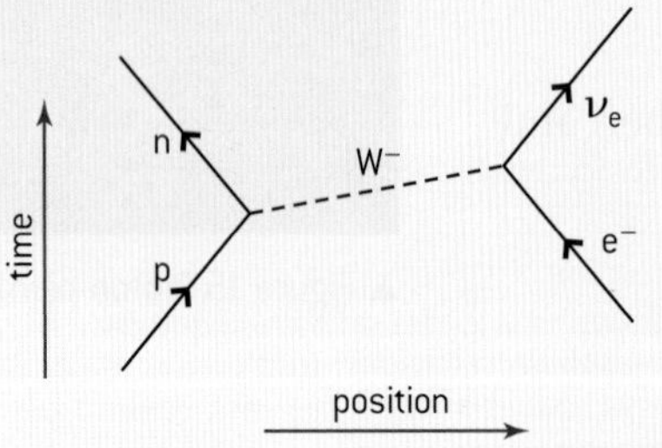

Figure 17 Feynman diagram for an proton–electron collision.

In this case the proton and the electron collide and produce a neutron and an electron neutrino. This interaction is mediated by the W^- boson.

c)

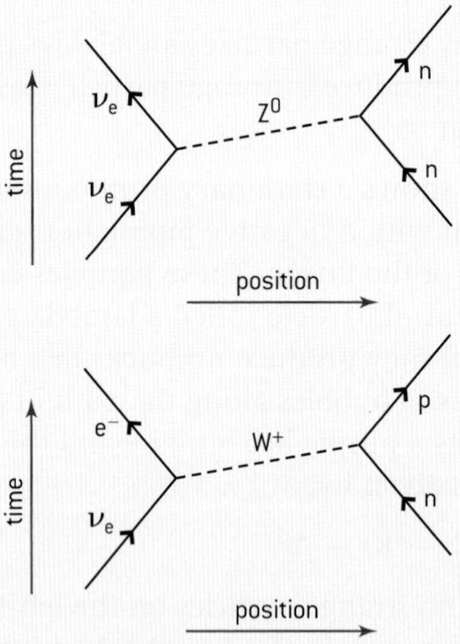

Figure 18 Feynman diagrams for the two types of neutron-electron neutrino collision.

The most likely collision between a neutron and an electron neutrino is one in which the Z^0 boson mediates the collision and the neutrino effectively bounces off the electron – this is known as a **neutral current interaction**. The electron neutrino can occasionally also interact through the W boson by changing a neutron into a proton. These are the **charged current interactions**.

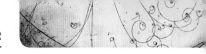

The Higgs boson

The Standard Model has been very successful at tying together theory and experimental results. The original theory predicted that leptons and quarks should have zero mass – this clearly did not agree with the experimental results for leptons and bosons in which finite masses of the particles had been already measured. To solve this problem the British physicist Peter Higgs introduced a theory that explained the mass of particles including the W and Z bosons. By introducing the *Higgs mechanism*, the equations of the Standard Model were changed in such a way as to allow these particles to have mass. According to this theory, particles gain mass by interacting with the Higgs field that permeates all space. The theory about the Higgs mechanism could be tested experimentally because it predicted the existence of a new particle not previously seen. This particle, called the **Higgs particle**, is a boson-like force mediator; however, it does not, in this case, mediate any force. The mass of this particle is very large and, therefore, requires a great deal of energy to be produced. With the opening of the large hadron collider (LHC) at the European Centre for Nuclear Research (CERN) in Geneva, a particle accelerator became available that was capable of providing sufficient energy to produce the Higgs boson.

In July 2012, scientists both at CERN and Fermilab announced that they had established the existence of a "Higgs-like" boson and, by March 14 2013, this particle was tentatively confirmed to be positively charged and to have zero spin; these are two of the properties of a Higgs boson. The discovery of the Higgs boson leads to the adaptation of the diagram of the Standard Model as shown in figure 19.

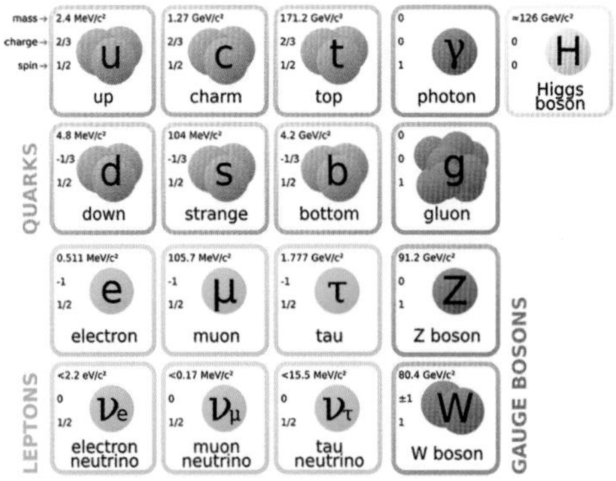

▲ Figure 19 The Standard Model including the Higgs boson.

Nature of science

International collaboration

The history of particle physics is one of international collaboration and the particle research facility at CERN is an excellent example of how the international science community can co-operate effectively. CERN is run by 20 European member states with four nations waiting to join; in addition there are more than 50 countries with agreements with CERN. The CERN website claims that half the world's particle physicists and over ten thousand scientists from more than a hundred countries do research there. With funds at a premium it makes financial sense to pool resources and work internationally. In terms of international cooperation, the work undertaken by particle physicists has much in common with the ideals of the IB.

Questions

1 *(IB)*

a) Light is emitted from a gas discharge tube. Outline briefly how the visible line spectrum of this light can be obtained.

The table below gives information relating to three of the wavelengths in the line spectrum of atomic hydrogen.

Wavelength / 10^{-9} m	Photon energy / 10^{-19} J
1880	1.06
656	3.03
486	4.09

b) Deduce that the photon energy for the wavelength of 486×10^{-9} m is 4.09×10^{-19} J.

The diagram below shows two of the energy levels of the hydrogen atom, using data from the table above. An electron transition between these levels is also shown.

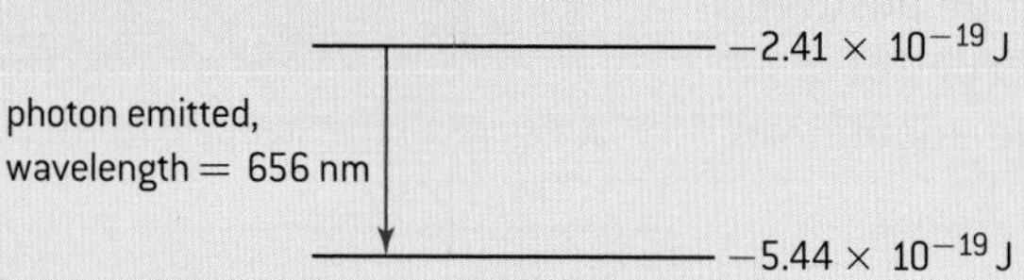

c) (i) On a copy of the diagram above, construct the other energy level needed to produce the energy changes shown in the table above.

(ii) Draw arrows to represent the energy changes for the two other wavelengths shown in the table above.

(9 marks)

2 *(IB)* Diagram 1 below shows part of the emission line spectrum of atomic hydrogen. The wavelengths of the principal lines in the visible region of the spectrum are shown.

diagram 1

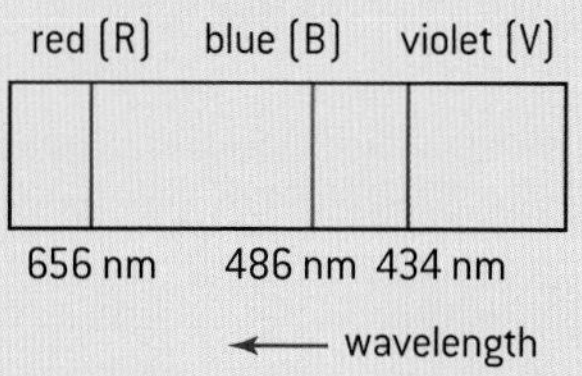

diagram 2

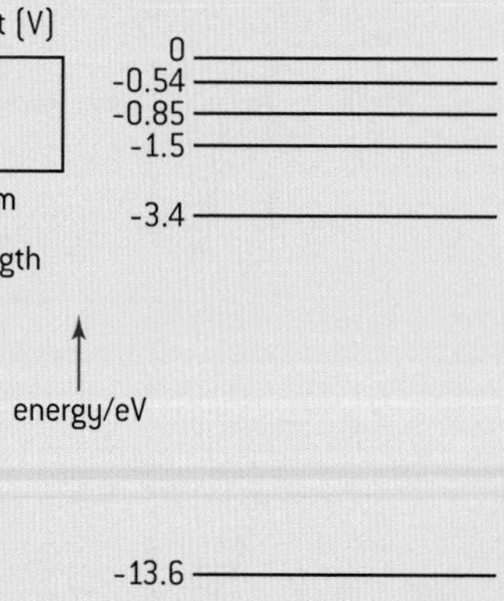

Diagram 2 shows some of the principal energy levels of atomic hydrogen.

a) Show, by calculation, that the energy of a photon of red light of wavelength 656 nm is 1.9 eV.

b) On a copy of diagram 2, draw arrows to represent:

(i) the electron transition that gives rise to the red line (label this arrow R)

(ii) a possible electron transition that gives rise to the blue line (label this arrow B).

(4 marks)

3 *(IB)* A nucleus of the isotope xenon, Xe-131, is produced when a nucleus of the radioactive isotope iodine I-131 decays.

a) Explain the term *isotopes*.

b) Fill in the boxes on a copy of the equation below in order to complete the nuclear reaction equation for this decay.

$${}^{131}_{\square}\mathrm{I} \rightarrow {}^{131}_{54}\mathrm{Xe} + \beta^- + \square$$

c) The activity A of a freshly prepared sample of the iodine isotope is 3.2×10^5 Bq. The variation of the activity A with time t is shown below.

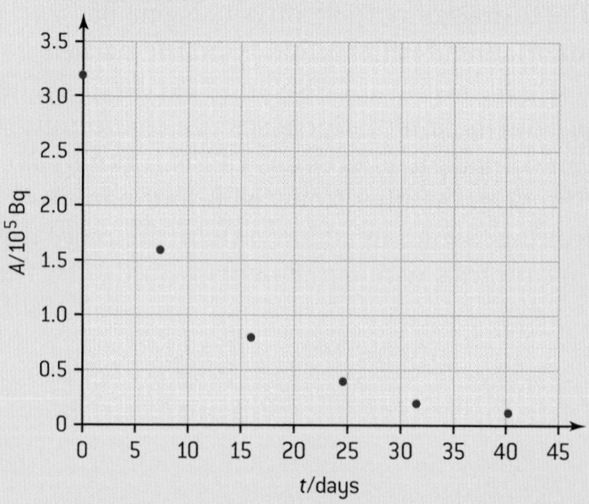

On a copy of this graph, draw a best-fit line for the data points.

d) Use the graph to estimate the half-life of I-131.

(8 marks)

4 *(IB)* One isotope of potassium is potassium-42 ($^{42}_{19}K$). Nuclei of this isotope undergo radioactive decay with a half-life of 12.5 hours to form nuclei of calcium.

a) Complete a copy of the nuclear reaction equation for this decay process.

$$^{42}_{19}K \rightarrow {}_{20}Ca +$$

b) The graph below shows the variation with time of the number N of potassium-42 nuclei in a particular sample.

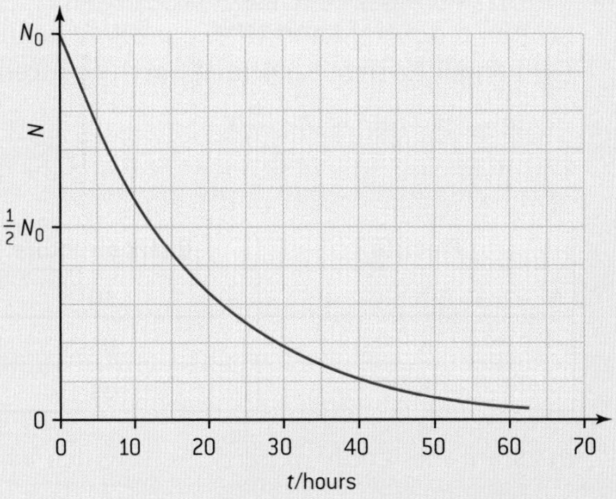

The isotope of calcium formed in this decay is stable.

On a copy of the graph above, draw a line to show the variation with time *t* of the number of calcium nuclei in the sample.

c) Use the graph in (c), or otherwise, to determine the time at which the ratio

$$\frac{\text{number of calcium nuclei in sample}}{\text{number of potassium-42 nuclei in sample}}$$

is equal to 7.0.

(7 marks)

5 *(IB)*

a) Explain what is meant by a *nucleon*.

b) Define what is meant by the *binding energy* of a nucleus.

The plot below shows the variation with nucleon number of the binding energy per nucleon.

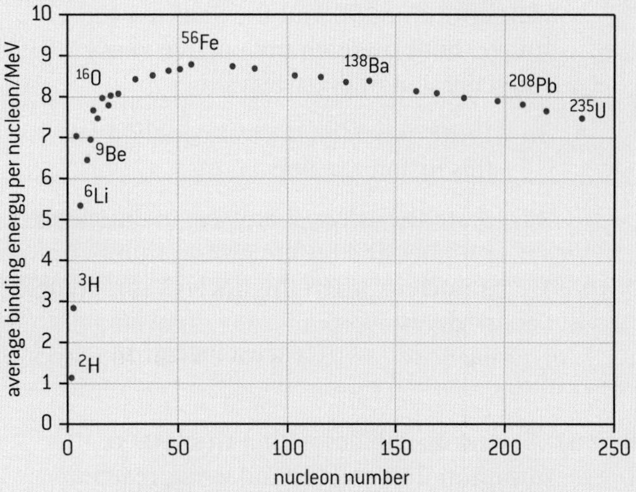

c) With reference to the graph, explain why energy can be released in both the fission and the fusion processes.

(5 marks)

6 *(IB)*

a) Use the following data to deduce that the binding energy per nucleon of the isotope $^{3}_{2}He$ is 2.2 MeV.

nuclear mass of $^{3}_{2}He$ = 3.016 03u

mass of proton = 1.007 28u

mass of neutron = 1.008 67u

b) In the nuclear reaction $^{2}_{1}H + {}^{2}_{1}H \rightarrow {}^{3}_{2}He + {}^{1}_{0}n$ energy is released.

(i) State the name of this type of reaction.

(ii) Sketch the general form of the relationship between the binding energy per nucleon and the nucleon number.

(iii) With reference to your graph, explain why energy is released in the nuclear reaction above.

(9 marks)

7 **a)** Distinguish between *nuclear fission* and *radioactive decay*.

b) A nucleus of uranium-235 ($^{235}_{92}U$) may absorb a neutron and then undergo fission to produce nuclei of strontium-90 ($^{90}_{38}Sr$) and xenon-142 ($^{142}_{54}Xe$) and some neutrons.

The strontium-90 and the xenon-142 nuclei both undergo radioactive decay with the emission of β^- particles.

(i) Write down the nuclear equation for this fission reaction.

(ii) State the effect, if any, on the nucleon number and on the proton number of a nucleus when the nucleus undergoes β^- decay.

(6 marks)

8 a) A neutron collides with a **nucleus** of uranium-235 and the following reaction takes place.

$$^{235}_{92}U + ^{1}_{0}n \rightarrow ^{96}_{37}Rb + ^{138}_{55}Cs + 2^{1}_{0}n$$

State the name of this type of reaction.

b) Using the data below, calculate the energy, in MeV, that is released in the reaction.

mass of $^{235}_{92}U$ nucleus $= 235.0439u$

mass of $^{96}_{37}Rb$ nucleus $= 95.9342u$

mass of $^{138}_{55}Cs$ nucleus $= 137.9112u$

mass of $^{1}_{0}n$ nucleus $= 1.0087u$

c) Suggest the importance of the two neutrons released in the reaction.

d) The rest mass of each neutron accounts for about 2 MeV of the energy released in the reaction. Explain what accounts for the remainder of the energy released.

(9 marks)

9 The diagram below illustrates a proton decaying into a neutron by beta positive (β^+) decay.

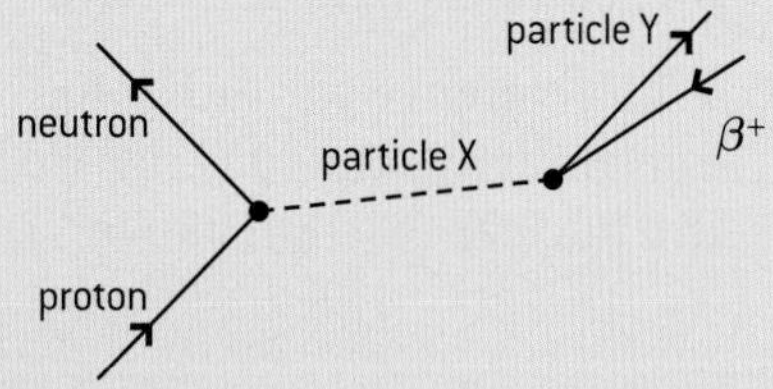

State the name of:

a) the force involved in this decay

b) the particle X

c) the particle Y involved in the decay.

(3 marks)

10 a) Possible particle reactions are given below. In each case apply the conservation laws to determine whether or not the reactions violate any of them.

(i) $\mu^- \rightarrow e^- + \gamma$

(ii) $p + n \rightarrow p + \pi^0$

(iii) $p \rightarrow \pi^+ + \pi^-$

b) State the name of an exchange particle involved in the weak interaction.

(10 marks)

11 *(IB)* When a negative kaon (K^-) collides with a proton, a neutral kaon (K^0), a positive kaon (K^+) and a further particle (X) are produced.

$$K^- + p \rightarrow K^0 + K^+ + X$$

The quark structure of kaons is shown in the table.

Particle	Quark structure
K^-	$s\bar{u}$
K^+	$u\bar{s}$
K^0	$d\bar{s}$

a) State the family of particles to which kaons belong.

b) State the quark structure of the proton.

c) The quark structure of particle X is sss. Show that the reaction is consistent with the theory that hadrons are composed of quarks.

(4 marks)

12 The Feynman diagram below represents a β^- decay via the weak interaction process. Time is shown as upwards. The wiggly line represents a virtual exchange particle.

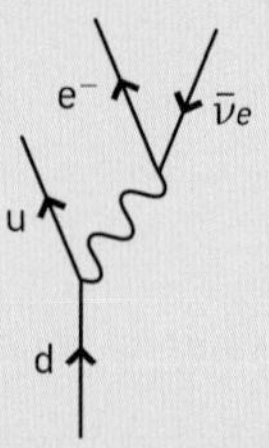

a) State what is meant by *virtual exchange particle.*

b) Determine whether the virtual particle in the process represented by the Feynman diagram is a W^+, a W^-, or a Z^0 boson.

(4 marks)

Introduction

In Topic 2 we looked at the principles behind the transfer of energy from one form to another. We now look in detail at sources that provide the energy we use every day. The provision of energy is a global issue. On the one hand, fossil fuel reserves are limited and these fuels can be a source of pollution and greenhouse gases, yet they are a convenient and energy-rich resource. The development of renewable energy sources continues but they are not yet at a point where they can provide all that we require. Political rhetoric and emotion often obscure scientific assessments about energy resources. Everyone – not just scientists – needs a clear understanding of the issues involved in order to make sound judgements about the future of our energy provision.

8.1 Energy sources

Understanding

- Primary energy sources
- Renewable and non-renewable energy sources
- Electricity as a secondary and versatile form of energy
- Sankey diagrams
- Specific energy and energy density of fuel sources

Nature of science

We rely on our ability to harness energy. Our large-scale production of electricity has revolutionized society. However, we increasingly recognize that such production comes at a price and that alternative sources are now required. There are elements of risk in our continued widespread use of fossil fuels: risk to the planet and risk to the supplies themselves. Society has to make important decisions about the future of energy supply on the planet.

Applications and skills

- Describing the basic features of fossil fuel power stations, nuclear power stations, wind generators, pumped storage hydroelectric systems, and solar power cells
- Describing the differences between photovoltaic cells and solar heating panels
- Solving problems relevant to energy transformations in the context of these generating systems
- Discussing safety issues and risks associated with the production of nuclear power
- Sketching and interpreting Sankey diagrams
- Solving specific energy and energy density problems

Equations

- $\text{power} = \dfrac{\text{energy}}{\text{time}}$
- wind power equation $= \frac{1}{2}A\rho v^3$

Primary and secondary energy

We use many different types of energy and energy source for our heating and cooking, transport, and for the myriad other tasks we undertake in our daily lives. There is a distinction between two basic types of energy source we use: primary sources and secondary sources.

A **primary source** is one that has not been transformed or converted before use by the consumer, so a fossil fuel – coal, for example – burnt directly in a furnace to convert chemical potential energy into the internal energy of the water and surroundings is an example of a primary source. Another example is the kinetic energy in the wind that can be used to generate electricity (a secondary source) or to do mechanical work such as in a windmill (a device used, for example, to grind corn or to pump up water from underground).

The definition of a **secondary source** of energy is one that results from the transformation of a primary source. The electrical energy we use is generated from the conversion of a primary source of energy. This makes electrical energy our most important secondary source. Another developing secondary source is hydrogen, although this is, at the moment, much less important than electricity. Hydrogen makes a useful fuel because it burns with oxygen releasing relatively large amounts of energy (you will know this if you have ever observed hydrogen exploding with oxygen in the lab). The product of this reaction (water) has the advantage that it is not a pollutant. However, hydrogen does not exist in large quantities in the atmosphere. So energy from a primary source would have to be used to form hydrogen from hydrocarbons, or by separating water into hydrogen and oxygen. The hydrogen could then be transported to wherever it is to be used as a source of energy.

Renewable and non-renewable energy sources

The primary sources can themselves be further divided into two groups: **renewable** and **non-renewable**. Renewable sources, such as biomass, can be replenished in relatively short times (on the scale of a human lifetime), whereas others such as wind and water sources are continually generated from the Sun's energy. Non-renewable sources, on the other hand, can be replaced but only over very long geological times.

A good way to classify renewable and non-renewable resources is by the rates at which they are being consumed and replaced. Coal and oil, both non-renewable, are produced when vegetable matter buried deep below ground is converted through the effects of pressure and high temperature. The time scale for production is hundreds of millions of years (the deposits of coal on Earth were formed from vegetation that lived and died during the Carboniferous geological period, roughly 300 million years ago). There are mechanisms active today that are beginning the process of creating fossil fuels in suitable wetland areas of the planet, but our present rate of usage of these fossil fuels is far greater than the rate at which they are being formed.

On the other hand, renewable fuels such as biomass use biological materials such as trees that were only recently alive. Such sources can be grown to maturity relatively quickly and then used for energy conversion. The rate of usage of the fuels can be similar to the rate at which they are being grown.

There are further advantages in the use of biomass and other renewable sources, where the material has been produced recently. When these renewable resources are converted, they will release carbon dioxide (one of the greenhouse gases) back into the atmosphere. But this is "new carbon" that was taken from the atmosphere and trapped in the biomass material relatively recently. The conversion of fossil fuels releases carbon dioxide into the atmosphere that was fixed in the fossil fuels hundreds of millions of years ago. The carbon dioxide content of the atmosphere was more than 150% greater in the Carboniferous period than it is today, and thus the burning of fossil fuels increases the overall amount of this greenhouse gas in the present-day atmosphere.

Types of energy sources

In Topic 2 there was a list of some of the important energies available to us. Some were mechanical in origin. Other energies were related to the properties of bulk materials and atomic nuclei. Of particular importance are the nuclear reactions, both fission and fusion, that you met in Topic 7.

Primary sources

The table below gives an indication of many of the primary sources that are used in the world today – although not all sources can be found in all locations. The use of geothermal energy, for example, requires that the geology of the location has hot rocks suitably placed below the surface.

Energy sources			
		source	**Energy form**
Non-renewable sources	Nuclear fuels	uranium-235	nuclear
	Fossil fuels	crude oil	chemical potential
		coal	
		natural gas	
Renewable sources		Sun	radiant (solar)
		water	kinetic
		wind	kinetic
		biomass	chemical potential
		geothermal	internal

Not all the primary sources in the table are necessarily used to provide electricity as a secondary source, it depends on local circumstances. A water wheel using flowing water in a river can be used to grind corn in a farming community rather than be harnessed to an electrical generator. A solar furnace may be used in an African village to boil water or to cook, while photovoltaic cells may produce the electrical energy required by the community. In some situations, this is often a better solution than that of changing all the solar energy to an electrical form that has to be reconverted subsequently.

Some energy is always degraded into an internal form in a conversion. Nevertheless, many of the world's primary sources are used to provide electricity using a power station where the secondary source output is in the form of electrical energy.

Primary energy use

Data for the present usage of various energy sources are readily available from various sources on the Internet. Figure 1 shows two examples of data released by the US Department of Energy.

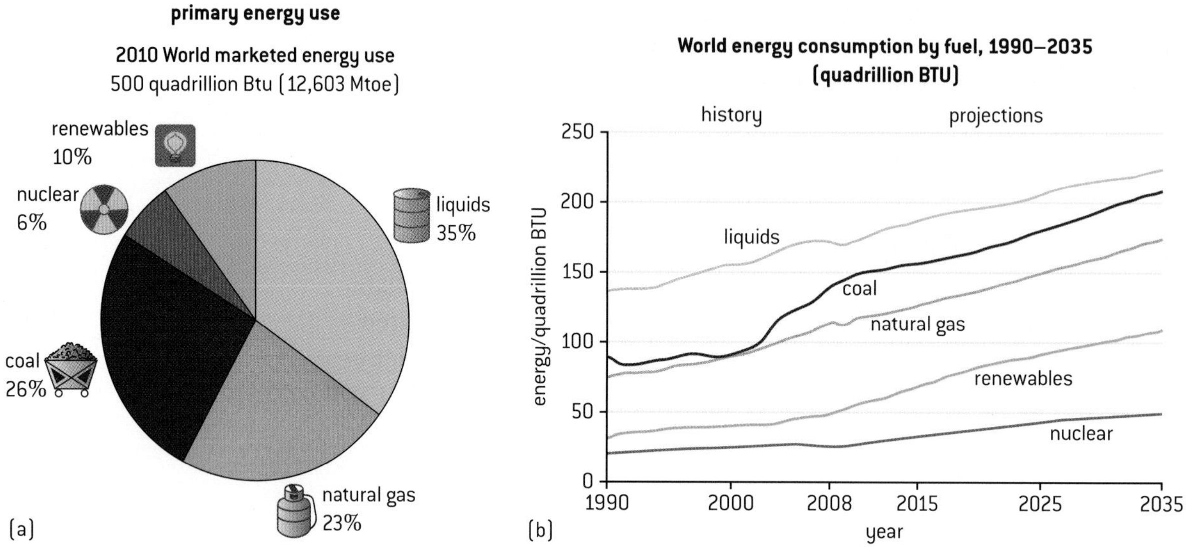

▲ Figure 1 Total world energy usage (US Energy Information Administration, report #DOE/EIA-0484 (2010)).

The first example (figure 1(a)) is a chart that shows the various energy sectors that account for the total world usage of energy sold on the open market in 2010. There are two units here that are common in energy data. When you carry out your own investigations of global energy usage, you will almost certainly come across these non-SI units.

- The British Thermal Unit (BTU) is still used in energy resource work; it was historically very important. The BTU is defined as the energy required to raise one British pound of water (about 0.5 kg) through 1 °F (a change of roughly 0.6 K) and is equivalent to about 1000 J.
- Mtoe stands for "million tonnes of oil equivalent". One tonne of oil equivalent is the energy released when one tonne (1000 kg) of crude oil is burnt; this is roughly 42 GJ, leading to a value of 5×10^{20} J for 2010 usage total.

The US chart uses an unusual multiplier – the energy total for 2010 is given as 500 quadrillion. The quadrillion is either 10^{15} or 10^{24} depending on the definition used. In this particular case it is 10^{15} and the total world energy usage in 2010 was about 5×10^{17} BTU, which also leads to about 5×10^{20} J.

The second chart (figure 1(b)) gives a projection from 1990 up to the year 2035 for world energy usage. Adding the various contributions to the total indicates that the US Department of Energy predicts a world total energy usage in 2035 of about 7.5×10^{20} J; a 50% increase on the 2010 figure. One feature of this graph (in the light of the enhanced greenhouse effect that we shall discuss later) is that the relative proportions of the various energy sources do not appear to be changing greatly over the timescale of the prediction.

Investigate!

Where are we today?

- Textbooks are of necessity out-of-date! They represent the position on the day that part of the book was completed. So the examples here are to give you some insight into the type and quantity of data that are available to you. But they are not to be regarded either as the last word or information that you must memorize for the examination. All the resources shown here were accessed on the Internet without difficulty.
- You should search for the latest tables and graphs using the source information printed with the graphs. (This is known as a bibliographic reference and it is essential to quote this when you use other people's data.) Search for the latest data and discuss it in class. Divide up the jobs so that students bring different pieces of data to the discussion. Ask yourselves: What are the trends now? Has the position changed significantly since this book was written?
- Science does not stand still, and this area of environmental science is moving as quickly as any other discipline. You need to have accurate data if you are to make informed judgments.
- Predictions are based on assumptions. They cannot predict critical events that might alter the situation e.g. a disaster triggered by a tsunami.

Specific energy and energy density

Much of the extraction of fossil fuels involves hard and dangerous work in mines or on oilrigs whether at sea or on land. On the face of it, the effort and risk of mining fossil fuels does not seem to be justified when there are other sources of energy available. So why are fossil fuels still extracted? The answer becomes more obvious when we look at the energy available from the fossil fuel itself. There are two ways to measure this: specific energy and energy density.

The word "specific" has the clear scientific meaning of "per unit mass", or (in SI) "per kg". So, **specific energy** indicates the number of joules that can be released by each kilogram of the fuel. Typical values for a particular fuel can vary widely, coal, for example, has a different composition and density depending on where it comes from, and even depending on the location of the sample in the mining seam.

Density is a familiar concept; it is the amount of quantity possessed by one cubic metre of a substance. **Energy density** is the number of joules that can be released from 1 m^3 of a fuel.

Fuel	Specific energy/ $MJ\ kg^{-1}$	Energy density/ $MJ\ m^{-3}$
Wood	16	1×10^4
Coal	20–60	$(20-60) \times 10^6$
Gasoline (petrol)	45	35×10^6
Natural gas at atmospheric pressure	55	3.5×10^4
Uranium (nuclear fission)	8×10^7	1.5×10^{15}
Deuterium/tritium (nuclear fusion)	3×10^8	6×10^{15}
Water falling through 100 m in a hydroelectric plant	10^{-3}	10^3

The table shows comparisons between some common fuels (and in the case of fusion, the possible energy yields if fusion should become commercially viable). You should look at these and other values in detail for yourself. Notice the wide range of values that appear in this table. Explore data for other common fuels. As you learn about different fuels, find out data for their specific energy and energy density and add these values to your own table.

Worked examples

1 A fossil-fuel power station has an efficiency of 25% and generates 1200 MW of useful electrical power. The specific energy of the fossil fuel is 52 MJ kg^{-1}. Calculate the mass of fuel consumed each second.

Solution

If 1200 MW of power is developed then, including the efficiency figure, $\frac{1200 \times 100}{25} = 4800$ MW of energy needs to be supplied by the fossil fuel.

The specific energy is 52 MJ kg^{-1}, so the mass of fuel required is $\frac{4800}{52} = 92$ kg s^{-1}. (That is roughly 1 tonne every 10 s, or one railcar full of coal every 2 minutes.)

2 When a camping stove that burns gasoline (petrol) is used, 70% of the energy from the fuel reaches the cooking pot. The energy density of the gasoline is 35 GJ m^{-3}.

a) Calculate the volume of gasoline needed to raise the temperature of 1 litre of water from 10 °C to 100 °C. Assume that the heat capacity of the pot is negligible. The specific heat capacity of the water is 4.2 kJ kg^{-1} K^{-1}.

b) Estimate the volume of fuel that a student should purchase for a weekend camping expedition.

Solution

a) 1 litre of water has a mass of 1 kg so the energy required to heat the water is 4200 × 1 × 90 which is 0.38 MJ.

Allowing for the inefficiency, 0.54 MJ of energy is required and this is a volume of fuel of $\frac{0.54 \times 10^6}{35 \times 10^9} = 1.6 \times 10^{-4}$ m^3 or about 200 ml.

b) Assume that 2 litres of water are required for each meal, and that there will be 5 cooked meals during the weekend. So 2 litres of fuel should be more than enough.

Thermal power stations

A thermal power station is one in which a primary source of energy is converted first into internal energy and then to electrical energy. The primary fuel can include nuclear fuel, fossil fuel, biomass or other fuel that can produce internal energy. (In principle, a secondary source could be used to provide the initial internal energy, but this would not be sensible as it incurs substantial extra losses.) We will discuss the different types of primary conversion later. However, once the primary energy has been converted to internal energy, all thermal power stations use a common approach to the conversion of internal into electrical energy: the energy is used to heat water producing steam at high temperatures and high pressures. Figure 2 shows what happens.

Energy from the primary fuel heats water in a pressure vessel to create steam. This steam is superheated. This means that its temperature is well above the familiar 100 °C boiling point that we are used to at atmospheric pressure. To attain such high temperatures the steam has to be at high pressure, hundreds of times more than atmospheric pressure. At these high pressures the water in the vessel does not boil in the

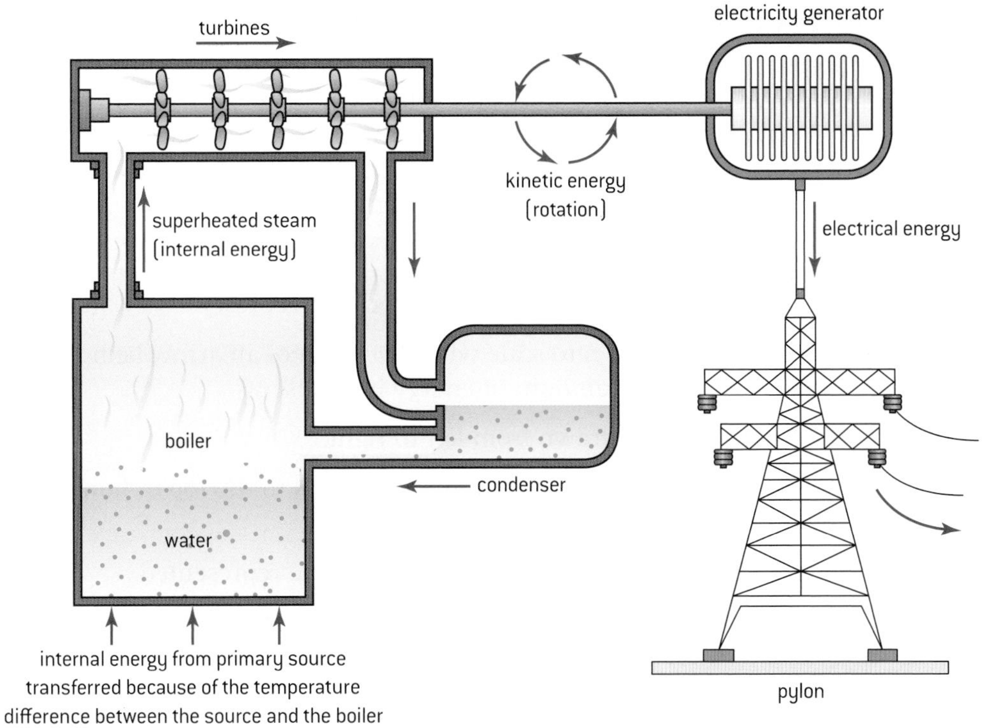

▲ Figure 2 Energy conversions inside a power station.

way that is familiar to us, it goes straight into the steam phase without forming the bubbles in the liquid that you see in a cooking pot on a stove. However, we will continue to use the term "boiler" for the vessel where the water is converted to steam even though the idea of boiling is technically incorrect.

The high-pressure steam is then directed to a turbine. Turbines can be thought of as "reverse" fans where steam blows the blades around (whereas in a fan the blades turn to move the gas). There is usually more than one set of blades and each set is mounted on a common axle connected to an alternating current (ac) electricity generator. In the generator, the electrical energy is produced when coils of wire, turned by the turbine, rotate in a magnetic field. The energy that is generated is sent, via a network of electrical cables, to the consumers. You can find the details of the physics of electrical energy generation and its subsequent transmission in Topic 11.

There are really three energy transfers going on in this process: primary energy to the internal energy of water, this internal energy to the kinetic energy of the turbine, and kinetic energy of the turbine to electrical energy in the generator. It is easy to forget the kinetic energy phase and to say that the internal energy goes straight to electrical.

Of course, the original internal energy is produced in different ways in different types of thermal power station. In fossil and biomass stations, there is a straightforward combustion process where material is set alight and burnt. In nuclear stations, the process has to be more complicated. We shall look at the differences between stations in the initial conversion of the energy later.

Tip

Do not confuse the roles of turbine and dynamo in a power station. The energy conversions they carry out are quite different.

Sankey diagrams

Different types of thermal power station have different energy losses in their processes and different overall efficiencies. The **Sankey diagram** is a visual representation of the flow of the energy in a device or in a process (although in other subjects outside the sciences, the Sankey diagram is also used to show the flow of material).

There are some rules to remember about the Sankey diagram:

- Each energy source and loss in the process is represented by an arrow.
- The diagram is drawn to scale with the width of an arrow being proportional to the amount of energy it represents.
- The energy flow is drawn from left to right.
- When energy is lost from the system it moves to the top or bottom of the diagram.
- Power transfers as well as energy flows can be represented.

Here is an example to demonstrate the use of a Sankey diagram.

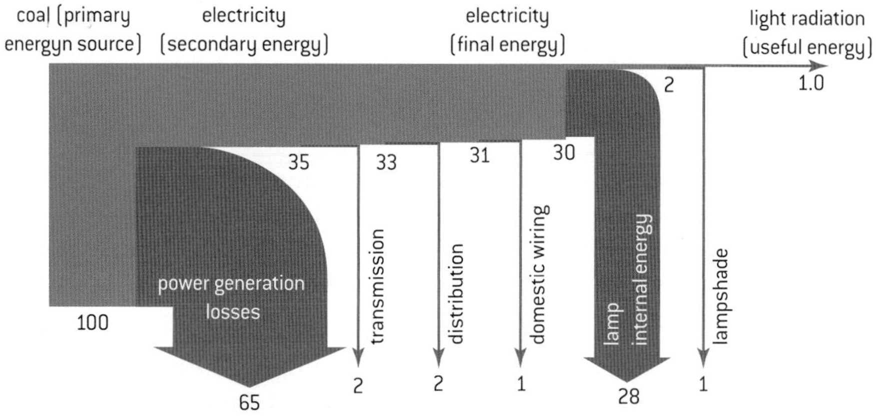

▲ Figure 3 Sankey diagram for a lamp.

It shows the flows of energy that begin with the conversion of chemical energy from fossil fuels and end with light energy emitted from a filament lamp. The red arrows represent energy that is transferred from the system in the form of energy as a result of temperature differences. It is important to recognize that in any process where there is an energy transformation, this energy is "lost" and is no longer available to perform a useful job. This is **degraded energy** and there is always a loss of energy like this in all energy transfers.

Of the original primary energy (100% shown in blue at the bottom of the diagram), only 35% appears as useful secondary energy. The remaining 65% (shown in red) is lost to the surroundings in the process, shown by an arrow that points downwards off the chart. The secondary energy is then shown with the losses involved in the transmission and distribution of the electricity, and to losses in the house wiring. In the lamp itself, most of the energy (28% of the original) is transferred to the internal energy of the surroundings. Finally only 1% of the original primary energy is left as light energy for illumination.

A Sankey diagram is a useful way to visualize the energy consumption of nations. There are many examples of this available on the Internet. Figure 4 shows the energy flows associated with the Canadian economy. Look on the Internet and you will probably find the Sankey diagram for your own national energy demand.

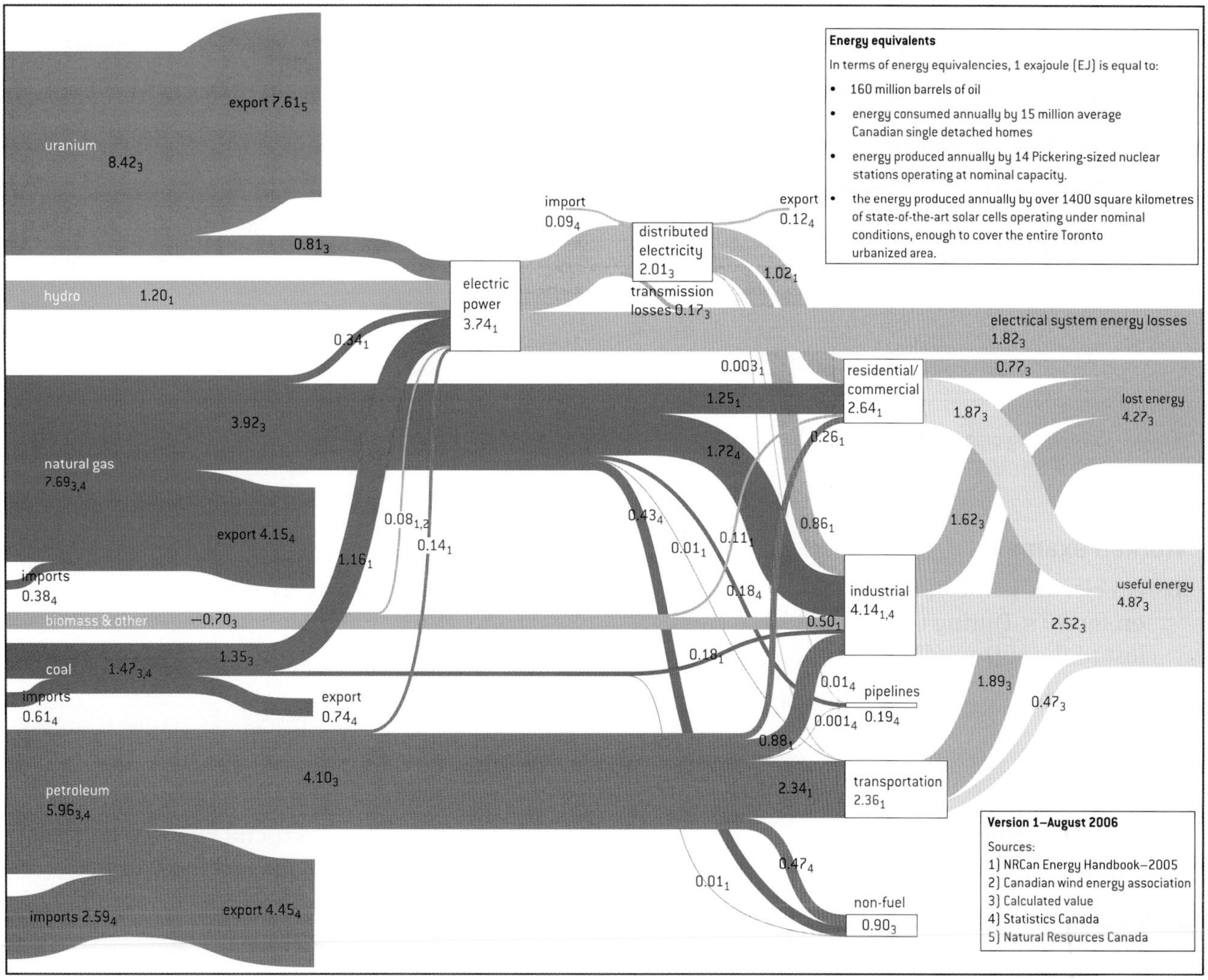

▲ Figure 4 The energy flow in Canada in 2003. The values are in exajoule. (http://ww2.nrcan.ge.ca/es/oerd).

Nature of science

Sankey diagram in another context

Here is a completely different use of the Sankey diagram constructed in a historical context. It shows the progress of Napoleon's army to and from his Russian campaign 1812–1813. The width of the strips gives the number of men in the army: brown in the journey to Moscow, black on the return. The graph below the diagram gives the average temperature experienced by the army on its return journey in Réaumur degrees. The Réaumur temperature scale is named after the French scientist who suggested a scale that had fixed points of 0 at the freezing point of water and 80 at the boiling point of water.

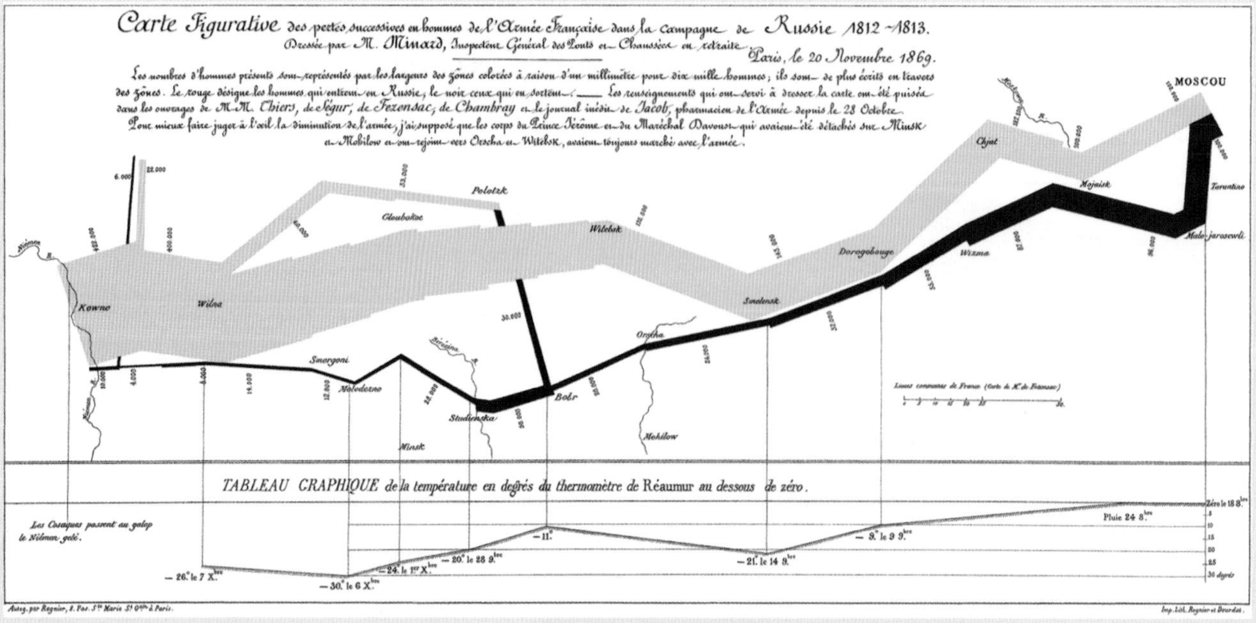

▲ Figure 5 A Sankey diagram showing the change in size of Napoleon's army during his Russian campaign.

Worked examples

1 An electric kettle of rating 2.0 kW is switched on for 90 s. During this time 20 kJ of energy is lost to the surroundings from the kettle.

Draw a Sankey diagram for this energy transfer.

Solution

The energy supplied in 90 s is
$2 \times 1000 \times 90 = 180$ kJ.

20 000 J is lost to the surroundings; this is 11% of the total.

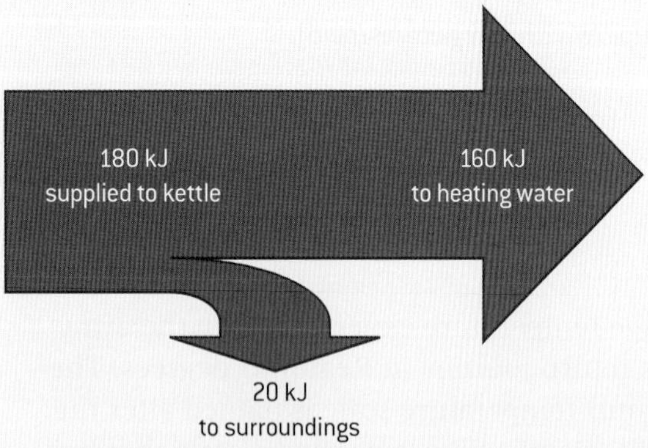

2 In a petrol-powered car 34% of the energy in the fuel is converted into the kinetic energy of the car. Heating the exhaust gases accounts for 12% of the energy lost from the fuel. The remainder of the energy is wasted in the engine, the gearbox and the wheels.

Use these data to sketch a Sankey diagram for the car.

Solution

Of the 100% of the original fuel energy, 12% is lost in the exhaust, and 34% is useful energy. This leaves 54% energy lost in the engine and the transmission.

A convenient way to draw the diagram is on squared paper. Use a convenient scale:
$10\% \equiv 1$ large square is a reasonable scale here.

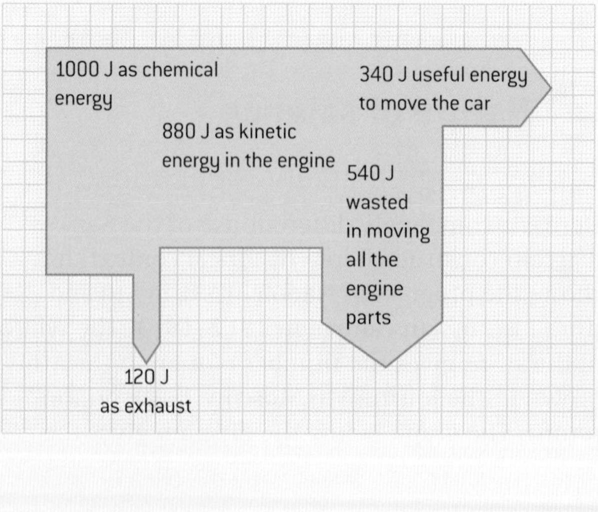

Primary sources used in power stations

This section compares the different ways in which the initial energy required by a thermal power station can be generated.

Fossil fuels

Modern fossil-fuel power stations can be very large and can convert significant amounts of power. The largest in the world at the time of writing has a maximum power output of about 6 GW.

The exact process required before the fuel is burnt differs slightly depending on the fuel used, whether coal, oil, or natural gas. The gas and oil can be burnt readily in a combustion chamber that is thermally connected to the boiler. In the case of coal, some pre-treatment is normally required. Often the coal is crushed into a fine powder before being blown into the furnace where it is burnt.

There are obvious disadvantages to the burning of fossil fuels. Some of these are environmental, but other disadvantages can be seen as a misuse of these special materials:

- The materials have taken a very large time to accumulate and will not be replaced for equally long times.
- The burning of the fuels releases into the atmosphere large quantities of carbon dioxide that have been locked in the coal, oil, and gas for millions of years. This has a major impact on the response of the atmospheric system to the radiation incident on it from the Sun (the greenhouse and enhanced greenhouse effects).
- Fossil fuels have significant uses in the chemical industry for the production of plastics, medicines, and other important products.
- It makes sense to locate power stations as close as possible to places where fossil fuels are recovered; however, this is not always possible and, in some locations, large-scale transportation of the fuels is still required. A need for transport leads to an overall reduction in the efficiency of the process because energy has to be expended in moving the fuels to the power stations.

Nuclear fuel

Sub-topic 7.2 dealt with the principles that lie behind nuclear fission. It explained the origin of the energy released from the nucleus when fission occurs and showed you how to calculate the energy available per fission.

In this course we will only consider so-called "thermal fission reactors", but there are other types in frequent use. A particularly common variety of the thermal reactors is the pressurized water reactor (PWR). Uranium-235 is the nuclide used in these reactors. As with all our other examples of power stations, the aim is to take the energy released in the nuclear fission and use this to create high-pressure steam to turn turbines connected to an electrical generator. However, the energy is not gained quite so easily as in the case of the fossil fuels.

Figure 6 shows a schematic of a PWR with the final output of steam to, and the return pipe from, the turbines on the right-hand side of the diagram. The remainder of the power station is as in figure 2 on p313.

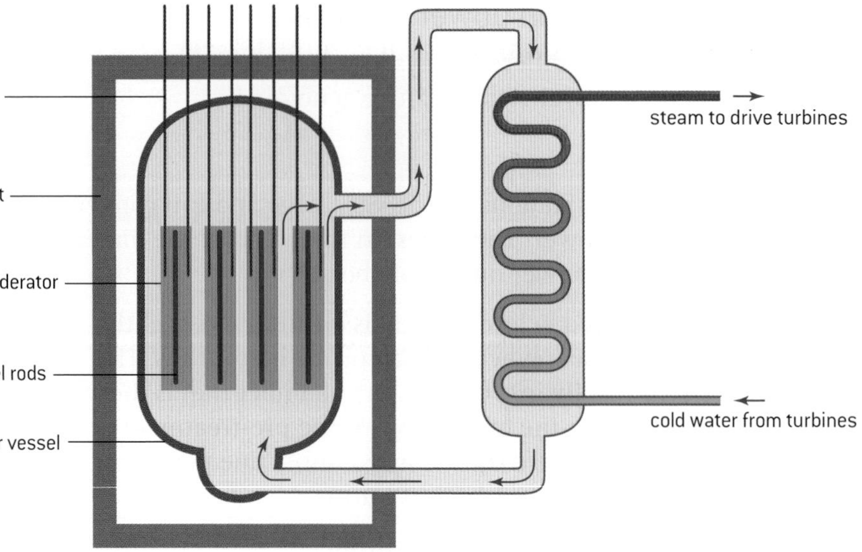

▲ Figure 6 Basic features of a pressurised water reactor (PWR).

Uranium is mined as an ore in various parts of the world, including Australia, Canada, and Kazakhstan (which together produce about 60% of the world's ore every year). The US, Russia, and parts of Africa also produce smaller amounts of uranium ore. About 99% of the ore as it comes directly from the ground is made up of uranium-238, with the remainder being U-235; it is the U-235 not U-238 that is required for the fission process. This means that an initial extraction process is required to boost the ratio of U-235 : U-238. The fuel needs to contain about 3% U-235 before it can be used in a reactor. This is because U-238 is a good absorber of neutrons and too much U-238 in the fuel will prevent the fission reaction becoming self-sustaining. The fuel with its boosted proportions of U-235 is said to be **enriched**.

Fission product	energy/MeV
fission fragments	160
decay of fission fragments	21
emitted gamma rays	7
emitted neutrons	5

The enriched material is then formed into fuel rods – long cylinders of uranium that are inserted into the core of the reactor. Most of the energy (about 200 MeV or 3×10^{-11} J per fission) is released in the form of kinetic energy of the fission fragments and neutrons emitted during the fission. Immediately after emission, the neutrons are moving at very high speeds of the order of 10^4 km s^{-1}. However, in order for them to be as effective as possible in causing further fissions to sustain the reaction they need to be moving with kinetic energies much lower than this, of the order of 0.025 eV (with a speed of about 2 km s^{-1}). This slower speed is typical of the speeds that neutrons have when they are in equilibrium with matter at about room temperatures. Neutrons with these typical speeds are known as **thermal neutrons**.

The requirement of removing the kinetic energy from the neutrons is not only that the neutrons can stimulate further fissions effectively, but that their energy can be efficiently transferred to the later stages of the power station. The removal of energy is achieved through the use of a **moderator**, so-called because it moderates (slows down) the speeds of the neutrons.

Typical moderators for the PWR type include water and carbon in the form of graphite. The transfer of energy is achieved when a fast-moving neutron strikes a moderator atom inelastically, transferring energy to the atom and losing energy itself. After a series of such collisions the neutron will have lost enough kinetic energy for it to be moving at thermal speeds and to have a high chance of causing further fission.

A further problem is that U-238 is very effective at absorbing high-speed neutrons, so if the slowing down is carried out in the presence of the U-238 then few free neutrons will remain at the end of it. So reactor designers have moderators close to, but not part of, the fuel rods; the neutrons slow down in the presence of moderator but away from the U-238. The fuel rods and the moderating material are kept separate and neutrons move from one to the other at random. The reactor vessel and its contents are designed to facilitate this.

The criteria for a material to be a good moderator include not being a good absorber of neutrons (absorption would lower the reaction rate and possibly stop the reaction altogether) and being inert in the extreme conditions of the reactor. Some reactor types, for example, use deuterium (^2_1H) in the form of deuterium oxide (D_2O, called "heavy water") rather than graphite.

You should be able to recognize that the best moderator of all ought to be a hydrogen atom (a single proton in the nucleus) because the maximum energy can be transferred when a neutron strikes a proton. However, hydrogen is a very good absorber of neutrons and it cannot be used as a moderator in this way.

There is a need to regulate the power output from the reactor and to shut down its operation if necessary. This is achieved through the use of **control rods**. These are rods, often made of boron or some other element that absorbs neutrons very well, that can be lowered into the reactor. When the control rods are inserted a long way into the reactor, many neutrons are absorbed in the rods and fewer neutrons will be available for subsequent fissions; the rate of the reaction will drop. By raising and lowering the rods, the reactor operators can keep the energy output of the reactor (and therefore the power station) under control.

The last part of the nuclear power station that needs consideration is the mechanism for conveying the internal energy from inside the reactor to the turbines. This is known as the **heat exchanger**. The energy exchange cannot be carried out directly as in the fossil-fuel stations. There needs to be a closed-system heat exchanger that collects energy from the moderator and other hot regions of the reactor, and delivers it to the water. The turbine steam cannot be piped directly through the reactor vessel because there is a chance that radioactive material could be transferred outside the reactor vessel. Using a closed system prevents this.

The pressurized water reactor is given its name because it transfers the energy from moderator and fuel rods to the boiler using a closed water system under pressure. Water is not the only substance available for this. In the Advanced Gas-cooled Reactors (AGR) used in the UK, carbon dioxide gas is used rather than water, but the principle of transferring energy safely through a closed system is the same.

Nature of science

Fast breeder reactors

A remarkable design for nuclear reactors comes with the development of the fast breeder reactor. Plutonium-239 (Pu-239) is the fissionable material in this case just as uranium-235 is used in PWRs. However, a blanket of uranium-238 surrounds the plutonium core. Uranium-238 does not easily fission and is a good absorber of neutrons (remember that its presence is undesirable in a PWR). This U-238 absorbs neutrons lost from the reactor core and is transmuted into Pu-239 – the fuel of the fast breeder reactor! The reactor is making its own fuel and generating energy at the same time. It has been reported that, under the right conditions, a fast breeder reactor can produce 5 kg of fissionable plutonium for every 4 kg used in fission. This is a good way to convert the large stockpiles of virtually useless uranium-238 into something of value.

There are drawbacks of course: large amounts of high-level radioactive waste from the fast breeder reactor need to be dealt with and, in the wrong hands, the plutonium can be used for nuclear weapons.

Investigate!

Running your own reactor

- No schools are likely to have their own nuclear reactor for students to use! However you can still investigate the operation of a nuclear power station.
- There are a number of simulations available on the Internet that will give you virtual control of the station. A suitable starting point is to enter "nuclear reactor applet" into a search engine.

Worked examples

1 When one uranium-235 nucleus undergoes fission, 3.2×10^{-11} J of energy is released. The density of uranium-235 is 1.9×10^{4} kg m^{-3}.

Calculate, for uranium-235:

a) the specific energy

b) the energy density.

Solution

a) The mass of the atom is 235u (ignoring the mass of the electrons in the atom).

This is equivalent to a mass of $1.7 \times 10^{-27} \times 235 = 4.0 \times 10^{-25}$ kg.

So the specific energy of uranium-235 is $\frac{3.2 \times 10^{-11}}{4.0 \times 10^{-25}} = 8.0 \times 10^{13}$ J kg^{-1}

b) 1 kg of uranium-235 has a volume of $\frac{1}{1.9 \times 10^{4}}$ $\approx 5 \times 10^{-5}$ m^3. Therefore the energy density of pure uranium-235 $= 8.0 \times 10^{13} \div 5 \times 10^{-5}$ $= 1.6 \times 10^{18}$ J m^{-3}

2 Explain what will happen in a pressured water reactor if the moderator is removed.

Solution

The role of the moderator is to remove kinetic energy from neutrons so that there is a high probability that further fissions will occur. When neutrons are moving at high speeds, there is a very high probability that uranium-238 nuclei will absorb them without fission occurring. So the removal of the moderator will mean that neutrons are no longer slowed down, and will be absorbed by U-238. The fission reaction will either stop or its rate will be reduced.

3 When a moving neutron strikes a stationary carbon-12 atom head-on, the neutron loses about 30% of its kinetic energy. Estimate the number of collisions that would be required for a 1 MeV neutron to be slowed down to 0.1 eV.

Solution

After one collision the remaining kinetic energy of the neutron will be 0.7×1 MeV.

After two collisions the energy of the neutron will be $0.7 \times 0.7 \times 1$ MeV

$= 0.7^2 \times 1$ MeV

After n collisions the energy of the neutron will be $0.7^n \times 1$ MeV.

So $0.7^n = 0.1 \times 10^{-6}$,

or $n = \frac{-7}{\log_{10} 0.7} = 45.2$

So 46 collisions are required to reduce the neutron speed by this factor of 10^7.

(In practice, about 100 are required; you might want to consider why the actual number is larger than the estimate.)

Safety issues in nuclear power

There needs to be a range of safety measures provided at the site of a nuclear reactor to protect the work force, the community beyond the power station, and the environment.

- The reactor vessel is made of thick steel to withstand the high temperatures and pressures present in the reactor. This has the benefit of also absorbing alpha and beta radiations together with some of the gamma rays and stray neutrons.
- The vessel itself is encased in layers of very thick reinforced concrete that also absorb neutrons and gamma rays.
- There are emergency safety mechanisms that operate quickly to shut the reactor down in the event of an accident.
- The fuel rods are inserted into and removed from the core using robots so that human operators do not come into contact with the spent fuel rods, which become highly radioactive during their time in the reactor.

The issue of the disposal of the waste produced in the nuclear industry is much debated in many parts of the world. Some of the waste will remain radioactive for a very long time – but, as you know from Topic 7, this implies that its activity will be quite low during this long period. There are however other problems involving the chemical toxicity of this waste material which mean that it is vital to keep it separate from biological material and thus the food chain. The technology required to achieve this is still developing.

At the end of the life of a reactor (of the order of 25–50 years at the moment), the reactor plant has to be decommissioned. This involves removing all the fuel rods and other high-activity waste products and enclosing the reactor vessel and its concrete shield in a larger shell of concrete. It is then necessary to leave the structure alone for up to a century to allow the activity of the structure to drop to a level similar to that of the local background. Such long-term treatment is expensive and it is important to factor these major costs into the price of the electricity as it is being produced during the lifetime of the power station.

TOK

Of all the scientific issues of our time, perhaps nuclear energy invokes the greatest emotional response in both scientists and non-scientists alike. It is vital to carry out accurate risk assessments for all energy sources, not just nuclear. Is it possible we could forget the location of the waste sites in 50, 100, or 1000 years? Are human errors part of the equation? How can such assessments be carried out in an emotionally charged debate?

Nature of science

Society and nuclear power

The use of nuclear power has been growing throughout the world since the late 1940s. However, society has never been truly comfortable with its presence and use. There are many reasons for this, including a lack of public understanding of the fission process (both advantages and disadvantages), public reaction to accidents that have occurred periodically over the years, and a fear of radioactive materials. Major accidents have included the Chernobyl incident, the event at Three Mile Island, and the Fukushima accident of 2011 that was triggered by a tsunami and possibly also an associated earthquake.

At the time of writing there has been a withdrawal of approval for nuclear plans by some governments as a result of public opinion changes after the Fukushima accident. Continuing decisions about how we generate energy will be required by society as time goes on, and as our energy-generation technology improves.

When you debate these issues, ensure that you understand the scientific facts and statistics that surround the issue of nuclear power.

It is important also to understand the meaning of risk, not just in the context of the nuclear power industry, but also in relation to things we do every day. Do some research for yourself. Find out the risks to your health of:

- living within 20 km of a nuclear power station
- flying once from Europe to a country on the Pacific Rim
- smoking 20 cigarettes every day
- driving 1000 km in a car.

Try to find numerical estimates of the risk, not just written statements.

Wind generators

Wind generators can be used successfully in many parts of the world. Even though the wind blows erratically at most locations, this can be countered by the provision of separate wind farms, each with large numbers of individual turbines all connected to the electrical power grid.

There are two principal designs of generator: horizontal-axis and vertical-axis. In both cases a rotor is mounted on an axle that is either horizontal or vertical, hence the names. The rotor is rotated by the wind and, through a gearbox, this turns an electrical generator. The electrical energy is fed either to a storage system (but this increases the expense) or to the electrical grid.

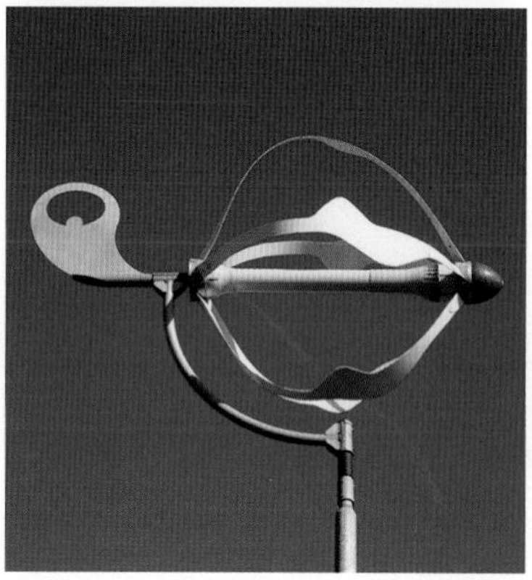

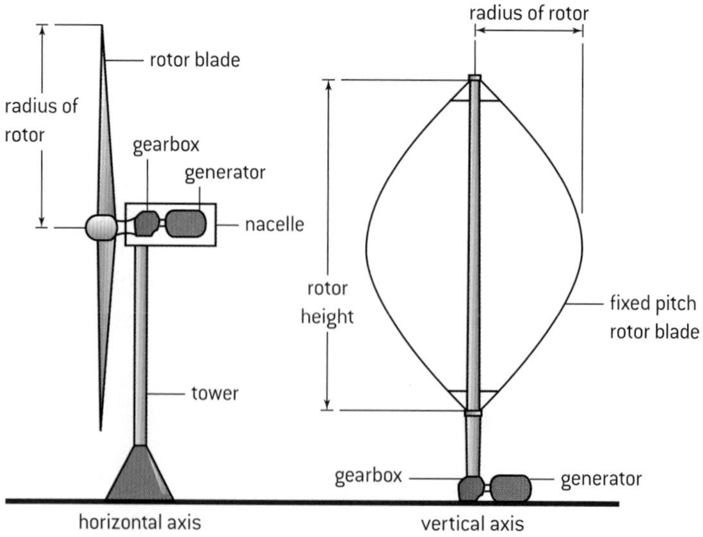

▲ Figure 7 Wind turbine configurations.

The horizontal-axis machine can be steered into the wind. The vertical-axis type does not have to be steered into the wind and therefore its generator can be placed off-axis.

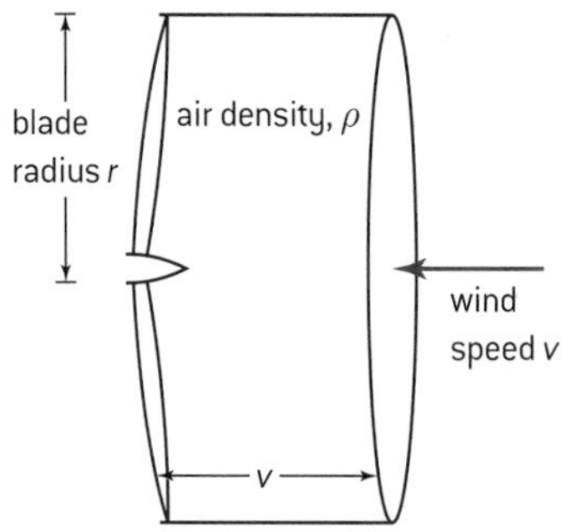

▲ Figure 8 Volume of air entering a wind turbine in one second.

It is possible to estimate the maximum power available from a horizontal-axis wind turbine of blade area A. In one second, the volume of air moving through the turbine is vA when the speed of the wind is v (figure 8).

The mass of air moving through the turbine every second is ρvA where ρ is the density of the air, and the kinetic energy of the air arriving at the turbine in one second is

$$\frac{1}{2}(mass) \times (speed)^2 = \frac{1}{2}(\rho vA)v^2 = \frac{1}{2}\rho Av^3$$

If a wind turbine has a blade radius r then the area A swept out by the blades is πr^2 and the maximum theoretical kinetic energy arriving at the turbine every second (and hence the maximum theoretical power) is

$$\frac{1}{2}\rho\pi r^2 v^3$$

This is a *maximum theoretical value of the available power* as there are a number of assumptions in the proof. In particular, it is assumed that *all* the kinetic energy of the wind can be used. Of course, the wind has to move out of the back of the turbine and so must have some kinetic energy remaining after being slowed down. Also, if the turbine is part of a wind farm then the presence of other nearby turbines disturbs the flow of the air and leads to a reduction in the energy from turbines at the rear of the array.

The turbine power equation suggests that a high wind speed and a long blade (large A or r) will give the best energy yields. However, increasing the radius of the blade also increases its mass and this means that the rotors will not rotate at low wind speeds. The blade radius has to be a compromise that depends on the exact location of the wind farm.

Many wind farms are placed off-shore; the wind speeds are higher than over land. This is because the sea is relatively smooth compared with the buildings, hills and so on that are found on land. However, installation and subsequent maintenance for off-shore arrays are more expensive because of the access issues.

Another ideal place for wind farms is at the top of a hill. The effect of the land shape is to constrict the flow of the wind into a more confined volume so that in consequence the wind speed rises as the air moves up the hill. Wind speeds therefore tend to be greater at the top of hills and, because the power output of the turbine is proportional to v^3, even a small increase in average wind speed is advantageous.

Some people object to both on- and off-shore arrays on the grounds of visual pollution. There are also suggestions that wind farms compromise animal habitats in some places and that the turbines are noisy for those who live close by.

The factors for and against wind being used as an energy source are summarized in this table:

Advantages	Disadvantages
No energy costs	Variable output on a daily or seasonal basis
No chemical pollution	Site availability can be limited in some countries
Capital costs can be high but reduce with economies of scale	Noise pollution
Easy to maintain on land; not so easy off-shore	Visual pollution
	Ecological impact

Worked examples

1 A wind turbine produces a power P at a particular wind speed. If the efficiency of the wind turbine remains constant, estimate the power produced by the turbine:

a) when the wind speed doubles

b) when the radius of the blade length halves.

Solution

The equation for the kinetic energy arriving at the wind turbine every second is $\frac{1}{2}\rho\pi r^2v^3$.

a) When the wind speed v doubles, v^3 increases by a factor of 8, so the power output will be $8P$.

b) When the radius of the blade halves, r^2 will go down by a factor of 4 and (if nothing else changes) the output will be $\frac{P}{4}$.

2 A wind turbine with blades of length 25 m is situated in a region where the average wind speed is 11 m s^{-1}.

a) Calculate the maximum possible output of the wind turbine if the density of air is 1.3 kg m^{-3}.

b) Outline why your estimate will be the maximum possible output of the turbine.

Solution

a) Using the wind turbine equation:

the kinetic energy arriving at the wind turbine every second is $\frac{1}{2}\rho\pi r^2v^3$,

this will be the maximum power output and is $\frac{1}{2} \times 1.3 \times \pi \times 25^2 \times 11^3 = 1.7$ MW.

b) Mechanical and electrical inefficiencies in the wind turbine have not been considered. The calculation assumes that all the kinetic energy of the wind can be utilized; this is not possible as some kinetic energy of the air will remain as it leaves the wind turbine.

Pumped storage

There are a number of ways in which water can be used as a primary energy resource. These include:

- pumped storage plants
- hydroelectric plants
- tidal barrage
- tidal flow systems
- wave energy.

All these sources use one of two methods:

- The gravitational potential energy of water held at a level above a reservoir is converted to electrical energy as the water is allowed to fall to the lower level (used in hydroelectric (figure 9(a)), pumped storage, and tidal barrage).

- The kinetic energy of moving water is transferred to electrical energy as the water flows or as waves move (river or tidal flow or wave systems). Figure 9(b) shows a picture of the Canadian Beauharnois run-of-the-river power station on the St-Laurent river that can generate 1.9 GW of power.

In this topic we focus on the **pumped storage system**.

The wind farms and nuclear power stations we have discussed so far are known as **baseload** stations. They run 24 hours a day, 7 days a week converting energy all the time. However, the demand that consumers make for energy is variable and cannot always be predicted. From time to time the demand exceeds the output of the baseload stations. Pumped storage is one way to make up for this deficit.

(a)

(b)

▲ Figure 9 Water as a primary energy resource.
(a) A hydroelectric plant in Thailand.
(b) A run-of-the-river plant in Canada.

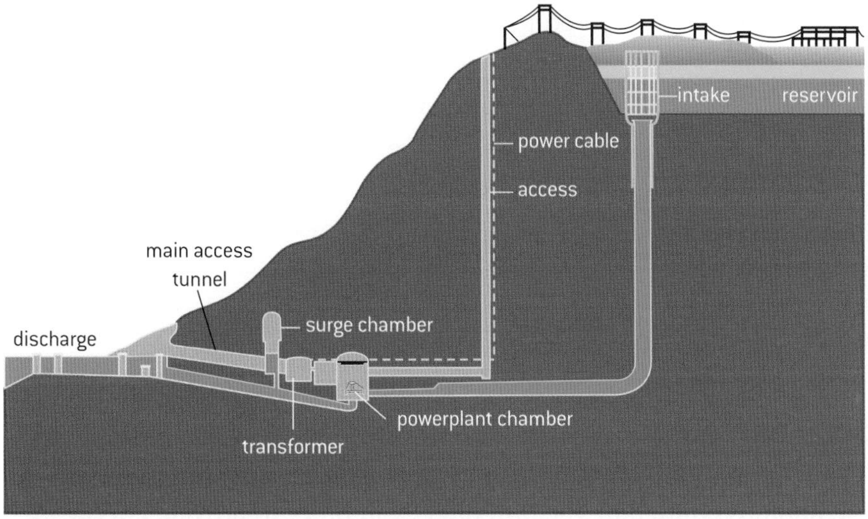

▲ Figure 10 A pumped-storage generating station.

A pumped storage system (Figure 10) involves the use of two water reservoirs – sometimes a natural feature such as a lake, sometimes a man-made lake or an excavated cavern inside a mountain. These reservoirs are connected by pipes. When demand is high, water is allowed to run through the pipes from the upper reservoir to the lower via water turbines. When demand is low, and electrical energy is cheap, the turbines operate in reverse to pump water back from the lower to the upper reservoir.

Some pumped storage systems can go from zero to full output in tens of seconds. The larger systems take longer to come up to full power, however, substantial outputs are usually achieved in only a few minutes from switch on.

For a pumped storage system that operates through a height difference of Δh, the gravitational potential energy available $= mg\Delta h$ where m is the mass of water that moves through the generator and g is the gravitational field strength.

So the maximum power P available from the water is equal to the rate at which energy is converted in the machine and is

$$P = \frac{m}{t}g\Delta h = \left(\frac{V}{t}\rho\right)g\Delta h$$

where t is the time for mass m to move through the generator, V is the volume of water moving through the generator in time t and ρ is the density of the water.

Worked examples

1 Water from a pumped storage system falls through a vertical distance of 260 m to a turbine at a rate of 600 kg s^{-1}. The density of water is 1000 kg m^{-3}. The overall efficiency of the system is 65%.

Calculate the power output of the system.

Solution

In one second the gravitational potential energy lost by the system is $mg\Delta h = 600 \times g \times 260 =$ 1.5 MJ.

The efficiency is 65%.

Output power $= 1.5 \times 10^6 \times \frac{65}{100} = 0.99$ MW

2 In a tidal barrage system water is retained behind a dam of height h. Show that the gravitational potential energy available from the water stored behind the dam is proportional to h^2.

Solution

Assume that the cross-sectional area of the dam is A and that the cross-section is rectangular.

The volume of water held by the dam is Ah.

The mass of the water held by the dam is ρAh where ρ is the density of water.

When the dam empties completely the centre of mass of the water falls through a distance $\frac{h}{2}$(because the centre of mass is half way up the height of the dam).

The gravitational potential energy of the water is $mgh = (\rho Ah) \times g \times \frac{h}{2} = \frac{1}{2}\rho Ah^2$.

Thus, gpe $\propto h^2$

Solar energy

Although most energy is ultimately derived from the Sun, two systems use photons emitted by the Sun directly in order to provide energy in both large- and small-scale installations.

Solar heating panels

Solar heating panels is a technique for heating water using the Sun's energy.

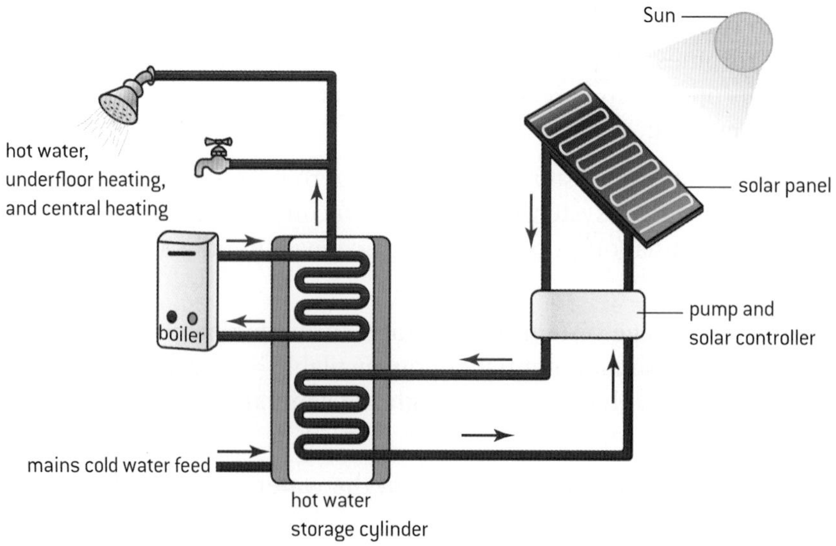

▲ Figure 11 Solar thermal-domestic hot water system.

A **solar heating panel** contains a pipe, embedded in a black plate, through which a glycol–water mixture is circulated by a pump (glycol has a low freezing point, necessary in cold countries). The liquid heats up as infra-red radiation falls on the panel. The pump circulates the liquid to the hot-water storage cylinder in the building. A heat-exchanger system transfers the energy to the water in the storage cylinder. A pump is needed because the glycol–water mixture becomes less dense as it heats up and would therefore move to the top of the panel and not heat the water in the cylinder. A controller unit is required to prevent the system pumping hot water from the cylinder to the panel during the winter when the panel is cold.

▲ Figure 12 A domestic photovoltaic panel system.

Solar photovoltaic panels

The first solar "photocells" were developed around the middle of the nineteenth century by Alexandre-Edmond Becquerel (the father of Henri the discoverer of radioactivity). For a long time, the use of solar cells, based on the element selenium, was restricted to photography. With the advent of semiconductor technology, it was possible to produce **photovoltaic cells** (as they are properly called, abbreviated to PV) to power everything from calculators to satellites. In many parts of the world, solar panels are mounted on the roofs of houses. These panels not only supply energy to the house, but excess energy converted during sunny days is often sold to the local electricity supply company.

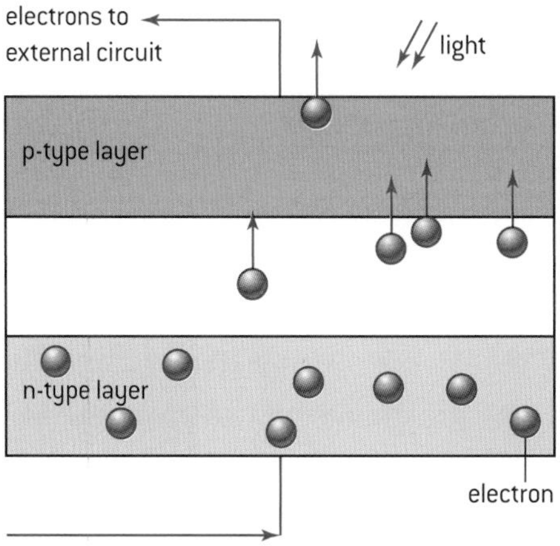

▲ Figure 13 Cross-section of a photovoltaic cell.

The photovoltaic materials in the panel convert electromagnetic radiation from the Sun into electrical energy. A full explanation of the way in which this happens goes beyond the IB syllabus, but a simplified explanation is as follows.

The photovoltaic cell consists of a single crystal of semiconductor that has been doped so that one face is p-type semiconductor and the opposite face is n-type. These terms n-type and p-type indicate the most significant charge carriers in the substance (electrons in n-type, positive "holes" – an absence of electrons – in p-type). Normally there is equilibrium between the charge carriers in both halves of the cell. However, when energy in the form of photons falls on the photovoltaic cell, then the equilibrium is disturbed, electrons are released and gain energy to move from the n-region to the p-region and hence around the external circuit. The electrons transfer this energy to the external circuit in the usual way and do useful work.

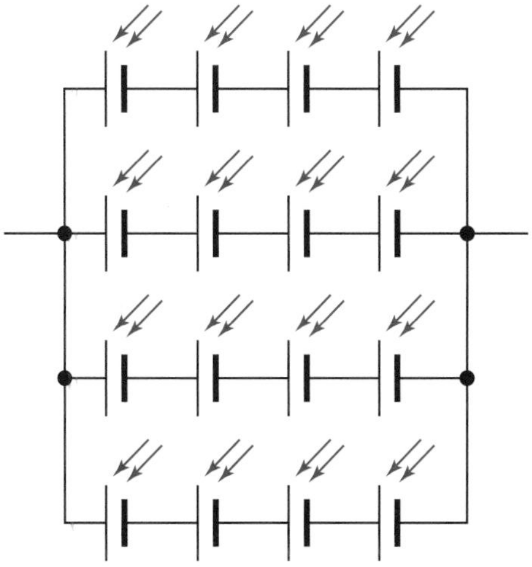

▲ Figure 14 Connecting photovoltaic cells together.

One single cell has a small emf of about 1 V (this is determined by the nature of the semiconductor) and so banks of cells are manufactured in order to produce usable currents on both a domestic and commercial scale. Many cells connected in series would give large emfs but also large internal resistances; the compromise usually adopted is to connect the cells in a combination of both series and parallel.

The efficiencies of present-day solar cells are about 20% or a little higher. However, extensive research and development is being carried out in many countries and it is likely that these efficiencies will rise significantly over the next few years.

The advantages of both solar heating panels and photovoltaic cells are that maintenance costs are low, and that there are no fuel costs. Individual households can use them. Disadvantages include high initial cost, and the relatively large inefficiencies (so large areas of cells are needed). It goes without saying that the outputs of both types of cell are variable, and depend on both season and weather. We will look at the reasons for these variations in Sub-topic 8.2.

The mathematics of both photovoltaic cells and solar heating panels is straightforward. At a particular location and time of day, the Sun radiates a certain power per square metre I (also known as the intensity of the radiation) to the panels. Panels have an area A, so the power arriving at the surface of the panel will be IA. Panels have an efficiency η which is the fraction of the energy arriving that is converted into internal energy (of the liquid in the heating panels) or electrical energy (photovoltaics). So the total power converted by the panel is ηIA.

Worked examples

1 A house requires an average power of 4.0 kW in order to heat water. The average solar intensity at the Earth's surface at the house is 650 W m^{-2}. Calculate the minimum surface area of solar heating panels required to heat the water if the efficiency of conversion of the panel is 22%.

Solution

4000 W are required, each 1 m^2 of panel can produce $650 \times \frac{22}{100} = 140$ W

$$\text{Area required} = \frac{4000}{140} = 28 \text{ m}^2$$

2 Identify the energy changes in photovoltaic cells and in solar heating panels.

Solution

A solar heating cell absorbs radiant energy and converts it to the internal energy of the working fluid.

A photovoltaic cell absorbs photons and converts their energy to electrical energy.

8.2 Thermal energy transfer

Understanding

- → Conduction, convection, and thermal radiation
- → Black-body radiation
- → Albedo and emissivity
- → Solar constant
- → Greenhouse effect
- → Energy balance in the surface–atmosphere system

Nature of science

The study of the Earth's climate illustrates the importance of modelling in science. The kinetic theory for an ideal gas is a good model for the way that real gases actually behave. Scientists model the Earth's climate in an attempt to understand the implications of the release of greenhouse gases for global warming.

The climate is a much more complex system than a simple gas. Issues for scientists include: the availability of data for the planet as a whole, and greater computing power means that more sophisticated models can be tested. Collaboration between research groups means that debate about the accuracy of the models can take place.

Applications and skills

- → Sketching and interpreting graphs showing the variation of intensity with wavelength for bodies emitting thermal radiation at different temperatures
- → Solving problems involving Stefan–Boltzmann and Wien's laws
- → Describing the effects of the Earth's atmosphere on the mean surface temperature
- → Solving albedo, emissivity, solar constant, and the average temperature of the Earth problems

Equations

- → Stefan-Boltzmann equation: $P = e\sigma AT^4$
- → Wien's Law: $\lambda_{max}\,(\text{metres}) = \dfrac{2.90 \times 10^{-3}}{T(\text{kelvin})}$
- → intensity equation: $I = \dfrac{\text{power}}{A}$
- → $\text{albedo} = \dfrac{\text{total scattered power}}{\text{total incident power}}$

Introduction

We considered some of the energy resources in use today in Sub-topic 8.1. In the course of that sub-topic some of the pollution and atmospheric effects of the resources were mentioned. In this sub-topic we look in more detail at the physics of the Earth's atmosphere and the demands that our present need for energy are making on it. After a review of the ways in which energy is transferred due to differences in temperature, we discuss the Sun's radiation and its effect on the atmosphere. Finally, we look at how changes in the atmosphere modify the climate.

Thermal energy transfer

Any object with a temperature above absolute zero possesses internal energy due to the motion of its atoms and molecules. The higher the temperature of the object, the greater the internal energy associated with the molecules. Topic 3 showed that in a gas the macroscopic quantity that we call absolute temperature is equivalent to the average of the kinetic energy of the molecules. Given the opportunity, energy will spontaneously transfer from a region at a high temperature to a region at a low temperature. "Heat", we say rather loosely, "flows from hot to cold".

In this sub-topic we look at the ways in which energy can flow due to differences in temperature. There are three principal methods: conduction, convection, and thermal radiation. All are important to us both on an individual level and in global terms.

Thermal conduction

There are similarities between electrical conduction and thermal conduction to the extent that this section is labelled thermal conduction for correctness. However, for the rest of our discussion we will use the term "conduction", taking the word to mean thermal conduction.

We all know something of practical conduction from everyday experience. Burning a hand on a camping stove, plunging a very hot metal into cold water which then boils, or melting ice in the hand all give experience of energy moving by conduction from a hot source to a cold sink.

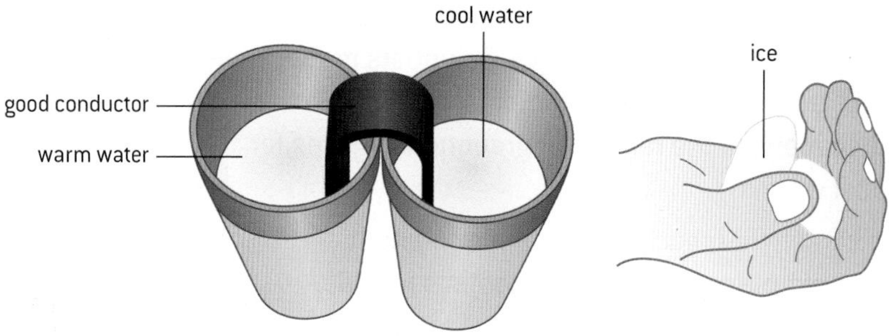

▲ Figure 1 Laboratory conduction demonstrations.

Metals are good thermal conductors, just as they are also good electrical conductors. Poor thermal conductors such as glass or some plastics also conduct electricity poorly. This suggests that there are similarities between the mechanisms at work in both types of conduction. However, it should be noted that there are still considerable differences in scale between the very best metal conductors (copper, gold) and the worst metals (brass, aluminium). There are many lab experiments that you may have seen designed to help students recognize the different thermal properties of good and poor conductors.

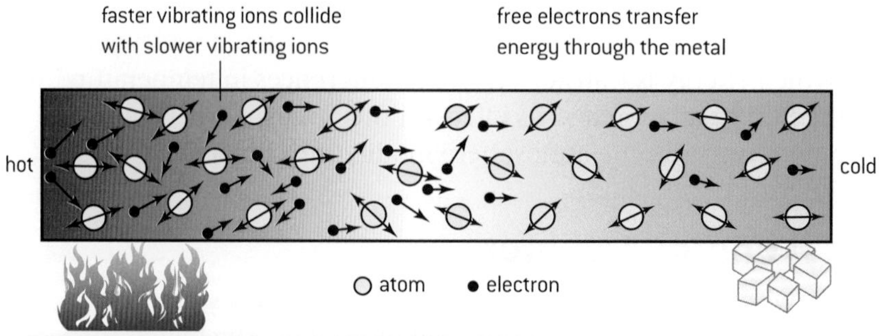

▲ Figure 2 How conduction occurs at an atomic level.

In conduction processes, energy flows through the bulk of the material without any large-scale relative movement of the atoms that make up the solid. Conduction (electrical and thermal) is known as a transport phenomenon. Two mechanisms contribute to thermal conduction:

- **Atomic vibration** occurs in all solids, metal and non-metal alike. At all temperatures above 0 K, the ions in the solid have an internal energy. So they are vibrating about their average fixed position in the solid. The higher the temperature, the greater is the average energy, and therefore the higher their mean speed. Imagine a bar heated at one end and cooled at the other (see figure 2). At the hot end, the ions vibrate at a large amplitude and with a large average speed. At the cold end the amplitude is lower and the speed is smaller. At the position where the bar is heated the ions vibrate with increasing amplitude and collide with their nearest neighbours. This transfers internal energy and the amplitude of the neighbours will increase; this process continues until the bar reaches thermal equilibrium. In this case the energy supplied to each ion is equal to that transferred by the ion to its neighbours in the bar or the surroundings. Each region of the bar will now be at the same uniform temperature.
- Conduction can occur in gases and liquids as well as solids, but, because the inter-atomic connections are weaker and the atoms (particularly in the gases) are farther apart in fluids, conduction is much less important in many gases and liquids than is convection.
- Although thermal conduction by atomic vibration is universal in solids, there are other conduction processes that vary in importance depending on the type of solid under discussion. As we saw in Topic 5, electrical conductors have a covalent (or metallic) bonding that releases **free electrons** into what is essentially an electron gas filling the whole of the interior of the solid. These free electrons are in thermal equilibrium with the positive ions that make up the atomic lattice of the solid. The electrons can interact with each other and the energy from the high-temperature end of the solid "diffuses" along the solid by interactions between these electrons. When an electron interacts with an atom, then energy is transferred back into the atomic lattice to change the vibrational state of the atom. This free-electron mechanism for conduction depends critically on the numbers of free electrons available to the solid. Good electrical conductors, where there are many charge carriers (free electrons) available per unit volume, are likely also to be good thermal conductors. For example, in copper there is one free electron per atom. You should be able to use Avogadro's number, the density of copper and its relative atomic mass, to show that there are 8.4×10^{28} electrons in every cubic metre of copper.

Nature of science

Thermal and electrical resistivity

Analogies are often used in science to aid our understanding of phenomena. Electrical and thermal conduction are closely linked and so it ought to be possible to transfer some ideas from one to the other. There is an analogy between thermal and electrical effects when thermal conduction is compared with electrical conduction for a wire of cross-section area A and length Δl:

Thermal

The rate of transfer of internal energy Q is proportional to temperature gradient:

$$\frac{\Delta Q}{\Delta t} = -\frac{A}{k}\left(\frac{\Delta T}{\Delta l}\right)$$

where k is the thermal resistivity of the material.

Electrical

The rate of transfer of charge q (the current) is proportional to potential gradient:

$$\frac{\Delta Q}{\Delta t} = -\frac{A}{\rho}\left(\frac{\Delta V}{\Delta l}\right)$$

where ρ is the electrical resistivity of the material.

The thermal resistivity k of a material corresponds to the electrical resistivity ρ and the temperature gradient corresponds to the electric potential gradient. The rate at which both internal energy and charge are carried through the wire is related to the presence of free electrons in the metal. It is more than a coincidence of equations, physics at the atomic scale is involved. Are good electrical conductors also good thermal conductors? Compare the values of ρ and k for different metals.

Convection

Convection is the movement of groups of atoms or molecules within fluids (liquids and gases) that arises through variations in density. Unlike conduction, which involves the microscopic transfer of energy, convection is a bulk property. Convection cannot take place in solids. An understanding of convection is important in many areas of physics, astrophysics and geology. In some hot countries, houses are designed to take advantage of natural convection to cool down the house in hot weather.

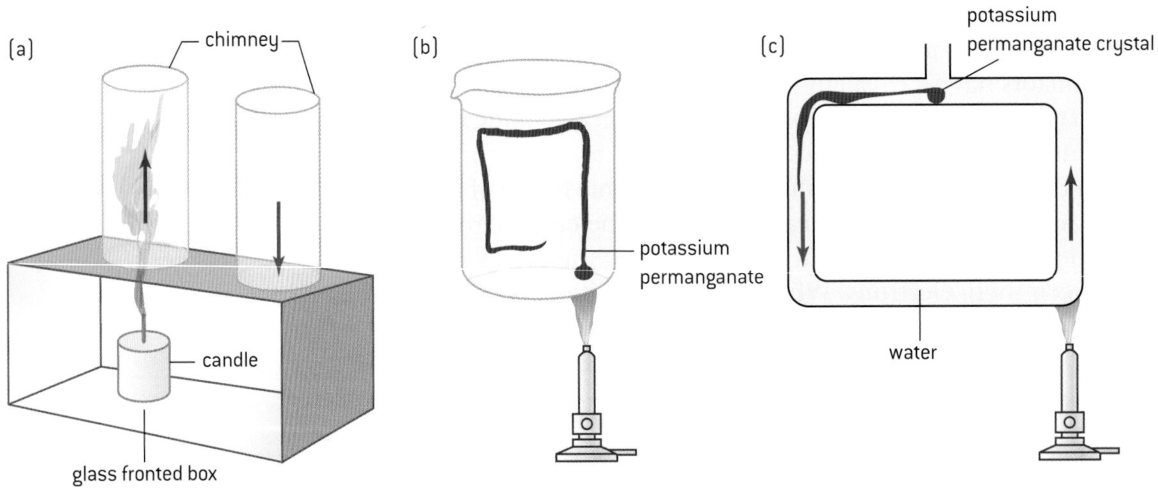

▲ Figure 3 Convection currents.

Figure 3 shows three lab experiments that involve convection. In all three cases, energy is supplied to a fluid. Look at the glass-fronted box (a) first. A candle heats the air underneath a tube (a chimney) that leads out of the box. The air molecules immediately above the flame move further apart decreasing the air density in this region. With a smaller density these molecules experience an upthrust and move up through the chimney.

This movement of air reduces the pressure slightly which pulls cooler air down the other chimney. Further heating of the air above the flame leads to a continuous current of cold air down the right-hand chimney and hot air up the left-hand tube. This is a **convection current**.

Similar currents can be demonstrated in liquids. Figure 3(b) shows a small crystal of a soluble dye (potassium permanganate) placed at the bottom of a beaker of water. When the base of the beaker is heated gently near to the crystal, water at the base heats, expands, becomes less dense, and rises. This also leads to a convection current as in figure 3(c) where a glass tube, in the shape of a rectangle, again with a small soluble coloured crystal, can sustain a convection current that moves all around the tube.

This is the mechanism by which all the water heated in a saucepan on a stove eventually reaches a uniform temperature.

Examples of convection

There are many examples of convection in action. Figures 4 and 5 show examples from the natural world; there are many others.

Sea breezes

If you live by an ocean you will have noticed that the direction of the breeze changes during a 24-hour period. During the day, breezes blow on-shore from the ocean, at night the direction is reversed and the breeze blows off-shore from the land to the sea.

Convection effects explain this. During the day the land is warmer than the sea and warm air rises over the land mass, pulling in cooler air from above the ocean. At night the land cools down much more quickly than the sea (which has a temperature that varies much less) and now the warmer air rises from the sea so the wind blows off-shore. (You might like to use your knowledge of specific heat capacity to explain why the sea temperature varies much less than that of the land.)

A similar effect occurs in the front range of the Rocky Mountains in the USA. The east-facing hills warm up first and the high-pressure region on the plains means that the wind blows towards the mountains. Later in the day, the east-facing slopes cool down first and the effect is reversed.

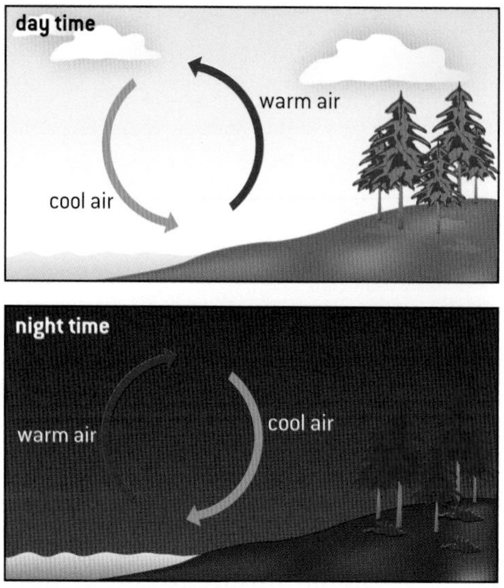

▲ Figure 4 Sea breezes.

Convection in the Earth

At the bottom of the Atlantic Ocean, and elsewhere on the planet, new crust is being created. This is due to convection effects that are occurring below the surface. The Earth's core is at a high temperature and this drives convection effects in the part of the planet known as the upper mantle. Two convection currents operate and drive material in the same direction. Material is upwelling at the top of these currents to reach the surface of the Earth at the bottom of the ocean. This creates new land that is forcing the Americas, Europe, and Africa apart at the rate of a few centimetres every year. In other parts of the world convection currents are pulling material back down below the surface (subduction). These convection currents have, over time, given rise to the continental drift that has shaped the continents that we know today.

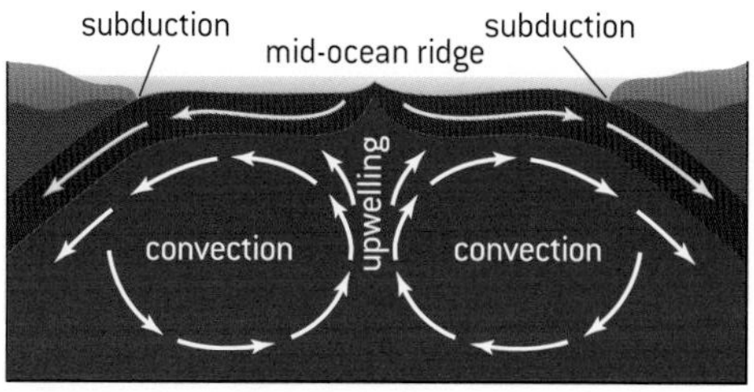

▲ Figure 5 Convection currents in the Earth's mantle.

Why the winds blow

The complete theory of why and how the winds blow would occupy a large part of this book, but essentially the winds are driven by uneven heating of the Earth's surface by the Sun. This differential heating can be due to many causes including geographical factors and the presence of cloud. However, where the land or the sea heat up, the air just above them rises and creates an area of low pressure. Conversely, where the air is falling a high-pressure zone is set up. The air moves from the high- to the low-pressure area and this is what we call a wind. There is a further interaction of the wind velocity with the rotation of the Earth (through an effect known as the Coriolis force). This leads to rotation of the air masses such that air circulates clockwise around a high-pressure region in the northern hemisphere but anti-clockwise around a high-pressure area in the southern hemisphere.

Nature of science

Modelling convection – using a scientific analogy

Faced with a hot cup of morning coffee and little time to drink it, most of us blow across the liquid surface to cool it more quickly. This causes a more rapid loss of energy than doing nothing and therefore the temperature of the liquid drops more quickly too. This is an example of **forced convection** – when the convection cooling is aided by a draught of air. Doing nothing and allowing the convection currents to set up by themselves is **natural convection**.

Newton stated an empirical law for cooling under conditions of forced convection. He suggested that the rate of change of the temperature of the cooling body $\frac{d\theta}{dt}$ was proportional to the temperature difference between the temperature of the cooling body θ and the temperature of the surroundings θ_s.

In symbols,

$$\frac{d\theta}{dt} \propto (\theta - \theta_s)$$

The key to understanding this equation is to realize that it is about the temperature *difference* between the hot object and its surroundings; we call this the temperature excess. Newton's law of cooling leads to a half-life behaviour in just the same way that radioactive half-life follows from the radioactive equation

$$\frac{dN}{dt} \propto N$$

where N is the number of radioactive atoms in a sample.

Using radioactivity as an analogy, there is a cooling half-life so that the time for the temperature excess over the surroundings to halve is always the same for a particular situation of hot object and surroundings.

This is another analogy that helps us to understand science by linking apparently different phenomena.

Worked examples

1 Explain the role played by convection in the flight of a hot-air balloon.

Solution

The air in the gas canopy is heated from below and as a result its temperature increases. The hot air in the balloon expands and its density decreases below that of the cold air outside the gas envelope. There is therefore an upward force on the balloon. If this exceeds the weight of the balloon (plus basket and occupants) then the balloon will accelerate upwards.

2 Suggest two reasons why covering the liquid surface of a cup of hot chocolate with marshmallows will slow down the loss of energy from the chocolate.

Solution

The marshmallows, having air trapped in them, are poor conductors so they allow only a small flow of energy through them. The upper surface of marshmallow will be at a lower temperature than the lower surface. This reduces the amount of convection occurring at the surface as the convection currents that are set up will not be so strongly driven as the differential densities will not be so great.

Thermal radiation

Basics

Thermal radiation is the transfer of energy by means of electromagnetic radiation. Electromagnetic radiation is unique as a wave in that it does not need a medium in order to move (propagate). We receive energy from the Sun even though it has passed through about 150 million km of vacuum in order to reach us. Radiation therefore differs from conduction and convection, both of which require a bulk material to carry the energy from place to place.

Thermal radiation has its origins in the thermal motion of particles of matter. All atoms and molecules at a temperature greater than absolute zero are in motion. Atoms contain charged particles and when these charges are accelerated they emit photons. It is these photons that are the thermal radiation.

Investigate!

Black and white surfaces

- Take two identical tin cans and cut out a lid for each one from polystyrene. Have a hole in each lid for a thermometer. Paint one can completely with matt black paint, paint the other shiny white.

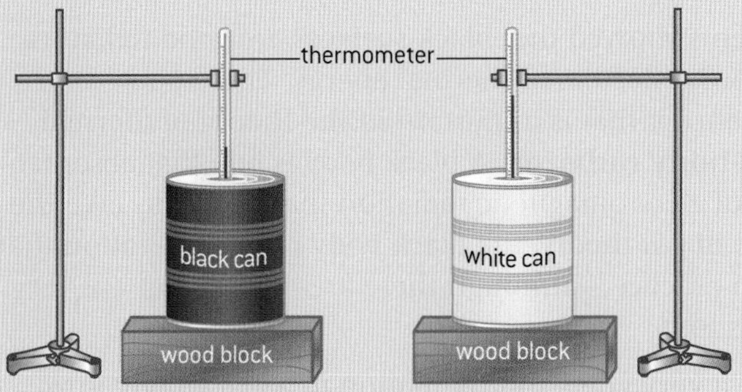

▲ Figure 6 Comparing emission of radiation from two surfaces.

- Fill both cans with the same volume of hot water at the same temperature, replace the lids and place the thermometers in the water.

  Keep the cans apart so that radiation from one is not incident on the other.
- Collect data to enable you to plot a graph to show how the temperature of the water in each can varies with time. This is called a cooling curve.
- You could also consider doing the experiment in reverse, beginning with cold water and using a radiant heater to provide energy for the cans. In this case, you must make sure that the heater is the same distance from the surface of each can and that the shiny can is unable to reflect radiation to the black one.

Experiments such as the one in *Investigate!* suggest that black surfaces are very good at radiating and absorbing energy. The opposite is true for white or shiny surfaces; they reflect energy rather than absorb it and are poor at radiating energy. This is why dispensers of hot drinks are often shiny – it helps them to retain the energy.

Nature of science

Radiator or not?

In many parts of the world, houses need to be heated during part or all of the year. One way to achieve this is to circulate hot water from a boiler through a thin hollow panel often known as a "radiator". But is this the appropriate term?

The outside metal surface of the panel becomes hot because energy is conducted from the hot water through the metal.

The air near the surface of the panel becomes hotter and less dense; it rises, setting up a convection current in the room.

There is some thermal radiation from the surface but as its temperature is not very different from that of the room, the net radiation is quite low – certainly lower than the contributions from convection.

Should the radiator be called a radiator?

Making a saucepan

We all need pans to cook our food. What is the best strategy for designing a saucepan?

The pan will be placed on a flat hot surface heated either by flame or radiant energy from an electrically heated filament or plate. The energy conducts through the base and heats the contents of the saucepan. The base of the pan needs to be a good conductor to allow a large energy flux into the pan. The walls of the saucepan need to withstand the maximum temperature at which the pan will be used but should not lose energy if possible. Giving them a shiny silver finish helps this.

The handle of the pan needs to be a poor conductor so that the pan can be lifted easily and harmlessly. Don't make it solid, make it strong and easy to hold but as thin as possible (think how electrical resistance varies with the shape and thickness of a conductor).

Conclusion: a good pan will have a thick copper base (a good conductor), sides, and handle of stainless steel (a poor conductor) and the overall finish will be polished and silvery.

Black-body radiation

The simple experiments showed that black surfaces are good radiators and absorbers but poor reflectors of thermal energy. These poor reflectors lead to a concept that is important in the theory of thermal radiation: the **black-body radiator**. A black body is one that absorbs all the wavelengths of electromagnetic radiation that fall on it. Like the ideal gas that we use in gas theory, the black body is an idealization that cannot be realized in practice – although there are objects that are very close approximations to it.

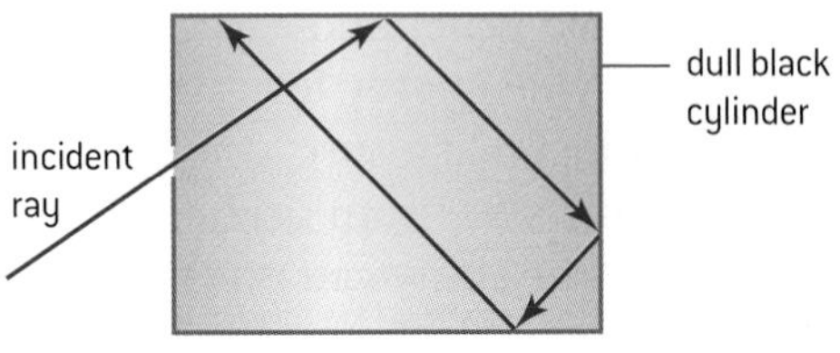

▲ Figure 7 A black-body enclosure.

One way to produce a very good approximation to a black body is to make a small hole in the wall of an enclosed container (a cavity) and to paint the interior of the container matt black. The container viewed through the hole will look very black inside.

Some of the first experiments into the physics of the black body were made by Lummer and Pringsheim in 1899 using a porcelain enclosure made from fired clay. When such enclosures are heated to high temperatures, radiation emerges from the cavity. The radiation appears coloured depending on the temperature of the enclosure. At low temperatures the radiation is in the infra-red region, but as the temperature rises, the colour emitted is first red, then yellow, eventually becoming white if the temperature is high enough. The intensity of the radiation coming from the hole or cavity is higher when the cavity is at a higher temperature. The emission from the hole is not dependent on the material from which the cavity is made unlike the emission from the surface of a container.

This can be seen in the picture of the interior of a steel furnace (see figure 8). In the centre of the furnace at its very hottest point, the colour appears white, at the edges the colour is yellow. At the entrance to the furnace where the temperature is very much lower, the colour is a dull red.

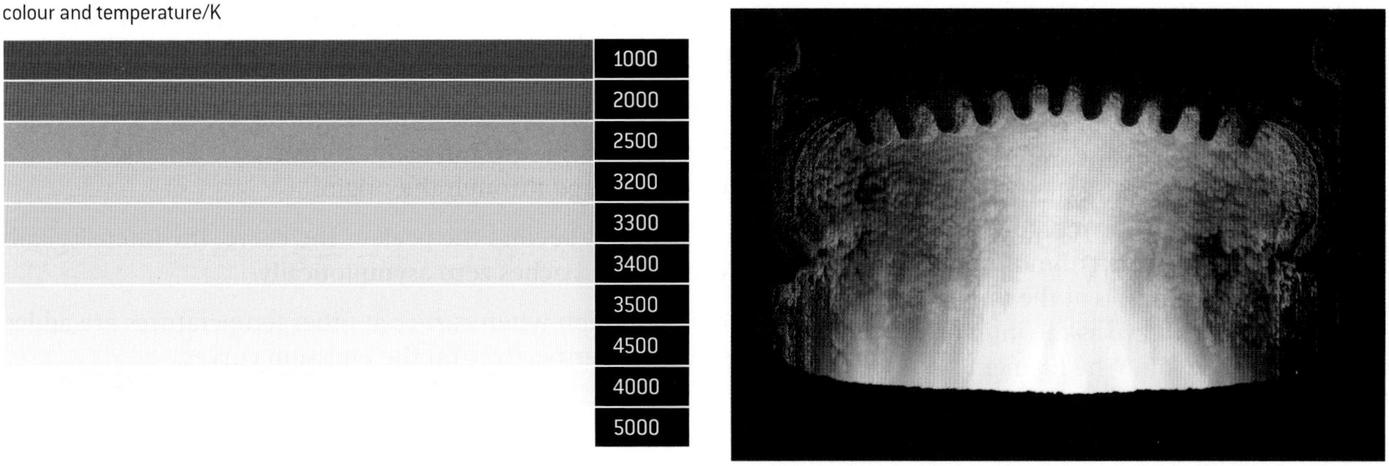

▲ Figure 8 Interior of furnace.

The emission spectrum from a black body

Although there is a predominant colour to the radiation emitted from a black-body radiator, this does not mean that only one wavelength emerges. To study the whole of the radiation that the black body emits, an instrument called a spectrometer is used. It measures the intensity of the radiation at a particular wavelength. Intensity is the power emitted per square metre.

As an equation this is written:

$$I = \frac{P}{A}$$

where I is the intensity, P is the power emitted, and A is the area on which the power is incident. The units of intensity are W m^{-2} or $\text{J s}^{-1}\text{ m}^{-2}$.

A typical intensity–wavelength graph is shown in figure 9 for a black body at the temperature of the visible surface of the Sun, about 6000 K. The Sun can be considered as a near-perfect black-body radiator. The

Nature of science

The potter's kiln

A potter needs to know the temperature of the inside of a kiln while the clay is being "fired" to transform it into porcelain. Some potters simply look into the kiln through a small hole. They can tell by experience what the temperature is by seeing the radiating colour of the pots inside. Other potters use an instrument called a pyrometer. A tungsten filament is placed at the entrance to the kiln between the kiln interior and the potter's eye. An electric current is supplied to the filament and this is increased until the filament disappears by merging into the background. At this point it is at the same temperature as the interior of the kiln. The filament system will have previously been calibrated so that the current required for the filament to disappear can be equated to the filament temperature.

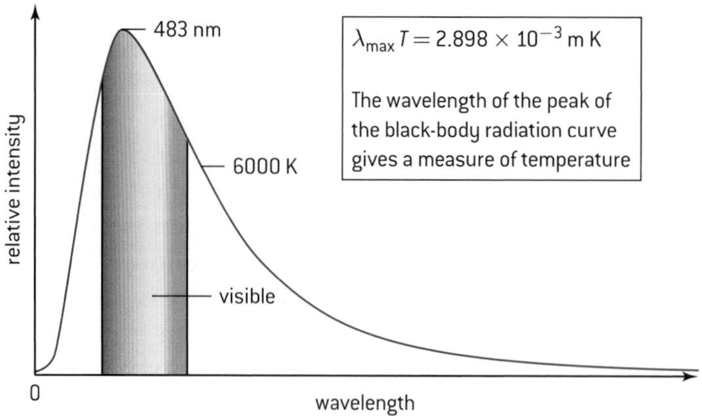

▲ Figure 9 Intensity against wavelength for a black body at the temperature of the Sun.

graph shows how the relative intensity of the radiation varies with the wavelength of the radiation at which the intensity is measured.

There are a number of important features to this graph:

- There is a peak value at about 500 nm (somewhere between green and blue light to our eyes). (Is it a coincidence that the human eye has a maximum sensitivity in this region?)
- There are significant radiations at all visible wavelengths.
- There is a steep rise from zero intensity–notice that the line does not go through the origin.
- At large wavelengths, beyond the peak of the curve, the intensity falls to low levels and approaches zero asymptotically.

Figure 10 shows the graph when curves at other temperatures are added and this gives further perspectives on the emission curves.

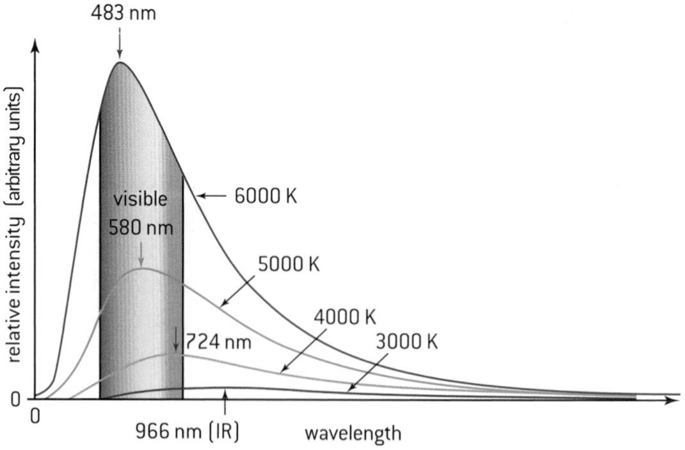

▲ Figure 10 Intensity against wavelength for black bodies at four temperatures.

This diagram shows four curves all at different temperatures. As before the units are *arbitrary*, meaning that the graph shows relative and not absolute changes between the curves at the four temperatures.

This family of curves tells us that, as temperature increases:

- the overall intensity at each wavelength increases (because the curve is higher)
- the total power emitted per square metre increases (because the total area under the curve is greater)
- the curves skew towards shorter wavelengths (higher frequencies)
- the peak of the curve moves to shorter wavelengths.

The next step is to focus on the exact changes between these curves.

Wien's displacement law

In 1893 Wilhelm Wien was able to deduce the way in which the shape of the graph depends on temperature. He showed that the height of the curve and the overall width depends on temperature alone. His full law allows predictions about the height of any point on the curve but we will only use it to predict the peak of the intensity curve.

Wien's displacement law states that the wavelength at which the intensity is a maximum λ_{max} in metres is related to the absolute temperature of the black body T by

$$\lambda_{max} = \frac{b}{T}$$

where b is known as Wien's displacement constant. It has the value 2.9×10^{-3} m K.

Tip

Notice that the unit for *b* is metre kelvin and must be written with a space between the symbols, take care not to write it as mK which means millikelvin and is something quite different!

Stefan–Boltzmann law

The scientists Stefan and Boltzmann independently derived an equation that predicts the total power radiated from a black body at a particular temperature. The law applies across all the wavelengths that are radiated by the body. Stefan derived the law empirically in 1879 and Boltzmann produced the same law theoretically five years later.

The **Stefan–Boltzmann law** states that the total power P radiated by a black body is given by

$$P = \sigma A T^4$$

where A is the total surface area of the black body and T is its absolute temperature. The constant, σ, is known as the Stefan–Boltzmann constant and has the value 5.7×10^{-8} W m^{-2} K^{-4}.

The law refers to the power radiated by the object, but this is the same as the energy radiated per second. It is easy to show that the energy radiated each second by one square metre of a black body is σT^4. This variant of the full law is known as Stefan's law.

Grey bodies and emissivity

In practice objects can be very close to a black body in behaviour but not quite 100% perfect in the way they behave. They are often called **grey** objects to account for this. A grey object at a particular temperature will emit less energy per second than a perfect black body of the same dimensions at the same temperature. The quantity known as **emissivity**, e, is the measure of the ratio between these two powers:

$$e = \frac{\text{power emitted by a radiating object}}{\text{power emitted by a black body of the same dimensions at the same temperature}}$$

Being a ratio, emissivity has no units.

For a real material, the power emitted can be written as

$$P = e\sigma AT^4$$

using the same symbols as before. A perfect black body has an emissivity value of 1. An object that completely reflects radiation without any absorption at all has an emissivity of 0. All real objects have an emissivity somewhere these two values. Typical values of emissivity for some substances are shown in the table. Emissivity values are a function of the wavelength of the radiation. It is surprising that snow and ice, although apparently white and reflective, are such effective emitters (and absorbers) in the infra-red.

Material	Emissivity
water	0.6–0.7
snow	0.9
ice	0.98
soil	0.4–0.95
coal	0.95

Nature of science

Building a theory

By the end of the 19th century, the graph of radiation intensity emitted by a black body as a function of wavelength was well known. Wien's equation fitted the experiments but only at short wavelengths. Rayleigh attempted to develop a new theory on the basis of classical physics. He suggested that charges oscillating inside the cavity produce standing electromagnetic waves as they bounce backwards and forwards between the cavity walls. Standing waves that escape from the cavity produce the observed black-body spectrum. Rayleigh's model fits the observations at long wavelengths but predicts an "ultraviolet catastrophe" of an infinitely large intensity at short wavelengths. Max Planck varied Rayleigh's theory slightly. He proposed that the standing waves could not carry all *possible* energies but only certain quantities of energy E given by nhf where n is an integer, h is a constant (Planck's constant) and f is the frequency of the allowed energy. Planck's model fitted the experimental results at all wavelengths and thus, in 1900, a new branch of physics was born: *quantum physics*. Planck limited his theory to the space inside the cavity, he believed that the radiation was continuous outside.

Later, Einstein proposed that the photons outside the cavity also had discrete amounts of energy. Planck was the scientific referee for Einstein's paper and it is to Planck's credit that he recognised the value of Einstein's work and accepted the paper for publication even though it overturned some of this own ideas.

Worked examples

1 The Sun has a surface temperature of 5800 K and a radius of 7.0×10^8 m. Calculate the total energy radiated from the Sun in one hour.

Solution

$P = \sigma AT^4$

Surface area of Sun $= 4\pi r^2 = 4 \times 3.14 \times (7 \times 10^8)^2$

$= 6.2 \times 10^{18}$

So power $= 5.7 \times 10^{-8} \times 6.2 \times 10^{18} \times 5800^4$

$= 4.0 \times 10^{26}$ W

In one hour there are 3600 s, so the energy radiated in one hour is 1.4×10^{30} J.

2 A metal filament used as a pyrometer in a kiln has a length of 0.050 m and a radius of 1.2×10^{-3} m. Determine the temperature of the filament at which it radiates a power of 48 W.

Solution

The surface area of the filament is $2\pi rh = (2\pi \times 1.2 \times 10^{-3}) \times 0.050 = 3.8 \times 10^{-4}$ m^2

So the power determines the temperature as

$$48 = 5.7 \times 10^{-8} \times 3.8 \times 10^{-4} \times T^4$$

$$T = \sqrt[4]{\frac{48}{5.7 \times 10^{-8} \times 3.8 \times 10^{-4}}} = 1200 \text{ K.}$$

3 A spherical black body has an absolute temperature T_1 and surface area A. Its surroundings are kept at a lower temperature T_2.

Determine the net power lost by the body.

Solution

The power emitted by the body is; $\sigma A T_1^{\,4}$

the power absorbed from the surroundings is $\sigma A T_2^{\,4}$.

So the net power lost is $\sigma A(T_1^{\,4} - T_2^{\,4})$.

Note that this is not the same as

$\sigma A(T_1 - T_2)^4$.

Sun and the solar constant

The Sun emits very large amounts of energy as a result of its nuclear fusion reactions. Because the Earth is small and a long way from the Sun, only a small fraction of this arrives at the top of the Earth's atmosphere. A black body at the temperature of the Sun has just under half of its radiation in our visible region, roughly the same amount in the infra-red, and 10% in the ultraviolet. It is the overall difference between this incoming radiation and the radiation that is subsequently emitted from the Earth that determines the energy gained by the Earth from the Sun. This energy is used by plants in photosynthesis and it drives the changes in the world's oceans and atmospheres; it is crucial to life on this planet.

The amount of energy that arrives at the top of the atmosphere is known as the **solar constant**. A precise definition is that *the solar constant is the amount of solar radiation across all wavelengths that is incident in one second on one square metre at the mean distance of the Earth from the Sun on a plane perpendicular to the line joining the centre of the Sun and the centre of the Earth.*

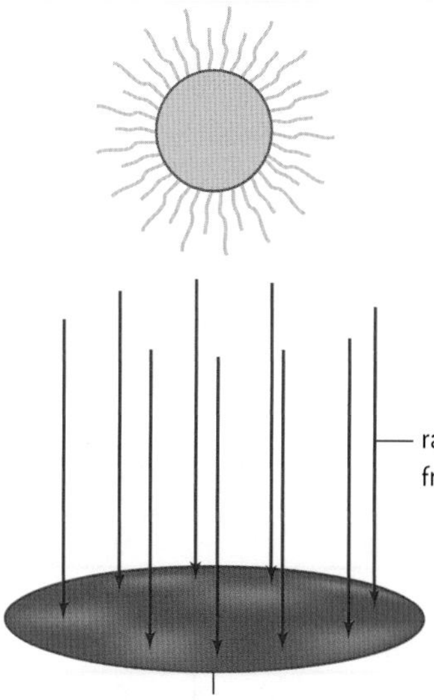

▲ Figure 11 Defining the solar constant.

The energy from the Sun is spread over an imaginary sphere that has a radius equal to the Earth–Sun distance. The Earth is roughly 1.5×10^{11} m from the Sun and so the surface area of this sphere is 2.8×10^{23} m^2.

The Sun emits 4×10^{26} J in one second. The energy incident in one second on one square metre at the distance of the Earth from the Sun is $\frac{4.0 \times 10^{26}}{2.8 \times 10^{23}} = 1400$ J. The answer is quoted to 2 s.f., a reasonable precision for this estimate and represents about 5×10^{-10} of the entire output of the Sun.

The value of the solar constant varies periodically for a number of reasons:

- The output of the Sun varies by about 0.1% during its principal 11-year sunspot cycle.
- The Earth's orbit is elliptical with the Earth slightly closer to the Sun in January compared to July; this accounts for a difference of about

7% in the solar constant. (Note that this difference is *not* the reason for summer in the Southern Hemisphere – the seasons occur because the axis of rotation of the Earth is not perpendicular to the plane of its orbit around the Sun.)

- Other longer-period cycles are believed to occur in the Sun with periods ranging from roughly hundreds to thousands of years.

Energy balance in the Earth surface–atmosphere system

The solar constant is the power incident on the top of the atmosphere. It is not the power that arrives at ground level. As the radiation from the Sun enters and travels through the atmosphere, it is subject to losses that reduce the energy arriving at the Earth's surface. Radiation is absorbed and also scattered by the atmosphere. The degree to which this absorption and scattering occurs depends on the position of the Sun in the sky at a particular place. When the Sun is lower in the sky (as at dawn and sunset), its radiation has to pass through a greater thickness of atmosphere and thus more scattering and absorption takes place. This gives rise to the colours in the sky at dawn and dusk.

Even when the energy arrives at ground level, it is not necessarily going to remain there. The surface of the Earth is not a black body and therefore it will reflect some of the energy back up towards the atmosphere. The extent to which a particular surface can reflect energy is known as its **albedo** (from the Latin word for "whiteness"). It is given the symbol a:

$$a = \frac{\text{energy reflected by a given surface in a given time}}{\text{total energy incident on the surface in the same time}}$$

Like emissivity, albedo has no units, it varies from 0 for a surface that does not reflect any energy (a black body) to 1 for a surface that absorbs no radiation at all. Unless stated otherwise, the albedo in the Earth system is normally quoted for visible light (which as we saw earlier accounts for nearly a half of all radiation at the surface).

The average annual albedo for the whole Earth is about 0.35, meaning that on average about 35% of the Sun's rays that reach the ground are reflected back into the atmosphere.

This figure of 0.35 is, however, very much an average because albedo varies depending on a number of factors:

- It varies daily and with the seasons, depending on the amount and type of cloud cover (thin clouds have albedo values of 0.3–0.4, thick cumulo-nimbus cloud can approach values of 0.9).
- It depends on the terrain and the material of the surface. The table gives typical albedo values for some common land and water surfaces.

Surface	Albedo
Ocean	0.06
Fresh snow	0.85
Sea ice	0.60
Ice	0.90
Urban areas	0.15
Desert soils	0.40
Pine forest	0.15
Deciduous forest	0.25

The importance of albedo will be familiar to anyone who lives where snow is common in winter. Fresh snow has a high albedo and reflects most of the radiation that is incident on it – the snow will stay in place for a long time without melting if the temperature remains cold. However, sprinkle some earth or soot on the snow and, when the sun shines, the snow will soon disappear because the dark material on its surface absorbs energy. The radiation provides the latent heat needed to melt the snow. Albedo effects help to explain why satellites (including the International Space Station) in orbit around the Earth can take pictures of the Earth's cloud cover and surface in the visible spectrum.

Worked examples

1 Four habitats on the Earth are: forest, grassland (savannah), the sea, an ice cap.

Discuss which of these have the greatest and least albedo.

Solution

A material with a high albedo reflects the incident visible radiation. Ice is a good reflector and consequently has a high albedo. On the other hand, the sea is a good absorber and has a low albedo.

2 The data give details of a model of the energy balance of the Earth. Use the data to calculate the albedo of the Earth that is predicted by this model.

Data

Incident intensity from the Sun	$= 340\ \text{W m}^{-2}$
Reflected intensity at surface	$= 100\ \text{W m}^{-2}$
Radiated intensity from surface	$= 240\ \text{W m}^{-2}$
Re-radiated intensity from atmosphere back to surface	$= 2\ \text{W m}^{-2}$

Solution

The definition of albedo is clear.

It is $\frac{\text{power reflected by a given surface}}{\text{total power incident on the surface}}$

So in this case the value is $\frac{100}{340} = 0.29$

The greenhouse effect and temperature balance

The Earth and the Moon are the same average distance from the Sun, yet the average temperature of the Moon is 255 K, while that of the Earth is about 290 K. The discrepancy is due to the Earth having an atmosphere while the Moon has none.

The difference is due to a phenomenon known as the **greenhouse effect** in which certain gases in the Earth's atmosphere trap energy within the Earth system and produce a consequent rise in the average temperature of the Earth. The most important gases that cause the effect include carbon dioxide (CO_2), water vapour (H_2O), methane (CH_4), and nitrous oxide (dinitrogen monoxide; N_2O), all of which occur naturally in the atmosphere. Ozone (O_3), which has natural and man-made sources, makes a contribution to the greenhouse effect.

It is important to distinguish between:

- the "natural" greenhouse effect that is due to the naturally occurring levels of the gases responsible, and
- the enhanced greenhouse effect in which increased concentrations of the gases, possibly occurring as a result of human-derived processes, lead to further increases in the Earth's average temperature and therefore to climate change.

The principal gases in the atmosphere are nitrogen, N_2, and oxygen, O_2, (respectively, 70% and 20% by weight). Both of these gases are made up of tightly bound molecules and, because of this, do not absorb energy from sunlight. They make little contribution to the natural greenhouse effect. The 1% of the atmosphere that is made up of the CO_2, H_2O, CH_4 and N_2O has a much greater effect.

The molecular structure of greenhouse-gas molecules means that they absorb ultraviolet and infra-red radiation from the Sun as it travels through the atmosphere. Visible light on the other hand is not so readily absorbed by these gases and passes through the atmosphere to be absorbed by the land and water at the surface. As a result the temperature of the surface rises. The Earth then re-radiates just like any other hot object. The temperature of the Earth's surface is far lower than that of the Sun, so the wavelengths radiated from the Earth will peak in the long-wavelength infra-red. The absorbed radiation had, of course, mostly been in the visible region of the electromagnetic spectrum. So, just as gases in the atmosphere absorbed the Sun's infra-red on the way in, now they absorb energy in the infra-red from the Earth on the way out. The atmosphere then re-radiates the energy yet again, this time in all directions meaning that some returns to Earth. This energy has been trapped in the system that consists of the surface of the Earth and the atmosphere.

The whole system is a dynamic equilibrium reaching a state where the total energy incident on the system from the Sun equals the energy total being radiated away by the Earth. In order to reach this state, the temperature of the Earth has to rise and, as it does so, the amount of energy it radiates must also rise by the Stefan–Boltzmann law. Eventually, the Earth's temperature will be such that the balance of incoming and outgoing energies is attained. Of course, this balance was established over billions of years and was steadily changing as the composition of the atmosphere and the albedo changed with changes in vegetation, continental drift, and geological processes.

Nature of science

Other worlds, other atmospheres

The dynamic equilibrium in our climate has been very important for the evolution of life on Earth. Venus and Mars evolved very differently from Earth.

Venus has similar dimensions to the Earth but is closer to the Sun with a very high albedo at about 0.76. It's atmosphere is almost entirely carbon dioxide. In consequence, the surface temperature reaches a 730 K and a runaway greenhouse effect acts. Venus and Mars are clear reminders to us of the fragility of a planetary climate.

Why greenhouse gases absorb energy

Ultraviolet and long-wavelength infra-red radiations are absorbed by the atmosphere.

Photons in the ultraviolet region of the electromagnetic spectrum are energetic and have enough energy to break the bonds within the gas molecules. This leads to the production of ionic materials in the atmosphere. A good example is the reaction that leads to the production of ozone from the oxygen atoms formed when oxygen molecules are split apart by ultraviolet photons.

The energies of infra-red photons are much smaller than those of ultraviolet and are not sufficient to break molecules apart. When the frequency of a photon matches a vibrational state in a greenhouse gas molecule then an effect called **resonance** occurs. We will look in detail at the vibrational states and resonance in carbon dioxide, but similar effects occur in all the greenhouse gas molecules.

In a carbon dioxide molecule, the oxygen atoms at each end are attached by double bonds to the carbon in a linear arrangement. The bonds resemble springs in their behaviour.

The molecule has four vibrational modes as shown in figure 12. The first of these modes – a linear symmetric stretching does not cause infrared absorption, but the remaining three motions do. Each one has

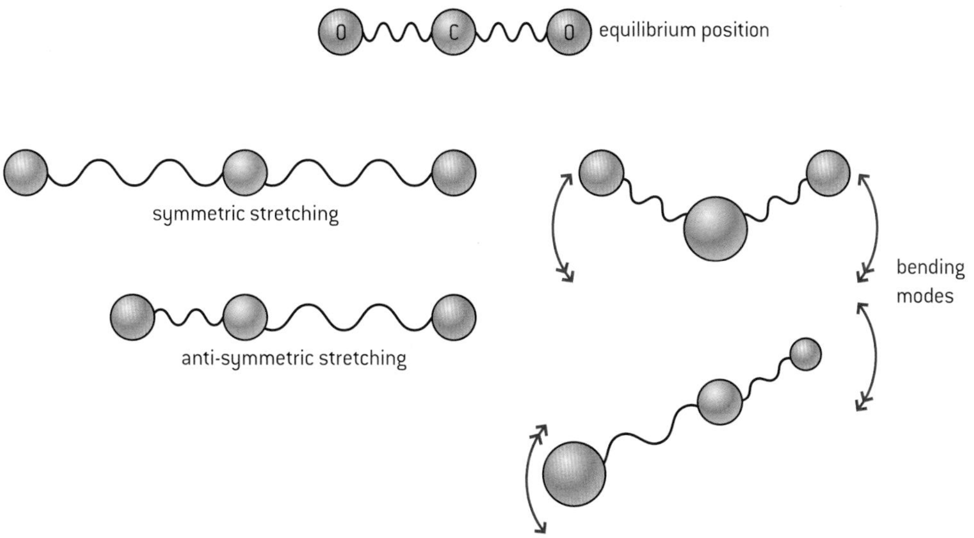

▲ Figure 12 Vibrational states in the carbon dioxide molecule.

a characteristic frequency. If the frequency of the radiation matches this, then the molecule will be stimulated into vibrating with the appropriate mode and the energy of the vibration will come from the incident radiation. This leads to vibrational absorption at wavelengths of 2.7 μm, 4.3 μm and 15 μm.

These effects of these absorptions can be clearly seen in figure 13 which shows part of the absorption spectrum of carbon dioxide. In this diagram a peak indicates a wavelength at which significant absorption occurs.

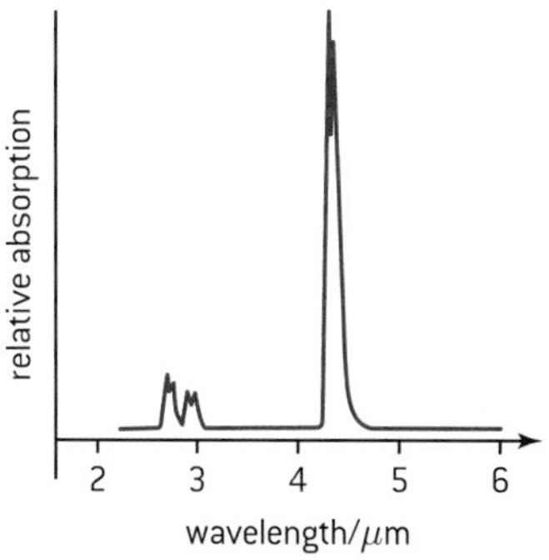

▲ Figure 13 Part of the absorption spectrum for carbon dioxide.

Modelling the climate balance

We said earlier that about 1400 J falls on each square metre of the upper atmosphere each second: the solar constant. We use the physics introduced in this topic to see what the consequences of this are for the Earth's surface–atmospheric system.

The full 1400 J does not of course reach the surface. Of the total, about 25% of the incident energy is reflected by the clouds and by particles in the atmosphere, about 25% is absorbed by the atmosphere, and about 6% is reflected at the surface.

The incoming radiation falls on the portion of the Earth's surface which is normal to the Sun's radiation – i.e. a circle of area equal to $\pi \times (\textit{radius of Earth})^2$, as only one side of the Earth faces the Sun at any one time. However this radiation has to be averaged over the whole of the surface which is $4\pi \times (\textit{radius of Earth})^2$. So the mean power arriving at each square metre is $\frac{1400}{4} = 350$ W.

The albedo now has to be taken into account to give an effective mean power at one square metre of the surface of

$$(1 - a) \times 350$$

For the average Earth value for a of about 0.3, the mean power absorbed by the surface per square metre is 245 W.

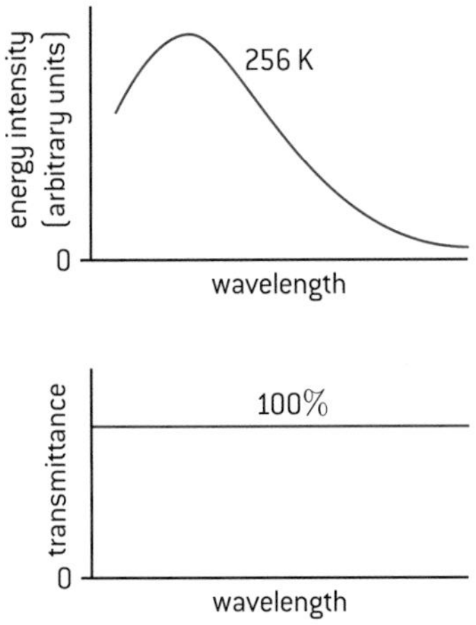

▲ Figure 14 Intensity and transmittance for a completely transparent atmosphere.

The knowledge of this emitted power allows a prediction of the temperature of a black body T that will emit 245 W m^{-2}. Using the Stefan–Boltzmann law:

$$245 = \sigma T^4$$

$$\text{So } T = \sqrt[4]{\frac{245}{5.67 \times 10^{-8}}} = 256\,\text{K}$$

This is very close to the value for the Moon, which has no atmosphere. We need to investigate why the mean temperature of the Earth is about 35 K higher than this.

We made the assumption that the Earth emits 245 W m^{-2} and that this energy leaves the surface and the atmosphere completely. This would be true for an atmosphere that is completely transparent at all wavelengths, but Earth's atmosphere is not transparent in this way.

Figure 14 shows the relative intensity–wavelength graph for a black body at 256 K. As expected, the area under this curve is 245 W m^{-2} and represents the predicted emission from the Earth. It shows the response of an atmosphere modelled as perfectly transparent at all wavelengths. (Technically, this graph shows the transmittance of the atmosphere as a function of wavelength, a value of 100% means that the particular wavelength is completely transmitted, 0 means that no energy is transmitted at this wavelength.) Not surprisingly all the black-body radiation leaves the Earth because the transmittance is 1 for all wavelengths in this model.

In fact the atmosphere absorbs energy in the infra-red and ultraviolet regions. A simple, but slightly more realistic model for this absorption will leave the transmittance at 100% for the visible wavelengths and change the transmittance to 0 for the absorbed wavelengths. Figure 15 shows how this leads to an increased surface temperature.

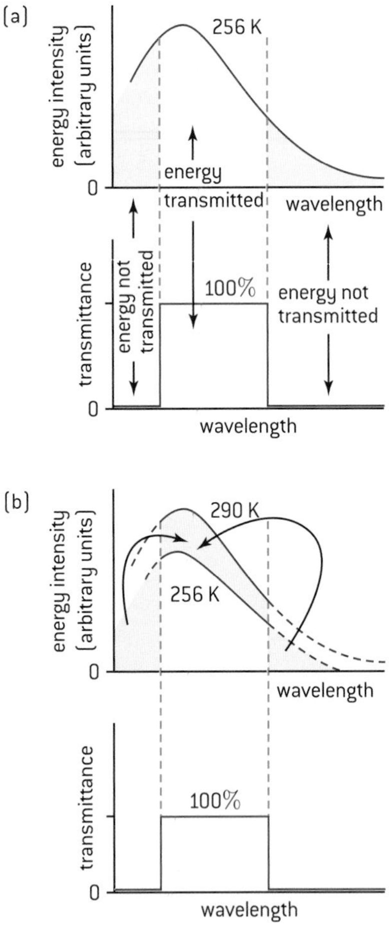

▲ Figure 15 Intensity and transmittance for an atmosphere opaque to infra-red and ultraviolet radiation.

When the transmittance graph is merged with the graph for black-body radiation to give the overall emission from the Earth into space, the area under the overall emission curve will be less than 245 W m^{-2} because the infra-red and ultraviolet wavelengths are now absorbed in the atmosphere and these energies are not lost (Figure 15(a)). This deficit will be re-radiated in all directions; so some returns to the surface.

In order to get the energy balance correct again, the temperature of the emission curve must be raised so that the area under the curve returns to 245 W m^{-2}. As the curve changes with the increase in temperature, the area under the curve increases too. The calculation of the temperature change required is difficult and not given here. However, for the emission from the surface to equal the incoming energy from the Sun, allowing for the absorption, the surface temperature must rise to about 290 K. The net effect is shown in Figure 15(b) with a shifted and raised emission curve compensating for the energy that cannot be transmitted through the atmosphere.

The suggestion that the atmosphere completely removes wavelengths above and below certain wavelengths is an over-simplification.

Figure 16 shows the complicated transmittance pattern in the infrared and indicates which absorbing molecules are responsible for which regions of absorption.

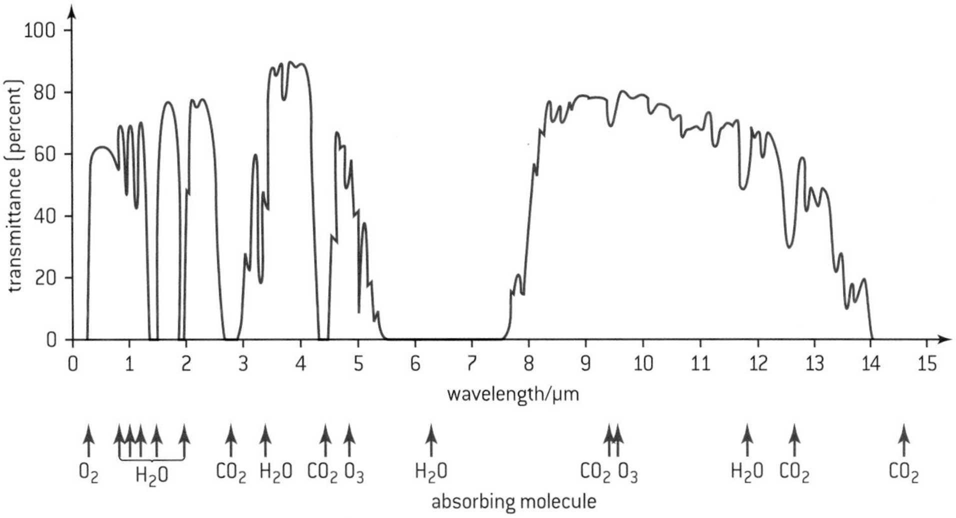

▲ Figure 16 Transmittance of the atmosphere in the infra-red.

The energy balance of the Earth

The surface–atmosphere energy balance system is very complex; figure 17 is a recent diagram showing the basic interactions and you should study it carefully.

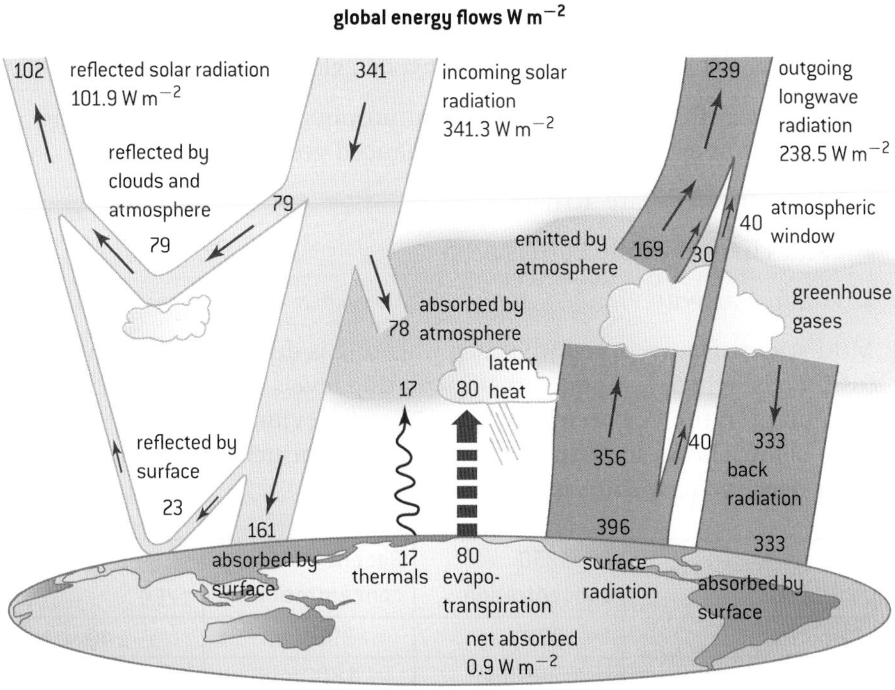

▲ Figure 17 Factors that make up the energy balance of the Earth (after Stephens and others. 2012. *An update on earth's energy balance in light of the latest global observations*. Nature Geoscience.).

Global warming

There is little doubt that climate change is occurring on the planet. We are seeing a significant warming that may lead to many changes to the sea level and in the climate in many parts of the world. The fact that there is change should not surprise us. We have recently (in geological terms) been through several Ice Ages and we are thought to be in an interstadial phase at the moment (interstadial means "between Ice Ages"). In the seventeenth century a "Little Ice Age" covered much of northern Europe and North America. The River Thames regularly froze and the citizens held fairs on the ice. In 1608, the Dutch painter Hendrick Avercamp painted a winter landscape showing the typical extent and thickness of the ice in Holland (figure 18).

▲ Figure 18 Winter landscape with skaters (1608), Hendrick Avercamp.

Many models have been suggested to explain global warming, they include:

- changes in the composition of the atmosphere (and specifically the greenhouse gases) leading to an enhanced greenhouse effect
- increased solar flare activity
- cyclic changes in the Earth's orbit
- volcanic activity.

Most scientists now accept that this warming is due to the burning of fossil fuels, which has gone on at increasing levels since the industrial revolution in the eighteenth century. There is evidence for this. The table below shows some of the changes in the principal greenhouse gases over the past 250 years.

Gas	Pre-1750 concentration	Recent concentration	% increase since 1750
Carbon dioxide	280 ppm	390 ppm	40
Methane	700 ppb	1800 ppb	160
Nitrous oxide	270 ppb	320 ppb	20
Ozone	25 ppb	34 ppb	40

ppm = parts per million; ppb = parts per billion

The recent values in this table have been collected directly in a number of parts of the world (there is a well-respected long-term study of the variation of carbon dioxide in Hawaii where recently the carbon dioxide levels exceeded 400 ppm for the first time for many thousands of years). The values quoted for pre-1750 are determined from a number of sources:

- Analysis of Antarctic ice cores. Cores are extracted from the ice in the Antarctic and these yield data for the composition of the atmosphere during the era when the snow originally fell on the continent. Cores can give data for times up to 400 000 years ago.
- Analysis of tree rings. Tree rings yield data for the temperature and length of the seasons and the rainfall going back sometimes hundreds of years.
- Analysis of water levels in sedimentary records from lake beds can be used to identify historical changes in water levels.

An enhanced greenhouse effect results from changes to the concentration of the greenhouse gases: as the amounts of these gases increase, more absorption occurs both when energy enters the system and also when the surface re-radiates. For example, in the transmittance–wavelength graph for a particular gas, when the concentration of the gas rises, the absorption peaks will increase too. The surface will need to increase its temperature in order to emit sufficient energy at sea level so that emission of energy by Earth from the top of the atmosphere will equal the incoming energy from the Sun.

Global warming is likely to lead to other mechanisms that will themselves make global warming increase at a greater rate:

- the ice and snow cover at the poles will melt, this will decrease the average albedo of Earth and increase the rate at which heat is absorbed by the surface.
- a higher water temperature in the oceans will reduce the extent to which CO_2 is dissolved in seawater leading to a further increase in atmospheric concentration of the gas.

Other human-related mechanisms such as deforestation can also drive global warming as the amount of carbon fixed in the plants is reduced.

This is a problem that has to be addressed at both an international and an individual level. The world needs greater efficiency in power production and a major review of fuel usage. We should encourage the use of non-fossil-fuel methods. As individuals we need to be aware of our personal impact on the planet, we should be conscious of our carbon footprint. Nations can capture and store carbon dioxide, and agree to increase the use of combined heating and power systems. What everyone agrees is that doing nothing is not an option.

Nature of science

An international perspective

There have been a number of international attempts to reach agreements over the ways forward for the planet. These have included:

- The Kyoto Protocol was originally adopted by many (but not all) countries in 1997 and later extended in 2012.
- The Intergovernmental Panel on Climate Change.
- Asia–Pacific Partnership on Clean Development and Climate.
- The various other United Nations Conventions on Climate Change, e.g. Cancùn, 2010.

Do some research on the Internet to find what is presently agreed between governments.

Questions

1 (IB)

a) A reactor produces 24 MW of power. The efficiency of the reactor is 32%. In the fission of one uranium-235 nucleus 3.2×10^{-11} J of energy is released.

Determine the mass of uranium-235 that undergoes fission in one year in this reactor.

b) During its normal operation, the following set of reactions takes place in the reactor.

$$^{1}_{0}n + ^{238}_{92}U \rightarrow ^{239}_{92}U$$
$$^{239}_{92}U \rightarrow ^{239}_{93}Np + ^{0}_{-1}e + \bar{\nu}$$
$$^{239}_{93}Np \rightarrow ^{239}_{94}Pu + ^{0}_{-1}e + \bar{\nu}$$

Comment on the international implications of the product of these reactions.

2 (IB)

The diagram shows a pumped storage power station used for the generation of electrical energy.

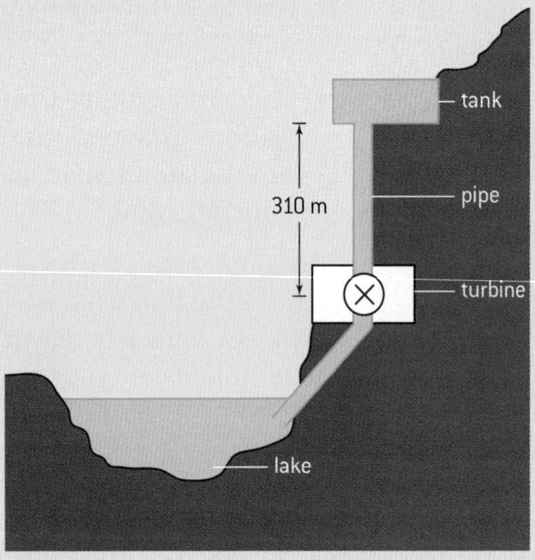

Water stored in the tank falls through a pipe to a lake through a turbine that is connected to an electricity generator.

The tank is 50 m deep and has a uniform area of 5.0×10^4 m^2. The height from the bottom of the tank to the turbine is 310 m.

The density of water is 1.0×10^3 kg m^{-3}.

a) Show that the maximum energy that can be delivered to the turbine by the falling water is about 8×10^{12} J.

b) The flow rate of water in the pipe is 400 m^3 s^{-1}. Calculate the power delivered by the falling water.

3 (IB)

The energy losses in a pumped storage power station are shown in the following table.

Source of energy loss	Percentage loss of energy
friction and turbulence of water in pipe	27
friction in turbine and ac generator	15
electrical heating losses	5

a) Calculate the overall efficiency of the conversion of the gravitational potential energy of water in the tank into electrical energy.

b) Sketch a Sankey diagram to represent the energy conversion in the power station.

4 (IB)

A nuclear power station uses uranium-235 (U-235) as fuel.

a) Outline:

(i) the processes and energy changes that occur through which the internal energy of the working fluid is increased

(ii) the role of the heat exchanger of the reactor and the turbine in the generation of electrical energy.

b) Identify **one** process in the power station where energy is degraded.

5 (IB)

The intensity of the Sun's radiation at the position of the Earth is approximately 1400 W m^{-2}.

Suggest why the average power received per unit area of the Earth is 350 W m^{-2}.

6 *(IB)*

The diagram shows a radiation entering or leaving the Earth's surface for a simplified model of the energy balance at the Earth's surface.

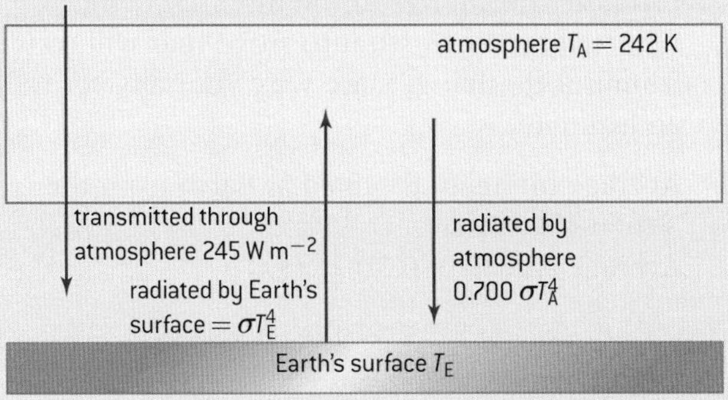

a) State the emissivity of the atmosphere.

b) Determine the intensity of the radiation radiated by the atmosphere towards the Earth's surface.

c) Calculate T_E.

7 a) Outline a mechanism by which part of the radiation radiated by the Earth's surface is absorbed by greenhouse gases in the atmosphere. Go on to suggest why the incoming solar radiation is not affected by the mechanism you outlined.

b) Carbon dioxide (CO_2) is a greenhouse gas. State **one** source and **one** sink (that removes CO_2) of this gas.

8 *(IB)*

The graph shows part of the absorption spectrum of nitrogen oxide (N_2O) in which the intensity of absorbed radiation A is plotted against frequency f.

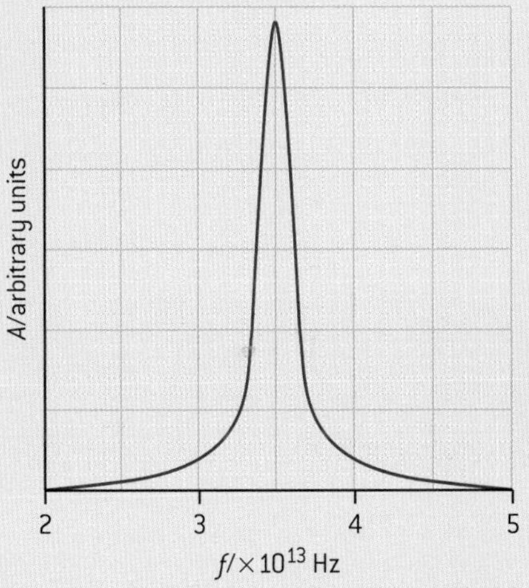

a) State the region of the electromagnetic spectrum to which the resonant frequency of nitrogen oxide belongs.

b) Using your answer to (a), explain why nitrogen oxide is classified as a greenhouse gas.

9 *(IB)*

The diagram shows a simple energy balance climate model in which the atmosphere and the surface of Earth are treated as two bodies each at constant temperature. The surface of the Earth receives both solar radiation and radiation emitted from the atmosphere. Assume that the Earth's surface and the atmosphere behave as black bodies.

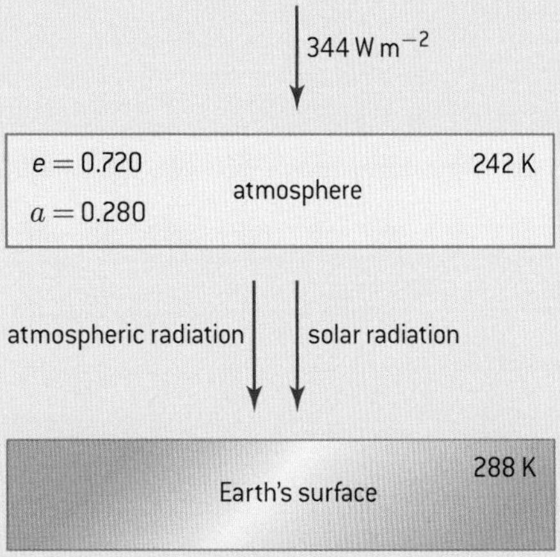

Data for this model:

average temperature of the atmosphere of Earth = 242 K

emissivity e of the atmosphere of Earth = 0.720

average albedo a of the atmosphere of Earth = 0.280

solar intensity at top of atmosphere = 344 W m^{-2}

average temperature of the surface of Earth = 288 K

a) Use the data to determine:

(i) the power radiated per unit area of the atmosphere

(ii) the solar power absorbed per unit area at the surface of the Earth.

b) It is suggested that, if the production of greenhouse gases were to stay at its present level, then the temperature of the Earth's atmosphere would eventually rise by 6 K.

Calculate the power per unit area that would then be

(i) radiated by the atmosphere

(ii) absorbed by the Earth's surface.

c) Estimate the increase in temperature of the Earth's surface.

10 *(IB)*

It has been estimated that doubling the amount of carbon dioxide in the Earth's atmosphere changes the albedo of the Earth by 0.01. Estimate the change in the intensity being reflected by the Earth into space that will result from this doubling. State why your answer is an estimate.

Average intensity received at Earth from the Sun = $340\ \mathrm{W\,m^{-2}}$
Average albedo = 0.30

9 WAVE PHENOMENA (AHL)

Introduction

In this topic we develop many of the concepts introduced in Topic 4. In general, a more mathematical approach is taken and we consider the usefulness of modelling using a spreadsheet – both for graphing and developing relationships through iteration.

9.1 Simple harmonic motion

Understanding

- The defining equation of SHM
- Energy changes

Nature of science

The importance of SHM

The equation for simple harmonic motion (SHM) can be solved analytically and numerically. Physicists use such solutions to help them to visualize the behaviour of the oscillator. The use of the equations is very powerful as any oscillation can be described in terms of a combination of harmonic oscillators using Fourier synthesis. The modelling of oscillators has applications in virtually all areas of physics including mechanics, electricity, waves and quantum physics. In this sub-topic we will model SHM using a simple spreadsheet and see how powerful this interpretation can be.

Applications and skills

- Solving problems involving acceleration, velocity and displacement during simple harmonic motion, both graphically and algebraically
- Describing the interchange of kinetic and potential energy during simple harmonic motion
- Solving problems involving energy transfer during simple harmonic motion, both graphically and algebraically

Equations

- angular velocity–period equation: $\omega = \frac{2\pi}{T}$
- defining equation for shm: $a = -\omega^2 x$
- displacement–time equations: $x = x_0 \sin \omega t$; $x = x_0 \cos \omega t$
- velocity–time equations: $v = \omega x_0 \cos \omega t$; $v = -\omega x_0 \sin \omega t$
- velocity-displacement equation: $v = \pm\omega \sqrt{(x_0^2 - x^2)}$
- kinetic energy equation: $E_K = \frac{1}{2} m\omega^2 (x_0^2 - x^2)$
- total energy equation: $E_T = \frac{1}{2} m\omega^2 x_0^2$
- period of simple pendulum: $T = 2\pi\sqrt{\frac{l}{g}}$
- period of mass–spring: $T = 2\pi\sqrt{\frac{m}{k}}$

Introduction

In this sub-topic we treat SHM more mathematically but restrict ourselves to two systems – the simple pendulum and of a mass oscillating on a spring. Each of these systems is isochronous and is usually lightly damped; this means that a large number of oscillations occur before the energy in the system is transferred to the internal energy of the system and the surrounding air.

Angular speed or frequency (ω)

In Sub-topic 6.1 we considered the angular speed ω in relation to circular motion; it is the rate of change of angle with time and is also called **angular frequency**. It is measured in radians per second (rad s^{-1}). This quantity is important when we deal with simple harmonic motion because there is a very close relationship between circular motion and SHM. This relationship can be demonstrated using the apparatus shown in figure 1.

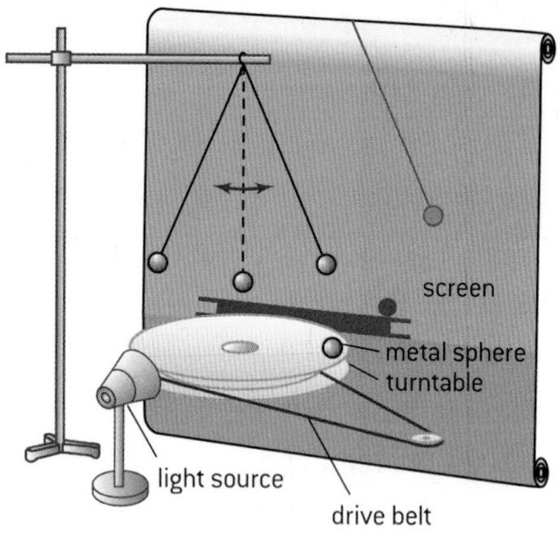

▲ Figure 1 Comparison of SHM and circular motion.

A metal sphere is mounted on a turntable that rotates at a constant angular speed. A simple pendulum is arranged so that it is in line with the sphere and oscillates with the same periodic time as that of the turntable – a little trial and error should give a good result here. The pendulum and the turntable are illuminated by a light that is projected onto a screen. The shadows projected, onto the screen, of the circular motion of the sphere and oscillatory motion of the pendulum show these motions to be identical.

Circular motion and SHM

In mathematical terms the demonstration in figure 1 is equivalent to projecting the two-dimensional motion of a point onto the single dimension of a line. Imagine a point P rotating around the perimeter of a circle with a constant angular speed ω. The radius of the circle r joins P with the centre of the circle O. At time $t = 0$ the radius is horizontal and at time t it has moved through an angle θ radians. For constant angular speed $\omega = \frac{\theta}{t}$, rearranging this gives $\theta = \omega t$. Projecting P onto the y-axis gives the vertical component of r as $r \sin \theta$. Projecting P onto the x-axis gives the horizontal component of r as $r \cos \theta$. The variation of the vertical component with time or angle is shown on the right of figure 2 and takes the form of a sine curve. Because the rate of rotation is constant, the angle θ or ωt is proportional to time. If we drew a graph of y against t the quantities 2π and π would be replaced by T and $\frac{T}{2}$ respectively.

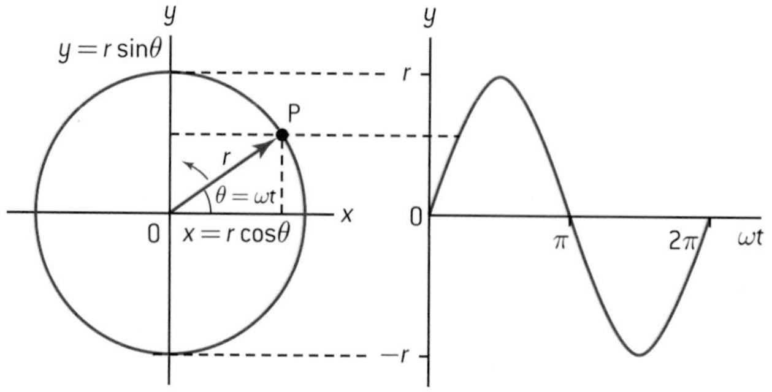

▲ Figure 2 Projection of circular motion on to a vertical line.

The equations of the projections are $y = y_0 \sin \omega t$ and $x = x_0 \cos \omega t$.

Here y_0 and x_0 are the maximum values of y and x, which in this case are identical to r. These are the amplitudes of the motion.

You may remember from Sub-topic 4.1 that in SHM the displacement, velocity and acceleration all vary sinusoidally with time. Thus, the projection of circular motion on the vertical or horizontal takes the same shape as SHM. This is very useful in analysing SHM.

The relationship between displacement, velocity, and acceleration

In Sub-topic 4.1 we saw that, starting from the displacement–time curve, we could derive the velocity–time and acceleration–time curves from the gradients. This can be done graphically, but another technique is to differentiate the equations with respect to time (differentiation is equivalent to finding the gradient).

We have seen from the comparison with circular motion that the displacement takes the form: $x = x_0 \sin \omega t$ or $x = x_0 \cos \omega t$ depending on when we start timing. It really doesn't matter whether the projection is onto the x-axis or the y-axis – therefore we can use x and y interchangeably.

Let's start with $x = x_0 \sin \omega t$

This means that the velocity is given by $v \left(= \frac{dx}{dt}\right) = x_0\, \omega \cos \omega t$, where x_0 is the amplitude and ω the angular frequency.

The maximum value that cosine can take is 1 so the maximum velocity is $v_0 = x_0\, \omega$ thus making the equation for the velocity at time t become $v = v_0 \cos \omega t$

We know that the acceleration will be the gradient of a velocity–time graph so we have

$$a = \frac{dv}{dt} = -v_0\omega \sin \omega t = -x_0\, \omega^2 \sin \omega t$$

As for cosine, the maximum value that sine can take is 1 so $a_0 = v_0\, \omega = x_0\, \omega^2$ giving $a = -a_0 \sin \omega t$

Comparing the equations $x = x_0 \sin \omega t$ and $a = -x_0\, \omega^2 \sin \omega t$ we can see a common factor of $x_0 \sin \omega t$

meaning that

$$a = -\omega^2 x$$

This may remind you, in Sub-topic 4.1, we saw that $a = -kx$ for SHM.

So the constant k must actually be ω^2 (the angular frequency squared). We will look at the significance of ω^2 very soon.

In examinations, you can be asked to find maximum values for velocity and acceleration – this makes the calculations easier because the maximum of sine and cosine are each 1 and, therefore, we don't need to include the sine or cosine term in our calculations. Thus the maximum velocity will be $v_0 = x_0\, \omega$ and the magnitude of the maximum acceleration will be $a_0 = x_0\, \omega^2$ (the direction of a_0 will be opposite that of x_0).

Modelling SHM with a spreadsheet

Much can be learned about SHM by using a spreadsheet to graph it. Figure 3 is a screen shot of part of a spreadsheet set up for this purpose

Note

- You will meet differential calculus in your Mathematics or Mathematics Studies course. This is not the place to teach you to differentiate – you will not be expected use calculus on your IB Physics course. However, for students studying physics, engineering and allied subjects at a higher level, calculus will form a major aspect of your course. In this case, we will differentiate the equations, but it is the results that are important not the method of obtaining them.

	A	B	C	D
1	time/s	displacement/m	velocity/m s^{-1}	acceleration/m s^{-2}
2	0.0	0.00	6.30	0.00
3	0.2	1.25	6.10	−1.98
4	0.4	2.41	5.52	−3.83
5	0.6	3.43	4.58	−5.45
6	0.8	4.23	3.36	−6.71
7	1.0	4.76	1.93	−7.56
8	1.2	4.99	0.37	−7.92
9	1.4	4.91	−1.21	−7.79
10	1.6	4.51	−2.71	−7.16
11	1.8	3.83	−4.05	−6.09
12	20.	2.91	−5.12	−4.62
13	2.2	1.81	−5.87	−2.87
14	2.4	0.59	−6.26	−0.93
15	2.6	−0.67	−6.24	1.06
16	2.8	−1.88	−5.84	2.99
17	3.0	−2.98	−5.06	4.73
18	3.2	−3.89	−3.96	6.17
19	3.4	−4.55	−2.62	7.22
20	3.6	−4.92	−1.11	7.81
21	3.8	−4.99	0.48	7.92
22	4.0	−4.73	2.03	7.52
23	4.2	−4.18	3.45	6.64
24	4.4	−3.37	4.66	5.35
25	4.6	−2.34	5.57	3.72
26	4.8	−1.17	6.13	1.85
27	5.0	0.08	6.30	−0.13
28	5.2	1.33	6.07	−2.11
29	5.4	2.49	5.46	−3.95
30	5.6	3.49	4.51	−5.54
31	5.8	4.27	3.27	−6.78
32	6.0	4.79	1.83	−7.60
33	6.2	5.0	0.26	−7.93
34	6.4	4.89	−1.31	−7.76
35	6.6	4.48	−2.81	−7.11
36	6.8	3.78	−4.13	−6.00
37	7.0	2.24	−5.18	−4.51
38	7.2	1.73	−5.91	−2.74
39	7.4	0.50	−6.27	−0.84
40	7.6	−0.75	−6.23	1.20
41	7.8	−1.96	−5.79	3.11
42	8.0	−3.05	−5.00	4.84
43	8.2	−3.94	−3.88	6.25
44	8.4	−4.58	−2.52	7.28
45	8.6	−4.94	−1.00	7.84
46	8.8	−4.98	0.58	7.90
47	9.0	−4.71	2.13	7.47
48	9.2	−4.14	3.54	6.57
49	9.4	−3.31	4.73	5.25
50	9.6	−2.27	5.62	3.60

	G	H	I	J
1	$t_1 =$	0	$x_0 =$	5
2	$\Delta t =$	0.2	$\omega =$	1.26

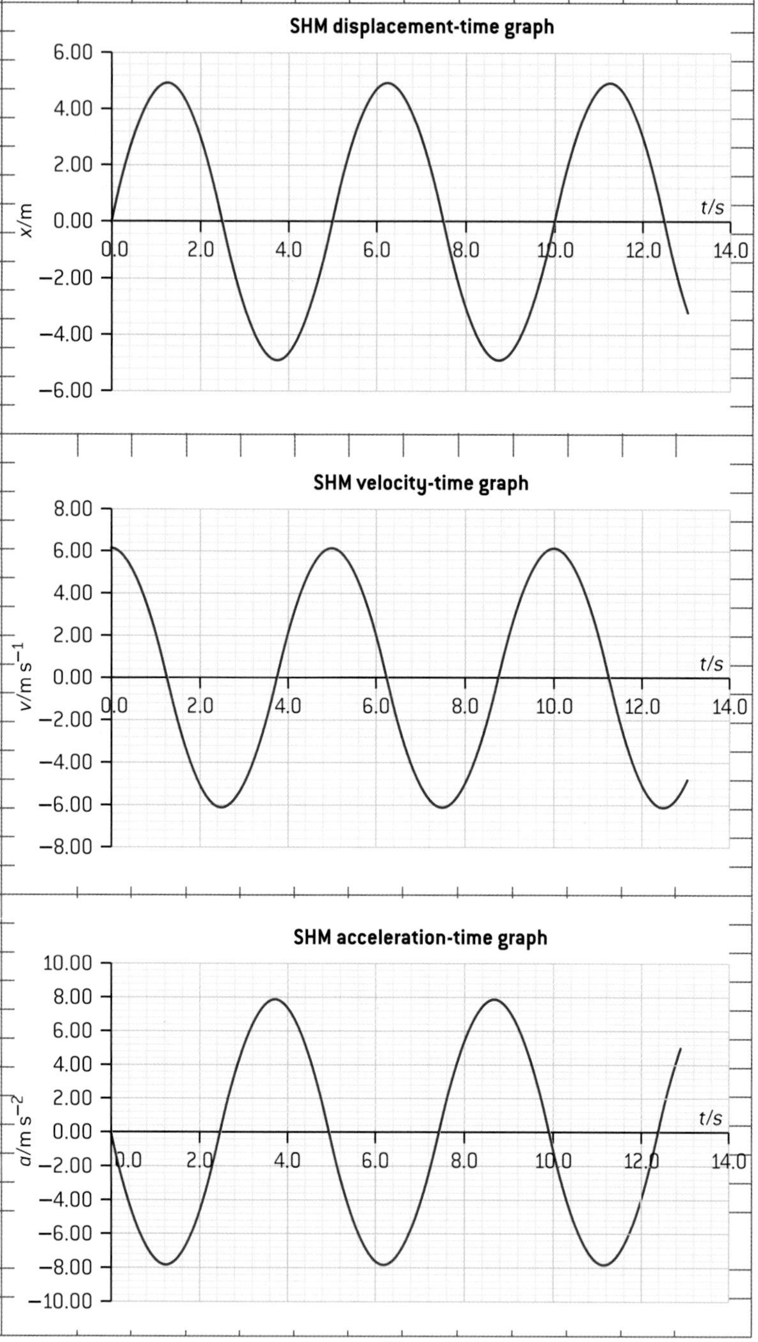

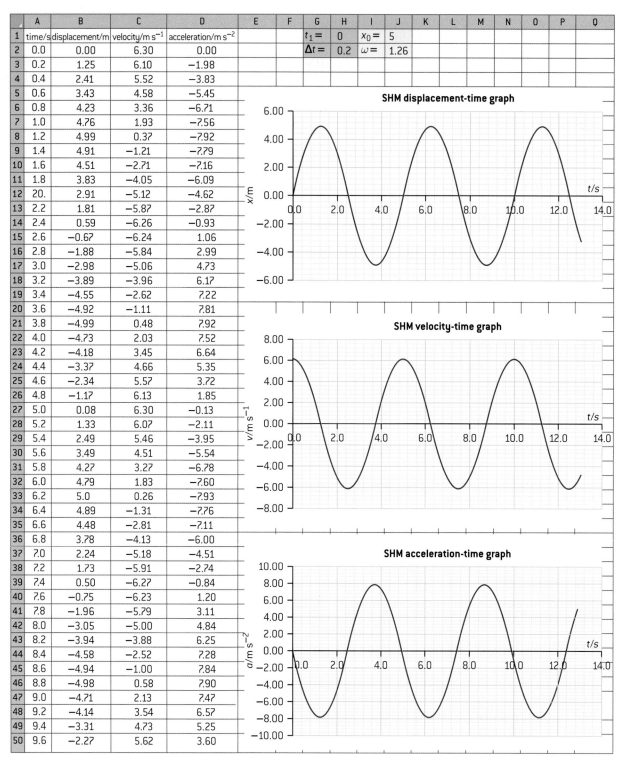

▲ Figure 3 Spreadsheet for SHM.

(as before this uses Microsoft Excel but other spreadsheets will have similar functions). Column A contains incremental times starting from zero (cell H1 is copied into A2 – this means that the starting time can be changed). Cell H2 determines the time increments (in this case 0.2 s) by adding the contents of H2 to each previous cell; the times can be increased down the column. The formula in cell A3, in this case, is =A2+$H$2 and the formula in cell A4 is =A3+$H$2, etc.

The times generated, together with chosen values of x_0 and ω, are used to generate the displacement, velocity and time curves. The values for x_0 and ω are inserted into cells J1 and J2 respectively – these can then be changed at will. The equation for the displacement, $x = x_0 \sin \omega t$ is converted into the formula =$J$1*SIN(A2*$J$2), which is copied into column B. The equation for the velocity, $v = x_0 \omega \cos \omega t$ is converted into the formula =$J$1*$J$2*COS(A2*$J$2) which is copied into column C. Finally, the equation for the acceleration, $a = -x_0 \omega^2 \sin \omega t$, is converted into the formula =(−1)*$J$1*($J$2)^2*SIN(A2*$J$2) and is copied into column D.

To produce the curves shown in figure 3, you now need to:

- insert a scatter chart
- choose the data series
- do a little formatting to improve the size and position of the chart and label the axes, etc.

Note

We have used a spreadsheet to show the solutions to the SHM equation. Later we will discuss how we can actually solve the SHM equation using iteration.

The SHM equation and ω^2

Re-visiting the equation $a = -\omega^2 x$ discussed earlier in this sub-topic, we see that it fits the definition of SHM (**motion in which the acceleration is proportional to the displacement from a fixed point and is always directed towards that fixed point**). The equation is the defining equation for SHM.

The sinusoidal graphs provide the solutions to this equation with respect to time. This may seem strange because there is no apparent time factor in $a = -\omega^2 x$. This is where ω^2 comes in. We saw with circular motion that ω represents the angular speed. Therefore ω^2 is simply the square of this measured in rad^2 s^{-2}. ω^2 has dimensions equivalent to $\left(\frac{1}{\text{time}}\right)^2$.

In circular motion we defined ω as being $\frac{\Delta\theta}{\Delta t}$ or simply $\frac{\theta}{t}$ when it is constant. For a complete revolution the angle will be 2π radians and the time will be the periodic time T. This means that $\omega = \frac{2\pi}{T}$ or, alternatively, $\omega = 2\pi f$; comparing this equation with $f = \frac{1}{T}$ explains why we can call ω the angular frequency.

Worked example

An object performs SHM with a period of 0.40 s and has amplitude of 0.20 m. The displacement is zero at time zero. Calculate:

a) the maximum velocity

b) the magnitude of the velocity after 0.10 s

c) the maximum acceleration of the object.

Solution

a) $v_0 = x_0 \omega = x_0 \times \frac{2\pi}{T}$

$v_0 = 0.20 \times \frac{2\pi}{0.40} = 3.1 \text{ m s}^{-1}$

b) As the displacement is zero at time zero this must be a sine or negative sine wave. Thus the velocity will be a cosine or negative cosine.

We are only asked to find the magnitude of the velocity so it doesn't matter which of the cosine curves the motion really takes.

Using $v = x_0 \omega \cos \omega t$

$$\omega t = \frac{2\pi t}{T} = \frac{(2\pi \times 0.10)}{0.40} \approx \frac{\pi}{2} \text{ rad}$$

So $v = 0.20 \times \frac{2\pi}{0.40} \times \cos \frac{\pi}{2} = 0$

This could have been done without the full calculation once we had decided that it was a cosine. 0.10 s represents the time for a quarter of a period (0.40 s); the value of any cosine at a quarter of a period (or $\frac{\pi}{2}$ radian) is zero.

c) $a_0 = x_0 \omega^2 = 0.20 \times \left(\frac{2\pi}{0.40}\right)^2 \approx 49 \text{ m s}^{-1}$

The velocity equation

The derivation of the SHM equation is for reference and you don't need to reproduce it – following it through, however, will help you to understand what is going on.

We have already seen that $v = x_0 \omega \cos \omega t$ and $x = x_0 \sin \omega t$ and you may know that $\sin^2 \theta + \cos^2 \theta = 1$.

This means that $\cos \theta = \pm\sqrt{1 - \sin^2 \theta}$ (don't forget that squaring either the positive or negative cosine will give $(+)\cos^2 \theta$. This means that $\pm$ must be included in the square root of this.

So $v = \pm x_0 \omega \sqrt{1 - \sin^2 \omega t}$

But $\sin \omega t = \frac{x}{x_0}$ so $\sin^2 \omega t = \left(\frac{x}{x_0}\right)^2$

$$v = \pm x_0 \omega \sqrt{1 - \left(\frac{x}{x_0}\right)^2} = \pm\omega\sqrt{x_0{}^2 - x_0{}^2\left(\frac{x}{x_0}\right)^2} = \pm\omega\sqrt{x_0{}^2 - x^2}$$

The equation $v = \pm\omega\sqrt{x_0{}^2 - x^2}$ is useful for finding the velocity at a particular position when you know the amplitude and period (or frequency or angular frequency) – you don't need to know the time being considered.

Worked example

An object oscillates simple harmonically with frequency 60 Hz and amplitude 25 mm. Calculate the velocity at a displacement of 8 mm.

Solution

$\omega = 2\pi f = 120\pi$ (there is no need to calculate the value here but do not leave π in the answer in an examination.

$v = \pm\omega\sqrt{x_0{}^2 - x^2}$

$= \pm 120\pi \sqrt{(25 \times 10^{-3})^2 - (8 \times 10^{-3})^2}$

$= \pm 8.9 \text{ m s}^{-1}$

Note

- $\pm$ tells us that the object could be going in either of the two opposite directions.
- Don't forget $x_0{}^2 - x^2 \neq (x_0 - x)^2$; it is a common mistake for students to equate these two expressions.

Simple harmonic systems

1. The simple pendulum

The simple pendulum represents a straightforward system that oscillates with SHM when its amplitude is small. When the pendulum bob (the mass suspended on the string) is displaced from the rest position there is a component of the bob's weight that tends to restore the bob to its normal rest or equilibrium position. A condition of a system oscillating simple harmonically is that there is a restoring force that is proportional to the displacement from the equilibrium position – this is, in effect, tying in with the equation defining SHM because $F = ma$ and $a = -\omega^2 x$ this means that $F = -m\omega^2 x$.

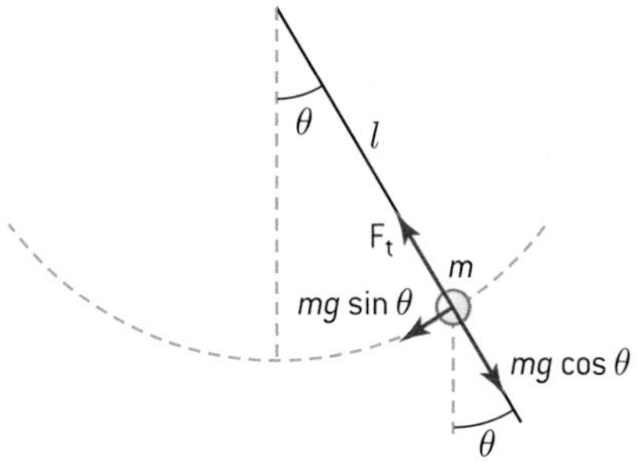

▲ Figure 4 Restoring force for simple pendulum.

Figure 4 shows the forces acting on a pendulum bob. The bob is in equilibrium along the radius when the tension in the string F_t equals the component of the weight in line with the string ($= mg\cos\theta$). The component of the weight perpendicular to this is not in equilibrium and provides the restoring force.

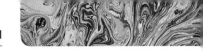

So the restoring force must be equal to the mass multiplied by the acceleration according to Newton's second law of motion:

$$mg \sin \theta = ma$$

For a small angle $\sin \theta \approx \theta \approx \frac{x}{l}$

rearranging gives $-m\left(\frac{g}{l}\right)x = ma$

(The minus sign is because the displacement (to the right) is in the opposite direction to the acceleration (to the left) in figure 4.

Cancelling m

$$a = -\left(\frac{g}{l}\right)x$$

this compares with the defining equation for SHM $a = -\omega^2 x$ leading to

$$\omega^2 = \frac{g}{l}$$

As the periodic time for SHM is given by $T = \frac{2\pi}{\omega}$, this shows that $T = 2\pi\sqrt{\frac{l}{g}}$.

Note

- The period of a simple pendulum is independent of the mass of the pendulum.
- The period of a simple pendulum is independent of the amplitude of the pendulum.
- l is the length of the pendulum from the point of suspension to the centre of mass of the bob.
- The equation only applies to small oscillations (swings making angle of less than 10° with the rest position).

Investigate!

1 Experimenting with the simple pendulum:

- attach a piece of thread of length about 2 metres to a pendulum bob (a lump of modelling clay would do for this)
- suspend the thread through a split cork to provide a stable point of suspension
- align a pin mounted in another piece of modelling clay with the rest position of the bob – to act as a reference point (sometimes called a fiducial marker)
- set the bob oscillating through a small angle and start timing as the bob passes the fiducial marker (you should consider why your measurements are likely to be more reliable when the bob is moving at its fastest)
- time thirty oscillations (remember that each oscillation is from **A** to **C** and back to **A**)
- measure the length of the thread from the point of suspension to the centre of mass of the bob (take this to be the centre of the bob)
- repeat the procedure and find an average period for the pendulum

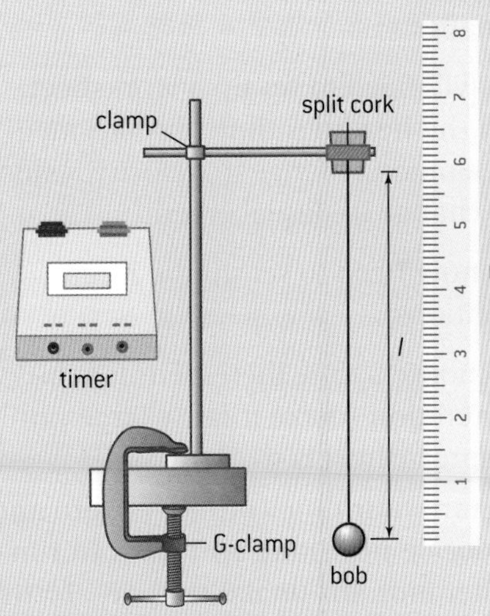

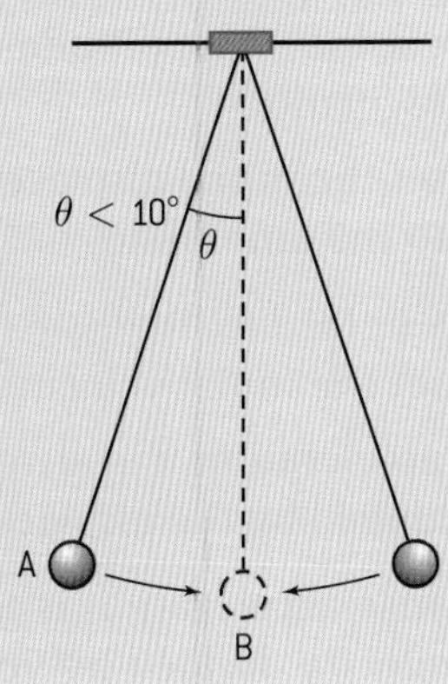

- repeat your measurements for a range of lengths
- first plot a graph of periodic time T against length l
- referring to the simple pendulum equation you may see that this graph should not be linear – what should you plot in order to linearize your results?
- how can you use the graph to calculate a value for the acceleration of freefall g?
- don't forget to do an error analysis and compare your result with the accepted value of g.

2 Use of iteration

Making use of the powerful iterative functionality of a spreadsheet can be a valuable learning aid. With iteration a graphical investigation is possible from basic principles and basic equations without knowing the solution and without advanced mathematics.

- you will need to set up a worksheet with the headings time, displacement, acceleration, change in velocity, velocity and change in displacement
- you now need to set up the initial conditions:
  1. choose a time increment for the iteration (make it fairly large until your spreadsheet is working well)
  2. start your time column at zero and decide how long you want to run the iteration for (make this, say, 10 time increments initially)
  3. choose an amplitude value for your oscillation
  4. set the initial displacement value equal to the amplitude
  5. choose a constant (k or ω^2)
  6. set the initial velocity to be zero
- the acceleration will always be $-k$ multiplied by the displacement
- because acceleration is the change in velocity divided by your time increment the change in velocity will be the acceleration multiplied by the time increment – when you add this to your previous velocity you will get the new velocity (actually it's the average velocity through the time increment)
- because velocity is the change in displacement divided by the time increment, you can find the change in displacement by multiplying your current velocity value by your chosen time increment
- the new value of the displacement will be the previous value added to the change in displacement
- This now feeds back into finding the next value of the acceleration and the cycle repeats and you generate your data for which you can plot displacement-time, velocity-time and acceleration-time graphs.

The following flow chart illustrates the iteration:

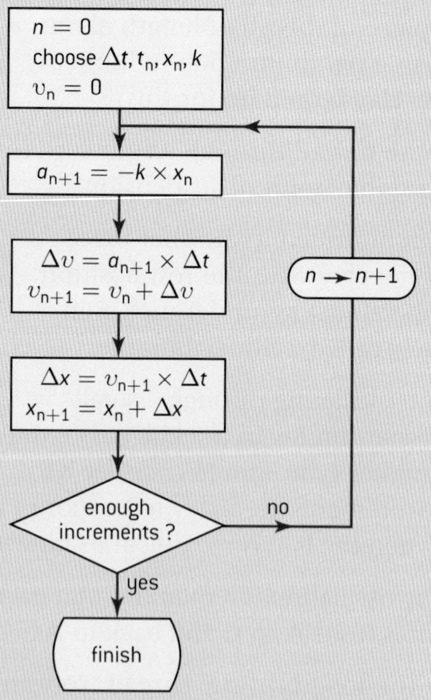

Subscripts n represent the current value of a variable and $n + 1$ the next value; at each loop the next value always replaces the previous one.

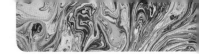

2. Mass–spring system

We have focused on a simple pendulum as being a very good approximation to SHM. A second system which also behaves well and gives largely undamped oscillations is a mass–spring system. We will consider a mass being oscillated horizontally by a spring (see figure 5); this is more straightforward than taking account of including the effects of gravity experienced in vertical motion. We will assume that the friction between the mass and the base is negligible. The mass, therefore, exchanges elastic potential energy (when it is fully extended and compressed) with kinetic energy (as it passes through the equilibrium position).

Note

- The period of the mass-spring system is independent of amplitude (for small oscillations).
- The period of the mass-spring system is independent of the acceleration of gravity.

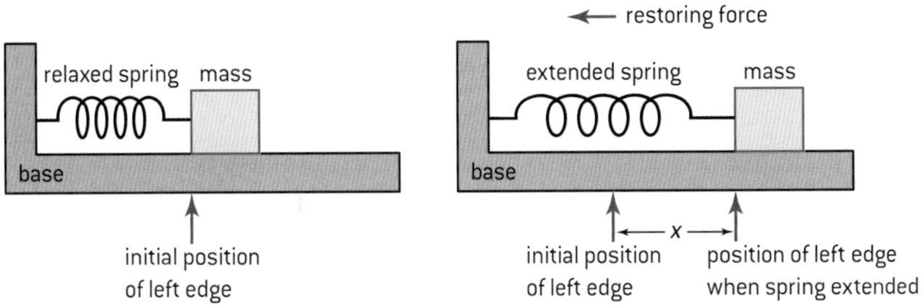

▲ Figure 5 Restoring force for mass-spring system.

When a spring (having spring constant k) is extended by x from its equilibrium position there will be a restoring force acting on the mass given by $F = -kx$ (the force is in the opposite direction to the extension).

Using Newton's second law

$$ma = -kx$$

which can be written as

$$a = -\left(\frac{k}{m}\right)x$$

this compares with the defining equation for SHM

$$a = -\omega^2 x$$

$$\omega^2 = \left(\frac{k}{m}\right)$$

As the periodic time for SHM is given by

$$T = \frac{2\pi}{\omega}$$

this shows that

$$T = 2\pi\sqrt{\frac{m}{k}}$$

When the spring is compressed the quantity x represents the compression of the spring. When the mass is to the left of the equilibrium position the compression is positive but the restoring force will be negative. This, therefore, leads to the same outcomes as for extensions.

Worked example

The mass–spring system is used in many common accelerometer designs. A mass is suspended by a pair of springs which displaces when acceleration occurs. An accelerometer contains a mass of 0.080 kg coupled to a spring with spring constant of 4.0 kN m^{-1}. The amplitude of the mass is 20 mm. Calculate:

a) the maximum acceleration

b) the natural frequency of the mass.

Solution

a) $a = -\left(\frac{k}{m}\right)x$

$$a = -\left(\frac{4.0 \times 10^3}{0.08}\right) 20 \times 10^{-3}$$

$$= 1000 \text{ m s}^{-2}$$

b) $T = 2\pi\sqrt{\frac{m}{k}}$ so $f = \frac{1}{T}$

$$= \frac{1}{2\pi}\sqrt{\frac{k}{m}}$$

$$f = \frac{1}{2\pi}\sqrt{\frac{4.0 \times 10^3}{0.08}} = 36 \text{ Hz}$$

Energy in SHM systems

We have seen that, in a simple pendulum, there is energy interchange between gravitational potential and kinetic; in a horizontal mass–spring system the interchange is between elastic potential and kinetic energy. Because each system involves kinetic energy we will focus on this form of energy and bear in mind that the potential energy will always be the difference between the total energy and the kinetic energy at a particular time. The total energy will be equal to the maximum kinetic energy.

From Topic 2 we know that the kinetic energy of an object of mass m, moving at velocity v, is given by

$$E_K = \frac{1}{2}mv^2$$

We also know that the equation for the velocity at a particular position is

$$v = \pm\omega\sqrt{x_0^2 - x^2}$$

Combining these equations gives the kinetic energy at displacement x:

$$E_K = \frac{1}{2}m\omega^2(x_0^2 - x^2)$$

This tells us that the maximum kinetic energy will be given by

$$E_{Kmax} = \frac{1}{2}m\omega^2 x_0^2$$

and this must be the total energy (when the potential energy is zero) so we can say

$$E_T = \frac{1}{2}m\omega^2 x_0^2$$

The potential energy at any position will be the difference between the total energy and the kinetic energy

so

$$E_P = E_T - E_K = \frac{1}{2}m\omega^2 x_0^2 - \frac{1}{2}m\omega^2(x_0^2 - x^2) = \frac{1}{2}m\omega^2 x^2$$

Figure 6 illustrates the variation with displacement of energy: the green line shows the total energy, the red the potential energy and the blue the kinetic energy. At any position the total energy is the sum of the kinetic energy and the potential energy – as we would expect.

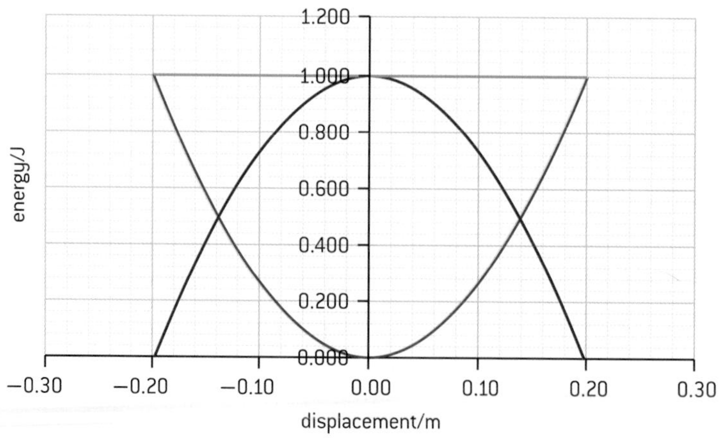

▲ Figure 6 Variation of energy with displacement.

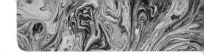

The variation of the potential energy and kinetic energy with displacement are both parabolas. With all the quantities in the total energy equation being constant, the total energy is, of course, constant for an undamped system.

In addition to looking at the variation of energy with displacement we should consider the variation of energy with time. Again, let us start with the kinetic energy.

The velocity varies with time according to the equation

$$v = x_0 \omega \cos \omega t$$

so the kinetic energy will be

$$E_K = \frac{1}{2}mv^2 = \frac{1}{2}m(x_0 \omega \cos \omega t)^2$$

or

$$E_K = \frac{1}{2}mx_0^2 \omega^2 \cos^2 \omega t$$

When the cosine term equals 1, this gives the maximum kinetic energy. The maximum kinetic energy occurs when the potential energy is zero and so is numerically equal to the total energy at that instant.

$$E_T = \frac{1}{2}mx_0^2 \omega^2$$

The potential energy will be the difference between the total energy and the kinetic energy so

$$\begin{aligned} E_P &= E_T - E_K \\ &= \frac{1}{2}mx_0^2 \omega^2 - \frac{1}{2}mx_0^2 \omega^2 \cos^2 \omega t \\ &= \frac{1}{2}mx_0^2 \omega^2 \sin^2 \omega t \end{aligned}$$

This relationships are shown on figure 7.

The green line represents the total energy, the red curve the potential energy and the blue curve the kinetic energy.

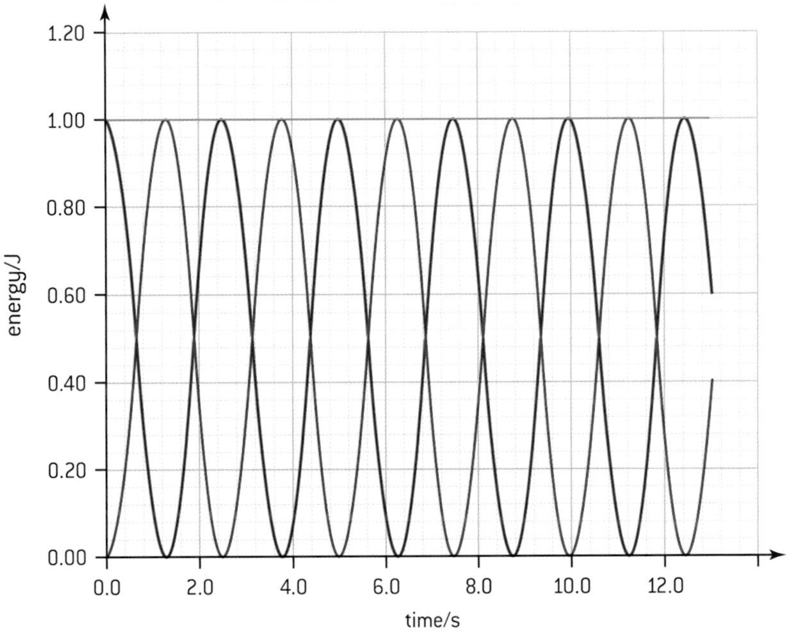

▲ Figure 7 Graph showing variation of energy with time.

Note

- The total energy is always the sum of the kinetic and potential energies.
- The graphs (unlike sine and cosine) never become negative.
- The period of the energy change is half that of the variation with time of displacement, velocity, or acceleration.
- The frequency of the energy change is twice that of the variation with time of displacement, velocity, or acceleration.

9.2 Single-slit diffraction

Understanding

→ The nature of single-slit diffraction

Nature of science

Development of theories

That rays "travel in straight lines" is one of the first theories of optics that students encounter. It comes as a surprise when this theory cannot explain the diffraction seen at the edges of shadows cast by small objects illuminated using point sources. Although partial shadows can be explained by considering light sources to be extended, this cannot account for the diffraction from a point source. The wave theory of diffraction and how diffraction can be explained in terms of wavefront propagation from secondary sources is a good example of how theories have been developed in order to explain a wider variety of phenomena.

Applications and skills

→ Describing the effect of slit width on the diffraction pattern

→ Determining the position of first interference minimum

→ Qualitatively describing single-slit diffraction patterns produced from white light and from a range of monochromatic light frequencies

Equations

→ *angle between first minimum and central maximum* $\theta = \frac{\lambda}{a}$

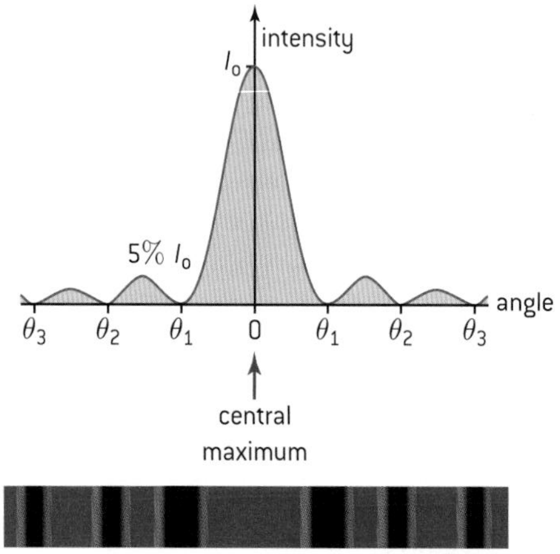

▲ Figure 1 Variation of intensity with angle for a diffraction pattern.

Introduction

In Sub-topic 4.4 we introduced diffraction and saw that when a wave passes through an aperture it spreads into the geometric shadow region. We also saw that the diffraction pattern consists of a series of bright and dark fringes. We will now consider the diffraction pattern in more detail.

Graph of intensity against angle

Figure 1 shows a single-slit diffraction pattern together with a graph of the variation of the intensity of the diffraction pattern with the angle measured from the straight-through position.

θ_1, θ_2, and θ_3 are the angles with the straight-through position made by the minima.

Note

- The central maximum has **twice the angular width** of the secondary maxima (each of these have the same angular width).
- The intensity falls off quite significantly from the principal maximum to the secondary maxima – the intensity of the first secondary maximum is approximately 5% of that of the principal maximum, the second is about 2% and the third is about 1% (figure 1 is not drawn to scale – the secondary maxima are all larger than a scale-diagram would show).
- No light reaches the centre of the minima but, in going towards the maxima, the intensity gradually increases – it is very difficult to decide the exact positions of the maxima and minima.
- The intensity is proportional to the square of the amplitude.

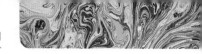

The single-slit equation

Returning to the work of Sub-topic 4.4 we saw how waves can superpose to give constructive interference (when they meet in phase) or destructive interference (when they meet anti-phase).

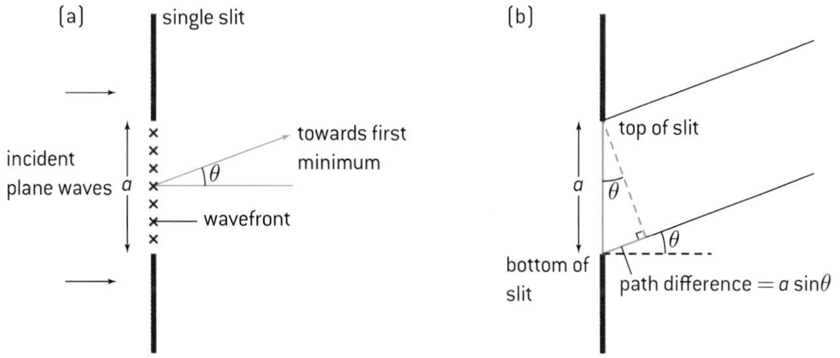

▲ Figure 2 Deriving the single slit equation.

Figure 2(a) shows a single-slit of width a. Each point on the wavefront within the slit behaves as a source of waves. These waves interfere when they meet beyond the slit. For the two waves coming from the edges of the slit making an angle θ with the straight through, there is a path difference of $a \sin \theta$. Waves from a point halfway along the slit will have a path difference of $\frac{a}{2} \sin \theta$ from the waves coming from each of the edges. When this path difference equals half a wavelength, the waves from halfway along the slit will interfere destructively with the waves coming from the bottom edge of the slit. It follows that for each point in the bottom half of the slit there will be a point in the top half of the slit at a distance $\frac{a}{2} \sin \theta$ from it. This means that when $\frac{a}{2} \sin \theta = \frac{\lambda}{2}$, there is destructive interference between a wave coming from a point in the upper half of the slit and an equivalent wave from the lower half of the slit.

Because we are dealing with small angles we can approximate $\sin \theta$ to θ and, by cancelling the factor of two, we arrive at the equation

$$\theta = \frac{\lambda}{a}$$

The position of other minima (shown in figure 1) will be given by $\theta = \frac{n\lambda}{a}$ (where $n = 2, 3, 4, \ldots$) but you will not be examined on this relationship.

Note

- This is the equation for the angle of the first **minimum**.
- It follows from the equation $\theta = \frac{n\lambda}{a}$ that the minima are not actually separated by equal distances. However, for values of θ that are less than about 10°, it is a good approximation to consider the minima to be equally spaced.
- The principal or central maximum occurs because the pairs of waves from the top and bottom halves of the slit will travel the same distance and have no path difference.
- The angular width of the principal maximum (from first minimum on one side to first minimum on the other) is 2θ.

TOK

Small angle approximation

We say that diffraction is most effective when the aperature is of the same order of magnitude as the width of the slit; we can demonstrate this effectively using a ripple tank. In the equation $\theta = \frac{\lambda}{a}$ if we make $\lambda \approx a$, then $\theta = 1$ radian (or 57.3°) or, if you avoid the small angle approximation, $\sin \theta = 1$ and $\theta = 90°$. Examine how closely diffraction in a ripple tanks agrees with small angle approximation prediction. With poor agreement (as this should show) why do we continue to use the equation $\theta = \frac{\lambda}{a}$?

Worked example

a) Explain, by reference to waves, the diffraction of light at a single slit.

b) Light from a helium–neon laser passes through a narrow slit and is incident on a screen 3.5 m from the slit. The graph below shows the variation with distance x along the screen of intensity I of the light on the screen.

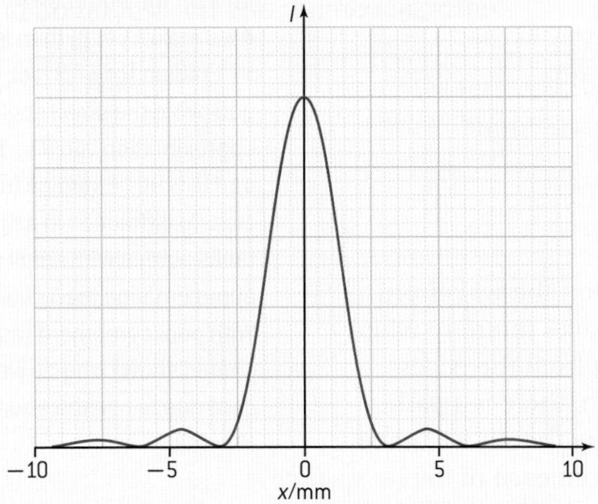

(i) The wavelength of the light emitted by the laser is 630 nm. Determine the width of the slit.

(ii) State **two** changes to the intensity distribution of the central maximum when the single slit is replaced by one of greater width.

Solution

a) The wavefront within the slit behaves as a series of point sources which spread circular wave fronts from them. These secondary waves superpose in front of the slit and the superposition of the waves produces the diffraction pattern.

b) (i) In this case the graph is drawn for distance not angle, so we need to calculate the angle θ in order to be able to use $\theta = \frac{\lambda}{a}$.

$\theta = \frac{s}{D} = \frac{3.0 \times 10^{-3}}{3.5}$

$= 0.86 \times 10^{-3}$ rad with s being the distance from the centre of the principal maximum to the first minimum. It is more reliable to measure 2θ from the first minimum on one side of the principal maximum to the first minimum on the other side of the principal maximum. D is the distance from the slit to the screen.

$a = \frac{\lambda}{\theta} = \frac{630 \times 10^{-9}}{0.86 \times 10^{-3}}$

$= 0.73 \times 10^{-3}$ m or 0.73 mm

(ii) With a wider slit more light is able to pass through. This will result in an increase in the intensity of the beam and so the peaks will all be higher.

a in the equation $\theta = \frac{\lambda}{a}$ increases but λ remains constant; the angle θ will decrease which means the principal maximum will become narrower and the minima will move closer together.

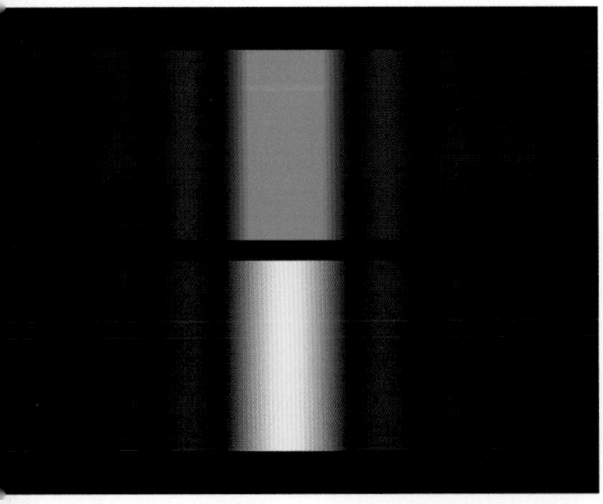

▲ Figure 3 Single slit with monochromatic and white light.

Single slit with monochromatic and white light

Figure 3 shows two images of the light emerging from a single-slit. The upper image is obtained using monochromatic (one frequency) green light and the lower image is obtained using white light. Both the angular width of the central maximum **and** the angular separation of successive secondary maxima depend on the wavelength of the light – this is the reason why the secondary maxima produced by the diffraction of white light are coloured. You will see that, for the secondary maxima, the blue light is less deviated than the other colours as it has the shortest wavelength. The edges of the principal maximum are coloured rather than pure white. This is because the principal maxima for the colours at the blue end of the visible spectrum are less spread than the colours at the red end; the edges are therefore a combination of red, orange, and yellow and have an orange hue.

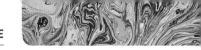

Investigate!

Diffraction at a single slit

- Using a laser pointer, shine the light on the gap between the jaws of a pair of digital or vernier callipers (forming a slit) and project it onto a white screen. The screen needs to be at least 3 m from the callipers.
- Adjust the callipers so that the image is clear.
- Measure the separation, s, of the first minima on the screen using a metre ruler.

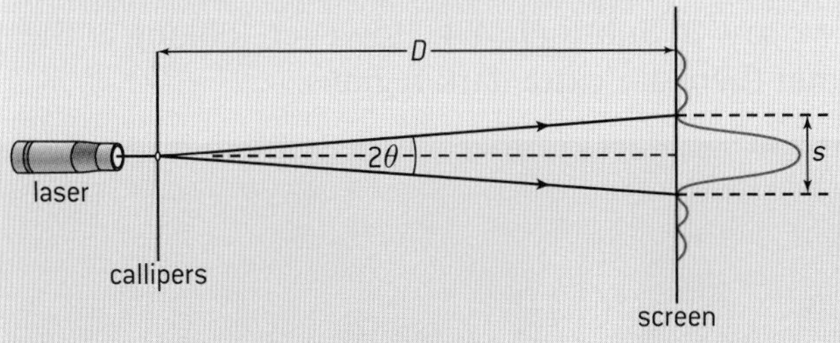

▲ Figure 4 Investigating laser light passing through slit.

- Measure the distance between the callipers and the screen, D, using a tape-measure.
- Calculate the wavelength of the light using $\lambda \approx a\theta$ as we know that $2\theta \approx \frac{s}{D}$
- Both the dependence of the width of the central maximum and the separation of the maxima on the wavelength of light can be investigated by using laser pointers with different colours.
- Research the wavelength ranges of the laser light and then suggest how you could modify the experiment to check the calibration of the callipers at small jaw separation.

We will return to diffraction when we look at resolution in Sub-topic 9.4.

9.3 Interference

Understanding

→ Young's double-slit experiment
→ Modulation of two-slit interference pattern by one-slit diffraction effect
→ Multiple slit and diffraction grating interference patterns
→ Thin film interference

Nature of science

Thin film interference

The observation of colour is not simply a question of colour pigmentation. Certain mollusc and beetle "shells", butterfly wings, and the feathers of hummingbirds and kingfishers can produce beautiful colours as a result of thin film interference known as "iridescence". The observed colour changes depend on the angle of illumination or viewing, and are caused by multiple reflections from the surfaces of films with thicknesses similar to the wavelength of light. These natural aesthetics require the analysis of the physics of interference.

Applications and skills

→ Qualitatively describing two-slit interference patterns, including modulation by one-slit diffraction effect
→ Investigating Young's double-slit experimentally
→ Sketching and interpreting intensity graphs of double-slit interference patterns
→ Solving problems involving the diffraction grating equation
→ Describing conditions necessary for constructive and destructive interference from thin films, including phase change at interface and effect of refractive index
→ Solving problems involving interference from thin films

Equations

fringe separation for double slit: $s = \frac{\lambda D}{d}$

diffraction grating equation: $n\lambda = d\sin\theta$

light reflection by parallel-sided thin film:

constructive interference $2dn = (m + \frac{1}{2})\lambda$

destructive interference $2dn = m\lambda$

Introduction

In Sub-topic 4.4 we considered interference of the waves emitted by two coherent sources. We saw that interference is a property of all waves, and we considered the equal fringe spacing in the Young double-slit experiment. In this sub-topic we focus on the intensity variation in a double-slit experiment and see how interference is achieved using multiple slits. We then look at a second way of achieving interference, using division of amplitude instead of division of wavefront as with double-slit interference.

Intensity variation with the double-slit

Figure 1 shows the image of the light from a helium–neon laser that has passed through a double slit. The alternate red and dark fringes are equally spaced as we saw in Sub-topic 4.4. However, looking at the image closely we see that there are extra dark regions.

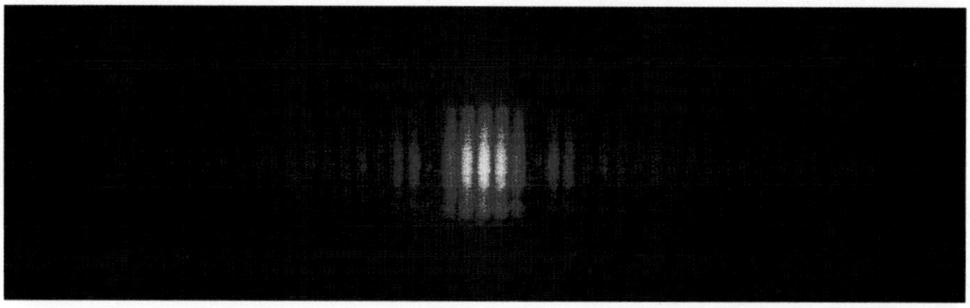

▲ Figure 1 Double-slit diffraction pattern for light from He–Ne laser.

We know from Sub-topic 9.2 that a single slit produces a diffraction pattern with a very intense principal maximum and much less intense secondary maxima. A double slit is, of course, two single slits so each of the slits produces a diffraction pattern and the waves from the two slits interfere. The two effects mean that the intensity of the interference pattern is not constant, but is modified by the diffraction pattern to produce the intensity.

Figure 2(a) below shows how the relative intensity would vary for a double-slit interference pattern without any modification due to diffraction. By using relative intensity we avoid the need to think about the actual intensity values and units. Figure 2(b) shows the variation of relative intensity with angle for a single slit. Figure 2(c) shows the superposition of the two effects so that the single-slit diffraction pattern behaves as the envelope of the interference pattern. Shaping a pattern in this way is called *modulation* and is important in the theory of AM (amplitude modulation) radio. You will note that the fringe spacing does not change between figures 2(a) and (c) but that the bright fringes occurring between 11° and 14° in figure 2(a) are reduced to a much lower intensity. An interference maximum coinciding with a diffraction minimum is suppressed and does not appear in the overall pattern.

We saw in Sub-topic 4.4 that the fringe separation s for light of wavelength λ is given by

$$s = \frac{\lambda D}{d}$$

where D is the distance from the double slit to the screen and d is the separation of the slits. The value of s used in this equation is for the interference pattern not the modulated pattern caused by diffraction.

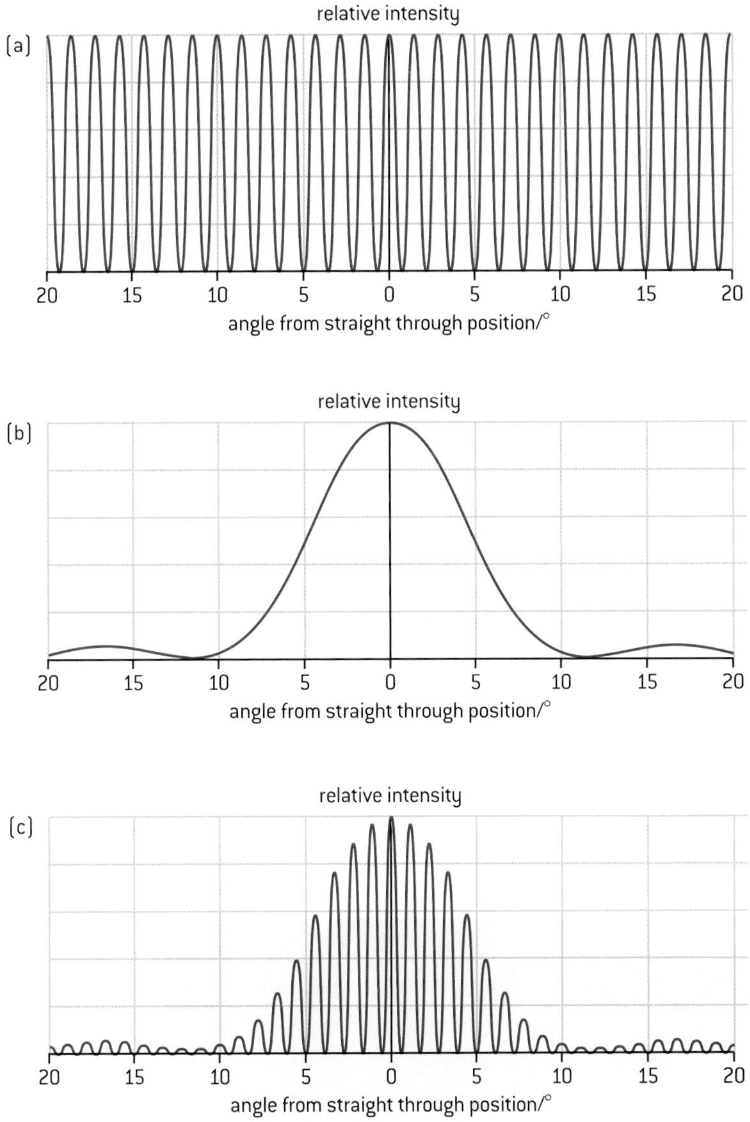

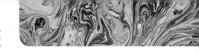

▲ Figure 2 Combination of diffraction and interference.

Multiple-slit interference

We have now seen the interference patterns produced both by a single slit and a double slit. What happens if there are more than two slits? The answer is that the bright fringes, which come from constructive interference of the light waves from different slits, remain in the same positions as for a double slit but the pattern becomes sharper. The bright fringes are narrower and their intensity is proportional to the square of the number of slits. Why is this?

At the centre of the principal maximum the waves reaching the screen from all of the slits are in phase and so it is very bright here. By moving to a position close to the centre of this maximum the path difference between the light from two adjacent slits has changed by such a small amount that it hardly affects the interference and so it is still bright. Slits further away from each other will have more likelihood of a greater path difference and therefore meeting out of phase.

Imagine having three slits: at the principal maximum the path difference between the waves will be small as their paths are nearly parallel. Each of the three wave trains will meet nearly in phase at positions around the principal maximum peak, which makes it fairly wide.

Now imagine having 100 slits: for every slit there will a second slit somewhere that transmits a wave with a half wavelength path difference from the first. When the waves interfere at a position close to the centre of the maximum, destructive interference occurs and reduces the overall intensity. This will be true for the waves coming from the two slits adjacent to this pair, the two next to those and so on. Increasing the number of slits will give destructive interference close to the centre of the maxima – this reduces the width of the central maximum (and the other maxima too). The extra energy that is needed to increase the intensity of the maxima must come from the regions that are now darker, so the maxima are more intense.

The mathematics of modulation of waves is quite complex and for the purpose of the IB Physics course you will simply need to recognize that the modulation is happening and remember the effect of increasing the number of slits.

Figure 3 shows the interference patterns obtained when red laser light passes through various numbers of slits of identical width. Figures 3 and 4 show that the single-slit diffraction pattern always acts as an envelope for the multiple-slit interference patterns.

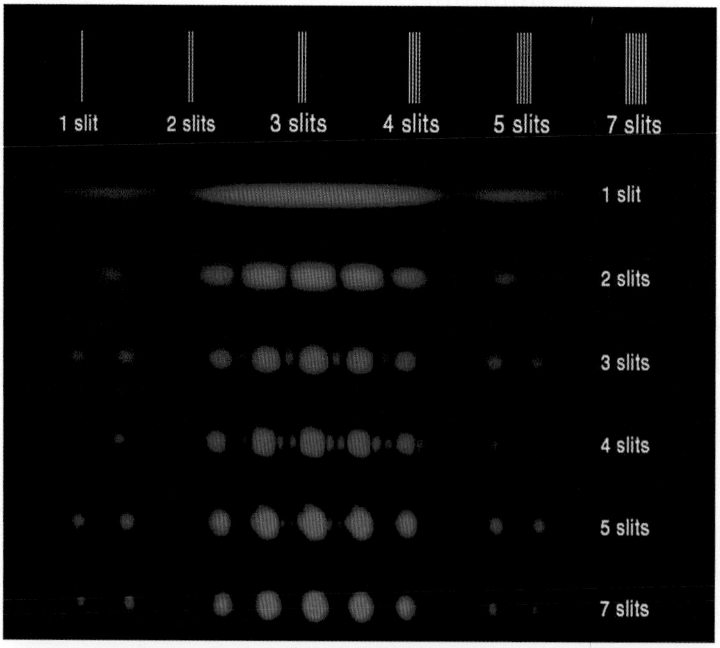

▲ Figure 3 The effect of increasing the number of slits on an interference pattern.

In figure 4 the relative intensities have been drawn the same size in order to show the increased sharpness; however, for two, three, and five slits the actual intensities are in the ratio 1 : 9 : 25.

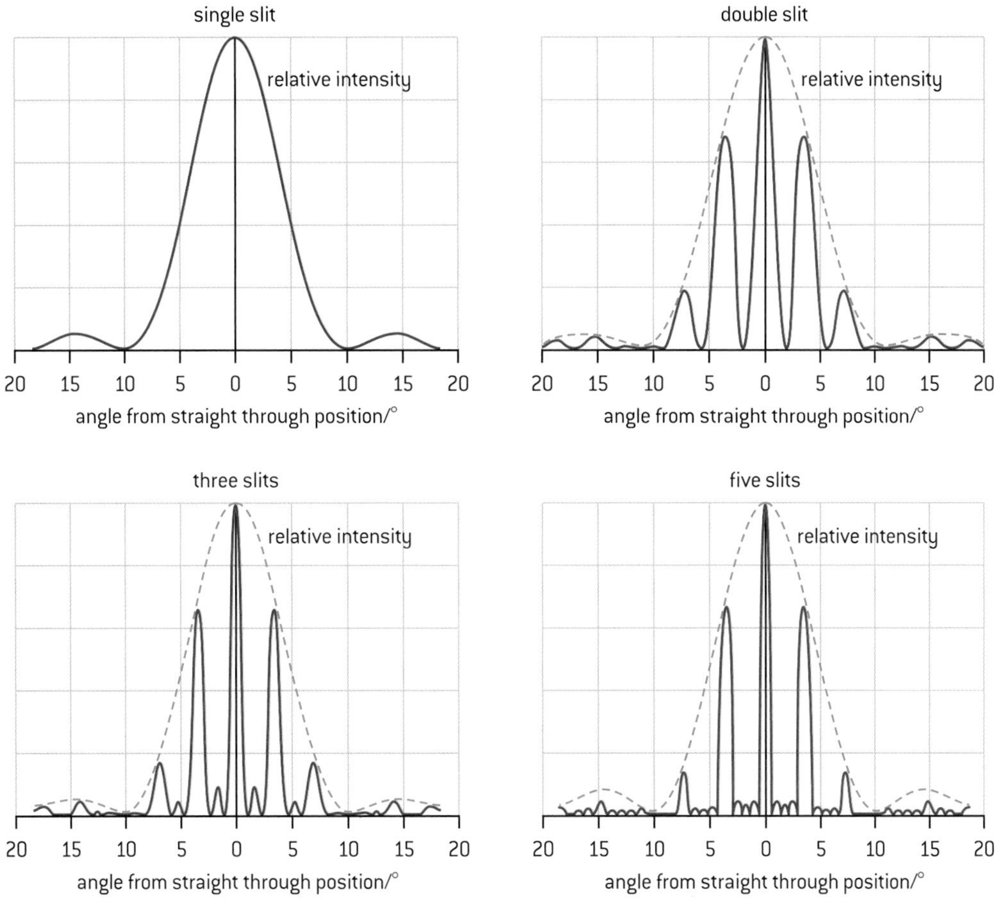

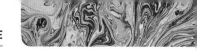

▲ Figure 4 The effect of increasing the number of slits on variation of intensity.

The diffraction grating

A diffraction grating is a natural consequence of the effect on the interference pattern when the number of slits is increased. Diffraction gratings are used to produce optical spectra. A grating contains a large number of parallel, equally spaced slits or "lines" (normally etched in glass or plastic) – typically there are 600 lines per millimetre (see figure 5). When light is incident on a grating it produces interference maxima at angles θ given by

$$n\lambda = d\sin\theta$$

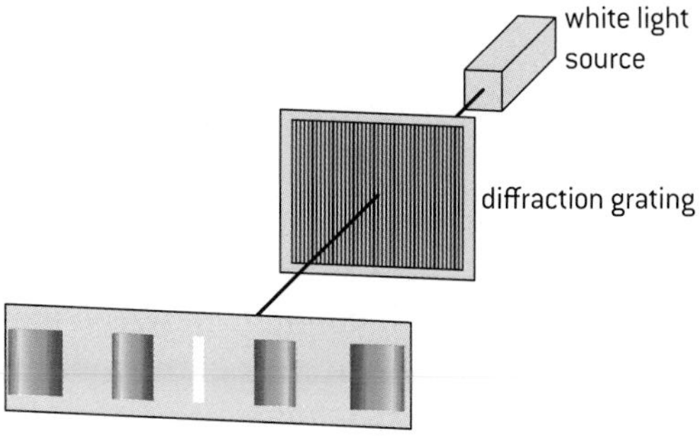

▲ Figure 5 Diffraction grating.

The spacing between the slits is small, which makes the angle θ large for a fixed wavelength of light and n. This means that we cannot use the small angle approximation for relating the wavelength to the position of the maxima as we did for a double-slit. Figure 6 shows a section of a diffraction grating in which three consecutive slits deviate the incident waves towards a maximum. The slits are so narrow and the screen is so far away from the grating that the angles made by the diffracted waves are virtually identical. Providing the path difference between waves coming from the same part of successive slits is an integral number of wavelengths, the waves will reach the screen in phase and give a maximum. In the case of a diffraction grating the distance d is taken as the length of both the transparent and opaque sections of the slit (they are taken to be of equal width). From figure 6 we see that the path difference between those waves coming from A and B and the path between waves coming B and C will each be $d\sin\theta$. This means

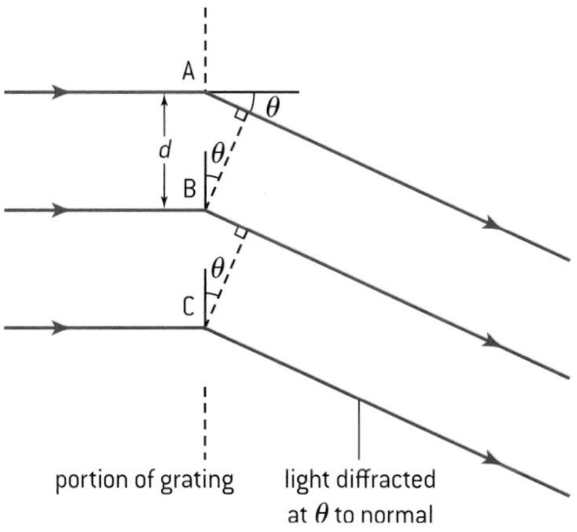

▲ Figure 6 Deriving the diffraction grating equation.

▲ Figure 7 Dispersing white light with a diffraction grating.

that $n\lambda = d\sin\theta$ (where n is the "order" of the maximum and is zero for the central maximum, 1 for the first maximum on each side of the centre, etc.).

We can see that the angular positions of the interference maxima depend on the grating spacing, d. The shape of the diffraction envelope, however, is determined by the width of the clear spaces (as for a single slit). The maxima must lie within the envelope of the single slit diffraction pattern if they are to be relatively intense; this sets an upper limit on how wide the transparent portions of the grating can be.

One of the uses of a diffraction grating is to disperse white light into its component colours: this is because different wavelengths produce maxima at different angles. Figure 7 shows that light of greater wavelength (for any given order) is deviated by a larger angle. This is in line with what we would predict from using $n\lambda = d\sin\theta$. Each successive visible spectrum repeats the order of the colours of the previous one but becomes less intense and more spread out.

Grating spacing and number of lines per mm

It is usual for the "number of lines per mm" or (N) to be quoted for a diffraction grating, rather than the spacing. N must be converted into d in order to use the diffraction grating equation $n\lambda = d\sin\theta$. This is straightforward because $d = \frac{1}{N}$ but N must first be converted to the number of lines per metre by multiplying by 1000 before taking the reciprocal.

Nature of science

Appropriate wavelengths for effective dispersion

For a diffraction grating to produce an observable pattern, the grating spacing must be comparable to the wavelength of the waves. The wavelength of visible light is between approximately 400–700 nm. A grating with 600 lines per mm has a spacing of approximately 2000 nm or four times the wavelength of green light. This spacing produces very clear images.

The wavelengths of low-energy X-rays is around 10^{-10} m or 0.1 nm. The 600 lines per millimetre optical grating will not produce any observable maximum ... therefore, what will diffract X-rays? The spacing of ions in crystals is of the same order of magnitude as X-rays and the regular lattice shape of a crystal can perform the same task with X-rays as a diffraction grating does with visible light. The diffracted patterns are now commonly detected with charge-coupled device (CCD) detectors.

Worked example

A diffraction grating having 600 lines per millimetre is illuminated with a parallel beam of monochromatic light, which is normal to the grating. This produces a second-order maximum which is observed at 42.5° to the straight-through direction. Calculate the wavelength of the light.

Solution

$N = 6.00 \times 10^5$ lines per metre.

$d = \frac{1}{N} = \frac{1}{6.00 \times 10^5} = 1.67 \times 10^{-6}$ m.

$n\lambda = d\sin\theta$ so $\lambda = \frac{d\sin\theta}{n} = \frac{1.67 \times 10^{-6} \times \sin 42.5}{2}$ (don't forget to have your calculator in degrees)

$\lambda = 5.64 \times 10^{-7} = 564$ nm (all data is given to three significant figures so the answer should also be to this precision).

Nature of science

Using a spectrometer

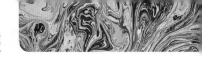

A spectrometer is a useful, if expensive, piece of laboratory equipment that allows the wavelengths of light emitted by sources to be analysed. The instrument consists of a collimator, turntable and telescope. A diffraction grating or other light disperser is placed on the turntable, which is carefully levelled. The collimator uses lenses to produce a parallel beam of light from a source – this light beam is then incident on the grating that disperses it. The end of the collimator furthest from the grating has a vertical adjustable slit that serves as the source – when we talk about spectral lines we infer that a spectrometer is being used because the "lines" are the dispersed images of the collimator slit. Before measurements are made the telescope is focused on the collimator slit – this means that the dispersed light will also be in focus. It is usual for sources that are said to be "monochromatic" to actually emit a variety of different wavelengths. By using the spectrometer, the angle θ in the diffraction grating equation can be measured. This is usually done by measuring the angle between the two first-order images on either side of the straight-through position i.e. 2θ. With a knowledge of the grating spacing d, the wavelength λ can be determined.

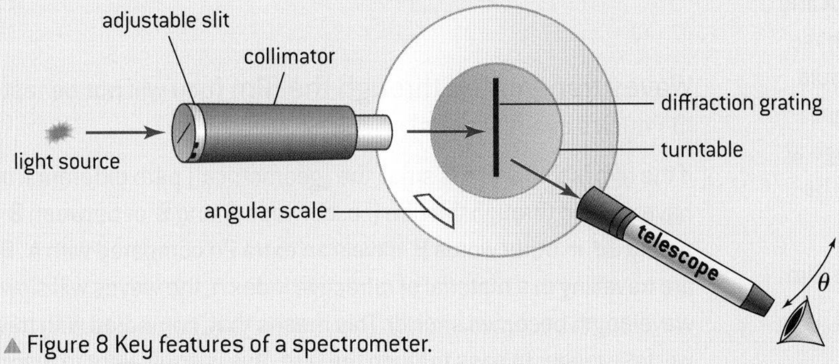

▲ Figure 8 Key features of a spectrometer.

Interference by division of amplitude

It was mentioned in the introduction to this sub-topic that there are two ways of providing coherent sources that are able to interfere. Young's double slit and multiple slits all derive their interfering waves by taking waves from different parts of the same wavefront. Because the interfering waves have all come from the same wavefront they will be in phase with each other. Wherever the waves meet they will interfere and a fringe pattern can be obtained anywhere in front of the sources (the slits). Since this interference can be found anywhere the fringes are said to be "non-localized".

Division of amplitude is a method of achieving interference using two waves that have come from the same point on a wavefront. Each wave has a portion of the amplitude of the original wave. In order to achieve interference by division of amplitude, the source of light must come from a much bigger source than the slit used for division of wavefront interference. The image produced will, however, be "localized" to one place instead of being found anywhere in front of the sources.

Thin film interference

Figure 9 shows a wave incident at an angle θ to the normal to the surface of a film of transparent material (such as low-density oil or detergent) having refractive index n. This diagram is not drawn to scale

– θ and d are both very small so that incident wave is effectively normal to the surface. The incident wave partially reflects at the top surface of the film and partially refracts into the film. This refracted wave, on reaching the lower surface of the film, again partially reflects (remaining in the film) and partially refracts into the air below the film. This process can occur several times for the same incident wave.

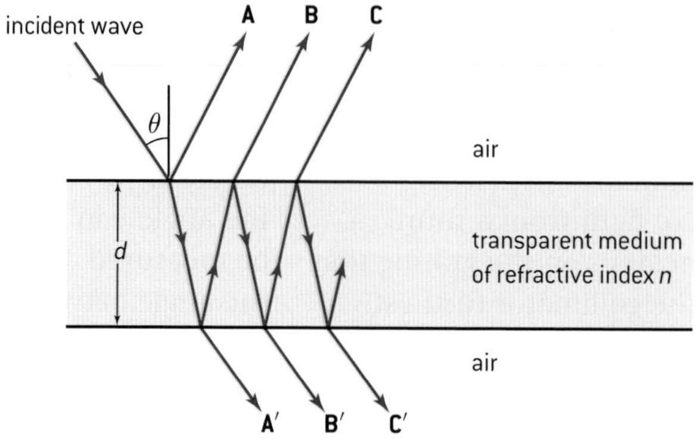

▲ Figure 9 Interference at parallel-sided thin film.

Waves reflected by the film *(you may be tested on this in IB Physics examinations)*

In this case **A** has been reflected from the top surface of the film and, because the reflection is at an optically denser medium, there is a phase change of π radian (equivalent to half a wavelength). The wave **B** travels an optical distance of $2dn$ before it refracts back into the air. Thus the optical path difference between **A** and **B** will be $2dn$. If there had been no phase change then this optical distance would equal $m\lambda$ for constructive interference. However, because of **A**'s phase change at the top surface the overall effect will be destructive interference.

Thus for the light reflecting from the film when

$$2dn = m\lambda$$

there will be destructive interference and when

$$2dn = (m + \tfrac{1}{2})\lambda$$

there will be constructive interference.

Waves transmitted through the film (you will **not** be tested on this in IB Physics examinations)

If the film is thin and θ is small the (geometrical) path difference between waves that have passed through the film (i.e. between **A'** and **B'** or between **B'** and **C'**) will be very nearly $2d$, in other words **B'** travels an extra $2d$ compared with **A'**. Because the waves are travelling in a material of refractive index n, the waves will slow down and the wavelength becomes shorter. This means that, compared with travelling in air, they will take longer to pass through the film. This is equivalent to them travelling through a thicker film at their normal speed. The optical path difference will, therefore, be $2dn$. If this distance is equal to $m\lambda$ where $m = 0, 1, 2$ etc. (we are using m to avoid confusion with the refractive index n) then there will be constructive interference. If the optical path difference is equal to an odd number of half wavelengths, $(m + \tfrac{1}{2})\lambda$, then there will be destructive interference.

Nature of science

Coating of lenses

When light is incident on a lens some of it will be reflected and some transmitted. The reflected light is effectively wasted, reducing the intensity of any image formed (by the eye using corrective lenses, in a camera or in a telescope). By coating the lens with a transparent material of quarter of a wavelength thickness, the light reflected by the coating and the light reflected by the lens can be made to interfere destructively and thus eliminate the reflection altogether. Magnesium fluoride is often used for the coating and is optically denser than air but less dense than glass (having a refractive index of 1.38). When it is used to coat a lens the waves reflecting from both the magnesium fluoride and the glass will undergo a phase change of π radian ... so the phase changes effectively cancel each other out. The optical path difference between the waves will be equal to the refractive index of magnesium fluoride multiplied by twice the thickness of the coating. For destructive interference the optical path difference is $2nd = \frac{\lambda}{2}$ so the thickness of the coating is given by $d = \frac{\lambda}{4n}$. With white light there will be a range of wavelengths and so it is impossible to match the thickness of the coating to all of these wavelengths. If the thickness is matched to green light, then red and blue will still be reflected giving the lens the appearance of being magenta (= red + blue). By using multiple layers of a material of low refractive index and one of higher refractive index, it is possible to reduce the amount of light reflected to as little as 0.1% for a chosen wavelength.

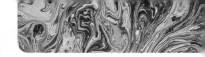

Worked example

a) Name the wave phenomenon that is responsible for the formation of regions of different colour when white light is reflected from a thin film of oil floating on water.

b) A film of oil of refractive index 1.45 floats on a layer of water of refractive index 1.33 and is illuminated by white light at normal incidence.

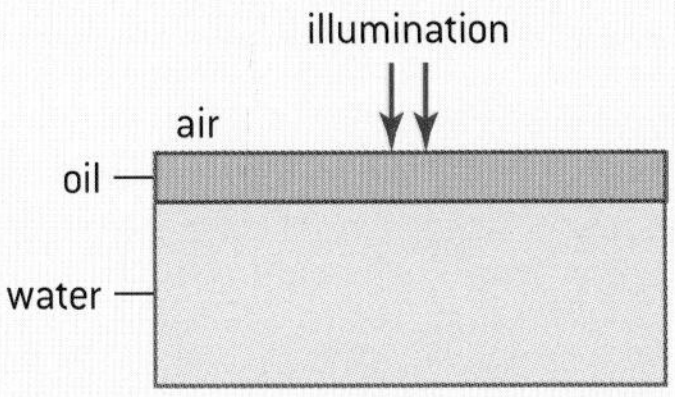

When viewed at near normal incidence a particular region of the film looks red, with an average wavelength of about 650 nm.

(i) Explain the significance of the refractive indices of oil and water with regard to observing the red colour.

(ii) Calculate the minimum film thickness.

Solution

a) Although there is reflection involved, the colours come about because of interference.

b) (i) n is the refractive index of the oil. Because the waves are travelling in oil, they move more slowly than they would do in air and so the effective path difference between the waves reflected at the air–oil interface and the waves at the oil–water interface is longer by a factor of 1.45. The wave reflected at the air–oil interface undergoes a phase change equivalent to $\frac{\lambda}{2}$ (oil is denser than air). The waves reflected at the oil–water interface undergo no phase change on reflection at the less dense medium. So for the bright constructive red interference $2dn = (m + \frac{1}{2})\lambda$

if $m = 0$,

$2dn = \frac{\lambda}{2}$ or $\lambda = 4dn$

(ii) $d = \frac{650}{(4 \times 1.45)} = 110$ nm

Nature of science

Vertical soap films

Figure 10 shows the colours of light transmitted by a vertical soap film. Over a few seconds the film drains and becomes thinner at the top and thicker at the bottom. When the film is illuminated with white light, the reflected light appears as a series of horizontal coloured bands. The bands move downwards as the film drains and the top becomes thinner. The top of the film appears black just before the film breaks. It has now become too thin for there to be a path difference between the waves coming from the two surfaces of the film. The phase change that occurs for the light reflected by the surface of the film closest to the source means that, for all colours, there is cancellation and so no light can be seen.

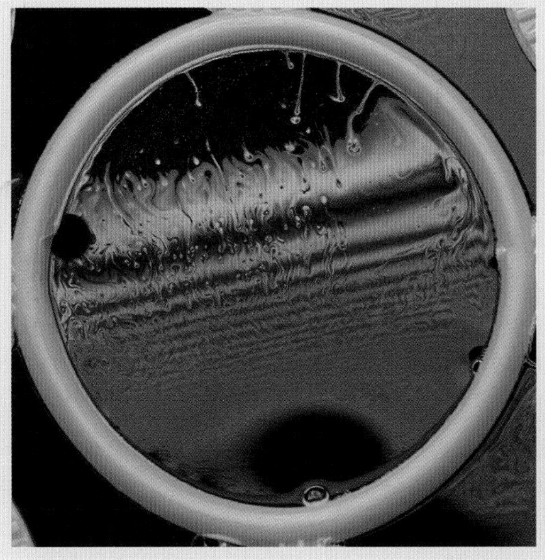

▲ Figure 10 Thin film interference in a vertical soap film.

TOK

The aesthetics of physics

The colour in the soap film makes a beautiful image. Does physics need to rely upon the arts to be aesthetic? Herman Bondi the Anglo-Austrian mathematician and cosmologist implied that Einstein believed otherwise when he wrote:

> *"What I remember most clearly was that when I put down a suggestion that seemed to me cogent and reasonable, Einstein did not in the least contest this, but he only said, 'Oh, how ugly.' As soon as an equation seemed to him to be ugly, he really rather lost interest in it and could not understand why somebody else was willing to spend much time on it. He was quite convinced that beauty was a guiding principle in the search for important results in theoretical physics."*
>
> *— H. Bondi*

How does a mathematical equation convey beauty? Is the beauty in the mathematics itself or what the mathematics represents?

9.4 Resolution

Understanding

- → The size of a diffracting aperture
- → The resolution of simple monochromatic two-source systems

Applications and skills

- → Solving problems involving the Rayleigh criterion for light emitted by two sources diffracted at a single slit
- → Describing diffraction grating resolution

Equations

- → Rayleigh's criterion: $\theta = 1.22\frac{\lambda}{b}$
- → resolvance of a diffraction grating: $R = \frac{\lambda}{\Delta\lambda} = mN$

Nature of science

How far apart are atoms?

When we view objects, we are limited by the wavelength of the light used to make an observation – shorter wavelengths such as X-rays, gamma rays and fast-moving electrons will improve the resolution that is achievable. However, even using the shortest wavelengths obtainable does not allow us to locate the exact position of objects on the atomic and sub-atomic scale. The process of making an observation using these waves disturbs the system and increases the uncertainty with which we can locate an object. Heisenberg's uncertainty principle places a limit on how close we can be to finding the exact position of objects on the quantum scale.

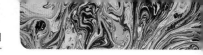

Introduction

Resolution is the ability of an imaging system to be able to produce two separate distinguishable images of two separate objects. The imaging system could be an observer's eye, a camera, a radio telescope, etc. Whether objects can be resolved will depend on the wavelength coming from the objects, how close they are to each other and how far away they are from the observer.

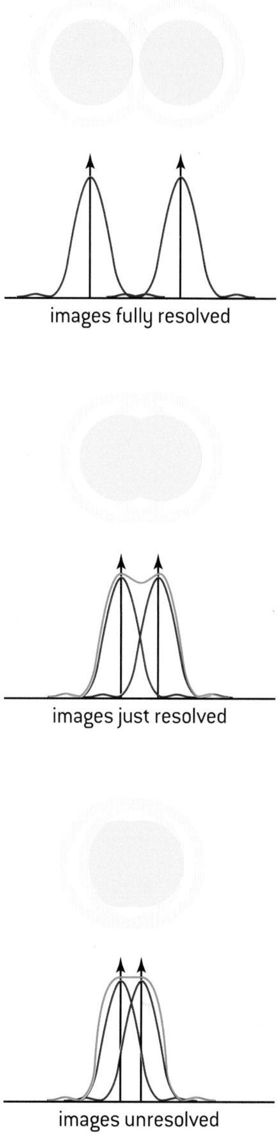

▲ Figure 1 Diffraction intensity patterns of two objects viewed through a circular aperture.

Diffraction and resolution

We have seen that when light passes through an aperture a diffraction pattern is formed. For an optical system the aperture could be the pupil of the observer's eye or the objective lens of a telescope. When there are two sources of light two diffraction patterns will be formed by the system.

How close can these patterns be for us to still recognize that there are two sources?

Two objects observed through an aperture will produce two diffracted images which may or may not overlap. In the late nineteenth century the English physicist, John William Strutt, 3rd Baron Rayleigh, proposed what is now known as the ***Rayleigh criterion*** for resolution of images. This states that **two sources are resolved if the principal maximum from one diffraction pattern is no closer than the first minimum of the other pattern.**

The limit to resolution is when the principal maximum of the diffraction pattern from one source lies on the first minimum diffraction pattern from the second source (and vice versa). Diffracted images further apart than this limit will be resolved and those closer will be unresolved.

Figure 1 shows the diffraction intensity patterns produced by two objects the same distance apart but viewed through a circular aperture from different distances. The variation of intensity with angle for each of the diffraction patterns is shown below the image. According to the Rayleigh criterion, the uppermost pair of images are fully resolved because the principal maximum of each diffraction pattern lies further from the other than the first minimum. The central images are just resolved since the principal maximum of one diffraction pattern is at the same position as the first minimum of the second diffraction pattern. The bottom images are unresolved as the principal maximum of one diffraction pattern lies closer to the second pattern than its first minimum.

Nature of science

Other criteria for resolution?

Rayleigh's criterion is not the only one used in optics. Many astronomers believe that they can resolve better than Rayleigh predicts. C M Sparrow developed another criterion for telescopes that leads to an angular separation at resolution about half that of Rayleigh. His criterion is that the two diffraction patterns when added together give a constant amplitude in the regions of the two central maxima.

Resolution equation

For single slits we saw in Sub-topic 9.2 that the first minimum occurs when the angle with the straight-through position is given by $\theta = \frac{\lambda}{a}$ where λ is the wavelength of the waves and b is the slit width. With a circular aperture the equation is modified by a factor of 1.22, but derivation of this factor is beyond the scope of the IB Physics course. So we have

$$\theta = 1.22\frac{\lambda}{a}$$

in this case a is the diameter of the circular aperture (or very commonly the diameter of the lens or mirror forming the image).

The pupil of the eye has a diameter of about 3 mm and, taking visible light to have a wavelength in the order of 6×10^{-7} m, the minimum angle of resolution for the eye is $\theta = \frac{1.22 \times 6 \times 10^{-7}}{3 \times 10^{-3}} \approx 2 \times 10^{-4}$ rad. Optical defects in the eye mean that this limit is probably a little small.

The primary mirror of the Hubble Space Telescope has a diameter of 2.4 m. For this telescope, the minimum angle of resolution is $\theta = \frac{1.22 \times 6 \times 10^{-7}}{2.4} \approx 3 \times 10^{-7}$ rad. This is a factor of about a thousand smaller than that achieved by the unaided eye – which means that the Hubble Space Telescope is much better at resolving images. As this telescope is in orbit above the atmosphere, it avoids the atmospheric distortion which degrades images achieved by Earth-based telescopes. These concepts are covered in more detail in Option C (Imaging).

Worked example

A student observes two distant point sources of light. The wavelength of each source is 550 nm. The angular separation between these two sources is 2.5×10^{-4} radians subtended at the pupil of a student's eye.

a) State the Rayleigh criterion for the two images on the retina to be just resolved.

b) Estimate the diameter of the circular aperture of the eye if the two images are just to be resolved.

Solution

a) The images will be just resolved when the diffraction pattern from one of the point sources has its central maximum at the same position as the first minimum of the diffraction pattern of the other point source.

b) $\theta = 1.22\frac{\lambda}{b} => b = 1.22\frac{\lambda}{\theta} = \frac{1.22 \times 550 \times 10^{-9}}{2.5 \times 10^{-4}}$

$= 2.7 \times 10^{-3}\text{ m} \approx 3\text{ mm}$

Note

- You may wish to support your answer by drawing the two intensity–angle curves if you think your answer may not be clear.
- It is common for students to miss out the word "diffraction" in their answers – this is crucial to score all marks.

Nature of science

Diffraction and the satellite dish

When radiation is emitted from a transmitting satellite dish the waves diffract from the dish. The diameter of the dish behaves as an aperture. The angle θ made by the first minimum with the straight-through is related to diameter b by the equation

$$\theta = 1.22\frac{\lambda}{b}$$

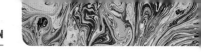

When the wavelength increases so does the angle through which the waves become diffracted. This means that the diffracted beam now covers a larger area. However, increasing the diameter of a dish narrows the beam and it covers a smaller area. In satellite communications the **footprint** is the portion of the Earth's surface over which the satellite dish delivers a specified amount of signal power. The footprint will be less than the region covered by the principal maximum because the signal will be too weak to be useful at the edges of the principal maximum. Figure 2 shows the footprint of a communications satellite.

- Small values of the dish diameter will give a large footprint but the intensity may be quite low since the energy is spread over a large area.
- The footprint of a satellite has social and political implications ranging from unwarranted observation to sharing television programmes.

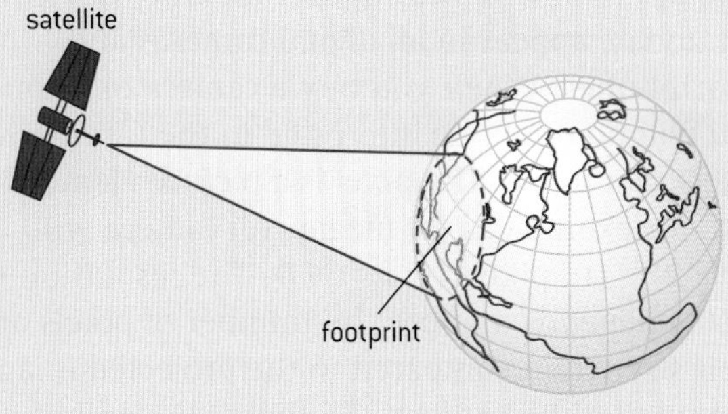

▲ Figure 2 Satellite footprint.

Resolvance of diffraction gratings

We have seen that diffraction gratings are used to disperse light of different colours. Such gratings are usually used with spectrometers to allow the angular dispersion for each of the colours to be measured. When using a spectrometer, knowing the number of lines per millimetre on the diffraction grating and measuring the angles that the wavelengths of light are deviated through allows the wavelengths of the colours to be determined. We have seen that there is a limit to how close two objects can be before their diffraction images are indistinguishable – the same is true of the different wavelengths that can be resolved using a particular diffraction grating.

When we previously looked at interference patterns for multiple slits in Sub-topic 9.3, we saw that increasing the number of slits improves the sharpness of the maxima formed. Figure 3 shows that, when light of the same wavelength is viewed from the same distance x, the angular dispersion θ_D for the principal maximum with the double slit is larger than angular dispersion θ_G for the diffraction grating. A sharper principal maximum is one with less angular dispersion. With wider maxima there is more overlap of images from different sources and lower resolution.

Using this argument we see that, when beams of light are incident on a diffraction grating, a wider beam covers more lines (a greater number of slits) and will produce sharper images and better resolution.

The resolvance R for a diffraction grating (or other device used to separate the wavelengths of light such a multiple slits) is defined as the ratio of the wavelength λ of the light to the smallest difference in wavelength that can be resolved by the grating $\Delta\lambda$.

The resolvance is also equal to Nm where N is the total number of slits illuminated by the incident beam and m is the order of the diffraction.

$$R = \frac{\lambda}{\Delta\lambda} = Nm$$

The larger the resolvance, the better a device can resolve.

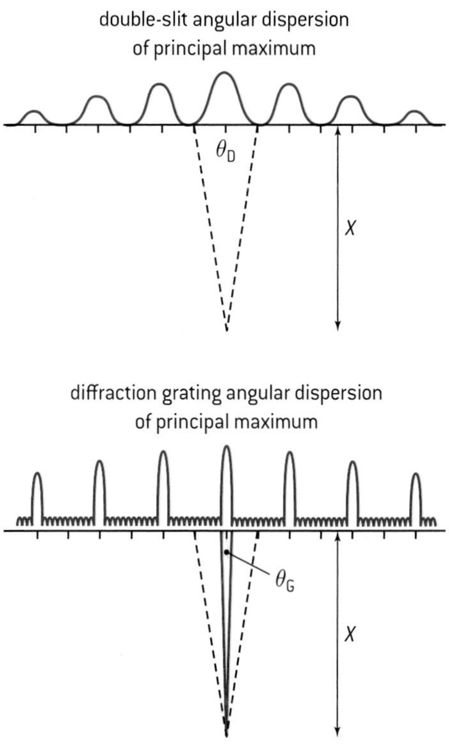

▲ Figure 3 Angular dispersion of principal maximum for diffraction grating compared with a double-slit.

Nature of science

Resolution in a CCD

Charge-coupled devices (CCDs) were originally developed for use in computer memory devices but, today, appear in all digital cameras and smartphones. When you buy a camera or phone you will no doubt be interested in the number of pixels that it has. The pixel is a picture element and, for example, a 20 megapixel camera will have 2×10^7 pixels on its CCD. The resolution of a CCD depends on both the number of pixels and their size when compared to the projected image. The smaller the camera, the more convenient it is to carry, and (at present) cameras with pixels of dimensions as small as 2.7 μm × 2.7 μm are mass-produced. In general CCD images are resolved better with larger numbers of smaller pixels. Figure 4 shows two images of the words "IB Physics" – the top uses a small number of large pixels, while the lower one is far better resolved by using a large number of small pixels. The upper image is said to be "pixellated".

▲ Figure 4 Images with a small number of large pixels and a large number of small pixels.

Worked example

Two lines in the emission spectrum of sodium have wavelengths of 589.0 nm and 589.6 nm respectively. Calculate the number of lines per millimetre needed in a diffraction grating if the lines are to be resolved in the second-order spectrum with a beam of width 0.10 mm.

Solution

$R = \frac{\lambda}{\Delta\lambda} = \frac{589.0}{0.6}$ (both values are in nanometres so this factor cancels)

You would be equally justified in using 589.6 (or 589.3 – the mean value) in the numerator here.

$R = 981.7$ (no units)

Thus $981.7 = Nm = N \times 2$

$\therefore N = 498.8$ lines

This is in a beam of width 0.10×10^{-3} m so, in 1 mm, there needs to be $4988 \approx 5000$ lines.

Since diffraction gratings are not normally made with 4988 lines mm^{-1}, the sensible choice is to use one with 5000 lines mm^{-1}!

9.5 The Doppler effect

Understanding

- The Doppler effect for sound waves and light waves

Nature of science

From water waves to the expansion of the universe

In his 1842 paper, *Über das farbige Licht der Doppelsterne* (Concerning the coloured light of the double stars), Doppler used the analogy of the measurement of the frequency of water waves to reason that the effect that bears his name should apply to all waves. Three years later, Buys Ballot verified Doppler's hypothesis for sound using stationary and moving groups of trumpeters. During his lifetime Doppler's hypothesis had no practical application; one hundred years on it has far-reaching implications for cosmology, meteorology and medicine.

Applications and skills

- Sketching and interpreting the Doppler effect when there is relative motion between source and observer
- Describing situations where the Doppler effect can be utilized
- Solving problems involving the change in frequency or wavelength observed due to the Doppler effect to determine the velocity of the source/observer

Equations

Doppler equation

- for a moving source:

$$f' = f\left(\frac{v}{v \pm u_s}\right)$$

- for a moving observer:

$$f' = f\left(\frac{v \pm u_o}{v}\right)$$

- for electromagnetic radiation:

$$\frac{\Delta f}{f} = \frac{\Delta\lambda}{\lambda} \approx \frac{v}{c}$$

Introduction

When there is relative motion between a source of waves and an observer, the observed frequency of the waves is different to the frequency of the source of waves. The apparent change in pitch of an approaching vehicle engine and a sounding siren are common examples of this effect. The Doppler effect has wide-ranging implications in both atomic physics and astronomy.

The Doppler effect with sound waves

Although we use general equations for this effect, we build up the equations under different conditions before combining them. In the following derivations (that need *not* be learned but will help you to understand the equations and how to answer questions on this topic) the letter s refers to the source of the waves (the object giving out the sound) and the letter o refers to the observer; f will always be the frequency of the source and f' the apparent frequency as measured by the observer.

Note

It is only the component of the wave in the source–observer direction that is used; perpendicular components of the motion do not alter the apparent frequency of the wave.

1. Moving source and stationary observer

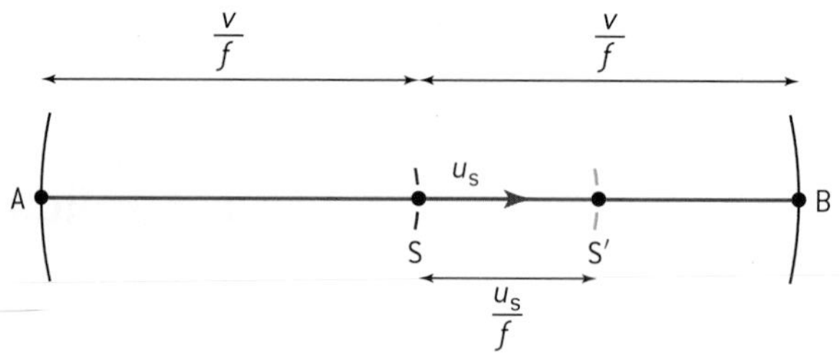

▲ Figure 1 The Doppler effect for a moving source and stationary observer.

At time $t = 0$ the source is at position S and it emits a wave that travels outwards in all directions, with a velocity v as shown in figure 1. At time $t = T$ (i.e. one period later) the wave will have moved a distance equivalent to one wavelength or (using $v = f\lambda$) a distance $= \frac{v}{f}$ to reach positions A and B. When the source is moving to the right with a velocity u_s, in time T it will have travelled to S′, a distance $u_sT(= \frac{u_s}{f})$. To a stationary observer positioned at B it will appear that the previously emitted crest has reached B but the source that emitted it has moved forwards and now is at S′. Therefore, to the observer at B, the apparent wavelength (λ') is the distance S′B $= \frac{v}{f} - \frac{u_s}{f}$

$$\therefore \quad \lambda' = \frac{v - u_s}{f}$$

Thus, the wavelength appears to be squashed to a smaller value. This means that the observer at B will hear a sound of a higher frequency than would be heard from a stationary source – the sound waves travel at speed v (which is unchanged by the motion of the source) so

$$f' = \frac{v}{\lambda'} = f\left(\frac{v}{v - u_s}\right)$$

To an observer at A the wavelength would have appeared to be stretched to a longer value given by S′A meaning that

$$\lambda' = \frac{v + u_s}{f}$$

and the observed frequency would be lower than that of a stationary source, meaning that

$$f' = \frac{v}{\lambda'} = f\left(\frac{v}{v + u_s}\right)$$

The two equations for f' can be combined into a single equation

$$f' = f\left(\frac{v}{v \pm u_s}\right)$$

in which the $\pm$ sign is changed to $-$ for a source moving towards a stationary observer and to $+$ for a source moving away from a stationary observer.

2. Moving observer and stationary source

In this case the source remains stationary but the observer at B moves towards the source with a velocity u_o. The source emits crests at a frequency f but the observer, moving towards the source, encounters the crests more often, in other words at a higher frequency f'. Relative to the observer, the waves are travelling with a velocity $v + u_o$ and the frequency is f'. The wavelength of the crests does not appear to have changed and will be $\lambda = \frac{v}{f}$. The wave equation ($v = f\lambda$) applied by the observer becomes $v + u_o = f'\frac{v}{f}$ meaning that the frequency measured by the observer will be

$$f' = f\left(\frac{v + u_o}{v}\right)$$

When the observer moves away from the source the wave speed appears to be $v - u_o$ and so

$$f' = f\left(\frac{v - u_o}{v}\right)$$

Again the two equations for f' can be combined into a single equation

$$f' = f\left(\frac{v \pm u_o}{v}\right)$$

in which the $\pm$ sign is changed to $-$ for an observer moving away from a stationary source and to $+$ for an observer moving towards a stationary source.

Worked example

A stationary loudspeaker emits sound of frequency of 2.00 kHz. A student attaches the loudspeaker to a string and swings the loudspeaker in a horizontal circle at a speed of 15 m s^{-1}. The speed of sound in air is 330 m s^{-1}.

An observer listens to the sound at a close, but safe, distance from the student.

a) Explain why the sound heard by the observer changes regularly.

b) Determine the maximum frequency of the sound heard by the observer.

Solution

a) As the loudspeaker approaches the observer, the frequency appears higher than the stationary frequency because the apparent wavelength is shorter – meaning that the wavefronts are compressed. As it moves away from the observer, the frequency appears lower than the stationary frequency because the apparent wavelength is stretched. The overall effect is, therefore, a continuous rise and fall of pitch heard by the observer.

b) The maximum observed frequency occurs when the speaker is approaching the observer so:

$f' = f\left(\frac{v}{v - u_s}\right) = 2000\left(\frac{330}{330 - 15}\right) = 2100$ Hz (this is 2 s.f. precision in line with 15 m s^{-1})

The Doppler effect with light

The Doppler effect occurs not only with sound but also with light (and other electromagnetic waves); in which case the frequency and colour of the light differs from that emitted by the source. There is a significant difference in the application of Doppler effect for sound and light waves. Sound is a mechanical wave and requires a medium through which to travel; electromagnetic waves need no medium. Additionally, one of the assumptions or postulates of special relativity is that the velocity of light waves is constant in all inertial reference frames – this means that, when measured by an observer who is not accelerating, the observer will measure the speed of light to be 3.00×10^8 m s^{-1} irrespective of whether the observer moves towards the source or away from it and, therefore, it is impossible to distinguish between the motion of a source and an observer. This is not true for sound waves, as we have seen.

Although the Doppler effect equations for light and sound are derived on completely different principles, providing the speed of the source (or observer) is much less than the speed of light, the equations give approximately the correct results for light or sound. The quantities u_s and u_o in the Doppler equations for sound have no significance for light, and so we need to deal with the relative velocity v between the source and observer. Thus, in either of the equations

$$f' = f\left(\frac{v}{v \pm u_s}\right)$$

or

$$f' = f\left(\frac{v \pm u_o}{v}\right)$$

the wave speed is that of electromagnetic waves (c) and one of the velocities u_s or u_o is made zero while the other is replaced by the relative velocity v.

The first of these equations becomes

$$f' = f\left(\frac{c}{c + v}\right) = f\left(\frac{1}{1 + \frac{v}{c}}\right) = f\left(1 + \frac{v}{c}\right)^{-1}$$

This can be expanded using the binomial theorem to approximate to

$$f' = f\left(1 - \frac{v}{c}\right) \approx f - f\frac{v}{c}$$

(ignoring all the terms after the second in the expansion).

This can be written as

$$f - f' \approx f\frac{v}{c}$$

or

$$\Delta f \approx f\frac{v}{c}$$

This equation is equivalent to

$$\Delta\lambda \approx \lambda\frac{v}{c}$$

Note

- Substituting values into the second equation will give the same resulting relationship – you may like to try this but remember you do not need to know any of these derivations.
- The equation is only valid when $c \gg v$ and so cannot usually be used with sound (where the wave speed is $\approx$ 300 m s^{-1} – unless the source or observer is moving much more slowly than this speed).

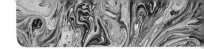

Worked example

As the Sun rotates, light waves received on Earth from opposite ends of a diameter show equal but opposite Doppler shifts. The speed of the edge of the Sun relative to the Earth is 1.90 km s^{-1}. What wavelength shift should be expected in the helium line having wavelength 587.5618 nm?

Solution

Using $\Delta f = f\frac{v}{c}$ this is equivalent to $\Delta\lambda \approx \lambda\frac{v}{c}$.

$1.90 \text{ km s}^{-1} = 1.90 \times 10^3 \text{ m s}^{-1}$

$$\Delta\lambda = 587.5618 \times \frac{1.90 \times 10^3}{3.00 \times 10^8} = 0.003\,7 \text{ nm}$$

Since one edge will approach the Earth – the shift from this edge will be a decrease in wavelength (blue shift) and that of the other edge (receding) will be an increase in wavelength (red shift).

Nature of science

Applications of the Doppler effect

1. Astronomy

The Doppler effect is of particular interest in astronomy – it has been used to provide evidence about the motion of the objects throughout the universe.

The Doppler effect was originally studied in the visible part of the electromagnetic spectrum. Today, it is applied to the entire electromagnetic spectrum. Astronomers use Doppler shifts to calculate the speeds of stars and galaxies with respect to the Earth. When an astronomical body emits light there is a characteristic spectrum that corresponds to emissions from the elements in the body. By comparing the position of the spectral lines for these elements with those emitted by the same elements on the Earth, it can be seen that the lines remain in the same position relative to each other but shifted either to longer or shorter wavelengths (corresponding to lower or higher frequencies). Figure 2 shows an unshifted absorption spectrum imaged from an Earth-bound source together with the same spectral lines from distant astronomical objects. The middle image is red-shifted indicating that the source is moving away from the Earth. The lower image is blue-shifted showing the source to be local to the Earth and moving towards us.

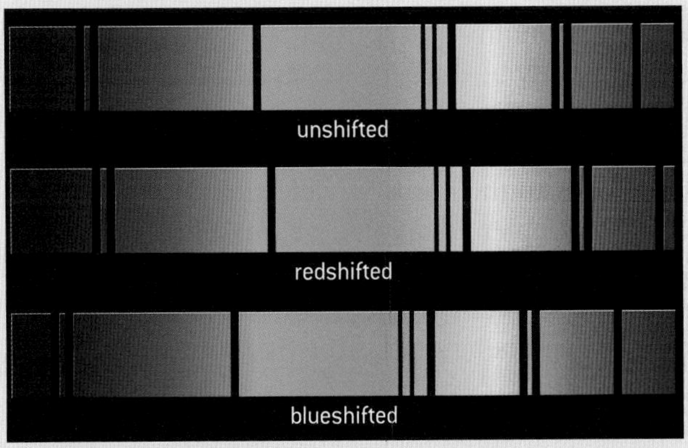

▲ Figure 2 The Doppler shifted absorption spectra.

Frequency or wavelength shifts can occur for reasons other than relative motion. Electromagnetic waves moving close to an object with a very strong gravitational field can be red-shifted – this, unsurprisingly, is known as *gravitational red-shift* and is discussed further in Option A. The cosmological red-shift arises from the expansion of space following the Big Bang and is what we currently detect as the cosmic microwave background radiation (this is further discussed in Option D).

2. Radar

Radar is an acronym for "radio detection and ranging". Although it was developed for tracking aircraft during the Second World War, the technique has wide-ranging uses today, including:

- weather forecasting
- ground-penetrating radar for locating geological and archaeological artefacts
- providing bearings
- radar astronomy

- use in salvaging
- collision avoidance at sea and in the air.

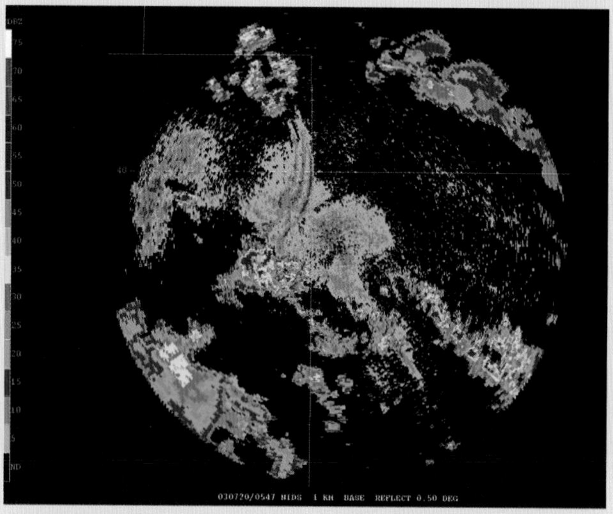

▲ Figure 3 Radar screen used in weather forecasting.

Radar astronomy differs from radio astronomy as it can only be used for Moon and planets close to the Earth. This depends on microwaves being transmitted to the object which then reflects them back to the Earth for detection. In order to use the Doppler equations, we must recognize that the moving object first of all behaves as a moving observer and then, when it reflects the microwaves, behaves as a moving source.

This means, for microwave sensing, the equation $\Delta f \approx f\frac{v}{c}$ is adapted to become $\Delta f \approx 2f\frac{v}{c}$.

3. Measuring the rate of blood flow

The Doppler effect can be used to measure the speed of blood flow in blood vessels in the body. In this case ultrasound (i.e. longitudinal mechanical waves of frequency above

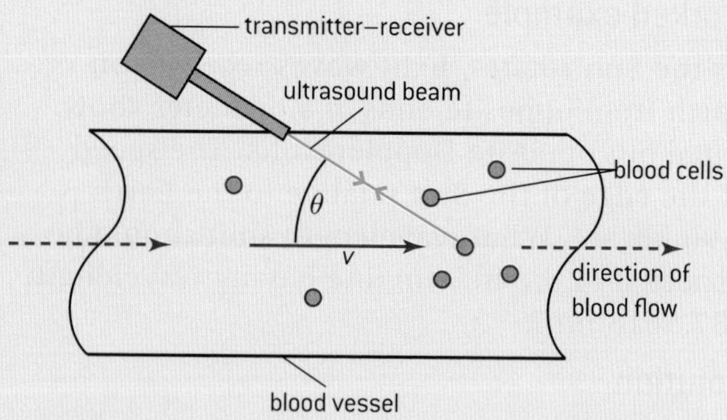

▲ Figure 4 The Doppler effect used to measure the speed of blood cells.

approximately 20 kHz) is transmitted towards a blood vessel. The change in frequency of the beam reflected by a blood cell is detected by the receiver. The speed of sound and ultrasound is around 1500 m s^{-1} in body tissue so the equation $\Delta f \approx f\frac{v}{c}$ is appropriate for blood cells moving at speeds of little more than 1 m s^{-1}. As the blood does not flow in the direction of the transmitter–receiver, there needs to be a factor that will give the necessary component of the blood velocity in a direction parallel to the transmitter – receiver. As with radar, there will need to be a factor of two included in the equation – the shift is being caused by the echo from a moving reflector. As can be seen from figure 4, the equation for the rate of blood flow will be $\Delta f \approx 2f\frac{v\cos\theta}{c}$.

The Doppler effect flowmeter may be used in preference to "in-line" flowmeters because it is non-invasive and its presence does not affect the rate of flow of fluids. The fact that it will remain outside the vessel that is carrying the fluid means it will not suffer corrosion from contact with the fluid.

Worked example

Microwaves of wavelength 150 mm are transmitted from a source to an aircraft approaching the source. The shift in frequency of the reflected microwaves is 5.00 kHz. Calculate the speed of the aircraft relative to the source.

Solution

The data gives a wavelength and a change of frequency so we need to find the frequency of the microwaves.

Using

$c = f\lambda$ gives $f = \frac{c}{\lambda} = \frac{3.00 \times 10^8}{150 \times 10^{-3}} = 2.00 \times 10^9$ Hz.

Using the Doppler shift equation for radar

$\Delta f \approx 2f\frac{v}{c} \Rightarrow v \approx \frac{\Delta f c}{2f} = \frac{5.00 \times 10^3 \times 3.00 \times 10^8}{2 \times 2.00 \times 10^9}$
$= 375$ m s^{-1}

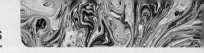

Questions

1 *(IB)*

The variation with displacement x of the acceleration a of a vibrating object is shown below.

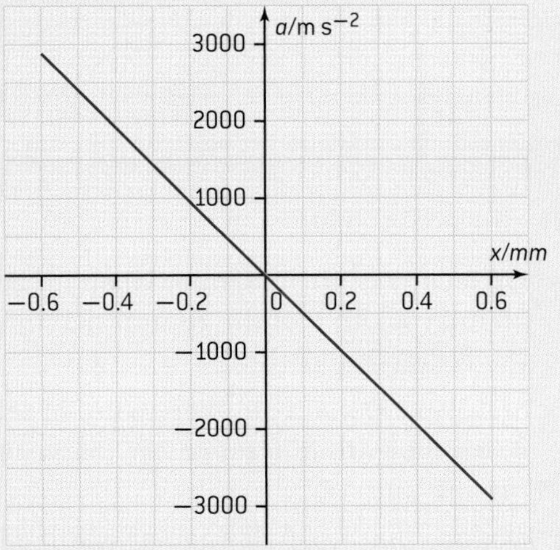

a) State and explain **two** reasons why the graph indicates that the object is executing simple harmonic motion.

b) Use data from the graph to show that the frequency of oscillation is 350 Hz.

c) State the amplitude of the vibrations.

(9 marks)

2 *(IB)*

a) A pendulum consists of a bob that is suspended from a rigid support by a light inextensible string. The pendulum bob is moved to one side and then released. The sketch graph shows how the displacement of the pendulum bob undergoing simple harmonic motion varies with time over one time period.

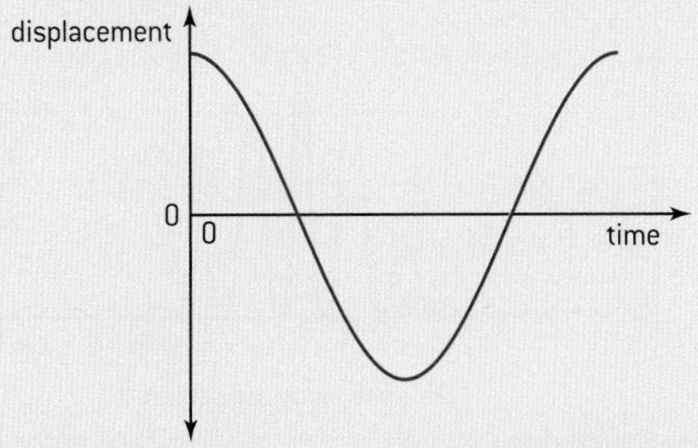

Copy the sketch graph and on it clearly label

(i) a point at which the acceleration of the pendulum bob is a maximum.

(ii) a point at which the speed of the pendulum bob is a maximum.

b) Explain why the magnitude of the tension in the string at the midpoint of the oscillation is greater than the weight of the pendulum bob.

c) The pendulum bob is moved to one side until its centre is 25 mm above its rest position and then released.

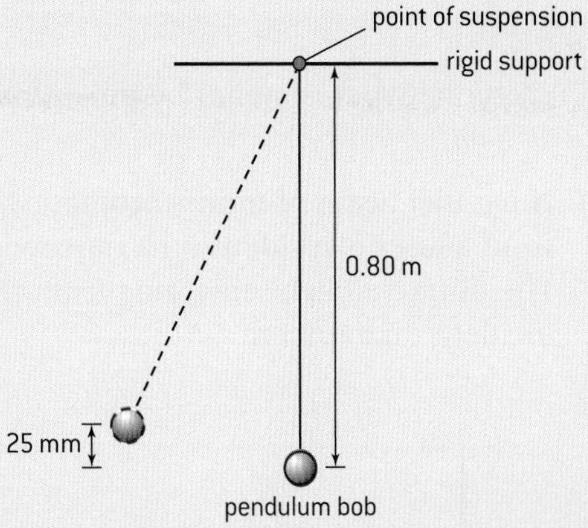

(i) Show that the speed of the pendulum bob at the midpoint of the oscillation is 0.70 m s^{-1}.

(ii) The mass of the pendulum bob is 0.057 kg. The centre of the pendulum bob is 0.80 m below the support. Calculate the magnitude of the tension in the string when the pendulum bob is vertically below the point of suspension.

(10 marks)

3 *(IB)*

a) A particle of mass m attached to a light spring is executing simple harmonic motion in a **horizontal direction**.

State the condition (relating to the net force acting on the particle) that is necessary for it to execute simple harmonic motion.

b) The graph shows how the kinetic energy E_K of the particle in (a) varies with the displacement x of the particle from equilibrium.

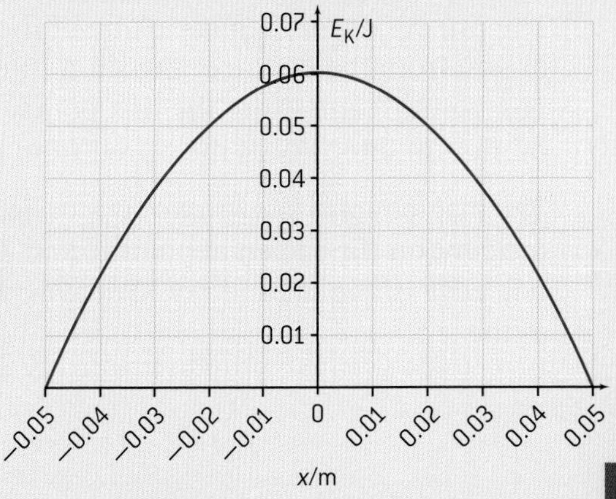

(i) On a copy of the axes above, sketch a graph to show how the potential energy of the particle varies with the displacement x.

(ii) The mass of the particle is 0.30 kg. Use data from the graph to show that the frequency f of oscillation of the particle is 2.0 Hz.

(8 marks)

4 *(IB)*

a) Describe what is meant by the *diffraction of light.*

b) A parallel beam of monochromatic light from a laser is incident on a narrow slit. The diffracted light emerging from the slit is incident on a screen.

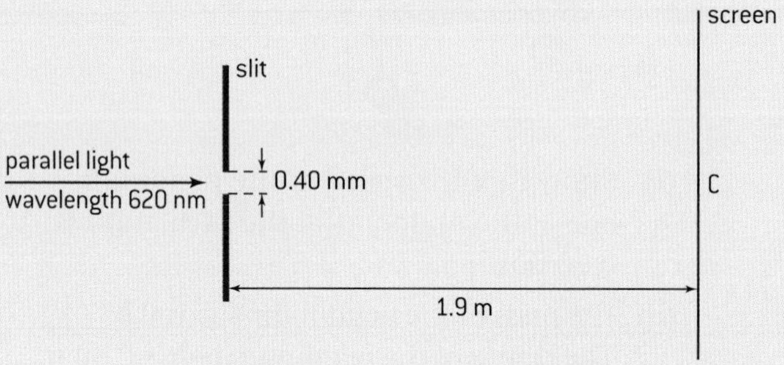

The centre of the diffraction pattern produced on the screen is at C. Sketch a graph to show how the intensity I of the light on the screen varies with the distance d from C.

c) The slit width is 0.40 mm and it is 1.9 m from the screen. The wavelength of the light is 620 nm. Determine the width of the central maximum on the screen.

(8 marks)

5 *(IB)*

Plane wavefronts of monochromatic light are incident on a narrow, rectangular slit whose width b is comparable to the wavelength λ of the light. After passing through the slit, the light is brought to a focus on a screen.

The line XY, normal to the plane of the slit, is drawn from the centre of the slit to the screen. The points P and Q are the first points of minimum intensity as measured from point Y.

The diagram also shows two rays of light incident on the screen at point P. Ray ZP leaves one edge of the slit and ray XP leaves the centre of the slit.

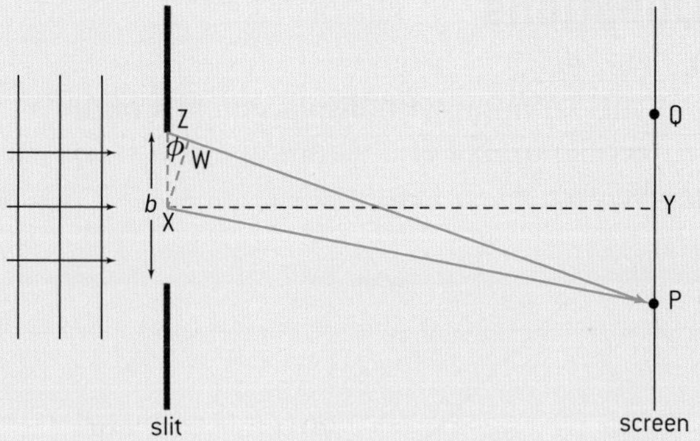

The angle ϕ is small.

a) On a copy of the diagram, label the half angular width θ of the central maximum of the diffraction pattern.

b) State and explain an expression, in terms of λ, for the path difference ZW between the rays ZP and XP.

c) Deduce that the half angular width θ is given by the expression

$$\theta = \frac{\lambda}{b}$$

d) In a certain demonstration of single slit diffraction, $\lambda = 450$ nm, $b = 0.15$ mm and the screen is a long way from the slits.

Calculate the angular width of the central maximum of the diffraction pattern on the screen.

(8 marks)

6 *(IB)*

Monochromatic parallel light is incident on two slits of equal width and close together. After passing through the slits, the light is brought to a focus on a screen. The diagram below shows the intensity distribution of the light on the screen.

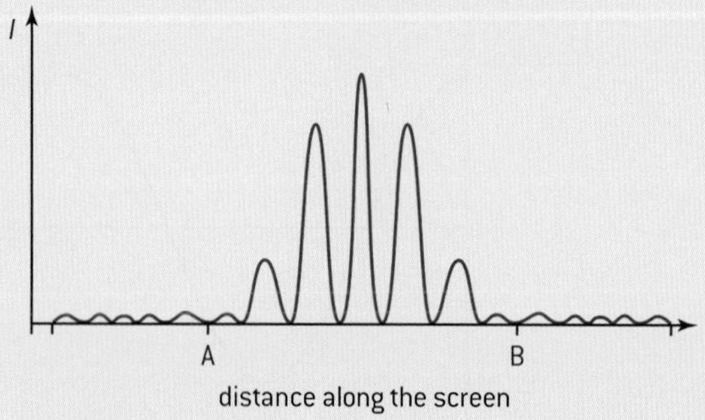

a) Light from the same source is incident on many slits of the same width as the widths of the slits above. On a copy of the diagram, draw a possible new intensity distribution of the light between the points A and B on the screen.

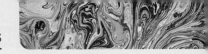

A parallel beam of light of wavelength 450 nm is incident at right angles on a diffraction grating. The slit spacing of the diffraction grating is 1.25×10^{-6} m.

b) Determine the angle between the central maximum and first order principal maximum formed by the grating.

(4 marks)

7 *(IB)*
Light of wavelength 590 nm is incident normally on a diffraction grating, as shown below.

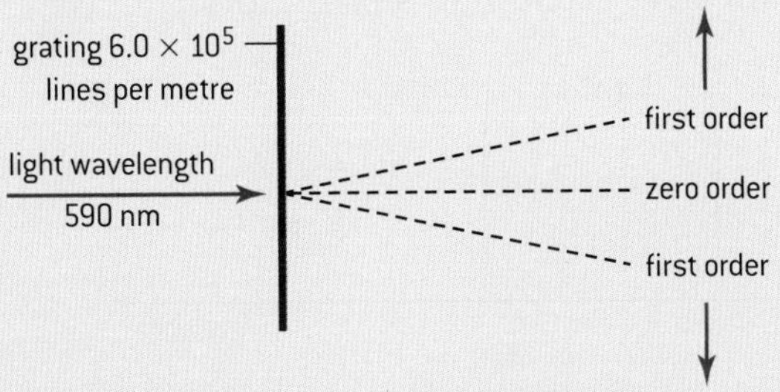

The grating has 6.0×10^5 lines per metre.

a) Determine the **total** number of orders of diffracted light, including the zero order, that can be observed.

b) The incident light is replaced by a beam of light consisting of two wavelengths, 590 nm and 589 nm.

State **two** observable differences between a first-order spectrum and a second-order spectrum of the diffracted light.

(6 marks)

8 *(IB)*
Monochromatic light is incident on a thin film of transparent plastic as shown below.

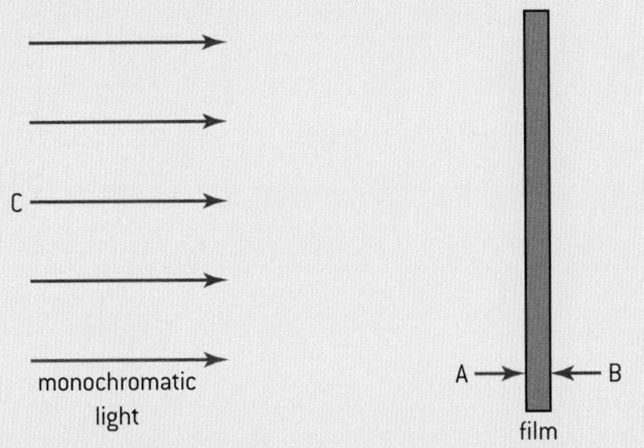

The plastic film is in air.

Light is partially reflected at both surface A and surface B of the film.

a) State the phase change that occurs when light is reflected from

(i) surface A

(ii) surface B.

The light incident on the plastic has a wavelength of 620 nm. The refractive index of the plastic is 1.4.

b) Calculate the minimum thickness of the film needed for the light reflected from surface A and surface B to undergo destructive interference.

(5 marks)

9 *(IB)*

The two point sources A and B emit light of the same frequency. The light is incident on a rectangular narrow slit and, after passing through the slit, is brought to a focus on the screen.

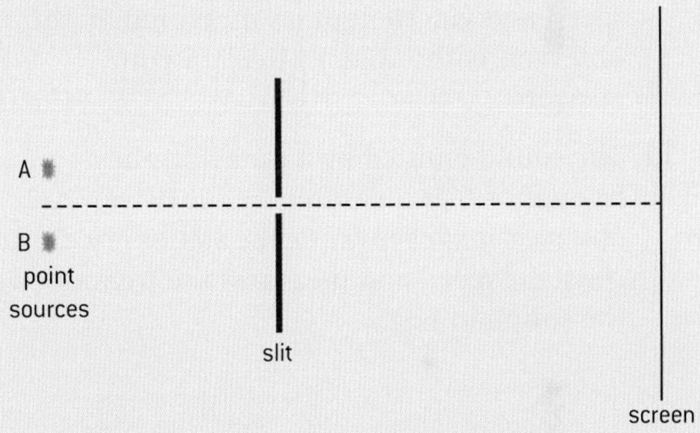

a) B is covered. Sketch a graph to show how the intensity I of the light from A varies with distance along the screen. Label the curve you have drawn A.

b) B is now uncovered. The images of A and B on the screen are just resolved. Using your axes, sketch a graph to show how the intensity I of the light from B varies with distance along the screen. Label this curve B.

c) The bright star Sirius A is accompanied by a much fainter star, Sirius B. The mean distance of the stars from Earth is 8.1×10^{16} m. Under ideal atmospheric conditions, a telescope with an objective lens of diameter 25 cm can just resolve the stars as two separate images.

Assuming that the average wavelength emitted by the stars is 500 nm, estimate the apparent, linear separation of the two stars.

(6 marks)

10 *(IB)*

a) Explain what is meant by the term *resolvance* with regards to a diffraction grating.

b) A grating with a resolvance of 2000 is used in an attempt to separate the red lines in the spectra of hydrogen and deuterium.

(i) The incident beam has a width of 0.2 mm. For the first order spectrum, how many lines per mm must the grating have?

(ii) Explain whether or not the grating is capable of resolving the hydrogen lines which have wavelengths 656.3 nm and 656.1 nm.

(6 marks)

11 A source of sound approaches a stationary observer. The speed of the emitted sound and its wavelength, measured at the source, are v and λ respectively. Compare the wave speed and wavelength, as measured by the observer, with v and λ . Explain your answers. (4 marks)

12 The sound emitted by a car's horn has frequency f, as measured by the driver. An observer moves towards the stationary car at constant speed and measures the frequency of the sound to be f'.

a) Explain, using a diagram, any difference between f' and f.

b) The frequency f is 3.00×10^2 Hz. An observer moves towards the stationary car at a constant speed of 15.0 m s^{-1}. Calculate the observed frequency f' of the sound. The speed of sound in air is 3.30×10^2 m s^{-1}.

(5 marks)

13 *(IB)*

The wavelength diagram shown below represents three lines in the emission spectrum sample of calcium in a laboratory.

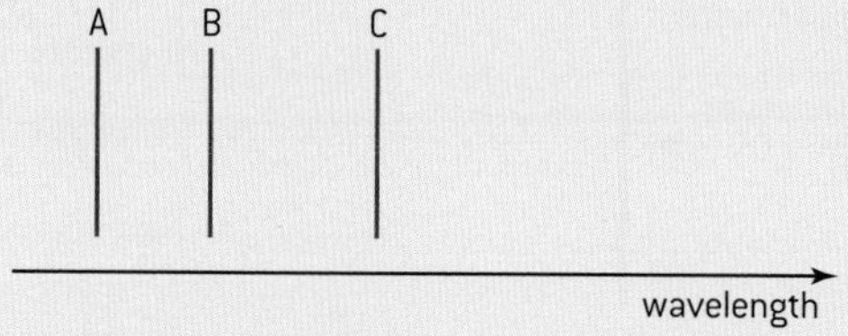

A distant star is known to be moving directly away from the Earth at a speed of 0.1c. The light emitted from the star contains the emission spectra of calcium. Copy the diagram and sketch the emission spectrum of the star as observed in the laboratory. Label the lines that correspond to A, B, and C with the letters A*, B*, and C*. Numerical values of the wavelengths are **not** required. (3 marks)

10 FIELDS (AHL)

Introduction

Earlier in the book you were introduced to gravitation and electrostatics. At that stage, these were treated as separate sets of ideas, but there are strong similarities between the concepts in both topics. In this topic we are going to discuss these similarities and develop the concepts of gravitation and electrostatics in parallel.

10.1 Describing fields

Understanding

- → Gravitational fields
- → Electrostatic fields
- → Electric potential and gravitational potential
- → Field lines
- → Equipotential surfaces

Applications and skills

- → Representing sources of mass and charge, lines of electric and gravitational force, and field patterns using an appropriate symbolism
- → Mapping fields using potential
- → Describing the connection between equipotential surfaces and field lines

Equations

- → work–electric potential equation: $W = Q\Delta V_e$
- → work–gravitational potential equation: $W = m\Delta V_g$

Nature of science

Our everyday experience of forces is that they are direct and observable. They act directly on objects and the response of the object is calculable. Acceptance of the field concept means acceptance of action at a distance. One object can influence another without the need for contact between them. The paradigm shift from one world view to another is difficult. It took a significant effort in the history of science and demands a leap in conceptual understanding from physicists.

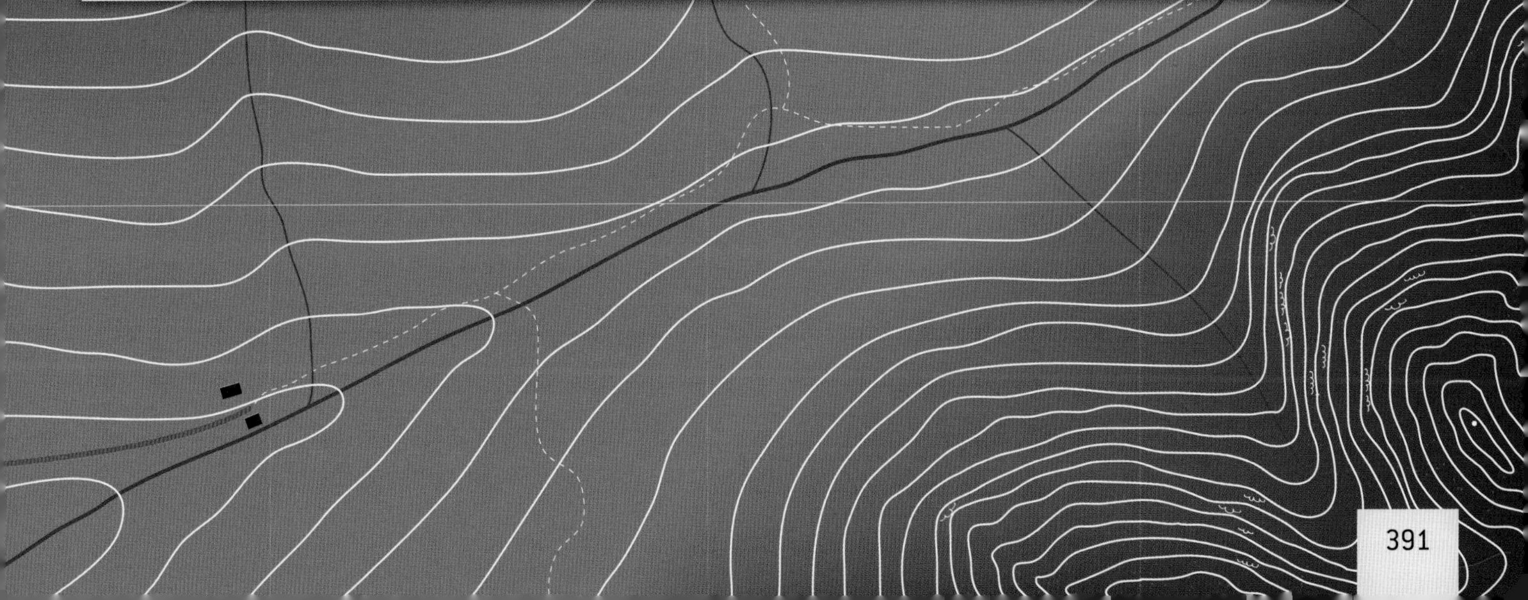

Nature of science

Scalar and vector fields

Strictly, the fields here are vector fields as the forces have magnitude and direction. Other fields are different in that the quantities they represent only have magnitude – they are scalar fields. The temperature associated with each point around a camp fire is an example of a scalar field.

Nature of science

Action-at-a-distance

The *action-at-a-distance* envisaged here is imagined to act simultaneously even though it does not. Topic 12 teaches you that contemporary models of physics explain interactions using the concept of *exchange particles* that have a finite (and calculable) lifetime.

The parts of this chapter devoted to electric fields have a blue background or a blue line in the margin; the sections dealing with gravity have a green background or a green line in the margin. You can concentrate on one type of field by sticking to one colour.

Before we introduce new ideas, here is a review of those ideas that we considered in the earlier topics.

Fields

A field is said to exist when one object can exert a force on another object at a distance.

Electric fields	***Gravity fields***
An electrostatic force exists between two charged objects. There are two types of charge: • negative, corresponding to a surplus of electrons in the object • positive, corresponding to a deficit of electrons in the object.	A gravitational force exists between two objects that both have mass. A gravitational field is associated with each mass. Any other mass in this field has a gravitational field acting on it.
When the two objects have the same sign of charge, then the force between them is repulsive. When the objects have opposite signs, then the force between them is attractive.	Gravity is always an attractive force. Repulsion between masses is never observed.
All electric charges give rise to an electric field. An electrostatic force acts on a charge that is in the field of another charge.	

Field strength

Electric fields	***Gravity fields***
The definition of field strength is similar in both types of field. It arises from a "thought experiment" involving the measurement of force acting on a test object. Near the surface of the Earth, the gravitational field strength is 9.8 N kg^{-1} and, on a clear day, the electric field strength is about 100 N C^{-1}. Both fields point downwards. In both gravity and electrostatics there is a problem with carrying out the measurement of field strength practically. The presence of a test object will distort and alter the field in which it is placed since the test object carries its own field.	
electric field strength $= \frac{\text{force acting on positive test charge}}{\text{magnitude of test charge}}$ The direction of the field is the same as the direction of the force acting on a positive charge. We have to pay particular attention to direction in electrostatics because of the presence of two signs of charge. The unit of electric field strength is N C^{-1}.	gravitational field strength $= \frac{\text{force acting on test mass}}{\text{magnitude of test mass}}$ Gravitational force is always attractive. This means that the direction of field strength is always towards the mass that gives rise to the field. The unit of gravitational field strength is N kg^{-1}.

Energy ideas so far

Electric fields	*Gravity fields*
Potential difference In Topic 5 electric potential difference was used in electrostatics and current electricity as a measure of energy transfer. $V = \frac{W}{Q}$ V is the electric potential difference and W is the work done on a positive test charge of size Q. V is measured in volts, which is equivalent to joule coulomb^{-1}.	**Potential energy** In Topic 2 gravitational potential energy was used as the measure of energy transfer when a mass m is moved in a gravitational field. $\Delta E_p = mg\Delta h$ Where ΔE_p is the change in gravitational potential energy, m is the mass, g is the gravitational field strength, and Δh is the change in vertical height. ΔE_p is measured in joules. This equation applies when g is effectively constant over the height change being considered. Remember that g has to be the local value of the field strength – near the Earth's surface 9.8 N kg^{-1} is the value to use, but further out a smaller value would be required.
The electric potential difference refers to work done per unit charge, gravitational potential energy refers to work done = (*force* (*weight*) × *distance moved*).	

Now read on ...

The recap of earlier work at the start of this topic should have reminded you of the similarities and differences between gravitational and electric fields.

The differences include:

- Gravity is most noticeable when masses are very large – the size of planets, stars, and galaxies.
- Electric forces largely control the behaviour of atoms and molecules but, because there is usually a complete balance between numbers of positive and negative charges, the effects are not easily noticed over large distances.
- Gravitational forces on the other hand are perceived to act over astronomical distances.

Despite these differences, the two fields share an approach that allows the study of one to inform the other. Studying the two types of field in parallel will help you understand the links. If you need to concentrate on either gravity or electric fields, look at the relevant background colour or line in the margin.

Field lines

Field lines help us to visualize the shapes of electric fields that arise from static charges. They also aid the study of the magnetic fields associated with moving charges. Magnetic fields and their consequences are discussed further in Topic 11.

If an object in a field has the relevant field property (mass for gravity, charge for electricity, etc.), then it will be influenced by the field. In a Topic 5 *Investigate!*, small particles (semolina grains) were suspended in a fluid. This experiment gives an impression of the field shape between charged parallel plates and other arrangements of charge. However the experiments are tricky to carry out and give no quantitative data. Here is another qualitative experiment that gives more insight into the behaviour of electrostatic fields.

Investigate!

Field between parallel plates

- A small piece of foil that has been charged can be used to detect the presence of an electric field.
- The detector is made from a rod of insulator – a plastic ruler or strip of polythene are ideal. Attached to the rod is a small strip of foil: thin aluminium or gold foil or "Dutch" metal are suitable. The dimensions of the foil need to be about 4 cm × 1 cm and the foil can be attached to the rod using adhesive tape.
- Set up two vertical parallel metal plates connected to the terminals of a power supply that can deliver about 1 kV to the plates. Take care when carrying out this experiment (use the protective resistor in series with the supply if necessary). Begin with the plates separated by a distance about one-third of the length of their smaller side.
- Touch the foil briefly to one of the plates. This will charge the foil. You should now see the foil bend away from the plate it touched.
- The angle of bend in the foil indicates the strength of the electric field. Explore the space between the plates and outside them too. Notice where the field starts to become weaker as the detector moves outside the plate region. Does the force indicated by the detector vary inside the plate region or is it constant?
- Turn off the supply and change the spacing between the plates. Does having a larger separation produce a larger or a smaller field?
- Change the pd between the plates. Does this affect the strength of the field?
- An alternative way to carry out the experiment is to use a candle flame in the space between the plates. What do you notice about the shape of the flame when the field is turned on? Can you explain this in terms of the charged ions in the flame?
- The foil detector itself can also be used to explore the field around a charged metal sphere such as the dome of a van der Graaf generator.

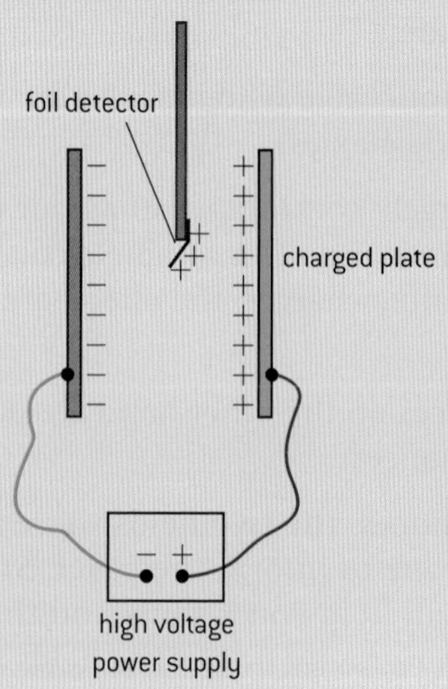

▲ Figure 1 Electric field detector.

Experiments such as this suggest that the field between two parallel plates:

- is uniform in the region between the plates
- becomes weaker at the edges (these are known as **edge effects**) as the field changes from the between-the-plates value to the outside-the-plates value (often zero). You should be able to use the properties of the field lines from Topic 5 to explain why there can be no abrupt change in field strength.

  A close study of the field shows field lines like those in figure 2. At the edge, the field lines curve outwards as the field gradually weakens from the large value between the plates to the much weaker field well away from them. For the purposes of this course, you should assume that this curving begins at the end of the plate (although in reality it begins a little way in as you may have seen in your experiment with the field detector).

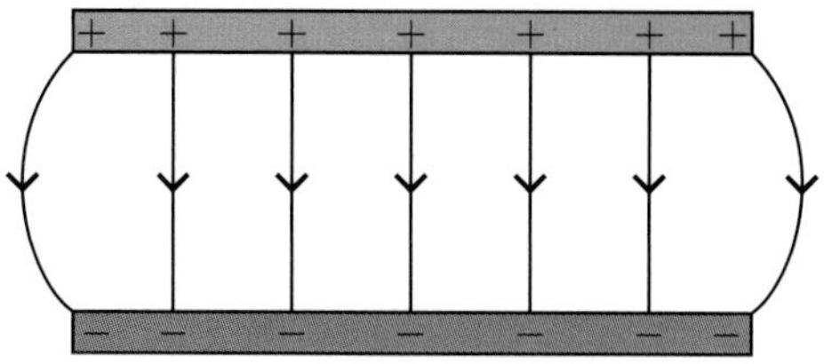

▲ Figure 2 Electric field line pattern for parallel plates.

Try to predict the way in which the shape of the field lines might change if a small conductive sphere is introduced in the middle of the space between the two plates. Look back at some of the field pattern results from Topic 5 and decide how the field detector used here would respond to the fields there.

Sometimes you will see or hear the field lines called "lines of force". It is easy to see why. The field detector shows that forces act on charges in a field. The lines of force originate at positive charge (by definition) and end at negative charge as indicated by the arrow attached to the line. If the line is curved, the tangent at a given point gives the direction of the electric force on a positive charge. The density of the lines (how close they are) shows the strength of the force.

Linking field lines with electric potential

The *Investigate!* indicates that the strength of the electric field depends on the potential difference between them. Another experiment will allow you to investigate this in more detail.

Investigate!

Measuring potentials in two dimensions

- This experiment gives a valuable insight into the way potential varies between two charged parallel plates.
- You need a sheet of a paper that has a uniform graphite coating on one side, one type is called "Teledeltos" paper. In addition, you will need a 6 V power supply (domestic batteries are suitable), two strips of copper foil, and two bulldog clips. Finally, you will need a high-resistance voltmeter (a digital meter or an oscilloscope will be suitable) and connecting leads.

- Connect the circuit shown in the diagram. Take particular care with the connections between the copper strips and the paper. One way to improve these is to paint the connection between copper and paper using a liquid consisting of a colloidal suspension of graphite in water.
- Press the voltage probe onto the paper. There should be a potential difference between the lead and the 0 V strip.
- Choose a suitable value for the voltage (say 3 V) and explore the region between the copper strips. Mark a number of points where the voltage is 3 V and draw on the paper joining the points up. Repeat for other values of voltage. Is there a consistent pattern to a line that represents a particular voltage?
- Try other configurations of plates. One important arrangement is a point charge, which can be simulated with a single point at 6 V and a circle of copper foil outside it at 0 V. One way to create the point is to use a sewing needle or drawing pin (thumb tack). You will need the colloidal graphite to make a good connection between the point and the paper.
- An alternative way to carry out the experiment, this time in three-dimensions, is to use the same circuit with a tank of copper(II) sulfate solution and copper plates.

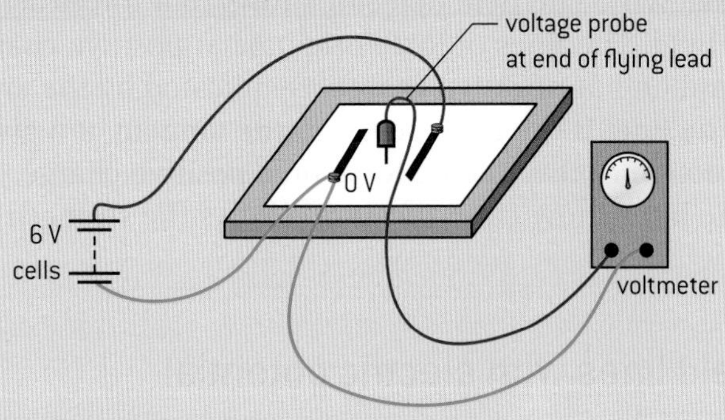

▲ Figure 3 Equipotentials in two-dimensions.

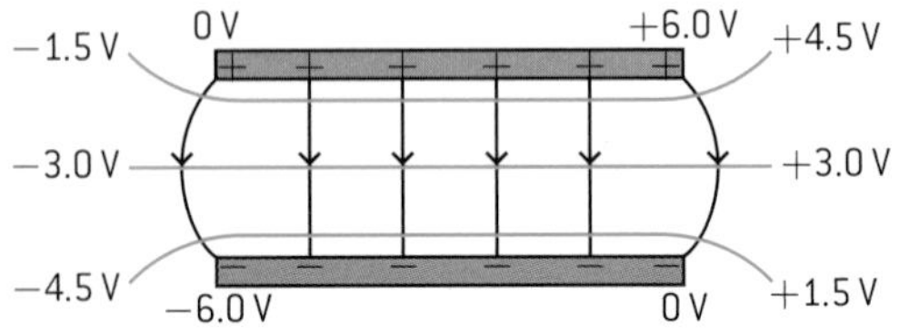

▲ Figure 4 Lines of equal potential.

The lines of equal potential difference between two parallel plates drawn in this experiment resemble those shown in figure 4. These lines are called **equipotentials** because they represent points on the two-dimensional paper where the voltage is always the same. When a charge moves from one point to another along an equipotential line, no work is done.

Figure 4 also shows the electric field lines between the plates. You will see a simple relationship between the field lines and the equipotentials. The angle between them is always 90°. If you know either the shape of the equipotential lines or the shape of the field lines then you can deduce the shape of the other. The same applies to the relationship between rays and wavefronts in optics.

Tip

Think about why the field line and the equipotential must be at 90° to each other. As a hint, remember the definition of work done in terms of force and distance moved from Topic 2.

You will have noticed that the word "difference" has disappeared and that we are using the plain word "potential". As long as we always refer to some agreed reference point (in this case the 0 V copper strip on the paper), the absolute values of potential measured on the voltmeter are identical to the values of potential difference measured across the terminals of the meter.

To illustrate this, the parallel-plate equipotentials in figure 4 are re-labelled on different sides of the diagram taking a different point in the circuit to be zero. You will see that the potential differences are unchanged, but the change of reference position shifts all the potential values by a fixed amount. This definition of a reference position for potential will be considered in the next sub-topic.

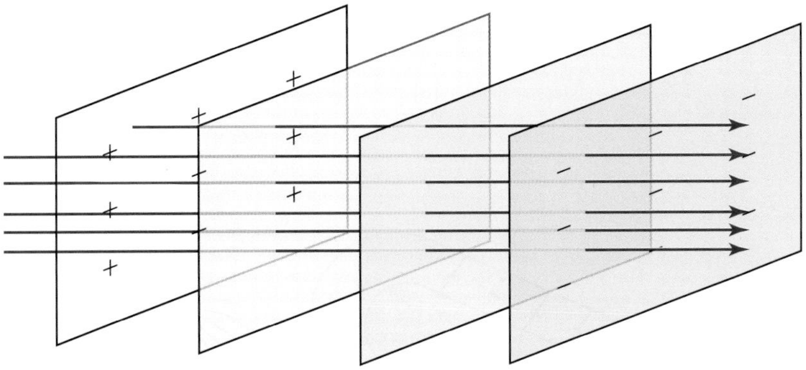

Figure 5 Equipotentials in three dimensions.

The conducting-paper experiment is two-dimensional. The foil detector allowed you to explore a three-dimensional field. The equipotentials between the charged plates in three dimensions are in the form of sheets parallel to the plates themselves always at 90° to the field lines. So another consequence of extending to three dimensions is that it is possible to have equipotential surfaces and even volumes. An example of this is a solid conductor: if there is a potential difference between any part of the conductor and another, then charge will flow until the potential difference has become zero. All parts of the conductor must therefore be at the same potential as each other – the whole conductor is an equipotential volume. Consequently the field lines must emerge from the volume at 90° – whatever shape it is.

In summary, equipotential surfaces or volumes:

- link points having the same potential
- are regions where charges can move without work being done on or by the charge
- are cut by field lines at 90°
- should be referred to a zero of potential
- do not have direction (potential, like any energy, is a scalar quantity)
- can never cross or meet another equipotential that has a different value.

Tip

Calculate the change in potential going from −4.5 V to −3.0 V as well as from +1.5 V to +3.0 V. In both cases, we go from a low potential point to a higher potential point so the potential difference should be positive and the same in both cases (+1.5 V). Remember differences are calculated from final state minus initial state.

Equipotentials and field lines in other situations

Figure 6 shows the equipotentials and electric field lines for a number of common situations.

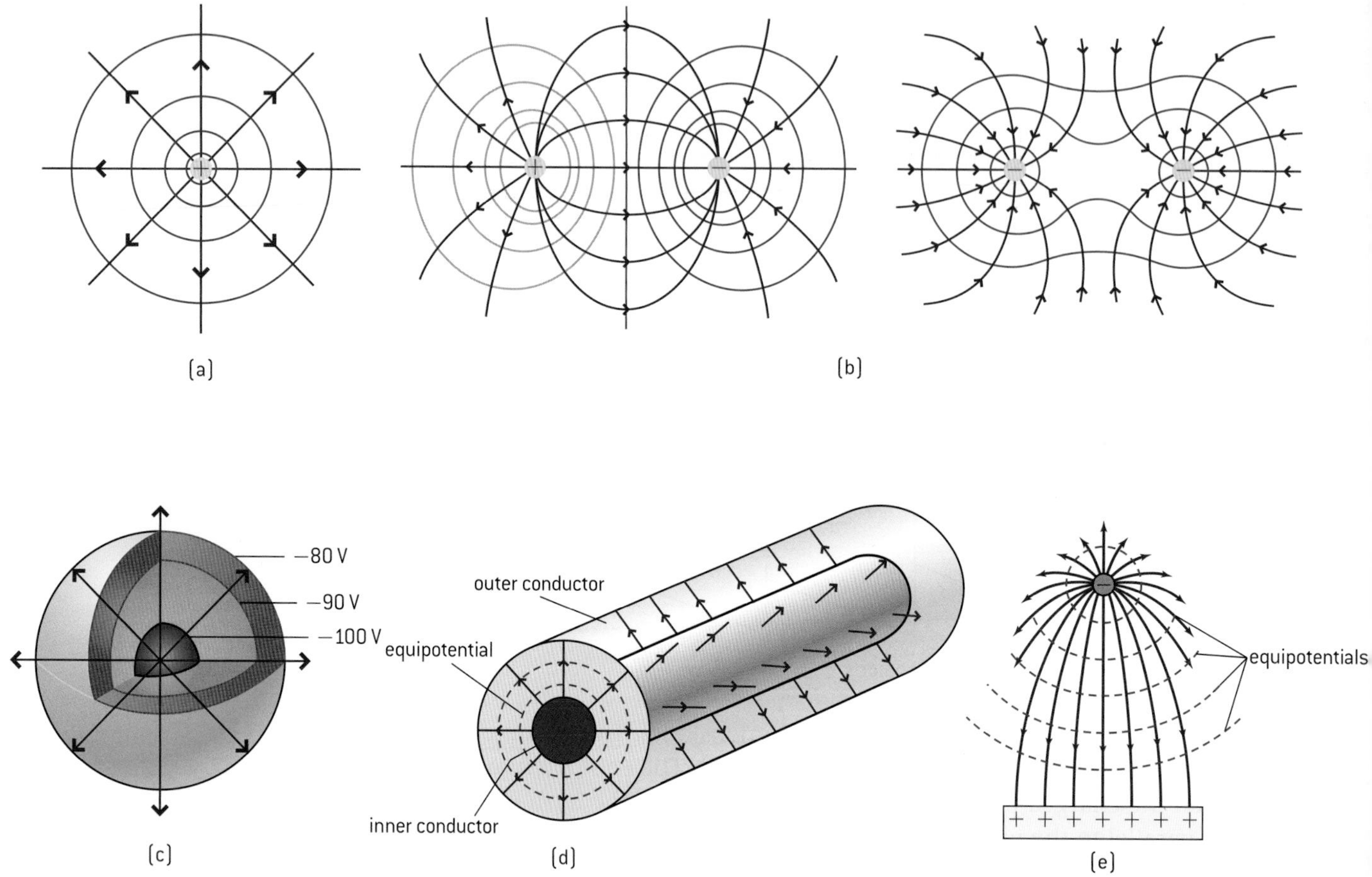

▲ Figure 6 Equipotentials around charge arrangements.

(a) Field due to a single point charge

Earlier we looked at the field due to a single point charge.

In the case of a positive charge, the field lines radiate out from the point and the field in this case is called a **radial field**. For a negative charge the only change is that the arrows now point inwards to the charge.

What are the equipotentials for this case? The key lies in the 90° relationship between field line and equipotential.

Figure 6(a) shows the arrangement of equipotentials around a single point charge. They are a series of concentric shells centred on the point charge. This arrangement is shown as a two-dimensional arrangement, but it is important always to think in three dimensions. Two-dimensional diagrams are however easier to draw and perfectly acceptable in an examination.

(b) Field due to two point charges of the same and opposite sign

These arrangements bear *some* similarities to the magnetic field patterns in the space between two bar magnets; they demonstrate the use of the rules for the electric field lines.

(c) Field due to a charged sphere

A hollow or solid conducting sphere is at a single potential.The field outside the sphere is identical to that of a point charge of the same magnitude placed at the centre of the sphere. This time however, there are no field lines inside the sphere.

(d) Field due to a co-axial conductor

Co-axial conductors are commonly used in the leads that connect a domestic satellite dish to a TV. There is a central conductor with an earthed cylinder outside it, separated and spaced by an insulator. The symmetry of the arrangement is different from that of a sphere and so the symmetry of field pattern changes too.

(e) Field between a point charge and a charged plate

The key to this diagram is to remember that the field lines are radial to the point charge when very close to it and the lines must be at 90° to the surface of the plate.

Field and equipotential in gravitation

We did not use the concept of gravitational field lines in Topic 6. However, field lines can be used to describe gravitational fields. The concepts of field lines and equipotentials transfer over to gravity without difficulty.

Although there is no easy experiment that can make the gravitational field lines visible (unlike magnetic or electric field lines), the direction can be determined simply by hanging a weight on a piece of string! The direction of the string gives the local direction of the field pointing towards the centre of the planet.

The arrangement of gravitational field lines that surround a point mass or a spherical planet is similar to the pattern for electric field lines around a negative point charge. Again, the field is radial for both cases with the field lines directed towards the mass – because gravitational force can only be attractive.

All that has been written about equipotentials in electric fields also applies to gravitational equipotentials. Imagine a mass positioned at a vertical height above the Earth's surface, say 10 m. If the mass moves to another location which is also 10 m above the surface, then no work has to be done (other than against the non-conservative forces of friction) to move it. The mass begins and ends its journey on the same equipotential and no work has been transferred. These two points lie on the surface of a sphere whose centre is at the centre of the planet.

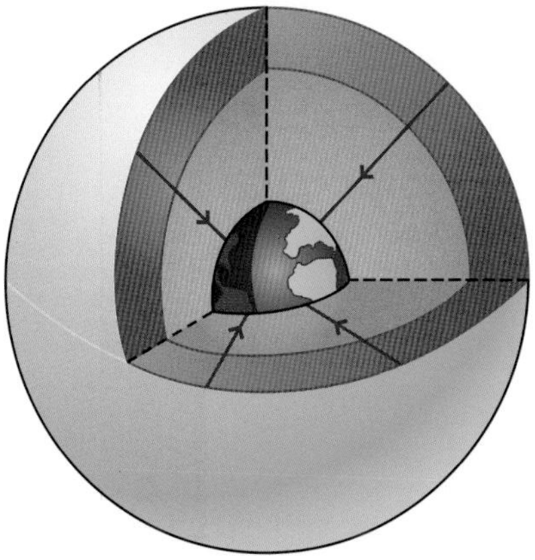

▲ Figure 7 Field lines and equipotentials around a planet.

There is another way to interpret gravitational equipotentials and field lines. It is possible that you regularly use equipotentials without even realizing it!

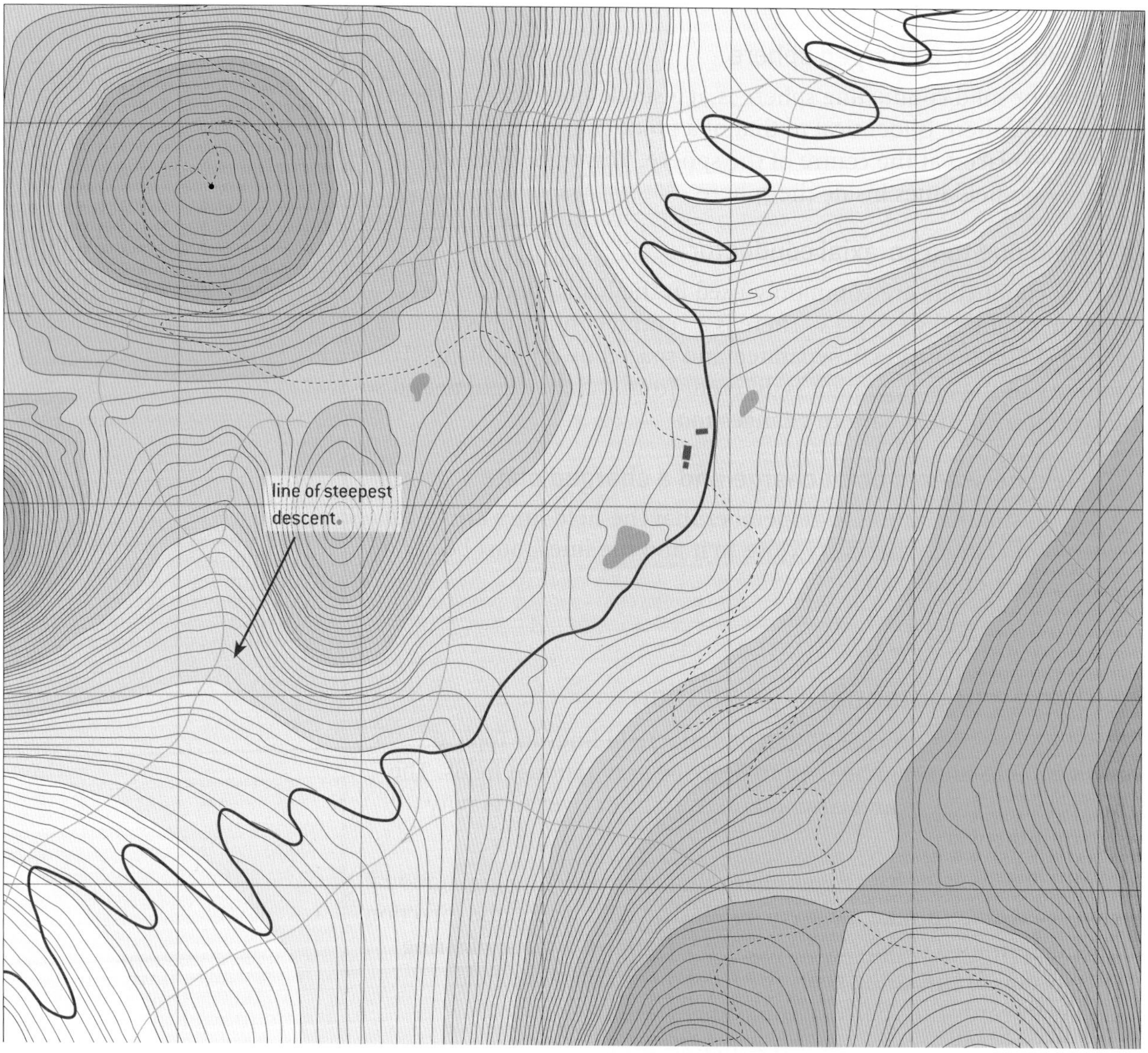

▲ Figure 8 Contour map (contour heights have been removed for clarity).

The contour lines on this map are equipotentials. If you walk along a contour line then your vertical height does not change, and you will neither gain nor lose gravitational potential energy. Mountain walkers have to be experts in using contour maps, not just to know where they are, but to choose where it is safe to go. Looking across the contours where they are closest together gives the line of steepest descent. This gives the direction of the gravitational field down the slope at that point.

Just as for electric fields, a gravitational field is uniform and directed downwards at 90° normal to the surface when close to the surface of the object.

Worked examples

1 An electrostatic device has electric field lines as shown in the diagram.

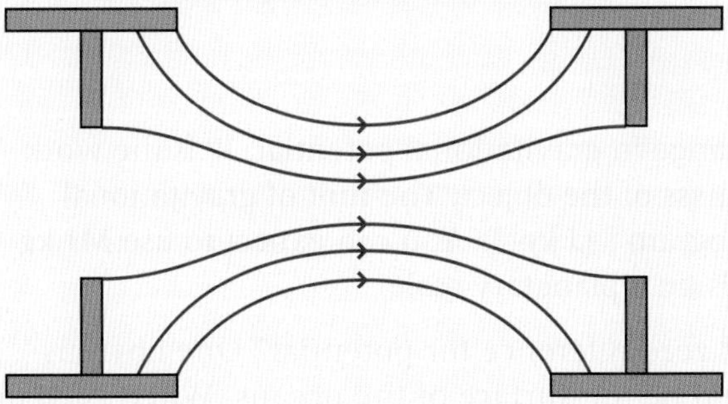

Draw the diagram and add to it **five** possible equipotential lines for the arrangement.

Solution

The equipotential lines must always meet the field lines at 90°. One possible line is a vertical line in the centre of the diagram. On each side of this the lines must bend to meet the lines appropriately.

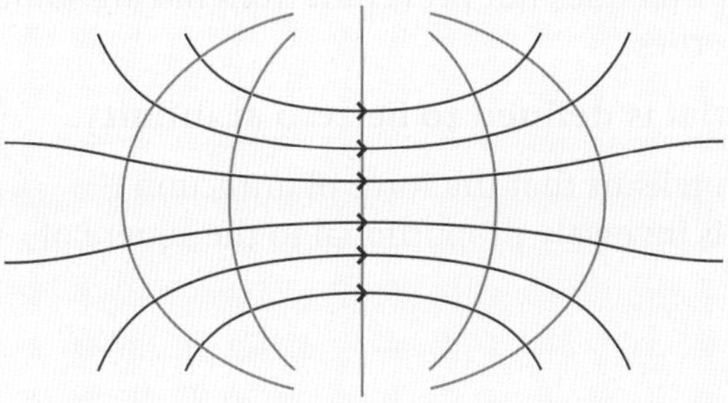

2 The diagram shows electric equipotential lines for an electric field. The values of the equipotential are shown. Explain where the electric field strength has its greatest magnitude.

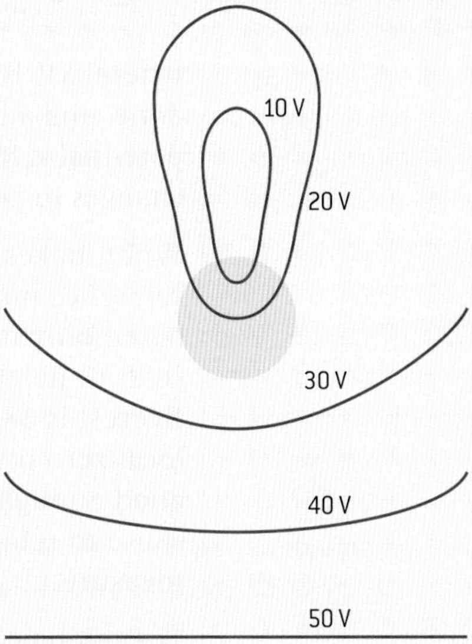

Solution

The work done in moving between equipotentials is the same between each equipotential. The *work done* is equal to *force* × *distance*. So the force on a test charge is greatest where the distance between lines is least. This is in the region around the base of the 20 V equipotential. Providing that potential change is the same between neighbouring equipotential lines, then the closer the equipotential lines are to each other, the stronger the electric field strength will be.

Potential at a point

Gravitational potential

So far the terms *potential* and *potential difference* have been used almost interchangeably by saying that *potential* needs a reference point. But this needs clarification. We need to define *potential* carefully for both gravitation and electrostatics and, in particular, we need to decide on the reference point we will use for our zero.

It is easier to begin with gravity:

Gravitational potential difference is the work done in moving a unit mass (1 kg in our unit system) between two points.

To calculate the change in gravitational potential of an object when it leaves one point and arrives at another, we need to know both the gravitational energy needed to achieve the move and also the mass of the object. Then the change in the gravitational potential of the arrival point relative to the departure point is $\frac{\text{work done to move the object}}{\text{mass of the object}}$.

In symbols

$$\Delta V_g = \frac{W}{m}$$

where ΔV_g is the change in gravitational potential, W is the work done, and m is the mass of the object. The unit of gravitational potential is joule kilogram^{-1} (J kg^{-1}). You can expect to use MJ kg^{-1} for changes in potentials on a planetary scale.

What makes a good zero reference for potential? One obvious reference point could be the surface of the oceans. When travel is based only on Earth this is a reasonable reference to choose. We refer to heights as being above or below sea level so that Mount Everest is 8848 m asl (above sea level). Despite being an appropriate local zero of potential for Earth-based activities, sea level is not good enough as soon as astronauts leave the Earth. It makes no sense to refer planet Mars to the Earth sea level – we need another measure.

The reference point that makes most sense – although it might seem implausible on first meeting – is "infinity". Infinity is not a real place, it is in our imagination, but that does not prevent it from having some interesting and useful properties!

Gravitational potential is defined to be zero at infinity.

Newton's law of gravitation tells us that the force of attraction F between two point masses is inversely proportional to the square of their separation r

$$F \propto \frac{1}{r^2}$$

The definition of gravitational potential reflects the fact that there is no interaction between two masses when they are separated by this imaginary infinite distance. Newton's law predicts that if the separation between two masses increases by one thousand times, then the attractive force goes down by a factor of one million. So if r is very large, then the force becomes very small indeed. At "infinity" we know that the force is zero even though we cannot go there to check.

Now we can study what happens as the two masses begin to approach each other from this infinite separation. Imagine that you are one of the masses and that the second mass (originally at infinite distance) is moved a little closer to you. This mass is now gravitationally attracted to you (and vice versa). Work would now need to be done on the system to push the mass away from you against the force of attraction back to its original infinitely distant starting point.

Energy has to be added to the system to return the mass to infinity. On arrival at infinity the mass will have returned to zero potential (because it is at infinity) and therefore whenever the masses are closer than infinity, the system of the two masses has a **negative potential**. The more closely they approach each other, the more negative the stored energy becomes because we have to put increasing amounts of energy back into the system to return the masses to an infinite separation (figure 9).

Gravitational potential at a point is defined to be equal to the work done per unit mass (kilogram) in moving a test mass from infinity to the point in question.

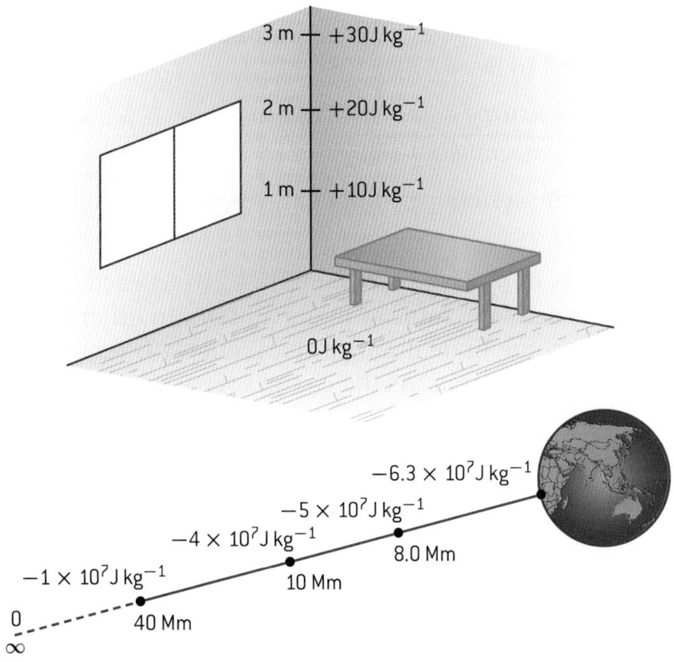

Figure 9 School lab and the Earth.

On the Earth, sea level corresponds to a gravitational potential of -6.25×10^7 J kg^{-1}. (We will calculate this in the next section.) Near the surface, raising one kilogram through one metre from sea level requires 9.8 (N) × 1 (m) of energy, in other words, about 10 J of energy for each kilogram raised. So the potential becomes more positive by 10 J kg^{-1} for each metre raised. You could label the wall of your school laboratory in potential values!

Comparing the gravitational potential at sea level with the gravitational potential at the top of Mount Everest shows that the potential at the top of the mountain is 86 710 J kg^{-1} greater than at sea level. This makes the gravitational potential at the top of Everest equal to -6.24×10^7 J kg^{-1}. Not apparently a large change, but then infinity is a good deal further away!

If we know the change in potential, then we can use it to calculate the total gravitational potential energy required for a change in height. For a mountaineer and kit of total mass 120 kg, the gravitational potential energy required to scale Mount Everest is about 10^7 J (that is 86 710 J kg^{-1} × 120 kg) starting at sea level.

Tip

At infinity, the gravitational potential is **maximum** and equal to zero, by the convention. At separations less than this the masses are bound to each other gravitationally and so have **negative** gravitational potential.

Returning to the idea of our contour maps, we could re-label all the contours changing the values from heights in metres to potentials in joule kg^{-1}. The shape would be the same. Instead of knowing the difference in height, we now know the amount of energy that will be used or released when each kilogram of the mass moves from one place on the map to another.

Electrical potential

The situation needs to be modified for electric potential. Again, the zero of potential is defined to be at infinity and the change in electric potential ΔV_e is defined to be

$$\Delta V_e = \frac{\textit{work done in moving charge between two points}}{\textit{magnitude of charge}} = \frac{W}{Q}$$

The energy stored in the system per unit charge is what we call the potential of the system.

If we have two charges of opposite sign then they attract each other, and the situation is exactly the same as in gravitation. In other words, when the charges are some finite distance apart, energy is required to move them to infinite separation. As the charges separate, the electric potential will gradually climb to zero.

The situation changes when the two charges have the same sign: now both charges repel each other.

Imagine two negative charges held at some finite distance apart. Energy was stored in the system when the charges were brought together from infinity. If they are released, the two charges will fly apart (to infinity where the repulsive force becomes zero). The system does work on the charges. There must have been a positive amount of energy stored in the system before the charges were allowed to separate. The potential is still defined to be zero at infinity, but this time as one charge is moved from infinity towards the other charge, the overall value of the potential rises above the zero level to become positive. Work has been done on the charge.

The electric potential at a point is the work done in bringing a unit positive test charge from infinity to the point.

Tip

These charges are not bound to each other. They have a high electric potential energy. When returned to infinity they will have zero electric potential energy and this value is a **minimum.**

Tip

Don't forget there are three separate points in this definition:

- work done in moving unit charge
- the test charge is positive
- the test charge moves from infinity to the point

10.2 Fields at work

Understanding

- Potential and potential energy
- Potential gradient
- Potential difference
- Escape speed
- Orbital motion, orbital speed, and orbital energy
- Forces and inverse-square law behaviour

Nature of science

Electric charge is invisible to our senses, yet we accept its existence. We are aware of it through the force laws that it obeys and via the field properties that we assign to it. In the same way, the motion of planets and satellites on a larger scale also requires that scientists develop ways to visualize and then report their theories to non-scientists.

Applications and skills

- Determining the potential energy of a point mass and the potential energy of a point charge
- Solving problems involving potential energy
- Determining the potential inside a charged sphere
- Solving problems involving the speed required for an object to go into orbit around a planet and for an object to escape the gravitational field of a planet
- Solving problems involving orbital energy of charged particles in circular orbital motion and masses in circular orbital motion
- Solving problems involving forces on charges and masses in radial and uniform fields

Equations

potential energy equations:	$V_e = \frac{kQ}{r}$	$V_g = -\frac{GM}{r}$
field strength equations:	$E = -\frac{\Delta V_e}{\Delta r}$	$g = -\frac{\Delta V_g}{\Delta r}$
relation between field strength and potential:	$E_p = qV_e = \frac{kQq}{r}$	$E_p = mV_g = -\frac{GMm}{r}$
force laws:	$F_E = k\frac{q_1 q_2}{r^2}$	$F_G = G\frac{m_1 m_2}{r^2}$

- escape speed: $v_{esc} = \sqrt{\frac{2GM}{r}}$
- orbital speed: $v_{orbit} = \sqrt{\frac{GM}{r}}$

Introduction

In the second part of this topic we look at the underlying mathematics of fields and how it applies to gravitational and electric fields.

Parts of this chapter devoted to electric fields have a blue background or margin line; the sections dealing with gravity have a green background or margin line. You can concentrate on one type of field by sticking to one colour.

Forces and inverse-square law behaviour

Electric fields	Gravity fields
Both fields obey an inverse-square law in which the force between two objects is inversely proportional to the distance between them squared.	
In a vacuum: $$F_E = +\frac{kq_1 q_2}{r^2}$$ where F_E is the force between two point charges q_1 and q_2 and r is the distance between them. This is known as Coulomb's law. The constant k in the equation is $\frac{1}{4\pi\varepsilon_0}$ where ε_0 is known as the permittivity of a vacuum or the permittivity of free space. If the field is in a medium other than a vacuum then the ε_0 in the equation is replaced by ε, the permittivity of the medium. The sign in the equation is positive. The sign of the overall result of a calculation indicates the direction in which the force acts. Negative indicates attraction between the charges; positive indicates repulsion.	$$F_G = \frac{Gm_1 m_2}{r^2}$$ where F_G is the force between two point masses m_1 and m_2 and r is the distance between them. This is known as Newton's law of gravitation. The constant in the equation is G known as the universal gravitational constant. The value of G is a universal constant.

Nature of science

Inverse-square laws

Both fields obey the inverse-square law in which the force depends on $\frac{1}{\text{distance}^2}$. Sub-topic 4.4 showed how the inverse-square law also arises in the context of radiation from the geometry of space.

The value of n in $\frac{1}{r^n}$ has been tested a number of times since the inverse-square law behaviour of electric and gravitational fields was suggested, and n is known to be very close to 2 (for Coulomb's law, within 10^{-16} of 2)

One consequence of a force law being inverse-square is that the force between the charges or masses becomes weaker as the distance increases, but never becomes zero at any real distance. We said earlier that the force was zero when the objects were separated by an infinite distance. There is no actual infinity point in space, but there is one in our imagination. We use infinity as a useful concept for our energy ideas. In one sense, infinity can be pushed further away as the instruments that measure field strength become more precise!

Tip

Charges q_1 and q_2 can be positive or negative. A positive force F_E indicates a *repulsive* force.

A negative force F_E indicates an attractive force.

The masses m_1 and m_2 are always positive, and so in this notation a positive force F_G always indicates an *attractive* force.

We saw in Topic 5 that, close to any surface, the field is uniform. The argument was that, locally, the surface appears flat and that there is cancellation of the components of field parallel to the surface due to each charge. This leaves only components perpendicular to the surface to contribute to the field.

The electric fields between parallel plates or the gravitational or electric fields close to a curved surface are uniform and are useful in developing ideas about fields.

Investigate!

Charged parallel plates

- This experiment illustrates how the field strength and other factors are connected for parallel plates and will help you to understand these factors.
- Set up a pair of parallel plates, a 5 kV power supply, a ***well-insulated*** flying lead, and a coulombmeter (a meter that will measure charge directly in coulombs) as shown in the circuit. Later you will need to replace the parallel plates with ones that have different areas. You will also need to record the distance between the plates.
- Set and record a suitable voltage V on the power supply. Your teacher will tell you the value to use. Take great care at all times with the high voltages used in this experiment.
- Measure and record the distance d between the parallel plates.
- Touch the flying lead to the right-hand plate briefly and then remove it from the plate. This charges the plates.
- Zero the coulombmeter and then touch its probe to the right-hand plate that you just charged. Record the reading Q on the meter. You may wish to repeat this measurement as a check.
- Change the distance between the plates without changing the setting on the power supply.
- Repeat the measurements of distance apart and charge stored. Don't forget to switch off the power supply before you measure the separation of the plates.
- Carry out another experiment where the power supply emf and the plate separation are unchanged but plates of different areas are used. Record the plate areas.
- Carry out another experiment where the emf V of the power supply is changed but the plate area and distance are not changed.
- Plot your results as Q versus V, Q versus A, and Q versus d. where Q is the charge on the plates, V is the potential of the plate, A is the area of the plates, and d is the separation of the plates.

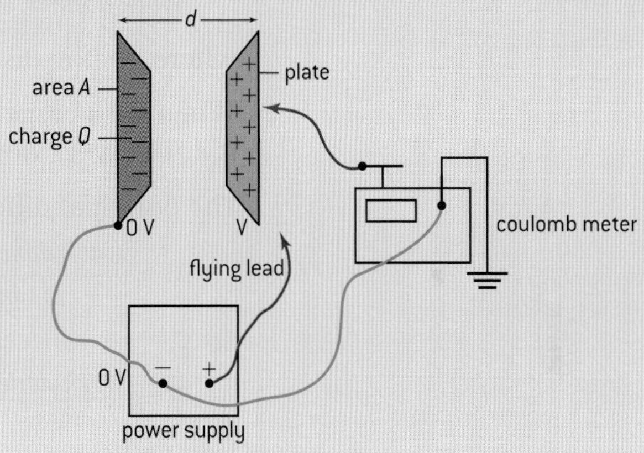

▲ Figure 1 How the charge stored on parallel plates varies.

The results of the experiment show that

- $Q \propto V$
- $Q \propto A$
- $Q \propto \frac{1}{d}$.

These results can be expressed as

$$Q \propto \frac{VA}{d}$$

The constant in the equation turns out to be ε_0; in principle careful measurements in this experiment could provide this answer too.

So

$$Q = \varepsilon_0 \frac{VA}{d}$$

where

$$\varepsilon_0 = 8.854 \times 10^{-12}\ \text{m}^{-3}\ \text{kg}^{-1}\ \text{s}^4\ \text{A}^2.$$

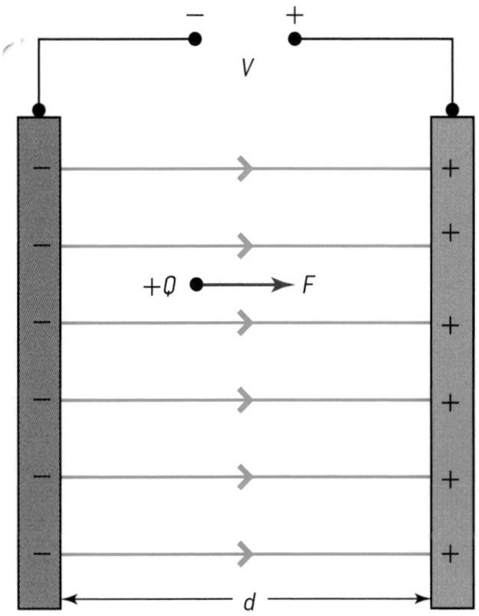

▲ Figure 2 Field strength between parallel plates.

Electric field strength and potential gradient

The uniform electric field between two charged plates provides us with another way to think about electric field strength.

In Topic 5 (page 177) we mentioned that electric field strength could be measured as the change in potential divided by the change in distance (we call this the potential gradient). We can now show that this result is correct.

In figure 2, a positive charge that has a size Q is in a field between two charged plates. This field has a strength E and the force F that acts on the charge is therefore (from the definition of electric field strength given in Sub-topic 5.1)

$$F = EQ$$

It is easy to determine the work done by the field on the charge when the charge is moved at constant speed because the field is uniform and the force is constant. As usual, the work done is *force × distance moved* and the lines of force are in the same direction as the distance moved so there is no component of distance to worry about

$$\textit{Work done} = EQ \times x$$

where x is the distance moved.

When the charge goes from one plate to the other the distance moved is d. The work done on the charge is EQd.

The potential difference V between the plates is (see Topic 5.1)

$$V = \frac{\textit{work done in moving a charge}}{\textit{magnitude of charge}}$$

So

$$V = \frac{\textit{work done}}{Q} = \frac{EQd}{Q}$$

leading to

$$E = \frac{V}{d}$$

This shows that electric field strength can be written and calculated in two ways:

- $\frac{\textit{force acting on charge}}{\textit{magnitude of charge}}$, this gives the unit N C^{-1}
- $\frac{\textit{potential change}}{\textit{distance moved}}$, this gives the unit V m^{-1}.

So as well as two equations, there are two possible units for electric field strength and both are correct.

Writing the equations in full for the uniform field between parallel plates

$$E = \frac{V}{d} = \frac{F}{Q}$$

In practice, it is easier to measure a potential difference with a voltmeter in the laboratory than to measure a force. This is why an electric field strength is commonly expressed in volt metre^{-1}.

Worked examples

1 A pair of parallel plates with a potential difference between them of 5.0 kV are separated by 120 mm. Calculate:

a) the electric field strength between the plates

b) the electric force acting on a doubly ionized oxygen ion between the plates.

Solution

a) $E = \frac{V}{d} = \frac{5000}{0.12} = 4.2 \times 10^4\ \text{V m}^{-1}$

b) The charge of the ion is $2e = +3.2 \times 10^{-19}$ C.

$F = QE = 3.2 \times 10^{-19} \times 4.2 \times 10^4 = 1.34 \times 10^{-14}$ N

The force is directed towards the negative plate.

2 A pair of parallel plates are separated by 80 mm. A droplet with a charge of 11.2×10^{-19} C is in the field.

a) Calculate the potential difference required to produce a force of 3.6×10^{-14} N on the droplet.

b) The plates are now moved closer to each other with no change to the potential difference. The force on the droplet changes to 1.4×10^{-13} N. Calculate the new separation of the plates.

Solution

a) $E = \frac{F}{Q} = \frac{3.6 \times 10^{-14}}{11.2 \times 10^{-19}} = 3.2 \times 10^{-4}\ \text{N C}^{-1}$

$V = Ed = 3.2 \times 10^4 \times 0.080 = 2600$ V

b) The force changes by a factor of

$\frac{1.4 \times 10^{-13}}{3.6 \times 10^{-14}} = 3.9$

The separation decreases by this factor, to

$\frac{80}{3.9} = 21$ mm.

Electric field strength and surface charge density

The expression

$$Q = \varepsilon_0 \frac{VA}{d}$$

rearranges to

$$\frac{Q}{A} = \varepsilon_0 \frac{V}{d}$$

$\frac{Q}{A}$ is σ, the surface charge density on the plate. We say that for the uniform field between the plates, the electric field strength E can be written as $\frac{V}{d}$.

So between the plates

$$E = \frac{\sigma}{\varepsilon_0}$$

Remember, this equation is for the field between *two* parallel plates. Each plate contributes half the field, so the electric field E close to the surface of any conductor is equal to

$$E = \frac{\sigma}{2\varepsilon_0}$$

where σ is the charge per unit area on the surface and ε_0 has its usual meaning.

This is dimensionally correct because the units of σ are C m^{-2} and the units of ε_0 are C^2 N^{-1} m^{-2}, $\frac{\sigma}{\varepsilon_0}$ is $\frac{\text{C m}^{-2}}{\text{C}^2\ \text{N}^{-1}\ \text{m}^{-2}}$ which simplifies to N C^{-1}, the same units as for E.

You will be given this equation in an examination if you need to use it.

Tip

Generally, the term *gradient* refers to a change per unit of distance. The term *rate* refers to a change per unit to time.

(Sometimes gradient is used a word for slope)

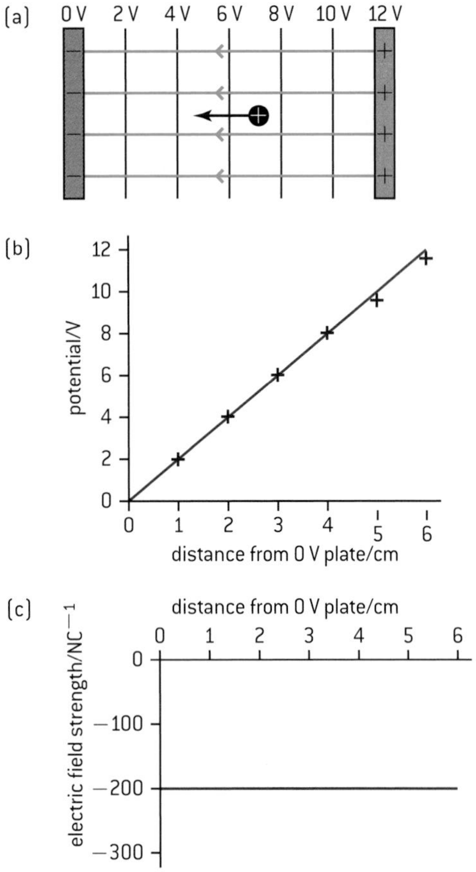

Figure 3 Electric field strength and potential.

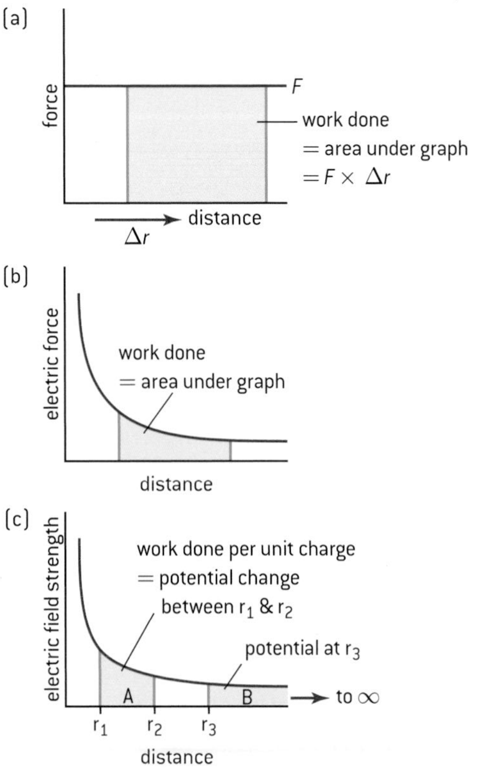

Figure 4 Work done and work done per unit charge.

Graphical interpretations of electric field strength and potential

Figure 3(a) is a reminder of the field pattern between charged parallel plates.

The equipotentials are equally spaced. Figure 3(b) shows the data from measurements of the potential at a series of points between the plates plotted as a graph of potential against distance from the zero potential plate. It is a straight line because the electric field is uniform. A graph of electric field strength against distance from the 0 V plate is a straight line parallel to the distance axis (Figure 3(c)).

The gradient of the potential–distance graph is $\frac{\text{change in } V}{\text{change in } x}$ and because it is a straight line, this is equal to $\frac{V}{d}$, in other words, E.

This leads to:

$$E\text{, electric field strength} = -\text{potential gradient} = -\frac{\Delta V_e}{\Delta x}$$

(Δ, as usual, stands for "change in").

The minus sign that appears in the equation needs an explanation. It means that the direction of the vector electric field is always opposite to the variation of the potential (ΔV_e) of a positive charge. So, travel in the opposite direction to that of the field means moving to a position of higher potential and thus a positive gain in potential. In Figure 3(a) the motion of the positive charge is to the left, ΔV_e is negative (final state minus initial state: 0 V − 12 V); so, according to the equation, E will be positive and this tells us that the motion is in the direction of the field. In other words, going *upstream* in the field (against the field) means going to higher potential so a gain in potential (positive change in ΔV_e). Going downstream, ΔV_e is negative and E is positive (product of two minus signs) indicating the motion is in the direction of field.

You may be wondering whether the argument changes if the moving charge is an electron. If an electron moves from 0 V towards the 12 V plate (a position of higher potential) its potential energy is reduced because the potential energy change ΔE_p is given by $e\Delta V$ as usual but here e is negative and ΔV is positive. The electron will accelerate in this direction if free to do so gaining kinetic energy from the field at the expense of the electrical potential energy.

In numerical terms (figures 3(b) and (c)), the gradient $\frac{\Delta V}{\Delta x}$ is +200 V m^{-1} and so the electric field strength is −200 N C^{-1}.

The equation $E = -\frac{\Delta V}{\Delta x}$ can be re-written as $E \times \Delta x = -\Delta V$, and in this form has a graphical interpretation.

A graph of force against distance for a constant force is shown in figure 4(a). The work done when this force moves an object through a distance Δr is equal to the area under the graph. In symbols

$$F \times \Delta r = W$$

When an electric force varies with distance then the work done is still the area under the graph (Figure 4(b)). Electric field strength (Figure 4(c)) is the force *per unit charge*, and so the area under a graph of electric field strength against distance is equal to the work done *per unit charge* – in other words the change in potential.

Figure 4(c) shows the variation of electric field strength against distance for a single point positive charge – this is an inverse-square variation. Two areas are shown on this graph, one area (A) that shows the *potential change* between r_1 and r_2, the other (B) shows the *potential* at r_3 – in this case the area included goes all the way to infinity.

The equation

$$E = -\frac{\Delta V_e}{\Delta r}$$

can be written in calculus form as

$$E = -\frac{dV_e}{dr}$$

This new form can be used to take the equation for the field strength

$$E = \frac{kQ}{r^2}$$

Q here is the charge creating the field (not a charge that would react to the field) and show that V_e is given by

$$V_e = \frac{Q}{4\pi\varepsilon_0 r} \quad \text{or} \quad \frac{kQ}{r}$$

using the constant $k = \frac{1}{4\pi\varepsilon_0}$.

This is the expression for the potential V at a distance r from a point charge. It predicts that the closer we are to a (positive) charge, the greater (more positive) is the potential. Conversely, the closer we are to a negative charge, the more negative is the potential. These predictions correspond to the conclusions we reached when we considered figure 3.

Potential is the work done in moving a positive unit charge (from infinity) to a particular point, so the work required to move a charge of size q from infinity to the point will be: $V_e \times q$. This is the potential energy that charge q possesses due to its position in the field that is giving rise to the potential.

If a charge q is in a field that arises from another single point charge Q then its potential *energy* $E_p = qV_e$, or

$$E_p = \frac{kqQ}{r} = \frac{qQ}{4\pi\varepsilon_0 r}$$

Graphs of field strength and potential against distance for the field due to a single positive point charge are shown in figure 5.

Tip

Notice the relative shapes of the curves: $\frac{1}{r^2}$ is more sharply curving than the $\frac{1}{r}$ of the potential. Notice also the links between the two graphs.

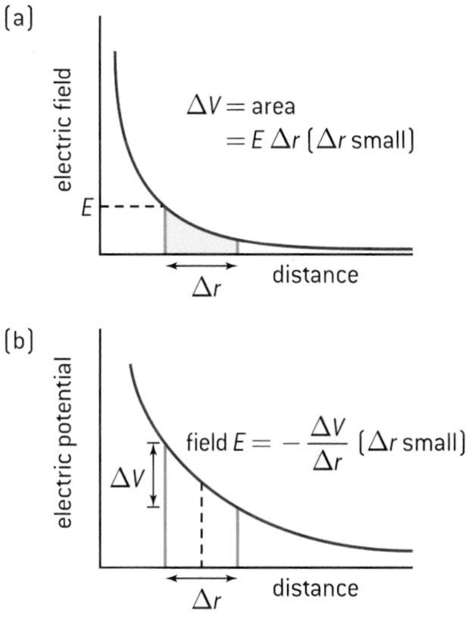

Figure 5 Electric field strength and potential against distance for a positive point charge.

Worked examples

1 Two parallel metal plates have a potential difference between them of 1500 V and are separated by 0.25 m. Calculate the electric field strength between the plates.

Solution

$$\textit{Electric field strength} = \frac{\textit{potential difference}}{\textit{separation}}$$

$$= \frac{1500}{0.25} = 6 \text{ kV m}^{-1} \ (\textit{or} \text{ kN C}^{-1})$$

2 A point charge has a magnitude of −0.48 nC. Calculate the potential 1.5 m from this charge.

Solution

$$V_e = \frac{Q}{4\pi\varepsilon_0 r}$$

Substituting

$$V_e = \frac{-0.48 \times 10^{-9}}{4\pi\varepsilon_0 \times 1.5} = 2.9 \text{ V}$$

Nature of science

Linking electrostatics and gravity

The conclusion reached here arises because of the inverse-square law. It also applies to a *hollow* planet. We can show, in a similar way, that the gravitational field strength is zero. This helps us later to show how the gravitational field strength varies inside a solid planet of uniform density.

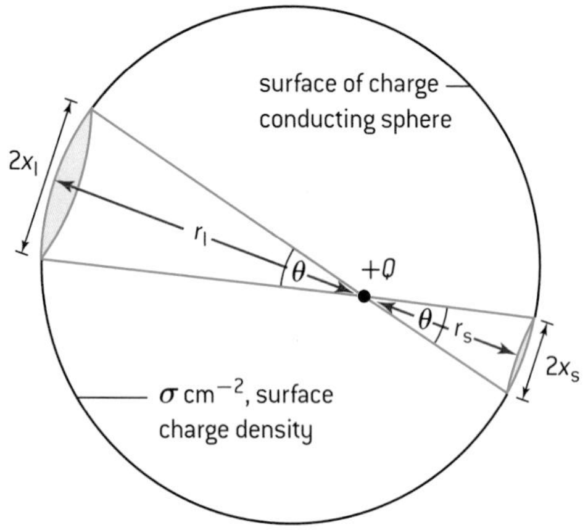

▲ Figure 6 Field inside a charged conductor.

Potential inside a hollow conducting charged sphere

So far we have considered point charges and the parallel–plate arrangement in detail. Now we turn to another important charge configuration: the hollow conducting charged sphere.

Outside a charged conducting sphere, the field is indistinguishable from that of the point charge. The argument is that because the field lines leave the surface of the sphere at 90° these lines must be radial, so an observer outside the sphere cannot distinguish between a charged sphere and a point charge with the same magnitude of charge placed at the centre of the sphere.

The field inside the sphere

Because the sphere is a conductor, all the surplus charge must reside on the outside of the sphere. This follows because:

- the charges will move until they are as far apart as possible
- the charges will move until they are all in equilibrium (which in practice means that they have to be equidistant on the surface).

To find the field strength inside the conductor we need to compute the force that acts on a positive test charge Q placed anywhere inside the conductor.

We choose a point inside the conductor at random and place our test charge there. This is shown in figure 6. The next step is to find the force acting on this test charge. To make this easier we consider two cones that meet at the test charge. These cones have the same solid angle at the apex. One of these cones is large and the other is small. The test charge is close to the conductor surface on the side of the small cone but not so close on the side of the large cone.

Call the distance from the test charge to the sphere r_s for the small cone and r_l for the large cone. The cone radii are x_s and x_l for the small and large cones respectively.

The surface charge density on the cone is σ and is the same over the whole sphere.

Therefore

	For the small cone	For the large cone
Area of the end of the cone	πx_s^2	πx_l^2
Charge on the end of the cone	$\sigma \pi x_s^2$	$\sigma \pi x_l^2$
Distance of area from test charge	r_s	r_l
Force on test charge due to area	$\frac{kQ\sigma \pi x_s^2}{r_s^2}$	$\frac{kQ\sigma \pi x_l^2}{r_l^2}$
	Directed away from the surface	Directed away from the surface

The geometry of the cones is such that

$$\tan\left(\frac{\theta}{2}\right) = \frac{x_s}{r_s} = \frac{x_l}{r_l}$$

and therefore

$$\frac{kQ\sigma\pi x_s^2}{r_s^2} \text{ is equal to } \frac{kQ\sigma\pi x_1^2}{r_1^2}$$

The forces acting on the test charge due to the two charge areas are equal in size and opposite in direction. Remarkably, they completely cancel out leaving no net force due to these two areas of charge.

We chose the angle of the cones arbitrarily so this result applies for any angle we could have chosen. Similarly, we chose the position for the test charge at random. Therefore this proof must apply for any pair of cones and any test point inside the sphere. There is therefore no net force on the test charge anywhere inside the sphere and consequently there is no electric field either. This proof relies on the inverse-square law and the fact that the area of the ends of the cones also depends on the (*distance of the test charge from the area*)^2.

Because the electric field is zero the potential inside the sphere must be a constant since the gradient $\frac{\Delta V}{\Delta r}$ is zero inside the sphere. V being constant must mean that ΔV is zero so no work is required to move a charge at constant speed inside the charged sphere. The potential inside is equal to the potential at the surface of the sphere.

So we can now plot both the electric field strength and the electric potential for a charged conducting sphere and these graphs are shown in figure 7.

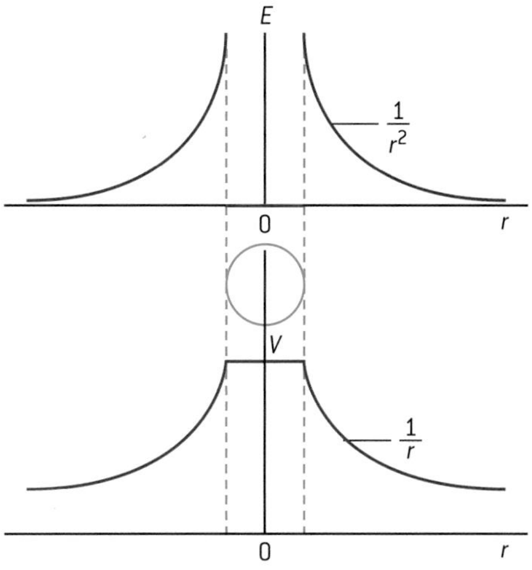

Figure 7 Field and potential inside and outside the conductor.

Nature of science

Potential and potential energy

The connection between field strength and potential gradient is universal and applies to all fields based on an inverse-square law. It tells us about the fundamental relationship between force and the work the force does and it tells us this in a way that is independent of the mass (or the charge) of the test object that is moved into the field. Thus, field strength is a concept that represents force but with the mass or charge term of the test object removed. Of course, the mass or charge element that gives rise to the field in the first place is still important and remains in E and g.

So

Mass/charge independent		*Mass/charge dependent*
Field strength	↔	**force**
Potential	↔	**potential energy**

Gravitational potential

Gravitation

As with electric fields

$$g = -\frac{\Delta V_g}{\Delta r}$$

where g is the gravitational field strength, V_g is the gravitational potential, and r is distance.

For the case of the field due to a single point mass M with a point mass m a distance r from it, the potential V_g at r is

$$V_g = -\frac{GM}{r}$$

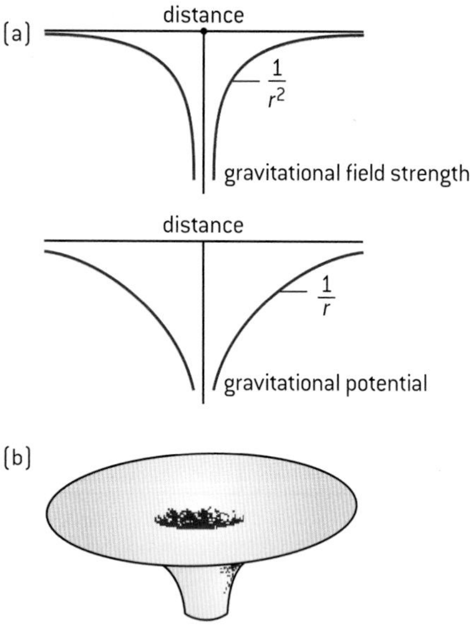

▲ Figure 8 Gravitational field strength and potential variation with distance.

The potential energy of m at r from mass M is

$$E_p = mV_g = -\frac{GMm}{r}$$

These equations lead to graphs similar to those for electric fields, which are illustrated in figure 8(a).

The forces are always attractive for gravity and we saw that this has the consequence that the potential is always negative with a maximum of zero at infinity. The graphs in figure 8 are plotted to reflect this.

The graphs should be shown in three-dimensions, but are plotted in two dimensions for clarity. When thinking of the plotted surface of this negative $\frac{1}{r}$ potential in 3-D, treat it as a potential well (figure 8(b)). This well will trap any particle that has insufficient kinetic energy to escape the mass. Later when we discuss how spacecraft leave the Earth you can imagine that a spaceship without enough energy can only climb up so far. From this point the spaceship can circle around the mass (that is orbit the mass), but climb no higher.

Worked examples

1 At a point X the gravitational potential is -8 J kg^{-1}. At a point Y the gravitational potential is -3 J kg^{-1}. Calculate the change in gravitational potential energy when a mass of size 4 kg moves from X to Y.

Solution

A change in potential is calculated from final state minus initial state so change here is $(-3) - (-8) = +5$ J kg^{-1}. So energy required is $+20$ J, moving from a lower point in the potential well to a higher point (climbing toward infinity where the value is maximum and equal to zero).

2 Which diagram shows the gravitational equipotential surfaces due to two spheres of equal mass?

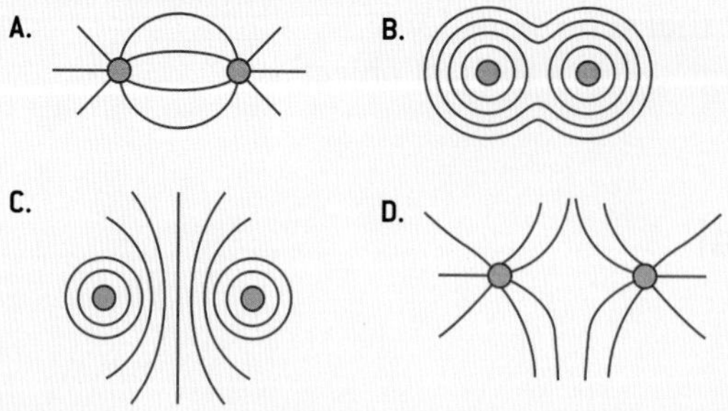

Solution

The spheres attract and so the gravitational field lines joining them resemble the pattern between two unlike charges (this pattern is similar to that in A). The equipotential surfaces must cut these field lines at 90° so the best fit to the appropriate pattern is C.

3 Calculate the potential at the surface of the Earth. The radius of the Earth is 6.4 Mm and the mass of the Earth is 6.0×10^{24} kg.

Solution

$$V_g = -\frac{GM}{r}$$

$$V_g = -\frac{6.7 \times 10^{-11} \times 6.0 \times 10^{24}}{6.4 \times 10^6}$$

$$= -6.3 \times 10^7 \text{ J kg}^{-1}$$

Potential inside a planet

In a charged conducting sphere the mobile charges move to the outside of the sphere and this results in zero field inside. This is not the same for the gravitational field of a solid massive sphere ("massive" in this context means "having mass" not "having a large mass").

This case was mentioned in Topic 6 in the context of a tunnel drilled from the surface to the centre of the Earth. The essential physics is that if we are at some intermediate depth between surface and centre we can think of the Earth as having two parts: the solid sphere beneath our feet and the spherical shell "above our head".

The inner solid sphere, a miniature Earth with a smaller radius, behaves as a normal Earth but with a different gravitational field strength given by

$$g' = \frac{G \times \text{reduced mass of the Earth}}{\text{radius of small Earth}^2}$$

The outer spherical shell does not give rise to a gravitational field. The argument is similar to that which we used to prove that the electric field inside a charged conductor was zero.

The *reduced mass of the Earth* is equal to $\frac{4}{3}\pi\rho r'^3$ where ρ is the density of the Earth (assumed constant), so

$$g' = \frac{4\pi G\rho r'}{3}$$

where r' is the reduced radius of the "small Earth"

The gravitational field is proportional to the distance from the centre of the Earth until we reach the surface. This is an unexpected result. The inverse-square law actually gives rise to a linear relationship – because the $\frac{1}{r^2}$ behaviour of the force cancels with the r^3 variation of the mass of the *reduced Earth.*

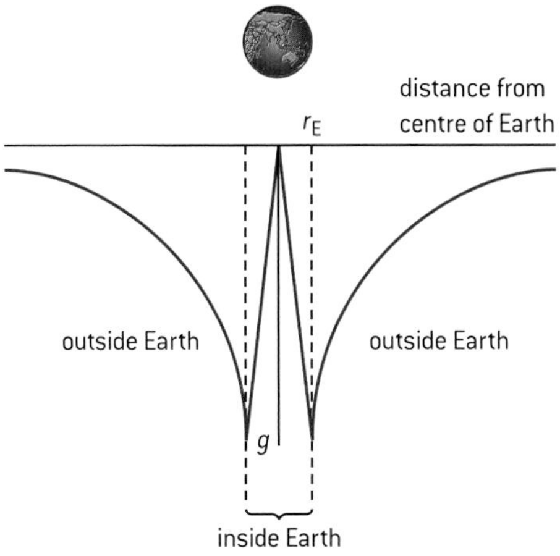

▲ Figure 9 Gravitational field inside the Earth.

Nature of science

Delivering the mail!

The acceleration g' is always directed toward the centre of Earth and is proportional to r. So a vehicle inside a tunnel drilled right through the Earth will perform simple harmonic motion if friction is neglected (see Topic 9). The vehicle will stop momentarily when arriving at the *antipode* 84 minutes after release. This is also the period of a satellite that goes round the Earth in low-earth orbit (technically, just skimming the surface).

Get the timing right and mail bags could be exchanged at the surface!

A chordal tunnel dug between Vancouver and Montreal or between New York and Rio de Janeiro or between London and Johannesburg (*not* passing through the center of the Earth) also allows the mail to be delivered. Use some of the ideas and equations in Topic 9 to assess the engineering issues with this plan!

Leaving the Earth

Picture a spacecraft stationary on its launch pad. A space launch from Earth may put a satellite into Earth orbit. Or it may be a mission to explore a region far away from Earth requiring the craft to escape Earth's gravity and eventually come under the influence of other planets and stars.

What minimum energy is needed to allow the craft to escape the Earth?

Orbiting

In Topic 6 there are some basic ideas about circular orbits of a satellite about a planet. We showed that the linear orbital speed v of a satellite about a planet is

$$v_{orbit} = \sqrt{\frac{GM}{r}}$$

and that the orbital time T for this orbit is

$$T_{\text{orbit}} = \sqrt{\frac{4\pi^2 r^3}{GM}}$$

here M is the mass of the planet and r is the radius of the orbit.

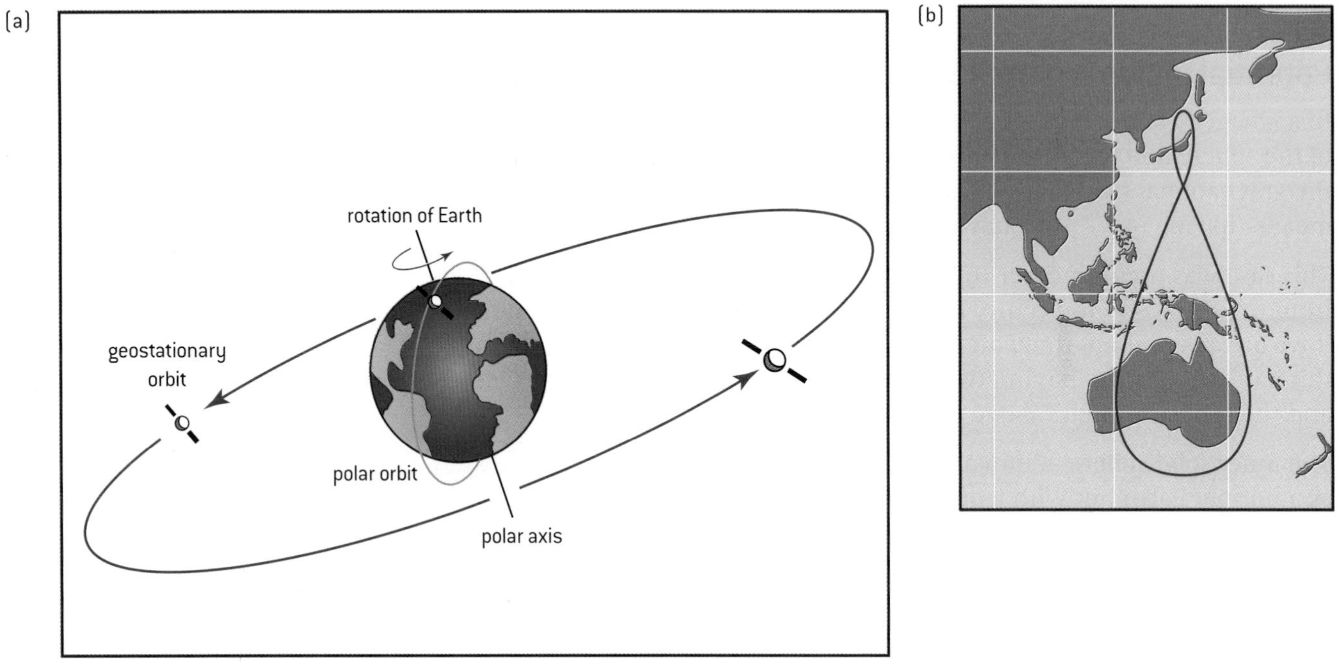

Figure 10 Polar and geostationary orbits (not to scale) and the track of a geosynchronous orbit.

Worked example

Calculate the orbital time for a satellite in polar orbit.

Solution

The orbital time for these satellites is obtained by using the appropriate value for r in the T equation. The radius of the orbit is the orbital height (100 km) plus the radius of the Earth (6.4 Mm), so $r = 6.5 \times 10^6$ m.

$$T = \sqrt{\frac{4 \times \pi^2 \times (6.5 \times 10^6)^3}{6.67 \times 10^{-11} \times 6.0 \times 10^{24}}}$$

$= 5200$ s (about 87 min)

Although satellites can be put into orbits of any radius (provided that the radius is not so great that a nearby astronomical body can disturb it), there are two types of orbit that are of particular importance.

The **polar orbit** is used for satellites close to the Earth's surface (close in this context means about 100 km above the surface). The orbit is called "polar" because the satellites in this orbit are often put into orbit over the poles. Imagine you are viewing the Earth and the orbiting satellites from some way away in space. You will see the satellite orbiting in one plane (intersecting the centre of the Earth) with Earth rotating beneath it. In the course of 24 hours, the satellite will view every point on the Earth.

For the orbital radius of a polar orbit, the linear speed of the satellite is 7800 m s^{-1} or 28 000 km per hour.

Geosynchronous satellites orbit at much greater distances from the Earth and have orbital times equal to one sidereal day which is roughly 24 hours. This means that the geosynchronous satellite can be made to stay in the same area of sky and typically follows a figure-of-eight orbit above a region of the planet. A typical track in the southern hemisphere is shown in figure 10(b). This track shows where the satellite is overhead during the 24 hour cycle. Because the plane of the orbit coincides with the centre of the Earth but is not through the Equator, the overhead position wanders. To see this, use a globe imagining the satellite going around its orbit as the Earth rotates below it at the same rate.

A **geostationary orbit** on the other hand is a special case of the geosynchronous orbit. In this case the satellite is placed in orbit above the plane of the Equator and will not appear to move if viewed from the surface. This has the advantage that a receiving antenna (aerial) or satellite dish on Earth also does not have to track the transmitting antenna on the satellite. A TV satellite dish pointing towards the south (equator) in the northern hemisphere will be always aligned with the geostationary satellite. These satellites are generally used for communication purposes and for collecting whole-disk images of the Earth for weather forecasting purposes.

Worked examples

a) A geostationary satellite has a orbital time of 24 hours. Calculate the distance of the orbit from the surface of the Earth.

b) Calculate the gravitational field strength at the orbital radius of a geostationary satellite.

Solution

a) 24 hours is 86 400 s.

Rearranging the orbital time equation

$$r = \sqrt[3]{\frac{T^2\,GM}{4\pi^2}}$$

$$r = \sqrt[3]{\frac{86\,400^2 \times 6.67 \times 10^{-11} \times 6.0 \times 10^{24}}{4\pi^2}}$$

$$= 42 \times 10^6 \text{ m} = 42\,000 \text{ km}$$

This is (42 000 – 6400) = 36 000 km above the surface (the subtraction is to find the distance from the surface).

b)

$$g = \frac{GM_E}{(r_E + r)^2}$$

$$= \frac{6.7 \times 10^{-11} \times 6.0 \times 10^{24}}{(4.2 \times 10^7)^2}$$

$$= 0.23 \text{ N kg}^{-1}$$

Escaping the Earth

The total energy of a satellite is made up of the gravitational potential energy and the kinetic energy (ignoring any energy transferred to the internal energy of the atmosphere by the satellite). To escape from the surface of the Earth, work must be done on the satellite to take it to infinity. For an unpowered projectile this is simply equal to the kinetic energy – meaning that the total of the (negative) gravitational potential energy and the (positive) kinetic energy must add up to zero. Thus to reach infinity

gravitational potential energy + kinetic energy = 0

and therefore

$$-\frac{GM_E\, m_S}{R_E} + \frac{1}{2} m_S v_E^2 = 0$$

where R_E is the distance of the satellite from the centre of the Earth, v_E is the escape speed, and M_E and m_s are the masses of the Earth and the satellite respectively. The gravitational potential energy is negative because this is a bound system. Kinetic energy is always positive. (It is important to keep track of the signs in this proof.)

To escape the Earth's gravitational field completely the total energy of the satellite must be (at least) zero. For the case where it is exactly zero (for the satellite to just reach infinity)

$$v_E = \sqrt{\frac{2GM_E}{R_E}}$$
$$= \sqrt{2gR_E}$$

where g is the gravitational field strength at the surface. This speed must be about 11 200 m s^{-1} (400 00 km $hour^{-1}$) for something to escape.

It is important to be clear about the true meaning of escape speed. This is the speed at which an *unpowered* object, something like a bullet, would have to be travelling to leave the Earth from the surface. In theory, a rocket with enough fuel can leave the Earth at any speed. All that is required is to supply the 63 MJ for each kilogram of the mass of the rocket (the gravitational potential of the Earth's field at the surface is −62.5 MJ kg^{-1}). However, in practice it is best to reach the escape speed as soon as possible.

Similarly, if a spacecraft begins its journey from a parking orbit, then less fuel will be required from there because part of the energy has already been supplied to reach the orbit.

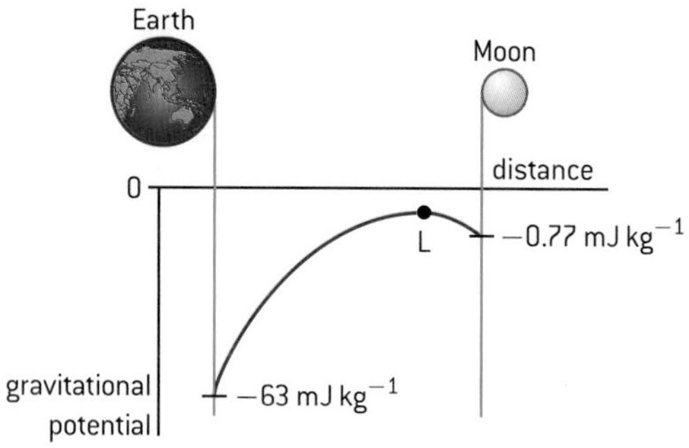

▲ Figure 11 Gravitational potential between the Earth and the Moon (not to scale).

The graph of potential against distance helps here. The graph in figure 11 shows the variation in gravitational potential from the Earth to the Moon. Before takeoff, a spacecraft sits in a potential well on the Earth's surface. When the rockets are fired the spacecraft gains speed and moves away from the Earth. It needs enough energy to reach the maximum of the potential at point L (this is known as the Lagrangian point and it is also the point where the gravitational field strengths of the Earth and Moon are equal and opposite). Once at L the spacecraft can fall down the potential hill to arrive at the Moon.

Orbit shapes

We have now identified two speeds: orbital speed close to the surface (7.8 km s^{-1}) and the escape speed (11.2 km s^{-1}). At speeds between these values the spacecraft will achieve different orbits. These are shown in figure 12.

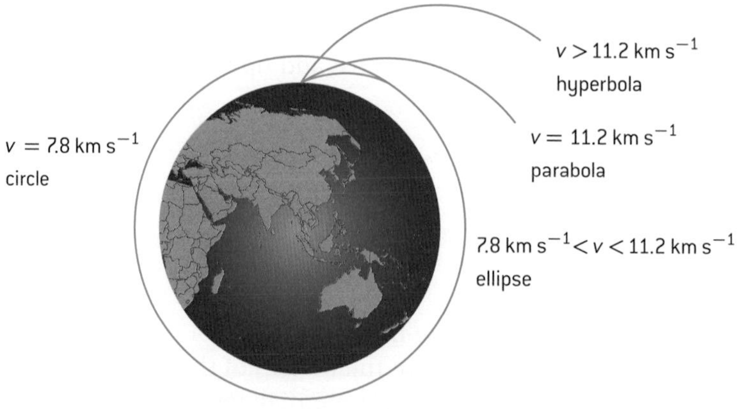

▲ Figure 12 Orbital shapes of satellites.

- At speeds less than 7.8 km s^{-1} the craft will return to Earth.
- At a speed of 7.8 km s^{-1} the craft will have a circular path.
- At speeds between 7.8 km s^{-1} and 11.2 km s^{-1} the orbit will be an ellipse.
- At a speed of 11.2 km s^{-1} the craft will escape following a parabolic path.
- At speeds greater than 11.2 km s^{-1} the craft will escape the Earth along a hyperbolic path. (It will not however necessarily escape the Sun and leave the Solar System.)

You will only have to consider circular orbits for the examination.

Nature of science

Using the Earth's rotation and the planets

Rotation

Many launch sites for rockets are close to the Equator. This is because the Equator is where the linear speed of the Earth's surface is greatest. This equatorial velocity is about 500 m s^{-1} from west to east meaning that a rocket taking off to the east has to attain a relative speed of $11.2 - 0.5 = 10.7$ km s^{-1}, a significant saving in fuel.

Slingshots

Another technique much used in missions to distant planets is to use "slingshot" techniques in which a spacecraft is attracted towards a planet and then swings around it to shoot off in a different direction possibly at a greater speed. The interaction is effectively a collision between the craft and the planet in which momentum is conserved but the craft gains some energy from the orbital energy of the planet.

The spacecraft *Voyager 2* was launched in 1977 and in the course of its journey has passed Jupiter, Saturn, Uranus, and Neptune, using each planet to change its trajectory achieving a "Grand Tour" of the planets. The spacecraft is still travelling and is now a considerable distance from Earth (it has its own Twitterfeed if you wish to know where it is today). It is expected to lose all its electrical energy sometime in 2025 and will then continue, powered down, through space carrying messages from the people of Earth to any other life form that may encounter it.

Gases in the atmosphere

The Earth's atmosphere only contains small amounts of the lighter gases such as hydrogen and helium. All gases in the atmosphere are at the same average temperature T and thus have the same mean kinetic energy $\overline{E}_K$ for the molecules (because, from Topic 3, $\overline{E}_K \propto T$).

Kinetic energy is $\frac{1}{2}mv^2$, therefore the mean speed of the molecules of a particular element with mass m is proportional to $\frac{1}{\sqrt{m}}$. Calculations show that the lightest gases in the atmosphere have mean speeds that are close to (but less than) escape speed. But even so, the very fastest of these molecules can escape the atmosphere so that eventually almost all the light gases disappear. The Moon has no atmosphere, because its escape speed is lower than that of the Earth and ultimately almost all its gas molecules leave after release. The low-density atmosphere of Mars is slowly leaking away into space.

Charges moving in magnetic and electric fields

Charges moving in magnetic fields

We saw in Topic 5 that a force acts on a charge moving in a magnetic field.

In Figure 13 an electron moves to the right into a uniform magnetic field in which the field lines are oriented at 90° to the direction in which the electron moves (the field lines are out of the plane of the paper in the figure).

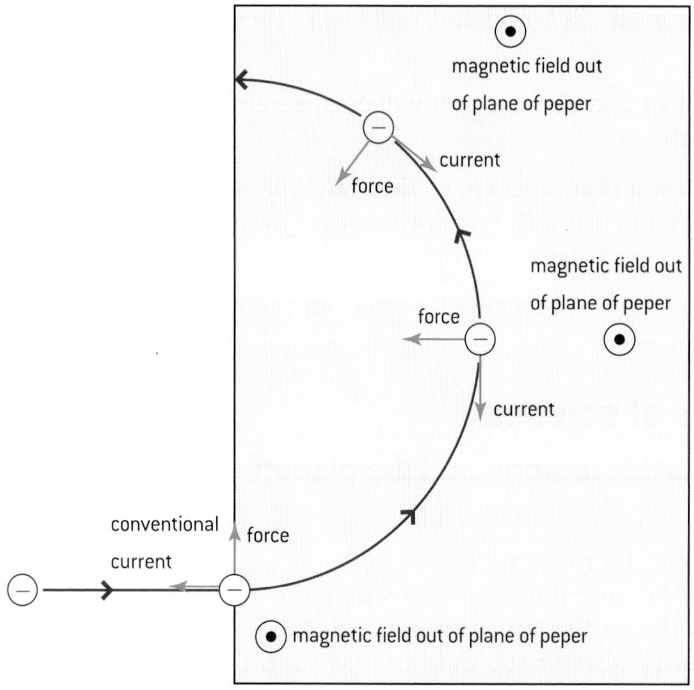

▲ Figure 13 Force acting on an electron in a uniform magnetic field.

Fleming's left-hand rule predicts the effect on the electron. The force will be at right angles to both the velocity and the direction of the magnetic field. In using Fleming's rule, remember that it applies to conventional current, and that here the electron is moving to the right – so the conventional current is initially to the left.

The electron accelerates in response to the force, and its direction of motion must change. The direction of this change is such that the electron will still travel at right angles to the field and the magnetic force will continue to be at right angles to the electron's new direction. This is exactly the condition required for the electron to move in a circle. The magnetic force acting on the electron is providing the centripetal force for the electron. As the electron continues in a circle so the magnetic force direction alters as shown in figure 13.

The force acting on the electron depends on its charge q, its speed v, and the magnetic field strength B. The centripetal force will lead to a circular orbit of radius r and

$$\frac{m_e v^2}{r} = Bqv$$

So the radius of the circle is (written in a number of ways)

$$r = \frac{m_e v}{Bq} = \frac{p}{Bq} = \frac{(2m_e E_k)^{\frac{1}{2}}}{Bq}$$

where p is the momentum and m_e is the mass of the electron. It is possible to use observations of this electron motion to measure the specific charge on the electron. This is the charge per unit mass measured in C kg^{-1}. A fine-beam tube is used in which a beam of

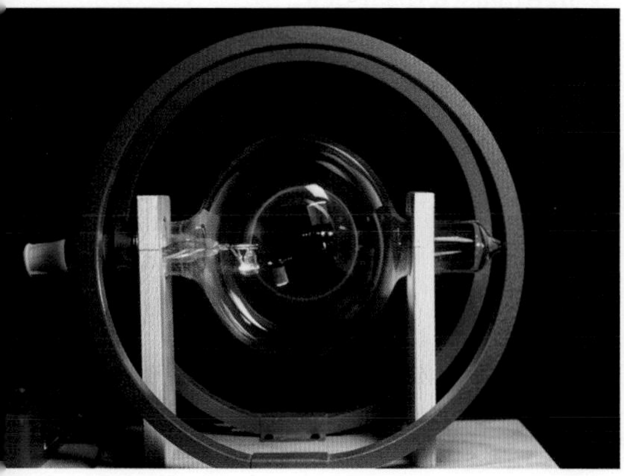

▲ Figure 14 Electrons moving in a circle demonstrated in a fine-beam tube.

electrons is fired through a gas at very low pressure (which means that the electrons do not collide with too many gas atoms). When a uniform magnetic field is applied at right angles to the beam direction, the electrons move in a circle and their path is shown by the emission of visible light from atoms excited by collisions with electrons along the path.

The electrons of mass m_e are accelerated using a potential difference V before entering the field. Their energy is

$$\frac{1}{2} m_e v^2 = qV$$

So

$$v = \sqrt{\frac{2qV}{m_e}}$$

which with

$$v = \frac{Bqr}{m_e} \text{ gives}$$

$$\frac{q}{m_e} = \frac{2V}{B^2 r^2}$$

In a case where the electron beam is moving into a magnetic field that is not at 90° to the beam direction, then it is necessary to take components of the electron velocity both perpendicular to the field and parallel to it. The perpendicular component leads to a circular motion exactly as before. The only difference will be that

$$r = \frac{m_e v \sin\theta}{Bq}$$

where θ is the angle between the field direction and the beam direction. So the radius of the circle will be smaller than the perpendicular case.

The parallel component of velocity does not lead to a circular motion. The electrons will continue to move at the component speed ($v \cos\theta$) in this direction. The overall result is that as the beam enters the field it will begin to move in a helical path.

Large particle accelerators can use magnetic fields to steer the charged particles as they are accelerated to higher and higher speeds. The acceleration is carried out using electric fields.

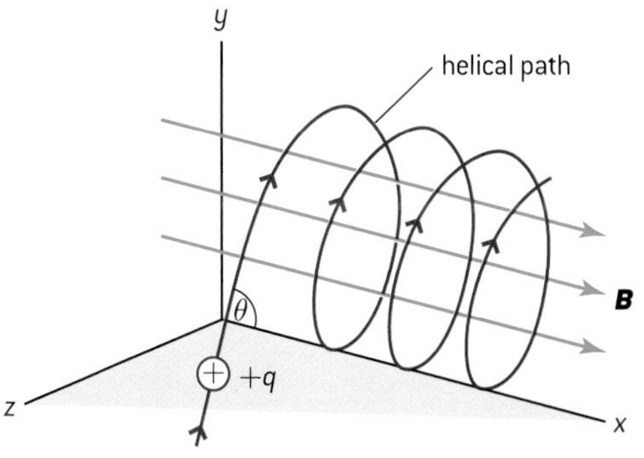

Figure 15 A positively-charged particle moving in a magnetic field not at 90° to its direction of motion.

Charges moving in electric fields

The situation is different for electrons moving into the uniform electric field produced by a pair of charged parallel plates.

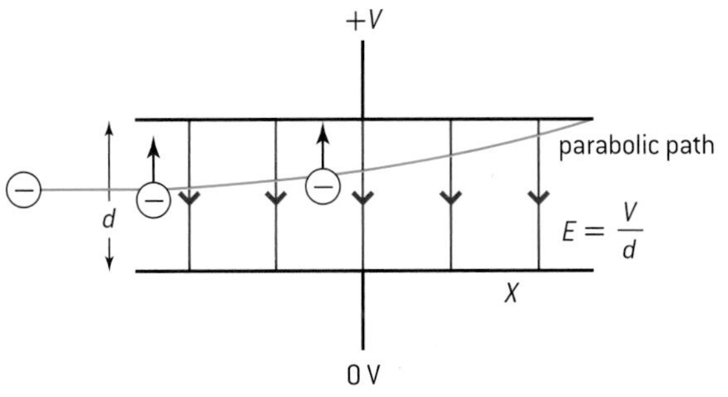

Figure 16 Motion of an electron in a uniform electric field.

The force on the electron now acts parallel to the field lines. In the case shown in the figure 16 the electron is accelerated vertically. Because the force is constant in both magnitude and direction (uniform field) the acceleration will also be constant. Mechanics equations from Topic 2 are used to analyse the situation:

$E = \frac{V}{d}$ as usual for a uniform field, and

$$a_{\text{vertical}} = \frac{F}{m_e} = \frac{qE}{m_e} = \frac{qV}{m_e d}$$

where m_e is the mass of the electron.

There is no horizontal acceleration and the horizontal component of the velocity does not change. The time t taken to travel between the plates of length x is therefore

$$t = \frac{x}{v_{\text{horizontal}}}$$

So the vertical component of the velocity as the electron leaves the plates is (using $v = u + at$)

$$v_{\text{vertical}} = a_{\text{vertical}} \times t = \frac{qV}{m_e d} \times \frac{x}{v_{\text{horizontal}}}$$

To obtain the final speed with which the electron leaves the field, it is necessary to combine the two components in the usual way as

$$\sqrt{v^2_{\text{vertical}} + v^2_{\text{horizontal}}}$$

The deflection s of the electron while in the field is proportional to $t^2 \left(s = \frac{1}{2}at^2\right)$. This form is equivalent to the equation for a parabola, so the trajectory of the electron is parabolic not circular unlike the case with a magnetic field. This derivation ignores the effects of gravity on the electron. Is this justified?

Charge moving in magnetic and electric fields

Despite the fact that magnetic fields lead to circular orbits and electric fields lead to parabolic trajectories, it is possible to use them in a particular configuration to do a useful job. The trick is to combine the magnetic and electric fields at right angles to each other.

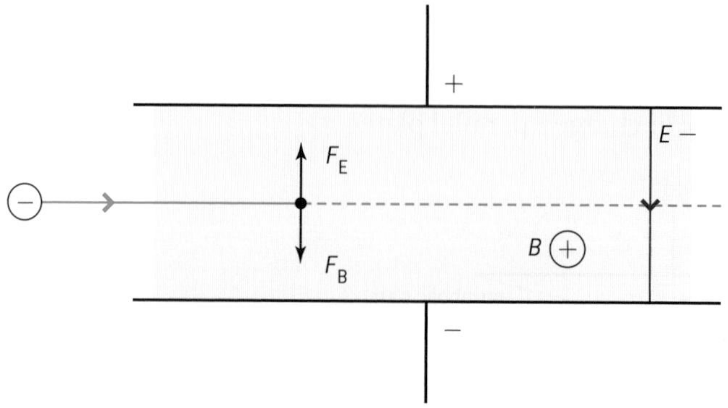

Figure 17 Crossed electric and magnetic fields can cancel out.

For the arrangement shown in the diagram, the electric force is vertically upwards and the magnetic force is downwards. When these forces are equal then there will be no net force acting on the charged particle.

For this to be the case

$$F_E = F_B; \qquad qE = Bqv$$

leading to

$$v = \frac{E}{B}$$

This means that for a particular ratio of $E : B$ there is one speed at which the forces will be balanced. Charged particles travelling more slowly than this speed will have a larger electric force than magnetic force, and the net force will be upwards on the diagram. For faster electrons the magnetic force will be larger and the electron will be deflected downwards. This provides a way to filter the speeds of charged particles and the arrangement is known as a **velocity selector**.

Nature of science

Bainbridge mass spectrometer

The Bainbridge mass spectrometer is a good example of a number of aspects of electric and magnetic fields being brought together to perform a useful job.

There are two parts to the instrument: a velocity selector with crossed electric and magnetic fields and a deflection chamber with just a magnetic field (in this diagram, into the paper). It is left as an exercise for the reader to see that the motion of equally charged ions of the same speed but different mass travel in circles of different radii in the deflection chamber. The ions arrive at the photographic plate or electronic sensor at different positions and, by measuring these positions, their mass (and relative abundance) can be determined.

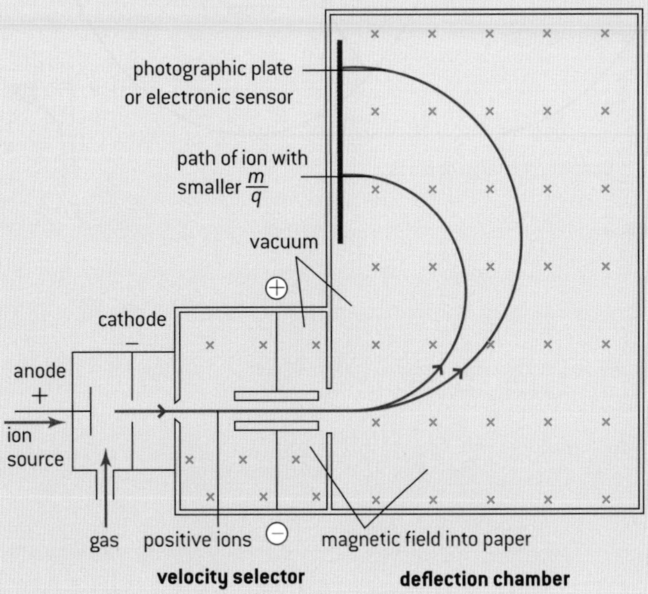

▲ Figure 18 Bainbridge mass spectrometer.

Concept map for field theory

Figure 19 gives a visual summary of all the relationships presented in this topic. It applies to both gravitational and electric fields. Both sets of equations are represented on it. The diagram shows the cycle of force ⟶ field strength ⟶ potential ⟶ potential energy ⟶ force and the relationships between them.

Comparison of gravitational and electric fields

The table below sets out some of the important similarities between electric and gravitational forces and summarizes the contents of Topics 5, 6, and 10.

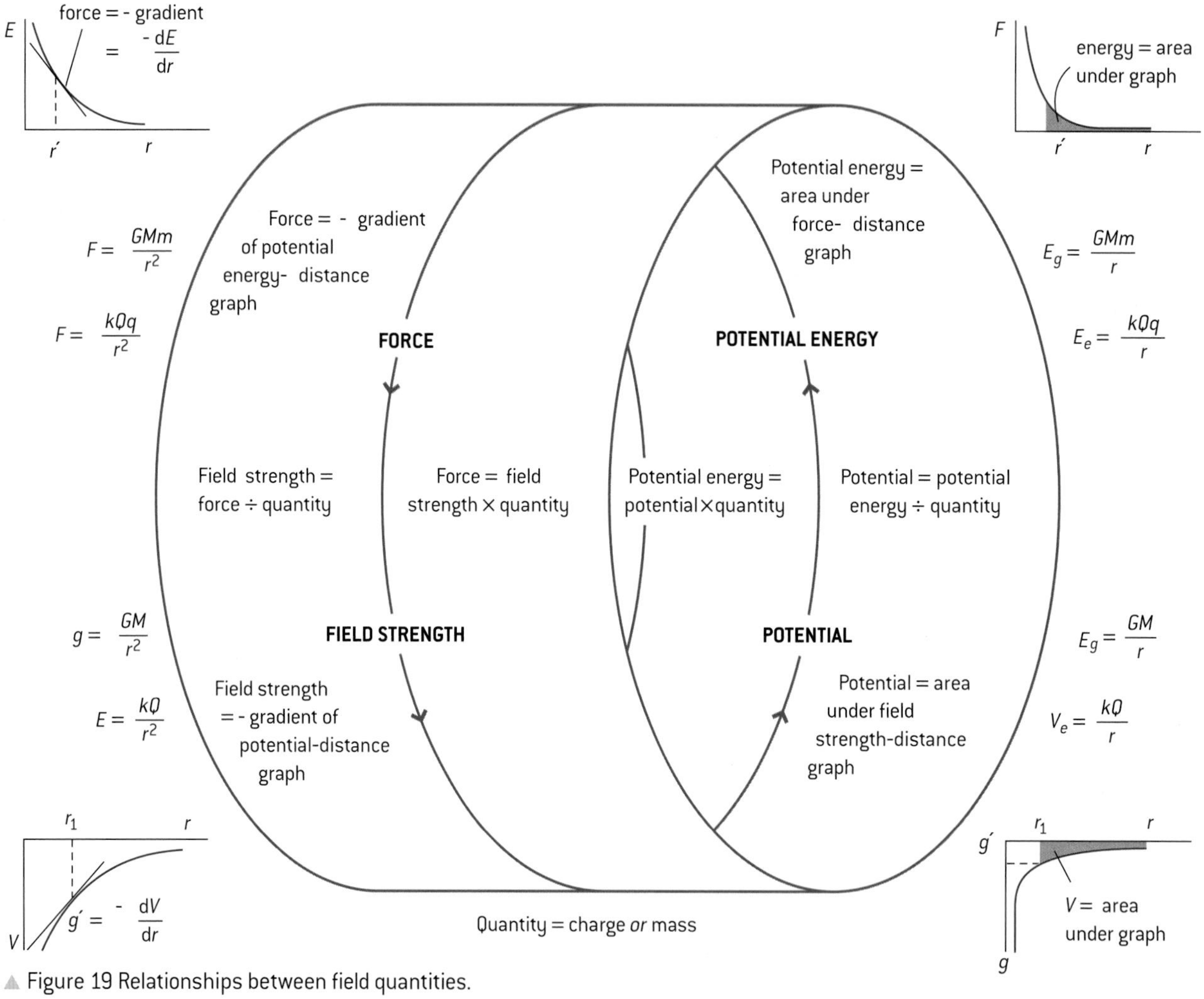

▲ Figure 19 Relationships between field quantities.

		Electric	Gravitational
Force law		$F = \frac{Qq}{4\pi\varepsilon_0 r^2}$ (Coulomb's law)	$F = \frac{GMm}{r^2}$ (Newton's law)
Field strength	Modification when not in a vacuum	Replace ε_0 with ε	No change
	Definition	$E = \frac{F}{q}$	$g = \frac{F}{m}$
	Unit	$N\,C^{-1}$ or $V\,m^{-1}$	$N\,kg^{-1}$ or $m\,s^{-2}$
	Distance *r* from a point object	$E = \frac{Q}{4\pi\varepsilon_0 r^2}$	$g = \frac{GM}{r^2}$
	At *r* from centre of sphere of radius R, $r \geq R$	$E = \frac{Q}{4\pi\varepsilon_0 r^2}$	$g' = \frac{GM}{r^2}$
	At *r* from centre of sphere of radius R, $r < R$	$E = 0$	$g' = \frac{4\pi G\rho r'}{3}$
Potential	Definition	Electric potential energy per unit charge	Gravitational potential energy per unit mass
	Unit	$V \equiv J\,C^{-1}$	$J\,kg^{-1}$
	For two point charges or masses	$V_p = \frac{Q}{4\pi\varepsilon_0 r}$	$V_g = -\frac{GM}{r}$
Differences value at infinity: Attractive: zero and maximum Repulsive: zero and minimum		• Between charges • Attractive and repulsive depending on charge sign	• Between masses • Only attractive
	Constant in force law	$\frac{1}{4\pi\varepsilon_0}$ (= Coulomb constant k)	G

Questions

1 (*IB*) The mass of Earth is M_E and the radius of Earth is R_E. At the surface of Earth the gravitational field strength is g.

A spherical planet of uniform density has a mass of 3 M_E and a radius 2 R_E. Calculate the gravitational field strength at the surface of the planet. (1 mark)

2 (*IB*) A spacecraft travels away from a planet in a straight line with its rockets switched off. At one instant the speed of the spacecraft is 5400 m s^{-1} when the time $t = 0$. When $t = 600$ s, the speed is 5100 m s^{-1}. Calculate the average gravitational field strength acting on the spacecraft during this time interval. (1 mark)

3 (*IB*)

a) State, in terms of electrons, the difference between a conductor and an insulator.

b) Suggest why there must be an electric field inside a current-carrying conductor.

c) The magnitude of the electric field strength inside a conductor is 55 N C^{-1}. Calculate the force on a free electron in the conductor.

d) The electric force between two point charges is a fundamental force that applies to charges whereas gravity is the gravitational force that applies to two masses. State **one** similarity between these two forces and **two** other differences.

e) The force on a mass of 1.0 kg falling freely near the surface of Jupiter is 25 N. The radius of Jupiter is 7.0×10^7 m.

(i) State the value of the magnitude of the gravitational field strength at the surface of Jupiter.

(ii) Calculate the mass of Jupiter.

(13 marks)

4 An astronaut in orbit around Earth is said to be "weightless". Explain why this is.

5 (*IB*) This question is about the gravitational field of Mars.

a) Define the *gravitational potential energy* of a mass at a point.

b) The graph shows the variation of the gravitational potential V with distance r from the centre of Mars. R is the radius of Mars which is 3.3 Mm. (Values of V inside the planet are not shown.)

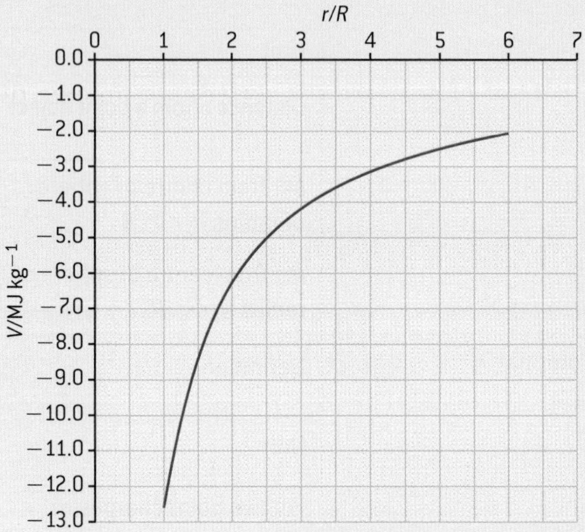

A rocket of mass 12 Mg lifts off from the surface of Mars.

(i) Calculate the change in gravitational potential energy of the rocket at a distance 4R from the centre of Mars.

(ii) Determine the magnitude of the gravitational field strength at a distance 4R from the centre of Mars.

c) Determine the magnitude of the gravitational field strength at the surface of Mars.

d) The gravitational potential at the surface of Earth is −63 MJ kg^{-1}. Without any further calculation, compare the escape speed required to leave the surface of Earth with that of the escape speed required to leave the surface of Mars.

(10 marks)

6 (*IB*) A small sphere X of mass M is placed a distance d from a point mass. The gravitational force on sphere X is 90 N. Sphere X is removed and a second sphere Y of mass 4M is placed a distance 3d from the same point mass. Calculate the gravitational force on sphere Y.

11 ELECTROMAGNETIC INDUCTION (AHL)

Introduction

The physics of electromagnetic induction has profound implications for the way we generate electrical energy and therefore for the way we live. Every year throughout the world about 10^{20} J of energy are converted into an electrical form using electromagnetism. This vast conversion of energy enables us to feed our ever-growing appetite for technologies that require electricity. It also raises considerable issues about the sustainability of the energy sources used for the conversion.

11.1 Electromagnetic induction

Understandings

- → Electromotive force (emf)
- → Magnetic flux and magnetic flux linkage
- → Faraday's law of induction
- → Lenz's law

Nature of science

Much of the physics in this sub-topic was discovered through painstaking experimentation by a handful of scientists. Pre-eminent among them was Michael Faraday who observed currents induced in a coil when magnetic fields were varied nearby. These observations led him to the laws of electromagnetic induction that we use in our large-scale generation of electrical energy.

Applications and skills

- → Describing the production of an induced emf by a changing magnetic flux and within a uniform magnetic field
- → Solving problems involving magnetic flux, magnetic flux linkage, and Faraday's law
- → Explaining Lenz's law through the conservation of energy

Equations

- → Flux: $\Phi = BA\cos\theta$
- → Faraday's / Neumann's equation: $\varepsilon = -N\frac{\Delta\Phi}{\Delta t}$
- → emf induced in moving rod: $\varepsilon = Bvl$
- → in side of coil with N turns: $\varepsilon = BvlN$

Inducing an emf

In Topic 5 we saw that when an electric charge moves in a magnetic field, then a force acts on the charge (and therefore on the conductor in which it is moving). In a reverse sense, a movement or change in a magnetic field relative to a stationary charge gives rise to an electric current. This phenomenon is called **electromagnetic induction**.

Investigate!

Making a current

- Begin with a magnet and a coil or a solenoid of wire. Arrange the coil horizontally and connect it to a galvanometer (a form of sensitive ammeter, with the zero in the middle of the scale). You can use a laboratory coil, or you can wind your own from suitable metal wire using a cylinder as a former.
- Move the bar magnet so that its north-seeking pole approaches one end of the coil and observe the effect on the meter. Record the direction of the current as indicated by the meter and the peak value shown.
- Repeat the movement, moving the bar magnet with its south-seeking pole towards the coil.
- Move the bar magnet away from the coil.
- Change the speed with which you move the magnet.
- Compare the current directions for each case and also the size of the current produced.
- Now try moving the coil and keeping the magnet still. Does this change your observation?
- Now try moving the coil and the magnet at the same speed and in the same direction. What is the size of the current now? If your coil allows it, you might also try making the magnet enter the coil at an angle rather than along its axis. You could also try moving the magnet completely through the coil and out of the other side. Try to interpret this complex situation when you have understood the simpler cases.
- Relate the direction of current flow in the coil to the magnetic poles produced at the ends of the coil using the ideas in Topic 5. (Figure 1(b) reminds you of the rule.)

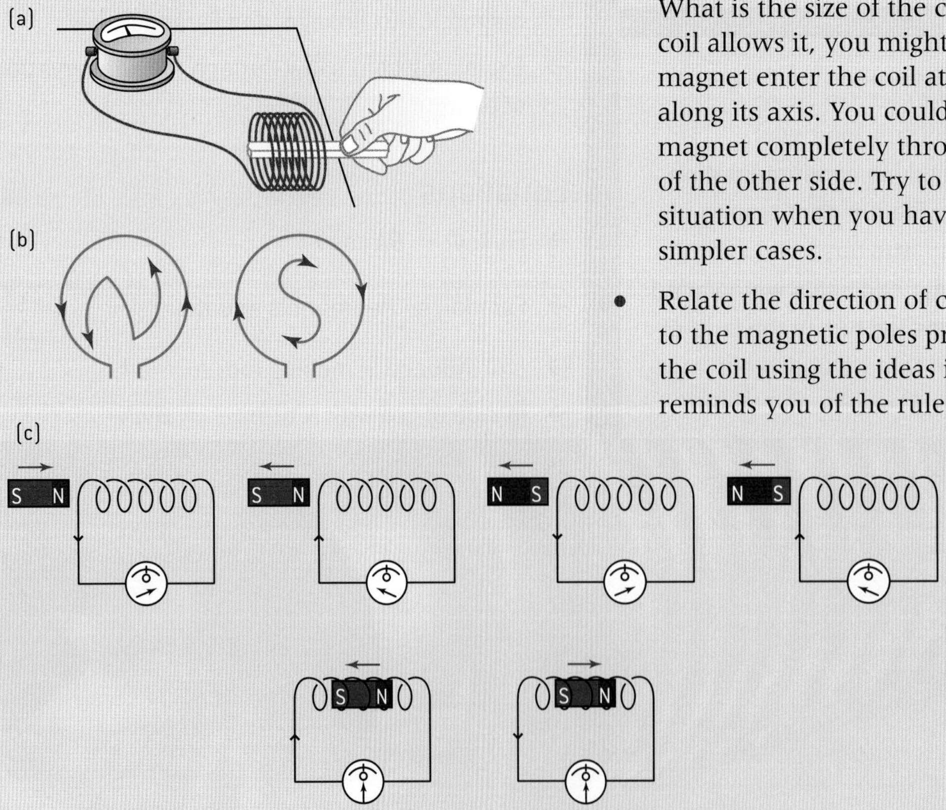

▲ Figure 1 (a) the experiment, (b) the N and S rule, and (c) typical results.

The results for this simple experiment are shown in figure 1(c). You should focus both on the direction of the conventional current flow and on the magnitude of the current. You are observing results similar to those made by Faraday in the 19th century.

A number of conclusions emerge from these simple experiments:

- The current only appears when there is relative motion between the coil and the magnet; either or both can move to produce the effect. However, if both coil and magnet move with no relative motion between them then no current occurs. Only movement of the wire in the coil relative to the field is important.
- When the north-seeking pole of the magnet is inserted into the coil, the current in the coil tends to reduce the magnet's motion by producing another north-seeking pole at the magnet end of the coil. Push a south-seeking pole in and another south pole appears at the magnet end of the coil. It is as though the system acts to repel the bar magnet and to reduce its movement. The system appears to oppose any change in the magnetic flux; the greater the rate of change, the greater is this opposition. We will look at this again later.
- In the same way, when a magnetic pole is moved away from the coil, the opposite pole is formed by the current in the coil in an attempt to attract the magnet and reduce its speed of motion.
- Moving the coil at greater speeds relative to the magnet increases the sizes of the currents. The effect is at a maximum when the axis of the magnet between its poles is perpendicular to the area of cross-section of the coil.

The keys to understanding these effects lie in what we said earlier in Topic 5 about the motion of charges in a magnetic field and what we know about the internal structure of the conducting wire that makes up the coil.

Figure 2(a) shows electrons in a metal rod that is moving through a uniform, unchanging magnetic field. Free electrons in the wire are moving upwards with the rod in an external magnetic field that acts into the page. A force acts on each electron to the right. This force is equivalent in direction to that which would act on a positive charge moving down the page (a conventional current downwards). This direction is perpendicular to both the field and the direction of motion of the rod and is determined using Fleming's left-hand rule.

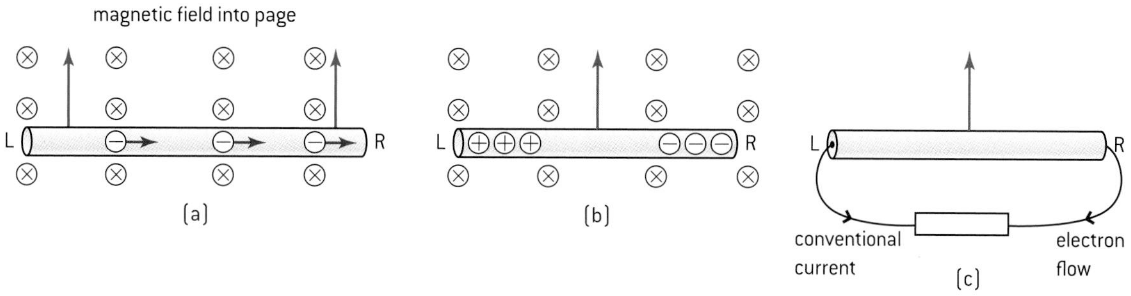

Figure 2 Electrons forced to move in a magnetic field experience a force.

In figure 2(b), with no external connection between L and R the electrons accumulate at the right-hand end (R) of the rod making it negatively charged, and a lack of electrons at the left-hand end (L) makes it positive. A potential difference exists between L and R, L being at the higher potential. When there is no external connection between L and R, no current will circulate. Charges will accumulate at the end of the rod, that is electrons will move to one end leaving the other end positive. Without a current in a closed circuit no work is required, no transformation of energy takes place.

If the circuit is closed externally between L and R (figure 2(c)), a flow of electrons will occur as shown. Inside the rod the conventional current flows from R to L. The electrons flow out of the right-hand end of the rod and this is a conventional current in the external circuit from L to R. A current has been generated, or induced, in the circuit.

The system is acting to move the electrons through the resistor and, because this is a transformation of energy into an electric form, the source of the energy is termed an electromotive force (emf). As usual, we can identify the amount of energy transferred for each coulomb of charge that moves around the circuit and we use the term **induced emf** to show that the emf arises from an induction effect. (The word "induction" is another term used in the early days of the study of electromagnetism.)

Nature of science

Electromagnetic force

Perhaps you can now see why the term electromagnetic force arose in the early days of electromagnetic induction – and why it still persists. Some physicists object that no force acts when the term emf is applied to electric cells, batteries, piezoelectric devices, and so on – therefore, they say, emf is not a good expression. It is true that it is difficult to see how the word "force" can apply in the case of a cell. But in the case of electromagnetic induction, it is clear that a force is acting on the electrons in the conductor that is being moved and so the term emf continues to be used in physics. The fact that we often use the abbreviation emf rather than the full expression is a reminder that we should not focus on the term "force" but rather on the units of emf: $J\ C^{-1}$.

Lenz's law

An important aspect of electromagnetic induction is that the induced emf exists *whether the charge flows in a complete circuit or not*. In the case shown in figure 2(b) the circuit is incomplete. Electrons flow along the rod until an excess of them sets up an electric field which repels further electrons, stopping further flow. It is only when the circuit is complete that charge flows. An induced emf is *always* generated by the system and the induced current will exist *only* if there is a complete circuit.

We can look at the system in terms of a possible direction rule. Fleming's left-hand rule was able to predict the force on, and therefore a flow of, the electrons. This flow of electrons is equivalent to a conventional current

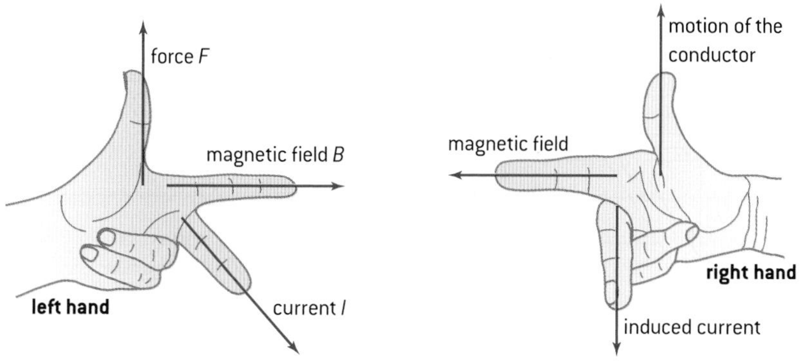

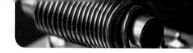

Fleming's left-hand rule for motor effects

Fleming's right hand rule for induction effects

▲ Figure 3 Fleming's rules.

acting in the opposite direction. You have two choices: either use Fleming's left-hand rule to work out the force direction from first principles and let this lead you to the conventional current direction (the argument is given above), or you can use another rule (Fleming's right-hand rule), which uses the symmetry between our left and right hands to give the relationship between the motion of the conductor, the direction of the field and the direction of the induced conventional current. It's your choice!

The observations you made in the *Investigate!* can be interpreted by looking closely at the directions of current in the coil relative to the movement of field and coil. A rule that describes this was summed up by the German scientist Heinrich Lenz in 1833. He stated that

the direction of the induced current is such as to oppose the change that created the current.

Check your experimental notes (or the diagrams in figure 1 that sum them up) and see if your results confirm this law.

In fact, Lenz's law is little more than the conservation of energy. Suppose that, rather than opposing the induced effect, the change were to enhance it. This would imply an attraction instead of a repulsion between magnet and coil; the magnet would be pulled into the coil, accelerating as it goes. This would increase the speed and lead to an even greater acceleration. The magnet would move faster and faster into the coil, gaining kinetic energy from nowhere. Conservation of energy tells us this cannot happen.

Another way to look at the consequences of Lenz's law is to realize that you cannot do work without having some opposition. The induced current in the coil is such that the induced field produced by this current opposes the motion of the magnet you are holding. If the circuit is open, there is no current, no opposition, and no electric energy produced. If you move the magnet very fast you will clearly feel the opposite force acting on you!

Electrons had not been discovered in Lenz's time and we can see how his law arises from first principles. In Topic 5 we saw that when a current flows within a magnetic field the current produces a magnetic field which distorts the original field pattern. The result was that a force acted on (and could accelerate) the current-carrying conductor. This was the effect that led us to the basis of an electric motor.

Nature of science

Another rule

Another possibility is to use the right hand with its thumb in the direction of the current and fingers in the direction of the magnetic field. By pushing you get the direction of the force on a positive charge or a conventional current. If an electron is the moving charge, point the thumb in the opposite direction to get the force in opposite direction This convention can also be used to get the direction of the magnetic field due a conventional current e.g. by curling the fingers around a wire with current.

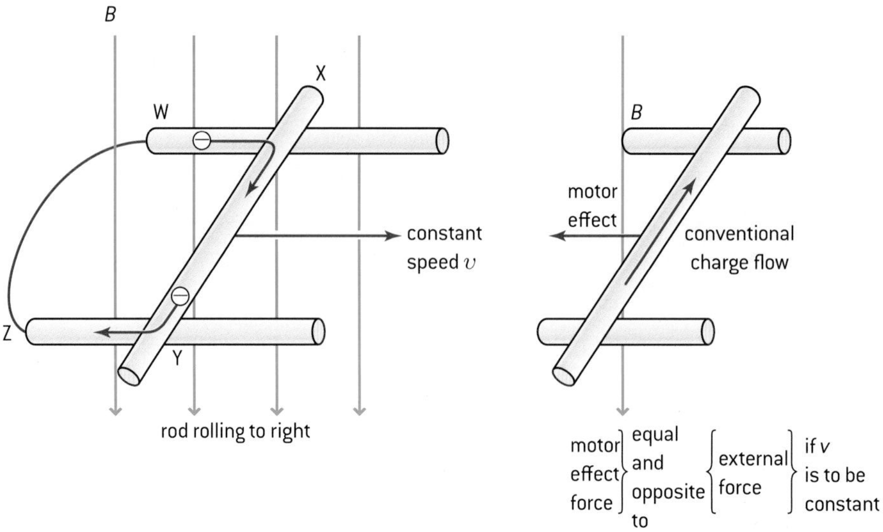

Figure 4 Conducting rod rolling in a magnetic field.

In electromagnetic induction, there is a current in the conductor as a result of the motion of the conductor. Figure 4 extends the example of the moving rod that is now rolling to the right at a constant speed through the magnetic field on a pair of rails. The rails conduct and form part of a complete electrical circuit WXYZ. (Rolling means that we do not have to worry about friction between the rod and the rails.) Charges, driven by the induced emf, flow around the circuit giving rise to an induced current in the direction shown. This induced current interacts with the uniform magnetic field to give rise to a force - the motor effect force that we discussed in Topic 5 (page 234). If you now use Fleming's left-hand rule you will see that the induced current leads to a motor effect (a force) acting to the left in figure 4, that is, opposite to the direction in which the conductor is moving. If the rod is to move at a constant speed then (by Newton's first law of motion) an external force must be exerted on it. This is where the energy conversion comes in. The work done by the external force to keep the conductor moving at a constant speed appears as electrical energy in the conductor.

Magnetic flux and flux density

We can use the physics from Topic 5 to extend these qualitative ideas. The magnitude of this magnetic force is BIl where B is the magnetic field strength, I is the induced current in the rod, and l is the length of the rod. Fleming's left-hand rule shows that the magnetic force arising from the induced current opposes the original force. In other words, the opposing magnetic force is to the left if the original applied force is to the right. The net force is zero and the rod moves at constant speed. It is possible for work to be done because of the opposition to the motion presented by the external magnetic field that produces a magnetic force on the induced current. No opposition, no work possible (or needed!).

From Newton's first law, to keep the rod in figure 4 moving at constant velocity, a constant force equal to BIl must act on the rod to the right.

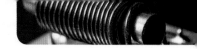

The energy we have to supply therefore in a time Δt is

force $\times$ *distance moved*

which is

$BIl \times \Delta x$

where Δx is the distance moved to the right by the rod.

The induced emf is equal to the energy per coulomb supplied to the system. In other words

$$\frac{\text{energy supplied}}{\text{charge moved}} = \varepsilon = \frac{BIl \times \Delta x}{Q} = \frac{BIl\Delta x}{I\Delta t}$$

and therefore

$\varepsilon = Blv$

where v is the speed of the rod.

Cancelling and rearranging gives

$$\varepsilon = \frac{B\,\Delta A}{\Delta t} = B \times \text{rate of change of area}$$

This is because the area that is swept through by the rod in a time t is $l\Delta x = \Delta A$ (because it is the amount by which the area *changes* in time Δt)

So, in words,

induced emf = magnetic flux density × rate of change of area

This introduces you to an alternative term in magnetism for what was earlier called the magnetic field strength – **magnetic flux density**.

The magnetic field strength is numerically equivalent to magnetic flux density.

In Topic 5 we used the term "magnetic field strength" because in that topic we were concerned with basic ideas of field. In both electrostatics and gravity, the term field strength has a meaning of $\frac{\text{force}}{\text{mass}}$ or $\frac{\text{force}}{\text{charge}}$ depending on the context. We defined magnetic field strength in Topic 5 as

$$\frac{\text{force}}{\text{current} \times \text{length}}$$

However, this definition does not take account of the old but helpful view of magnetic field lines as lines directed from a north-seeking to a south-seeking pole with a line density that is a measure of the strength of the field. It is this visualization of a field in terms of lines (close together when the field is strong, and well-separated when the field is weak) that links magnetic field strength to magnetic flux density.

When lines of force (field lines) are close together, then the magnetic field strength is large. There will be many lines through a given area. We say also that the magnetic flux density is large. The total number of lines per square metre is a measure of the magnetic flux density and therefore the total number of lines in a given area is a measure of the magnetic flux.

Flux is an old English word that has the meaning of "flow" and one way to think about flux is to imagine a windsock used to show the direction and speed of wind at an airfield. If the wind is strong then the flux

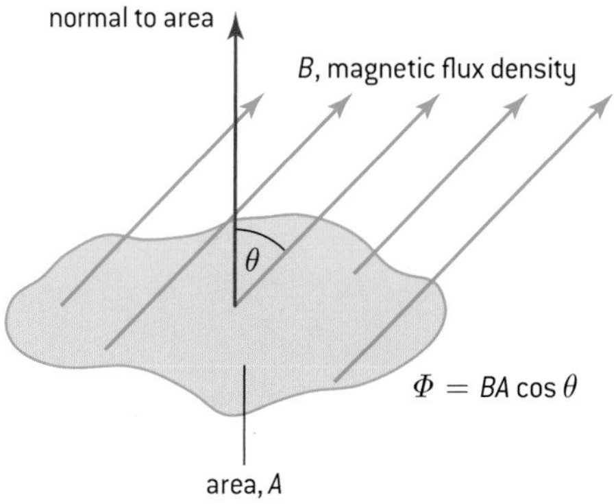

▲ Figure 5 Flux and flux density.

density is high. The flux is the number of streamlines going through the sock. If the wind has the same speed for two windsocks of different sizes then the windsock with the larger opening will have a larger flux even though the flux density is the same.

If magnetic flux density is defined as the number of flux lines per unit area, then flux itself must be equivalent to

flux density × *area*

so that, in symbols,

$$\Phi = B \times A$$

This equation assumes that B and A are at right angles. Figure 5 shows the relationship between area and flux density when this is not the case. In this case, we need to consider the component of the field normal to the plane. A normal to the area is constructed and this normal makes an angle θ with the lines of flux. So,

$\Phi = BA\cos\theta$ so that when $\theta = 0$ (area normal to field lines) $\Phi = BA$, and when $\theta = 90°$, $\Phi = 0$.

To sum up:

- Magnetic flux density B is related to the number of field lines per unit area. It is a vector quantity.
- Magnetic flux is equal to BA (also written as Φ). It is a scalar quantity.
- The equation $\Phi = BA\cos\theta$ is used if the area is not normal to the lines.

The unit of flux is the weber (Wb) and is defined in terms of the emf induced when a magnetic field changes. The equation $\varepsilon = B\frac{\Delta A}{\Delta t}$ can be re-written as $\varepsilon = \frac{\Delta(BA)}{\Delta t}$ or $\frac{\textit{change in flux}}{\textit{time taken for change}}$ and so

a rate of change of flux of one weber per second induces an emf of one volt across a conductor

For a particular conductor, knowledge of a magnetic field in terms of the magnetic flux and the rate at which it changes allows a direct calculation of the magnitude of the induced emf that will appear.

We can now make a direct link between flux density and field strength.

The magnetic flux density $\left(\frac{\text{flux}}{\text{area over which it acts}}\right.$, measured in weber metre$^{-2}\left.\right)$ is numerically equal to the magnetic field strength $\left(\frac{\text{force}}{\text{current} \times \text{length}}\right)$.

One tesla (T) ≡ one weber per square metre (Wb m^{-2})

We therefore also have a link between changes in the magnetic field strength and the induced emf.

Magnetic flux linkage

Finally, one more quantity appears. Our derivation of $\varepsilon = B\frac{\Delta A}{\Delta t}$ above used a single rod rolling along two rails. This is equivalent to a single rectangular coil of wire that is gradually increasing in area. Imagine this single turn of coil gradually increasing its area to include more and more field lines. As before the emf across the ends of the coil will be equal to the rate of change of area multiplied by the magnetic flux density. If there are N turns of wire in the coil then the induced emf will be N times greater so that $= NB\frac{\Delta A}{\Delta t} = \frac{\Delta(N\Phi)}{\Delta t}$. $N\Phi$ is known as the **magnetic flux linkage**.

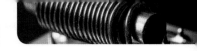

The unit of flux linkage is often written as **weber turns,** although this is entirely equivalent to weber because the number of turns is simply a number.

The relationships between these interlinked quantities can be shown as:

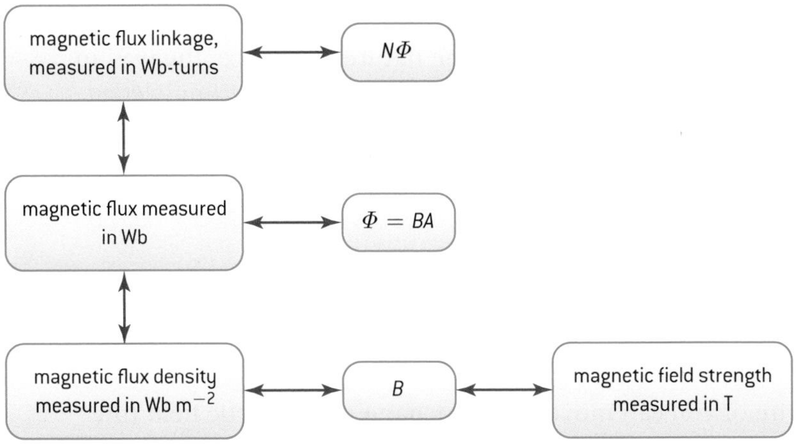

Magnetic induction is summed up in a law devised by Faraday himself that is known as Faraday's law, it states that

> **the induced emf in a circuit is equal to the rate of change of magnetic flux linkage through the circuit.**

In our usual notation this is written algebraically as

$$\varepsilon = -\frac{N\Delta\Phi}{\Delta t}$$

The negative sign is added to include Lenz's law. In its full mathematical form, the equation is also known as Neumann's equation. This equation combines Faraday's and Lenz's ideas. It reminds us that, (as we shall see in the next section) magnetic flux can be changed in three different ways: by changing the area of cross-section with time $\left(\frac{\Delta A}{\Delta t}\right)$, by changing θ with time $\left[\frac{\Delta(\cos\theta)}{\Delta t}\right]$ or by changing magnetic flux density with time $\left(\frac{\Delta B}{\Delta t}\right)$.

Nature of science

Cutting lines of force

Faraday first introduced the field-line model though his model is not quite the same as our modern interpretation of a magnetic field. He considered the lines of force to be at the edges of "tubes of force", like elongated elastic bands. At the time, an invisible "aether", thought to have elastic properties, was considered to fill space. Later on, Faraday and others took the concept of the field line further by suggesting that it was the action of the conductor "cutting" the tubes of force that led to an induced emf. This is a helpful way to think of the process, although it conceals the link between a charge being moved in a magnetic field and the magnetic force that acts on the charge as a result. But we need always to remember that Faraday and the others did not know of the existence of the electron, and that they were very familiar with the ideas of field lines. In the nineteenth century, Maxwell refined these models by including both electrostatic and electromagnetic forces in one set of equations.

This illustrates two things about the nature of science: the way in which scientists allow a discovery to illuminate prior knowledge in a different way, and the power of the visual image to help us to comprehend a phenomenon.

Changing fields and moving coils

An emf can be electromagnetically induced in a number of ways that, on the face of it, appear different from each other:

- A wire or coil can move in an unchanging magnetic field (the example of the rolling rod above).
- The magnetic field can change in strength but the conductor does not move or change.
- A coil can change its size or orientation in an unchanging magnetic field.
- Combinations of these changes can occur.

We will look at these cases in the context of a rectangular coil interacting with a uniform (but possibly changing) magnetic field.

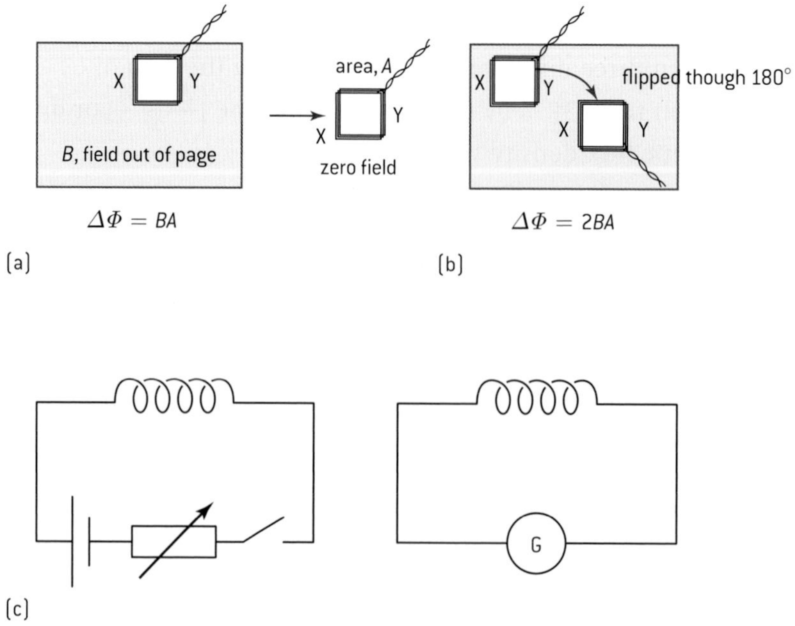

▲ Figure 6 Moving coils and changing fields.

Case 1: Straight wire moving in a uniform field

This is the case of the rolling rod above. The change in area per second is lv, the length l of the rod multiplied by v, the speed of the rod.

The induced emf is therefore $\varepsilon = Bvl$ when the wire moves at 90° to the field lines. If the wire motion is not at 90° to the field then, as usual, the component of field at 90° to the direction should be used.

Case 2: Coil moving

The coil can move as shown in figure 6(a) from one position in a magnetic field to another position where the field may be different. If the coil begins and ends in positions where the field is identical, then there is no change in the flux linkage and there is no induced emf. Although the coil is cutting lines, the same number is being cut on opposite sides of the coil. Two emfs are induced but in opposite senses and they therefore cancel out.

If the coil moves from a position where the flux is Φ to a position where the flux is zero, the change in flux linkage is $N\Phi$ and the induced emf ε is

$$\varepsilon = \frac{N\Phi}{\textit{time taken for change}}$$

An interesting variant of this occurs where a coil in a field is flipped through 180° (figure 6(b)). To visualize this, look at things from the point of view of the coil. The field lines appear to reverse their direction through the coil, and so the change in flux is $\Phi - (-\Phi)$, in other words, 2Φ. So the emf induced will be equal to $\frac{2N\Phi}{\textit{time taken to rotate coil}}$.

When a coil rotates in a field the emf produced instantaneously depends on the rate of change of the flux linkage, and this in turn depends on the angle the coil makes instantaneously with the field. If the coil rotates at a constant angular speed, then the emf output varies in a sinusoidal way. This is the basis of an alternating current (ac) generator as we shall see later.

Case 3: Magnetic field changes

Sometimes the field changes from one value to another – it gets stronger or weaker – but the coil does not move. The act of cutting field lines is not so obvious here.

Suppose the field is being turned on from zero. Before the field begins to change there are no field lines inside the coil. You can think of the lines as moving from the outside into the area bounded by the coil. The change stops when the flux density is at its final value. In so doing the lines must have cut through the stationary coil.

Again, the solution is not difficult. Now $\varepsilon = -\frac{N\Delta\Phi}{\Delta t}$ becomes $\varepsilon = -NA\frac{\Delta B}{\Delta t}$ because only B is changing. You need to know the rate at which the field is changing with time (or the total change and the total time over which it happens). Another example is the case of two coils face-to-face as in figure 6(c). One coil is connected to a galvanometer alone, the other coil is connected to a circuit with a cell, a variable resistor and a switch. In what way will you expect the galvanometer reading to change when the switch is closed and remains closed? When the switch is opened? When the switch is closed and the resistance in the circuit is varied?

Worked examples

1 The graph shows the variation of magnetic flux with time through a coil of 500 turns.

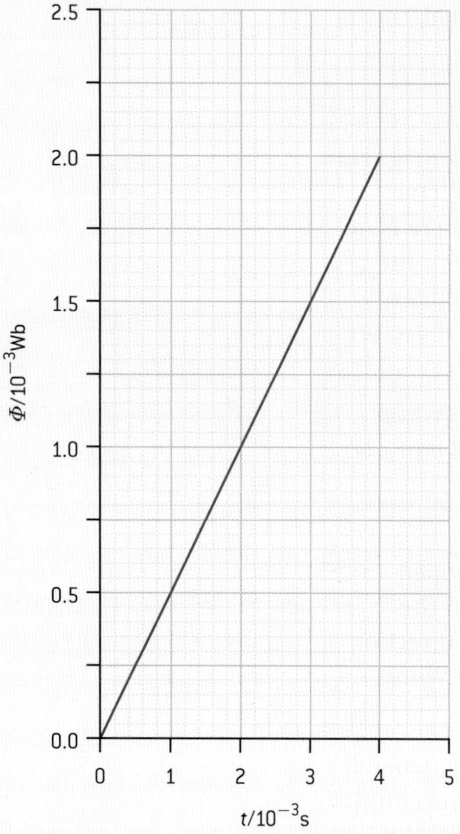

Calculate the magnitude of the emf induced in the coil.

Solution

The change in flux is 2×10^{-3} Wb and this occurs in a time of 4.0 ms. the rate of change of flux is the gradient of this graph (as always) – as the flux is proportional to the time we can use any corresponding values of ϕ and t.

$$\frac{\Delta \Phi}{\Delta t} = \frac{2\ \text{mWb}}{4\ \text{ms}} = 0.5\text{V}$$

thus the induced emf $= 500 \times 0.5\ \text{V} = 250\ \text{V}$

2 A small cylindrical magnet and an aluminum cylinder (which is non-magnetic) of similar shape and mass are dropped from rest down a vertical copper tube of length 1.5 m.

a) Show that the aluminum cylinder will take about 0.5 s to reach the bottom of the tube.

b) The magnet takes 5 s to reach the bottom of the tube. Explain why the objects take different times to reach the bottom.

Solution

a) Use a kinematic equation, e.g. $s = ut + \frac{1}{2}at^2$. Then $1.5 = \frac{1}{2} \times 9.8 \times t^2$

$t = 0.55$ s.

b) As the magnet falls the copper tube experiences a changing magnetic flux, and as a result an emf is induced in the walls of the tube. This emf results in a current in the tube. The current leads to another magnetic field that opposes the motion of the magnet by Lenz's law. There is an upward force on the magnet so that its acceleration is less than the value for free fall. In the case of the aluminium cylinder no current arises and it falls with the usual acceleration.

Nature of science

Applications of electromagnetic induction

There are many applications of electromagnetic induction over and above the generation of electrical energy. They include electromagnetic braking, which is used in large commercial road vehicles; the use of an induction coil to generate the large pds required to provide the spark that ignites the petrol–air mixture in a car engine; and the generation of the signal in geophones and metal detectors. In each of these examples, a changing magnetic field leads to the generation of an emf and demonstrates the physics developed in this sub-topic.

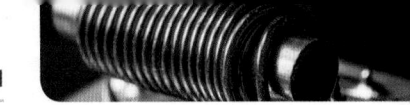

11.2 Power generation and transmission

Understanding

- Alternating current (ac) generators
- Average power and root mean square (rms) values of current and voltage
- Transformers
- Diode bridges
- Half-wave and full-wave rectification

Nature of science

The provision of abundant, cheap electrical energy has been one of the ways in which an abstract science and its technological development have directly affected the lives of many people on the planet. Who could have imagined the enormous impact that Faraday's discoveries would make? ...certainly not Faraday himself! This is an example of research in pure science leading to great practical applications.

Applications and skills

- Explaining the operation of a basic ac generator, including the effect of changing the generator frequency
- Solving problems involving the average power in an ac circuit
- Solving problems involving step-up and step-down transformers
- Describing the use of transformers in ac electrical power distribution
- Investigating a diode bridge rectification circuit experimentally
- Qualitatively describing the effect of adding a capacitor to a diode bridge rectification circuit

Equations

- rms and peak values:
  $I_{rms} = \frac{I_0}{\sqrt{2}}$
- potential difference: $V_{rms} = \frac{V_0}{\sqrt{2}}$
- resistance: $R = \frac{V_0}{I_0} = \frac{V_{rms}}{I_{rms}}$
- maximum power: $P_{max} = I_0 V_0$
- average power: $P_{average} = \frac{1}{2} I_0 V_0$
- transformer equation: $\frac{\varepsilon_p}{\varepsilon_s} = \frac{N_p}{N_s} = \frac{I_s}{I_p}$

Introduction

We have seen that a rod rolling along two parallel rails generates induced emf and induced current; this is hardly a sensible way to generate electrical energy on a large scale. The practicalities of generation were solved by scientists from the 1830s onwards, first for direct current, and later for alternating current.

Alternating current (ac) generators

In this IB course we focus on the ac generator because it is commonly used for the generation of energy. Such a generator consists of a coil with a large number of turns; the coil rotates relative to a magnetic field.

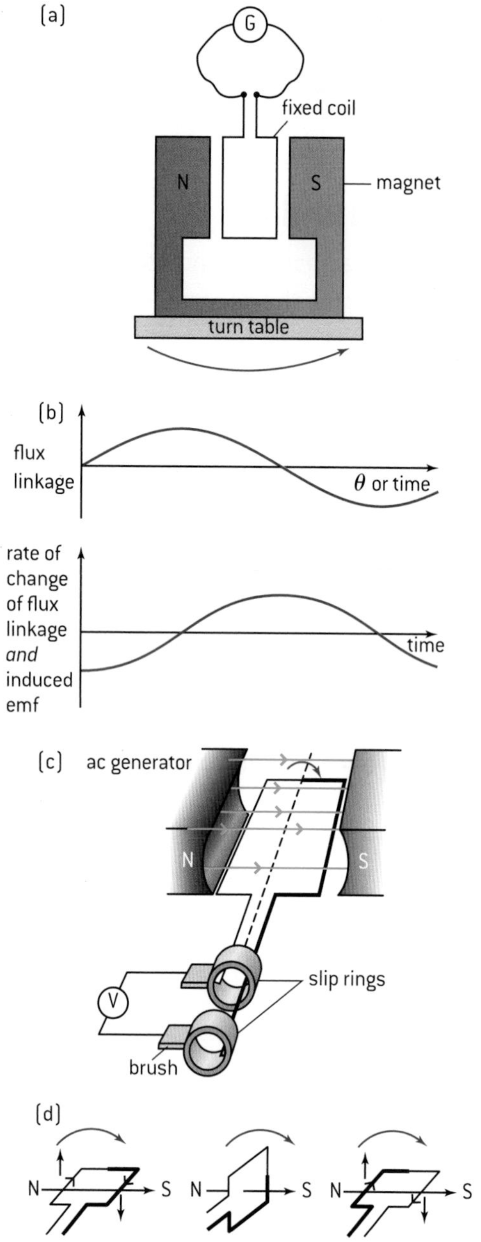

▲ Figure 1 Basic ac generator.

For the moment we will imagine a fixed coil placed between the poles of a U-shaped magnet that stands at the centre of a rotating turntable (figure 1(a)). The turntable can turn at different angular speeds and the coil can have different numbers of turns and cross-sectional areas. The coil is connected to a galvanometer or to a data logger that registers the emf across the coil. When the magnetic flux in the coil is maximum, the emf induced (current) is minimum (0) and vice-versa. Changing the speed of the turntable changes the frequency of the emf as well as its amplitude. Increasing the number of turns or increasing the area of the coil will increase the amplitude of the emf but leave the frequency unchanged if the turntable speed does not change.

Magnetic field lines cut the coil as the turntable rotates and an induced emf is generated. However the emf varies as the turntable rotates. Figure 1(b) shows how the flux linkage varies with θ, the angle between the normal to the coil and field lines. When θ is equal to 90°, the field lines lie in the plane of the coil and so the flux through the coil is zero ($\cos 90° = 0$ in equation $\phi = BA\cos\theta$). When θ is equal to 0° the field lines are at 90° to the plane of the coil and the flux through the coil is a maximum. When the turntable rotation speed is constant, this graph also shows how the flux varies with time. The graph is a sine curve.

The emf induced in the coil is equal to $-\frac{\Delta\phi}{\Delta t}$, in other words the negative of the gradient of the flux-time graph. Figure 1(b) also shows how the induced emf varies with time; this graph can be obtained either by differentiation or by a consideration of the gradient of the flux-time graph. When the normal to the coil and the field lines are parallel (θ is 90°) then the emf is zero because the coil does not – instantaneously – cut the field lines at all.

However, while some ac generators have a rotating magnetic field, others have fixed magnets and a rotating coil (Figure 1(c)). The principle is however the same. The direction of charge flow in the coils varies with the position of the coil. The use of a direction rule shows this.

When the left-hand wire is moving upwards as shown in Figure 1(d), the conventional current direction in this wire is to the back of the coil. The right-hand wire is moving downwards at the same instant and the current in this wire is towards the front of the coil. Charge flows clockwise (looking from above) in the coil and out into the external circuit.

Half a cycle later the sides of the coil will have exchanged position. Charge is again flowing clockwise, but because the coil has rotated, the current (so far as the meter is concerned) is in the opposite direction. Figure 1(d) shows this too.

If there were wires permanently connecting the coil to the meter they would quickly become twisted. Energy needs to be extracted from the generator without this happening. Slip rings are used for this. The ends of the coil terminate in two rings of metal that rotate with the coil about the same axis. Two stationary brushes, connected to the external part of the circuit, press onto the rotating rings and charge flows out into the circuit through these connections (Figure 1(c)).

The essential requirements for an ac generator are therefore:

- a rotating coil
- a magnetic field
- relative movement between the coil and the magnetic field
- a suitable connection to the outside world.

Real-life generators are more sophisticated than our simple models and an Internet search will allow you to see many different types of ac generator.

For real generators, there is another issue that arises because an induced current is generated in the coils. As we saw in Sub-topic 11.1, any moving conductor carrying an induced current in a magnetic field has two forces acting on it: the force that moves it, and an opposite force that arises because of the induced current. This also applies to the rotating current-carrying coil in the ac generator. Fleming's left-hand rule and Lenz's law show that this force opposes whatever is turning the coil.

If a generator coil is being rotated clockwise by an external agent, then the magnetic forces that arise from the induced current interacting with the magnetic field will exert a turning force anticlockwise on the coil. Some of the energy provided by the external agent turning the coil has to be used to overcome this anticlockwise magnetic effect. This reduces the induced current that can be made available to the external circuit. However, remember that with no opposition, no work is done and no energy will be transferred from the agent (doing the turning) to the coil (and its associated electrical circuit).

This is easily demonstrated using a bicycle dynamo (a device similar to our first rotating-magnet generator) (see figure 2). In this type of dynamo a permanent magnet is rotated in the gap inside a coil. When the lamp is switched off so that no induced current is produced, the dynamo is relatively easy to turn (remember, there will still be an induced emf across the terminals). When the dynamo supplies current and lights the lamp, more effort is required to rotate the dynamo at the same speed since an opposing magnetic force will act on the current (coil). This is Lenz's law in action.

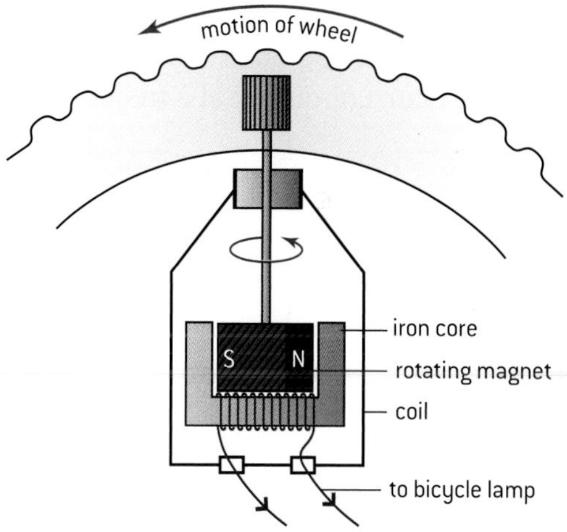

▲ Figure 2 Bicycle dynamo.

Nature of science

Modelling an ac generator

Although the following derivation will not be required in the examination, it shows you how Faraday's law can be used to model the behaviour of a simple ac generator.

The coil has an average length l and an average width of w with N turns (these dimensions are shown in figure 3(a)). The area of the coil A is therefore $l \times w$.

At the instant when the normal to the plane of the coil is at an angle θ to the magnetic field, the flux linkage through the coil is $N \times \Phi$, which is $N \times BA\cos\theta$. The coil spins at a constant angular speed ω In time t the angle swept is θ ($= \omega t$).

Therefore the flux linkage varies with t as $NB \cos \omega t$ and a graph showing how flux linkage varies with time has a cosine shape with maximum and minimum values of NBA at $t = 0$ and $-NBA$ when the coil is halfway through one cycle.

The value of the induced emf at any instant is equal to the (−) rate of change of the flux linkage ($N\Phi$) and this is the the (−) gradient of the flux linkage–time graph. So the graph of induced emf is a (negative) sine curve with an equation of which has maximum and minimum values for the emf of $\varepsilon_0 = \pm BAN\omega$.

A supply in which the current and voltage vary as a sine wave is an **alternating supply.** Although ω is used for convenience in the mathematics of the emf induced in the generator, for everyday purposes we use the frequency. This has the same meaning as elsewhere in physics: the number of cycles that occur each second. A generator that rotates 50 times in one second has an alternating current output at a frequency of 50 Hz.

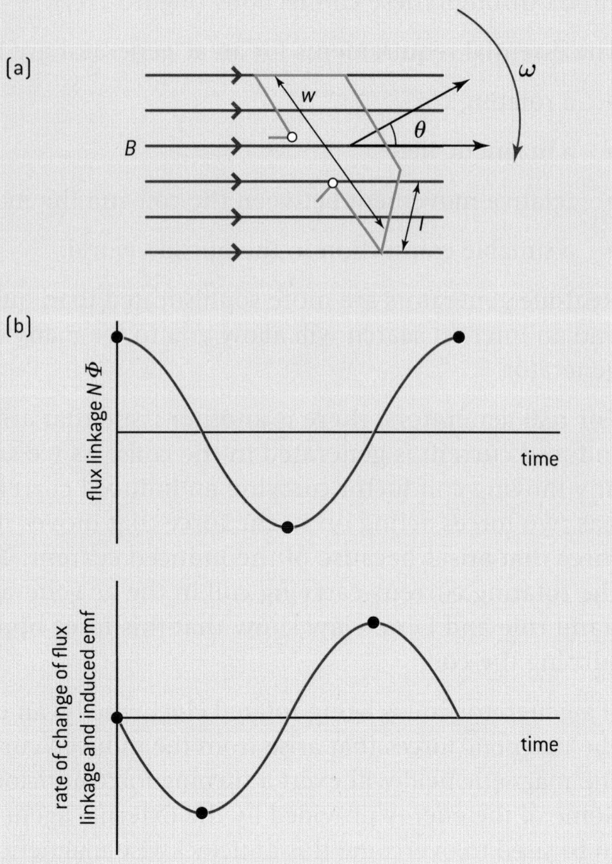

▲ Figure 3

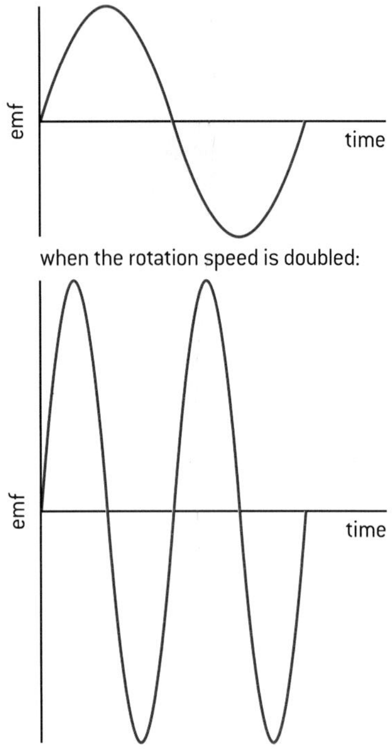

▲ Figure 4 Increasing the rotation speed of the coil.

The model above shows that the output emf of a generator is sinusoidal and that one complete rotation of the coil through 360° gives one cycle of the alternating current.

Figure 4 shows the effect on the emf of changing the angular speed of the coil (without changing any other feature of the coil or field). If the angular speed of the coil is increased then:

- the coil will take a shorter time to complete one cycle and so there will be more cycles every second hence the frequency increases.
- the time between maximum and minimum flux linkage will decrease and therefore (as the flux linkage is constant) the rate of change increases and hence the peak emf increases.

Other ways to increase the output of the emf, but without changing frequency, include increasing:

- the magnetic field strength (B)
- the number of turns on the rotating coil (N)
- the larger coil area (A).

Worked example

A coil rotates at a constant rate in a uniform magnetic field. The variation of the emf E with angle θ between the coil and the field direction is shown.

Copy the graph below and, on the same axes, add the emf that will be produced when the rate of rotation of the coil is doubled. Explain your answer.

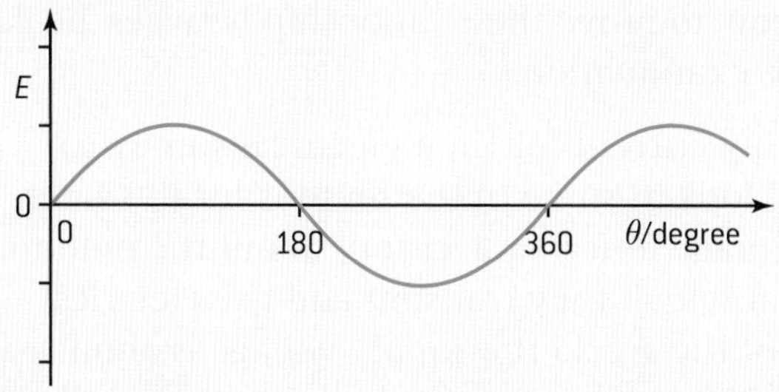

Solution

The rate of change of the flux linkage will double, so the magnitude of the peak emf will also double. This is because Faraday's law states that the induced emf is proportional to the rate of change of flux linkage. However, the coil now rotates in half the time, so the time for one cycle will be halved. On the same scales, the new graph is:

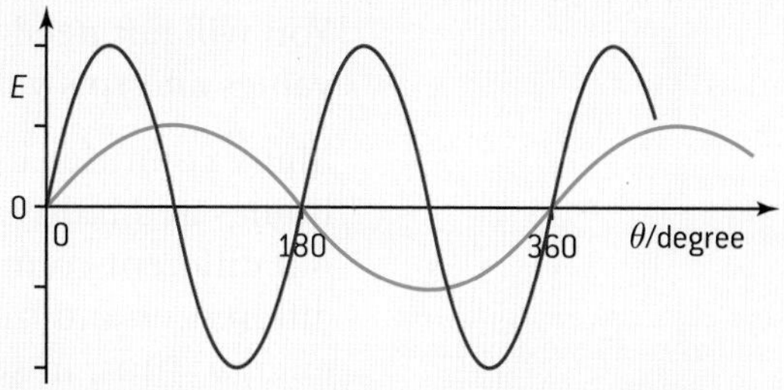

Measuring alternating currents and voltages

The current and voltage of an alternating supply change constantly throughout one cycle. Measuring these quantities is not straightforward because, with positive and negative half-cycles, the average value for current or voltage over one cycle is zero.

One way around this problem is to consider the power supplied to a resistance R connected to the generator. The instantaneous power dissipated in the resistance is I^2R where I is the instantaneous current.

Figure 5 shows both the current–time graph and the power dissipated–time graph with the same time axes. Notice the difference between the two:

- The power–time graph is always positive (which we would expect because the power is I^2R and a number squared is always positive).
- The power graph cycles at twice the frequency of the current.

To see this, suppose that the time period of the ac generator is very large taking 10 s to turn once through one cycle. If you watch a filament lamp supplied with an ac supply of such a low frequency you will see the lamp flash on and off *twice* in each cycle. The lamp is on when the emf is near its maximum (positive) and minimum (negative) values.When we look at a filament lamp powered by the AC mains, persistence of vision prevents us seeing its flashing like this because it is switched on and off at 100 or 120 times per second (twice the normal mains frequency of 50 or 60 Hz).

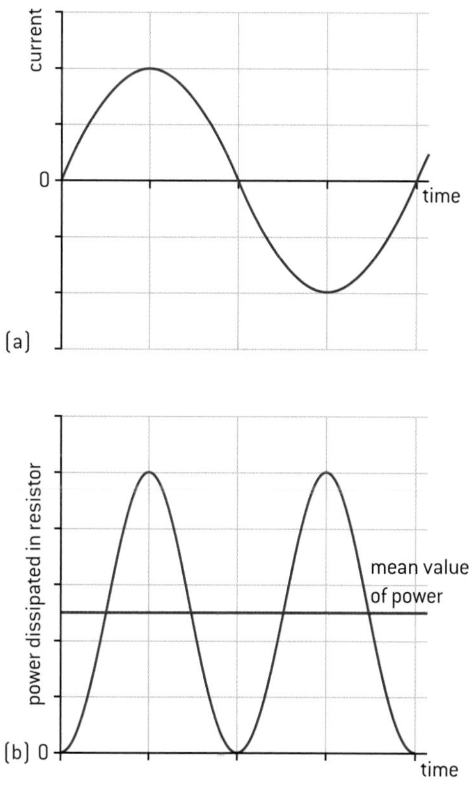

▲ Figure 5 Current–time and power–time graphs for ac.

Alternating values are measured using the equivalent direct current that delivers the same power as the ac over one cycle. A lamp supplied from a dc supply would have the same brightness as the average brightness of our ac lamp flashing on and off twice a cycle. This equivalent dc current is that which gives the average value of the power in the power–time graph. Because the average value of a $\sin^2$ graph is halfway between the peak and zero values and the curve is symmetrical about this line (shown on figure 5(b)), the areas above and below the average line are the same.

So the mean power that an ac circuit supplies is $\frac{1}{2}\ I_0^2 R$ where I_0 is the peak value of the current. The dc current required to give this power is $\sqrt{\frac{1}{2}\ I_0^2}$ which is $\frac{I_0}{\sqrt{2}}$.

This value is known as the **root mean square (rms) current** $\boldsymbol{I_{rms}} = \frac{\boldsymbol{I_0}}{\sqrt{\boldsymbol{2}}}$. In a similar way $V_{rms} = \frac{V_0}{\sqrt{2}}$. The power P dissipated in a resistance $= I_{rms} V_{rms} = \frac{I_0}{\sqrt{2}} \frac{V_0}{\sqrt{2}} = \frac{I_0 V_0}{2}$

with the usual equivalents:

$$P = I_{rms} V_{rms} = \frac{1}{2}\ I_{rms}^2\ R = \frac{1}{2}\ \frac{V_{rms}^2}{R}$$

You will not need to know how to prove the relationship between peak values and rms values for the examination.

Many countries use alternating current for their electrical supply to homes and industry. We will look at some of the reasons for this later, but different countries have made differing decisions about the potential differences and frequencies at which they transmit and use electrical energy. Thus, in some parts of the world the supply voltage is about 100 V; in others it is roughly 250 V. Likewise, frequencies are usually either 50 Hz or 60 Hz.

Worked examples

1 The diagram below shows the variation with time t of the emf E generated in a rotating coil.

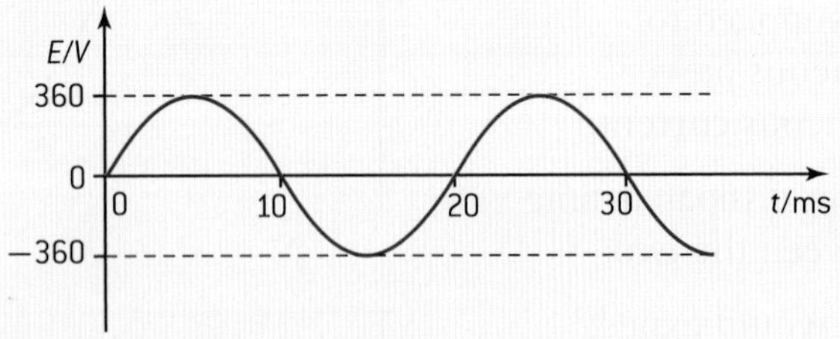

Calculate:

a) the rms value of the emf

b) the frequency of rotation of the coil.

Solution

a) the peak value of the emf is 360 V, so the rms value is $\frac{360}{\sqrt{2}} = 255$ V

b) $f = \frac{1}{T} = \frac{1}{0.02} = 50$ Hz

2 A resistor is connected in series with an alternating current supply of negligible internal resistance. The peak value of the supply voltage is 140 V and the peak value of the current in the resistor is 9.5 A. Calculate the average power dissipation in the resistor.

Solution

The average power $= \frac{\textit{peak current} \times \textit{peak pd}}{2}$

$$= \frac{1}{2} \times 140 \times 9.5 = 670 \text{ W}$$

Transformers

One of the reasons why ac supplies are so common is because a device called a transformer can be used to change alternating supplies from one pd to another. Transformers come in many shapes and sizes ranging from devices that convert voltages at powers of many megawatts down to the small devices used to power domestic devices that need a low-voltage supply.

A transformer consists of three parts:

- an input (or primary) coil
- an output (or secondary) coil
- an iron core on which both coils are wound.

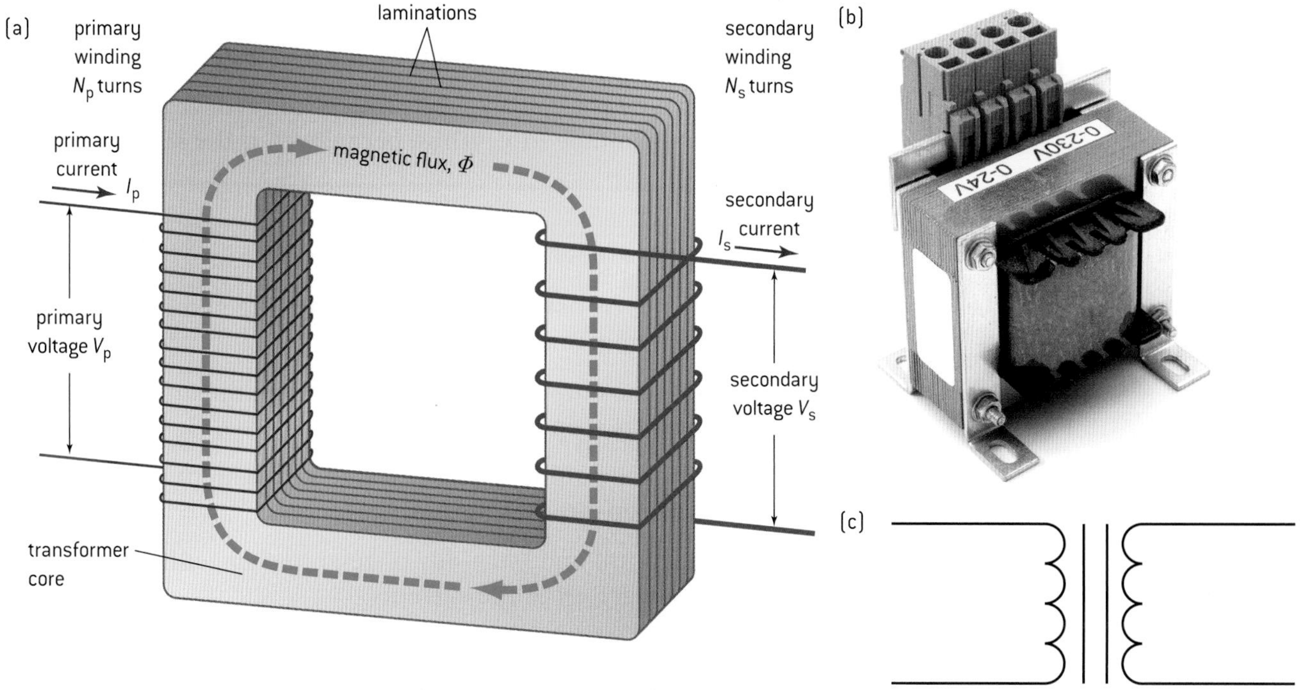

▲ Figure 6 Transformers in theory and practice.

Figure 6 shows a schematic diagram of a transformer (a), a real-life transformer (b), and also the transformer symbol used in IB examinations (c).

The operation of a transformer, like the ac generator, relies on electromagnetic induction.

- Alternating current is supplied to the primary coil.
- A magnetic field is produced by the current in the primary coil and this field links around a core made from a magnetic material, usually soft iron. (The basic ideas behind the production of this field were covered in Topic 5.)
- Because the primary current is alternating, the magnetic field in the core alternates at the same frequency also. The field goes first in one direction around the core and then reverses its direction.
- The transformer is designed so that the changing flux also links the secondary coil.
- The secondary coil has a changing field inside it and an induced alternating emf appears at its terminals. When the coil is connected to an external load, charge will flow in the circuit of the secondary coil and its load.
- Energy has been transferred from the primary to the secondary circuit through the core.

Suppose that an alternating pd with a peak value of V_p is applied to the primary coil and that this results in a flux of Φ in the core. The flux linked to the secondary coil of N_s turns is therefore $N_s\Phi$ and the induced emf in the secondary coil is $V_s = N_s \frac{\Delta\Phi}{\Delta t}$. There is also a flux linkage to the primary coil even though the primary current was originally responsible for setting up the field in the first place. This gives rise to $\varepsilon_p = N_p \frac{\Delta\Phi}{\Delta t}$

where ε_p and N_p are the induced emf and the number of turns in the primary coil. This induced primary emf will oppose the applied pd.

If we assume that the resistance of the primary coil is negligible, then ε_p will be equal in magnitude to V_p. Thus, because $\frac{\Delta\Phi}{\Delta t}$ is the same for both coils $\frac{\varepsilon_p}{N_p} = \frac{V_p}{N_p} = \frac{V_s}{N_s}$ and

$$\frac{\varepsilon_p}{\varepsilon_s} = \frac{V_p}{V_s} = \frac{N_p}{N_s}$$

This is known as the **transformer rule**. It relates the number of turns on the coils to the input and output voltages:

- When $N_s > N_p$ the output voltage is greater than the input voltage. This is known as a **step-up** transformer.
- When $N_s < N_p$ the output voltage is less than the input voltage. This is known as a **step-down** transformer.
- The terms "step-up" and "step-down" refer to changes in the alternating voltages *not* to the currents.

You may wonder how zero current in the primary coil can lead to any energy transfer to the secondary. The answer is: it cannot. The equation applies to the case where there is no current in the secondary (in other words it has not yet been connected to a load). Once again Lenz's law has a part to play. When the secondary coil supplies current (because a resistor is now connected across its terminals), then because charge flows through this secondary coil, another magnetic field is set up in the coil and through the core. This magnetic field tries to oppose the changes occurring in the system and so tends to reduce the flux in the core. This in turn reduces the opposing emf in the primary (often called a "back emf" for this reason) and so now there will be an overall current in the primary that allows energy to be transferred.

Electrical engineers use a more complex theory of the transformer than the one presented here to take account of this and other factors. But for our simple theory, which assumes that the transformer loses no energy, the energy entering the primary is equal to the energy leaving the secondary so

$$I_p V_p = I_s V_s$$

where I_p and I_s are the currents in the primary and secondary circuits respectively. This is a second transformer equation that you should know be able to use.

In fact, many transformers have an efficiency that is close to 100% as energy losses can be reduced by good design. The **efficiency of a transformer** is

$$\frac{\text{energy supplied by secondary coil}}{\text{energy supplied to primary coil}} \times 100\%$$

Ways to improve the efficiency include:

Laminating the core

- Iron is a good conductor and the changing flux in the transformer produces currents flowing inside the iron core itself. These are known as **eddy currents**. To prevent these, transformer designers use thin layers of insulating material that separate layers (sheets) of iron in the core (shown in Figure 6(a)). This has the result that although the magnetic properties of the iron are largely unaltered, the electrical resistance of the core is significantly increased. The currents are forced to travel along longer paths within the layers hence increasing electrical resistance and reducing the current. Laminations reduce the energy losses that result from a reduction in the amount of flux and from a rise in temperature of the iron that would occur if the eddy currents were large.

Choosing the core material

- The magnetic material of the core is a "soft" magnetic material. It can be magnetized and demagnetized very readily and also maintain high fluxes too. These are all desirable properties for the core.

Choosing the wire in the coils

- Low-resistance wires are used in the primary and secondary coils as high resistances would lead to heating losses (called **joule heating**) in the coils.

Core design

- It is important that flux is not allowed to leak out of the core. As much flux as possible should link both coils so that the maximum rate of change of flux linkage occurs.

> **Note**
>
> **Hard and soft magnetic materials**
>
> We sometimes talk about magnetic materials being "soft" or "hard". A soft material, such as iron, can be easily magnetised by another magnet or a current-carrying coil. When the magnetic influence is removed, however, the iron loses all or most of its magnetism easily. Such materials are excellent for the cores of transformers because they respond well to the variations in magnetic field. Hard materials such as some iron alloys (a steel) do not magnetise easily but are good at retaining the magnetism. These materials are best for the manufacture of permanent magnets such as the ones you use in the laboratory.

Worked example

A transformer steps down a mains voltage of 120 V to 5 V for a tablet computer. The computer requires 1.0 W of power to operate correctly. There are 2300 turns on the primary coil of the transformer.

a) Calculate the number of turns on the secondary coil.

b) Calculate the current in the secondary coil when it is operating correctly.

c) The input current to the primary is 0.0090 A. Calculate the efficiency of the transformer.

Solution

a) $\frac{V_p}{V_s} = \frac{N_p}{N_s}$ so $N_s = \frac{V_s \times N_p}{V_p} = \frac{5 \times 2300}{120} = 96$ turns

b) Current in the secondary $I_s = \frac{\text{tablet power}}{V_s} = \frac{1.0}{5}$ $= 0.20$A

c) Efficiency $= \frac{\text{power supplied by secondary coil}}{\text{power supplied to primary coil}}$ $= \frac{5 \times 0.2}{120 \times 0.0090} = 0.93$ (or 93%)

Investigate!

Transformer action

- In this experiment you are asked to investigate the basic ideas behind the transformer. You will require an ac low-voltage power supply to provide an input, an oscilloscope to view the output of your coils, and the coils themselves. One coil should have significantly less turns than the other, typical values might be 240 turns on one coil and 1100 turns on the other. You will also need additional apparatus for part of the experiment. (Some coil kits have special laminated iron cores that clip together to give a continuous magnetic circuit,)

- Connect the coil with fewer turns to the supply (this is the primary circuit) and the other coil to the oscilloscope (this is the secondary). Set the coils close so that the secondary coil is linked to the flux produced by the primary. Look closely at the output from the secondary coil and compare the relative sizes of the input and output and also their phases.
- Try the following changes to see how they affect the output:
  - Alter the separation and orientation of the coils (reverse one coil relative to the other and observe the difference). Alter the frequency of the primary current. Link the coils by inserting a piece of iron or several iron nails through the centre of both coils.
  - With the coils linked with iron compare the output voltages when the 240 turn coil is the primary and, later, when the 110 turn coil is the primary. If you have access to other coils with different numbers of turns collect data to complete the table.

Number of primary turns N_p	Primary pd, V_p/V	Number of secondary turms N_s	Secondary pd, V_s/V	$\frac{V_p}{V_s}$	$\frac{N_p}{N_s}$

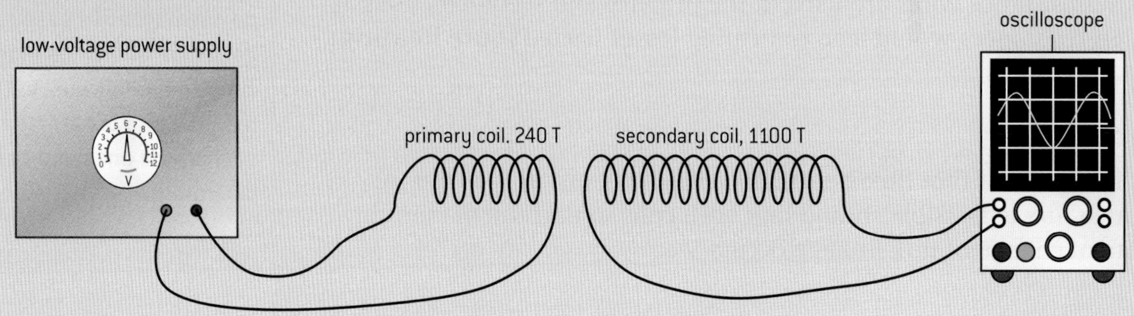

▲ Figure 7

Transformers in action

Many countries have an electrical grid system so that each separate community within the country does not have to provide its own energy. In the event of power failures in one part of the grid system, energy can be diverted where needed. When energy has to be sent over long distances it is advantageous to send it at very high voltages. This is because transmitting at high voltages and low currents helps to reduce the energy lost in the power transmission lines that are used in the grid.

A laboratory example will help here:

Figure 8 shows two alternative ways to transmit electrical energy from one point in a school laboratory to another. In both circuits a power supply acts as the "generator" providing energy at an alternating potential difference of 12 V. This energy is transmitted through two transmission lines (leads strung between two retort stands). In one case (Figure 8(a)), the energy is tramsitted at 12 V, but in the other (Figure 8(b)) a step-up transformer with a turns ratio of 1:10 is used to transmit the energy at 120 V. At the other end of this transmission line, a step-down transformer returns the pd supply to the voltage level required by the lamps in the circuit.

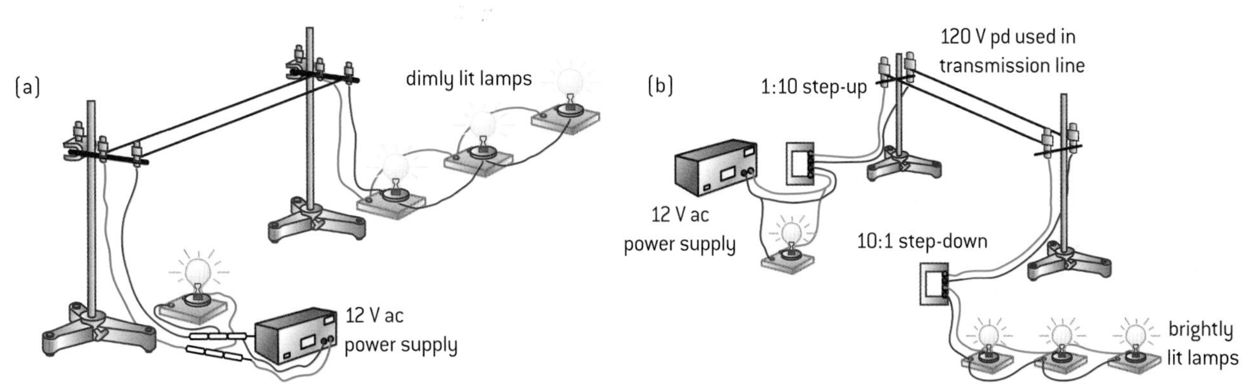

▲ Figure 8 Transmission at high voltages.

A numerical calculation will help you to compare the two cases: In each case, there are three lamps each rated at 0.3 A, 12V (a total power requirement of 3.6 W) and a transmission line of total resistance 1.5 ohms.

Without the transformers, the total current required by lamps will be 0.9 A and power loss in transmission line = $I^2R = 0.81 \times 1.5 = 1.2$ W.

With the voltage stepped up to 120 V along the transmission line, the total current will be stepped down to 0.09 A during transmission. This means the power loss in the transmission line = $I^2R = 0.081 \times 1.5 = 0.012$ W.

So, stepping *up* the voltage by a factor of 10 reduces the current by a factor of 10 and therefore reduces the power lost during transmission by a factor of 100 (i.e. 10^2). This is an important saving for electricity supply companies (and their customers!). Transformers play a crucial role in increasing the efficiency of transmission.

In practice, grid systems are usually like the one shown in figure 9. Numbers are not given on this diagram as they vary from country to country. Find out the details of the grid voltages used where you live.

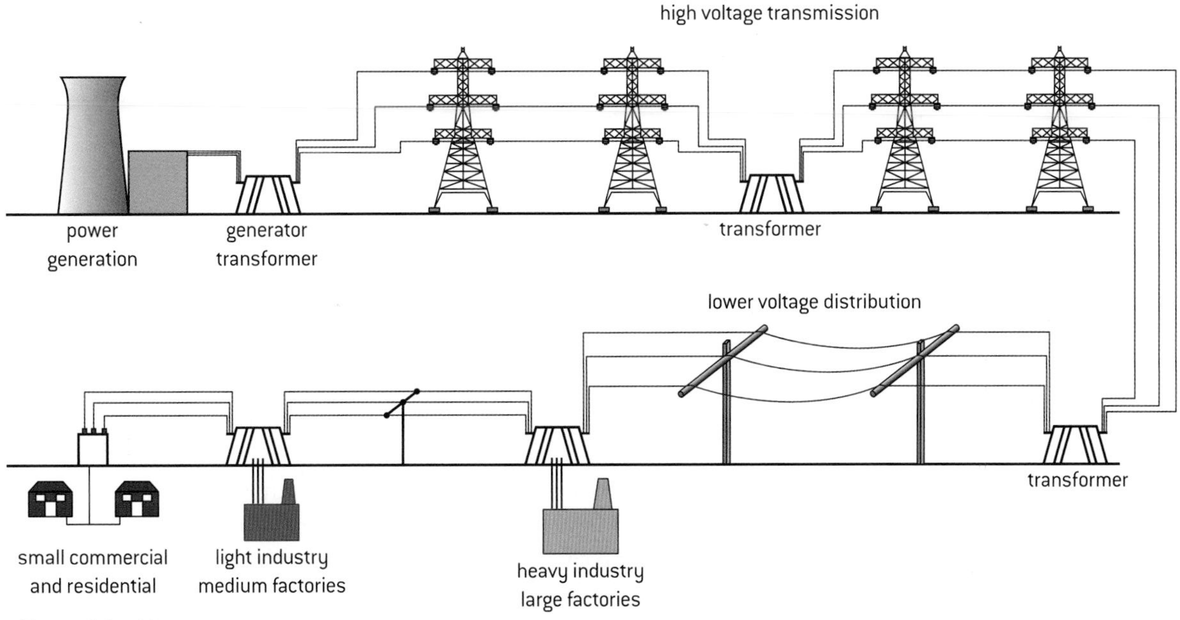

▲ Figure 9 A grid system.

Nature of science

HVDC and international collaboration

The argument that high transmission voltages lead to better efficiency does not apply simply to ac. The argument is still valid for dc but historically the transformation between voltages was easier using ac. However, things are now changing. As the physics and engineering of electrical transmission improve, so it is advantageous to use high-voltage direct-current transmission (HVDC).

Although the cost of the equipment to convert between two dc voltages is greater than the cost of a transformer, there are other factors in the equation. Countries use different supply frequencies and this is a major problem when feeding electricity from one country into the grid of another. Using undersea cables over long distances with ac also involves larger currents than might be expected as the cables have capacitance and induction effects. Additional currents are required to move charges every cycle.

Some of the longest HVDC transmission paths in the world include the 2400 km long connection between the Amazon Basin and south-eastern Brazil that carries about 3 GW of electrical power, and the Xiangjiaba–Shanghai system in China that carries 6.4 GW over a distance of 2000 km.

Governments work together to maintain electricity supplies. There are many examples of electrical links between countries provided so that one nation can supply energy to another nation during times of shortage. These are not necessarily times of crisis. There are short-term fluctuations in the demand for electricity. Sometimes energy is fed from country to country at one time of the day and then back again later. Examples of such links include the electrical links from the Netherlands to the UK and the HVDC link between Italy and Greece.

Rectifying ac

The convenience of ac for domestic distribution is clear, but it raises the question of how devices that can only operate on dc, including computers and electronic equipment, can be made to function when connected to an ac supply.

The process of converting an ac supply into dc is called **rectification**. A device that carries this out is known as a **rectifier**. We consider two varieties of rectifier in this course: half wave and full wave. In half wave, only half of each cycle of the current is used whereas in full wave a more complex circuit leads to the use of both halves of the ac cycle.

Half-wave rectification

Figure 10(a) shows the basic circuit. A single diode is connected in series with the secondary terminals of a transformer and the load resistor. The load resistor represents the part of the circuit that is being supplied with the rectified current.

Diodes are devices that only allow charge to flow through them in one direction (the symbol for the diode has an arrowhead that points in the direction of conventional current allowed by the device). We say that the diode is *forward biased* when it is conducting. When no charge can move through it (because the pd across the device has the wrong polarity) we say that the diode is reverse biased. The load resistor will only have a pd across it for about half a cycle (figure 10(b)). The waveform is not quite half a cycle wide because the diode does not conduct from exactly 0 V; it requires a small forward pd for conduction to begin.

The way in which the circuit operates is straightforward. When terminal X on the transformer is positive with respect to terminal Y, then there will be a current through the diode in its forward conducting direction. When the ac switches so that X is negative with respect to Y, then the diode is reverse biased and there is no conduction or current.

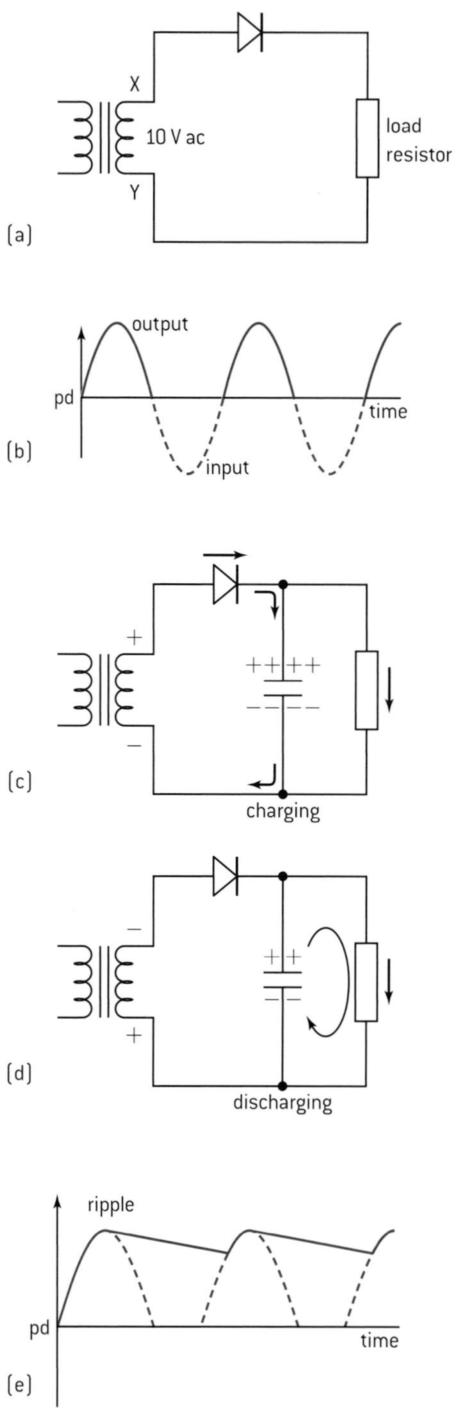

Figure 10 Half-wave rectifier.

An obvious problem is that although the charge now flows in only one direction, the current is not constant. We need a way to smooth the waveform so that it more closely resembles the constant value of a dc supply. One way to achieve this is to use a resistor and capacitor connected in parallel between the diode and the load (figure 10(c)). During the first half-cycle, the capacitor charges up and the potential difference across it approaches the peak of the transformer output emf. When the current is zero in the second (negative) half-cycle, the capacitor discharges through the resistor at a rate determined by the time constant of the circuit (Figure 10(d)). If the time constant is much larger than the time for half a cycle, the amount of charge released by the capacitor (and therefore the pd across it) will be small and the pd will not change very much. When the diode conducts again in the next half-cycle, the charge stored on the capacitor will be "topped up". From now on there will be a discharge–charge cycle with the pd across the capacitor varying much less than in the basic circuit (Figure 10(e)). When a capacitor is used in this way, it is often referred to as a smoothing or "reservoir" capacitor. The small variation in the output pd is known as the "ripple" voltage.

The choice of capacitor and resistor values depends on the application in use. A large time constant provides good smoothing, but at the cost of a more expensive capacitor and a trade-off in the shape of the waveform due to the large charging currents that are drawn from the transformer.

Full-wave rectification

For some applications, the amount of ripple in a half-wave rectifier cannot be tolerated. In such cases, full-wave rectifiers are used.

Figure 11(a) below shows one way to achieve full-wave rectification. This arrangement uses two diodes and requires a centre-tap on the transformer. This means that there is a connection half way along the length of the wire that has been wound to make the secondary coil. This can be clearly seen at Y on the transformer symbol in the diagram. Also in the diagram is the resistor–capacitor pair that will smooth the rectified wave.

The best way to understand how the circuit works is to imagine that terminal Y is always at zero potential. Then, for half the time X will be positive and Z will be negative relative to point Y. For the other half of the cycle, these polarities reverse. When X is positive relative to Y, diode D_1 will conduct. Notice that only half of the secondary coil connected to D_1 takes part in conduction at any time. One half cycle later, D_1 will not conduct because X will be negative and Z will be positive relative to Y. D_2 conducts during this half cycle. Current is again supplied to a capacitor–resistor combination during both halves of the cycle leading to full-wave rectification.

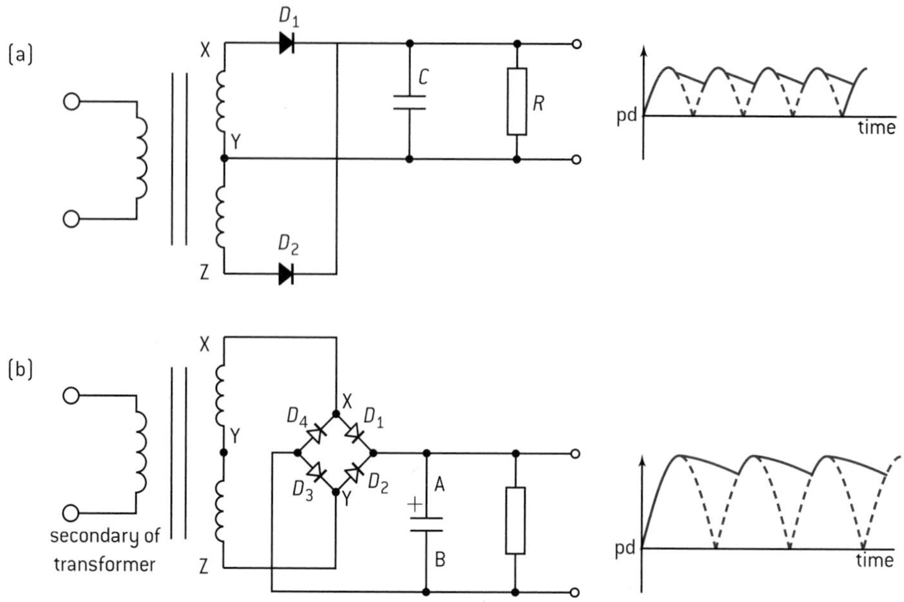

Figure 11 Full-wave rectification.

In order to achieve a particular value of peak pd, twice as many turns are required on the secondary compared to the half-wave arrangement. This disadvantage can be overcome using a diode bridge (figure 11(b)).

The full secondary coil is now used and the diode arrangement allows the whole of the coil to supply current throughout the cycle. The disadvantage is the need for four diodes and a more complex circuit arrangement.

When X is positive relative to Z then the junction between D_1 and D_4 is positive relative to the junction between D_2 and D_3. Of D_1 and D_4, D_1 is the one that conducts so that point A of the capacitor will be positive too. Similarly, D_3 conducts and point B of the capacitor becomes negative. The capacitor charges and current is supplied to the rest of the circuit. When the polarity of the secondary coil switches, X becomes negative relative to Z and the conducting diodes are now D_2 and D_4. The connections are arranged so that the direction of the conventional current in D_4 is away from B and the direction in D_2 is towards A. This is the same as in the previous half cycle and the polarity of charge delivered to the capacitor is unchanged.

The best way to learn this arrangement is not necessarily to memorize the orientation of the diodes but by understanding how they function in order to provide a consistent pattern of polarity at the capacitor during the full cycle.

These examples of rectifying circuits are said to be passive, meaning that none of the devices amplify or modify the waveforms. Some modern power supplies for computers are active devices. Switched-mode power supplies change the frequency of the supplied ac to a much higher value to allow the signal to be modified by an electronic circuit. The advantage is that less energy is wasted in resistance but this is achieved only at the expense of much greater complexity in the electronic circuits.

Nature of science

The war of the currents

In the 1880s a commercial battle broke out between the direct-current distribution system developed by Edison and the alternating-current system of Westinghouse. Both companies were attempting to corner the US market. In those days, electricity was principally used to provide energy for filament lamps. Edison's system had the advantage that batteries could be used as a backup if the power failed (which was a frequent event when electrical distribution began). On the other hand, Westinghouse's ac system could be transmitted with lower energy losses using larger and fewer distribution stations. Westinghouse made full use of scientists from around the world in developing ac distribution, including Nikola Tesla who had made much progress in transformer design.

The full story of the battle is complex, but more and more companies (principally the newly formed General Electric Company) began to follow Westinghouse's lead and eventually dc supplies largely disappeared. However, as late as the 1980s, direct current was generated in the UK to supply the dc powered printing presses in London. In the US, the hotel where Tesla lived used dc into the 1960s. Consolidated Edison (as Edison's company became known) shut down its last dc supply on 14 November 2007.

Wheatstone and Wien bridge circuits

The four-diode rectifier arrangement is one of a class of circuits known as **bridge circuits**.

The **Wheatstone bridge** (figure 12) was popularized by Sir Charles Wheatstone in 1833 and is a completely resistive arrangement that can be used to estimate the resistance of an unknown resistor. It is normally used in a dc context.

The working of the circuit is straightforward (figure 12). Three known resistors R_1, R_2 and R_3, are connected in the circuit with a fourth unknown resistor R_X. One of the known resistors is variable and is adjusted until the current in a galvanometer that "bridges" the pairs of resistors is zero. This galvanometer is usually a centre-zero meter (that shows negative and positive values with the 0 marking in the centre of the scale) with high sensitivity.

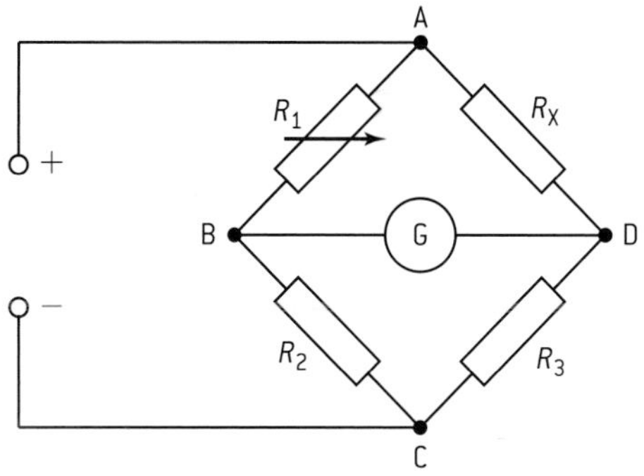

Figure 12 The Wheatstone bridge.

When the current in the galvanometer is 0, the bridge is said to be balanced. No charge is flowing through the galvanometer at the balance point because the potentials at B and D are identical so there is no potential difference across BD. This means that the potential differences across R_1 and R_X are the same as each other, and the potential differences across R_2 and R_3 are also identical.

$$V_1 = I_1 R_1 = I_2 R_X \qquad V_2 = I_1 R_2 = I_2 R_3$$

So $\frac{I_1}{I_2} = \frac{R_X}{R_1} = \frac{R_3}{R_2}$

therefore

$$R_X = \frac{R_1 \times R_3}{R_2}$$

Knowledge of three of the resistors means that that R_X can be calculated. For accurate work, the bridge requires that the known resistors have accurately measured values and that the ammeter is sufficiently sensitive to give precise results.

Investigate!

Using the Wheatstone bridge

- The details of this experiment will depend on the equipment that you have in your school. Your teacher will advise you on this.
- Set up the circuit you are to use. Choose R_1 so that it is similar in value to the unknown resistor.

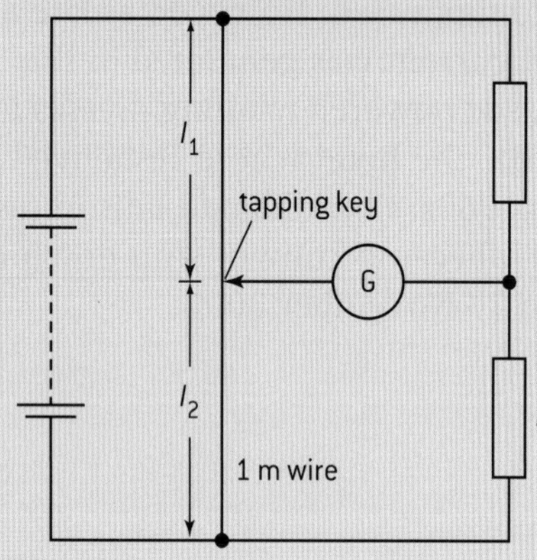

▲ Figure 13

- Set the ratio $\frac{R_3}{R_2}$ so that the current in the ammeter (the galvanometer) is zero.
- One common way to carry out the experiment is to use a long (1 m) straight wire of uniform cross-section perhaps taped along a metre ruler. The resistance per unit length of such a wire should be constant along its length. Figure 13 shows the way the wire is connected in the circuit.
- The way to determine the ratio $\frac{R_3}{R_2}$ is to find the point on the metre wire at which the current in the ammeter is zero. Then the ratio of the lengths of the two parts of the wire is equal to the ratio of their resistances.
- The equation for the bridge becomes
  $R_3 = R_1 \times \frac{l_1}{l_2}$
  where l_1 and l_1 are the relevant lengths on the metre wire.

Nature of science

Unbalanced Wheatstone bridges

It is possible to make use of the Wheatstone bridge when it is out of balance. Suppose the unknown resistor R_X is replaced by a thermistor. At a particular temperature the bridge can be made to balance with a particular combination of three fixed resistances and the thermistor itself. When the temperature changes from this starting point, the thermistor resistance will also change, increasing if the temperature drops, and becoming smaller if the temperature rises. These changes will take the bridge out of balance with a potential difference appearing across the meter that bridges the resistor pairs. The whole circuit can be calibrated so that a known temperature change at the thermistor gives rise to a known out-of-balance voltage across the meter. Such a circuit can be made to be extremely sensitive if the sensitivity of the meter and the resistor values are chosen carefully.

Other sensors can be chosen too: for example, a light-dependent resistor for changes in light intensity, and a length of resistance wire under tension (a strain gauge) for changes in length.

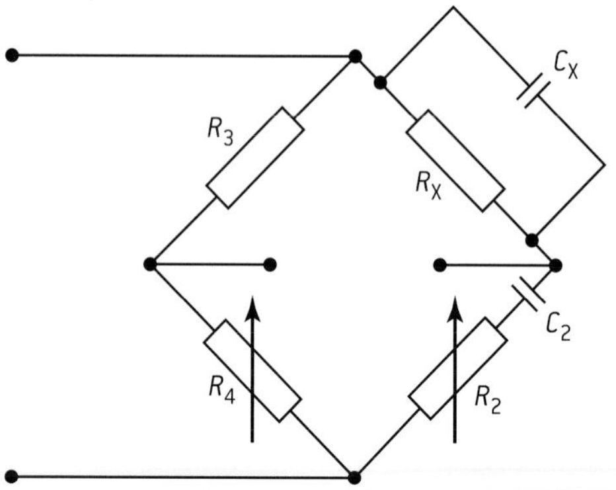

▲ Figure 14 Wien bridge.

The **Wien bridge** circuit is a modification of the Wheatstone arrangement to allow the identification of resistance and capacitance values for an unknown component. This bridge operates with an alternating power supply.

The bridge is modified by the addition of a capacitor in series with R_2. Again the current in the centre arm of the bridge is minimized. R_2, C_2 and the frequency of the supply need to be adjusted for the minimum current. Knowing the values of the components allows the operator to calculate the value of R_X and C_X.

The theory for this bridge will not be tested in the examination.

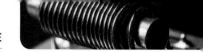

11.3 Capacitance

Understanding

- Capacitance
- Dielectric materials
- Capacitors in series and parallel
- Resistor–capacitor (RC) series circuits
- Time constant

Nature of science

This sub-topic contains many important links and analogies. Some are straightforward, the link between the energy stored in a spring and the energy stored in a capacitor, for example. Others, however, reflect the importance of the over-arching ideas of exponential growth and decay that underpin many areas of sciences and social sciences. Rates of reaction in chemistry, changes in living populations in biology, and radioactive decay in physics are all related by the important idea that the rate of change depends on the total instantaneous number and a constant probability of change. This is one of the many ways in which scientists use mathematics to model reality.

Applications and skills

- Describing the effect of different dielectric materials on capacitance
- Solving problems involving parallel-plate capacitors
- Investigating combinations of capacitors in series or parallel circuits
- Determining the energy stored in a charged capacitor
- Describing the nature of the exponential discharge of a capacitor
- Solving problems involving the discharge of a capacitor through a fixed resistor
- Solving problems involving the time constant of an RC circuit for charge, voltage, and current

Equations

- definition of capacitance: $C = \frac{Q}{V}$
- combining capacitors in parallel: $C_{\text{parallel}} = C_1 + C_2 + \ldots$
- series: $\frac{1}{C_{\text{series}}} = \frac{1}{C_1} + \frac{1}{C_2} + \ldots$
- capacitance of a parallel-plate capacitor: $C = \varepsilon \frac{A}{d}$
- energy stored in a capacitor: $E = \frac{1}{2}CV^2$
- time constant: $\tau = RC$
- exponential discharge charge: $Q = Q_0 e^{-\frac{t}{\tau}}$
- current: $I = I_0 e^{-\frac{t}{\tau}}$
- potential difference: $V = V_0 e^{-\frac{t}{\tau}}$

Introduction

In Topic 10, a pair of charged parallel plates was used to produce a uniform field. A charge was transferred to the plates using a power supply. In this topic we look in detail at this transfer of charge onto, and from, the plates.

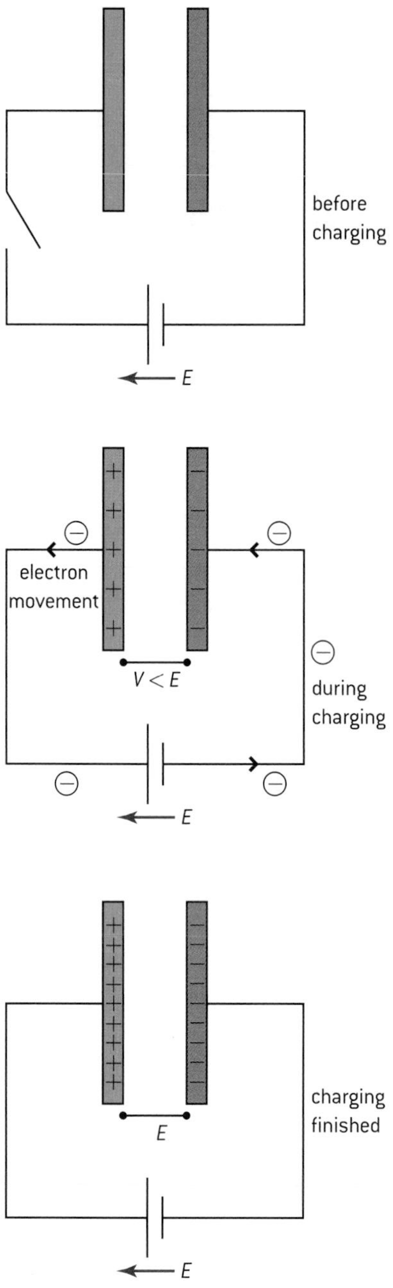

▲ Figure 1 Charging a capacitor.

Capacitors in theory

The arrangement of parallel plates in which the two plates are separated by an insulator is called a **capacitor**. The insulator might be a vacuum, it could also be air or another gas providing that a spark (and therefore charge) cannot jump between the plates. It could also be a non-conducting material such as a plastic.

Figure 1 shows an arrangement of two parallel plates (in a vacuum for simplicity) connected to a cell. The plates are initially uncharged.

When the switch is closed, electrons begin to flow. There is no current between the plates because of the insulation between them. Electrons are removed from the plate connected to the positive terminal of the cell and are transferred to the plate connected to the negative.

Charge is being separated by the system and stored. This requires energy and, not surprisingly, this is provided by the cell, the only source of emf in the circuit. Energy is being stored on the plates as the electrons arrive there.

To understand this, imagine the first electron that moves through the cell from the positive plate in the diagram to the negative one. The cell will not need much energy to do this as the plates were initially uncharged. However, the second electron to move will find it slightly more difficult (figure 1). This is because the first electron is now on the right-hand plate and repels this second electron. The cell has to use more energy to move the second electron. The energy required from the cell increases with each subsequent electron until finally there is insufficient potential energy available in the cell to move any more electrons.

At this point, if we were to disconnect the cell and measure the potential difference across the capacitor then we would find that the capacitor pd is equal to the emf of the cell. Had this been done earlier in the charging, then the unconnected capacitor pd would be less than the cell emf. The capacitor pd cannot exceed the emf of the power supply.

The charge Q on the capacitor is stored at a potential difference V. The ratio of the charge stored to the potential difference between the plates is defined to be the **capacitance** of the capacitor.

$$\text{capacitance, } C = \frac{\text{charge stored on one plate}}{\text{potential difference between plates}} = \frac{Q}{V}$$

The charge stored on one plate is the same as the charge transferred through the cell and moved from one plate to the other.

The unit of capacitance is the coulomb per volt, or farad (symbol: F) and is named after Michael Faraday, the English physicist, who developed so much of our understanding about current electricity and magnetism. In fundamental units the farad is equivalent to $A^2 S^4 kg^{-1} m^{-2}$.

One farad is a very large unit of capacitance since 1 C of charge is a very large amount of charge and you will usually use capacitances measured in millifarads (mF), microfarads (μF) and picofarads (pF). Try not to confuse mF and μF.

Tip

Notice that, even though one plate has charge $+Q$ and the other $-Q$, the charge stored is Q not $2Q$. This is because it is only one group of charges (electrons) that are being moved between the plates.

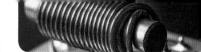

Worked examples

1 A pair of parallel plates store 2.5×10^{-6} C of charge at a potential difference of 15 V. Calculate the capacitance of this capacitor.

Solution

$$C = \frac{Q}{V} = \frac{2.5 \times 10^{-6}}{15} = 1.7 \times 10^{-7}\ \text{F} = 170\ \text{nF}.$$

2 Calculate the potential difference across a capacitor of capacitance 0.15 μF when it stores a charge of 7.8×10^{-8} C.

Solution

$$V = \frac{Q}{C} = \frac{7.8 \times 10^{-8}}{1.5 \times 10^{-7}} = 0.52\ \text{V}$$

Nature of science

Changing names

The name capacitance has not always been used. In the late nineteenth century and even later it was called *capacity* in English (in French *"capacité d'un condensateur" and in Spanish "la capacidad de un condensador"*)

These were in many ways obvious names, and the idea of the capacity of a container can provide a model to help you understand capacitance ideas.

Imagine two containers of water the same height, one narrow, the other wide.

Both containers, initially empty, are filled with water at the same rate (the same volume of water every second). Obviously, the narrower container overflows first. Thinking in terms of the top of both containers having the same gravitational potential, the liquid in the narrow container reaches this potential first.

Put another way, when both containers are full, for the same potential (height of the water surface above the table is equivalent to the potential difference of the capacitor) the wider container will hold more liquid (equivalent to more charge) than the other.

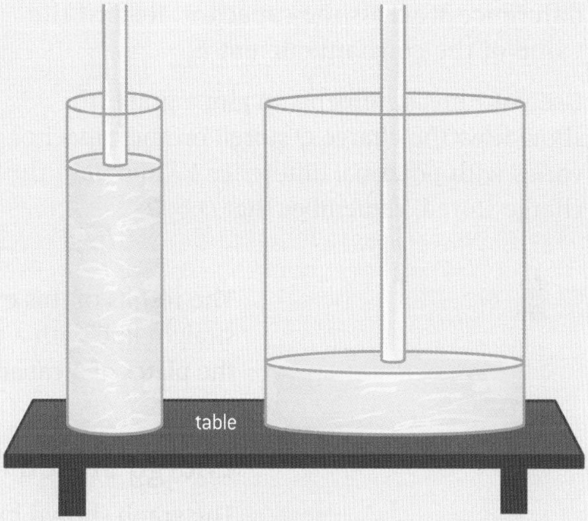

▲ Figure 2 Water container analogy.

This is another link within physics that leads to analogous relationships.

Investigate!

Charging a capacitor with a constant current

- It is unusual for a capacitor to charge or discharge with a constant current. But this exercise will help you to understand the charging process for any capacitor.
- Set up the circuit. A value of 470 μF is suitable for this experiment.
- You also require a clock with a second hand. You may wish to use a data logger with a voltage sensor in place of the voltmeter and the clock (the data logger can make both measurements for you).
- Begin by shorting out the capacitor with the flying lead so that it is completely discharged. The voltmeter should indicate zero.
- In the experiment you need to alter the resistance of the variable resistor to keep the

charging current at a constant value. If the resistance remains at a constant value, the current will fall, reaching zero when the capacitor is fully charged so you will need to decrease the resistance as the experiment proceeds.

- Set the variable resistor to its maximum value. Close the switch with the flying lead connected across the capacitor. Remove the flying lead, start the clock, and record measurements of time elapsed *t* and potential difference *V* across the capacitor. Record the value of the constant current *I*.
- Use your measurements to plot a graph to show how the charge *Q* stored on the capacitor varies with potential difference *V*. To obtain the charge stored, remember that $Q = It$.

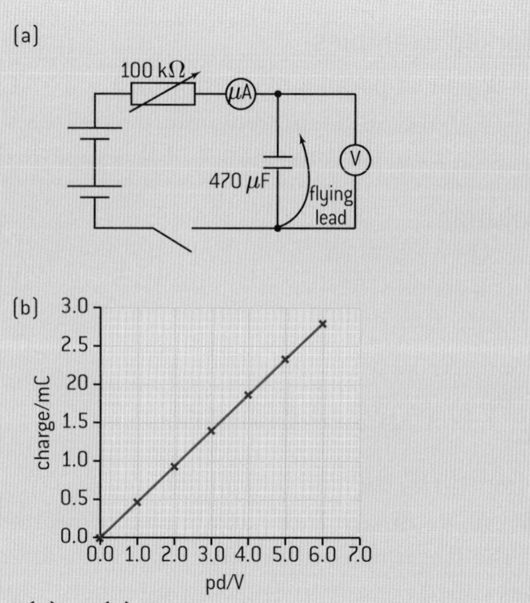

▲ Figure 3(a) and (b) Charging a capacitor with constant current.

The results of this experiment are shown in figure 3(b). The graph is a straight line with a gradient that is equal to *C*. The potential difference across the plates of a capacitor is directly proportional to the charge it carries.

Energy stored in a capacitor

The graph shown in figure 3(b) can give more information than the gradient. The potential difference *V* is the energy per unit of charge stored on the capacitor and *Q* is the charge, so the product of these two quantities is the energy stored and this quantity is equal to the area under the graph.

Algebraically, this area is

$$\frac{1}{2}(\text{base} \times \text{height}) = \frac{1}{2} \times Q \times V$$

There are two additional ways that this expression for the energy can be written:

$$\text{Energy stored on capacitor} = \frac{1}{2}QV = \frac{1}{2}\frac{Q^2}{C} = \frac{1}{2}CV^2$$

This is an idea that is analogous to the potential energy stored in a stretched string where there was also a factor of $\frac{1}{2}$. (We looked at this in Topic 2.)

Worked examples

1 Calculate the energy that can be stored on a 5000 μF capacitor that is charged to a potential difference of 25 V.

Solution

$$\text{Energy stored} = \frac{1}{2}CV^2 = 0.5 \times 5 \times 10^{-3} \times 25^2 = 1.6\ \text{J}$$

2 A capacitor of value 100 mF stores an energy of 250 J. Calculate the pd across the capacitor.

Solution

$$\text{Energy stored } V = \sqrt{\frac{2 \times \text{energy}}{C}} = \sqrt{\frac{500}{0.1}} = 71\ \text{V}$$

Capacitance of a parallel-plate capacitor

In Topic 10 we saw that for a pair of parallel plates

$$\frac{Q}{A} = \varepsilon_0 \frac{V}{d}$$

where Q is the charge stored, A is the area of overlap of the plates, V is the potential difference between the plates, and d is the separation of the plates.

Rearranging this expression gives

$$\frac{Q}{V} = \varepsilon_0 \frac{A}{d}$$

and $\frac{Q}{V}$ is the capacitance C of the two-plate system.

Finally,

$$C = \varepsilon_0 \frac{A}{d}$$

This enables us to calculate the capacitance of a capacitor given the area of overlap of the plates and their separation. However, in real capacitors, edge effects will reduce this value. Earlier we mentioned the edge effects, the field distortions at the edge of the plates that lead to differences between the theoretical results here and what happens in practice with real capacitors. You saw some of the results of edge effects when you considered electric field-line shapes at the edges of plates in Topic 10.

If the gap between the plates is filled with an insulator of permittivity ε, this becomes

$$C = \varepsilon \frac{A}{d}$$

Worked example

1 Two parallel plates both have an area $0.015\ \text{m}^2$ and are placed 2.0×10^{-3} m apart. Calculate the capacitance of this arrangement.

Solution

$$C = \varepsilon_0 \frac{A}{d}$$

$$= 8.9 \times 10^{-12} \times \frac{0.015}{2 \times 10^{-3}}$$

$$= 6.7 \times 10^{-11}\ \text{F}$$

Capacitors in practice

Capacitors are used extensively in electronic circuits. They store charge and can be used to reduce the effects of fluctuations in a circuit. They are also the basis of timing circuits. This means that electric component designers need to be able to design capacitors of various sizes and capacitance values.

The equation $C = \varepsilon_0 \frac{A}{d}$ gives the designers a basis for this. It is evident that there are three ways to increase the capacitance of the plates:

- Make the plate overlap area larger because $C \propto A$
- Set the plates closer together because $C \propto \frac{1}{d}$
- Change the value of the constant in the equation (in other words change the permittivity from that of a vacuum to some other value).

Increasing the area of plate overlap increases capacitance because more charge can be stored for a given potential difference between the plates and therefore $\frac{Q}{V}$ increases.

The second way to increase C is to move the plates closer together. To understand the mechanism here, imagine the plates charged and

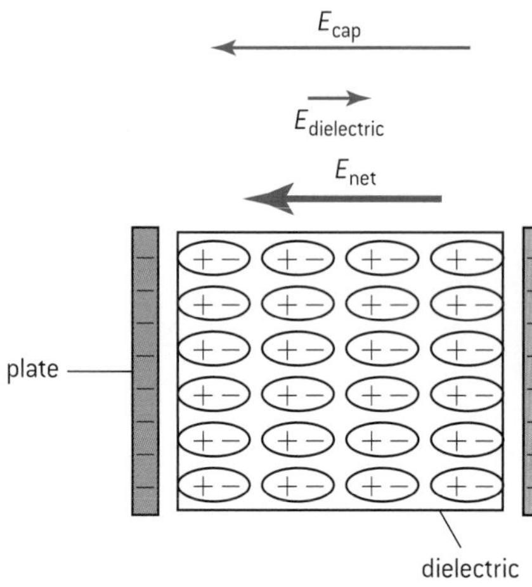

▲ Figure 4 How a dielectric increases capacitance.

Dielectric strength of dielectrics

If the potential difference across a capacitor is increased, a maximum field strength will be reached after which the dielectric becomes a conductor. A spark will then jump between the plates. This maximum field is called the *dielectric strength* of the dielectric. For air this is about 1 kV mm^{-1} (the exact value depends on temperature, pressure and how dry the air is); for teflon the dielectric strength is 60 kV mm^{-1}. If during thunderstorms the electric field strength in air between charged clouds or between a charged cloud and the ground exceeds this critical value, then a lightning strike may occur.

isolated, i.e. not connected to a power supply. The charge on the plates cannot change as there is no route along which the charges can move between them. The plates attract because they are positive and negative. If they are allowed to move closer together at a constant speed (without touching) then they will do work on whoever is moving them. This energy must come from somewhere, the only possible source is the capacitor itself and so the potential difference must drop and once again $\frac{Q}{V}$ increases.

The third way to change capacitance is to insert a material between the plates that replaces the air (or vacuum) that we have so far imagined to fill the space. This means that the new material can do two things for the capacitor designer: it can be used to separate the plates by a fixed amount, and also change the properties of the capacitor at the same time.

In Topic 5 we mentioned that materials have their own permittivity which is greater than that of a vacuum. To treat this mathematically, when a material of permittivity ε is present we replace ε_0 in the equations where it occurs with ε alone. Some typical values for ε were also given in Topic 5 for a number of different materials.

A particularly useful thing happens when a **dielectric material** is inserted between capacitor plates to fill the whole space between them. A dielectric is an electrical insulator that is polarized when placed in an electric field. The origin of the polarization is in the molecules of the dielectric substance. Each molecule is slightly more positive at one end than the other. This means that when a molecule is in an electric field, it responds either by moving slightly or by rotating so that the more positive end of the molecule moves in the direction of the electric field. If the molecules are in a solid and the solid is packed between the plates of a capacitor then this reduces the electric field strength between the plates.

When the dielectric is in place (figure 4), the molecules inside it respond to the field produced by the capacitor. Notice the field directions carefully. In the diagram the original field E_{cap} of the capacitor is from right to left, but the dielectric field $E_{dielectric}$ (indicated by the charges at the surface of the dielectric) is from left to right. Therefore the net field between the plates is equal to the capacitor field minus the dielectric field. The overall field E_{net} in the dielectric is reduced and because $E_{net} = E_{cap} - E_{dielectric} = \frac{V}{d}$ (V is the potential difference) then V decreases too as d is fixed. The insertion of the dielectric reduces the potential difference between the plates because some of the stored energy of the capacitor has been used to align the dielectric molecules. The overall charge stored is unchanged so $\frac{Q}{V}$ is increased and hence so is the capacitance of the capacitor. Dielectrics increase the capacitance of a capacitor. Another way to describe the action of the dielectric is by saying that the presence of the dielectric raises the potential of the negative plate and lowers the potential of the positive plate; this reduces the potential *difference* between the plates.

This explanation of the dielectric effect is a simplified one. There are other reasons for the increase in capacitance, but in many cases the explanation can be given in terms of potential change as here.

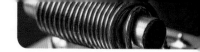

Figure 5(a) shows some typical designs used for practical capacitors. A very common type of capacitor is the electrolytic capacitor shown as a cutaway in figure 5(b). The designer here uses the dielectric advantageously by using it to space the two metal foils apart. The capacitor layers are then rolled up together (rather like rolling up three carpets initially on top of each other). This type of capacitor must be connected into a circuit correctly. The dielectric material is a chemical (the electrolyte in the diagram) that is a good dielectric when the electric field direction is correct. In this design the layers are very thin (good because d is small) and the plate area is large (even better) giving some of the largest values of capacitance possible in a given volume.

Materials used for the dielectric include; paper, mica (a mineral that can be cut into thin layers), Teflon, plastics, ceramics, and the oxides of various metals such as aluminium.

The table shows how good some of these materials can be at improving the ability of a capacitor to store charge at a given voltage. For each material the number given is the ratio of the permittivity of the material to that of a vacuum, in symbols $\frac{\varepsilon}{\varepsilon_0}$. (This ratio is called the relative permittivity or sometimes the "dielectric constant", but you will not be asked about this in the examination.)

Material	$\frac{\varepsilon}{\varepsilon_0}$
vacuum	1
air	$1.000\,54 \cong 1$
paper	4
mica	5
polystyrene	3
ceramic	100–15 000
aluminium oxide	9–11
teflon	2.1
paraffin	2.3
water (pure)	80

Tip

If you are using electrolytic capacitors in the laboratory, take care that the polarity of the capacitor in the circuit is correct. If the polarity is reversed, the dielectric can become hot and cause the capacitor to explode.

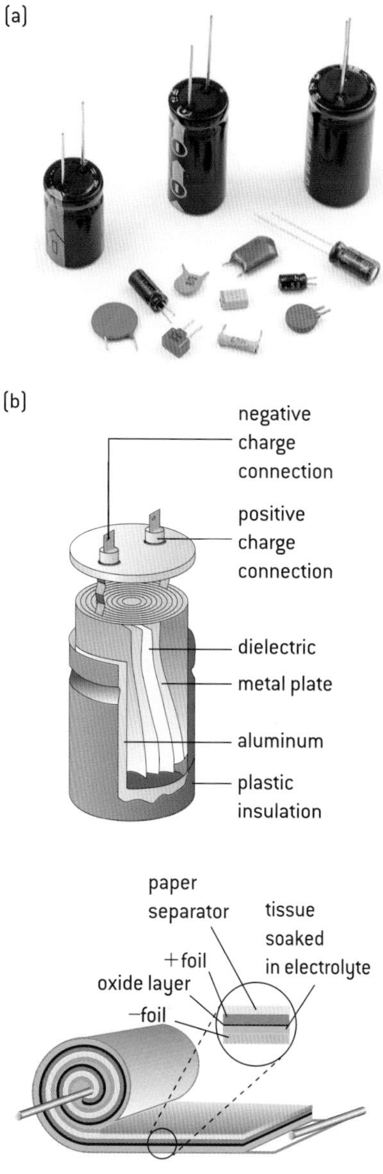

▲ Figure 5 Practical capacitors.

Worked example

1 In a laboratory experiment, two parallel plates, each of area 100 cm², are separated by 1.5 cm. Calculate the capacitance of the arrangement if the gap between the plates is filled with:

a) air

b) polystyrene.

Solution

a) Convert the centimetre units to metres: area $= 100 \times 10^{-4} = 10^{-2}$ m²; separation = 0.015 m

$$C = \varepsilon_0 \frac{A}{d} = 8.9 \times 10^{-12} \times \frac{10^{-2}}{0.015} = 5.9 \times 10^{-12}\ \text{F}$$

b) Polystyrene has a dielectric constant of 3, which means that $\varepsilon = 3 \times 8.9 \times 10^{-12}$. Capacitance is now 1.8×10^{-11} F.

2 Calculate the area of overlap of two capacitor plates separated by a thickness of 0.010 m of air. The capacitance is 1 nF.

Solution

$$A = \frac{Cd}{\varepsilon} = \frac{1 \times 10^{-9} \times 0.01}{8.9 \times 10^{-12}} = 1.1\ \text{m}$$

Combining capacitors in parallel and series

A further way to modify capacitance values is to combine two or more capacitors together in much the same way that resistors were combined in Topic 5. Like resistors, capacitors can be connected in parallel and in series.

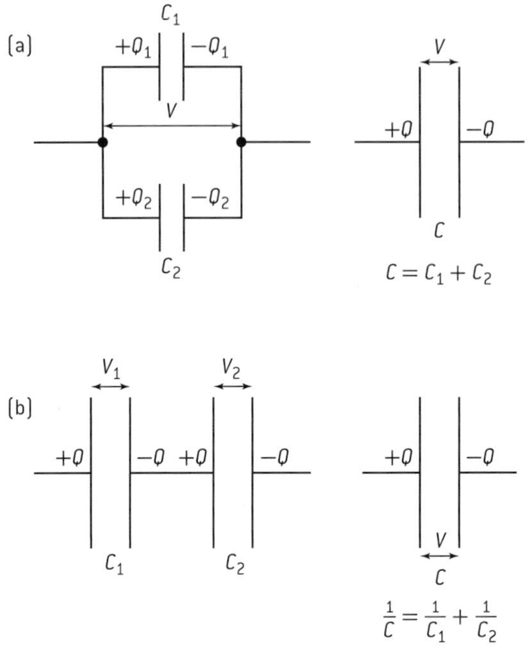

▲ Figure 6 Capacitors in parallel and series.

Parallel

Parallel capacitors have the same potential difference across them when connected as in figure 6(a). The total charge stored is $Q_1 + Q_2$ (as shown on the diagram) and these charges are $Q_1 = VC_1$ and $Q_2 = VC_2$. The single capacitor that is equivalent to the two parallel ones has a charge of $Q = VC$.

So

$$Q = Q_1 + Q_2 \text{ (conservation of charge)}$$

and therefore

$$VC = VC_1 + VC_2$$

Cancelling the V terms gives

$$\mathbf{C = C_1 + C_2}$$

When two capacitors are connected in parallel, the total capacitance is equal to the sum of the capacitances.

Series

Capacitors in series (figure 6(b)) store the same amount of charge Q as each other because the same current charges both capacitors (an example of Kirchhoff's first law in action). Kirchhoff's second law tells us that the potential differences across the capacitors V_1 and V_2 must add up to give the emf of the cell V.

Thus

$$V = V_1 + V_2 \text{ (conservation of energy)}$$

Worked example

Calculate the capacitance of the network below.

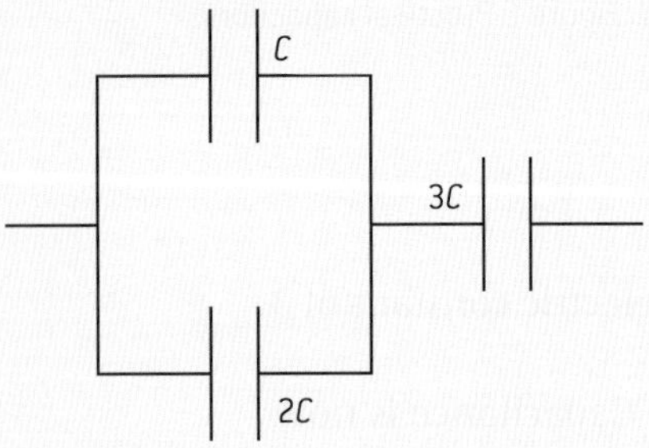

Solution

The capacitors in parallel have capacitance of $C + 2C = 3C$

Combining the parallel capacitors and $3C$ gives:

$\frac{1}{C_{total}} = \frac{1}{3C} + \frac{1}{3C} = \frac{2}{3C}$ so $C_{total} = 1.5C$.

Using the definition of capacitance

$$\frac{Q}{C} = \frac{Q}{C_1} + \frac{Q}{C_2}$$

Q cancels to give

$$\frac{\mathbf{1}}{\boldsymbol{C}} = \frac{\mathbf{1}}{\boldsymbol{C_1}} + \frac{\mathbf{1}}{\boldsymbol{C_2}}$$

So this time, **the reciprocal of the total capacitance is equal to the sum of the reciprocals of each capacitance**.

This is a reversal of the equations for combining resistors and can be a convenient way to remember the equations.

Discharging and charging a capacitor

In an earlier *Investigate!* a capacitor was charged with a constant current. This is an unusual situation that required continuous changes to the total resistance in the circuit to achieve a constant flow of charge. When the total resistance is constant then the charging current varies with time. This section examines the nature of this variation.

Investigate!

Discharging a capacitor

- Set up the circuit shown in figure 7(a). This circuit has two functions: to charge the capacitor and then to discharge it with the power supply disconnected from the circuit. Suitable values for the components are: capacitance, 100 μF; resistance 470 kΩ.
- Charge the capacitor by connecting the flying lead to point *X*. Then begin the discharge by disconnecting the flying lead.
- Record the variation of the potential difference across the capacitor with time. The capacitance and resistance values are designed to allow you to carry out the experiment by yourself. However, you may find it even easier with two people, one recording the data. Alternatively use a voltage sensor together with a data logger to collect and display the data.
- Plot the graph of potential difference against time for the discharge.
- Your results will probably resemble those shown in figure 7(b).

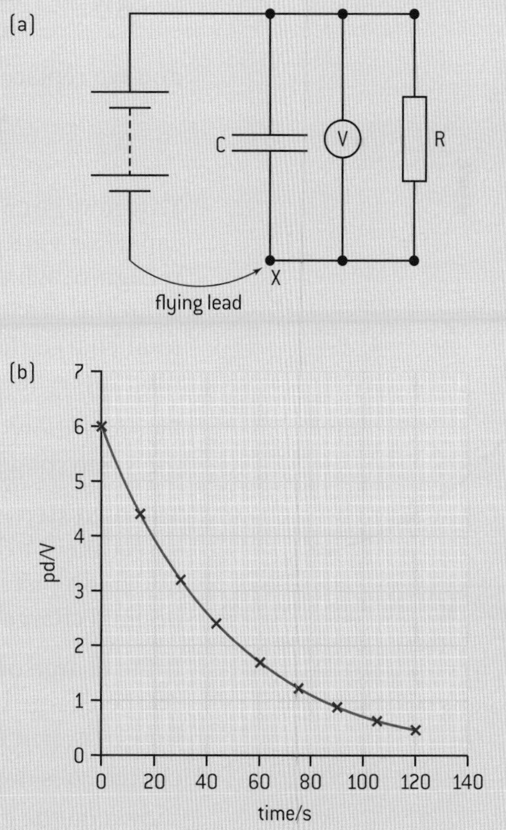

▲ Figure 7 Circuit for capacitor discharge and results.

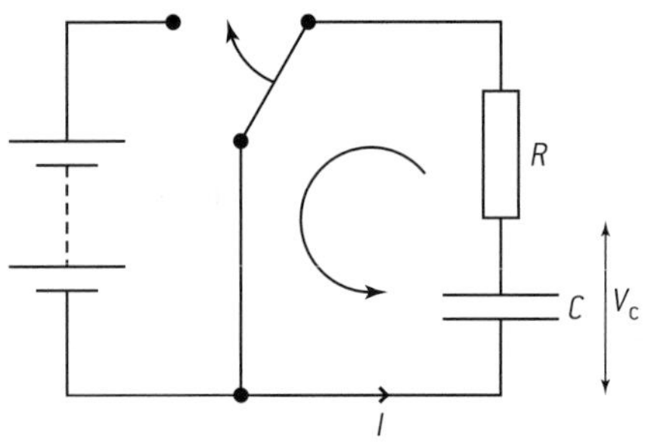

Figure 8 Discharging circuit.

Discharging

You can think of the capacitor as taking the place of a power source and, at any instant, the current I and the pd V_C across the capacitor are related by

$$V_C = IR$$

Here both V_C and I change with time. This equation is obtained by applying Kirchhoff's second law to the circuit loop in Figure 8.

Therefore

$$\frac{\Delta Q}{\Delta t} = \frac{V_C}{R}$$

where Q is the charge on the capacitor and t is the time that has elapsed since discharging began.

So, because

$$V_C = \frac{Q}{C}$$

$$\frac{\Delta Q}{\Delta t} = -\frac{Q}{RC}$$

A negative sign has appeared in the right-hand side of the expression. As the capacitor discharges, the charge on the capacitor falls and in each Δt the change in the charge is a *negative* value.

Rearranging gives

$$\Delta Q = -\frac{Q\Delta t}{RC}$$

and we replace RC by the constant τ where $\tau = R \times C$ so that

$$\Delta Q = -\frac{Q\Delta t}{\tau}$$

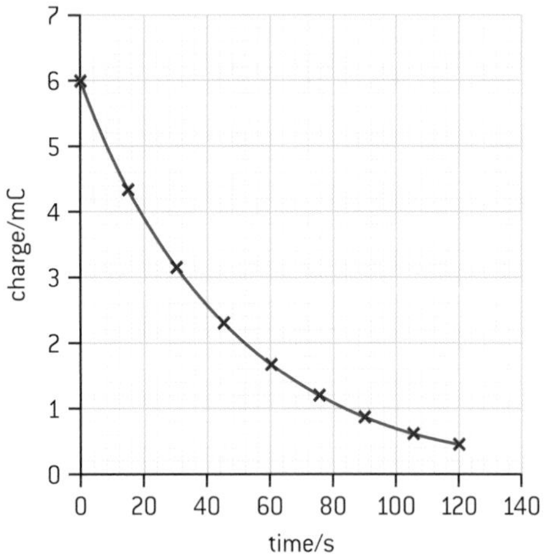

Figure 9 Charge versus time for a discharging capacitor.

This new equation allows us to analyse how the charge on the capacitor varies with time during the discharge. The first step is to recognize what the equation says. It predicts that the *loss of charge* from the capacitor in a time interval Δt will be equal to $\frac{1}{\tau}$ of the total charge that was stored on the capacitor at the beginning of the time interval.

Figure 9 shows what this means in terms of a graph of Q against t. It has the same shape as figure 7(b) because $C = \frac{Q}{V}$ and so $Q \propto V$.

The next step is to model the discharge using $\Delta Q = \frac{Q\Delta t}{\tau}$. We will use a spreadsheet to do this and begin by first concentrating on the part of the Q–t graph just after the discharge begins (at time $t = 0$) at which time the original charge on the capacitor is Q_0.

If the charge on the capacitor did not change, then after a time interval Δt the charge would still be Q_0 and the graph would be parallel to the time axis. But this is not what happens, the equation tells us that the charge goes down (remember the minus sign!) by $\frac{Q_0 \Delta t}{\tau}$. This change is shown on the graph.

- The charge remaining on the capacitor after Δt will be $Q_0 - \frac{Q_0 \Delta t}{\tau}$.
- The graph line has a gradient of $-\frac{Q_0}{\tau}$ during the first time interval Δt.

What happens next? A second time interval begins and charge continues to move off the capacitor plates. But at the start of this interval the charge and, therefore, the pd across the capacitor are less than when t was 0. Call the new charge stored Q_1. The change in charge this time is $\Delta Q = -\frac{Q_1 \Delta t}{\tau}$ because the charge is no longer Q_0. Because Q_1 is less than at the start, ΔQ will also be less over this second time interval. The gradient of the graph becomes less and has a new value of $\frac{Q_1}{\tau}$.

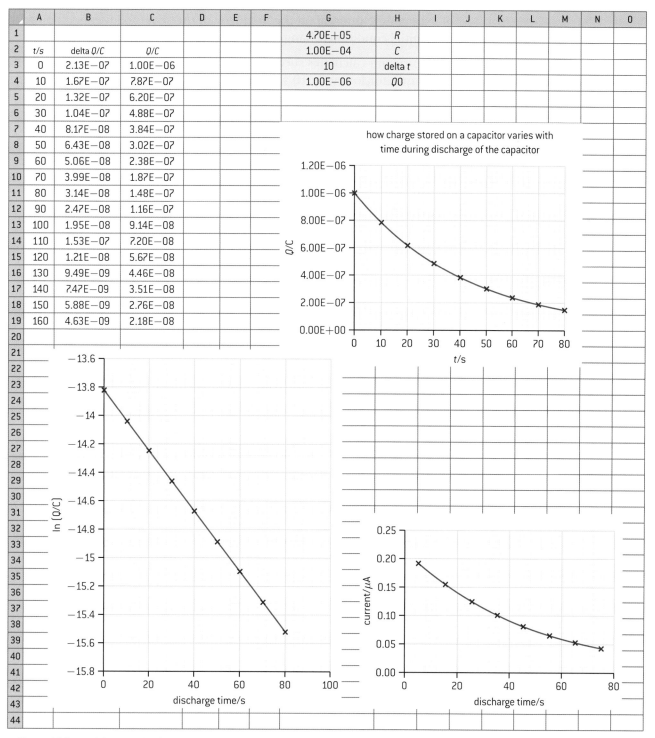

	A	B	C	D	E	F	G	H
1							4.70E+05	*R*
2	t/s	delta Q/C	Q/C				1.00E−04	*C*
3	0	2.13E−07	1.00E−06				10	delta t
4	10	1.67E−07	7.87E−07				1.00E−06	Q0
5	20	1.32E−07	6.20E−07					
6	30	1.04E−07	4.88E−07					
7	40	8.17E−08	3.84E−07					
8	50	6.43E−08	3.02E−07					
9	60	5.06E−08	2.38E−07					
10	70	3.99E−08	1.87E−07					
11	80	3.14E−08	1.48E−07					
12	90	2.47E−08	1.16E−07					
13	100	1.95E−08	9.14E−08					
14	110	1.53E−07	7.20E−08					
15	120	1.21E−08	5.67E−08					
16	130	9.49E−09	4.46E−08					
17	140	7.47E−09	3.51E−08					
18	150	5.88E−09	2.76E−08					
19	160	4.63E−09	2.18E−08					

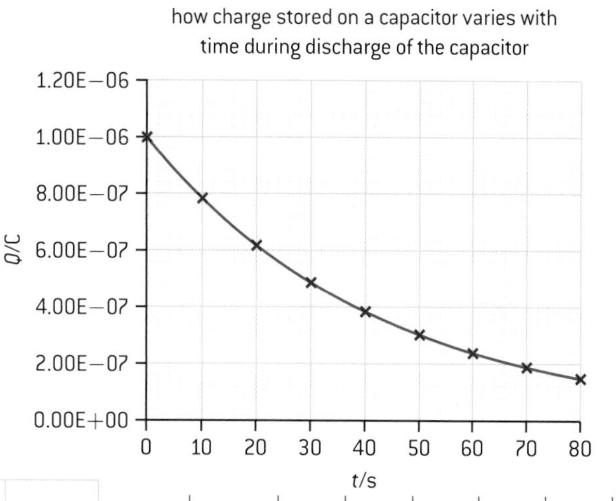

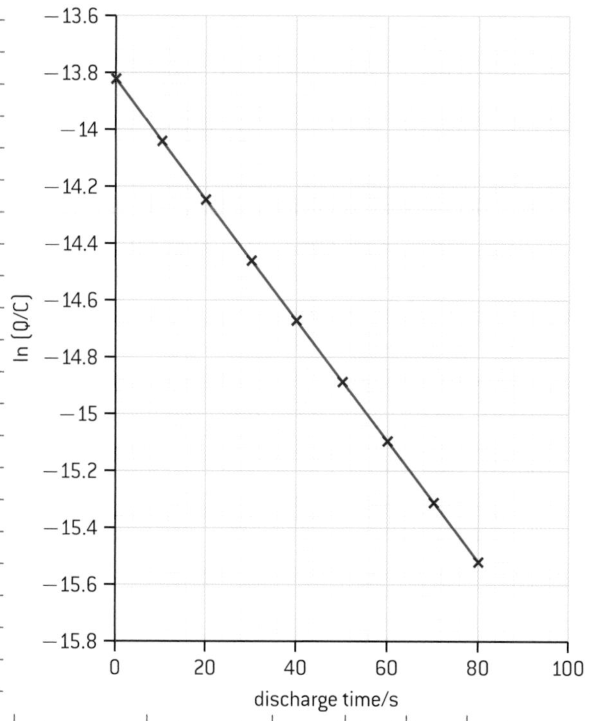

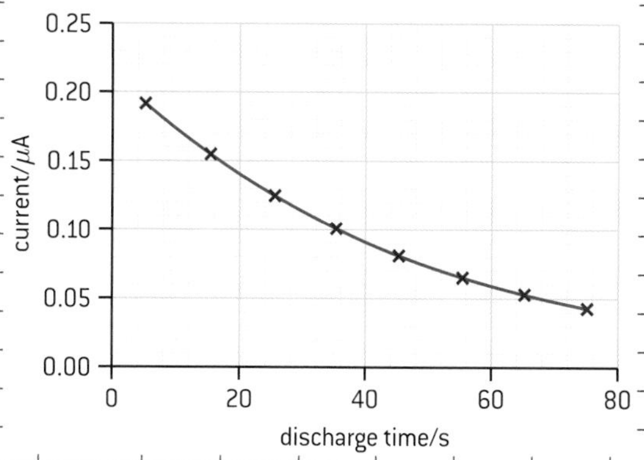

Figure 10 Spreadsheet model for capacitor discharge.

In each successive Δt the change in the charge will be less than in the previous time interval and as time goes on the graph line will curve.

- Column A shows the time increasing in steps of 10 s (this is a deliberately large time increment, 1 s or 0.1 s would be better).
- Column B shows the calculation of ΔQ using $\Delta Q = -\frac{Q\Delta t}{RC}$.
- Column C shows the new value of Q at the end of the time interval.
- The value of Q in column C is then applied as the initial charge in the next time interval on the next row of the sheet following a loop with initial conditions changing at each cycle.

The program calculates successive values for the charge that remains on the capacitor for every 10 s of the discharge. The program then plots a graph of charge of the capacitor against time.

(If you want to produce your own version of this spreadsheet, the formulae are:

- in cell B3 and below **=C3*$G$3/($G$1*$G$2)**
- in cell C9 and below **=C3-B3**
- the symbol $ is required to ensure that the spreadsheet uses the same row and column for C, R and Δt when the columns are copied downwards from row to row – if these were not there then G3 would become G4 then G5 then G6 and so on as the rows are copied down the sheet.

At the top left-hand corner of the spreadsheet are the constants required in the solution: R, C, Δt and Q_0. R and C have the same value as in the *Investigate!* so this solution should match up with the results you obtained then.

Look closely at this graph (which only has the first 80 s of the discharge plotted) and compare it to your own results for the discharging experiment. The graphs have interesting properties. Examine the time it takes the charge on the capacitor to halve. The initial charge at $t = 0$ is 1×10^{-6} C (written as 1.00E-6 in Excel notation) and half this value (5.00E-7) occurs at about $t = 29$ s. Compare this with the time taken for the charge to halve from 5.00E-7 to 2.50E-7. This is from $t = 29$ s to $t = 58$ s – another time of 29 s. The next halving from 2.50E-7 to 1.25E-7 takes another 29 s. For this combination of R and C it always takes 29 s for the charge to halve. If you used values or 100 μF and 470 kΩ in your experiment check that your results give a similar result.

This behaviour is similar to that shown by decaying radioactive nuclei. The activity of the radioactive substance halves in one half-life. For a capacitor, half the charge leaves the capacitor during a half-life. Such behaviour is characteristic of **exponential decay**.

The equation

$$\frac{\Delta Q}{Q} = -\frac{1}{\tau}\Delta t$$

has a similar form to the radioactivity decay equation

$$\frac{\Delta N}{N} = -\lambda \Delta t$$

(N is the number of nuclei present at a given instant and λ is the probability of decay) and therefore we expect them to share the same solution

$$N = N_0 \, e^{-\lambda t}$$

and

$$Q = Q_0 \, e^{-\frac{t}{\tau}}$$

The capacitor discharge equation can also be written (by taking natural logs of both sides of the expression)

$$\log_e Q = \log_e Q_0 - \frac{t}{\lambda}$$

A graph of $\log_e Q$ against t should be a straight line with gradient $-\frac{1}{\lambda}$. The $\log_e Q - t$ graph for the spreadsheet results is shown in figure 10. It has an intercept on the y-axis of -13.8 and a gradient of $-0.021\ \text{s}^{-1}$. This tells us that $\log_e (Q_0/\text{C}) = -13.8$ (remember that the units are written inside the brackets to make the whole log number unitless).

Therefore $Q_0 = e^{-13.8} = 1.0\ \mu\text{C}$ which was the original charge the model gave the capacitor. R was 470 000 Ω and C was 100 μF in the model, so $RC = \frac{1}{\text{Gk}\tau} = 47$ s and the model predicts $\text{Gk}\tau$ to be $0.021\ \text{s}^{-1}$. This is also confirmed by the gradient.

The computer spreadsheet allows us to model one other aspect of capacitor discharge. As the charge flows, discharging the capacitor, there is a current in the circuit. The current is equal to the change in charge per second over each 10 s time interval. This can be modelled by assigning the average current during a time interval to the value of the current at a time halfway through the time interval (i.e. at 5 s for the first 0–10 s interval, at 15 s for the second 10–20 s interval and so on).

Another graph in figure 10 shows that the current – just like charge and potential difference – decays exponentially with a half-life behaviour.

The product $R \times C = \tau$ is known as the **time constant** for the circuit. The name is fully justified because the units of RC are

$$\text{ohm} \times \text{farad} = \frac{\text{volt}}{\text{ampere}} \times \frac{\text{coulomb}}{\text{volt}} = \frac{\text{coulomb}}{\text{ampere}} = \frac{\text{ampere} \times \text{second}}{\text{ampere}}$$

$$= \text{second}.$$

The capacitor time constant is *not* the same as a half-life. After a time of RC since the start, Q will be

$$Q = Q_0 \, e^{-1}$$

because $t = RC = \tau$

So

$$\frac{Q}{Q_0} = \frac{1}{e} = 0.37$$

During each time interval τ the charge drops to 37% of its value at the start of the interval. After 2τ this will be $(0.37)^2$ or 13.7%. After 5τ the charge remaining on the capacitor will be less than 1% of its initial value.

Tip

To obtain the instantaneous current in the resistor from the graph of charge on the capacitor against time, remember that the rate at which charge leaves the capacitor is also the rate at which charge flows through the resistor. In other words, the gradient of the tangent of the Q-t graph will give you the value of the current in the resistor.

Tip

There are three equations that give the variation of charge, current and potential difference with time for a discharge circuit:

$$Q = Q_0 e^{-\frac{t}{\tau}}$$

$$I = I_0 e^{-\frac{t}{\tau}}$$

$$V = V_0 e^{-\frac{t}{\tau}}$$

Worked examples

1 A 220 μF capacitor charged to 30 V discharges through a 330 kΩ resistor.

a) Calculate the time taken for the capacitor to discharge to 10 V.

b) Calculate the charge moved from the capacitor in this time.

Solution

a) $V = V_0 e^{-\frac{t}{RC}}$ so $10 = 30e^{-\frac{t}{330 \times 10^3 \times 220 \times 10^{-6}}}$.
Therefore $\ln 0.33 = -\frac{t}{72.6}$ and $t = 80$ s.

b) $\Delta Q = C\Delta V$. So $\Delta Q = 220 \times 10^{-6} \times 20 = 4.4$ mC.

2 Calculate the number of multiples of the time constant that are required for a capacitor to lose 90% of its charge.

Solution

$Q = Q_0 e^{-\frac{t}{RC}}$

For this case $0.1 = e^{-\frac{t}{RC}}$, therefore $-\frac{t}{RC} = \ln 0.1 = 2.3$, making $t = 2.3RC$.

2.3 time constants are required to lose 90% of charge.

Charging

Again we approach the solution of these equations through an experiment followed by making a spreadsheet model.

Investigate!

Charging a capacitor

- This is very similar to the experiment during which you varied the total resistance in a circuit to achieve a constant current. This time, however, the resistance does not change. As the potential difference at which the capacitor is storing charge increases, the current in the circuit will decrease.
- Suitable values for the fixed resistor and capacitor are, as before, 100 μF and 470 kΩ.
- Make sure the capacitor is discharged, by short-circuiting it with the flying lead, and then disconnect the flying lead to begin the experiment. Wait for a data logger to do its job or take manual readings of the voltage across the capacitor every 5 s until the pd becomes constant.
- Plot a graph of pd against time and examine it closely. To what extent does the behaviour resemble the discharge case?

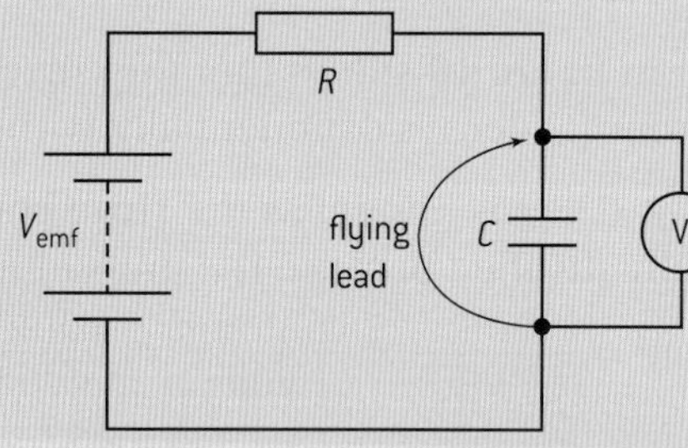

▲ Figure 11

A model for charging

The capacitor is in series with a resistor of resistance R and a cell of emf V_{emf}. When the circuit is switched on with the capacitor uncharged, charge begins to flow.

Using Kirchhoff's second law

$$V_{emf} = V_C + V_R$$

and therefore

$$V_R = V_{emf} - V_C = IR$$

so

$$I = \frac{(V_{\text{emf}} - V_C)}{R}$$

In the time interval Δt, the change in the charge on the capacitor ΔQ is related to the change in the potential difference between the plates ΔV by

$$I = \frac{\Delta Q}{\Delta t} = \frac{C\Delta V}{\Delta t}$$

therefore (rearranging)

$$\Delta V = \frac{(V_{\text{emf}} - V_C) \times \Delta t}{RC}$$

This expression can be integrated mathematically, but a numerical solution illustrates more of the physics of the charging.

	A	B	C	D	E	F	G	H	I	J	K	L
1					R	4.70E+05						
2	time/s	delta *V*/C	*Vc*/C		C	1.00E−04						
3	0	0	0		delta *t*	10						
4	10	1.28E+00	1.28E+00		Vemf	6						
5	20	1.00E+00	2.28E+00									
6	30	7.91E-01	3.07E+00									
7	40	6.23E-01	3.70E+00									
8	50	4.90E-01	4.19E+00									
9	60	3.86E-01	4.57E+00									
10	70	3.04E-01	4.88E+00									
11	80	2.39E-01	5.11E+00									
12	90	1.88E-01	5.30E+00									
13	100	1.48E-01	5.45E+00									
14	110	1.17E-01	5.57E+00									
15	120	9.19E-02	5.66E+00									
16	130	7.23E-02	5.73E+00									
17	140	5.69E-02	5.79E+00									
18	150	4.48E-02	5.83E+00									
19	160	3.53E-02	5.87E+00									
20	170	2.78E-02	5.90E+00									
21	180	2.19E-02	5.92E+00									
22	190	1.72E-02	5.94E+00									
23	200	1.36E-02	5.95E+00									
24	210	1.07E-02	5.96E+00									
25	220	8.40E-03	5.97E+00									
26	230	6.61E-03	5.98E+00									
27	240	5.21E-03	5.98E+00									
28	250	4.10E-03	5.98E+00									
29	260	3.23E-03	5.99E+00									
30	270	2.54E-03	5.99E+00									
31	280	2.00E-03	5.99E+00									
32	290	1.57E-03	5.99E+00									
33	300	1.24E-03	6.00E+00									
34												

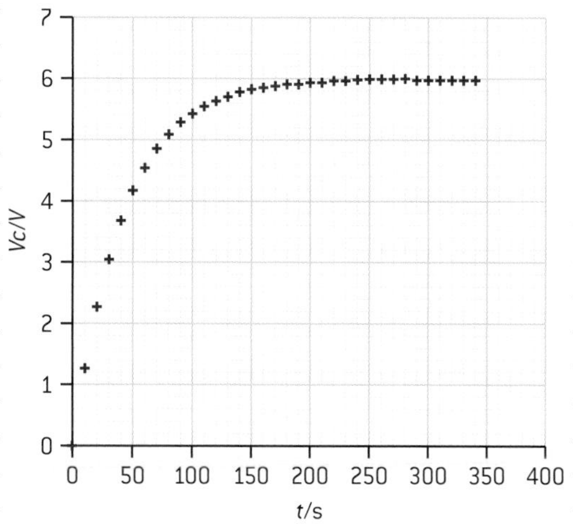

Figure 12 Spreadsheet model for capacitor charging.

The three columns of the spreadsheet model are:

- time is increased row by row by an amount Δt; the formula for **A4** is **=A3+$F$3** (as before the $ signs force the spreadsheet to use the same row and column each time).
- deltaV_c is the change in V_c since the last time; it is given by the equation above and **B4** translates into **=($F$4-B3)*$F$3/($F$1*$F$2)** in the spreadsheet language. Look carefully at the contents of each cell and you should see why this is correct.
- V_c is the new value of V_c that incorporates the deltaV_c from the previous line; it is a simple addition **=C3+B4**

The spreadsheet is copied for about 50 rows (not all shown here) and then the values of t and V_c are used to plot a graph.

Does this graph have similarities to the one you plotted from experimental data?

What are the similarities between charge and discharge?

TOK

Exponential decay

The similarity between radioactivity and capacitor discharge is shared with many other phenomena throughout the whole of science and extends into economic theory. In radioactivity the reason for the behaviour is that each nucleus of a particular isotope has an identical probability of decay per second. Therefore the number decaying in one second is directly proportional to the number of nuclei that remain undecayed. In capacitor discharge, the potential difference (which is proportional to the charge) between the plates is related to the discharge current through $V = IR$ so $\frac{Q}{C} = R\frac{\Delta Q}{\Delta t}$. We can use our water analogy again, this time for water flowing from the bottom of a large tank. When the tank is full the pressure ("potential") from the weight of water at the bottom is large and the flow rate is large too. When the tank is nearly empty the pressure is lower and the flow rate smaller.

There are other examples of exponential changes in the natural world. Sea animals with shells grow at a rate that depends on the mass of food they can eat and this mass depends on their mouth size. So the shell grows exponentially and the shape of the shell can be in the form of an exponential curve. Population growth is an exponential change if there are no predators to remove a species.

Why do so many academic areas have this linking relationship?

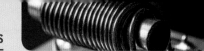

Questions

1 (*IB*) A bar magnet is suspended above a coil of wire by means of a spring, as shown below.

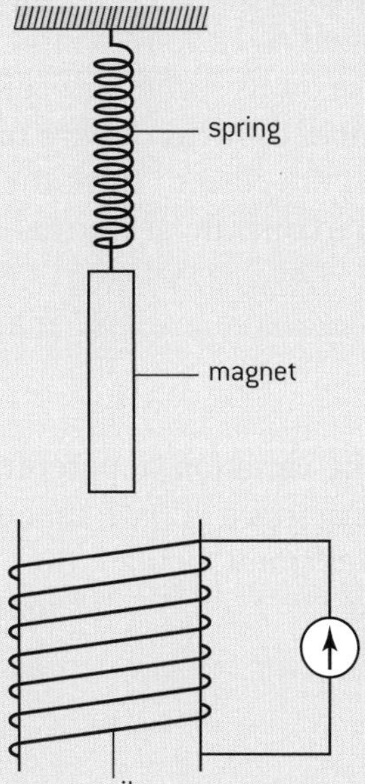

The ends of the coil are connected to a sensitive high-resistance voltmeter. The bar magnet is pulled down so that its north pole is level with the top of the coil. The magnet is released and the variation with time t of the velocity v of the magnet is shown below.

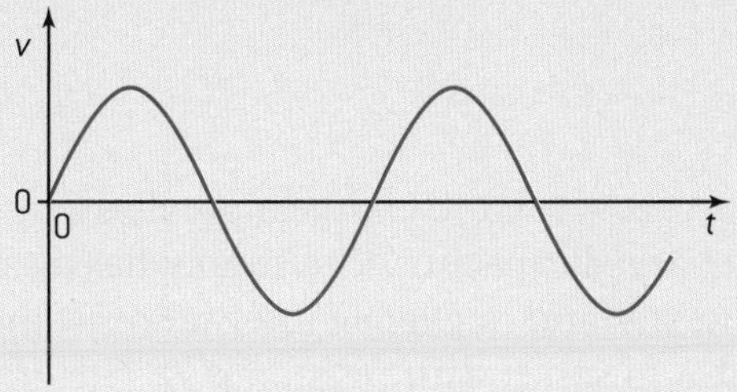

a) Copy the diagram and on it:

(i) mark with the letter M, one point in the motion where the reading of the voltmeter is a maximum

(ii) mark with the letter Z, one point where the reading on the voltmeter is zero.

b) Explain, in terms of changes in flux linkage, why the reading on the voltmeter is alternating.

(4 marks)

2 A uniform magnetic field of strength B completely links a coil of area S. The field makes an angle ϕ to the plane of the coil.

State the magnitude of the magnetic flux linking the coil.

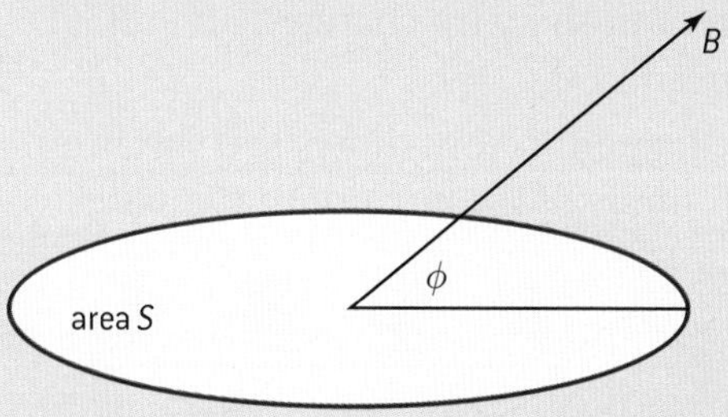

3 (*IB*)

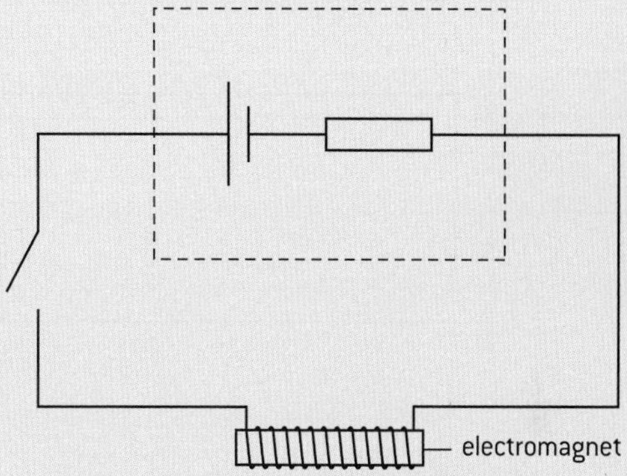

The current in the circuit is switched on.

a) State Faraday's law of electromagnetic induction and use the law to explain why an emf is induced in the coil of the electromagnet.

b) State Lenz's law and use the law to predict the direction of the induced emf in part **a)**.

c) Magnetic energy is stored in the electromagnet. State and explain, with reference to the induced emf, the origin of this energy.

(8 marks)

4 (*IB*) A small coil is placed with its plane parallel to a long straight current-carrying wire, as shown below.

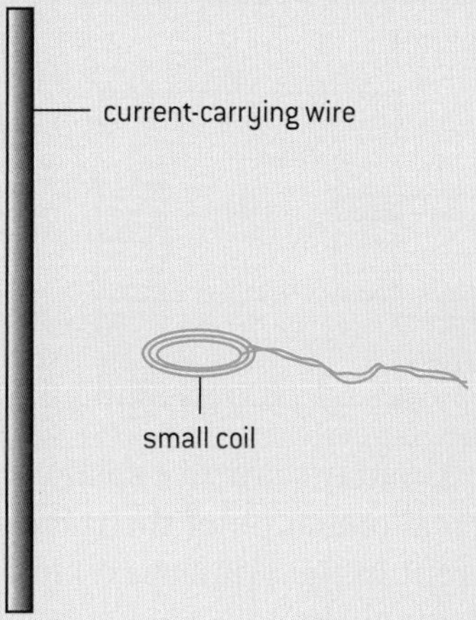

a) Use Faraday's law of electromagnetic induction to explain why, when the current in the wire changes, an emf is induced in the coil.

The diagram below shows the variation with time t of the current in the wire.

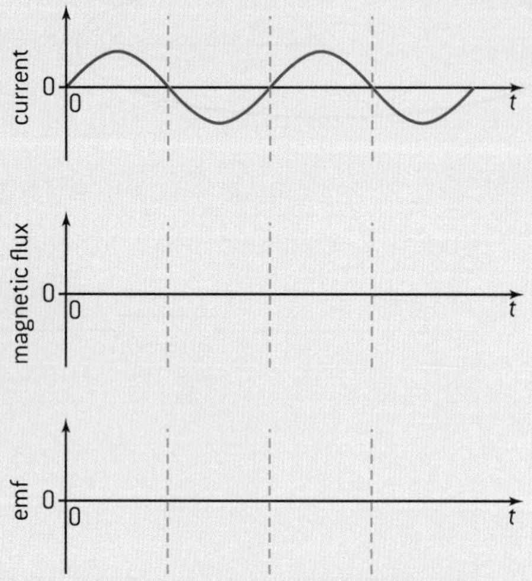

b) **(i)** Copy the diagrams and sketch, on the axes, graphs to show the variation with time t of the magnetic flux in the coil.

(ii) Sketch, on your axes, a graph to show the variation with time t of the emf induced in the coil.

(iii) State and explain the effect on the maximum emf induced in the coil when the coil is further away from the wire.

c) Such a coil may be used to measure large alternating currents in a high-voltage cable. Identify **one** advantage and **one** disadvantage of this method.

(8 marks)

5 (*IB*) The diagram below shows an ideal transformer.

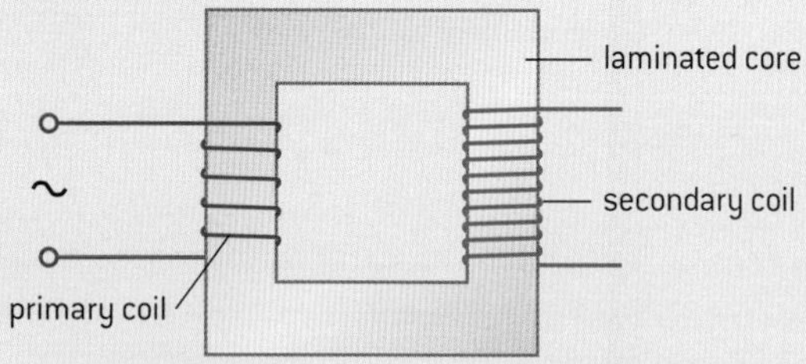

a) Use Faraday's law to explain why, for normal operation of the transformer, the current in the primary coil must vary continuously.

b) Outline why the core is laminated.

c) The primary coil of an ideal transformer is connected to an alternating supply rated at 230 V. The transformer is designed to provide power for a lamp rated as 12 V, 42 W and has 450 turns of wire on its secondary coil. Determine the number of turns of wire on the primary coil and the current from the supply for the lamp to operate at normal brightness.

(7 marks)

6 (*IB*) The graph shows the variation of potential difference V with time t across a 220 μF capacitor discharging through a resistor. Calculate the resistance of the resistor. (3 marks)

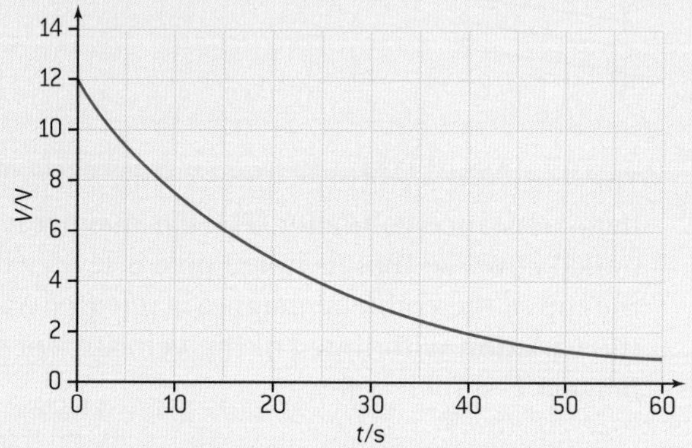

7 **a)** A circuit is used to charge a previously uncharged capacitor. The supply has an emf of 12.0 V and negligible internal resistance and is in series with resistance R. The graph shows how the potential difference across the capacitor varies with time after the switch is closed.

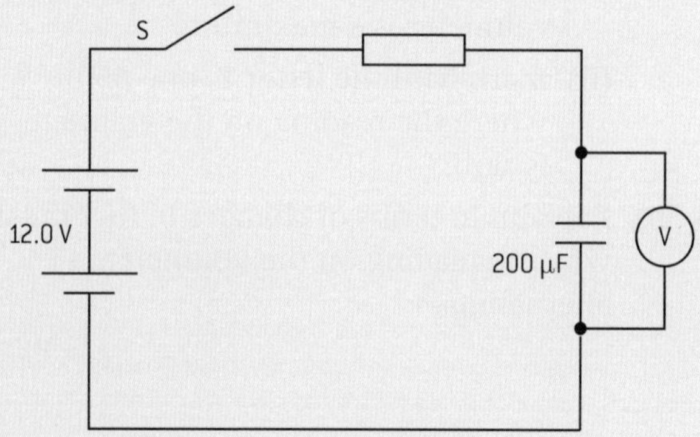

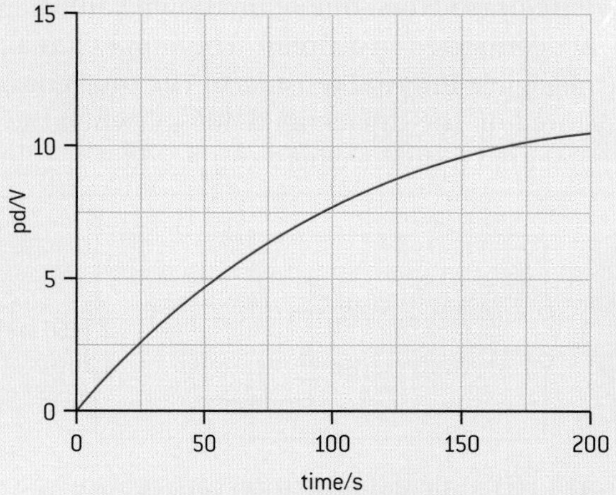

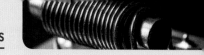

(i) Determine the time taken for the potential difference across the capacitor to reach half the maximum value.

(ii) Calculate *R*.

(iii) Calculate the initial charging current.

b) A 100 μF capacitor is added in series with the 200 μF capacitor.

(i) Calculate the effective capacitance of the combination.

(ii) Draw a graph showing how the potential difference across the combination varies with time when the combination is charged.

(10 marks)

8 **a)** (i) A capacitor has a capacitance of 1 μF. Outline what this tells you about the capacitor.

(ii) Sketch a graph to show how the charge on the 1 μF capacitor varies with the potential difference across it over a range of 6 V.

(iii) Explain how you could determine the energy stored by the capacitor for a potential difference of 6 V.

b) A 0.047 F capacitor is charged to a pd of 20 V, disconnected from the supply and connected to a small motor without discharging. The motor can then lift an object of mass 0.25 kg through a height of 0.90 m before the capacitor is fully discharged.

Calculate:

(i) the initial energy stored by the capacitor

(ii) the efficiency of the system.

(10 marks)

9 A capacitor stores a charge of 30 μC when the pd across it is 15 V. Calculate the energy stored by the capacitor when the pd across it is 10 V. (2 marks)

10 In a timer, an alarm sounds after a time controlled by a discharge circuit powered by a 9.0 V cell of negligible internal resistance. The alarm time is varied using resistor R.

The capacitor is charged by moving the two-way switch to position S_1. The timing starts when the switch is moved to S_2.

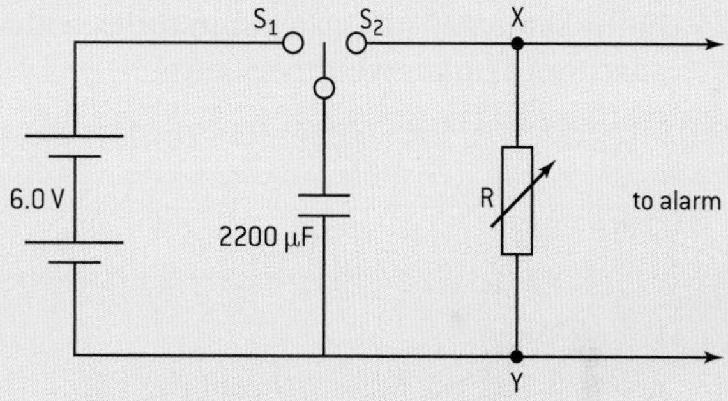

An alarm rings when the potential difference across R reaches 3.0 V.

a) In one setting the time constant of the circuit when the capacitor is discharging is 3.0 minutes. Sketch a graph to show how the potential difference across R varies with time for two time constants of the discharge.

b) Use your graph to state the time at which the alarm will sound.

c) Calculate the resistance of the variable resistor when the time constant is 3.0 minutes.

d) Determine the maximum value of the resistance **R** that is needed for the timer to operate for up to 5 minutes.

e) State how a capacitor could be connected to the circuit to increase the range of the timer.

(11 marks)

11

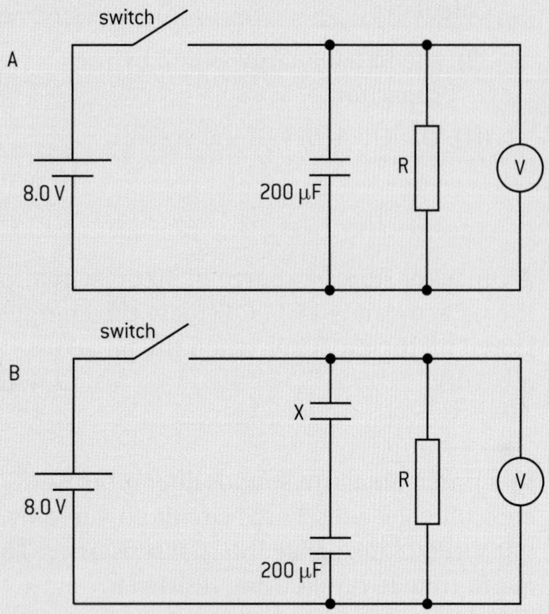

The diagram shows two graphs of the variation of p.d. with time for the discharge through resistor of value R of (i) a 220 μF capacitor and (ii) the same 220 μF capacitor in series with a capacitor of unknown capacitance.

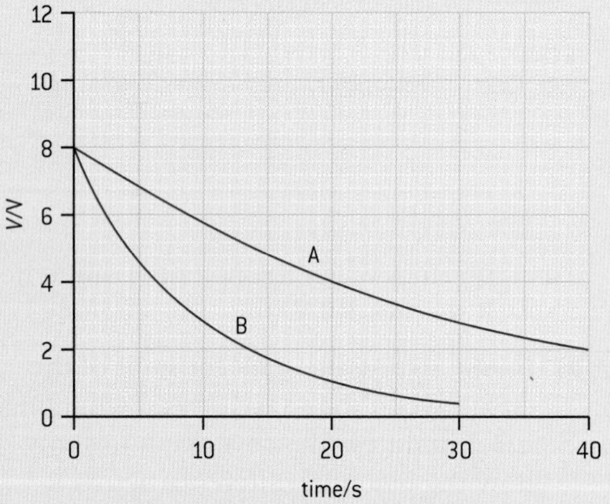

a) Outline why the p.d. in experiment B falls more rapidly than in experiment A.

b) **(i)** Determine R.

(ii) Determine the capacitance of the unknown capacitor.

(8 marks)

12 Three capacitors, one of unknown value, are connected in a circuit. The supply has a negligible internal resistance. The total charge stored on the capacitors is 400 μC when the potential difference between A and B is 12.0 V.

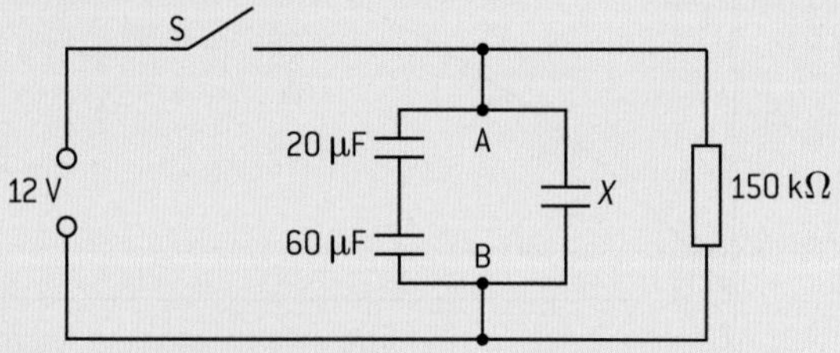

a) **(i)** Calculate the total capacitance of the circuit.

(ii) Calculate X.

(iii) Calculate the potential difference across the 20 μF capacitor.

b) The supply is disconnected and the capacitors discharge through the 150 Ω resistor.

Calculate the:

(i) time taken for the potential difference between A and B to fall to 6.0 V

(ii) potential difference between A and B 8.0 s after opening the switch.

(10 marks)

13 A parallel plate capacitor is made from two circular metal plates with an air gap of 1.8 mm between them. The capacitance was found to be 2.3×10^{-11} F.

Calculate:

a) the diameter of the plates

b) the energy stored when the potential difference between the plates is 6.0 V.

(5 marks)

12 QUANTUM AND NUCLEAR PHYSICS (AHL)

Introduction

This topic develops material we first met in Topic 7. The physics included here was ground-breaking when it was first proposed (mostly in the early to mid 20th century); there were many highly respected physicists who totally rejected much of the theory. Today these ideas are seen as being fairly uncontroversial but there are still those who believe that reality should be modelled by theories that are simpler or more universal. Time will, no doubt, tell whether we have currently got it right!

12.1 The interaction of matter with radiation

Understandings

- Photons
- The photoelectric effect
- Matter waves
- Pair production and pair annihilation
- Quantization of angular momentum in the Bohr model for hydrogen
- The wave function
- The uncertainty principle for energy and time, and position and momentum
- Tunnelling, potential barrier, and factors affecting tunnelling probability

Nature of science

Scientists' increasing dependence on quantum phenomena

Much of what is described by quantum mechanics depends upon probability and seems in many ways counter-intuitive. Einstein, who remained unconvinced about quantum mechanical interpretations of matter, said that he did not … "believe that God plays dice".

Applications and skills

- Discussing the photoelectric effect experiment and explaining which features of the experiment cannot be explained by the classical wave theory of light
- Solving photoelectric problems both graphically and algebraically
- Discussing experimental evidence for matter waves, including an experiment in which the wave nature of electrons is evident
- Stating order of magnitude estimates from the uncertainty principle

Equations

- Planck relationship: $E = hf$
- Einstein photoelectric equation: $E_{max} = hf - \Phi$
- Bohr orbit energies: $E = -\frac{13.6}{n^2} eV$
- quantization of angular momentum: $mvr = \frac{nh}{2\pi}$
- Probability density: $P(r) = |\psi|^2 \Delta V$
- Heisenberg relationships: position–momentum $\Delta x \Delta p \geq \frac{h}{4\pi}$
- energy–time: $\Delta E \Delta t \geq \frac{h}{4\pi}$

Introduction

In Topic 4 we treated light as a wave, but in Topic 7 we needed to modify our views of this in order to describe energy changes that happen in atoms and to explain atomic spectra. In Sub-topic 7.3 we looked at the Rutherford model of the atom with a nucleus surrounded by orbiting electrons. We said that this model did not obey the laws of classical physics but had much to commend it in terms of agreement with experiment. We now consider the further leap of faith that needed to be made in order to reconcile the Rutherford's experimental results with theory; this leap of faith meant abandoning aspects of classical physics and our everyday experiences and to accept many outlooks which go against our intuition. We will now consider the photoelectric effect and go on to look at Bohr's old quantum theory and its development into the modern interpretation of the quantum theory; facets of physics that would mean that physicists could never be truly certain of anything again!

Nature of science

The ultraviolet catastrophe

In the later part of the nineteenth century physicists were attempting to explain the radiation emitted by a black body – a perfect emitter and absorber of radiation. The radiation emitted by a black body was modelled using classical physics by the Rayleigh–Jeans law, whereby the intensity is proportional to the square of the frequency. This theory worked well for the visible and infrared parts of the spectrum, where it matched the practical curve, but it failed in the ultraviolet region by implying that ultraviolet radiation would be emitted with an infinite intensity – something that clearly was not possible (see figure 1). In 1900, the German physicist Max Planck suggested that the ultraviolet catastrophe would be rectified if electrons oscillating in the atoms of hot bodies were to have energies that were quantized in integral values of hf – where f is the frequency of the electrons and h is Planck's constant. The theory that he developed worked well at interpreting black-body radiation, but it was not well-received by the physics community because he was unable to justify why the energies of electrons should be quantized. It was not until Einstein used similar assumptions to those of Planck in order to explain the photoelectric effect that physicists began to take Planck's ideas seriously.

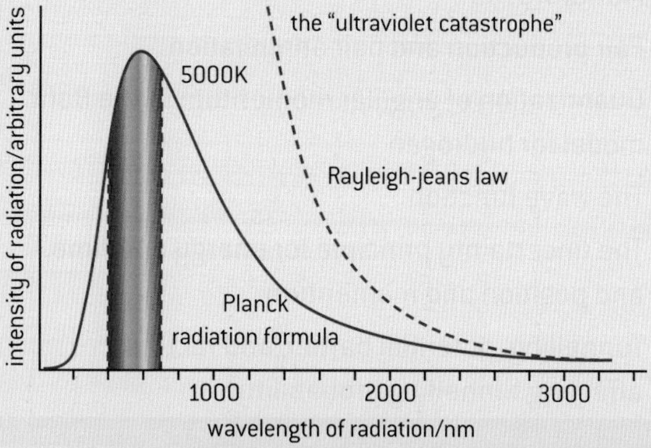

▲ Figure 1 The ultraviolet catastrophe – the experimental curve and one based on classical physics.

The photoelectric effect

Demonstration of the photoelectric effect

The photoelectric effect can be demonstrated using a gold-leaf electroscope (or a coulombmeter). A freshly cleaned sheet of zinc should be mounted on the electroscope plate and the sheet charged negatively by connecting it to a high negative potential (of around – 3 kV). When a range of electromagnetic radiation from infra-red to ultraviolet is incident on the sheet the divergence of the leaf only falls when ultraviolet radiation is used; the leaf then immediately collapses. This is because the zinc sheet and electroscope leaf are discharging by the emission of electrons.

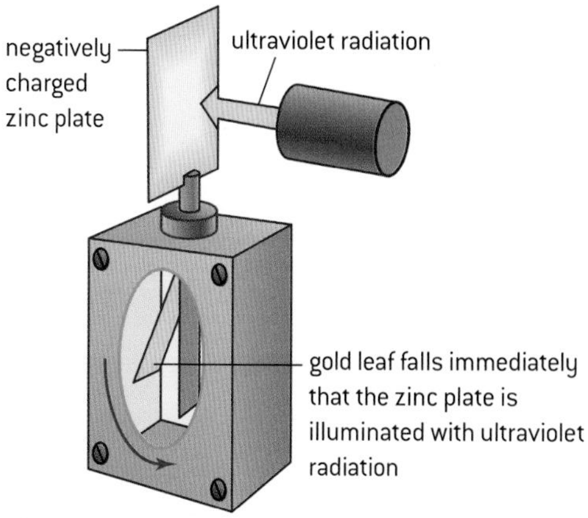

▲ Figure 2 Gold-leaf electroscope demonstration of the photoelectric effect.

Explanation of the photoelectric effect

Einstein explained the photoelectric effect in the following way:

- Light can be considered to consist of photons, each of energy $= hf$
- Each photon can only interact with a single electron.
- There is a minimum photon frequency – called the **threshold frequency** (f_0) below which no electron can be emitted.
- Energy is needed to do the work to overcome the attractive forces that act on the electron within the metal – this energy is called the **work function** (Φ).
- Any further energy supplied by a photon becomes the kinetic energy of the emitted electron (often called a photoelectron).
- Increasing the intensity of light simply increases the number of photons incident per second.

Explaining observations from the gold leaf experiment

The zinc sheet has a certain work function and photons must have a greater energy than this to be able to emit electrons. The ultraviolet radiation has the highest frequency of the radiation used and so it is these photons that are able to free the electrons from the metal.

a) **With very intense visible or infra-red light incident on the sheet the leaf remains diverged.**

Increasing the intensity only increases the number of photons incident per second; with long wavelength (or low frequency) none of the photons have enough energy to liberate electrons so the leaf remains diverged.

b) **If low intensity ultraviolet is used, the leaf still falls immediately.**

One photon interacts with one electron so no time is needed to build up the energy to release an electron; as soon as the electron absorbs a sufficiently energetic photon it will be ejected from the metal.

c) **Placing a sheet of glass between the ultraviolet source and the zinc prevents the leaf from falling.**

Glass will only transmit low energy visible photons; it absorbs the higher energy ultraviolet ones so none reach the zinc.

d) **If the zinc sheet is charged positively the leaf remains diverged for all wavelengths of radiation.**

Raising the potential of the metal presents a much deeper potential well for the electrons to escape from, meaning that the photons no longer have sufficient energy to liberate the electrons. This is similar

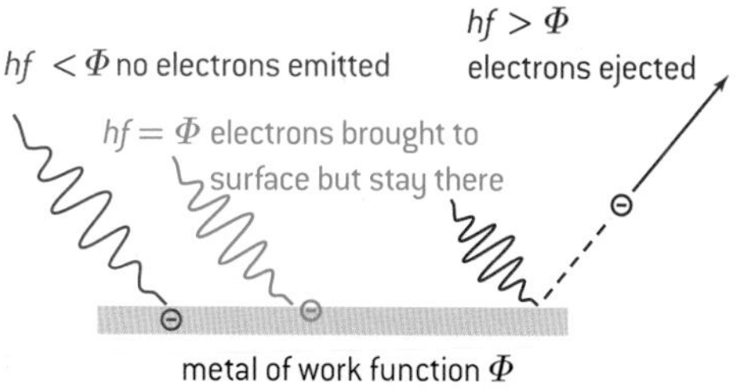

▲ Figure 3 Photoelectric emission and the work function.

to replacing the zinc with a metal of greater work function. This is analogous to someone being trapped in a well with vertical sides and only being able to jump out in one leap.

Einstein's photoelectric equation

We are now in a position to look at how Einstein expressed the photoelectric effect in a single equation. His equation takes the form:

$$E_{max} = hf - \Phi$$

In this equation E_{max} represents the maximum kinetic energy of the emitted electron, hf is the energy of the incident photon, and Φ is the work function of the metal. Each of the terms in the equation represents a quantity of energy that could be measured in joule. However, it is quite common to express the energies in electronvolts simply because this avoids using very small powers of ten. The kinetic energy is expressed as a maximum value because the work function is defined as the minimum energy required to liberate an electron – those embedded further inside the metal will take more energy than this to be liberated.

Worked examples

1 The work function for aluminium is 6.6×10^{-19} J.

a) Calculate the photoelectric threshold frequency for aluminium.

b) Calculate the maximum kinetic energy of the electrons emitted when photons of frequency 1.2×10^{15} Hz are incident on the aluminium surface.

Solution

The work function for aluminium is 6.6×10^{-19} J.

a) Using Einstein's equation $E_{max} = hf - \Phi$ at the threshold value $E_{max} = 0$ so $\Phi = hf_0$ where f_0 is the threshold frequency.

$$f_0 = \frac{\Phi}{h} = \frac{6.6 \times 10^{-19}}{6.63 \times 10^{-34}}$$

$$= 1.0 \times 10^{15} \text{ Hz}$$

This is to two significant figures in line with the work function value.

b) Using $E_{max} = hf - \Phi$

$$\Rightarrow E_{max} = 6.63 \times 10^{-34} \times 1.2 \times 10^{15} - 6.6 \times 10^{-19}$$

$$= 1.4 \times 10^{-19} \text{ J}$$

2 Photons incident on a metal surface have energy 2.2 eV.

a) Calculate the frequency of this radiation.

b) Calculate the maximum speed with which a photoelectron may be emitted from a potassium surface by photons of this energy.

The work function for potassium is 1.5 eV.

Solution

a) First the energy needs to be converted into joules:

$$2.2 \text{ eV} = 2.2 \times 1.6 \times 10^{-19} = 3.5 \times 10^{-19} \text{ J}$$

Then $E = hf \Rightarrow f = \frac{E}{h} = \frac{3.5 \times 10^{-19}}{6.63 \times 10^{-34}}$

$$= 5.3 \times 10^{14} \text{ Hz}$$

b) Again using $E_{max} = hf - \Phi$ but sticking to electronvolts to start with:

$$E_{max} = 2.2 - 1.5 = 0.7 \text{ eV}$$

So the maximum kinetic energy = 0.7 eV = $0.7 \times 1.6 \times 10^{-19}$ J = 1.12×10^{-19} J

This means that $\frac{1}{2}m_e v^2 = 1.12 \times 10^{-19}$ J

So $v = \sqrt{\frac{2 \times 1.12 \times 10^{-19}}{m_e}}$

Looking up the value for the mass of electron in the data booklet gives

$$m_e = 9.11 \times 10^{-31} \text{ kg}$$

$$v = \sqrt{\frac{2 \times 1.12 \times 10^{-19}}{9.11 \times 10^{-31}}}$$

$$= 5.0 \times 10^5 \text{ m s}^{-1}$$

Investigate!

Millikan's photoelectric experiment

In 1916, the American physicist Robert Millikan designed an elegant experiment with which to test Einstein's photoelectric equation. We can use a photocell for essentially the same experiment with an arrangement as shown in figure 4.

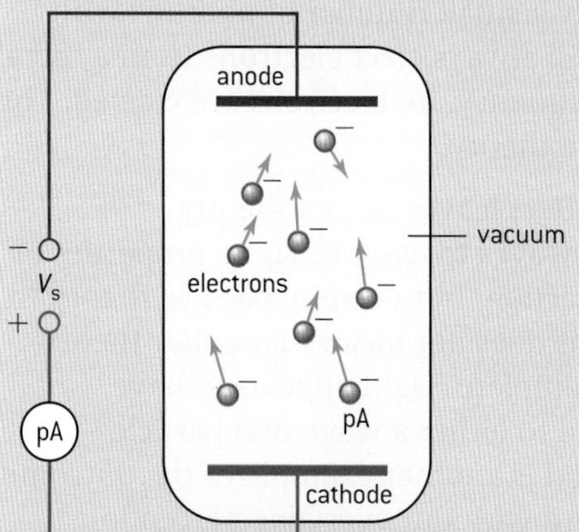

▲ Figure 4 Photocell for measuring the Planck constant.

- A variety of coloured filters are used with a white light source to allow incident photons of different frequencies to fall on the cathode of the photocell.
- The electrons are emitted from the photocell and travel across the vacuum towards the anode.
- In this way the electrons complete the circuit and a small current of a few picoamps registers on the picoammeter.
- The maximum kinetic energy is obtained by adjusting the voltage of the anode with respect to the cathode using a potential divider (not shown on the diagram) – the anode is actually made negative!
- When the potential difference across the tube is just sufficient to prevent electrons from crossing the tube, the maximum kinetic energy of the emitted electrons is equal to eV_s (where V_s is the stopping potential and e is the elementary charge on an electron).
- Einstein's equation $E_{max} = hf - \Phi$ can now be rearranged to give $eV_s = hf - hf_0$ where f_0 is the threshold frequency.
- The filters are usually provided with a range of transmitted wavelengths which gives a value for their uncertainties. This means Einstein's equation needs further rearrangement into

$$eV_s = \frac{hc}{\lambda} - \frac{hc}{\lambda_0}$$

where λ_0 is the threshold wavelength and c is the speed of electromagnetic waves in a vacuum.

- Dividing throughout by e gives

$$V_s = \frac{hc}{e\lambda} - \frac{hc}{e\lambda_0} = \frac{hc}{e}\left(\frac{1}{\lambda} - \frac{1}{\lambda_0}\right)$$

This is of the form:

$$y = mx + c$$

- A graph of V_s against $\frac{1}{\lambda}$:
  - will be of gradient $\frac{hc}{e}$ and
  - have an intercept on the $\frac{1}{\lambda}$ axis equal to $\frac{1}{\lambda_0}$
  - have an intercept on the V_s axis equal to $-\frac{hc}{e\lambda_0}$ (as shown in figure 5)

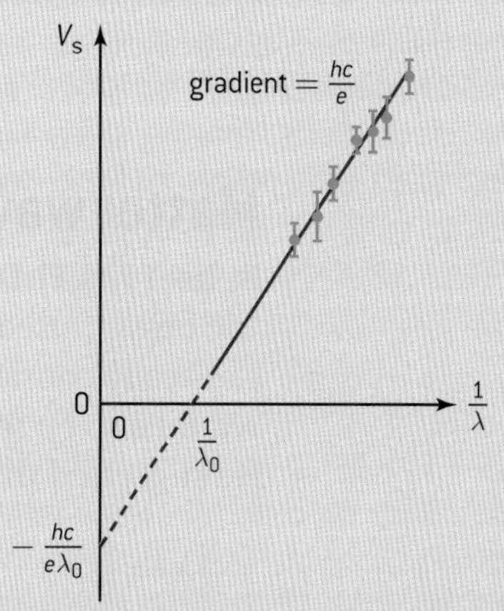

▲ Figure 5 Planck's constant graph.

Note

Different metal cathodes will have different work functions so they will always give graphs of the same gradient as figure 5 but they will produce parallel lines with different intercepts.

The wave theory and the photoelectric effect

The reason that the wave theory fails to account for the photoelectric effect is explained by the instantaneous nature of the emission that occurs when light falls onto a metal surface. Waves provide a continuous supply of energy, the intensity of which is proportional to the square of the wave amplitude. According to classical wave theory, when low intensity electromagnetic radiation of any frequency is incident on a metal surface, given sufficient time, enough energy should eventually accumulate to allow an electron to escape from its potential well. However we find that actually a small number of photons with frequencies above the threshold will always eject electrons from a metal surface immediately; below this frequency, no electrons are ejected. This is completely contradictory to wave theory.

The outcome of the photoelectric effect leaves us in a slightly uncomfortable position of needing waves to describe some properties of light, such as interference and diffraction, but needing particle theory to explain the photoelectric effect. This does not mean that either theory is "wrong" rather that both are incomplete. Light appears to have characteristics that can be attributed to either a wave or a particle – we call this **wave–particle duality** and is another example of the outcomes of experiments failing to be in accord with our everyday perception.

Worked example

Photons of frequency of 1.2×10^{15} Hz are incident on a metal of work function of 1.8 eV.

a) Calculate the energy transferred to the metal by each photon.

b) Calculate the maximum kinetic energy of the emitted electrons in electronvolts.

c) Determine the stopping potential for the electrons.

Solution

a) Using $E = hf$ to calculate the photon energy in joules (we are not asked for this to be in eV so it is fine to leave it in joules).

$E = 6.63 \times 10^{-34} \times 1.2 \times 10^{15} = 8.0 \times 10^{-19}$ J

b) We now need to work in electronvolts so the photon energy $= \frac{8.0 \times 10^{-19}}{1.6 \times 10^{-19}} = 5.0$ eV

With $\Phi = 1.8$ eV this means the maximum kinetic energy is $(5.0 - 1.8) = 3.2$ eV

c) Working in electronvolts now really comes into its own since the stopping potential will be 3.2 V.

Matter waves

In his 1924 PhD thesis, the French physicist, Louis de Broglie (pronounced "de broy"), used the ideas of symmetry to suggest that if something classically considered to be a wave had particle-like properties, the opposite would also be true. Matter could, therefore, also have wave-like properties. He suggested that the wavelength λ associated with a particle is given by

$$\lambda = \frac{h}{p}$$

Here h is the Planck constant and p is the momentum of the particle ($= mv$). This wavelength is known as the **de Broglie wavelength**.

De Broglie used ideas from the special theory of relativity and the photoelectric effect in order to derive this equation for light (you don't need to learn this derivation).

The total energy of an object (from the special theory of relativity) is the total of the rest energy and kinetic energy, given by $E = \sqrt{p^2c^2 + m_0^{\ 2}c^4}$ (the first term in the equation is a kinetic energy term and the second term is the rest mass energy term). For a photon the second term is zero (photons have no rest mass) so $E = \sqrt{p^2c^2}$ or $E = pc$.

As we have seen from the photoelectric effect $E = \frac{hc}{\lambda}$ equating these $pc = \frac{hc}{\lambda}$ or $\lambda = \frac{h}{p}$

With this equation derived for light, de Broglie simply speculated that the same thing would be true for matter, and very soon he was shown to be correct!

Electron diffraction

In 1925, two American physicists, Clinton Davisson and Lester Germer, demonstrated de Broglie's hypothesis experimentally by observing interference maxima when a beam of electrons was reflected by a nickel crystal. In 1928, the British physicist George Thomson independently repeated Davisson and Germer's work at the University of Aberdeen.

Figure 6(a) and (b) shows a laboratory arrangement for demonstrating the effect using the transmission of electrons through a thin slice of crystal.

Electrons from a heated cathode pass through a thin film of carbon atoms (the crystal). If the electrons behaved like particles they would be only slightly deviated by collisions with the carbon atoms and would form a bright region in the centre of the screen.

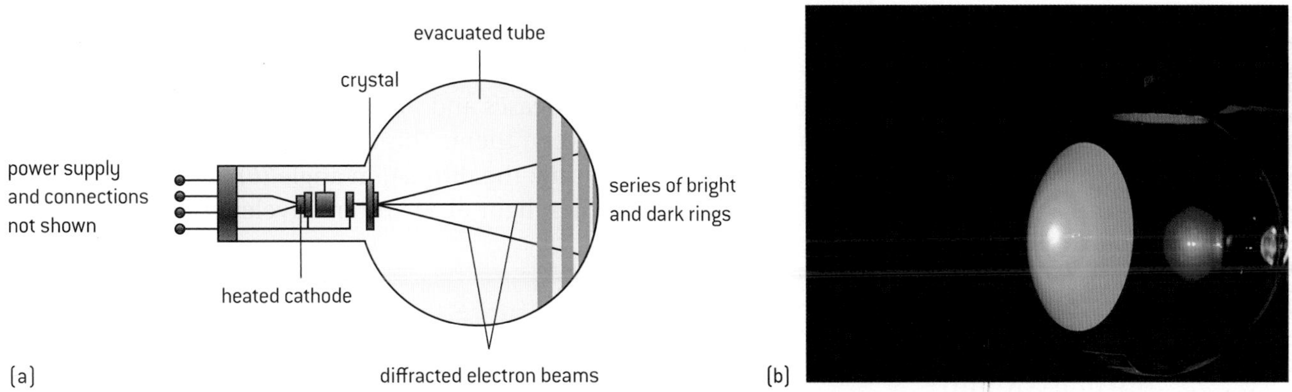

▲ Figure 6 Electron diffraction tube.

The bright rings indicate where the electrons land on the screen. Where there is a bright glow there is a high probability of electrons reaching that point – where there is darkness there is a low probability of the electrons reaching that point. The same pattern builds up slowly, even if there are only a few electrons travelling in the tube at any one time. This pattern is very similar to the interference pattern that is obtained with light using a diffraction grating and can be explained by assuming that electrons behave in a similar way to waves:

When electrons are accelerated through a potential difference they gain kinetic energy

$$eV\left(= \frac{1}{2}mv^2\right)$$

Assuming that the accelerated electrons do not travel close to the speed of light then the momentum of the electrons is given by $p = mv$, meaning that $p^2 = (mv)^2$

so $\frac{1}{2}mv^2 = \frac{p^2}{2m} = eV$

and $p = \sqrt{2meV}$

using the de Broglie relationship

$$\lambda = \frac{h}{p}$$

this gives $\lambda = \frac{h}{\sqrt{2meV}}$

Worked example

Calculate the de Broglie wavelength of electrons accelerated through a potential difference of 3.00 kV.

Solution

In an IB question on this you would be asked to develop this equation step by step and then do this calculation.

$$\lambda = \frac{h}{\sqrt{2meV}} = \frac{6.63 \times 10^{-34}}{\sqrt{2 \times 9.110 \times 10^{-31} \times 1.60 \times 10^{-19} \times 3000}}$$

$$= 2.24 \times 10^{-11}\ \text{m}$$

This is similar to the wavelength of the X-rays used to form diffraction patterns when they are incident on crystals. Increasing the accelerating voltage increases the energy and momentum of the electrons. The wavelength, therefore, decreases and so produces smaller diameter rings with smaller spacing between them. This is analogous to light passing through a diffraction grating: the diffraction angle (θ) in the equation $n\lambda = d\sin\theta$ is reduced when light of a shorter wavelength is used.

Nature of science

Wave–particle duality

From our discussion of the photoelectric effect and electron diffraction we have seen that both electrons and photons sometimes behave as waves and sometimes as particles. This gives rise to questions such as "is matter a wave or a particle?" and "is light a wave or a particle?". The answer to each of these questions is "neither": matter is matter and light is light, therefore they are neither waves nor particles. However, in each case we need the wave model to explain some properties of each and we need the particle model to explain other properties. Nobody knows the mechanisms by which electrons and photons behave, because they do not enter into the realms of everyday experience. We interpret their behaviour by using mathematics (of increasing complexity), but it is impossible to unravel the properties of waves or electrons by relating these to everyday occurrences. That individual photons can pass through individual slits and produce a pattern consisting of regions of high photon densities and regions of low photon densities makes no sense when we think of the interference of particles – yet this is what happens in the quantum world.

The Bohr model

In order to interpret the scattering of alpha particles as discussed in Sub-topic 7.3, Niels Bohr proposed a model of an atom in which electrons could only occupy orbits of certain radii. His model was based on the following three assumptions:

1 **Electrons in an atom exist in stationary states.**

- Contradicting classical physics, electrons could remain in these orbits without emitting any electromagnetic radiation.

2 **Electrons may move from one stationary state to another by absorbing or emitting a quantum of electromagnetic radiation**

- If an electron absorbs a quantum of radiation it can move from one stationary state to another of greater energy; when an electron moves from a stationary state of higher energy to one of lower energy it emits a quantum of radiation.
- The difference in energy between the stationary states is given by $\Delta E = hf$.

3 **The angular momentum of an electron in a stationary state is quantized in integral values of $\frac{h}{2\pi}$**

- This can be represented mathematically by:

$$mvr = \frac{nh}{2\pi}$$

- Angular momentum is the (vector) product of the momentum of a particle and the radius of its orbit – so, for a particle in a circular orbit, the angular momentum will be constant.
- This assumption is equivalent to suggesting that an integral number of de Broglie-type wavelengths fits the electron's orbital:

$$\lambda = \frac{h}{p} \quad \therefore \quad p(= mv) = \frac{h}{\lambda}$$

The circumference of an orbit (of radius r) $= 2\pi r$

When the number of complete waves fitting this orbit is n then each

$$\lambda = \frac{2\pi r}{n}$$

Equating the two values for λ gives

$$\frac{h}{p} = \frac{2\pi r}{n}$$

or, rearranging and substituting mv for p, gives

$$mvr = \frac{nh}{2\pi}$$

This pattern is shown in figure 7 and corresponds to standing waves.

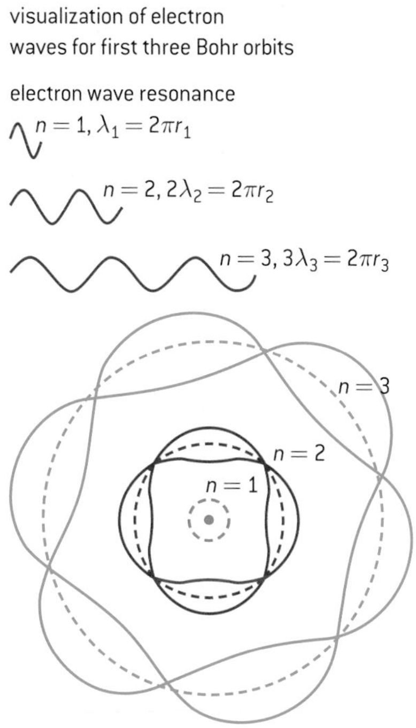

▲ Figure 7 Standing waves in an atom for $n = 2$ and $n = 3$.

Energies in the Bohr orbits

One of the triumphs of the Bohr atom was that it produced an equation that agreed with the experimental equation for the spectrum of the hydrogen atom. By measuring the total kinetic and potential energy of the hydrogen atom we find that, for an electron in the nth energy level (where $n = 1$ represents the ground state, $n = 2$ the first excited state, etc. and is called the **principal quantum number**), the total energy E in electronvolts at each level is given by:

$$E = -\frac{13.6}{n^2}$$

See figure 8.

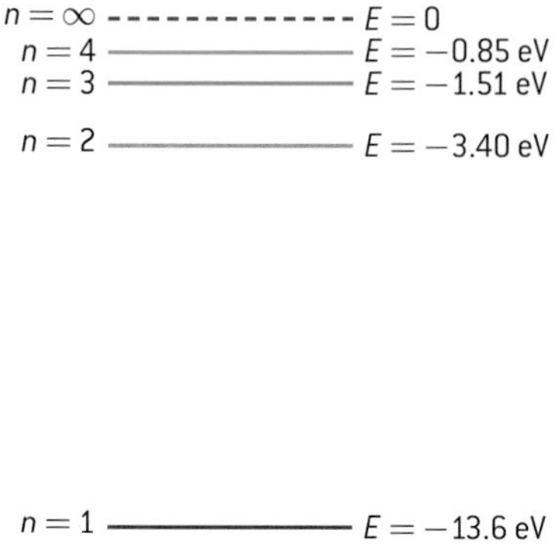

▲ Figure 8 Energy levels in a hydrogen atom.

This total energy is negative because the electron is bound to the nucleus and energy must be supplied to the system in order to completely separate the electron from the proton. Bohr went on to modify this equation so that it could accommodate other hydrogen-like (i.e. one electron) systems such as singly ionized helium and doubly ionized lithium, etc. The model failed, however, to be extended to more complicated systems of atoms and it could not explain why certain allowed transitions were more likely to occur than others. Despite its flaws, the Bohr model proved to instigate a more fundamental approach to the atom; this is now called **quantum mechanics**.

Worked example

In his theory of the hydrogen atom, Bohr refers to *stable electron orbits.*

a) State the Bohr postulate that determines which stable orbits are allowed.

b) Describe how the existence of such orbits accounts for the emission line spectrum of atomic hydrogen.

The Bohr model of the hydrogen atom can be extended to singly ionized helium atoms. The model leads to the following expression for the energy E_n of the electron in an orbit specified by the integer n.

$$E_n = \frac{k}{n^2}$$

where k is a constant.

In the spectrum of singly ionized helium, the line corresponding to a wavelength of 362 nm arises from electron transitions between the orbit $n = 3$ to the orbit $n = 2$.

c) Deduce the value of k.

Solution

a) The angular momentum (mvr) of an electron in a stationary state is quantized in integral values of $\frac{h}{2\pi}$

b) *There are many things that can be said here but some of the key points are:*

- When in a stable orbit an electron does not emit radiation.
- When an electron drops to a lower energy level or orbit it emits a photon.
- The frequency of the emitted photon is proportional to the difference in energy between the two levels.
- A transition between two levels results in a spectral line of a single wavelength being emitted.

c) $E = \frac{hc}{\lambda} = \frac{k}{n^2}$ so the difference in energy between the two levels will be

$$= k\left(\frac{1}{4} - \frac{1}{9}\right) = 0.139\,k$$

$$k = \frac{hc}{0.139\lambda} = \frac{6.63 \times 10^{-34} \times 3 \times 10^{8}}{362 \times 10^{-9} \times 0.139}$$

$$= 3.95 \times 10^{-18}\ \text{J}$$

Schrödinger's equation

Wave–particle duality explains a bright interference fringe as being the place where there is a high probability of finding a particle. The position of particles is described mathematically by probability waves. As with classical waves, probability waves superpose with one another to produce the expected interference pattern. A low-intensity beam of photons incident on a single slit one at a time will build up a distribution that is identical to the expected diffraction pattern, provided that we wait a sufficiently long time. A similar pattern is obtained by firing electrons at a slit of suitably small width (see figure 10). From this distribution we can predict the places where electrons will not reach but we are unable to predict where an individual electron will be detected – although, statistically, there will be approximately 22 times the number of electrons arriving at the area around the principal diffraction maximum compared to the number around the secondary maximum.

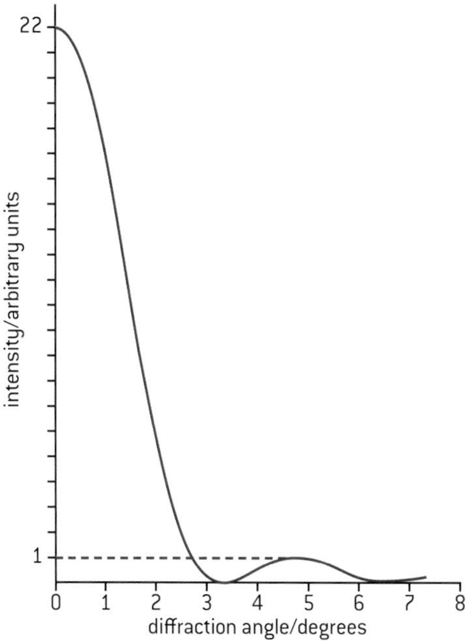

▲ Figure 9 Diffraction pattern intensity distribution.

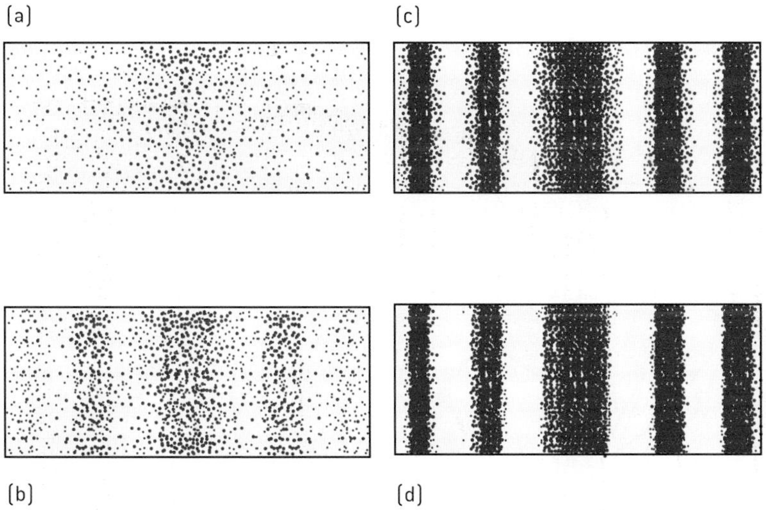

▲ Figure 10 Diffraction pattern being built up by individual electrons.

The concept of probability was developed into the quantum mechanical model of the atom in 1926 by the Austrian (and later Austrian–Irish) physicist, Erwin Schrödinger. **Schrödinger's wave function** ψ describes the quantum state of particles. His wave equation has many similarities to a classical wave equation but it is not a derived equation. It has, however, been thoroughly verified experimentally and its solution for the single electron hydrogen atom agrees with the Bohr relationship $E = -\frac{13.6}{n^2}$.

The wave function is not a directly observable quantity but its amplitude is very significant. With light waves we observe the intensity, not the amplitude, and we have seen that the intensity is proportional to the square of the amplitude. For the wave function, where the square of the amplitude is a maximum there is the greatest probability of finding a photon. When the square of the amplitude is zero there is zero probability of finding the photon. The quantity ψ may be thought of as the amplitude of the de Broglie wave corresponding to a particle (although it does not have any physical significance); however, the square of the amplitude of the wave function $|\psi|^2$ is proportional to the probability per unit volume of finding the particle – this is known as the probability density. Mathematically we write this as:

$$P(r) = |\psi|^2 \Delta V$$

Here $P(r)$ is the probability of finding a particle a distance r from a chosen origin and ΔV is the volume being considered.

For double slit interference, in terms of the probability wave, the wave function is considered to be such that a single photon or electron passes through both slits and be everywhere on the screen until it is observed or measured. When this happens, the wave function collapses to the classical case and the particle is detected. This is known as the **Copenhagen interpretation**, so named by the German physicist, Werner Heisenberg. It relates to the interpretation of quantum mechanics used by Heisenberg, Bohr and their co-workers between 1924 and 1927. It can be summarized as **nothing is real unless it is observed**. So matter or light can be considered to be a wave or a particle. If it behaves like a particle then it *is* a particle. If it behaves like a wave, then it *is* a wave.

"We've agreed to count it as both a wave and a particle for tax purposes."

▲ Figure 11 The Copenhagen interpretation.

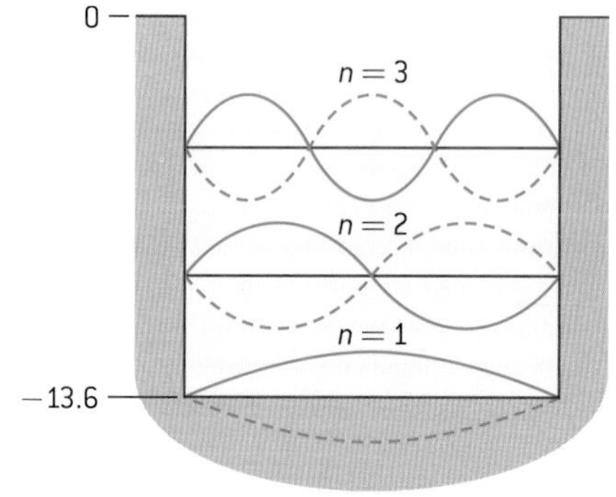

▲ Figure 12 Electron standing waves in a potential well.

In the simplified (one-dimensional) version of the hydrogen atom, as shown in figure 12, an electron would be detected somewhere between the nucleus and the outside edge of the atom – these are shown by the edges of a potential well. A more realistic model would show the potential varying as the inverse of the distance from the nucleus $\left(V \propto -\frac{1}{r}\right)$.

Within the well, the electron energy must be such that the wave function has nodes at the sides. In the electron wave model the probabilities of finding an electron within the nucleus or outside the atom are both zero so the wave amplitude is zero at these points. The electron is most likely to be found (highest probability) where the amplitude is maximum; this is midway between the nodes.

Worked example

The graph below shows the variation with distance r from the nucleus of the square of the wave function, Ψ^2, of an electron in the hydrogen atom according to the Schrödinger theory. The nucleus is assumed to be a single point.

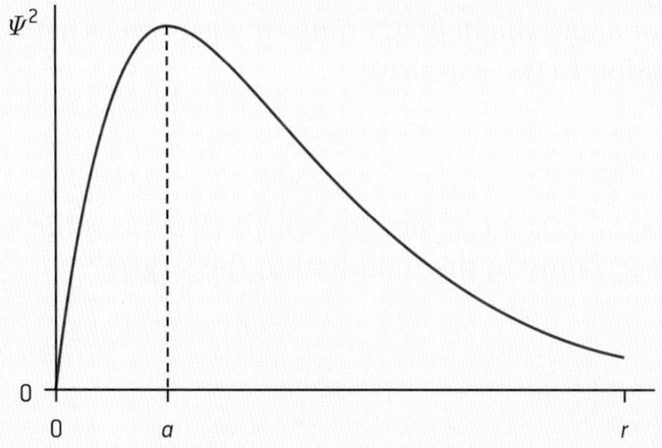

State where the electron is most likely to be found. Explain whether the electron could be found either in the nucleus or at positions corresponding to the largest value of r shown on the graph.

Solution

The position around a has the highest value of the square of the wave function. This is the position where there is the highest probability of finding the electron. Because the square of the wave function is zero at the position of the nucleus, the electron definitely cannot be found there but there is a finite chance of finding the electron at large values of r since the graph has not fallen to zero. The electron's probability cloud would be densest at a, but it will still have a little density at large values of r.

The Heisenberg uncertainty principle

We have seen that when a quantum is diffracted it is only possible to predict its subsequent path in terms of the probability of the wave function. The outcome of this experiment is in line with Heisenberg's uncertainty principle.

This is usually written as $\Delta x \Delta p \geq \frac{h}{4\pi}$ and places a limit on how precisely we are able to know the position and momentum of something in the quantum mechanical realm.

If we wish to know where an electron is positioned at a given time, the principle tells us that there is an uncertainty (given by Δx) with which we can know that position. If Δx is very small, then the uncertainty (Δp) in knowing the momentum of the electron is very large. Therefore, it is not possible to precisely determine the position of the electron and its momentum at the same time; the product of the two uncertainties will always be greater than or equal to $\frac{h}{4\pi}$.

If we imagine the wave function of a free electron (one that is not in any field that would change its motion) to be a sine wave then we can measure its wavelength perfectly. Since we know its wavelength perfectly, we also know its momentum perfectly ($p = \frac{h}{\lambda}$). This implies that the position of the electron has an infinite uncertainty and is spread out over all of space.

In order to detect a quantum particle it would be necessary to use something that has a comparable size to that particle. Using radiation with which to detect a nucleus would need a wavelength of $\approx 10^{-15}$ m. We know from the de Broglie relationship ($\lambda = \frac{h}{p}$) that the shorter the

Note

In the limit when one of the quantities is known perfectly the other quantity has an infinite uncertainty.

wavelength is, the greater the momentum so, with a wavelength of 10^{-15} m, the momentum will be $\frac{6.63 \times 10^{-34}}{10^{-15}} \approx 10^{-18}$ N s.

On a nuclear level this is a very large momentum and would mean that any radiation of this wavelength would impart energy to the nucleus which would then make its position effectively immeasurable.

In dealing with electrons diffracting through a narrow gap, perhaps of size $\approx 10^{-18}$ m, the uncertainty principle also applies. In passing through the gap the uncertainty of the electron's position in the gap will be $\pm$ half the gap width. This then puts a limit on the precision with which we can know the component of the momentum of the electron parallel to the gap (and therefore its wavelength). Thus $\Delta p \approx \frac{h}{4\pi \times 0.5 \times 10^{-18}} \approx \pm 1 \times 10^{-16}$ N s and $\Delta\lambda \approx \frac{h}{\Delta p} \approx \pm 10^{-17}$ m (or approximately ten times the gap size. That is a large uncertainty in relation to the gap size!)

Worked example

The diagrams show the variation, with distance x, of the wave function ψ of four different electrons. The scale on the horizontal axis in all four diagrams is the same. For which electron is the uncertainty in the momentum the largest?

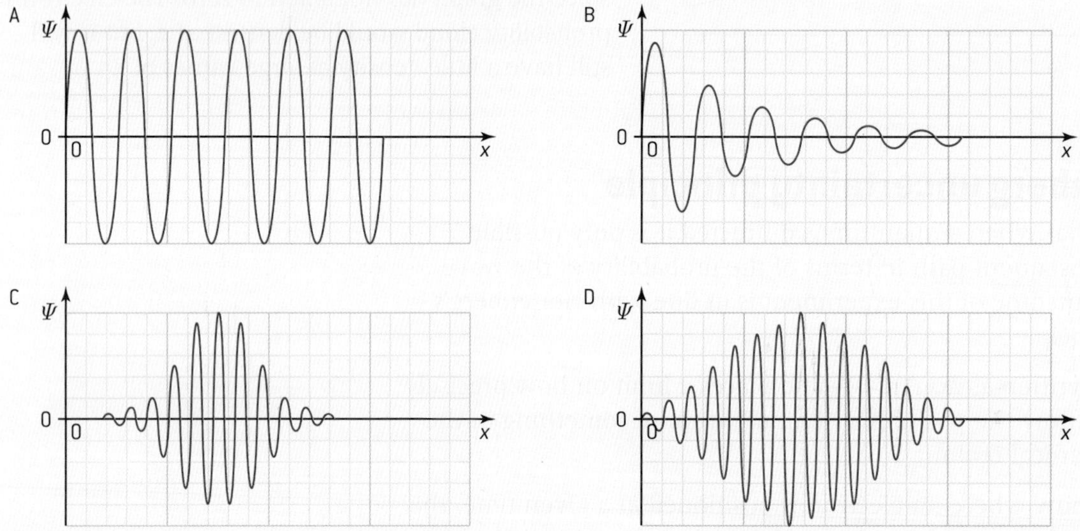

Solution

The square of the wave function is proportional to the probability of finding the electron. Each of B, C and D have positions where the square of the wave function is greatest (close to zero in B and centrally in C and D). This means that there is a high probability of finding the electron at one position and, therefore, there is low uncertainty in (the Heisenberg) position – but, as a consequence, there will be large uncertainty in momentum. For A there are many positions where the electron could be found with equal probability – at each of the maxima or minima there is large uncertainty in position but low uncertainty in momentum. Because C has the position where the electron is most well-defined it must have the largest uncertainty in its momentum.

Pair production and annihilation

Close to an atomic nucleus, where the electric field is very strong, a photon of the right energy can turn into a particle along with its antiparticle. This could be an electron and a positron or a proton and an antiproton. The outcome will always be a particle and an antiparticle in order to conserve charge, lepton number, baryon number and strangeness. The particle and antiparticle are said to be a "pair" and the effect is known as **pair production**. The antiparticle will have a mass equal to that of the particle meaning that the photon must have enough energy to create the masses of the two particles. The minimum energy needed to do this is given by:

$$E = 2mc^2$$

where m is the (rest) mass of the particle/antiparticle and c is the speed of electromagnetic waves in a vacuum. Figure 13 shows a Feynman diagram for the production of an electron–positron pair.

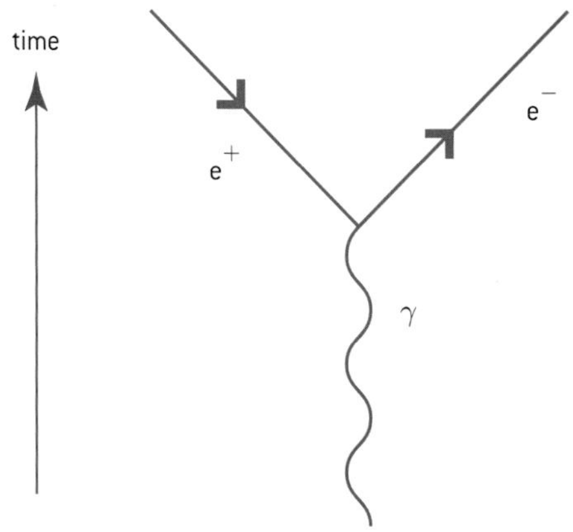

▲ Figure 13 Feynman diagram of electron–positron pair production.

The gamma ray photon (γ) must have an energy of at least 1.02 MeV (which is twice the rest energy of an electron). Any photon energy in excess of this amount is converted into the kinetic energy of the electron–positron pair and the original electron. Pair production can also occur in the vicinity of an orbital electron but in this case more energy will be needed as the orbital electron itself gains considerable momentum and kinetic energy. Figure 14 shows pair production taking place near an atomic electron. The photon is non-ionizing and leaves no track but the newly formed electron and positron can be seen spiralling in opposite directions in the applied magnetic field. The recoiling electron gains a great deal of kinetic energy and this is why it hardly bends in the magnetic field – although it can be seen to bend in the same direction as the other electron. Theory shows that the threshold energy needed for this type of pair production is $4mc^2$ (= 2.04 MeV).

The equation for this interaction is:

$$\gamma + e^- \rightarrow e^- + e^- + e^+$$

When a particle meets its antiparticle they **annihilate**, forming two photons. The total energy of the photons is equal to the total mass-energy of the annihilating particles. Sometimes a pair of particles annihilate but then one of the photons produces another pair of particles. The positron that is formed in the interaction, as shown in figure 14, quickly disappears as it is re-converted into photons in the process of annihilation with another electron in matter.

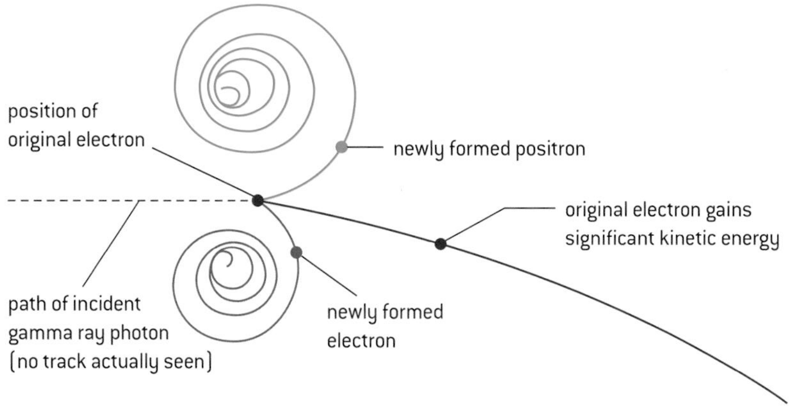

▲ Figure 14 Bubble chamber tracks of electron–positron pair production.

Pair production and the Heisenberg uncertainty principle

There are pairs of variables, other than x and p, to which the uncertainty principle also applies – of these energy and time are two important **conjugate variables**. This version of the uncertainty principle is written as:

$$\Delta E \, \Delta t \geq \frac{h}{4\pi}$$

Some interesting things can happen in the quantum world. It is found experimentally that the threshold energy required for the production of an electron–positron pair can be much less than the expected 1.02 MeV in the presence of a heavy nucleus. Imagine a 10 eV photon in the vicinity of a heavy nucleus: the low energy photon produces an electron–positron pair. A very short while later the electron and the positron collide to produce two 5 eV photons. This looks like a huge violation of the conservation of mass-energy but it is allowable under the uncertainty principle. During the lifetime of the composite electron–positron pair there is uncertainty regarding the total energy. If the uncertainty was equal to 1.02 MeV, what would be the limit on the uncertainty of the lifetime of the pair? The time Δt can be calculated by the energy–time formulation of the uncertainty principle:

$$\Delta t = \frac{h}{\Delta E 4\pi} = \frac{6.63 \times 10^{-34}}{1.02 \times 10^{6} \times 1.6 \times 10^{-19} \times 4\pi} = 3.2 \times 10^{-22}\ \text{s}$$

Here we have converted the energy in MeV into J by multiplying by the electronic charge.

This lifetime is so short that a measurement of the energy of the pair would have an uncertainty of at least 1.02 MeV and the experiment would not be able to detect the violation of the law of conservation of energy. If we cannot perform an experiment to detect a violation of the conservation law, then quantum mechanics says there is some probability of the process occurring. Another way of explaining this example is to say that nature will cheat if it can get away with it!

This example has been verified experimentally but the theory behind it is beyond that covered in the IB Diploma physics course.

Quantum Tunnelling

According to quantum mechanics, a particle's wave function has a finite probability of being everywhere in the universe at the same time. The probability may be infinitesimally small away from the effect that we would expect from classical physics but it is, nevertheless, finite. This means that, for example, an electron in the ground state of a hydrogen atom could escape the attraction of the nucleus with less than the expected 13.6 eV. In agreement with the uncertainty principle a particle can effectively "borrow" energy from its surroundings, pass through a barrier and then pay the energy back, providing it does not take too long.

Figure 15 shows a situation in which the wave function of the electron extends across a physical barrier giving the electron a finite possibility of it existing there (ignoring any modification of the probability curve by the presence of the barrier).

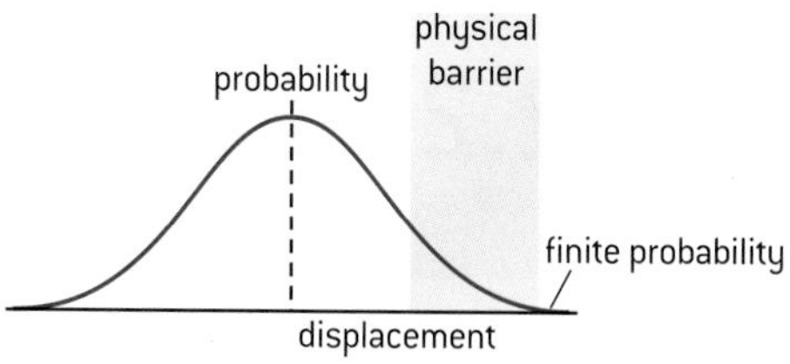

▲ Figure 15 Quantum tunnelling to pass through a physical barrier.

Quantum tunnelling is responsible for the relatively low temperature fusion that occurs in main sequence stars such as the Sun. The repulsive forces between a pair of protons that are to fuse means that they require kinetic energies of just over 1 MeV – this requires them to be at a temperature of around 10^{10} K. This is much higher than the temperature of the core of the Sun (which is around 3×10^7 K). As the result of the high pressures and quantum tunnelling there is a small chance that hydrogen atoms can fuse at a temperature below that expected. Because of the immense numbers of atoms in the Sun, even with a very low probability of quantum tunnelling fusion occurring, there is still a great deal going on. In the case of the Sun there is estimated to be in excess of four million tonnes of hydrogen fusing in this way in every second.

The scanning tunnelling microscope (STM), invented in 1981 by IBM Zurich, has revolutionized the study of material surfaces and has been used to manipulate individual strands of DNA – thus offering the potential to repair genetic damage. STMs use the currents generated when electrons tunnel into a surface in order to map out the structure of the surface. Quantum tunnelling is currently used in quantum tunnelling composites (QTCs). QTCs are the basis of touch screen technology. This technology has applications in smartphones, computer tablets, cameras and monitors. The entanglement of quantum particles is starting to bear fruit in photon teleportation, quantum cryptography and computing.

Worked example

The graph shows the variation with distance x of the wave function Ψ of an electron at a particular instant of time. The electron is confined within a region of length 2.0×10^{-10} m.

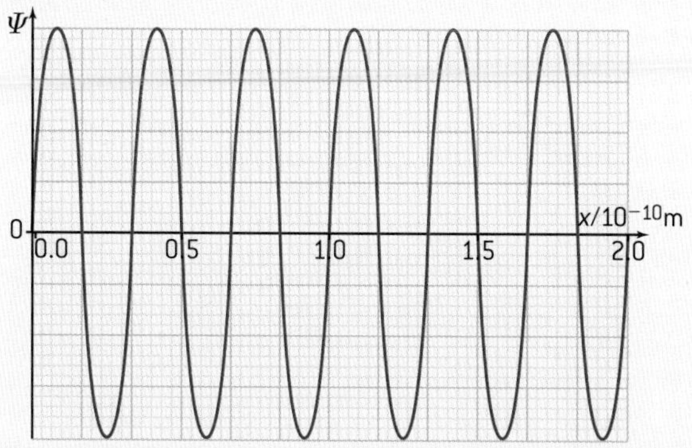

a) State what is meant by the *wave function* of an electron.

b) Using data from the graph estimate, for this electron:

(i) its momentum

(ii) the uncertainty in its momentum.

Solution

a) The wave function is a property of the electron everywhere in space. The square of the wave function is proportional to the probability of finding the electron somewhere.

b) (i) the (de Broglie) wavelength of the electron is

$$\frac{2 \times 10^{-10}}{6} = 3.3 \times 10^{-11}\ \text{m}$$

and so

$$p = \left(\frac{h}{\lambda} = \frac{6.63 \times 10^{-34}}{3.3 \times 10^{-11}} =\right) 2.0 \times 10^{-23}\ \text{Ns}$$

(ii) The electron is confined to 2.0×10^{-10} m, this means the uncertainty in its position $(\Delta x) = 2.0 \times 10^{-10}$ m.

Using the uncertainty principle this means that the uncertainty in momentum will be at least:

$$\Delta p = \left(\frac{h}{4\pi\Delta x} = \frac{6.63 \times 10^{-34}}{4\pi \times 2.0 \times 10^{-10}} =\right)$$

2.6×10^{-25} Ns

12.2 Nuclear physics

Understanding

- Rutherford scattering and nuclear radius
- Nuclear energy levels
- The neutrino
- The law of radioactive decay and the decay constant

Nature of science

Why we need particle accelerators

When Rutherford, Geiger and Marsden performed their scattering experiments they needed to use naturally occurring sources of alpha particles with relatively low energies of between 3 and 7 MeV. This is not enough energy for them to penetrate the electrostatic potential energy barrier around the nucleus. Although there are cosmic rays reaching the Earth with energies a million times greater than can be produced in a particle accelerator (10^{21} eV compared with 10^{15} eV on Earth), these are unpredictable and cannot be used. In CERN's large hadron collider (LHC) the energy available has been raised to 7 TeV. By using particle accelerators, detectors and sophisticated computers the number of known particles has increased from the proton, neutron and electron to the enormous number possible in the standard model. If our knowledge of the atom is to progress still further, particle accelerators are the most likely way forward. Plans are already in place to construct a very large hadron collider (VLHC) with a 240 km circumference chamber and beam energy of at least 50 TeV.

Applications and skills

- Describing a scattering experiment including location of minimum intensity for the diffracted particles based on their de Broglie wavelength
- Explaining deviations from Rutherford scattering in high energy experiments
- Describing experimental evidence for nuclear energy levels
- Solving problems involving the radioactive decay law for arbitrary time intervals
- Explaining the methods for measuring short and long half-lives

Equations

- relationship between radius of nucleus and nucleon number: $R = R_0 A^{\frac{1}{3}}$
- decay equation for number of nuclei at time t: $N = N_0 e^{-\lambda t}$
- decay equation for activity at time t: $A = \lambda N_0 e^{-\lambda t}$
- angle of electron diffraction first minimum: $\sin\theta \approx \frac{\lambda}{D}$

Introduction

In Sub-topic 7.3 we looked at the Rutherford model of the atom with a nucleus surrounded by orbiting electrons. We will now look more closely at the implications of the alpha scattering experiment and see how similar experiments have provided a much better understanding of the nucleus. We will then consider a more mathematical approach to the radioactive decay of nuclei.

Rutherford scattering and the nuclear radius

In Sub-topic 7.3 we saw that the main results of the alpha scattering experiment were that:

- most of the alpha particles passed through the gold leaf undeflected
- some alpha particles were deflected through very wide angles
- some alpha particles rebounded in the opposite direction.

The interpretations of these results were:

- most of the atom is empty space
- the atom contains small dense regions of electric charge
- these small dense regions are positively charged.

We will now look at the analysis of this ground-breaking experiment in some detail.

The method of closest approach

Alpha particles that backscatter are those colliding head-on with a gold nucleus. As only about 1 in 8000 alpha particles are scattered through large angles it must mean that the probability of a head-on collision is very small and that the nucleus occupies a very small portion of the total atomic volume ... but how much?

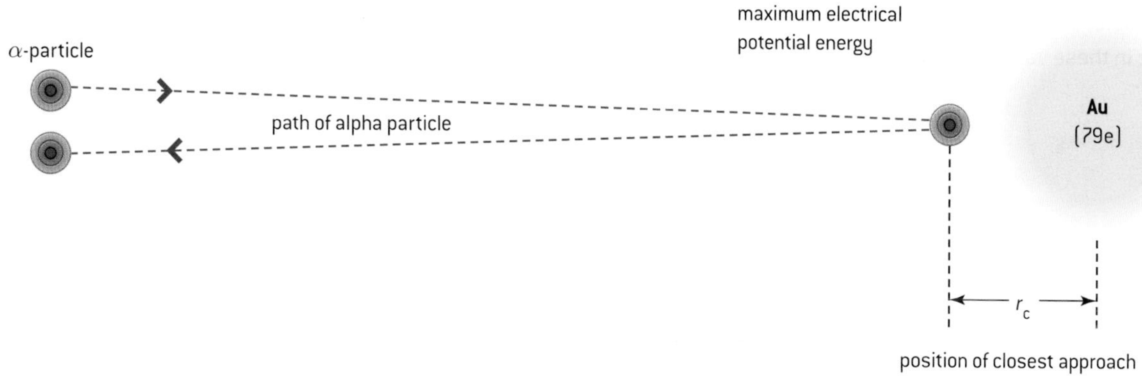

▲ Figure 1 Method of closest approach.

Figure 1 shows an alpha particle that is incident head-on with a gold nucleus. As the alpha particle becomes closer to the nucleus its kinetic energy falls and its electrical potential energy increases. When the alpha particle is at its closest to the nucleus, its kinetic energy has fallen to zero and it has momentarily stopped moving. Taking an alpha particle of kinetic energy E_α at the position of closest approach (= distance r_c from the nucleus) we can equate its kinetic energy to the electrical potential energy giving:

$$E_\alpha = \frac{kZe \times 2e}{r_c}$$

where k is the Coulomb constant, Z is the proton number of gold (making Ze the charge on the gold nucleus) and $2e$ represents the alpha particle charge.

So for an alpha particle

$$r_c = \frac{k2Ze^2}{E_\alpha}$$

In Rutherford's experiment $E_\alpha = 7.68$ MeV and $Z = 79$, this gives a value for r_c of

$$r_c = \frac{8.99 \times 10^9 \times 2 \times 79 \times (1.60 \times 10^{-19})^2}{(7.68 \times 10^6 \times 1.60 \times 10^{-19}} = 2.96 \times 10^{-14}\ \text{m}$$

This derivation is an approximation because we have treated the gold nucleus as being a point mass. If the alpha particle were to have penetrated the nucleus, then the Coulomb force would not have been applicable as we now know that the strong nuclear force is dominant within the nucleus. Rutherford went on to obtain an expression for the number of alpha particles scattered through a variety of angles and he also arrived at values that were in agreement with the upper limit of the gold nucleus being approximately 3×10^{-14} m. However, further experiments using more energetic alpha particles have approached the nucleus closer than this value. At this separation the strong nuclear force becomes dominant and the alpha particle must have penetrated the nucleus. As the volume V of a nucleus must be proportional to the number of its nucleons, we would expect $V \propto A$ (where A is the nucleon number) and so the nuclear radius R would be expected to be $\propto A^{\frac{1}{3}}$. This gives

$$R = R_0 A^{\frac{1}{3}}$$

Here R_0 is called the Fermi radius and has a value of 1.2×10^{-15} m and is measured experimentally.

Note

You should not confuse the Fermi radius (which is related to a nucleus) with the Bohr radius (which is related to an atom): there is a factor of 10^{-5} difference in these values.

Nuclear density

If we imagine the nucleus to be spherical, it follows that its volume can be calculated using the following equation:

$$V = \frac{4}{3}\pi R^3 = \frac{4}{3}\pi A R_0^{\ 3}$$

The density of nuclear material will be given by

$$\rho = \frac{M}{V} = \frac{Au}{\frac{4}{3}\pi A R_0^{\ 3}} = \frac{3u}{4\pi R_0^{\ 3}}$$

where u is the uniform atomic mass and Au is the total mass of a nucleus of nucleon number A.

As each of the quantities in the equation is a constant, it implies that the density of any nucleus is independent of the number of nucleons in the nucleus.

Substituting values into this equation gives:

$$\rho = \frac{3 \times 1.66 \times 10^{-27}}{4\pi \times (1.2 \times 10^{-15})^3} = 2.3 \times 10^{17}\ \text{kg m}^{-3}$$

This is an incredibly dense material – a volume of 1 cm^3 would have a mass of 200 000 tonne! In nature the only object that has a nuclear density of this value is a **neutron star** – a type of stellar remnant resulting from the gravitational collapse of a massive star, which is composed almost entirely of neutrons.

Deviations from Rutherford scattering

The scattering experiments performed by Rutherford, Geiger, and Marsden were limited by the energies of the alpha parties emitted by the radioactive sources available to them. When their experiments are repeated using more energetic, accelerated alpha particles it is found that, at these higher energies, the Rutherford scattering relationship does not agree with experimental results. At higher energies the alpha particles were able to approach the target nucleus so closely that the strong nuclear attractive force overcomes the electrostatic repulsion. Figure 2 shows how the strong nuclear force and the repulsive coulomb force (between protons) vary with distance.

The method of closest approach gives an approximation of the size of a nucleus. More reliable values for the size of a nucleus can be found using electron diffraction.

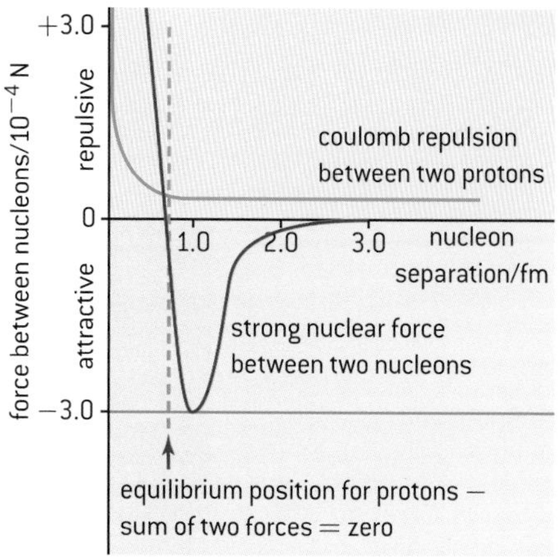

▲ Figure 2 Variation of the strong nuclear force and coulomb force with distance.

Electron diffraction

As electrons are leptons (and not hadrons) they are not affected by the strong nuclear force but are affected by the charge distribution of the nucleus. High-energy electrons have a short de Broglie wavelength of the order of 10^{-15} m. As this is also the order of magnitude of the size of a nucleus, it means that diffraction analogous to that observed with light incident on a narrow slit or small object can be observed.

For light incident on a small circular object of diameter D, the angle θ that the first diffraction minimum makes with the straight-through position ($\theta = 0°$) is given by

$$\sin\theta \approx \frac{\lambda}{D}$$ where λ is the wavelength of the light

The elastic scattering of high energy electrons by a nucleus produces a similar effect. With the arrangement shown in figure 3(a), the intensity of the diffracted beam is seen to be a maximum in the straight-through position, falling to a minimum before slightly increasing again. The minimum differs from light because it never reaches zero for scattered electrons. Again, the relationship can be approximated by

$$\sin\theta \approx \frac{\lambda}{D}$$

Here D is the nuclear diameter and λ is the de Broglie wavelength of the electrons.

Note

With angles greater than 10° the small angle approximation that $\sin\theta \approx \theta$ cannot be applied to the electron scattering.

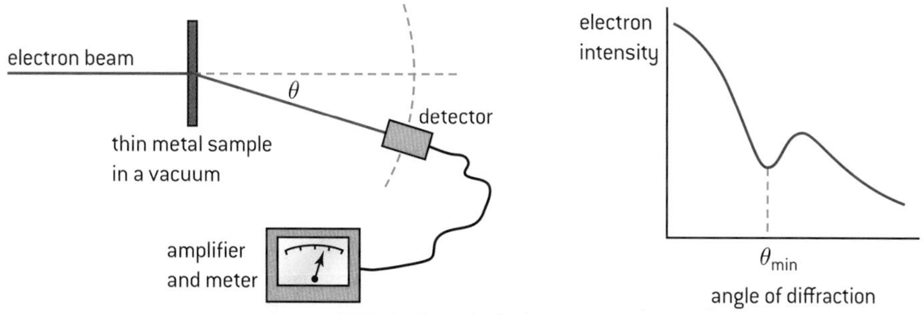

(a) outline of experiment

(b) typical results

▲ Figure 3 Experimental arrangement and results.

To achieve an appropriate de Broglie wavelength the electrons used in this scattering experiment need energies in the region of 400 MeV. At these energies, electrons are travelling close enough to the speed of light to mean that relativistic corrections should be applied to both the electron momentum and the associated wavelengths. With the electron rest energy of approximately 0.5 MeV, the total energy can be taken as 400 MeV without any serious error consideration.

The wavelength of an electron is given by

$$\lambda = \frac{hc}{E}$$

For an electron with energy 400 MeV

$$\lambda = \frac{6.63 \times 10^{-34} \times 3.00 \times 10^{8}}{400 \times 10^{6} \times 1.60 \times 10^{-19}} = 3.1 \times 10^{-15}\ \text{m}$$

This is the same order of magnitude as the size of a nucleus.

Worked example

The nuclear radius of calcium-40 has an accepted value of 4.54 fm. It is being investigated using a beam of electrons of energies 420 MeV.

a) How closely does the value of the radius of calcium-40, obtained using the relationship $=R_0 A^{\frac{1}{3}}$, agree with the accepted value?

b) Calculate the de Broglie wavelength of an electron having energy 420 MeV.

c) Determine the angle that the first minimum in the diffraction pattern makes with the straight-through direction.

d) Explain why 50 MeV electrons would be unlikely to provide reliable results in this experiment.

Solution

a) A is the nucleon number that, for calcium-40, is 40. Thus $A^{\frac{1}{3}} = \sqrt[3]{40} = 3.42$ which gives a value for R of $3.42 \times R_0 = 3.42 \times 1.2$ fm $=$ 4.10 fm.

There is an error of 0.44 fm between the accepted value and that obtained using the relationship given (4.54 – 4.10). As a percentage this is $= \frac{0.44}{4.54} = 9.7\% \approx 10\%$.

b) $\lambda = \frac{hc}{E} = \frac{6.63 \times 10^{-34} \times 3.00 \times 10^{8}}{420 \times 10^{6} \times 1.60 \times 10^{-19}}$

$= 2.9 \times 10^{-15}$ m

c) Using $\sin\theta \approx \frac{\lambda}{D} \Rightarrow \theta \approx \sin^{-1}\left(\frac{\lambda}{D}\right)$

Thus, as the radius is 4.54 fm, the diameter is 9.08 fm giving

$$\theta \approx \sin^{-1}\frac{2.9 \times 10^{-15}}{9.08 \times 10^{-15}} = 18.6°$$

d) As the de Broglie wavelength is given by $\lambda = \frac{hc}{E}$, 50 MeV electrons would have a de Broglie wavelength of $\frac{420}{50} = 8.4$ times greater than that of 420 MeV electrons. This means that

$$\theta \approx \sin^{-1}\left(\frac{2.4 \times 10^{-14}}{9.08 \times 10^{-15}}\right)$$

This is not calculable as $\sin\theta$ cannot be greater than one.

The de Broglie wavelength is too long to be diffracted by a calcium nucleus.

Using electrons of higher energies

When electrons of much greater energies than 420 MeV are used in scattering experiments, something very different happens. The collisions are no longer elastic (the bombarding electrons lose kinetic energy). This energy is "converted" into mass as several mesons are emitted from the nucleus. At still higher energies, the electrons penetrate deeper into the nucleus and scatter off the quarks within protons and neutrons – this is known as deep inelastic scattering and provides direct evidence for the quark model of nucleons.

Energy levels in the nucleus

Much of our evidence for the nucleus having energy levels comes from the radioactive decay of nuclides. The emission of gamma radiation is analogous to the emission of photons by electrons undergoing energy level transitions. The emission of alpha or beta particles by radioactive parent nuclei often leaves the daughter nucleus in an excited state. The daughter nucleus then emits one or more gamma ray photons as it reaches the ground state. Figure 4 shows some of the decay routes of americium-241. Each nucleus emits an alpha particle having one of a number of possible energies (three of which are shown) to become a nucleus of neptunium-237. Depending on the energy of the emitted alpha particle, the neptunium nucleus can be in the ground state or an excited state. From this it will decay into the ground state by emitting a single gamma photon or, when it decays in two steps, two photons.

From the differences in the energies of the alpha particles we can see that the energy level E_1 will be (5.545 – 5.486) = 0.059 MeV above the ground state (E_0) and energy level E_2 will be (5.545 – 5.443) = 0.102 MeV above the ground state. From the differences in the energy levels we can calculate the energies of each of the three gamma ray photons that could be emitted – they will be 0.102 MeV, 0.059 MeV, and 0.043 MeV.

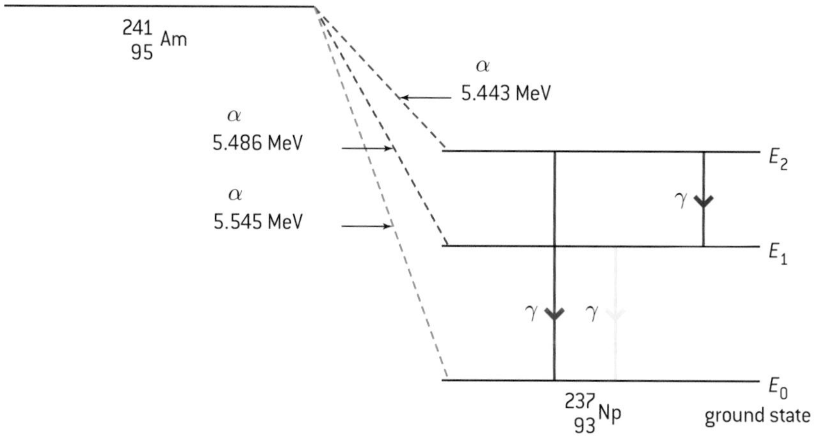

▲ Figure 4 Decay of americium-241.

As we see from the decay of americium, the energies of the alpha particles are also quantized and provide evidence for the nucleus having energy levels. The mechanism by which the alpha particle leaves the nucleus is more complex than that producing the emission of the gamma ray photons. Alpha particles form as clusters of two protons and two neutrons inside the nucleus well before they are emitted as alpha particles. The nucleons are in random motion within the nucleus but their kinetic energies are much smaller than those needed to escape from the nucleus. This is because the strong nuclear force provides a potential energy barrier which the alpha particle needs to overcome before it can escape from the nucleus (when the electrostatic repulsion will ensure that it accelerates away from the nucleus). From a classical mechanics point of view the alpha particle simply should not leave the nucleus (see figure 5).

From the quantum mechanical standpoint, the wave function for the alpha particle is not localized to the nucleus and allows an overlap with the potential energy barrier provided by the strong nuclear force. This means that there is a finite but very small probability of observing the alpha particle outside the nucleus. Although the probability is extremely small, some alpha particles will tunnel out of the nucleus. Experimentally it is found that, with a higher potential barrier and greater thickness to cross, a nucleus will have a longer lifetime. This explains the very long half-lives of uranium and polonium. The Russian-born American physicist, George Gamow, was the first person to describe alpha decay in terms of quantum tunnelling (discussed in Sub-topic 12.1). When the wave function is at its maximum the probability of tunnelling is greatest, meaning that alpha particles with specific energies are most likely to be emitted.

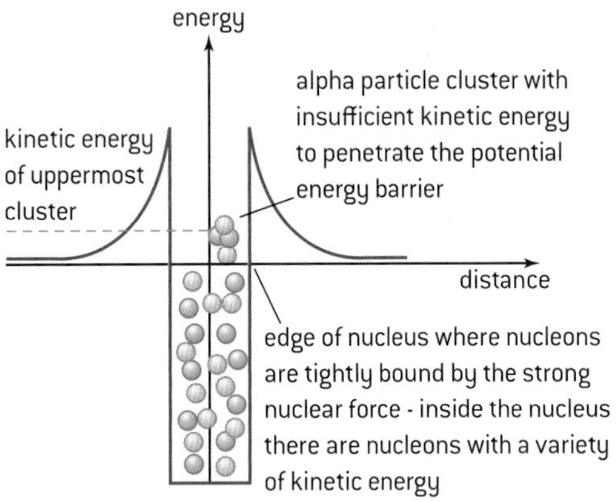

▲ Figure 5 Classical mechanics view of alpha decay.

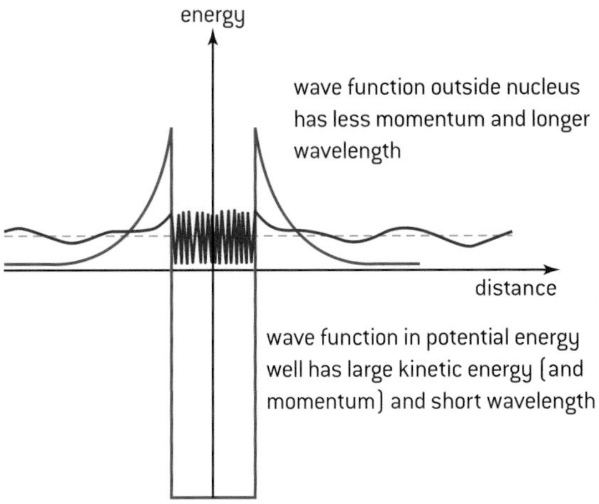

▲ Figure 6 Quantum tunnelling view of alpha decay.

Negative beta decay

We looked at beta decay in Sub-topic 7.1 and returned to it in Sub-topic 7.3 when it was explained that an anti-neutrino accompanies the electron emitted in negative beta decay. The first theory of beta decay was proposed in 1934 by Fermi; at this time the existence of quarks was unknown and neutrinos were hypothetical. Experiments show that beta particles emitted by a source have a continuous energy spectrum and are not of discrete single energy as are alpha particles and gamma photons. Figure 7 shows a typical negative beta-energy spectrum.

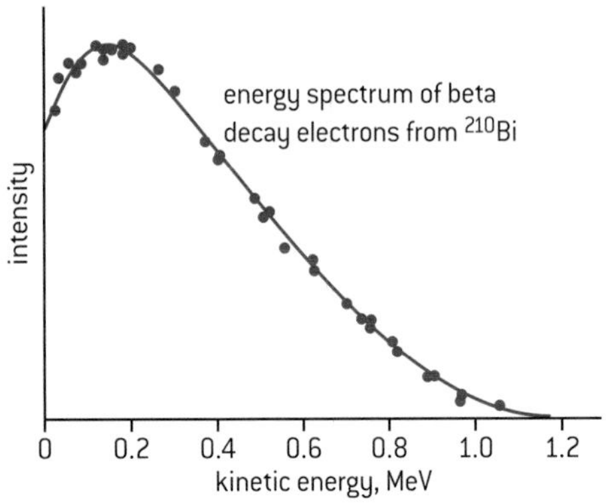

▲ Figure 7 Negative beta-energy spectrum.

Possible explanations for this spectrum were that mass–energy and momentum were not conserved in beta decay. These were very unlikely solutions since both of these principles are considered to be fundamental to physics. Pauli suggested that, if a third particle was to be emitted in the decay, not only would this solve the mass–energy and momentum problems but it would also allow spin angular momentum to be conserved in the emission. The emission (of what has proved to be) an electron antineutrino meant that for a particular nucleus the energy would be shared between the electron (the beta particle) and the antineutrino.

TOK

Awe and wonder or turn-off?

Rumour has it that Murray Gell-Mann, when searching for a name to call what is now known as the "quark", had a epiphany upon seeing the word "quark" in the novel *Finnegan's Wake* by the Irish novelist James Joyce. The poet, John Updike, on reading about neutrinos chose to write the following:

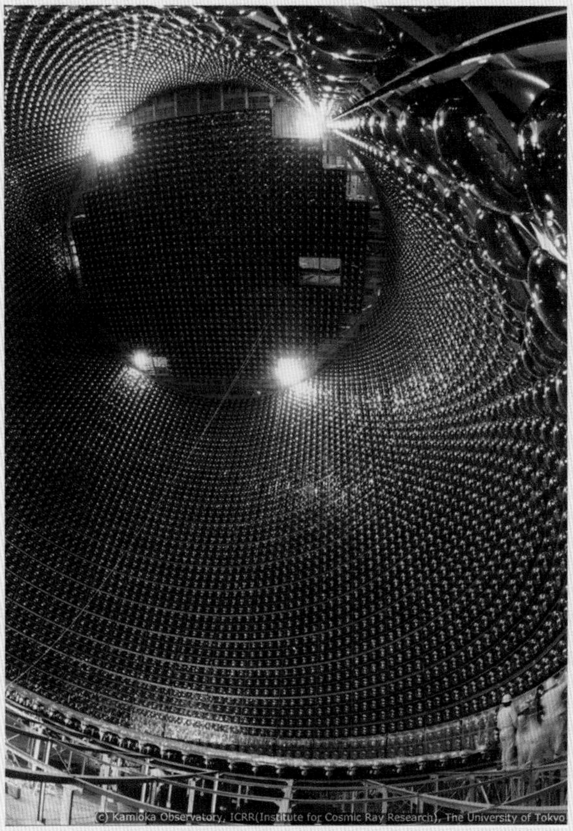

▲ Figure 8 Super-Kamioka Neutrino Detection Experiment Mount Kamioka near Hida, Japan.

Cosmic Gall

Neutrinos, they are very small.
They have no charge and have no mass
And do not interact at all.
The earth is just a silly ball
To them, through which they simply pass,
Like dustmaids down a draughty hall
Or photons through a sheet of glass.
They snub the most exquisite gas,
Ignore the most substantial wall,
Cold shoulder steel and sounding brass,
Insult the stallion in his stall,
And scorning barriers of class,
Infiltrate you and me! Like tall
And painless guillotines, they fall
Down through our heads into the grass.
At night they enter at Nepal
And pierce the lover and his lass
From underneath the bed – you call
It wonderful; I call it crass.

Telephone Poles and Other Poems, John Updike, (Knopf, 1960)

Updike clearly was not a believer in the neutrino. Does he have a strong case for his disbelief? Is the concept of something as challenging as the neutrino inspirational or is it too fanciful to be credible?

The law of radioactive decay

As we saw in Sub-topic 7.1, radioactive decay is a **random** and **unpredictable** process. There is no way of telling which nucleus in a sample of material will decay next. What we do know is that the more radioactive nuclei present, the greater the probability of some decaying. With a sample of many millions of nuclei the rules of statistics can be applied with a virtual certainty.

The **probability that an individual nucleus will decay in a given time interval** (of one second, one minute, one hour, etc.) is known as the **decay constant, λ**. The units for λ are time^{-1} (s^{-1}, minute^{-1}, h^{-1}, etc.).

The activity of a sample ***A* is the number of nuclei decaying in a second** – it is measured in becquerel (Bq). In a sample of N undecayed nuclei, the activity will be equal to the number of nuclei present multiplied by the probability that one will decay in a second. In equation form:

$$A = \lambda N$$

N will decrease with time; this relationship is often written with a minus sign (although this is not the case on the IB Diploma Programme physics syllabus)

$$A = -\lambda N$$

Using calculus notation, this is can be written as:

$$\frac{dN}{dt} = -\lambda N$$

This is the relationship that gives an exponential decay and is analogous to capacitor discharge as seen in Sub-topic 11.3 with the λ constant being analogous to τ^{-1}.

In general, when the rate of change of a quantity is proportional to the amount of the quantity left to change, an exponential relationship will always be obtained. We will now show this for radioactive decay – the process will involve integral calculus and so you will not be tested on this in examinations.

Rearranging the equation

$$\frac{dN}{dt} = -\lambda N$$

gives

$$\frac{dN}{N} = -\lambda dt$$

this can be integrated from time $t = 0$ to time $t = t$ when the number of undecayed nuclei will fall from $N = N_0$ to $N = N$

$$\int_{N_0}^{N} \frac{dN}{N} = -\int_{0}^{t} \lambda dt$$

Integrating this gives

$$[\ln N]_{N_0}^{N} = -\lambda t$$

So

$$\ln N - \ln N_0 = -\lambda t$$

or

$$\ln\left(\frac{N}{N_0}\right) = -\lambda t$$

Raising both sides to the power of e gives

$$\frac{N}{N_0} = e^{-\lambda t} \text{ or } N = N_0 e^{-\lambda t}$$

As A is proportional to N, this equation can be written in terms of the activity to give

$$A = A_0 e^{-\lambda t}$$

Here A_0 is the activity of sample of radioactive material at time $t = 0$ and this can also be written as

$$A = \lambda N_0 e^{-\lambda t}$$

Note

Again, we can make use iteration to avoid using calculus when solving the decay equation. The following flow chart shows an algorithm for doing this:

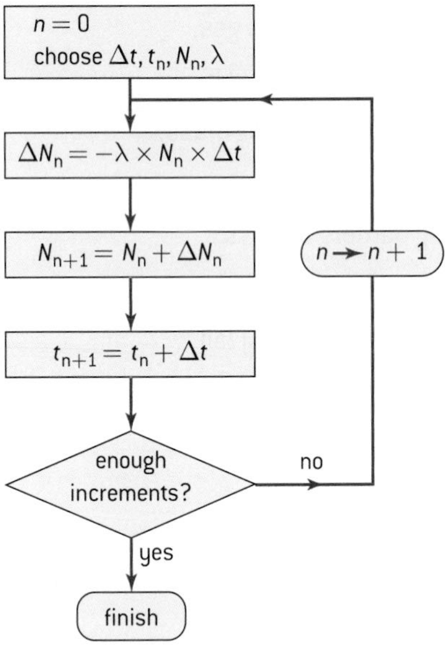

Worked example

Radium-226 emits alpha particles. The decay constant is $1.35 \times 10^{-11}\ s^{-1}$

What mass of radium 226 is needed to give an activity of 2200 Bq?

Solution

$$A = \lambda N \text{ so } N = \frac{A}{\lambda} = \frac{2200}{1.35 \times 10^{-11}} = 1.63 \times 10^{14}$$

1 mol of a substance contains 6.02×10^{23} particles so this quantity amounts to

$$\frac{1.63 \times 10^{14}}{6.02 \times 10^{23}} = 2.7 \times 10^{-10} \text{ mol}$$

As a mol of radium 226 has an approximate mass of 226 g the sample has a mass of

$$2.7 \times 10^{-10} \times 226 \times 10^{-3} \text{ kg} = 6.1 \times 10^{-11} \text{ kg}$$

Decay constant and half-life

We dealt with integral numbers of half-lives in Sub-topic 7.1. Now we consider cases where the number of half-lives is not a whole number.

The half-life is the time that it takes for the number of radioactive nuclei to halve. So, in this time, N falls from N_0 to $\frac{N_0}{2}$.

Substituting these values into $N = N_0 e^{-\lambda t}$ gives $\frac{N_0}{2} = N_0 e^{-\lambda t_{\frac{1}{2}}}$ where $t_{1/2}$ is the half-life.

When we rearrange this and take logs to base e we get

$$\ln\left(\frac{\frac{N_0}{2}}{N_0}\right) = \ln\left(\frac{1}{2}\right) = -\lambda t_{1/2}$$

Calculating log to base e of $\frac{1}{2}$ gives

$$t_{1/2} = \frac{-0.693}{-\lambda} = \frac{0.693}{\lambda}$$

Worked example

A laboratory prepares a 10 μg sample of caesium-134. The half-life of caesium-134 is approximately 2.1 years.

a) Determine, in s^{-1}, the decay constant for this isotope of caesium.

b) Calculate the initial activity of the sample.

c) Calculate the activity of the sample after 10.0 years.

Solution

a) $$\lambda = \frac{0.693}{t_{1/2}} = \frac{0.693}{2.1 \times 365 \times 24 \times 3600}$$

$$= 1.05 \times 10^{-8}\ s^{-1}\ (= 0.33\ y^{-1})$$

b) 1 mol of ^{134}Cs has a mass of approximately 134 g so 10 μg comprises of $\frac{10 \times 10^{-6}}{134}$ mol or $= 7.5 \times 10^{-8}$ mol.

This means that there are $7.5 \times 10^{-8} \times 6.02 \times 10^{23} = 4.49 \times 10^{16}$ atoms of Cs

$$A = \lambda N = 1.05 \times 10^{-8} \times 4.49 \times 10^{16} = 4.7 \times 10^{8} \text{ Bq}$$

c) $A = A_0 e^{-\lambda t} = 4.7 \times 10^{8} e^{-0.33 \times 10}$ (working in years)

$A = 1.7 \times 10^{7}$ Bq

Note

A second way to tackle this is to calculate the number of half-lives that have elapsed $= \frac{10}{2.1} = 4.76$

The amount that will be left is then $0.5^{4.76} = 0.037$, the activity will then be $0.037 \times 4.7 \times 10^{8} = 1.7 \times 10^{7}$ Bq

Investigate!

Measuring short half-lives

The G–M tube was discussed in Sub-topic 7.1. This is a very important instrument used to measure half-lives of radioactive materials. For those with fairly short half-lives (from a few seconds to a few hours) it is a straightforward process to measure the count rate using a G–M tube and counter or a data logger.

- Measure the background count rate with your apparatus with no radioactive sources in the vicinity.
- Position the source close to the window of the G-M tube so that almost none of the radiation is absorbed by the air.
- Take readings of the count rate at appropriate time intervals until the count rate is the same as the background count.
- Subtract the background count rate from your readings to give the corrected count rate (R); assuming that the radiation is emitted equally in all directions, the count rate shown on the counter will be proportional to the activity of the source.
- Plot a graph of the natural log of the corrected count rate, $\ln(R)$, against time.
- As $R \propto A$ we can write $R = R_0 e^{-\lambda t}$

  Taking natural logs of this gives $\ln R = \ln R_0 - \lambda t$

  So a graph of $\ln R$ (on the y-axis) against t will be of gradient $-\lambda$ and intercept on the $\ln R$ axis of $\ln R_0$ as shown in figure 9.
- Note the notation for the unit of a log quantity – logs have no units but R does so the unit is bracketed to R. The corrected count rate is in counts per second, as "count" has no unit this is equivalent to s^{-1}.
- What is the advantage of plotting this log graph when compared with plotting count rate against time?
- How do you use the gradient to calculate the half-life of the radioactive nuclide?
- Compare your value with the accepted value to give you an idea of the uncertainty.
- What is the difference between the activity of source and the count rate?

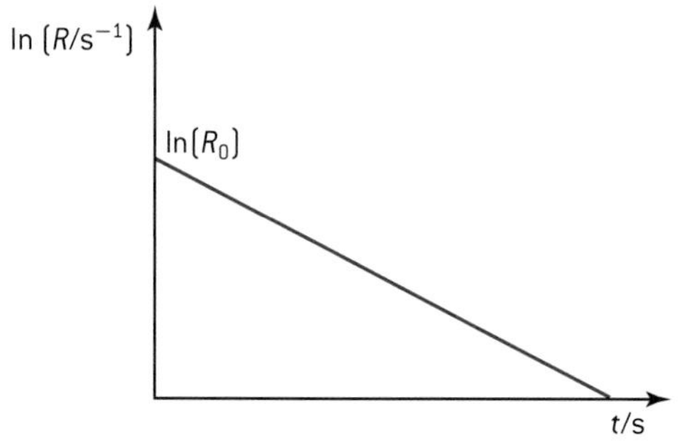

▲ Figure 9 Graph of natural log of the corrected count rate against time.

Measuring long half-lives

Some nuclides have very long half-lives, for example uranium-238 has one of just under 4.5 billion years. When a radioactive nuclide has a half-life that is long compared to the time interval over which radioactive decay observations are possible, there is no apparent rate of decay and it is not possible to measure the half-life in the manner suggested using a G–M tube.

In these cases a pure sample of the nuclide in a known chemical form needs to be separated, its mass measured and then a count rate taken. From this reading the activity can be calculated by multiplying the count rate by the ratio:

$$\frac{\text{area of sphere of radius equal to the position of the G–M tube window}}{\text{area of G–M tube window}}$$

The decay constant is then determined from the mass of the specimen using the method shown in the first worked example in this section.

Investigate!

Decay in height of water column

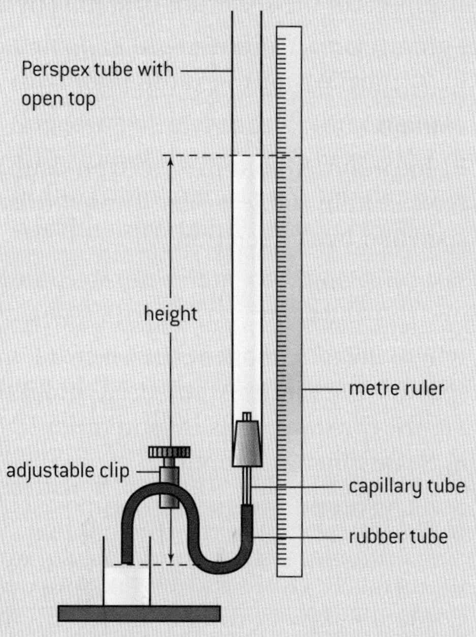

▲ Figure 10 Decay in height of a water column.

This is an analogue of radioactive decay using a long vertical tube of water, which is connected to a capillary tube.

- Make sure the clip is closed, then fill the long tube almost to the brim with water.
- Undo the clip completely to allow the water to flow through the capillary tube – you will need to repeat the experiment with the same setting so, if the clip is not fully undone, you will need to count how many turns you make from the clip being fully tight.
- As the water level in the long tube reaches a previously chosen point near the top of the tube, start measuring the height at regular time intervals – you will need to perform a trial experiment to allow you to judge what a sensible time interval should be – one that will give you a minimum of six points on your graph.
- Record your results in a table.
- Repeat the readings twice more – always starting the clock when the water level is at the same mark on the scale.
- Find the average of the readings for each time interval.
- Plot graphs of height against time and the natural log of height against time.
- From each of these graphs calculate the half-life and decay constant of the water.
- Determine which of the two values is the most reliable.
- Consider how this experiment models radioactive decay and in which ways it is different.

Questions

1 When light is incident on a metal surface, electrons may be ejected. The following graph shows the variation with frequency f of the maximum kinetic energy E_{max} of the ejected electrons.

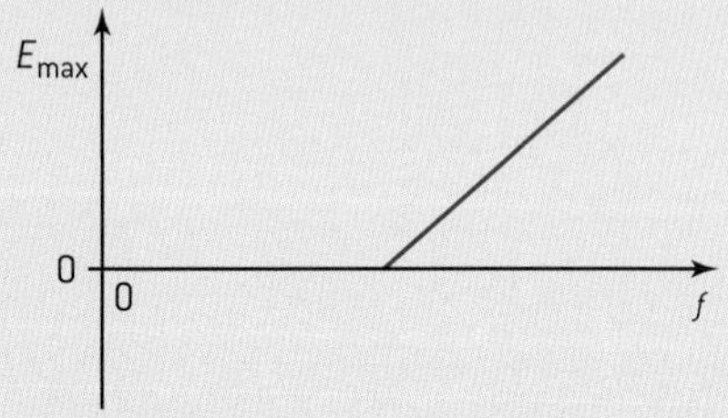

a) Sketch the graph obtained for the variation with frequency of the maximum kinetic energy of the emitted electrons when a different metal with a lower threshold frequency is used.

b) Explain why you have drawn the graph in this way.

(4 marks)

2 *(IB)*

In order to demonstrate the photoelectric effect, the apparatus shown below is used.

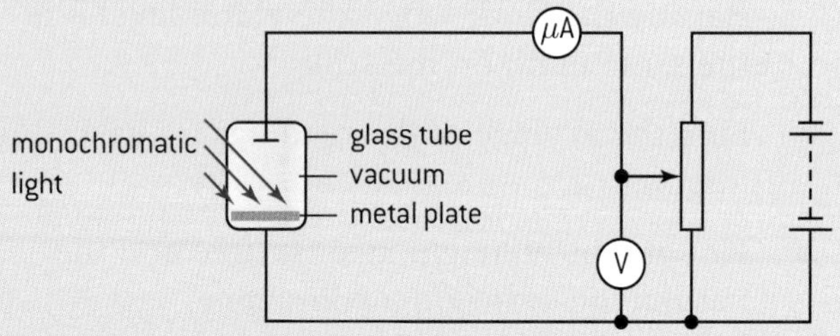

Monochromatic light is incident on the metal plate. The potentiometer is adjusted to give the minimum voltage at which there is zero reading on the microammeter.

a) State and explain what change, if any, will occur in the reading of the microammeter when:

(i) the intensity of the incident light is increased but the frequency remains unchanged

(ii) the frequency of the light is increased at constant intensity.

b) For light of wavelength 540 nm, the minimum reading on the voltmeter for zero current is 1.9 V.

(i) State the connection between photon energy and the energy of the emitted electron.

(ii) Using your answer to (i) calculate the work function of the surface of the metal plate.

(8 marks)

3 *(IB)*

In 1913 Niels Bohr developed a model of the hydrogen atom which successfully explained many aspects of the spectrum of atomic hydrogen.

a) State **one** aspect of the spectrum of atomic hydrogen that Bohr's model did not explain.

Bohr proposed that the electron could only have certain stable orbits. These orbits are specified by the relation

$$mvr = \frac{nh}{2\pi} \quad \text{with } n = 1, 2, 3\ldots$$

where m is the mass of the electron, v its speed, r the radius of the orbit and h the Planck constant. This is sometimes known as Bohr's first assumption.

b) State a second assumption proposed by Bohr.

By using Newton's second law and Coulomb's law in combination with the first assumption, it can be shown that

$$r = \frac{n^2h^2}{4\pi^2mke^2} \quad \text{where } k = \frac{1}{4\pi\varepsilon_0}.$$

It can also be shown that the total energy E_n of the electron in a stable orbit is given by

$$E_n = -\frac{ke^2}{2r}.$$

c) Using these two expressions, deduce that the total energy E_n may be given as

$$E_n = -\frac{K}{n^2} \quad \text{where } K \text{ is a constant.}$$

d) State and explain what physical quantity is represented by the constant K.

e) Outline how the Schrödinger model of the hydrogen atom leads to the concept of energy levels.

(10 marks)

4 *(IB)*

A beam of electrons is incident normally to the plane of a narrow slit as shown below.

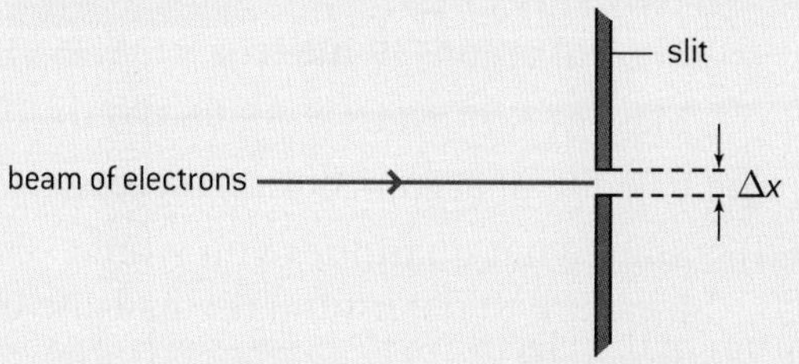

The slit has width Δx equal to 0.01 mm.

As an electron passes through the slit, there is an uncertainty Δx in its position.

a) Calculate the minimum uncertainty Δp in the momentum of the electron.

b) Suggest, by reference to the original direction of the electron beam, the direction of the component of the momentum that has the uncertainty Δp.

(3 marks)

5 *(IB)*

An α-particle approaches a nucleus of palladium. The initial kinetic energy of the α-particle is 3.8 MeV. The particle is brought to rest at point P, a distance d from the centre of the palladium nucleus. It then reverses its incident path.

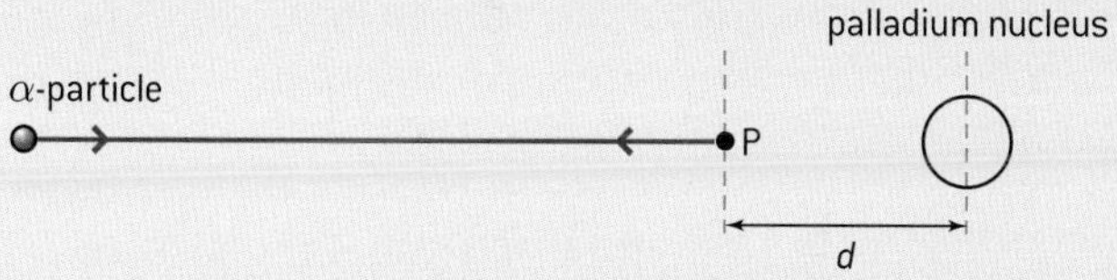

a) Calculate the value, in joules, of the electric potential energy of the α-particle at point P. Explain your working.

b) The proton number of palladium is 46. Calculate the distance d.

c) Gold has a proton number of 79.

Explain whether the distance of closest approach of this α-particle to a gold nucleus would be greater or smaller than your answer in (b).

d) The radius (in metre) R of a nucleus with nucleon number A is given by

$$R = 1.2 \times 10^{-15} A^{\frac{1}{3}}.$$

(i) State in terms of the unified atomic mass unit u, the approximate mass of a nucleus of mass number A.

(ii) The volume of a sphere of radius R is given by $v = \frac{4\pi R^3}{3}$. Deduce that the density of all nuclei is approximately 2×10^{17} kg m^{-3}.

(8 marks)

6 *(IB)*

A nucleus of the nuclide xenon, Xe-131, is produced when a nucleus of the radioactive nuclide iodine, I-131 decays.

a) Explain the term *nuclide*.

b) Complete the nuclear reaction equation for this decay.

$$^{131}_{\square}\text{I} \rightarrow ^{131}_{54}\text{Xe} + \beta^- + \square$$

c) The activity A of a freshly prepared sample of I-131 is 6.4×10^5 Bq and its half-life is 8.0 days.

(i) Sketch a graph to show the variation of the activity of this sample over a time of 25 days.

(ii) Determine the decay constant of the isotope I-131 (in day^{-1}).

The sample is to be used to treat a growth in the thyroid of a patient. The isotope should not be used until its activity is equal to 0.5×10^5 Bq.

(iii) Calculate the time it takes for the activity of a freshly prepared sample to be reduced to an activity of 0.5×10^5 Bq

(11 marks)

7 *(IB)*

a) A stable isotope of argon has a nucleon number of 36 and a radioactive isotope of argon has a nucleon number of 39.

(i) State what is meant by a *nucleon*.

(ii) Outline the quark structure of nucleons.

(iii) Suggest, in terms of the number of nucleons and the forces between them, why argon-36 is stable and argon-39 is radioactive.

b) Argon-39 undergoes β^- decay to an isotope of potassium (K). The nuclear reaction equation for this decay is

$${}^{39}_{18}\text{Ar} \rightarrow \text{K} + \beta^- + x$$

(i) State the proton number and the nucleon number of the potassium nucleus and identify the particle x.

(ii) The existence of the particle x was postulated some years before it was actually detected. Explain the reason, based on the nature of β^- energy spectra, for postulating its existence.

(iii) Use the following data to determine the maximum energy, in J, of the β^- particle in the decay of a sample of argon-39.

Mass of argon-39 nucleus = 38.96431 u

Mass of K nucleus = 38.96370 u

c) The half-life of argon-39 is 270 years.

(i) State what quantities you would measure to determine the half-life of argon-39.

(ii) Explain how you would calculate the half-life using the quantities you have stated in (i).

(21 marks)

8 *(IB)*

a) Outline a method for the measurement of the half-life of a radioactive isotope having a half-life of approximately 10^9 years.

b) A radioactive isotope has a half-life $T_{1/2}$. Determine the fraction of this isotope that remains in a particular sample of the isotope after a time of 1.6 $T_{1/2}$.

(5 marks)

A RELATIVITY

Introduction

Earlier we described the rules of classical mechanics as developed by Newton and others. These rules provided the basis for physics for 300 years. Ultimately, however, Newtonian and Galilean mechanics struggles: it cannot deal with things that move very fast or with objects that are very small. The insights of Einstein and others around the beginning of the twentieth century enabled physicists to change their understanding of time and space – this type of change is called a paradigm shift. This topic describes our present view of what we now called spacetime.

A.1 The beginnings of relativity

Understandings

- → Reference frames
- → Galilean relativity and Newton's postulates concerning time and space
- → Maxwell and the constancy of the speed of light
- → Forces on a charge or current

Nature of science

Einstein's great insight was to realize that the speed of light is constant for all inertial observers. This has enormous consequences for our understanding of space and time. A paradigm shift occurred and the Newtonian view of time was overturned. There are many other examples of paradigm shift in science but perhaps none quite as profound as this.

Applications and skills

- → Using the Galilean transformation equations
- → Determining whether a force on a charge or current is electric or magnetic in a given frame of reference
- → Determining the nature of the fields observed by different observers

Equations

- → Galilean transformation equations: $x' = x - vt$
- → $u' = u - v$

Reference frames

A reference frame allows us to refer to the position of a particle. Reference frames consist of an origin together with a set of axes. In this course we generally use the Cartesian reference frame, in which position is defined using three distances measured along axes that are at 90° to each other. The axes of a three-dimensional graph make up a Cartesian reference frame.

Other frames are, of course, available. Sailors use latitude and longitude; these together with the distance of an object from the centre of the Earth constitute a different reference frame. Astronomers usually use angles when defining the position of a star that is being observed; they only need two angles because the distance to the star (for observational purposes) is irrelevant.

For some of the frames of reference in use, one or more of Newton's laws of motion do not hold. Our own frame on the surface of a rotating planet shows this straight away. Everything off-planet appears to be spinning around us. An object that is at rest relative to us on planet Earth is not moving at a constant velocity at all. This has consequences that we have already seen in earlier topics; the fictitious centrifugal force and the Coriolis force used to "explain" the movement of weather systems are two cases in point.

For Newton's first law to be valid we need to be careful about the nature of the reference frame in which we use the law. We define an **inertial frame of reference** as a frame in which an object obeys Newton's first law: so that it travels at a constant velocity because no external force acts on it.

Do inertial frames exist? The best way to find one is to take a spaceship out into deep space, well away from the gravitational effects of planets and stars, and then turn off the engines. No forces act from outside or inside the spacecraft and this will be a true inertial frame of reference.

Galilean relativity and Newton's postulates

Even though we have to take some trouble to reach one, there are an infinite number of inertial frames of reference in the universe and there are a number of ways in which we can move between them.

- The first obvious way is to step sideways from one frame to another. This is known as a *translation* (figure 1(a)).
- The set of axes can be rotated to form another set. This is known as a *rotation* (figure 1(b)). We shall not consider these in detail in this course.
- One frame can move relative to another frame with a constant relative velocity. This is known as a *boost*. (figure 1(c))

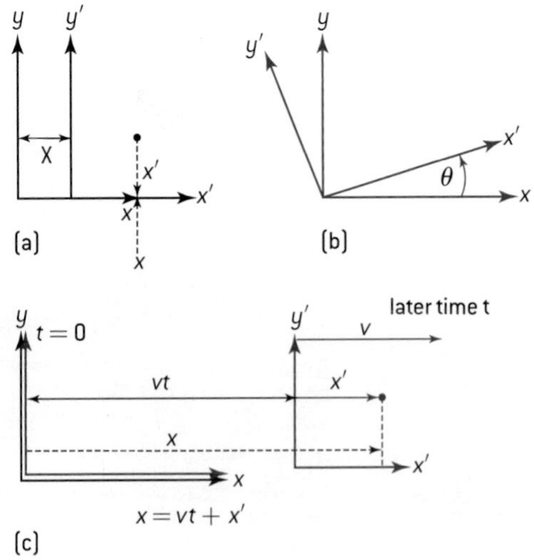

▲ Figure 1 Translations, rotations, and boosts.

It is easy to see that if an object moves with constant velocity in one reference frame then under any of these three conditions it will be measured as having a constant (but different) velocity in the other reference frame.

The principle of relativity is often associated with Albert Einstein. In fact, Galileo was probably the first person to discuss the principle. He describes how, in a large sailing ship, butterflies in a cabin with no windows would be observed to fly at random whether the ship were moving at constant velocity or not. An observer in the cabin could not deduce by observing the butterflies whether the ship was moving. And the butterflies would certainly not be pinned against the back wall of the cabin.

This principle of relativity as described by Galileo tells us about the nature of our universe.

- The translation rule is equivalent to saying that there is no special place in the universe; position is relative.
- The rotation rule says that there is no special direction; direction is relative.
- The boost rule says that there is no special velocity. Stationary is not an absolute condition and any object can be described as stationary with respect to another object. (As we shall see, the disagreement between Galileo's ideas and those of Einstein is crucial here.)

Thinking, for a translation, in terms only of the *x*-direction (that is, in one dimension), if the distance between the origins of two inertial frames S' and S' is X then a position x in S is related to position x' in frame S' by

$$x' = x - X$$

For a boost, if one inertial frame moves relative to the other by a constant relative velocity v along the x-axis, then the distance between the origins of the reference frames must be changing by v every second and this distance is vt where t is the time since the frames coincided.

Imagine that the origins of the two reference frames S and S' (Figure 1(c)) were at the same position (that is, they were coincident) at time $t = 0$. Technically, we say that clocks in the frames were adjusted so that $x = x' = 0$ when $t = t' = 0$. At a later time t, the origins of S and S' will be separated by vt where v is the velocity of frame S' relative to frame S. Therefore a position x in frame S will be related to position x' in S' by

$$x = x' + vt \text{ and } x' = x - vt$$

The velocities also transform in an obvious way. If the velocity in S is u and the velocity in S' is u', then

$$u' = u - v$$

This set of equations that link two reference frames by their relative velocity are known as the **Galilean transformations**.

Newton developed Galileo's ideas further in his *Principa Mathematica* by suggesting two important postulates (a postulate is an assertion or assumption that is not proved and acts as the starting point for a proof):

- Newton treated space and time as fixed and absolute. This is implied in our use of t in both equations above (t' does not appear, only t). A time interval between two events described in frame S is identical to the time interval between the same two events as described in frame S'. The evidence of our senses seems to confirm this (but remember that we do not travel close to the speed of light in everyday life).
- Newton recognized that two observers in separate inertial frames must make the same observations of the world. In other words, they will both arrive at the same physical laws that describe the universe.

Nature of science

The way Newton put it.

These postulates were expressed somewhat differently by Newton in the *Principa Mathematica* – the book (in Latin) that he wrote to publish some of his discoveries. In translation his postulates were:

> " I. Absolute, true, and mathematical time, of itself, and from its own nature, flows equably without relation to anything external .
>
> II. Absolute space, in its own nature, without any relation to anything external, remains always similar and immovable."

(Translation by Mottes (1971), revised by Cajorio, University of California Press.)

Worked example

1 In a laboratory an electron travels at a speed of 2×10^4 m s^{-1} relative to the laboratory and another electron travels at 4×10^4 m s^{-1} relative to the laboratory. Use a Galilean transformation to calculate the speed of one electron in the frame of the other when they are travelling **a)** in opposite directions **b)** in the same direction.

Solution

a) The closing speed of the electrons is $2 \times 10^4 - (-4 \times 10^4) = 6 \times 10^4$ m s^{-1}

b) The relative speed of one electron relative to the other is $2 \times 10^4 - (4 \times 10^4) = 2 \times 10^4$ m s^{-1}

TOK

Changing perspectives

In Topic 2 we began by describing Newton's second law of motion in a simple way as:

force = mass × acceleration

Later we showed that this was better expressed as

force = rate of change of momentum

A similar change of expression is possible here: Newton's first law is usually given as a variant of "Every object continues in its state of rest or uniform motion unless net external forces act on it" (and you should continue to use this or a similar wording in your own work). But this is not the only possibility. A succinct and interesting way to express Newton's first law is as "Inertial frames exist".

To what extent do concise forms of scientific laws help or hinder our understanding?

Charges and currents – a puzzle

The choice of inertial frame can make a radical difference to the perception of a situation.

Consider a positively-charged particle moving initially at velocity v some distance away from and parallel to a wire carrying a current. In the wire the electrons are also moving with speed v. The positive charges in the wire are stationary. We will use two inertial reference frames in this example. One frame is at rest relative to the positive charges in the wire. The other frame is at rest relative to the moving charge q.

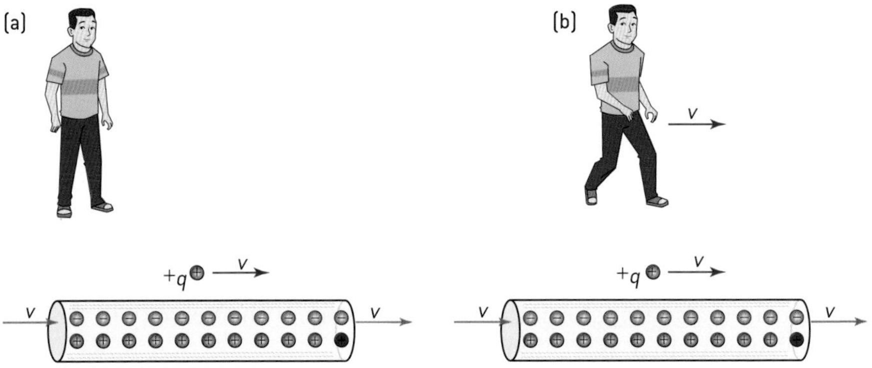

▲ Figure 2 Moving charges in reference frames.

To an observer at rest (figure 2(a)) with respect to the stationary positive charges the situation in the wire seems clear. The numbers of negative and positive charges in the wire are equal and therefore the lone moving charge $+q$ does not experience an electrostatic (electric) force. The movement of the electrons in the wire to the right (a conventional current to the left) gives rise to circular magnetic field lines centred on the wire and going into the page above the wire. The single moving charge is therefore moving perpendicularly with respect to this field. Fleming's left-hand rule indicates that there is a magnetic force acting perpendicularly outwards on the charge and therefore the charge will be accelerated in this direction. The stationary observer detects a magnetic repulsive force acting between the charge and the current-carrying wire.

What is the situation from the point of view of the moving charge q? An observer moving with the charge (figure 2(b)) sees the single charge $+q$ as stationary; electrons in the wire that also appear to be stationary, but the positive charges in the wire appear to be moving to the left at speed v. We could imagine that the moving positive charges lead to a magnetic field around the wire just as before – but this does not help! To the moving observer the charge $+q$ is stationary relative to the resulting magnetic field and therefore no magnetic force should arise.

What happens is that (as we shall see later) the relative movement of the positive charges in the wire leads to a contraction in the spacing of these charges as perceived by the observer moving with $+q$. There are more positive charges than negative charges per unit length as detected by the moving observer and so there is effectively a net repulsive force acting on $+q$ (we shall see how this change arises from length contraction in a later sub-topic). Now the observer moving with q explains the force acting on q as electrostatic in origin, not magnetic as before.

However, both observers report a force and in both cases the force acts outwards from the wire. The physical result is the same even though the explanations differ. A mathematical analysis from the standpoint of the two inertial frames also confirms that the magnitude of the force is identical in both cases.

Another situation is that of two point electric charges moving in parallel directions at the same speed as each other. This case (figure 3) is not identical to that of the charge moving near a current-carrying wire. There is no balance of positive and negative charges to complicate matters this time. Again, we consider the situation from the standpoint of two separate inertial frames: that of an observer stationary relative to the point charges, and that of an observer moving at a different velocity relative to the charges. Observations made in the frame of the point charges are easier to understand (figure 3(a)): the observer moving with the charges sees two positive charges that repel and no magnetic attraction between the two. This observer describes the repulsion as purely electrostatic.

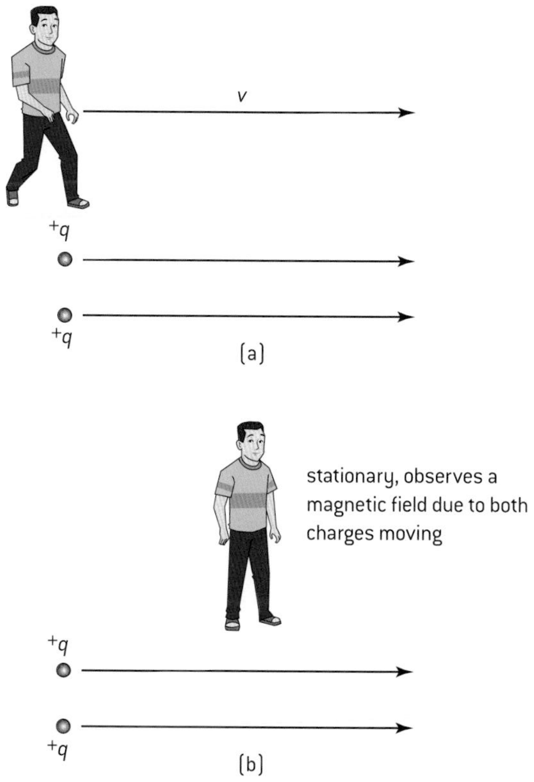

Figure 3 Two point charges moving parallel.

For an observer no longer stationary relative to the point charges (figure 3(b)), the situation is changed. From this point of view, the repulsive electric field is increased (through relativistic length contraction). There is also a magnetic field that was not apparent when the observer was stationary relative to the charges. This is because a moving charge gives rise to a magnetic field. Each charge now appears to be moving within the magnetic field due to the other charge and consequently there is an attraction that the observer describes as magnetic in origin. There is both an increased (electrostatic) repulsion and a new (magnetic) attraction compared with the stationary observer frame. Again, a mathematical analysis shows that the force between the charges is identical for all observer inertial frames of reference. This is what we expect given that all physical systems observed in inertial frames must obey the same laws. As Einstein himself said:

> *"What led me ... to the special theory of relativity was the conviction that the electromagnetic force acting on a body in motion in a magnetic field was nothing else but an electric field."*

Maxwell and electromagnetism

In 1861 James Maxwell established the connection between electrostatics, electromagnetic induction, and the speed of light. He developed four equations that between them describe the whole of electrical and magnetic theory and lead to the recognition that light is a form of electromagnetic radiation. The four equations incorporate the value of the speed of light travelling in a vacuum (free space) in a fundamental way. The conclusion that must be drawn from Maxwell's equations is that, if observers in different inertial frames make observations of the speed of light then, if they are to agree about physical laws, they must observe identical values for the speed of light. But the Galilean transformations predict a different result from this. The Galileo predicts that the speed of light differs in different frames by the magnitude of the relative velocity between the frames.

So the inescapable conclusion that follows from Maxwell (same speed of light for all observers) is directly contrary to the assumption of absolute time and absolute space as postulated by Newton, and as embodied in the Galilean transformations. Physics had reached an impasse; it required the genius of Maxwell to recognize the problem. It required another genius, Einstein, to move the subject forward again half a century later.

TOK

Validating a paradigm shift

The development of Einstein's relativity ideas required a shift in the view that scientists took of the physical rules that govern the universe. How do scientists ensure that the need to shift perspectives is valid?

TOK

Maxwell's advances

Einstein wrote:

> *The precise formulation of the time–space laws was the work of Maxwell. Imagine his feelings when ...equations he had formulated proved to him that electromagnetic fields spread in the form of polarised waves, and at the speed of light!... it took physicists some decades to grasp the full significance of Maxwell's discovery, so bold was the leap that his genius forced upon the conceptions of his fellow workers.*
>
> (Albert Einstein, *Science*, May 24, 1940)

To what extent do you think this is true?

A.2 Lorentz transformations

Understanding

- The two postulates of special relativity
- Clock synchronization
- The Lorentz transformations
- Velocity addition
- Invariant quantities (spacetime interval, proper time, proper length, and rest mass)
- Time dilation
- Length contraction
- The muon decay experiment

Nature of science

Einstein's theory of relativity stems from two postulates. He deduced the rest of the theory mathematically. This is an example of pure deductive science at work.

Applications and skills

- Solving problems involving velocity addition
- Solving problems involving time dilation and length contraction
- Solving problems involving the muon decay experiment

Equations

- Lorentz transformation equations:
$$\gamma = \frac{1}{\sqrt{1 - \frac{v^2}{c^2}}}$$
- $x' = \gamma(x - vt)$; $\Delta x' = \gamma(\Delta x - v\Delta t)$
- $t' = \gamma\left(t - \frac{vx}{c^2}\right)$; $\Delta t' = \gamma\left(t - \frac{v\Delta x}{c^2}\right)$
- $u' = \dfrac{u - v}{1 - \frac{uv}{c^2}}$
- $\Delta t = \gamma \Delta t_0$

The two postulates of special relativity

Newton's two postulates of space and time from Sub-topic A.1 require an absolute time that is the same for all observers in inertial frames. This implies, through the Galilean transformation, that a moving observer will observe a different value for the speed of light in free space from that of a stationary observer.

Maxwell's discoveries began to prompt serious questions about the peculiar behaviour of light and the validity of Galilean transformations. In 1887, two US scientists, Michelson and Morley, using very precise apparatus showed experimentally that any such difference between moving and stationary observers was below the measurement limits of their experiment. Scientists such as Lorentz, Fitzgerald, and Poincaré attempted to explain both Maxwell's conclusions and the results of the Michelson-Morley experiment.

Einstein's great leap forward was to recognize that if Maxwell's four electromagnetic equations were to be true in all inertial frames (which had to be the case) then some modifications of Newton's postulates were required. He was able to show that some equations that had already been developed by Lorentz as a way of avoiding the problems of the null result of Michelson and Morley could, alternatively and more properly, be derived assuming only Einstein's own modifications of Newton's postulates.

Einstein's two postulates are:

- The laws of physics are the same in all inertial frames of reference (Newton's first postulate too).
- The speed of light in free space (a vacuum) is the same in all inertial frames of reference (replacing the concept of absolute time and space, Newton's second postulate).

The Lorentz transformation

Earlier we saw that, for boosts, the Galilean transformations lead to the equations $x' = x - X$, $x' = x - vt$, and $u' = u - v$. There is an additional new transformation that arises from Newton's postulate that time is absolute. This is represented as $\Delta t' = \Delta t$.

To make the Maxwell's equations consistent for all inertial reference frames, Lorentz introduced a factor γ given by

$$\gamma = \frac{1}{\sqrt{1 - \frac{v^2}{c^2}}}$$

v is (as usual) the speed of one inertial frame relative to the other and c is the speed of light in free space, γ is called the **Lorentz factor.** As v tends to c, γ tends to infinity. Lorentz then used this factor to modify the Galilean expressions so that

$$x' = \gamma(x - vt) \text{ and } \Delta t' = \gamma \Delta t$$

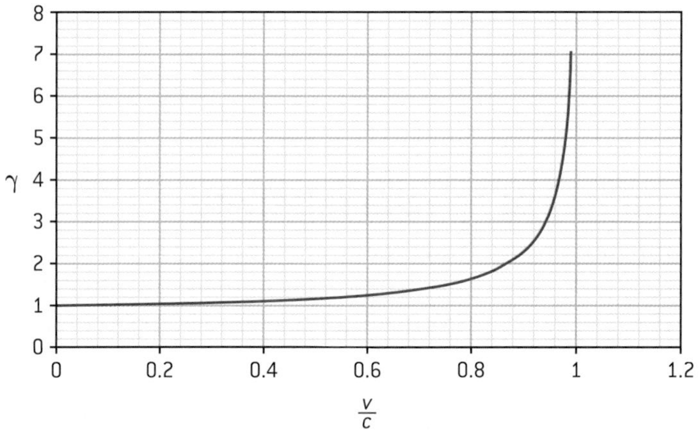

▲ Figure 1 How the Lorentz factor varies with speed.

Figure 1 shows how the Lorentz factor changes with speed. The scale on the x-axis is in the ratio of $\frac{v}{c}$ and shows us that, for speeds up to 20% of that of light, the Lorentz factor remains close to 1. If the Galilean transformations were true for all speeds then the graph would show a horizontal line at value 1 for all values of $\frac{v}{c}$.

Strictly, the equation involving x and x' is used to translate from a *position* in one frame to the *position* in the other frame at the same instant in time. In the Galilean transformation, lengths measured within one frame transform without change into the same length in any other frame because length Δx is the difference between two positions, and

$$\Delta x = x_1 - x_2 = (x_1' - vt) - (x_2' - vt) = x_1' - x_2' = \Delta x'.$$

In the Lorentz transformation this equality of Δx and $\Delta x'$ is no longer true and

$$\Delta x' = \gamma(\Delta x - v\Delta t)$$

This is a crucial result because it tells us that if one inertial reference frame is moving at a constant velocity relative to another, then an observer in one frame making a length measurement of an object in the other frame will not agree with the measurement made by an observer in the other frame. Space is no longer absolute.

The expression $x' = \gamma(x - vt')$ gives the position x of the object as observed in the moving reference frame. Sometimes we observe the position in the moving frame x' and we need the value of x in the stationary frame. This is given by $x = \gamma(x' + vt)$ and is known as the **inverse Lorentz transformation**.

The Lorentz transformation for time is

$$t' = \gamma\left(t - \frac{v}{c^2}x\right)$$

with an inverse transformation of $t = \gamma\left(t' + \frac{-v}{c^2}x'\right)$

Like space, time has also lost the property of being absolute. Time measured in different frames differs when there is relative velocity between the frames. Also, terms in x now appear in the time equations and terms in t in the expressions for x' and x.

We have assumed so far that there is no relative motion between the frames in the y or z directions. If this is true then there will be no relativistic changes in these directions either (this can be proved formally). The assumption of no motion in directions y and z and the previous expressions lead to the complete set of Lorentz transformations which are here compared with their Galilean equivalents.

Galilean	Lorentz	Inverse Lorentz
$x' = x - vt$	$x' = \gamma\left(x - \frac{v}{c}ct\right)$	$x = \gamma\left(x' + \frac{v}{c}ct'\right)$
$y' = y$ $z' = z$	$y' = y$ $z' = z$	$y = y'$ $z = z'$
$t' = t$	$t = \gamma\left(t - \frac{-v}{c^2}x\right)$ $ct' = \gamma\left(ct - \frac{v}{c}x\right)$	$t = \gamma\left(t + \frac{-v}{c^2}x\right)$ $ct = \gamma\left(ct' + \frac{v}{c}x'\right)$

In each case, if $v \ll c$ then $\gamma \approx 1$ and the Lorentz equations reduce to the Galilean equations with which we are already familiar.

An additional change in the Lorentz equations in this table is the expression of time using ct rather than t alone (this gives the time equation the dimensions of distance). A second change is to include the speed of light, c, twice in the distance equations. These changes make the equations appear more symmetric and help to explain why later in this topic we use axes of ct against x to draw spacetime diagrams.

Worked examples

1 Calculate the Lorentz factor for an object travelling at 2.7×10^8 m s^{-1}.

Solution

This speed is $0.9c$.

$$\gamma = \frac{1}{\sqrt{1-\frac{v^2}{c^2}}} = \frac{1}{\sqrt{(1-0.9^2)}} = 2.3$$

2 Clocks in two frames S' and S are adjusted so that when $x = x' = 0$, $t = t' = 0$. Frame S' as a speed of $0.8c$ relative to S.

Event 1 occurs at $x_1 = 50$ m, $y_1 = 0$, $z_1 = 0$, and $t_1 = 0.3\ \mu$s.

Event 1 occurs at $x_2 = 80$ m, $y_2 = 0$, $z_2 = 0$, and $t_2 = 0.4\ \mu$s.

Calculate, as measured in S',

a) The distance between x_1 and x_2

b) The time interval between t_1 and t_2.

Solution

a) $\gamma = \frac{5}{3} = 1.67$

From the Lorentz transformation

$$x_2' - x_1' = \gamma((x_2 - x_1) - v(t_2 - t_1))$$

Substituting gives

$$x_2' - x_1' = 1.67((80 - 50) - 0.8 \times 3 \times 10^8 (4 \times 10^{-7}) - (3 \times 10^{-7}))$$

$$x_2' - x_1' = 10 \text{ m}$$

b) $$t_2' - t_1' = \gamma\left((t_2 - t_1) - \frac{v}{c}\frac{(x_2 - x_1)}{c}\right)$$

$$= 1.67\left((t_2 - t_1) - \frac{v}{c}\frac{(t_2 - t_1)}{c}\right)$$

$$t_2' - t_1' = 30 \text{ ns (2 s.f.)}$$

Velocity addition

An object is moving in frame A with a constant velocity u_A. Frame A itself is moving with a constant velocity v with respect to frame B. What do the Lorentz equations have to say about the velocity u_B of the object when it is viewed by an observer in frame B?

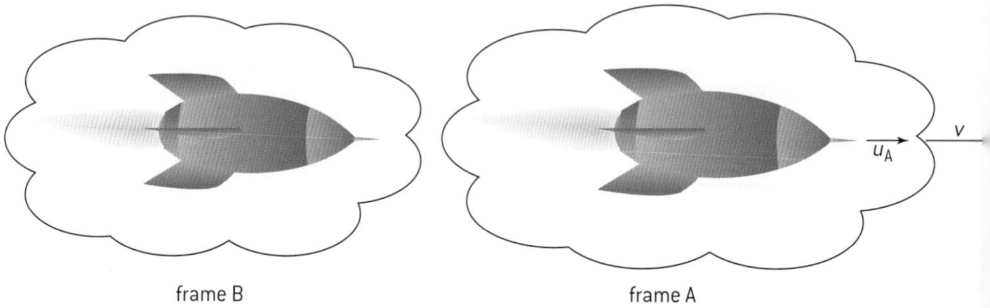

▲ Figure 2 Relativistic relative velocities.

Galilean relativity has no problem with this, the answer is simple: $u_B = u_A + v$. But this cannot be correct from a relativistic view. Suppose the rocket in frame A is moving at the speed of light. Then if v is positive u_B will exceed c; this is not allowed by Einstein's second postulate.

We need to use the Lorentz equations. The speed u_A is equal to $\frac{x_A}{t_A}$ when viewed in frame A (here x_A and t_A correspond to x' and t' in the earlier transformation equations).

Similarly $u_B = \frac{x_B}{t_B}$ (x and t from before)

so

$$u_B = \frac{x_B}{t_B} = \frac{\gamma(x_A + vt_A)}{\gamma\left(t_A + \frac{vx_A}{c^2}\right)}$$

using the Lorentz transformations. The positive signs are there because we are expressing x_B and t_B in terms of x_A and t_A rather than the other way round as we did before.

You should satisfy yourself that substituting $x_A = u_A t_A$ into this expression gives

$$u_B = \frac{u_A + v}{1 + \frac{u_A v}{c^2}}$$

This is the case where an observer *at rest in frame B* is measuring the speed of the object.

The inverse case of an observer in frame A (sitting in the frame A rocket in figure 2) measuring the speed of the object in frame B gives a similar expression with a change in sign.

$$u_A = \frac{u_B - v}{1 - \frac{u_B v}{c^2}}$$

A way to get the signs correct for a particular combination of frame velocities is to begin with the Galilean transformation (that is, when $uv \ll c$) and see what signs you would expect in this case. Then remember that the sign matches top and bottom of the equation.

Invariant quantities

Spacetime interval

In developing his theory of special relativity, Einstein realized that absolute time and absolute space are not invariant (unchanging) properties when moving from one inertial reference frame to another. However, not all quantities change when moving between inertial frames.

In the previous section we chose to express time in the Lorentz transformations not as plain t but as the product of the speed of light and time, ct. This quantity ct has the dimensions of length and this leads us to an invariant quantity known as the ***spacetime interval*** Δs that is defined for motion in the x direction as $\Delta s^2 = (c\Delta t)^2 - \Delta x^2$ (sometimes called the **invariant interval**). (You may see this defined in some books as $\Delta x^2 - (c\Delta t)^2$, in other words, as the negative of our definition.)

The spacetime interval is invariant because in another frame of reference

$$c\Delta t = c(t_2 - t_1) = \gamma\left(ct_2' + \frac{v}{c}x_2'\right) - \gamma\left(ct_1' + \frac{v}{c}x_1'\right) = \gamma\left(c\Delta t' + \frac{v\Delta x'}{c}\right)$$

and

$$\Delta x = x_2 - x_1 = \gamma\left(x_2' + \frac{v}{c}ct_2'\right) - \gamma\left(x_1' + \frac{v}{c}ct_1'\right) = \gamma(\Delta x' + v\Delta t')$$

therefore

$$\Delta s^2 = \gamma^2\left(c\Delta t' + \frac{v\Delta x'}{c}\right)^2 - \gamma^2(\Delta x' + v\Delta t')^2$$

$$= \gamma^2(c^2 - v^2)\Delta t'^2 - \gamma^2\left(1 - \frac{v^2}{c^2}\right)\Delta x'^2$$

$$= (c\Delta t')^2 - \Delta x'^2$$

This is obviously identical to the original definition using the same quantities (and no others) measured in the new frame.

Worked example

1 Jean and Phillipe are in separate frames of reference, neither of which is accelerating. Jean observes a spacecraft moving to his right at 0.8c. Phillippe observes a spacecraft moving to his left at 0.9c. Calculate the velocity of Phllippe's frame of reference relative to Jean's.

Solution

Suppose Jean is the observer at rest and Phillippe is moving relative to Jean with speed v.

$u_A = \frac{u_B - v}{1 - \frac{u_B v}{c^2}}$ can be re-arranged to give $v = \frac{u_B - u_A}{1 - \frac{u_B u_A}{c^2}}$

This gives the relative velocity of the reference frame in terms of the known individual speeds in the reference frames.

So $v = \frac{(0.8c - (-0.9c))}{1 + \frac{0.72}{c^2}}$

$= \frac{1.7c}{1.72} = 0.989c.$

In three dimensions the spacetime interval becomes $\Delta s^2 = (c\Delta t)^2 - \Delta x^2 - \Delta y^2 - \Delta z^2$.

TOK

So what about time travel?

The spacetime interval has a bearing on the cause and effect relationship between two objects or events. Intervals can be classified as space-like, time-like, or light-like depending on the value of Δs^2 for the two events.

Space-like intervals

If Δs^2 is < 0 (i.e. negative) then $(\Delta x^2 + \Delta y^2 + \Delta z^2) > (c\Delta t)^2$. This means that the distance between the events is too great for light (or anything travelling slower) from one event to have any effect on the other. They do not occur in each other's past or future and although there is a reference frame in which they occur at the same *time*, there is no reference frame in which the events can occur at the same *place*.

Time-like intervals

In this case, Δs^2 is > 0 (ie positive) and $(c\Delta t)^2 > (\Delta x^2 + \Delta y^2 + \Delta z^2)$. Now there is sufficient time for there to be a cause and effect relationship between the events because the time part of the spacetime interval is greater than the spatial separation.

Light-like intervals

The intermediate case is where $(c\Delta t)^2 = (\Delta x^2 + \Delta y^2 + \Delta z^2)$ and therefore $\Delta s^2 = 0$. The spatial distance and the time interval are exactly the same. Such events are linked by, for example, a photon travelling at the speed of light.

Time travel has always been a fascination of science-fiction authors. The spacetime interval can tell us the extent to which two events in space and time can affect each other.

To what extent do fictional works that you know mirror scientific truth?

Rest mass

A second invariant quantity is the rest mass m_0 of a particle. This is an important quantity and is defined as the mass of a particle in the frame in which the particle is at rest. This will be discussed at more length in a later sub-topic, as the concept of mass is bound up with what we mean by energy.

Proper time

Some of the most dramatic differences between our everyday perceptions of space and time and the predictions made by special relativity concern the time and length differences that arise between frames of reference moving relative to each other.

Figure 3 shows a simple **light clock** that consists of two mirrors facing each other across a room. An observer sits at rest in the room and watches light reflect between the mirrors. The distance across the room is L and so the time taken for the light to return to a mirror is $t = \frac{2L}{c}$.

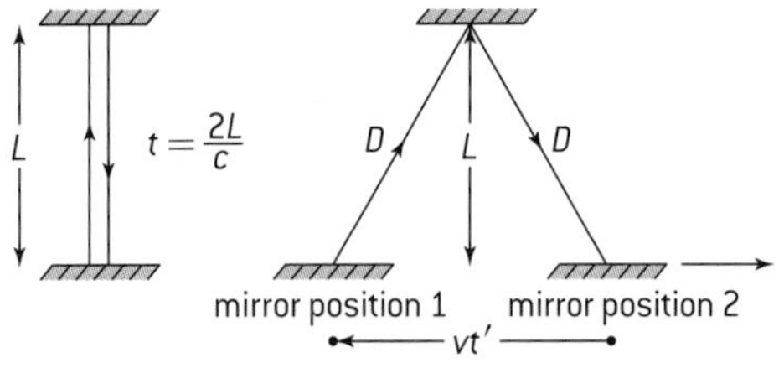

▲ Figure 3 A light clock from stationary and moving frames of reference.

Another observer moves to the left parallel to the mirrors at constant velocity v relative to the mirrors, watching the reflections. The right-hand diagram shows the bottom mirror at two positions as seen by the moving observer: when the light leaves the bottom mirror and when it returns. In the frame of this observer, the light appears to travel to the right at an angle to the direction of motion (but, of course, at the same speed of light). The distance travelled by the light is now $2D$ and the time observed between reflections at the same mirror is now $t' = \frac{2D}{c}$.

The distance travelled horizontally by the moving observer in time t' is vt' and by an application of Pythagoras' theorem:

$$D^2 = L^2 + \frac{v^2t'^2}{4^2}$$

$$D = \frac{ct'}{2}$$

Rearranging for t' gives

$$t' = \frac{\frac{2L}{c}}{\sqrt{1 - \frac{v^2}{c^2}}}$$

and so

$$t' = \frac{t}{\sqrt{1 - \frac{v^2}{c^2}}}$$

which reduces to $t' = \gamma t$.

This result also follows directly from the Lorentz transformations: Two events (the light leaving from, and returning to, the bottom mirror) occur at t_1 and t_2. These events occur at the same place (the mirror) so we do not need to include the x terms in our proof (because $x_1 = x_2 = x$). The time interval between these events is $\Delta t = t_2 - t_1$.

In the observer frame therefore

$$\Delta t' = t_2' - t_1' = \gamma\left(t_2 - \frac{vx}{c^2} - \left(t_1 - \frac{vx}{c^2}\right)\right) = \gamma \Delta t$$

The same result as before and

time interval in the observer frame = time in the mirror frame × γ

$$\Delta t = \gamma \Delta t_0$$

The quantity Δt_0, the time interval in the stationary frame, is known as the **proper time**. Its definition is the time interval between two events measured in the reference frame where both events occur at the same position. Proper time is our third invariant quantity. Alternatively, it is the shortest possible measured time interval between two events.

The result shows that time measured in a moving frame is *always* longer than the time measured in a frame that is stationary relative to a clock. The effect is known as *time dilation* ("dilated" means "expanded"). If the moving observer in our example here also has a clock then an observer stationary with respect to the mirror frame observes the moving clock running slower than the mirror clock. The situation is symmetric. We will discuss this point further in the next section.

Worked example

1 Tom is flying a plane at 0.9c. The landing lights of the plane flash every 2 s as measured in the reference frame of the plane. Sam watches the plane go by. Calculate the time between flashes as observed by Sam.

Solution

When $v = 0.5c$, $\gamma = 2.3$ (this was calculated in an earlier example). The proper time is 2 s.

The time between flashes for Sam $= \gamma t = 2.3 \times 2 = 4.6$ s.

The time is dilated in Sam's frame of reference.

Proper length

The length of an object also changes when observed in frames moving relative to the object. When discussing proper time we had to be careful to specify that the positions at which time was measured were the same for both measurements. This time, the length L of an object (where $L = x_2 - x_1$) must have x_1 and x_2 measured at the same time so that t_1 and t_2 in the Lorentz equations are both equal to t.

In reference frame S, x_1 and x_2 represent the ends of an object of length L_0. *This is the frame in which the object is at rest.* In S' these ends become x_1' and x_2' with a length L'. The reference frame of S moves at speed v relative to S'.

In S

$$L_0 = x_2 - x_1$$

which in S' using the Lorentz transformations

$$= \frac{x_2' + vt_2'}{\sqrt{1 - \frac{v^2}{c^2}}} - \frac{x_1' + vt_1'}{\sqrt{1 - \frac{v^2}{c^2}}}$$

The ends of the rod are measured at the same time and so t_1' and t_2' are equal.

So

$$L_0 = \frac{L'}{\sqrt{1 - \frac{v^2}{c^2}}}$$

and

$$L' = \frac{L_0}{\gamma}$$

This leads to the fourth invariant quantity (L_0) called the **proper length** defined as the length of an object measured by an observer at rest relative to the object. By implication the two measurement events have to be made at the same time. The proper length can also be regarded as the longest measured length that can be determined for an object.

There is direct experimental evidence for time dilation and length contraction (which are two sides of the same coin). Muons are particles that can be created either in high-energy accelerators or (more cheaply!) in the upper atmosphere when cosmic rays strike air molecules. These muons have very short mean lifetimes of about 2.2 μs. When travelling at 0.98c, the distance the muon will travel in one mean lifetime is roughly 660 m – far less than the height of 10 km above the Earth's surface where they are created. On a Newtonian basis very few muons would be expected to reach the surface as the time to reach it is about 15 mean lifetimes. Yet a considerable number of muons are detected at the surface. This is due to time dilation (or length contraction, whichever viewpoint you choose). At a speed of 0.98c, $\gamma = \frac{1}{\sqrt{1 - 0.98^2}} = 5$. So, in the reference frame of the Earth, the mean lifetime becomes 11 μs. The time to travel 10 km at 0.98c is 33 μs so that a significant number of muons will remain undecayed at the surface.

In the frame of reference of the muon, the 10 km (as measured by an observer on the Earth) from atmosphere to Earth's surface is only $\frac{10\text{ km}}{\gamma}$ (to the muon). This is 2 km in the muon's rest frame corresponding to a travel time of about 3 mean lifetimes allowing many more muons to reach the surface than Galilean relativity would suggest.

Depending on the observer's viewpoint, either time dilation or length contraction can be used to explain the observed large number of muons at the surface.

Clock synchronization

The problem of synchronizing the clocks we have used in our discussions is an important one. Suppose an observer is standing close to a clock and simultaneously viewing another clock 1 km away. Because the speed of light is invariant, the distant clock – even if originally synchronized – will appear to register a time $\frac{10^4}{3 \times 10^8} = 30\ \mu\text{s}$ later than the nearby clock. However, if the two clocks are in the same inertial frame, they will be synchronized (tick at the same rate).

One way to achieve this synchronization is to synchronize both clocks when they are close together and then move one to its final position, but to do it very slowly so that the two clocks continue (approximately) to share a reference frame with each other. Another method would be to have both clocks at their final positions and to use a third clock moving slowly between them to transfer the times from one to another.

Worked example

1 Clare and Phillippe fly identical spacecraft that are 16 m long in their own frame of reference. Clare's spacecraft is travelling at a speed of $0.5c$ relative to Phillippe's. Calculate the length of

a) Clare's aircraft according to Phillippe

b) Phillippe's aircraft according to Clare.

Solution

$\gamma = 1.15$ for this relative speed.

a) The length of Clare's aircraft is $\frac{16}{1.15}$ m according to Phillippe, this is 13.9 m.

b) Because the situation is symmetrical Clare will also think that Phillippe's spacecraft is 13.9 m long.

Nature of science

GPS – a study in relativity

Satellite navigation units (satnavs) in cars and other domestic devices that use the global positioning system are now very common. They can pinpoint their position to within a few metres.

At the time of writing there is a network of 24 satellites orbiting at about 2×10^7 m above the Earth with orbital periods of about 12 hours. The orbits are such that at least four satellites are above the horizon at any point on the surface at all times. Inside the satellite is an atomic clock that is accurate to about 10^{-9} s and transmits a signal to the receivers on or above the Earth.

The receivers triangulate the signals from satellites above the horizon to arrive at a positional fix to within metres within a few seconds. Wait a little longer with some special receivers and this precision can rise to orders of millimetres. It is now routine for a satnav in a moving vehicle to show its speed and heading in real time.

The design of both the satellite transmitters and the GPS receivers need to take account of relativity. The atomic clocks are adjusted so that once in orbit they run at the same rate as Earth-bound clocks. The GPS receivers have microcomputers that carry out the required calculations to make the relativistic corrections.

A.3 Spacetime diagrams

Understanding

- → Spacetime diagrams
- → Worldlines
- → The twin paradox

Nature of science

The spacetime (or Minkowski) diagram is a perfect example of a picture adding value to conceptual understanding. At first, Einstein did not realize the full potential of these diagrams but came to see that they had considerable value in making some of the difficult ideas of relativity – including simultaneity – more obvious.

Applications and skills

- → Representing events on a spacetime diagram as points
- → Representing the positions of a moving particle on a spacetime diagram by a curve (the worldline)
- → Representing more than one inertial reference frame on the same spacetime diagram
- → Determining the angle between a worldline and the time axis for specific speed on a spacetime diagram
- → Solving problems on simultaneity and kinematics using spacetime diagrams
- → Representing time dilation and length contraction on spacetime diagrams
- → Describing the twin paradox
- → Resolving of the twin paradox through spacetime diagrams

Equations

- → angle of worldline to ct axis $\theta = \tan^{-1}\left(\frac{v}{c}\right)$

Spacetime diagrams and worldlines

In 1908, Minkowski introduced a way to visualize the concept of spacetime. This will help you to understand many of the ideas and concepts that arise in special relativity.

Physicists are well used to graphs as ways to visualise data. Minkowski attempted to represent the four-dimensional nature of spacetime using a graphical picture known as a **spacetime diagram** or sometimes as a Minkowski diagram.

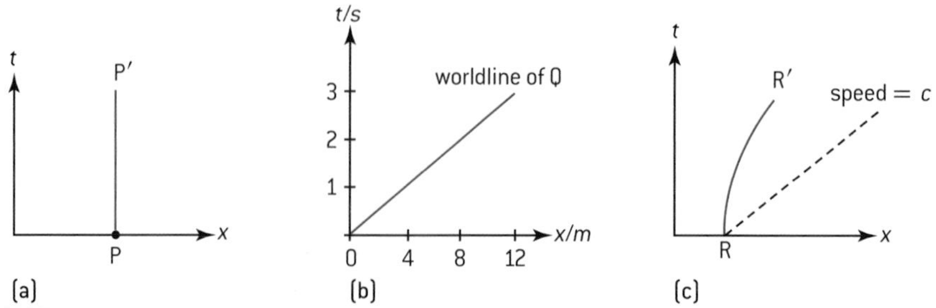

▲ Figure 1 Spacetime diagrams.

Spacetime diagrams show the position of an object in one dimension (x) at a time (t) in an inertial frame. The axes themselves constitute the inertial frame. The diagram resembles (but should not be confused with) the ordinary distance–time graphs with which you are familiar in mechanics, except that time is plotted on the y-axis and position on the x-axis.

Figure 1(a) shows the spacetime diagram for a particle that is stationary with respect to the inertial frame that is represented by the diagram. At $t = 0$ the particle P is on the x-axis. As time goes on, because the object is stationary, it does not change its position (x) in the reference frame. Time is, of course, increasing. Line PP′ shows the **trajectory** of the particle through spacetime and is known as the **worldline** of the particle.

Figure 1(b) shows a different particle Q moving at a constant velocity in the reference frame of this spacetime diagram. At $t = 0$, Q is at the origin of the diagram ($x = 0$) and it is moving at 4 m s^{-1} to the right. Each second after the origin time, Q is 4 m further to the right and so its worldline in the spacetime diagram is a line at an angle to the axes. It should be easy for you to see that if a further particle R were to be accelerating relative to the reference frame of the diagram, then the worldline of R would be a curve (figure 1(c)).

There must be a limit to the R worldline because nothing can exceed the speed of light in free space. So the gradient of the dotted line on figure 1(c) shows the maximum limiting speed of R. This dotted line also represents the world line of a photon in the diagram (the minimum gradient of RR′).

Returning to particle Q which is moving with constant velocity in the reference frame of P, we could think of Q as being stationary in its own reference frame and in this event the spacetime diagram for Q in its own frame would be identical in shape to figure 1(a). We can combine these two separate spacetime diagrams for different inertial frames moving at constant speed relative to each other and this combination of axes will be of particular help later in this sub-topic when we resolve some of the paradoxes that special relativity appears to create.

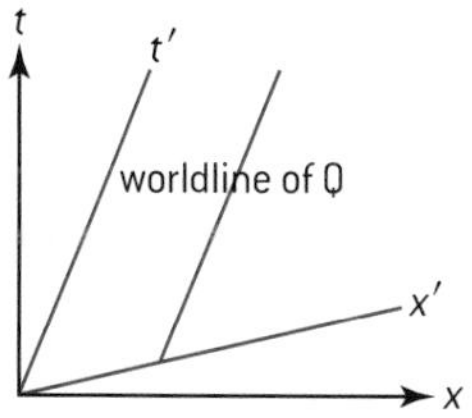

▲ Figure 2 Two inertial frames – one spacetime diagram.

Figure 2 shows the combined spacetime diagram for Q and P drawn for the reference frame of P. Notice what has happened to the Q axes ($t' - x'$), they have swung away from t and x and become closed up together. The Q x'-axis is now at the same angle as the previous Q worldline in the P reference frame. Q is, of course, moving along a worldline parallel to the time axis in the Q reference frame.

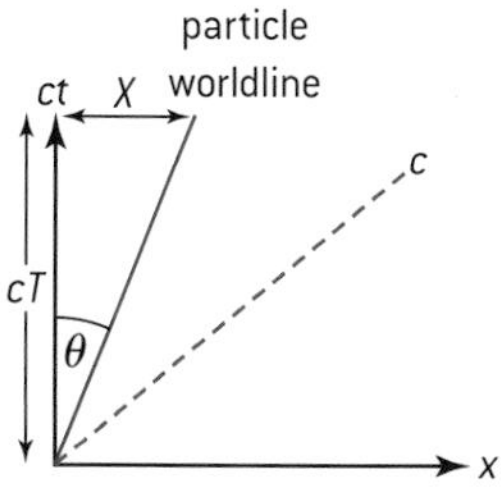

▲ Figure 3 Definition of θ.

We have so far used t and x for our axes, but this is not the only convention used. Sometimes you will see time plotted in units of ct. This is convenient because it means that the limiting line that represents the speed of light will have a gradient of 1 on the spacetime diagram. You can expect to see both conventions used in IB examinations.

An additional convention is that sometimes physicists define c to be equal to 1 so that, in calculations, large values for the answers do not trouble them. Equally, expect to see speeds quoted as, for example, 0.95c meaning 95% of the speed of light in free space (2.85×10^8 m s^{-1}).

Some simple geometry (Figure 3) shows that when we are using *ct-x* axes, then the angle θ between the worldline of a particle and the *ct* axis is given by

$$\tan\theta = \frac{opposite}{adjacent} = \frac{X}{cT} = \frac{1}{c} \times \frac{X}{T} = \frac{v}{c}$$

or

$$\theta = \tan^{-1}\left(\frac{v}{c}\right)$$

When $v = c$, then $\theta = 45°$ ($\tan 45° = 1$) and the worldline for a photon starting at the origin of the spacetime diagram is a line at 45° to both *ct* and *x* axes.

Simultaneity

There are significant changes to our ideas about the order in which things happen or whether two events happen simultaneously under special relativity. This is because the speed of light is always observed to have the same value by observers in different frames.

The classic "thought" experiment to illustrate this is the example of a train carriage moving at constant velocity past an observer standing on a station platform (Figure 4). A person in the carriage (Jack) switches on a lamp that hangs from the centre of the ceiling. Jack observes that the light from the lamp reaches the two end walls of the carriage (R and L) at the same moment.

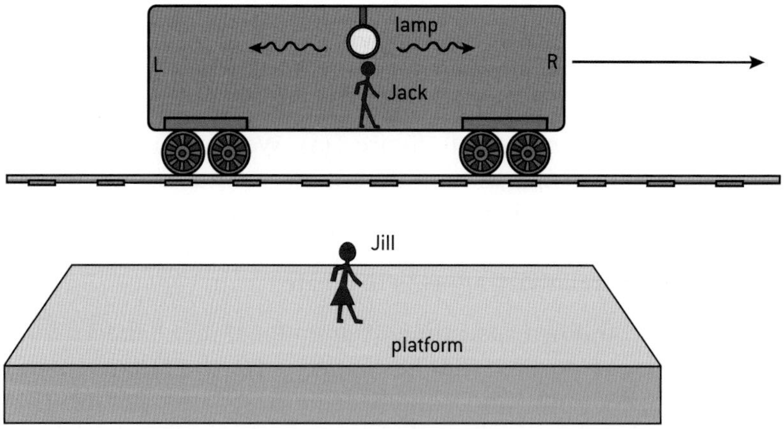

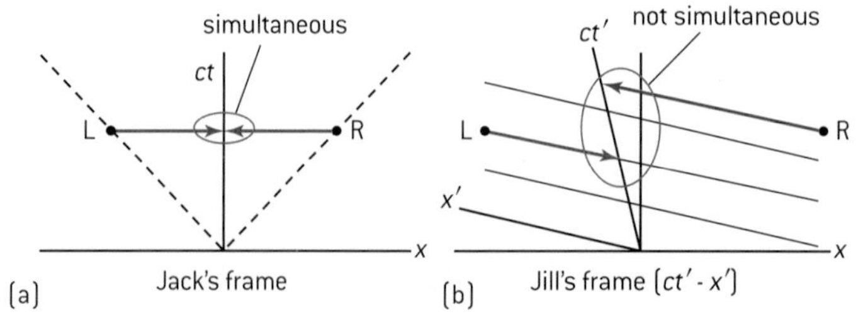

▲ Figure 4 Simultaneity at work in a spacetime diagram.

Jill who is on the platform, does not agree with this observation. The light from the lamp moves to both ends of the carriage at the same speed *c*. However, in the time it takes the light to get to the ends, the left-hand end of the carriage has moved towards the light and the right-hand end has moved away. The consequence is that the light (according to Jill) hits the left-hand end first (event L) before the right (event R).

This result becomes clear in a spacetime diagram. In the reference frame of Jack (*ct*-*x*) the events R and L occur at the same instant because they are on a line parallel to the *x* axis and are at the same *ct* coordinate (figure 4(a)). In Jill's frame plotted on figure 4(b) (*ct'*-*x'*) it is clear that L occurs before R when you look these events in terms of the *ct'* axis.

There is a danger of confusion here because it is possible to misunderstand and think that the loss of simultaneity is to do with the transmission of the information. In other words, that this difference of opinion between Jack and Jill arises because the light travels through different distances from the ends of the carriage to their eyes. That is not the explanation of what is happening. The lack of simultaneity arises because the speed of light is always constant even if a particular observer is moving relative to the light source. As far as Jack is concerned, he is *always* midway between the carriage endwalls. As far as Jill is concerned, once the photons have left the lamp, then they travel at *c* and the carriage will continue to move while the photons themselves are in transit.

One of the reasons for this confusion is the use of the term "observer", which is a very common one in books and articles about relativity. We often think of an observer as being located at one point in the inertial reference frame. This is not the true meaning. It is better to think of the observer as being in overall charge of an (infinitely) large number of clocks and rulers that are located throughout the observer's frame. Jack (the stationary observer in this case) can take a reading at the instant when the light hits the end wall of the carriage without having to worry about the time taken for this information to travel from the carriage to his position. Another way is to think about the observer as being a whole team of observers with each one able to make measurements of his or her immediate region of the reference frame.

This can be explained in terms of the Lorentz transformations. Imagine that two events are simultaneous in one frame of reference. This will mean that for both events the time coordinate will be *ct* for any value of *x*. However, in a frame moving at *v* relative to the first frame, the Lorentz transformation shows that:

$$t' = \frac{t - \frac{vx}{c^2}}{\sqrt{1 - \frac{v^2}{c^2}}}$$

Therefore unless *x* is the same for both events then *t'* cannot be the same for both events – they will not be simultaneous.

Conversely, this also tells us that if two events *are* simultaneous in the second, moving frame then they must be occurring at the same position (*x*).

Time dilation and length contraction re-visited

Figure 5(a) shows the spacetime diagram for an observer in frame B who is viewing an automobile moving at a constant velocity *v* relative to B. The diagram shows the stationary frame A for the automobile with the B spacetime axes included. We need to be quite clear about what we mean by the term *time interval* here. It is the time between two events measured at the same place in the reference frame (in other words, the proper time). The spacetime diagram should show that a measurement in any other frame leads to a time interval greater than the proper time.

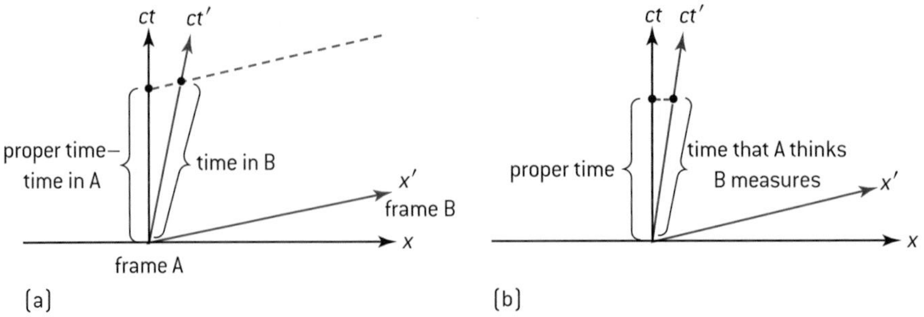

▲ Figure 5 Measuring times in two reference frames.

At $t = 0$ the origins of both reference frames coincide and the automobile is at the origin in both frames. The automobile is stationary in frame A and its worldline lies along the time axis as usual. Frame B is moving at constant velocity and its axes are modified as usual in frame A. The event that marks the end of the proper time interval is transformed along the x' axis to meet the ct' axis in order to obtain the time that B will measure at what B thinks is the same instant in the moving frame (figure 5(a)).

The question of what A (in the stationary frame) thinks is the same instant is different (figure 5(b)). However, whichever view we take of the measurement in frame B (whether from the A or B standpoint), the time measured in B is always greater than the proper time because the length of the line in the spacetime diagram is always along the hypotenuse of the triangle, whereas the proper time is one of the other sides of the triangle.

A spacetime diagram also helps understanding of length contraction. To see why distance measurements change in a moving frame, again we must understand what is being measured. We need to consider the distance between two points at the same instant in time as judged by the two observers in different frames.

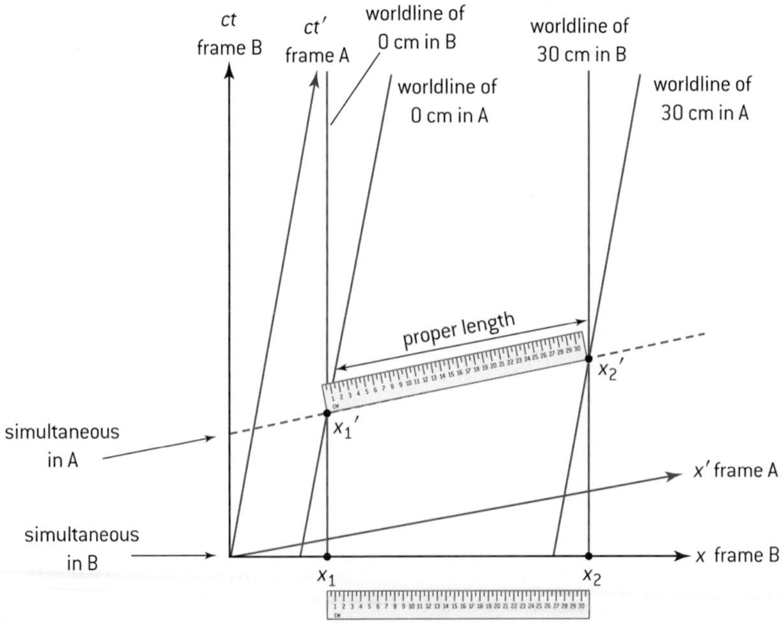

▲ Figure 6 Measuring lengths in two reference planes.

Figure 6 shows a ruler stationary in frame A. The ruler lies along the x'-axis because this is where we will determine the proper length. The rules for measuring proper length indicate that the two position determinations must be made simultaneously. The other frame B is considered to be the stationary frame for the purposes of the spacetime diagram. Thus, A must be moving relative to B and as a result has the axes $ct'-x'$. The diagram shows what happens. The proper length of the ruler is the distance measured between the worldlines and parallel to the x' axis. The equivalent simultaneous measurement in B will always be shorter than that in A – the length is contracted.

The twin paradox

Many of the ideas we have discussed so far in the study of special relativity come together in a series of paradoxes. The twin paradox is the most famous of these and can be very simply stated:

Mark and Maria are twins. Mark decides to go on a journey to a distant star at a high speed with Lorentz factor equal to γ. After reaching the star taking a time T in Maria's frame of reference. Mark returns with his journey also taking a time T (to Maria) to do so. At the end of his journey Maria has aged $2T$ but she is amazed to find that Mark has only aged by $\frac{2T}{\gamma}$.

This is nothing more than time dilation – so where does the paradox arise? Think about Mark's experience. He sits in his spacecraft – his frame – and watches Maria and the Earth move away at high speed (with the same γ as before) so why is Maria not younger than him on his return? We would expect some symmetry between the two frames.

The answer to the paradox is that there is no symmetry at all between the two cases. Maria has remained in an inertial frame throughout Mark's journey. Mark has not. He needs to accelerate four times during the journey: at the start of the trip, when he slows down at the star, when he accelerates back up to top speed and finally when he decelerates to arrive at the Earth. Moving out of an inertial frame of reference even once breaks the symmetry and as a consequence Mark and Maria age at different rates relative to each other over the whole journey.

The spacetime diagram (figure 7(a)) shows what happens. The frame for Maria is $ct-x$, the frame for Mark is $ct'-x'$. Mark leaves Maria at the origin of her reference frame and she remains here throughout. Of course, she moves along the ct axis as time increases. Meanwhile, Mark moves along his worldline which is at his origin $x' = 0$ or (in Maria's frame) at $x = vt$ where v is Mark's speed relative to Maria. Mark reaches the star at event P. We can draw two lines of simultaneity; one for Mark and one for Maria. In Maria's frame she thinks Mark arrives at the star at time Q. R is the time that Mark thinks Maria observes when he arrives at the star. They disagree about the simultaneity of events Q and R as we should expect. At this stage, both Mark and Maria think that the other is younger by a factor of γ – as predicted by the time dilation result from earlier.

There is a problem in thinking about the return journey because if it is to happen at all, Mark has to change speed. It is not necessary for him to do this, though. We can imagine that as soon as he reaches the star, he synchronizes his clock with another clock on a spacecraft that belongs

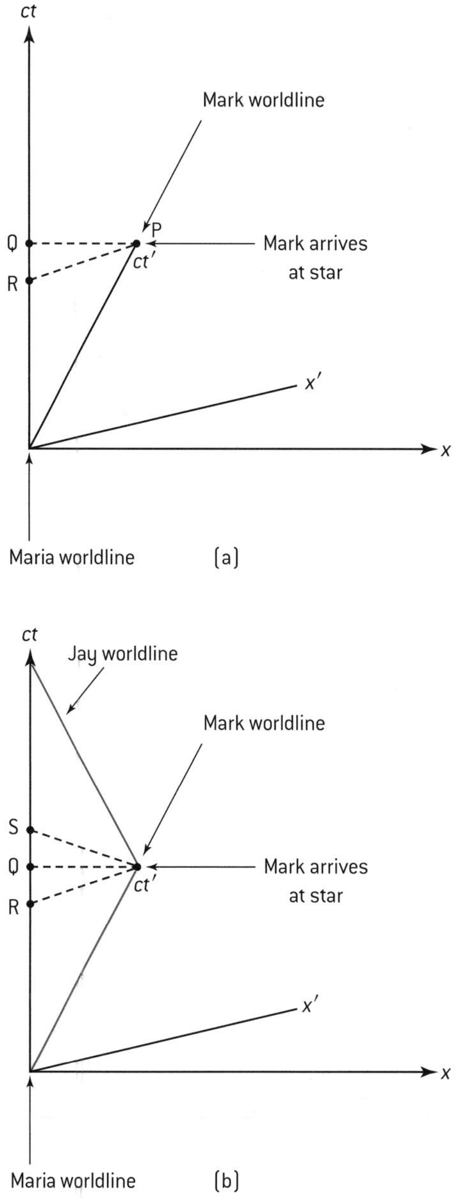

▲ Figure 7

to Jay. Jay is already on his way to Earth (and therefore Maria) at the same speed that Mark had when approaching the star. Figure 7(b) shows the added worldline for Jay. As Jay leaves the star, he thinks that Maria is at S. If Mark were to slow down at the star, turn round, and go back to the Earth, the acceleration procedure would make Maria appear to rapidly age from Q to S.

Worked example

In the distant future a network of four warning beacons W, X, Y and Z is set up to warn spaceship commanders of the approach lanes for planet Earth. The beacons flash in sequence. The spacetime diagram shows the reference frame in which the beacons are at rest and one cycle of the sequence. The worldline for a spaceship is also shown.

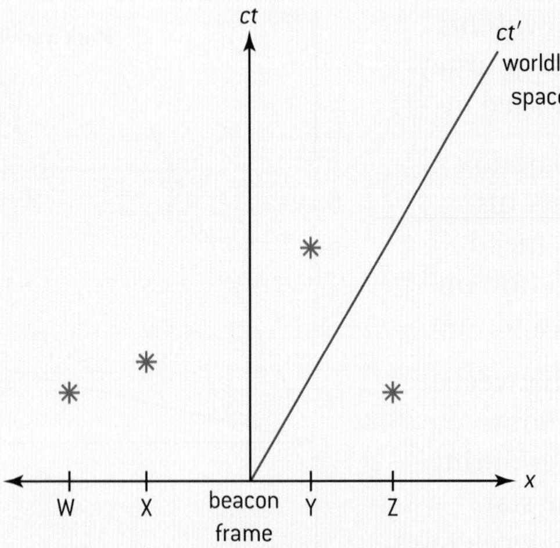

a) Determine the order in which the four beacons flash according to:

(i) an observer stationary in the frame of the beacons

(ii) an observer on the spaceship

b) Determine the order in which the observer on the spaceship sees the beacons flash.

c) Calculate the speed of the spaceship.

Solution

a) (i) The spacetime diagram in the frame of the beacons indicates the chronological order in which the beacons flash: W and Z simultaneously, then X, and then Y.

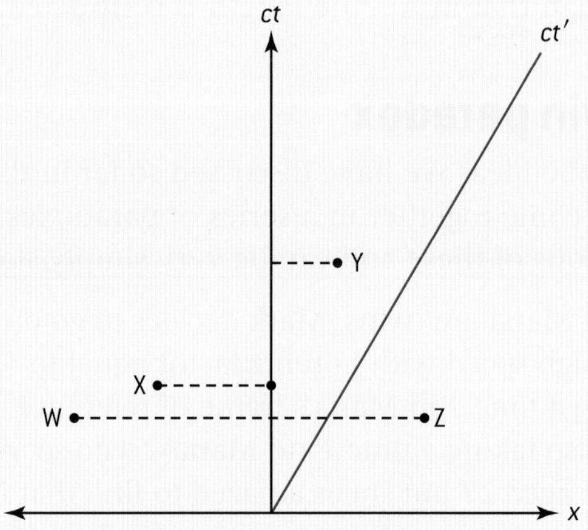

(ii) The order of the flashes in the spaceship frame has to be obtained from constructing lines parallel to the x' axis.

In this frame, Z is observed to flash first, W and X then flash simultaneously, finally Y flashes.

b) To decide on the arrival of the light from the beacons it is necessary to add the photon worldlines to the diagram. These are lines that begin at the beacon flash and travel at 45° to the axes.

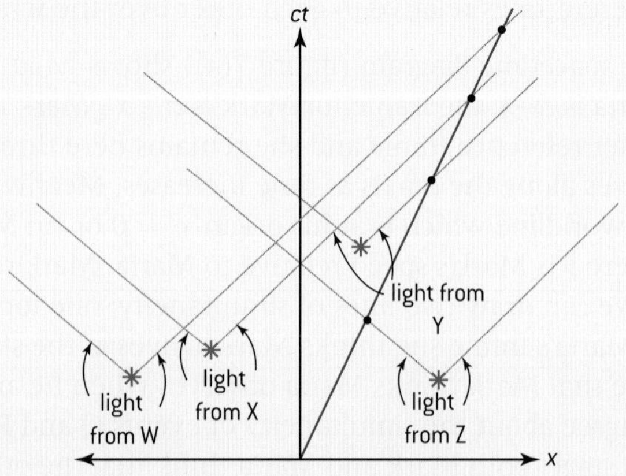

The intersection of the photon worldline with the ct' axis gives the arrival time at the spacecraft. The order is Z, Y, X, W.

c) $\tan\theta = \frac{v}{c} = 0.58$

$v = 0.58c$

A.4 Relativistic mechanics (AHL)

Understanding

- Total energy and rest energy
- Relativistic momentum
- Particle acceleration
- Electric charge as an invariant quantity
- Photons
- MeV c^{-2} as the unit of mass and MeV c^{-1} as the unit of momentum

Applications and skills

- Describing the laws of conservation of momentum and conservation of energy within special relativity
- Determining the potential difference necessary to accelerate a particle to a given speed or energy
- Solving problems involving relativistic energy and momentum conservation in collisions and particle decays

Equations

- total relativistic energy $E = \gamma m_0 c^2$
- rest-mass energy $E_0 = m_0 c^2$
- relativistic kinetic energy $E_K = (\gamma - 1)m_0 c^2$
- relativistic momentum $p = \gamma m_0 v$
- energy-momentum relation $E^2 = p^2 c^2 + m_0^2 c^4$
- acceleration through pd V $qV = \Delta E_K$

Nature of science

A further paradigm shift occurred in physics when Einstein realized that some conservation laws (momentum and energy) broke down as inviolate laws of physics unless modifications were made to them. This led him (amongst other things) to formulate one of the most famous equations in the whole of physics.

So far we have dealt almost exclusively with the basic concepts of space and time and we have shown that they change their nature when we have regard to the relative motion of inertial frames. Now we have to identify and explain the changes that need to take place in the other laws of physics for them to be invariant for all inertial observers.

Total energy and rest energy

In a previous sub-topic we mentioned that the rest mass m_0 of a particle was an invariant quantity. Einstein, in one of his famous 1905 papers, proved that when an object loses energy its mass changes and by so doing he suggested that energy and mass are related. This led him to possibly the most well-known physics equation in the world – and possibly the most misunderstood, $E = mc^2$.

When a particle is viewed from its rest frame then its mass will be observed as the rest mass m_0. This means that the rest energy of the particle is $E_0 = m_0c^2$. Like the rest mass, this is an invariant quantity for a particular particle.

Some physicists also use the concept of relativistic mass, meaning the mass observed when the particle is in a reference frame moving relative to the observer. However, the consequences of energy and mass having an equivalence means that it is not necessary to use both relativistic mass and energy; we will only use the term **total energy** to signify both these ideas. The total energy can easily be converted into an equivalent mass if required using Einstein's equation (but we will not do so in this course).

The total energy E of a particle is equal to the sum of the rest energy E_0 and the kinetic energy E_K:

$$E = E_K + E_0$$

ignoring potential energy and its changes, and energy dissipation. This leads to a set of equations relating the energies and masses:

$$E = m_0 \gamma c^2$$

and therefore

$$E_K = mc^2 - m_0 c^2 = m_0 (\gamma - 1) c^2$$

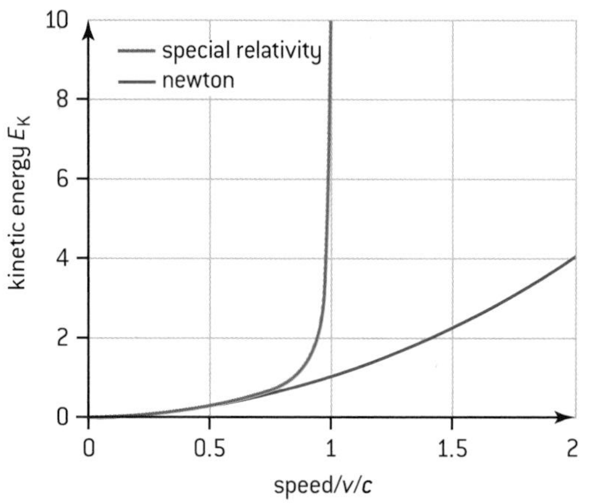

▲ Figure 1 Relativistic kinetic energy against speed.

Figure 1 shows the shape of the graph of E_K against speed (expressed as a fraction of c).

When the speed is zero, the energy is the rest energy. At high speeds approaching the speed of light, the curve becomes asymptotic to the line $\frac{v}{c} = 1$.

Relativistic momentum

Momentum must still be conserved within the special theory. However in order to achieve this, the expression for momentum must incorporate γ to change from the Newtonian mu to $m_0 \gamma u$.

In fact, both momentum and energy are jointly conserved within the theory with the proviso that energy has to include the whole bundle of energies associated with a particle including its rest energy. The momentum p of the particle is expressed as $p = m\gamma u$ and (as before) the total energy is $E = m_0 \gamma c^2$.

Combining these equations and eliminating u leads to an expression for the total energy:

$$E^2 = p^2c^2 + m_0^2c^4 = (pc)^2 + (m_0c^2)^2$$

This is known as the **energy–momentum relation.** It has an important property. If the equation is rearranged as

$$(m_0c^2)^2 = E^2 - (pc)^2$$

then the quantity on the left-hand side is the invariant mass – it does not change between inertial frames. Therefore the right-hand side must also be invariant, so we can write

$$(m_0c^2)^2 = E^2 - (pc)^2 = E'^2 - (p'c)^2$$

The knowledge that $(E^2 - p^2c^2)$ in the frame is invariant allows us to, for example, move easily between the particle frame in a particle accelerator and the lab frame.

Many tests of these equations were made from the time of Einstein's first suggestions of the relationships. The equations were always verified but some of the determinations were indirect and had many sources of error. Perhaps the most direct and convincing experiment was that developed by Bertozzi in 1964. He accelerated electrons to a high speed and measured their speed as they travelled through a vacuum. Immediately after passing through the speed-measuring apparatus, the electrons were absorbed by an aluminium disc, thus transferring their energy to the internal energy of the disc. The temperature of the disc increased as a result and the energy of the incident electrons could be measured directly. Bertozzi's results are shown in figure 2. This was a direct and convincing verification of Einstein's theory.

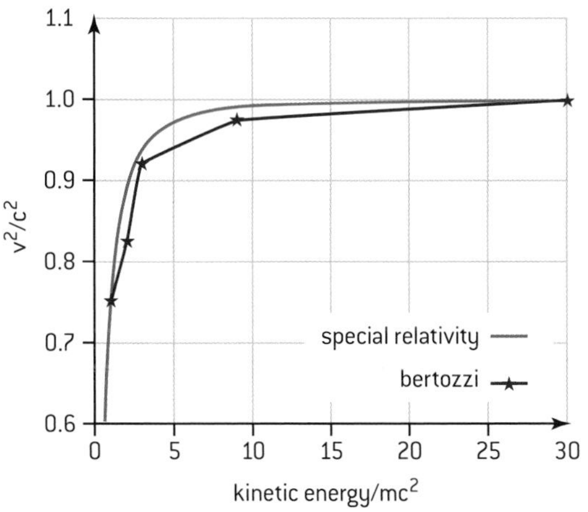

▲ Figure 2 Results of Bertozzi's experiment.

Worked example

1 Calculate the speed at which a particle must travel for its total energy to equal five times its rest mass energy.

Solution

$E = \gamma mc^2 = 5mc^2$

As $\gamma = 5 = \dfrac{1}{\sqrt{1 - \dfrac{v^2}{c^2}}} \rightarrow$

$1 - \dfrac{v^2}{c^2} = \dfrac{1}{25}$ and $\dfrac{v^2}{c^2} = \dfrac{24}{25}$

Therefore, $v = 0.98c$.

Nature of science

Operating a particle accelerator

Although one of the most direct verifications of special theory was only carried out in the 1960s, it was clear that the theory was the appropriate one to use. Cyclotrons had already been designed and built, and in these particle accelerators the circulating particles gain considerable energy. The energy that they gain (in our frame of reference) is found not to vary with $\frac{1}{2}mv^2$ and a relativistic correction for this is required. Essentially, the particles become more massive than would be expected from a Newtonian consideration. As cyclotrons were developed that could supply more and more energy there came a point where a synchronization mechanism was required to allow for the relativistic changes. This led to the development of synchrocyclotrons and other high-energy accelerators.

Particle acceleration

The obvious way to accelerate a charged particle is to place it in an electric field. As we saw in Topic 5 this leads to an electric force on the particle and to a gain in energy that can be expressed in terms of the charge on the particle and the potential difference through which it moves. The key here is that charge q is our fifth invariant quantity so the term γ does not enter into our specification of charge in a moving frame. Thus the charge of the particle during the acceleration does not change and we can see directly (as in earlier parts of this book) that

$$qV = \Delta E_K$$

where ΔE_K is the change in kinetic energy and V is the potential difference.

This leads to a new set of units that are extensively used in particle and relativistic physics. Rather than use kg for mass and kg m s^{-1} for momentum it is much easier to think and work in terms of the energy equivalent of these units.

Thus, for mass we use eV c^{-2} (or, more commonly, multiples of this, MeV c^{-2} and GeV c^{-2}) and for momentum MeV c^{-1} and GeV c^{-1}. What has effectively happened is that c in the energy–momentum relation has been made equal to 1. It is as though the equation has been written:

$$E^2 = p^2 + m_0^{\,2}$$

One of the worked examples below is a repeat of an earlier example to show you how this change in units works.

Photons

Photons have zero mass. What does this mean for their properties within the special theory?

The starting point is the energy–momentum relation once again, but this time m_0 is zero and so the right-most term disappears leaving

$$E^2 = p^2c^2$$

So the momentum of a photon is $p = \frac{E}{c} = \frac{hf}{c} = \frac{h}{\lambda}$

TOK

Conservation laws

This discussion of conservation rules that have needed to be modified in the light of the special theory calls into question the nature of all conservation rules. Do the laws of natural science differ from the laws that exist in other branches of knowledge, for example, in economic theory?

Worked examples

1 A particle of charge $+e$ has a rest mass of 9.1×10^{-31} kg. It is accelerated from rest to a speed of 2.4×10^{8} m s^{-1}.

Calculate, for this particle:

a) the rest energy in MeV

b) the total energy after acceleration

c) the kinetic energy after acceleration

d) the potential difference through which it must be accelerated to reach this speed.

Solution

a) The rest energy $= m_0c^2 = 8.2 \times 10^{-14}$ J; this is $\frac{8.2 \times 10^{-14}}{1.6 \times 10^{-19}} = 0.51$ MeV.

b) The speed is equivalent to $\gamma = 1.7$.

The total energy $= m_0\gamma c^2 = 1.7 \times 8.2 \times 10^{-14} = 1.4 \times 10^{-13}$ J $= 0.87$ MeV

c) The kinetic energy $= (1.7 - 1) \times 8.2 \times 10^{-14} = 5.5 \times 10^{-14}$ J $=$ $(0.87 - 0.51)$ MeV $= 0.36$ MeV

d) To attain a kinetic energy of 360 keV must require a pd of 360 kV.

2 Calculate the momentum of a photon of visible light of wavelength 560 nm.

Solution

$$p = \frac{h}{\lambda}$$

SO $p = \dfrac{6.6 \times 10^{-34}}{5.6 \times 10^{-7}} = 1.2 \times 10^{-27}$ kg m s^{-1}.

3 A small insect with a mass of 1.5×10^{-3} kg flies at 0.48 m s^{-1}. Calculate the momentum and kinetic energy of the insect.

Solution

This is a non-relativistic solution as the speed is much less than that of light.

1 eV $\equiv 1.6 \times 10^{-19}$ J and 1 kg $\equiv 9 \times 10^{16}$ J $= 9 \times 10^{4}$ TJ.

Kinetic energy $= \dfrac{1}{2} \times 1.5 \times 10^{-3} \times 0.48^2 = 1.7 \times 10^{-4}$ J

$= \dfrac{1.7 \times 10^{-4}}{1.6 \times 10^{-19}} = 1.1 \times 10^{15}$ eV $= 1.1 \times 10^{3}$ TeV

Momentum $= 1.5 \times 10^{-3} \times 0.48 = 7.2 \times 10^{-4}$ kg m s^{-1} $= 7.2 \times 10^{-4} \times 3 \times 10^{8} \div 1.6 \times 10^{-19} = 1.35 \times 10^{24}$ eV c^{-1} $= 1.4 \times 10^{12}$ TeV c^{-1}.

A.5 General relativity (AHL)

Understanding

- The equivalence principle
- The bending of light
- Gravitational redshift and the Pound-Rebka-Snider experiment
- Schwarzschild black holes
- Event horizons
- Time dilation near a black hole
- Applications of general relativity to the universe as a whole

Nature of science

After publishing his special theory, which applied to non-accelerated reference frames, Einstein tackled the general theory incorporating the effects of acceleration and gravity. This required intuition, imagination and creative thinking on his part. He needed to modify his ideas of spacetime in the light of the effects of mass and the curvature that it produces. In this way he was responsible for yet another paradigm shift in physics.

Applications and skills

- Using the equivalence principle to deduce and explain light bending near massive objects
- Using the equivalence principle to deduce and explain gravitational time dilation
- Calculating gravitational frequency shifts
- Describing an experiment in which gravitational redshift is observed and measured
- Calculating the Schwarzschild radius of a black hole
- Applying the formula for gravitational time dilation near the event horizon of a black hole

Equations

- Gravitational frequency shift: $\frac{\Delta f}{f} = \frac{g\Delta h}{c^2}$
- Schwarzchild radius: $R_s = \frac{2GM}{c^2}$
- gravitational time dilation $\Delta t = \frac{\Delta t_0}{\sqrt{1 - \frac{R_s}{r}}}$

The equivalence principle

In Topic 2 we introduced the idea that there are two types of mass: inertial and gravitational and we suggested that these are equivalent. Mass that is gravitationally attracted is taken to be the same as the mass that responds to a force by accelerating. This equivalence had been recognized since the time of Galileo but was first discussed in detail by the German physicist Mach at the end of the nineteenth century (having been touched on by various philosophers before him). Einstein and others named the central idea: Mach's principle. Mach had rejected Newton's view of absolute time and space, taking a relational view of the universe in which any motion can only be seen with respect to other objects in the universe. Thus we cannot say merely that an object is rotating but must refer to the axis about which it rotates. Mach's writings, although controversial amongst scientists, partly led to Einstein developing the general theory of relativity that he published in 1916. In this he proposed a principle of equivalence:

Einstein's principle of equivalence states that gravitational effects cannot be distinguished from inertial effects.

A thought experiment helps to explain the equivalence principle; the experiment involves two observers and an elevator (lift). The elevator is a long way from any other mass so that it is essentially a gravity-free zone.

One of the observers (X) is in the elevator and carries an object that has mass; X cannot see out of the elevator. Another observer (Y) is outside the elevator and not connected to it in any way but can view what happens inside.

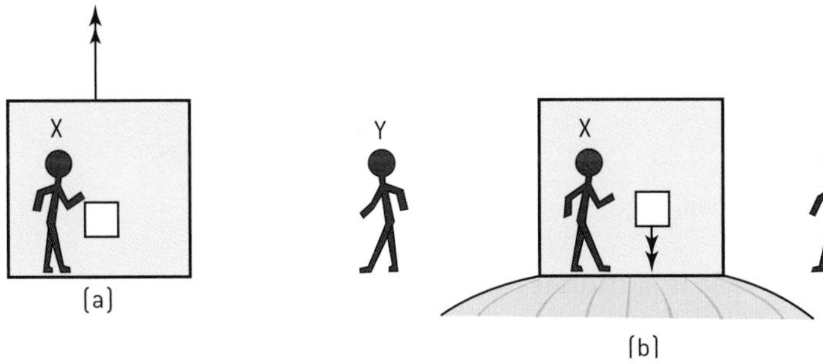

▲ Figure 1 The elevator "thought" experiment.

Initially the elevator is moving at constant velocity. X releases the object and it stays exactly where it is placed. (Remember that there are no nearby masses or planets to attract the mass.)

X repeats the experiment (Figure 1(a)) but this time, as X releases the object, the elevator begins to accelerate at 9.8 m s^{-2} in the direction of the elevator roof (we will call this "upwards" for brevity). X will think that the object accelerates downwards at 9.8 m s^{-2}. Y observes from outside that the object stays where it is in space and that the lift is accelerating upwards around it.

Consider the same experiment repeated (Figure 1(b)) with the elevator stationary relative to and close to the surface of the Earth. When X releases the object it will be accelerated downwards at 9.8 m s^{-2}. Y will explain this acceleration as due to the attraction of the Earth.

The important point here is that in both experiments the object accelerated downwards, in one case due to an inertial (acceleration) effect and in the other due to the effects of gravity. But to observer X in the elevator these two cases are identical and cannot be distinguished. It is this inability to decide what causes the acceleration that lies at the heart of the principle of equivalence. There is in principle no experiment that the observer in the lift can carry out to decide which effect is which.

In figure 1(b) we can either regard the lift as accelerating within a universe (i.e. everything else) that is stationary or we can view the lift as stationary in the frame of observer X with the rest of the universe accelerating around it. According to Einstein there is no absolute motion, only relative motion exists. This means that the special status held by observers in inertial frames of reference in the special theory no longer exists in the general theory and – accelerated or not – all observers have the same status and obey the same laws. However, it may be that some frames allow the laws to be stated in a more straightforward way.

Gravitational redshift

The general theory predicts that gravity can affect the motion of light itself. The equivalence principle helps us understand why.

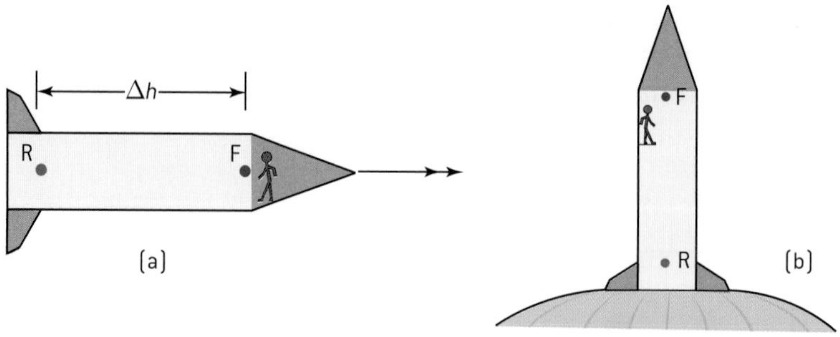

▲ Figure 2 Equivalence in action.

Imagine a spaceship that is stationary (or moving at constant velocity) relative to the rest of the universe. Inside the ship are an observer and two identical light sources, one light source at the rear of the ship (lamp R) and one close to the observer at the front (lamp F).

The spaceship begins to accelerate Figure 2(a)). At the instant that the acceleration begins, both light sources start to emit light of the same frequency. (It helps to imagine that the light emission begins as the spaceship starts to accelerate, but this is not an essential part of the argument.)

The light from R takes a finite time $\Delta t = \frac{\Delta h}{c}$ to reach the observer (where Δh is the distance between the light sources). During this time the velocity of the spaceship and observer will have changed by an amount $\Delta v = g\Delta t$, where g is the acceleration. However, the light in transit from R to the observer will not change its speed because (as usual) the speed of light is independent of the observer. The observer will observe a longer wavelength for the light from R compared to the light from F as lamp R will appear to have been Doppler redshifted.

This can be made quantitative. Merging the two equations above gives

$$\Delta v = g\Delta t = \frac{g\Delta h}{c}$$

Δv is the change in speed of the observer over the time period, Δt is relative to the speed of light source R when it emitted the observed light, and so the frequency shift Δf observed is

$$\frac{\Delta f}{f} = \frac{\Delta v}{c} = \frac{g\Delta h}{c^2}$$

where f is the frequency of the emitted light.

The equivalence principle predicts that the effects in a gravitational field should be indistinguishable from an inertial system, so we do not need necessarily to say more. However, it is instructive to consider the problem from another different (but ultimately identical) standpoint.

Figure 2(b) shows the spaceship at rest relative to, and sitting on, the surface of the Earth. Again lamp R emits light to the observer. Each emitted light photon has energy hf when it leaves the source. To reach the observer some of this energy has to be traded off against the gravitational field as gravitational potential energy. This loss of photon energy (reduction in f) is equivalent to an increase in the wavelength λ of the light ($f\lambda = c$) and so again, the light from R is redshifted compared to the light from lamp F.

These discussions of frequency and wavelength shifts lead to the idea of **gravitational time dilation.** We can regard the arrival of the wave as a series of ticks of a clock. A redshift means that the clock is observed to tick more slowly than the original. So the observer at the front of the rocket thinks that lamp R is "ticking" more slowly than lamp F and the observer at the top of a mountain on Earth thinks that time is running more slowly for an observer at sea level.

This effect is gravitational time dilation. It is different from the time dilation effect observed when inertial frames are moving relative to each other.

We mentioned the GPS system earlier (p 521) in the context of special relativity, which predicts that the clocks on the satellites fall behind ground clocks by about 7 μs every 24 hours due to their relative motion with respect to the surface. General relativity, on the other hand, predicts that the satellite clocks should advance compared with the surface clocks by about 45 μs in the same time period. The net result is that the clocks in the GPS satellite gain on clocks back on Earth by about 38 μs every day. This factor swamps the 20 ns accuracy required of the Earth-bound GPS receivers. If relativistic effects are not taken into account, then the errors in a position become serious after about 100 s and accumulate at a rate of tens of kilometres every day. This would be completely unacceptable for any navigational systems. The GPS receivers in our cars and on our mobile phones are constantly carrying out relativistic corrections to adjust for the unavoidable time changes due to relativistic effects.

Worked examples

1 The frequency of a line in the emission spectrum of sodium is measured in a frame of reference in which the sodium source is stationary and well away from gravitational influences. Calculate the fractional frequency shift that will be measured by an observer also stationary with respect to the sodium source but placed 1 km above the source close to the surface of Earth.

Solution

The fractional frequency shift is given by

$$\frac{\Delta f}{f} = \frac{g\Delta h}{c^2}$$

so the wavelength shift is $\frac{\Delta \lambda}{\lambda} = \frac{g\Delta h}{c^2} = \frac{9.8 \times 1000}{9 \times 10^{16}} \approx 1.1 \times 10^{-13}$

This is a redshift so the wavelength is increased by this fractional amount.

2 Calculate the difference in time per day due to gravitational redshift of a clock at the top of Everest compared to a clock at sea level. Mount Everest is 8800 m above sea level. Assume that the acceleration due to gravity is constant over this height difference.

Solution

$\frac{\Delta T}{T} = \frac{\Delta f}{f}$ where T is the time for one day.

This means that $\frac{\Delta T}{T} = \frac{g\Delta h}{c^2}$ and $\Delta T = \frac{gT\Delta h}{c^2} = \frac{9.8 \times 86\,400 \times 8800}{(3 \times 10^8)^2}$ this is 80 ns every day. The observer on the mountain thinks that the sea-level clock is running more slowly so the mountain clock appears to gain time.

Nature of science

Tidal forces, and a more precise definition

The earlier explanation of the elevator "thought" experiment was too simplistic. The equivalence of gravity and acceleration is true only for uniform gravitational fields. Real gravitational fields will almost always be non-uniform and inhomogeneous.

At the surface of the Earth with human-sized experiments, the field is very close to uniform and nearby objects appear to fall in parallel directions with the same acceleration. But two objects in a very large elevator will converge towards the centre of the Earth as they fall. An observer in the elevator sees the two objects apparently moving closer together – and will wonder why.

This is known as a tidal effect (figure 3) and means that a precise statement of the principle of equivalence should include a clause stating that it applies when tidal effects can be neglected. Alternatively, we can say that over an infinitesimally small spacetime region the laws of physics in general relativity are equivalent to those under special relativity.

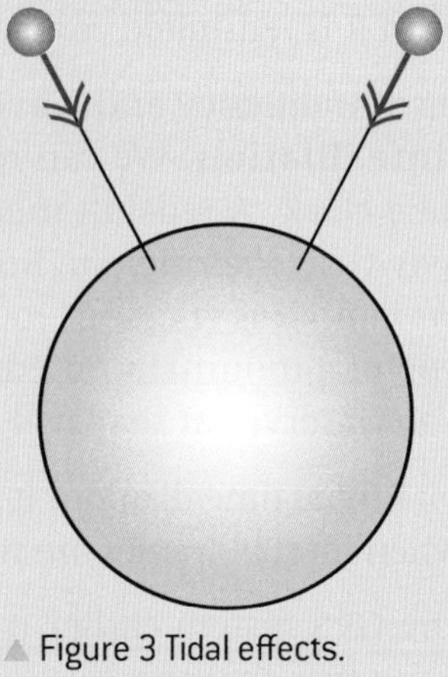

▲ Figure 3 Tidal effects.

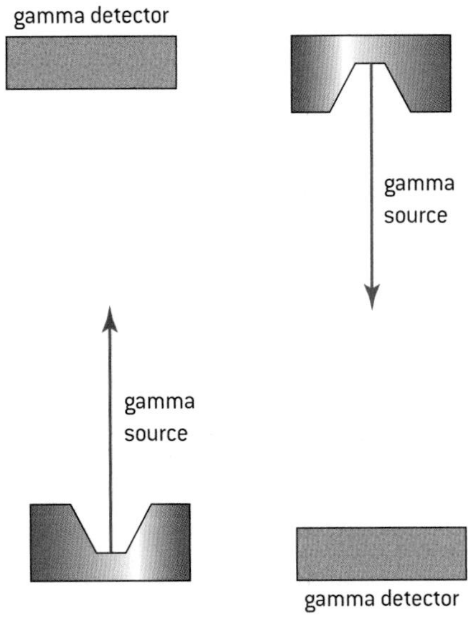

▲ Figure 4 Pound–Rebka–Snider experiment.

Although the need for relativistic corrections in the GPS satellite navigation system indicates the truth of Einstein's general theory, there have been formal scientific tests made over the past century to verify that it is correct. They include experiments some of which were suggested by Einstein himself:

- The precession of the perihelion of Mercury (the point in its orbit where Mercury is closest to the Sun)
- Deflection of light by the Sun
- Gravitational redshift of light

One of the most recent of these is the experiment devised by Pound, Rebka and Snider in 1959 to measure gravitational redshift. This experiment was the last of Einstein's suggestions to be attempted using the (then) recently discovered technique of Mossbauer spectroscopy. This technique allows small shifts in the frequencies of gamma rays to be measured.

In a very precise experiment, Pound and his co-workers fired a beam of gamma rays vertically upwards towards a detector placed about 22 m above the gamma source. They repeated the experiment with the gamma source firing downwards to the detector. If the upward and downward fractional changes in energy of the gammas are compared then

$$\left(\frac{\Delta E}{E}\right)_{\text{upwards}} - \left(\frac{\Delta E}{E}\right)_{\text{downwards}} = \frac{2g\Delta h}{c^2}$$

For the Pound–Rebka–Snider experiment, $\Delta h = 22.6$ m and $g = 9.81$ m s^{-2} which leads to a theoretical difference in the fractional energies of 4.9×10^{-15}. They measured the difference to be $(5.1 \pm 0.5) \times 10^{-15}$ which compares well with the theoretical value. This was a convincing verification of the general theory.

If a single planet orbits a star, then Newtonian gravitation and mechanics predict that the orbit of the planet will follow the same elliptical path forever. The presence of other nearby planets and moons, however, disturbs this motion. In practice, planets in the solar system orbit the Sun in an ellipse that rotates gradually around the Sun; this is known as precession. The rate at which this rotation occurs can be predicted very accurately on the basis of Newtonian mechanics.

Observations made in 1859 showed that the precession of Mercury is faster than predicted by Newtonian mechanics. The change to the precession rate is small but much larger than the error in its measurement; there is no doubt that the precession prediction is incorrect. One of the first successes of Einstein's general theory was that it correctly predicted the value of the observed precession rate of Mercury.

Einstein's third suggestion was that a massive star would deflect light from its straight-line path. Shortly after Einstein published his theory, the English physicist Eddington travelled to the west coast of Africa with colleagues to observe stars during the 1919 total eclipse of the Sun. This enabled them to confirm that the mass of the Sun deflects light according to the general theory. This will be discussed in more detail in the next section.

A fourth effect was tested by Irwin Shapiro in 1964. Einstein had predicted that the time for a radar (microwave) pulse of radiation to go past the Sun and return to Earth after reflection from Venus and Mercury would take longer than expected because of the effect of the Sun's gravity. Shapiro measured this time and made a similar measurement when the signal did not travel close to the Sun (because the planets had moved on in their orbits). He found that the delay existed as Einstein had predicted and that its magnitude was as expected. Similar experiments have been repeated in various ways since the 1960s, and always indicate a time delay as predicted by Einstein.

The bending of light

So far we have discussed the general theory in terms of the equivalence it suggests between gravitational and inertial effects and the changes it predicts for astronomical observations. The general theory, of course, offers much more than this.

For Newtonian gravitation, the gravity field is a model of reality that can be analysed using the law of gravitation. In the general theory, Einstein constructed a set of ten equations known as the Einstein field equations. These equations indicate that gravity is the effect observed when spacetime is curved (distorted) by the presence of mass and energy.

An analogy for spacetime curvature is that of a rubber sheet. The sheet represents a two-dimensional space in a three-dimensional spacetime continuum. (This means that it is only an analogy to the real world because one dimension has been suppressed.) In the absence of energy or mass, the sheet is flat, horizontal and undistorted. However, if a mass is placed on the sheet, then the sheet deforms under the influence of the mass as shown in figure 5.

Figure 6 A primary star and lensing effect.

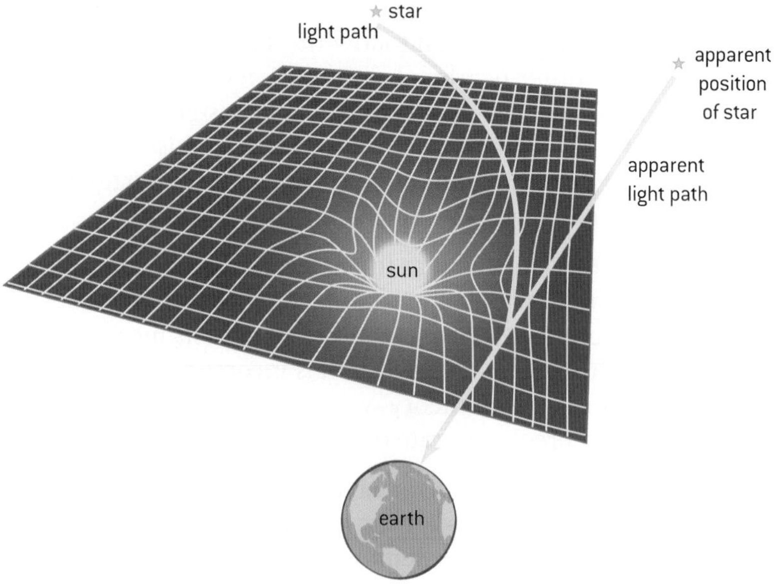

Figure 5 Rubber sheet analogy for gravity.

The effect of this deformation on the passage of light from a distant star can be seen in figure 5. Light comes from the real location of the star but, as it passes near the Sun, the direction of the light is altered by the warping of spacetime. After leaving the vicinity of the Sun the light appears to have come from a different direction. In 1919 Eddington was able to measure this apparent shift in the star position when the Sun moves close to the line between the Earth and the star. Measurements were made in Brazil and Africa (both places where the eclipse was total) and they confirmed the predictions made by the general theory. However the data were hard to collect and it is only recently that re-workings of the data have confirmed that Eddington's conclusion was not affected by observational errors and confirmation bias.

Of course, the **deflection** effect is happening in the four dimensions of spacetime, not the three of our analogy. This means that a three-dimensional (to us) gravitational lensing can be observed. Figure 6 is a striking image taken by the Faint Object Camera of a European Space Agency satellite showing the primary image of a star and four additional images of it formed by this lensing effect.

The Einstein field equations do not always have exact solutions (in the sense that two simultaneous equations with two unknowns have an exact solution). For example, at present, the equations cannot provide an exact solution for the spacetime of two binary stars – one of the commonest star arrangements. Physicists usually make simplifying approximations when studying the implications of the equations. If, for example, the assumptions of low speeds and very weak gravity are made, then the field equations can be manipulated to give Newton's law of gravitation which, therefore, proves to be a special case of general relativity.

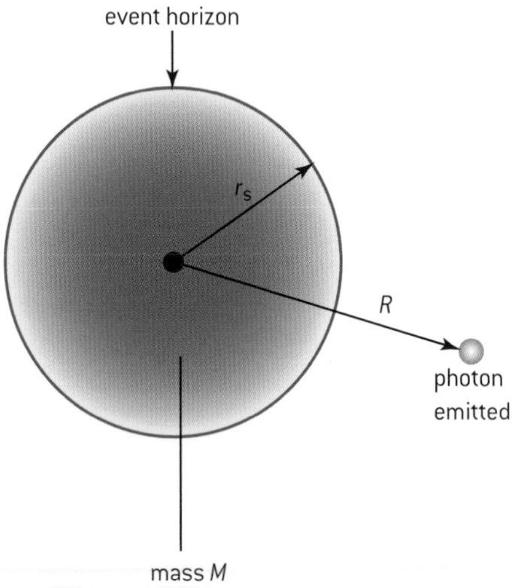

Figure 7 The Schwarzschild radius.

Schwarzschild black holes

Within a few weeks of the publication of the general theory, Karl Schwarzchild was able to produce one of the first exact solutions of the Einstein field equations. He assumed the presence in spacetime of a spherical, non-rotating, uncharged mass and was able to derive equations for the gravitational field that surrounds such an object.

Using Schwarzchild's equations, the photon redshift that results from the effect of the gravitational field of the mass can be derived and is given (approximately) by

$$\frac{\Delta\lambda}{\lambda} = \frac{R_s}{2r} = \frac{gM}{rc^2}$$

where λ is the emitted wavelength shifted by $\delta\lambda$ and r is the distance from the centre of the mass to the point at which the photon is emitted and the constant $R_s = \frac{2GM}{c^2}$. This approximate equation assumes that r is much greater than R_s. Notice the similarity between this equation and the earlier expression for the fractional change in frequency

$$\frac{\Delta f}{f} = \frac{\Delta v}{c} = \frac{g\Delta h}{c^2}$$

The constant in the expression, R_s, has the dimensions of length and is known as the **Schwarzschild radius**; it is equal to $\frac{2GM}{c^2}$.

Outside the Schwarzchild radius ($r > R_s$) gravity obeys the usual rules, but inside the spherical region defined R_s where $r < R_s$ the normal structure of spacetime does not apply.

The change between the two regimes where $r = R_s$ is known as an **event horizon**. Events happening inside the event horizon cannot affect an observer outside the horizon. Although the event horizon is not a true boundary (observers are able to pass through it from outside to inside), the event horizon represents the surface where the gravitational pull is so large that nothing can escape – not even light. A different interpretation of the event horizon is that it is the surface at which the speed needed to escape from the mass becomes equal to the speed of light. Light emitted from inside the event horizon cannot escape.

The strong gravitational field (extreme warping of spacetime) associated with the black hole, attracts nearby mass to it. Mass will appear to collapse towards the centre of the black hole.

Clocks in a strong gravitational field run more slowly than in the absence of gravity. The same phenomenon is observed near an event horizon. As a clock moves towards the event horizon from the outside, an external observer will see the clock slow down with the clock never quite passing through the horizon itself. The light emitted from the clock will also be increasingly gravitationally redshifted as the clock approaches the Schwarzschild radius (we already know that this is equivalent to time dilation). We might expect the situation in the frame of the clock to be different and, indeed, the clock (in its frame) will pass through the horizon in a finite amount of proper time.

The proper time interval Δt_0 for an observer at distance R_0 within the gravitational field of the mass that has a Schwarzchild radius of R_s is related to the time interval Δt_d measured by a distant observer by

$$\Delta t_0 = \Delta t_d\sqrt{1 - \frac{R_s}{R_0}}$$

This equation assumes that the object giving rise to the event horizon does not rotate.

Worked example

Calculate the Schwarzchild radius for:

a) the Earth (mass $= 6 \times 10^{24}$ kg)

b) the Sun (mass $= 2.0 \times 10^{30}$ kg).

Solution

a) The Schwarzchild radius $= \frac{2GM}{c^2}$

$= \frac{2 \times 6.67 \times 10^{-11} \times 6 \times 10^{24}}{9 \times 10^{16}}$

$= 8.9$ mm

b) A similar substitution for the mass of the Sun gives: 3.0 km.

Worked example

Ticks separated by intervals of 1.0 s are emitted by a clock that is 2×10^5 m from the event horizon of a black hole of mass 3×10^{31} kg. Calculate the number of ticks detected by an observer in a distant rocket in a period of 10 minutes in the frame of reference of the rocket.

Solution

The Schwarzchild radius is $\frac{2 \times 6.67 \times 10^{-11} \times 3 \times 10^{31}}{9 \times 10^{16}} = 4.5 \times 10^4$ m

$\Delta t_p = \Delta t_0\sqrt{1 - \frac{R_s}{R_0}}$ and when the measured time interval on the rocket ship is 600 s the proper time at the horizon is

$600 \times \sqrt{1 - \frac{4.5 \times 10^4}{2.0 \times 10^{31}}}$

$= 529$ s. There will be 592 ticks.

This applies to all bodies that have mass and the worked examples show typical Schwarzchild radii for planetary and solar bodies. In theory, any object that can be compressed sufficiently for all its mass to be inside the event horizon will demonstrate these effects. The crucial point is whether this compression is possible or not. For almost all objects it is not, because the gravitational self-attraction of a planet's mass is insufficient to overcome the repulsion of the electrons in the atomic shells of atoms. However, for very dense objects, the mass can fit inside the Schwarzchild radius and this object then becomes a black hole. A star needs, typically, to be about three times the mass of the Sun for this to occur.

Black holes are massive, extremely dense objects from which matter and radiation cannot escape. Spacetime inside the event horizon is so warped that any path taken by light inside the Schwarzchild radius will curve farther into the black hole.

Nature of science

Links between classical and quantum physics

For many years after Einstein's discovery it was thought that nothing could escape a black hole. It is now recognized that this is only true for black holes under a classical theory. Under quantum theory a black hole can be shown to radiate in a similar way to a black body. This result was unexpected and led to other connections between black holes and the classical study of thermodynamics.

The work of the physicist Stephen Hawking and others has led to an understanding of the thermodynamics and mechanics of black holes, so that it is now recognized that black holes (or rather the strong gravity field near them) can lead to the emission of Hawking radiation. This would in principle allow a black hole to be observed and studied. In practice, the radiation emitted by mass being accelerated towards the event horizon will swamp the Hawking radiation. This emission of this radiation implies that black holes can evaporate over time leading to a dynamic process of hole creation and disappearance,

Applications of general relativity to the universe as a whole

There are many topics in cosmology that rely on general relativity. Some of these have been solved; some remain the subject of active research. Others remain as tantalizing theoretical predictions.

- Studies of the Einstein field equations lead to a number of different solutions (of which Schwarzchild's was one of the first) in particular the solutions of Friedmann, Lemaitre, Robertson and Walker (FLRW). These allow the behaviour of the universe over its lifetime to be modelled and the equations have been highly successful in their application. Many aspects of the early universe are illustrated in the FLRW solutions including the large-scale structure of the universe, the way in which chemical elements were created, and the presence of the cosmic background radiation. Additionally, knowledge of the rate at which the universe is expanding allows the total mass of the universe to be estimated.

The answer the equations produce does not agree with the amount of matter that we can actually see around us in the universe. Physicists now suggest the presence of an (at present) unobservable dark matter and dark energy.

- There is a suggestion that single super-massive black holes can be found at the centre of galaxies. The mass of these objects can range up to a mass of several billion times that of our Sun. Such black holes will have been influential in the formation of galaxies and other larger structures in the universe. As interstellar dust and other materials fall into the galactic black holes, a number of artefacts appear, many of which are predicted by the general theory. These include the emission of jets of very energetic particles at speeds close to that of light that can in principle be observed. Astronomers look carefully for evidence of black holes at the centre of galaxies. There is a strong candidate in the Milky Way with an object (Sagittarius A*) that has a diameter similar to the radius of Uranus but a mass about 4 million times greater than that of the Sun.
- Binary black holes, normally in orbit around each other, can merge. The general theory suggests that the resulting event should lead to the emission of gravitational waves. Other events in the universe may also cause gravitational-wave emission. There are experiments in operation and others being devised that will, it is hoped, detect the effects of these gravitational waves and provide further verification of some of Einstein's predictions.

TOK

The unnecessary constant

One form of the Einstein field equations is

$$R_{\mu v} - \frac{1}{2} g_{\mu v} R + g_{\mu v} \Lambda = \frac{8\pi G}{c^4} T_{\mu v}$$

(this is *not* going to be tested in the examination!).

R and g tell us about the curvature of spacetime, and T is concerned with the matter and energy in the universe. The constants G and c have their usual meaning.

The reason for including this equation is to draw your attention to the third term on the left-hand side of the equation and the constant Λ (capital lambda) that it contains. Einstein called Λ the *cosmological constant*.

In the early part of the twentieth century when Einstein was working on the general theory, the scientific view was that the universe was stationary and unchanging. Einstein realized that his theory predicted a universe that was neither static nor unchanging – according to the equations it should be growing larger. Einstein add*ed* the Λ term to adjust the theory to predict a stationary state. A few years later the astronomer Edwin Hubble found strong evidence that the universe is expanding. At this point, Einstein realized that his inclusion of Λ had been unnecessary. He later described putting the constant into the equation as "my greatest blunder".

What other examples are there of scientists whose work was initially rejected (either by themselves or others) but which was later accepted?

Questions

1 Explain what is meant by an *inertial frame of reference*. (1 mark)

2 An electron is travelling parallel to a metal wire that carries an electric current. Discuss the nature of the force on the electron in terms of the frame of reference of:

a) a proton stationary with respect to the wire

b) the moving electron.

3 Two electrons are travelling directly towards one another. Each has a speed of $0.002c$ relative to a stationary observer. Calculate the relative velocity of approach, as measured in the frame of reference of one of the electrons according to the Galilean transformation. (2 mark)

4 *(IB)*

a) Define:

(i) *proper length*

(ii) *proper time*

Muons are accelerated to a speed of $0.95c$ as measured in the reference frame of the laboratory. They are counted by detector 1 and any muons that do not decay are counted by detector 2. The distance between detector 1 and 2 is 1370 m.

Half the number of muons pass through detector 2 as pass through detector 1 in the same given time.

b) Determine, the half-life of the muons

(i) in the laboratory frame

(ii) in the rest frame of the muons.

(iii) Detemine the separation of the counters in the muon rest frame.

c) Use your answers in (b) to explain what is meant by the terms *time dilation* and *length contraction*. (11 marks)

5 *(IB)*
The radioactive decay of a particular nuclide involves the release of a β-particle. A beta-particle detector is placed 0.37 m from the actinium source, as measured in the laboratory reference frame. The Lorentz factor of the beta particle after release is 4.9.

a) Calculate, for the laboratory reference frame:

(i) the speed of the β-particle

(ii) the time taken for the β-particle to reach the detector.

b) The events described in (a) can be described in the β-particle's frame of reference.

For the frame of the beta particle:

(i) state the speed of the detector

(ii) calculate the distance travelled by the detector. (7 marks)

6 *(IB)*

a) Explain what is meant by an *inertial frame of reference*.

b) An observer in reference frame A measures the relativistic mass and the length of an object that is at rest in their reference frame. The observer also measures the time interval between two events that take place at one point in the reference frame. The relativistic mass and length of the object and the time interval are also measured by a second observer in reference frame B. B is moving at constant velocity relative to A.

c) **(i)** State whether the observer in frame B measures the quantities as being larger, the same size or smaller than when measured in frame A.

(ii) Compare the density of the object measured in frame B with the same measurement made in A. (7 marks)

7 Some students are marooned on a planet carrying out field work when it becomes clear that the star of the planet system is about to become a supernova. A spaceship is despatched to rescue them and the students can be beamed

aboard the spaceship without the need for it to change speed. The spacetime diagram shows the frame of the star and planet and the frame of the spaceship. The star and the planet do not move relative to each other. The star becomes a supernova at spacetime point S.

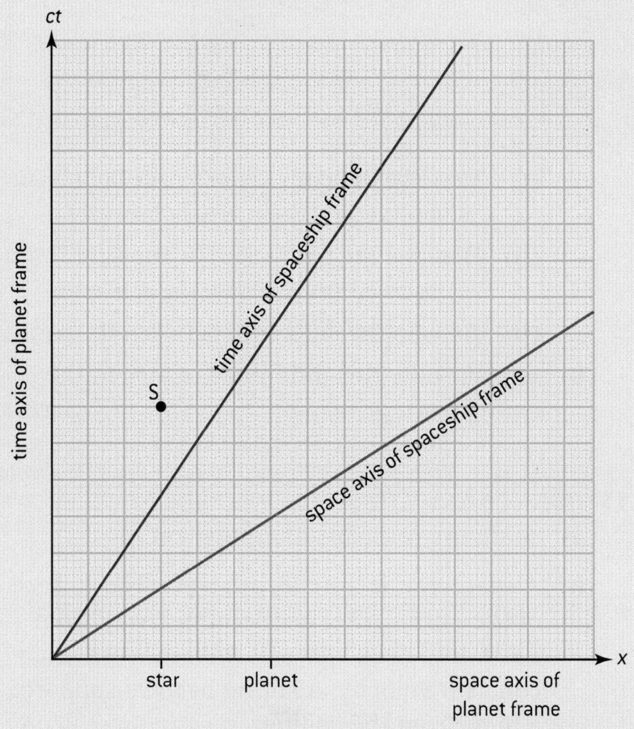

a) Copy the diagram to scale and on it identify:

(i) the spacetime point at which the spaceship arrives at the planet

(ii) the spacetime point of the star when the spaceship arrives at the planet.

b) Discuss whether the spaceship arrives at the planet before or after the supernova:

(i) in the frame of the spaceship

(ii) in the frame of the planet.

c) Discuss when the IB students see the supernova. (9 marks)

8 Two athletes compete in a race. Athlete S is slow and is awarded a handicap so that he has less far to run than athlete F. The spacetime diagram for the race is shown for the frame of reference of the referee who is stationary relative to the finishing lines.

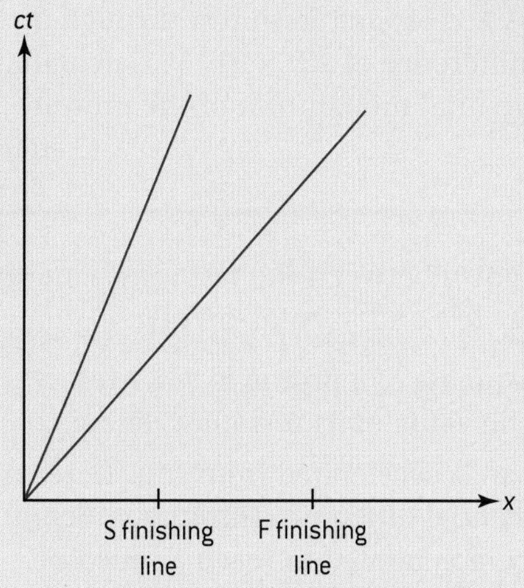

a) Discuss who wins the race in the reference frame of:

(i) F

(ii) S

(iii) the referee.

b) Discuss whether there is agreement about the result of the race. (10 marks)

9 *(IB)*

a) Define *rest mass.*

b) An electron of rest mass m_0 is accelerated through a potential difference V. Explain why, for large values of V, the equation $\frac{1}{2}m_0v^2 = eV$ cannot be used to determine the speed v of the accelerated electron.

c) Determine the mass equivalence of the change in kinetic energy of the electron when V is 5 MV. (7 marks)

10 *(IB)*

a) A charged particle of rest mass m_0 and carrying charge e, is accelerated from rest through a potential difference V. Deduce that

$$\gamma = 1 + \frac{eV}{m_0c^2}$$

where γ is the Lorentz factor and c is the speed of light in free space.

b) Calculate the speed attained by a proton accelerated from rest through a potential difference of 500 MV. (5 marks)

11 A proton is accelerated from rest through a potential difference of 2.0×10^9 V. Calculate, in MeV c^{-1} the final momentum of the proton. (3 marks)

12 *(IB)*

a) Distinguish between the rest mass–energy of a particle and its total energy.

b) The rest mass of a proton is 938 MeV c^{-2}. State the value of its rest mass–energy.

c) A proton is accelerated from rest. Determine the potential difference through which it must be accelerated to reach a speed of 0.98*c*, as measured by a stationary observer in the laboratory reference frame. (9 marks)

13 *(IB)*

a) In both the special and general theories of relativity, Einstein introduced the idea of *spacetime*.

Describe, with a diagram, what is meant by *spacetime* with reference to a particle that is far from other masses and is moving with a constant velocity.

b) Explain how the general theory of relativity accounts for the gravitational attraction between the Earth and an orbiting satellite.

c) Describe what is meant by a *black hole*.

d) Estimate the radius of the Sun necessary for it to become a black hole. (12 marks)

14 *(IB)*

a) A spaceship in a gravity-free region of space accelerates uniformly with respect to an inertial observer, in a direction perpendicular to its base. A narrow beam of light is initially directed parallel to the base.

Diagram 1: View with respect to inertial observer **Diagram 2:** View with respect to observer in ship

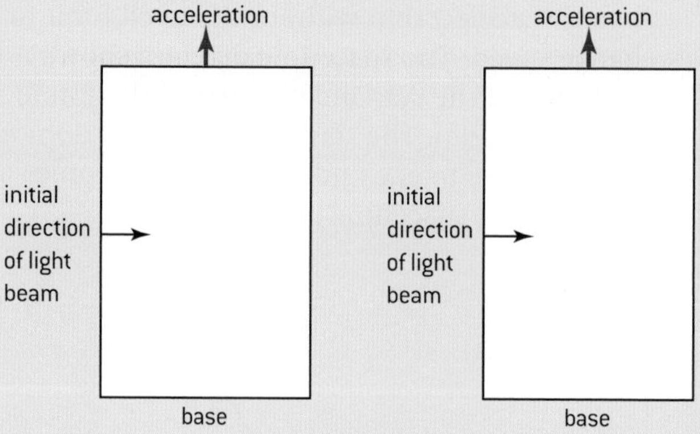

(i) Describe the path taken by the light beam as observed by the inertial observer.

(ii) Describe the path taken by the light beam as observed from within the spaceship.

(iii) Explain the difference between the paths.

b) Explain how this difference relates to the equivalence principle. (7 marks)

15 *(IB)*

a) State the principle of equivalence.

b) A spacecraft is initially at rest on the surface of the Earth. After accelerating away from Earth into deep space, it then moves with a constant velocity. A spring balance supports a mass from the ceiling.

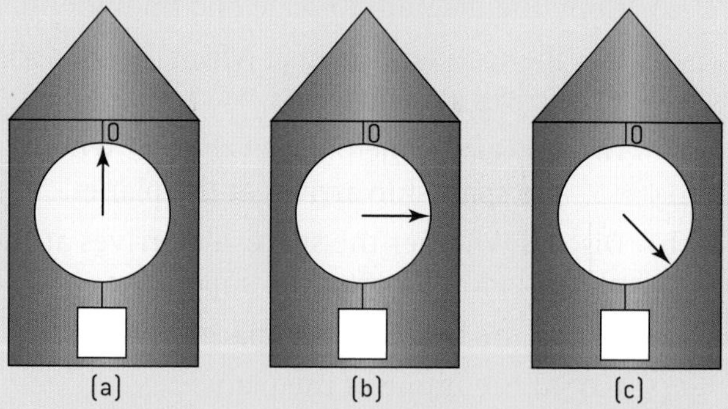

The diagrams show the readings on the spring balance at different stages of the motion.

Identify, with an explanation, the reading that will be obtained when the spaceship is:

(i) at rest on the Earth's surface

(ii) moving away from Earth with acceleration

(iii) moving at constant velocity in deep space.

c) The spacecraft now accelerates in deep space such that the acceleration equals that of free fall at the Earth's surface.

State and explain which reading would be observed on the spring balance. (10 marks)

16 *(IB)*
The gravitational field of a black hole warps spacetime.

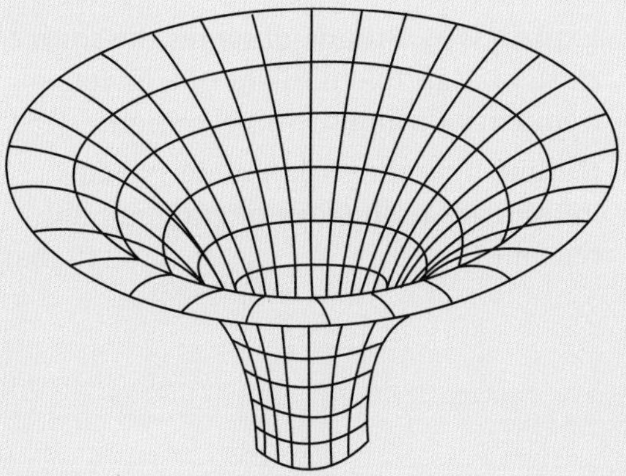

a) (i) Describe what is meant by the *centre* and the *surface* of a black hole.

(ii) Define the *Schwarzchild radius.*

(iii) Calculate the Schwarzchild radius for a star that has a mass 10 times that of the Sun (solar mass $= 2.0 \times 10^{30}$ kg).

b) In 1979, astronomers discovered "two" very distant quasars separated by a small angle. Examination of the images showed that they were identical. Outline how these observations give support to the theory of general relativity. (9 marks)

17 One prediction from the principle of equivalence is the effect known as gravitational lensing.

a) State the *principle of equivalence.*

b) Use the principle to explain *gravitational lensing.* (5 marks)

18 *(IB)*
On 29 March 1919, an experiment by Eddington carried out during a total eclipse provided evidence to support Einstein's general theory. The diagram below (not to scale) shows the relative position of the Sun, Earth and a star S on this date.

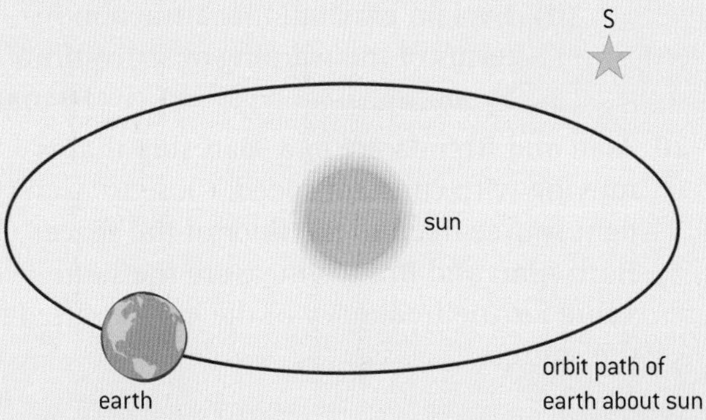

Eddington measured the apparent position of the star and six months later, he again measured the position of the star from Earth.

a) State why the experiment was carried out during a total solar eclipse.

b) Explain why the position of the star was measured again six months later.

c) Copy the diagram and draw the path of a ray of light from S to the Earth as suggested by the general theory.

d) Explain how Einstein's theory accounts for the path of the ray.

e) Label the apparent position of the star as seen from Earth. (6 marks)

19 *(IB)*

a) (i) Describe, with reference to spacetime, the nature of a black hole.

(ii) Define, with reference to (a)(i), the Schwarzchild radius.

(iii) A star that has a mass of 4.0×10^{31} kg evolves into a black hole.

Calculate the Schwarzchild radius of the black hole, stating any assumption that you make.

b) A spacecraft approaches the black hole in (a)(iii). If it were to continue to travel in a straight line it would pass within 1000 km of the black hole.

(i) Suggest the effect the black hole would have on the motion of the spacecraft.

(ii) Explain gravitational attraction in terms of the warping of space-time by matter. (10 marks)

20 Alan and Brenda are in a spaceship that is moving with constant speed. Close to Alan is a light source fixed to the floor of the spaceship. Both Alan and Brenda measure the same value for the frequency of the light emitted by the source.

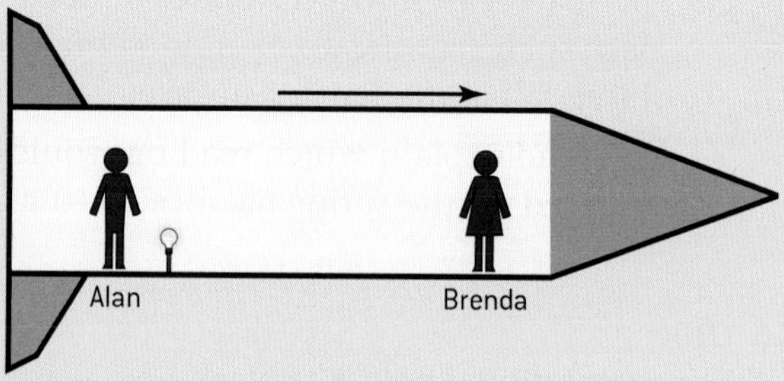

The spaceship begins to accelerate.

a) Explain why Brenda observes the source close to Alan to emit light of a lower frequency during the acceleration.

b) Outline how the situation described in (a) leads to the idea of gravitational redshift. (5 marks)

B ENGINEERING PHYSICS

Introduction

Engineering physics covers some of the key topics that you would expect to meet on an undergraduate engineering course. These are quite diverse topics covering rotational motion, thermodynamics, fluid mechanics, and forced vibrations and resonance. The mathematical content of these topics is naturally restricted to be in line with a pre-university course; however, those students with a strong mathematical inclination will find much in these sub-topics to stimulate further research.

B.1 Rigid bodies and rotational dynamics

Understanding

- Torque
- Moment of inertia
- Rotational and translational equilibrium
- Angular acceleration
- Equations of rotational motion for uniform angular acceleration
- Newton's second law applied to angular motion
- Conservation of angular momentum

Nature of science

Extended objects

The work covered until now has almost invariably used the model of an object being represented by a single particle of zero dimensions. This model works well for many situations but falls short when forces are not applied along a line passing through the centre of mass of the object. In such cases an extended object can still be modelled; however, objects of different shape will not always move in the same way when forces are applied to them. Using simplifying models, engineers were able to design and manufacture sophisticated machines that had significant impact during the industrial revolution of the late eighteenth to early nineteenth century.

Applications and skills

- Calculating torque for single forces and couples
- Solving problems involving moment of inertia, torque, and angular acceleration
- Solving problems in which objects are in both rotational and translational equilibrium
- Solving problems using rotational quantities analogous to linear quantities
- Sketching and interpreting graphs of rotational motion
- Solving problems involving rolling without slipping

Equations

- Torque equation: $\Gamma = Fr\sin\theta$
- Moment of inertia: $I = \sum mr^2$
- *Newton's second law for rotational motion:* $\Gamma = I\alpha$
- Relationship between angular frequency and frequency: $\omega = 2\pi f$
- Equations of motion for constant angular acceleration:
- $\omega_f = \omega_i + \alpha t$
- $\omega_f^2 = \omega_i^2 + 2\alpha\theta$
- $\theta = \omega_i t + \frac{1}{2}\alpha t^2$
- Angular momentum: $L = I\omega$
- Rotational kinetic energy: $E_{K_{rot}} = \frac{1}{2}I\omega^2$

Introduction

This sub-topic is very closely related to the linear mechanics and circular motion that you studied in Topics 2 and 6. There are quantities that are directly analogous to the linear quantities of displacement, velocity, acceleration, force, etc. With the knowledge that you have already absorbed and an understanding that a rigid body is an extension of a point object you will be able to solve most rotational dynamics problems.

Note

- The angular acceleration is different from both centripetal acceleration (a_c) and tangential acceleration (a_t) – this is the rate of change of the linear tangential velocity with time.
- As an object moving in a circle undergoes angular acceleration, its tangential acceleration increases as does its centripetal acceleration.
- The tangential acceleration of the body is related to the angular acceleration and the radius of the circle (r) by the relationship:

$$a_t = r\alpha$$

This is because $v_t = \omega r$

This means that $\alpha = \frac{\Delta v}{r\Delta t} = \frac{a_t}{r}$ where v_t is the tangential velocity of the object. Thus the linear quantity (velocity or acceleration) is the angular quantity multiplied by the radius.

Uniform motion in a circle

You will remember that for a body moving around a circle with a constant linear speed (or angular speed) there needs to be a means of providing a centripetal force. This can be a contact force, such as tension or friction (or a component of these forces), or a force provided by a field, such as Newton's law of gravitation or Coulomb's law.

In all cases the force (or component) providing the centripetal force will be given by:

$$F = \frac{mv^2}{r} = m\omega^2 r$$

These terms were defined in Topic 6. Remember that ω is known as the angular velocity or angular frequency and is related to frequency f by the equation:

$$\omega = 2\pi f$$

Angular acceleration

Let us consider an object moving in the circle so that it is given an angular acceleration and its angular speed increases. We define **angular acceleration (α) as the rate of change of angular speed with time.**

$$\alpha = \frac{\Delta\omega}{\Delta t}$$

α is measured in radian per second squared (rad s^{-2})

Worked example

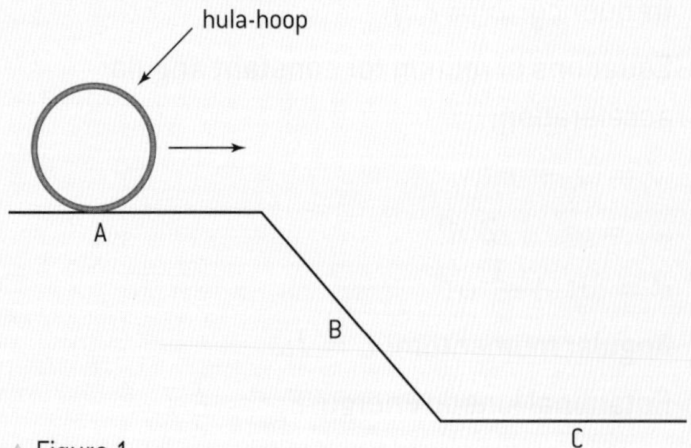

Figure 1

The diagram shows a hula-hoop (a large plastic ring) rolling with constant angular speed along a horizontal surface prior to rolling down a uniform inclined plane. When it reaches a second horizontal surface it, again, moves with a constant angular speed.

Sketch graphs to show the variation with time of a) the angular velocity and b) the angular acceleration of the hula-hoop.

Solution

a) The graph shows the hula-hoop travelling with constant angular velocity along A and C. It has a greater value along C since it has now undergone angular acceleration. As B is of constant gradient, the angular acceleration is constant here.

b) The second graph shows zero angular acceleration throughout A and C and a constant angular acceleration along B. *You should compare these graphs with those for a point object moving along a frictionless surface.*

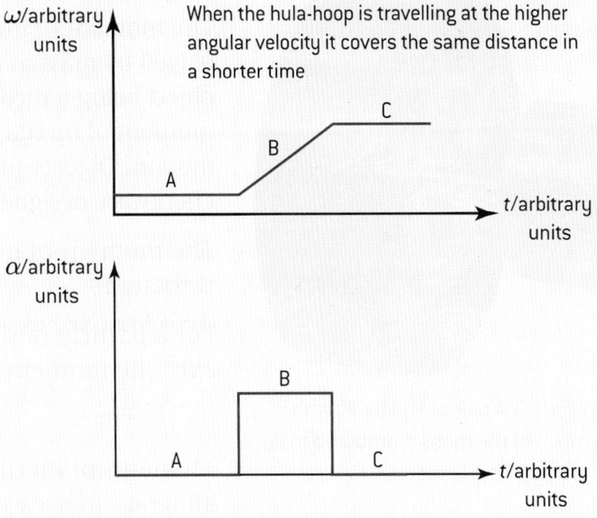

ω/arbitrary units
When the hula-hoop is travelling at the higher angular velocity it covers the same distance in a shorter time
A
B
C
t/arbitrary units
α/arbitrary units
A
B
C
t/arbitrary units

▲ Figure 2

Equations of motion

In Topic 2 we met the four equations of motion under constant linear acceleration:

$$v = u + at$$

$$s = ut + \frac{1}{2}at^2$$

$$v^2 = u^2 + 2as$$

$$s = \frac{(v + u)t}{2}$$

In rotational dynamics there are four analogous equations that apply to a body moving with constant angular acceleration:

$$\omega_f = \omega_i + \alpha t$$

$$\theta = \omega_i t + \frac{1}{2}\alpha t^2$$

$$\omega_f^2 = \omega_i^2 + 2\alpha\theta$$

$$\theta = \frac{(\omega_i + \omega_f)t}{2}$$

Here ω_i = initial velocity (in rad s^{-1}), ω_f = final velocity (in rad s^{-1}), θ = angular displacement (in rad) or sometimes just "angle", α = angular acceleration (in rad s^{-2}), t = time taken for the change of angular speed (in s).

Note

The IB Physics Data Booklet does not include the last of these rotational equations – it is probably the most straightforward of the equations because it simply equates two ways of expressing the average angular speed.

Worked example

A wheel is rotated from rest with an angular acceleration of 8.0 rad s^{-2}. It accelerates for 5.0 s.

Determine:

a) the angular speed

b) the number of revolutions that the wheel has rotated through.

Solution

a) $\omega_f = \omega_i + \alpha t \Rightarrow \omega_f = 0 + 8.0 \times 5.0$
$= 40.0$ rad s^{-1}

b) $\theta = \frac{(\omega_i + \omega_f)t}{2} = \frac{(0 + 40.0)5.0}{2} = 100$ rad

Each revolution makes an angle of 2π radian so the number of revolutions $= \frac{100}{2\pi}$

$= 15.9$ revolutions.

Moment of inertia

▲ Figure 3 A pair of flywheels with much of the mass distributed to be around the perimeter.

The moment of inertia ,I, of a body is the rotational equivalent of the role played by mass in linear dynamics. In a similar way to the inertial mass of an object being a measure of its opposition to a change in its linear motion, the moment of inertia of an object is its resistance to a change in its rotational motion. Objects such as flywheels that need to retain their rotational kinetic energy are designed to have large moments of inertia as shown in figure 3.

The moment of inertia of an object depends on the axis about which it is rotated.

For a particle (a single point) of mass m rotating at a distance r about an axis, the moment of inertia is given by:

$$I = mr^2$$

Moments of inertia are scalar quantities and are measured in units of kilogram metres squared (kg m^2).

For an object consisting of more than one point mass, the moment of inertia about a given axis can be calculated by adding the moments of inertia for each point mass.

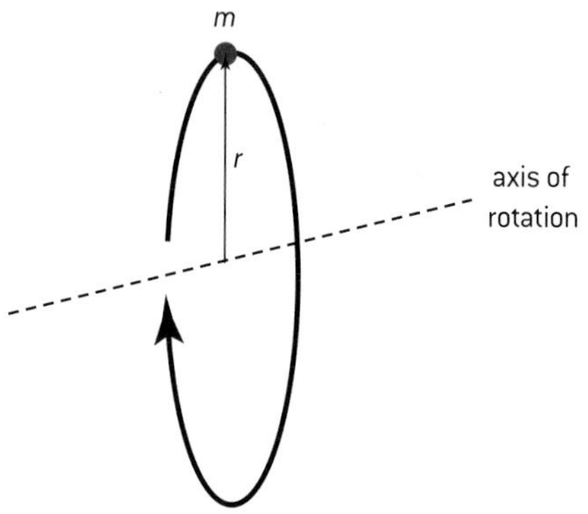

▲ Figure 4 Moment of inertia of a point mass.

$$I = \sum mr^2$$

This summation is a mathematical abbreviation for $m_1r_1^2 + m_2r_2^2 + m_3r_1^3 + \ldots + m_nr_n^2$ when the object is made up of n point masses.

The moment of inertia of a simple pendulum of length l and mass m is given by

$$I_{\text{pendulum}} = ml^2$$

while that of a simple dumb-bell consisting of two masses m connected by a light rod of length l, when rotating about the centre of the rod, will be

$$I_{\text{dumb-bell}} = m\left(\frac{l}{2}\right)^2 + m\left(\frac{l}{2}\right)^2 = 2m\left(\frac{l}{2}\right)^2 = \frac{1}{2}ml^2$$

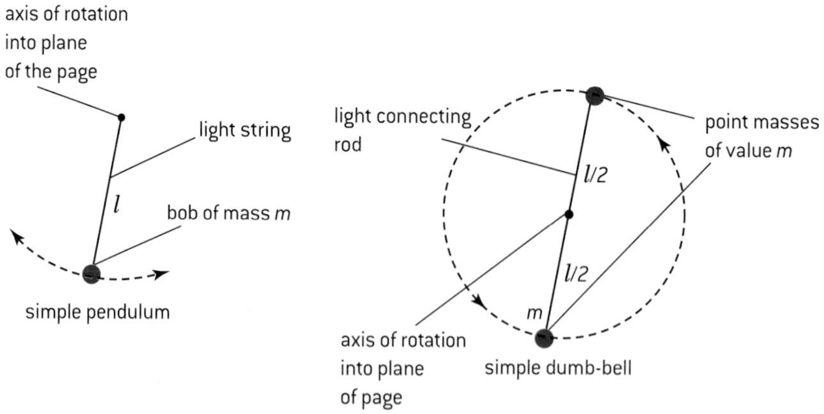

▲ Figure 5 Moment of inertias of a simple pendulum and dumb-bell.

In each of these cases we assume that the mass is a point mass and that the string and the connecting rod are so light that their masses can be ignored.

For a more complicated structure we need to use integral calculus in order to derive formulae for moments of inertia. In IB Diploma Programme physics questions, any equations that need to be used for moments of inertia will be provided in the question.

Note

With linear motion there is just one "single" mass value for an object. In rotational motion, the moment of inertia is also a function of the position of the axis of rotation. This means that there are an infinite range of possible moments of inertia for any one object.

Torque

Torque ,Γ, can be defined as shown in figure 6. Consider force F acting at point P on an object. The direction of F is such that it makes an angle θ to the radius of the circle in which the object rotates. The torque will be given by:

$$\Gamma = Fr\sin\theta$$

This is sometime called the force multiplied by the "arm of the lever".

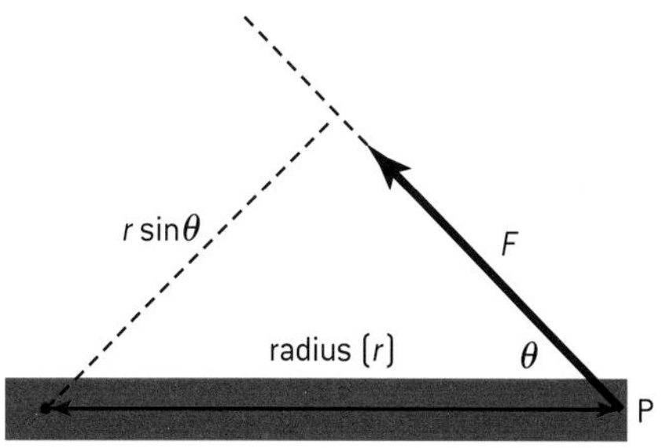

▲ Figure 6 Definition of torque.

Torques are also called "moments", but to avoid confusion between "moment" and "momentum" we advise sticking to the term torque. They are (pseudo) vector quantities with the direction perpendicular to the plane of the circle in which the object rotates. Imagine your right hand gripping the axis about which the object rotates so that the fingers curl round the axis, when your thumb is "up" it will be pointing in the direction of the torque (as shown in figure 7). This rule also applies to the directions of angular displacement ,θ, angular velocity ,ω, and angular acceleration ,α, which, like torque, are all vector quantities.

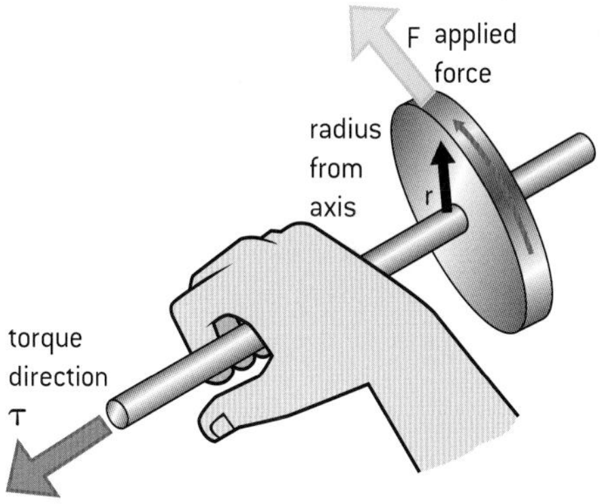

▲ Figure 7 Direction of torque.

The maximum torque that a force can apply to a body is when the force is perpendicular to the arm of the lever. In this case $\theta = 90°$ and so $\sin\theta = 1$. This means that

$$\Gamma = Fr$$

Couples

***A* couple consists of a pair of equal and opposite forces that do not act in the same straight line** (see figure 7). This combination of forces produces a torque that causes an object to undergo angular acceleration without having any translational acceleration.

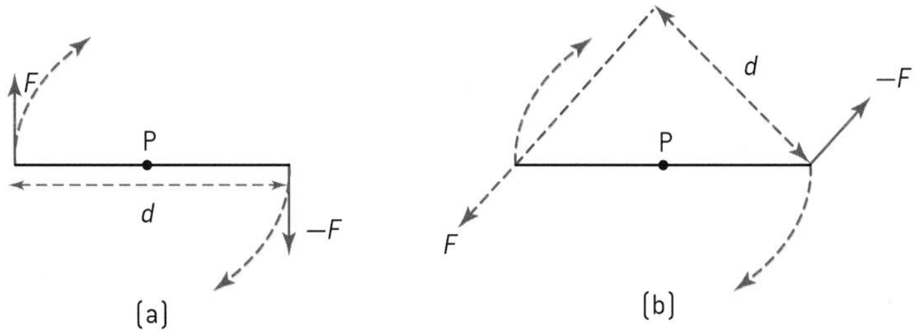

▲ Figure 8 Two examples of couples.

The torque of the couple about the pivot P in both figure 8(a) and (b) is equal to Fd (that is the product of one of the forces and the perpendicular distance between them).

- A single resultant force acting through the centre of mass of an object will produce translational acceleration. This means that the object will move in a straight line. An object in free-fall is an example of this as (ignoring air resistance) the pull of gravity is the only force acting on it.
- The same force, but acting through a point displaced from the centre of mass of the object, will produce a combination of both linear and angular acceleration. This means that the object will move in a helical path.

Note

Torque is measured in units of newton metre (N m) but don't confuse this with the unit of work or energy (the joule). In the case of torque the force is perpendicular to the direction in which the object moves but in the case of work the force and direction moved are the same. Torque is a vector quantity while work is a scalar.

Figure 9 shows a force acting at the top of an object – the effect of this force can be split into the (i) same force acting at the centre of mass of the object (producing translation) and (ii) a couple (producing rotation). The couple consists of $\frac{1}{2}F$ at the top and $-\frac{1}{2}F$ at the bottom of the object.

This is a general principle that can be applied to any force not acting through the centre of mass of an object.

A couple acting on an object will produce angular acceleration with no linear motion at all. This means that the object will move in a circular path (as shown in figure 8).

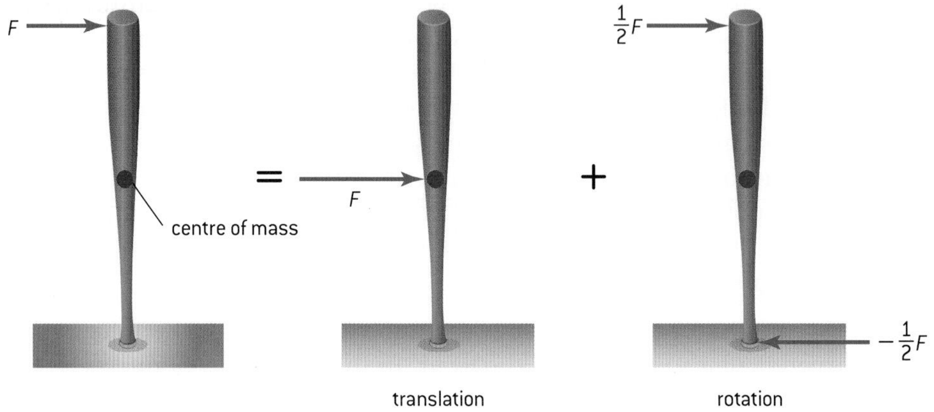

▲ Figure 9 Force acting through point displaced from centre of mass.

Newton's first law for angular motion – rotational equilibrium

We saw in Topic 2 that a body in (translational) equilibrium does not accelerate but remains in a state of rest or travels with uniform velocity. This means that there can be no resultant force acting on the body in line with Newton's first law of motion. For an object to be in rotational equilibrium there can be no external resultant torque acting on it – it will then remain in a state of rest or continue to rotate with constant angular velocity.

For rotational motion Newton's first law may be stated as:

An object continues to remain stationary or to move at a constant angular velocity unless an external torque acts on it.

When the object is in rotational equilibrium:

total clockwise torque = total anticlockwise torque.

This statement is often called the "principle of moments".

Newton's second law for angular motion – angular acceleration

Again, in Topic 2 we considered the simpler statement of Newton's second law of (translational) motion equating force to the product of mass and acceleration. The rotational equivalent of Newton's second law relates the angular acceleration and torque on a body of moment of inertia.

$$\Gamma = I\alpha$$

In IB Diploma Programme Physics questions, the axis to which the torque is applied and the axis about which the moment of inertia is taken will always be the same (although there are quite simple ways of dealing with situations when this is not the case).

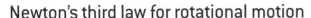

Newton's third law for rotational motion

You can probably guess that this version of the law says that action torque and reaction torque are equal and opposite – this pair of torques, like action and reaction forces for linear motion, act on different bodies. If body A applies a torque to body B, then body B applies an equal and opposite torque to body A.

Angular momentum

Linear momentum (p) is a vector quantity defined as being the product of mass and velocity. Angular momentum (L) is the rotational equivalent to this and is defined as being the product of a body's moment of inertia and its angular velocity; this, too, is a vector.

$$L = I\omega$$

Angular momentum is measured in units of kg m^2 rad s^{-1}.

Worked example

A couple, consisting of two 4.0 N forces, acts tangentially on a wheel of diameter 0.60 m. The wheel starts from rest and makes one complete rotation in 2.0 s. Calculate:

a) the angular acceleration

b) the moment of inertia of the wheel.

Solution

a) $\theta = \omega_i t + \frac{1}{2}\alpha t^2 \Rightarrow 2\pi = 0 + \frac{1}{2}\alpha \times 2.0^2$

$\alpha = \frac{4\pi}{4.0} = \pi = 3.14 \text{ rad s}^{-2}$

b) torque of couple $\Gamma = Fd = 4.0 \times 0.60 = 2.4$ N m

as $\Gamma = I\alpha \Rightarrow I = \frac{\Gamma}{\alpha} = \frac{2.4}{3.14} = 0.76$ kg m^2

The conservation of angular momentum

In linear dynamics we find the total (linear) momentum of a system remains constant providing no external forces act on it. In rotational dynamics, the total angular momentum of a system remains constant providing no external torque acts on it.

Figure 10 shows a small mass being gently dropped onto a freely spinning disc. The addition of the mass increases the combined moment of inertia of the disc and the mass and so the angular velocity of the system now falls in order to conserve the angular momentum.

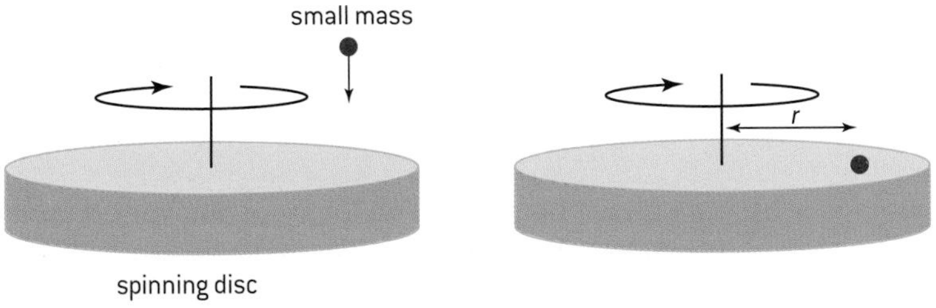

▲ Figure 10 An example of the conservation of angular momentum.

The conservation of angular momentum has many applications in physics, such as the speeding up of an ice skater when pirouetting with decreasing angular momentum, or a gymnast putting in turns or twists by changing the distribution of mass.

Rotational kinetic energy

You probably feel confident about the analogies between linear and rotational dynamics, so it should not surprise you that the relationship for angular kinetic energy is

$$E_{K_{rot}} = \frac{1}{2} I\omega^2$$

This, like all energies, is a scalar quantity and could, in principle, be found by adding the (translational) kinetic energies of all the particles making up a rotating object.

Two more useful rotational analogies are given in the table below.

Quantity	Linear equation	Angular equation
work	$W = Fs$	$W = \Gamma\theta$
power	$P = Fv$	$P = \Gamma\omega$

Rolling and sliding

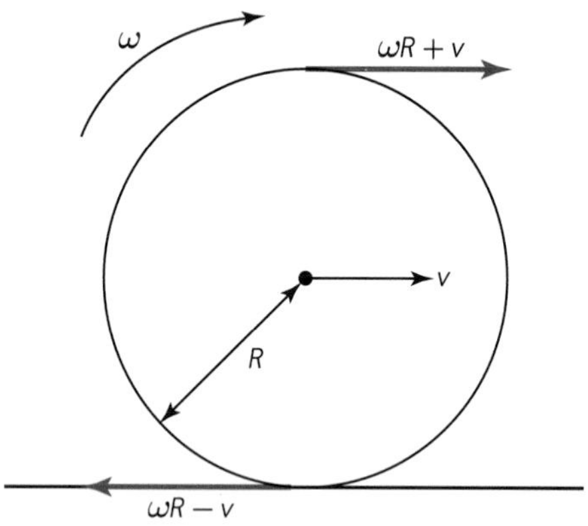

▲ Figure 11 Disc rolling on a flat surface.

If an object makes a perfectly frictionless contact with a surface it is impossible for the object to roll – it simply slides. When there is friction the object can roll; as the point of contact between the rolling body and the surface along which it rolls is instanteously stationary, the coefficient of static friction should be used in calculations involving rolling. The point of contact must be stationary because it does not slide. Figure 11 shows a disc of radius R rolling along a flat surface such that its centre of mass has a velocity v. Each point on the perimeter of the disc will have a tangential velocity $= \omega R$.

This means that the top of the disc will have total velocity of $\omega R + v$ and the point of contact with the surface will have a velocity of $\omega R - v$. As the disc is not slipping, the bottom of the disc has zero instantaneous velocity and so $\omega R = v$. This means that the top of the disc will have an instantaneous velocity $= 2v$.

The total kinetic energy of a body that is rolling without slipping will be $= \frac{1}{2}I\omega^2 + \frac{1}{2}mv^2$. When an object rolls down a slope so that it loses a vertical height h, the loss of gravitational potential energy will become the total kinetic energy. This gives us

$$mgh = \frac{1}{2}I\omega^2 + \frac{1}{2}mv^2$$

Worked example

A solid ball, of radius of 45 mm, rolls down an inclined plane of length 2.5 m. The sphere takes a time of 6.0 s to roll down the plane. Assume that the ball does not slip.

The moment of inertia, I, of a solid ball of mass M and radius R is given by $I = \frac{2}{5}MR^2$

a) Calculate the velocity of the sphere as it reaches the end of the plane.

b) Calculate the angular velocity of the sphere as it reaches the end of the plane.

c) Determine the angle of inclination of the plane.

d) Comment on whether the assumption that the ball does not slip was appropriate in this instance.

Solution

a) For the ball starting at rest rolling along the slope for time t the (translational) equations of motion give:

$$s = \frac{1}{2}at^2 \text{ and } v = at$$

Substituting the values for s and t

$$a = \frac{2s}{t^2} = \frac{2 \times 2.5}{6.0^2} = 0.14 \text{ m s}^{-2}$$

$$v = at = 0.14 \times 6.0 = 0.84 \text{ m s}^{-1}$$

b) The angular speed $\omega = \frac{v}{r} = \frac{0.84}{45 \times 10^{-3}} = 18.7 \approx$ 19 rad s^{-1}

c) from the equations $mgh = \frac{1}{2} I\omega^2 + \frac{1}{2} mv^2$ and $v = \omega R$

$$mgh = \frac{1}{2} v^2\left(\frac{I}{R^2} + m\right)$$

substituting for the moment of inertia equation we have:

$$Mgh = \frac{1}{2} v^2\left(\frac{\frac{2}{5}MR^2}{R^2} + M\right)$$

Cancelling M and R^2 gives

$$gh = \frac{1}{2} v^2\left(\frac{2}{5} + 1\right) = \frac{7}{10} v^2$$

So $h = \frac{7}{10} \times \frac{v^2}{g}$ where h is the height of the end of the inclined plane.

$$h = \frac{7}{10} \times \frac{0.84^2}{9.81} = 0.05 \text{ m}$$

$$\sin\theta = \frac{0.05}{2.5} \quad \therefore \quad \theta = \sin^{-1}\left(\frac{0.05}{2.5}\right) = 1.1°$$

d) This angle is very small and so there is no problem with assuming that the ball rolls without slipping.

Investigate!

Measuring the moment of inertia of a flywheel

A flywheel is mounted so that it can rotate on a horizontal axle (the axle forming part of the flywheel). A mass, suspended by a string, is wound round the axle of a flywheel ensuring that it does not overlap. When the mass is released, the gravitational potential energy of mass is converted into the linear kinetic energy of the mass + the rotational kinetic energy of the flywheel + work done against the frictional forces (all these values are taken when the string loses contact with the axle).

$$mgh = \frac{1}{2}mv^2 + \frac{1}{2}I\omega^2 + n_1W$$

The symbols here have their usual meaning with n_1 being the number of turns of string and W the work done against frictional forces during each revolution. When the string disengages, the rotational kinetic energy makes a further n_2 rotations before it comes to rest – so the rotational kinetic energy of the flywheel must be equal to n_2W or

$$\frac{1}{2}I\omega^2 = n_2W \text{ making } W = \frac{I\omega^2}{2n_2}$$

This means $mgh = \frac{1}{2}mv^2 + \frac{1}{2}I\omega^2 + \frac{n_1I\omega^2}{2n_2}$

or $\quad mgh = \frac{1}{2}mv^2 + \frac{1}{2}I\omega^2\left(1 + \frac{n_1}{n_2}\right)$

In addition to this $\omega = vr$ and $\frac{v}{2} = \frac{h}{t}$

- Measure the radius r of the axle of the system using vernier or digital callipers.

- Attach a mass hanger to one end of the piece of string and make a loop at the other end so that, as the mass hanger just touches the floor, the loop slips off the peg in the axle.
- Loop the string over the peg, wind up the string for a whole number of revolutions n_1 and measure the height h of hanger above the floor. Make sure the windings do not overlap.

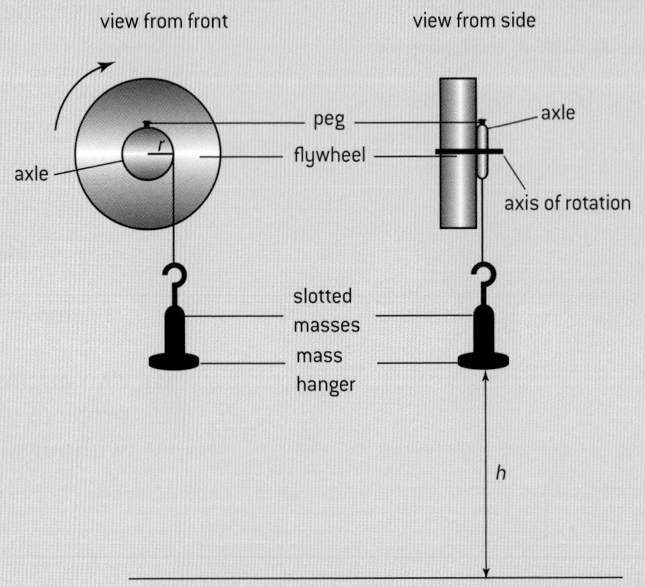

▲ Figure 12 Measuring the moment of inertia of a flywheel.

- Adjust the number of slotted masses on the hanger.
- Measure the time t from the moment of release until the hanger reaches the floor.
- Count the number of revolutions n_2 the flywheel makes after the string comes off the peg until it comes to rest.
- Repeat several times and calculate mean values from which a value for the moment of inertia of the flywheel and axle I can be deduced.
- Consider how the data could be used graphically to find a value for I.

B.2 Thermodynamics

Understandings

- The first law of thermodynamics
- The second law of thermodynamics
- Entropy
- Cyclic processes and pV diagrams
- Isovolumetric, isobaric, isothermal, and adiabatic processes
- Carnot cycle
- Thermal efficiency

Nature of science

Different viewpoints of the second law of thermodynamics

Thermodynamics is an area of physics which, through different scientific eras, has been shaped by a combination of practice and theory. When Sadi Carnot wrote his treatise about heat engines, he was a believer in the caloric standpoint – yet his ideas were sufficiently developed to influence the development of machines in the industrial revolution. Clausius, Boltzmann, Kelvin, and Gibbs were all responsible for different statements of the second law of thermodynamics – a law that has a fundamental impact on whether or not a process, allowed by the first law of thermodynamics, can actually occur.

Applications and skills

- Describing the first law of thermodynamics as a statement of conservation of energy
- Explaining sign convention used when stating the first law of thermodynamics as $Q = \Delta U + W$
- Solving problems involving the first law of thermodynamics
- Describing the second law of thermodynamics in Clausius form, Kelvin form and as a consequence of entropy
- Describing examples of processes in terms of entropy change
- Solving problems involving entropy changes
- Sketching and interpreting cyclic processes
- Solving problems for adiabatic processes for monatomic gases using $pV^{\frac{5}{3}} = \text{constant}$
- Solving problems involving thermal efficiency

Equations

- First law of thermodynamics: $Q = \Delta U + W$
- Internal energy: $U = \frac{3}{2}nRT$
- Entropy change: $\Delta S = \frac{\Delta Q}{T}$
- Equation of state for adiabatic change: $pV^{\frac{5}{3}} = \text{constant}$ (for monatomic gases)
- Work done when volume changes at constant pressure: $W = p\Delta V$
- Thermal efficiency: $\eta = \frac{\text{useful work done}}{\text{energy input}}$
- Carnot efficiency: $\eta_{\text{Carnot}} = 1 - \frac{T_{\text{cold}}}{T_{\text{hot}}}$

Introduction

When using thermodynamics, there is a convention that we talk about the body that we are interested in as being the **system**. Everything else that may have an impact on the system is known as the **surroundings**. The system is separated from the surroundings by a **boundary** or **wall**. Everything including the system and the surroundings is called the **universe**.

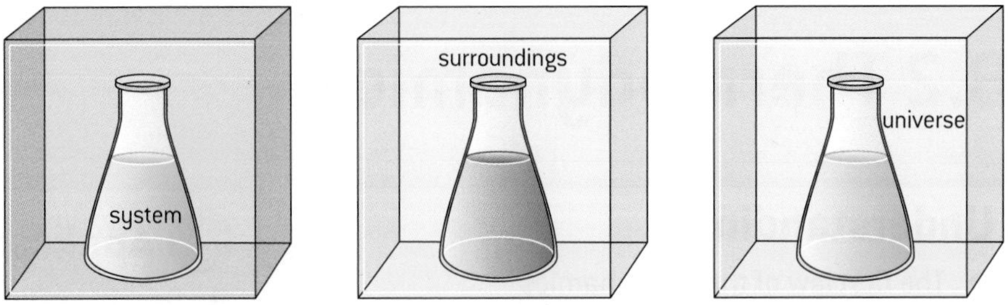

▲ Figure 1 Thermodynamic system and surroundings.

As we saw in Topic 3, James Joule showed that work done on a system or energy transferred to the system because of temperature differences result in the same outcome: the internal energy of that system increases.

The nature of the way the system changes and how that energy reveals itself depends on whether or not the phase of the substance changes. When there is no change of state the most apparent effect is the increase in the mean random kinetic energy of the particles; when there is a change of state the increase in the potential energy is the most significant effect.

Remember that *the internal energy of a system is the total of the potential energy and the random kinetic energy of all the particles making up the system.*

We often simplify discussion by considering the system to be an ideal gas. In this case the internal energy is entirely kinetic and we can use the relationship derived in Sub-topic 3.2. This showed that, for n moles of an ideal gas, the internal energy U is related to the absolute temperature T by the equation $U = \frac{3}{2} nRT$, where R is the universal molar gas constant.

The first law of thermodynamics

The internal energy of a system may change as any combination of (i) doing work on the system (or allowing the system to do work on the surroundings) and (ii) transferring energy to or from the system as a result of a difference in temperature. Saying this amounts to stating the conservation of energy. There are a variety of versions of the equation for the first law of thermodynamics and each has its merits – they will all be self-consistent and understandable but you will need to make sure that you read the definition of each of the terms in the equation. The preferred version of the equation for the IB Physics syllabus is written as:

$$Q = \Delta U + W$$

When each of these quantities is **positive**:

- Q represents the energy transferred **from** the surroundings to the system because the surroundings are at a higher temperature than the system
- ΔU represents the **increase** in the internal energy of the system (this is not simply a change, it is an increase)
- W represents the work done **by** the system as it expands and pushes back the surroundings.

When each of these quantities is **negative**:

- Q represents the energy transferred **from** the system to the surroundings because the system is at a higher temperature than the surroundings

- ΔU represents the **decrease** in the internal energy of the system
- W represents the work done **on** the system as the surroundings compresses it.

Worked example

A system's internal energy falls by 200 J as a result of energy transfer and work being done. The system does 500 J of work on the surroundings.

a) State and explain whether the system is at a higher or lower temperature than its surroundings.

b) Calculate the amount of energy transferred causing the reduction in internal energy.

Solution

a) The system must gain energy in order to be able to do this amount of work and so its temperature must be below that of the surroundings.

b) Using the first law of thermodynamics $Q = \Delta U + W$, ΔU must be negative and W must be positive.

$Q = -200 + 500 = 300\text{J}$

So 300 J are transferred to the system from the surroundings.

Nature of science

Human metabolism and the first law

Let us consider a human body as a thermodynamic system. When we eat, our internal energy increases because we are taking in food in the form of chemical potential energy. When there is no energy transfer because of temperature differences this must be a process which is related to work – although we will not discuss the biochemistry here.

So, using the first law with no energy transfer:

$$0 = \Delta U + W$$

As ΔU is positive this means that W is negative.

The chemical energy in the food we eat does three main things: it allows us to do work on our surroundings, it allows us to transfer energy to the (usually) cooler surroundings and the remainder is stored in our bodies (as fat). With a well-adjusted, balanced diet the build up of fat and change in internal energy is zero and the food we eat allows us to stay warm and do work and be active. So the ideal system is to make the net change in the internal energy of our bodies zero – eat too much and we will build up fat (ΔU is positive), eat too little and we will lose weight (ΔU is negative). In reality, human metabolism is much more complicated than this, but the first law of thermodynamics does represent a model that has many more applications than the gases we will now focus on.

Using the first law for ideal gases

When discussing the changes that can be made to the state of the gas, it is usual to consider an ideal gas enclosed in a cylinder by a moveable piston. The gas represents the system, the cylinder and the piston represent the boundaries or walls. Everything else becomes the surroundings.

Calculating the work done in an isobaric change

An **isobaric** change is one which occurs at **constant pressure**. Consider an ideal gas at a pressure p enclosed in a cylinder of cross-sectional area A. When the gas expands it pushes the piston a distance Δx so that the volume

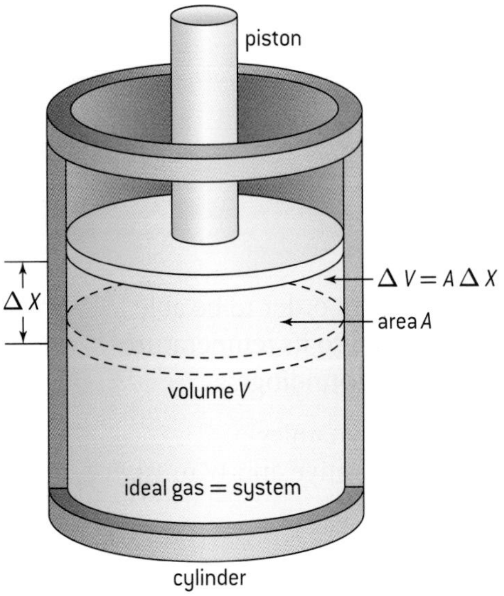

▲ Figure 2 Work done by an ideal gas.

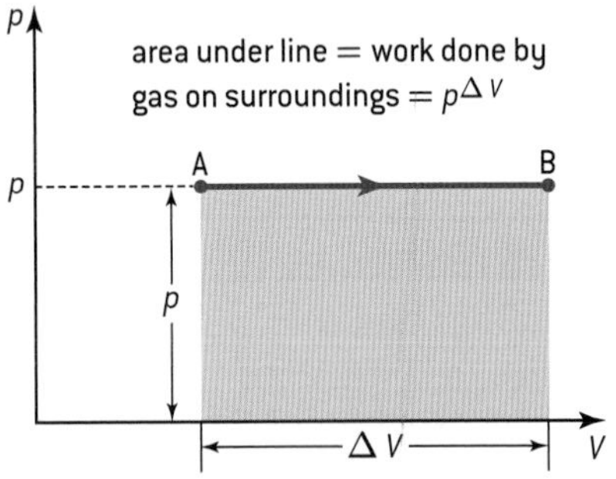

▲ Figure 3 Work done in an isobaric change.

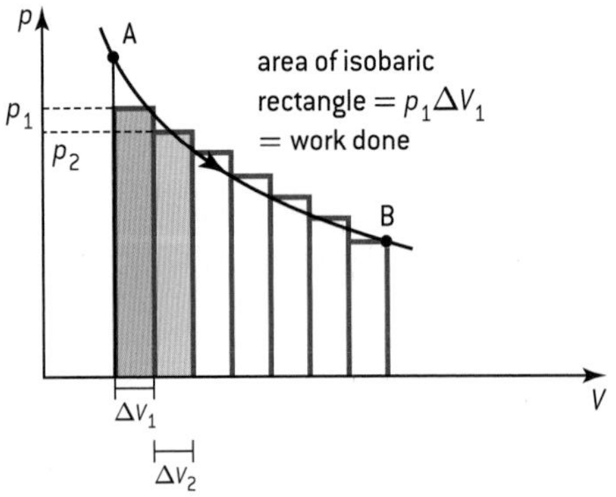

▲ Figure 4 Work done in a non-isobaric change.

Note

Isothermal changes normally take place very slowly and the boundary between the system and the surrounding must be a good conductor of energy.

of the gas increases by ΔV ($= A\Delta x$). When energy is supplied to the gas from the surroundings the pressure can remain constant at value at p. The work done by the gas on the surroundings during this expansion will be W. The force F of the gas on piston $= pA$. This means that the work done during expansion $= F\Delta x = pA\Delta x = p\Delta V$.

$$W = p\Delta V$$

This will be a positive quantity because the system is doing work on the surroundings. When the gas is compressed W will be negative.

Figure 3 shows a p–V graph for an isobaric change. The area under the graph is equal to the work done. The arrow on the line connecting the end points AB shows that the gas is expanding and doing work on the surroundings – an arrow in the opposite direction would show a compression.

Applying the first law of thermodynamics to this, we get $Q = \Delta U + p\Delta V$. This could result in any number of possibilities but they must be consistent with this equation; Q must be positive for an expansion and negative for a compression. For example, supplying 20 J to the gas could increase the internal energy by 19 J and allow the gas to do 1 J or work or increase the internal energy by 19.1 and do 0.9 J of work etc.

The equation of state for an isobaric change is $\frac{V}{T} =$ constant.

Work done for non-isobaric changes

When changes do not occur at constant pressure we can still calculate the work done from the area under a p–V graph. We can make the assumption that the pressure will be unchanged over a small change in volume and, therefore, we approximate the overall change to a series of small isobaric changes as shown in figure 4.

The area of the first constant pressure rectangle $= p_1\Delta V_1 = W_1$

The area of the second rectangle would be $p_2\Delta V_2 = W_2$, etc.

Therefore, the total work done $= \Sigma_n W_n = \Sigma_n p_n \Delta V_n$ for n rectangles – this is the area under the curve. So for any p–V graph the area under the curve will give the work done – this depends on the path taken and not just on the end points.

Isothermal changes

Isothermal changes are those that occur **resulting in the internal energy of the system staying constant**. The internal energy of an ideal gas consists of the sum of the mean random kinetic energies of the particles of gas. The mean kinetic energy is proportional to the temperature of the gas. An isothermal change, therefore, means that there is **no change in the temperature of the system**.

Using the first law of thermodynamics $Q = \Delta U + W$ with $\Delta U = 0$, this leaves $Q = W$. As there is no change in internal energy, the energy transferred to the system because of a temperature difference between the system and the surroundings Q will allow the system to do work W on the surroundings. As $W = p\Delta V$ this can only mean that the gas is expanding – the direction arrow on the graph should go from A to B.

The other possibility is that $-Q = -W$ so the energy transferred from the system to the surroundings is equal to the work done on the system by the surroundings. In this case the gas is being compressed – the direction arrow on the graph should go from B to A.

We saw in Topic 3 that when the temperature does not change $pV =$ constant. This is the equation of the line on a $p-V$ graph for an isothermal change. The lines are known as isotherms. As shown in figure 6, changes at higher temperatures will always produce isotherms that are further from the origin than those at lower temperatures. For a given volume the pressure will always increase in moving from a low temperature isotherm to one at higher temperature.

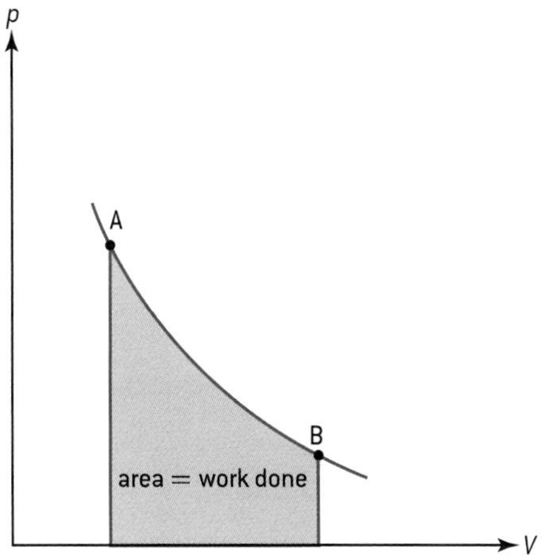

▲ Figure 5 Work done in an isothermal change.

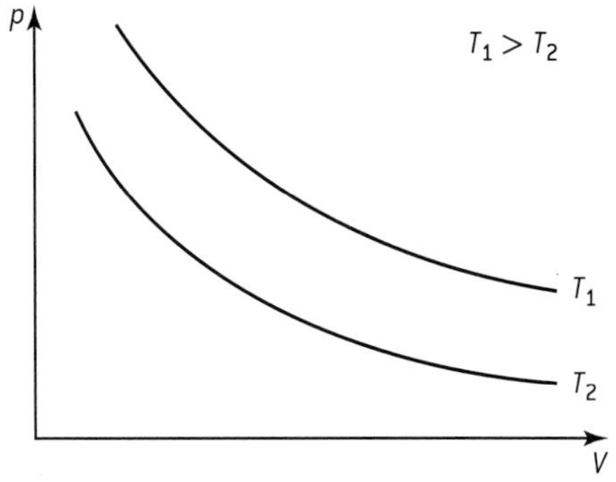

▲ Figure 6 Isotherms at different temperatures.

Adiabatic changes

This is the name given to a change in which **no energy is transferred** between the system and the surroundings. This does not mean that the system and the surroundings are always at the same temperature, although that could be so, but it is more likely that there is a well-insulated barrier between them. Adiabatic changes usually happen very quickly, which means that there is no time for the energy to transfer.

Applying the first law of thermodynamics to a system undergoing an adiabatic change gives:

$Q = \Delta U + W$ but, as $Q = 0$, this can mean either $\Delta U = -W$ (an increase in internal energy occurs because of work being done on the system) or $-\Delta U = W$ (a decrease in internal energy occurs because the system is doing work on the surroundings). For an ideal monatomic gas the equation for an adiabatic change takes the form:

$$pV^{\frac{5}{3}} = \text{constant}$$

This can be written as $p_1V_1^{\frac{5}{3}} = p_2V_2^{\frac{5}{3}}$ when there is a constant temperature.

As the gas is an ideal gas the equation

$$\frac{p_1V_1}{T_1} = \frac{p_2V_2}{T_2}$$

also applies to an adiabatic change.

Dividing $p_1V_1^{\frac{5}{3}} = p_2V_2^{\frac{5}{3}}$ by $\frac{p_1V_1}{T_1} = \frac{p_2V_2}{T_2}$

gives

$$T_1V_1^{\frac{2}{3}} = T_2V_2^{\frac{2}{3}}$$

In the $p-V$ equation for an adiabatic change V is raised to a power of greater than one. This means that for an adiabatic change the line will be steeper than that for an isothermal change – as shown in figure 7.

The area under the adiabatic change, as for all other changes, will be the work done. The direction of the arrow will determine whether work is done on or by the gas.

Note

You don't need to derive the equations for adiabatic changes. The exponent for V does vary if the gas molecules are diatomic (two atom molecules) or polyatomic (three or more molecules in the atom). The value is actually the ratio of the principal molar specific heats but, as this is not going to be examined, we leave you to do further research on this.

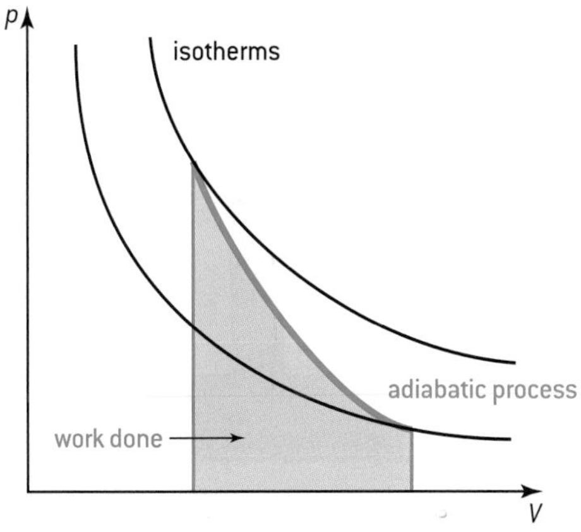

▲ Figure 7 Adiabatic and isothermal changes.

Isovolumetric changes

These changes occur at constant volume and, therefore, mean that **no work can be done** by or on the system.

$$Q = \Delta U + W = \Delta U + p\Delta V$$

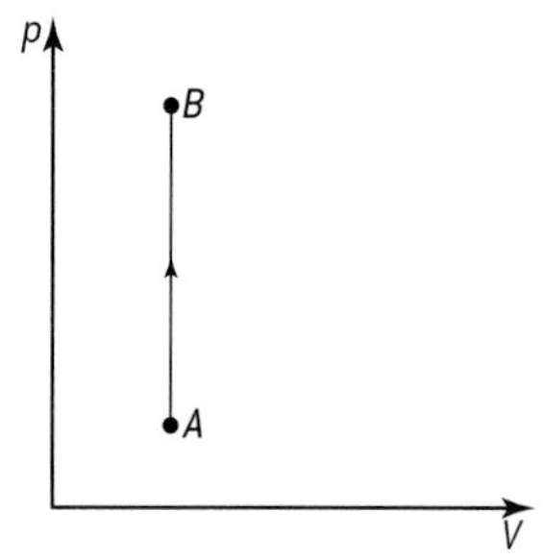

▲ Figure 8 Isovolumetric change.

With no change in the volume, the first law of thermodynamics becomes $Q = \Delta U$ (when the energy transferred to the system increases the internal energy) or $-Q = -\Delta U$ (when the energy transferred from the system decreases the internal energy).

Figure 8 shows an isovolumetric change on a p–V graph. In this case the change shows an increase in temperature (moving to a higher isotherm on an isothermal graph).

The equation of state for an isovolumetric change is $\frac{p}{T} = $ constant.

Worked example

a) Distinguish between an *isothermal* process and an *adiabatic* process as applied to an ideal gas.

b) An ideal gas is held in a cylinder by a moveable piston and energy is supplied to the gas such that the gas expands at a constant pressure of 1.5×10^5 Pa. The initial volume of the cylinder is 0.040 m^3 and its final volume is 0.12 m^3. The total energy supplied to the gas during the process is 7.5×10^3 J.

(i) State and explain the type of change that the gas undergoes.

(ii) Determine the work done by the gas.

(iii) Calculate the change in internal energy of the gas.

Solution

a) An *isothermal* process is one that takes place at constant temperature (and constant internal energy) so there is an interchange of energy transferred because of the temperature difference between the gas and the surroundings and work done one on the other.

An *adiabatic* process is one in which there is no energy exchanged between the system and the surroundings. This means that changes in the internal energy (and hence temperature) occurs because of work done by or on the ideal gas.

b) (i) As the pressure does not change the gas undergoes an isobaric expansion.

(ii) Work done = area under the p–V graph for the change $(= p\Delta V) = 1.5 \times 10^5 (0.12 - 0.04) = 1.2 \times 10^4$ J

(iii) Using the first law of thermodynamics $Q = \Delta U + W$

$\Delta U = Q - W$ because only 7.5×10^3 J is supplied to the gas and 12.0×10^3 J of work is done, the internal energy must fall by 4.5×10^3 J

Cycles and engines

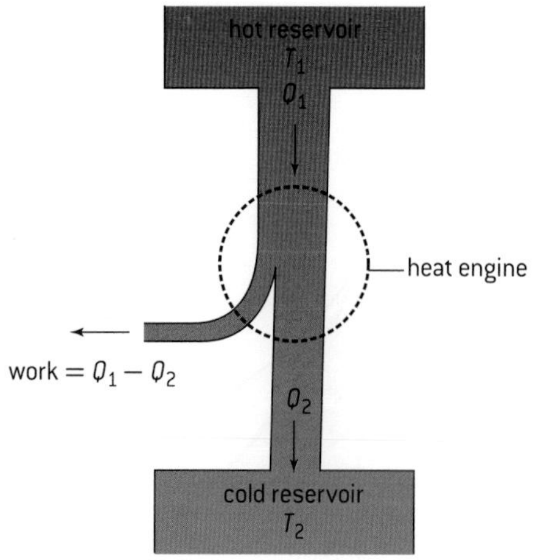

▲ Figure 9 The principle of a heat engine.

Work can be converted into internal energy effectively through frictional forces. Work done by friction is usually undesirable because it increases the temperature of the system (good) and the surroundings (bad). The reverse process of continuously converting energy into work is more difficult to achieve, but it can be done using a **heat engine** that operates through a cycle of changes. In 1824 the French physicist and engineer, Sadi Carnot, published the first description of a heat engine; in this he described what has come to be known as the "Carnot cycle". The principle of a heat engine is to take in energy at a high temperature, reject energy at a low temperature and use the remainder of the energy to do work on the system as illustrated in figure 9.

We can think of the system as being a gas enclosed in a cylinder with a frictionless moveable piston. When energy Q_1 is supplied to the gas from a hot reservoir at temperature T_1, the gas expands and moves the piston, doing work. This will stop as soon as the gas pressure is equal to that of the surroundings. The gas now needs to be returned to its original state before it can do further work. This can only happen if some of the energy (Q_2) that was initially absorbed is rejected to a cold reservoir at lower temperature T_2.

The thermal efficiency η of the heat engine will be given by:

$$\eta = \frac{\text{useful work done}}{\text{energy input}} = \frac{W}{Q_1} = \frac{Q_1 - Q_2}{Q_1}$$

In the Carnot cycle we imagine a completely friction-free engine that is able to take a gas through a cycle of two isothermal and two adiabatic changes as shown in figure 10.

Starting at point A where the gas is at its highest temperature T_1, it expands isothermally to B by absorbing energy Q_1. The internal energy of the gas does not change so all the energy absorbed is doing work.

The gas now expands adiabatically to C where the temperature falls to T_2. No energy is now being absorbed but the gas still does work on the surroundings by losing some internal energy.

During the expansion ABC the area under the two expansion curves gives the work done on the surroundings.

At C the gas now needs work being done on it so it is compressed isothermally to D and rejects energy Q_2. The internal energy does not change so the work done on the gas is all rejected as energy.

Finally, the gas is further compressed adiabatically from D back to A. The work done on the gas is all used to increase the internal energy in returning the gas to temperature T_1.

During the compression CDA the area under the two compression curves gives the work done by the surroundings on the gas. **The area enclosed by the curve is the net work done by the gas on the surroundings in one cycle.**

The Carnot heat engine is said to be reversible. This is a theoretical concept in which, at any part in the cycle, the system can be returned to a previous state without any energy transference – this must be done infinitely slowly and means that the system returns exactly to its initial state at the end of the cycle. It can be shown that, for the Carnot cycle (and all reversible heat engines), that the thermal efficiency is given by

$$\eta = \frac{T_1 - T_2}{T_1} = 1 - \frac{T_2}{T_1}$$

This is written in the IB Physics Syllabus as

$$\eta_{\text{Carnot}} = 1 - \frac{T_{\text{cold}}}{T_{\text{hot}}}$$

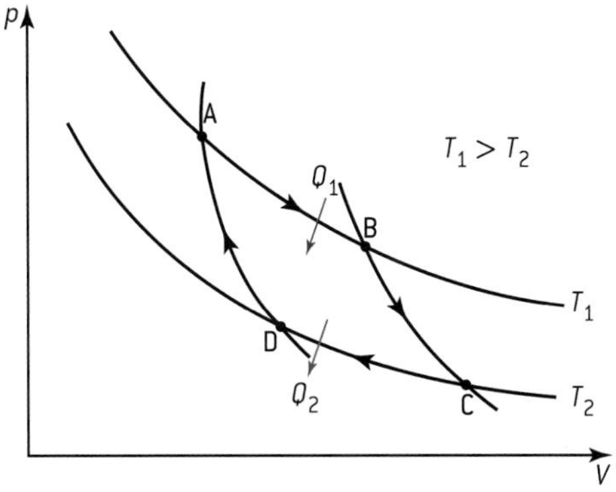

▲ Figure 10 The Carnot cycle.

Note

- Temperatures used here must be in kelvin.
- This is the equation for the maximum efficiency of a heat engine.
- The maximum efficiency is increased by raising the temperature of the hot reservoir and/or by lowering the temperature of the cold reservoir.
- The maximum efficiency can never equal 100% as this would mean that the cold reservoir was at absolute zero or else the hot reservoir was at an infinitely high temperature – neither of these requirements is possible.
- In practice thermodynamic cycles can be achieved but none will be more efficient than the Carnot cycle.

Worked example

A quantity of an ideal gas is used as the working substance of a heat engine. The cycle of operation of the engine is shown in the p–V graph opposite. Change CA is isothermal.

The temperature of the gas at A is 300 K.

a) During the change AB the change in internal energy of the gas is 7.2 kJ.

(i) Calculate the temperature, at B, of the gas.

(ii) Determine the amount of energy transferred during change AB.

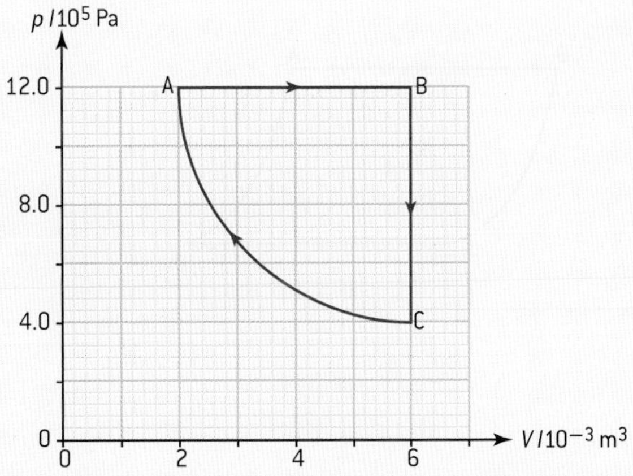

b) State why, for the change BC, the change in the internal energy of the gas is numerically the same as that in AB.

c) Calculate:

(i) the net work done in one cycle

(ii) the efficiency.

Solution

a) (i) For an ideal gas $\frac{pV}{T}$ = constant

As pressure is constant $\frac{V_A}{T_A} = \frac{V_B}{T_B} => T_B$

$= \frac{V_B \times T_A}{V_A} = \frac{6.0 \times 10^{-3} \times 300}{2.0 \times 10^{-3}} = 900\text{ K}$

(ii) $W = p\Delta V$

$= 12.0 \times 10^5 \times (6.0 \times 10^{-3} - 2.0 \times 10^{-3})$

$= 4.8\text{ kJ}$

The increase in internal energy = 7.2 kJ so the energy transferred to the gas must equal 4.8 + 7.2 = 12.0 kJ.

b) The gas undergoes the same change in temperature and, as the gas is ideal, this means that the change in internal energy depends solely on the temperature.

c) (i) We need to find the area enclosed by the cycle.

Each large square is equivalent to $1.0 \times 10^{-3} \times 2.0 \times 10^5\text{ J} = 200\text{ J}$

Estimate that there are 14 large squares, making a total 2800 J or 2.8 kJ.

(ii) $\eta = \frac{W}{Q_1} = \frac{2.8}{12} = 0.23$ or 23%

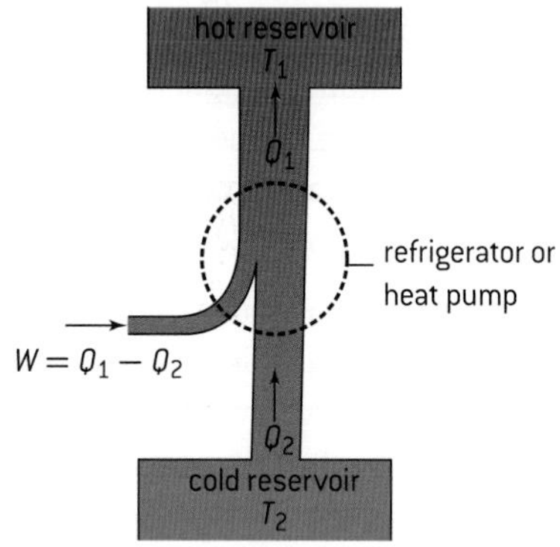

▲ Figure 11 The principle of a heat pump and refrigerator.

Heat pumps and refrigerators

Heat pumps and refrigerators act in a similar manner to a heat engine working in reverse. They take in energy Q_2 at a low temperature T_2, do work W on the working substance and reject energy Q_1 at the high temperature T_1. Although they are very similar in their working the heat pump is designed to add energy to the high temperature reservoir (for example, the room being heated by extracting energy from the ground) whilst the refrigerator is designed to remove energy from the low temperature reservoir (the cool box). Air conditioning units are another example of heat pumps.

Worked example

The diagram shows the relationship between the pressure p and the volume V of the working substance of a refrigerator for one cycle of its operation. The working substance is a volatile liquid which is made to vaporize and condense.

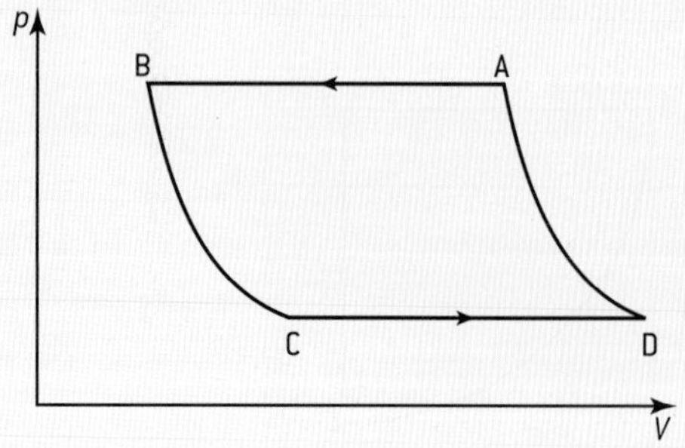

a) The working substance at point C of the cycle is entirely in the liquid phase.

Suggest the reason why both the changes from CD and AB are isothermal, isobaric changes.

b) State during which process of the cycle energy is absorbed from the cold reservoir and during which process energy is transferred to the hot reservoir.

c) State how the value of the work done during one cycle may be determined from the pV diagram.

Solution

a) Both changes are isobaric and isothermal because there is no pressure or temperature change. Each of the changes occurs because of a change of phase of the working substance. From C to D the liquid vaporizes and from A to B the vapour condenses.

b) Energy is absorbed during C to D as the liquid needs energy to vaporize and it is ejected during A to B in order for the vapour to condense.

c) The area enclosed by the cycle will always indicate the net amount of work – in this case it is work done on the system (working substance).

The second law of thermodynamics

Although this is a fundamental law that has its origins in practical experiences, the second law of thermodynamics can be stated in a number of different ways. The first law of thermodynamics equates work to energy, the second law deals with the circumstances in which energy can be converted to work. Each of the statements is equivalent to the others and communicates an expression of the impracticability of reversing real thermodynamic processes.

The Clausius version of the second law can be stated as:

It is impossible to transfer energy from a body at a lower temperature to one at higher temperature without doing work on the system.

The Kelvin (or Kelvin-Planck) version states:

It is impossible to extract energy from a hot reservoir and transfer this entirely into work.

If the second law was not true it would be possible to power ships by energy extracted from the sea – this cannot be done because there needs to be a cold reservoir into which the difference between the energy extracted and the work done would be rejected. When, after a time, we return to full cup of coffee we do not expect to find it at a higher temperature than when it was made. Energy passes from the hot coffee to the cooler room until the two bodies are at the same temperature – if we wanted the coffee to heat up we would have to transfer energy to it using a heating coil or else do work on it by stirring it rapidly!

The internal energy of an object is related to the random motion of the molecules of the object. By trying to convert this internal energy into work we are trying to convert random motion into something more ordered. It is impossible to do this because we cannot take control over the individual motion of a colossal number of molecules.

Entropy

A third version of the second law of thermodynamics involves the concept of **entropy** S. This quantity can be defined (for a reversible change) in terms of the equation

$$\Delta S = \frac{\Delta Q}{T}$$

ΔS is the increase in entropy

ΔQ is the energy absorbed by the system

T is the temperature in kelvin at which this occurs.

Entropy is a scalar quantity and has units of joule per kelvin ($J\ K^{-1}$)

Note

- T is always positive so when energy is absorbed by a system and ΔQ is positive there will be a positive change in entropy – an increase. When energy is rejected by the system the entropy will decrease.
- For an adiabatic change $\Delta Q = 0$ and so $\Delta S = 0$
- A substance taken through a complete reversible cycle will undergo no change in entropy as $\eta = 1 - \frac{T_2}{T_1} = 1 - \frac{Q_2}{Q_1}$ this means $\frac{Q_1}{T_1} = \frac{Q_2}{T_2}$
- All heat engines reject energy to the surroundings and generate an overall increase in entropy of the universe.

Worked example

0.20 kg of ice at 0 °C melts. The specific latent heat of fusion of water is $3.3 \times 10^5\ J\ kg^{-1}$. Calculate the change in entropy of the ice as it melts.

Solution

Energy needed to melt the ice $= mL = 0.20 \times 3.3 \times 10^5 = 6.6 \times 10^4$ J

$$\Delta S = \frac{\Delta Q}{T} = \frac{6.6 \times 10^4}{273} = 2.4 \times 10^2\ J\ K^{-1}$$

Entropy as a measure of disorder

We have now seen that for real processes the entropy of the universe increases – in fact this is another way of stating the second law of thermodynamics first suggested by Boltzmann.

Real processes always degrade the energy, i.e. change the energy from being localized to being more spread out. If some hot water is mixed with cold water in a completely insulated container there is no loss of energy, however, the opportunity to use the energy to do work is now restricted by having cold water. It is not a sensible proposition to separate the most energetic molecules to produce some hot water and some cold water from the mixture – the energy has changed from the localized situation in the hot water molecules to a situation where it is spread-out amongst all of the molecules.

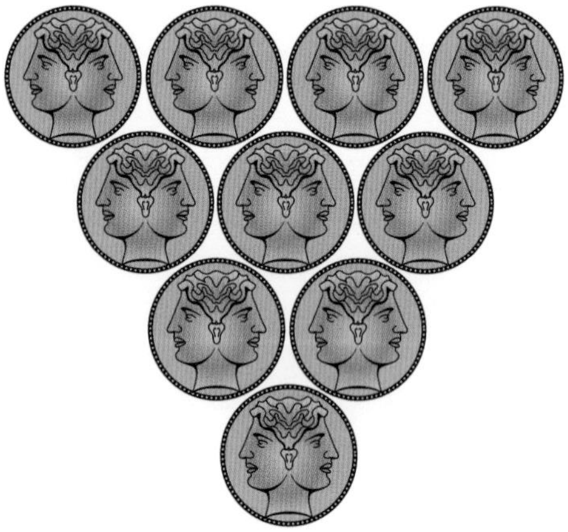

▲ Figure 12 Ten coins all "heads up".

Imagine having 10 coins all placed heads up on the table. The coins are picked up and shaken before being returned to the table, without looking to see where they are placed; some coins will, therefore, have heads up but others will have heads down. The coins have moved from an ordered state to a disordered state. There is a small likelihood $\left(\left(\frac{1}{2}\right)^{10} = \frac{1}{1024}\right)$ that the coins would be replaced with all 10 heads up. By increasing the number of coins from 10 to 100 it decreases the likelihood of them all being heads up to 1 in 1.3×10^{30}, and, this really isn't likely to happen! The coin experiment mirrors nature in that a system does not naturally become more ordered. We have now seen that the entropy of a system naturally increases with the disorder of the system. This is not coincidental since it can be shown that *entropy is a measure of the disorder of a system.*

TOK

The arrow of time

Many scientists have discussed increasing entropy as representing the "arrow of time". Because the disorder of a large system will increase with time, finding such a system with increased order would be equivalent to time going backwards. Perpetually increasing entropy is contrary to Newton's laws of motion in which the change of state of a system is equally predictable whether we go forwards or backwards in time.

Does this statement of the Second law prevent the possibility of time travel?

Worked example

a) State what is meant by an *increase in entropy* of a system.

b) State, in terms of entropy, the second law of thermodynamics.

c) When a chicken develops inside an egg, the entropy of the egg and its contents decreases. Explain how this observation is consistent with the second law of thermodynamics.

Solution

a) When the entropy increases, there is an increase in the degree of disorder in the system.

b) The total entropy of the universe increases.

c) Entropy of the surroundings must increase more than the decrease of entropy in the developing egg. The energy generated by the biochemical processes within the egg becomes more spread out as a consequence of some passing into the surroundings.

B.3 Fluids and fluid dynamics (AHL)

Understandings

- Density and pressure
- Buoyancy and Archimedes' principle
- Pascal's principle
- Hydrostatic equilibrium
- The ideal fluid
- Streamlines
- The continuity equation
- The Bernoulli equation and the Bernoulli effect
- Stokes' law and viscosity
- Laminar and turbulent flow and the Reynolds number

Applications and skills

- Determining buoyancy forces using Archimedes' principle
- Solving problems involving pressure, density and Pascal's principle
- Solving problems using the Bernoulli equation and the continuity equation
- Explaining situations involving the Bernoulli effect
- Describing the frictional drag force exerted on small spherical objects in laminar fluid flow
- Solving problems involving Stokes' law
- Determining the Reynolds number in simple situations

Equations

- Buoyancy force: $B = \rho_f V_f g$
- Pressure in a fluid: $p = p_0 + \rho_f g d$
- Continuity equation: $Av =$ constant
- The Bernoulli equation: $\frac{1}{2}\rho v^2 + \rho g z + p =$ constant
- Stokes' law: $F_D = 6\pi\eta r v$
- Reynolds number: $R = \frac{v r \rho}{\eta}$

Nature of science

Fluids in motion

The study of the transportation of mass is very important in medicine and engineering. Knowledge of how the velocity and pressure changes throughout a moving fluid is vital in our understanding of many physical processes, including how blood flows around the body, how aircraft fly, and how jet engines operate. The visualization of fluid flow can be extremely appealing and has been used extensively in the visual arts.

Introduction

The gaseous and liquid states of matter are jointly known as **fluids** – substances that can flow and take the shape of their containers. There are important differences between gases and liquids:

- It is easy to compress a gas but liquids, like solids, are almost incompressible.
- Gases are not restricted by a surface but liquids are – you can have half a cup of coffee but not half a cup of air! The most energetic molecules may, however, be able to break through the surface to form a vapour above the liquid.

Although this topic is called fluids and fluid dynamics, it is almost entirely restricted to idealized fluids – those that cannot be compressed, are non-viscous, and flow in a steady manner. These properties are best represented by a low viscosity liquid, such as water, and not by a 'sticky' one, such as oil (which has much internal friction).

Static fluids – density and pressure

The density ρ of a substance is given by the ratio of the mass m of the substance to the volume V of the substance:

$$\rho = \frac{m}{V}$$

Pressure is the ratio of the perpendicular contact force acting on a surface to the area of the surface:

$$p = \frac{F}{A}$$

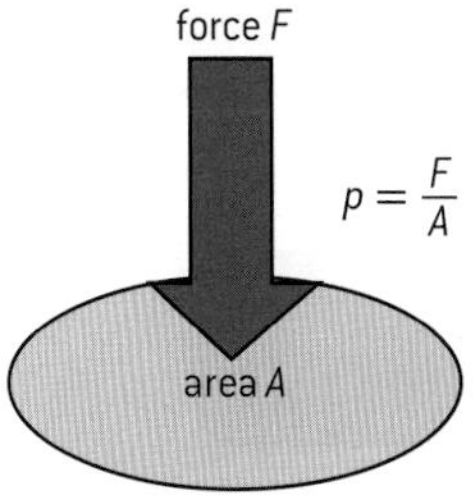

▲ Figure 1 Definition of pressure.

Although the direction of force is at right angles to the surface on which it acts, pressure is a scalar quantity and acts in all directions; with the force in newton and the area in metres squared, pressure is measured in pascal (Pa). In a fluid, pressure is found to increase with depth and, at a given depth, the pressure is found to be equal in all directions. This results in a perpendicular force acting on any surface at a given depth. If this was not the case, any pressure differences would cause the liquid to flow until the pressure was constant.

Consider a cylinder of height h and base area A in a fluid of density ρ as shown in figure 2 – the top of the cylinder is at the surface of the fluid. The forces acting on the bottom of the cylinder are the weight W of the column of liquid above it acting downwards and the pressure force pA from the liquid acting upwards. These are in equilibrium so

$$W = pA$$

W = mass of cylinder × gravitational field strength = mg, this means that

$$W = Ah\rho g = pA$$

and therefore

$$p = h\rho g$$

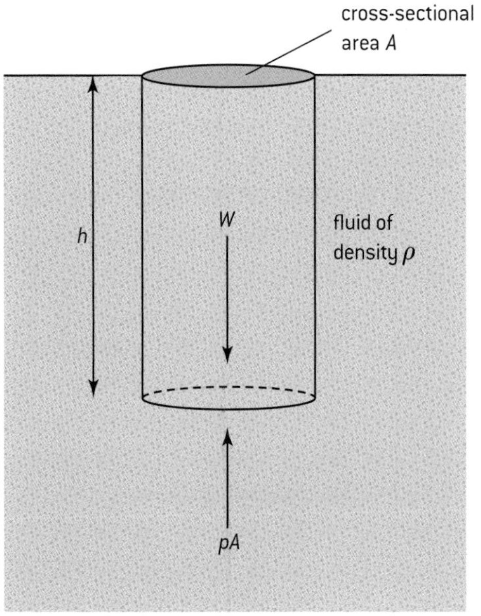

▲ Figure 2 Pressure in a fluid.

In a liquid that is not in a sealed container the atmospheric pressure p_0 will also act on the liquid surface. This makes the total pressure p at depth h become $h\rho g + p_0$

$$p = h\rho g + p_0$$

Nature of science

Measuring density of immiscible liquids using a U-tube

When two liquids that do not mix are added to the limbs of a U-tube, their levels settle in a way that is dependent on their relative densities. At the level X–Y in figure 3 the pressure will be constant, so

$$p_0 + h_A\rho_A g = p_0 + h_B\rho_B g$$

Cancelling p_0 and g gives

$$h_A\rho_A = h_B\rho_B$$

When we know the density of one liquid we are then able to calculate that of the second liquid by taking a ratio of the heights. This experiment is an example of hydrostatic equilibrium.

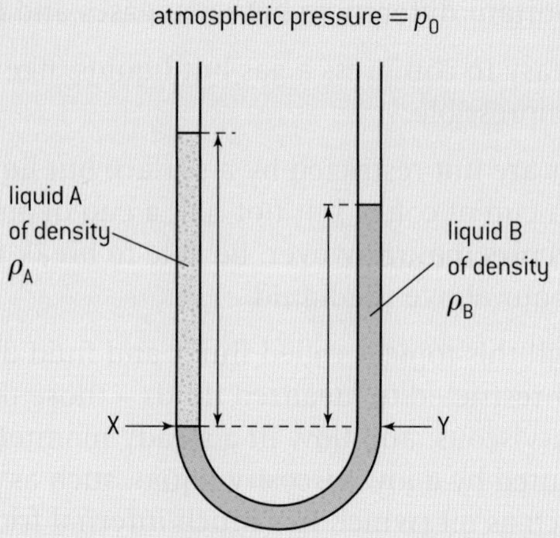

▲ Figure 3 Hydrostatic equilibrium to measure density.

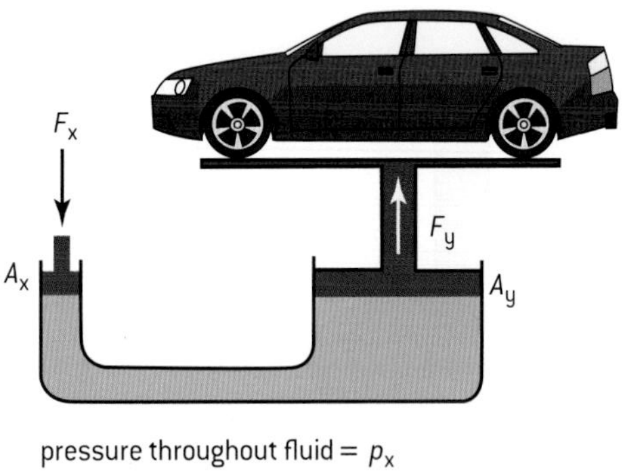

▲ Figure 4 Hydraulic jack.

Pascal's principle

This principle says that **the pressure applied at one point in an enclosed fluid under equilibrium conditions is transmitted equally to all parts of the fluid.** The principle allows hydraulic systems to operate. For example, the hydraulic jack is used to raise heavy objects such as cars.

A small force F_x applied to a piston of small cross-sectional area A_x produces a pressure P_x in a liquid such as oil. This pressure is communicated through the liquid so that it acts on a second, larger piston of cross-sectional area A_y. This will then produce a larger force F_y so that

$$p_x = \frac{F_x}{A_x} = \frac{F_y}{A_y}$$

The device cannot amplify the energy, so the distance moved by F_x must be far more than that moved by F_y in order to conserve energy (work being force × distance).

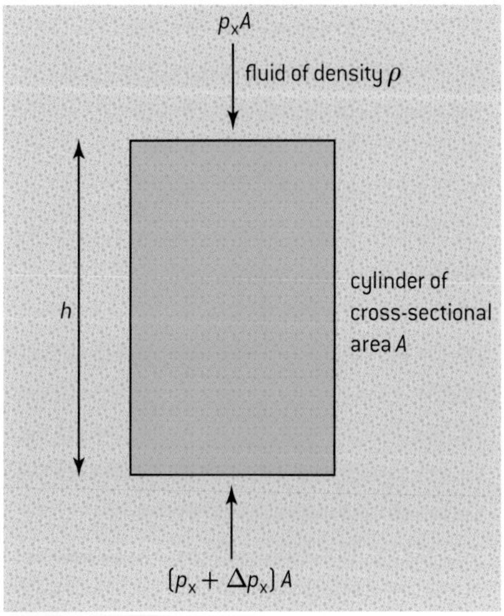

▲ Figure 5 Archimedes' principle.

Archimedes' principle

When you dive into a swimming pool you quickly feel the buoyancy force of the water acting on you. The magnitude of this upward force is expressed by Archimedes' principle; this says that **for an object wholly or partially immersed in a fluid there will be an upward buoyancy force acting on the object which is equal to the weight of fluid that the object displaces.** The buoyancy force is often known as the **upthrust**.

We can deduce this relationship by considering a cylinder of height h and cross-sectional area A fully immersed in a liquid of density ρ. The pressure is p_x at the top and $(p_x + \Delta p_x)$ at the bottom. The latter is higher because of the extra depth of fluid. The forces on the top and bottom of the cylinder will be p_xA and $(p_x + \Delta p_x)A$. We ignore the atmospheric pressure because it acts on both the top and the bottom of the cylinder.

The buoyancy force B will, therefore, be the difference between these forces

$$B = (p_x + \Delta p_x)A - p_x A = \Delta p_x A$$

$$\Delta p_x = h\rho g \text{ making } B = h\rho g A$$

$$V = hA$$

And so $B = \rho V g$, which is the weight of the fluid displaced by the cylinder.

To emphasize that this applies to a fluid, the IB Physics data booklet shows this equation as

$$B = \rho_f V_f g$$

Note

When an object is floating:

- The buoyancy force acting on the object must equal the weight of the object.
- The buoyancy force will equal the weight of fluid that is displaced by the submerged portion of the floating object.

Worked example

A block of wood of mass 50 kg is completely immersed in water.

The density of the wood is 1.1×10^3 kg m^{-3} and that of the water 1.0×10^3 kg m^{-3}

a) Calculate the resultant force on the block.

b) When the block is cut into planks and made into a boat it floats, displacing 0.15 m^3 of water when empty. Calculate the buoyancy force acting on the boat.

Solution

a) The weight of the block $= mg = 50 \times 9.81 = 4.90 \times 10^2$ N

Buoyancy force = weight of water displaced $= B = \rho_f V_f g$

$$V_f = \text{volume of block} = \frac{m}{\rho} = \frac{50}{1.1 \times 10^3} = 0.045 \text{ m}^3$$

$$B = \rho_f V_f g = 1.1 \times 10^3 \times 0.045 \times 9.81 = 4.86 \times 10^2 \text{ N}$$

The overall force is, therefore,
$4.90 \times 10^2 - 4.86 \times 10^2 = 4$ N downwards

b) The weight of water displaced by the boat
$= 0.15 \times 1.0 \times 10^3 \times 9.81 = 1.5 \times 10^3$ N

So $B = 1.5 \times 10^3$N (around three times the original value)

Fluid dynamics

There are many instances of fluid flow. Examples include the smoke from a fire, hot water in a central heating system and wind spinning around in a tornado. Fluid dynamics also deals with solid objects moving through stationary fluids. An ideal fluid offers no resistance either to a solid moving through it or it moving through or around a solid object. An ideal fluid is said to be **non-viscous**. We will start our discussion of fluid dynamics by considering a non-viscous, incompressible fluid in a stream tube.

Streamline (or laminar) flow

Ideal fluids flow in a very predictable way. We can represent the paths taken by particles within a fluid using the concept of **streamlines**. In streamline flow, the motion of a particle passing a particular point is identical to the motion of all the particles that preceded it at that point. Streamlines will be close together when the particles move quickly and further apart when they move more slowly. A group of streamlines is known as a stream tube. Fluid never crosses the surface of a **stream tube**. Figure 6 shows the streamlines in a tube that narrows and widens again.

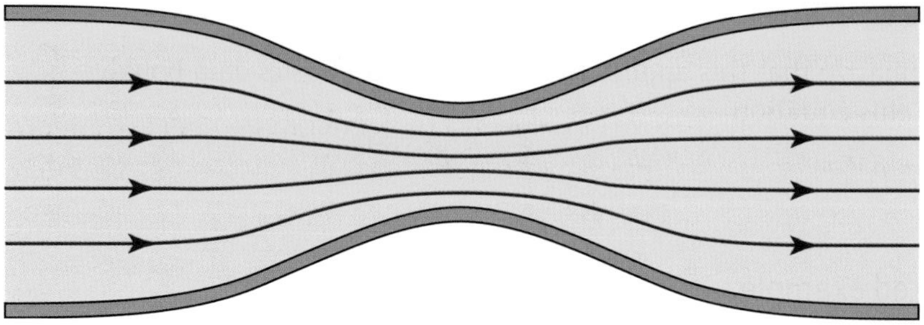

▲ Figure 6 Streamlines in a tube that narrows.

The continuity equation

With **steady flow** the mass of fluid entering one end of a stream tube must be equal to that leaving the other end. The stream tube may be a physical tube or pipe or it may be a series of streamlines within the fluid.

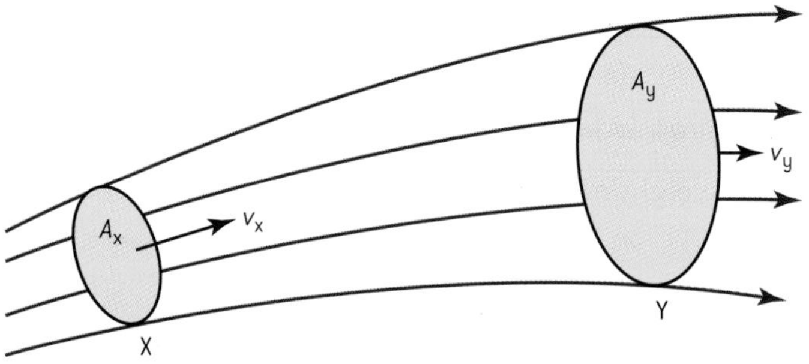

▲ Figure 7 Streamlines at the boundary of a stream tube.

For a length of stream tube the fluid enters at X through an area of cross-section A_x at velocity v_x and leaves at Y through an area of cross-section A_y with velocity v_y.

In a short time, Δt, the fluid leaving X will travel a distance $v_x\Delta t$ – thus meaning that a volume $A_x v_x \Delta t$ and a mass $A_x v_x \rho \Delta t$ enter the tube in this time.

In the same time a mass $A_y v_y \rho \Delta t$ will leave the stream tube through Y.

As there cannot be any discontinuities in an ideal liquid these masses must be equal so

$A_x v_x \rho \Delta t = A_y v_y \rho \Delta t$ meaning that when $\rho \Delta t$ is cancelled from both sides $A_x v_x = A_y v_y$

We see from this that

$$Av = \text{constant}$$

This is known as the **continuity equation** – and the product Av is known as the **volume flow rate** (or, sometimes, just **flow rate**). Volume flow rate is measured in units of $m^3\ s^{-1}$.

The Bernoulli equation

This equation is a generalized equation which deals with how a fluid is able to both speed up and rise to a higher level as it passes through a stream tube. The equation is derived from the conservation of energy and, although the derivation is shown below, you will not need to repeat this in an IB Physics examination. In your data booklet the Bernoulli equation is written as

$$\frac{1}{2}\rho v^2 + \rho g z + p = \text{constant}$$

Here ρ is the fluid density, v is its speed, g the gravitational field strength, z the height above a chosen level, and p the pressure at that height. You will notice in the equation that, if ρ was replaced by m in each of the first two terms, we would have kinetic energy and potential energy.

As density ρ is equal to $\frac{m}{V}$, multiplying the Bernoulli equation by V would give

$$\frac{1}{2}mv^2 + mgz + pV = \text{different constant}$$

This has now turned the Bernoulli equation into the conservation of energy – because it implies that the kinetic energy + gravitational potential energy (z is height here) + work done (remember the first law of thermodynamics) remains the same. The equation in the Bernoulli form expresses each of the terms as an energy density (energy per unit volume) and is measured in J m^{-3}. This equation also tells us that, for any point in a continuous steady flow, the total of all the quantities will remain the same: if the pressure changes then one or more of the other terms must also change.

Derivation of the Bernoulli equation

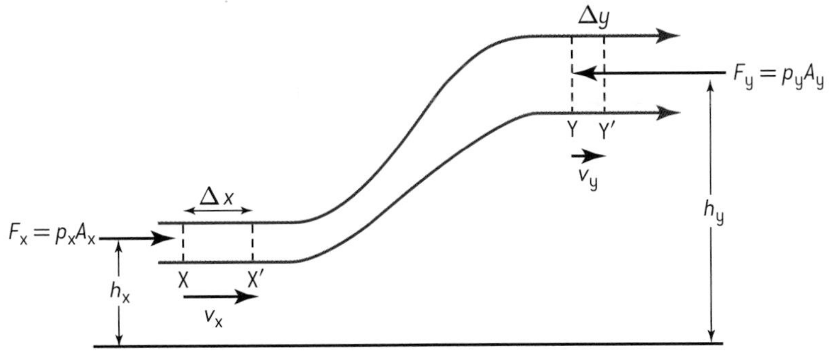

▲ Figure 8 The Bernoulli equation.

Consider a fluid entering and leaving a pipe that becomes wider and higher. It is obvious that the pressure at the inlet must be greater than the pressure at the outlet. When the inlet pressure is p_x the force F_x at the inlet is p_xA_x and that at the outlet is p_yA_y.

In a short time, Δt, the fluid entering the pipe at X moves a short distance Δx from X to X′ at velocity v_x. This means that the work done on the fluid is $F_x\,\Delta x = p_xA_x\,\Delta x$.

In the same time, the same mass of fluid moves a distance Δy in moving from Y to Y′ and the work done will be $F_y\,\Delta y = p_yA_y\Delta y$.

The net work done will be $p_x A_x \Delta x - p_y A_y \Delta y$

As this happens in the same time, Δt, and because equal masses must mean equal volumes for an ideal fluid of constant density, this gives

$$A_x \Delta x = A_y \Delta y = V \text{ thus net work done} = V(p_x - p_y)$$

the change of kinetic energy for this mass $= \frac{1}{2} m(v_y^2 - v_x^2)$

and the change in potential energy $= mg(h_y - h_x)$

Overall, for the conservation of energy:

work done = gain in kinetic energy + gain in gravitational potential energy

so

$$V(p_x - p_y) = \frac{1}{2}m(v_y^2 - v_x^2) + mg(h_y - h_x)$$

Dividing by V and collecting terms for x and y we get

$$p_x + \frac{1}{2}\rho v_x^2 + \rho g h_x = p_y + \frac{1}{2}\rho v_y^2 + \rho g h_y$$

This is the Bernoulli equation.

In the IB Physics data booklet the symbol z is used for heights.

Applications of the Bernoulli equation

From the Bernoulli equation it follows that, when a fluid speeds up, there must be a decrease in either the pressure or the gravitational potential energy or both of these. When the flow is horizontal there can be no change in the gravitational potential energy, so there must be a reduction in pressure. Aerofoils, used on aircraft wings and mounted on racing cars, are shaped so that the air flows faster over the more curved surface than over the flatter surface – in most cases it is the aerofoil that moves, but the relative effect is identical. The faster moving air produces a lower pressure and less force acts on that section of the aerofoil – this causes the aerofoil to be forced upwards (a lift) or downwards (a down thrust) depending on which way it is positioned. The effect of the shape of the aerofoil causes the streamlines to be closer together on the curved side and to be further apart on the flatter side.

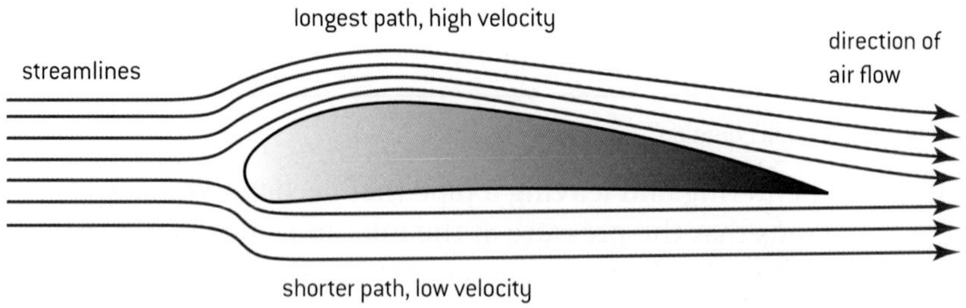

▲ Figure 9 The aerofoil.

Venturi tubes

When a tube narrows, the speed of the fluid increases at the narrow part of the tube before returning to the original value at the wider part. This means that the pressure is higher at the edges of the tube and lower in the centre. This is the principle of the Venturi gauge, which is used for measuring fluid speeds as demonstrated in figure 10. The higher air pressure has a greater effect on the right limb of the manometer. This pressure difference can be calibrated to measure relative fluid speeds and flow rates.

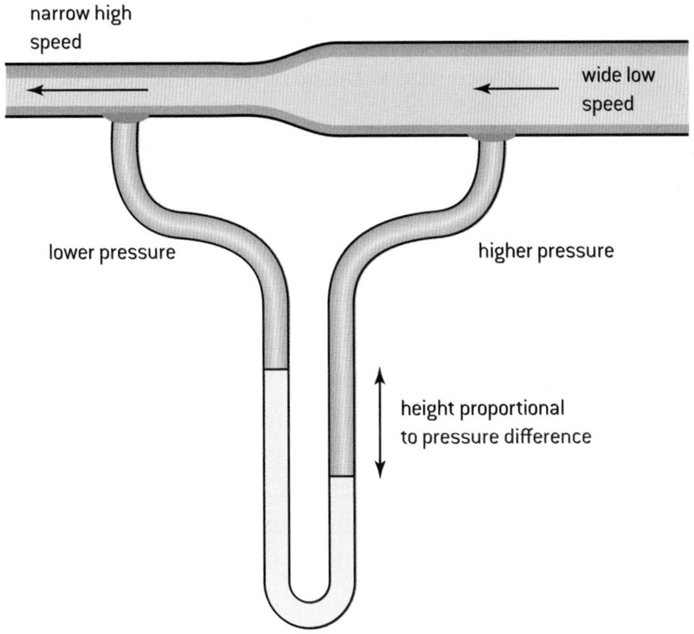

▲ Figure 10 Venturi gauge.

Pitot static tubes

Pitot static tubes are used for measuring the velocity of a fluid of density ρ. The tubes X and Y must be on the same streamline. The opening of tube X is perpendicular to the direction of fluid flow but the opening of tube Y is parallel to the flow. The fluid can then flow into the opening of tube Y. The two tubes must be far enough apart for tube X not to affect the velocity at Y. When there is steady flow, no fluid particles can move from one streamline to another and so none enter tube X. Tube X measures the static pressure p_x. Particles will enter tube Y where their kinetic energy will be brought to zero.

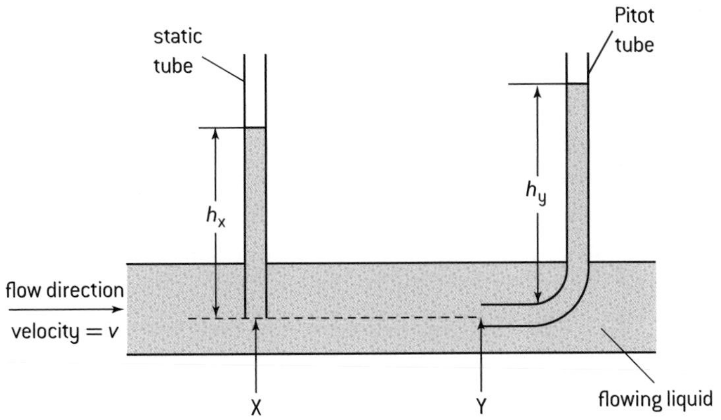

▲ Figure 11 Pitot static tubes.

Applying the Bernoulli equation to the streamline along X–Y gives

$$p_x + \frac{1}{2}\rho v_x^2 = p_y$$

which gives

$$v_x = \sqrt{\frac{2}{\rho}(p_y - p_x)}$$

As $p_x = h_x\rho g + p_0$ and $p_y = h_y\rho g + p_0$ the velocity equation becomes

$$v_x = \sqrt{2g(h_y - h_x)}$$

For an idealized fluid this will be the velocity along the streamline X–Y. This arrangement is the one used for measuring the velocity of liquids, but it can be adapted to measure the flow of gases. A Pitot tube is installed on aircraft wing to measure the speed of the aircraft relative to the air.

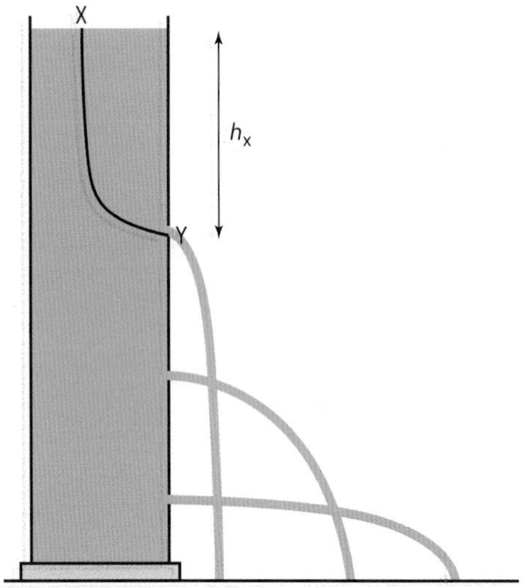

▲ Figure 12 Leaking can demonstration.

Flow out of a container

You may be familiar with the leaking can demonstration. Figure 12 shows liquid pressure increasing with depth. This is another application of the Bernoulli equation. The line X–Y represents a streamline joining water at the surface to a hole in the side of the can. Each of these two points will be open to the atmosphere and so they will both be at atmospheric pressure. This means that $p_x = p_y = p_0$.

The Bernoulli equation becomes:

$$\frac{1}{2}\rho v_x^2 + \rho g h_x = \frac{1}{2}\rho v_y^2 + \rho g h_y$$

Here h_x is the height above Y and $h_y = 0$. If we assume the container is wide enough and the holes small enough for the velocity of the water surface to approximate to zero, the equation reduces to

$$\rho g h_x = \frac{1}{2}\rho v_y^2$$

Cancelling ρ and re-arranging the equation gives

$$v_y = \sqrt{2gh_x}$$

This confirms that the velocity is related to the height of water above opening. Viscosity and turbulence have been neglected.

Worked example

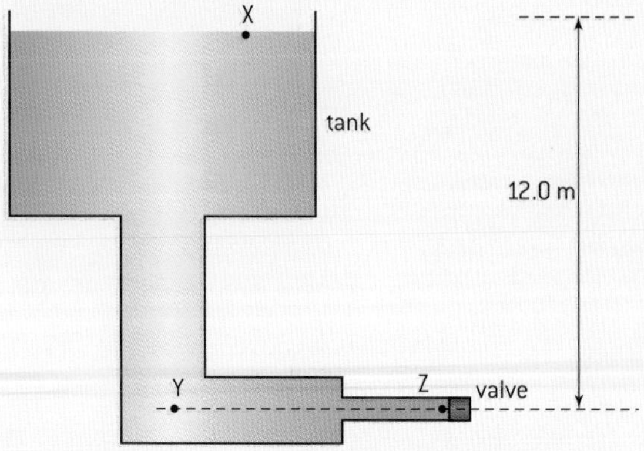

The diagram shows a large water tank, open to the atmosphere. The tank feeds a pipe of cross-sectional area 4.0×10^{-3} m^2 which, in turn, feeds a narrow tube of cross-sectional areas 5.0×10^{-4} m^2. The narrow tube is sealed with a closed valve. The water surface in the tank is 12.0 m above the centre of the two tubes.

density of water = 1.00×10^3 kg m^{-3}

atmospheric pressure = 1.0×10^5 Pa

a) Calculate the pressure at Y.

b) Calculate the pressure at Z.

c) The valve is now opened so that water leaves the system. The pressure at the open valve is now atmospheric.

(i) Determine the velocity of the water at Z.

(ii) Calculate the velocity of the water at Y.

(iii) Calculate the pressure at Y.

Solution

a) and **b)** The pressure at Y and Z will be equal because they are at the same level. The pressure will be $p = p_0 + \rho_f gd$
$= 1.0 \times 10^5 + (1.00 \times 10^3 \times 9.81 \times 12.0)$
$= 2.18 \times 10^5 \text{ Pa} \approx 2.2 \times 10^5 \text{ Pa}$

c) (i) Assuming the surface area is large enough for the kinetic energy at X to be zero we have shown that
$v = \sqrt{2gh} = \sqrt{2 \times 9.81 \times 12.0}$
$= 15.3 \text{ m s}^{-1}$

> **Note**
>
> That this is not an equation given in the IB Physics data booklet. If you were tested on it, the question would need to be structured in order to lead you through the development of this equation.

(ii) Using the continuity equation

$A_y v_y = A_z v_z$

$4.0 \times 10^{-3} \times v_y = 5.0 \times 10^{-4} \times 15.3$

$v_y = 1.9 \text{ m s}^{-1}$

(iii) Using the Bernoulli equation between Y and Z (they are on the same level so there will be no difference in gravitational potential energy).

$p_z + \frac{1}{2}\rho v_z^2 = p_y + \frac{1}{2}\rho v_y^2$

and as $p_z = p_0$ then $p_y = p_0 + \frac{1}{2}\rho(v_z^2 - v_y^2)$

$p_y = 1.0 \times 10^5 + \frac{1}{2} \times 1.00 \times 10^3 (15.3^2 - 1.9^2)$

$= 2.2 \times 10^5 \text{ Pa}$

Viscosity of fluids

We have seen that ideal fluids are non-viscous. Real fluids do, of course, have viscosity with some fluids being more viscous than others. For a viscous fluid in laminar flow, each layer impedes the motion of its neighbouring layers. In a pipe, the layers adjacent to the inner walls of the tube are stationary while the layers in the centre of it travel at the highest speed.

The viscosity of fluids is highly temperature dependent, with most liquids becoming less viscous at higher temperatures and gases becoming more viscous at higher temperatures. The viscosity of a fluid is measured in terms of its coefficient of viscosity η (at a given temperature). This is a quantity measured in units of pascal second (Pa s).

Stokes' law

When a sphere, of radius r, falls slowly through a viscous fluid it pulls cylindrical layers with it. Under these conditions of streamline flow, the sphere experiences a viscous drag force F_D. This is given by

$$F_D = 6\pi\eta rv$$

where v is the velocity of the sphere and η is the coefficient of viscosity of the fluid. If the sphere is contained in a tube, the tube would need to be very wide for Stokes' equation to apply.

When a small metal ball is released in oil, the ball will initially accelerate until the drag force (plus the buoyancy force) is equal to the weight of the ball – the ball will then travel at its terminal speed so

$$6\pi\eta rv_t + \frac{4}{3}\pi r^3 \rho g = mg$$

or

$$6\pi\eta r v_t + \frac{4}{3}\pi r^3 \rho g = \frac{4}{3}\pi r^3 \sigma g$$

and

$$v_t = \frac{2r^2 g(\sigma - \rho)}{9\eta}$$

Here v_t is the ball's terminal speed, ρ is the density of the fluid and σ is the density of the ball.

Investigate!

Measuring the coefficient of viscosity of oil

Using the apparatus shown in figure 13 you can investigate Stokes' equation.

- Use six or more balls of different radii (but made of the same material) – small balls should reach the terminal velocity in a few centimetres if you use a viscous liquid such as engine oil or glycerol.
- You will need to know the densities of the liquid and the material of the balls.
- You will need to make sure that the balls reach their terminal velocity – use a long, wide transparent cylinder.
- If you use steel balls, a magnet can be used to remove them from the oil.
- Think about ways of releasing the balls consistently.
- Think how you are going to measure the diameter of balls accurately.

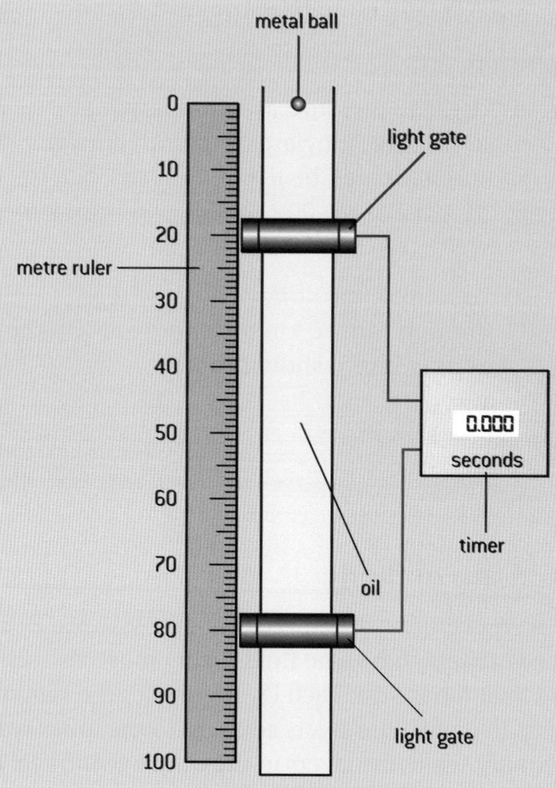

▲ Figure 13 Stokes' law investigation.

Turbulent flow

At low velocities fluids flow steadily and in layers that do not mix – because the fluid is viscous the layers travelling closest to the walls will move slowest as shown in figure 14. When the fluid velocity is increased or obstacles project into the fluid the flow becomes turbulent. The flow is no longer laminar and the particles in the different layers mix with each other. This causes the smooth streamlines, seen during laminar flow, to break up and form eddies and vortices.

It is not easy to predict when the rate of flow is sufficiently high to cause the onset of turbulence. A quantity, known as the Reynolds number R, gives a practical "rule of thumb" that can be used to predict whether the flow is great enough to become turbulent. This is a dimensionless quantity which is calculated from the equation:

$$R = \frac{vr\rho}{\eta}$$

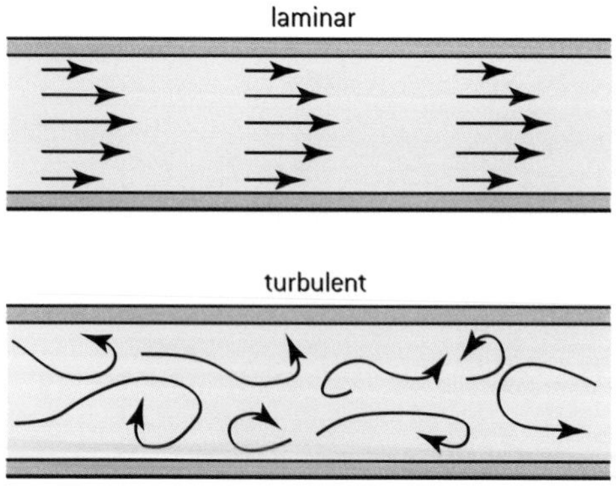

▲ Figure 14 Laminar and turbulent flow.

For a fluid v is velocity, r the radius of the pipe (or other dimension for a river, length of a plate etc.), ρ the density and η the coefficient of viscosity. Using this definition a Reynolds number that is less than 1000 is taken to represent laminar flow. If the Reynolds number is greater than 1000, it does not mean that the flow will be turbulent because there is a transition stage between the fluid being laminar and becoming turbulent. It is generally accepted that, using this definition of the Reynolds number, a value of above 2000 will be turbulent.

Worked example

Oil flows through a pipe of diameter 30 mm with a velocity of 2.5 m s^{-1}. The oil has viscosity 0.30 Pa s and density of 890 kg m^{-3}

a) Determine whether or not this flow is laminar.

b) Calculate the maximum velocity at which the oil will flow through the pipe and still remain laminar.

Solution

Coefficient of viscosity is often just called "viscosity".

a) $R = \frac{vr\rho}{\eta} = \frac{2.5 \times 15 \times 10^{-3} \times 890}{0.30} = 111$

This value is below 1000 so the flow is laminar.

b) $v = \frac{R\eta}{r\rho} = \frac{1000 \times 0.30}{15 \times 10^{-3} \times 890} = 22 \text{ m s}^{-1}$

B.4 Forced vibrations and resonance (AHL)

Understandings

- → Natural frequency of vibration
- → Q factor and damping
- → Periodic stimulus and the driving frequency
- → Resonance

Nature of science

Risk assessment and resonance

Resonance can be a useful phenomenon and, as discussed in this sub-topic, it is utilized in many walks of life. Despite this, the possibility of one system interacting with another has its drawbacks – loud music with heavy bass notes or the noise during construction work can carry over long distances and can, therefore, impinge upon people's lives. Most nations promote the inclusion of resonance hazards during risk assessment applicable when carrying out building or maintenance work. For example, devising safety measures such as incorporating sound insulation in wall cavities or installing double or triple glazing to shield from the effects of unwanted sounds from outside.

Applications and skills

- → Qualitatively and quantitatively describing examples of under-, over-, and critically-damped oscillations
- → Graphically describing the variation of the amplitude of vibration with driving frequency of an object close to its natural frequency of vibration
- → Describing the phase relationship between driving frequency and forced oscillations
- → Solving problems involving Q factor
- → Describing the useful and destructive effects of resonance

Equations

Quality factor equations

- → $Q = 2\pi \dfrac{\text{energy stored}}{\text{energy dissipated per cycle}}$
- → $Q = 2\pi \times \text{resonant frequency} \times \dfrac{\text{energy stored}}{\text{power loss}}$

Introduction

All mechanical systems, and some electrical systems, will vibrate when they are set in motion. We have seen this in Sub-topics 4.1 and 9.1 when studying simple harmonic motion. More intricate examples of oscillations, that are not simple harmonic, include the motion of a tall building in the wind, the shaking of an unbalanced washing machine as its drum spins, or the vibrations caused by a car engine misfiring. In some instances, the vibrations are useful but, often, the function of the system will benefit from being damped – when the amplitude of oscillation is restricted. Dampers are increasingly being incorporated in tall buildings to restrict potential earthquake damage.

Free vibrations

When a mechanical system is displaced from its rest position and allowed to vibrate, without any external forces being applied, it will oscillate at its natural frequency f_0. Such vibrations are called **free vibrations**.

Oscillations and damping

At a given time the amplitude of an oscillation depends on how much the system is damped as well as the size of the driving force. This happens when an oscillating system experiences a resistive force which causes the amplitude of the object oscillating to decay. The resistive force acts in the opposite direction to the motion of the system and it increases with the speed of the oscillation. This means that the damping force is a maximum when the system passes through its equilibrium position as it will have a maximum speed at this point (and it will be zero at the maximum displacement at which point the system momentarily stops). In opposing the damping force, the system must do work and this reduces the energy that it stores – causing the amplitude to decay. The resistive nature of the force slows the oscillator down which increases the time period (although this effect is negligible when a system is lightly damped).

Figure 1 shows how the displacement of an oscillator varies with time when there is light damping. This is also known as **under-damping**. The envelope of the curve takes an exponential shape when the damping or resistive force is proportional to the speed; this means the ratio of the amplitudes at half-period intervals is a constant value $\left(\frac{0.25}{0.21} \approx \frac{0.21}{0.18} \approx \frac{0.18}{0.15} \text{ etc.}\right)$.

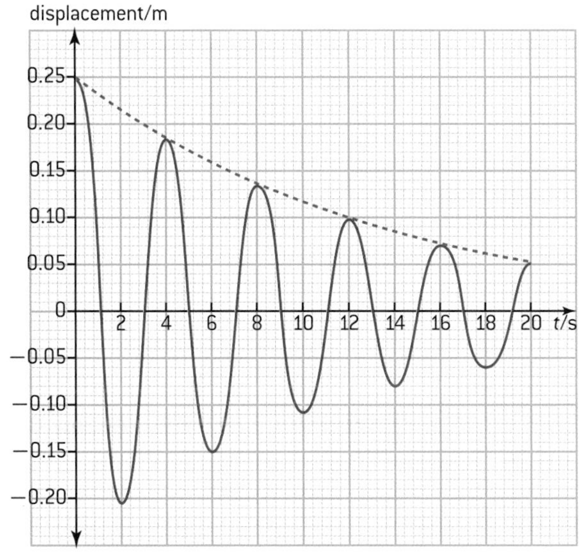

▲ Figure 1 Under-damped vibration.

Although the amplitude decreases with time, the frequency and time period of the oscillator are approximately constant.

Heavier damping may completely stop a system from oscillating. When the system is displaced and it returns to its equilibrium position without overshooting it, the system is "heavily damped". A system returning to its equilibrium position in the shortest possible time is said to be **critically damped**, while a system taking longer than this to return to the equilibrium position is **over-damped**. Critical damping is necessary in many mechanical systems – using this type of damping in car suspension systems avoids oscillations on uneven surfaces; such oscillations may lead to a loss of control and a very uncomfortable ride. For a similar reason, fire doors in buildings are often fitted with automatic closers that are critically damped.

Investigate!

Iteratation with damped SHM

- The equation for a velocity dependent SHM can be written as $ma = -kx - bv$

  (the symbols having their usual meaning and b being the damping factor ... a number between 0 and 1).
- The term bv should now be subtracted from each acceleration value in the spreadsheet iteration discussed in the SHM spreadsheet investigate! in sub-topic 9.1(the second box in the flow chart is modified to $a_n + 1 = -kx_n - bv_n$).
- You should try out different values for b in order to judge the impact that it has on the damping of the shm.
- An example of this spreadsheet is provided on the webpage.

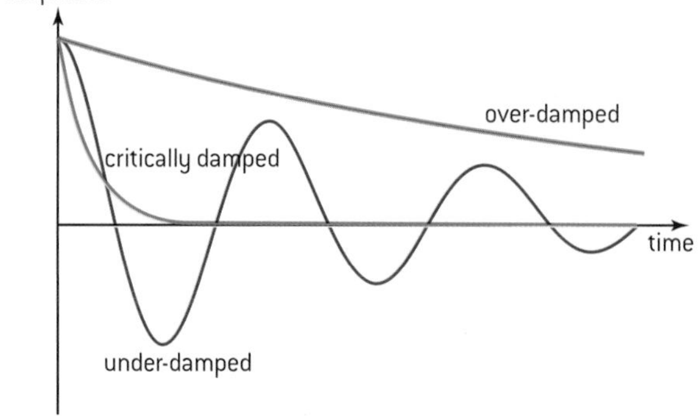

▲ Figure 2 Variation of amplitude with time for different degrees of damping.

These curves can be modelled by adding a velocity-dependent damping term to the simple harmonic motion equation.

▲ Figure3 Child being pushed on swing.

Forced vibrations

When an external force acts on a mechanical system, the force may have its own frequency of vibration, which may affect the motion of the mechanical system. A simple example of this is a child on a swing. When the child is left to his own devices he will only be able to swing at a particular frequency which is a property of the child–swing system. Pushing the child provides an external force that causes the motion of the swing to change. The amplitude of the swing will only increase if the pushes are applied at a rate matching the swing frequency. When the applied frequency does not match the swing frequency the amplitude might decrease. **Forced vibrations** are those that occur when a regularly changing external force is applied to a system resulting in the system vibrating at the same frequency as the force.

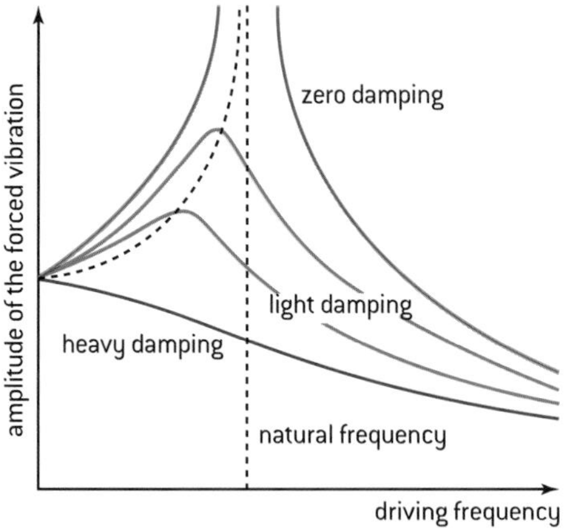

▲ Figure 4 Effect of damping on resonance.

Resonance

When a mechanical system is forced to oscillate by a driving force that has the same frequency as the natural frequency of the mechanical system, it will vibrate with maximum amplitude. This is called **resonance**. When the frequency of the driving force becomes closer to the natural frequency of the system the amplitude of the oscillation will be greater. This can be seen in figure 4, which shows the variation with the driving frequency of the system's amplitude. The degree of damping alters the system's amplitude response. Resonance occurs in many physical systems ranging from tuning radios to the tidal effects of the Moon and the function of lasers.

Note

- For zero damping the amplitude would be infinite, however, this could not happen in practice because infinite amplitude would require an infinite amount of energy to be supplied!
- As the damping increases the peak moves slightly to the left of the natural frequency.
- Increased damping reduces the sharpness of the resonance peak.

Q factor

The Q or "quality" factor is a criterion by which the sharpness of resonance can be assessed. It is defined by the relationships:

$$Q = 2\pi \frac{\text{energy stored}}{\text{energy dissipated per cycle}}$$

$$= 2\pi \times \text{resonant frequency} \times \frac{\text{energy stored}}{\text{power loss}}$$

This is an arbitrary definition with the 2π being included so that using the equation with a real system becomes simplified! In the equations for many rotating or oscillating systems the 2π factor will cancel ...making estimation of Q factors straightforward. The two definitions are equivalent relationships:

writing E_s for energy stored, E_d for the energy dissipated per cycle, f_0 for the resonant frequency, T_0 for the resonant period, P for power loss (= power dissipated) and E for the energy lost in time t.

The second relationship becomes

$$Q = 2\pi f_0 \frac{E_s}{P} = 2\pi \frac{1}{T_0} \frac{E_s}{\frac{E}{t}}$$

$$= 2\pi \frac{t}{T_0} \frac{E_s}{E}$$

The factor $\frac{t}{T_0} = \frac{\text{time considered}}{\text{time for one oscillation}} = n$ (number of oscillations)

So $Q = 2\pi n \frac{E_s}{E}$

$$= 2\pi \frac{E_s}{\frac{E}{n}} = 2\pi \frac{E_s}{E_d}$$

or in words $Q = 2\pi \frac{\text{energy stored}}{\text{energy dissipated per cycle}}$

This is the first relationship.

The Q factor is a numerical quantity and has no unit. A system with a high Q factor is lightly damped and will continue vibrating for many oscillations as the energy dissipated per cycle will be small. As a rule of thumb, the Q factor is approximately the number of oscillations that the system will make before its amplitude decays to zero (without further energy input). The larger the Q factor, the sharper the resonance peak on the graph of amplitude against driving frequency. A high quality factor means a low loss of energy.

For light damping $Q = m\omega/b$ where b is the damping factor relating the resistive force F_r to the speed v: $F_r = -bv$

Some typical values of Q factors are:

Oscillator	Q factor
critically damped door	0.5
loaded test tube oscillating in water	10
mass on spring	50
simple pendulum	200
oscillating quartz crystal	30 000

Worked example

An electrical pendulum clock has a period of 1.0 s. An electrical power supply of 25 mW maintains its constant amplitude. As the pendulum passes its equilibrium position it has kinetic energy of 40 mJ.

a) Explain how these quantities apply to the Q factor relationship.

b) Calculate the Q factor for the pendulum clock.

Solution

a) The pendulum has a frequency of 1.0 Hz.

As it is storing 40 mJ, the rate of energy supplied must equal the rate at which energy is lost i.e., the power is supplied at a rate of 25 mW.

b) $Q = 2\pi \times \text{resonant frequency} \times \frac{\text{energy stored}}{\text{power loss}}$

$$= 2\pi \times 1 \times \frac{40 \times 10^{-3}}{25 \times 10^{-3}}$$

$$= 10$$

The Q factor is an especially important quantity for electrical oscillations transmitting radio waves. When selecting radio and television stations it is essential that the transmitting and receiving circuits are tuned to the same resonance frequency.

Barton's pendulums

Figure 5 shows an arrangement of pendulums that can be used to investigate resonance. The apparatus is known as Barton's pendulums – named after the British physicist, Edwin Barton.

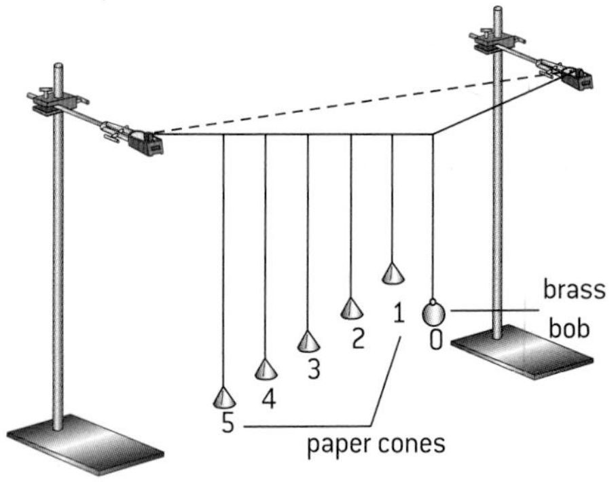

▲ Figure 5 Barton's pendulums.

As the diagram shows, the set up consists of a number of paper cone pendulums of varying lengths. All are suspended from the same string as the brass bob that acts as the "driver" pendulum. When the driver pendulum is displaced from its rest position and released, it forces all the paper cone pendulums to oscillate with the same frequency, but with different amplitudes.

This is an example of forced oscillations. The "cone" that has the same length as the driver pendulum has the greatest amplitude because it has the same natural frequency as the driver pendulum. In calculating the period using the equation for the period of a simple pendulum, the "equivalent length" of each pendulum should be taken from the centre of the bob to the horizontal broken line shown on the diagram.

Careful observation shows that:

- cone 3 (which has the same length as the driver pendulum) always lags behind the driver pendulum by one quarter of period (equivalent to 90° or $\frac{\pi}{2}$ radian)
- the shorter cones (1 and 2) are almost in phase with the driver pendulum
- the longer pendulums (4 and 5) are almost in anti-phase (180° or π radian out of phase) with the driver pendulum.

Figure 6 shows the variation, with forcing frequency, of the phase lag for a pair of pendulums. The length of the driver pendulum string is adjusted – when it is shorter than the driven pendulum the forcing frequency is higher and the driven pendulum will lag behind the driver by π radian, etc.

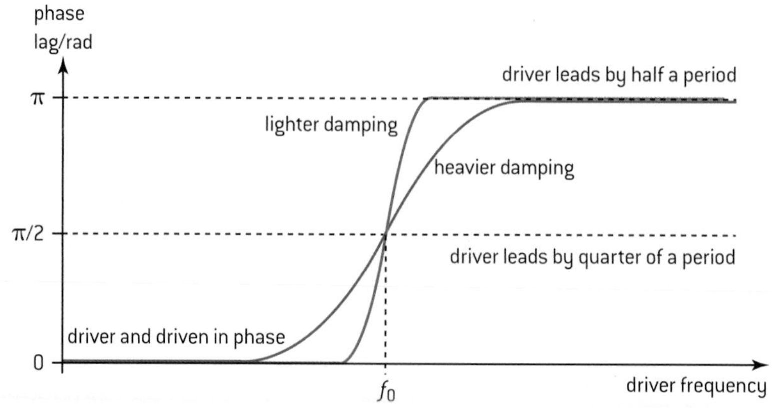

▲ Figure 6 Phase relationship for the displacement of a forced vibration.

Nature of science

Examples of resonance

In nature, resonance is a very common phenomenon having examples in virtually all areas of physics. Further examples include:

- The human voice uses resonance to produce loud sounds from a relatively weak source – the vocal cord.
- The sound box of musical instruments amplifies the energy, causing air in the box to resonate at the same frequency.
- Ozone in the stratosphere has natural frequencies of vibration that match the frequency of ultraviolet and absorb this radiation.
- Microwave cookers emit electromagnetic waves that match the natural frequency of water molecules in food.
- The optical cavities in lasers set up standing waves for light in order to produce coherent beams.
- MRI (magnetic resonance imaging) scans use resonating protons in atoms to provide vital information about body cells.

Resonance can have drawbacks as well as benefits, for example:

- Vibrations in machinery can cause vibrations in nearby objects – overtaking mirrors on lorries that are idling can be seen to vibrate with large amplitude.
- Vibration set up by soldiers marching across bridges has meant that they are told to "break step". In June 2000, London's Millennium Footbridge (figure 7) was found to sway alarmingly from the resonances set by pedestrians crossing it. Fitting dampers solved the problem but failed to prevent the bridge being known as "the wobbly bridge".
- The Tacoma Narrows Bridge in Washington State, USA, collapsed in November 1940 as a result of cross winds matching its natural frequency.

▲ Figure 7 The London Millennium Footbridge.

- Feedback can be heard at rock concerts – a loud howling sound is produced when microphones or pickups are too close to loudspeakers, giving an uncontrolled amplification of the sound.

Investigate!

Resonance of a hacksaw blade

Apparatus similar to that shown in figure 8 can be used to investigate the resonance of a hacksaw blade.

- An electromagnet is powered by a signal generator that behaves as a variable frequency ac supply.
- The frequency can be varied by adjusting the signal generator.
- The clamped hacksaw blade will have a natural frequency that depends on the length projecting.
- The amplitude of the end of the blade can be determined using a vertically clamped millimetre scale (not shown).
- This apparatus can be used to measure the variation of amplitude with (a) supply frequency f (b) projecting length L.

- The general form of the relationship may be $x_0 = kf^n$ or $x_0 = kL^n$
- This suggests a log–log graph should generate a straight–line graph.

  Melde's string offers a similar investigation when a long metallic wire carries a current in a magnetic field (see Sub-topic 4.5):
- An alternating small current is set up in the wire.
- The wire is placed between the poles of a strong U-shaped magnet and is clamped at both ends. The magnet is positioned somewhere near one end of the wire.
- By changing the tension in the wire the natural frequency of the wire will approach the frequency of the AC current.
- When resonance is reached, standing waves will appear along the wire, the harmonic depending on the conditions.
- Take care because the wire can become red-hot!

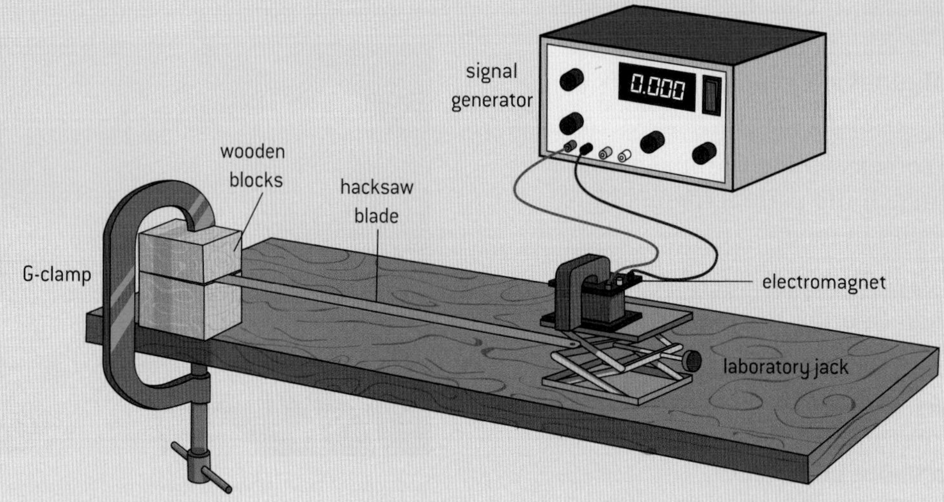

▲ Figure 8 Resonance of a hacksaw blade.

Questions

1 An object of mass 3.0 kg is attached to a string which is wrapped round a thin uniform disc of radius 0.25 m and mass 7.0 kg. The disc rotates about a horizontal axis passing through its centre. Assume the rotation of the disc is frictionless.

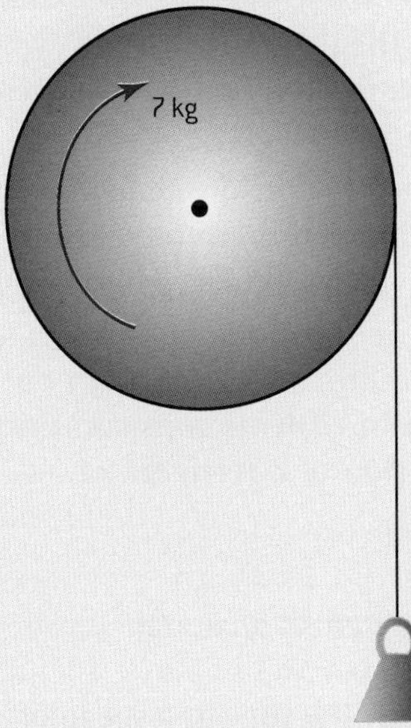

Calculate:

a) the angular acceleration of the disc when the mass is released from rest

b) the angular acceleration if the force were applied by pulling the cord with a constant force of 30 N.

The moment of inertia I of a thin uniform disc of mass M and radius R is given by $I = \frac{1}{2}MR^2$ (5 marks)

2 A skater has moment of inertia 2.85 kg m^2 with her arms outstretched. She rotates with an initial angular speed of 2.0 rad s^{-1} as shown in the diagram. By bringing in her arms the skater's moment of inertia reduces to 1.5 kg m^2.

For the skater, calculate:

a) her final angular speed

b) the change in her rotational kinetic energy. (5 marks)

3 The angular speed of a rotating disc is increased from 20 rad s^{-1} to 85 rad s^{-1} in a time of 6.0 s by a constant torque.
The disc has mass of 8.0 kg and radius is 0.35 m. Calculate:

a) the work done by the torque in this time

b) the average power applied to the disc during this time. (The moment of inertia of a disc is given by the equation in question 1.) (7 marks)

4 (*IB*)
The graph shows the variation with volume of the pressure of a fixed mass of gas when it is compressed adiabatically and also when the same sample of gas is compressed isothermally.

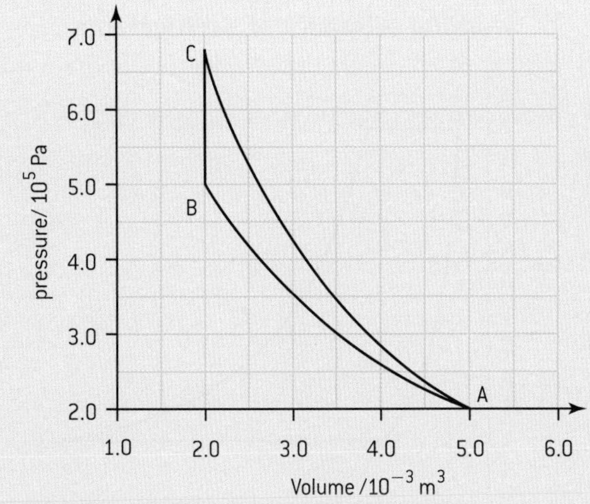

a) State and explain whether line AB or AC represents the isothermal compression.

b) On a copy of the graph, shade the area that represents the difference in work done in the adiabatic change and in the isothermal change.

c) Determine the difference in work done, as identified in (b).

d) Use the first law of thermodynamics to explain the change in temperature during the adiabatic compression. (9 marks)

5 *(IB)*

An ideal gas at an initial pressure of 4.0×10^5 Pa is expanded isothermally from a volume of 3.0 m^3 to a volume of 5.0 m^3.

a) Calculate the final pressure of the gas.

b) On graph paper, sketch a graph to show the variation with volume V of the pressure p during this expansion.

c) Use the sketch graph in (b) to:

(i) estimate the work done by the gas during this process

(ii) explain why less work would be done if the gas were to expand adiabatically from the same initial state to the same final volume. (7 marks)

6 *(IB)*

a) State what is meant by *isobaric change*.

b) The diagram below shows the pressure–volume (p–V) changes for one cycle of the working substance of a refrigerator.

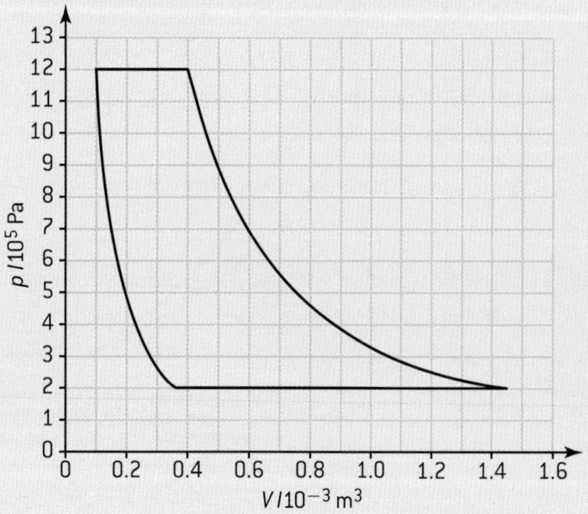

On a copy of the diagram above:

(i) draw arrows to show the direction of the changes

(ii) label with the letter A an isobaric change

(iii) label with the letter B the change during which energy is transferred to the working substance because of a temperature difference.

c) Use data from the diagram in (b) to estimate the work done during one cycle of the working substance.

d) (i) By reference to entropy change, state the second law of thermodynamics.

(ii) The cycle of the working substance in (b) reduces the temperature inside the refrigerator. Explain how your statement in (d)(i) is consistent with the operation of a refrigerator. (11 marks)

7 A horizontal tube of cross-sectional area 3.0×10^{-4} m^2 narrows to 1.6×10^{-4} m^2. Water flows through the tube at a speed of 0.40 m s^{-1} at the wider part.

a) Calculate:

(i) the volume rate of flow in m^3 s^{-1} through the tube

(ii) the speed of the water in the narrower part of the tube.

b) Explain how the flow of water in the pipe is accelerated. (6 marks)

8 The diagram shows a venturi meter which is used to measure the flow rate of low-density liquid through a horizontal pipe. When the liquid is flowing, the mercury manometer is used to measure the difference between the pressure of the liquid in the pipe at the entrance to the venturi meter and that in the throat of the meter:

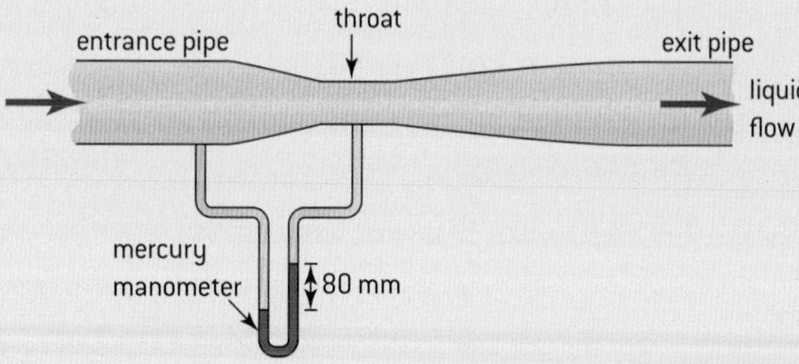

a) In one measurement the difference in levels of the mercury in the manometer arms is 80 mm. Calculate the pressure difference between the liquid in the entrance to the venturi meter and that in the throat, expressing your answer in pascals.

density of mercury $= 1.4 \times 10^4$ kg m^{-3}

b) the corss-sectional areas of the pipe and venturi meter throat are 4.0×10^{-2} m^2 and 1.0×10^{-2} m^2 respectively.

Determine the ratio of the speed of the liquid flowing in the throat of the venturi meter to its speed in the entrance pipe.

c) The liquid has a density of 8.0×10^2 kg m^{-3}.

(i) Use the Bernoulli equation to estimate the speed of the liquid in the horizontal pipe.

(ii) Hence estimate the flow rate of the liquid in kg s^{-1}.

(10 marks)

9 **a)** (i) Use the continuity equation to show that an incompressible liquid moving from a wider pipe to a narrower pipe must increase its velocity.

(ii) State Bernoulli's relation for the flow of an incompressible inviscid fluid along a horizontal stream line.

b) Figure 1 shows a cross-section through a simple laboratory filter pump.

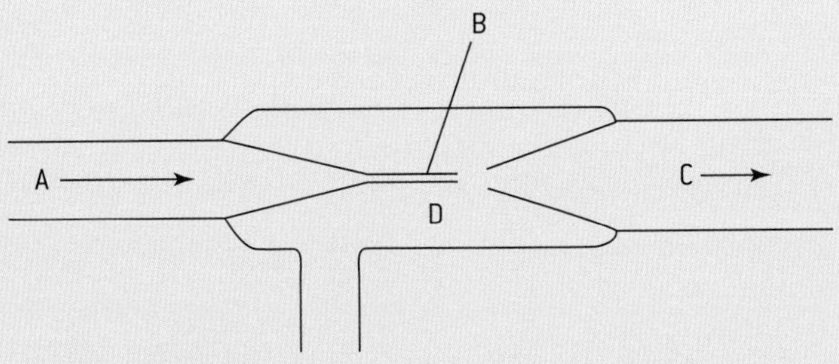

▲ Figure 1

The Pressure at C is at atmospheric pressure (100 kPa) and the water at C is moving very slowly.

The Pressure at D is 45 kPa. Nozzle B has a diameter of 2.0 mm. Calculate

(i) Explain how a flow of water from A to C through the pump produces a partial vacuum at D.

(ii) Calculate the velocity of the water emerging from the nozzle B.

(iii) Calculate the rate, in m^3 s^{-1}, at which water flows through the pump. (density of water = 1000 kg m^{-3})

(13 marks)

10 A spring for which the extension is directly proportional to the weight hung on it is suspended vertically from a fixed support. When a weight of 2.0 N is attached to the end of the spring the spring extends by 50 mm. A mass of 0.50 kg is attached to the lower end of the unloaded spring. The mass is pulled down a distance of 20 mm from the equilibrium position and then released.

a) (i) Show that the time period of the simple harmonic vibrations is 0.70 s.

(ii) On a sheet of graph paper sketch the displacement of the mass against time, starting from the moment of release and continuing for two oscillations. Show appropriate time and distance scales on the axes.

b) The mass–spring system described in part (a) is attached to a support that can be made to vibrate vertically with a small amplitude. Describe the motion of the mass–spring system with reference to frequency and amplitude when the support is driven at a frequency of

(i) 0.5 Hz (ii) 1.4 Hz

(8 marks)

11 The Millennium Footbridge in London was discovered to oscillate when large numbers of pedestrians were walking across it.

a) What name is given to this kind of physical phenomenon?

b) Explain the conditions that would cause this phenomenon become particularly hazardous?

c) Suggest **two** measures which engineers might adopt in order to reduce the size of the oscillations of a bridge. (7 marks)

12 a) A *forced vibration* could *show resonance*.

Explain what is meant by:

(i) forced vibrations

(ii) resonance.

b) (i) Explain what is meant by *damping*.

(ii) What effect does damping have on resonance? (5 marks)

C IMAGING

Introduction

Data collection in science often depends on using our senses to make an observation. Sometimes the image we perceive needs to be enhanced in some way. For example, doctors may need to "see" inside the human body in order to make a diagnosis. This topic deals with the physics of imaging in both visual and non-visual contexts.

C.1 Introduction to imaging

Understanding

- Thin lenses
- Converging and diverging lenses
- Converging and diverging mirrors
- Ray diagrams
- Real and virtual images
- Linear and angular magnification
- Spherical and chromatic aberrations

Nature of science

- Virtual images cannot be formed directly on a screen. The technique of ray tracing allows the position and size of a virtual image or a virtual object to be inferred. This is an example of deductive logic in action.

Applications and skills

- Describing how a curved transparent interface modifies the shape of an incident wavefront
- Identifying the principal axis, focal point, and focal length of a simple converging or diverging lens on a scaled diagram
- Solving problems involving not more than two lenses by constructing scaled ray diagrams
- Solving problems involving not more than two curved mirrors by constructing scaled ray diagrams
- Solving problems involving the thin lens equation, linear magnification, and angular magnification
- Explaining spherical and chromatic aberrations and describing ways to reduce their effects on images

Equations

- Thin lens equation: $\frac{1}{f} = \frac{1}{v} + \frac{1}{u}$
- Power of a lens: $P = \frac{1}{f}$
- Linear magnification: $m = \frac{h_i}{h_o} = -\frac{v}{u}$
- Angular magnification: $M = \frac{\theta_i}{\theta_o}$
- Magnification of a magnifying glass: $M_{nearpoint} = \frac{D}{f} + 1$; $M_{infinity} = \frac{D}{f}$

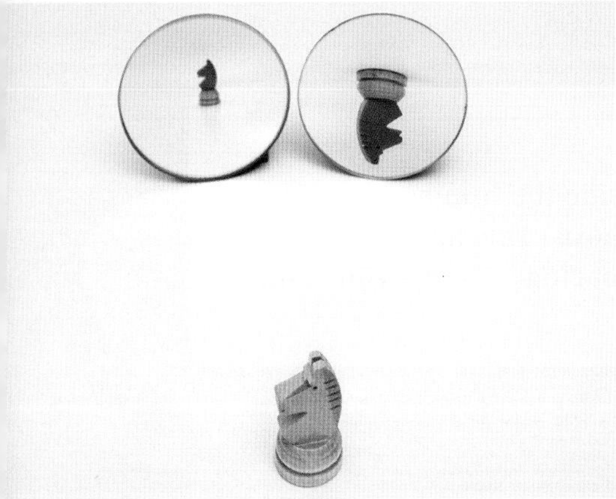

▲ Figure 1 Concave and convex mirrors and their reflections.

Converging and diverging mirrors

Topic 4 introduced the rules of reflection at plane surfaces. In this section we look at what happens when surfaces are no longer flat. A curved reflecting surface allows the rays to be manipulated, leading to images of various types formed by mirrors. The shapes of convex and concave surfaces are shown in figure 2.

Mirrors made with these shapes are commonly found in a number of situations: as mirrors used for makeup or shaving where a magnified image is needed, or as a way for seeing a wide angle of view.

We will mainly consider mirror surfaces that have a spherical profile. Figure 2 shows what happens when light is incident on both convex and concave reflectors that have been formed from the surface of a hollow sphere.

The line that goes through the centre of the mirror surface at 90° to the surface is the **principal axis**.

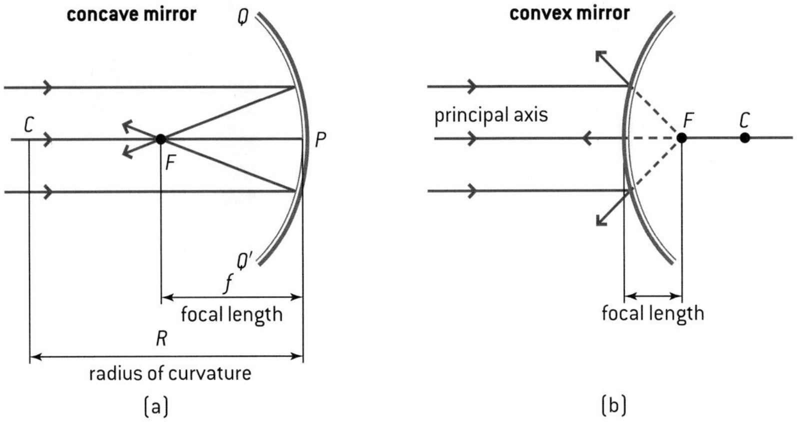

▲ Figure 2

When rays that are parallel and close to the principal axis are incident on a concave mirror then they all reflect through a point known as the **principal focus** F. F is also known as the **focal point** of the mirror. It is easy to see why this has to be the case. The second law of reflection tells us that the angle of incidence at the mirror is equal to the angle of reflection. The surface of the mirror is part of a sphere (drawn as a circle in the two-dimensional diagram) and therefore the line joining the centre of the sphere C to the mirror at the point where the light ray is incident is the normal to the surface. This defines the angle of incidence i.

The angle of reflection r must equal i. If the incident ray is close to the principal axis then, after reflection, the ray will cross the principal axis at a point F that is half way from the **pole** P of the mirror to the centre of the sphere. The distance from F to the pole is known as the **focal length** f.

When the distance between the principal axis and the incident ray is small, f is half the radius of curvature of the mirror. We shall look at what happens when incident rays lie well away from the principal axis later. For now, our assumption is always that the distance between rays and the axis is small.

When a convex mirror is used (figure 2(b)) the position is very similar, except that this time the reflected ray appears to have come *from* the focus. There is the same relationship between f and the radius of the mirror as before.

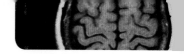

Investigate!

Images from a mirror

These experiments are intended to introduce you to the images formed by a mirror. The mirrors that will work best for these experiments have a focal length of about 15–25 cm. You need to know the approximate focal length of each mirror before you begin.

(1) Concave

Begin by taking the concave mirror and looking into it with an approximate distance between your eye and the mirror that is (a) about double the focal length, (b) less than the focal length. What do you notice about the image you see? Is the image the right way up or upside down? Can you describe what happens when the eye is exactly at the focal point?

Now set the mirror on the bench using suitable stands so that the light from an illuminated object is incident on the mirror and then reflected to form an image on a screen. Make sure that the object–mirror distance u is greater than the focal length. Find out the range of u for which the image is larger, smaller, and the same size as the object.

(2) Convex

This time simply look into the mirror. Where is the image formed? Is it similar to the image formed by a plane mirror, what are the differences?

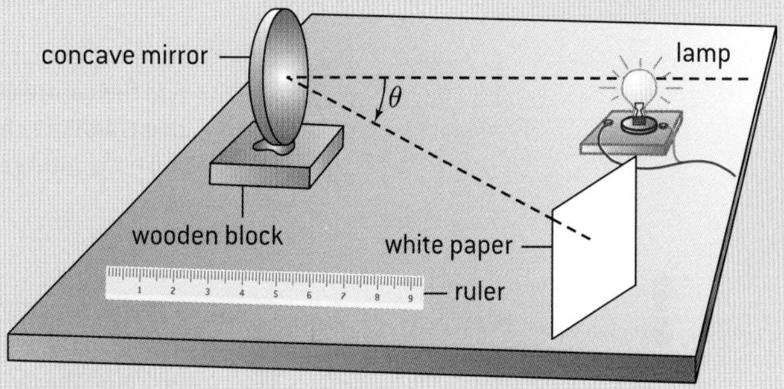

▲ Figure 3.

To explain the positions of images formed by curved mirrors we use the technique of **drawing a ray diagram**. A ray diagram establishes the relationship between an object and its image that results when rays from the object are reflected in the mirror.

Using the idea that light is reversible we can identify **predictable rays** that we will use to construct a scaled ray diagram. Reversibility means that when light rays are reflected back along their original path they will trace out the incident path but in the reverse direction.

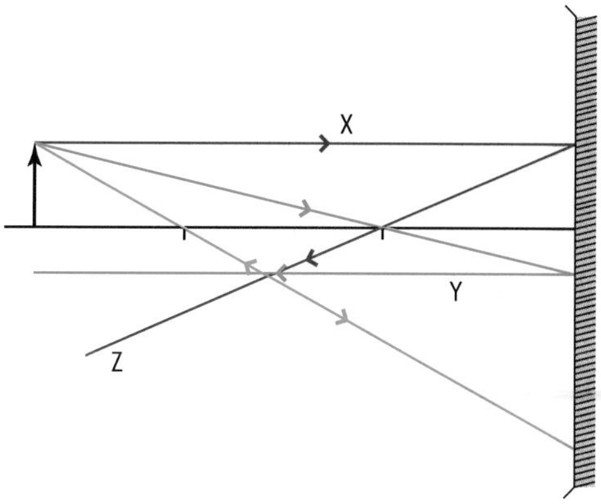

▲ Figure 4 Predictable rays.

The predictable rays we will use are:

- **X:** initially parallel to the principal axis and then going, after reflection, through the focal point.
- **Y:** initially through the focal point and then going, after reflection, parallel to the principal axis.
- **Z:** a ray through the centre of curvature that goes after reflection back along its original path.

Using these rays we can determine the position and the nature (size and type) of the image formed by a curved mirror. Remember that X, Y, and Z are close to the principal axis and the mirror is almost flat at the centre.

The drawing technique uses several conventions:

- The mirror is drawn in two dimensions as a straight line at 90° to the principal axis. Only the top and bottom of the mirror are curved to indicate whether the mirror is convex or concave.
- The vertical scale and the horizontal scale need not be the same. Objects and mirrors are sometimes only centimetres high but can

be separated by distances of metres. Using the same scale in both directions might compress the diagram in one dimension.

- The object is represented by a vertical arrow drawn to show the largest dimension of the object to scale.
- At least two predictable rays are drawn from the top of the object arrow. The position of the top of the image of the arrowhead will be shown by the direction of the rays after reflection.
- Arrows are drawn to show the direction in which the rays travel.

The step-by-step construction of one ray diagram each for a concave and a convex mirror is shown in sequence in figure 5.

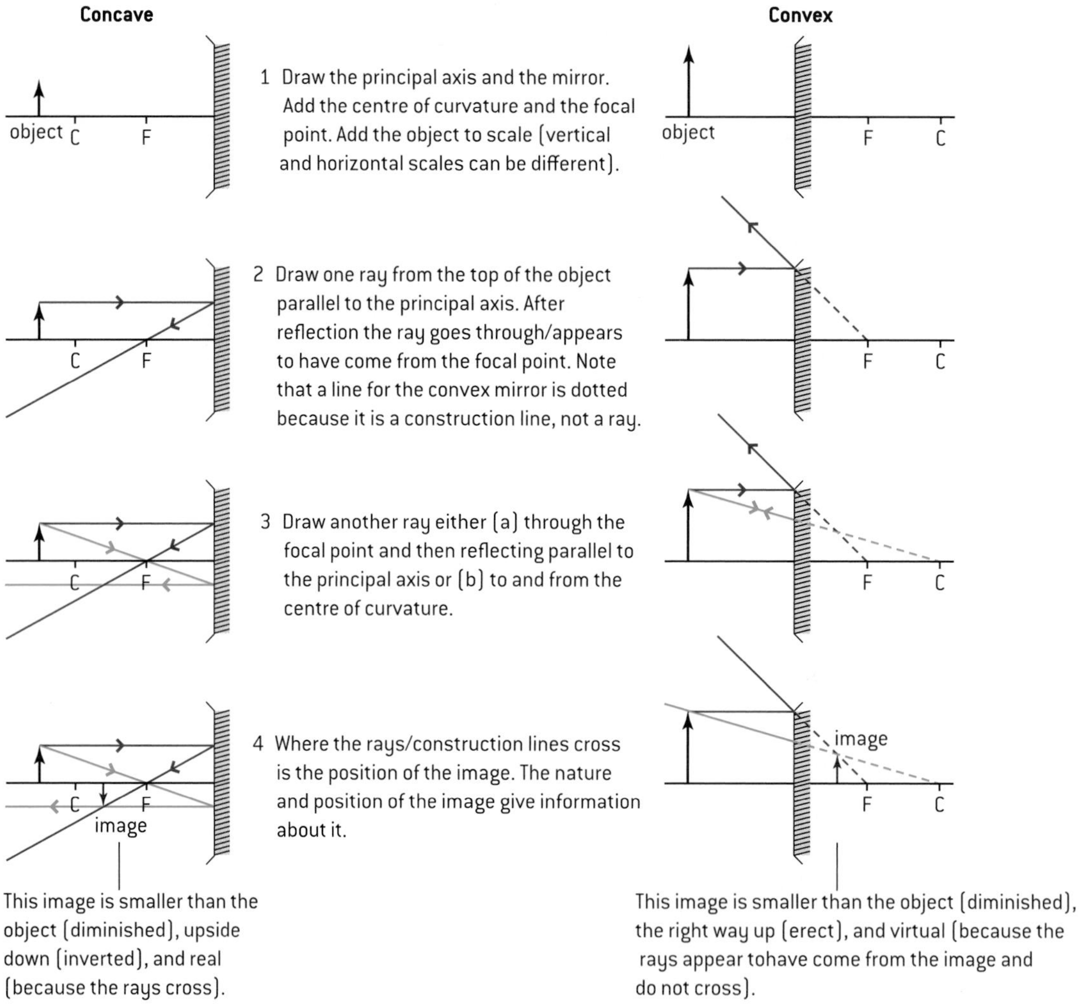
Concave
Convex
object
C
F
1 Draw the principal axis and the mirror. Add the centre of curvature and the focal point. Add the object to scale (vertical and horizontal scales can be different).
2 Draw one ray from the top of the object parallel to the principal axis. After reflection the ray goes through/appears to have come from the focal point. Note that a line for the convex mirror is dotted because it is a construction line, not a ray.
3 Draw another ray either (a) through the focal point and then reflecting parallel to the principal axis or (b) to and from the centre of curvature.
4 Where the rays/construction lines cross is the position of the image. The nature and position of the image give information about it.
image
This image is smaller than the object (diminished), upside down (inverted), and real (because the rays cross).
This image is smaller than the object (diminished), the right way up (erect), and virtual (because the rays appear tohave come from the image and do not cross).

▲ Figure 5 Step-by-step instructions for drawing concave mirror and convex mirror ray drawing.

The concave mirror has a number of separate cases each of which depends on the position of the object. If the rays from the top of the object converge to a single point then this marks the top of the arrowhead. This is now below the principal axis (whereas the object was above) because the image is an inversion of the object (turned upside down). This point where the rays converge is still called the "top" of the image.

The rays in these cases form a **real image**. Rays that intersect (converge) in this way can be used to form an image on a screen. This is what was happening when you saw a "picture" forming on the screen in the *Investigate!*

However, when the object is closer than the focal point, the final ray directions diverge (move apart) and do not cross. When these diverging rays are constructed back (notice that construction lines are always drawn as broken lines in the worked examples as they are not rays), they form a different type of image known as a **virtual image**. This type of image cannot be formed on a screen but it can be seen or manipulated using another mirror or a lens. Virtual images in the *Investigate!* were formed by the lens in your eye.

A convex mirror by itself can never form a real image from a real object. As the rays always diverge, the convex mirror is called a **diverging mirror**. For the concave mirror, unless the object is closer to the lens than the focal point, the rays converge. Concave mirrors are usually called **converging mirrors** – in some books they are referred to as converging (concave) mirrors.

Magnification

Because the diagrams for both types of mirrors are scaled, the final scale sizes of the object and image can be used to measure the **linear magnification** m of the mirror system.

$$m = \frac{\text{height of image}}{\text{height of object}} = \frac{h_i}{h_o}$$

Geometry shows that m is also equal to $\frac{v}{u}$, where u and v are the distances from the mirror pole to the object and the image respectively.

In some mirror examples that we shall see later, the object and image heights cannot be easily identified. In these cases **angular magnification** M is used. $M = \frac{\theta_i}{\theta_o}$ where θ_i and θ_o are the angles subtended at the mirror by the rays from the image of the top of the image and the rays from the top of the object, respectively.

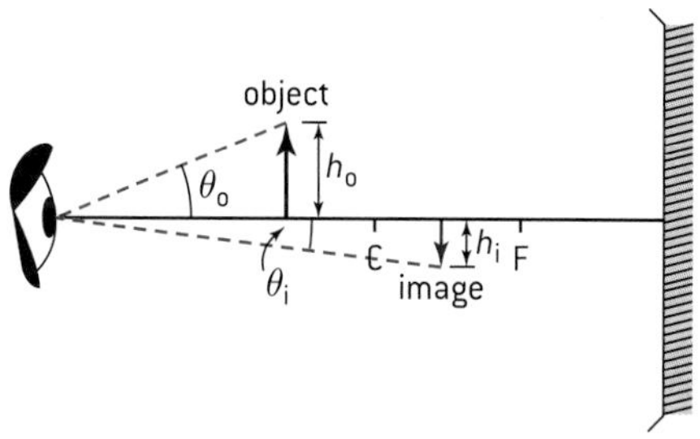

▲ Figure 6 Magnification by a mirror, linear and angular.

Aberrations in a mirror

So far we have assumed that the rays from the object are travelling close and approximately parallel to the principal axis. When this is not the case then the spherical mirror introduces distortions (aberrations) into the images.

The principal reason for this is that at large distances from the pole of the mirror the rays no longer reflect through the focal point. When rays parallel to the principal axis are incident over the whole mirror surface then they form a pattern known as a **caustic curve**. You may have noticed this pattern forming when strong sunlight strikes the reflecting surface of a circular cup of coffee or another opaque liquid. The inner surface of the cup reflects the light onto the liquid surface. The position of focal point and the shape of the caustic can be clearly seen.

To reduce this problem in cases where a large mirror surface is needed, a parabolic surface can be used. Rays parallel to the principal axis of a parabolic mirror always pass through the focus and no caustic forms in these circumstances (figure 7(b)). An example is the large mirror used for an astronomical telescope.

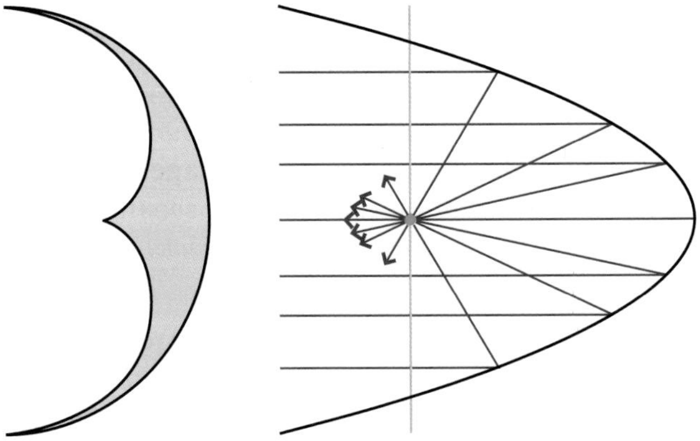

▲ Figure 7 Spherical mirror distortions and the parabolic mirror.

The reversibility of light tells us that a parabolic mirror shape can also be used to generate a parallel beam of light if a light source is placed at the focus. This is often used to produce a searchlight beam that diverges very little even at large distances from the mirror.

In some situations, two or more mirrors can be combined in optical instruments such as telescopes. We will give examples of these in a later section.

Worked example

1 Construct a ray diagram for a concave mirror when:

a) the object is between F and C

b) the object is at C

c) the object is at F

d) the object is between the pole of the mirror and F.

Solution

a)

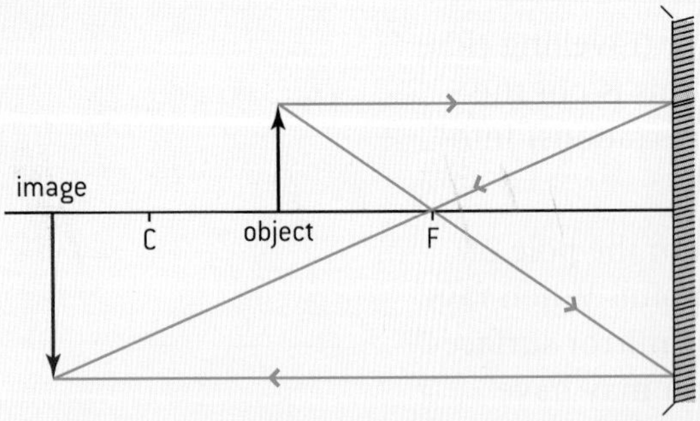

b)

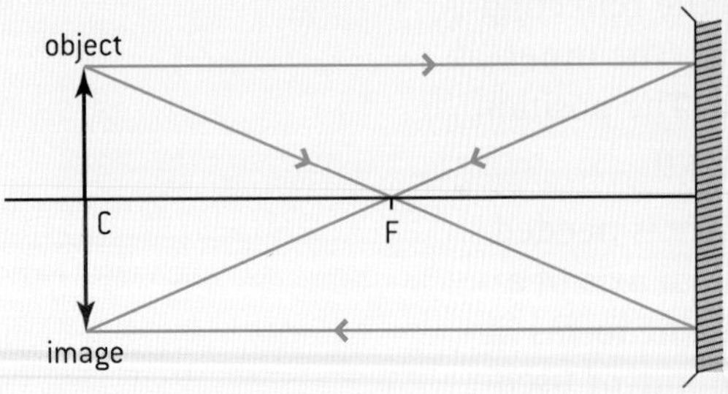

c)

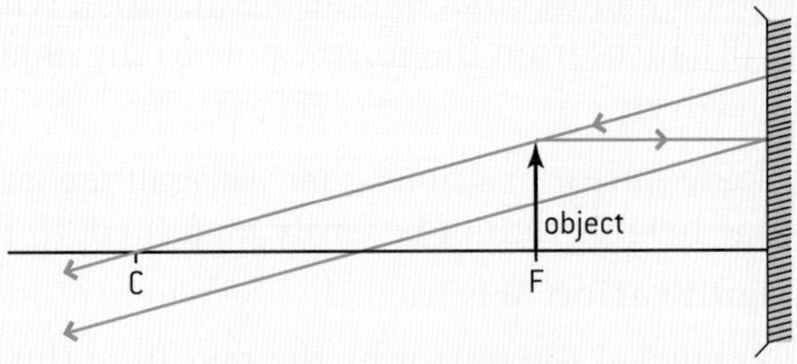

d)

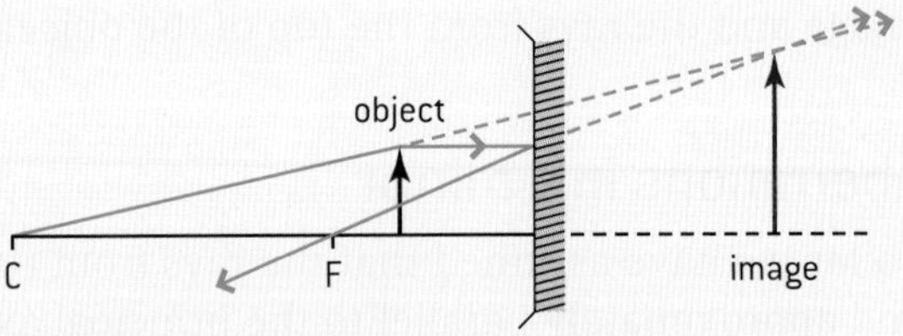

2 Construct a ray diagram for a convex mirror when the object is between the pole of the mirror and the focal distance.

Solution

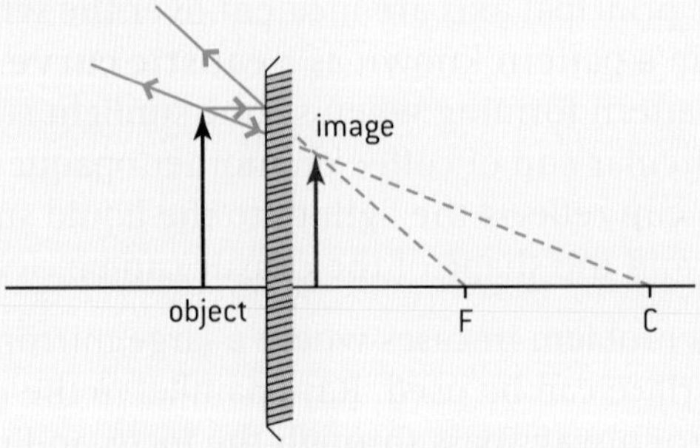

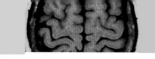

Converging and diverging thin lenses

A mirror has only one surface and the light always travels in the air. When light enters a transparent medium it is refracted and the speed and wavelength of the light change.

This change of speed makes the analysis of light passing through a curved interface into a different medium more complex than the reflection case. Lenses usually have two interfaces through which the light enters and leaves. Before considering this double refraction in detail we will look at the effect of a single interface between two different media.

In this example, the light travels slower in the second medium than the first. Figure 8(a) shows parallel rays of light as they travel through a convex surface from one medium to another of greater optical density. Parallel plane wave fronts are associated with these rays and are also shown. The centre of these wavefronts meets the curved convex surface earlier than the outer edges of the wave front. As the wave enters the medium it slows down and the wavelength of the light becomes smaller. This means that the centre of the wave is travelling at a slower speed in the second medium than the wave edges that have not yet reached the interface.

The overall effect on the wave when it has fully entered the medium is that what was a parallel wave has become curved. The rays are no longer parallel either. They meet at a focal point.

One interpretation of the action of a curved surface is that it adds curvature to wavefronts (in the convex case) or removes curvature (in the concave case). The curvature of a surface can be defined in general terms as $\frac{1}{R}$ where R is the radius of the surface. The smaller the radius of the surface, the more curved it is.

Now we can see what happens when a real lens is used to modify rays (figure 8(b)). There are two curved interfaces and it is the cumulative effect of both interfaces that determine the properties of the lens (figure 9).

We use a similar model to that of the mirrors to define the parts of the lens (figure 10). The **optical centre** of the lens is the point in the lens that does not deviate rays of light passing through it. As with the mirror we define a **principal axis**, **focal point** and **focal length**. Rays parallel to the principal axis pass through the focal point after refraction for a convex lens. For a concave lens, the parallel rays appear to have come from the focal point.

We shall assume that the convex or concave lenses have lens thicknesses that can be ignored compared to the distances between object and image. The **thin-lens theory** that uses this assumption is more straightforward than when the thickness of the lens is included.

We will also consider only spherical lenses, that is, those with surfaces shaped so as to form part of the surface of a sphere. Many modern lenses are designed to have an **aspherical** shape (non-spherical), such lenses, for example, spectacle lenses, can be made thinner and lighter than their spherical counterparts.

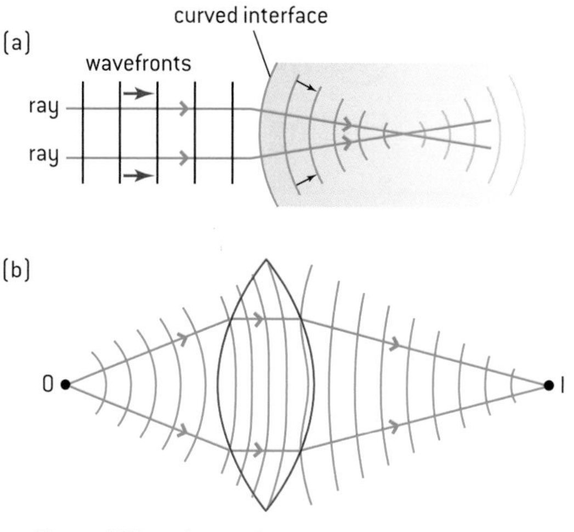

▲ Figure 8 Wave fronts through a medium.

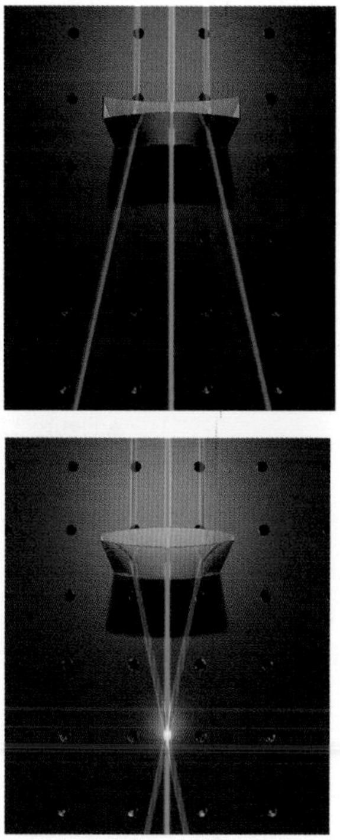

▲ Figure 9 Rays through convex and concave lenses.

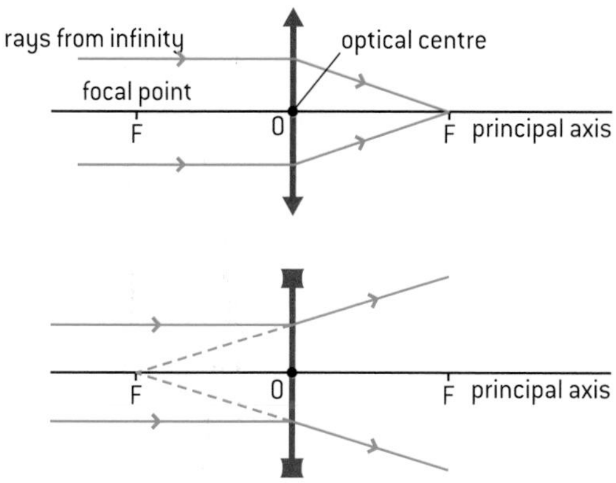

▲ Figure 10 Definitions of the parts of a lens.

Investigate!

Finding an approximate value for the focal length of a convex lens

- We have used the idea of rays that travel parallel to the principal axis. These can be obtained (approximately) by using rays that have come from a distant object. The diagram shows how these arise. Two rays that leave the same place on an object and then travel a large distance compared with the diameter of the lens have to be travelling very close to parallel if they are both to enter the lens.
- Hold your lens near a window so that rays from a distant object outside (more than about 20 m away will do) form an image on a sheet of card (called a screen) held behind the lens.
- The distance from the card to the lens is, approximately, the focal length. Get someone to measure this distance while you hold the lens and the card. Notice that the image is upside-down.

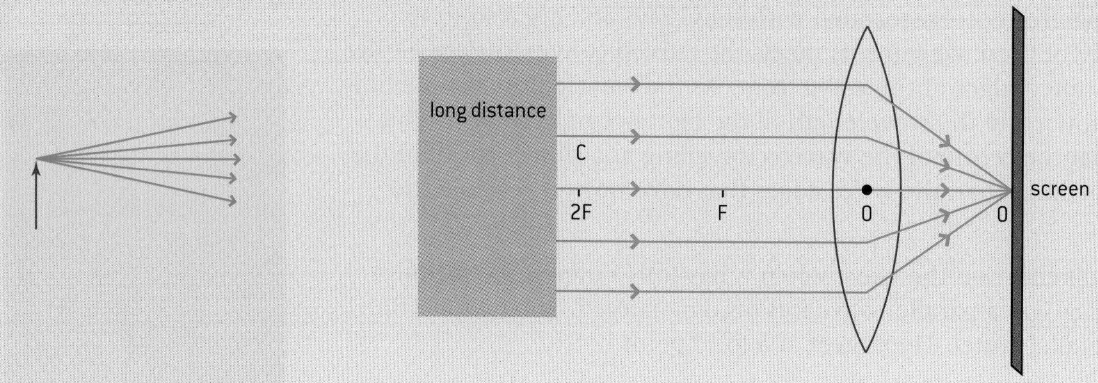

▲ Figure 11

Investigate!

The relationship between object and image for a thin convex lens

- For this experiment you will need a convex lens of approximate focal length 0.15 m, an illuminated object, a screen, and a ruler to measure the distance between the object and the lens (the object distance) and the lens and the screen (the image distance).
- Set the object about 0.4 m from the lens. Move the screen until there is a clearly focused picture of the object on the screen – you should expect it to be upside–down.
- Measure the object distance u and the image distance v and record them.
- Repeat the procedure for a range of object distances (do not attempt to have the object distance closer than the focal length).
- Plot your data on a graph of:
  $\frac{1}{v}$ against $\frac{1}{u}$

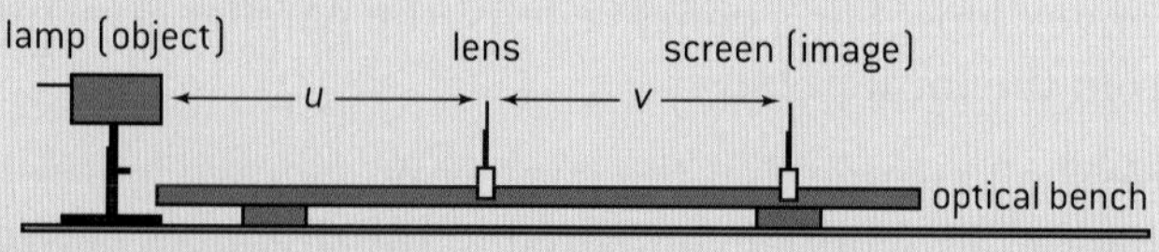

▲ Figure 12

The results of this experiment can be summed up using scaled ray diagrams.

Ray diagrams for both convex and concave lenses are constructed on a similar basis to those for mirrors. Additional rules are that:

- The lens is drawn as a vertical line to emphasize that it is thin. Symbols at the top and bottom of the lens indicate whether it is convex or concave.

- The focal length should be shown on *both* sides of the lens.
- There are three predictable rays:
  - A ray from the top of the object parallel to the principal axis goes through the focal point after refraction by the convex lens or, in the case of a concave lens, deviates so that it appears to have passed through the virtual focus.
  - A ray from the top of the object through the optical centre of the lens does not deviate.
  - A ray from the top of the object through the focal point travels parallel to the principal axis after refraction.
  - Only two of these rays are needed to complete the diagram and show the position size and nature of the image. The third can be used as a check.
- Rays should have an arrow to show the direction in which they travel.

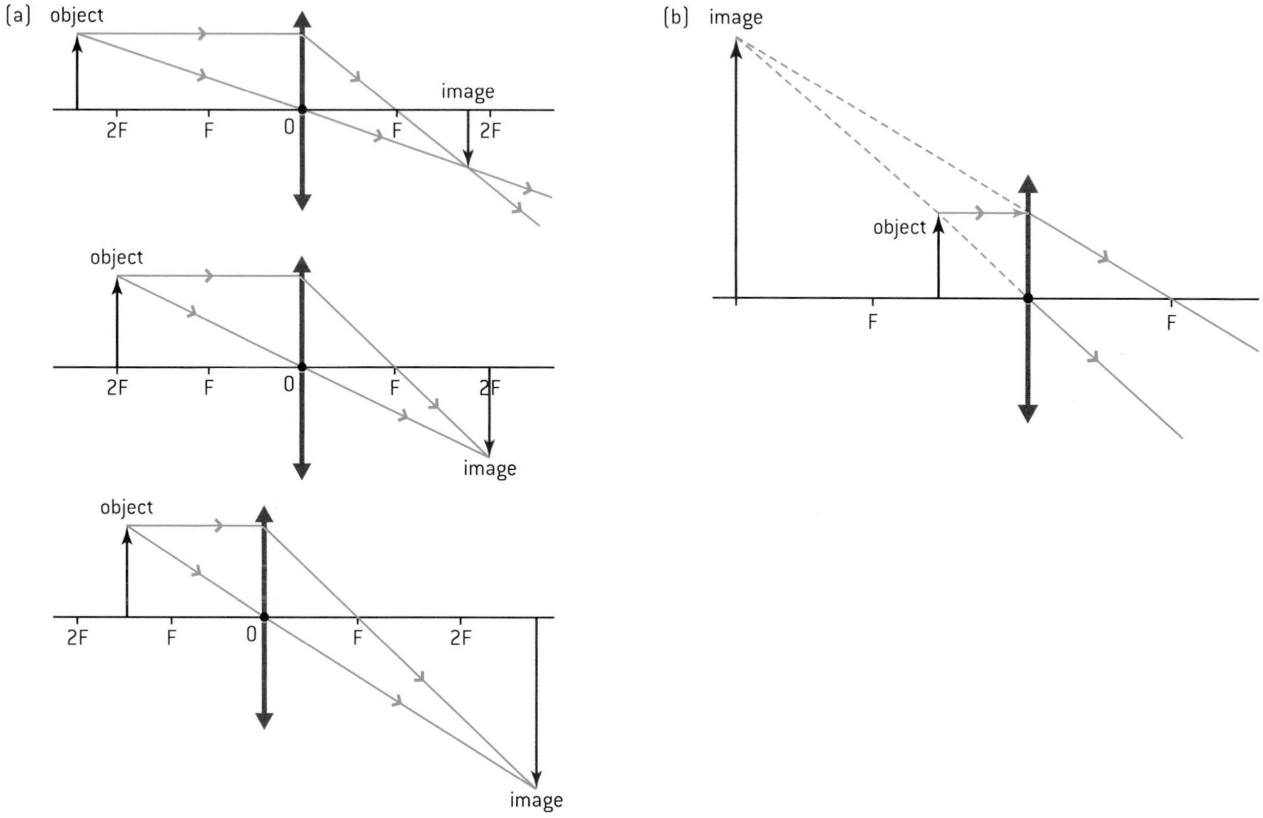

▲ Figure 13 Ray diagrams for a convex (converging) lens.

Figure 13(a) shows the ray diagrams that match some of the cases you tested in the *Investigate!*

When the object is closer to the lens that the focal point, a real image cannot be formed and a virtual image results (figure 13(b)). This is the ray diagram for a **magnifying glass**.

Diverging lenses have fewer ray diagrams than the converging lenses, the final image is always virtual. Such images cannot be formed on a screen and an additional converging lens (such as the lens in the eye) is required to form a real image that can be projected (figure 14).

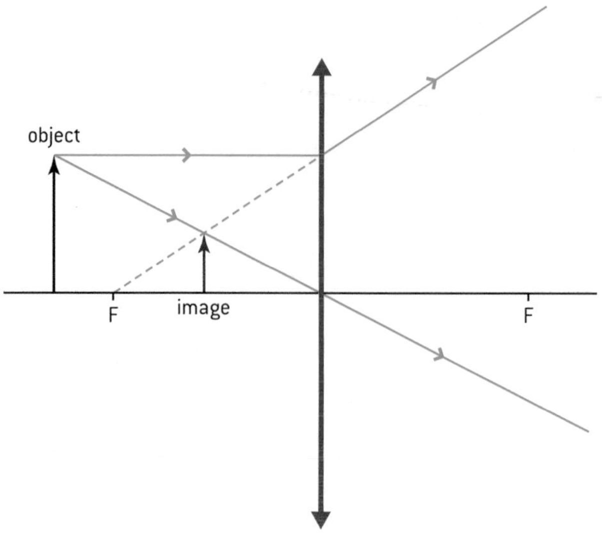

▲ Figure 14 Ray diagram for a concave (diverging) lens.

The lens equation

The *Investigate!* shows that the position of the image depends both on the position of the object and the focal length of the lens. The lens equation quantifies this relationship.

In curvature terms, a particular lens of focal length f adds the same amount of curvature to any wavefront passing through it. The amount added is $\frac{1}{f}$. This is because a lens with a small f gives more curvature to incident parallel rays than does a lens where f is large (the small f lens bends the rays more). The quantities $\frac{1}{u}$ and $\frac{1}{v}$ determine the curvature of the object and image wavefronts. Figure 15 shows how the symbols are defined and relate to each other. Therefore we can write:

curvature of wavefronts leaving lens = curvature of wavefronts entering lens + curvature added by lens

so

$$\frac{1}{v} = \frac{1}{u} + \frac{1}{f}$$

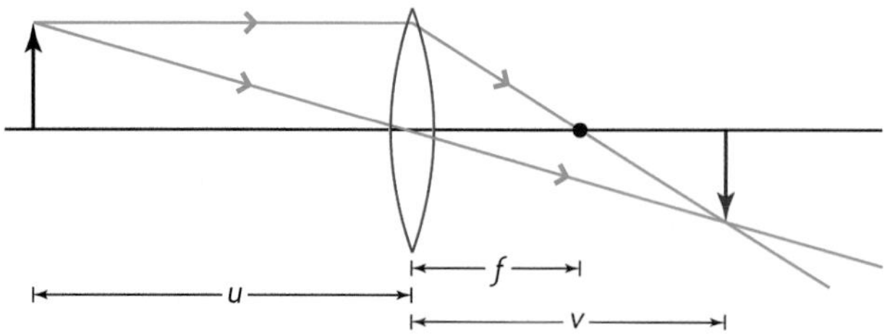

▲ Figure 15 The lens equation.

However, the wavefronts from the object are spreading out when they reach the lens whereas the wavefronts that have left the lens to form the image are converging. The curvatures of the object wavefronts are in the opposite sense to those of the image. To allow for this we should consider the u value to be negative with respect to v. This leads to the lens equation:

$$\frac{1}{u} + \frac{1}{v} = \frac{1}{f}$$

Some care needs to be taken with the signs given to u, v and f. The rule that applies when using the lens equation in this form is:

- real objects and images are treated as positive
- virtual objects and images are treated as negative
- the focal length of a converging lens is positive
- the focal length of a diverging lens is negative
- this is known as the **real-is-positive** sign convention; you may see texts that use other conventions.

The ability or **power** of the lens to add (or subtract) curvature to a wavefront is:

$$\text{power, } P \text{ (in dioptre)} = \frac{1}{f(\text{in metre})}.$$

A "strong" lens has a small focal length and adds a large curvature to any wavefront passing through it. The power of a lens is measured in **dioptres** (not an SI unit). A converging lens with a focal length of 50 cm (0.5 m) has a power of +2D; the lens is said to be a 'positive lens'. A diverging lens of focal length –12.5 cm has a power of –8D; the lens is said to be a 'negative lens'. If you wear eye glasses or contact lenses you may have seen these units used on your lens prescription.

The definitions of linear and angular magnification are identical to those for curved mirrors.

$$m = \frac{\text{height of image}}{\text{height of object}} = -\frac{v}{u}$$

$$M = \frac{\theta_i}{\theta_o}$$

where the θ_o and θ_i are the angles subtended by the object and the image at the lens.

Note

In fact, although this will not be tested in the examination, the lens equation also applies to mirrors. In this case f is equal to $1/R$.

Worked examples

1 An object of height 5 cm is placed 12 cm from a converging lens of power +10 D.

Calculate:

a) the focal length of the lens

b) the position of the image

c) The nature of the image.

Solution

a) $f = \frac{1}{D} = 0.10 \text{ m} = 10 \text{ cm}$

b) $\frac{1}{u} + \frac{1}{v} = \frac{1}{f}$ so $\frac{1}{v} = \frac{1}{f} - \frac{1}{u}$

$\frac{1}{v} = \frac{1}{10} - \frac{1}{12} = \frac{2}{120}$;

$v = +60$ cm from the lens

c) The positive sign tells us that the image is real, magnified by $\frac{60}{12} = 5$ times and is therefore 25 cm high. The image is upside–down and on the opposite side of the lens to the object.

2 An object of height 5 cm is placed 8.0 cm from a converging lens of focal length +10 cm. Calculate the position and nature of the image.

Solution

$\frac{1}{v} = \frac{1}{f} - \frac{1}{u} = \frac{1}{10} - \frac{1}{8} = \frac{-2}{80}$; $v = -40$ cm

The image is formed 40 cm from the lens; it is virtual, magnified 4 times, and is the right way up. It forms on the same side as the object. This is the magnifying glass arrangement.

3 Calculate the position and nature of an image placed 12 cm from a diverging lens of focal length −24 cm.

Solution

$$\frac{1}{v} = \frac{1}{f} - \frac{1}{u} = \frac{1}{-24} - \frac{1}{+12} = -\frac{36}{24 \times 12} = \frac{-1}{8}$$

The image is formed 8 cm from the lens on the same side as the object. It is diminished with a magnification of $(-)\frac{8}{12} = 0.67$. The image is upright and virtual.

4 a) A converging lens of focal length +15 cm is 25 cm from an object. Calculate the position of the image.

b) Another converging lens also of focal length +25 cm is placed 18 cm from the lens on the image side. Calculate the new position of the image.

Solution

a) $\frac{1}{v} = \frac{1}{f} - \frac{1}{u} = \frac{1}{15} - \frac{1}{25}$; $v = 37.5$ cm.

b) The new u is $37.5 - 18 = 19.5$ cm from the lens. This is a virtual object for the second lens and so will have a negative sign in the lens equation.

$$\frac{1}{v} = \frac{1}{25} - \frac{1}{-19.5}; \; v = +11 \text{ cm}$$

A real image forms 11 cm from the second ($f = 25$ cm) lens.

Although not asked for in the question, the ray diagram is given as an example of a more complicated arrangement.

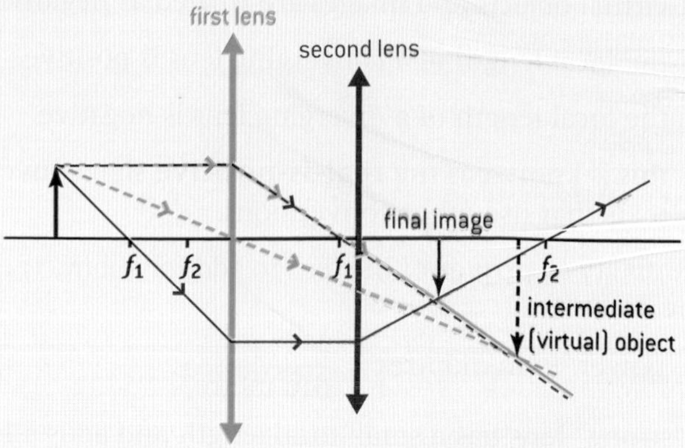

The first lens and its image is drawn first (green lines) the rays beyond the position of the second will not form so are drawn dashed as construction lines.

Then the second lens is added together with its focal points (red lines). There are a number of possible predictable rays that can be drawn. One shown here is the ray that is aimed at the optical centre of the second lens – this ray will not deviate as it goes through the lens and must also go through the top of the virtual image. The other ray for the second lens comes from the top of the original object, through f_1 and then must go parallel to the principal axis after being refracted by the lens. After passing through the second lens, it must go through f_2 and defines the top of the final real image.

The simple magnifying glass

A single convex lens used as a magnifying glass is the oldest optical instrument recorded. Around 2400 years ago, Aristophanes magnified small objects using spherical flasks filled with water like those used by present-day chemists.

We have already drawn the ray diagram for the magnifying glass, but because of its importance, both alone and in conjunction with other lenses, it merits a section of its own.

The human eye has a **near point** distance D that is the closest distance at which the eye can focus on an object without strain. For a normal eye this distance is taken to be 25 cm (although it varies greatly from individual to individual). Therefore, the best we can do to see the fine detail in an object is to hold it 25 cm away from the eye. The magnifying glass helps us to improve on this.

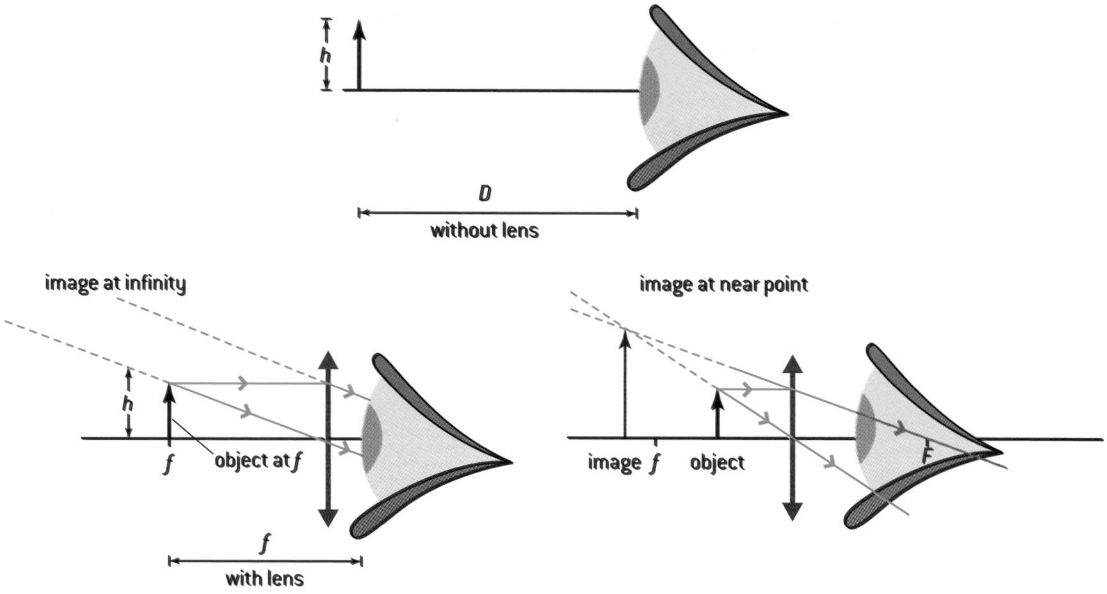

▲ Figure 16 Image formed by a magnifying glass at near point and infinity

One way (and the least tiring method for prolonged viewing) is to position the object at the focus of the converging lens and therefore form the image at infinity where it can be viewed with relaxed eye muscles (figure 16). If f for the lens is smaller than 25 cm the object will need to be closer than the near point with the eye muscles relaxed. In such circumstances, the eye should be as close to the lens as possible.

In this case, the angle subtended by the unaided eye is the size of the image $\frac{h}{D}$ whereas the angle subtended by the aided eye is $\frac{h}{f}$ so the angular magnification M is

$$\frac{\frac{h}{f}}{\frac{h}{D}} = \frac{D}{f}.$$

So a magnifying glass with focal length 0.1 m will give a magnification of ×2.5. However this is not the best that the magnifying glass can do.

Another way to use the lens, but more tiring because the muscles are not relaxed, is to place the eye close to the lens as before and to adjust the position of the object so that the image forms at the near point. The object can now be much closer to the eye than would be comfortable without the lens.

The angular magnification now changes to $M = \frac{\frac{h'}{u}}{\frac{h}{D}}$ where u, h' and h are defined in figure 17.

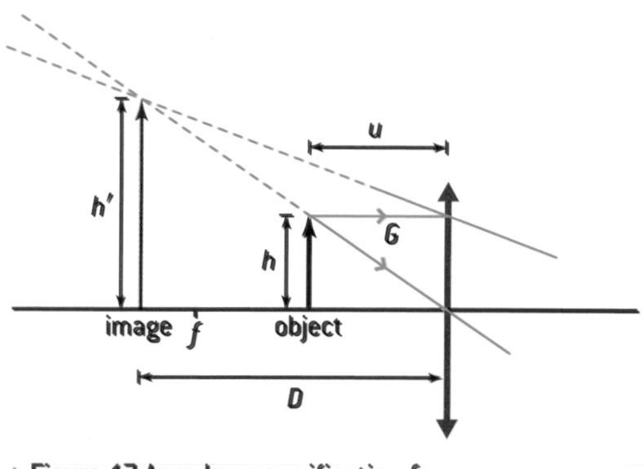

▲ Figure 17 Angular magnification for a magnifying glass.

The lens equation gives

$$\frac{1}{f} = \frac{1}{-D} + \frac{1}{u}$$

So $\frac{D-u}{u} = \frac{D}{f}$ or $\frac{D}{u} - 1 = \frac{D}{f}$

So for this setting of the magnifying glass

$$M = \frac{D}{f} + 1$$

Note

You may see that some books give the equation as $M = \frac{D}{f} - 1$. This results when the sign of D is not introduced during the proof but is used in the final calculation. Using the numerical examples above together with the equation with the negative sign, when the image is at infinity is M × −2.5 and when the image is at the near point it is ×−3.5. So the magnitude of the magnification still remains larger for the second case.

Worked example

A magnifying glass has a focal length of 7.5 cm. It is used by a person with a near point of 25 cm to view an object. Calculate the angular magnification of the object when the image is at:

a) infinity

b) the near point.

Solution

a) $M = \frac{D}{f}$; $M = \frac{25}{7.5} = 3.3$

b) M = infinity angular magnification + 1 = 4.3

For a magnifying glass with focal length 0.10 m and a near point of 0.25 m, the magnification is ×3.5.

Spherical and chromatic aberrations

Chromatic aberrations

The absolute refractive indices of the materials used for making lenses vary with wavelength. Different colours travel through the lenses at different speeds and thus form focal points at different distances from the lens.

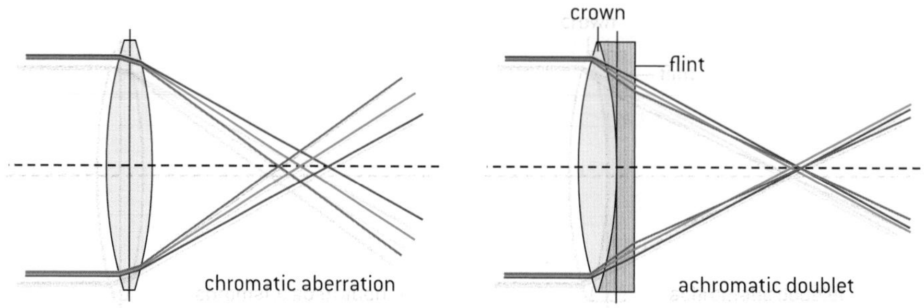

▲ Figure 18 Chromatic aberration and a way to correct it.

This problem is known as **chromatic aberration** and it leads to the appearance of colour fringes around the image. The best place to view the image is at the intersection of the green rays. This is the point known as the **circle of least confusion**. A circle drawn around all the rays here is at its smallest.

One common way to correct for chromatic aberrations is to use a **doublet lens**, one of the lenses is converging with positive power and the other diverging with a negative and smaller magnitude power. The lenses produce chromatic aberrations in opposite senses. Over the whole visible range, the colour performance of the combined lens is better than with one lens alone.

Spherical aberrations

Like mirrors, lenses also suffer from aberrations caused through the geometry of the spherical lens.

As figure 19 shows, rays that are far from the optical centre are brought to a focus closer to the convex lens than those near the lens centre. This is **spherical aberration** and leads to an object consisting of a square grid distorting as shown. This is known as barrel distortion.

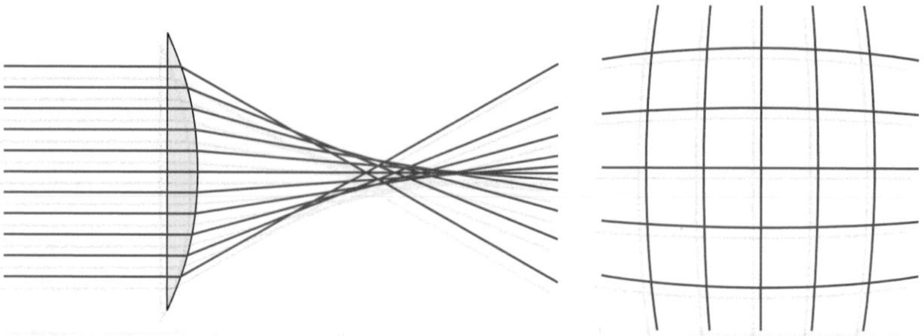

▲ Figure 19 Rays that produce spherical aberration and the effect on a square grid.

The easiest way to cure spherical aberration is to reduce the aperture (diameter) of the lens perhaps by putting an obstacle with a hole cut in the centre over the lens. Of course, this will reduce the amount of light energy arriving at the image position and will make the image appear less bright.

Nature of science

Same travel times

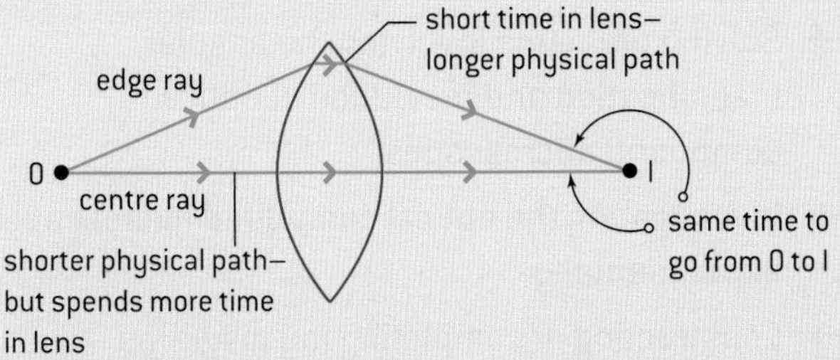

▲ Figure 20

There is another interpretation of how lenses work. We have regarded the lens as refracting light rays or as adding curvature to or subtracting curvature from a wavefront.

Think about two rays going through a converging lens (figure 20). One goes through the centre of the lens where it is thickest. The other goes through the edge of the lens where there is not much glass. The image forms at the place on the other side of the lens where the time taken by the light to take these two paths is the same.

The ray through the centre travels farther than the other ray in the glass, and because the speed of the light is slower in the glass, this ray spends longer in the glass than the edge ray. The edge ray has a longer overall path. What is special about the image position is that it is the (unique) place where all rays from the object take the same time to reach their respective places on the image.

This has to be the case given our wavefront interpretation. The lens keeps the wave together so that all parts meet at the same time at the same place on the image. Taking the same time and having the same effective path distance is what makes a lens a useful thing to manipulate rays. This is an example of Fermat's principle of least time.

TOK

Sign conventions – what is their effect?

There are two common sign conventions used in optics: the real-is-positive convention used here and the "New Cartesian' convention. Earlier we used a conventional current of positive charges in electricity. To what extent, if any, do sign conventions affect our understanding and use of science?

C.2 Imaging instrumentation

Understanding

- Optical compound microscopes
- Simple optical astronomical refracting telescopes
- Simple optical astronomical reflecting telescopes
- Single-dish radio telescopes
- Radio interferometry telescopes
- Satellite-borne telescopes

Nature of science

Optical instruments have been developed over the centuries to improve our ability to observe very distant objects in the sky and very small objects on Earth. Instruments to improve our vision were used in the time of the Egyptians. Authors described combinations of lenses in mediaeval Europe and it is certain that the knowledge of optics had been known in the Arab world since at least 1000 years BP. This line of improvement in instrumentation is clear in optics.

Applications and skills

- Constructing and interpreting ray diagrams of optical compound microscopes at normal adjustment
- Solving problems involving the angular magnification and resolution of optical compound microscopes
- Investigating the optical compound microscope experimentally
- Constructing or completing ray diagrams of simple optical astronomical refracting telescopes at normal adjustment
- Solving problems involving the angular magnification of simple optical astronomical telescopes
- Investigating the performance of a simple optical astronomical refracting telescope experimentally
- Describing the comparative performance of Earth-based telescopes and satellite-borne telescopes

Equations

- Magnification of astronomical telescope $M = \frac{f_o}{f_e}$

Optical compound microscope

Although the convex lens used as a magnifying glass provides magnification, small focal lengths are required to produce large magnifications ($M = \frac{D}{f}$). Lenses with small focal length have large surface curvatures and this can give rise to so much aberration that features in the image cannot be seen clearly.

The compound microscope is designed to produce a virtual magnified image of a small object. The term "compound" means that it is composed of more than one lens: an **objective lens**, close to the object, with a very short focal length f_o, and an **eyepiece lens** into

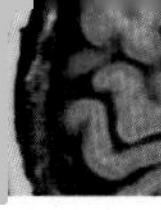

which the user looks As with all optical systems, you should try to understand the function of each component in the microscope:

- The objective forms a real, highly magnified image of the object at a position that is closer to the eyepiece than f_e.
- The eyepiece then acts as a magnifying glass the objective's real image as its own object.
- The final image, viewed by the observer's eye, is virtual and very highly magnified.
- The final image should not be closer to the eye than the near point D and when the image is formed here then the microscope is said to be in **normal adjustment**. Many experienced microscope users do not focus the microscope in full normal adjustment, however. It is often much more convenient for the final image to be focused at the plane of the bench on which the microscope stands. A notebook next to the microscope on the bench will then also be in focus for note keeping or drawing and can be viewed by the other open eye.

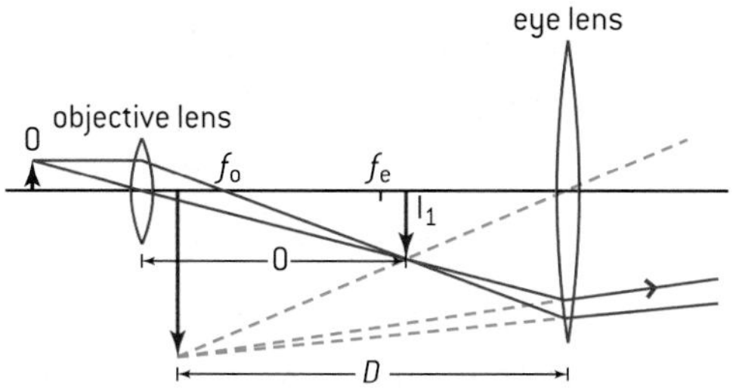

▲ Figure 1 Compound microscope.

The ray diagram (see figure 1) shows how the microscope in normal adjustment forms its image. When learning about microscopes and telescopes, try to understand the role of the separate elements as this will make it easier for you to draw the ray diagrams in examinations. It is a good idea to devise a strategy for drawing the diagram as it is easy to end up with a final image that is too large for the paper or off the page. Figure 3 (on page 611) shows one route to achieve a good sketch.

Notice that the vertical size of the lenses does not necessarily reflect their actual sizes in the instrument itself. The eyepiece and objective apertures are usually very small in diameter, but are shown large on the diagram so that it appears that the rays actually go through them. In practice, with real microscopes the lenses are small to minimise the aberrations and so that the lenses can be ground very accurately. As usual, if the ray diagram is drawn to scale, then measurements can be taken of the object and image scaled sizes to establish the overall magnification.

For the objective lens the magnification is equal to $\frac{L}{f_o}$ where L is the length in the microscope tube between the objective focal point and the eyepiece focal point (so that the total physical length of the microscope tube is $f_o + L + f_e$). The angular magnification of the eyepiece is $\frac{D}{f_e}$ as it is a magnifying glass (remember that D is the near point distance). The angular magnification of the complete microscope is given by the product of the magnification of the two lenses acting separately and this is approximately

$$\frac{DL}{f_o f_e}$$

Investigate!

Make a microscope

- Take two converging lenses, with focal lengths perhaps 5 cm and 15 cm. Determine their focal lengths by using rays of light from a distant object and measuring the approximate focal length between the lens and a piece of card.

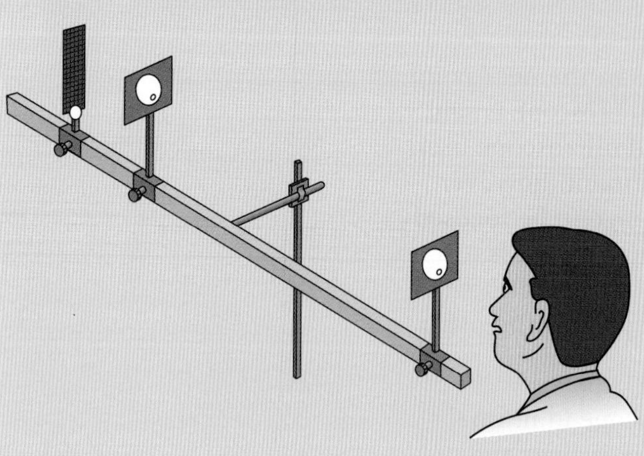

▲ Figure 2

- Fix the short focal length (objective) lens securely to the end of a short ruler (length 30 cm or so) using modelling clay or similar material. Set up the ruler so that it is a short distance from an object – a well-illuminated piece of graph paper makes a good object for this experiment.
- Fix the other (eyepiece) lens so that when you look into it, you see a magnified virtual object.
- By comparing the size of the grid on the object with the size of the grid on the image, estimate the magnification of the microscope.
- Does your microscope approach the theoretical value of the magnification? The approximations assume that f_o and f_e are very small compared with D and L. Is this true in this case?

Resolution of a microscope

All light that goes through an aperture such as a lens is diffracted. Point objects become image disks as a result. If the image disks of two adjacent point objects overlap, it is difficult to tell them apart and the two images are not resolved. This is a significant problem in microscopy.

Rayleigh suggested that two images were just resolved when the minimum of the diffraction pattern of one image coincided with the central maximum of the pattern of the other. This means that when the images are just resolved, the angle θ subtended at the eye by rays from each image is given by:

$$\sin \theta = 1.22 \frac{\lambda}{d}$$

where λ is the wavelength of the light and d is the diameter of the aperture (in this case circular). For the microscope the aperture diameter is the effective diameter of the lenses. This is known as the Rayleigh criterion.

You can find more discussion of the Rayleigh criterion in Topic 9.

The criterion is often modified for use with microscopes and the effective aperture is replaced by N the numerical aperture of the lens. This is equal to $n \sin \theta$ where n is the refractive index of the medium in which the lens is placed ($n = 1$ in the case of most microscopes that are in the air) and θ is the half angle of the maximum cone of light that can enter the lens. The criterion becomes:

$$\sin \theta = 1.22 \frac{\lambda}{N}$$

(You will not need to use this equation in the examination.)

Nature of science

Electron microscopes

Electron microscopes are used to image the smallest objects. The electrons are accelerated and as a result have extremely small wavelengths, much less that that of light.

Glass lenses cannot be used to focus the electrons but magnetic and electrostatic fields can, and these are used to bend the paths of the electrons to produce a focusing effect. Look at some web sites that explain how the various types of electron microscope work. Try to understand how the fields replace the lenses of the compound optical microscope.

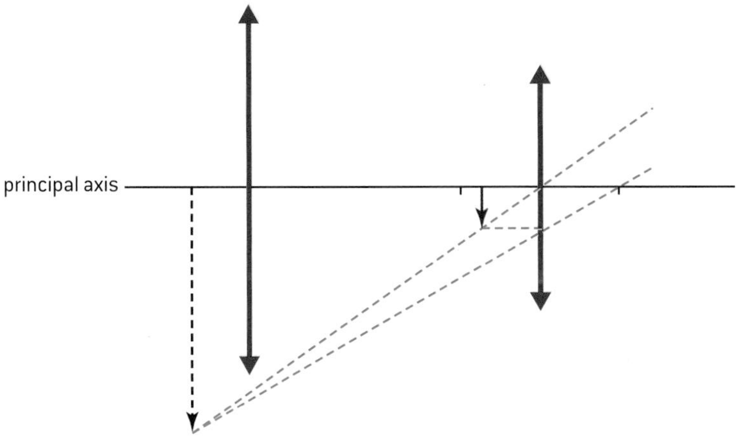

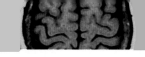

First, draw the principal axis. Add the lenses a scaled distance apart if you have been given dimensions. Then draw the intermediate image – use an intermediate image height that will allow the final image to be on the page.

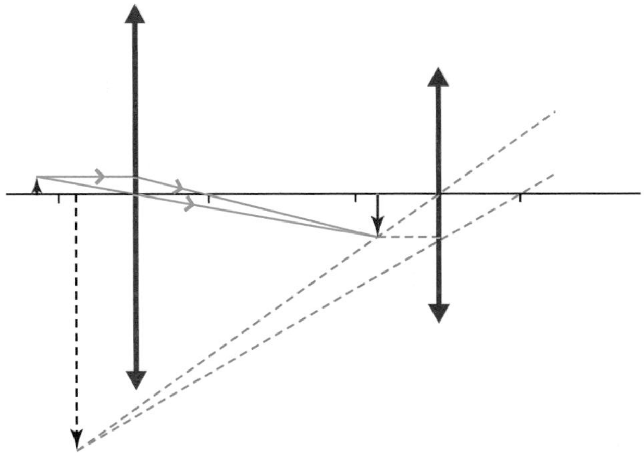

Work backwards from the intermediate image to form the object to the left of the objective. Either choose the focal lengths yourself or use the scaled values if they have been provided. Don't draw rays to the right of the intermediate image yet.

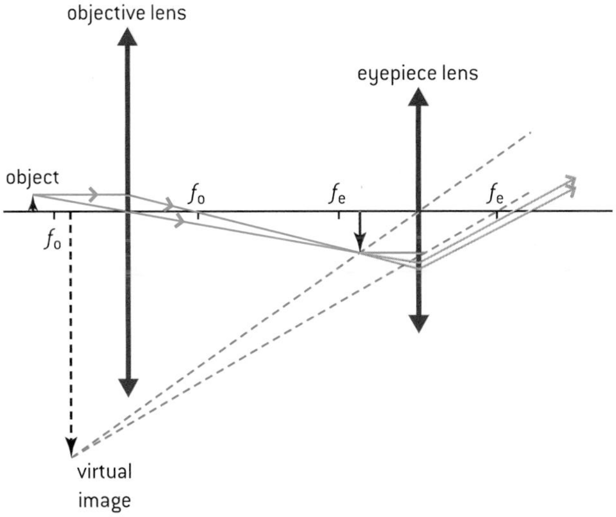

Finally, draw the rays to the right of the intermediate image. These do not deviate between the lenses, but after going through the eyepiece they must appear to have come from the top of the virtual image. To complete the diagram add the arrows to all rays (not construction lines) and label all focal points and both lenses.

▲ Figure 3 Strategy for drawing the compound microscope diagram.

The equations suggest that the resolution of a microscope can be improved by:

- using a short wavelength for the light (sometimes microscopists use ultraviolet radiation for taking photographs of specimens to improve the resolution)
- using as a wide an aperture for the objective as is consistent with reducing aberrations (the design of the final eye piece lens is fixed by the average size of the eye)

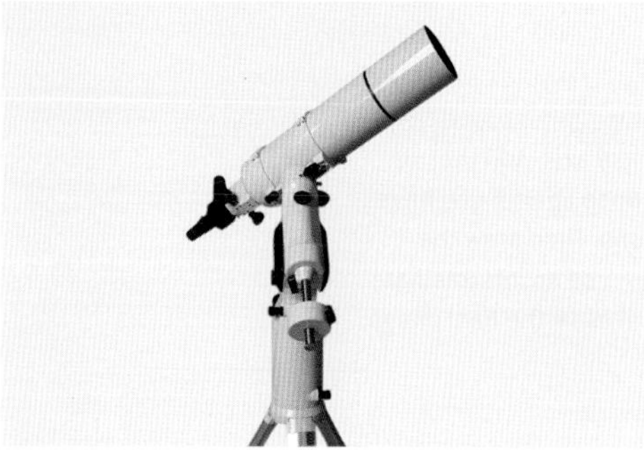

▲ Figure 4 An astronomical reflector.

- using a liquid of high refractive index between the specimen and the objective lens. The liquid objective allows a wider cone of light rays to be collected by the lens and this extra information improves the resolution of the image. Typically an oil of refractive index around 1.6 is used in special microscopes that have oil-immersion objectives.

Astronomical refracting telescopes

Objects in the night sky have always fascinated astronomers, and telescopes for viewing the sky were, like the microscope, developed early in the history of science.

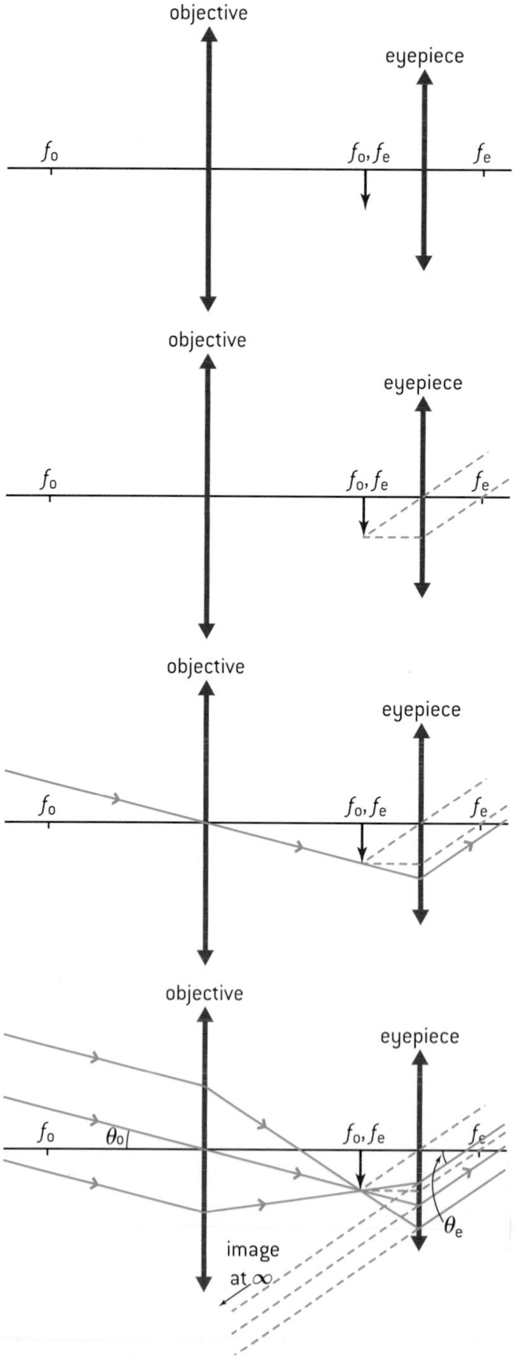

Begin by drawing the principal axis, the lenses and the common focal points. Add an intermediate image. Do not make it too small – allow enough room for the construction later.

Draw construction lines to show you the final direction of the rays when they have been refracted by the eyepiece. These are construction lines so should be dashed.

The first ray should be the ray that ends at the top of the intermediate image and goes through the optical centre. This ray is not deviated at the objective. After passing through the eyepiece the ray must travel parallel to the construction lines.

Choose two more rays parallel to the first ray before they enter the objective. After passing through the objective they must go to the top of the intermediate image. After passing through the eyepiece they must go parallel to the first ray (and the construction lines). Finally, add further construction lines to show the direction of the final image. Then label the diagram and add the ray arrows

▲ Figure 5 Telescope ray diagram and drawing sequence.

Also like the microscope, the astronomical refracting telescope uses two converging lenses. The **objective lens** has a long focal length f_o and an **eyepiece lens** that has a short focal length f_e. The telescope is designed primarily to focus the light from very distant objects and in its normal adjustment is set up for this purpose only.

Light from infinity (that is, parallel rays) from the object are focused by the objective lens at its focal point. The focal points of the objective and eyepiece lenses are coincident (at the same place) inside the telescope and therefore the eyepiece treats the image formed by the objective as though it is a real object placed at f_e. We know from earlier that this leads to parallel rays emerging from the eyepiece and these then enter the eye of the observer.

The angle subtended by the rays from the top of the image and the principal axis is much greater than the angle subtended by the rays from the top of the object. The image is magnified. The ray directions show that the viewed image is also upside–down, but this is not an issue for astronomers (maps of the Moon are often inverted to take account of this). The angular magnification of the telescope follows directly by looking at the incident and emerging rays and the angles, θ_f and θ_e, they make with the principal axis

$$M = \frac{\theta_o}{\theta_e} = \frac{f_o}{f_e}$$

Note

In astronomy the term "power" is sometimes used in place of "angular magnification". The maximum practical power for any given astronomical telescope is limited by the need to produce a high resolution image. This in itself is determined by the atmospheric conditions.

Investigate!

Making a telescope

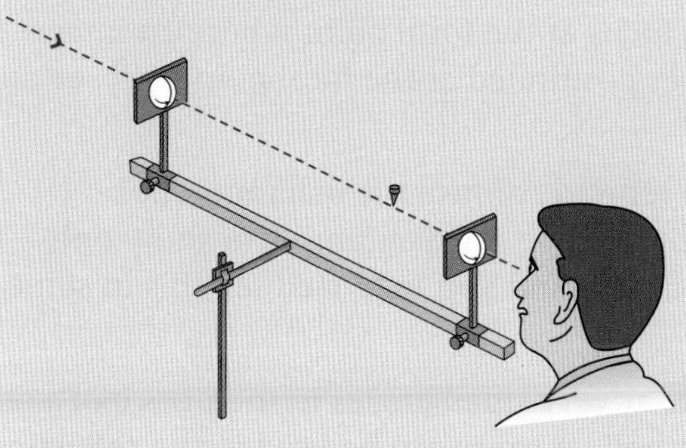

▲ Figure 6

- The basic method is similar to that of the microscope in that the instrument is constructed on a ruler or optical bench.
- This time a metre ruler is required together with a long focal length objective (f_o about 50 cm or longer) and a short focal length eyepiece (f_e about 10–15 cm). The combined total of $f_o + f_e$ should not exceed the length of the ruler.
- Measure and record the focal lengths of both lenses.
- Fix the eyepiece lens at the end of the ruler with modelling clay as before.
- Holding or clamping the ruler horizontally, fix the objective lens a distance equal to $f_o + f_e$ from the eyepiece while looking through the eyepiece. You may need to move the objective forward and backwards along the ruler a short distance to get the optimum focus for your own eye.
- View a distant object (say a brick wall) and estimate how much larger one brick is when viewed through the telescope compared with using your naked eye.
- Does your telescope approach the theoretical angular magnification give above?

Refracting telescopes are frequently used by amateur astronomers and also by professional astronomers for some applications. However, the largest refracting telescopes made have objective apertures up to 2 m and this represents the limit of the technology for grinding the glass lens. It is also difficult to support the lens given that it needs to be at the end of a long tube. Mirrors can be made much larger and therefore collect more light energy meaning that more distant objects can be viewed.

Worked example

1 A small object is placed 30 mm from the objective lens of a compound microscope in normal adjustment. A real intermediate image is formed 150 mm from the objective lens. The eyepiece lens has a focal length of 75 mm and forms a virtual image at the near point. For the observer, the near point distance is 300 mm. Calculate the overall magnification of the telescope.

Solution

Magnification of the objective $= \frac{150}{30} = 5\times$.

For the eyepiece $\frac{1}{75} = \frac{1}{u} + \frac{1}{-300}$; $\frac{1}{u} = \frac{5}{300}$; $u = 60$ mm. So magnification of eyepiece $= \frac{300}{60} = 5\times$.

Overall magnification $= 5 \times 5 = 25\times$

2 An astronomical telescope in normal adjustment has an objective focal length of 150 cm and an eyepiece focal length of 5.0 cm. Calculate:

a) the angular magnification of the telescope

b) the overall length of the telescope.

Solution

a) $M = \frac{f_o}{f_e} = \frac{150}{5} = 30\times$ (angular magnification and linear are the same for the telescope in normal adjustment)

b) Overall length $= f_o + f_e = 1.55$ m

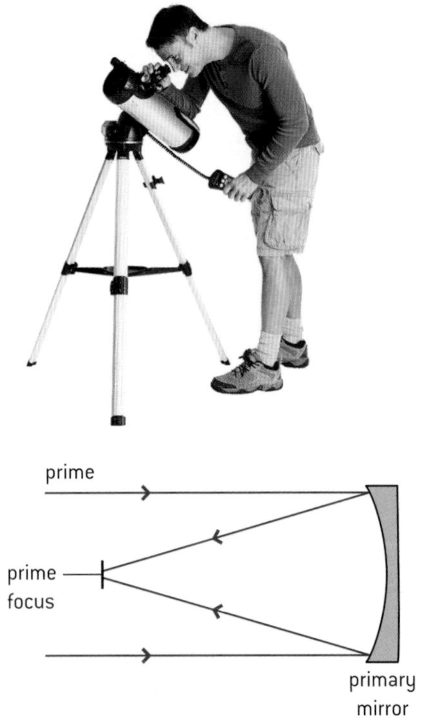

▲ Figure 7 Picture of astronomical telescope and basic ray diagram.

Astronomical reflecting telescopes

The first known reflecting telescope was completed by Newton in 1668. The basis of the telescope is straightforward.

Rays from the top and bottom of a distant object are reflected by a concave mirror and come to a focus at the focal point where an image of the object forms. An observer positioned at the focal point will see an image in the mirror.

Rays from the top and bottom tend to confuse the diagram. The convention in many reflecting telescope ray diagrams is to show only the rays parallel to the principal axis and you may well see this in some books.

Because the rays are parallel or at very small angles to the principal axis, either spherical mirrors or parabolic mirrors can be used in the reflecting telescopes. For astronomical use, parabolic is chosen so that the spherical aberrations discussed in the earlier section are eliminated.

As with compound microscopes, the angular resolution of a reflecting telescope is given by

$$\sin\theta = 1.22\frac{\lambda}{d}$$

where d is the diameter of the aperture. So a large aperture for the mirror is an advantage as this increases the resolving power of the instrument.

There are many advantages of reflecting telescopes over their refracting equivalents:

- There is one surface which is usually ground to shape and then coated with a reflective metallic surface. The rays of light do not have to pass through a number of layers of glass. The telescope can therefore in principle have no **chromatic aberration**.

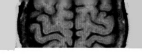

- The mirror only has to be supported from one side. This makes engineering large reflecting telescopes more straightforward.
- With only one surface to grind, the telescopes are cheaper for a comparable quality of image.
- Only one surface has to be made perfect.

There are also some disadvantages however:

- The mirror surface is vulnerable to damage and needs to be cleaned (unless covered – which introduces aberration).
- The optics can easily get out of alignment if the support for the mirror shifts in some way.

In addition, this simple design means that the observer's head or a camera blocks some of the light travelling along the tube thereby reducing some of the benefits of using a large mirror.

For this reason, various arrangements are used to transfer the image from inside to outside the main tube of the telescope. There are a number of different design variants for achieving this, some of which use lenses and other devices to produce a high-quality image. In this course we will examine only two simple systems: the newtonian telescope and the cassegrain telescope. Both are named for their inventors.

Newtonian mounting

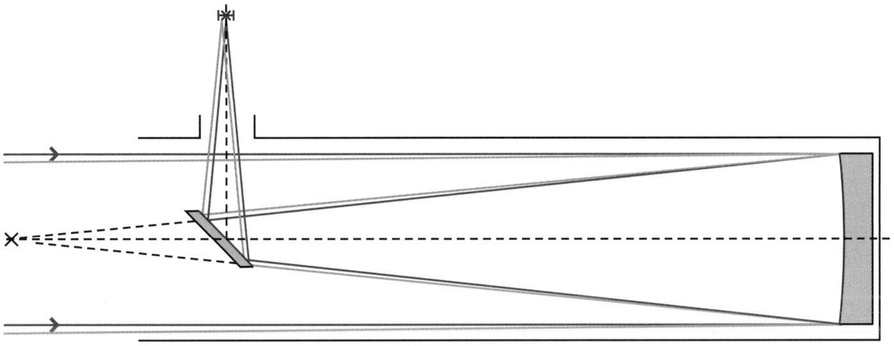

newtonian mounting

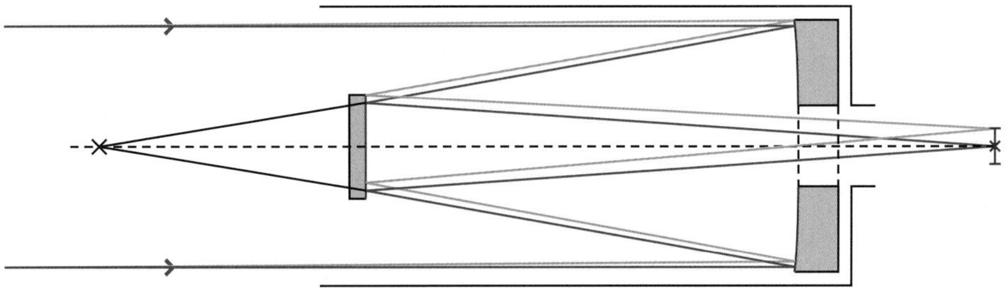

cassegrain mounting

▲ Figure 8 Newtonian and cassegrain mirror systems.

In this system a small secondary plane mirror diverts the rays from inside to outside the tube. Plane mirrors of high quality can be produced that introduce little or no distortion to the image. This is one of the cheapest designs of telescope and is very popular with amateur astronomers, some of whom grind their own mirrors and build their telescopes from scratch.

Cassegrain mounting

Replacing the plane mirror with a hyperbolic secondary mirror enables the rays to be sent through a hole in the primary mirror surface. This means the observer can be in line with the telescope and can look in the observing direction. A principal advantage of this mounting is that the focal length of the telescope can be effectively longer than the tube itself. However, the secondary mirror inside the tube cuts off some of the incoming light, and the principal mirror has lost some of its reflecting surface. In general, all telescope types have specific observational advantages and disadvantages.

Radio telescopes

Many of the principles of the reflecting telescopes are carried over into the single-dish radio telescopes. These instruments are intended to collect electromagnetic signals in the radio region that originate from astronomical objects.

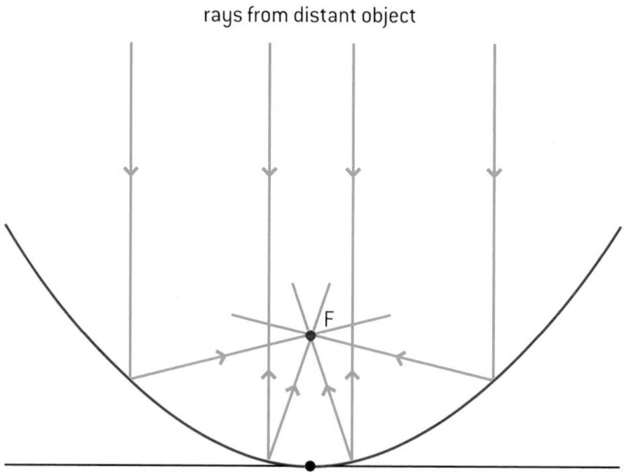

▲ Figure 9 Radio telescopes.

A parabolic dish has the property that all rays parallel to the principal axis are brought to a focus at the same point. If a radio antenna (aerial) is placed at this point then it will collect all the focused radio waves that arrive simultaneously. The larger the area of the dish, the greater the power collected, so in principle the designer should aim for as large an area as possible. Another advantage of a large dish diameter is that, as with the optical reflecting telescopes, the resolution of the telescope improves (θ in the Rayleigh criterion equation decreases) as the diameter becomes larger.

▲ Figure 10 The Arecibo Observatory, Puerto Rico.

Resolution is a particularly important factor as the wavelengths used by the telescopes are those of the radio segment of the electromagnetic spectrum. These wavelengths are of the order of centimetres to tens of metres and are much larger than those of the visible light used with optical instruments (order of 10^{-7} m). So a telescope working at radio wavelengths needs to have an aperture at least 100 000 times greater than its optical counterpart to have the same resolution.

This raises engineering problems for a dish that has to be steered to point at particular objects in the sky. There is the difficulty of moving the dish and problems associated with the dish deforming from the ideal parabolic shape under its own weight. Some radio telescopes get over this problem by building the dish into a cavity in the ground.

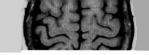

The disadvantage is that the dish can no longer be steered. Arecibo Observatory in Puerto Rico is a good example of this (figure 10).

Some flexibility can be engineered into the system by moving the antenna, which in this telescope is suspended at the focus and can be moved on a suspended railway track that can be seen in the photograph.

The Arecibo telescope is 300 m across but a larger telescope of 500 m diameter, also in a natural crater, is being constructed in Pingtang County, Guizhou Province, south-west China. The Chinese telescope is designed so that the panels that make up the dish are moveable to allow the principal axis of the telescope to be steered within limits. By contrast, the largest dish telescope in the world at present is the Green Bank 100 m diameter telescope in West Virginia USA (figure 11).

▲ Figure 11 The Green Bank Telescope, West Virginia, USA.

Interferometer telescopes

To overcome the inherent design problems of a dish and the inability to steer a telescope built in a crater, a recent trend is to construct interferometer telescopes.

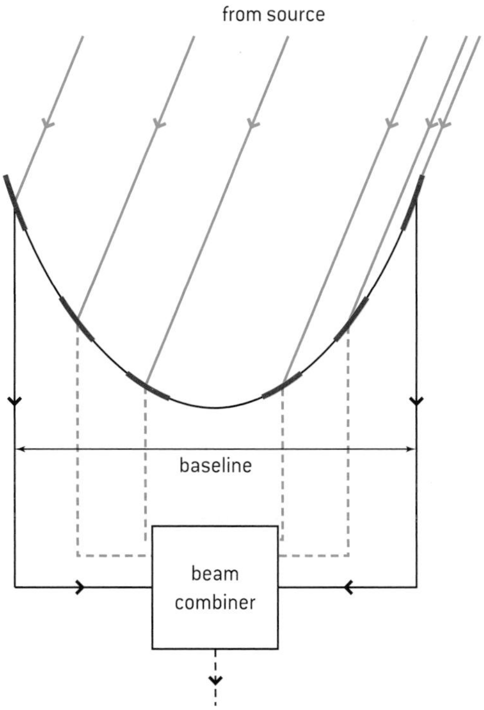

▲ Figure 12 Interferometer telescope.

Such telescopes come in various formats with two or more radio telescopes combined, but the basic idea behind all of the arrangements is that signals from a source are combined in the individual components of the telescope to produce a total signal. The signals can be combined in such a way that the whole telescope is equivalent to a single dish with an effective diameter equal to a baseline B. The baseline is the dimension across the individual dishes that make up the telescope. The resolution of such an array is $\sin\theta = \frac{\lambda}{B}$.

One format that is frequently used is a series of small steerable dishes. These are individually of low cost and do not have the engineering problems associated with large dishes. The signals are combined using computers to give a final signal.

Another approach is to use a very long baseline. Signals from steerable dishes situated in different countries can be combined to give a network, the baseline of which can approach the size of the planet. For example the European Very-long-baseline interferometry Network (EVN) is a combination of 12 telescopes and can, itself, be combined with interferometers in the UK and the USA to make a global network that achieves a very high resolution.

There are plans to build a square kilometre array (SKA) that will combine telescopes in Australia, South Africa and other countries in the southern hemisphere with a baseline of the order of 3000 km. The telescope is due to be completed by 2024.

Other interferometer formats are possible and these at present include linear arrays of two long antennae or arrangements of antennae built in the shape of a plus sign.

Nature of science

Discoveries in radio astromony

Not everyone has to be a professional. Some branches of science are known for the number of discoveries made by amateurs. Astronomy is one of these.

The first radio telescope was constructed in 1931 by a radio engineer named Karl Jansky who worked in the Bell Telephone Laboratories in the US. He studied sources of static noise in radio signals and discovered that some of the static originated outside the Earth. His telescope was a mobile collection of radio antennae (aerials). The first true dish radio telescope was built by a North American amateur radio enthusiast named Grote Reber (call sign W9GFZ) who was inspired by Jansky's work. He spent considerable time (from 1933–38) constructing a series of parabolic dish reflectors that were eventually sensitive enough to reproduce Jansky's results. Reber and Jansky were the first true pioneers of radio astronomy.

Today, amateur astronomers have excellent telescopes coupled to digital cameras. This makes it possible for them to contribute to science in a significant way. Professional astronomers cannot view all areas of the sky night after night. For example, amateur astronomers discover deep sky supernovae in distant galaxies and alert professional observers who then make observations using larger telescopes.

Earth and satellite-borne telescopes

Recently, high-quality telescopes have been placed in satellite orbit above the Earth or on spacecraft that are aimed away from the Earth into the Solar System.

Observational astronomy carried out on the surface is subject to a number of limitations, as described overleaf:

- When we look at the night sky with the naked eye, the stars appear to twinkle. This is because our observations are made through tens of kilometres of air. The atmosphere introduces brightness variations and position errors into the images of the stars through local short-term changes in the air density caused by heating and convection effects. A common way to reduce this problem is to build optical telescopes on mountain tops in places where the atmosphere is relatively stable.

Some recent telescopes use up-to-date techniques to remove light variation using adaptive optics. When viewing a distant dim star the optics make reference also to light from a nearby bright star that is assumed to have a constant intensity and position. Parts of the telescope mirror surfaces are then moved rapidly to correct for the distortion of the light, which is assumed to be the same for both reference and observed stars. However, observations that are made from a satellite platform above the atmosphere do not need these corrections.

- There is an increasing problem of electromagnetic pollution from artificial sources of radiation associated with cities on Earth. Optical telescopes need to be constructed in increasingly remote areas, and radio telescopes also need to be placed in places where the radio spectrum is relatively uncluttered.
- Stars emit radiation right across the electromagnetic spectrum, and all this information is of value to astronomers. Many wavelengths from the stars are absorbed by the atmosphere (figure 13) and cannot be detected at the Earth's surface. The only way to measure these is through the use of a telescope or sensor mounted on a satellite. X-rays are a particularly important region of the spectrum to astronomers and orbiting satellites are almost the only way to study emissions at these wavelengths from stars (figure 13).

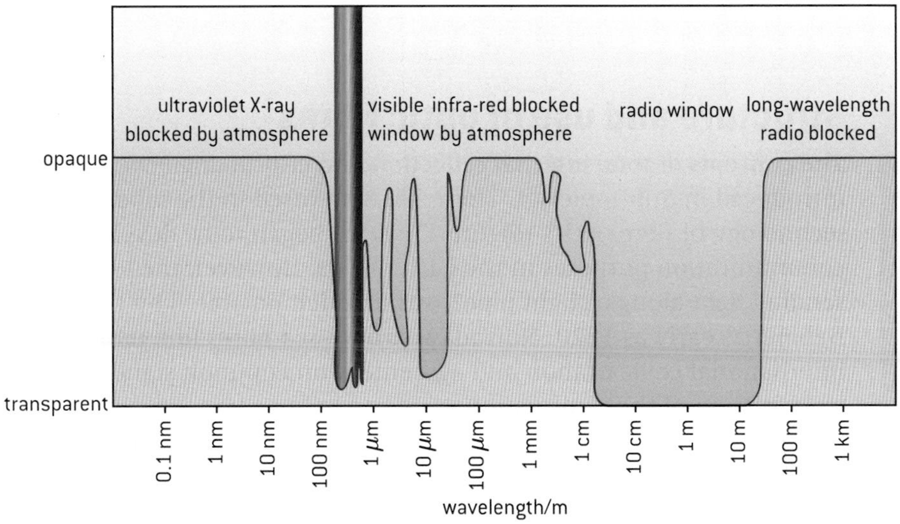

▲ Figure 13 Absorption of the electromagnetic spectrum by the atmosphere.

- An Earth-based telescope, whether radio or optical, can either be fixed or steerable. A fixed telescope can only view the parts of the sky that move in front of it. A steerable instrument can view (within the limits of the local horizon) the whole of the hemisphere on which it is centred. Similarly, some satellite telescopes are fixed so that they observe certain sections of the sky. Other sky-survey satellites allow the entire sky to be mapped over the period of the satellite's lifetime.

The expense of placing an observing satellite into orbit and the costs of building surface-bound telescopes is so great that international collaboration is common in astronomy with research groups from around the world booking observation time on the instruments.

TOK

What do we see?

Any optical or radio instrument extends our senses beyond their "design limits". This can be in terms of wavelengths to which we are not normally sensitive, or in terms of magnifying beyond what we would normally see. To what extent do the images we see or interpret represent true reality?

C.3 Fibre optics

Understandings

- → Structure of optic fibres
- → Step-index fibres and graded-index fibres
- → Total internal reflection and critical angle
- → Waveguide and material dispersion in optic fibres
- → Attenuation and the decibel (dB) scale

Nature of science

Modern communications rely heavily on the use of fibre optics. A relatively simple piece of science has led through applied science to a transformation of all types of communications across the globe.

Applications and skills

- → Solving problems involving total internal reflection and critical angle in the context of fibre optics
- → Describing how waveguide and material dispersion can lead to attenuation and how this can be accounted for
- → Solving problems involving attenuation
- → Describing the advantages of fibre optics over twisted pair and coaxial cables

Equations

- → critical angle equation $n = \frac{1}{\sin c}$
- → attenuation (dB) $= 10 \log \frac{I}{I_0}$

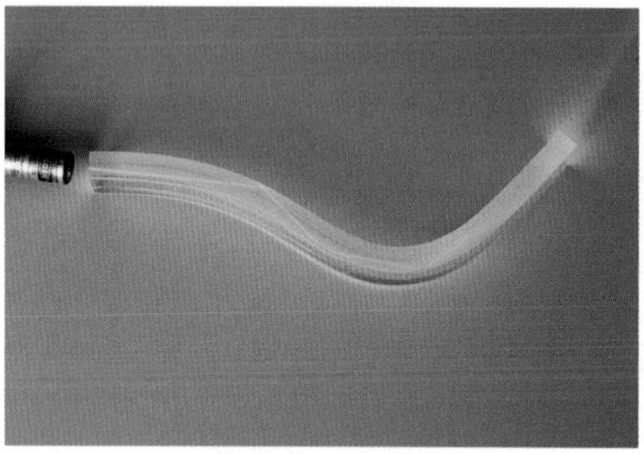

Structure and use of optic fibres

The concepts of total internal reflection and critical angle were introduced in Sub-topic 4.4. These ideas are used in the modern technology of fibre optics (Figure 1) which began to be developed for communication purposes in the early 1970s. However, the idea of sending light along a "light pipe" was first demonstrated by Colladon and Babinet as early as 1820. Nowadays, under-sea fibres link nations with international collaboration and agreement on common standards for the transmission of the information.

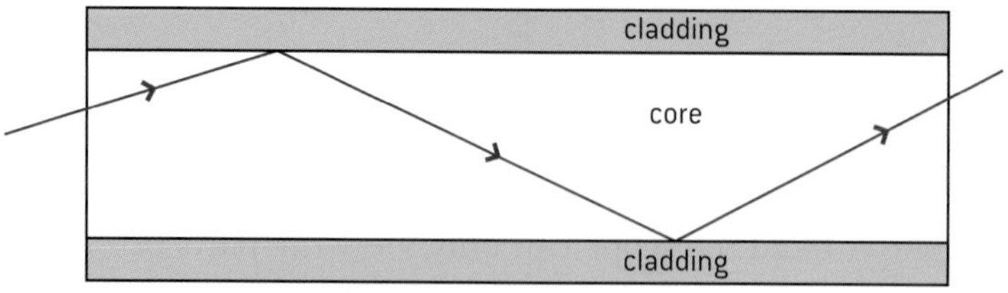

▲ Figure 1 Total internal reflection in a fibre.

Communications and the advantages of optic fibres

The principal methods using physical links for communication (as opposed to the radiation of electromagnetic waves as in radio) are:

- **Twisted pair.** In this technique developed by Alexander Bell, the wires connecting one operator with another were twisted together (hence the name). External electrical signals can induce an emf in both wires and this emf is to some extent cancelled out by the twisting. However, the noise cancellation is not necessarily complete and if two pairs of twisted-pair cables are next to each other then signals from one pair can be induced in the other cable pair giving cross-talk and a lack of security. Twisted-pair cables are generally used for low-frequency applications.

- **Coaxial cable.** This cable is constructed with a central core conductor insulated from a metallic shield that is earthed at zero potential relative to the changes in the signal voltage (but which carries the return current). The cable is completed with a tough protective outer cover. Such cable is generally used for radio-frequency signals. You may well have seen an example of it carrying the (very weak) signals from your tv antenna or satellite dish to the tv itself. This type of cable rejects outside electrical noise as the earth shield acts as the surface of a conducting shell. Little noise from outside can distort the weak signal travelling along the cable and equally the signal itself cannot radiate significantly beyond the earthed conductor. However, the cable is bulky and expensive.
- **Optic fibres.** In this technique the signal to be transmitted down the fibre modulates an electromagnetic wave (usually at visible or infra-red wavelengths). This modulated wave is shone along a very thin glass fibre. The thickness of the fibre is so small that the light strikes the walls at angles (much) greater than the critical angle and so the light propagates along the fibre. Electrical noise does not affect the passage of the light through the fibre. There are wavelength windows in optic fibre glasses at which the loss of signal with distance (the attenuation of the glass) is very low. The fibre is surrounded by a cladding material with a refractive index *smaller* than that of the core in order to ensure that the critical angle condition is maintained.

  Digital signals are transmitted along the optic fibres. Such signals consist of a sequence of changes between two states; these might be on/off, variations between two fixed frequencies, or abrupt changes in the phase of the signal. For our discussions we will assume a simple model where the light is either on or off, so that a series of light pulses is transmitted down the cable.

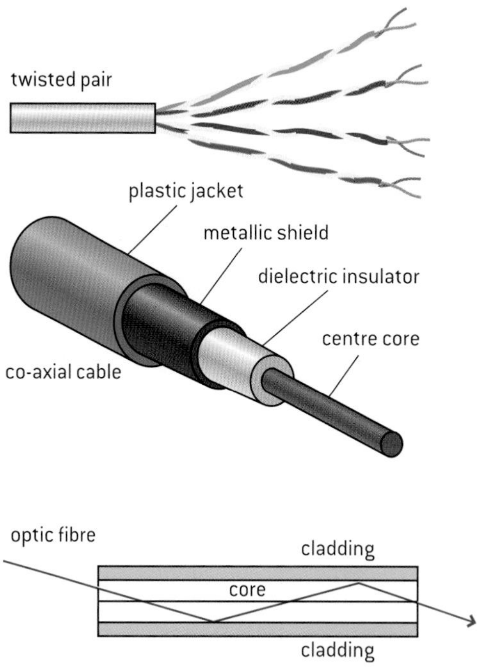

▲ Figure 2 Physical communication links.

The advantage of optic fibres over conductors are:

- Immunity to electromagnetic interference unlike most electrical cables.
- Low attenuation loss over long distances in glass compared to metal conductors.
- Broader bandwidth (total range of usable frequencies) in the glass compared to wires. This means that one fibre can carry millions of telephone conversations.
- Glass is an insulator and does not suffer from the inherent problems of a electrical-cable telephone system where electrical connections to earth (shorts) are a problem.
- Very small diameter so that compact bundles of fibres can be constructed thus increasing the capacity of the system even more for the same physical space as an electrical cable.

- Greater security to signal detection from outside the fibre ("fibre tapping"). Some optical fibres have been designed with a dual core and are said to be completely secure to outside tapping.

The main types of fibre are:

- **Step-index** fibre in which there is an abrupt change in the refractive index at the interface between core and cladding.
- **Graded-index** fibre in which there is a gradual reduction in the refractive index from the centre to the outside of the core.
- **Single-mode** fibre, which has a diameter of only a few times the wavelength of the light used in it. At such small diameters, the rules of geometrical optics that we use for our discussion of optic fibres break down and wave theories are required. We are not going to consider this type of fibre further in this course.

We will look at the behaviour of step- and graded-index fibres in more detail later.

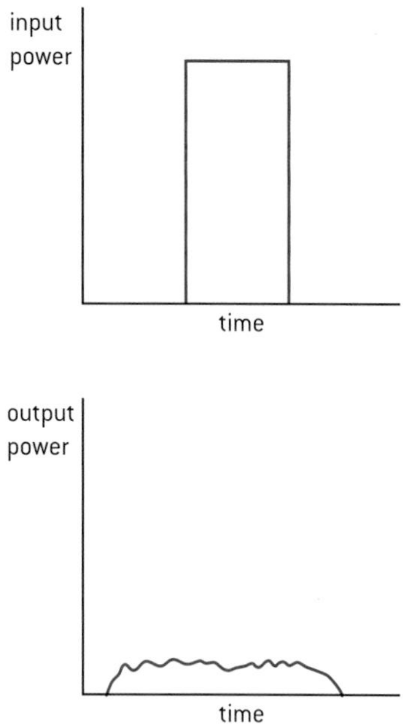

▲ Figure 3 Attenuation with time in an optical fibre. The time axes are to the same scale.

Attenuation and dispersion

Optic fibres have very low loss. If sea water had the same optical properties as glass used in fibres, then the details at the bottom of the 11 km deep Mariana Trench, off the coast of Japan, would be clearly visible from the ocean surface. Nevertheless, eventually the signal weakens (attenuates) so that it needs to be amplified before it can continue its journey. The device that carries out this re-amplification is known as a **repeater**.

At the point where the signal enters the cable, the pulse will have an abrupt on/off change in intensity.

Figure 3 shows how the power of the pulse varies with time before and after passing along the cable: the power is reduced (by a very large factor) so that the total energy in the pulse (the area under the power–time curve) is reduced also. This is **attenuation**; energy has been lost to the cable. Separate **dispersion** effects change the shape and cause the pulse to "spread out" as it travels along the fibre. If two pulses that were initially separate overlap through dispersion, then the receiving system cannot disentangle them from each other. Dispersion imposes an upper limit on the rate at which a particular fibre can transmit information.

The repeater needs to do two things: re-shape the pulse into its original square format and also boost the amplitude of the signal. We will discuss the reasons for the loss of amplitude and the dispersion later, but for the moment we concentrate on how attenuation is measured.

The signal amplitude can fall by many orders of magnitude before it needs to be boosted. To deal easily with large ratio changes a logarithmic scale called the **Bel scale** is used. In this scale, the attenuation in bels (B) is defined to be the logarithm to base 10 of the ratio of the intensity (or power) of a signal to a reference level of the signal.

So

$$\text{attenuation in bel} = \log_{10} \frac{I}{I_0}$$

Note

This definition uses the intensity (or power) of the signal. The intensity I of the signal is related to its amplitude A by $I \propto A^2$. So

$$\text{attenuation} = 10 \log_{10} \frac{I}{I_0}$$

translates to

$$\text{attenuation} = 10 \log_{10} \frac{A^2}{A_0^2}$$

This can be re-written as

$$\text{attenuation} = 20 \log_{10} \frac{A}{A_0}$$

where I is the attenuated (output) power level of the signal and I_0 is the input intensity level (the subscript is a zero not a letter 'o'). This means that 10 B is equal to a power ratio $\left(\frac{I}{I_0}\right)$ of 10^{10}, which is a large ratio. As power ratios tend to be smaller than this, it is more usual to quote power ratios in **decibels**, where

$$\text{attenuation in decibel (dB)} = 10 \log_{10} \frac{I}{I_0}$$

A change in power of 10 dB is equivalent to a power ratio of 10 times. A change of 3 dB is equivalent to a change by a factor of 2 in the power.

The table shows some typical values for attenuations in optic fibres, twisted cable, and coaxial cable.

Link	Attenuation per unit length dB/100 m
Coaxial cable	15–300 for frequencies up to about 1 GHz depending on signal frequency and cable design
Twisted pair	5–50 for frequencies up to about 300 MHz
Optic fibre	0. 02 at 10^{14} Hz

Worked example

A signal of power 7.5 mW is input to a fibre that has an attenuation loss of 3.0 dB km^{-1}. The signal needs to be amplified when the power has fallen to 1.5×10^{-18} W. Calculate the distance required between amplifiers in the system.

Solution

Power loss $= 10 \log_{10}\left(\frac{7.5 \times 10^{-3}}{1.5 \times 10^{-18}}\right) = 157$ dB

The fibre loses 3 dB every kilometre so another amplifier will be required in $\frac{157}{3} = 52$ km.

The frequencies most used for fibre optics are in the range 0.8–1.5 μm. Attenuations in these ranges arise from two principal causes:

- **Absorption**
  This loss arises from the chemical composition of the glass and from any impurities that remain in it after manufacture. Glass absorbs some infra-red wavelengths strongly and these wavelengths have to be avoided for transmission.
- **Scattering**
  Rayleigh scattering (named after the British physicist Lord Rayleigh, who developed the ideas of image resolution) is the main scattering loss. It is caused by small variations in refractive index of the glass introduced during manufacture. Rayleigh scattering accounts for about 95% of the attenuation.

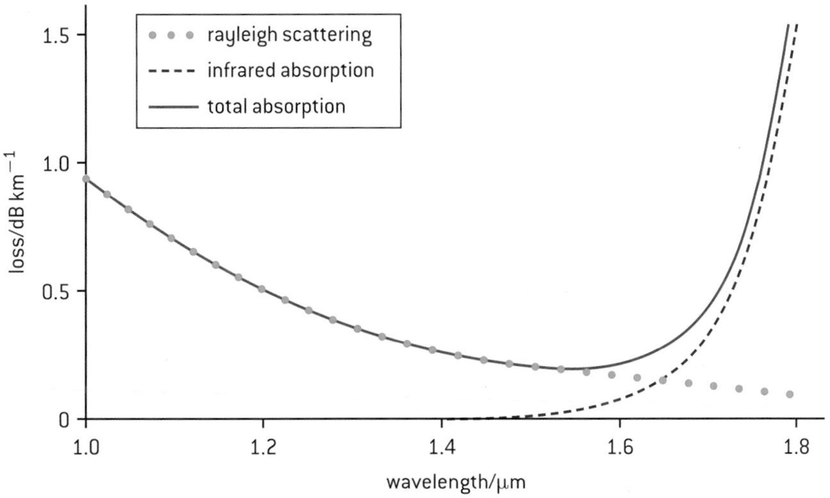

▲ Figure 4 Attenuation mechanisms in optic fibres.

TOK

Grace Hopper and her nanoseconds

The US computer scientist Grace Hopper was famous as one of the first people to construct a compiling language for computers. She was also famous for her lectures on computer science. To illustrate the problem of communication at high speed she would hand out lengths of wire 30 cm long. This is the distance travelled by light in a vacuum in one nanosecond. She used this to link the speeds and sizes of computers in the minds of her audience.

To what extent does the finite speed of light put limitations on our ability to carry out computations faster and faster?

▲ Figure 5

Light in the core interacts with the molecules of the glass. If the light continues in the general forward direction there is no attenuation, but if the light is scattered in directions other than from which it came, then attenuation occurs. This depends very strongly on the size of the small variations in density, etc. in the glass and also on the wavelength λ of the light. The loss due to Rayleigh scattering is proportional to the λ^{-4} and is greatest at short wavelengths.

The graph (figure 4) shows how the two effects of absorption and scattering contribute to the overall attenuation of a typical fibre for wavelengths greater than 1 um.

Nature of science

Why is the sky blue?

Rayleigh scattering is the reason why the sky is blue. Because the blue and violet light from the Sun have shorter wavelengths than the other colours, they are preferentially scattered out of the direct beam by the gas molecules in the air compared to longer wavelengths. We see the Sun as having a yellow disk with a blue sky. At sunset, the Sun's rays pass through a thicker atmosphere and the presence of particles in the air scatters longer wavelengths too, so the Sun and the clouds now appear red. In fact, this is a simplistic explanation of a complex phenomenon. If you want to know more, investigate both Rayleigh and Mie scattering.

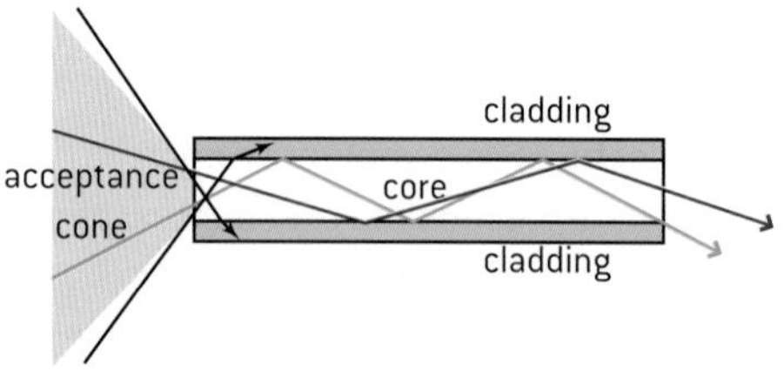

▲ Figure 6 Dispersion.

Two other effects contribute to distortions of the light as it passes through the fibre:

- **Material dispersion**
  This is similar to the chromatic aberration we met earlier in Sub-topic C.1. Both occur because the refractive index of the material used for the lens or fibre depends on wavelength. Refractive index is equal to the ratio of: $\frac{\text{wave speed in vacuun}}{\text{wave speed in transmitting medium}}$.

  The refractive index of glass decreases as the wavelength increases and so the wave speed in the glass also increases with increasing wavelength (red light travels faster than violet).

  Using light with a wide range of colours (a large bandwidth) means that the long wavelength light will reach the end of the fibre ahead of the shorter wavelengths leading to a spreading of the pulse. The answer is to restrict the wavelengths but this, at the same time, means that fewer wavelengths can be used and so there may be fewer channels available for communication.

- **Waveguide (or modal) dispersion**
  Even if monochromatic light is used, a large diameter step-index optic fibre will still be affected by waveguide dispersion. Rays of light that propagate along the fibre can travel by different routes depending on their initial angle of incidence at the end of the fibre.

  Compare a ray that travels along the central axis of the fibre with one that is at a large angle to the axis and reflects many times.

Although the scale of this drawing (figure 7) is unrealistic (because, to scale, it has far too large a diameter for any real optic fibre), you should see that the large-angle ray takes a much longer path than the central ray to travel from one end of the fibre to the other. The large-angle ray will arrive later and the time difference will appear as an increase in the pulse width.

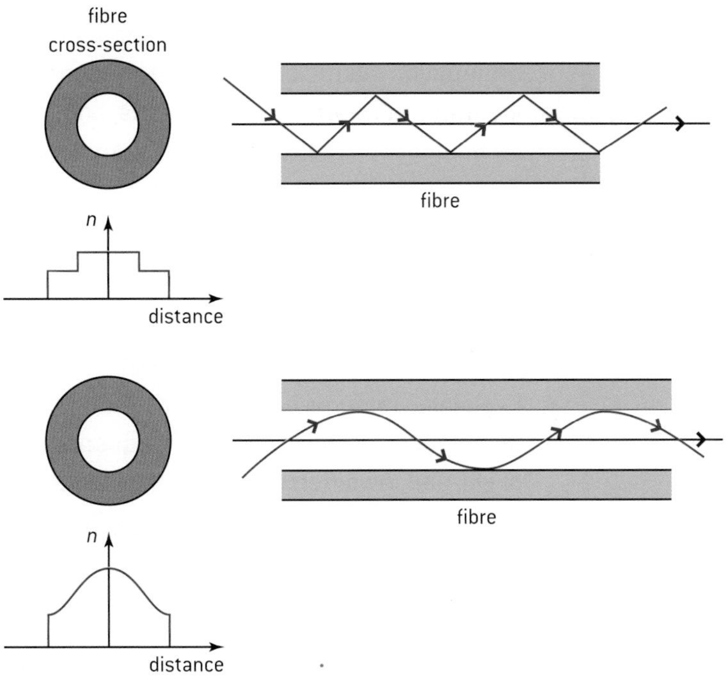

▲ Figure 7 Graded-index fibres.

To make a correction for this effect, graded-index fibres are used. As described earlier the refractive index of the core is not constant. It has a high index in the centre of the core and a decreasing index towards the core–cladding interface. The speed of the light is slowest in the centre with the speed increasing towards the cylinder wall. This means that large-angle rays travel more quickly than central axis rays during the periods when they are close to the wall. Combining this with a smaller overall diameter of core reduces (but does not eliminate) the waveguide dispersion effect.

C.4 Imaging the body (AHL)

Understanding

- Detection and recording of X-ray images in medical contexts
- Generation and detection of ultrasound in medical contexts
- Medical imaging techniques (magnetic resonance imaging) involving nuclear magnetic resonance (NMR)

Nature of science

Decisions made by a physician can involve an assessment of risk. The doctor will use some imaging techniques in the knowledge that the ionizing radiations involved can harm the patient. The real question is whether the techniques lead to a possible overall benefit.

Applications and skills

- Explaining features of X-ray imaging, including attenuation coefficient, half-value thickness, linear/mass absorption coefficients, and techniques for improvements of sharpness and contrast
- Solving X-ray attenuation problems
- Solving problems involving ultrasound acoustic impedance, speed of ultrasound through tissue and air, and relative intensity levels
- Explaining features of medical ultrasound techniques, including choice of frequency, use of gel, and the difference between A and B scans
- Explaining the use of gradient fields in NMR
- Explaining the origin of the relaxation of proton spin and consequent emission of signal in NMR
- Discussing the advantages and disadvantages of ultrasound and NMR scanning methods, including a simple assessment of risk in these medical procedures

Equations

- Attenuation (dB): $L_I = 10 \log \frac{I_1}{I_0}$
- Intensity: $I = I_0 e^{-\mu x}$
- Linear absorption: $\mu x_{1/2} = \ln 2$
- Acoustic impedence: $Z = \rho c$

Introduction

Medicine and physics combine in the world of medical diagnosis and treatment. Doctors have come to rely on technology from developments in physics. There has been a revolution in the methods that enable a doctor to visualize the interior of a patient's body.

X-ray images in medicine

Rontgen discovered X-rays in 1895 and, within one month of the publication of his original scientific paper, the radiation had been used for medical purposes. Nowadays, X-radiation is used both diagnostically and therapeutically. Here we concentrate on the use of X-rays to generate an image of the interior of the body that will inform doctors in making their diagnoses.

The basic principles behind the imaging are that X-rays are shone through part or all of the body. Materials such as bone absorb some of the X-rays or scatter them out of the beam, so that a reduced intensity emerges from the patient. Other, softer materials such as muscle and tissue do not absorb or scatter the rays so well. The X-rays are then incident on a photographic plate or a sensor. The plate is developed or a computer computes the data from the sensor, in both cases an image on film or monitor is produced. Where the intensity of the X-rays is high (not absorbed by the tissue), the plate is exposed (darkened) or the sensor detects the arrival of many photons. When significant absorption has occurred, fewer photons leave the body and the plate is not exposed to the same degree. As X-rays image are traditionally viewed as negatives, this means that the photograph is darkened where the rays are not absorbed and remains transparent (white) where the radiation has been absorbed by the body.

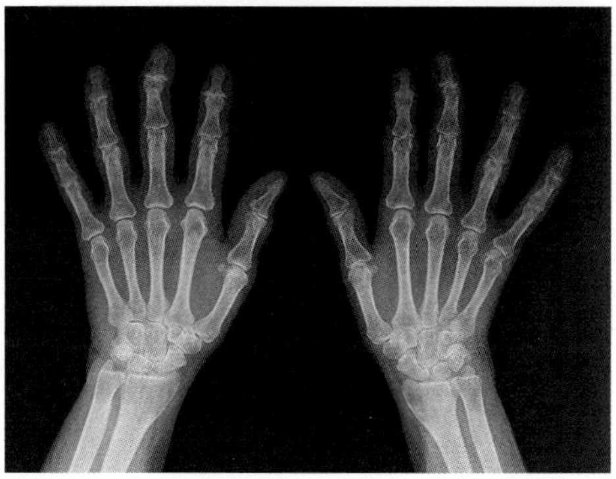
▲ Figure 1 X-ray images.

The X-rays themselves are produced when electrons travelling at high speed are decelerated in a heavy-metal target such as tungsten. The electrons are first accelerated in a vacuum through tens of thousands of volts in an electric field and then strike the target. On colliding with the target, electrons are slowed down rapidly and as a result lose energy to internal energy of the target (99% of the incident energy) and as energy in the form of X-ray photons (1%). Given the large amount of internal energy, the X-ray tubes need to be cooled and the anode target is often rotated to avoid hot spots developing on it. Details of the generation of the X-rays are not required for the examination.

X-rays are attenuated as they pass through material, and this means that some of the photons are removed. Others change direction (so that they can no longer be considered part of the beam). The principal mechanisms for removing photons or changing their direction are:

- **Coherent scattering**
  A process similar to the scattering mechanisms described in the previous sub-topic on fibre optics. It is the predominant mechanism for low-energy X-ray photons up to energies of about 30 keV.
- **Photoelectric effect**
  This is identical to the effect described in Topic 12. The incoming electron has enough energy to remove an inner-shell electron from an atom. When other electrons from higher energy states lose energy to occupy the inner shell, a photon of light is emitted. This mechanism is most important in the energy range 30–100 keV. The photoelectric scattering is particularly important as it provides contrast on the image between tissue and bone.
- **Compton scattering**
  In this mechanism, which occurs at energies generally greater than those used in diagnostic X-rays, the high-energy X-ray photon ejects an outer-shell electron from an atom and, as a result, a photon of lower energy moves off in a different direction to the original photon. This scattering mode is of principal importance in therapeutic X-ray medicine where X-rays of high energy are involved.
- **Pair production**
  As explained in Topic 12, at high energies in excess of 1 MeV, a photon can interact with a nucleus to produce an electron–positron pair.

Attenuation also occurs when the intensity of the X-ray beam decreases with distance from the X-ray tube as the beam diverges.

For the photons in a monochromatic beam of X-rays, the chance of an individual photon being scattered or absorbed is related to the probability of this photon interacting with an atom in the material being X-rayed. Suppose a photon has a 10% chance of removal by a particular thickness of material (figure 2).

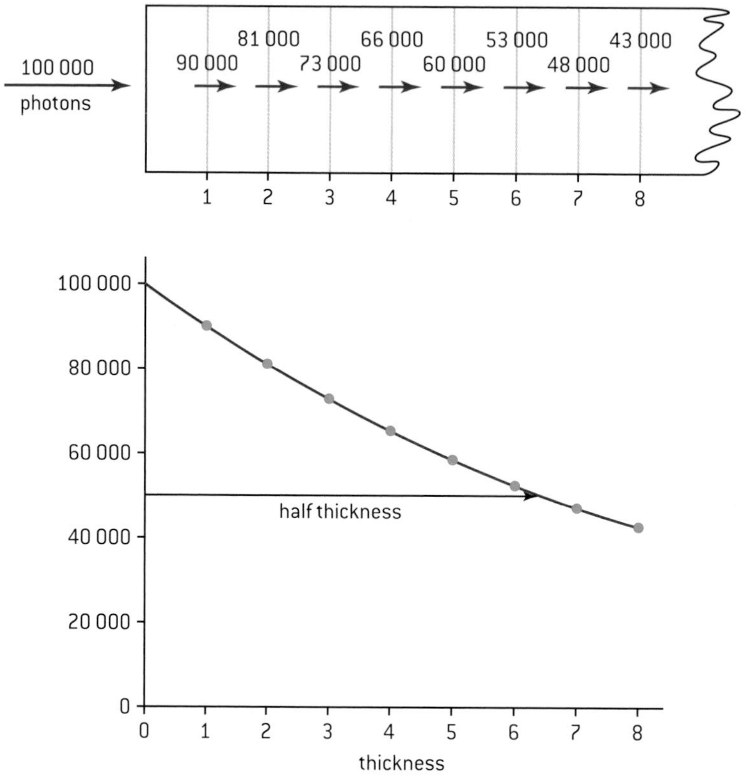

▲ Figure 2 Absorption effects lead to a half thickness.

Then if 100 000 photons are incident on material of this thickness, 90 000 (90% of 100 000) will remain in the beam when it leaves the material. If the material were twice as thick, then 81 000 (90% of 90 000) will remain after this further thickness. This is an identical argument to that used in radioactive decay (except that there the argument related the time taken for the ensemble of atoms to decay to the probability of decay for one atom). However, this similarity means that, like radioactivity, the intensity of the X-rays obeys an exponential relationship and we can state that:

$$I = I_0 e^{-\mu_l x}$$

where I is the intensity of the attenuated beam, I_0 is the intensity of the incident beam, x is the thickness of the material and μ_l is the **linear absorption coefficient** measured in units of metre^{-1}.

Although the linear absorption coefficient follows directly from the way the absorption probability is defined, it is not the most convenient absorption coefficient to use as it depends on the density of the absorbing material. As an example, compare water vapour (steam) with ice. Ice will scatter more effectively than steam because it has a greater density and therefore the X-ray photons are more likely to encounter a water molecule every centimetre of travel in ice than in steam. It is more convenient to quote a single **mass absorption coefficient** which is density independent and depends only on the element or compound that is absorbing the X-rays.

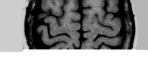

The relationship between the linear absorption coefficient μ_l and the mass absorption coefficient μ_m is

$$\mu_m = \frac{\mu_l}{\rho}$$

where ρ is the density of the material. The units of mass absorption coefficient are $m^2\ kg^{-1}$.

As a consequence, when the mass absorption coefficient is used, the appropriate intensity equation is

$$I = I_0\, e^{-\mu_m \rho x}$$

This equation is only strictly correct when the beam is monochromatic because values of the absorption coefficient vary with the energy of the X-ray photons. Highly penetrating radiation has a short wavelength (about 0.01 nm) and the absorption coefficients are small at these wavelengths; such radiation said to be **hard**. Long wavelength photons (about 1 nm) are termed **soft X-rays** and the absorption coefficients are much larger.

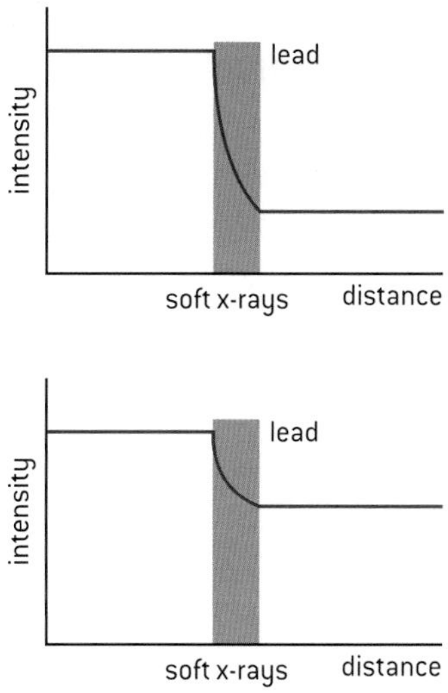

▲ Figure 3

In the same way that radioactive materials have a defined half-life, so for X-rays the thickness of material required to halve the intensity is called the half thickness $x_{\frac{1}{2}}$.

Again, following radioactive decay:

$$x_{1/2} = \frac{\ln 2}{\mu_l} \text{ or } x_{1/2} = \frac{\ln 2}{\rho \mu_m}$$

The value of μ_l may be determined from the gradient of a graph of ln I against x.

Frequently the beam will pass through a number of layers of different materials all with different thicknesses and absorption coefficients. This is easy to treat mathematically if the layers have parallel plane interfaces.

Suppose that layer 1 has a thickness x' and a linear absorption coefficient μ_l' and that layer 2 has a thickness x'' and a linear absorption coefficient μ_l''.

Then the intensity when the beam leaves layer 1, I_1 is

$$I_1 = I_0\, e^{-\mu_l' x'}$$

and the intensity when the beam leaves layer 2, I_2 is

$$I_2 = I_1\, e^{-\mu_l'' x''}$$

Therefore

$$I_2 = I_0\, e^{-\mu_l' x'}\, e^{-\mu_l'' x''}$$

or

$$I_2 = I_0\, e^{-(\mu_l' x' + \mu_l'' x'')}$$

For each layer add the product of the linear absorption coefficient and the layer thickness and then use the exponential function to find the final intensity of the beam.

When the mass absorption coefficients are known the equation becomes:

$$I_2 = I_0\, em^{-(\rho' \mu_m' x' + \rho'' \mu_m'' x'')}$$

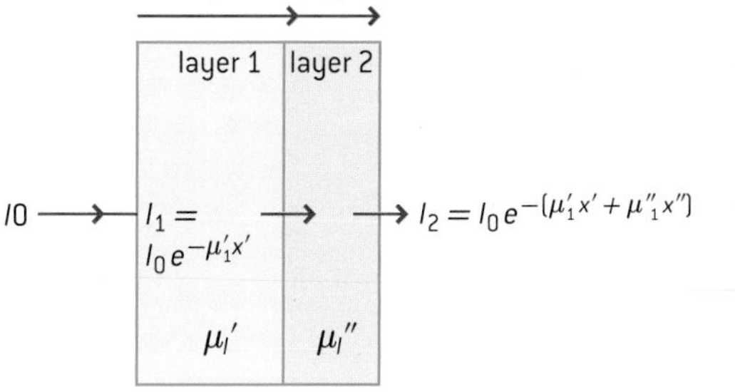

▲ Figure 4

Worked examples

1 A sample of a metal of density 4000 kg m^{-3} has a half thickness of 12 mm with a particular wavelength of radiation. Calculate, for the metal, the:

a) linear absorption coefficient

b) mass absorption coefficient.

Solution

a) $\mu_l = \frac{\ln 2}{x_{1/2}} = 58 \text{ m}^{-1}$

b) $\mu_m = \frac{\mu_l}{\rho} = \frac{58}{4000} = 1.4 \times 10^{-2} \text{ m}^2 \text{ kg}^{-1}$

2 Calculate the thickness of muscle tissue that will reduce the intensity of a certain X-radiation by a factor of 10^3, assuming that the linear absorption coefficient for muscle is 0.035 cm^{-1}.

Solution

$$I_1 = I_0 e^{-\mu_l x}$$

So $0.001 = e^{-0.035 \times x}$

$\ln 0.001 = -0.035 \times x$

$$x = \frac{6.9}{0.035} = 197 \text{ cm}$$

About 2 m of muscle tissue is required to reduce the intensity by 1000×.

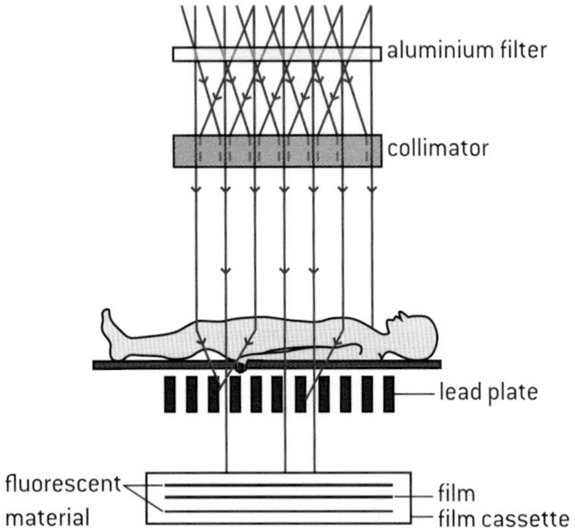

▲ Figure 5 Contrast techniques in X-ray imaging.

Improving the image

Collimation

Figure 5 shows a common arrangement used to take diagnostic X-ray images. A number of features are used to enhance both the sharpness and contrast of the image formed on the photographic plate:

- The X-ray beam is filtered by passing it through a thin plate of aluminium. This selectively removes low-energy photons because the absorption coefficient is much greater at low energies. These low-energy photons would simply be absorbed by the patient (adding to the radiation dose) and would serve no useful imaging purpose. The plate also reduces the intensity of the beam somewhat, but overall the penetrating power of the beam increases and the X-rays become harder.
- The beam as it leaves the tube is very divergent. It is collimated by passing through a series of channels in lead plates. Off-axis rays are absorbed by the lead and do not reach the patient. This produces a narrower, more parallel beam. This is beneficial because photons scattered from the off-axis angles tend to blur the photographic image.
- As for the X-rays that reach the patient: some do not interact, some are absorbed and disappear from the system, and others are scattered. Again, the scattering leads to blurring on the image. A grid system of lead plates below the patient, arranged parallel to the beam, is used to absorb the scattered rays but allows the on-axis rays to reach the imaging system.
- The X-rays finally arrive at the film cassette and this contains two fluorescent screens, one each side of the film. As the X-rays interact with these screens they cause the materials in the screens to fluoresce and emit light that improves the blackening of the photographic negative. Many of the X-rays would not contribute to the image without this arrangement.

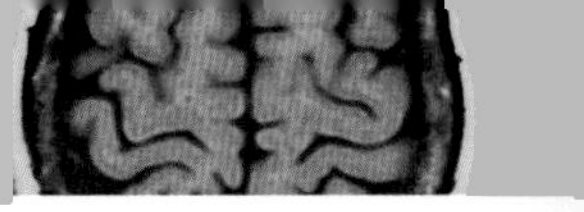

Contrast

Some tissues in the body are hard to distinguish on the photographs without enhancements to the contrast. Heavy elements with large absorption coefficients can be introduced into the body to help with this. Barium and bismuth are examples of elements used. Patients with stomach disorders may be asked to drink barium sulfate ($BaSO_4$), which coats the lining of the stomach and makes its outline very clear on the image. Iodine can be introduced intravenously to produce clear images of the cardiovascular system.

Other ways to enhance detection include using charge-coupled sensors in place of film together with computer systems to form electronic images viewed on monitors.

The advantage of X-ray imaging is that it is a quick and inexpensive technique, costing far less than an MRI or CT scan. The X-ray machines can be highly portable, meaning that patients may not have to travel to a central location for a scan.

However, unlike ultrasound and MRI techniques, X-rays involve ionizing radiation and this presents a risk to both the patient and the radiographer. Techniques to reduce the X-ray intensity and exposure time are constantly being improved to reduce the risks. Nevertheless this exposure needs to be kept in perspective. The average chest or dental X-ray provides a much smaller dose of radiation to the body than a commercial intercontinental flight at 12 km above the Earth.

Computed tomography

Computer (computerized) tomography (CT scan), also known as computed axial tomography (CAT scan) was introduced during the 1970s. It gives a much greater range of grey scales to the image and provides an axial scan – an image of a slice through the patient.

X-rays are incident on the area of the patient under investigation and the scattered and direct photons are detected by a series of small detectors placed in an arc around the patient. After the first exposure the X-ray tube and the detectors move around the patient taking exposure after exposure until an entire range covering 360° around the body has been made (figure 6).

The detection method uses scintillation counters that respond to the presence of a photon by emitting a flash of light, which is then detected by a photomultiplier that produces an enhanced electric current for each flash that occurs.

A computer builds up the information from each detector and each exposure to produce a complete image of the slice of the patient up to a few tens of millimetres thick. If further slices are required, the patient can be moved under computer control to a new position and the process is repeated.

The sensitivity of the system is very high, and many features can be seen and interpreted by doctors. A disadvantage of the CT scan, however, is that the cumulative dose to the patient is high, and there is a greater risk of damage to the patient than with a normal X-ray. However, exposures from the scanners are dropping all the time as improved detection techniques are developed.

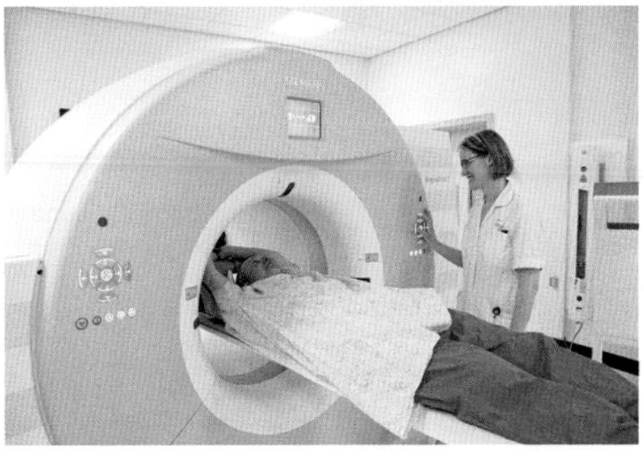

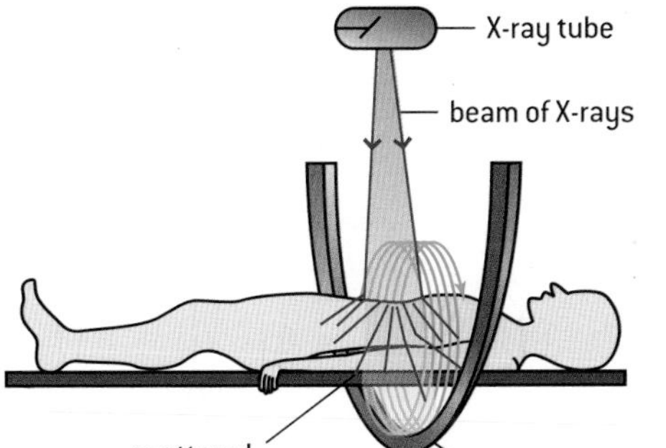

▲ Figure 6 CT scanning.

Ultrasound in medicine

Generation

Ultrasound is a sound wave generated at a frequency above that at which humans hear. The lower frequency limit of ultrasound is taken to be 20 kHz. The range of frequencies used in medicine is from 2 to 20 MHz.

Like electromagnetic radiation, ultrasound waves can be absorbed by matter, and reflected and refracted when they meet a boundary between two media. The usual rules for reflection and refraction of light at an interface also apply to ultrasound.

Brothers Pierre and Paul-Jacques Curie first observed the piezoelectric effect that is used to generate the ultrasound. They found that when a quartz crystal is deformed, it produces a small emf between opposite faces of the crystal. Conversely, applying a potential difference across the crystal causes it to deform and, under the right circumstances, to vibrate at high frequencies. This property of piezoelectricity was found to be shared by other materials besides quartz including some synthetic ceramic materials. These newer materials are now used in preference to quartz for producing ultrasound.

To make the scan a piezoelectric transducer (which can both emit and receive the ultrasound) is placed in contact with the skin. A gel is used between the transducer and skin to prevent a large loss of energy that would occur at the transducer–air and air–skin interfaces. A single pulse of the ultrasound is transmitted into the tissues and the system then waits for the reflected wave (an echo) to return from each interface inside the patient. There are a number of ways to display the scan but the simplest display (known as an A-scan) is to show a graph of reflected signal strength against time. Figure 7(a) shows a typical A-scan with the various echo returns labelled. A knowledge of the speed of the ultrasound and the time for return enables the size and location of an organ to be determined (remembering that, like all echoes, the ultrasound has to travel to the interface and return).

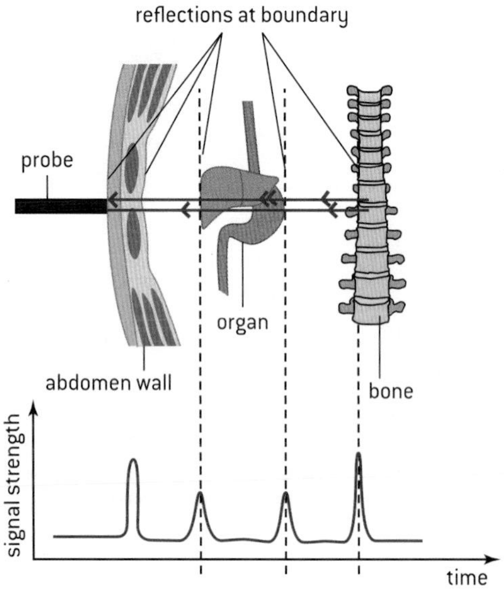

a) A scan

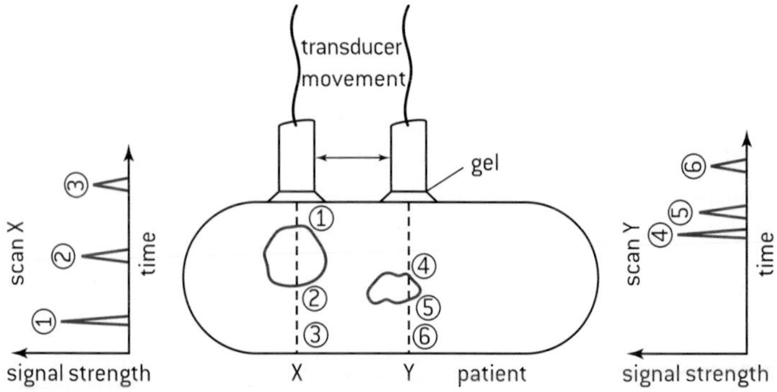

b) B scan

▲ Figure 7 A and B scans.

A more complex type of scan is known as the B-scan (figure 7(b)). This requires a computer to combine a series of scans produced by the transducer. The operator rocks the transducer backwards and forwards to illuminate the internal surfaces and the computer builds up a slice image (in a similar way to the CT scan) through the patient by combining the signals from a whole series of A-scans.

The transmitted ultrasound that passes through the patient is not used in ultrasound diagnostic techniques, so it is of major importance to minimize the absorption of the ultrasound by tissues and to maximize the energy in the returned echo.

To compare the ultrasound performance of different types of material and tissue we use the acoustic impedance of the medium. This is a number that indicates how easy it is to transmit ultrasound through a particular material. The **acoustic impedance Z** of a material is found to depend on the speed of sound c in the material and the density ρ of the material:

$$Z = \rho c$$

The SI unit for acoustic impedance is kg m^{-2} s^{-1} ; this is sometimes abbreviated to the non-SI unit the rayl (named for Lord Rayleigh) but we will use the SI unit here.

The table gives some typical values for speeds, densities and Z values for various tissues, though it should be remembered that c and therefore Z depend on the ultrasound frequency being used.

Medium	Speed / m s^{-1}	Density / kg m^{-3}	Acoustic impedance / 10^6 kg m^{-2} s^{-1}
Bone	4100	1900	7.8
Soft tissue	1500	1050	1.6
Liver	1550	1070	1.7
Muscle	1600	1080	1.7
Water	1480	1000	1.5
Air at 15°C	340	1.21	4.1×10^{-4}

The proportion of the incident wave energy that is reflected at an interface depends on differences between the acoustic impedances of the two media. It can be shown that the ratio of the initial intensity I_0 to the reflected intensity I_r is:

$$\frac{I_r}{I_0} = \frac{(Z_2 - Z_1)^2}{(Z_2 + Z_1)^2}$$

where Z_1 and Z_2 are the acoustic impedances of the first material and the second material respectively.

The resolution of the image is of great importance in ultrasound imaging. The obvious approach might be to aim for the highest frequency possible as this will give the shortest wavelength and the best resolution. However, the attenuation of the wave also increases markedly with frequency as does the resolution itself (typically about 2 mm) if the frequency is taken beyond a certain maximum. Generally the doctor has to accept a compromise, but frequencies in the range 2–5 MHz are used for most applications.

Not all medical ultrasound use leads to an image in the conventional sense. There are many other diagnostic and therapeutic uses for ultrasound. These include the detection of blood flow and measurement of its speed using Doppler shifts, and recent innovations where microbubbles of gas are introduced to enhance the image of blood vessels (in a similar way to the injection of contrast enhancers in X-radiography).

The advantages of ultrasound include:

- It is a non-invasive technique.
- It is relatively quick and inexpensive.
- There are no known harmful effects.
- It is of particular value in imaging soft tissues.

However:

- Image resolution can be limited.
- Ultrasound does not transmit through bone.
- Ultrasound cannot image the lungs and the digestive system as these contain gas which strongly reflects at the interface with tissue.

Nuclear magnetic resonance (NMR) in medicine

Magnetic resonance imaging (MRI scans) are medical diagnostic tools that use the phenomenon of nuclear magnetic resonance (NMR).

In use, a patient is placed in a strong uniform magnetic field produced by an electromagnet. Other magnetic fields around the patient are varied and signals emitted from the tissues are detected, measured and transformed into an image using a computer. The images produced have good resolution and good contrast for parts of the body that contain large proportions of water (and hence hydrogen nuclei). The resolution for NMR techniques depends on the resonance frequency of the signal, so the resolution is proportional to the magnetic field strength. At the time of writing, a typical resolution for NMR imaging is 2 mm. MRI is a technique preferred for diagnosis of brain and central nervous system disorders.

An explanation of MRI comes in two parts: the basic NMR effect itself and then a description of the way it is modified for medical use.

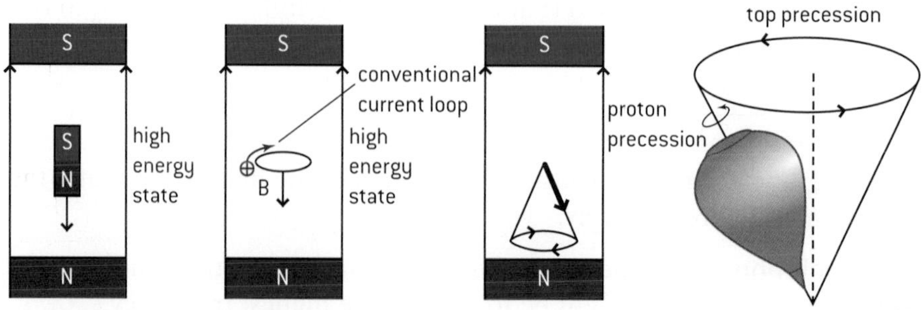

▲ Figure 8 Larmor precession.

The basic NMR effect

- Protons possess the properties of charge and spin and this leads them to behave as small magnets. The situation is analogous to charges moving around a circular loop of wire producing a magnetic field that acts through the centre of the loop.

- Under normal conditions, hydrogen-rich materials such as water have the proton spins arranged randomly to give no overall magnetic field. However, when a strong magnetic field is imposed on the material the spins of the protons, and thus their magnetic fields, line up with the imposed magnetic field.
- Normally the protons line up in the lowest energy state, but if a radio-frequency (rf) field of an appropriate frequency is applied to the system, some of the protons will flip into the opposite high energy state. This is analogous to a bar magnet flipping into a state where the north-seeking pole of the bar magnet is next to the north-seeking pole of the field source (see figure 8).
- The appropriate frequency to cause the flip is known as the **Larmor frequency** and, crucially for MRI, is directly proportional to the strength of the magnetic field in which the material is placed. The Larmor frequency in Hz is $4.26 \times 10^7\ B$ where B is the magnetic field strength in T.
- What happens to the spinning protons when the rf is applied can be thought of in terms of a spinning child's top. A top precesses around its spin axis in response to the Earth's gravitational field. The protons can be thought of as similarly precessing at the Larmor frequency.
- As the protons precess, the changing direction of the magnetic field that they produce induces an emf in a coil of wire nearby.
- When the rf field is switched off, the protons in the high-energy state relax (return) to the low-energy state.

MRI modifications

- The process is essentially the same as the basic phenomenon. The patient is placed in a strong uniform magnetic field. However an additional **gradient field** is added to the strong field.

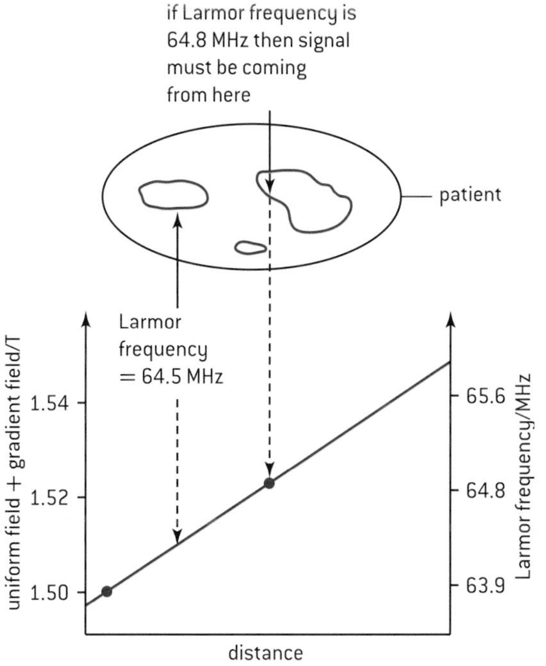

▲ Figure 9 A specific Larmor frequency comes from a unique slice in the patient.

TOK

Henry Thoreau said "It's not what you look at that matters, it's what you see." To what extent do you agree with this comment in the context of imaging inside the body?

- The gradient field is designed so that the total field (uniform + gradient) varies linearly across the patient. The Larmor frequency itself depends linearly on B and therefore the Larmor frequency varies predictably with position in the body.
- The rf field is switched on and proton precession occurs. The relaxation of the protons after the rf field has been removed leads to an electromagnetic signal that contains information about the number of protons emitting at each Larmor frequency. This information is recovered from the signal allowing a computer to plot the information spatially on an image.

MRi produces excellent images for diagnostic purpose without exposing the patient to radiation (whether from X-rays or radio-isotopes). The energy of the photons in MRI are well below the 1 eV levels that correspond to molecular bonds and also are below the energies of X-ray photons. However, MRi is not entirely without risk. Factors involved in the risk include:

- ***Strong magnetic fields***. Some patients are unsuitable for an MRI scan if they have, for example, a knee or hip joint replacement that would distort the magnetic field or a heart pacemaker that could be severely affected by currents induced in it when the strong magnetic fields change. There is no known risk from a strong magnetic field by itself.
- ***Radio-frequency (rf) fields***. These can lead to local heating in the tissues of the patient.
- ***Noise***. The large changes in magnetic field strength within the scanner cause parts of it to attempt to change shape during the scan. This can give rise to high intensities of sound that patients can find disturbing.
- ***Claustrophobia***. A strong uniform field is difficult to produce and can only be maintained over a small volume of space. This means that, typically, the patient is scanned while in a small-diameter tunnel, which some people find uncomfortable.

Additionally, there may be elements of discomfort for some patients in that a scan can take up to 90 minutes to complete; lying still in a confined space may prove difficult for some. In particular, young children and babies need to be sedated as they cannot understand the need to remain still.

Questions

1 Four rays of light from O are incident on a thin concave (diverging) lens. The *focal points* of the lens are labelled F. The lens is represented by the straight line XY.

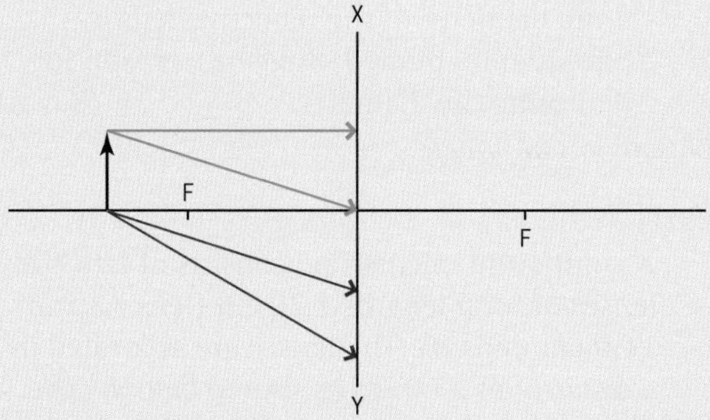

a) Define the term *focal point*.

b) On a copy of the diagram:

(i) complete the four rays to locate the position of the image formed by the lens

(ii) show where the eye must be placed in order to view the image.

c) State and explain whether the image is real **or** virtual.

d) The focal length of the lens is 50.0 cm. Calculate the linear magnification of an object placed 75.0 cm from the lens.

e) Half of the lens is now covered such that only rays on one side of the principal axis are incident on the lens. Describe the effects, if any, that this will have on the linear magnification and the appearance of the image. (14 marks)

2 *(IB)*

a) The diagram shows a small object O represented by an arrow placed in front of a converging lens. The focal points of the lens are labelled F.

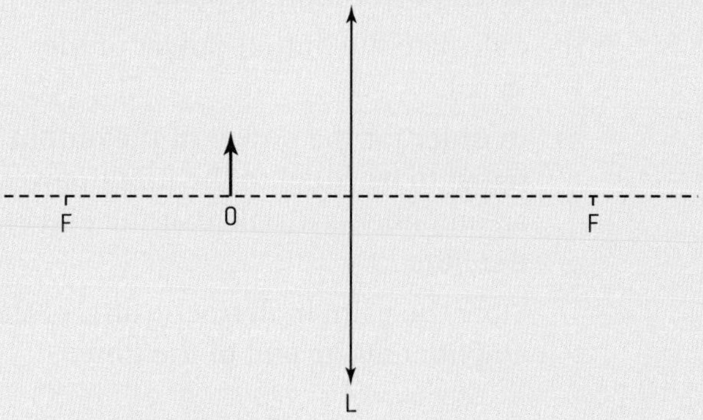

(i) On a copy of the diagram, draw rays to locate the position of the image of the object formed by the lens.

(ii) Explain whether the image is real or virtual.

b) A convex lens of focal length 62.5 cm is used to view an insect of length 8.0 mm that is crawling on a table. The lens is held 50 mm above the table.

(i) Calculate the distance of the image from the lens.

(ii) Calculate the length of the image of the insect. (8 marks)

3 *(IB)*

a) A parallel beam of light is incident on a convex lens of focal length 18 cm. The light is focused at point X as shown below.

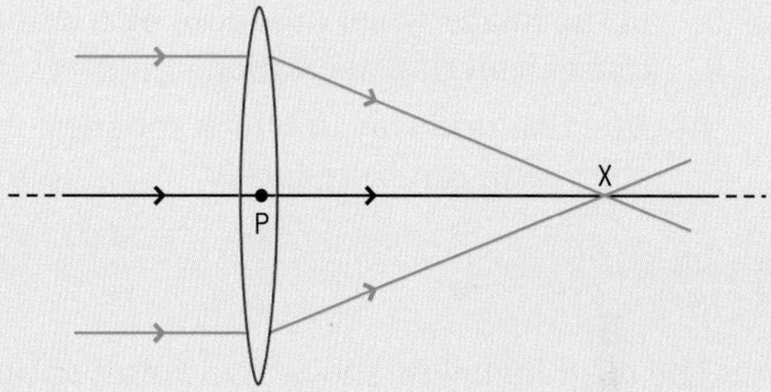

State the value of the distance PX.

b) A diverging lens of focal length 24 cm is now placed 12 cm from the convex lens as shown below.

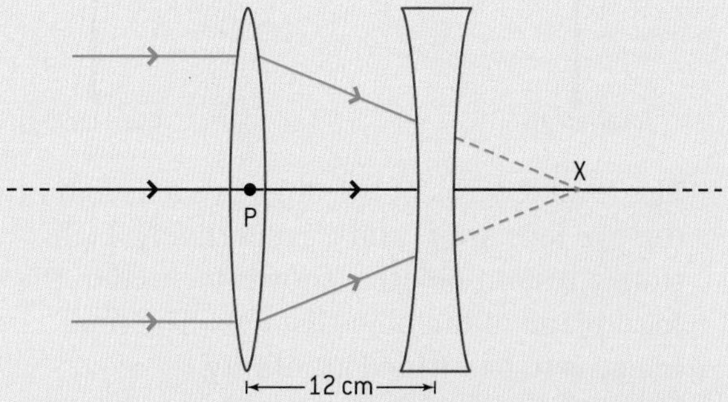

c) (i) Explain why point X acts as a virtual object for the diverging lens.

(ii) Calculate the position of the image as produced by the diverging lens.

d) A lens combination, such as a diverging and a convex lens, is referred to as a telephoto lens. Suggest why a telephoto lens is considered to have a longer focal length than that of a single convex lens. (7 marks)

4 *(IB)*

The diagram below shows the image of a square grid produced by a lens that does not cause spherical aberration.

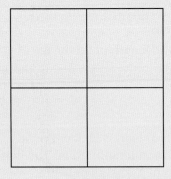

a) On a copy of the diagram, draw the shape of the image when produced by a lens that causes spherical aberration.

b) Describe **one** way in which spherical aberration can be reduced. (4 marks)

5 *(IB)*

The diagram below shows two lenses arranged so as to form an astronomical telescope. The two lenses are represented as straight lines.

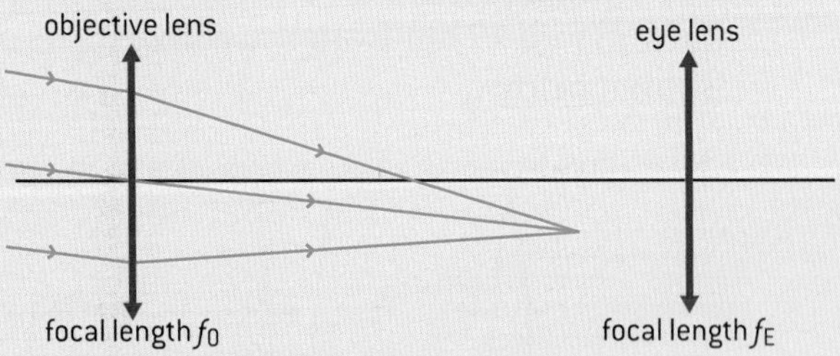

The focal lengths of the objective lens and of the eye lens are f_O and f_E respectively. Light from a distant object is shown focused in the focal plane of the objective lens. The final image is to be formed at infinity.

a) Complete the ray diagram to show the formation of the final image.

b) (i) State what is meant by *angular magnification*.

(ii) Using the completed ray diagram above, derive an expression in terms of f_O and f_E for the angular magnification of an astronomical telescope. Assume that the final image is at infinity.

c) When specifying an astronomical telescope, the diameter of the objective lens is frequently quoted. Suggest a reason for quoting the diameter. (8 marks)

6 *(IB)*

A compound microscope consists of two convex lenses of focal lengths 1.20 cm (lens A) and 11.0 cm (lens B). The lenses are separated by a distance of 23.0 cm as shown below. (The diagram is not drawn to scale.)

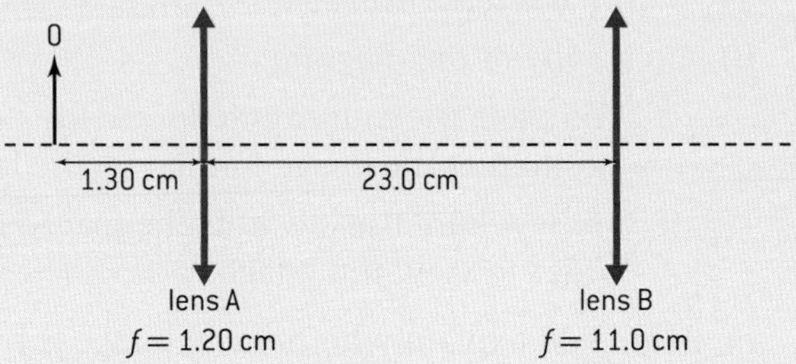

An object O is placed 1.30 cm from lens A. An image of O is formed 15.6 cm from A.

a) This image forms an object for lens B. Calculate the object distance for lens B.

b) Calculate the distance from lens B of the image as produced by lens B.

c) Calculate the magnification of the microscope. (5 marks)

7 *(IB)*

a) State **one** cause of attenuation and **one** cause of dispersion in an optical fibre.

b) An optical fibre of length 5.4 km has an attenuation per unit length of 2.8 dB km^{-1}. The signal power input is 80 mW.

(i) Calculate the output power of the signal.

(ii) In order for the power of the output signal to be equal to the input power, an amplifier is installed at the end of the fibre.

State the gain, in decibels (dB), of the amplifier at the end of the fibre.

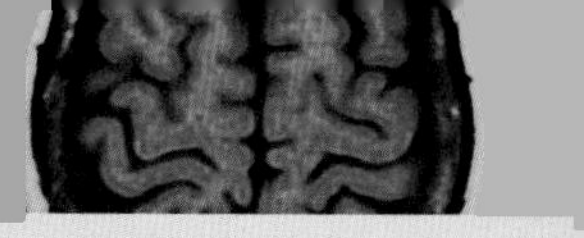

c) The signal to noise ratio (SNR), in dB, is defined as

$\text{SNR} = 10 \log \frac{P_{signal}}{P_{noise}}$ where P_{signal} and P_{noise} are the powers of the signal and noise respectively.

The SNR of the signal in (b) before amplification was 20 dB. Calculate the SNR after amplification. (7 marks)

8 *(IB)*

The variation with time t of the input power to an optic fibre is shown in diagram 1.

The variation with time t of the output power from the optic fibre is shown in diagram 2.

diagram 1

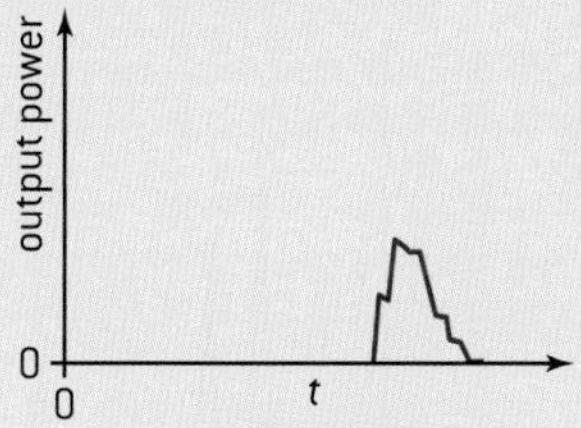

diagram 2

The scales are the same on both diagrams.

a) State and explain the feature of the graphs that shows that there is:

(i) attenuation of the signal

(ii) signal noise.

b) The duration (time width) of the signal increases as it travels along the optic fibre.

(i) State **two** reasons for this increased time duration.

(ii) Suggest why this increase in the width of the pulse sets a limit on the frequency of pulses that can be transmitted along an uninterrupted length of optic fibre. (7 marks)

9 *(IB)*

a) State what is meant by *X-ray quality*.

A parallel beam of X-rays of intensity I_0 is incident on a material of thickness x as shown below. The intensity of the emergent beam is I.

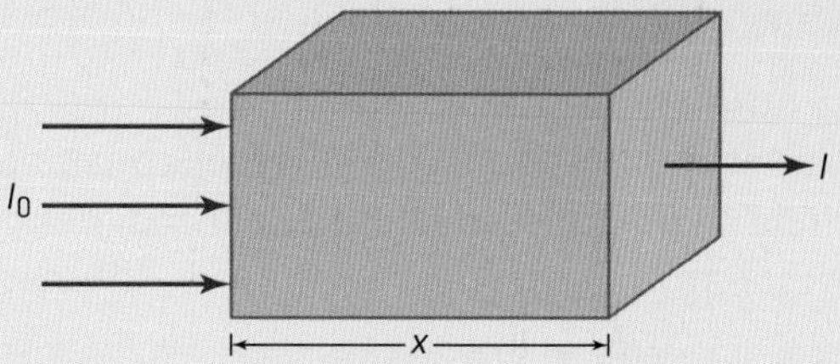

b) Define *half-value thickness*.

c) Draw a sketch-graph to show how intensity I varies with distance x.

d) Annotate your sketch-graph to show the half-value thickness $x_{1/2}$.

e) State the name of one of the mechanisms responsible for the attenuation of diagnostic X-rays in matter. (6 marks)

10 a) State and explain **one** situation, in each case, where the following diagnostic techniques would be used.

(i) X-rays

(ii) Ultrasound

(iii) Nuclear magnetic resonance

b) Apart from health hazards, explain why different means of diagnosis are needed. (6 marks)

11 Beam energies of about 30 keV are used for diagnostic X-rays. This results in good contrast on the radiogram because the most important attenuation mechanism is not simple scattering.

a) Outline the most important attenuation mechanism that is taking place at this energy.

b) Explain what is meant by:

(i) *attenuation coefficient*

(ii) *half-value thickness.*

c) The attenuation coefficient at 30 keV varies with the atomic number Z as

attenuation coefficient $\propto Z^3$

The data given below list average values of the atomic number Z for different biological materials.

Biological material	Atomic number Z
fat	5.9
muscle	7.4
bone	13.9

(i) Calculate the ratio:

$$\frac{\text{attenuation coefficient for bone}}{\text{attenuation coefficient for muscle}}$$

(ii) Suggest why X-rays of 30 keV energy are useful for diagnosing a broken bone, but a different technique must be used for examining a fat-muscle boundary.

(13 marks)

12 *(IB)*

a) State and explain which imaging technique is normally used:

(i) to detect a broken bone

(ii) to examine the growth of a fetus.

The graph below shows the variation of the intensity I of a parallel beam of X-rays after it has been transmitted through a thickness x of lead.

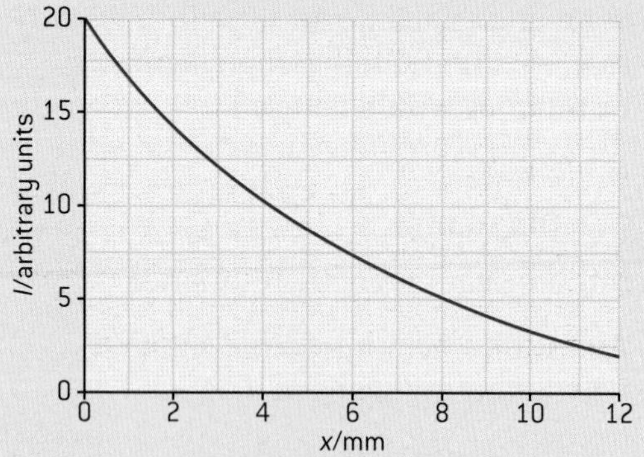

b) (i) Use the graph to estimate the half-value thickness $x_{1/2}$ for this beam in lead.

(ii) Determine the thickness of lead required to reduce the intensity transmitted to 20% of its initial value.

(iii) A second metal has a half-value thickness $x_{1/2}$ for this radiation of 8 mm. Calculate what thickness of this metal is required to reduce the intensity of the transmitted beam by 80%.

(11 marks)

13 The attenuation of X-rays depends not only on the nature of the material through which they travel but also on the photon energy. For photons with energy of about 30 keV, the *half-value* thickness of muscle is about 50 mm and for photons of energy 5 keV, it is about 10 mm.

Explain which photon energy would be most suitable for obtaining a sharp picture of a broken leg. (2 marks)

14 a) State a typical value for the frequency of ultrasound used in medical scanning.

b) The diagram below shows an ultrasound transmitter and receiver placed in contact with the skin.

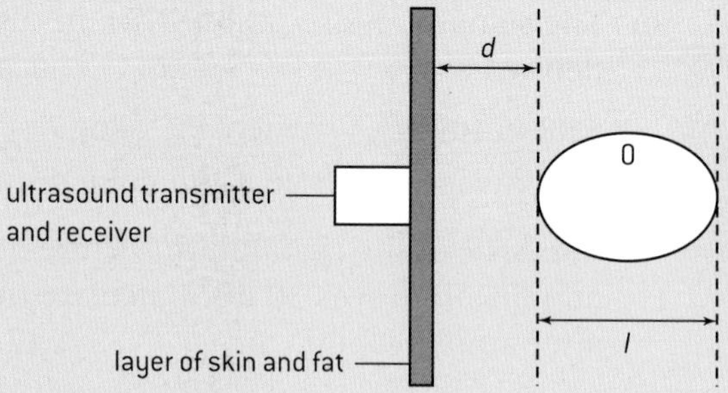

The scan is to estimate the depth d of the organ labelled O and also to find its length, l.

The pulse strength of reflected pulses is plotted against time t where t is the time elapsed between the pulse being transmitted and the time that the pulse is received.

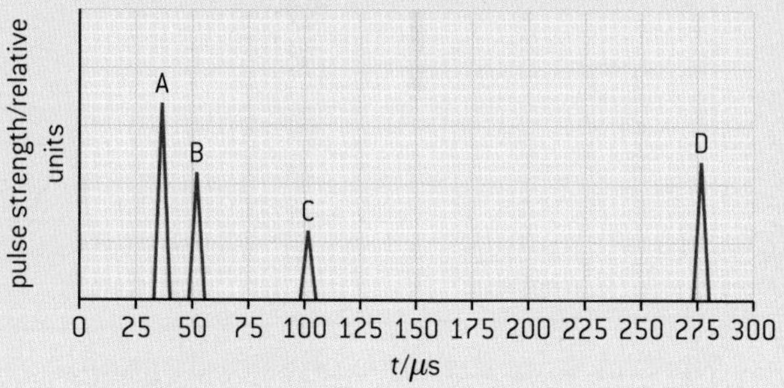

(i) Identify the origin of the reflected pulses A, B and C and D.

(ii) The mean speed in tissue and muscle of the ultrasound used in this scan is 1.5×10^3 m s^{-1}. Using data from the above graph, estimate the depth d of the organ beneath the skin and the length l of the organ O.

c) The above scan is known as an A-scan. State **one** way in which a B-scan differs from an A-scan.

d) State **one** advantage and **one** disadvantage of using ultrasound as opposed to using X-rays in medical diagnosis. (10 marks)

15 State and explain the use of:

a) a barium meal in X-ray diagnosis

b) a gel on the skin during ultrasound imaging

c) a non-uniform magnetic field superimposed on a much larger constant field in diagnosis using nuclear magnetic resonance. (7 marks)

D ASTROPHYSICS

Introduction

Astrophysics probes some of the most fundamental questions that humanity has sought to answer since the dawn of civilization. It links the experimental discipline of astronomy with the theoretical understanding of everything in the universe – cosmology. It offers insights into the universe and provides answers, albeit tentatively, to its size, age, and content. Astrophysicists can theorize on the life cycles of stars and gain an appreciation of how the universe looked at the dawn of time. The weakest of the four fundamental forces, gravity, comes into its own on an astronomic scale. It provides the mechanism to attach planets to stars, stars to other stars (in galaxies), and galaxies to other galaxies (in clusters and super clusters).

D.1 Stellar quantities

Understandings

- → Objects in the universe
- → The nature of stars
- → Astronomical distances
- → Stellar parallax and its limitations
- → Luminosity and apparent brightness

Nature of science

When we look upwards, away from the Earth, on a dark clear night we see many points of light in the sky. Some of these, such as the planets, artificial satellites, and aircraft, are visible because they are reflecting light from the Sun. Others, such as stars and galaxies, are visible because of the light that they emit. The light from distant stars and galaxies has travelled truly astronomical distances to reach us and we are able to construct a historical account of space from analysing this. However, the light from the different sources will have travelled varying distances to reach us and have been emitted at a range of times. This is analogous to looking at a family photograph containing a mixture of many generations of a family at different stages in their lives. It is a credit to humanity that we are able to conjecture so much about space when we can only justify our reasoning with circumstantial evidence. In this way astrophysics mirrors forensic science – we can never obtain evidence at the actual time an event happens.

Applications and skills

- → Identifying objects in the universe
- → Qualitatively describing the equilibrium between pressure and gravitation in stars
- → Using the astronomical unit (AU), light year (ly) and parsec (pc)
- → Describing the method to determine distance to stars through stellar parallax
- → Solving problems involving luminosity, apparent brightness, and distance

Equations

- → parsec definition: $d\ (\text{parsec}) = \dfrac{1}{p\ (\text{arc-second})}$
- → luminosity equation: $L = \sigma A T^4$
- → apparent brightness equation: $b = \dfrac{L}{4\pi d^2}$

Introduction

In this first sub-topic we rapidly move outwards from our solar system to the remainder of our galaxy and beyond. In doing this we briefly examine some of the objects that make up the universe. We then consider the range of units that we use for astronomical measurements. We end by considering how we can estimate the distance and luminosity of relatively near stars by treating them as black-body radiators and using their apparent brightness.

Objects that make up the universe

The solar system

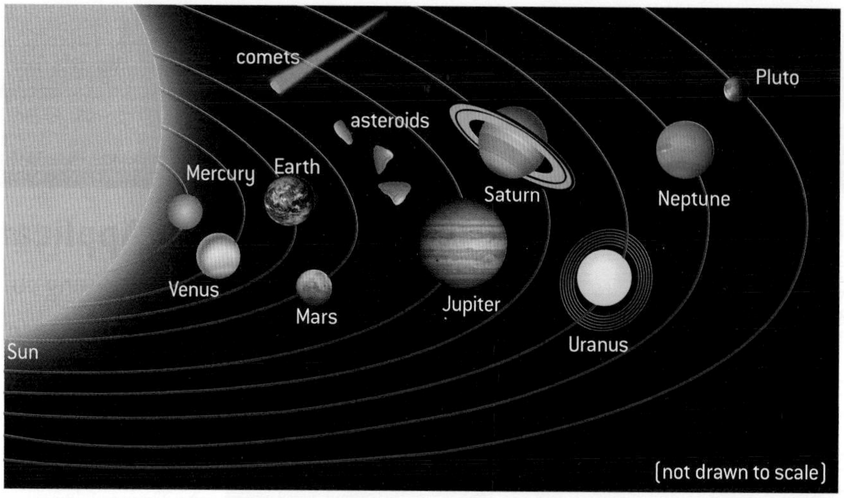

▲ Figure 1 The solar system.

The solar system is a collection of planets, moons, asteroids, comets, and other rocky objects travelling in elliptical orbits around the Sun under the influence of its gravity. The Sun is a star formed, we believe, from a giant cloud of molecular hydrogen gas that gravitated together, forming clumps of matter that collapsed and heated up. A gas disc around the young, spinning Sun evolved into the planets. It is thought that the planets were formed about 4.6×10^9 years ago (see figure 1).

The high temperature close to the Sun permitted only those compounds with high condensation temperatures to remain solid, gradually accreting (sticking together) particles to form the four terrestrial planets: Mercury, Venus, Earth, and Mars. Further away from the Sun the "Gas Giants" or "jovian" planets comprising of Jupiter, Saturn, Uranus, and Neptune were formed from cores of rock and metal and an abundance of ice. The huge quantity of ice meant that these planets became very large and produced strong gravitational fields that captured the slow moving hydrogen and helium. Pluto used to be called "the ninth planet" but, in 2006, it was downgraded to a "dwarf planet". The planets move in elliptical orbits round the Sun with only Mercury occupying a plane significantly different to that of the other planets. Further out from the Sun, beyond Neptune, is the Kuiper belt. This is similar to the asteroid belt but much larger; it is the source of short-period comets and contains dwarf planets (including Pluto). The Kuiper belt is set to be the next frontier of exploration in our solar system.

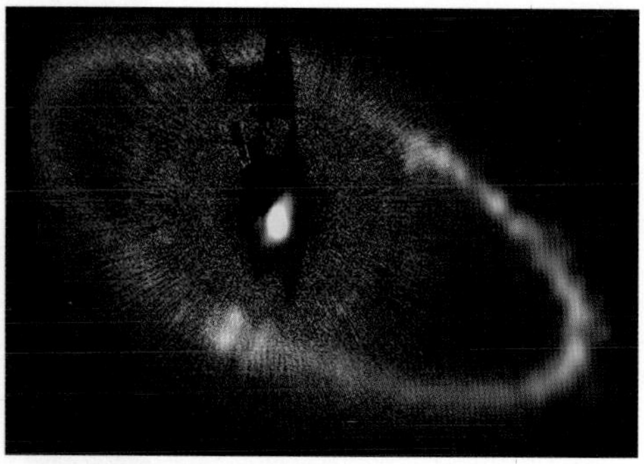

▲ Figure 2 Thermal emission from the young star Fomalhaut and the debris disc surrounding it.

Six of the planets have moons orbiting them: Mars has two moons, while Jupiter has at least 50 acknowledged moons with several provisional ones that might be asteroids captured by its gravitational field. About 4.5 billion years ago, the Earth's moon is believed to have been formed from material ejected when a collision occurred between a Mars-size object and the Earth.

Asteroids are rocky objects orbiting the Sun – with millions of them contained in solar orbit in the asteroid belt situated between Mars and Jupiter. Those of size less than 300 km have irregular shape because their gravity is too weak to compress them into spheres. Some of the asteroids, such as Ceres with a diameter of about 10^6 m, are large enough to be considered as "minor planets".

Comets are irregular objects a few kilometres across comprising frozen gases, rock, and dust. Observable comets travel around the Sun in sharply elliptical orbits with periods ranging from a few years to thousands of years. As they draw near to the Sun the gases in the comet are vaporized, forming the distinctive comet tail that can be millions of kilometres long and always points away from the Sun.

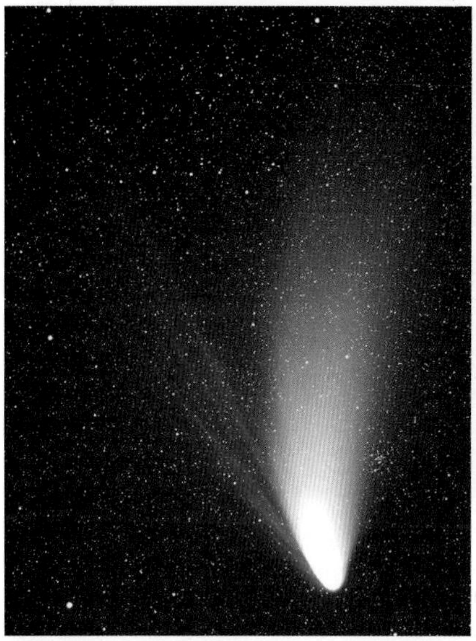

▲ Figure 3 Comet Hale-Bopp.

Stars

Like the Sun, all stars initially form when gravity causes the gas in a nebula to condense. As the atoms move towards one another, they lose gravitational potential energy that is converted into kinetic energy. This raises the temperature of the atoms which then form a protostar. When the mass of the protostar is large enough, the temperature and pressure at the centre will be sufficient for hydrogen to fuse into helium, with the release of very large amounts of energy – the star has "ignited". Ignition produces emission of radiation from the core, producing a radiation pressure that opposes the inward gravitational forces. When this is balanced the star is in a state of hydrostatic equilibrium and will remain stable for up to billions of years because it is on the "main sequence". As the hydrogen is used up the star will eventually undergo changes that will move it from the main sequence. During these changes the colour of the star alters as its surface temperature rises or falls and it will change size accordingly. The original mass of material in the star determines how the star will change during its lifetime.

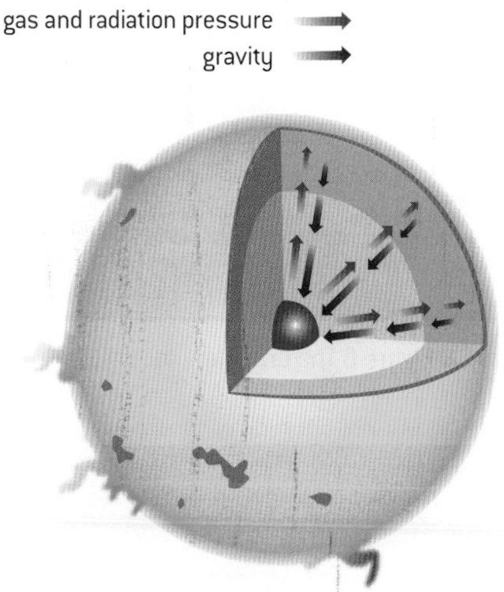

▲ Figure 4 Hydrostatic equilibrium.

Groups of stars

Despite the difficulties in assessing whether stars exist singly or in groups of two or more, it is thought that around fifty per cent of the stars nearest to the Sun are part of a star system comprising two or more stars. **Binary stars** consist of two stars that rotate about a common centre of mass. They are important in astrophysics because their interactions allow us to measure properties that we have no other way of investigating. For example, careful measurement of the motion of the stars in a binary system allows their masses to be estimated.

A **stellar cluster** is a group of stars that are positioned closely enough to be held together by gravity. Some clusters contain only a few dozen stars while others may contain millions. All of the stars in a star cluster were formed at the same time from the same nebula. The Pleiades (figure 5) is a stellar cluster of about 500 stars that can be seen with the naked eye;

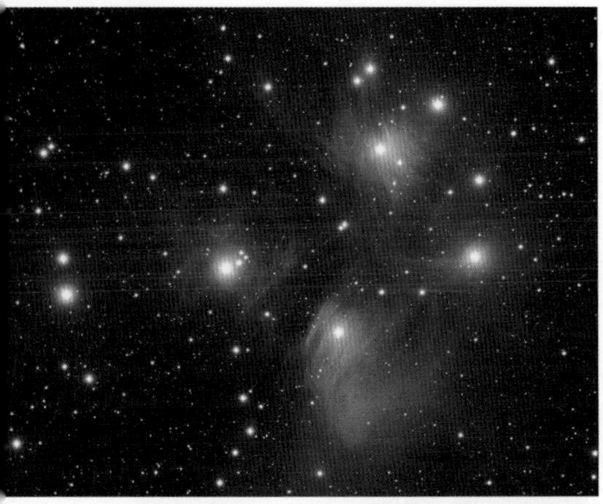

▲ Figure 5 The Pleiades.

▲ Figure 6 The Crab Nebula.

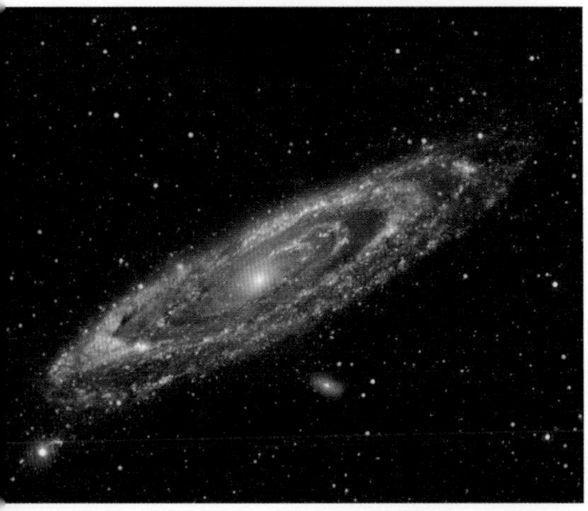

▲ Figure 7 The Andromeda galaxy with two smaller satellite galaxies.

this is an example of an **open cluster**. Open clusters consist of up to several hundred stars that are younger than ten billion years and may still contain some gas and dust. They are located within our galaxy, the Milky Way, and so lie within a single plane. **Globular clusters** contain many more stars and are older than eleven billion years and, therefore, contain very little gas and dust. There are 150 known globular clusters lying just outside the Milky Way in its galactic halo. Globular clusters are essentially spherically shaped.

A **constellation**, unlike a stellar cluster, is a pattern formed by stars that are in the same general direction when viewed from the Earth. They are more significant historically than physically as many ancient societies attributed them with religious importance. Today, the patterns made by the constellations are helpful to astronomers in locating areas of the sky for telescopic study. Naturally, some of the stars in a constellation are much closer to the Earth than others. Because of proper motion they will appear very different in, say, ten thousand years time. Such stars are not held together by gravity.

Nebulae

Regions of intergalactic cloud of dust and gas are called *nebulae* (singular is *nebula*). As all stars are "born" out of nebulae, these regions are known as stellar nurseries. There are two different origins of nebulae. The first origin of nebulae occurred in the "matter era" around 380 000 years after the Big Bang. Dust and gas clouds were formed when nuclei captured electrons electrostatically and produced the hydrogen atoms that gravitated together. The second origin of nebulae is from the matter which has been ejected from a supernova explosion. The Crab Nebula, shown in figure 6, is a remnant of such a supernova. Other nebulae can form in the final, red giant, stage of a low mass star such as the Sun.

Galaxies

A galaxy is a creation of stars, gas, and dust held together by gravity and containing billions of stars. The Milky Way contains about 3×10^{11} stars and, probably, at least this number of planets. Some galaxies exist in isolation but the majority of them occur in groups known as clusters that have anything from a few dozen to a few thousand members. The Milky Way is part of a cluster of about 30 galaxies called the "Local Group" which includes Andromeda (figure 7) and Triangulum. Regular clusters consist of a concentrated core and are spherical in shape. Irregular clusters also exist, with no apparent shape and a lower concentration of galaxies within them. Since the launch of the Hubble Space Telescope it has been observed that even larger structures, called superclusters, form a network of sheets and filaments; approximately 90% of galaxies can be found within these. In between the clusters there are voids that are apparently empty of galaxies.

The Milky Way and Andromeda are members of the most common class of galaxies – spiral galaxies. These are characterized by having a disc-shape with spiral arms spreading out from a central galactic bulge that contains the greatest density of stars. It is increasingly speculated that, at the centre of the galactic bulge, there is a black hole. The spiral arms contain many young blue stars and a great deal of dust and gas. Other galaxies are elliptical in shape, being ovoid or spherical – these contain much less gas and dust than spiral galaxies; they are thought to have been formed from collisions between spiral galaxies. Irregular galaxies are shapeless and may have been stretched by the presence of other massive galaxies – the Milky Way appears to be having this effect on some nearby dwarf galaxies.

Astronomical distances

Resulting from the huge distances involved in astronomical measurements, some unique, non-SI units have been developed. This avoids using large powers of ten and allows astrophysicists to gain a feel for relative sizes and distances.

The light year (ly): The speed of light is one of the most fundamental constants in physics; all inertial observers measure light as travelling at the same speed (see Option A for a background to this). This property can be used to define a set of units based on the speed of light. For example, the distance travelled by light in one minute is called a *light minute*. As it takes light approximately 8 minutes to travel from the Sun to the Earth, the distance between them is 8 light minutes. The light year (ly) is a more commonly used unit and is the distance travelled by light in one year.

$$1\text{ ly} = 9.46 \times 10^{15}\text{m}$$

The astronomical unit (AU): This is the average distance between the Sun and the Earth. It is really only useful when dealing with the distances of planets from the Sun.

$$1\text{ AU} = 1.50 \times 10^{11}\text{ m} \approx 8\text{ light minutes}$$

The parsec (pc): This is the most commonly used unit of distance in astrophysics.

$$1\text{ pc} = 3.26\text{ ly} = 3.09 \times 10^{16}\text{ m}$$

Distances between nearby stars are measured in pc, while distances between distant stars within a galaxy will be in kiloparsecs (kpc), and those between galaxies in megaparsecs (Mpc) or gigaparsecs (Gpc).

The distances used in astronomy are truly enormous, and this has meant that a variety of indirect methods have been developed for their measurement. The method used to measure the distance of an astronomical object from the Earth is dependent on its proximity.

Stellar parallax

On Earth, surveyors measure distances by using the method of triangulation. A known or measured length is taken as a baseline. The angles that a distant object makes at either end of the baseline are then measured (using a theodolite). From these angles the distance of the object can be calculated. When measuring the distance of a star from the Earth (in parsec) a similar technique is employed. Parallax is based on the fact that nearby objects will appear to cross distant objects when viewed from different positions. This can be seen from inside a moving car when fence posts by the roadside appear to speed by but distant hills hardly move. As the Earth orbits the Sun, the stars that are quite close to us appear to move across the distant "fixed" stars as shown in figure 8.

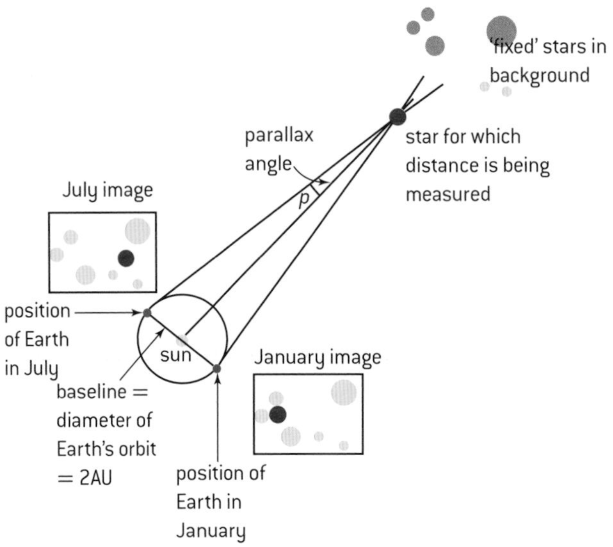

▲ Figure 8 Stellar parallax.

When the image of a nearby star is recorded in both January and July the star appears to have moved across the fixed stars in the background. Using the equation

$$d = \frac{1}{p}$$

gives the distance d in parsecs when p is the parallax angle in arcsecond.

This simple relationship is used for defining the parsec: when a star is at a distance of 1 pc from the Earth the parallax angle given by the equation will be one arcsecond.

In a circle there are 360 degrees. In every degree there are 60 arcminutes and 60 arcseconds in every arcminute. Thus 1 arcsecond is very small being just $\frac{1}{3600}$ of a degree.

There is a limit to the distance that can be measured using stellar parallax – parallax angles of less than 0.01 arcsecond are difficult to measure from the surface of the Earth because of the absorption and scattering of light by the atmosphere. Turbulence in the atmosphere also limits the resolution because it causes stars to "twinkle". Using the parallax equation, gives a maximum range of $d = \frac{1}{0.01} = 100$ pc.

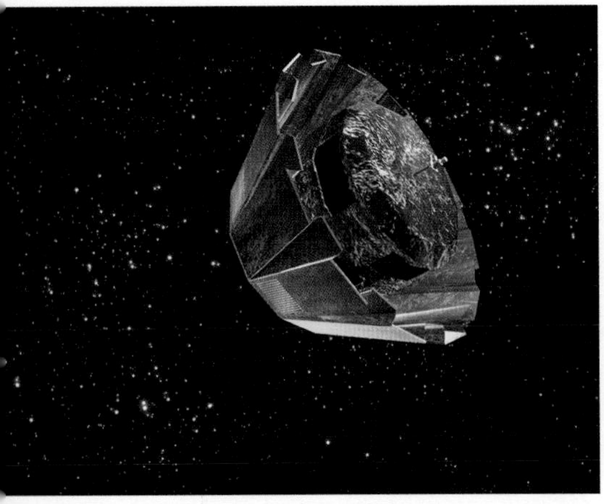

▲ Figure 9 The Gaia mission.

In 1989, the satellite Hipparcos (an acronym for High Precision Parallax Collecting Satellite) was launched by the European Space Agency (ESA). Being outside the atmosphere, Hipparcos was able to measure the parallaxes of 118 000 stars with an accuracy of 0.001 arcsecsond (to distances 1000 pc); its mission was completed in 1993. Gaia (figure 9), Hipparcos's successor, was launched in 2013 and is charged with the task of producing an accurate three-dimensional map showing the positions of about a billion stars in the Milky Way. This is about one per cent of the total number of stars in the galaxy! Gaia is able to resolve a parallax angle of 10 microarcsecond measuring stars at a distance of 100 000 pc. Amateur astronomers taking images will be working at approximately 1 arcsecond per pixel.

Luminosity and apparent brightness of stars

In Sub-topic 4.3 we looked at the intensity of a wave at a point distance r from the source. Intensity was defined as the power emitted by a source divided by the area of the sphere over which the energy is spread equally.

$$I = \frac{P}{4\pi r^2}$$

For stars P is called the luminosity (L) of the star and represents the total energy emitted by the star per second in watt. I in this context is the apparent brightness (b) and is measured in watts per square metre (W m^{-2}); the distance from the star is usually denoted by d. This version of the equation is written as:

$$b = \frac{L}{4\pi d^2}$$

The luminosity of a star is a very important quantity in establishing the nature of a star. Hence, the equation is particularly useful because, although we are not in a position to measure the luminosity of the star, we can measure its apparent brightness (for example, by using a telescope and a charge-coupled device). If we also calculate the distance of the star we can then work out its luminosity. Alternatively, when we know the luminosity of a star (because it is similar to other stars of known luminosity) we are able to use the measurement of its apparent brightness to estimate the star's distance.

Worked example

a) For a star, state the meaning of the following terms:

(i) *Luminosity*

(ii) *Apparent brightness.*

b) The spectrum and temperature of a certain star are used to determine its luminosity to be approximately 6.0×10^{31} W. The apparent brightness of the star is 1.9×10^{-9} Wm^{-2}. These data can be used to determine the distance of the star from Earth. Calculate the distance of the star from Earth in parsec.

c) Distances to some stars can be measured by using the method of stellar parallax.

(i) Outline this method.

(ii) Modern techniques enable the measurement from Earth's surface of stellar parallax angles as small as 5.0×10^{-3} arcsecond. Calculate the maximum distance that can be measured using the method of stellar parallax.

Solution

a) (i) The luminosity is the total power emitted by the star.

(ii) The apparent brightness is the incident power per unit area received at the surface of the Earth.

b) Using $b = \frac{L}{4\pi d^2}$ we rearrange to get

$$d = \sqrt{\frac{L}{4\pi b}} = \sqrt{\frac{6.0 \times 10^{31}}{4\pi \times 1.9 \times 10^{-9}}} = 5.0 \times 10^{19}\ \text{m}$$

As 1 pc = 3.26 ly and 1 ly = 9.46×10^{15} m then 1 pc = $3.26 \times 9.46 \times 10^{15}$ m

$= 3.08 \times 10^{16}$ m.

5.0 $\times 10^{19}$ m is $\frac{5.0 \times 10^{19}}{3.08 \times 10^{16}}$ or 1623 pc $\approx$1600 pc

c) (i) The angular position of the star against the background of fixed stars is measured at six month intervals. The distance d is then found using the relationship $d = \frac{1}{p}$ (this is shown in figure 6.)

(ii) $d = \frac{1}{5 \times 10^{-3}} = 200$ pc.

Worked example

Some data for the variable star Betelgeuse are given below.

Average apparent brightness $= 1.6 \times 10^{-7}$ Wm^{-2}

Radius = 790 solar radii

Earth–Betelgeuse separation = 138 pc

The luminosity of the Sun is 3.8×10^{26} W and it has a surface temperature of 5800 K.

a) Calculate the distance between the Earth and Betelgeuse in metres.

b) Determine, in terms of the luminosity of the Sun, the luminosity of Betelgeuse.

c) Calculate the surface temperature of Betelgeuse.

Solution

a) As 1 pc $= 3.1 \times 10^{16}$ m,

$138 \text{ pc} = 138 \times 3.1 \times 10^{16}$

$= 4.3 \times 10^{18}$ m

b) $b = \frac{L}{4\pi d^2} \therefore L = 4\pi d^2 b$

$= 4\pi[4.3 \times 10^{18}]^2 \times 1.6 \times 10^{-7}$

$= 3.7 \times 10^{31}$ W

Dividing by the luminosity of the Sun gives

$\frac{3.7 \times 10^{31}}{3.8 \times 10^{26}} = 9.7 \times 10^4.$

So Betelgeuse has a luminosity of $9.7 \times 10^4\ L_{Sun}$

c) As $L = \sigma 4\pi R^2 T^4$, by taking ratios we get

$$\frac{L_{Sun}}{L_{Betelgueuse}} = \frac{\sigma 4\pi R_{Sun}{}^2 T_{Sun}{}^4}{\sigma 4\pi R_{Betelgeuse}{}^2 T_{Betelgeuse}{}^4}$$

$$\frac{T_{Betelgeuse}}{T_{Sun}} = \sqrt[4]{\frac{L_{Betelgeuse}\, R_{Sun}{}^2}{L_{Sun}\, R_{Betelgeuse}{}^2}}$$

$$= \sqrt[4]{\frac{9.8 \times 10^4}{790^2}} = 0.63$$

$$T_{Betelgeuse} = 0.63 \times 5800 \text{ K}$$

$$= 3700 \text{ K}$$

Black-body radiation and stars

In Sub-topic 8.2 we considered black bodies as theoretical objects that absorb all the radiation that is incident upon them. Because there is no reflection or re-emission they appear completely black – as their name suggests. Such bodies would also behave as perfect emitters of radiation, emitting the maximum amount of radiation possible at their temperature. All objects at temperatures above absolute zero emit black-body radiation. This type of radiation consists of every wavelength possible but containing different amounts of energy at each wavelength for a particular temperature. Although stars are not perfect black-bodies they are capable of emitting and absorbing all wavelengths of electromagnetic radiation. Figure 10 shows the black-body radiation curves for the Sun, the very hot star Spica, and the cold star Antares. Because each of the stars will produce different intensities, the curves have been normalized by dividing the intensity emitted at a given wavelength by the maximum intensity that the star yields – this means the vertical scale has no unit and the maximum value it can take is 1.00. The maximum intensity of radiation emitted by the Sun has a wavelength of just over 500 nm making it appear yellow; the peak intensity for Spica is in the ultraviolet region, but there is sufficient intensity in the blue region for it to appear blue; the peak for Antares is in the near infra-red but, with plenty of red light emitted, it appears to be red to the naked eye.

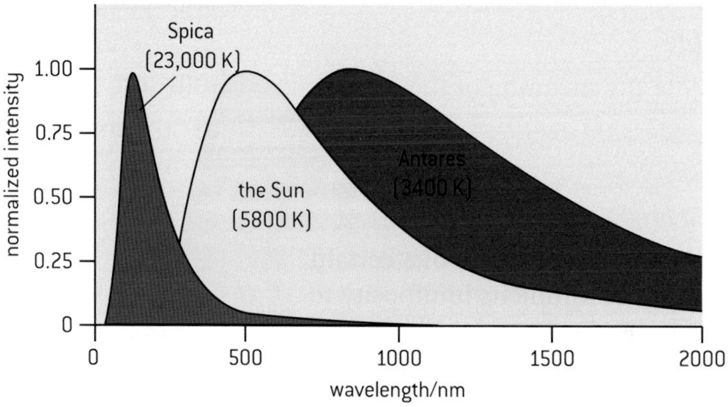

▲ Figure 10 Black-body radiation curves for three stars.

For a star the Stefan–Boltzmann law is written as

$$L = \sigma A T^4$$

where L is the luminosity in watt, A the surface area of the star in square metres, and T the temperature in kelvin. σ is the Stefan–Boltzmann constant $= 5.67 \times 10^{-8}$ Wm^{-2} K^{-4}.

When we assume that a star is spherical we can use this equation in the form:

$$L = \sigma 4\pi R^2 T^4$$

where R is the radius of the star.

We can see that the luminosity of a star depends on its temperature and its size (measured here by its surface area). In the next sub-topic we will see how the balance between temperature and surface star size is used to categorize star type.

D.2 Stellar characteristics and stellar evolution

Understandings

- → Stellar spectra
- → Hertzsprung–Russell (HR) diagram
- → Mass–luminosity relation for main sequence stars
- → Cepheid variables
- → Stellar evolution on HR diagrams
- → Red giants, white dwarfs, neutron stars, and black holes
- → Chandrasekhar and Oppenheimer–Volkoff limits

Nature of science

In 1859, the physicist Gustav Kirchhoff and chemist Robert Bunsen worked at the University of Heidelberg in Germany. Having developed the first spectroscope, they repeated Foucault's experiment of passing sunlight through a bright sodium flame to find the absorption lines seen in the solar spectrum. They then repeated their experiment with other alkali metals and were able to show that the solar absorption spectra were the reverse of emission spectra. Kirchhoff went on to provide strong evidence for the presence of iron, magnesium, sodium, nickel, and chromium in the atmosphere of the Sun. Their experiments paved the way for our understanding of many of the properties of stars.

Applications and skills

- → Explaining how surface temperature may be obtained from a star's spectrum
- → Explaining how the chemical composition of a star may be determined from the star's spectrum
- → Sketching and interpreting HR diagrams
- → Identifying the main regions of the HR diagram and describing the main properties of stars in these regions
- → Applying the mass–luminosity relation
- → Describing the reason for the variation of Cepheid variables
- → Determining distance using data on Cepheid variables
- → Sketching and interpreting evolutionary paths of stars on an HR diagram
- → Describing the evolution of stars off the main sequence
- → Describing the role of mass in stellar evolution

Equations

- → Wien's law: $\lambda_{max} T = 2.9 \times 10^{-3}$ mK
- → luminosity-mass relationship: $L \propto M^{3.5}$

Introduction

The magnitude of the distance between stars is unimaginable with the nearest star (ignoring the Sun) being so distant that, even with the fastest rocket ever built, it would take almost a hundred thousand years for us to reach it. The curiosity of the human race knows no bounds and we have striven to find out all we can about stars – their life cycles and their distances from the Earth. In this sub-topic we investigate how it is possible to estimate the distance of stars more than 100 pc from the Earth, and explore the lives of stars of different mass, radius, temperature, and colour.

Stellar spectra

In Sub-topic 7.1 we considered emission and absorption spectra occurring as electrons make transitions between energy levels. Such spectra provide important information about the chemical composition, density, surface temperature, rotational and translational velocities of stars. When we observe the spectrum of stars using a spectrometer we find that nearly all stars show a continuous spectrum which is crossed by dark absorption lines; some stars also show bright emission lines. Seeing absorption lines across a continuous stellar spectrum tells us that the stars have a hot dense region (to produce the continuous spectrum) surrounded by cooler, low-density gas (to produce the absorption lines). In general, the density and temperature of a star decreases with distance from its centre. Because its temperature is so high, a star's core has to be composed of high-pressure gases and not of molten rock, unlike the cores of some planets.

Composition of stars

Absorption of certain wavelengths is apparent when we observe the intensity–wavelength relationship for stars. The smooth theoretical black-body curve is modified by absorption dips as can be seen in figure 1. The graph shows the variation with wavelength of the intensity for the Sun and Vega (which is much hotter than the Sun having a surface temperature of 9600 K). In the case of Vega, the cooler hydrogen in the star's outer layers (the photosphere) absorbs the photons emitted by hydrogen. The clear pattern between the wavelengths absorbed and those of the visible part of the hydrogen absorption spectrum shown in figure 2 strongly suggests that Vega's photosphere is almost entirely made up of hydrogen.

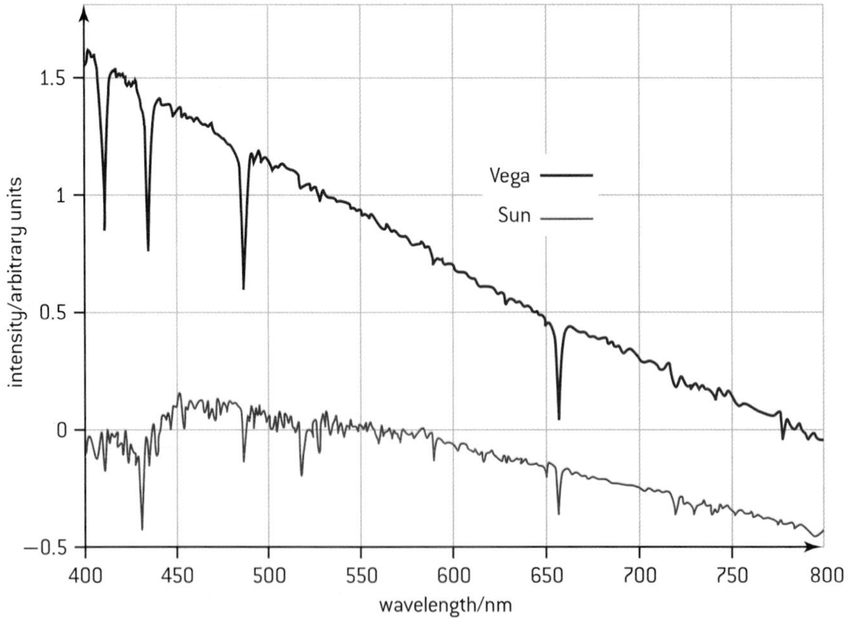

▲ Figure 1 Intensity–wavelength relation for Vega and the Sun.

The absorption lines correspond to those of hydrogen shown in the spectrum of figure 2.

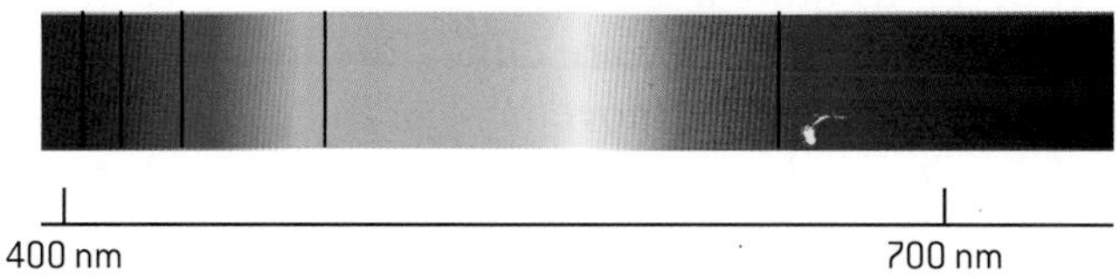
hydrogen absorption spectrum
400 nm
700 nm

▲ Figure 2 Atomic hydrogen absorption spectrum.

It also implies that the transitions occur in agreement with the Balmer series from level $n = 2$ to higher levels. On the other hand, the Sun has some of the visible hydrogen lines. Because the Sun is cooler than Vega, many of its hydrogen atoms are in the ground state producing absorption lines that are in the ultraviolet region of the spectrum (corresponding to transitions from level $n = 1$ in the Lyman series). Stars that are even hotter than Vega tend not to produce hydrogen absorption lines in the visible spectrum because their high temperatures mean that the hydrogen in the photosphere is ionized (and therefore has no electron to become excited by the absorption of a photon).

Nature of science

Spectral classes

Stars were originally categorized in terms of the strength of the hydrogen absorption lines – with the stars producing the darkest absorption lines being called type A and those with successively weaker lines type B and C etc. Within the last century astronomers recognized that the line strength depended on temperature of the stars rather than their composition. Subsequently, the whole system of classification was revised with the result that many categories were abandoned and many others reordered – as shown in the table below. This form of star classification will not be included in your IB Diploma Programme Physics examinations, but is included here for historical interest and to indicate how many scientists have a reluctance to abandon a well-loved system!

Main sequence star properties

Colour	Class	Solar masses	Solar diameters	Temperature/K	Prominent lines
bluest	O	20–100	12–25	40 000	ionized helium
bluish	B	4–20	4–12	18 000	neutral helium, neutral hydrogen
blue-white	A	2–4	1.5–4	10 000	neutral hydrogen
white	F	1.05–2	1.1–1.5	7000	neutral hydrogen, ionized calcium
yellow-white	G	0.8–1.05	0.85–1.1	5500	neutral hydrogen, strongest ionized calcium
orange	K	0.5–0.8	0.6–0.85	4000	neutral metals (calcium, iron), ionized calcium
red	M	0.08–0.5	0.1–0.6	3000	molecules and neutral metals

Using the absorption spectrum to determine the chemical elements in a particular star is not easy as the spectra of many elements (differently ionized) are superimposed on one another. Although the lines present in an absorption spectrum tell us quite a lot about the temperature of the photosphere (as described above), they are difficult to interpret in terms of the abundance of elements. The movement of the gas atoms in the star causes the photons of light emitted to undergo both red and blue Doppler shifts. In hotter stars, the atoms move faster than in cooler stars and, therefore, the Doppler broadening is more pronounced. The rotation of the stars themselves means that the light reaching us will come from different parts of the star – one edge moving towards us, the other moving away from us and the central region rotationally stationary. This, of course, adds to the thermal Doppler broadening.

Wien's displacement law and star temperature

By treating a star as a black body it is possible to estimate its surface temperature using Wien's law – as discussed in Sub-topic 8.2. Stars range in surface temperature from approximately 2000 K to 40 000 K. The temperature T of the star in kelvin is given by:

$$T = \frac{2.9 \times 10^{-3}\ \text{m K}}{\lambda_{max}}$$

where λ_{max} is the wavelength in metres at which the black-body radiation is of *maximum intensity* and T is in kelvin.

Worked example

a) Explain the term *black-body radiation.*

The diagram is a sketch graph of the black-body radiation spectrum of a certain star.

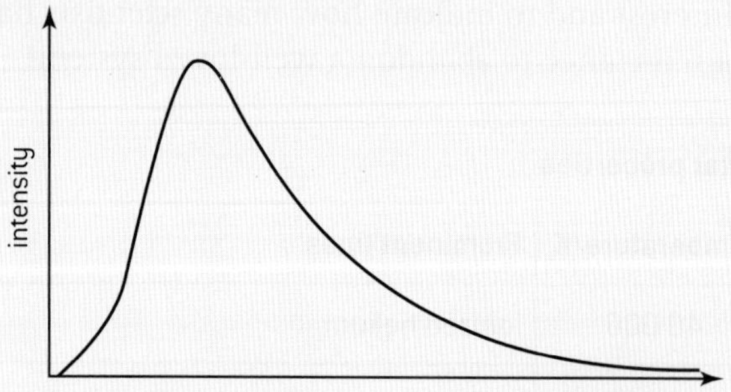

b) Copy the graph and label its horizontal axis.

c) On your graph, sketch the black-body radiation spectrum for a star that has a lower surface temperature and lower apparent brightness than this star.

d) The star Betelgeuse in the Orion constellation emits radiation approximating to that emitted by a black-body radiator with a maximum intensity at a wavelength of 0.97 μm.

Calculate the surface temperature of Betelgeuse.

Solution

a) Black-body radiation is that emitted by a theoretical perfect emitter for a given temperature. It includes all wavelengths of electromagnetic waves from zero to infinity.

b) and **c)**

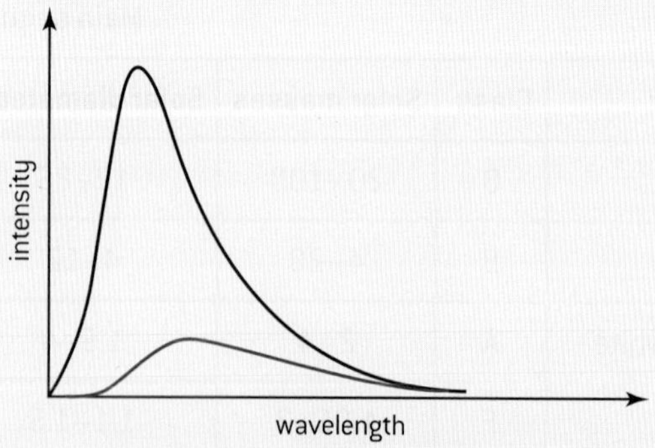

The red line intensity should be consistently lower and the maximum shown shifted to a longer wavelength.

d) $T = \frac{2.9 \times 10^{-3}\ \text{m K}}{\lambda_{max}} = \frac{2.9 \times 10^{-3}}{0.97 \times 10^{-6}}$

$= 2.99 \times 10^{3}\ \text{K} \approx 3000\text{K}$

Cepheid variables

Cepheid variables are extremely luminous stars that undergo regular and predictable changes in luminosity. Because they are so luminous it means that very distant Cepheids can be observed from the Earth. In 1784, the periodic pulsation of the supergiant star, Delta Cephei, was discovered by the English amateur astronomer, John Goodricke. This star has a period of about 5.4 days – as seen in figure 3.

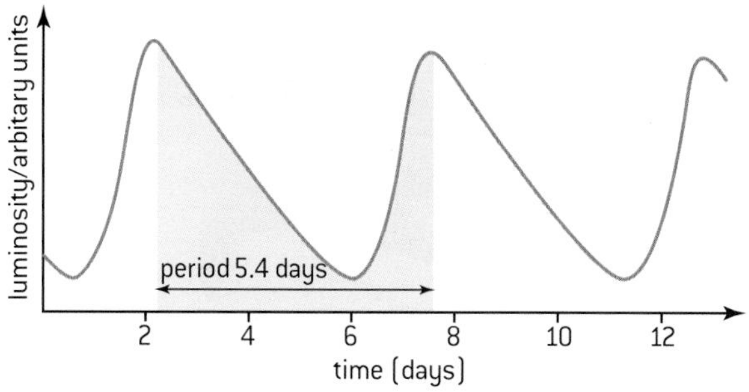

▲ Figure 3 Luminosity–time relationship for Delta Cephei.

In 1908, the American astronomer, Henrietta Leavitt, was working at the Harvard College Observatory. She published an article describing the linear relationship between the luminosity and period of pulsation for 25 stars, all practically the same distance from the Earth, in the Small Magellanic Cloud – a galaxy that orbits the Milky Way. This relationship allows us to estimate the luminosity of a given star by measuring its period of pulsation. It is now known that there are many more of these variable stars – collectively known as *Cepheids*. The period of these stars varies between twelve hours and a hundred days. Although the period is regular it is not sinusoidal and it takes less time for the star to brighten than it does to fade.

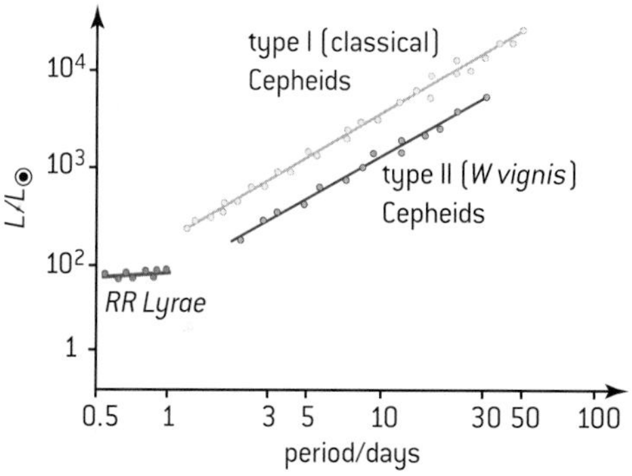

▲ Figure 4 Relative luminosity – period relationships for Cepheid stars.

Figure 4 shows the relative luminosity–period relationship for Cepheid stars. The relative luminosity is the ratio of the luminosity of the star to that of the Sun. The Sun's luminosity is conventionally written as $L_{\odot}$

Cepheid stars are stars that have completed the hydrogen burning phase and moved off the main sequence (see later for an explanation of this). The variation in luminosity occurs because the outer layers within the star expand and contract periodically. This is shown diagrammatically in figure 5. If a layer of an element loses hydrostatic equilibrium (between the gas and radiation pressure and that due to gravity) and is pulled inwards by gravity (1), the layer becomes compressed and less transparent to radiation (2); this means that the temperature inside the layer increases, building up the internal pressure (3) and causing the layer to be pushed outwards (4). During expansion the layer cools, becoming less dense (5) and more transparent, allowing radiation to escape and letting the pressure inside fall (6). Subsequently the layer falls inwards under gravity (1) and the cycle repeats causing the pulsation of the radiation emitted by the star.

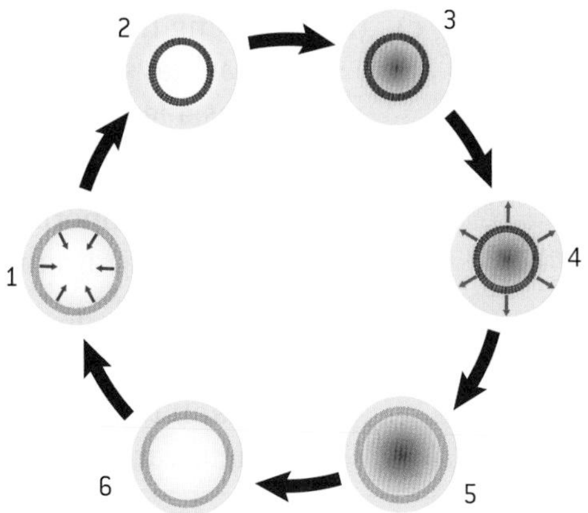

▲ Figure 5 Cycle in a Cepheid star.

Cepheid variable stars are known as "standard candles" because they allow us to measure the distances to the galaxies containing Cepheid variable stars. The distances, d, of the Cepheids can be calculated from the apparent brightness–luminosity equation:

$$b = \frac{L}{4\pi d^2}.$$

The apparent brightness is measured using a telescope and CCD and the luminosity is calculated from a measurement of the period of the Cepheid.

TOK

Women in science, technology, and engineering

Over the ages there have been few women scientists, and physics has been no exception. Astronomy, however, is one of the areas in which women have shone. Many women have made significant contributions. For example, in astrophysics, we have the German–British astronomer Caroline Herschel (who discovered several comets), Agnes Clerk, Maria Mitchell, Annie Jump Cannon, and Henrietta Leavitt. Even today, there continues to be fewer women following careers in science, and fewer opportunities for them to gain promotion to senior roles. Why should women be so under- represented in science, technology, and engineering?

Worked example

a) Define (i) *luminosity* (ii) *apparent brightness.*

b) State the mechanism for the variation in the luminosity of the Cepheid variable.

The variation with time t, of the apparent brightness b, of a Cepheid variable is shown below.

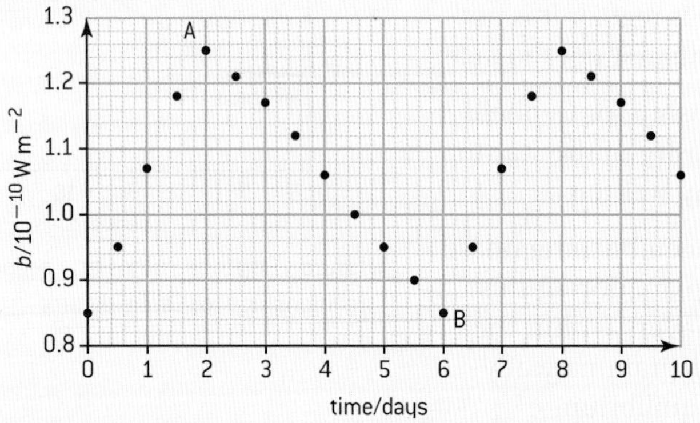

Two points in the cycle of the star have been marked A and B.

c) (i) Assuming that the surface temperature of the star stays constant, deduce whether the star has a larger radius after two days or after six days.

(ii) Explain the importance of Cepheid variables for estimating distances to galaxies.

d) (i) The maximum luminosity of this Cepheid variable is 7.2×10^{29} W. Use data from the graph to determine the distance of the Cepheid variable.

(ii) Cepheids are sometimes referred to as "standard candles". Explain what is meant by this.

Solution

a) (i) Luminosity is the total power radiated by a star.

(ii) Apparent brightness is the power from a star received by an observer on the Earth per unit area of the observer's instrument of observation.

b) Outer layers of the star expand and contract periodically due to interactions of the elements in a layer with the radiation emitted.

c) (i) The radius is larger after two days (point A) because, at this time the luminosity is higher and so the star's surface area is larger.

(ii) Cepheid variables show a regular relationship between period of variation of the luminosity and the luminosity. By measuring the period the luminosity can be calculated and, by using the equation $b = \frac{L}{4\pi d^2}$, the distances to the galaxy can be measured. This assumes that the galaxy contains the Cepheid star.

d) (i) $b = \frac{L}{4\pi d^2}$ thus $1.25 \times 10^{-10} = \frac{7.2 \times 10^{29}}{4\pi d^2}$

$$d = \sqrt{\frac{7.2 \times 10^{29}}{4\pi \times 1.25 \times 10^{-10}}}$$

$d = 2.14 \times 10^{19}$ m

(ii) A *standard candle* is a light source of known luminosity. Measuring the period of a Cepheid allows its luminosity to be estimated. From this, other stars in the same galaxy can be compared to this known luminosity.

Hertzsprung-Russell (HR) diagram

We saw in Sub-topic D.1 that the luminosity of a star is proportional both to its temperature to the fourth power and to its radius squared. Clearly, large hot stars are the most inherently bright, but how do other combinations of temperature and radius compare to these?

In the early 1900s, two astronomers, Ejnar Hertzsprung in Denmark and Henry Norris Russell in America, independently devised a pictorial way of illustrating the different types of star. By plotting a scattergram of the luminosities of stars against the stars' temperatures, clear patterns emerged. These scattergrams are now known as **Hertzsprung-Russell** (HR) diagrams. In general, cooler red stars tend to be of relatively low luminosity, while hotter blue stars tend to be of high luminosity. With high temperature conventionally drawn to the left of the horizontal axis, the majority of stars create a diagonal stripe which goes from top left to bottom right – this is known as the *main sequence*. A small number of stars do not follow the main sequence pattern but, instead, form island groups above and below the main sequence. The vertical axis is commonly modified to show the ratios of star luminosity to that of the Sun (denoted as $L_{\odot}$) as shown in figure 6 – in this case the axis is logarithmic and has no unit. The temperature axis is also logarithmic and doubles with every division from right to left.

The HR diagram shows the position of many stars of different ages; during the lifetime of a star its position will move on the diagram as its temperature and luminosity changes. We know from black-body radiation that the luminosity depends on the size of a star and its temperature. Small, cool stars will be dim and be positioned to the bottom right of the diagram – from Wien's law we know they will be red. Large, hot stars will be of high luminosity and blue or blue-white in colour thus placing them at the top left of the diagram. A modification of the HR diagram to include the different star classes is shown in figure 7.

Main sequence stars are ordinary stars, like the Sun, that produce energy from the fusion of hydrogen and other light nuclei such as helium and carbon. Nearly 90% of all stars fit into this category.

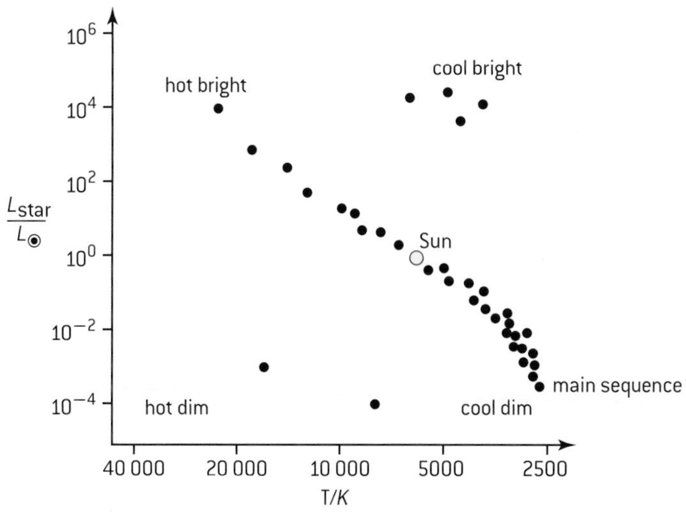

▲ Figure 6 Hertzsprung–Russell diagram.

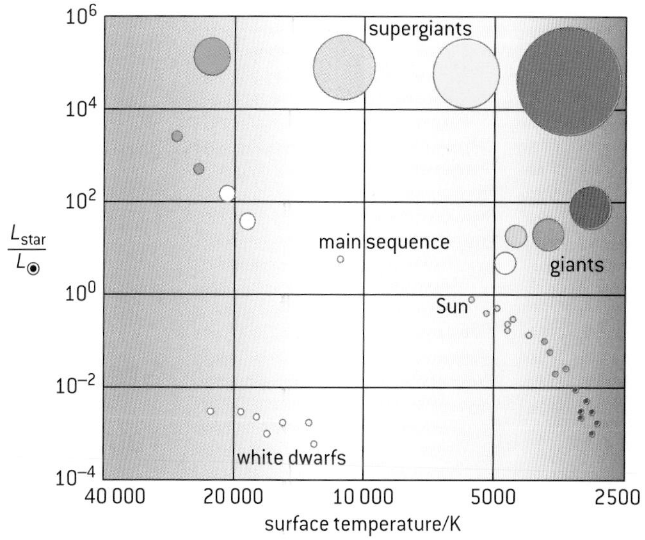

▲ Figure 7 Hertzsprung–Russell diagram showing the different classes of stars.

Red giants are cooler than the Sun and so emit less energy per square metre of surface. However, they have a higher luminosity, emitting up to 100 times more energy per second than the Sun. This means that they must have a much greater surface area to be able to emit such large energies. They, therefore, have a much larger diameter than the Sun – making them "giant" stars.

Supergiant stars are gigantic and very bright. A supergiant emitting 100 000 times the energy per second and at the same temperature of the Sun must have a surface area 100 000 times larger. This leads to a diameter that is over 300 times the diameter of the Sun. Only about 1% of stars are giants and supergiants.

White dwarfs are the remnants of old stars and constitute about 9% of all stars. Although they were very hot when they finally stopped producing energy, they have a relatively low luminosity showing them to have a small surface area. These very small, hot stars are very dense and take billions of years to cool down.

Mass–luminosity relation for main sequence stars

Not all main sequence stars are the same as the Sun – some are smaller and cooler while others are larger and hotter. High mass stars have shorter lifetimes – a star with a mass of 10 times the solar mass might only live 10 million years compared to the expected lifetime of around 10 billion years for the Sun. Observations of thousands of main sequence stars have shown there to be a relationship between the luminosity and the mass. For such stars this takes the form

$$L \propto M^{3.5}$$

where L is the luminosity in W (or multiples of the Sun's luminosity, $L_{\odot}$) and M is the mass in kg (or multiples of the Sun's mass, $M_{\odot}$).

Because mass is raised to a positive power greater than one, this means that even a slight difference in the masses of stars results in a large difference in their luminosities. For example, a main sequence star of 10 times the mass of the Sun has a luminosity of $(10)^{3.5} \approx 3200$ times that of the Sun.

For a star to be stable it needs to be in hydrostatic equilibrium, where the pressure due to the gravitational attraction of inner shells is equalled by the thermal and radiation pressure acting outwards. For a stable star of higher mass there will be greater gravitational compression and so the core temperature will be higher. Higher temperatures make the fusion between nuclei in the core more probable giving a greater rate of nuclear reaction and emission of more energy; thus increasing the luminosity. The mass of a star is fundamental to the star's lifetime – those with greater mass have far shorter lives.

Stellar evolution

Formation of a star

We saw in Sub-topic D.1 that the initial process in the formation of a star is the gravitational attraction of hydrogen nuclei. The loss of potential energy leads to an increase in the gas temperature. The gas becomes denser and, when the protostar has sufficient mass, the temperature becomes high enough for nuclear fusion to commence. The star moves onto the main sequence where it remains for as long as its hydrogen is being fused into helium – this time occupies most of a star's life. Eventually when most the hydrogen in the core has fused into helium the star moves off the main sequence.

The fate of stars

All stars collapse when most of the hydrogen nuclei have fused into helium. Gravity now outweighs the radiation pressure and the star shrinks in size and heats up. The hydrogen in the layer surrounding the shrunken core is now able to fuse, raising the temperature of the outer layers which makes them expand, forming a giant star. Fusion of the hydrogen adds more helium to the core which continues to shrink and heat up, forming heavier elements including carbon and oxygen. The very massive stars will continue to undergo fusion until iron and nickel (the most stable elements) are formed. What happens at this stage depends on the mass of the star.

A. Sun-like stars

For stars like the Sun of moderate mass (up to about 4 solar masses) the core temperature will not be high enough to allow the fusion of carbon. This means that, when the helium is used up, the core will continue to shrink while still emitting radiation. This "blows away" outer layers forming a planetary nebula around the star. When the remnant of the core has shrunk to about the size of the Earth it consists of carbon and oxygen ions surrounded by free electrons. It is prevented from further shrinking by **an electron degeneracy pressure**. Pauli's exclusion principle prevents two electrons from being in the same quantum state and this means that the electrons provide a repulsive force that prevents gravity from further collapsing the star. The star is left to cool over billions of years as **a white dwarf**. Such stars are of very high density of about 10^9 kg m^{-3}. Figure 8 shows Sirius B, the companion star of Sirius A, and the first white dwarf to be identified.

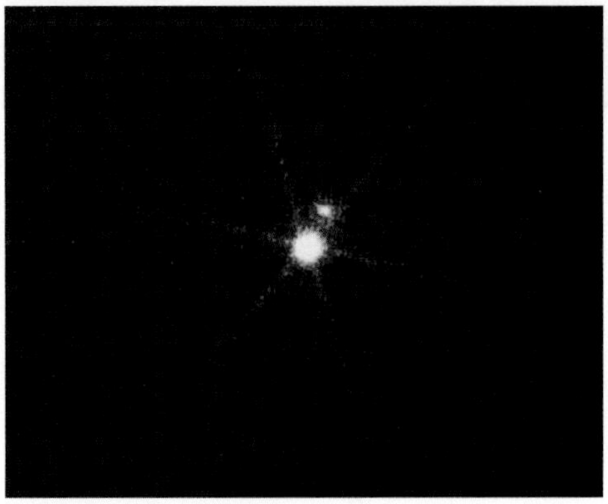

▲ Figure 8 Sirius A and B.

The probable future for the Sun is shown as the purple line on the HR diagram figure 9.

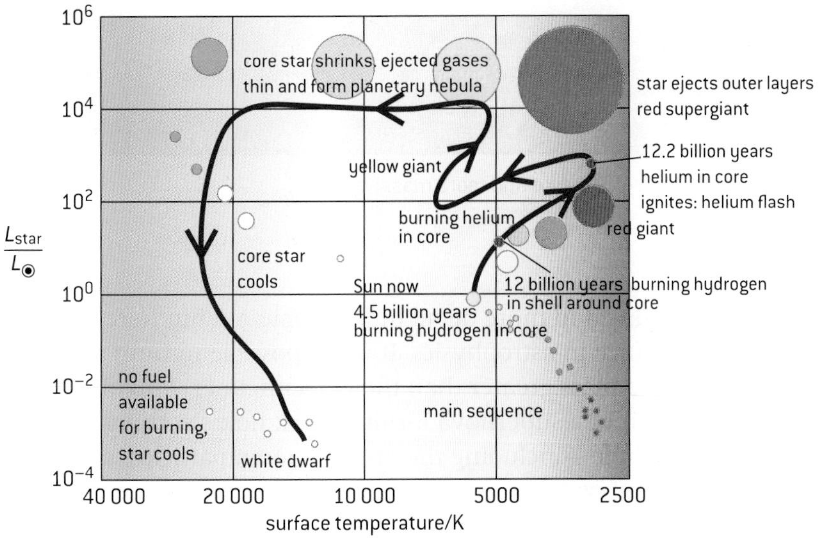

▲ Figure 9 Hertzsprung–Russell diagram showing the Sun's path.

B. Larger stars

The core of a star that is much bigger than the Sun undergoes a different evolutionary path from that of a Sun-like star. When such stars are in the red giant phase, the core is so large that the resulting high temperature causes the fusion of nuclei to create elements heavier than carbon. The giant phase ends with the star having layers of elements with proton numbers that decrease from the core to the outside (much like layers in an onion – see Sub-topic D.4, figure 8). The dense core causes gravitational contraction which, as for lighter stars, is opposed by electron degeneracy pressure. Even with this pressure, massive stars cannot stabilize. **The Chandrasekhar limit** stipulates that it is impossible for a white dwarf to have a mass of more than 1.4 times the mass of the Sun. When the mass of the core reaches this value the electrons combine with protons to form neutrons – emitting neutrinos in the process. The star collapses with neutrons coming as close to each other as in a nucleus. The outer layers of the star rush in towards the core but bounce off it in a huge explosion – a supernova. This blows off the outer layers and leaves the remnant core as a neutron star. In a quantum mechanical process similar to that of electron degeneracy, the neutrons provide a **neutron degeneracy pressure** that resists further gravitational collapse. The **Oppenheimer–Volkoff limit** places an upper value on a neutron star for which neutron degeneracy is able to resist further collapse into a black hole. This value is currently estimated at between 1.5 and 3 solar masses.

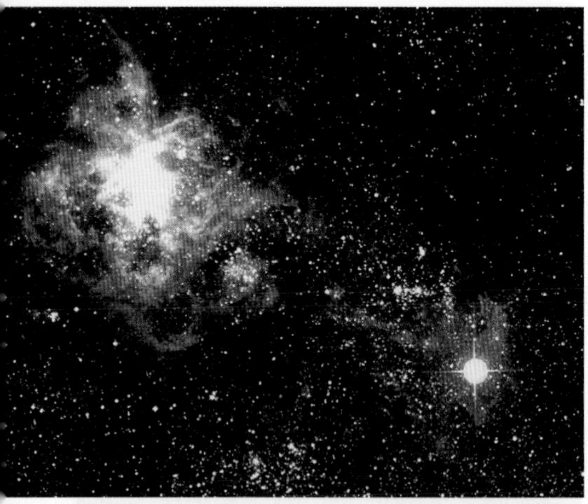

▲ Figure 10 Supernova 1987A with the right hand image the region of the sky taken just before the event

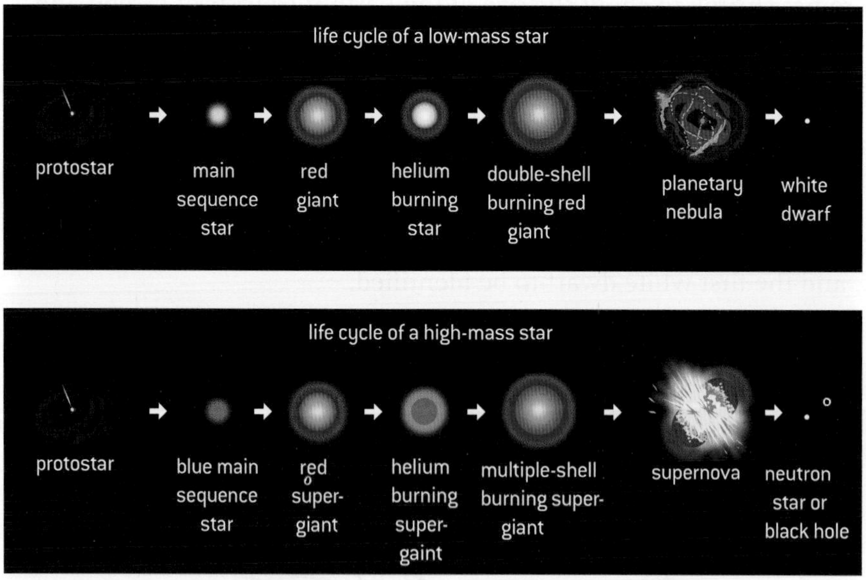

▲ Figure 11 The evolution of Sun-like and more massive stars.

Black holes

Black holes are discussed in more detail in Sub-topic A.5 but here we discuss their importance in astrophysics. It is not possible to form a neutron star having a mass greater than the Oppenheimer–Volkoff limit – instead the remnant of a supernova forms a black hole. Nothing can escape from a black hole – including the fastest known particles, photons. For this reason it is impossible to see a black hole directly but their existence can be strongly inferred by the following.

- The X-rays emitted by matter spiralling towards the edge of a black hole and heating up. X-ray space telescopes, such as NASA's Chandra, have observed such characteristic radiation.
- Giant jets of matter have been observed to be emitted by the cores of some galaxies. It is suggested that only spinning black holes are sufficiently powerful to produce such jets.
- The unimaginably strong gravitational fields have been seen to influence stars in the vicinity, causing them to effectively spiral. A black hole has been detected in the centre of the Milky Way and it has been suggested that there is a black hole at the centre of every galaxy.

Worked example

A partially completed Hertsprung–Russell (HR) diagram is shown below.

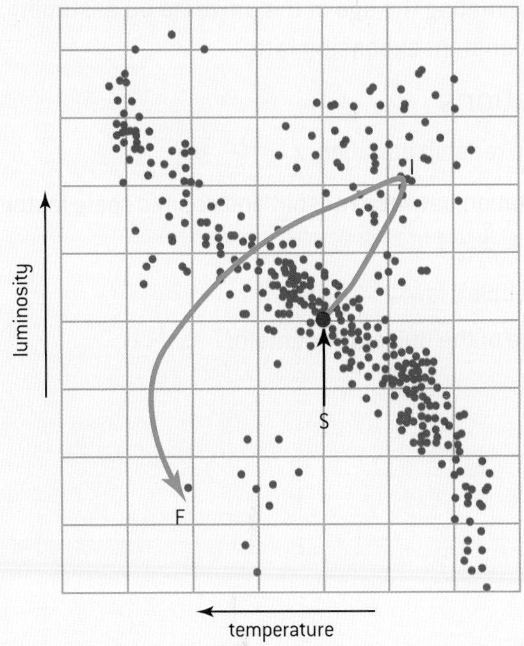

The line indicates the evolutionary path of the Sun from its present position, S, to its final position, F. An intermediate stage in the Sun's evolution is labelled by I.

a) State the condition for the Sun to move from position S.

b) State and explain the change in the luminosity of the Sun that occurs between positions S and I.

c) Explain, by reference to the Chandrasekhar limit, why the final stage of the evolutionary path of the Sun is at F.

d) On the diagram, draw the evolutionary path of a main sequence star that has a mass of 30 solar masses.

Solution

a) Most of the Sun's hydrogen has fused into helium.

b) Both the luminosity and the surface area increase as the Sun moves from S to I.

c) White dwarfs are found in region F of the HR diagram. Main sequence stars that end up with a mass under the Chandrasekhar limit of 1.4 solar masses will become white dwarfs.

d) The path must start on the main sequence above the Sun. This should lead to the super red giant region above I and either stop there or curve downwards towards and below white dwarf in the region between F and S.

D.3 Cosmology

Understandings

- The Big Bang model
- Cosmic microwave background (CMB) radiation
- Hubble's law
- The accelerating universe and redshift (Z)
- The cosmic scale factor (R)

Nature of science

Cosmology and particle physics

When we look at astronomical objects we see them as they were in the past. The Sun is always viewed as it was 8 minutes earlier and the most distant galaxies appear as they were more than 10 billion (10^{10}) years ago. The further away the galaxy, the further back in time we are looking. Cosmology is a way of studying the history of the universe. Particle physicists study the universe in a different way, but they have the same objectives. They use particle accelerators, such as the Large Hadron Collider (LHC) or the Relativistic Heavy Ion Collider (RHIC), to attempt to recreate events that mimic conditions in the very early universe.

Applications and skills

- Describing both space and time as originating with the Big Bang
- Describing the characteristics of the CMB radiation
- Explaining how the CMB radiation is evidence for a hot big bang
- Solving problems involving z, R, and Hubble's law
- Estimating the age of the universe by assuming a constant expansion rate

Equations

- the redshift equation: $z = \frac{\Delta\lambda}{\lambda_0} \approx \frac{v}{c}$
- relation between redshift and cosmic scale factor: $z = \frac{R}{R_0} - 1$
- Hubble's law: $v = H_0 d$
- age of the universe estimate: $T \approx \frac{1}{H_0}$

Introduction

Sir Isaac Newton believed that the universe was infinite and static. In his model of the universe he argued that the stars would exert equal gravitational attractions on each other in all directions, and this would provide a state of equilibrium. In 1823, the German astronomer, Heinrich Olbers, suggested that Newton's view of an infinite universe conflicted with observation; when we look up into the night sky we see darkness but, in an infinite universe, we should be able to see a star in every direction and, therefore, the night sky should be uniformly bright on a cloudless night. In 1848, the America author, Edgar Allan Poe, suggested that the universe, although infinite, was simply not old enough to have allowed the light to travel from the most far-flung regions to reach us. Although Poe was a naive scientist, he had suggested a theory that was a forerunner of the Big Bang model. In this sub-topic we will consider how the (inflationary) Big Bang model has developed into the most probable explanation of the beginning of the universe.

The redshift and Hubble's law

In the 1920s, the American astronomer, Edwin Hubble, was working at the Mount Wilson Observatory in California. He set out to find an additional way of gauging the distance of remote galaxies that would supplement using Cepheid stars. Following on from the work of others Hubble started to compare the spectra of distant galaxies with their Earth-bound equivalents. He found that the spectra from the galaxies invariably appeared to be redshifted in line with the Doppler effect. Such consistent results could only mean that all the galaxies were moving away from the Earth. Figure 1 shows the typical absorption spectra for a range of sources.

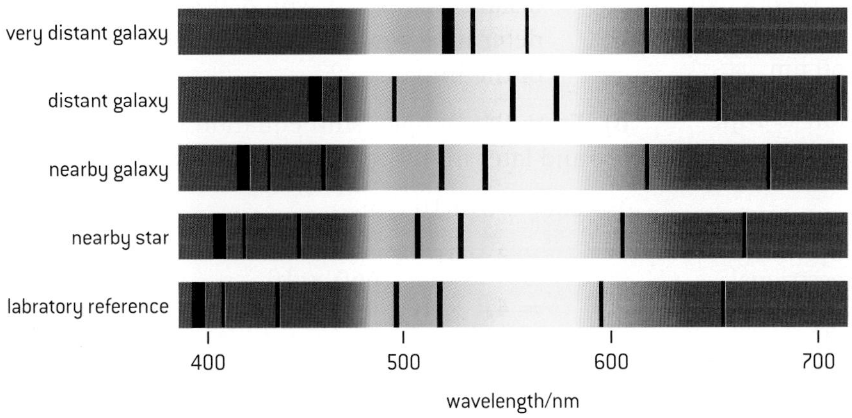

▲ Figure 1 Redshifted absorption spectra.

For optical spectra the wavelengths are moved **towards** the red end of the spectrum. The shift applies to all waves in the spectrum, so the absorption lines in the spectrum can be seen to have shifted.

In addition to recognizing the consistent redshift, Hubble showed that the further away the galaxy the greater the redshift. To do this he used the standard candles available to him, Cepheid variables. Although his data had large uncertainties (see figure 2) he suggested that the recessional speed of a galaxy is proportional to its distance from Earth. Hubble's law is written as

$$v = H_0 d$$

where v is the velocity of recession and d is the distance of the galaxy (both measured from Earth), H_0 is the Hubble constant.

v is usually being measured in km s^{-1} and d in Mpc, H_0 is usually measured in km s^{-1} Mpc^{-1}.

From the gradient of the graph of Hubble's data you will see that it gives a value for H_0 of about 500 km s^{-1} Mpc^{-1}. Now that we have more reliable data we can see that Hubble's intuition was valid but, by using 1355 galaxies, data give a modern value for H_0 that is much closer to 70 km s^{-1} Mpc^{-1}. However, it can be seen in figure 3 that there is still uncertainty in the value of the Hubble constant.

Note

- In the case of infra-red or microwaves the radiation already has wavelengths that are longer than red light. For these, the term redshift is ambiguous because "redshift" changes their wavelengths to even longer wavelengths and not towards the wavelength of red light.
- Stars within the Milky Way and, therefore, relatively close to the Earth might actually be moving towards us and could show a blueshift – this is simply a local phenomenon.

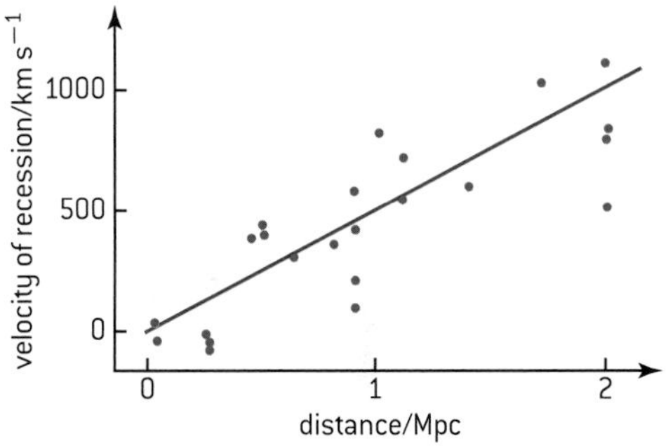

▲ Figure 2 Recreation of Hubble's original 1929 data.

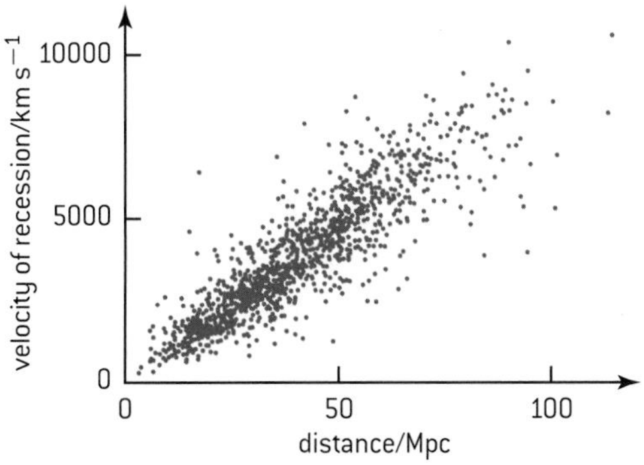

▲ Figure 3 More recent velocity–distance plot.

Worked example

a) The value of the Hubble constant H_0 is accepted by some astronomers to be in the range 60 km s^{-1} Mpc^{-1} to 90 km s^{-1} Mpc^{-1}.

(i) State and explain why it is difficult to determine a precise value of H_0.

(ii) State **one** reason why it would be desirable to have a precise value of H_0.

b) The line spectrum of the light from the quasar 3C 273 contains a spectral line of wavelength 750 nm. The wavelength of the same line, measured in the laboratory, is 660 nm.

Using a value of H_0 equal to 70 km s^{-1} Mpc^{-1}, estimate the distance of the quasar from Earth.

Solution

a) (i) The Hubble constant is the constant of proportionality between the recessional velocity of galaxies and their distance from Earth. The further galaxies are away (from Earth) the more difficult it is to accurately determine how far away they are. This is because of the difficulty of both locating a standard candle, such as finding a Cepheid variable within the galaxy, and the difficulties of accurately measuring its luminosity.

(ii) Having a precise value of H_0 would allow us to gain an accurate value of the rate of expansion of the universe and to determine an accurate value to distant galaxies. It would also allow us to determine a more reliable value for the age of the universe.

b) From the Doppler shift equation (see Topic 9 and later in this topic) $\frac{\Delta\lambda}{\lambda_0} \approx \frac{v}{c}$

$\Delta\lambda = 90 \times 10^{-9}$ m

$$v = 3 \times 10^8 \times \frac{90 \times 10^{-9}}{660 \times 10^{-9}}$$
$$= 4.1 \times 10^7 \text{ m s}^{-1}$$

$$d = \frac{v}{H_0} = \frac{4.1 \times 10^4}{70} = 590 \text{ Mpc}$$

> **TOK**
>
> **Values for H_0**
>
> There is still much debate about the value of the Hubble constant. Using better data has brought Hubble's original value down by a factor of almost 10 but the nature of measurement of intergalactic distances means that there are inevitable uncertainties. In December 2012, NASA's Wilkinson Microwave Anisotropy Probe (WMAP) gave a value for H_0 of (69.32 ± 0.80) km s^{-1} Mpc^{-1} but, in March 2013, the Planck Mission provided a value of (67.80 ± 0.77) km s^{-1} Mpc^{-1}. With two very credible scientific projects providing values that barely overlap at their extreme values, should we believe scientists' claims?

The Big Bang model and the age of the universe

Hubble's conclusion that the galaxies are moving further apart provides compelling evidence that they were once much closer together. According to the Standard Model, about 13.7 billion years ago the universe occupied a space smaller than the size of an atom. At that instant the entire universe exploded in a Big Bang, undergoing an immense expansion in which both time and space came into being. Starting off at a temperature of 10^{32} K, the universe rapidly cooled so that one second after the Big Bang it had fallen to 10^{10} K. In the time since the Big Bang, the universe has continued to cool to 2.7 K. In this time there has been an expansion of the fabric of space – the universe has not been expanding into a vacuum that was already there. As the galaxies move apart the space already between them becomes stretched. This is what we mean by "expansion" and why we believe the redshift occurs. The space through which the electromagnetic radiation travels is expanding and it stretches out the wavelength of the light. The further away the source of the light, the greater space becomes stretched, resulting in a more stretched-out wavelength and increasing the **cosmological redshift** – not to be confused with local redshift due to the Doppler effect.

Cosmologists have spent a great deal of time considering what happened between the time that the Big Bang occurred and the present day. One of the key questions is "how did the universe appear in the distant past?" Hubble's law certainly suggests that the galaxies were closer together than they are now and, logically, it follows that there must have been a point in time when they were all in the same place – that time being the Big Bang. It is possible to estimate the age of the universe using Hubble's law.

Assuming that Hubble's law has held true for all galaxies at all times, the light from the most distant star (at the edge of the observable universe) has taken the age of the universe to travel to us. If the light was emitted immediately after the time of the Big Bang, the space between the galaxy and the Earth must have expanded at slightly less than the speed of light for the light to have just reached us. This makes the recessional speed of the galaxy (almost) that of the speed of light, c. The distance that light has travelled from the galaxy $= c \times T$ where T is the age of the universe

$$v = H_0 d$$

$$c \approx H_0 cT$$

$$T \approx \frac{1}{H_0}$$

Using a value for H_0 of 70 km s^{-1} Mpc^{-1}

from the IB Physics data booklet

1 ly $= 9.46 \times 10^{15}$ m and 1 pc $= 3.26$ ly

so 1 Mpc $= 10^6 \times 9.46 \times 10^{15} \times 3.26$

$$\approx 3.1 \times 10^{22} \text{ m}$$

$$\frac{1}{H_0} = \frac{3.1 \times 10^{22}}{70 \times 10^3} = 4.4 \times 10^{17} \text{ s}$$

$$\approx 1.4 \times 10^{10} \text{ yr (or 14 billion years)}$$

The derivation of the age of the universe equation assumes that the galaxy and the Earth are moving at a relative constant speed of c and that there is nothing in their way to slow them down. In Sub-topic D.5 we consider a range of possibilities for the continued expansion of the universe.

The importance of the cosmic microwave background (CMB)

Until the 1960s there were two competing theories of the origin of the universe. One was the Big Bang theory and the other was known as the *steady state* theory. One aspect of the Big Bang theory is that it suggested a very high temperature early universe that cooled as the universe expanded. In 1948, Gamow, Alpher, and Herman predicted that the universe should show the spectrum of a black-body emitter at a temperature of about 3 K. In the Big Bang model, at approximately 4×10^5 years after the formation of the universe, the temperature had cooled to about 3000 K and the charged ion matter was able to attract electrons to form neutral atoms. This meant that space had become transparent to electromagnetic radiation, allowing radiation to escape in all directions (previously, when matter was ionic, it had been opaque to radiation). The expansion of the universe has meant that each of the photons emitted at this time has been shifted to a longer wavelength that now peaks at around 7 cm – in the microwave region of the spectrum. At earlier times the photons would have been much more energetic and of far shorter wavelengths, peaking in the visible or ultraviolet region of the electromagnetic spectrum.

The CMB in the sky looks essentially the same in all directions (it is "isotropic") and does not vary with the time of day; this provides compelling support for the Big Bang model. With the discovery of CMB, the advocates of the steady state theory were forced to concede to the strength of evidence. The Wilkinson Microwave Anisotropy

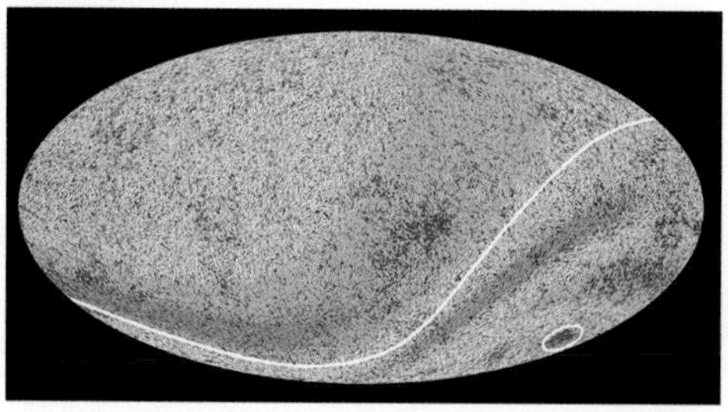

▲ Figure 4 The CMB sky.

Probe (WMAP) was able to show that, on a finer scale, there are some fluctuations in the isotropy of the CMB; this indicates the seeds of galaxies in the early universe.

Figure 4 is an image produced by the European Space Agency's Planck mission team showing an all-sky image of the infant universe created from 15.5 months of data. The radiation is essentially uniform but shows tiny variations in temperature (<0.2 mK) – the false colours in the image have been enhanced to show these minute fluctuations. The temperature differences show where galaxies are likely to have had a higher probability of forming. The effect of the Milky Way has been removed from the image.

Nature of science

The discovery of CMB

Discoveries in science are sometimes made by serendipity. The predictions of cosmologists were confirmed by two radio engineers, Arno Penzias and Robert Wilson, when setting up a new microwave receiver at the Bell Laboratories in the USA. They verified that a 7.4 cm wavelength signal, being picked up by their aerial, was not a fault with their electronics (as they first thought) but was coming from space and was equally strong in all directions. To support the Big Bang theory, however, it was necessary to look for the distribution of energy for other wavelengths, some of which did not penetrate the Earth's atmosphere. The Cosmic Background Explorer (COBE) satellite was launched in 1989 and enabled a team of scientists from Berkeley to show that the radiation was black-body radiation at a temperature of 2.7 K.

The redshift equation and the cosmic scale factor

We met the Doppler Effect for electromagnetic radiation in Sub-topic 9.5. Although the cause of the redshift is the stretching of space rather than a constantly moving source, the electromagnetic Doppler equation holds true and can be used in astrophysics where the redshift ratio $\frac{\Delta\lambda}{\lambda_0}$ is denoted by the symbol z, giving

$$z = \frac{\Delta\lambda}{\lambda_0} \approx \frac{v}{c}$$

Because CMB suggests that the universe is essentially isotropic and homogenous at any point in space at a chosen (proper) time after the Big Bang, it is essentially true that the density of matter should be the same throughout the universe. Soon after the Big Bang the density would have been greater and at later times smaller. The expansion of the universe can be considered to be a rescaling of it. As the universe expands, all distances are stretched with the cosmic scale factor R. In other words, if the radiation had wavelength λ_0 when it was emitted but λ when it was detected, the cosmic scale factor would have changed from R_0 to R. This means that space has stretched by an amount ΔR in the time that the wavelength has stretched by the amount $\Delta\lambda$. Hubble's law holds because, rather than galaxies receding from one another, space is expanding; this results in the redshift being a Hubble redshift as opposed to a Doppler redshift.

$$\frac{\Delta\lambda}{\lambda_0} \approx \frac{v}{c}$$

and

$$z = \frac{\Delta\lambda}{\lambda_0} = \frac{\Delta R}{R_0} = \frac{R - R_0}{R_0} = \frac{R}{R_0} - 1$$

Worked example

A distant quasar is detected to have a redshift of value = 5.6.

a) Calculate the speed at which the quasar is currently moving relative to the Earth.

b) Estimate the ratio of the current size of the universe to its size when the quasar the emitted photons that were detected.

Solution

a) $z = \frac{\Delta\lambda}{\lambda_0} \approx \frac{v}{c} = 5.6$

$v = 5.6c$

b) $\frac{R}{R_0} - 1 = 5.6 \therefore \frac{R}{R_0} = 6.6$

$\frac{R_0}{R} = \frac{1}{6.6} = 0.15$ so the universe was approximately 15% of its current size.

Type Ia supernovae and the accelerating universe

In the late 1990s, Type Ia supernovae (discussed in more detail in Sub-topic D.4) were found to offer key evidence regarding the expansion of the universe. By using Type Ia supernovae as standard candles to estimate galactic distances up to around 1000 Mpc and measuring their redshifts, strong evidence was obtained suggesting the universe might currently be undergoing an accelerated expansion. The universe is known to contain a significant amount of ordinary matter that has a tendency to slow down its expansion. Acceleration, therefore, would require some sort of invisible energy source and, although none has been directly observed, it has been named "dark energy". Figure 5 shows a NASA/WMAP graphic illustrating the evolution of the universe including the latter day acceleration.

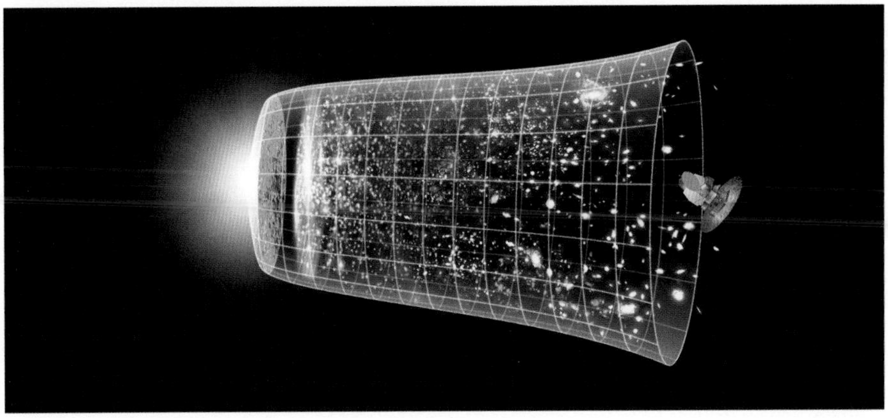

▲ Figure 5 Evolution of the accelerating universe.

Note

- The exception to universal expansion is any gravitationally bound system such as the stars within a galaxy or a galaxy within a cluster.
- The expansion of spacetime is not bound by the speed of light. This is because expansion of spacetime does not correspond to material objects moving within a single reference frame.
- With the distant universe expanding faster than the speed of light, there is an effective "event horizon". This means that, because the space between the Earth and the distant galaxies is expanding faster than the speed of light, it will never be possible for us to see photons emitted from this distance.

D.4 Stellar processes AHL

Understandings

- The Jeans criterion
- Nuclear fusion
- Nucleosynthesis off the main sequence
- Type Ia and II supernovae

Nature of science

As a carbon-based life form we are composed of much more than hydrogen and helium. Because supernovae are responsible for the nucleosynthesis and distribution of the heavy elements we can justifiably say that we are all stars – or at least made of the material produced in stars. The iconic Canadian composer/musician/artist Joni Mitchell wrote the song "Woodstock" in 1969. One of the lyrics in this song echoed the view of the great cosmologist, Carl Sagan: "We are stardust...". Sadly Joni missed the mark at being a physicist with a later song that was called "You Turn Me On, I'm A Radio"!

Applications and skills

- Applying the Jeans criterion to star formation
- Describing the different types of nuclear fusion reactions taking place off the main sequence
- Applying the mass–luminosity relation to compare lifetimes on the main sequence relative to that of our Sun
- Describing the formation of elements in stars that are heavier than iron, including the required increases in temperature
- Qualitatively describing the s and r processes for neutron capture
- Distinguishing between Type Ia and II supernovae

Introduction

This sub-topic builds on some of the ideas and concepts that we have met previously. We discuss how interstellar material can aggregate to form a star and how fusion in stars synthesizes all the naturally occurring elements. We look at the two main types of supernova and see how Type Ia can be used as a standard candle to allow us to measure the distance of galaxies as far away as 1000 Mpc.

The Jeans Criterion for star formation

We have seen in Sub-topic D.1 how stars form out of nebulae – interstellar clouds of dust, hydrogen, helium, and heavier elements such as in the Helix nebula shown in figure 1.

Such clouds might exist for many millions of years in a relatively constant state (although they will be constantly losing and gaining gas from the region around them). Eventually something happens to disturb the calm – this could be a collision with another cloud or a shockwave given out by a supernova exploding in the vicinity of the cloud. The result is that the cloud can become unstable and could collapse. By examining the spectra of such gas clouds we know that they are often very cold – sometimes just a few kelvin above absolute zero.

▲ Figure 1 The Helix Nebula.

Whether or not the gravitational attraction of the gas is sufficient for star formation depends on how quickly the gas temperature rises to prevent the gas freefalling into the centre of the cloud. Compressions in the gas travel at the speed of sound. With small quantities of gas, the compressions pass through it quickly enough to prevent it from collapsing – the gas oscillates and stabilizes. With large quantities of gas the compressions travel too slowly to prevent collapse. Whether there is sufficient mass for gravity to overcome radiation pressure is given by Jeans' criterion (named after the British astrophysicist Sir James Jeans, who developed the first workable theory of star formation). The total energy of a gas cloud is a combination of positive kinetic energy and negative gravitational potential energy. At infinite separation of the gas particles the total energy is kinetic and they are not bound together. When they are close and the potential energy dominates the particles are bound (the total energy being negative). The Jeans criterion for the gas to collapse is simply that the magnitude of the potential energy must be greater than the kinetic energy. This depends on the temperature and the particle density in the cloud. *A cold, dense gas is far more likely to collapse than a hot low-density gas (which will have too much kinetic energy and too little gravitational potential energy).*

Typically in a cool cloud, the density of gas is about 3×10^7 particles per cubic metre and the local temperature will be around 100 K. Under these circumstances the "Jeans mass" must be nearly four hundred thousand solar masses to facilitate star formation ($M_J > 400\ 000\ M_\odot$). Given such mass, it is no surprise that stars form in clusters in these regions – individual stars have a mass between 0.1 $M_\odot$ and 150 $M_\odot$. In even cooler regions of space, with higher gas densities of 10^{11} particles per cubic metre and temperatures of 10 K, the Jeans mass falls to approximately 50 $M_\odot$. With these masses, short-lived giant stars are formed. The atmospheric number density close to the surface of the Earth is of the order of 10^{25} molecules per cubic metre. But, with the high temperature and low mass of gas, you need not worry that a star will form in the atmosphere!

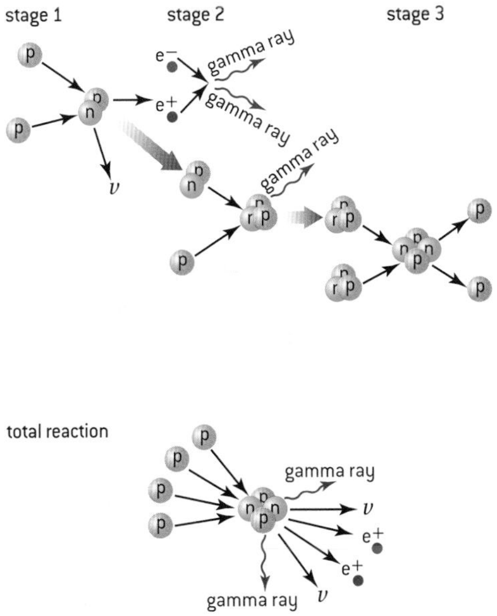

▲ Figure 2 The proton–proton chain.

Nuclear fusion

We have seen that the energy generated by a star is the result of thermonuclear fusion reactions that take place in the core of the star. The process that occurs during the main sequence is known as "hydrogen burning"; here hydrogen fuses into helium. For Sun-like stars the process advances through the **proton–proton chain** but stars of greater than four solar masses undergo a series of reactions known as the **CNO cycle**.

The proton–proton chain has three stages:

- $^1_1H + ^1_1H \rightarrow ^2_1H + ^0_1e + ^0_0\nu$ [two protons (hydrogen-1 nuclei) fuse into a hydrogen-2 nucleus (plus a positron and a neutrino)]
- $^1_1H + ^2_1H \rightarrow ^3_2He + \gamma$ [a third proton fuses with the hydrogen-2 to form a helium-3 nucleus + a gamma-ray photon]
- $^3_2He + ^3_2He \rightarrow ^4_2He + ^1_1H + ^1_1H$ [two helium-3 nuclei fuse to produce helium-4 and two hydrogen-1nuclei]

Thus, in order to produce a helium nucleus, four hydrogen nuclei are used in total (six are used in the fusion reactions and two are generated).

The CNO process occurs in the larger stars with a minimum core temperature of 2×10^7 K and involves six stages:

- $^1_1H + ^{12}_6C \rightarrow ^{13}_7N + \gamma$ [proton fuses with carbon-12 to give unstable nitrogen-13 + a gamma-ray photon]
- $^{13}_7N \rightarrow ^{13}_6C + ^0_1e + ^0_0\nu$ [nitrogen-13 undergoes positron decay into carbon-13]
- $^1_1H + ^{13}_6C \rightarrow ^{14}_7N + \gamma$ [carbon-13 fuses with proton to give nitrogen-14 + a gamma-ray photon]
- $^1_1H + ^{14}_7N \rightarrow ^{15}_8O + \gamma$ [nitrogen-14 fuses with proton to give unstable oxygen-15 + a gamma-ray photon]
- $^{15}_8O \rightarrow ^{15}_7N + ^0_1e + ^0_0\nu$ [oxygen-15 undergoes positron decay into nitrogen-15]
- $^1_1H + ^{15}_7N \rightarrow ^{12}_6C + ^4_2He$ [nitrogen-15 fuses with proton to give carbon-12 (again) and helium-4]

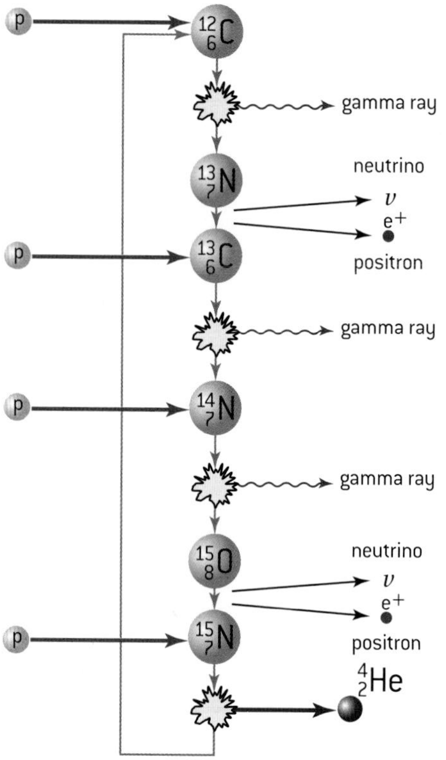

▲ Figure 3 CNO chain.

Again, four protons are used to undergo the fusion process; carbon-12 is both one of the fuels and one of the products. Two positrons, two neutrinos and three gamma-ray photons are also emitted in the overall process. The fusing of hydrogen into helium takes up the majority of a star's lifetime and is the reason why there are far more main sequence stars than those in other phases of their life-cycle.

> **Note**
> The CNO cycle is part of the main sequence and no heavy elements are synthesized in this process.

Fusion after the main sequence

We saw in Sub-topic D.2 that, once the hydrogen in a star's core is used up, the core converts into helium. The lack of radiation pressure causes the core to shrink and heat up. The hot core instigates the fusion and expansion of the hydrogen surrounding it, causing the red giant phase. Eventually, the core's rising temperature becomes high enough to make the star move off the main sequence. Helium now fuses into the unstable beryllium which then fuses with a further helium nucleus to produce carbon and then oxygen:

- $^4_2He + ^4_2He \rightarrow ^8_4Be$ [two helium nuclei fuse to produce unstable beryllium-8]
- $^4_2He + ^8_4Be \rightarrow ^{12}_6C$ [a helium nucleus quickly fuses with beryllium-8 to produce carbon-12]
- $^4_2He + ^{12}_6C \rightarrow ^{16}_8O$ [a further helium nucleus fuses with carbon-12 to produce oxygen-16]

You can see from these reactions that **nucleosynthesis** (the production of different nuclides by the fusion of nuclei) is not a simple process. Eventually, both the carbon and oxygen produced will undergo fusion and form nuclei of silicon, magnesium, sodium, and so on until iron-56 is reached. This element represents one of the most stable of all the elements (nickel-62 is the most stable nuclide but is far less abundant in stars than iron-56). These nuclides have the highest binding energy per nucleon. Energy cannot be released by further fusions of these elements to produce even heavier nuclides, instead energy will need to be **taken in** to allow fusion to occur.

If nickel-62 is the most stable nuclide then how are heavier nuclides made? This is where neutron capture comes in. Neutrons, being uncharged, do not experience electrostatic repulsion and can approach so close to nuclei that the strong nuclear force is able to capture them. The capture of a neutron increases the nucleon number by one and so does not produce a new element, just a heavier isotope of the original element (isotopes of X in figure 4). The newly created isotope will be excited and decay to a less energetic sibling by emitting a gamma-ray photon. The neutron in the newly formed nucleus might be stable or it could decay into a proton, an electron and antineutrino (by negative beta decay). This raises the proton number of the nucleus by one and produces the nucleus of a new element (Y in figure 4). The new nuclide will initially be excited and releases a gamma-ray photon in decaying into a less energetic nuclide. The half-life of beta decay depends solely on the nature of the particular parent nuclide. Whether or not there is sufficient time for a nucleus to capture a further neutron depends on the density of neutrons bombarding the nuclei.

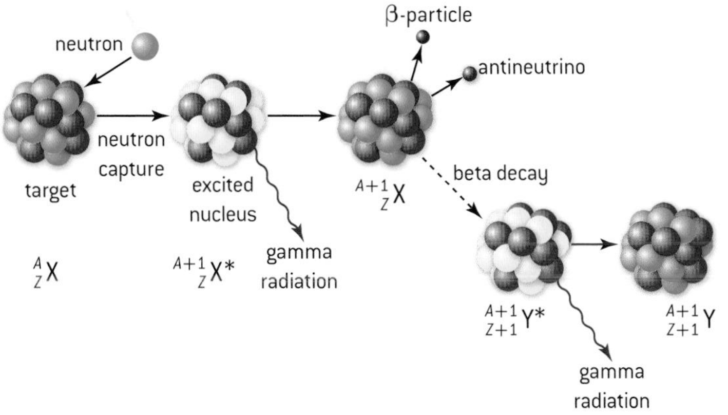

▲ Figure 4 Neutron capture.

In a massive star, heavy nuclides (up to bismuth-209) can be produced by slow neutron capture or the "**s-process**". These stars provide a fairly small neutron flux as a by-product of carbon, oxygen, and silicon burning. This means that there is time for nuclides to undergo beta decay before further neutron captures build up their nucleon number – producing successively heavier isotopes of the original element.

In rapid neutron capture, or the "**r-process**", there is insufficient time for beta decay to occur so successively heavier isotopes are built up very quickly, one neutron at a time. Type II supernovae produce a very high neutron flux and form nuclides heavier than bismuth-209 in a matter of minutes (and well before there is any likelihood of beta decay occurring). This is something

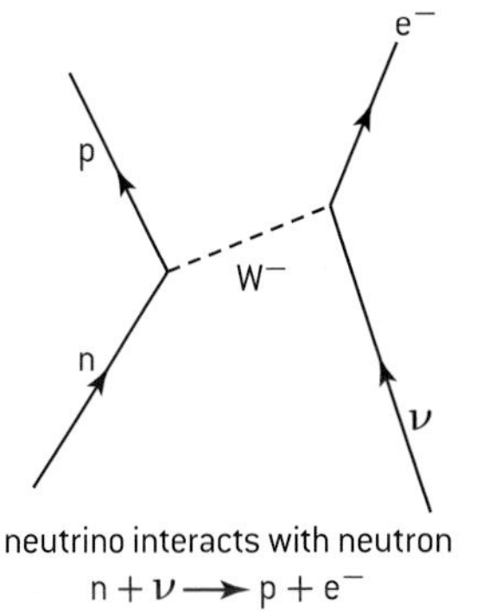

▲ Figure 5 Neutrino interacts with neutron.

that does not appear to happen in the massive stars. There is also a high neutrino flux in a supernova and this has the effect of causing neutrons to convert into protons through the weak interaction, forming new elements. This weak interaction is shown in the Feynman diagram, figure 5.

Worked example

a) Suggest why nuclear fusion processes inside stars can only synthesize elements with a nucleon number less than 63.

b) Outline how heavier elements could be produced by stars.

Solution

a) Energy is released when the binding energy per nucleon increases. The binding energy per nucleon is a maximum at nucleon number 62 (nickel) and so further fusions would require energy to be supplied. This means that, in a star producing heavier elements, fusion is no longer energetically favourable.

b) With a strong neutron flux, nuclei can absorb neutrons. This can either be a slow process in massive stars, which are capable of producing nuclei no more massive than bismuth-209, or it can be a rapid process in a supernova, that is able to produce still more massive nuclei.

Lifetimes of main sequence stars

It might appear that the more matter a star contains the longer it can exist. However, this takes no account of stars having different luminosities and, in fact, the more massive a star is, the shorter its lifetime! Massive stars need higher core temperatures and pressures to prevent them from collapsing under gravity. This means that fusion proceeds at a faster rate than in stars with lower mass. Therefore, massive stars use up their core hydrogen more quickly and spend less time on the main sequence than stars of lower mass.

In Sub-topic D.2 we saw that the luminosity of a main sequence star is related to its mass by the relationship:

$$L \propto M^{3.5}$$

Luminosity is the total energy E released by the star per unit time while hydrogen is being fused or

$$L = \frac{E}{t}$$

While fusion occurs, the energy emitted is accompanied by a loss of mass. This will amount to a proportion κ of the total star mass during its lifetime – so a star of mass M loses a mass κM. Using Einstein's mass–energy relationship $E = mc^2$, this makes the energy emitted during the hydrogen burning phase of the star's life $E = \kappa Mc^2$. Its average luminosity will be given by:

$$L = \frac{\kappa Mc^2}{\tau}$$

where τ is the lifetime of the star.

This can be rearranged to give:

$$\tau = \frac{\kappa Mc^2}{L}$$

and because

$$L \propto M^{3.5}$$

we can deduce that

$$\tau \propto \frac{M}{M^{3.5}}$$

or

$$\tau \propto M^{-2.5}$$

It is useful to compare the mean lifetime of a star with that of the Sun and so, using the normal notation for solar quantities, we have:

$$\frac{\tau}{\tau_{\odot}} = \left(\frac{M}{M_{\odot}}\right)^{-2.5}$$

The Sun is expected to have a lifetime of approximately ten billion years ($= 10^{10}$ years). A bigger star of, say, 10 solar masses would start with ten times the Sun's hydrogen but it would have a much shorter lifetime given by

$$\frac{\tau_{\text{star}}}{10^{10}} = \left(\frac{10M_{\odot}}{M_{\odot}}\right)^{-2.5} = 3.2 \times 10^{-3}$$

This gives $\tau_{\text{star}} = 3.2 \times 10^{7}$ years. We can see from this that although it has ten times the Sun's mass it only lives for 0.3% of the lifetime of the Sun.

Note

You will not be tested on this relationship in your IB Physics examinations, but it is included here to help explain the basic ideas, as these might be tested.

Figure 6 is a HR diagram showing the masses and lifetimes of a number of stars. This HR diagram is for interest only and there is no need to try to learn it; however, the various regions should make sense to you.

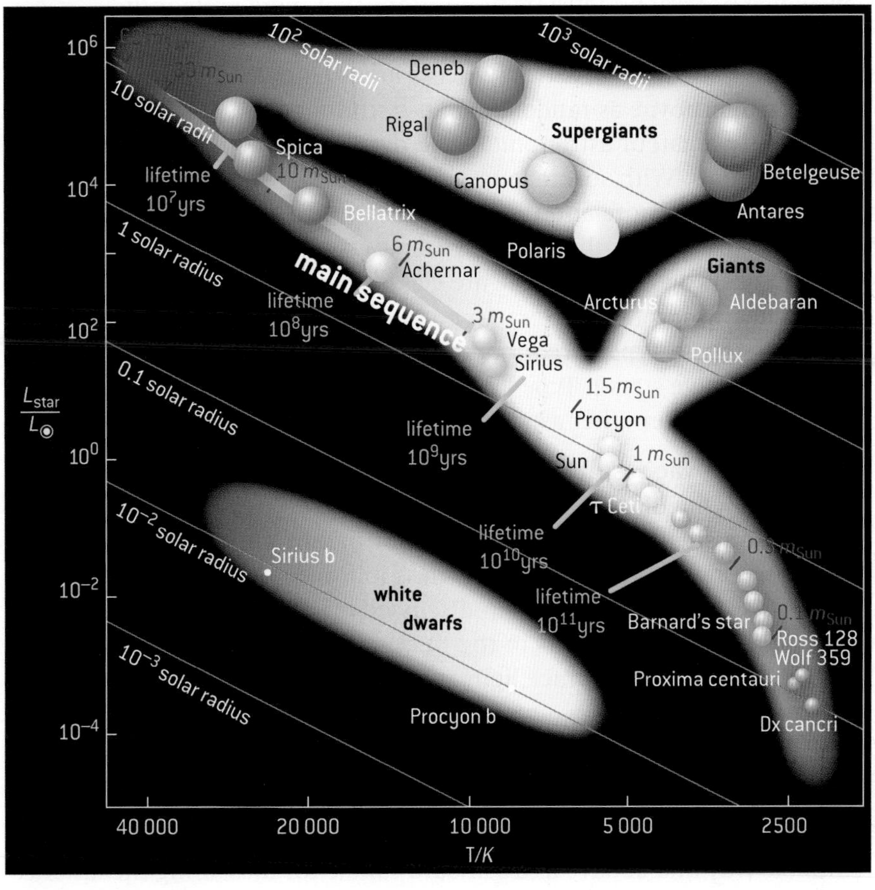

▲ Figure 6 HR diagram.

Worked example

a) Explain why a star having a mass of 50 times the solar mass would be expected to have a lifetime of many times less than that of the Sun.

b) By referring to the mass–luminosity relationship, suggest why more massive stars will have shorter lifetimes.

Solution

a) The more massive stars will have much more nuclear material (initially hydrogen). Massive stars have greater gravity so equilibrium is reached at a higher temperature at which the outward pressure due to radiation and the hot gas will balance the inward gravitational pressure. This means that fusion proceeds at a faster rate than in stars with lower mass – meaning that the nuclear fuel becomes used up far more rapidly.

b) As the luminosity of the star is the energy used per second, stars with greater luminosity are at higher temperatures and will use up their fuel in shorter periods of time. The luminosity of a star is related to its mass by the relationship $L \propto M^{3.5}$. Therefore, increasing the mass raises the luminosity by a much larger factor which in turn means the temperature is much higher. At the higher temperature the fuel will be used in a much shorter time.

Supernovae

We now delve into supernovae a little more because of their importance to astrophysics. Supernovae are very rare events in any given galaxy but, because there are billions of galaxies, they are detected quite regularly. They appear as very bright stars in positions that were previously unremarkable in brightness. The advent of high-powered automated space telescopes has meant that astrophysicists are no longer dependent on observing supernovae as random events within or close to the Milky Way. Until recently amateur astronomers had discovered more supernovae than the professionals. With several automated surveys, such as the Catalina Real-Time Survey, more and more supernovae are being detected so that the number detected in the last ten years is greater than the total detected before ten years ago. Supernovae are classified as being Type I or Type II in terms of their absorption spectra (Type I have no hydrogen line but Type II do. This is because Type I are produced by old, low-mass stars and Type II by young, massive stars.). Type I supernovae are subcategorized as Ia, Ib, Ic, etc. depending on other aspects of their spectra.

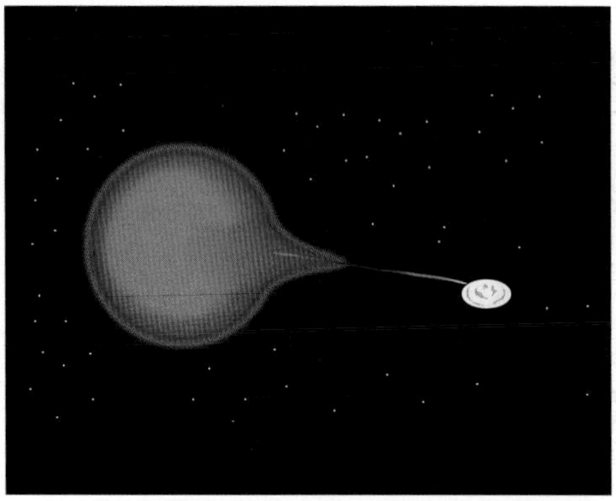

▲ Figure 7 An artist's impression of the accretion of stellar matter leading to a Type Ia supernova. The white dwarf is the star on the right, while the left-hand star is a red giant.

Type Ia supernovae result from accretion of matter between two stars in a binary star. One of the stars is a white dwarf and the other is either a giant star or a smaller white dwarf. The formation of these supernovae show up as a rapid increase in brightness followed by a gradual tapering off. Type II supernovae have been discussed briefly in Sub-topic D.2 and consist of single massive stars in the final stages of their evolution. These classes of supernovae produce light curves with different characteristics as shown in figure 9.

Type Ia supernovae

These are very useful to astrophysicists as they always emit light in a predictable way and behave as a standard candle for measuring the distance of the galaxy in which the supernova occurs. Given the immense density of the material within a white dwarf, the gravitational field is unimaginably strong and attracts matter from the companion star. When the mass of the growing white dwarf exceeds the Chandrasekhar limit of 1.4 solar masses, the star collapses under gravity. The fusion of carbon and oxygen into nickel generates such radiation pressure that the star is blown apart, reaching a luminosity of 10^{10} times that of the Sun. Because the chain reaction always occurs at this mass we know how bright the supernova actually is and, by comparing it with the apparent brightness observed on Earth, we can estimate the distance of the supernova's galaxy up to distances of 1000 Mpc. After the explosion the ejected material continues to expand in a shell around the remnant for thousands of years until it mixes with the interstellar material – giving the potential to form a new generation of stars.

Type II supernovae

After approximately 10 million years (for stars of 8–10 solar masses) all the hydrogen in the core has converted into helium and hydrogen fusion can now only continue in a shell around the helium core. The core undergoes gravitational collapse until its temperature is high enough for fusion of helium into carbon and oxygen. This phase lasts for about a million years until the core's helium is exhausted; it will then contract again under gravity, causing it to heat up and allowing the fusion of carbon into heavier elements. It takes about 10 thousand years until the carbon is exhausted. This pattern continues, with each heavier element lasting for successively shorter lengths of time, until silicon is fused into iron-56 – taking a few days (see figure 8). At this point the star is not in hydrostatic equilibrium because there is now little radiation pressure to oppose gravity. On reaching the Chandrasekhar limit of 1.4 $M_{\odot}$, electron degeneracy pressure is insufficient to oppose the collapse and the star implodes producing neutrons and neutrinos. The implosion is opposed by a neutron degeneracy pressure that causes an outward shock wave. This passes through the outer layers of the star causing fusion reactions to occur. Although this process lasts just a few hours it results in the heavy elements being formed. As the shock wave reaches the edge of the star, the temperature rises almost instantly to 20 000 K and the star explodes, blowing material off as a supernova.

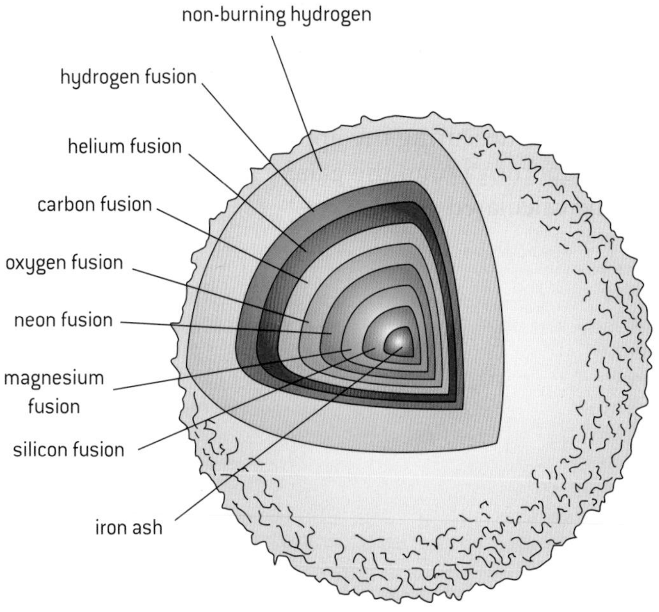

▲ Figure 8 The onion model of a massive star before it goes supernova.

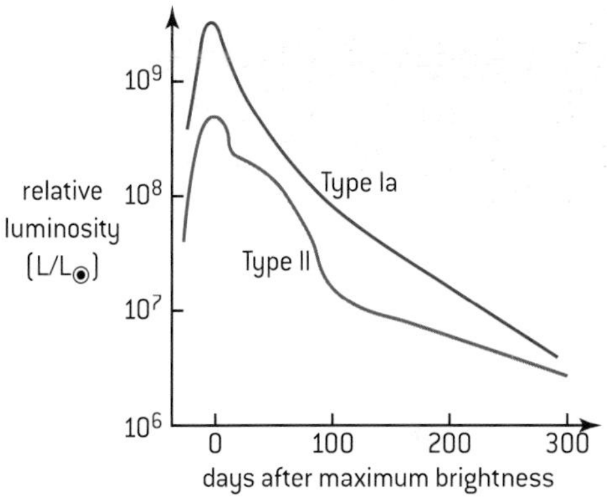

▲ Figure 9 Light curves for class Ia and II supernovae.

Type Ia and Type II supernovae can be distinguished by observers on Earth from the manner in which the stars emit light. The Type Ia emits light up to 10^{10} times the luminosity of the Sun; this rapidly reaches a maximum and then gradually tails off over six months or so. The Type II emits light up to 10^{9} times the luminosity of the Sun; however, the burst falls a little before reaching a slight plateau where it stays for some days before falling away more rapidly. The light curves for these supernovae are shown in figure 9.

Worked example

a) Outline the difference between a Type Ia and a Type II supernova.

b) (i) What is meant by a standard candle?

(ii) Explain how a Type Ia supernova can be used as a standard candle.

Solution

a) A Type Ia supernova results from a white dwarf in a binary star system accreting material from its companion giant or smaller white dwarf star. Eventually, when the mass of the white dwarf passes the Chandrasekhar limit, the star collapses and is blown apart as a supernova by the huge temperature generated. Type II supernovae occur in stars having masses between 8 and 50 times the mass of the Sun. When the core has changed into inert iron and nickel, with no further fusion occurring, gravity collapses the core. Again, on reaching the Chandrasekhar limit of 1.4 $M_{\odot}$, electron degeneracy pressure is insufficient to oppose the collapse and the star implodes producing neutrons and neutrinos. The implosion is opposed by a neutron degeneracy pressure that causes an outward shock wave blowing away the surrounding material as a supernova.

b) (i) A standard candle is a star of known luminosity that, when compared with its apparent brightness, can be used to calculate its distance.

(ii) As all Type Ia supernova occur when the mass reaches the Chandrasekhar mass they are all (essentially) of the same peak luminosity. By measuring the apparent brightness, the distance of the supernova (and the galaxy in which it is a member) can be calculated (from $b = \frac{L}{4\pi d^2}$).

D.5 Further cosmology AHL

Understanding

- The cosmological principle
- Rotation curves and the mass of galaxies
- Dark matter
- Fluctuations in the CMB
- The cosmological origin of redshift
- Critical density
- Dark energy

Nature of science

How constant is the Hubble constant?

...or when is a constant not a constant? The Hubble constant H_0 is a quantity that we have used in Sub-topic D.3 to estimate the age of the universe. This is a very important cosmological quantity, indicating the rate of expansion of the universe. The current value is thought to be around 70 km s^{-1} Mpc^{-1}, but it has not always been this value and will not be in the future. So it is a constant in space but not in time and it would be more appropriately named the "Hubble parameter". The zero subscript is used to indicate that we are talking about the present value of the constant – in general use we should omit the subscript. In the IB Physics course, H_0 is used to indicate all values of the Hubble constant.

Applications and skills

- Describing the cosmological principle and its role in models of the universe
- Describing rotation curves as evidence for dark matter
- Deriving rotational velocity from Newtonian gravitation
- Describing and interpreting the observed anisotropies in the CMB
- Deriving critical density from Newtonian gravitation
- Sketching and interpreting graphs showing the variation of the cosmic scale factor with time
- Describing qualitatively the cosmic scale factor in models with and without dark energy

Equations

- velocity of rotating galaxies: $v = \sqrt{\frac{4\pi G\rho}{3}}\,r$
- critical density of universe: $\rho_c = \frac{3H^2}{8\pi G}$

Introduction

In Sub-topic D.3 we saw that the Big Bang should correspond to the simultaneous appearance of space and time (spacetime). Hubble's law and the expansion of the universe tell us that the observable universe is certainly larger than it was in the past and that it can be traced back to something smaller than an atom, containing all the matter and energy currently in the universe. There is no special place in the universe that would be considered to be the source of the Big Bang, it expanded everywhere in an identical manner. It is not possible to use the Big Bang model to speculate about what is beyond the observable universe – it should *not* be thought of as expanding into some sort of vacuous void. In this sub-topic we will consider ways in which the universe might continue to expand and look at possible models for flat, open, and closed universes.

Nature of science

Philosophy or cosmology?

Although there is substantial evidence leading us to believe in the Big Bang model nobody actually knows what instigated the Big Bang. We have explained that this was the beginning of space and time and so we cannot ask "what happened before the Big Bang?" There are a number of theories that reflect on why the Big Bang occurred – such as fluctuations in gravity or quantum fluctuations but these theories stimulate other questions such as "what caused this?" As of yet these theories cannot be put to the test.

The Cosmological Principle

Buoyed up by the success of his general theory of relativity in 1915, Einstein sought to extend this theory to explain the dynamics of the universe or "cosmos". In order to make headway, because he recognized that this would be a complex matter, he made two simplifying assumptions – these have subsequently been shown to be essentially true on a large scale:

- the universe is homogenous
- the universe is isotropic

The first of these requirements simply says that the universe is the same everywhere – which, when we ignore the lumpiness of galaxies, it is. The second requirement is that the universe looks the same in all directions. Although this may not seem too different from the homogeneity idea it actually is! It really says that we are not in any special place in the universe – and ties in with the theory of relativity that says there is no special universal reference frame. Imagine that we are positioned towards the edge of a closed universe and we look outward – there would be a limited number of galaxies to send photons to us. If we look inward there would be an immense number of galaxies to send us photons. The two situations would appear very different. The isotropicity says that this isn't the case. Jointly, these two prerequisites of Einstein's theory are known as the "Cosmological Principle". These assumptions underpin the Big Bang cosmology and lead to specific predictions for observable properties of the universe.

Figure 1 shows an image produced by the Automated Plate Measurement (APM) Galaxy survey of around 3 million galaxies in the Southern Hemisphere sky. The image shows short-range patterns but, in line with the cosmological principal, on a large scale the image shows no special region or place that is different from any other.

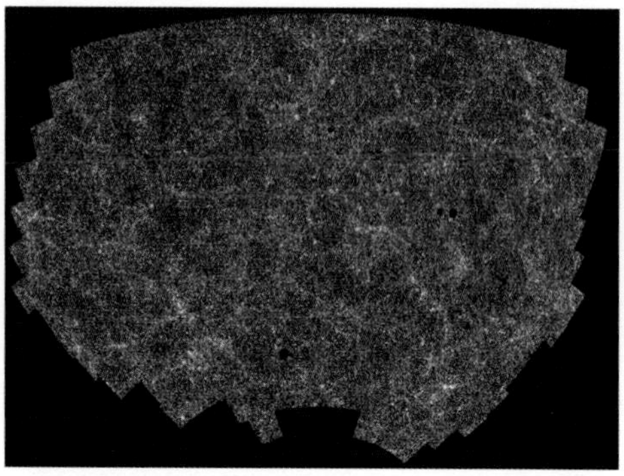

▲ Figure 1 APM Galaxy survey image.

Using the cosmological principle and the general theory of relativity it can be shown that matter can only distort spacetime in one of three ways. This is conventionally shown diagrammatically by visualizing the impact of the third dimension on a flat surface; however, the fourth dimension of time is also involved and this makes visualization even more complex.

- The flat surface can be positively curved into a spherical shape of a finite size. This means that, by travelling around the surface of the sphere, you could return to your original position or, by travelling through the universe, you could return to your original position in spacetime.

- The flat surface can be negatively curved like the shape of a saddle and have an infinite size. In this universe you would never return to the same point in spacetime.
- The surface could also remain flat and infinite as given in our everyday experiences. Again, you would never return to the same position in spacetime.

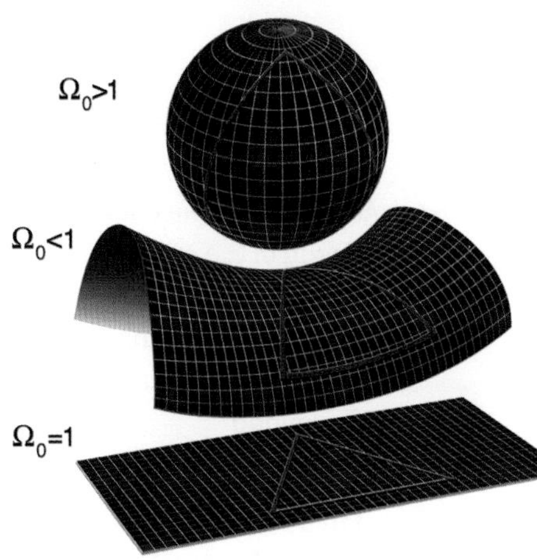

▲ Figure 2 The visualizations of closed, open, and flat universes.

Knowledge of the amount of matter within the universe is essential when determining which model is applicable. There is a critical density (ρ_c) of matter that would keep the universe flat and infinite – this density would provide a gravitational force large enough to prevent the universe running away but just too little to pull it back to its initial state. With less than the critical density the universe would be open and infinite. With greater density than the critical value the universe would be closed and finite – with gravity pulling all matter back to the initial state of spacetime. The critical density of matter appears to be no greater than ten particles per cubic metre and current research suggests that the average density is very close to this critical value.

The implications of the density of intergalactic matter

As can be seen in figure 5 in Sub-topic D.3, theory suggests that, after the initial inflationary period following the Big Bang, the rate of expansion of the universe has been slowing down. Instrumental to the fate of the universe is the uncertainty about how much matter is available to provide a strong enough gravitational force to reverse the expansion and cause a gravitational collapse. As discussed in Sub-topic D.3, data from Type Ia supernovae has suggested that the universe may actually be undergoing an accelerated expansion caused by mysterious "dark energy".

We can derive a relationship for the critical density using Newtonian mechanics:

Imagine a homogenous sphere of gas of radius r and density ρ. A galaxy of mass m at the surface of the sphere will be moving with a recessional speed v away from the centre of the sphere along a radius as shown in figure 3.

By Hubble's law the velocity of the galaxy is given by:

$$v = H_0 r$$

The total energy of the galaxy is the sum of its kinetic energy and its gravitational potential energy (relative to the centre of mass of the sphere of gas).

$$E_T = E_K + E_P$$

$$E_T = \frac{1}{2} mv^2 - G\frac{Mm}{r}$$

remembering that potential energy is always negative for objects separated by less than infinity.

The mass M is that of the sphere of gas is given by

$$M = \frac{4}{3}\pi r^3 \rho$$

$$E_T = \frac{1}{2} m(H_0 r)^2 - G\frac{\frac{4}{3}\pi r^3 \rho m}{r}$$

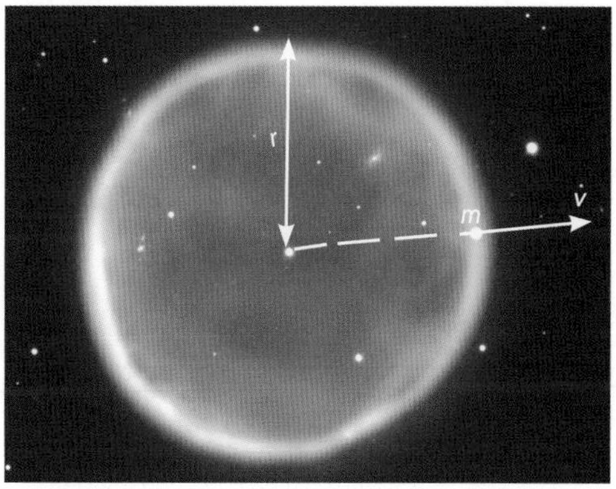

▲ Figure 3 Critical density for the universe.

Note

- This equation contains nothing but constants. Any value for the critical density of the material of the universe is dependent on how precisely the Hubble constant can be determined.
- A more rigorous derivation of this equation requires the use of general relativity.

The galaxy will continue to move providing that it has sufficient kinetic energy, thus making E_T positive. In the limit $E_T = 0$ this gives

$$\frac{1}{2}m(H_0 r)^2 = G\frac{\frac{4}{3}\pi r^3 \rho_c m}{r}$$

where ρ_c is the critical density of matter.

Simplifying this equation gives

$$\rho_c = \frac{3H_0^2}{8\pi G}$$

The cosmic scale factor and time

The ratio of the actual density of matter in the universe (ρ) to the critical density is called the density parameter (indicated in figure 2) and is given the symbol Ω_0.

$$\Omega_0 = \frac{\rho}{\rho_c}$$

There are three possibilities (shown in figure 4) for the fate of the universe, depending on the density parameter of the universe:

1. If $\Omega_0 = 1$ (or $\rho = \rho_c$) the density must equal the critical density and must be the value for a flat universe in which there is just enough matter for the universe to continue to expand to a maximum limit. However, the rate of expansion would decrease with time. This is thought to be the least likely option.
2. If $\Omega_0 < 1$ (or $\rho < \rho_c$) the universe would be open and would continue to expand forever.
3. If $\Omega_0 > 1$ (or $\rho > \rho_c$) then the universe would be closed. It would eventually stop expanding and would then collapse and end with a "Big Crunch".

An accelerated expansion of the universe (shown by the red line on figure 4) might be explained by the presence of dark energy. This offers an interesting and, increasingly likely, prospect.

In Sub-topic D.3 we considered the cosmic scale factor (R). This is essentially the relative size, or "radius", of the universe. Figure 4 shows how R varies with time for the different density parameters. Each of the models gives an Ω_0 value that is based on the total matter in the universe. An explanation for the accelerated universe depends on the concept of the (currently) hypothetical dark energy outweighing the gravitational effects of baryonic and dark matter.

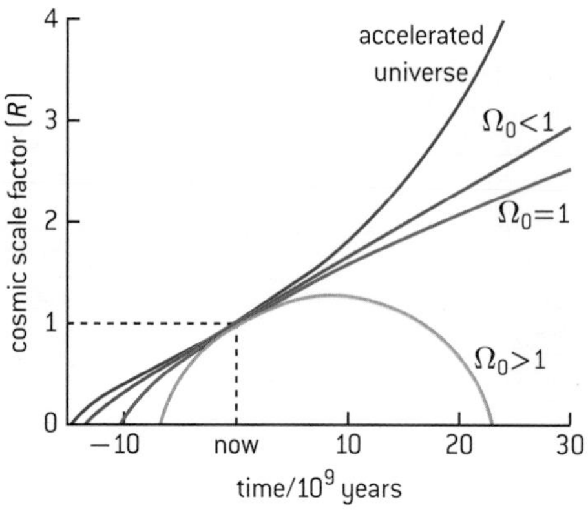

▲ Figure 4 Variation of R for different density parameters.

The cosmic scale factor and temperature

The wavelength of the radiation emitted by a galaxy will always be in line with the cosmic scale factor (R). So, as space expands, the wavelength will expand with it. We know from Wien's law that the product of the maximum intensity wavelength and the temperature is a constant. Assuming that the spectrum of a black body retains its shape during the expansion this means that Wien's law has been valid from the earliest times following the Big Bang. Thus, the wavelength and the cosmic scale factor are both inversely proportional to the absolute temperature.

$$T \propto \frac{1}{R} \text{ (and } T \propto \frac{1}{\lambda}\text{)}$$

Worked example

The diagram below shows the variation of the cosmic scale factor R of the universe with time t. The diagram is based on a closed model of the universe. The point $t = T$ is the present time.

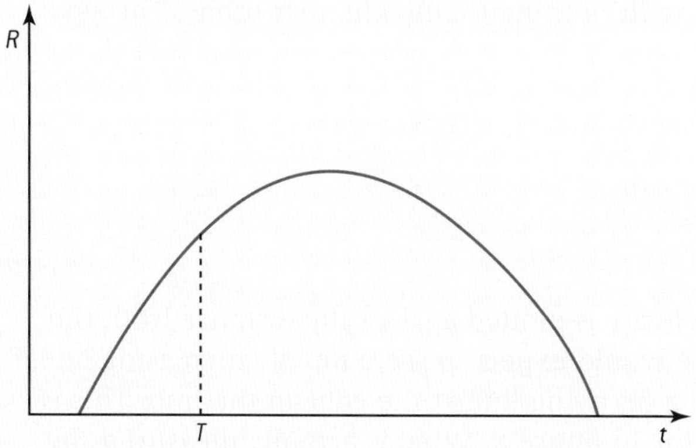

a) Explain what is meant by a *closed universe*.

b) On a copy of the diagram, draw the variation based on an open model of the universe.

c) Explain, by reference to your answer to b), why the predicted age of the universe depends upon the model of the universe chosen.

d) (i) What evidence suggests that the expansion of the universe is accelerating?

(ii) What is believed to be the cause of the acceleration?

Solution

a) A closed universe is one that will stop expanding at some future time. It will then start to contract due to gravity.

b)

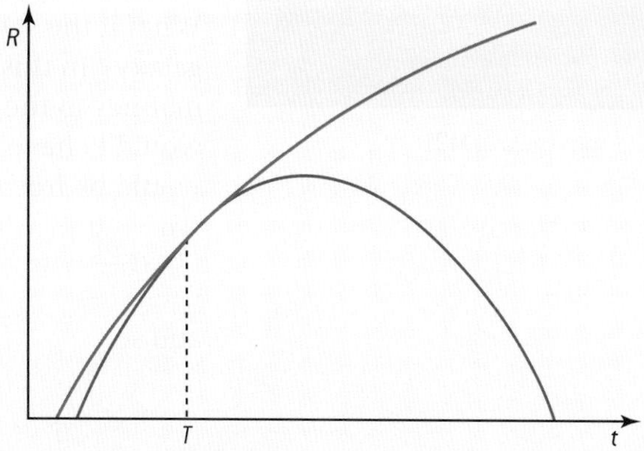

The graph should start at an early time (indicating an older universe) and touch the closed universe line at T. It should show curvature but not flatten out as a flat universe would do.

c) We only know the data for the present time so all curves will cross at T. By tracing the curve back to the time axis, we obtain the time for the Big Bang. This extrapolation will give a different time for the different models.

d) (i) The redshift from distant type Ia supernovae has suggested that the expansion of the universe is now accelerating.

(ii) The cause of this is thought to be dark energy – something of unknown mechanism but opposing the gravitational attraction of matter (both dark and baryonic).

Evidence for dark matter

Let us imagine a star of mass m near the centre of a spiral galaxy of total mass M. In this region the average density of matter is ρ. The star moves in a circular orbit with an orbital velocity v and radius r. By equating Newton's law of gravitation to the centripetal force we obtain

$$G\frac{Mm}{r^2} = \frac{mv^2}{r}$$

Cancelling m and r

$$G\frac{M}{r} = v^2$$

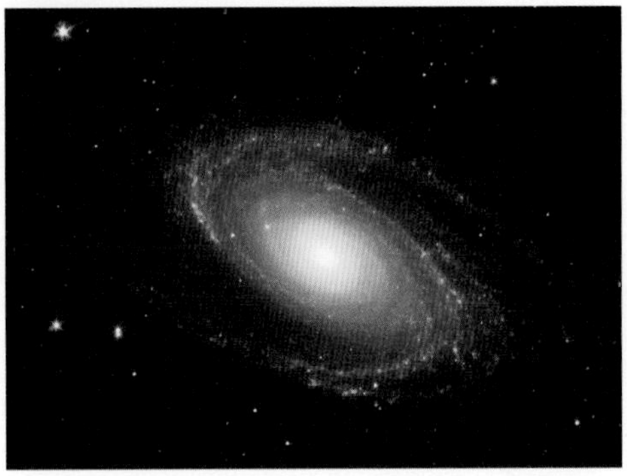

▲ Figure 5 The spiral galaxy M81.

In terms of the density and taking the central hub to be spherical, this gives

$$G\frac{\frac{4}{3}\pi r^3\rho}{r} = v^2$$

This means that $v = \sqrt{\frac{G4\pi\rho}{3}}\,r$ or $v = \text{constant} \times r$

From this we can see that the velocity is directly proportional to the radius.

What if the star is in one of the less densely populated arms of the galaxy? In this case we would expect the star to behave in a similar manner to the way in which planets rotate about the Sun. The galaxy would behave as if its total mass was concentrated at its centre; the stars would be free to move with nothing to impede their orbits. This gives

$$G\frac{M}{r} = v^2$$

and so

$$v \propto \frac{1}{\sqrt{r}}$$

When the rotational velocity is plotted against the distance from the centre of the galaxy, we would expect to see a rapidly increasing linear section that changes to a decaying line at the edge of the hub. This is shown by the broken line in figure 6. What is actually measured (by measuring the speed from the redshifts of the rotating stars) is the upper observed line. This is surprising because this "flat" rotation curve shows that the speed of stars, far out into the region beyond the arms of the galaxy, are moving with essentially the same speed as those well inside the galaxy. One explanation for this effect is the presence of **dark matter** forming a halo around the outer rim of the galaxy (as shown in figure 6). This matter is not normal "luminous" or "baryonic" matter and emits no radiation and, therefore, its presence cannot be detected.

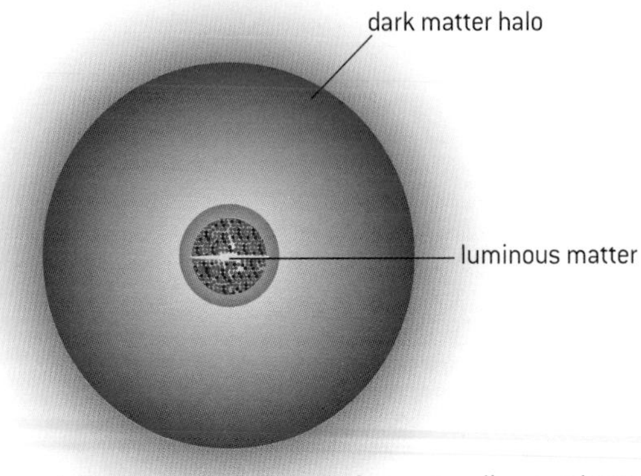

▲ Figure 6 Dark matter halo surrounding a galaxy.

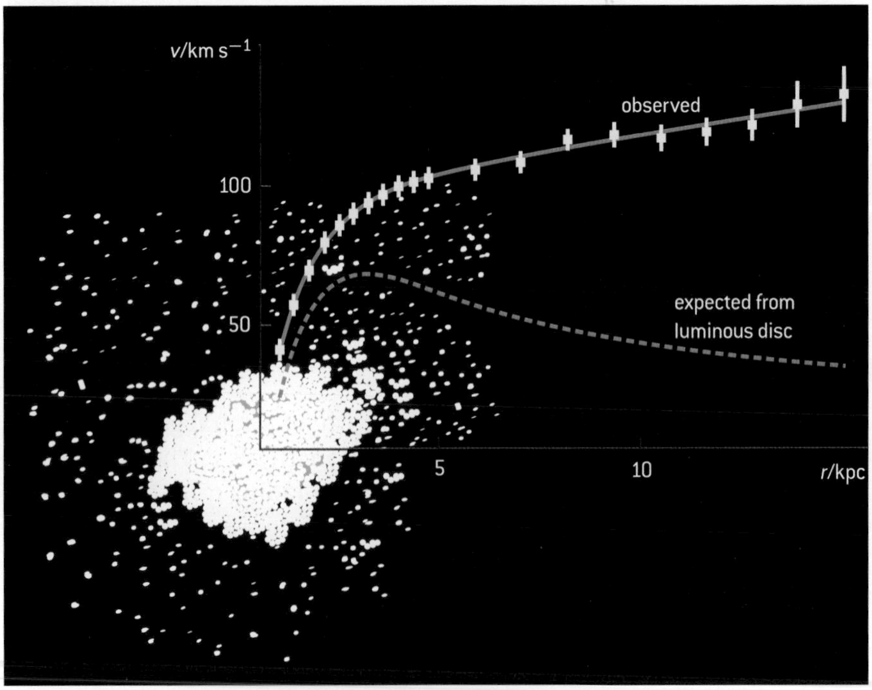

▲ Figure 7 The rotation curve for the spiral galaxy M33.

In figure 8, the experimental curve has been modelled by assuming that the halo adds sufficient mass to that of the galactic disc. This maintains the high rotational speeds well away from the galactic centre.

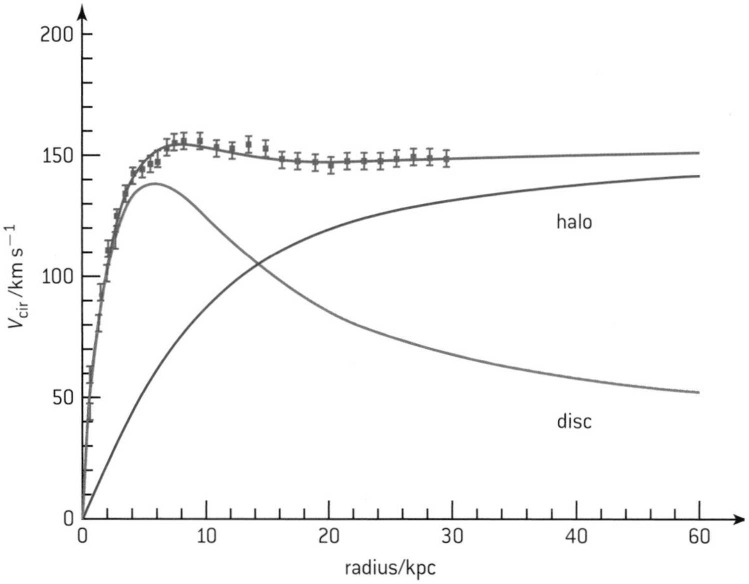

▲ Figure 8 distribution of dark matter in NGC 3198.

Other evidence for the presence of dark matter comes from:

- the velocities of galaxies orbiting each other in clusters – these galaxies emit far less light than they ought to in relation to the amount of mass suggested by their velocities
- the gravitational lensing effect of radiation from distant objects (such as quasars) – because the radiation passes through a cluster of galaxies it becomes much more distorted than would be expected by the luminous mass of the cluster
- the X-ray images of elliptical galaxies show the presence of haloes of hot gas extending well outside the galaxy. For this gas to be bound to the galaxy, the galaxy must have a mass far greater than that observed – up to 90% of the total mass of these galaxies is likely to be dark matter.

At the moment no one knows the nature of dark matter but there are some candidates:

- **MACHOs** are MAssive Compact Halo Objects that include black holes, neutron stars, and small stars such as brown dwarfs. These are all high density (compact) stars at the end of their lives and might be hidden by being a long way from any luminous objects. They are detected by gravitational lensing, but it is questionable whether or not there are sufficient numbers of MACHOs to be able to provide the amount of dark matter thought to be in the universe.
- **WIMPs** are Weakly Interacting Massive Particles – subatomic particles that are not made up of ordinary matter (they are non-baryonic). They are weakly interacting because they pass through ordinary, baryonic, matter with very little effect. Massive does not mean "big", it means that these particle have mass (albeit very small mass). To produce the amount of mass needed to make up the dark matter there would need to be

unimaginably large quantities of WIMPs. In 1998, neutrinos with very small mass were discovered and these are possible candidates for dark matter; other than this the theory depends on hypothetical particles called axions and neutralinos that are yet to be discovered experimentally.

Dark energy

In 1998, observations by the Hubble Space Telescope (HST) of a very distant supernova showed that the universe was expanding more slowly than it is today. Although nobody has a definitive explanation of this phenomenon its explanation is called "dark energy". ESA's Planck mission has provided data that suggest around 68% of the universe consists of dark energy (while 27% is dark matter leaving only 5% as normal "baryonic" matter – see figure 9).

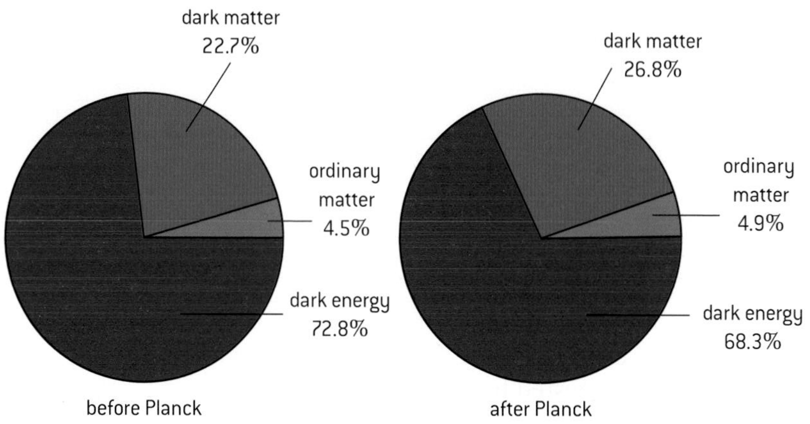

▲ Figure 9 The mass/energy recipe for the universe.

It has been suggested that dark energy is a property of space and so, with the expansion of the universe as space expands, so too does the amount of dark energy, i.e. more dark energy coming into existence along with more space. This form of energy would subsequently cause the expansion of the universe to accelerate. Nobody knows if this model is viable. It is possible that an explanation for the accelerated expansion of the universe requires a new theory of gravity or a modification of Einstein's theory – such a theory would still need to be able to account for all the phenomena that are, at the moment, correctly predicted by the current theory.

Nature of science

The Dark Energy Survey (DES)

With dark energy being one of the most up-to-the-minute research topics in the whole of science, it is unsurprising that international collaboration is prolific. The DES is a survey with the aim of gaining the best possible data for the rate of expansion of the universe. This is being carried out by observing around 3000 distant supernovae, the most distant of which exploded when the universe was about half its current size. Using the Victor M. Blanco Telescope at Cerro Tololo Inter-American Observatory (CTIO) in Chile, 120 collaborators from 23 institutions in 5 countries are using a specially developed camera to obtain images in the near infra-red part of the visible spectrum. The survey will take five years to complete and will add to the sky-based research of missions such as WMAP and Planck.

Anisotropies in the CMB

In Sub-topic D.3 we discussed the importance of the cosmic microwave background with regard to the Big Bang. In this model, the universe came into being almost 13.8 billion years ago when its density and temperature were both very high (this is often referred to as the *hot Big Bang*). Since then the universe has both expanded and cooled. The Planck satellite image in Sub-topic D.3, figure 4 showed that, although the CMB is essentially isotropic, there are minute temperature fluctuations; these variations are called **anisotropies**. In the early 1990s, the Cosmic Background Explorer (COBE) satellite provided the first evidence of these anisotropies but, with the launch of NASA's WMAP in 2001 and the ESA's Planck satellite in 2009, the resolution has been improved dramatically. Both these missions have shown significant low-level temperature fluctuations. It is thought that these fluctuations appear as the result of tiny, random variations in density, implanted during cosmic inflation – the period of accelerated expansion that occurred immediately after the Big Bang. When the universe was 380 000 years old and became transparent, the radiation emitted from the Big Bang was released and it has travelled outwards through space and time, including towards the Earth. This radiation was in the red part of the electromagnetic spectrum when it was released, but its wavelength has now been stretched with the expansion of the universe so that it corresponds to microwave radiation. The pattern shown in the variation demonstrates the differences present on the release of the radiation: fluctuations that would later grow into galaxies and galaxy clusters under the influence of gravity.

The information extracted from the Planck sky map shows isotropy on a large scale but with a lack of symmetry in the average temperatures in opposite hemispheres of the sky. The Standard Model suggests that the universe should be isotropic but, given these differences, this appears not to be the case. There is also a cold spot (circled in Sub-topic D.3, figure 4) extending over a patch of sky and this is larger than WMAP had previously shown. The concepts of dark matter and dark energy have already been added to the Standard Model as additional parameters. The evidence from the CMB anisotropies may require further tweaks to the theory or even a major re-think because of this.

The Planck data also identify the Hubble constant to be 67.15 km s^{-1} Mpc^{-1} (significantly less than the current standard value in astronomy of around 100 km s^{-1} Mpc^{-1}). The data imply that the age of the universe is 13.82 billion years. Over the next few years this value may well be modified because of additional data being collected by this mission and from orbiting telescopes – including the James Webb space telescope (figure 10) and the joint NASA/ESA Euclid mission. There has, arguably, never been a more productive time in the history of cosmology.

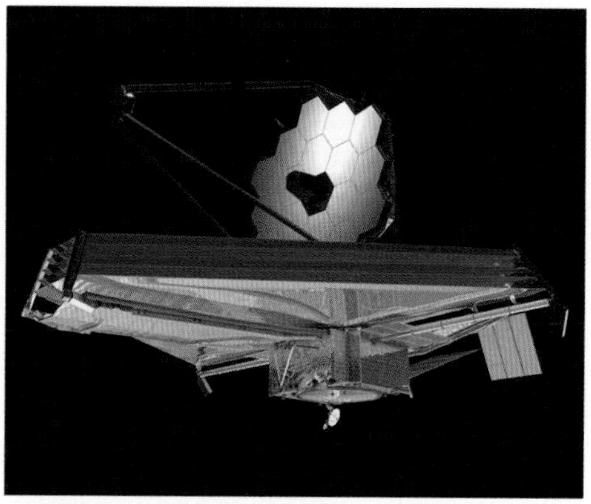

▲ Figure 10 The James Webb space telescope – artist's impression.

Questions

1 *(IB)*

a) The star Wolf 359 has a parallax angle of 0.419 arcsecond.

(i) Describe how this parallax angle is measured.

(ii) Calculate the distance in light-year from Earth to Wolf 359.

(iii) State why the method of parallax can only be used for stars at a distance less than a few hundred parsecond from Earth.

b) The ratio

$$\frac{\text{apparent brightness of Wolf 359}}{\text{apparent brightness of the Sun}} \text{ is } 3.7 \times 10^{-15}.$$

Show that the ratio

$$\frac{\text{luminosity of Wolf 359}}{\text{luminosity of the Sun}} \text{ is } 8.9 \times 10^{-4}.$$

(11 marks)

2 The average intensity of the Sun's radiation at the surface of the Earth is 1.37×10^3 Wm^{-2}. Calculate (a) the luminosity and (b) the surface temperature of the Sun.

The mean separation of the Earth and the Sun = 1.50×10^{11} m, radius of the Sun = 6.96×10^8 m, Stefan–Boltzmann constant = 5.67×10^{-8} Wm^{-2} K^{-4}. (4 marks)

3 *(IB)*

The diagram below is a flow chart that shows the stages of evolution of a main sequence star such as the Sun. (Mass of the Sun, the solar mass = $M_\odot$)

main sequence star mass $\approx M_\odot$ → red giant nebula → planetary nebula → white dwarf

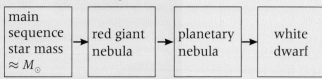

a) Copy nad complete the boxes below to show the stages of evolution of a main sequence star that has a mass greater than $8M_\odot$.

main sequence star mass $>8M_\odot$ → [ ] → [ ] → [ ]

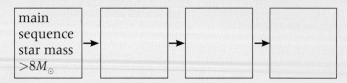

b) Outline why:

(i) white dwarf stars cannot have a greater mass than $1.4M_\odot$

(ii) it is possible for a main sequence star with a mass equal to $8M_\odot$ to evolve into a white dwarf. (6 marks)

4 *(IB)*

a) Define *luminosity.*

b) The sketch-graph below shows the intensity spectrum for a black body at a temperature of 6000 K.

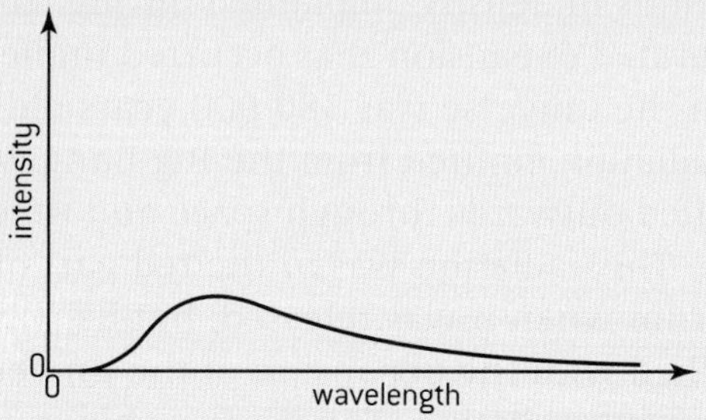

On a copy of the axes, draw a sketch-graph showing the intensity spectrum for a black body at 8000 K.

c) A sketch of a Hertzsprung–Russell diagram is shown below.

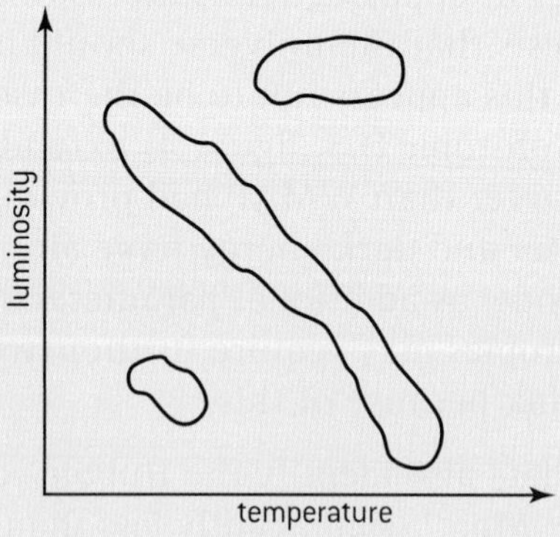

Copy the diagram above and identify the:

(i) main sequence (label this M)

(ii) red giant region (label this R)

(iii) white dwarf region (label this W).

d) In a Hertzsprung–Russell diagram, luminosity is plotted against temperature. Explain why the diagram alone does not enable the luminosity of a particular star to be determined from its temperature. (8 marks)

5 *(IB)*

The diagram below shows the grid of a Hertzsprung–Russell (HR) diagram on which the positions of the Sun and four other stars A, B, C and D are shown.

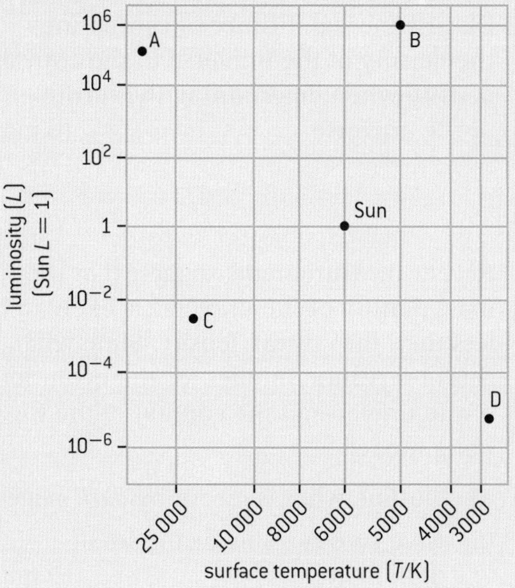

a) Name the type of stars shown by A, B, C, and D.

b) Explain, using information from the HR diagram and without making any calculations, how astronomers can deduce that star B is larger than star A.

c) Using the following data and information from the HR diagram, show that star B is at a distance of about 700 pc from Earth.

Apparent brightness of the Sun = 1.4×10^3 W m^{-2}

Apparent brightness of star B = 7.0×10^{-8} W m^{-2}

Mean distance of the Sun from Earth = 1.0 AU

1 parsec = 2.1×10^5 AU (11 marks)

6 *(IB)*

a) State what is meant by *cosmic microwave background radiation.*

b) Describe how the cosmic microwave background radiation provides evidence for the expanding universe. (5 marks)

7 *(IB)*

a) In an observation of a distant galaxy, spectral lines are recorded. Spectral lines at these wavelengths cannot be produced in the laboratory. Explain this phenomenon.

b) Describe how Hubble's law is used to determine the distance from the Earth to distant galaxies.

c) Explain why Hubble's law is not used to measure distances to nearby stars or nearby galaxies (such as Andromeda). (6 marks)

8 *(IB)*

One of the most intense radio sources is the Galaxy NGC5128. Long exposure photographs show it to be a giant elliptical galaxy crossed by a band of dark dust. It lies about 1.5×10^7 light years away from Earth.

a) Describe any differences between this galaxy and the Milky Way.

Hubble's law predicts that NGC5128 is moving away from Earth.

b) (i) State Hubble's law.

(ii) State and explain what experimental measurements need to be taken in order to determine the Hubble constant.

c) A possible value for the Hubble constant is 68 km s^{-1} Mpc^{-1}. Use this value to estimate:

(i) the recession speed of NGC5128

(ii) the age of the universe. (10 marks)

9 **a)** Describe what is meant by a *nebula*.

b) Explain how the Jeans criterion applies to star formation. (3 marks)

10 Outline how hydrogen is fused into helium in:

a) stars of mass similar to that of the Sun

b) stars of mass greater than ten solar masses. (6 marks)

11 a) (i) Explain what is meant by *neutron capture*.

(ii) Write a nuclear equation to show nuclide A capturing a neutron to become nuclide B.

b) Outline the difference between s and r processes in nucleosynthesis. (8 marks)

12 Explain why the lifetimes of more massive main sequence stars are shorter than those of less massive ones. (4 marks)

13 Briefly explain the roles of electron degeneracy, neutron degeneracy and the Chandrasekhar limit in the evolution of a star that goes supernova. (6 marks)

14 *(IB)*

a) Explain the significance of the *critical density* of matter in the universe with respect to the possible fate of the universe.

The critical density ρ_c of matter in the universe is given by the expression:

$$\rho_c = \frac{3H_0^2}{8\pi G}$$

where H_0 is the Hubble constant and G is the gravitational constant.

An estimate of H_0 is $2.7 \times 10^{-18}\ s^{-1}$.

b) (i) Calculate a value for ρ_c.

(ii) Using your value for H_0, determine the equivalent number of nucleons per unit volume at this critical density.

(5 marks)

15 *(IB)*

a) Describe the observational evidence in support of an expanding universe.

b) Explain what is meant by the term *critical density* of the universe.

c) Discuss the significance of comparing the density of the universe to the critical density when determining the future of the universe. (6 marks)

16 *(IB)*

a) Recent measurements suggest that the mass density of the universe is likely to be less than the critical density. State what this observation implies for the evolution of the universe in the context of the Big Bang model.

b) (i) Outline what is meant by *dark matter*.

(ii) Give **two** possible examples of dark matter. (5 marks)

INTERNAL ASSESSMENT

Introduction

In this chapter you will discover the important role of experimental work in physics. It guides you through the expectations and requirements of an independent investigation called the internal assessment (IA). The investigation itself is likely to occur late in your second year, so you do not need to read this chapter until your teacher advises you to.

Advice on the internal assessment

Understanding

- theory and experiment
- internal assessment requirements
- internal assessment guidance
- internal assessment criteria

Nature of science

Empirical evidence is a key to objectivity in science. Evidence is obtained by observation, and the details of observation are embedded in experimental work. Theory and experiment are two sides of the same coin of scientific knowledge.

Applications and skills

- to appreciate the interrelationship of theory and experiment
- ability to plan your internal assessment
- understand teacher guidance
- appreciate the formal requirements of an internal assessment
- to be critically aware of academic honesty

Theory and experiment

The sciences use a wide variety of methodologies and there is no single agreed scientific method. However, all sciences are based on evidence obtained by experiment. Evidence is used to develop theories, which then form laws. Theories and laws are used to make predictions that can be tested in experiments. Science moves in a cycle that moves between theory and experiment. Observations inform theory. However, the refinement of a theory and improvements in instrumentation re-focus on more observation. Experimentation allows us to have confidence that a theory is not merely pure speculation.

Consider a famous analogy used by Albert Einstein and Leopold Infeld, of a man trying to understand the mechanism of a pocket watch. In the following quote, Einstein illustrates that our scientific knowledge can be tested against reality. He shows that we can confirm or deny a theory by experiment, but we can never know reality itself. There is a continual dance between theory and experiment.

> "Physical concepts are free creations of the human mind, and are not, however it may seem, uniquely determined by the external world. In our endeavor to understand reality we are somewhat like a man trying to understand the mechanism of a closed watch. He sees the face and the moving hands, even hears it ticking, but he has no way of opening the case if he is ingenious he may form some picture of the mechanism which could be responsible for all the things he observes, but he may never be quite sure his picture is the only one which could explain his observations. He will never be able to compare his picture with the real mechanism and he cannot even imagine that possibility of the meaning of such a comparison. But he certainly believes that, as his knowledge increases, his picture of reality will become simpler and simpler and will explain a wider and wider range of his sensuous impressions. He may also believe in the existence of the ideal limit of knowledge and that it is approached by the human mind. He may call this ideal limit the objective truth."
>
> *—Albert Einstein and Leopold Infeld, "The Evolution of Physics."*

The internal assessment requirements

Experimental work is not only an essential part of the dynamic of scientific knowledge it also plays a key role in the teaching and learning of physics. Experimental work should be an integral and regular part of your physics lessons consisting of demonstrations, hands-on group work, and individual investigations. It may include computer simulations, mathematical models, and online databases resources. It is only natural then that time should be allocated to you in order to formulate, design, and implement your own physics experiment. You will produce a single investigation that is called an internal assessment. This means that your teacher will assess your report using IB criteria, and the IB will externally moderate your teacher's assessment.

Your investigation will consist of:

- selecting an appropriate topic
- researching the scientific content of your topic
- defining a workable research question
- adapting or designing a methodology
- obtaining, processing, and analysing data
- appreciating errors, uncertainties, and limits of data
- writing a scientific report 6–12 A4 pages long
- receiving continued guidance from your teacher.

Planning and guidance

After your teacher introduces the idea of an internal assessment investigation, you will have an opportunity to discuss your investigation topic with your teacher. Through dialogue with your teacher you can select an appropriate topic, define a workable research question, and begin by doing research into what is already known about your topic. You will not be penalized for seeking your teacher's advice.

It is your **teacher's responsibility** to provide you with a clear understanding of the IA expectations, rules, and requirements.

Your teacher will:

- provide you with continued guidance at all stages of your work
- help you focus on a topic, then a research question, and then an appropriate methodology
- provide quidance as you work and read a draft of your report, making general suggestions for improvements or completeness.

Your teacher will not, however, edit your report nor give you a tentative grade or achieving level for your report until it is finally completed. Once your report is completed and formally submitted you are not allowed to make any changes.

Your teacher is responsible for your guidance, making sure you understand the IA expectations, and that your work is your own.

As the student it is **your responsibility** to appreciate the meaning of academic honesty, especially authenticity and the respect of intellectual property. You are also responsible for initiating your research question with the teacher, seeking help when in doubt, and demonstrating independence of thought and initiative in the design and implementation of your investigation. You are also responsible for meeting the deadlines set by your teacher.

The internal assessment report

There is no prescribed format for your investigation report. However, the IA criteria encourage a logical and justified approach, one that demonstrates personal involvement and exhibits sound scientific work.

The style and form of your report for the IA investigation should model a scientific journal article. You should be familiar with a number of high school level physics journal articles. For example, journals like the British "Physics Education" publication (http://iopscience.iop.org) or the American "The Physics Teacher" publication (http://tpt.aapt.org) often have articles that are appropriate for high school work. Moreover, many of these articles can provide good ideas for an investigation.

There is no prescribed narrative mode, and your teacher will direct you to the style that they wish you to use. However, because a report describes what you have done, it is reasonable to write in the past tense. Descriptions are always clearer to understand if you avoid the use of pronouns (usually 'it') and refer specifically to the relevant noun ('the wire', 'the ammeter', 'a digital caliper', etc.).

Academic honesty

The IB learner profile (page iii) describes IB students as aspiring to develop many qualities, including that of being "principled". This means that you act with integrity and honesty, with a strong sense of fairness and justice, and that you take responsibility for your actions and their consequences. The IA is your responsibility, and it is your work. Plagiarism and copying others' work is not permissible. You must clearly distinguish between your own words and thoughts and those of others by the use of quotation marks (or other methods like indentation) followed by an appropriate citation that denotes an entry in the bibliography.

Although the IB does not prescribe referencing style or in-text citation, certain styles may prove most commonly used; you are free to choose a style that is appropriate. It is expected that the minimum information included is: name of author, date of publication, title of source, and page numbers as applicable.

Types of investigations

After you have covered a number of physics syllabus topics and performed a number of hands-on experiments in class, you will be required to research, design, perform, and write up your own investigation. This project, known as an **internal assessment**, will count for 20% of your grade, so it is important that you make the most of your opportunity to do well. You will have 10 hours of class time, be able to consult with your teacher at all stages of your work, and research and write your report out of class. Your IA investigation may not be used as part of a physics extended essay.

The variety and range of possible investigations is large, you could choose from:

- **Traditional hands-on experimental work.** You may want to measure the acceleration of gravity using Atwood's approach, or determine the gas constant using standard Boyle's law equipment.
- **Database investigations.** You may obtain data from scientific websites and process and analyse the information for your investigation. Perhaps you find a pattern in the ebb and flow of the ocean tides, or process information on global warming, or use an astronomical database to confirm Kepler's law.
- **Spreadsheet.** You can make use of a spreadsheet with data from any type of investigation. You can process the data, graph the results, even design a simple model to compare textbook theory with your experimental values.
- **Simulations.** It may not be feasible to perform some investigations in the classroom, but you may be able to find a computer simulation. The data from a simulation could then be processed and presented in such a way that something new is revealed. Perhaps you might determine the universal gravitation constant through a simulation (an experiment too sensitive to perform in most school laboratories). Or you might investigate the effect of air resistance on projectile motion.

Combinations of the above are also possible. The subject matter of your investigation is up to you. It may be something within the syllabus, something you have already studied or are about to study, or it can be within or outside the syllabus. The depth of understanding should be, however, commensurate with the course you are taking. This means that your knowledge of IB Physics (either SL or HL) will be sufficient to achieve maximum marks when assessed.

The assessment criteria

Your IA consists of a single investigation with a report 6–12 pages long. The report should have an academic and scholarly presentation, and demonstrate scientific rigor commensurate with the course. There is the expectation of personal involvement, an understanding of physics, and that the study is set within a known academic context. This means you need to research your topic and find out what is already known about it.

There are six assessment criteria, ranging in weight from 8–25% of the total possible marks. Each criterion reflects a different aspect of your overall investigation.

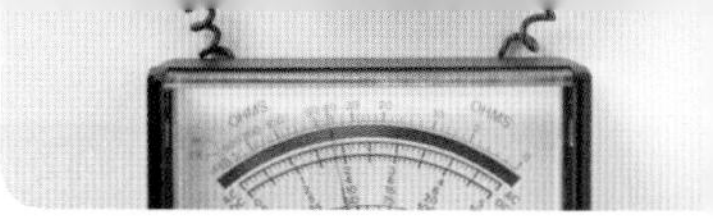

Criterion	Points	Weight
Personal engagement	0–2	8%
Exploration	0–6	25%
Analysis	0–6	25%
Evaluation	0–6	25%
Communication	0–4	17%
Total	**0–24**	**100%**

The IA grade will count for 20% of your total physics grade. The criteria are the same for standard and higher level students. We will now consider each criterion in detail.

PERSONAL ENGAGEMENT. *This criterion assesses the extent to which you engage with the investigation and make it your own. Personal engagement may be recognized in different attributes and skills. These include thinking independently and/or creatively, addressing personal interests, and presenting scientific ideas in your own way.*

For maximum marks under the personal engagement criterion, you must provide clear evidence that you have contributed significant thinking, initiative, or insight to your investigation: that you take the responsibility for ownership of your investigation. Your research question could be based upon something covered in class or an extension of your own interest.

For example, you may be a keen music student and your teacher may have demonstrated resonance with a wine glass whilst studying sound. You could have been fascinated with this phenomenon and decide to design and perform an investigation on the resonance of a wine glass. Personal significance, interest, and curiosity are expressed here.

You may also demonstrate personal engagement by showing personal input and initiative in the design, implementation, or presentation of the investigation. Perhaps you designed an improved method for measuring the timing of a bouncing ball or devised an interesting method for the analysis of data. You are not to simply perform a cookbook-like experiment.

The key here is to be involved in your investigation, to contribute something that makes it your own.

EXPLORATION. *This criterion assesses the extent to which you establish the scientific context for your work, state a clear and focused research question, and use concepts and techniques appropriate to the course you are studying. Where appropriate, this criterion also assesses awareness of safety, environmental, and ethical considerations.*

For maximum marks under the exploration criterion, your topic must be appropriately identified and you must describe a relevant and fully focused research question. Background information about your investigation must be appropriate and relevant, and the methodology must be suitable to address your research question. Moreover, for maximum marks, your research must identify significant factors that may influence the relevance, reliability, and sufficiency of your data. Finally, your work must be safe and it must demonstrate a full awareness of relevant environmental and ethical issues.

The key here is your ability to select, develop, and apply appropriate methodology and produce a scientific work.

ANALYSIS. *This criterion assesses the extent to which your report provides evidence that you have selected, processed, analysed, and interpreted the data in ways that are relevant to the research question and can support a conclusion.*

For maximum marks under the analysis criterion, your investigation must include sufficient raw data to support a detailed and valid conclusion to your research question. Your processing of the data must be carried out with sufficient accuracy. Moreover, your analysis must take a full and appropriate account of experimental uncertainty. Finally, for maximum marks, you must correctly interpret your data, so that completely valid and detailed conclusions to the research questions can be deduced.

The key here is to make an appropriate and justified analysis of your data that is focused on your research question.

EVALUATION. *This criterion assesses the extent to which your report provides evidence of evaluation of the investigation and results with regard to the research question and the wider world.*

For maximum marks under the evaluation criterion, you must describe a detailed and justified conclusion that is entirely relevant to the research question, and fully supported by your analysis of the data presented. You should make a comparison to the accepted scientific context if relevant. The strengths and weakness of your investigation, such as the limitations of data and sources of uncertainty, must be discussed and you must provide evidence of a clear understanding of the methodological issues involved in establishing your conclusion. Finally, to earn maximum marks for evaluation, you must discuss realistic and relevant improvements and possible extensions to your investigation.

The key here is different from the analysis criterion. The focus of evaluation is to incorporate the methodology and to set the results within a a wider scientific context while making reference to your research topic.

COMMUNICATION. *This criterion assesses whether the investigation is presented and reported in a way that supports effective communication of the investigation's focus, process, and outcomes.*

For maximum marks under the communication criterion, your report must be clear and easy to follow. Although your writing does not have to be perfect, any mistakes or errors should not hamper the understanding, focus, process, and outcomes of your investigation. Your report must be well structured and focused on the necessary information, the process and outcomes must be presented in a logical and coherent way. Your text must be relevant and avoid wandering off onto tangential issues. Your use of specific physics terminology and conventions must be appropriate and correct. Graphs, tables, and images must all be well presented. Your lab report should be 6–12 pages long. Excessive length (beyond 12 pages) will be penalized under the communication criterion.

The key here is to demonstrate a concise, logical, and articulate report, one that is easy to follow and is written in a scientific context. This is not an assessed criterion but nevertheless is likely to be key to the fulfilment of a successful IA. In conclusion, the IA represents a unique opportunity for you to take ownership of your physics learning by investigating something that matters to you. It is a chance for you to work independently and to follow your own scientific instincts. True, you should heed the advice and experience of your teacher and be guided so that you don't go off down a blind alley; however you should be prepared to research your topic independently and approach your teacher full of ideas and suggestions. Experience suggests that those students brim full of proposals are likely to be successful, providing they stick to the physics skills and principles that have been encouraged to develop throughout the course.

INDEX

Page numbers in *italics* refer to questions.